rs

in
Bergen, Department of Informatics
, 5020 Bergen, Norway
@ii.uib.no

ds
Latvia, Faculty of Computing
, 1586 Riga, Latvia
freivalds@lu.lv

wska
Oxford, Department of Computer Science
ing, Parks Road, Oxford OX1 3QD, UK
kwiatkowska@cs.ox.ac.uk

itute of Science, Faculty of Mathematics and Computer Science
Rehovot, Israel
peleg@weizmann.ac.il

3   e-ISSN 1611-3349
2-39211-5   e-ISBN 978-3-642-39212-2
78-3-642-39212-2
lberg Dordrecht London New York

gress Control Number: 2013941217

ssification (1998): F.2, F.1, C.2, H.3-4, G.2, I.2, I.3.5, E.1

ry: SL 1 – Theoretical Computer Science and General Issues

Lecture Notes in Computer Science   7966

Commenced Publication in 1973
Founding and Former Series Editors:
Gerhard Goos, Juris Hartmanis, and Jan van Leeuwe

Advanced Research in Computing and Software Science

Subline of Lectures Notes in Computer Science

Subline Series Editors

Subline Advisory Board

Fedor V. Fomin   Rūsiņš Fre
Marta Kwiatkowska   Davi(

Automata, L
and Progran

40th International Colloqu
Riga, Latvia, July 8-12, 2(
Proceedings, Part II

Volume Edit

Fedor V. Fom
University of
Postboks 780
E-mail: fomin

Rūsiņš Freiva
University of
Raina bulv. 19
E-mail: rusins

Marta Kwiatk
University of
Wolfson Build
E-mail: marta.

David Peleg
Weizmann Inst
POB 26, 76100
E-mail: david,

ISSN 0302-97
ISBN 978-3-64
DOI 10.1007/9
Springer Heide

Library of Con

CR Subject Cla

LNCS Sublibra

Typesetting: Camera

Printed on acid-free

Springer is part of S

 Springer

Preface

ICALP, the International Colloquium on Automata, Languages and Programming, is arguably the most well-known series of scientific conferences on Theoretical Computer Science in Europe. The first ICALP was held in Paris, France, during July 3–7, 1972, with 51 talks. The same year EATCS, the European Association for Theoretical Computer Science, was established. Since then ICALP has been the flagship conference of EATCS.

ICALP 2013 was the 40th conference in this series (there was no ICALP in 1973). For the first time, ICALP entered the territory of the former Soviet Union. It was held in Riga, Latvia, during on July 8–12, 2013, in the University of Latvia. This year the program of ICALP was organized in three tracks: Track A (Algorithms, Complexity and Games), Track B (Logic, Semantics, Automata and Theory of Programming), and Track C (Foundations of Networked Computation: Models, Algorithms and Information Management).

In response to the Call for Papers, 436 papers were submitted; 14 papers were later withdrawn. The three Program Committees worked hard to select 71 papers for Track A (out of 249 papers submitted), 33 papers for Track B (out of 113 papers), and 20 papers for Track C (out of 60 papers). The average acceptance rate was 29%. The selection was based on originality, quality, and relevance to theoretical computer science. The quality of the submitted papers was very high indeed. The Program Committees acknowledge that many rejected papers deserved publication but regrettably it was impossible to extend the conference beyond 5 days.

The best paper awards were given to Mark Bun and Justin Thaler for their paper "Dual Lower Bounds for Approximate Degree and Markov-Bernstein Inequalities" (in Track A), to John Fearnley and Marcin Jurdziński for the paper "Reachability in Two-Clock Timed Automata is PSPACE-complete" (in Track B), and to Dariusz Dereniowski, Yann Disser, Adrian Kosowski, Dominik Pająk, and Przemysław Uznański for the paper "Fast Collaborative Graph Exploration" (in Track C). The best student paper awards were given to Radu Curticapean for the paper "Counting matchings of size k is #W[1]-hard" (in Track A) and to Nicolas Basset for the paper "A maximal entropy stochastic process for a timed automaton" (in Track B).

ICALP 2013 contained a special EATCS lecture on the occasion of the 40th ICALP given by:

– Jon Kleinberg, Cornell University

 Invited talks were delivered by:

– Susanne Albers, Humboldt University
– Orna Kupferman, Hebrew University
– Dániel Marx, Hungarian Academy of Sciences

- Paul Spirakis, University of Patras
- Peter Widmayer, ETH Zürich

The main conference was preceded by a series of workshops on Sunday July 7, 2013 (i.e., one day before ICALP 2013). The list of workshops consisted of:

- Workshop on Automata, Logic, Formal languages, and Algebra (ALFA 2013)
- International Workshop on Approximation, Parameterized and Exact Algorithms (APEX 2013)
- Quantitative Models: Expressiveness and Analysis
- Foundations of Network Science (FONES)
- Learning Theory and Complexity
- 7th Workshop on Membrane Computing and Biologically Inspired Process Calculi (MeCBIC 2013)
- Workshop on Quantum and Classical Complexity

We sincerely thank our sponsors, members of the committees, referees, and the many colleagues who anonymously spent much time and effort to make ICALP 2013 happen.

May 2013

Fedor V. Fomin
Rūsiņš Freivalds
Marta Kwiatkowska
David Peleg

Organization

Program Committee

Track A

Andris Ambainis	University of Latvia, Latvia
Edith Elkind	Nanyang Technological University, Singapore
Leah Epstein	University of Haifa, Israel
Rolf Fagerberg	University of Southern Denmark, Denmark
Fedor Fomin	University of Bergen, Norway (Chair)
Pierre Fraigniaud	CNRS and University Paris Diderot, France
Fabrizio Grandoni	Dalle Molle Institute, Switzerland
Joachim Gudmundsson	University of Sydney, Australia
Kazuo Iwama	Kyoto University, Japan
Valentine Kabanets	Simon Fraser University, Canada
Stavros Kolliopoulos	National and Kapodistrian University of Athens, Greece
Daniel Král'	University of Warwick, UK
Daniel Lokshtanov	University of California, San Diego, USA
Konstantin Makarychev	Microsoft Research, Redmond, USA
Peter Bro Miltersen	Aarhus University, Denmark
Ilan Newman	University of Haifa, Israel
Konstantinos Panagiotou	Ludwig Maximilians University, Munich, Germany
Alexander Razborov	University of Chicago, USA
Saket Saurabh	The Institute of Mathematical Sciences, India
David Steurer	Microsoft Research, New England, USA
Kunal Talwar	Microsoft Research, Silicon Valley, USA
Dimitrios Thilikos	National and Kapodistrian University of Athens, Greece
Virginia Vassilevska Williams	University of California, Berkeley, and Stanford, USA
Gerhard Woeginger	Eindhoven University of Technology, The Netherlands

Track B

Christel Baier	TU Dresden, Germany
Chiara Bodei	University of Pisa, Italy
Mikołaj Bojańczyk	University of Warsaw, Poland
Patricia Bouyer-Decitre	CNRS/ENS Cachan, France
Vassilis Christophides	University of Crete, Greece
Yuxin Deng	Shanghai Jiao-Tong University, China
Marcelo Fiore	University of Cambridge, UK

Patrice Godefroid	Microsoft Research, Redmond, USA
Andy Gordon	MSR Cambridge and University of Edinburgh, UK
Alexey Gotsman	Madrid Institute for Advanced Studies (IMDEA), Spain
Masami Hagiya	University of Tokyo, Japan
Michael Huth	Imperial College London, UK
Stephan Kreutzer	TU Berlin, Germany
Antonín Kučera	Masaryk University, Brno, Czech Republic
Viktor Kuncak	EPFL, Lausanne, Switzerland
Marta Kwiatkowska	University of Oxford, UK (Chair)
Leonid Libkin	University of Edinburgh, UK
Rupak Majumdar	Max Planck Institute, Kaiserslautern, Germany
Jerzy Marcinkowski	University of Wroclaw, Poland
Annabelle McIver	Macquarie University, Australia
Catuscia Palamidessi	INRIA Saclay and LIX, Ecole Polytechnique, Paris, France
Frank Pfenning	Carnegie Mellon University, USA
André Platzer	Carnegie Mellon University, USA
Jean-François Raskin	UL Brussels, Belgium
Jan Rutten	CWI and Radboud University Nijmegen, The Netherlands
Peter Selinger	Dalhousie University, Canada
Andreas Winter	University of Bristol, UK

Track C

James Aspnes	Yale Univerity, USA
Ioannis Caragiannis	University of Patras, Greece
Xavier Defago	JAIST, Japan
Josep Diaz	UPC, Barcelona, Spain
Stefan Dobrev	Slovak Academy of Sciences, Bratislava, Slovak Republic
Michele Flammini	University of L'Aquila, Italy
Leszek Gąsieniec	University of Liverpool, UK
Cyril Gavoille	Univerity of Bordeaux, France
David Kempe	University of Southern California, USA
Valerie King	University of Victoria, Canada
Amos Korman	CNRS, Paris, France
Miroslaw Kutylowski	Wroclaw University of Technology, Poland
Dahlia Malkhi	Microsoft Research, Silicon Valley, USA
Luca Moscardelli	University of Chieti, Pescara, Italy
Thomas Moscibroda	Microsoft Research Asia and Tsinghua University, China
Marina Papatriantafilou	Chalmers University of Technology, Goteborg, Sweden

David Peleg   Weizmann Institute of Science, Israel (Chair)
Yvonne Anne Pignolet   ETH Zurich, Switzerland
Sergio Rajsbaum   UNAM, Mexico
Liam Roditty   Bar-Ilan University, Israel
José Rolim   University of Geneva, Switzerland
Christian Scheideler   Technical University of Munich, Germany
Jennifer Welch   Texas A&M University, USA

Organizing Committee

(all from University of Latvia, Latvia)

Andris Ambainis
Kaspars Balodis
Juris Borzovs   (Organizing Chair)
Rūsiņš Freivalds   (Conference Chair)
Marats Golovkins
Nikolay Nahimov
Jeļena Poļakova
Alexander Rivosh
Agnis Škuškovniks   (Organizing Deputy Chair)
Juris Smotrovs
Abuzer Yakaryılmaz

Sponsoring Institutions

QuBalt
University of Latvia

Additional Reviewers

Aaronson, Scott
Aceto, Luca
Adamaszek, Anna
Afshani, Peyman
Agrawal, Manindra
Ahn, Kook Jin
Aichholzer, Oswin
Albers, Susanne
Allouche, Jean-Paul
Alur, Rajeev
Alvarez, Carme
Amano, Kazuyuki
Andoni, Alexandr

Arvind, V.
Askalidis, Georgios
Atserias, Albert
Aumüller, Martin
Avigdor-Elgrabli, Noa
Avis, David
Badanidiyuru,
    Ashwinkumar
Bae, Sang Won
Balmau, Oana
Bampis, Evripidis
Bansal, Nikhil
Barcelo, Pablo

Barman, Siddharth
Barto, Libor
Belovs, Aleksandrs
Bendlin, Rikke
Benoit, Anne
Benzaken, Veronique
Berman, Itay
Bertrand, Nathalie
Berwanger, Dietmar
Bianchi, Giuseppe
Biedl, Therese
Bilò, Davide
Bilò, Vittorio

Björklund, Andreas
Björklund, Henrik
Blais, Eric
Bläser, Markus
Blum, William
Bodirsky, Manuel
Bodlaender, Hans L.
Bohy, Aaron
Boker, Udi
Bollig, Benedikt
Bonnet, François
Bonsangue, Marcello
Bortolussi, Luca
Boularias, Abdeslam
Bourhis, Pierre
Boyar, Joan
Boyle, Elette
Brandes, Philipp
Brandstadt, Andreas
Braverman, Mark
Braverman, Vladimir
Brazdil, Tomas
Bringmann, Karl
Brodal, Gerth Stølting
Brody, Joshua
Bulatov, Andrei
Byrka, Jarek
Cabello, Sergio
Cachin, Christian
Carbone, Marco
Carton, Olivier
Cerny, Pavol
Cervesato, Iliano
Chakrabarty, Deeparnab
Chakraborty, Sourav
Chaloulos, Konstantinos
Chan, Siu On
Charatonik, Witold
Chase, Melissa
Chattopadhyay, Arkadev
Chávez, Edgar
Chawla, Sanjay
Chechik, Shiri
Chen, Jie
Chen, Kevin

Chen, Ning
Chen, Taolue
Chen, Xin
Chester, Sean
Chistikov, Dmitry
Chitnis, Rajesh
Chmelík, Martin
Chrobak, Marek
Cicalese, Ferdinando
Clark, Alex
Conchinha, Bruno
Cormode, Graham
Corradini, Andrea
Crescenzi, Pierluigi
Currie, James
Cygan, Marek
Czerwiński, Wojciech
Czumaj, Artur
Dal Lago, Ugo
Datta, Samir
Daum, Marcus
Davies, Rowan
Dawar, Anuj
de Gouw, Stijn
de Groote, Philippe
de Haan, Robert
de Wolf, Ronald
Dell, Holger
Deshpande, Amit
Devanur, Nikhil
Devroye, Luc
Díaz-Báñez, José-Miguel
Dinitz, Michael
Dittmann, Christoph
Doerr, Benjamin
Doerr, Carola
Dorrigiv, Reza
Dósa, György
Doty, David
Doyen, Laurent
Dregi, Markus
Drucker, Andrew
Dräger, Klaus
Ducas, Léo
Dunkelman, Orr

Ďuriš, Pavol
Dutta, Kunal
Dvořák, Zdeněk
Dziembowski, Stefan
Ebtekar, Aram
Eidenbenz, Raphael
Eikel, Martina
Eisenbrand, Friedrich
Elbassioni, Khaled
Elkin, Michael
Emek, Yuval
Ene, Alina
Englert, Matthias
Eppstein, David
Erlebach, Thomas
Escoffier, Bruno
Esmaeilsabzali, Shahram
Faenza, Yuri
Fanelli, Angelo
Faust, Sebastian
Favrholdt, Lene
Fehnker, Ansgar
Feige, Uri
Feldman, Moran
Feng, Yuan
Fenner, Stephen
Feret, Jérôme
Ferrari, Gianluigi
Fertin, Guillaume
Fijalkow, Nathanaël
Filmus, Yuval
Fineman, Jeremy
Fischer, Eldar
Fischlin, Marc
Fisher, Jasmin
Floderus, Peter
Fountoulakis, Nikolaos
Frandsen, Gudmund
Frati, Fabrizio
Friedrich, Tobias
Frieze, Alan
Friggstad, Zachary
Fu, Hongfei
Függer, Matthias
Fujioka, Kaoru

Funke, Stefan
Gagie, Travis
Gairing, Martin
Galletta, Letterio
Gamarnik, David
Ganesh, Vijay
Ganian, Robert
Garg, Jugal
Gärtner, Bernd
Gasarch, William
Gaspers, Serge
Gauwin, Olivier
Gavinsky, Dmitry
Gay, Simon
Geeraerts, Gilles
Gemulla, Rainer
Georgiadis, Giorgos
Ghorbal, Khalil
Giannopoulos, Panos
Gimbert, Hugo
Giotis, Ioannis
Gmyr, Robert
Goasdoué, François
Gogacz, Tomasz
Golas, Ulrike
Goldberg, Andrew
Goldhirsh, Yonatan
Göller, Stefan
Golovach, Petr
Goncharov, Sergey
Gopalan, Parikshit
Gorbunov, Sergey
Gorry, Thomas
Gottlieb, Lee-Ad
Gourvès, Laurent
Goyal, Navin
Graça, Daniel
Grenet, Bruno
Guo, Alan
Gupta, Anupam
Gupta, Sushmita
Gurvich, Vladimir
Gutwenger, Carsten
Habib, Michel
Hadzilacos, Vassos

Hahn, Ernst Moritz
Hähnle, Nicolai
Hajiaghayi,
  Mohammadtaghi
Halldórsson,
  Magnús M.
Hansen, Kristoffer
  Arnsfelt
Hanzlik, Lucjan
Harsha, Prahladh
Hassin, Refael
Hasuo, Ichiro
Hayman, Jonathan
He, Chaodong
He, Shan
Heggernes, Pınar
Heindel, Tobias
Hellwig, Matthias
Hirai, Yoichi
Hitchcock, John M.
Hliněný, Petr
Hoeksma, Ruben
Höfner, Peter
Hon, Wing-Kai
Horiyama, Takashi
Huang, Chien-Chung
Huber, Anna
Hüllmann, Martina
Hur, Chung-Kil
Ibsen-Jensen, Rasmus
Ilcinkas, David
Im, Sungjin
Imai, Katsunobu
Imreh, Csanád
Indyk, Piotr
Ishii, Toshimasa
Ito, Hiro
Ito, Tsuyoshi
Itoh, Toshiya
Iván, Szabolcs
Iwamoto, Chuzo
Jager, Tibor
Jain, Rahul
Jančar, Petr
Jansen, Bart

Jansen, Klaus
Jarry, Aubin
Jeż, Artur
Jeż, Łukasz
Jonsson, Bengt
Jordán, Tibor
Jurdziński, Tomasz
Jürjens, Jan
Kaibel, Volker
Kamiński, Marcin
Kammer, Frank
Kanellopoulos,
  Panagiotis
Kannan, Sampath
Kaplan, Haim
Kapralov, Michael
Karakostas, George
Karanikolas, Nikos
Karavelas, Menelaos I.
Karhumäki, Juhani
Kärkkäinen, Juha
Kartzow, Alexander
Kawahara, Jun
Kayal, Neeraj
Keller, Nathan
Keller, Orgad
Kellerer, Hans
Kemper, Stephanie
Kerenidis, Iordanis
Khot, Subhash
Kiayias, Aggelos
Kiefer, Stefan
Kik, Marcin
Kim, Eun Jung
Kissinger, Alexander
Klauck, Hartmut
Klein, Karsten
Kliemann, Lasse
Klíma, Ondřej
Klin, Bartek
Klonowski, Marek
Kluczniak, Kamil
Kniesburges, Sebastian
Kobayashi, Koji
Kobayashi, Yusuke

Nonner, Tim
Nordström, Jakob
Novotný, Petr
Nowotka, Dirk
Nutov, Zeev
O'Donnell, Ryan
Obdrzalek, Jan
Ogierman, Adrian
Okamoto, Yoshio
Okawa, Satoshi
Oliehoek, Frans
Ollinger, Nicolas
Ölveczky, Peter
Onak, Krzysztof
Ono, Hirotaka
Ostrovsky, Rafail
Otachi, Yota
Ott, Sebastian
Oualhadj, Youssouf
Oveis Gharan, Shayan
Paes Leme, Renato
Pagh, Rasmus
Paluch, Katarzyna
Pandey, Omkant
Pandurangan, Gopal
Pandya, Paritosh
Panigrahi, Debmalya
Pankratov, Denis
Panolan, Fahad
Paparas, Dimitris
Pardubská, Dana
Pascual, Fanny
Pasechnik, Dmitrii
Passmore, Grant
Paul, Christophe
Paulusma, Daniël
Peikert, Chris
Perdrix, Simon
Petig, Thomas
Pferschy, Ulrich
Phillips, Jeff
Picaronny, Claudine
Pierrakos, George
Pietrzak, Krzysztof
Piliouras, Georgios

Pilipczuk, Marcin
Pilipczuk, Michał
Pinkas, Benny
Piskac, Ruzica
Polák, Libor
Policriti, Alberto
Porat, Ely
Pottier, François
Pouly, Amaury
Prabhakar, Pavithra
Prabhakaran, Manoj M.
Pratikakis, Polyvios
Pratt-Hartmann, Ian
Price, Eric
Puglisi, Simon
Quaas, Karin
Rabehaja, Tahiry
Rabinovich, Roman
Rabinovich, Yuri
Räcke, Harald
Radzik, Tomasz
Raghunathan, Ananth
Rajaraman, Rajmohan
Raman, Venkatesh
Ramanujan, M.S.
Ranzato, Francesco
Raptopoulos,
   Christoforos
Ravi, R.
Rawitz, Dror
Raz, Ran
Razenshteyn, Ilya
Regev, Oded
Řehák, Vojtěch
Rémy, Didier
Restivo, Antonio
Rettinger, Robert
Reutenauer, Christophe
Reyzin, Lev
Richerby, David
Rigo, Michel
Röglin, Heiko
Ron, Dana
Rosén, Adi
Rotbart, Noy

Rote, Günter
Roth, Aaron
Rouselakis, Yannis
Russo, Claudio
Rusu, Irena
Sadakane, Kunihiko
Saei, Reza
Saha, Barna
Sakai, Yoshifumi
Sakurada, Hideki
Salem, Iosif
Salinger, Alejandro
Sanders, Peter
Sankowski, Piotr
Sankur, Ocan
Santhanam, Rahul
Saptharishi, Ramprasad
Saraf, Shubhangi
Sassolas, Mathieu
Satti, Srinivasa Rao
Sau, Ignasi
Sauerwald, Thomas
Sawa, Zdeněk
Sağlam, Mert
Schäfer, Andreas
Schindelhauer, Christian
Schmid, Stefan
Schneider, Johannes
Schnoebelen, Philippe
Schröder, Matthias
Schubert, Aleksy
Schumacher, André
Schwartz, Roy
Schweitzer, Pascal
Schwentick, Thomas
Scott, Elizabeth
Sebő, András
Sedgewick, Bob
Segev, Danny
Seki, Shinnosuke
Sénizergues, Géraud
Sereni, Jean-Sébastien
Serna, Maria
Seshadhri, C.
Seto, Kazuhisa

Severini, Simone
Sgall, Jiří
Shapira, Asaf
Shavit, Nir
Sherstov, Alexander
Shi, Yaoyun
Shpilka, Amir
Siddharthan, Rahul
Sidiropoulos, Anastasios
Silva, Alexandra
Silveira, Rodrigo
Silvestri, Riccardo
Simmons, Robert
Simon, Hans
Singh, Mohit
Skrzypczak, Michał
Sloane, Tony
Sly, Allan
Soltanolkotabi, Mahdi
Soto, José A.
Špalek, Robert
Spirakis, Paul
Srba, Jiří
Srinathan, Kannan
Srinivasan, Srikanth
Stapleton, Gem
Staton, Sam
Stauffer, Alexandre
Stenman, Jari
Storandt, Sabine
Strauss, Martin
Strichman, Ofer
Strothmann, Thim
Subramani, K.
Suri, Subhash
Sutner, Klaus
Svensson, Ola
Svitkina, Zoya
Szegedy, Mario
Sznajder, Nathalie
Talmage, Edward
Tamir, Tami
Tan, Li-Yang
Tanabe, Yoshinori

Tanaka, Keisuke
Tancer, Martin
Tang, Bangsheng
Tassa, Tamir
Telikepalli, Kavitha
Tentes, Aris
Terauchi, Tachio
Tessaro, Stefano
Thachuk, Chris
Thaler, Justin
Thapper, Johan
Tirthapura, Srikanta
Todinca, Ioan
Toninho, Bernardo
Tonoyan, Tigran
Torenvliet, Leen
Toruńczyk, Szymon
Torán, Jacobo
Triandopoulos, Nikos
Tudor, Valentin
Tůma, Vojtěch
Tzameret, Iddo
Tzevelekos, Nikos
Ueckerdt, Torsten
Uehara, Ryuhei
Ueno, Kenya
Valencia, Frank
Valiant, Gregory
van Dam, Wim
van Den Heuvel, Jan
van Leeuwen, Erik Jan
van Melkebeek, Dieter
van Rooij, Johan M.M.
van Stee, Rob
Varacca, Daniele
Variyam, Vinod
Vatshelle, Martin
Veanes, Margus
Végh, László
Vereshchagin, Nikolay
Vergnaud, Damien
Verschae, José
Viderman, Michael
Vidick, Thomas

Vijayaraghavan,
   Aravindan
Villanger, Yngve
Visconti, Ivan
Vishnoi, Nisheeth
Vondrák, Jan
Vrgoč, Domagoj
Vrťo, Imrich
Walukiewicz, Igor
Wan, Andrew
Wang, Haitao
Wang, Yajun
Watanabe, Osamu
Watrous, John
Watson, Thomas
Weimann, Oren
Weinstein, Omri
Wieczorek, Piotr
Wiese, Andreas
Wiesner, Karoline
Williams, Ryan
Wirth, Tony
Wojtczak, Dominik
Wollan, Paul
Wong, Prudence W.H.
Wood, David
Wootters, Mary
Worrell, James
Wulff-Nilsen, Christian
Xiao, David
Yamamoto, Mitsuharu
Yaroslavtsev, Grigory
Yehudayof, Amir
Yoshida, Yuichi
Zadimoghaddam,
   Morteza
Zawadzki, Erik
Zetzsche, Georg
Zhang, Qin
Zikas, Vassilis
Zimmermann, Martin
Živný, Stanislav
Zwick, Uri

Table of Contents – Part II

Track C – Foundations of Networked Computation

Table of Contents – Part I

Track A – Algorithms, Complexity and Games

Algorithms, Networks, and Social Phenomena

Jon Kleinberg

Cornell University
Ithaca NY USA
http://www.cs.cornell.edu/home/kleinber/

Abstract. We consider the development of computational models for systems involving social networks and large human audiences. In particular, we focus on the spread of information and behavior through such systems, and the ways in which these processes are affected by the underlying network structure.

Keywords: social networks, random graphs, contagion.

Overview

A major development over the past two decades has been the way in which networked computation has brought together people and information at a global scale. In addition to its societal consequences, this move toward massive connectivity has led to a range of new challenges for the field of computing; many of these challenges are based directly on the need for new models of computation.

We focus here on some of the modeling issues that arise in the design of computing systems involving large human audiences — these include social networking and social media sites such as Facebook, Google Plus, Twitter, and YouTube, sites supporting commerce and economic exchange such as Amazon and eBay, and sites for organizing the collective creation of knowledge such as Wikipedia. The interactions on these sites are extensively mediated by algorithms, and in thinking about the design issues that come into play, we need to think in particular about the feedback loops created by interactions among the large groups of people that populate these systems — in the ways they respond to incentives [21,23,27], form social networks [9,12,20] and share information [25].

Within this broad space of questions, we consider models for the spread of information and behavior through large social and economic networks — it has become clear that this type of person-to-person transmission is a basic "transport mechanism" for such networks [14]. Among the issues informing this investigation are recent theoretical models of such processes [4,6,8,11,13,19,26,30], as well as incentive mechanisms for propagating information [1,2,5,15,24], techniques for reconstructing the trajectory of information spreading through a network given incomplete observations [7,10,18], and empirical results indicating the importance of network structure [3,16,17,22] — and in particular network neighborhood structure [28,29] — for understanding the ways in which information will propagate at a local level.

F.V. Fomin et al. (Eds.): ICALP 2013, Part II, LNCS 7966, pp. 1–3, 2013.

References

1. Arcaute, E., Kirsch, A., Kumar, R., Liben-Nowell, D., Vassilvitskii, S.: On threshold behavior in query incentive networks. In: Proc. 8th ACM Conference on Electronic Commerce, pp. 66–74 (2007)
2. Babaioff, M., Dobzinski, S., Oren, S., Zohar, A.: On bitcoin and red balloons. In: Proc. ACM Conference on Electronic Commerce, pp. 56–73 (2012)
3. Backstrom, L., Huttenlocher, D., Kleinberg, J., Lan, X.: Group formation in large social networks: Membership, growth, and evolution. In: Proc. 12th ACM SIGKDD International Conference on Knowledge Discovery and Data Mining (2006)
4. Blume, L., Easley, D., Kleinberg, J., Kleinberg, R., Tardos, É.: Which networks are least susceptible to cascading failures? In: Proc. 52nd IEEE Symposium on Foundations of Computer Science (2011)
5. Cebrián, M., Coviello, L., Vattani, A., Voulgaris, P.: Finding red balloons with split contracts: robustness to individuals' selfishness. In: Proc. 44th ACM Symposium on Theory of Computing. pp. 775–788 (2012)
6. Centola, D., Macy, M.: Complex contagions and the weakness of long ties. American Journal of Sociology 113, 702–734 (2007)
7. Chierichetti, F., Kleinberg, J.M., Liben-Nowell, D.: Reconstructing patterns of information diffusion from incomplete observations. In: Proc. 24th Advances in Neural Information Processing Systems. pp. 792–800 (2011)
8. Dodds, P., Watts, D.: Universal behavior in a generalized model of contagion. Physical Review Letters 92, 218701 (2004)
9. Easley, D., Kleinberg, J.: Networks, Crowds, and Markets: Reasoning about a Highly Connected World. Cambridge University Press (2010)
10. Golub, B., Jackson, M.O.: Using selection bias to explain the observed structure of internet diffusions. Proc. Natl. Acad. Sci. USA 107(24), 10833–10836 (2010)
11. Granovetter, M.: Threshold models of collective behavior. American Journal of Sociology 83, 1420–1443 (1978)
12. Jackson, M.O.: Social and Economic Networks. Princeton University Press (2008)
13. Kempe, D., Kleinberg, J., Tardos, É.: Maximizing the spread of influence in a social network. In: Proc. 9th ACM SIGKDD International Conference on Knowledge Discovery and Data Mining, pp. 137–146 (2003)
14. Kleinberg, J.: Cascading behavior in networks: Algorithmic and economic issues. In: Nisan, N., Roughgarden, T., Tardos, É., Vazirani, V. (eds.) Algorithmic Game Theory, pp. 613–632. Cambridge University Press (2007)
15. Kleinberg, J., Raghavan, P.: Query incentive networks. In: Proc. 46th IEEE Symposium on Foundations of Computer Science, pp. 132–141 (2005)
16. Kossinets, G., Watts, D.: Empirical analysis of an evolving social network. Science 311, 88–90 (2006)
17. Leskovec, J., Adamic, L., Huberman, B.: The dynamics of viral marketing. ACM Transactions on the Web 1(1) (May 2007)
18. Liben-Nowell, D., Kleinberg, J.: Tracing information flow on a global scale using Internet chain-letter data. Proc. Natl. Acad. Sci. USA 105(12), 4633–4638 (2008)
19. Mossel, E., Roch, S.: On the submodularity of influence in social networks. In: Proc. 39th ACM Symposium on Theory of Computing (2007)
20. Newman, M.E.J.: Networks: An Introduction. Oxford University Press (2010)
21. Nisan, N., Roughgarden, T., Tardos, É., Vazirani, V.: Algorithmic Game Theory. Cambridge University Press (2007)

22. Onnela, J.P., Saramaki, J., Hyvonen, J., Szabo, G., Lazer, D., Kaski, K., Kertesz, J., Barabasi, A.L.: Structure and tie strengths in mobile communication networks. Proc. Natl. Acad. Sci. USA 104, 7332–7336 (2007)
23. Papadimitriou, C.H.: Algorithms, games, and the internet. In: Proc. 33rd ACM Symposium on Theory of Computing, pp. 749–753 (2001)
24. Pickard, G., Pan, W., Rahwan, I., Cebrian, M., Crane, R., Madan, A., Pentland, A.: Time-critical social mobilization. Science 334(6055), 509–512 (2011)
25. Rogers, E.: Diffusion of Innovations, 4th edn. Free Press (1995)
26. Schelling, T.: Micromotives and Macrobehavior. Norton (1978)
27. Shoham, Y., Leyton-Brown, K.: Multiagent Systems: Algorithmic, Game-Theoretic, and Logical Foundations. Cambridge University Press (2009)
28. Ugander, J., Backstrom, L., Kleinberg, J.: Subgraph frequencies: Mapping the empirical and extremal geography of large graph collections. In: Proc. 22nd International World Wide Web Conference (2013)
29. Ugander, J., Backstrom, L., Marlow, C., Kleinberg, J.: Structural diversity in social contagion. Proc. Natl. Acad. Sci. USA 109(16), 5962–5966 (2012)
30. Watts, D.J.: A simple model of global cascades on random networks. Proc. Natl. Acad. Sci. USA 99(9), 5766–5771 (2002)

Recent Advances for a Classical Scheduling Problem

Susanne Albers

Department of Computer Science, Humboldt-Universität zu Berlin
`albers@informatik.hu-berlin.de`

Abstract. We revisit classical online makespan minimization which has been studied since the 1960s. In this problem a sequence of jobs has to be scheduled on m identical machines so as to minimize the makespan of the constructed schedule. Recent research has focused on settings in which an online algorithm is given extra information or power while processing a job sequence. In this paper we review the various models of resource augmentation and survey important results.

1 Introduction

Makespan minimization on parallel machines is a fundamental and extensively studied scheduling problem with a considerable body of literature published over the last forty years. Consider a sequence $\sigma = J_1, \ldots, J_n$ of jobs that have to be scheduled non-preemptively on m identical parallel machines. Each job J_t is specified by an individual processing time p_t, $1 \le t \le n$. The goal is to minimize the makespan, i.e. the maximum completion time of any job in the constructed schedule. In the offline variant of the problem the entire job sequence σ is known in advance. In the online variant the jobs arrive one by one as elements of a list. Whenever a job J_t arrives, its processing time p_t is known. The job has to be scheduled immediately on one of the machines without knowledge of any future jobs $J_{t'}$, with $t' > t$.

Already in 1966 Graham [25] presented the elegant *List* algorithm. This strategy, which can be used for the online setting, assigns each job of σ to a machine currently having the smallest load. Graham proved that *List* is $(2 - 1/m)$-competitive. An online algorithm A is called c-competitive if, for every job sequence, the makespan of A's schedule is at most c times the makespan of an optimal schedule [35]. In 1987 Hochbaum and Shmoys [26] devised a famous polynomial time approximation scheme for the offline problem, which is NP-hard [23].

Over the past 20-years research on makespan minimization has focused on the online problem. Deterministic algorithms achieving a competitive ratio smaller than $2 - 1/m$ were developed in [2,12,21,22,28]. The best deterministic strategy currently known has a competitiveness of 1.9201 [21]. Lower bounds on the performance of deterministic algorithms were presented in [2,12,13,20,24,31,32]. The strongest result implies that no deterministic online strategy can achieve a

F.V. Fomin et al. (Eds.): ICALP 2013, Part II, LNCS 7966, pp. 4–14, 2013.
© Springer-Verlag Berlin Heidelberg 2013

competitive ratio smaller than 1.88 [31]. Hence the remaining gap between the known upper and lower bounds is quite small.

Very few results have been developed for randomized online algorithms. For $m = 2$ machines, Bartal et al. [12] presented an algorithm that attains an optimal competitive ratio of $4/3$. Currently, no randomized algorithm is known whose competitiveness is provably below the deterministic lower bound of 1.88, for all values of m. A lower bound of $e/(e-1) \approx 1.581$ on the competitive ratio of any randomized online strategy, for general m, was given in [14,34]. The ratio of $e/(e-1)$ is also the best performance guarantee that can be achieved by deterministic online algorithms if job preemption is allowed [15].

Recent research on makespan minimization has investigated scenarios where the online constraint is relaxed. More precisely, an online algorithm is given additional information or extra power in processing a job sequence σ. The study of such settings is motivated by the fact that the competitiveness of deterministic online strategies is relatively high, compared to *List*'s initial performance guarantee of $2 - 1/m$. Furthermore, with respect to the foundations of online algorithms, it is interesting to gain insight into the value of various forms of resource augmentation. Generally, in the area of scheduling the standard type of resource augmentation is *extra speed*, i.e. an online algorithm is given faster machines than an offline algorithm that constructs optimal schedules. We refer the reader to [6,27,30] and references therein for a selection of work in this direction. However, for online makespan minimization, faster processors do not give particularly interesting results. Obviously, the decrease in the algorithms' competitive ratios is inversely proportional to the increase in speed.

For online makespan minimization the following scientifically more challenging types of resource augmentation have been explored. The problem scenarios are generally well motivated from a practical point of view.

- *Known total processing time:* Consider a setting in which an online algorithm knows the sum $\sum_{t=1}^{n} p_t$ of the job processing times of σ. The access to such a piece of information can be justified as follows. In a parallel server system there usually exist fairly accurate estimates on the workload that arrives over a given time horizon. Furthermore, in a shop floor a scheduler typically accepts orders (tasks) of a targeted volume for a given time period, say a day or a week.
- *Availability of a reordering buffer:* In this setting an online algorithm has a buffer of limited size that may be used to partially reorder the job sequence. Whenever a job arrives, it is inserted into the buffer; then one job of the buffer is removed and assigned in the current schedule.
- *Job migration:* Assume that at any time an online algorithm may perform reassignments, i.e. jobs already scheduled on machines may be removed and transferred to other machines. Job migration is a well-known and widely used technique to balance load in parallel and distributed systems.

In this paper we survey the results known for these relaxed online scenarios. It turns out that usually significantly improved competitive ratios can be achieved. Unless otherwise stated, all algorithms considered in this paper are deterministic.

Throughout this paper let $M_1, \ldots, M_m$ denote the m machines. Moreover at any time the *load* of a machine is the sum of the processing times of the jobs currently assigned to that machine.

2   Known Total Processing Time

In this section we consider the scenario that an online algorithm knows the sum $S = \sum_{t=1}^n p_t$ of the job processing times, for the incoming sequence σ. The problem was first studied by Kellerer et al. [29] who concentrated on $m = 2$ machines and gave an algorithm that achieves an optimal competitive ratio of $4/3$. The setting with a general number m of machines was investigated in [5,10,16]. Angelelli et al. [10] gave a strategy that attains a competitiveness of $(1 + \sqrt{6})/2 \approx 1.725$. The best algorithm currently known was developed by Cheng et al. [16] and is 1.6-competitive. Both the algorithms by Angelelli et al. and Cheng et al. work with job classes, i.e. jobs are classified according to their processing times. For each class, specific scheduling rules apply. The algorithm by Cheng et al. [16] is quite involved, as we shall see below. A simple algorithm not resorting to job classes was presented by Albers and Hellwig [5]. However, the algorithm is only 1.75-competitive and hence does not achieve the best possible competitive ratio. We proceed to describe the 1.6-competitive algorithm by Cheng et al. [16], which we call *ALG(P)*.

Description of *ALG(P)*: The job assignment rules essentially work with small, medium and large jobs. In order to keep track of the machines containing these jobs, a slightly more refined classification is needed. W.l.o.g. assume that $\sum_{t=1}^n p_t = m$ so that 1 is a lower bound on the optimum makespan. A job J_t, $1 \leq t \leq n$, is

- tiny if $p_t \in (0, 0.3]$ • little if $p_t \in (0.3, 0.6]$ • medium if $p_t \in (0.6, 0.8]$,
  - big if $p_t \in (0.8, 0.9]$ • very big if $p_t > 0.9$.

Tiny and little jobs are also called small. Big and very big jobs are large. At any given time let ℓ_j denote the load of M_j, $1 \leq j \leq m$. Machine M_j is

- empty if $\ell_j = 0$ • little if $\ell_j \in (0.3, 0.6]$ • small if $\ell_j \in (0, 0.6]$.

A machine M_j is called medium if it only contains one medium job. Finally, M_j is said to be nearly full if contains a large as well as small jobs and $\ell_j \leq 1.1$. *ALG(P)* works in two phases.

Phase 1: The first phase proceeds as long as (1) there are empty machines and (2) twice the total number of empty and medium machines is greater than the number of little machines. Throughout the phase *ALG(P)* maintains a lower bound L on the optimum makespan. Initially, $L := 1$. During the phase, for each new job J_t, the algorithm sets $L := \max\{L, p_t\}$. Then the job is scheduled as follows.

- J_t *is very big.* Assign J_t to an M_j with the largest load $\ell_j \leq 0.6$.
- J_t *is big.* If there are more than $m/2$ empty machines, assign J_t to an empty machine. Otherwise use the scheduling rule for very big jobs.
- J_t *is medium.* Assign J_t to an empty machine.
- J_t *is small.* Execute the first possible assignment rule: (a) If there exists a small machine M_j such that $\ell_j + p_t \leq 0.6$, assign J_t to it. (b) If there is a nearly full machine M_j such that $\ell_j + p_t \leq 1.6L$, assign J_t to it. (c) If there exist more than $m/2$ empty machines, then let $\ell^* = 0.9$; otherwise let $\ell^* = 0.8$. If there exist machines M_j such that $\ell_j > \ell^*$ and $\ell_j + p_t \leq 1.6L$ then, among these machines, assign J_t to one having the largest load. (d) Assign J_t to an empty machine.

Phase 2: If at the end of Phase 1 there are no empty machines, then in Phase 2 jobs are generally scheduled according to a *Best Fit* strategy. More specifically, a job J_t is assigned to a machine M_j having the largest load ℓ_j such that $\ell_j + p_t \leq 1.6L$. Here L is updated as $L := \max\{L, p_t, 2p^*\}$, where p^* is the processing time of the $(m+1)$-st largest job seen so far. If at the end of Phase 1 there exist empty machines, then the job assignment is more involved. First $ALG(P)$ creates batches of three machines. Each batch consists of two little machines as well as one medium or one empty machine. Each batch either receives only small jobs or only medium and large jobs. At any time there exists only one open batch to receive small jobs. Similarly, there exists one open batch to receive medium and large jobs. While batches are open or can be opened, jobs are either scheduled on an empty machines or using the *Best Fit* policy. Once the batches are exhausted, jobs are assigned using *Best Fit* to the remaining machines. We refer the reader to [16] for an exact definition of the scheduling rules.

Theorem 1. [16] *ALG(P) is 1.6-competitive, for general m.*

Lower bounds on the best possible competitive ratio of deterministic strategies were given in [5,10,16]. Cheng et al. [16] showed a lower bound of 1.5. Angelelli et al. [10] gave an improved bound of 1.565, as $m \to \infty$. The best lower bound currently known was presented in [5].

Theorem 2. [5] *Let A be a deterministic online algorithm that knows the total processing time of σ. If A is c-competitive, then $c \geq 1.585$, as $m \to \infty$.*

Hence the gap between the best known upper and lower bounds is very small. Nonetheless, an interesting open problem is to determine the exact competitiveness that can be achieved in the setting where the sum of the job processing times is known.

Further results have been developed for the special case of $m = 2$ machines. Two papers by Angelelli et al. [7,8] assume that an online algorithm additionally knows an upper bound on the maximum job processing time. A setting with $m = 2$ uniform machines is addressed in [9].

Azar and Regev [11] studied a related problem. Here an online algorithm even knows the value of the optimum makespan, for the incoming job sequence. In a scheduling environment it is probably unrealistic that the value of an optimal solution is known. However, the proposed setting represents an interesting bin packing problem, which Azar and Regev [11] coined *bin stretching*. Now machines correspond to bins and jobs correspond to items. Consider a sequence σ of items that can be feasibly packed into m unit-size bins. The goal of an online algorithm is to pack σ into m bins so as to stretch the size of the bins as least as possible. Azar and Regev [11] gave a 1.625-competitive algorithm, but the algorithm $ALG(P)$ is also 1.6-competitive for bin stretching.

Corollary 1. [16] *ALG(P) is 1.6-competitive for bin stretching.*

Azar and Regev [11] showed a lower bound.

Theorem 3. [11] *Let A be a deterministic online algorithm for bin stretching. If A is c-competitive, then $c \geq 4/3$.*

An open problem is to tighten the gap between the upper and the lower bounds. Bin stretching with $m = 2$ bins was addressed by Epstein [19].

3   Reordering Buffer

Makespan minimization with a reordering buffer was proposed by Kellerer et al. [29]. The papers by Kellerer et al. [29] and Zhang [36] focus on $m = 2$ machines and present algorithms that achieve an optimal competitive ratio of $4/3$. The algorithms use a buffer of size 2. In fact the strategies work with a buffer of size 1 because in [29,36] a slightly different policy is used when a new job arrives. Upon the arrival of a job, either this job or the one residing in the buffer may be assigned to the machines. In the latter case, the new job is inserted into the buffer.

The problem with a general number m of machines was investigated by Englert et al. [18]. They developed various algorithms that use a buffer of size $O(m)$. All the algorithms consist of two phases, an *iteration phase* and a *final phase*. Let k denote the size of the buffer. In the iteration phase the first $k - 1$ jobs of σ are inserted into the buffer. Subsequently, while jobs arrive, the incoming job is placed in the buffer. Then a job with the smallest processing time is removed from the buffer and assigned in the schedule. In the final phase, the remaining $k - 1$ jobs in the buffer are scheduled.

In the following we describe the algorithm by Englert et al. [18] that achieves the smallest competitiveness. We refer to this strategy as $ALG(B)$. For the definition of the algorithm and its competitiveness we introduce a function $f_m(\alpha)$. For any machine number $m \geq 2$ and real-valued $\alpha > 1$, let

$$f_m(\alpha) = (\alpha - 1)(H_{m-1} - H_{\lceil (1-1/\alpha)m \rceil - 1}) + \lceil (1 - 1/\alpha)m \rceil \alpha/m. \qquad (1)$$

Here $H_k = \sum_{i=1}^{k} 1/i$ denotes the k-th Harmonic number, for any integer $k \geq 1$. We set $H_0 = 0$. For any fixed $m \geq 2$, let α_m be the value satisfying $f_m(\alpha) = 1$. The paper by Albers and Hellwig [3] formally shows that α_m is well-defined.

Algorithm $ALG(B)$ is exactly α_m-competitive. The sequence $(\alpha_m)_{m\geq2}$ is non-decreasing with $\alpha_2 = 4/3$ and $\lim_{m\to\infty} \alpha_m = W_{-1}(-1/e^2)/(1 + W_{-1}(-1/e^2)) \approx$ 1.4659. Here W_{-1} denotes the lower branch of the Lambert W function.

Description of $ALG(B)$: The algorithm works with a buffer of size $k = 3m$. During the iteration phase $ALG(B)$ maintains a load profile on the m machines $M_1, \ldots, M_m$. Let

$$\beta(j) = \begin{cases} (\alpha_m - 1)\frac{m}{m-j} & \text{if } j \leq \lfloor m/\alpha_m \rfloor \\ \alpha_m & \text{otherwise.} \end{cases}$$

Consider any step during the iteration phase. As mentioned above, the algorithm removes a job with the smallest processing time from the buffer. Let p denote the respective processing time. Furthermore, let L be the total load on the m machines prior to the assignment. The job is scheduled on a machine M_j with a load of at most

$$\beta(j)(L/m + p) - p.$$

Englert et al. [18] prove that such a machine always exists. In the final phase $ALG(B)$ first constructs a virtual schedule on $M_1', \ldots, M_m'$ empty machines. More specifically, the $k - 1$ jobs from the buffer are considered in non-increasing order of processing time. Each job is assigned to a machine of $M_1', \ldots, M_m'$ with the smallest current load. The process stops when the makespan of the virtual schedule is at least three times the processing time of the last job assigned. This last job is removed again from the virtual schedule. Then the machines $M_1', \ldots, M_m'$ are renumbered in order of non-increasing load. The jobs residing on M_j' are assigned to M_j in the real schedule, $1 \leq j \leq m$. In a last step each of the remaining jobs is placed on a least loaded machine in the current schedule.

Theorem 4. [18] $ALG(B)$ is α_m-competitive, for any $m \geq 2$.

Englert et al. [18] prove that the competitiveness of α_m is best possible using a buffer whose size does not depend on the job sequence σ.

Theorem 5. [18] No deterministic online algorithm can achieve a competitive ratio smaller than α_m, for any $m \geq 2$, with a buffer whose size does not depend on σ.

Englert et al. [18] also present algorithms that use a smaller buffer. In particular they give a $(1 + \alpha_m/2)$-competitive algorithm that works with a buffer of size $m + 1$. Moreover, they analyze an extended *List* algorithm that always assigns jobs to a least loaded machine. The strategy attains a competitiveness of $2 - 1/(m - k + 1)$ with a buffer of size $k \in [1, (m + 1)/2]$.

The paper by Englert et al. [18] also considers online makespan minimization on related machines and gives a $(2 + \epsilon)$-competitive algorithm, for any $\epsilon > 0$, that uses a buffer of size m. Dósa and Epstein [17] studied online makespan minimization on identical machines assuming that job preemption is allowed and showed that the competitiveness is $4/3$.

4   Job Migration

To the best of our knowledge makespan minimization with job migration was first addressed by Aggarwal et al. [1]. However the authors consider an offline setting. An algorithm is given a schedule, in which all jobs are already assigned, and a budget. The algorithm may perform job migrations up to the given budget. Aggarwal et al. [1]. design strategies that perform well with respect to the best possible solution that can be constructed with the budget. In this article we are interested in online makespan minimization with job migration. Two models of migration have been investigated. (1) An online algorithm may migrate a certain volume of jobs. (2) An online algorithm may migrate a limited number of jobs. In the next sections we address these two settings.

4.1   Migrating a Bounded Job Volume

Sanders et al. [33] studied the scenario in which, upon the arrival of a job J_t, jobs of total processing time βp_t may be migrated. Here β is called the migration factor. Sanders et al. [33] devised various algorithms. A strength of the strategies is that they are robust, i.e. after the arrival of each job the makespan of the constructed schedule is at most c times the optimum makespan, for the prefix of the job sequence seen so far. As usual c denotes the targeted competitive ratio. On the other hand, over time, the algorithms migrate a total job volume of $\beta \sum_{t=1}^{n} p_t$, which depends on the input sequence σ.

In the remainder of this section we present some of the results by Sanders et al. [33] in more detail. The authors first show a simple strategy that is $(3/2 - 1/(2m))$-competitive; the migration factor is upper bounded by 2. We describe an elegant algorithm, which we call $ALG(M1)$, that is 3/2-competitive using a migration factor of 4/3.

Description of $ALG(M1)$: Upon the arrival of a new job J_t, the algorithm checks $m+1$ options and chooses the one that minimizes the resulting makespan, where ties are broken in favor of Option 0. *Option 0*: Assign J_t to a least loaded machine. *Option j* $(1 \leq j \leq m)$: Consider machine M_j. Ignore the job with the largest processing time on M_j and take the remaining jobs on this machine in order of non-increasing processing time. Remove a job unless the total processing volume of removed jobs would exceed $4/3 \cdot p_j$. Schedule J_t on M_j. Assign the removed jobs repeatedly to a least loaded machine.

Theorem 6. [33] *ALG(M1) is 3/2-competitive, where the migration factor is upper bounded by 4/3.*

Sanders et al. [33] show that a migration factor of 4/3 is essentially best possible for 3/2-competitive algorithms. More specifically, they consider schedules S whose makespan is at most 3/2 times the optimum makespan for the set of scheduled jobs and that additionally satisfy the following property: A removal of the largest job from each machine yields a schedule S' whose makespan is upper bounded by the optimum makespan for the set of jobs sequenced in S.

Sanders et al. [33] show that there exists a schedule S such that, for any $0 < \epsilon < 4/21$, upon the arrival of a new job, a migration factor of $4/3 - \epsilon$ is necessary to achieve a competitiveness of $3/2$.

Sanders et al. [33] also present a more involved algorithm that is $4/3$-competitive using a migration factor of $5/2$. Moreover, they construct a sophisticated online approximation scheme, where the migration factor depends exponentially on $1/\epsilon$.

Theorem 7. [18] *There exists a $(1 + \epsilon)$-competitive algorithm with a migration factor of $\beta(\epsilon) \in 2^{O(1/\epsilon \cdot \log^2(1/\epsilon))}$. The running time needed to update the schedule in response to the arrival of a new job is constant.*

Finally, Sanders et al. [33] show that no constant migration factor is sufficient to maintain truly optimal schedules.

Theorem 8. [33] *Any online algorithm that maintains optimal solutions uses a migration factor of $\Omega(m)$.*

4.2 Migrating a Limited Number of Jobs

Albers and Hellwig [3] investigated the setting where an online algorithm may perform a limited number of job reassignments. These job migrations may be executed at any time while a job sequence σ is processed but, quite intuitively, the best option is to perform migrations at the end of σ. The paper [3] presents algorithms that use a total of $O(m)$ migrations, independently of σ. The best algorithm is α_m-competitive, for any $m \geq 2$, where α_m is defined as in Section 3. Recall that α_m is the value satisfying $f_m(\alpha) = 1$, where $f_m(\alpha)$ is the function specified in (1). Again, there holds $\alpha_2 = 4/3$ and $\lim_{m \to \infty} \alpha_m = W_{-1}(-1/e^2)/(1 + W_{-1}(-1/e^2)) \approx 1.4659$. The algorithm uses at most $(\lceil (2 - \alpha_m)/(\alpha_m - 1)^2 \rceil + 4)m$ job migrations. For $m \geq 11$, this expression is at most $7m$. For smaller machine numbers it is $8m$ to $10m$. In the next paragraphs we describe the algorithm, which we call $ALG(M2)$.

Description of $ALG(M2)$: The algorithm operates in two phases, a *job arrival phase* and a *job migration phase*. In the job arrival phase all jobs of σ are assigned one by one to the machines. In this phase no job migrations are performed. Once σ is scheduled, the job migration phase starts. First the algorithm removes some jobs from the machines. Then these jobs are reassigned to other machines.

Job arrival phase: Let time t be the point in time when J_t has to be scheduled, $1 \leq t \leq n$. $ALG(M2)$ classifies jobs into small and large. To this end it maintains a lower bound L_t on the optimum makespan. Let p_t^{2m+1} denote the processing time of the $(2m + 1)$-st largest job in $J_1, \ldots, J_t$, provided that $2m + 1 \leq t$. If $2m + 1 > t$, let $p_t^{2m+1} = 0$. Obviously, when t jobs have arrived, the optimum makespan cannot be smaller than the average load $\frac{1}{m} \sum_{i=1}^{t} p_i$ on the m machines and, furthermore, cannot be smaller than three times the processing time of the $(2m + 1)$-st largest job. Hence let

$$L_t = \max\{\tfrac{1}{m} \textstyle\sum_{i=1}^{t} p_i, 3p_t^{2m+1}\}.$$

A job J_i, with $i \leq t$, is *small at time t* if $p_i \leq (\alpha_m - 1)L_t$; otherwise it is *large at time t*. Finally, let p_t^* be the total processing time of the jobs of $J_1, \ldots, J_t$ that are large at time t. Define $L_t^* = p_t^*/m$.

ALG(M2) maintains a load profile on the m machines $M_1, \ldots, M_m$ as far as small jobs are concerned. The profile is identical to the one maintained by *ALG(B)* in Section 3, except that here we restrict ourselves to small jobs. Again, let

$$\beta(j) = \begin{cases} (\alpha_m - 1)\frac{m}{m-j} & \text{if } j \leq \lfloor m/\alpha_m \rfloor \\ \alpha_m & \text{otherwise.} \end{cases}$$

For any machine M_j, $1 \leq j \leq m$, let $\ell_s(j,t)$ be the load on M_j caused by the jobs that are small at time t, prior to the assignment of J_t.

For $t = 1, \ldots, n$, each J_t is scheduled as follows. If J_t is small at time t, then it is scheduled on a machine with $\ell_s(j,t) \leq \beta(j)L_t^*$. Albers and Hellwig [3] show that such a machine always exists. If J_t is large at time t, then it is assigned to a machine having the smallest load among all machines. At the end of the job arrival phase let $L = L_n$ and $L^* = L_n^*$.

Job migration phase: The phase consists of a *job removal step*, followed by a *job reassignment step*. In the removal step *ALG(M2)* maintains a set R of removed jobs. While there exists a machine M_j whose load exceeds $\max\{\beta(j)L^*, (\alpha_m - 1)L\}$, *ALG(M2)* removes a job with the largest processing time currently residing on M_j and adds the job to R. In the job reassignment step the jobs of R are considered in non-increasing order of processing time. First *ALG(M2)* constructs sets $P_1, \ldots, P_m$ consisting of at most two jobs each. For $j = 1, \ldots, m$, the j-th largest job of R is inserted into P_j provided that the job is large at time n. Furthermore, for $j = m+1, \ldots, 2m$, the j-th largest job of R is inserted into P_{2m+1-j}, provided that the job is large at time n and twice its processing time is greater than the processing time of the job already contained in P_j. These sets $P_1, \ldots, P_m$ are renumbered in order of non-increasing total processing time. Then, for $j = 1, \ldots, m$, set P_j is assigned to a least loaded machine. Finally the jobs of $R \setminus (P_1 \cup \ldots, P_m)$ are scheduled one by one on a least loaded machine.

Theorem 9. [3] *ALG(M2) is α_m-competitive and uses at most $(\lceil (2-\alpha_m)/(\alpha_m - 1)^2 \rceil + 4)m$ job migrations.*

In fact Albers and Hellwig [3] prove that at most $(\lceil (2-\alpha_m)/(\alpha_m - 1)^2 \rceil + 4$ jobs are removed from each machine in the removal step. Furthermore, Albers and Hellwig show that the competitiveness of α_m is best possible using a number of migrations that does not depend on σ.

Theorem 10. [3] *Let $m \geq 2$. No deterministic online algorithm can achieve a competitive ratio smaller than α_m if $o(n)$ job migrations are allowed.*

Finally, Albers and Hellwig [3] give a family of algorithms that uses fewer migrations and, in particular, trades performance for migrations. The family is c-competitive, for any $5/3 \leq c \leq 2$. For $c = 5/3$, the strategy uses at most $4m$ job migrations. For $c = 1.75$, at most $2.5m$ migrations are required.

5 Conclusion

In this paper we have reviewed various models of resource augmentation for online makespan minimization. The performance guarantees are tight for the scenarios with a reordering buffer and job migration. Open problems remain in the setting where the total job processing time is known and in the bin stretching problem. The paper by Kellerer et al. [29] also proposed another augmented setting in which an online algorithm is allowed to construct several candidate schedules in parallel, the best of which is finally chosen. Kellerer et al. [29] present a 4/3-competitive algorithm for $m = 2$ machines. For a general number m of machines the problem is explored in [4].

References

1. Aggarwal, G., Motwani, R., Zhu, A.: The load rebalancing problem. Journal of Algorithms 60(1), 42–59 (2006)
2. Albers, S.: Better bounds for online scheduling. SIAM Journal on Computing 29, 459–473 (1999)
3. Albers, S., Hellwig, M.: On the value of job migration in online makespan minimization. In: Epstein, L., Ferragina, P. (eds.) ESA 2012. LNCS, vol. 7501, pp. 84–95. Springer, Heidelberg (2012)
4. Albers, S., Hellwig, M.: Online makespan minimization with parallel schedules, arXiv:1304.5625 (2013)
5. Albers, S., Hellwig, M.: Semi-online scheduling revisited. Theoretical Computer Science 443, 1–9 (2012)
6. Anand, S., Garg, N., Kumar, A.: Resource augmentation for weighted flow-time explained by dual fitting. In: Proc. 23rd Annual ACM-SIAM Symposium on Discrete Algorithms, pp. 1228–1241 (2012)
7. Angelelli, E., Speranza, M.G., Tuza, Z.: Semi-on-line scheduling on two parallel processors with an upper bound on the items. Algorithmica 37, 243–262 (2003)
8. Angelelli, E., Speranza, M.G., Tuza, Z.: New bounds and algorithms for on-line scheduling: two identical processors, known sum and upper bound on the tasks. Discrete Mathematics & Theoretical Computer Science 8, 1–16 (2006)
9. Angelelli, E., Speranza, M.G., Tuza, Z.: Semi-online scheduling on two uniform processors. Theoretical Computer Science 393, 211–219 (2008)
10. Angelelli, E., Nagy, A.B., Speranza, M.G., Tuza, Z.: The on-line multiprocessor scheduling problem with known sum of the tasks. Journal of Scheduling 7, 421–428 (2004)
11. Azar, Y., Regev, O.: On-line bin-stretching. Theoretical Computer Science 268, 17–41 (2001)
12. Bartal, Y., Fiat, A., Karloff, H., Vohra, R.: New algorithms for an ancient scheduling problem. Journal of Computer and System Sciences 51, 359–366 (1995)
13. Bartal, Y., Karloff, H., Rabani, Y.: A better lower bound for on-line scheduling. Infomation Processing Letters 50, 113–116 (1994)
14. Chen, B., van Vliet, A., Woeginger, G.J.: A lower bound for randomized on-line scheduling algorithms. Information Processing Letters 51, 219–222 (1994)
15. Chen, B., van Vliet, A., Woeginger, G.J.: A optimal algorithm for preemptive online scheduling. Operations Research Letters 18, 127–131 (1995)

16. Cheng, T.C.E., Kellerer, H., Kotov, V.: Semi-on-line multiprocessor scheduling with given total processing time. Theoretical Computer Science 337, 134–146 (2005)
17. Dósa, G., Epstein, L.: Preemptive online scheduling with reordering. In: Fiat, A., Sanders, P. (eds.) ESA 2009. LNCS, vol. 5757, pp. 456–467. Springer, Heidelberg (2009)
18. Englert, M., Özmen, D., Westermann, M.: The power of reordering for online minimum makespan scheduling. In: Proc. 49th Annual IEEE Symposium on Foundations of Computer Science, pp. 603–612 (2008)
19. Epstein, L.: Bin stretching revisited. Acta Informatica 39(2), 7–117 (2003)
20. Faigle, U., Kern, W., Turan, G.: On the performance of on-line algorithms for partition problems. Acta Cybernetica 9, 107–119 (1989)
21. Fleischer, R., Wahl, M.: Online scheduling revisited. Journal of Scheduling 3, 343–353 (2000)
22. Galambos, G., Woeginger, G.: An on-line scheduling heuristic with better worst case ratio than Graham's list scheduling. SIAM Journal on Computing 22, 349–355 (1993)
23. Garay, M.R., Johnson, D.S.: Computers and Intractability: A Guide to the Theory of NP-Completeness. W.H. Freeman and Company, New York (1979)
24. Gormley, T., Reingold, N., Torng, E., Westbrook, J.: Generating adversaries for request-answer games. In: Proc. 11th ACM-SIAM Symposium on Discrete Algorithms, pp. 564–565 (2000)
25. Graham, R.L.: Bounds for certain multi-processing anomalies. Bell System Technical Journal 45, 1563–1581 (1966)
26. Hochbaum, D.S., Shmoys, D.B.: Using dual approximation algorithms for scheduling problems theoretical and practical results. Journal of the ACM 34, 144–162 (1987)
27. Kalyanasundaram, B., Pruhs, K.: Speed is as powerful as clairvoyance. Journal of the ACM 47, 617–643 (2000)
28. Karger, D.R., Phillips, S.J., Torng, E.: A better algorithm for an ancient scheduling problem. Journal of Algorithms 20, 400–430 (1996)
29. Kellerer, H., Kotov, V., Speranza, M.G., Tuza, Z.: Semi on-line algorithms for the partition problem. Operations Research Letters 21, 235–242 (1997)
30. Pruhs, K., Sgall, J., Torng, E.: Online scheduling. In: Leung, J. (ed.) Handbook of Scheduling: Algorithms, Models, and Performance Analysis, ch. 15. CRC Press (2004)
31. Rudin III, J.F.: Improved bounds for the on-line scheduling problem. Ph.D. Thesis. The University of Texas at Dallas (May 2001)
32. Rudin III, J.F., Chandrasekaran, R.: Improved bounds for the online scheduling problem. SIAM Journal on Computing 32, 717–735 (2003)
33. Sanders, P., Sivadasan, N., Skutella, M.: Online scheduling with bounded migration. Mathematics of Operations Reseach 34(2), 481–498 (2009)
34. Sgall, J.: A lower bound for randomized on-line multiprocessor scheduling. Information Processing Letters 63, 51–55 (1997)
35. Sleator, D.D., Tarjan, R.E.: Amortized efficiency of list update and paging rules. Communications of the ACM 28, 202–208 (1985)
36. Zhang, G.: A simple semi on-line algorithm for $P2//C_{max}$ with a buffer. Information Processing Letters 61, 145–148 (1997)

Formalizing and Reasoning about Quality*

Shaull Almagor[1], Udi Boker[2], and Orna Kupferman[1]

[1] The Hebrew University, Jerusalem, Israel
[2] IST Austria, Klosterneuburg, Austria

Abstract. Traditional formal methods are based on a Boolean satisfaction no-tion: a reactive system satisfies, or not, a given specification. We generalize for-mal methods to also address the *quality* of systems. As an adequate specification formalism we introduce the linear temporal logic $LTL[\mathcal{F}]$. The satisfaction value of an $LTL[\mathcal{F}]$ formula is a number between 0 and 1, describing the quality of the satisfaction. The logic generalizes traditional LTL by augmenting it with a (parameterized) set $\mathcal{F}$ of arbitrary functions over the interval $[0, 1]$. For exam-ple, $\mathcal{F}$ may contain the maximum or minimum between the satisfaction values of subformulas, their product, and their average.

The classical decision problems in formal methods, such as satisfiability, model checking, and synthesis, are generalized to search and optimization problems in the quantitative setting. For example, model checking asks for the quality in which a specification is satisfied, and synthesis returns a system satisfying the specification with the highest quality. Reasoning about quality gives rise to other natural questions, like the distance between specifications. We formalize these basic questions and study them for $LTL[\mathcal{F}]$. By extending the automata-theoretic approach for LTL to a setting that takes quality into an account, we are able to solve the above problems and show that reasoning about $LTL[\mathcal{F}]$ has roughly the same complexity as reasoning about traditional LTL.

1 Introduction

One of the main obstacles to the development of complex computerized systems lies in ensuring their correctness. Efforts in this direction include *temporal-logic model check-ing* – given a mathematical model of the system and a temporal-logic formula that specifies a desired behavior of the system, decide whether the model satisfies the for-mula, and *synthesis* – given a temporal-logic formula that specifies a desired behavior, generate a system that satisfies the specification with respect to all environments [6].

Correctness is Boolean: a system can either satisfy its specification or not satisfy it. The richness of today's systems, however, justifies specification formalisms that are *multi-valued*. The multi-valued setting arises directly in systems in which components are multi-valued (c.f., probabilistic and weighted systems) and arises indirectly in ap-plications where multi values are used in order to model missing, hidden, or varying information (c.f., abstraction, query checking, and inconsistent viewpoints). As we elab-orate below, the multi-valued setting has been an active area of research in recent years.

* This work was supported in part by the Austrian Science Fund NFN RiSE (Rigorous Systems Engineering), by the ERC Advanced Grant QUAREM (Quantitative Reactive Modeling), and the ERC Grant QUALITY. The full version is available at the authors' URLs.

F.V. Fomin et al. (Eds.): ICALP 2013, Part II, LNCS 7966, pp. 15–27, 2013.
© Springer-Verlag Berlin Heidelberg 2013

No attempts, however, have been made to augment temporal logics with a quantitative layer that would enable the specification of the relative merits of different aspects of the specification and would enable to formalize the *quality* of a reactive system. Given the growing role that temporal logic plays in planning and robotics, and the criticality of quality in these applications [16], such an augmentation is of great importance also beyond the use of temporal logic in system design and verification.

In this paper we suggest a framework for formalizing and reasoning about quality. Our working assumption is that satisfying a specification is not a yes/no matter. Different ways of satisfying a specification should induce different levels of quality, which should be reflected in the output of the verification procedure. Consider for example the specification $G(req \to Fgrant)$. There should be a difference between a computation that satisfies it with grants generated soon after requests, one that satisfies it with long waits, one that satisfies it with several grants given to a single request, one that satisfies it vacuously (with no requests), and so on. Moreover, we may want to associate different levels of importance to different components of a specification, to express their mutual influence on the quality, and to formalize the fact that we have different levels of confidence about some of them.

Quality is a rather subjective issue. Technically, we can talk about the quality of satisfaction of specifications since there are different ways to satisfy specifications. We introduce and study the linear temporal logic $LTL[\mathcal{F}]$, which extends LTL with an arbitrary set $\mathcal{F}$ of functions over $[0,1]$. Using the functions in $\mathcal{F}$, a specifier can formally and easily prioritize the different ways of satisfaction. The logic $LTL[\mathcal{F}]$ is really a family of logics, each parameterized by a set $\mathcal{F} \subseteq \{f : [0,1]^k \to [0,1] \mid k \in \mathbb{N}\}$ of functions (of arbitrary arity) over $[0,1]$. For example, $\mathcal{F}$ may contain the $\min\{x,y\}$, $\max\{x,y\}$, and $1 - x$ functions, which are the standard quantitative analogues of the $\wedge$, $\vee$, and $\neg$ operators. As we discuss below, such extensions to LTL have already been studied in the context of quantitative verification [15]. The novelty of $LTL[\mathcal{F}]$, beyond its use in the specification of quality, is the ability to manipulate values by arbitrary functions. For example, $\mathcal{F}$ may contain the quantitative operator ∇_λ, for $\lambda \in [0,1]$, that tunes down the quality of a sub-specification. Formally, the quality of the satisfaction of the specification $\nabla_\lambda \varphi$ is the multiplication of the quality of the satisfaction of φ by λ. Another useful operator is the weighted-average function $\oplus_\lambda$. There, the quality described by the formula $\varphi \oplus_\lambda \psi$ is the weighted (according to λ) average between the quality of φ and that of ψ. This enables the quality of the system to be an interpolation of different aspects of it. As an example, consider the formula $G(req \to (grant \oplus_{\frac{3}{4}} Xgrant))$. The formula specifies the fact that we want requests to be granted immediately and the grant to hold for two transactions. When this always holds, the satisfaction value is 1. We are quite okay with grants that are given immediately and last for only one transaction, in which case the satisfaction value is $\frac{3}{4}$, and less content when grants arrive with a delay, in which case the satisfaction value is $\frac{1}{4}$.

An $LTL[\mathcal{F}]$ formula maps computations to a value in $[0,1]$. We accordingly generalize classical decision problems, such as model checking, satisfiability, synthesis, and equivalence, to their quantitative analogues, which are search or optimization problems. For example, the equivalence problem between two $LTL[\mathcal{F}]$ formulas φ_1 and φ_2 seeks the supremum of the difference in the satisfaction values of φ_1 and φ_2 over all computations. Of special interest is the extension of the synthesis problem. In conventional synthesis algorithms we are given a specification to a reactive system, typically by means of an LTL formula, and we transform it into a system that is guaranteed to satisfy the specification with respect to all environments [23]. Little attention has been paid to the quality of the systems that are automatically synthesized[1]. Current efforts to address the quality challenge are based on enriching the game that corresponds to synthesis to a weighted one [2,5]. Using $LTL[\mathcal{F}]$, we are able to embody quality within the specification, which is very convenient.

In the Boolean setting, the automata-theoretic approach has proven to be very useful in reasoning about LTL specifications. The approach is based on translating LTL formulas to nondeterministic Büchi automata on infinite words [25]. In the quantitative approach, it seems natural to translate formulas to *weighted automata* [21]. However, these extensively-studied models are complicated and many problems become undecidable for them [1,17]. We show that we can use the approach taken in [15], bound the number of possible satisfaction values of $LTL[\mathcal{F}]$ formulas, and use this bound in order to translate $LTL[\mathcal{F}]$ formulas to Boolean automata. From a technical point of view, the big challenge in our setting is to maintain the simplicity and the complexity of the algorithms for LTL, even though the number of possible values is exponential. We do so by restricting attention to feasible combinations of values assigned to the different subformulas of the specification. Essentially, our translation extends the construction of [25] by associating states of the automaton with functions that map each subformula to a satisfaction value. Using the automata-theoretic approach, we solve the basic problems for $LTL[\mathcal{F}]$ within the same complexity classes as the corresponding problems in the Boolean setting (as long as the functions in $\mathcal{F}$ are computable within these complexity classes; otherwise, they become the computational bottleneck). Our approach thus enjoys the fact that traditional automata-based algorithms are susceptible to well-known optimizations and symbolic implementations. It can also be easily implemented in existing tools.

Recall that our main contribution is the ability to address the issue of quality within the specification formalism. While we describe it with respect to Boolean systems, we show in Section 5 that our contribution can be generalized to reason about weighted systems, where the values of atomic propositions are taken from $[0, 1]$. We also extend $LTL[\mathcal{F}]$ to the branching temporal logic $CTL^\star[\mathcal{F}]$, which is the analogous extension of $CTL^\star$, and show that we can still solve decision and search problems. Finally, we define a fragment, $LTL^\triangledown$, of $LTL[\mathcal{F}]$ for which the number of different satisfaction values is linear in the length of the formula, leading to even simpler algorithms.

Related Work. In recent years, the quantitative setting has been an active area of research, providing many works on quantitative logics and automata [9,10,12,18].

Conceptually, our work aims at formalizing quality, having a different focus from each of the other works. Technically, the main difference between our setting and most

---

[1] Note that we do not refer here to the challenge of generating optimal (say, in terms of state space) systems, but rather to quality measures that refer to how the specification is satisfied.

of the other approaches is the source of quantitativeness: There, it stems from the nature of the system, whereas in our setting it stems from the richness of the new functional operators. For example, in *multi-valued systems*, the values of atomic propositions are taken from a finite domain [4,18]. In *fuzzy temporal logic* [22], the atomic propositions take values in $[0, 1]$. *Probabilistic temporal logic* is interpreted over Markov decision processes [8,20], and in the context of *real-valued signals* [11], quantitativeness stems from both time intervals and predicates over the value of atomic propositions.

Closer to our approach is [7], where CTL is augmented with discounting and weighted-average operators. Thus, a formula has a rich satisfaction value, even on Boolean systems. The motivation in [7] is to suggest a logic whose semantics is not too sensitive to small perturbations in the model. Accordingly, formulas are evaluated on weighted-system (as we do in Section 5) or on Markov-chains. We, on the other hand, aim at specifying quality of on-going behaviors. Hence, we work with the much stronger LTL and CTL* logics, and we augment them by arbitrary functions over $[0, 1]$.

A different approach, orthogonal to ours, is to stay with Boolean satisfaction values, while handling quantitative properties of the system, in particular ones that are based on unbounded accumulation [3]. The main challenge in these works is the border of decidability, whereas our technical challenge is to keep the simplicity of the algorithms known for LTL in spite of the exponential number of satisfaction values. Nonetheless, an interesting future research direction is to combine the two approaches.

2   Formalizing Quality

2.1   The Temporal Logic LTL[$\mathcal{F}$]

The linear temporal logic LTL[$\mathcal{F}$] generalizes LTL by replacing the Boolean operators of LTL with arbitrary functions over $[0, 1]$. The logic is actually a family of logics, each parameterized by a set $\mathcal{F}$ of functions.

Syntax. Let AP be a set of Boolean atomic propositions, and let $\mathcal{F} \subseteq \{f : [0, 1]^k \to [0, 1] \mid k \in \mathbb{N}\}$ be a set of functions over $[0, 1]$. Note that the functions in $\mathcal{F}$ may have different arities. An LTL[$\mathcal{F}$] formula is one of the following:

- True, False, or p, for $p \in AP$.
- $f(\varphi_1, ..., \varphi_k)$, $\mathsf{X}\varphi_1$, or $\varphi_1 \mathsf{U} \varphi_2$, for LTL[$\mathcal{F}$] formulas $\varphi_1, \ldots, \varphi_k$ and a function $f \in \mathcal{F}$.

Semantics. The semantics of LTL[$\mathcal{F}$] formulas is defined with respect to (finite or infinite) computations over AP. We use $(2^{AP})^\infty$ to denote $(2^{AP})^* \cup (2^{AP})^\omega$. A *computation* is a word $\pi = \pi_0, \pi_1, \ldots \in (2^{AP})^\infty$. We use π^i to denote the suffix $\pi_i, \pi_{i+1}, \ldots$. The semantics maps a computation π and an LTL[$\mathcal{F}$] formula φ to the *satisfaction value* of φ in π, denoted $[\![\pi, \varphi]\!]$. The satisfaction value is defined inductively as described in Table 1 below.[2]

---

[2] The observant reader may be concerned by our use of max and min where sup and inf are in order. In Lemma 1 we prove that there are only finitely many satisfaction values for a formula φ, thus the semantics is well defined.

Table 1. The semantics of LTL[$\mathcal{F}$]

Formula	Satisfaction value	Formula	Satisfaction value
$[\![\pi, \texttt{True}]\!]$	1	$[\![\pi, f(\varphi_1, ..., \varphi_k)]\!]$	$f([\![\pi, \varphi_1]\!], ..., [\![\pi, \varphi_k]\!])$
$[\![\pi, \texttt{False}]\!]$	0	$[\![\pi, \mathsf{X}\varphi_1]\!]$	$[\![\pi^1, \varphi_1]\!]$
$[\![\pi, p]\!]$	$\begin{matrix} 1 \text{ if } p \in \pi_0 \\ 0 \text{ if } p \notin \pi_0 \end{matrix}$	$[\![\pi, \varphi_1 \mathsf{U} \varphi_2]\!]$	$\max\limits_{0 \le i < \|\pi\|} \{\min\{[\![\pi^i, \varphi_2]\!], \min\limits_{0 \le j < i} [\![\pi^j, \varphi_1]\!]\}\}$

It is not hard to prove, by induction on the structure of the formula, that for every computation π and formula φ, it holds that $[\![\pi, \varphi]\!] \in [0, 1]$. We use the usual $\mathsf{F}\varphi_1 = \texttt{True}\mathsf{U}\varphi_1$ and $\mathsf{G}\varphi_1 = \neg(\texttt{True}\mathsf{U}(\neg\varphi_1))$ abbreviations.

The logic LTL coincides with the logic LTL[$\mathcal{F}$] for $\mathcal{F}$ that corresponds to the usual Boolean operators. For simplicity, we use the common such functions as abbreviation, as described below. In addition, we introduce notations for some useful functions. Let $x, y \in [0, 1]$. Then,

- $\neg x = 1 - x$
- $x \vee y = \max\{x, y\}$
- $x \wedge y = \min\{x, y\}$
- $\nabla_\lambda x = \lambda \cdot x$
- $x \oplus_\lambda y = \lambda \cdot x + (1 - \lambda) \cdot y$

To see that LTL indeed coincides with LTL[$\mathcal{F}$] for $\mathcal{F} = \{\neg, \vee, \wedge\}$, note that for this $\mathcal{F}$, all formulas are mapped to $\{0, 1\}$ in a way that agrees with the semantics of LTL.

Kripke Structures and Transducers. For a Kripke structure $\mathcal{K}$ and an LTL[$\mathcal{F}$] formula φ, we have that $[\![\mathcal{K}, \varphi]\!] = \min\{[\![\pi, \varphi]\!] : \pi \text{ is a computation of } \mathcal{K}\}$. That is, the value is induced by the path that admits the lowest satisfaction value. [3]

In the setting of open systems, the set of atomic propositions is partitioned into sets I and O of input and output signals. An (I, O)-transducer then models the computations generated (deterministically) by the system when it interacts with an environment that generates finite or infinite sequences of input signals.

Example 1. Consider a scheduler that receives requests and generates grants. Consider the LTL[$\mathcal{F}$] formula $\mathsf{G}(req \to \mathsf{F}(grant \oplus_{\frac{1}{2}} \mathsf{X}grant)) \wedge \neg(\nabla_{\frac{3}{4}} \mathsf{G}\neg req)$. The satisfaction value of the formula is 1 if every request is eventually granted, and the grant lasts for two consecutive steps. If a grant holds only for a single step, then the satisfaction value is reduced to $\frac{1}{2}$. In addition, if there are no requests, then the satisfaction value is at most $\frac{1}{4}$. This shows how we can embed vacuity tests in the formula.

2.2 The Basic Questions

In the Boolean setting, an LTL formula maps computations to $\{\texttt{True}, \texttt{False}\}$. In the quantitative setting, an LTL[$\mathcal{F}$] formula maps computations to $[0, 1]$. Classical decision problems, such as model checking, satisfiability, synthesis, and equivalence, are accordingly generalized to their quantitative analogues, which are search or optimization problems. Below we specify the basic questions with respect to LTL[$\mathcal{F}$]. While the

---

[3] Since a Kripke structure may have infinitely many computations, here too we should have a-priori used inf, and the use of min is justified by Lemma 1.

definition here focuses on LTL[$\mathcal{F}$], the questions can be asked with respect to arbitrary quantitative specification formalism, with the expected adjustments.

- The *satisfiability* problem gets as input an LTL[$\mathcal{F}$] formula φ and returns $\max\{[\![\pi, \varphi]\!] : \pi$ is a computation$\}$. Dually, the *validity* problem returns, given an LTL[$\mathcal{F}$] formula φ, the value $\min\{[\![\pi, \varphi]\!] : \pi$ is a computation$\}$. [4]
- The *implication* problem gets as input two LTL[$\mathcal{F}$] formulas φ_1 and φ_2 and returns $\max\{[\![\pi, \varphi_1]\!] - [\![\pi, \varphi_2]\!] : \pi$ is a computation$\}$. The symmetric version of implication, namely the *equivalence* problem, gets as input two LTL[$\mathcal{F}$] formulas φ_1 and φ_2 and returns $\max\{|[\![\pi, \varphi_1]\!] - [\![\pi, \varphi_2]\!]| : \pi$ is a computation$\}$.
- The *model-checking* problem is extended from the Boolean setting to find, given a system $\mathcal{K}$ and an LTL[$\mathcal{F}$] formula φ, the satisfaction value $[\![\mathcal{K}, \varphi]\!]$.
- The *realizability* problem gets as input an LTL formula over $I \cup O$, for sets I and O of input and output signals, and returns $\max\{[\![\mathcal{T}, \varphi]\!] : \mathcal{T}$ is an (I, O)-transducer$\}$. The *synthesis* problem is then to find a transducer that attains this value.

Decision Problems. The above questions are search and optimization problems. It is sometimes interesting to consider the decision problems they induce, when referring to a threshold. For example, the model-checking decision-problem is to decide, given a system $\mathcal{K}$, a formula φ, and a threshold t, whether $[\![\mathcal{K}, \varphi]\!] \geq t$. For some problems, there are natural thresholds to consider. For example, in the implication problem, asking whether $\max\{[\![\pi, \varphi_1]\!] - [\![\pi, \varphi_2]\!] : \pi$ is a computation$\} \geq 0$ amounts to asking whether for all computations π, we have that $[\![\pi, \varphi_1]\!] \geq [\![\pi, \varphi_2]\!]$, which indeed captures implication.

2.3   Properties of LTL[$\mathcal{F}$]

Bounding the Number of Satisfaction Values. For an LTL[$\mathcal{F}$] formula φ, let $V(\varphi) = \{[\![\pi, \varphi]\!] : \pi \in (2^{AP})^\infty\}$. That is, $V(\varphi)$ is the set of possible satisfaction values of φ in arbitrary computations. We first show that this set is finite for all LTL[$\mathcal{F}$] formulas.

Lemma 1. *For every* LTL[$\mathcal{F}$] *formula* φ, *we have that* $|V(\varphi)| \leq 2^{|\varphi|}$.

The good news that follows from Lemma 1 is that every LTL[$\mathcal{F}$] formula has only finitely many possible satisfaction values. This enabled us to replace the sup and inf operators in the semantics by max and min. It also implies that we can point to witnesses that exhibit the satisfaction values. However, Lemma 1 only gives an exponential bound to the number of satisfaction values. We now show that this exponential bound is tight.

Example 2. Consider the logic LTL[$\{\oplus\}$], augmenting LTL with the average function, where for every $x, y \in [0, 1]$ we have that $x \oplus y = \frac{1}{2}x + \frac{1}{2}y$. Let $n \in \mathbb{N}$ and consider the formula $\varphi_n = p_1 \oplus (p_2 \oplus (p_3 \oplus (p_4 \oplus ...p_n))...)$. The length of φ_n is in $O(n)$ and the nesting depth of $\oplus$ operators in it is n. For every computation π it holds that

$$[\![\pi, \varphi_n]\!] = \frac{1}{2}[\![\pi_0, p_1]\!] + \frac{1}{4}[\![\pi_0, p_2]\!] + ... + \frac{1}{2^{n-1}}[\![\pi_0, p_{n-1}]\!] + \frac{1}{2^{n-1}}[\![\pi_0, p_n]\!].$$

---

[4] Lemma 1 guarantees that max and min (rather than sup and inf) are defined.

Hence, every assignment $\pi_0 \subseteq \{p_1, ..., p_{n-1}\}$ to the first position in π induces a different satisfaction value for $[\![\pi, \varphi_n]\!]$, implying that there are 2^{n-1} different satisfaction values for φ_n.

A Boolean Look at LTL[$\mathcal{F}$]. LTL[$\mathcal{F}$] provides means to generalize LTL to a quantitative setting. Yet, one may consider a Boolean logic defined by LTL[$\mathcal{F}$] formulas and predicates. For example, having formulas of the form $\varphi_1 \geq \varphi_2$ or $\varphi_1 \geq v$, for LTL[$\mathcal{F}$] formulas φ_1 and φ_2, and a value $v \in [0, 1]$. It is then natural to compare the expressiveness and succinctness of such a logic with respect to LTL.

One may observe that the role the functions in $\mathcal{F}$ play in LTL[$\mathcal{F}$] is propositional, in the sense that the functions do not introduce new temporal operators. We formalize this intuition in the full version, showing that for every LTL[$\mathcal{F}$] formula φ and predicate $P \subseteq [0, 1]$, there exists an LTL formula $Bool(\varphi, P)$ equivalent to the assertion $\varphi \in P$. Formally, we have the following.

Theorem 1. *For every LTL[$\mathcal{F}$] formula φ and predicate $P \subseteq [0, 1]$, there exists an LTL formula $Bool(\varphi, P)$, of length at most exponential in φ, such that for every computation π, it holds that $[\![\pi, \varphi]\!] \in P$ iff $\pi \models Bool(\varphi, P)$.*

The translation described in the proof of Theorem 1 may involve an exponential blowup. We indeed conjecture that this blowup is unavoidable, implying that LTL[$\mathcal{F}$], when used as a Boolean formalism, is exponentially more succinct than LTL. Since very little is known about lower bounds for propositional formulas, we leave it as a conjecture. We demonstrate the succinctness with the following example.

Example 3. For $k \geq 1$, let $\oplus_{\frac{1}{k}}$ be the k-ary average operator. Consider the logic LTL[$\{\oplus_{\frac{1}{k}}\}$], for an even integer k, and consider the formula $\varphi_k = \oplus_{\frac{1}{k}}(p_1, \ldots, p_k)$, for the atomic propositions $p_1, \ldots, p_k$.

For every computation π, it holds that $[\![\pi, \varphi_k]\!] = \frac{|\{i: p_i \in \pi_0\}|}{k}$. Hence, $[\![\pi, \varphi_k]\!] = \frac{1}{2}$ iff exactly half of the atomic propositions $p_1, \ldots, p_k$ hold in π_0. We conjecture that the LTL formula $Bool(\varphi_k, \frac{1}{2})$ must be exponential in k. Intuitively, the formula has to refer to every subset of size $\frac{k}{2}$. A naive implementation of this involves $\binom{k}{k/2}$ clauses, which is exponential in k. The question whether this can be done with a formula that is polynomial in k is a long-standing open problem.

3 Translating LTL[$\mathcal{F}$] to Automata

The *automata-theoretic* approach uses the theory of automata as a unifying paradigm for system specification, verification, and synthesis [24,26]. In this section we describe an automata-theoretic framework for reasoning about LTL[$\mathcal{F}$] specifications. In order to explain our framework, let us recall first the translation of LTL formulas to nondeterministic generalized Büchi automata (NGBW), as introduced in [25]. We start with the definition of NGBWs. An NGBW is $\mathcal{A} = \langle \Sigma, Q, Q_0, \delta, \alpha \rangle$, where Σ is the input alphabet, Q is a finite set of states, $Q_0 \subseteq Q$ is a set of initial states, $\delta : Q \times \Sigma \to 2^Q$ is a transition function, and $\alpha \subseteq 2^Q$ is a set of sets of accepting states. The number of sets in α is the *index* of $\mathcal{A}$. A run $r = r_0, r_1, \cdots$ of $\mathcal{A}$ on a word $w = w_1 \cdot w_2 \cdots \in \Sigma^\omega$

is an infinite sequence of states such that $r_0 \in Q_0$, and for every $i \geq 0$, we have that $r_{i+1} \in \delta(r_i, w_{i+1})$. We denote by $\inf(r)$ the set of states that r visits infinitely often, that is $\inf(r) = \{q : r_i = q$ for infinitely many $i \in \mathbb{N}\}$. The run r is *accepting* if it visits all the sets in α infinitely often. Formally, for every set $F \in \alpha$ we have that $\inf(r) \cap F \neq \emptyset$. An automaton accepts a word if it has an accepting run on it. The language of an automaton $\mathcal{A}$, denoted $L(\mathcal{A})$, is the set of words that $\mathcal{A}$ accepts.

In the Vardi-Wolper translation of LTL formulas to NGBWs [25], each state of the automaton is associated with a set of formulas, and the NGBW accepts a computation from a state q iff the computation satisfies exactly all the formulas associated with q. The state space of the NGBW contains only states associated with maximal and consistent sets of formulas, the transitions are defined so that requirements imposed by temporal formulas are satisfied, and the acceptance condition is used in order to guarantee that requirements that involve the satisfaction of eventualities are not delayed forever.

In our construction here, each state of the NGBW assigns a satisfaction value to every subformula. Consistency then assures that the satisfaction values agree with the functions in $\mathcal{F}$. Similar adjustments are made to the transitions and the acceptance condition. The construction translates an LTL[$\mathcal{F}$] formula φ to an NGBW, while setting its initial states according to a required predicate $P \subseteq [0, 1]$. We then have that for every computation $\pi \in (2^{AP})^\omega$, the resulting NGBW accepts π iff $[\![\pi, \varphi]\!] \in P$.

We note that a similar approach is taken in [15], where LTL formulas are interpreted over quantitative systems. The important difference is that the values in our construction arise from the formula and the functions it involves, whereas in [15] they are induced by the values of the atomic propositions.

Theorem 2. *Let φ be an LTL[$\mathcal{F}$] formula and $P \subseteq [0, 1]$ be a predicate. There exists an NGBW $\mathcal{A}_{\varphi, P}$ such that for every computation $\pi \in (2^{AP})^\omega$, it holds that $[\![\pi, \varphi]\!] \in P$ iff $\mathcal{A}_{\varphi, P}$ accepts π. Furthermore, $\mathcal{A}_{\varphi, P}$ has at most $2^{(|\varphi|^2)}$ states and index at most $|\varphi|$.*

Proof. We define $\mathcal{A}_{\varphi, P} = \langle 2^{AP}, Q, \delta, Q_0, \alpha \rangle$ as follows. Let $cl(\varphi)$ be the set of φ's subformulas. Let C_φ be the collection of functions $g : cl(\varphi) \to [0, 1]$ such that for all $\psi \in cl(\varphi)$, we have that $g(\psi) \in V(\psi)$. For a function $g \in C_\varphi$, we say that g is *consistent* if for every $\psi \in cl(\varphi)$, the following holds.

- If $\psi = \texttt{True}$, then $g(\psi) = 1$, and if $\psi = \texttt{False}$ then $g(\psi) = 0$.
- If $\psi = p \in AP$, then $g(\psi) \in \{0, 1\}$.
- If $\psi = f(\psi_1, \ldots, \psi_k)$, then $g(\psi) = f(g(\psi_1), \ldots, g(\psi_k))$.

The state space Q of $\mathcal{A}_{\varphi, P}$ is the set of all consistent functions in C_φ. Then, $Q_0 = \{g \in Q : g(\varphi) \in P\}$ contains all states in which the value assigned to φ is in P.

We now define the transition function δ. For functions g, g' and a letter $\sigma \in \Sigma$, we have that $g' \in \delta(g, \sigma)$ iff the following hold.

- $\sigma = \{p \in AP : g(p) = 1\}$.
- For all $\mathsf{X}\psi_1 \in cl(\varphi)$ we have $g(\mathsf{X}\psi_1) = g'(\psi_1)$.
- For all $\psi_1 \mathsf{U} \psi_2 \in cl(\varphi)$ we have $g(\psi_1 \mathsf{U} \psi_2) = \max\{g(\psi_2), \min\{g(\psi_1), g'(\psi_1 \mathsf{U} \psi_2)\}\}$.

Finally, every formula $\psi_1 \mathsf{U} \psi_2$ contributes to α the set $F_{\psi_1 \mathsf{U} \psi_2} = \{g : g(\psi_2) = g(\psi_1 \mathsf{U} \psi_2)\}$.

Remark 1. The construction described in the proof of Theorem 2 is such that selecting the set of initial states allows us to specify any (propositional) condition regarding the sub-formulas of φ. A simple extension of this idea allows us to consider a set of formulas $\{\varphi_1, ..., \varphi_m\} = \Phi$ and a predicate $P \subseteq [0,1]^m$, and to construct an NGBW that accepts a computation π iff $\langle [\![\pi, \varphi_1]\!], ..., [\![\pi, \varphi_n]\!] \rangle \in P$. Indeed, the state space of the product consists of functions that map all the formulas in Φ to their satisfaction values, and we only have to choose as the initial states these functions g for which $\langle g(\varphi_1), ..., g(\varphi_n) \rangle \in P$. As we shall see in Section 4, this allows us to use the automata-theoretic approach also in order to examine relations between the satisfaction values of different formulas.

4 Solving the Basic Questions for LTL[$\mathcal{F}$]

In this section we solve the basic questions defined in Section 2.2. We show that they all can be solved for LTL[$\mathcal{F}$] with roughly the same complexity as for LTL. When we analyze complexity, we assume that the functions in $\mathcal{F}$ can be computed in a complexity that is subsumed by the complexity of the problem for LTL (PSPACE, except for 2EXPTIME for realizability), which is very reasonable. Otherwise, computing the functions becomes the computational bottleneck. A related technical observation is that, assuming the functions in $\mathcal{F}$ can be calculated in PSPACE, we can also enumerate in PSPACE the set $V(\varphi)$ of the possible satisfaction values of an LTL[$\mathcal{F}$] formula φ.

The questions in the quantitative setting are basically search problems, asking for the best or worst value. Since every LTL[$\mathcal{F}$] formula may only have exponentially many satisfaction values, one can reduce a search problem to a set of decision problems with respect to specific thresholds, remaining in PSPACE. Combining this with the construction of NGBWs described in Theorem 2 is the key to our algorithms.

We can now describe the algorithms in detail.

Satisfiability and Validity. We start with satisfiability and solve the decision version of the problem: given φ and a threshold v, decide whether there exists a computation π such that $[\![\pi, \varphi]\!] \geq v$. The latter can be solved by checking the nonemptiness of the NGBW $\mathcal{A}_{\varphi, P}$ with $P = [v, 1]$. Since the NGBW can be constructed on-the-fly, this can be done in PSPACE in the size of $|\varphi|$. The search version can be solved in PSPACE by iterating over the set of relevant thresholds.

We proceed to validity. It is not hard to see that for all φ and v, we have that $\forall \pi, [\![\pi, \varphi]\!] \geq v$ iff $\neg(\exists \pi, [\![\pi, \varphi]\!] < v)$. The latter can be solved by checking, in PSPACE, the nonemptiness of the NGBW $\mathcal{A}_{\varphi, P}$ with $P = [0, v)$. Since PSPACE is closed under complementation, we are done. In both cases, the nonemptiness algorithm can return the witness to the nonemptiness.

Implication and Equivalence. In the Boolean setting, implication can be reduced to validity, which is in turn reduced to satisfiability. Doing the same here is more sophisticated, but possible: we add to $\mathcal{F}$ the average and negation operators. It is not hard to verify that for every computation π, it holds that $[\![\pi, \varphi_1 \oplus_{\frac{1}{2}} \neg\varphi_2]\!] = \frac{1}{2}([\![\pi, \varphi_1]\!] - [\![\pi, \varphi_2]\!]) + \frac{1}{2}$. In particular, $\max\{[\![\pi, \varphi_1]\!] - [\![\pi, \varphi_2]\!] : \pi \text{ is a computation}\} = 2 \cdot \max\{[\![\pi, \varphi_1 \oplus_{\frac{1}{2}} \neg\varphi_2]\!] : \pi \text{ is a computation}\} - 1$. Thus, the problem reduces to the satisfiability of $\varphi_1 \oplus_{\frac{1}{2}} \neg\varphi_2$, which is solvable in PSPACE. Note that, alternatively, one can proceed as suggested in Remark 1 and reason about the composition of the NGBWs for φ_1 and φ_2. The solution to the equivalence problem is similar, by checking both directions of the implication.

Model Checking. The complement of the problem, namely whether there exists a computation π of $\mathcal{K}$ such that $[\![\pi, \varphi]\!] < v$, can be solved by taking the product of the NGBW $\mathcal{A}_{\varphi,(0,v]}$ from Theorem 2 with the system $\mathcal{K}$ and checking for emptiness on-the-fly. As in the Boolean case, this can be done in PSPACE. Moreover, in case the product is not empty, the algorithm returns a witness: a computation of $\mathcal{K}$ that satisfies φ with a low quality. We note that in the case of a single computation, motivated by multi-valued monitoring [11], one can label the computation in a bottom-up manner, as in CTL model checking, and the problem can be solved in polynomial time.

Realizability and Synthesis. Several algorithms are suggested in the literature for solving the LTL realizability problem [23]. Since they are all based on a translation of specifications to automata, we can adopt them. Here we describe an adoption of the Safraless algorithm of [19] and its extension to NGBWs. Given φ and v, the algorithm starts by constructing the NGBW $\mathcal{A}_{\varphi,[0,v)}$ and dualizing it to a universal generalized co-Büchi automaton (UGCW) $\tilde{\mathcal{A}}_{\varphi,[0,v)}$. Since dualization amounts to complementation, $\tilde{\mathcal{A}}_{\varphi,[0,v)}$ accepts exactly all computations π with $[\![\pi, \varphi]\!] \geq v$. Being universal, we can expand $\tilde{\mathcal{A}}_{\varphi,[0,v)}$ to a universal tree automaton $\mathcal{U}$ that accepts a tree with directions in 2^I and labels in 2^O if all its branches, which correspond to input sequences, are labeled by output sequences such that the composition of the input and output sequences is a computation accepted by $\tilde{\mathcal{A}}_{\varphi,[0,v)}$. Realizability then amounts to checking the nonemptiness of $\mathcal{U}$ and synthesis to finding a witness to its nonemptiness. Since φ only has an exponential number of satisfaction values, we can solve the realizability and synthesis search problems by repeating this procedure for all relevant values. Since the size of $\mathcal{A}_{\varphi,[0,v)}$ is single-exponential in $|\varphi|$, the complexity is the same as in the Boolean case, namely 2EXPTIME-complete.

5   Beyond LTL[$\mathcal{F}$]

The logic LTL[$\mathcal{F}$] that we introduce and study here is a first step in our effort to introduce reasoning about quality to formal methods. Future work includes stronger formalisms and algorithms. We distinguish between extensions that stay in the area of LTL[$\mathcal{F}$] and ones that jump to the (possibly undecidable) world of infinitely many satisfaction values. In the latter, we include efforts to extend LTL[$\mathcal{F}$] by temporal operators in which the future is discounted, and efforts to combine LTL[$\mathcal{F}$] with other qualitative aspects of systems [3]. In this section we describe two extensions of the first class: an extension of LTL[$\mathcal{F}$] to weighted systems and to a branching-time temporal logic. We also describe a computationally simple fragment of LTL[$\mathcal{F}$].

Weighted Systems. A *weighted Kripke structure* is a tuple $\mathcal{K} = \langle AP, S, I, \rho, L \rangle$, where AP, S, I, and ρ are as in Boolean Kripke structures, and $L : S \to [0, 1]^{AP}$ maps each state to a weighted assignment to the atomic propositions. Thus, the value $L(s)(p)$ of an atomic proposition $p \in AP$ in a state $s \in S$ is a value in $[0, 1]$. The semantics of LTL$[\mathcal{F}]$ with respect to a weighted computation coincides with the one for non-weighted systems, except that for an atomic proposition p, we have that $[\![\pi, p]\!] = L(\pi_0)(p)$.

It is not hard to extend the construction of $\mathcal{A}_{\varphi, P}$, as described in the proof of Theorem 2, to an alphabet W^{AP}, where W is a set of possible values for the atomic propositions. Indeed, we only have to adjust the transitions so that there is a transition from state g with letter $\sigma \in W^{AP}$ only if g agrees with σ on the values of the atomic propositions. Hence, in settings where the values for the atomic propositions are known, and in particular model checking, the solutions to the basic questions is similar to the ones described for LTL$[\mathcal{F}]$ with Boolean atomic propositions.

Formalizing Quality with Branching Temporal Logics. Formulas of LTL$[\mathcal{F}]$ specify ongoing behaviors of linear computations. A Kripke structure is not linear, and the way we interpret LTL$[\mathcal{F}]$ formulas with respect to it is universal. In *branching temporal logic* one can add universal and existential quantifiers to the syntax of the logic, and specifications can refer to the branching nature of the system [13].

The branching temporal logic CTL$^\star[\mathcal{F}]$ extends LTL$[\mathcal{F}]$ by the path quantifiers E and A. Formulas of the form Eφ and Aφ are referred to as *state formulas* and they are interpreted over states s in the structure with the semantics $[\![s, \mathsf{E}\varphi]\!] = \max\{[\![\pi, \varphi]\!] : \pi$ starts in $s\}$ and $[\![s, \mathsf{A}\varphi]\!] = \min\{[\![\pi, \varphi]\!] : \pi$ starts in $s\}$.

In [14], the authors describe a general technique for extending the scope of LTL model-checking algorithms to CTL*. The idea is to repeatedly consider an innermost state subformula, view it as an (existentially or universally quantified) LTL formula, apply LTL model checking in order to evaluate it in all states, and add a fresh atomic proposition that replaces this subformula and holds in exactly these states that satisfy it. This idea, together with our ability to model check systems with weighted atomic propositions, can be used also for model checking CTL$^\star[\mathcal{F}]$.

More challenging is the handling of the other basic problems. There, the solution involves a translation of CTL$^\star[\mathcal{F}]$ formulas to tree automata. Since the automata-theoretic approach for CTL* has the Vardi-Wolper construction at its heart, this is possible.

The Fragment LTL$^\triangledown$ *of* LTL$[\mathcal{F}]$. In the proof of Lemma 1, we have seen that a formula may take exponentially many satisfaction values. The proof crucially relies on the fact that the value of a function is a function of all its inputs. However, in the case of unary functions, or indeed functions that do not take many possible values, this bound can be lowered. Such an interesting fragment is the logic LTL$^\triangledown = \text{LTL}[\{\triangledown_\lambda, \blacktriangledown_\lambda\}_{\lambda \in [0,1]} \cup \{\vee, \neg\}]$, with the functions $\triangledown_\lambda(x) = \lambda \cdot x$ and $\blacktriangledown_\lambda(x) = \lambda \cdot x + (1 - \lambda)/2$.

This fragment is interesting in two aspects. First, computationally, an LTL$^\triangledown$ formula has only polynomially many satisfaction values. Moreover, for a predicate of the form $P = [v, 1]$ (resp. $P = (v, 1]$), the LTL formula $Bool(\varphi, P)$ can be shown to be of linear length in $|\varphi|$. This implies that solving threshold-problems for LTL$^\triangledown$ formulas can be done with tools that work with LTL with no additional complexity. Second, philosophically, an interesting question that arises when formalizing quality regards how the *lack* of quality in a component should be viewed. With quality between 0 and 1, we have that 1 stands for "good", 0 for "bad", and $\frac{1}{2}$ for "not good and not bad". While the $\triangledown_\lambda$ operator enables us to reduce the quality towards "badness", the $\blacktriangledown_\lambda$ operator enables us to do so towards "ambivalence".

References

1. Almagor, S., Boker, U., Kupferman, O.: What's decidable about weighted automata? In: Bultan, T., Hsiung, P.-A. (eds.) ATVA 2011. LNCS, vol. 6996, pp. 482–491. Springer, Heidelberg (2011)
2. Bloem, R., Chatterjee, K., Henzinger, T.A., Jobstmann, B.: Better Quality in Synthesis through Quantitative Objectives. In: Bouajjani, A., Maler, O. (eds.) CAV 2009. LNCS, vol. 5643, pp. 140–156. Springer, Heidelberg (2009)
3. Boker, U., Chatterjee, K., Henzinger, T.A., Kupferman, O.: Temporal specifications with accumulative values. In: 26th LICS, pp. 43–52 (2011)
4. Bruns, G., Godefroid, P.: Model checking with multi-valued logics. In: Díaz, J., Karhumäki, J., Lepistö, A., Sannella, D. (eds.) ICALP 2004. LNCS, vol. 3142, pp. 281–293. Springer, Heidelberg (2004)
5. Černý, P., Chatterjee, K., Henzinger, T.A., Radhakrishna, A., Singh, R.: Quantitative Synthesis for Concurrent Programs. In: Gopalakrishnan, G., Qadeer, S. (eds.) CAV 2011. LNCS, vol. 6806, pp. 243–259. Springer, Heidelberg (2011)
6. Clarke, E., Henzinger, T.A., Veith, H.: Handbook of Model Checking. Elsvier (2013)
7. de Alfaro, L., Faella, M., Henzinger, T.A., Majumdar, R., Stoelinga, M.: Model checking discounted temporal properties. TCS 345(1), 139–170 (2005)
8. Desharnais, J., Gupta, V., Jagadeesan, R., Panangaden, P.: Metrics for labelled markov processes. TCS 318(3), 323–354 (2004)
9. Droste, M., Kuich, W., Rahonis, G.: Multi-valued MSO logics over words and trees. Fundamenta Informaticae 84(3-4), 305–327 (2008)
10. Droste, M., Rahonis, G.: Weighted automata and weighted logics with discounting. TCS 410(37), 3481–3494 (2009)
11. Donzé, A., Maler, O., Bartocci, E., Nickovic, D., Grosu, R., Smolka, S.: On Temporal Logic and Signal Processing. In: Chakraborty, S., Mukund, M. (eds.) ATVA 2012. LNCS, vol. 7561, pp. 92–106. Springer, Heidelberg (2012)
12. Droste, M., Werner, K., Heiko, V.: Handbook of Weighted Automata. Springer (2009)
13. Emerson, E.A., Halpern, J.Y.: Sometimes and not never revisited: On branching versus linear time. Journal of the ACM 33(1), 151–178 (1986)
14. Emerson, E.A., Lei, C.L.: Modalities for model checking: Branching time logic strikes back. In: Proc. 12th POPL, pp. 84–96 (1985)
15. Faella, M., Legay, A., Stoelinga, M.: Model Checking Quantitative Linear Time Logic. TCS 220(3), 61–77 (2008)
16. Kress-Gazit, H., Fainekos, G.E., Pappas, G.J.: Temporal-Logic-Based Reactive Mission and Motion Planning. IEEE Trans. on Robotics 25(6), 1370–1381 (2009)
17. Krob, D.: The equality problem for rational series with multiplicities in the tropical semiring is undecidable. International Journal of Algebra and Computation 4(3), 405–425 (1994)
18. Kupferman, O., Lustig, Y.: Lattice automata. In: Cook, B., Podelski, A. (eds.) VMCAI 2007. LNCS, vol. 4349, pp. 199–213. Springer, Heidelberg (2007)
19. Kupferman, O., Vardi, M.Y.: Safraless decision procedures. In: Proc. 46th FOCS, pp. 531–540 (2005)

20. Kwiatkowska, M.Z.: Quantitative verification: models techniques and tools. In: FSE, pp. 449–458 (2007)
21. Mohri, M.: Finite-state transducers in language and speech processing. Computational Linguistics 23(2), 269–311 (1997)
22. Moon, S., Lee, K.H., Lee, D.: Fuzzy branching temporal logic. IEEE Transactions on Systems, Man, and Cybernetics, Part B 34(2), 1045–1055 (2004)
23. Pnueli, A., Rosner, R.: On the synthesis of a reactive module. In: Proc.16th POPL, pp. 179–190 (1989)
24. Thomas, W.: Automata on infinite objects. In: Handbook of Theoretical Computer Science, pp. 133–191 (1990)
25. Vardi, M.Y., Wolper, P.: An automata-theoretic approach to automatic program verification. In: Proc. 1st LICS, pp. 332–344 (1986)
26. Vardi, M.Y., Wolper, P.: Reasoning about infinite computations. I&C 115(1), 1–37 (1994)

The Square Root Phenomenon in Planar Graphs

Dániel Marx*

Computer and Automation Research Institute,
Hungarian Academy of Sciences (MTA SZTAKI)
Budapest, Hungary
dmarx@cs.bme.hu

Abstract. Most of the classical NP-hard problems remain NP-hard when restricted to planar graphs, and only exponential-time algorithms are known for the exact solution of these planar problems. However, in many cases, the exponential-time algorithms on planar graphs are significantly faster than the algorithms for general graphs: for example, 3-COLORING can be solved in time $2^{O(\sqrt{n})}$ in an n-vertex planar graph, whereas only $2^{O(n)}$-time algorithms are known for general graphs. For various planar problems, we often see a square root appearing in the running time of the best algorithms, e.g., the running time is often of the form $2^{O(\sqrt{n})}$, $n^{O(\sqrt{k})}$, or $2^{O(\sqrt{k})} \cdot n$. By now, we have a good understanding of why this square root appears. On the algorithmic side, most of these algorithms rely on the notion of treewidth and its relation to grid minors in planar graphs (but sometimes this connection is not obvious and takes some work to exploit). On the lower bound side, under a complexity assumption called Exponential Time Hypothesis (ETH), we can show that these algorithms are essentially best possible, and therefore the square root has to appear in the running time.

* Research supported by the European Research Council (ERC) grant 280152.

F.V. Fomin et al. (Eds.): ICALP 2013, Part II, LNCS 7966, p. 28, 2013.

A Guided Tour in Random Intersection Graphs[*]

Paul G. Spirakis[1,2], Sotiris Nikoletseas[1], and Christoforos Raptopoulos[1]

[1] Computer Technology Institute and Press "Diophantus" and
University of Patras, Greece
[2] Computer Science Department, University of Liverpool, United Kingdom
{spirakis,nikole}@cti.gr, raptopox@ceid.upatras.gr

1 Introduction and Motivation

Random graphs, introduced by P. Erdős and A. Rényi in 1959, still attract a huge amount of research in the communities of Theoretical Computer Science, Algorithms, Graph Theory, Discrete Mathematics and Statistical Physics. This continuing interest is due to the fact that, besides their mathematical beauty, such graphs are very important, since they can model interactions and faults in networks and also serve as typical inputs for an average case analysis of algorithms. The modeling effort concerning random graphs has to show a plethora of random graph models; some of them have quite elaborate definitions and are quite general, in the sense that they can simulate many other known distributions on graphs by carefully tuning their parameters.

In this tour, we consider a simple, yet general family of models, namely *Random Intersection Graphs (RIGs)*. In such models there is a universe $\mathcal{M}$ of *labels* and each one of n vertices selects a random subset of $\mathcal{M}$. Two vertices are connected if and only if their corresponding subsets of labels intersect.

Random intersection graphs may model several real-life applications quite accurately. In fact, there are practical situations where each communication agent (e.g. a wireless node) gets access only to some ports (statistically) out of a possible set of communication ports. When another agent also selects a communication port, then a communication link is implicitly established and this gives rise to communication graphs that look like random intersection graphs. Furthermore, random intersection graphs are relevant to and capture quite nicely social networking. Indeed, a social network is a structure made of nodes tied by one or more specific types of interdependency, such as values, visions, financial exchange, friends, conflicts, web links etc. Other applications may include oblivious resource sharing in a distributed setting, interactions of mobile agents traversing the web, social networking etc. Even epidemiological phenomena (like spread of disease between individuals with common characteristics in a population) tend to be more accurately captured by this "proximity-sensitive" family of random graphs.

---

[*] This research was partially supported by the EU IP Project MULTIPLEX contract number 317532.

F.V. Fomin et al. (Eds.): ICALP 2013, Part II, LNCS 7966, pp. 29–35, 2013.

1.1   A More Formal, First Acquaintance with RIGs

Random intersection graphs were introduced by M. Karoński, E.R. Sheinerman and K.B. Singer-Cohen [7] and K.B. Singer-Cohen [18]. The formal definition of the model is given below:

Definition 1 (Uniform Random Intersection Graph - $\mathcal{G}_{n,m,p}$ [7, 18]).
Consider a universe $\mathcal{M} = \{1, 2, \ldots, m\}$ *of labels and a set of n vertices V. Assign independently to each vertex* $v \in V$ *a subset S_v of $\mathcal{M}$, choosing each element* $i \in \mathcal{M}$ *independently with probability p and draw an edge between two vertices* $v \neq u$ *if and only if* $S_v \cap S_u \neq \emptyset$. *The resulting graph is an instance* $G_{n,m,p}$ *of the uniform random intersection graphs model.*

In this model we also denote by L_i the set of vertices that have chosen label $i \in M$. Given $G_{n,m,p}$, we will refer to $\{L_i, i \in \mathcal{M}\}$ as its *label representation*. It is often convenient to view the label representation as a bipartite graph with vertex set $V \cup \mathcal{M}$ and edge set $\{(v, i) : i \in S_v\} = \{(v, i) : v \in L_i\}$. We refer to this graph as the *bipartite random graph $B_{n,m,p}$ associated to $G_{n,m,p}$*. Notice that the associated bipartite graph is uniquely defined by the label representation.

It follows from the definition of the model the (unconditioned) probability that a specific edge exists is $1 - (1 - p^2)^m$. Therefore, if mp^2 goes to infinity with n, then this probability goes to 1. We can thus restrict the range of the parameters to the "interesting" range where $mp^2 = O(1)$ (i.e. the range of values for which the unconditioned probability that an edge exists does not go to 1). Furthermore, as is usual in the literature, we assume that the number of labels is some power of the number of vertices, i.e. $m = n^\alpha$, for some $\alpha > 0$.

It is worth mentioning that the edges in $G_{n,m,p}$ are not independent. In particular, there is a strictly positive dependence between the existence of two edges that share an endpoint (i.e. $\Pr(\exists\{u, v\}|\exists\{u, w\}) > \Pr(\exists\{u, v\}))$. This dependence is stronger the smaller the number of labels $\mathcal{M}$ includes, while it seems to fade away as the number of labels increases. In fact, by using a coupling technique, the authors in [4] prove the equivalence (measured in terms of total variation distance) of uniform random intersection graphs and Erdős-Rényi random graphs, when $m = n^\alpha, \alpha > 6$. This bound on the number of labels was improved in [16], by showing equivalence of sharp threshold functions among the two models for $\alpha \geq 3$. These results show that random intersection graphs are quite general and that known techniques for random graphs can be used in the analysis of uniform random intersection graphs with a large number of labels.

The similarity between uniform random intersection graphs and Erdős-Rényi random graphs vanishes as the number of labels m decreases below the number of vertices n (i.e. $m = n^\alpha$, for $\alpha \leq 1$). This dichotomy was initially pointed out in [18], through the investigation of connectivity of $G_{n,m,p}$. In particular, it was proved that the connectivity threshold for $\alpha > 1$ is $\sqrt{\frac{\ln n}{nm}}$, but it is $\frac{\ln n}{m}$ (i.e. quite larger) for $\alpha \leq 1$. Therefore, the mean number of edges just above connectivity is approximately $\frac{1}{2}n \ln n$ in the first case (which is equal to the mean number of edges just above the connectivity threshold for Erdős-Rényi random graphs), but it is larger by at least a factor of $\ln n$ in the second case. Other dichotomy results of similar flavor were pointed out in the investigation of the (unconditioned) vertex degree distribution by D. Stark [19], through the analysis of a suitable generating function, and in the investigation of the distribution of the number of isolated vertices by Y. Shang [17].

In this invited talk, we will focus on research related to both combinatorial and algorithmic properties of uniform random intersection graphs, but also of other, related models that are included in the family of random intersection graphs. In particular, we note that by selecting the label set of each vertex using a different distribution, we get random intersection graphs models whose statistical behavior can vary considerably from that of $G_{n,m,p}$. Two of these models are the following: (a) In the **General Random Intersection Graphs Model** $G_{n,m,p}$ [11], where $p = [p_1, p_2, \ldots, p_m]$, the label set S_v of a vertex v is formed by choosing independently each label i with probability p_i. (b) In the **Regular Random Intersection Graphs Model** $G_{n,m,\lambda}$ [6], where $\lambda \in \mathbb{N}$, the label set of a vertex is chosen independently, uniformly at random for the set of all subsets of $\mathcal{M}$ of cardinality λ.

2   Selected Combinatorial Problems

Below we provide a brief presentation of the main results on the topic obtained by our team. We also give a general description of the techniques used; some of these techniques highlight and take advantage of the intricacies and special structure of random intersection graphs, while others are adapted from the field of Erdős-Rényi random graphs.

2.1   Independent Sets

The problem of the existence and efficient construction of large independent sets in general random intersection graphs is considered in [11]. Concerning existence, exact formulae are derived for the expectation and variance of the number of independent sets of any size, by using a *vertex contraction technique*. This technique involves the characterization of the statistical behavior of an independent set of any size and highlights an *asymmetry* in the edge appearance rule of random intersection graphs. In particular, it is shown that the probability that any fixed label i is chosen by some vertex in a k-size S with no edges is exactly $\frac{kp_i}{1+(k-1)p_i}$. On the other hand, there is no closed formula for the respective probability when there is at least one edge between the k vertices (or even when the set S is complete)! The special structure of random intersection graphs is also used in the design of efficient algorithms for constructing quite large independent sets in uniform random intersection graphs. By analysis, it is proved that the approximation guarantees of algorithms using the label representation of random intersection graphs are superior to that of well known greedy algorithms for independent sets when applied to instances of $\mathcal{G}_{n,m,p}$.

2.2   Hamilton Cycles

In [15], the authors investigate the existence and efficient construction of *Hamilton cycles* in uniform random intersection graphs. In particular, for the case $m = n^\alpha, \alpha > 1$ the authors first prove a general result that allows one to apply (with the same probability of success) any algorithm that finds a Hamilton cycle with high probability in a $G_{n,M}$ random graph (i.e. a graph chosen equiprobably form the space of all graphs with M edges). The proof is done by using a simple coupling argument. A more complex coupling was given in [3], resulting in a more accurate characterization of the threshold function for Hamiltonicity in $G_{n,m,p}$ for the whole range of values of α. From an algorithmic perspective, the authors in [15] provide an expected polynomial time algorithm for the case where $m = O\left(\sqrt{\frac{n}{\ln n}}\right)$ and p is constant. For the more general case where $m = o\left(\frac{n}{\ln n}\right)$ they propose a *label exposure* greedy algorithm that succeeds in finding a Hamilton cycle in $G_{n,m,p}$ with high probability, even when the probability of label selection is just above the connectivity threshold.

2.3   Coloring

In [9], the problem of coloring the vertices of $G_{n,m,p}$ is investigated (see also [1]). For the case where the number of labels is less than the number of vertices and $mp \geq \ln^2 n$ (i.e. a factor $\ln n$ above the connectivity threshold of uniform random intersection graphs), a polynomial time algorithm is proposed for finding a *proper coloring* $G_{n,m,p}$. The algorithm is greedy-like and it is proved that it takes $O\left(\frac{n^2 mp^2}{\ln n}\right)$ time, while using $\Theta\left(\frac{nmp^2}{\ln n}\right)$ different colors. Furthermore, by using a one sided coupling to the regular random intersection graphs model $G_{n,m,\lambda}$ with $\lambda \sim mp$, and using an upper bound on its independence number from [13], it is shown that the number of colors used by the proposed algorithm is optimal up to constant factors.

To complement this result, the authors in [9] prove that when $mp < \beta \ln n$, for some small constant β, only np colors are needed in order to color $n - o(n)$ vertices of $G_{n,m,p}$ whp. This means that even for quite dense instances, using the same number of colors as those needed to properly color the clique induced by any label suffices to color almost all of the vertices of $G_{n,m,p}$. For the proof, the authors explore a combination of ideas from [5] and [8]. In particular, a martingale $\{X_t\}_{t \geq 0}$ is defined, so that X_n is equal to the maximum subset of vertices that can be properly colored using a predefined number of colors k. Then, by providing an appropriate lower bound on the probability that there is a sufficiently large subset of vertices that can be split in k independent sets of roughly the same size, and then using Azuma's Inequality for martingales, the authors provide a lower bound on $E[X_n]$ and also show that the actual value X_n is highly concentrated around its mean value.

Finally, due to the similarities that the $\mathcal{G}_{n,m,p}$ model has to the process of generating random hypergraphs, [9] includes a comparison of the problem of finding a proper coloring for $G_{n,m,p}$ to that of coloring hypergraphs so that no edge is monochromatic. In contrast to the first problem, it is proved that only two colors suffice for the second problem. Furthermore, by using the *method of conditional expectations* (see [14]) an algorithm can be derived that finds the desired coloring in polynomial time.

2.4   Maximum Cliques

In [12], the authors consider maximum cliques in the uniform random intersection graphs model $\mathcal{G}_{n,m,p}$. It is proved that, when the number of labels is not too large, we can use the label choices of the vertices to find a maximum clique in polynomial time (in the number of labels m and vertices n of the graph). Most of the analytical work in the paper is devoted in proving the *Single Label Clique Theorem*. Its proof includes a coupling to a graph model where edges appear independently and in which we can bound the size of the maximum clique by well known probabilistic techniques. The theorem states that when the number of labels is less than the number of vertices, any large enough clique in a random instance of $\mathcal{G}_{n,m,p}$ is formed by a single label. This statement may seem obvious when p is small, but it is hard to imagine that it still holds for *all* "interesting" values for p. Indeed, when $p = o\left(\sqrt{\frac{1}{nm}}\right)$, by slightly modifying an argument of [1], one can see that $G_{n,m,p}$ almost surely has no cycle of size $k \geq 3$ whose edges are formed by k distinct labels (alternatively, the intersection graph produced by reversing the roles of labels and vertices is a tree). On the other hand, for larger p a random instance of $\mathcal{G}_{n,m,p}$ is far from perfect[1] and the techniques of [1] do not apply. By using the Single Label Clique Theorem, a tight bound on the clique number of $G_{n,m,p}$ is proved, in the case where $m = n^{\alpha}, \alpha < 1$. A lower bound in the special case where mp^2 is constant, was given in [18]. We considerably broaden this range of values to also include vanishing values for mp^2 and also provide an asymptotically tight upper bound.

Finally, as yet another consequence of the Single Label Clique Theorem, the authors in [12] prove that the problem of inferring the complete information of label choices for each vertex from the resulting random intersection graph is *solvable* whp; namely, the maximum likelihood estimation method will provide a unique solution (up to permutations of the labels).[2] In particular, given values m, n and p, such that $m = n^{\alpha}, 0 < \alpha < 1$, and given a random instance of the $\mathcal{G}_{n,m,p}$ model, the label choices for each vertex are uniquely defined.

2.5   Expansion and Random Walks

The edge expansion and the cover time of uniform random intersection graphs is investigated in [10]. In particular, by using first moment arguments, the authors

---

[1] A *perfect graph* is a graph in which the chromatic number of every induced subgraph equals the size of the largest clique of that subgraph. Consequently, the clique number of a perfect graph is equal to its chromatic number.

[2] More precisely, if $\mathcal{B}$ is the set of different label choices that can give rise to a graph G, then the problem of inferring the complete information of label choices from G is *solvable* if there is some $B^* \in \mathcal{B}$ such that $\Pr(B^*|G) > \Pr(B|G)$, for all $\mathcal{B} \ni B \neq B^*$.

first prove that $G_{n,m,p}$ is an expander whp when the number of labels is less than the number of vertices, even when p is just above the connectivity threshold (i.e. $p = (1 + o(1))\tau_c$, where τ_c is the connectivity threshold). Second, the authors show that random walks on the vertices of random intersection graphs are whp *rapidly mixing* (in particular, the mixing time is logarithmic on n). The proof is based on upper bounding the second eigenvalue of the random walk on $G_{n,m,p}$ through coupling of the original Markov Chain describing the random walk to another Markov Chain on an associated random bipartite graph whose conductance properties are appropriate. Finally, the authors prove that the *cover time* of the random walk on $G_{n,m,p}$, when $m = n^\alpha, \alpha < 1$ and p is at least 5 times the connectivity threshold is $\Theta(n \log n)$, which is optimal up to a constant. The proof is based on a general theorem of Cooper and Frieze [2]; the authors prove that the degree and spectrum requirements of the theorem hold whp in the case of uniform random intersection graphs. The authors also claim that their proof also carries over to the case of smaller values for p, but the technical difficulty for proving the degree requirements of the theorem of [2] increases.

3   Epilogue

We discussed here recent progress on the Random Intersection Graphs (RIGs) Model. The topic is still new and many more properties await to be discovered especially for the General (non-Uniform) version of RIGs. Such graphs (and other new graph classes) are motivated by modern technology, and thus, some combinatorial results and algorithmic properties may become useful in order to understand and exploit emerging networks nowadays.

References

1. Behrisch, M., Taraz, A., Ueckerdt, M.: Coloring random intersection graphs and complex networks. SIAM J. Discrete Math. 23, 288–299 (2008)
2. Cooper, C., Frieze, A.: The Cover Time of Sparse Random Graphs. In: Random Structures and Algorithms, vol. 30, pp. 1–16. John Wiley & Sons, Inc. (2007)
3. Efthymiou, C., Spirakis, P.G.: Sharp thresholds for Hamiltonicity in random intersection graphs. Theor. Comput. Sci. 411(40-42), 3714–3730 (2010)
4. Fill, J.A., Sheinerman, E.R., Singer-Cohen, K.B.: Random intersection graphs when $m = \omega(n)$: an equivalence theorem relating the evolution of the $G(n, m, p)$ and $G(n, p)$ models. Random Struct. Algorithms 16(2), 156–176 (2000)
5. Frieze, A.: On the Independence Number of Random Graphs. Disc. Math. 81, 171–175 (1990)
6. Godehardt, E., Jaworski, J.: Two models of Random Intersection Graphs for Classification. In: Opitz, O., Schwaiger, M. (eds.). Studies in Classification, Data Analysis and Knowledge Organisation, pp. 67–82. Springer, Heidelberg (2002)
7. Karoński, M., Sheinerman, E.R., Singer-Cohen, K.B.: On Random Intersection Graphs: The Subgraph Problem. Combinatorics, Probability and Computing Journal 8, 131–159 (1999)
8. Luczak, T.: The chromatic number of random graphs. Combinatorica 11(1), 45–54 (2005)

9. Nikoletseas, S., Raptopoulos, C., Spirakis, P.G.: Coloring Non-sparse Random Intersection Graphs. In: Královič, R., Niwiński, D. (eds.) MFCS 2009. LNCS, vol. 5734, pp. 600–611. Springer, Heidelberg (2009)

10. Nikoletseas, S., Raptopoulos, C., Spirakis, P.G.: Expander properties and the cover time of random intersection graphs. Theor. Comput. Sci. 410(50), 5261–5272 (2009)

11. Nikoletseas, S., Raptopoulos, C., Spirakis, P.G.: Large independent sets in general random intersection graphs. Theor. Comput. Sci. 406, 215–224 (2008)

12. Nikoletseas, S., Raptopoulos, C., Spirakis, P.G.: Maximum Cliques in Graphs with Small Intersection Number and Random Intersection Graphs. In: Rovan, B., Sassone, V., Widmayer, P. (eds.) MFCS 2012. LNCS, vol. 7464, pp. 728–739. Springer, Heidelberg (2012)

13. Nikoletseas, S., Raptopoulos, C., Spirakis, P.G.: On the Independence Number and Hamiltonicity of Uniform Random Intersection Graphs. Theor. Comput. Sci. 412(48), 6750–6760 (2011)

14. Molloy, M., Reed, B.: Graph Colouring and the Probabilistic Method. Springer (2002)

15. Raptopoulos, C., Spirakis, P.G.: Simple and Efficient Greedy Algorithms for Hamilton Cycles in Random Intersection Graphs. In: Deng, X., Du, D.-Z. (eds.) ISAAC 2005. LNCS, vol. 3827, pp. 493–504. Springer, Heidelberg (2005)

16. Rybarczyk, K.: Equivalence of a random intersection graph and $G(n, p)$. Random Structures and Algorithms 38(1-2), 205–234 (2011)

17. Shang, Y.: On the Isolated Vertices and Connectivity in Random Intersection Graphs. International Journal of Combinatorics 2011, Article ID 872703 (2011), doi:10.1155/2011/872703

18. Singer-Cohen, K.B.: Random Intersection Graphs. PhD thesis, John Hopkins University (1995)

19. Stark, D.: The vertex degree distribution of random intersection graphs. Random Structures & Algorithms 24(3), 249–258 (2004)

To Be Uncertain Is Uncomfortable,
But to Be Certain Is Ridiculous*

Peter Widmayer

Institute of Theoretical Computer Science, ETH Zürich, Switzerland
widmayer@inf.ethz.ch

Traditionally, combinatorial optimization postulates that an input instance is given with absolute precision and certainty, and it aims at finding an optimum solution for the given instance. In contrast, real world input data are often uncertain, noisy, inaccurate. As a consequence, an optimum solution for a real world instance may not be meaningful or desired. While this unfortunate gap between theory and reality has been recognized for quite some time, it is far from understood, let alone resolved. We advocate to devote more attention to it, in order to develop algorithms that find meaningful solutions for uncertain inputs. We propose an approach towards this goal, and we show that this approach on the one hand creates a wealth of algorithmic problems, while on the other hand it appears to lead to good real world solutions.

This talk is about joint work with Joachim Buhmann, Matus Mihalak, and Rasto Sramek.

---

* Chinese proverb, sometimes also attributed to Goethe.

F.V. Fomin et al. (Eds.): ICALP 2013, Part II, LNCS 7966, p. 36, 2013.
© Springer-Verlag Berlin Heidelberg 2013

Decision Problems for Additive Regular Functions[*]

Rajeev Alur and Mukund Raghothaman

University of Pennsylvania
{alur,rmukund}@cis.upenn.edu

Abstract. Additive Cost Register Automata (ACRA) map strings to integers using a finite set of registers that are updated using assignments of the form "$x := y + c$" at every step. The corresponding class of *additive regular functions* has multiple equivalent characterizations, appealing closure properties, and a decidable equivalence problem. In this paper, we solve two decision problems for this model. First, we define the *register complexity* of an additive regular function to be the minimum number of registers that an ACRA needs to compute it. We characterize the register complexity by a necessary and sufficient condition regarding the largest subset of registers whose values can be made far apart from one another. We then use this condition to design a PSPACE algorithm to compute the register complexity of a given ACRA, and establish a matching lower bound. Our results also lead to a machine-independent characterization of the register complexity of additive regular functions. Second, we consider *two-player games over ACRAs*, where the objective of one of the players is to reach a target set while minimizing the cost. We show the corresponding decision problem to be EXPTIME-complete when the costs are non-negative integers, but undecidable when the costs are integers.

1 Introduction

Consider the following scenario: a customer frequents a coffee shop, and each time purchases a cup of coffee costing \$2. At any time, he may fill a survey, for which the store offers to give him a discount of \$1 for each of his purchases that month (including for purchases already made). We model this by the machine M_1 shown in figure 1. There are two states q_S and $q_{\neg S}$, indicating whether the customer has filled out the survey during the current month. There are three events to which the machine responds: C indicates the purchase of a cup of coffee, S indicates the completion of a survey, and $\#$ indicates the end of a month. The registers x and y track how much money the customer owes the establishment: in the state $q_{\neg S}$, the amount in x assumes that he will not fill out a survey that month, and the amount in y assumes that he will fill out a survey before the end of the month. At any time the customer wishes to settle his account, the machine outputs the amount of money owed, which is always the value in the register x.

---

[*] The full version of this paper is available on the arXiv (arXiv:1304.7029). This research was partially supported by the NSF Expeditions in Computing award 1138996.

F.V. Fomin et al. (Eds.): ICALP 2013, Part II, LNCS 7966, pp. 37–48, 2013.
© Springer-Verlag Berlin Heidelberg 2013

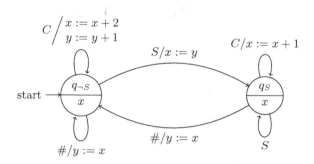

Fig. 1. ACRA M_1 models a customer in a coffee shop. It implements a function f_1 : $\{C, S, \#\}^* \to \mathbb{Z}$ mapping the purchase history of the customer to the amount he owes the store.

The automaton M_1 has a finite state space, and a finite set of integer-valued registers. On each transition, each register u is updated by an expression of the form "$u := v + c$", for some register v and constant $c \in \mathbb{Z}$. Which register will eventually contribute to the output is determined by the state after reading the entire input, and so the cost of an event depends not only on the past, but also on the future. Indeed, it can be shown that these machines are *closed under regular lookahead*, i.e. the register updates can be conditioned on regular properties of an as-yet-unseen suffix, for no gain in expressivity. The important limitation is that the register updates are test-free, and cannot examine the register contents.

The motivation behind the model is generalizing the idea of regular languages to quantitative properties of strings. A language $L \subseteq \Sigma^*$ is regular when it is accepted by a DFA. Regular languages are a robust class, permitting multiple equivalent representations such as regular expressions and as formulas in monadic second-order logic. Recently in [2], we proposed the model of regular functions: they are the MSO-definable transductions from strings to expression trees over some pre-defined grammar. The class of functions thus defined depends on the grammar allowed; the simplest is when the underlying domain is the set of integers $\mathbb{Z}$, and expressions involve constants and binary addition, and we call the resulting class *additive regular functions*. Additive regular functions have appealing closure properties, such as closure under linear combination, input reversal, and regular lookahead, and several analysis problems are efficiently decidable – such as containment, shortest paths and equivalence checking. The machine M_1 is an example of an *Additive Cost Register Automaton* (ACRA), and this class defines exactly the additive regular functions.

Observe that the machine M_1 has two registers, and it is not immediately clear how (if it is even possible) to reduce this number. This is the first question that this paper settles: Given an ACRA M, how do we determine the minimum number of registers needed by any ACRA to compute the function it defines, $[\![M]\!]$? We describe a property called register separation, and show that any equivalent ACRA needs at least k registers iff the registers of M are k-separable. It turns out that the registers of M_1 are 2-separable, and hence two registers are necessary. We then go on to show that determining k-separability is PSPACE-complete. Determining the register complexity is the natural analogue of the state minimization problem for DFAs [6].

The techniques used to analyse the register complexity allow us to state a result similar to the pumping lemma for regular languages: The register complexity of f is at least k iff for some m, we have strings $\sigma_0, \ldots, \sigma_m, \tau_1, \ldots, \tau_m$, suffixes w_1, $\ldots, w_k$, k distinct coefficient vectors $\mathbf{c}_1, \ldots, \mathbf{c}_k \in \mathbb{Z}^m$, and values $d_1, \ldots, d_k \in \mathbb{Z}$ so that for all vectors $\mathbf{x} \in \mathbb{N}^m$, $f\left(\sigma_0\tau_1^{x_1}\sigma_1\tau_2^{x_2}\ldots\sigma_m w_i\right) = \sum_j c_{ij}x_j + d_i$. Thus, depending on the suffix w_i, at least one of the cycles $\tau_1, \ldots, \tau_k$ contributes differently to the final cost.

Finally, we consider ACRAs with turn-based alternation. These are games where several objective functions are simultaneously computed, but only one of these objectives will eventually contribute to the output, based on the actions of both the system and its environment. Alternating ACRAs are thus related to multi-objective games and Pareto optimization [12], but are a distinct model because each run evaluates to a single value. We study the reachability problem in ACRA games: Given a budget k, is there a strategy for the system to reach an accepting state with cost at most k? We show that this problem is EXPTIME-complete when the incremental costs assume values from $\mathbb{N}$, and undecidable when the incremental costs are integer-valued.

Related Work. The traditional model of string-to-number transducers has been (non-deterministic) weighted automata (WA). Additive regular functions are equivalent to unambiguous weighted automata, and are therefore strictly sandwiched between weighted automata and deterministic WAs in expressiveness. Deterministic WAs are ACRAs with one register, and algorithms exist to compute the *state complexity* and for minimization [10]. Mohri [11] presents a comprehensive survey of the field. Recent work on the quantitative analysis of programs [5] also uses weighted automata, but does not deal with minimization or with notions of regularity. Data languages [7] are concerned with strings over a (possibly infinite) data domain $\mathbb{D}$. Recent models [3] have obtained Myhill-Nerode characterizations, and hence minimization algorithms, but the models are intended as acceptors, and not for computing more general functions. Turn-based weighted games [9] are ACRA games with a single register, and in this special setting, it is possible to solve non-negative optimal reachability in polynomial time. Of the techniques used in the paper, difference bound invariants are a standard tool. However when we need them, in section 3, we have to deal with disjunctions of such constraints, and show termination of invariant strengthening – to the best of our knowledge, the relevant problems have not been solved before.

Outline of the Paper. We define the automaton model in section 2. In section 3, we introduce the notion of separability, and establish its connection to register complexity. In section 4, we show that determining the register complexity

is PSPACE-complete. Finally, in section 5, we study ACRA reachability games –
in particular, that ACRA $(\mathbb{Z})$ games are undecidable, and that ACRA $(\mathbb{N})$ reachability games are EXPTIME-complete.

2   Additive Regular Functions

We will use additive cost register automata as the working definition of additive
regular functions, i.e. a function[1] $f : \Sigma^* \to \mathbb{Z}_\perp$ is regular iff it is implemented
by an ACRA. An ACRA is a deterministic finite state machine, supplemented
by a finite number of integer-valued registers. Each transition specifies, for each
register u, a test-free update of the form "$u := v + c$", for some register v, and
constant $c \in \mathbb{Z}$. Accepting states are labelled with output expressions of the form
"$v + c$".

Definition 1. *An ACRA is a tuple $M = (Q, \Sigma, V, \delta, \mu, q_0, F, \nu)$, where Q is a
finite non-empty set of states, Σ is a finite input alphabet, V is a finite set of
registers, $\delta : Q \times \Sigma \to Q$ is the state transition function, $\mu : Q \times \Sigma \times V \to V \times \mathbb{Z}$
is the register update function, $q_0 \in Q$ is the start state, $F \subseteq Q$ is the set of
accepting states, and $\nu : F \to V \times \mathbb{Z}$ is the output function.*

*The configuration of the machine is a pair $\gamma = (q, val)$, where q is the current
state, and $val : V \to \mathbb{Z}$ maps each register to its value. Define $(q, val) \to^a (q', val')$
iff $\delta(q, a) = q'$ and for each $u \in V$, if $\mu(q, a, u) = (v, c)$, then $val'(u) = val(v) + c$.*

*The machine M implements a function $[\![M]\!] : \Sigma^* \to \mathbb{Z}_\perp$ defined as follows.
For each $\sigma \in \Sigma^*$, let $(q_0, val_0) \to^\sigma (q_f, val_f)$, where $val_0(v) = 0$ for all v. If
$q_f \in F$ and $\nu(q_f) = (v, c)$, then $[\![M]\!](\sigma) = val_f(v) + c$. Otherwise $[\![M]\!](\sigma) = \perp$.*

We will write $val(u, \sigma)$ for the value of a register u after the machine has processed the string σ starting from the initial configuration. In the rest of this
section, we summarize some known results about ACRAs [2]:

Equivalent Characterizations. Additive regular functions are equivalent to
unambiguous weighted automata [11] over the tropical semiring. These are nondeterministic machines with a single counter. Each transition increments the
counter by an integer c, and accepting states have output increments, also integers. The unambiguous restriction requires that there be a single accepting path
for each string in the domain, thus the "min" operation of the tropical semiring
is unused. Recently, streaming tree transducers [1] have been proposed as the
regular model for string-to-tree transducers – ACRAs are equivalent in expressiveness to MSO-definable string-to-term transducers with binary addition as
the base grammar.

Closure Properties. What makes additive[2] regular functions interesting to
study is their robustness to various manipulations:

---

[1] By convention, we represent a partial function $f : A \to B$ as a total function
    $f : A \to B_\perp$, where $B_\perp = B \cup \{\perp\}$, and $\perp \notin B$ is the "undefined" value.
[2] We will often drop the adjective "additive", and refer simply to regular functions.

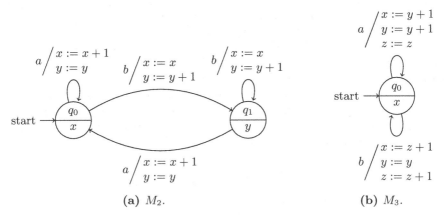

(a) M_2.

(b) M_3.

Fig. 2. ACRAs M_2 and M_3 operate over the input alphabet $\Sigma = \{a,b\}$. Both implement the function defined as $f_2(\epsilon) = 0$, and for all σ, $f_2(\sigma a) = |\sigma a|_a$, and $f_2(\sigma b) = |\sigma b|_b$. Here $|\sigma|_a$ is the number of occurrences of the symbol a in the string σ.

1. For all $c \in \mathbb{Z}$, if f_1 and f_2 are regular functions, then so are $f_1 + f_2$ and cf_1.
2. If f is a regular function, then f_{rev} defined as $f_{rev}(\sigma) = f(\sigma^{rev})$ is also regular.
3. If f_1 and f_2 are regular functions, and L is a regular language, then the function f defined as $f(\sigma) =$ if $\sigma \in L$, then $f_1(\sigma)$, else $f_2(\sigma)$ is also regular.
4. ACRAs are closed under regular lookahead, i.e. even if the machine were allowed to make decisions based on a regular property of the suffix rather than simply the next input symbol, there would be no increase in expressiveness.

Analysis Problems. Given ACRAs M_1 and M_2, equivalence-checking and the min-cost problem $(\min_{\sigma \in \Sigma^*} [\![M]\!](\sigma))$ can be solved in polynomial time. It follows then that containment (for all σ, $[\![M_1]\!](\sigma) \leq [\![M_2]\!](\sigma)$) also has a polynomial time algorithm.

3   Characterizing the Register Complexity

The register complexity of an additive regular function f is the minimum number of registers an ACRA needs to compute it. For example the register complexity of both $[\![M_1]\!]$ in figure 1 and $[\![M_2]\!]$ in figure 2a is 2. Computing the register complexity is the first problem we solve, and is the subject of this section and the next.

Definition 2. *Let $f : \Sigma^* \to \mathbb{Z}_\perp$ be a regular function. The register complexity of f is the smallest number k such that there is an ACRA M implementing f with only k registers.*

Informally, the registers of M are separable in some state q if their values can be pushed far apart. For example, consider the registers x and y of M_1 in the state q_0. For any constant c, there is a string $\sigma = C^c$ leading to q_0 so that $|val\,(x, \sigma) - val\,(y, \sigma)| \geq c$. In formalizing this idea, we need to distinguish registers that are live in a given state, i.e. those that can potentially contribute to the output. For example, M_1 could be augmented with a third register z tracking the length of the string processed. However, the value of z would be irrelevant to the computation of f_1. Informally, a register v is live[3] in a state q if for some suffix $\sigma \in \Sigma^*$, on processing σ starting from q, the initial value of v is what influences the final output.

Definition 3. *Let $M = (Q, \Sigma, V, \delta, \mu, q_0, \nu)$ be an ACRA. The registers of M are k-separable if there is some state q, and a subset $U \subseteq V$ so that*

1. $|U| = k$, *all registers $v \in U$ are live in q, and*
2. *for all $c \in \mathbb{Z}$, there is a string σ, such that $\delta\,(q_0, \sigma) = q$ and for all distinct $u, v \in U$, $|val\,(u, \sigma) - val\,(v, \sigma)| \geq c$.*

The registers of a machine M are not k-separable if at every state q, and subset U of k live registers, there is a constant c such that for all strings σ to q, $|val\,(u, \sigma) - val\,(v, \sigma)| < c$, for some distinct $u, v \in U$. Note that the specific registers which are close may depend on σ. For example, in the machine M_3 from figure 2b, if a string σ ends with an a, then x and y will have the same value, while if the last symbol was a b, then x and z are guaranteed to be equal.

Theorem 1. *Let $f : \Sigma^* \to \mathbb{Z}_\perp$ be a function defined by an ACRA M. Then the register complexity of f is at least k iff the registers of M are k-separable.*

We now sketch the proofs for each direction.

k-separability Implies a Lower Bound on the Register Complexity

Consider the machine M_1 from figure 1. Here $k = 2$, and the registers x and y are separated in the state $q_{\neg S}$. Let $\sigma_1 = \epsilon$, i.e. the empty string, and $\sigma_2 = S$ – these are suffixes which, when starting from $q_{\neg S}$, "extract" the values currently in x and y respectively.

Now suppose an equivalent counter-example machine M' is proposed with only one register v. At each state q' of M', observe the "effect" of processing suffixes σ_1 and σ_2. Each of these can be summarized by an expression of the form $v + c_{q'i}$ for $i \in \{1, 2\}$, the current value of register v, and $c_{q'i} \in \mathbb{Z}$. Thus, the outputs differ by no more than $|(v + c_{q'1}) - (v + c_{q'2})| \leq |c_{q'1}| + |c_{q'2}|$. Fix $n = \max_{q'} (|c_{q'1}| + |c_{q'2}|)$, and observe that for all σ, $|[\![M']\!]\,(\sigma\sigma_1) - [\![M']\!]\,(\sigma\sigma_2)| \leq n$. However, for $\sigma = C^{n+1}$, we know that $|f_1\,(\sigma\sigma_1) - f_1\,(\sigma\sigma_2)| > n$, so M' cannot be equivalent to M_1. This argument can be generalized to obtain:

Lemma 1. *Let M be an ACRA whose registers are k-separable. Then the register complexity of the implemented function f is at least k.*

---

[3] Live registers are formally defined in the full version of this paper.

Non-separability Permits Register Elimination

Consider an ACRA M whose registers are not k-separable. We can then state an invariant at each state q: there is a constant c_q such that for every subset $U \subseteq V$ of live registers with $|U| = k$, and for every string σ with $\delta(q_0, \sigma) = q$, there must exist distinct $u, v \in U$ with $|val(u, \sigma) - val(v, \sigma)| < c$. For example, with 3 registers x, y, z, this invariant would be $\exists c, |x - y| < c \vee |y - z| < c \vee |z - x| < c$. We will construct a machine M', where each state $q' = (q, C, \mathbf{v})$ has 3 components: the first component q is the state of the original machine, and C identifies some term (not necessarily unique) in the disjunction which is currently satisfied. Now for example, if we know that $|x - y| < c$, then it suffices to explicitly maintain the value of only one register, and the (bounded) difference can be stored in the state – this is the third component $\mathbf{v}$.

Since we need to track these register differences during the execution, the invariants must be inductive: if D_q and $D_{q'}$ are the invariants at states q and q' respectively, and $q \rightarrow^a q'$ is a transition in the machine, then it must be the case that $D_q \implies \text{WP}(D_{q'}, q, a)$. Here WP refers to the standard notion of the weakest precondition from program analysis: $\text{WP}(D_{q'}, q, a)$ is exactly that set of variable valuations val so that $(q, val) \rightarrow^a (q', val')$ for some $D_{q'}$-satisfying valuation val'.

The standard technique to make a collection of invariants inductive is strengthening: if $D_q \not\Longrightarrow \text{WP}(D_{q'}, q, a)$, then D_q is replaced with $D_q \wedge \text{WP}(D_{q'}, q, a)$, and this process is repeated at every pair of states until a fixpoint is reached. This procedure is seeded with the invariants asserting non-separability. However, before the result of this back-propagation can be used in our arguments, we must prove that the method terminates – this is the main technical problem solved in this section.

We now sketch a proof of this termination claim for a simpler class of invariants. Consider the class of difference-bound constraints – assertions of the form $C = \bigwedge_{u,v \in V} a_{uv} < u - v < b_{uv}$, where for each u, v, $a_{uv}, b_{uv} \in \mathbb{Z}$ or $a_{uv}, b_{uv} \in \{-\infty, \infty\}$. When in closed form[4], C induces an equivalence relation $\equiv_C$ over the registers: $u \equiv_C v$ iff $a_{uv}, b_{uv} \in \mathbb{Z}$. Let C and C' be some pair of constraints such that $C \not\Longrightarrow C'$. Then the assertion $C \wedge C'$ is strictly stronger than C. Either $C \wedge C'$ relates a strictly larger set of variables – $\equiv_C \subsetneq \equiv_{C \wedge C'}$ – or (if $\equiv_C = \equiv_{C \wedge C'}$) for some pair of registers u, v, the bounds $a'_{uv} < u - v < b'_{uv}$ imposed by $C \wedge C'$ are a strict subset of the bounds $a_{uv} < u - v < b_{uv}$ imposed by C. Observe that the first type of strengthening can happen at most $|V|^2$ times, while the second type of strengthening can happen only after a_{uv}, b_{uv} are established for a pair of registers u, v, and can then happen at most $b_{uv} - a_{uv}$ times. Thus the process of repeated invariant strengthening must terminate. This argument can be generalized to disjunctions of difference-bound constraints, and we conclude:

Lemma 2. *Consider an ACRA M whose registers are not k-separable. Then, we can effectively construct an equivalent machine M' with only $k - 1$ registers.*

---

[4] For all $u, v \in V$, $a_{uv} = -b_{vu}$, and for all $u, v, w \in V$, $a_{uv} + a_{vw} \leq a_{uw}$.

4   Computing the Register Complexity

4.1   Computing the Register Complexity Is in PSPACE

We reduce the problem of determining the register complexity of $[\![M]\!]$ to one of determining reachability in a directed "register separation" graph with $O\left(|Q|\, 2^{|V|^2}\right)$ nodes. The presence of an edge in this graph can be determined in polynomial space, and thus we have a PSPACE algorithm to determine the register complexity. Otherwise, if polynomial time algorithms are used for graph reachability and 1-counter 0-reachability, the procedure runs in time $O\left(c^3\, |Q|^4\, 2^{4|V|^2}\right)$, where c is the largest constant in the machine.

We first generalize the idea of register separation to that of separation relations: an arbitrary relation $\| \subseteq V \times V$ separates a state q if for every $c \in \mathbb{Z}$, there is a string σ so that $\delta\left(q_0, \sigma\right) = q$, and whenever $u \parallel v$, $|val\left(u, \sigma\right) - val\left(v, \sigma\right)| \geq c$. Thus, the registers of M are k-separable iff for some state q and some subset U of live registers at q, $|U| = k$ and $\{(u, v) \mid u, v \in U, u \neq v\}$ separates q.

Consider a string $\tau \in \Sigma^*$, so for some q, $\delta\left(q, \tau\right) = q$. Assume also that:

1. For every register u in the domain or range of $\|$, $\mu\left(q, \tau, u\right) = (u, c_u)$, for some $c_u \in \mathbb{Z}$, and
2. for some pair of registers x, y, $\mu\left(q, \tau, x\right) = (x, c)$ and $\mu\left(q, \tau, y\right) = (y, c')$ for distinct c, c'.

Thus, every pair of registers that is already separated is preserved during the cycle, and some new pair of registers is incremented differently. We call such strings τ "separation cycles" at q. They allow us to make conclusions of the form: If $\|$ separates q, then $\| \cup \{(x, y)\}$ also separates q.

Now consider a string $\sigma \in \Sigma^*$, such that for some q, q', $\delta\left(q, \sigma\right) = q'$. Pick arbitrary relations $\|$, $\|'$, and assume that whenever $u' \parallel' v'$, and $\mu\left(q, \sigma, u'\right) = (u, c_u)$, $\mu\left(q, \sigma, v'\right) = (v, c_v)$, we have $u \parallel v$. We can then conclude that if $\|$ separates q, then $\|'$ separates q' We call such strings σ "renaming edges" from $(q, \|)$ to $(q', \|')$.

We then show that if $\|$ separates q and $\|$ is non-empty, then there is a separation cycle-renaming edge sequence to $(q, \|)$ from some strictly smaller separation $(q', \|')$. Thus, separation at each node can be demonstrated by a sequence of separation cycles with renaming edges in between, and thus we reduce the problem to that of determining reachability in an exponentially large register separation graph. Finally, we show that each type of edge can be determined in PSPACE.

Theorem 2. *Given an ACRA M and a number k, there is a PSPACE procedure to determine whether its register complexity is at least k.*

4.2   Pumping Lemma for ACRAs

The following theorem is the interpretation of a path through the register separation graph. Given a regular function f of register complexity at least k, it guarantees the existence of m cycles $\tau_1, \ldots, \tau_m$, serially connected by strings σ_0, $\ldots, \sigma_m$, so that based on one of k suffixes $w_1, \ldots, w_k$, the cost paid on one of the cycles must differ. These cycles are actually the separation cycles discussed earlier, and intermediate strings σ_i correspond to the renaming edges. Consider for example, the function f_2 from figure 2, and let $\sigma_0 = \epsilon$, $\tau_1 = aab$, and $\sigma_1 = \epsilon$. We can increase the difference between the registers x and y to arbitrary amounts by pumping the cycle τ_1. Now if the suffixes are $w_1 = a$, and $w_2 = b$, then the choice of suffix determines the "cost" paid on each iteration of the cycle.

Theorem 3. *A regular function $f : \Sigma^* \to \mathbb{Z}_\perp$ has register complexity at least k iff there exist strings $\sigma_0, \ldots, \sigma_m, \tau_1, \ldots, \tau_m$, and suffixes $w_1, \ldots, w_k$, and k distinct coefficient vectors $\mathbf{c}_1, \ldots, \mathbf{c}_k \in \mathbb{Z}^m$, and values $d_1, \ldots, d_k \in \mathbb{Z}$ so that for all $x_1, \ldots, x_m \in \mathbb{N}$,*

$$f\left(\sigma_0 \tau_1^{x_1} \sigma_1 \tau_2^{x_2} \ldots \sigma_m w_i\right) = \sum_j c_{ij} x_j + d_i.$$

4.3 Computing the Register Complexity Is PSPACE-hard

We reduce the DFA intersection non-emptiness checking problem [8] to the problem of computing the register complexity. Let $A = (Q, \Sigma, \delta, q_0, \{q_f\})$ be a DFA. Consider a single-state ACRA M with input alphabet Σ. For each state $q \in Q$, M maintains a register v_q. On reading a symbol $a \in \Sigma$, M updates $v_q := v_{\delta(q,a)}$, for each q. Observe that this is simulating the DFA in reverse: if we start with a special tagged value in v_{q_f}, then after processing σ, that tag is in the register v_{q_0} iff σ^{rev} is accepted by A. Also observe that doing this in parallel for all the DFAs no longer requires an exponential product construction, but only as many registers as a linear function of the input size. We use this idea to construct in polynomial time an ACRA M whose registers are $(k + 2)$-separable iff there is a string $\sigma \in \Sigma^*$ which is simultaneously accepted by all the DFAs. Therefore:

Theorem 4. *Given an ACRA M and a number k, deciding whether the register complexity of $[\![M]\!]$ is at least k is PSPACE-hard.*

5   Games over ACRAs

We now study games played over ACRAs. We extend the model of ACRAs to allow alternation – in each state, a particular input symbol may be associated with multiple transitions. The system picks the input symbol to process, while the environment picks the specific transition associated with this input symbol. Accepting states are associated with output functions, and the system may choose to end the game in any accepting state. Given a budget k, we wish to decide whether the system has a winning strategy with worst-case cost no more than k. We show that ACRA games are undecidable when the incremental costs are integer-valued, and EXPTIME-complete when the incremental costs are from $\mathbb{D} = \mathbb{N}$.

Definition 4. *An   ACRA* ($\mathbb{D}$)   *reachability   game   G   is   a   tuple* $(Q, \Sigma, V, \delta, \mu, q_0, F, \nu)$, *where Q, Σ, and V are finite sets of states, input symbols and registers respectively,* $\delta \subseteq Q \times \Sigma \times Q$ *is the transition relation,* $\mu : \delta \times V \to V \times \mathbb{D}$ *is the register update function,* $q_0 \in Q$ *is the start state,* $F \subseteq Q$ *is the set of accepting states, and* $\nu : F \to V \times \mathbb{D}$ *is the output function.*

The game configuration is a tuple $\gamma = (q, val)$, *where* $q \in Q$ *is the current state, and* $val : V \to \mathbb{D}$ *is the current register valuation. A run* π *is a (possibly infinite) sequence of game configurations* $(q_1, val_1) \to^{a_1} (q_2, val_2) \to^{a_2} \cdots$ *with the property that*

1. *the transition* $q_i \to^{a_i} q_{i+1} \in \delta$ *for each i, and*
2. $val_{i+1}(u) = val_i(v) + c$, *where* $\mu(q_i \to^{a_i} q_{i+1}, u) = (v, c)$, *for each register u and for each transition i.*

A strategy is a function $\theta : Q^* \times Q \to \Sigma$ that maps a finite history $q_1 q_2 \ldots q_n$ to the next symbol $\theta(q_1 q_2 \ldots q_n)$.

Definition 5. *A run* π *is consistent with a strategy* θ *if for each i,* $\theta(q_1 q_2 \ldots q_i) = a_i$. θ *is winning from a configuration* (q, val) *with a budget of* $k \in \mathbb{D}$ *if for every consistent run* π *starting from* $(q_1, val_1) = (q, val)$, *for some* i, $q_i \in F$ *and* $\nu(q_i, val_i) \le k$.

For greater readability, we write tuples $(q, a, q') \in \delta$ as $q \to^a q'$. If $q \in F$, and val is a register valuation, we write $\nu(q, val)$ for the result $val(v) + c$, where $\nu(q) = (v, c)$. When we omit the starting configuration for winning strategies it is understood to mean the initial configuration (q_0, val_0) of the ACRA.

5.1   ACRA ($\mathbb{N}$) Reachability Games Can Be Solved in EXPTIME

Consider the simpler class of (unweighted) graph reachability games. These are played over a structure $G^f = (Q, \Sigma, \delta, q_0, F)$, where Q is the finite state space, and Σ is the input alphabet. $\delta \subseteq Q \times \Sigma \times Q$ is the state transition relation, $q_0 \in Q$ is the start state, and $F \subseteq Q$ is the set of accepting states. If the input symbol $a \in \Sigma$ is played in a state q, then the play may adversarially proceed to any state q' so that $(q, a, q') \in \delta$. The system can force a win if every run compatible with some strategy $\theta^f : Q^* \times Q \to \Sigma$ eventually reaches a state $q_f \in F$. Such games can be solved by a recursive back-propagation algorithm – corresponding to model checking the formula $\mu X \cdot (F \vee \bigvee_{a \in \Sigma} [a] X)$ – in time $O(|Q| |\Sigma|)$. In such games, whenever there is a winning strategy, there is a memoryless winning strategy θ_{small} which guarantees that no state is visited twice.

From every ACRA ($\mathbb{N}$) reachability game $G = (Q, \Sigma, V, \delta, \mu, q_0, F, \nu)$, we can project out an unweighted graph reachability game $G^f = (Q, \Sigma, \delta, q_0, F)$. Also, G^f has a winning strategy iff for some $k \in \mathbb{N}$, G has a k-winning strategy. Consider the cost of θ_{small} (computed for G^f) when used with G. Since no run ever visits the same state twice, θ_{small} is $c_0 |Q|$-winning, where c_0 is the largest constant appearing in G. We have thus established an upper-bound on the optimal reachability strategy, if it exists.

Given an upper-bound $k \in \mathbb{N}$, we would like to determine whether a winning strategy θ exists within this budget. Because the register increments are non-negative, once a register v achieves a value larger than k, it cannot contribute to the final output on any suffix σ permitted by the winning strategy. We thus convert G into an unweighted graph reachability G_k^f, where the value of each register is explicitly tracked in the state, as long as it is in the set $\{0, 1, \ldots, k\}$. This game can be solved for the optimal reachability strategy, and so we have:

Theorem 5. *The optimal strategy θ for an $ACRA\,(\mathbb{N})$ reachability game G can be computed in time $O\left(|Q|\,|\Sigma|\,2^{|V| \log c_0 |Q|}\right)$, where c_0 is the largest constant appearing in the description of G.*

Note that the optimal strategy in ACRA $(\mathbb{N})$ games need not be memoryless: the strategy may visit a state again with a different register valuation. However, the strategy θ constructed in the proof of the above theorem is memoryless given the pair (q, val) of the current state and register valuation.

5.2   Hardness of Solving ACRA $(\mathbb{D})$ Reachability Games

We reduce the halting problem for two-counter machines to the problem of solving an ACRA $(\mathbb{Z})$ reachability game. Informally, we construct a game G_M given a two-counter machine M so that the player has a 0-winning strategy through G_M iff M halts. This strategy encodes the execution of M, and the adversary verifies that the run is valid. A similar idea is used to show that deciding ACRA $(\mathbb{N})$ reachability games is EXPTIME-hard. The reduction in that case proceeds from the halting problem for linearly bounded alternating Turing machines [4]. Given such a machine M, we construct in polynomial time a game gadget G_M where the only strategy is to encode the runs of the Turing machine.

Theorem 6. *Determining whether there is a winning strategy with budget k in an $ACRA\,(\mathbb{N})$ reachability game is EXPTIME-hard.*

Theorem 7. *Determining whether there is a winning strategy with budget k in an $ACRA\,(\mathbb{Z})$ reachability game is undecidable.*

6   Conclusion

In this paper, we studied two decision problems for additive regular functions: determining the register complexity, and alternating reachability in ACRAs. The register complexity of an additive regular function f is the smallest number k so there is some ACRA implementing f with only k registers. We developed an abstract characterization of the register complexity as separability and showed that computing it is PSPACE-complete. We then studied the reachability problem in alternating ACRAs, and showed that it is undecidable for ACRA $(\mathbb{Z})$ and EXPTIME-complete for ACRA $(\mathbb{N})$ games. Future work includes proving similar characterizations and providing algorithms for register minimization in more general models such as streaming string transducers. String concatenation does not form a commutative monoid, and the present paper is restricted to unary operators (increment by constant), and so the technique does not immediately carry over. Another interesting question is to find a machine-independent characterization of regular functions $f : \Sigma^* \to \mathbb{Z}_\perp$. A third direction of work would be extending these ideas to trees and studying their connection to alternating ACRAs.

References

1. Alur, R., D'Antoni, L.: Streaming tree transducers. In: Czumaj, A., Mehlhorn, K., Pitts, A., Wattenhofer, R. (eds.) ICALP 2012, Part II. LNCS, vol. 7392, pp. 42–53. Springer, Heidelberg (2012)
2. Alur, R., D'Antoni, L., Deshmukh, J.V., Raghothaman, M., Yuan, Y.: Regular functions and cost register automata. To Appear in the 28th Annual Symposium on Logic in Computer Science (2013), Full version available at
   http://www.cis.upenn.edu/~alur/rca12.pdf
3. Bojanczyk, M., Klin, B., Lasota, S.: Automata with group actions. In: 26th Annual Symposium on Logic in Computer Science, pp. 355–364 (2011)
4. Chandra, A., Kozen, D., Stockmeyer, L.: Alternation. Journal of the ACM 28(1), 114–133 (1981)
5. Chatterjee, K., Doyen, L., Henzinger, T.A.: Quantitative Languages. In: Kaminski, M., Martini, S. (eds.) CSL 2008. LNCS, vol. 5213, pp. 385–400. Springer, Heidelberg (2008)
6. Hopcroft, J., Motwani, R., Ullman, J.: Introduction to Automata Theory, Languages, and Computation, 3rd edn. Prentice Hall (2006)
7. Kaminski, M., Francez, N.: Finite-memory automata. Theoretical Computer Science 134(2), 329–363 (1994)
8. Kozen, D.: Lower bounds for natural proof systems. In: 18th Annual Symposium on Foundations of Computer Science, pp. 254–266 (October 31-November 2, 1977)
9. Markey, N.: Weighted automata: Model checking and games. Lecture Notes (2008),
   http://www.lsv.ens-cachan.fr/ markey/Teaching/MPRI/2008-2009/
   MPRI-2.8b-4.pdf
10. Mohri, M.: Minimization algorithms for sequential transducers. Theoretical Computer Science 234, 177–201 (2000)
11. Mohri, M.: Weighted automata algorithms. In: Droste, M., Kuich, W., Vogler, H. (eds.) Handbook of Weighted Automata. Monographs in Theoretical Computer Science, pp. 213–254. Springer (2009)
12. Papadimitriou, C., Yannakakis, M.: Multiobjective query optimization. In: Proceedings of the 20th Symposium on Principles of Database Systems, PODS 2001, pp. 52–59. ACM (2001)

Beyond Differential Privacy: Composition Theorems and Relational Logic for f-divergences between Probabilistic Programs

Gilles Barthe and Federico Olmedo

IMDEA Software Institute, Madrid, Spain
{Gilles.Barthe,Federico.Olmedo}@imdea.org

Abstract. f-divergences form a class of measures of distance between probability distributions; they are widely used in areas such as information theory and signal processing. In this paper, we unveil a new connection between f-divergences and differential privacy, a confidentiality policy that provides strong privacy guarantees for private data-mining; specifically, we observe that the notion of α-distance used to characterize approximate differential privacy is an instance of the family of f-divergences. Building on this observation, we generalize to arbitrary f-divergences the sequential composition theorem of differential privacy. Then, we propose a relational program logic to prove upper bounds for the f-divergence between two probabilistic programs. Our results allow us to revisit the foundations of differential privacy under a new light, and to pave the way for applications that use different instances of f-divergences.

1 Introduction

Differential privacy [12] is a policy that provides strong privacy guarantees in private data analysis: informally, a randomized computation over a database D is differentially private if the private data of individuals contributing to D is protected against arbitrary adversaries with query access to D. Formally, let $\epsilon \geq 0$ and $0 \leq \delta \leq 1$: a randomized algorithm c is (ϵ, δ)-differentially private if its output distributions for any two neighbouring inputs x and y are (e^ϵ, δ)-close, i.e. for every event E:

$$\Pr c(x)E \leq e^\epsilon \Pr c(y)E + \delta$$

where $\Pr c(x)E$ denotes the probability of event E in the distribution obtained by running c on input x. One key property of differential privacy is the existence of sequential and parallel composition theorems, which allows building differentially private computations from smaller blocks. In this paper, we focus on the first theorem, which states that the sequential composition of an (ϵ_1, δ_1)-differentially private algorithm with an (ϵ_2, δ_2)-differentially private one yields an $(\epsilon_1 + \epsilon_2, \delta_1 + \delta_2)$-differentially private algorithm.

f-divergences [2,10] are convex functions that can be used to measure the distance between two distributions. The class of f-divergences includes many well-known notions of distance, such as statistical distance, Kullback-Leibler divergence (relative entropy),

F.V. Fomin et al. (Eds.): ICALP 2013, Part II, LNCS 7966, pp. 49–60, 2013.

or Hellinger distance. Over the years, f-divergences have found multiple applications in information theory, signal processing, pattern recognition, machine learning, and security. The practical motivation for this work is a recent application of f-divergences to cryptography: in [24], Steinberger uses Hellinger distance to improve the security analysis of key-alternating ciphers, a family of encryption schemes that encompasses the Advanced Encryption Standard AES.

Deductive Verification of Differentially Private Computations. In [6], we develop an approximate probabilistic Hoare logic, called apRHL, for reasoning about differential privacy of randomized computations. The logic manipulates judgments of the form:

$$c_1 \sim_{\alpha,\delta} c_2 : \Psi \Rightarrow \Phi$$

where c_1 and c_2 are probabilistic imperative programs, $\alpha \geq 1$, $0 \leq \delta \leq 1$ and Ψ and Φ are relations over states. As for its predecessor pRHL [5], the notion of valid judgment rests on a lifting operator that turns a relation R over states into a relation $\sim_R^{\alpha,\delta}$ over distributions of states: formally, the judgment above is valid iff for every pair of memories m_1 and m_2, $m_1 \ \Psi \ m_2$ implies $(\llbracket c_1 \rrbracket \ m_1) \sim_\Phi^{\alpha,\delta} (\llbracket c_2 \rrbracket \ m_2)$. The definition of the lifting operator originates from probabilistic process algebra [15], and has close connections with flow networks and the Kantorovich metric [11].

apRHL judgments characterize differential privacy, in the sense that c is (ϵ, δ)-differentially private iff the apRHL judgment $c \sim_{e^\epsilon, \delta} \Psi : c \Rightarrow \ \equiv$ is valid, where Ψ is a logical characterization of adjacency—for instance, two lists of the same length are adjacent if they differ in a single element.

Problem Statement and Contributions. The goal of this paper is to lay the theoretical foundations for tool-supported reasoning about f-divergences between probabilistic computations. To achieve this goal, we start from [6] and take the following steps:

1. as a preliminary observation, we prove that the notion of α-distance used to characterize differential privacy is in fact an f-divergence;
2. we define a notion of composability of f-divergences and generalize the sequential composition theorem of differential privacy to composable divergences;
3. we generalize the notion of lifting used in apRHL to composable f-divergences;
4. we define fpRHL, a probabilistic relational Hoare logic for f-divergences, and prove its soundness.

Related Work. The problem of computing the distance between two probabilistic computations has been addressed in different areas of computer science, including machine learning, stochastic systems, and security. We briefly point to some recent developments.

Methods for computing the distance between probabilistic automata have been studied by Cortes and co-authors [8,9]; their work, which is motivated by machine-learning applications, considers the Kullback-Leibler divergence as well as the L_p distance.

Approximate bisimulation for probabilistic automata has been studied, among others, by Segala and Turrini [23] and by Tracol, Desharnais and Zhioua [25]. The survey [1] provides a more extensive account of the field.

In the field of security, approximate probabilistic bisimulation is closely connected to quantitative information flow of probabilistic computations, which has been studied e.g. by Di Pierro, Hankin and Wiklicky [20]. More recently, the connections between quantitative information flow and differential privacy have been explored e.g. by Barthe and Köpf [4], and by Alvim, Andrés, Chatzikokolakis and Palamidessi [3]. Moreover, several language-based methods have been developed for guaranteeing differential privacy; these methods are based on runtime verification, such as PINQ [17] or Airavat [22], type systems [21,14], or deductive verification [7]. We refer to [19] for a survey of programming languages methods for differential privacy.

2 Mathematical Preliminaries

In this section we review the representation of distributions used in our development and recall the definition of f-divergences.

2.1 Probability Distributions

Throughout the presentation we consider distributions and sub-distributions over discrete sets only. A probability distribution (resp. sub-distribution) over a set A is an object $\mu : A \to [0, 1]$ such that $\sum_{a \in A} \mu(a) = 1$ (resp. $\sum_{a \in A} \mu(a) \leq 1$). We let $\mathcal{D}(A)$ (resp. $\mathcal{D}_{\leq 1}(A)$) be the set of distributions (resp. sub-distributions) over A.

Distributions are closed under convex combinations: given distributions $(\mu_i)_{i \in \mathbb{N}}$ in $\mathcal{D}(A)$ and weights $(w_i)_{i \in \mathbb{N}}$ such that $\sum_{i \in \mathbb{N}} w_i = 1$ and $w_i \geq 0$ for all $i \in \mathbb{N}$, the convex combination $\sum_{i \in \mathbb{N}} w_i \, \mu_i$ is also a distribution over A. Thus, given $\mu \in \mathcal{D}(A)$ and $M : A \to \mathcal{D}(B)$, we define the distribution bind $\mu \, M$ over B as $\sum_{a \in A} \mu(a) \, M(a)$. Likewise, sub-distributions are closed under convex combinations.

2.2 f-divergences

Let $\mathcal{F}$ be the set of non-negative convex functions $f : \mathbb{R}_0^+ \to \mathbb{R}_0^+$ such that f is continuous at 0 and $f(1) = 0$. Then each function in $\mathcal{F}$ induces a notion of distance between probability distributions as follows:

Definition 1 (f-divergence). *Given $f \in \mathcal{F}$, the f-divergence $\Delta_f(\mu_1, \mu_2)$ between two distributions μ_1 and μ_2 in $\mathcal{D}(A)$ is defined as:*

$$\Delta_f(\mu_1, \mu_2) \stackrel{def}{=} \sum_{a \in A} \mu_2(a) f\left(\frac{\mu_1(a)}{\mu_2(a)}\right)$$

The definition adopts the following conventions, which are used consistently throughout the paper:

$$0f\,(0/0) = 0 \qquad and \qquad 0f\,(t/0) = t \lim_{x \to 0^+} x f\,(1/x) \quad if\, t > 0$$

Moreover, if $\Delta_f(\mu_1, \mu_2) \leq \delta$ we say that μ_1 and μ_2 are (f, δ)-close.

f-divergence	f	Simplified Form
Statistical distance	$SD(t) = \frac{1}{2}|t - 1|$	$\sum_{a \in A} \frac{1}{2}|\mu_1(a) - \mu_2(a)|$
Kullback-Leibler[1]	$KL(t) = t\ln(t) - t + 1$	$\sum_{a \in A} \mu_1(a)\ln\left(\frac{\mu_1(a)}{\mu_2(a)}\right)$
Hellinger distance	$HD(t) = \frac{1}{2}(\sqrt{t} - 1)^2$	$\sum_{a \in A} \frac{1}{2}\left(\sqrt{\mu_1(a)} - \sqrt{\mu_2(a)}\right)^2$

Fig. 1. Examples of f-divergences

When defining f-divergences one usually allows f to take positive as well as negative values in $\mathbb{R}$. For technical reasons, however, we consider only non-negative functions. We now show that we can adopt this restriction without loss of generality.

Proposition 1. *Let $\mathcal{F}'$ be defined as $\mathcal{F}$, except that we allow $f \in \mathcal{F}'$ to take negative values. Then for every $f \in \mathcal{F}'$ there exists $g \in \mathcal{F}$ given by $g(t) = f(t) - f'_-(1)(t - 1)$, such that $\Delta_f = \Delta_g$. (Here f'_- denotes the left derivative of f, whose existence can be guaranteed from the convexity of f.)*

The class of f-divergences includes several popular instances; these include statistical distance, relative entropy (also known as Kullback-Leibler divergence), and Hellinger distance. In Figure 1 we summarize the convex function used to define each of them and we also include a simplified form, useful to compute the divergence. (In case of negative functions, we previously apply the transformation mentioned in Proposition 1, so that we are consistent with our definition of f-divergences.)

In general, Δ_f does not define a metric. The symmetry axiom might be violated and the triangle inequality holds only if f equals a non-negative multiple of the statistical distance. The identity of indiscernibles does not hold in general, either.

3   A Sequential Composition Theorem for f-divergences

In this section we show that the notion of α-distance used to capture differential privacy is an f-divergence. Then we define the composition of f-divergences and show that the sequential composition theorem of differential privacy generalizes to this setting.

3.1   An f-divergence for Approximate Differential Privacy

In [6] we introduced the concept of α-distance to succinctly capture the notion of differentially private computations. Given $\alpha \geq 1$, the α-distance between distributions μ_1 and μ_2 in $\mathcal{D}(A)$ is defined as

$$\Delta_\alpha(\mu_1, \mu_2) \overset{\text{def}}{=} \max_{E \subseteq A} d_\alpha(\mu_1(E), \mu_2(E))$$

---

[1] Rigorously speaking, the function used for defining the Kullback-Leibler divergence should be given by $f(t) = t\ln(t) + t - 1$ if $t > 0$ and $f(t) = 1$ if $t = 0$ to guarantee its continuity at 0.

where $d_\alpha(a, b) \stackrel{\text{def}}{=} \max\{a - \alpha b, 0\}$. (This definition slightly departs from that of [6], in the sense that we consider an asymmetric version of the α-distance. The original version, symmetric, corresponds to taking $d_\alpha(a, b) \stackrel{\text{def}}{=} \max\{a - \alpha b, b - \alpha a, 0\}$). Now we can recast the definition of differential privacy in terms of the α-distance and say that a randomized computation c is (ϵ, δ)-*differentially private* iff $\Delta_{e^\epsilon}(c(x), c(y)) \leq \delta$ for any two adjacent inputs x and y.

Our composition result of f-divergences builds on the observation that α-distance is an instance of the class of f-divergences.

Proposition 2. *For every* $\alpha \geq 1$, *the* α-*distance* $\Delta_\alpha(\mu_1, \mu_2)$ *coincides with the* f-*divergence* $\Delta_{\mathsf{AD}_\alpha}(\mu_1, \mu_2)$ *associated to function* $\mathsf{AD}_\alpha(t) \stackrel{\text{def}}{=} \max\{t - \alpha, 0\}$.

3.2 Composition

One key property of f-divergences is a monotonicity result referred to as the *data processing inequality* [18]. In our setting, it is captured by the following proposition:

Proposition 3. *Let* $\mu_1, \mu_2 \in \mathcal{D}(A)$, $M : A \to \mathcal{D}(B)$ *and* $f \in \mathcal{F}$. *Then*

$$\Delta_f(\text{bind } \mu_1 \ M, \text{bind } \mu_2 \ M) \leq \Delta_f(\mu_1, \mu_2)$$

In comparison, the sequential composition theorem for differential privacy [16] is captured by the following theorem.

Theorem 1. *Let* $\mu_1, \mu_2 \in \mathcal{D}(A)$, $M_1, M_2 : A \to \mathcal{D}(B)$ *and* $\alpha, \alpha' \geq 1$. *Then*

$$\Delta_{\alpha\alpha'}(\text{bind } \mu_1 \ M_1, \text{bind } \mu_2 \ M_2) \leq \Delta_\alpha(\mu_1, \mu_2) + \max_a \Delta_{\alpha'}(M_1(a), M_2(a))$$

Note that the data processing inequality for α-distance corresponds to the composition theorem for the degenerate case where M_1 and M_2 are equal. The goal of this paragraph is to generalize the sequential composition theorem to f-divergences. To this end, we first define a notion of composability between f-divergences.

Definition 2 (f-divergence composability). *Let* $f_1, f_2, f_3 \in \mathcal{F}$. *We say that* (f_1, f_2) *is* f_3-*composable iff for all* $\mu_1, \mu_2 \in \mathcal{D}(A)$ *and* $M_1, M_2 : A \to \mathcal{D}(B)$, *there exists* $\mu_3 \in \mathcal{D}(A)$ *such that*

$$\Delta_{f_3}(\text{bind } \mu_1 \ M_1, \text{bind } \mu_2 \ M_2) \leq \Delta_{f_1}(\mu_1, \mu_2) + \sum_{a \in A} \mu_3(a) \Delta_{f_2}(M_1(a), M_2(a))$$

Our notion of composability is connected to the notion of additive information measures from [13, Ch. 5]. To justify the connection, we first present an adaptation of their definition to our setting.

Definition 3 (f-divergence additivity). *Let* $f_1, f_2, f_3 \in \mathcal{F}$. *We say that* (f_1, f_2) *is* f_3-*additive iff for all distributions* $\mu_1, \mu_2 \in \mathcal{D}(A)$ *and* $\mu_1', \mu_2' \in \mathcal{D}(B)$,

$$\Delta_{f_3}(\mu_1 \times \mu_1', \mu_2 \times \mu_2') \leq \Delta_{f_1}(\mu_1, \mu_2) + \Delta_{f_2}(\mu_1', \mu_2')$$

Here, $\mu \times \mu'$ *denotes the product distribution of* μ *and* μ', *i.e.* $(\mu \times \mu')(a, b) \stackrel{\text{def}}{=} \mu(a)\mu'(b)$.

It is easily seen that composability entails additivity.

Proposition 4. *Let* $f_1, f_2, f_3 \in \mathcal{F}$ *such that* (f_1, f_2) *is* f_3-*composable. Then* (f_1, f_2) *is* f_3-*additive.*

The f-divergences from Figure 1 present good behaviour under composition. The statistical distance, Hellinger distance and the Kullback-Leibler divergence are composable w.r.t. themselves. Moreover, α-divergences are composable.

Proposition 5
- $(\mathsf{SD}, \mathsf{SD})$ *is* SD-*composable;*
- $(\mathsf{KL}, \mathsf{KL})$ *is* KL-*composable;*
- $(\mathsf{HD}, \mathsf{HD})$ *is* HD-*composable;*
- $(\mathsf{AD}_{\alpha_1}, \mathsf{AD}_{\alpha_2})$ *is* $\mathsf{AD}_{\alpha_1 \alpha_2}$-*composable for every* $\alpha_1, \alpha_2 \geq 1$.

The sequential composition theorem of differential privacy extends naturally to the class of composable divergences.

Theorem 2. *Let* $f_1, f_2, f_3 \in \mathcal{F}$. *If* (f_1, f_2) *is* f_3-*composable, then for all* $\mu_1, \mu_2 \in \mathcal{D}(A)$ *and all* $M_1, M_2 : A \to \mathcal{D}(B)$,

$$\Delta_{f_3}(\text{bind } \mu_1 \, M_1, \text{bind } \mu_2 \, M_2) \leq \Delta_{f_1}(\mu_1, \mu_2) + \max_a \Delta_{f_2}(M_1(a), M_2(a))$$

Theorem 2 will be the cornerstone for deriving the sequential composition rule of fpRHL. (As an intermediate step, we first show that the composition result extends to relation liftings.)

4   Lifting

The definition of valid apRHL judgment rests on the notion of lifting. As a last step before defining our relational logic, we extend the notion of lifting to f-divergences. One key difference between our definition and that of [6] is that the former uses two witnesses, rather than one. In the remainder, we let $\text{supp}\,(\mu)$ denote the set of elements $a \in A$ such that $\mu(a) > 0$. Moreover, given $\mu \in \mathcal{D}(A \times B)$, we define $\pi_1(\mu)$ and $\pi_2(\mu)$ by the clauses $\pi_1(\mu)(a) = \sum_{b \in B} \mu(a, b)$ and $\pi_2(\mu)(b) = \sum_{a \in A} \mu(a, b)$.

Definition 4 (Lifting). *Let* $f \in \mathcal{F}$ *and* $\delta \in \mathbb{R}_0^+$. *Then* (f, δ)-*lifting* $\sim_R^{f,\delta}$ *of a relation* $R \subseteq A \times B$ *is defined as follows: given* $\mu_1 \in \mathcal{D}(A)$ *and* $\mu_2 \in \mathcal{D}(B)$, $\mu_1 \sim_R^{f,\delta} \mu_2$ *iff there exist* $\mu_L, \mu_R \in \mathcal{D}(A \times B)$ *such that: i)* $\text{supp}\,(\mu_L) \subseteq R$; *ii)* $\text{supp}\,(\mu_R) \subseteq R$; *iii)* $\pi_1(\mu_L) = \mu_1$; *iv)* $\pi_2(\mu_R) = \mu_2$ *and v)* $\Delta_f(\mu_L, \mu_R) \leq \delta$. *The distributions* μ_L *and* μ_R *are called the left and right witnesses for the lifting, respectively.*

A pleasing consequence of our definition is that the witnesses for relating two distributions are themselves distributions, rather than sub-distributions; this is in contrast with our earlier definition from [6], where witnesses for the equality relation are necessarily sub-distributions. Moreover, our definition is logically equivalent to the original one from [15], provided $\delta = 0$, and f satisfies the identity of indiscernibles. In the case of statistical distance and α-distance, our definition also has a precise mathematical relationship with (an asymmetric variant of) the lifting used in [6].

Proposition 6. *Let* $\alpha \geq 1$, $\mu_1 \in \mathcal{D}(A)$ *and* $\mu_2 \in \mathcal{D}(B)$. *If* $\mu_1 \sim_R^{AD_\alpha, \delta} \mu_2$ *then there exists a sub-distribution* $\mu \in \mathcal{D}(A \times B)$ *such that: i)* $\mathrm{supp}(\mu) \subseteq R$; *ii)* $\pi_1(\mu) \leq \mu_1$; *iii)* $\pi_2(\mu) \leq \mu_2$ *and iv)* $\Delta_\alpha(\mu_1, \pi_1 \mu) \leq \delta$, *where* $\leq$ *denotes the natural pointwise order on the space of sub-distributions, i.e.* $\mu \leq \mu'$ *iff* $\mu(a) \leq \mu'(a)$ *for all* a.

We briefly review some key properties of liftings. The first result characterizes liftings over equivalence relations, and will be used to show that f-divergences can be characterized by our logic.

Proposition 7 (Lifting of equivalence relations). *Let* R *be an equivalence relation over* A *and let* $\mu_1, \mu_2 \in \mathcal{D}(A)$. *Then,*

$$\mu_1 \sim_R^{f,\delta} \mu_2 \iff \Delta_f(\mu_1/R, \mu_2/R) \leq \delta,$$

where μ/R *is a distribution over the quotient set* A/R, *defined as* $(\mu/R)([a]) \stackrel{def}{=} \mu([a])$. *In particular, if* R *is the equality relation* $\equiv$, *we have*

$$\mu_1 \sim_\equiv^{f,\delta} \mu_2 \iff \Delta_f(\mu_1, \mu_2) \leq \delta$$

Our next result allows deriving probability claims from lifting judgments. Given $R \subseteq A \times B$ we say that the subsets $A_0 \subseteq A$ and $B_0 \subseteq B$ are *R-equivalent*, and write $A_0 =_R B_0$, iff for every $a \in A$ and $b \in B$, $a \, R \, b$ implies $a \in A_0 \iff b \in B_0$.

Proposition 8 (Fundamental property of lifting). *Let* $\mu_1 \in \mathcal{D}(A)$, $\mu_2 \in \mathcal{D}(B)$, *and* $R \subseteq A \times B$. *Then, for any two events* $A_0 \subseteq A$ *and* $B_0 \subseteq B$,

$$\mu_1 \sim_R^{f,\delta} \mu_2 \wedge A_0 =_R B_0 \implies \mu_2(B_0) \, f\left(\frac{\mu_1(A_0)}{\mu_2(B_0)}\right) \leq \delta$$

Our final result generalizes the sequential composition theorem from the previous section to arbitrary liftings.

Proposition 9 (Lifting composition). *Let* $f_1, f_2, f_3 \in \mathcal{F}$ *such that* (f_1, f_2) *is* f_3-*composable. Moreover let* $\mu_1 \in \mathcal{D}(A)$, $\mu_2 \in \mathcal{D}(B)$, $M_1 : A \to \mathcal{D}(A')$ *and* $M_2 : B \to \mathcal{D}(B')$. *If* $\mu_1 \sim_{R_1}^{f_1, \delta_1} \mu_2$ *and* $M_1(a) \sim_{R_2}^{f_2, \delta_2} M_2(b)$ *for all* a *and* b *such that* $a \, R \, b$, *then*

$$(\mathrm{bind} \ \mu_1 \ M_1) \sim_{R_2}^{f_3, \delta_1 + \delta_2} (\mathrm{bind} \ \mu_2 \ M_2)$$

5 A Relational Logic for f-divergences

Building on the results of the previous section, we define a relational logic, called fpRHL, for proving upper bounds for the f-divergence between probabilistic computations written in a simple imperative language.

5.1 Programming Language

We consider programs written in a probabilistic imperative language pWHILE. The syntax of the programming language is defined inductively as follows:

$$
\begin{array}{lll}
\mathcal{C} ::= & \text{skip} & \text{nop} \\
\quad | & \mathcal{V} \leftarrow \mathcal{E} & \text{deterministic assignment} \\
\quad | & \mathcal{V} \xleftarrow{\$} \mathcal{DE} & \text{random assignment} \\
\quad | & \text{if } \mathcal{E} \text{ then } \mathcal{C} \text{ else } \mathcal{C} & \text{conditional} \\
\quad | & \text{while } \mathcal{E} \text{ do } \mathcal{C} & \text{while loop} \\
\quad | & \mathcal{C}; \mathcal{C} & \text{sequence}
\end{array}
$$

Here $\mathcal{V}$ is a set of variables, $\mathcal{E}$ is a set of deterministic expressions, and $\mathcal{DE}$ is a set of expressions that denote distributions from which values are sampled in random assignments. Program states or memories are mappings from variables to values. More precisely, memories map a variable v of type T to a value in its interpretation $[\![T]\!]$. We use $\mathcal{M}$ to denote the set of memories. Programs are interpreted as functions from initial memories to sub-distributions over memories. The semantics, which is given in Figure 2, is based on two evaluation functions $[\![\cdot]\!]_{\mathcal{E}}$ and $[\![\cdot]\!]_{\mathcal{DE}}$ for expressions and distribution expressions; these functions respectively map memories to values and memories to sub-distributions of values. Moreover, the definition uses the operator unit, which maps every $a \in A$ to the unique distribution over A that assigns probability 1 to a and probability 0 to every other element of A, and the null distribution μ_0, that assigns probability 0 to all elements of A. Note that the semantics of programs is a map from memories to sub-distributions over memories. Sub-distributions, rather than distributions, are used to model probabilistic non-termination. However, for the sake of simplicity, in the current development of the logic, we only consider programs that terminate with probability 1 on all inputs and leave the general case for future work.

$$
\begin{array}{lcl}
[\![\text{skip}]\!]\, m & = & \text{unit } m \\[4pt]
[\![c;\ c']\!]\, m & = & \text{bind } ([\![c]\!]\, m)\, [\![c']\!] \\[4pt]
[\![x \leftarrow e]\!]\, m & = & \text{unit } (m\, \{[\![e]\!]_{\mathcal{E}}\, m/x\}) \\[4pt]
[\![x \xleftarrow{\$} \mu]\!]\, m & = & \text{bind } ([\![\mu]\!]_{\mathcal{DE}}\, m)\, (\lambda v.\ \text{unit } (m\, \{v/x\})) \\[4pt]
[\![\text{if } e \text{ then } c_1 \text{ else } c_2]\!]\, m & = & \textit{if } ([\![e]\!]_{\mathcal{E}}\, m = \text{true})\ \textit{then } ([\![c_1]\!]\, m)\ \textit{else } ([\![c_2]\!]\, m) \\[4pt]
[\![\text{while } e \text{ do } c]\!]\, m & = & \lambda f.\ \sup_{n \in \mathbb{N}} ([\![[\text{while } e \text{ do } c]_n]\!]\, m\, f)
\end{array}
$$

$$
\text{where} \quad
\begin{array}{lcl}
[\text{while } e \text{ do } c]_0 & = & \textit{if } ([\![e]\!]_{\mathcal{E}}\, m = \text{true})\ \textit{then } (\text{unit } m)\ \textit{else } \mu_0 \\
[\text{while } e \text{ do } c]_{n+1} & = & \text{if } e \text{ then } c;\ [\text{while } e \text{ do } c]_n
\end{array}
$$

Fig. 2. Semantics of programs

5.2   Judgments

fpRHL judgments are of the form $c_1 \sim_{f,\delta} c_2 : \Psi \Rightarrow \Phi$, where c_1 and c_2 are programs, Ψ and Φ are relational assertions, $f \in \mathcal{F}$ and $\delta \in \mathbb{R}_0^+$. Relational assertions are first-order formulae over generalized expressions, i.e. expressions in which variables are tagged with a $\langle 1 \rangle$ or $\langle 2 \rangle$. Relational expressions are interpreted as formulae over pairs of memories, and the tag on a variable is used to indicate whether its interpretation should be taken in the first or second memory. For instance, the relational assertion $x\langle 1\rangle = x\langle 2\rangle$ states that the values of x coincide in the first and second memories. More generally, we use $\equiv$ to denote the relational assertion that states that the values of all variables coincide in the first and second memories.

An fpRHL judgment is valid iff for every pair of memories related by the pre-condition Ψ, the corresponding pair of output distributions is related by the (f, δ)-lifting of the post-condition Φ.

Definition 5 (Validity in fpRHL). *A judgment* $c_1 \sim_{f,\delta} c_2 : \Psi \Rightarrow \Phi$ *is valid, written* $\models c_1 \sim_{f,\delta} c_2 : \Psi \Rightarrow \Phi$, *iff*

$$\forall m_1, m_2 \bullet m_1 \, \Psi \, m_2 \implies (\llbracket c_1 \rrbracket \, m_1) \sim_\Phi^{f,\delta} (\llbracket c_2 \rrbracket \, m_2)$$

fpRHL judgments provide a characterization of f-divergence. Concretely, judgments with the identity relation as post-condition can be used to derive (f, δ)-closeness results.

Proposition 10. *If* $\models c_1 \sim_{f,\delta} c_2 : \Psi \Rightarrow \equiv$, *then for all memories* m_1, m_2,

$$m_1 \, \Psi \, m_2 \implies \Delta_f(\llbracket c_1 \rrbracket \, m_1, \llbracket c_2 \rrbracket \, m_2) \le \delta$$

Moreover, fpRHL characterizes continuity properties of probabilistic programs. We assume a continuity model in which programs are executed on random inputs, i.e. distributions of initial memories, and we use f-divergences as metrics to compare program inputs and outputs.

Proposition 11. *Let* $f_1, f_2, f_3 \in \mathcal{F}$ *such that* (f_1, f_2) *is* f_3-composable. *If we have* $\models c_1 \sim_{f_2,\delta_2} c_2 : \equiv \Rightarrow \equiv$, *then for any two distributions of initial memories* μ_1 *and* μ_2,

$$\Delta_{f_1}(\mu_1, \mu_2) \le \delta_1 \implies \Delta_{f_3}(\text{bind } \mu_1 \, \llbracket c_1 \rrbracket, \text{bind } \mu_2 \, \llbracket c_2 \rrbracket) \le \delta_1 + \delta_2$$

Finally, we can use judgments with arbitrary post-condictions to relate the probabilities of single events in two programs. This is used, e.g. in the context of game-based cryptographic proofs.

Proposition 12. *If* $\models c_1 \sim_{f,\delta} c_2 : \Psi \Rightarrow \Phi$, *then for all memories* m_1, m_2 *and events* E_1, E_2,

$$m_1 \, \Psi \, m_2 \wedge E_1 =_\Phi E_2 \implies (\llbracket c_2 \rrbracket \, m_2)(E_2) \, f\left(\frac{(\llbracket c_1 \rrbracket \, m_1)(E_1)}{(\llbracket c_2 \rrbracket \, m_2)(E_2)}\right) \le \delta$$

5.3 Proof System

Figure 3 presents a set of core rules for reasoning about the validity of an fpRHL judgment. All the rules are transpositions of rules from apRHL [6]. However, fpRHL rules do no directly generalize their counterparts from apRHL. This is because both logics admit symmetric and asymmetric versions, but apRHL and fpRHL are opposite variants: fpRHL is asymmetric and apRHL is symmetric. Refer to Section 5.4 for a discussion about the symmetric version of fpRHL.

$$\frac{\forall m_1, m_2 \bullet m_1 \; \Psi \; m_2 \implies (m_1 \{[\![e_1]\!] \, m_1/x_1\}) \; \Phi \; (m_2 \{[\![e_2]\!] \, m_2/x_2\})}{\vdash x_1 \leftarrow e_1 \sim_{f,0} x_2 \leftarrow e_2 : \Psi \Rightarrow \Phi} \text{[assn]}$$

$$\frac{\forall m_1, m_2 \bullet m_1 \; \Psi \; m_2 \implies \Delta_f([\![\mu_1]\!]_{\mathcal{DE}} \, m_1, [\![\mu_2]\!]_{\mathcal{DE}} \, m_2) \leq \delta}{\vdash x_1 \xleftarrow{\$} \mu_1 \sim_{f,\delta} x_2 \xleftarrow{\$} \mu_2 : \Psi \Rightarrow x_1\langle 1 \rangle = x_2\langle 2 \rangle} \text{[rand]}$$

$$\frac{\Psi \implies b\langle 1 \rangle \equiv b'\langle 2 \rangle}{\vdash c_1 \sim_{f,\delta} c_1' : \Psi \land b\langle 1 \rangle \Rightarrow \Phi \qquad \vdash c_2 \sim_{f,\delta} c_2' : \Psi \land \neg b\langle 1 \rangle \Rightarrow \Phi}{\vdash \text{if } b \text{ then } c_1 \text{ else } c_2 \sim_{f,\delta} \text{if } b' \text{ then } c_1' \text{ else } c_2' : \Psi \Rightarrow \Phi} \text{[cond]}$$

$$\frac{\begin{array}{c} (f_1, \ldots, f_n) \text{ composable and monotonic} \\ \Theta \stackrel{\text{def}}{=} b\langle 1 \rangle \equiv b'\langle 2 \rangle \qquad \Psi \land e\langle 1 \rangle \leq 0 \implies \neg b\langle 1 \rangle \\ \vdash c \sim_{f_1,\delta} c' : \Psi \land b\langle 1 \rangle \land b'\langle 2 \rangle \land e\langle 1 \rangle = k \Rightarrow \Psi \land \Theta \land e\langle 1 \rangle < k \end{array}}{\vdash \text{while } b \text{ do } c \sim_{f_n,n\delta} \text{while } b' \text{ do } c' : \Psi \land \Theta \land e\langle 1 \rangle \leq n \Rightarrow \Psi \land \neg b\langle 1 \rangle \land \neg b'\langle 2 \rangle} \text{[while]}$$

$$\frac{}{\vdash \text{skip} \sim_{f,0} \text{skip} : \Psi \Rightarrow \Psi} \text{[skip]} \qquad \frac{\begin{array}{c}(f_1, f_2) \text{ is } f_3\text{-composable} \\ \vdash c_1 \sim_{f_1,\delta_1} c_2 : \Psi \Rightarrow \Phi' \quad \vdash c_1' \sim_{f_2,\delta_2} c_2' : \Phi' \Rightarrow \Phi \end{array}}{\vdash c_1; c_1' \sim_{f_3,\delta_1+\delta_2} c_2; c_2' : \Psi \Rightarrow \Phi} \text{[seq]}$$

$$\frac{\vdash c_1 \sim_{f,\delta} c_2 : \Psi \land \Theta \Rightarrow \Phi}{\vdash c_1 \sim_{f,\delta} c_2 : \Psi \land \neg \Theta \Rightarrow \Phi}{\vdash c_1 \sim_{f,\delta} c_2 : \Psi \Rightarrow \Phi} \text{[case]} \qquad \frac{\vdash c_1 \sim_{f',\delta'} c_2 : \Psi' \Rightarrow \Phi'}{\Psi \Rightarrow \Psi' \quad \Phi' \Rightarrow \Phi \quad f \leq f' \quad \delta' \leq \delta}{\vdash c_1 \sim_{f,\delta} c_2 : \Psi \Rightarrow \Phi} \text{[weak]}$$

Fig. 3. Core proof rules

We briefly describe some main rules, and refer the reader to [6] for a longer description about each of them. Rule [seq] relates two sequential compositions and is a direct consequence from the lifting composition (see Proposition 9). Rule [while] relates two loops that terminate in lockstep. The bound depends on the maximal number of iterations of the loops, and we assume given a loop variant e that decreases at each iteration, and is initially upper bounded by some constant n. We briefly explain the side conditions: $(f_1, \ldots, f_n)$ is composable iff (f_i, f_1) is f_{i+1}-composable for every $1 \leq i < n$. Moreover, $(f_1, \ldots, f_n)$ is monotonic iff $f_i \leq f_{i+1}$ for $1 \leq i < n$. Note that the rule is given for $n \geq 2$; specialized rules exist for $n = 0$ and $n = 1$. This rule readily specializes to reason about (ϵ, δ)-differential privacy by taking $f_i = \text{AD}_{\alpha^i}$, where $\alpha = e^\epsilon$.

If an fpRHL judgment is derivable using the rules of Figure 3, then it is valid. Formally,

Proposition 13 (Soundness). *If* $\vdash c_1 \sim_{f,\delta} c_2 : \Psi \Rightarrow \Phi$ *then* $\models c_1 \sim_{f,\delta} c_2 : \Psi \Rightarrow \Phi$.

5.4   Symmetric Logic

One can also define a symmetric version of the logic by adding as an additional clause in the definition of the lift relation that $\Delta_f(\mu_R, \mu_L) \leq \delta$. An instance of this logic is the symmetric apRHL logic from [6]. All rules remain unchanged, except for the random sampling rule that now requires the additional inequality to be checked in the premise of the rule.

6 Conclusion

This paper makes two contributions: first, it unveils a connection between differential privacy and f-divergences. Second, it lays the foundations for reasoning about f-divergences between randomized computations. As future work, we intend to implement support for fpRHL in EasyCrypt [4], and formalize the results from [24]. We also intend to investigate the connection between our notion of lifting and flow networks.

Acknowledgments. This work was partially funded by the European Projects FP7-256980 NESSoS and FP7-229599 AMAROUT, Spanish project TIN2009-14599 DESAFIOS 10 and Madrid Regional project S2009TIC-1465 PROMETIDOS.

References

1. Abate, A.: Approximation metrics based on probabilistic bisimulations for general state-space markov processes: a survey. Electronic Notes in Theoretical Computer Sciences (2012) (in print)
2. Ali, S.M., Silvey, S.D.: A general class of coefficients of divergence of one distribution from another. Journal of the Royal Statistical Society. Series B (Methodological) 28(1), 131–142 (1966)
3. Alvim, M.S., Andrés, M.E., Chatzikokolakis, K., Palamidessi, C.: On the relation between differential privacy and Quantitative Information Flow. In: Aceto, L., Henzinger, M., Sgall, J. (eds.) ICALP 2011, Part II. LNCS, vol. 6756, pp. 60–76. Springer, Heidelberg (2011)
4. Barthe, G., Grégoire, B., Heraud, S., Béguelin, S.Z.: Computer-aided security proofs for the working cryptographer. In: Rogaway, P. (ed.) CRYPTO 2011. LNCS, vol. 6841, pp. 71–90. Springer, Heidelberg (2011)
5. Barthe, G., Grégoire, B., Zanella-Béguelin, S.: Formal certification of code-based cryptographic proofs. In: 36th ACM SIGPLAN-SIGACT Symposium on Principles of Programming Languages, POPL 2009, pp. 90–101. ACM, New York (2009)
6. Barthe, G., Köpf, B., Olmedo, F., Zanella-Béguelin, S.: Probabilistic relational reasoning for differential privacy. In: 39th ACM SIGPLAN-SIGACT Symposium on Principles of Programming Languages, POPL 2012, pp. 97–110. ACM, New York (2012)
7. Chaudhuri, S., Gulwani, S., Lublinerman, R., Navidpour, S.: Proving programs robust. In: 19th ACM SIGSOFT Symposium on the Foundations of Software Engineering and 13rd European Software Engineering Conference, ESEC/FSE 2011, pp. 102–112. ACM, New York (2011)
8. Cortes, C., Mohri, M., Rastogi, A.: Lp distance and equivalence of probabilistic automata. Int. J. Found. Comput. Sci. 18(4), 761–779 (2007)
9. Cortes, C., Mohri, M., Rastogi, A., Riley, M.: On the computation of the relative entropy of probabilistic automata. Int. J. Found. Comput. Sci. 19(1), 219–242 (2008)
10. Csiszár, I.: Eine informationstheoretische ungleichung und ihre anwendung auf den beweis der ergodizitat von markoffschen ketten. Publications of the Mathematical Institute of the Hungarian Academy of Science 8, 85–108 (1963)

11. Deng, Y., Du, W.: Logical, metric, and algorithmic characterisations of probabilistic bisimulation. Tech. Rep. CMU-CS-11-110, Carnegie Mellon University (March 2011)
12. Dwork, C.: Differential privacy. In: Bugliesi, M., Preneel, B., Sassone, V., Wegener, I. (eds.) ICALP 2006. LNCS, vol. 4052, pp. 1–12. Springer, Heidelberg (2006)
13. Ebanks, B., Sahoo, P., Sander, W.: Characterizations of Information Measures. World Scientific (1998)
14. Gaboardi, M., Haeberlen, A., Hsu, J., Narayan, A., Pierce, B.C.: Linear dependent types for differential privacy. In: 40th ACM SIGPLAN–SIGACT Symposium on Principles of Programming Languages, POPL 2013, pp. 357–370. ACM, New York (2013)
15. Jonsson, B., Yi, W., Larsen, K.G.: Probabilistic extensions of process algebras. In: Bergstra, J., Ponse, A., Smolka, S. (eds.) Handbook of Process Algebra, pp. 685–710. Elsevier, Amsterdam (2001)
16. McSherry, F.: Privacy integrated queries: an extensible platform for privacy-preserving data analysis. Commun. ACM 53(9), 89–97 (2010)
17. McSherry, F.D.: Privacy integrated queries: an extensible platform for privacy-preserving data analysis. In: 35th SIGMOD International Conference on Management of Data, SIGMOD 2009, pp. 19–30. ACM, New York (2009)
18. Pardo, M., Vajda, I.: About distances of discrete distributions satisfying the data processing theorem of information theory. IEEE Transactions on Information Theory 43(4), 1288–1293 (1997)
19. Pierce, B.C.: Differential privacy in the programming languages community. Invited Tutorial at DIMACS Workshop on Recent Work on Differential Privacy Across Computer Science (2012)
20. Di Pierro, A., Hankin, C., Wiklicky, H.: Measuring the confinement of probabilistic systems. Theor. Comput. Sci. 340(1), 3–56 (2005)
21. Reed, J., Pierce, B.C.: Distance makes the types grow stronger: a calculus for differential privacy. In: 15th ACM SIGPLAN International Conference on Functional programming, ICFP 2010, pp. 157–168. ACM, New York (2010)
22. Roy, I., Setty, S.T.V., Kilzer, A., Shmatikov, V., Witchel, E.: Airavat: security and privacy for MapReduce. In: 7th USENIX Conference on Networked Systems Design and Implementation, NSDI 2010, pp. 297–312. USENIX Association, Berkeley (2010)
23. Segala, R., Turrini, A.: Approximated computationally bounded simulation relations for probabilistic automata. In: 20th IEEE Computer Security Foundations Symposium, CSF 2007, pp. 140–156. IEEE Computer Society (2007)
24. Steinberger, J.: Improved security bounds for key-alternating ciphers via hellinger distance. Cryptology ePrint Archive, Report 2012/481 (2012), http://eprint.iacr.org/
25. Tracol, M., Desharnais, J., Zhioua, A.: Computing distances between probabilistic automata. In: Proceedings of QAPL. EPTCS, vol. 57, pp. 148–162 (2011)

A Maximal Entropy Stochastic Process
for a Timed Automaton[*][**]

Nicolas Basset[1,2]

[1] LIGM, University Paris-Est Marne-la-Vallée and CNRS, France
[2] LIAFA, University Paris Diderot and CNRS, France
nbasset@liafa.univ-paris-diderot.fr

Abstract. Several ways of assigning probabilities to runs of timed automata (TA) have been proposed recently. When only the TA is given, a
relevant question is to design a probability distribution which represents
in the best possible way the runs of the TA. This question does not seem
to have been studied yet. We give an answer to it using a maximal entropy
approach. We introduce our variant of stochastic model, the stochastic
process over runs which permits to simulate random runs of any given
length with a linear number of atomic operations. We adapt the notion of
Shannon (continuous) entropy to such processes. Our main contribution
is an explicit formula defining a process Y^* which maximizes the entropy.
This formula is an adaptation of the so-called Shannon-Parry measure to
the timed automata setting. The process Y^* has the nice property to be
ergodic. As a consequence it has the asymptotic equipartition property
and thus the random sampling w.r.t. Y^* is quasi uniform.

1 Introduction

Timed automata (TA) were introduced in the early 90's by Alur and Dill [4] and
then extensively studied, to model and verify the behaviours of real-time systems.
In this context of verification, several probability settings have been added to
TA (see references below). There are several reasons to add probabilities: this
permits (i) to reflect in a better way physical systems which behave randomly,
(ii) to reduce the size of the model by pruning the behaviors of null probability
[8], (iii) to resolve undeterminism when dealing with parallel composition [15,16].

In most of previous works on the subject (see e.g. [10,2,11,15]), probability
distributions on continuous and discrete transitions are given at the same time as
the timed settings. In these works, the choice of the probability functions is left
to the designer of the model. Whereas, she or he may want to provide only the
TA and ask the following question: what is the "best" choice of the probability
functions according to the TA given? Such a "best" choice must transform the

---

[*] An extended version of the present paper containing detailed proofs and examples
is available on-line http://hal.archives-ouvertes.fr/hal-00808909.
[**] The support of Agence Nationale de la Recherche under the project EQINOCS
(ANR-11-BS02-004) is gratefully acknowledged.

F.V. Fomin et al. (Eds.): ICALP 2013, Part II, LNCS 7966, pp. 61–73, 2013.

TA into a random generator of runs the least biased as possible, i.e it should generate the runs as uniformly as possible to cover with high probability the maximum of behaviours of the modeled system. More precisely the probability for a generated run to fall in a set should be proportional to the size (volume) of this set (see [16] for a same requirement in the context of job-shop scheduling). We formalize this question and propose an answer based on the notion of entropy of TA introduced in [6].

The theory developed by Shannon [21] and his followers permits to solve the analogous problem of quasi-uniform path generation in a finite graph. This problem can be formulated as follows: given a finite graph G, how can one find a stationary Markov chain on G which allows one to generate the paths in the most uniform manner? The answer is in two steps (see Chapter 1.8 of [19] and also section 13.3 of [18]): (i) There exists a stationary Markov chain on G with maximal entropy, the so called Shannon-Parry Markov chain; (ii) This stationary Markov chain allows to generate paths quasi uniformly.

In this article we lift this theory to the timed automata setting. We work with timed region graphs which are to timed automata what finite directed graphs are to finite state automata i.e. automata without labeling on edges and without initial and final states. We define stochastic processes over runs of timed region graphs (SPOR) and their (continuous) entropy. This generalization of Markov chains for TA has its own interest, it is up to our knowledge the first one which provides a continuous probability distribution on starting states. Such a SPOR permits to generate step by step random runs. As a main result we describe a maximal entropy SPOR which is stationary and ergodic and which generalizes the Shannon-Parry Markov chain to TA (Theorem 1). Concepts of maximal entropy, stationarity and ergodicity can be interesting by themselves, here we use them as the key hypotheses to ensure a quasi uniform sampling (Theorem 2). More precisely the result we prove is a variant of the so called Shannon-McMillan-Breiman theorem also known as asymptotic equipartition property (AEP).

Potential Applications. There are two kind of probabilistic model checking: (i) the almost sure model checking aiming to decide if a model satisfies a formula with probability one (e.g. [13,3]); (ii) the quantitative (probabilistic) model checking (e.g. [11,15]) aiming to compare the probability of a formula to be satisfied with some given threshold or to estimate directly this probability.

A first expected application of our results would be a "proportional" model checking. The inputs of the problem are: a timed region graph G, a formula φ, a threshold $\theta \in [0, 1]$. The question is whether the proportion of runs of G which satisfy φ is greater than θ or not. A recipe to address this problem would be as follows: (i) take as a probabilistic model $\mathcal{M}$ the timed region graph G together with the maximum entropy SPOR Y^* defined in our main theorem; (ii) run a quantitative (probabilistic) model checking algorithm on the inputs $\mathcal{M}, \varphi, \theta$ (the output of the algorithm is yes or no whether $\mathcal{M}$ satisfies φ with a probability greater than θ or not) (iii) use the same output for the proportional model checking problem.

A random simulation with a linear number of operations wrt. the length of the run can be achieved with our probabilistic setting. It would be interesting to incorporate the simulation of our maximal entropy process in a statistical model checking algorithms. Indeed random simulation is at the heart of such kind of quantitative model checking (see [15] and reference therein).

The concepts handled in this article such as stationary stochastic processes and their entropy, AEP, etc. come from information and coding theory (see [14]). Our work can be a basis for the probabilistic counterpart of the timed channel coding theory we have proposed in [5]. Another application in the same flavour would be a compression method of timed words accepted by a given deterministic TA.

Related Work. As mentioned above, this work generalizes the Shannon-Parry theory to the TA setting. Up to our knowledge this is the first time that a maximal entropy approach is used in the context of quantitative analysis of real-time systems.

Our models of stochastic real-time system can be related to numerous previous works. Almost-sure model checking for probabilistic real-time systems based on generalized semi Markov processes GSMPs was presented in [3] at the same time as the timed automata theory and by the same authors. This work was followed by [2,10] which address the problem of quantitative model checking for GSMPs under restricted hypotheses. The GSMPs have several differences with TA; roughly they behave as follows: in each location, clocks decrease until a clock is null, at this moment an action corresponding to this clock is fired, the other clocks are either reset, unchanged or purely canceled. Our probability setting is more inspired by [8,13,15] where probability densities are added directly on the TA. Here we add the new feature of an initial probability density function on states.

In [15], a probability distribution on the runs of a network of priced timed automaton is implicitly defined by a race between the components, each of them having its own probability. This allows a simulation of random runs in a non deterministic structure without state space explosion. There is no reason that the probability obtained approximates uniformness and thus it is quite incomparable to our objective.

Our techniques are based on the pioneering articles [6,7] on entropy of regular timed languages. In the latter article and in [5], an interpretation of the entropy of a timed language as information measure of the language was given.

2 Stochastic Processes on Timed Region Graphs

2.1 Timed Graphs and Their Runs

In this section we define a timed region graph which is the underlying structure of a timed automaton [4]. For technical reasons we consider only timed region graphs with bounded clocks. We will justify this assumption in section 3.1.

Timed Region Graphs. Let X be a finite set of variables called *clocks*. Clocks have non-negative values bounded by a constant M. A *rectangular constraint* has the form $x \sim c$ where $\sim \in \{\leq, <, =, >, \geq\}$, $x \in X$, $c \in \mathbb{N}$. A *diagonal constraint* has the form $x - y \sim c$ where $x, y \in X$. A guard is a finite conjunction of rectangular constraints. A *zone* is a set of clock vectors $\mathbf{x} \in [0, M]^X$ satisfying a finite conjunction of rectangular and diagonal constraints. A *region* is a zone which is minimal for inclusion (e.g. the set of points (x_1, x_2, x_3, x_4) which satisfy the constraints $0 = x_2 < x_3 - 4 = x_4 - 3 < x_1 - 2 < 1$). Regions of $[0, 1]^2$ are depicted in Fig 1.

As we work by analogy with finite graphs, we introduce timed region graphs which are roughly timed automata without labels on transitions and without initial and final states. Moreover we consider a state space decomposed in regions. Such a decomposition in regions are quite standard for timed automata and does not affect their behaviours (see e.g. [11,6]).

A timed region graph is a tuple $(X, Q, \mathbb{S}, \Delta)$ such that

- X is a finite set of clocks.
- Q is a finite set of *locations*.
- $\mathbb{S}$ is the set of *states* which are couples of a location and a clock vector ($\mathbb{S} \subseteq Q \times [0, M]^X$). It admits a region decomposition $\mathbb{S} = \cup_{q \in Q}\{q\} \times \mathbf{r}_q$ where for each $q \in Q$, $\mathbf{r}_q$ is a region.
- Δ is a finite set of *transitions*. Any transition $\delta \in \Delta$ goes from a *starting location* $\delta^- \in Q$ to an *ending location* $\delta^+ \in Q$; it has a set $\mathbf{r}(\delta)$ of clocks to reset when firing δ and a fleshy guard $\mathfrak{g}(\delta)$ to satisfy to fire it. Moreover, the set of clock vectors that satisfy $\mathfrak{g}(\delta)$ is projected on the region $\mathbf{r}_{\delta^+}$ when the clocks in $\mathbf{r}(\delta)$ are resets.

Runs of the Timed Region Graph. A *timed transition* is an element (t, δ) of $\mathbb{A} =_{def} [0, M] \times \Delta$. The *time delay* t represents the time before firing the transition δ.

Given a state $s = (q, \mathbf{x}) \in \mathbb{S}$ (i.e $\mathbf{x} \in \mathbf{r}_q$) and a timed transition $\alpha = (t, \delta) \in \mathbb{A}$ the *successor* of s by α is denoted by $s \triangleright \alpha$ and defined as follows. Let $\mathbf{x}'$ be the clock vector obtained from $\mathbf{x} + (t, \ldots, t)$ by resetting clocks in $\mathbf{r}(\delta)$ ($x_i' = 0$ if $i \in \mathbf{r}(\delta)$, $x_i' = x_i + t$ otherwise). If $\delta^- = q$ and $\mathbf{x} + (t, \ldots, t)$ satisfies the guard $\mathfrak{g}(\delta)$ then $\mathbf{x}' \in \mathbf{r}_{\delta^+}$ and $s \triangleright \alpha = (\delta^+, \mathbf{x}')$ else $s \triangleright \alpha = \bot$. Here and in the rest of the paper $\bot$ represents every undefined state.

We extend the successor action $\triangleright$ to words of timed transitions by induction: $s \triangleright \varepsilon = s$ and $s \triangleright (\alpha\alpha') = (s \triangleright \alpha) \triangleright \alpha'$ for all $s \in \mathbb{S}$, $\alpha \in \mathbb{A}$, $\alpha' \in \mathbb{A}^*$.

A *run* of the timed region graph $\mathcal{G}$ is a word $s_0\alpha_0 \cdots s_n\alpha_n \in (\mathbb{S} \times \mathbb{A})^{n+1}$ such that $s_{i+1} = s_i \triangleright \alpha_i \neq \bot$ for all $i \in \{0, \ldots, n-1\}$ and $s_n \triangleright \alpha_n \neq \bot$; its reduced version is $[s_0, \alpha_0 \ldots \alpha_n] \in \mathbb{S} \times \mathbb{A}^{n+1}$ (for all $i > 0$ the state s_i is determined by its preceding states and timed transition and thus is a redundant information). In the following we will use without distinction extended and reduced version of runs. We denote by $\mathcal{R}_n$ the set of runs of length n ($n \geq 1$).

Example 1. Let $\mathcal{G}^{ex1}$ be the timed region graph depicted on figure 1 with $\mathbf{r}_p$ and $\mathbf{r}_q$ the region described by the constraints $0 = y < x < 1$ and $0 = x < y < 1$

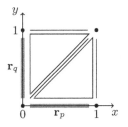
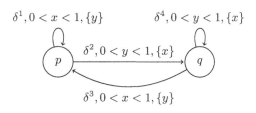

Fig. 1. The running example. Right: $\mathcal{G}^{\text{ex1}}$; left: Its state space (in gray).

respectively. Successor action is defined by $[p, (x, 0)] \triangleright (t, \delta^1) = [p, (x + t, 0)]$ and $[p, (x, 0)] \triangleright (t, \delta^2) = [q, (0, t)]$ if $x + t < 1$; $[q, (0, y)] \triangleright (t, \delta^3) = [p, (t, 0)]$ and $[q, (0, y)] \triangleright (t, \delta^4) = [q, (0, y + t)]$ if $y + t < 1$. An example of run of $\mathcal{G}^{\text{ex1}}$ is $(p, (0.5, 0))(0.4, \delta^1)(p, (0.9, 0))(0.8, \delta^2)(q, (0, 0.8))(0.1, \delta^3)(p, (0.1, 0))$.

Integrating over States and Runs; Volume of Runs. It is well known (see [4]) that a region is uniquely described by the integer parts of clocks and by an order on their fractional parts, e.g. in the region $\mathbf{r}^{\text{ex}}$ given by the constraints $0 = x_2 < x_3 - 4 = x_4 - 3 < x_1 - 2 < 1$, the integer parts are $\lfloor x_1 \rfloor = 2, \lfloor x_2 \rfloor = 0, \lfloor x_3 \rfloor = 4, \lfloor x_4 \rfloor = 3$ and fractional parts are ordered as follows $0 = \{x_2\} < \{x_3\} = \{x_4\} < \{x_1\} < 1$. We denote by $\gamma_1 < \gamma_2 < \cdots < \gamma_d$ the fractional parts different from 0 of clocks of a region $\mathbf{r}_q$ (d is called the dimension of the region). In our example the dimension of $\mathbf{r}^{\text{ex}}$ is 2 and $(\gamma_1, \gamma_2) = (x_3 - 4, x_1 - 2)$. We denote by Γ_q the simplex $\Gamma_q = \{\gamma \in \mathbb{R}^d \mid 0 < \gamma_1 < \gamma_2 < \cdots < \gamma_d < 1\}$. The mapping $\phi_{\mathbf{r}} : \mathbf{x} \mapsto \gamma$ is a natural bijection from the d dimensional region $\mathbf{r} \subset \mathbb{R}^{|X|}$ to $\Gamma_q \subset \mathbb{R}^d$. In the example the pre-image of a vector (γ_1, γ_2) is $(\gamma_2 + 2, 0, \gamma_1 + 4, \gamma_1 + 3)$.

Example 2 (Continuing example 1). The region $\mathbf{r}_p = \{(x, y) \mid 0 = y < x < 1\}$ is 1-dimensional, $\phi_{\mathbf{r}_p}(x, y) = x$ and $\phi_{\mathbf{r}_p}^{-1}(\gamma) = (\gamma, 0)$.

Now, we introduce simplified notation for sums of integrals over states, transitions and runs. We define the integral of an integrable[1] function $f : \mathbb{S} \to \mathbb{R}$ (over states):

$$\int_{\mathbb{S}} f(s)ds = \sum_{q \in Q} \int_{\Gamma_q} f(q, \phi_{\mathbf{r}_q}^{-1}(\gamma))d\gamma.$$

where $\int .d\gamma$ is the usual integral (w.r.t. the Lebesgue measure). We define the integral of an integrable function $f : \mathbb{A} \to \mathbb{R}$ (over timed transitions):

$$\int_{\mathbb{A}} f(\alpha)d\alpha = \sum_{\delta \in \Delta} \int_{[0,M]} f(t, \delta)dt$$

---

[1] A function $f : \mathbb{S} \to \mathbb{R}$ is integrable if for each $q \in Q$ the function $\gamma \mapsto f(q, \phi_{\mathbf{r}_q}^{-1}(\gamma))$ is Lebesgue integrable. A function $f : \mathbb{A} \to \mathbb{R}$ is integrable if for each $\delta \in \Delta$ the function $t \mapsto f(t, \delta)$ is Lebesgue integrable.

and the integral of an integrable function $f : \mathcal{R}_n \to \mathbb{R}$ (over runs) with the convention that $f[s, \boldsymbol{\alpha}] = 0$ if $s \rhd \alpha = \perp$:

$$\int_{\mathcal{R}_n} f[s, \boldsymbol{\alpha}]d[s, \boldsymbol{\alpha}] = \int_S \int_A \cdots \int_A f[s, \boldsymbol{\alpha}]d\alpha_1 \ldots d\alpha_n ds$$

To summarize, we take finite sums over finite discrete sets Q, $\varDelta$ and take integrals over dense sets Γ_q, $[0, M]$. More precisely, all the integrals we define have their corresponding measures[2] which are products of counting measures on discrete sets Σ, Q and Lebesgue measure over subsets of $\mathbb{R}^m$ for some $m \geq 0$ (e.g. Γ_q, $[0, M]$). We denote by $\mathfrak{B}(\mathbb{S})$ (resp. $\mathfrak{B}(\mathbb{A})$) the set of measurable subsets of $\mathbb{S}$ (resp. $\mathbb{A}$).

The volume of the set of n-length runs is defined by:

$$\mathtt{Vol}(\mathcal{R}_n) = \int_{\mathcal{R}_n} 1 d[s, \boldsymbol{\alpha}] = \int_S \int_{A^n} 1_{s \rhd \alpha \neq \perp} d\boldsymbol{\alpha} ds$$

Remark 1. The use of reduced version of runs is crucial when dealing with integrals (and densities in the following). Indeed the following integral on the extended version of runs is always null since variables are linked ($s_{i+1} = s_i \rhd \alpha_i$ for $i = 0..n-2$): $\int_A \int_S \cdots \int_A \int_S 1_{s_0 \alpha_0 \cdots s_{n-1} \alpha_{n-1} \in \mathcal{R}_n} ds_0 d\alpha_0 \ldots ds_{n-1} d\alpha_{n-1} = 0$.

2.2   SPOR on Timed Region Graphs

Let $(\Omega, \mathcal{F}, P)$ be a probability space. A *stochastic process over runs* (SPOR) of a timed region graph $\mathcal{G}$ is a sequence of random variables $(Y_n)_{n \in \mathbb{N}} = (S_n, A_n)_{n \in \mathbb{N}}$ such that:

C.1) For all $n \in \mathbb{N}$, $S_n : (\Omega, \mathcal{F}, P) \to (\mathbb{S}, \mathfrak{B}(\mathbb{S}))$ and $A_n : (\Omega, \mathcal{F}, P) \to (\mathbb{A}, \mathfrak{B}(\mathbb{A}))$.

C.2) The initial state S_0 has a probability density function (PDF) $p_0 : \mathbb{S} \to \mathbb{R}^+$ i.e. for every $\mathcal{S} \in \mathfrak{B}(\mathbb{S})$, $P(S_0 \in \mathcal{S}) = \int_{s \in \mathcal{S}} p_0(s)ds$ (in particular $P(S_0 \in \mathbb{S}) = \int_{s \in \mathbb{S}} p_0(s)ds = 1$).

C.3) Probability on every timed transition only depends on the current state: for every $n \in \mathbb{N}$, $\mathcal{A} \in \mathfrak{B}(\mathbb{A})$, for almost every[3] $s \in \mathbb{S}$, $y_0 \cdots y_n \in (\mathbb{S} \times \mathbb{A})^n$,

$$P(A_n \in \mathcal{A}|S_n = s, Y_n = y_n, \ldots, Y_0 = y_0) = P(A_n \in \mathcal{A}|S_n = s),$$

moreover this probability is given by a conditional PDF $p(.|s) : \mathbb{A} \to \mathbb{R}^+$ such that $P(A_n \in \mathcal{A}|S_n = s) = \int_{\alpha \in \mathcal{A}} p(\alpha|s)d\alpha$ and $p(\alpha|s) = 0$ if $s \rhd \alpha = \perp$ (in particular $P(A_n \in \mathbb{A}|S_n = s) = \int_{\alpha \in \mathbb{A}} p(\alpha|s)d\alpha = 1$).

C.4) States are updated deterministically knowing the previous state and transition: $S_{n+1} = S_n \rhd A_n$.

---

[2] We refer the reader to [12] for an introduction to measure and probability theory.

[3] A property *prop* (like "f is positive", "well defined"...) on a set B holds *almost everywhere* when the set where it is false has measure (volume) 0: $\int_B 1_{b \not\models prop} db = 0$.

For all $n \geq 1$, $Y_0 \cdots Y_{n-1}$ has a PDF $p_n : \mathcal{R}_n \to \mathbb{R}^+$ i.e. for every $R \in \mathfrak{B}(\mathcal{R}_n)$, $P(Y_0 \cdots Y_{n-1} \in R) = \int_{\mathcal{R}_n} p_n[s, \boldsymbol{\alpha}] 1_{[s,\boldsymbol{\alpha}] \in R} d[s, \boldsymbol{\alpha}]$. This PDF can be defined with the following chain rule:

$$p_n[s_0, \boldsymbol{\alpha}] = p_0(s_0) p(\alpha_0 | s_0) p(\alpha_1 | s_1) \ldots p(\alpha_{n-1} | s_{n-1})$$

where for each $j = 1..n - 1$ the state updates are defined by $s_j = s_{j-1} \triangleright \alpha_{j-1}$.

The SPOR $(Y_n)_{n \in \mathbb{N}}$ is called *stationary* whenever for all $i, n \in \mathbb{N}$, $Y_i \cdots Y_{i+n-1}$ has the same PDF as $Y_0 \cdots Y_{n-1}$ which is p_n.

Simulation According to a SPOR. Given a SPOR Y, a run $(s_0, \boldsymbol{\alpha}) \in \mathcal{R}_n$ can be chosen randomly w.r.t. Y with a linear number of the following operations: random pick according to p_0 or $p(.|s)$ and computing of a successor. Indeed it suffices to pick s_0 according to p_0 and for $i = 0..n - 1$ to pick α_i according to $p(.|s_i)$ and to make the update $s_{i+1} = s_i \triangleright \alpha_i$.

2.3   Entropy

In this sub-section, we define entropy for timed region graphs and SPOR. The first one is inspired by [6] and the second one by [21].

Entropy of a Timed Region Graph

Proposition-Definition 1. *Given a timed region graph $\mathcal{G}$, the following limit exists and defined the entropy of $\mathcal{G}$:*

$$\mathcal{H}(\mathcal{G}) = \lim_{n \to \infty} \frac{1}{n} \log_2(\mathrm{Vol}(\mathcal{R}_n)).$$

When $\mathcal{H}(\mathcal{G}) > -\infty$, the timed region graph is *thick*, the volume behaves w.r.t. n like an exponent: $\mathrm{Vol}(\mathcal{R}_n) \approx 2^{n\mathcal{H}}$. When $\mathcal{H}(\mathcal{G}) = -\infty$, the timed region graph is *thin*, the volume decays faster than any exponent: $\forall \rho > 0$, $\mathrm{Vol}(\mathcal{R}_n) << \rho^n$.

Entropy of a SPOR

Proposition-Definition 2. *If Y is a stationary SPOR, then*

$$-\frac{1}{n} \int_{\mathcal{R}_n} p_n[s, \boldsymbol{\alpha}] \log_2 p_n[s, \boldsymbol{\alpha}] d[s, \boldsymbol{\alpha}] \to_{n \to \infty} -\int_{\mathbb{S}} p_0(s) \int_{\mathbb{A}} p(\alpha|s) \log_2 p(\alpha|s) d\alpha ds.$$

This limit is called the entropy of Y, denoted by $H(Y)$.

Proposition 1. *Let $\mathcal{G}$ be a timed region graph and Y be a stationary SPOR on $\mathcal{G}$. Then the entropy of Y is upper bounded by that of $\mathcal{G}$: $H(Y) \leq \mathcal{H}(\mathcal{G})$.*

The main contribution of this article is a construction of a stationary SPOR for which the equality holds i.e. a timed analogue of the Shannon-Parry Markov Chain [21,20].

3   Maximal Entropy SPOR and Quasi Uniform Sampling

In this section $\mathcal{G}$ is a timed region graph satisfying the technical condition below (section 3.1). We present a stationary SPOR Y^* for which the upper bound on entropy is reached $H(Y^*) = \mathcal{H}(\mathcal{G})$ (Theorem 1). Another key property of this SPOR is the ergodicity we define now:

Ergodicity. Given a set of infinite runs $R \subseteq (\mathbb{S} \times \mathbb{A})^\omega$ and $i, j \in \mathbb{N}$, we denote by $R_i^{i+j} \subseteq (\mathbb{S} \times \mathbb{A})^{j+1}$ the set of finite runs $(s_i, \alpha_i) \cdots (s_{i+j}, \alpha_{i+j})$ that can occur between indices i and $i+j$ in an infinite run $(s_k, \alpha_k)_{k \in \mathbb{N}}$ of R. Let Y be a *stationary* SPOR then the sequence $P(Y_0 \cdots Y_n \in R_0^n)$ decreases and converges to a value called the probability of R and denoted by $P(R) = \lim_{n \to \infty} P(Y_0 \cdots Y_n \in R_0^n)$. The set R is *shift invariant* if for every $i, n \in \mathbb{N}$: $R_i^{i+n} = R_0^n$. A stochastic process is *ergodic* whenever it is stationary and every shift invariant set has probability 0 or 1. Definition of ergodicity for general probability measures can be found in [12].

3.1   Technical Assumptions

In this section we explain and justify several technical assumptions on the timed region graph $\mathcal{G}$ we make in the following.

Bounded Delays. If the delays were not bounded the sets of runs $\mathcal{R}_n$ would have infinite volumes and thus a quasi uniform random generation cannot be achieved.

Fleshy Transitions. We consider timed region graphs whose transitions are *fleshy* [6]: there is no constraints of the form $x = c$ in their guards. Non fleshy transitions yield a null volume and are thus useless. Delete them reduces the size of the timed region graph considered and ensures that every path has a positive volume (see [6,9] for more justifications and details).

Strong Connectivity of the Set of Locations. We will consider only timed region graph which are strongly connected i.e. locations are pairwise reachable. This condition (usual in the discrete case we generalize) is not restrictive since the set of locations can be decomposed in strongly connected components and then a maximal entropy SPOR can be designed for each components.

Thickness. In the maximal entropy approach we adopt, we need that the entropy is finite $\mathcal{H}(\mathcal{G}) > -\infty$. This is why we restrict our attention to *thick* timed region graph. The dichotomy between thin and thick timed region graphs was characterized precisely in [9] where it appears that thin timed region graph are degenerate. The key characterization of thickness is the existence of a forgetful cycle [9]. When the locations are strongly connected, existence of such a forgetful cycle ensures that the state space $\mathbb{S}$ is strongly connect i.e. for all $s, s' \in \mathbb{S}$ there exists $\alpha \in \mathbb{A}^*$ such that $s \triangleright \alpha = s'$.

Weak Progress Cycle Condition. In [6] the following assumption (known as the *progress cycle condition*) was made: for some positive integer constant D, on each path of D consecutive transitions, all the clocks are reset at least once.

Here we use a weaker condition: for a positive integer constant D, a timed region graph satisfies the D *weak progress condition* (D-WPC) if on each path of D consecutive transitions at most one clock is not reset during the entire path.

The timed region graph on figure 1 does not satisfy the progress cycle condition (e.g. x is not reset along δ^1) but satisfies the 1-WPC.

3.2   Main Theorems

Theorem 1. *There exists a positive real ρ and two functions $v, w : \mathbb{S} \mapsto \mathbb{R}$ positive almost everywhere such that the following equations define the PDF of an ergodic SPOR Y^* with maximal entropy: $H(Y^*) = \mathcal{H}(\mathcal{G})$.*

$$p_0^*(s) = w(s)v(s); \quad p^*(\alpha|s) = \frac{v(s \triangleright \alpha)}{\rho v(s)}. \tag{1}$$

Objects ρ, v, w are spectral attributes of an operator Ψ defined in the next section.

An ergodic SPOR satisfies an asymptotic equipartition property (AEP) (see [14] for classical AEP and [1] which deals with the case of non necessarily Markovian stochastic processes with density). Here we give our own AEP. It strongly relies on the pointwise ergodic theorem (see [12]) and on the Markovian property satisfied by every SPOR (conditions C.3 and C.4).

Theorem 2 (AEP for SPOR). *If Y is an ergodic SPOR then*

$$P[\{s_0\alpha_0s_1\alpha_1 \cdots \mid -(1/n)\log_2 p_n[s_0, \alpha_0 \cdots \alpha_n] \to_{n \to +\infty} H(Y)\}] = 1$$

This theorem applied to the maximal entropy SPOR Y^* means that long runs have a high probability to have a quasi uniform density:

$$p_n^*[s_0, \alpha_0 \cdots \alpha_n] \approx 2^{-nH(Y^*)} \approx 1/\text{Vol}(\mathcal{R}_n) \text{ (since } H(Y^*) = H(\mathcal{G})).$$

3.3   Operator Ψ and Its Spectral Attributes ρ, v, w

The maximal entropy SPOR is a lifting to the timed setting of the Shannon-Parry Markov chain of a finite strongly connected graph. The definition of this chain is based on the Perron-Frobenius theory applied to the adjacency matrix M of the graph. This theory ensures that there exists both a positive eigenvector v of M for the spectral radius[4] ρ (i.e. $Mv = \rho v$) and a positive eigenvector w of the transposed matrix $M^\top$ for ρ (i.e. $M^\top w = \rho w$). The initial probability distribution on the states Q of the Markov chain is given by $p_i = v_iw_i$ for $i \in Q$ and the transition probability matrix P is given by $P_{ij} = v_jM_{ij}/(\rho v_i)$ for $i, j \in Q$. The timed analogue of M is the operator Ψ introduced in [6]. To

---

[4] Recall from linear algebra (resp. spectral theory) that the spectrum of a matrix (resp. of an operator) Ψ is the set $\{\lambda \in \mathbb{C} \text{ s.t. } \Psi - \lambda Id \text{ is not invertible.}\}$. The spectral radius ρ of Ψ is the radius of the smallest disc centered in 0 which contains all the spectrum.

define ρ, v and w, we will use the theory of positive linear operators (see e.g. [17]) instead of the Perron-Frobenius theory used in the discrete case.

The operator Ψ of a timed region graph is defined by:

$$\forall f \in L_2(\mathbb{S}),\ \forall s \in \mathbb{S},\ \Psi f(s) = \int_A f(s \triangleright \alpha) d\alpha \ \text{(with } f(\perp) = 0), \qquad (2)$$

where $L_2(\mathbb{S})$ is the Hilbert space of square integrable functions from $\mathbb{S}$ to $\mathbb{R}$ with the scalar product $\langle f, g \rangle = \int_{\mathbb{S}} f(s)g(s)ds$ and associated norm $||f||_2 = \sqrt{\langle f, f \rangle}$.

Proposition 2. *The operator Ψ defined in (2) is a positive continuous linear operator on $L_2(\mathbb{S})$.*

The real ρ used in (1) is the spectral radius of Ψ.

Theorem 3 (adapted from [6] to $L_2(\mathbb{S})$). *The spectral radius ρ is a positive eigenvalue (i.e. $\rho > 0$ and $\exists v \in L_2(\mathbb{S})$ s.t. $\Psi v = \rho v$) and $\mathcal{H}(\mathcal{G}) = \log_2(\rho)$.*

The adjoint operator Ψ^* (acting also on $L_2(\mathbb{S})$) is the analogue of $M^\top$. It is formally defined by the equation:

$$\forall f, g \in L_2(\mathbb{S}),\ \langle \Psi f, g \rangle = \langle f, \Psi^* g \rangle. \qquad (3)$$

A more effective characterization of some power of Ψ^* and then of its eigenfunctions is ensured by proposition 3 below. The following theorem defines v, w used in the definition of the maximal entropy SPOR (1).

Theorem 4. *There exists a unique eigenfunction (up to a scalar constant) v of Ψ (resp. w of Ψ^*) for the eigenvalue ρ which is positive almost everywhere. Any non-negative eigenfunction of Ψ (resp. Ψ^*) is collinear to v (resp. w).*

Eigenfunctions v and w are chosen such that $\langle w, v \rangle = 1$. Operator Ψ and Ψ^* are easier to describe when elevated to some power greater than D the constant of weak progress cycle condition.

Proposition 3. *For every $n \geq D$ there exists a function $k_n \in L_2(\mathbb{S} \times \mathbb{S})$ such that: $\Psi^n(f)(s) = \int_{\mathbb{S}} k_n(s, s')f(s')ds'$ and $\Psi^{*n}(f)(s) = \int_{\mathbb{S}} k_n(s', s)f(s')ds'$.*

It is worth mentioning that for any $n \geq D$, the objects ρ, v (resp. w) are solutions of the eigenvalue problem $\int_{\mathbb{S}} k_n(s, s')v(s')ds' = \rho^n v(s)$ with v non negative (resp. $\int_{\mathbb{S}} k_n(s', s)w(s')ds' = \rho^n w(s)$ with w non negative); unicity of v (resp. w) up to a scalar constant is ensured by Theorem 4. Further computability issues for ρ, v and w are discussed in the conclusion.

Sketch of Proof of Theorem 4. The proof of Theorem 4 is based on theorem 11.1 condition e) of [17] which is a generalization of Perron-Frobenius Theorem to positive linear operators. The main hypothesis to prove is the irreducibility of Ψ whose analogue in the discrete case is the irreducibility of the adjacency matrix M of a finite graph i.e for all states i, j there exists $n \geq 1$ such that $M_{ij}^n > 0$ (this is equivalent to the strong connectivity of the graph). The following key lemma is a sufficient condition for the irreducibility of Ψ. It is based on the strong connectivity of the state space $\mathbb{S}$ ensured both by strong connectivity of the set of locations and by thickness (see section 3.1).

Lemma 1. *For every $q, q' \in Q$, there exists $n \geq D$ such that k_n (defined in Proposition 3) is positive almost everywhere on $(\{q\} \times \mathbf{r}_q) \times (\{q'\} \times \mathbf{r}_{q'})$.*

3.4 Running Example Completed

Example 3. Let us make (2) explicit on our running example.

$$\Psi f(p, (x, 0)) = \int f(p, (x+t, 0))\mathbf{1}_{0<x\leq x+t<1}dt + \int f(q, (0, t))\mathbf{1}_{0\leq t<1}dt$$
$$\Psi f(q, (0, y)) = \int f(p, (t, 0))\mathbf{1}_{0\leq t<1}dt \qquad + \int f(q, (0, y+t))\mathbf{1}_{0<y\leq y+t<1}dt$$

Integrals from left to right correspond to transitions δ^1, δ^2 for the first line and to δ^3, δ^4 for the second line.

We introduce the notations $v_p(\gamma) = v(p, (\gamma, 0))$ and $v_q(\gamma) = v(q, (0, \gamma))$. With these notations the eigenvalue equation $\rho v = \Psi v$ gives:

$$\rho v_p(\gamma) = \int_\gamma^1 v_p(\gamma')d\gamma' + \int_0^1 v_q(\gamma')d\gamma'; \quad \rho v_q(\gamma) = \int_0^1 v_p(\gamma')d\gamma' + \int_\gamma^1 v_q(\gamma')d\gamma'.$$

Similarly the eigenfunction w satisfies:

$$\rho w_p(\gamma) = \int_0^\gamma w_p(\gamma')d\gamma' + \int_0^1 w_q(\gamma')d\gamma'; \quad \rho w_q(\gamma) = \int_0^1 w_p(\gamma')d\gamma' + \int_0^\gamma w_q(\gamma')d\gamma'.$$

After some calculus we obtain that $\rho = 1/\ln(2)$; $v_p(\gamma) = v_q(\gamma) = C2^{-\gamma}$; $w_p(\gamma) = w_q(\gamma) = C'2^\gamma$ with C and C' two positive constants.

Finally the maximal entropy SPOR for $\mathcal{G}^{\text{ex1}}$ is given by:

$$p_0^*(p, (\gamma, 0)) = p_0^*(q, (0, \gamma)) = \frac{1}{2} \text{ for } \gamma \in (0, 1);$$

$$p^*(t, \delta^1|p, (\gamma, 0)) = p^*(t, \delta^4|q, (0, \gamma)) = \frac{2^{-t}}{\rho} \text{ for } \gamma \in (0, 1), t \in (0, 1);$$

$$p^*(t, \delta^2|p, (\gamma, 0)) = p^*(t, \delta^3|q, (0, \gamma)) = \frac{2^{\gamma-t}}{\rho} \text{ for } \gamma \in (0, 1), t \in (0, 1).$$

4   Conclusion and Perspectives

In this article, we have proved the existence of an ergodic stochastic process over runs of a timed region graph $\mathcal{G}$ with maximal entropy, provided $\mathcal{G}$ has finite entropy ($\mathcal{H} > -\infty$) and satisfies the D weak progress condition.

The next question is to know how simulation can be achieved in practice. Symbolic computation of ρ and v have been proposed in [6] for subclasses of deterministic TA. In the same article, an iterative procedure is also given to estimate the entropy $\mathcal{H} = \log_2(\rho)$. We think that approximations of ρ, v and w using an iterative procedure on Ψ and Ψ^* would give a SPOR with entropy as close to the maximum as we want. A challenging task for us is to determine an upper bound on the convergence rate of such an iterative procedure.

Connection with information theory is clear if we consider as in [5], a timed regular language as a source of timed words. A SPOR is in this approach a stochastic source of timed words. It would be very interesting to lift compression methods (see [19,14]) from untimed to timed setting.

Acknowledgements. I thank Eugene Asarin, Aldric Degorre and Dominique Perrin for sharing motivating discussions.

References

1. Algoet, P.H., Cover, T.M.: A sandwich proof of the Shannon-McMillan-Breiman theorem. The Annals of Probability 16(2), 899–909 (1988)
2. Alur, R., Bernadsky, M.: Bounded model checking for GSMP models of stochastic real-time systems. In: Hespanha, J.P., Tiwari, A. (eds.) HSCC 2006. LNCS, vol. 3927, pp. 19–33. Springer, Heidelberg (2006)
3. Alur, R., Courcoubetis, C., Dill, D.L.: Model-checking for probabilistic real-time systems. In: Leach Albert, J., Monien, B., Rodríguez-Artalejo, M. (eds.) ICALP 1991. LNCS, vol. 510, Springer, Heidelberg (1991)
4. Alur, R., Dill, D.L.: A theory of timed automata. Theoretical Computer Science 126, 183–235 (1994)
5. Asarin, E., Basset, N., Béal, M.-P., Degorre, A., Perrin, D.: Toward a timed theory of channel coding. In: Jurdziński, M., Ničković, D. (eds.) FORMATS 2012. LNCS, vol. 7595, pp. 27–42. Springer, Heidelberg (2012)
6. Asarin, E., Degorre, A.: Volume and entropy of regular timed languages: Analytic approach. In: Ouaknine, J., Vaandrager, F.W. (eds.) FORMATS 2009. LNCS, vol. 5813, pp. 13–27. Springer, Heidelberg (2009)
7. Asarin, E., Degorre, A.: Volume and entropy of regular timed languages: Discretization approach. In: Bravetti, M., Zavattaro, G. (eds.) CONCUR 2009. LNCS, vol. 5710, pp. 69–83. Springer, Heidelberg (2009)
8. Baier, C., Bertrand, N., Bouyer, P., Brihaye, T., Größer, M.: Probabilistic and topological semantics for timed automata. In: Arvind, V., Prasad, S. (eds.) FSTTCS 2007. LNCS, vol. 4855, pp. 179–191. Springer, Heidelberg (2007)
9. Basset, N., Asarin, E.: Thin and thick timed regular languages. In: Fahrenberg, U., Tripakis, S. (eds.) FORMATS 2011. LNCS, vol. 6919, pp. 113–128. Springer, Heidelberg (2011)
10. Bernadsky, M., Alur, R.: Symbolic analysis for GSMP models with one stateful clock. In: Bemporad, A., Bicchi, A., Buttazzo, G. (eds.) HSCC 2007. LNCS, vol. 4416, pp. 90–103. Springer, Heidelberg (2007)
11. Bertrand, N., Bouyer, P., Brihaye, T., Markey, N.: Quantitative model-checking of one-clock timed automata under probabilistic semantics. In: QEST, pp. 55–64. IEEE Computer Society (2008)
12. Billingsley, P.: Probability and measure, vol. 939. Wiley (2012)
13. Bouyer, P., Brihaye, T., Jurdziński, M., Menet, Q.: Almost-sure model-checking of reactive timed automata. QEST 2012, 138–147 (2012)
14. Cover, T.M., Thomas, J.A.: Elements of information theory, 2nd edn. Wiley (2006)
15. David, A., Larsen, K.G., Legay, A., Mikučionis, M., Poulsen, D.B., van Vliet, J., Wang, Z.: Statistical model checking for networks of priced timed automata. In: Fahrenberg, U., Tripakis, S. (eds.) FORMATS 2011. LNCS, vol. 6919, pp. 80–96. Springer, Heidelberg (2011)
16. Kempf, J.-F., Bozga, M., Maler, O.: As soon as probable: Optimal scheduling under stochastic uncertainty. In: Piterman, N., Smolka, S.A. (eds.) TACAS 2013 (ETAPS 2013). LNCS, vol. 7795, pp. 385–400. Springer, Heidelberg (2013)

17. Krasnosel'skij, M.A., Lifshits, E.A., Sobolev, A.V.: Positive Linear Systems: the Method of Positive Operators. Heldermann Verlag, Berlin (1989)
18. Lind, D., Marcus, B.: An Introduction to Symbolic Dynamics and Coding. Cambridge University Press (1995)
19. Lothaire, M.: Applied Combinatorics on Words (Encyclopedia of Mathematics and its Applications). Cambridge University Press, New York (2005)
20. Parry, W.: Intrinsic Markov chains. Transactions of the American Mathematical Society, 55–66 (1964)
21. Shannon, C.E.: A mathematical theory of communication. Bell Sys. Tech. J. 27, 379–423, 623–656 (1948)

Complexity of Two-Variable Logic on Finite Trees

Saguy Benaim, Michael Benedikt[1,*], Witold Charatonik[2,**], Emanuel Kieroński[2,***], Rastislav Lenhardt[1], Filip Mazowiecki[3,**], and James Worrell[1]

[1] University of Oxford
[2] University of Wrocław
[3] University of Warsaw

Abstract. Verification of properties expressed in the two-variable fragment of first-order logic FO^2 has been investigated in a number of contexts. The satisfiability problem for FO^2 over arbitrary structures is known to be NEXPTIME-complete, with satisfiable formulas having exponential-sized models. Over words, where FO^2 is known to have the same expressiveness as unary temporal logic, satisfiability is again NEXPTIME-complete. Over finite labelled ordered trees FO^2 has the same expressiveness as navigational XPath, a popular query language for XML documents. Prior work on XPath and FO^2 gives a 2EXPTIME bound for satisfiability of FO^2 over trees. This work contains a comprehensive analysis of the complexity of FO^2 on trees, and on the size and depth of models. We show that the exact complexity varies according to the vocabulary used, the presence or absence of a schema, and the encoding of labels on trees. We also look at a natural restriction of FO^2, its guarded version, GF^2. Our results depend on an analysis of types in models of FO^2 formulas, including techniques for controlling the number of distinct subtrees, the depth, and the size of a witness to satisfiability for FO^2 sentences over finite trees.

1 Introduction

The complexity of verifying properties over a class of structures depends on both the specification language and the type of structure. Stockmeyer [Sto74] showed that full first-order logic (FO) has non-elementary complexity even when applied to very restricted structures, such as words. The two-variable fragment, FO^2, is known to have better complexity. Grädel *et al.* [GKV97] showed that satisfiability over arbitrary relational vocabularies is NEXPTIME-complete, with satisfiable sentences having exponential-sized models. Over words, Etessami *et al.* showed that satisfiability remains NEXPTIME-complete, with satisfiable formulas again

---

* Supported by EPSRC grant EP/G004021/1.
** Supported by Polish NCN grant number DEC-2011/03/B/ST6/00346.
*** Supported by Polish Ministry of Science and Higher Education grant N N206 371339.

F.V. Fomin et al. (Eds.): ICALP 2013, Part II, LNCS 7966, pp. 74–88, 2013.

having exponential-sized models [EVW02]. Moreover the complexity bounds over words extend to bounds on a host of related verification problems [BLW12].

The NEXPTIME-completeness of FO^2 over both general structures and words raises the question of the impact of structural restrictions on the complexity of FO^2. Surprisingly the complexity of satisfiability for FO^2 over finite trees has not been not been investigated in detail. Marx and De Rijke [MdR04] showed that FO^2 over trees corresponds precisely to the navigational core of the XML query language XPath. From the work of Marx [Mar04] it follows that the satisfiability problem for XPath is complete for EXPTIME. Given that the translation from FO^2 to XPath in [MdR04] is exponential, this gives a 2EXPTIME bound on satisfiability for FO^2 over trees.

In this work we will consider the satisfiability problem for FO^2 over finite trees, and the corresponding question of the size and depth needed for witness models. In particular, we will consider:

- satisfiability in the presence of all navigational predicates: predicates for the parent/child relation, its transitive closure the descendant relation, the left/right sibling relation and its transitive closure;
- the impact on the complexity of limiting sentences to make use of predicates in a particular subset;
- satisfiability over general unranked trees, and satisfiability in the presence of a schema;
- satisfiability over trees where node labels are denoted with explicit unary labels versus the case where node labels are boolean combinations over a propositional alphabet;
- satisfiability of the full logic, versus the restriction to the case where quantification of variables must be guarded.

We will show that each of these variations affects the complexity of the problem. In the process, we will show that the tree case differs in a number of important ways from that of words. First, satisfiability is EXPSPACE-complete, unlike for general structures or words. Secondly, the basic technique for analyzing FO^2 on words [EVW02]—bounds on the number of quantifier-rank types that occur in a structure—is not useful for getting tight complexity bounds for FO^2 over trees. Instead we will use a combination of methods, including reductions to XPath, bounds on the number of subformula types, and a quotient construction that is based not only on types, but on a set of distinguished witness nodes. These techniques allow us to distinguish situations where satisfiable FO^2-formulas have models of (reasonably) small depth, and situations where they have models of small size.

Related Work. Two-variable logic on *data trees*, in which nodes are associated with values in an infinite set of data, has been studied by Bojańczyk et al. [BMSS09]. There the main result is decidability over the signature with data equality, the child relation, and the right sibling relation. Figueira [Fig12] considers two-variable logic with two successor relations, respectively corresponding to two different linear orders. This is quite different from considering the two successor relations derived from a tree order. Figueira's results were generalized in a recent work [CW13] where two-variable logic over structures that contain two forests of finite trees, even in presence of additional binary predicates and counting quantifiers, is proved to be decidable in NEXPTIME. However, these results are not comparable with ours because the logic in [CW13] is restricted to ranked trees, it cannot express the sibling relation and does not allow using the transitive descendant relation. The two-variable logic over two transitive relations is shown undecidable in [Kie05]. On the other hand the two-variable fragment with one transitive relation has been recently shown by Szwast and Tendera to be decidable [ST13]. The complexity of two-variable logic over ordinary trees is explicitly studied in [BK08]. Our results here show that the proof of the satisfiability result there (claiming NEXPTIME for full two-variable logic) is incorrect.

Organization. Section 2 gives preliminaries. Section 3 gives precise bounds for the satisfiability of full FO^2 on trees. Section 4 considers the case where the child predicate is absent, while Section 5 considers the case where the descendant predicate is absent. Section 6 considers restricting the logic to its guarded version. Section 7 gives conclusions.

2   Logics and Models

A *tree* is a finite directed acyclic graph whose edge relation is denoted $\downarrow$. We denote the transitive closure of the edge relation by $\downarrow_+$ and assume that $\downarrow_+$ is a partial order with only one minimum element—the root of the tree. We read $u \downarrow v$ as 'v is a child of u' and $u \downarrow_+ v$ as 'v is a descendant of u'. We assume also a *sibling* relation $\rightarrow$ whose transitive closure $\rightarrow^+$ restricts to a linear order on the children of each node. We read $u \rightarrow v$ as 'v is the right sibling of u'. Given a tree t and vertex v, SubTree(t, v) denotes the subtree of t rooted at v. We consider trees equipped with a family of unary predicates $P_1, P_2, \ldots$ on their vertices. We further say that a tree satisfies the *unary alphabet restriction* (UAR) if exactly one predicate P_i holds of each vertex.

We consider first-order logic with equality over signatures of the form $\tau_{un} \cup \tau_{bin}$, where τ_{un} consists of unary predicates and $\tau_{bin} \subseteq \{\downarrow, \downarrow_+, \rightarrow, \rightarrow^+\}$ is a subset of the *navigational relations*. Over such signatures we consider the two-variable subset of first-order logic FO^2 and its *guarded fragment* GF^2 (cf. [AvBN98]). The former comprises those formulas built using only variables x and y, while in the latter quantifiers are additionally relativised to atoms, i.e., universally quantified formulas have the form $\forall y \, (\alpha(x, y) \Rightarrow \varphi(x, y))$ and existentially quantified formulas have the form $\exists y \, (\alpha(x, y) \wedge \varphi(x, y))$, where $\alpha(x, y)$ is atomic. Atom $\alpha(x, y)$ is called a *guard*. Equalities $x = x$ or $x = y$ are also allowed as guards.

We write $FO^2[\tau_{bin}]$ or $GF^2[\tau_{bin}]$ to indicate that the only binary symbols that are allowed in formulas are those from τ_{bin} and that formulas are interpreted over trees. Although we allow equality in our upper bounds, it will not play any role in the lower bounds. When interpreting formulas of the logic in a tree, we will always assume the navigational relations are given their natural interpretation: $\downarrow$ as the child relation, $\downarrow_+$ the descendant relation, and so forth.

We consider also satisfiability for FO^2 over k-*ranked trees*, that is, trees where nodes have at most k children. Note that for k-ranked trees it is natural to consider signatures that include the relation $\downarrow_i$, connecting a node to its ith child for each $i \leq k$, either in place of or in addition to the predicates above. However we will not consider a separate signature for ranked trees, since it is easy to derive tight bounds for ranked trees for such signatures based on the techniques introduced here.

A *ranked tree schema* consists of a bottom-up tree automaton on trees of some rank k [Tho97]. A tree automaton takes trees labeled from a finite set Σ. We will thus identify the symbols in Σ with predicates P_i, and thus all trees satisfying the schema will satisfy the UAR.

We consider the following problems:

- Given an FO^2 sentence φ, determine if there is some tree (resp. k-ranked, UAR tree) that satisfies it.
- Given an FO^2 sentence φ and a schema S, determine whether φ is satisfied by some tree satisfying S. We consider the combined complexity in the formula and schema.

Some of our results will go through XPath, a common language used for querying XML documents viewed as trees. The navigational core of XPath is a modal language, analogous to unary temporal logic on trees, denoted NavXP. NavXP is built on binary modalities, referred to as *axis relations*. We will focus on the following axes: self, parent, child, descendant, descendant-or-self, ancestor-or-self, next-sibling, following-sibling, preceding-sibling, previous-sibling. In a tree t, we associate each axis a with a set R_a^t of pairs of nodes. R_{child}^t denotes the set of pairs of nodes (x, y) in t where y is a child of x, and similarly for the other axes (see [Mar04]).

NavXP consists of path expressions, which denote binary relations between nodes in a tree, and filters, denoting unary relations. Below we give the syntax (from [BK08]), using p to range over path expressions and q over filters. L ranges over symbols for each labelling of a node (i.e. for general trees, boolean combinations of predicates $P_1, P_2, \ldots$, for UAR trees a single predicate).

$$p ::= step \mid p/p \mid p \cup p \qquad\qquad step ::= axis \mid step[q]$$
$$q ::= p \mid lab() = L \mid q \wedge q \mid q \vee q \mid \neg q$$

where axis relations are given above.

The semantics of NavXP path expressions relative to a tree t is given by: 1. $[\![axis]\!] = R_{\text{axis}}^t$ 2. $[\![step[q]]\!] = \{(n, n') \in [\![step]\!] \ : \ n' \in [\![q]\!]\}$ 3. $[\![p_1/p_2]\!] = \{(n, n') \ : \ \exists w (n, w) \in [\![p_1]\!] \wedge (w, n') \in [\![p_2]\!]\}$ 4. $[\![p_1 \cup p_2]\!] = [\![p_1]\!] \cup [\![p_2]\!]$.

For filters we have: 1. $[\![lab() = L]\!] = \{n \ : \ n \text{ has label } L\}$ 2. $[\![p]\!] = \{n \ : \ \exists n' \ (n, n') \in [\![p]\!]\}$ 3. $[\![q_1 \wedge q_2]\!] = [\![q_1]\!] \cap [\![q_2]\!]$ 4. $[\![\neg q]\!](n) = \{n \ : \ n \notin [\![q]\!]\}$. A NavXP filter is said to hold of a tree t if it holds of the root under the above semantics.

As mentioned earlier, expressive equivalence of FO^2 and NavXP on trees, extending the translation to Unary Temporal Logic in the word case, is known:

Proposition 1 ([MdR04]). *There is an exponential translation from $FO^2[\downarrow_+]$ to NavXP with only the descendant and ancestor axes and from $FO^2[\downarrow, \downarrow_+, \rightarrow, \rightarrow^+]$ to NavXP with all axes.*

From the fact that NavXP has an exponential time satisfiability problem [Mar04] and the above proposition, we get the following (implicit in [MdR04]):

Corollary 1. *The satisfiability problem for $FO^2[\downarrow, \downarrow_+, \rightarrow, \rightarrow^+]$ is in 2EXPTIME.*

3   Satisfiability for Full FO^2 on Trees

Subformula Types and Exponential Depth Bounds. In the analysis of satisfiability of FO^2 for words of Etessami *et al.* [EVW02], a NEXPTIME bound is achieved by showing that any sentence with a finite model has a model of at most exponential size. The small-model property follows, roughly speaking, from the fact that *any* model realizes only exponentially many "quantifier-rank types"—maximal consistent sets of formulas of a given quantifier rank—and the fact that two nodes with the same quantifier-rank type can be identified.

In the case of trees, this approach breaks down in several places. It is easy to see that one cannot always obtain an exponential-sized model, since a sentence can enforce binary branching and exponential depth. Because there are doubly-exponentially many non-isomorphic small-depth subtrees, there can be doubly-exponentially many quantifier-rank types realized even along a single path in a tree: so quantifier-rank types can not be used even to show an exponential depth bound. We thus use *subformula types* of a given FO^2-formula φ (for short, φ-types), which are maximal consistent collections of subformulas of φ with one free variable. The φ-type of a node n in a tree, $Tp_\varphi(n)$, is defined as the set of subformulas of φ it satisfies. The number of φ-types is only exponential in $|\varphi|$, but subformula types are more delicate than quantifier-rank types. E.g. nodes with the same φ-type cannot always be identified without changing the truth of φ. Most of the upper bounds will be concerned with handling this issue, by adding additional conditions on nodes to be identified, and/or preserving additional parts of the tree.

Upper Bounds for FO^2 on Trees. We exhibit the issues arising and techniques used to solve them by giving an upper bound for the full logic, which improves on the 2EXPTIME bound one obtains via translation to NavXP.

Theorem 1. *The satisfiability problem for $FO^2[\downarrow, \downarrow_+, \rightarrow, \rightarrow^+]$ is in EXPSPACE.*

The key to the proof is to show an "exponential-depth property":

Lemma 1. *Every satisfiable $FO^2[\downarrow, \downarrow_+, \rightarrow, \rightarrow^+]$ sentence φ has a model t' whose depth is bounded by $2^{poly(|\varphi|)}$. The same bound holds for satisfiability with respect to UAR trees or ranked schemas. The outdegree of nodes can also be bounded by $2^{poly(|\varphi|)}$.*

We sketch the argument for the depth bound, leaving the similar proof for the branching bound to the full version. Given a tree t and nodes n_0 and n_1 in t with n_1 not an ancestor of n_0, the *overwrite* of n_0 by n_1 in t is the tree $t(n_1 \to n_0)$ formed by replacing the subtree of n_0 with the subtree of n_1 in t. Let F be the binary relation relating a node m in t to its copies in $t(n_1 \to n_0)$: n_1 and its descendants have a single copy if n_1 is a descendant of n_0, and two copies otherwise; nodes in $\mathsf{SubTree}(t, n_0)$ that are not in $\mathsf{SubTree}(t, n_1)$ have no copies, and other nodes have a single copy. In the case that n_1 is a descendant of n_0, F is a partial function. We say an equivalence relation $\equiv$ on nodes of a tree t is *globally φ-preserving* if for any equivalent nodes n_0, n_1 in t with $n_0 \notin \mathsf{SubTree}(t, n_1)$, the φ-type of a node n in t is the same as the φ-type of nodes in $F(n)$ within $t(n_1 \to n_0)$. We say it is *pathwise φ-preserving* if this holds for any node n_0, n_1 in t with n_1 a descendant of n_0. The *path-index* of an equivalence relation on t is the maximum of the number of equivalence classes represented on any path, while the *index* is the total number of classes.

We can not always overwrite a node with another having the same φ-type, but by adding additional information, we can get a pathwise φ-preserving relation with small path-index. For a node n, let $\mathsf{DescTypes}(n)$ be the set of φ-types of descendants of n, and $\mathsf{AncTypes}(n)$ the set of φ-types of ancestors of n. Let $\mathsf{IncompTypes}(n)$ be the φ-types of nodes n' that are neither descendants nor ancestors of n. Say $n_0 \equiv_{\mathsf{Full}} n_1$ if they agree on their φ-type, the set $\mathsf{DescTypes}$, and the set $\mathsf{IncompTypes}$.

Lemma 2. *The relation $\equiv_{\mathsf{Full}}$ is pathwise φ-preserving, and its path index is bounded by $2^{poly(|\varphi|)}$. Thus, there is a polynomial P such that for any tree t satisfying φ and root-to-leaf path p of length at least $2^{P(|\varphi|)}$, there are two nodes n_0, n_1 on p such that $t(n_1 \to n_0)$ still satisfies φ.*

Given Lemma 2, Lemma 1 follows by contracting all paths exceeding a given length until the depth of the tree is exponential in $|\varphi|$. In fact (e.g., for ranked trees) the equivalence classes of $\equiv_{\mathsf{Full}}$ can be used as the state set of a tree automaton A and then it can be arranged that A reaches the same state on n_0 as on n_1. The path index property implies that the automaton goes through only exponentially many states on any path of a tree. By taking the product of this automaton with a ranked schema, the corresponding depth bound relative to a schema follows.

We give a simple argument for the path index bound in Lemma 2. First, note that the total number of φ-types is exponential in $|\varphi|$. Now the sets $\mathsf{DescTypes}(n)$ either become smaller or stay the same as n varies down a path, and hence can only change exponentially often. Similarly the sets $\mathsf{IncompTypes}(n)$ grow bigger or stay the same, and thus can change only exponentially often. In intervals along a path where both of these sets are stable, the number of possibilities for the φ-type of a node is exponential. This gives the path index bound.

Theorem 1 follows from combining Lemma 1 with the following result on satisfiability of NavXP:

Theorem 2. *The satisfiability of a NavXP filter φ over trees of bounded depth b is in PSPACE (in b and $|\varphi|$).*

This result is a variant of a result from [BFG08] that finite satisfiability for the fragment of NavXP which contains only axis relations child, parent, next-sibling, preceding-sibling, previous-sibling and following-sibling is in PSPACE. Given Theorem 2 we complete the proof of Theorem 1 by translating an FO^2 sentence φ into an NavXP filter φ' with an exponential blow-up, using Proposition 1. By Lemma 1, the depth of a witness structure is bounded by an exponential in $|\varphi|$, and the EXPSPACE result follows.

Lower Bound. We now show a matching lower bound for the satisfiability problem.

Theorem 3. *The satisfiability problem for FO^2 on trees is EXPSPACE-hard, with hardness holding even when formulas are restricted to be in $GF^2[\downarrow_+]$.*

This is proved by coding the acceptance problem for an alternating exponential time machine. A tree node can be associated with an n-bit address, a path corresponds to one thread of the alternating computation, and the tree structure is used to code alternation. The equality and successor relations between the addresses associated to nodes x and y can be coded in $GF^2[\downarrow_+]$ using a standard argument—see [Kie02] for details, where it was shown that a restricted variant of the two-variable guarded fragment with some unary predicates and a single binary predicate that is interpreted as a transitive relation is EXPSPACE-hard. It is not hard to see that the proof presented there works fine (actually, it is even more natural) if we restrict the class of admissible structures to (finite) trees.

4   Satisfiability without Child

Unary Alphabet Restriction, Polynomial Alternation Bounds, and Polynomial Depth Bounds. The previous section showed EXPSPACE-completeness for satisfiability of $FO^2[\downarrow_+]$. However the EXPSPACE-hardness argument for $\downarrow_+$ makes use of multiple predicates holding at a given node, to code the address of a tape cell of an alternating exponential-time Turing Machine. It thus does not apply to satisfiability over UAR trees (as defined in Section 2) or to satisfiability with respect to a schema, since both cases restrict to a single alphabet symbol per node. In both cases we show that the complexity of satisfiability can be lowered to NEXPTIME, using distinct techniques for the case of ranked and unranked trees.

We start by noting that one always has at least NEXPTIME-hardness, even with UAR.

Theorem 4. *The satisfiability of $FO^2[\downarrow_+]$ UAR trees is NEXPTIME-hard, and similarly with respect to a ranked schema.*

The proof is a variation of the argument for NEXPTIME-hardness for words [EVW02], but this time using the frontier of a shallow but wide tree to code the tiling of an exponential grid.

We will prove a matching NEXPTIME upper bound for UAR trees and for satisfiability with respect to a ranked schema. To do this, we extend an idea introduced in the thesis of Weis [Wei11], working in the context of $\text{FO}^2[<]$ on UAR words: polynomial bounds on the number of times a formula changes its truth value while keeping the same symbol along a given path.

The following is a generalization of Lemma 2.1.10 of Weis [Wei11]. Consider an $\text{FO}^2[\downarrow_+]$ formula $\psi(x)$, a tree t satisfying the UAR, and fix a root-to-leaf path $p = p_1 \ldots p_{max(p)}$ in t. Given a label a, define an a-interval in p to be a set of the form $\{i : m_1 \le i < m_2; t, p_i \models a(x)\}$ for some m_1, m_2.

Lemma 3. *For every* $\text{FO}^2[\downarrow_+]$ *formula* $\psi(x)$, *UAR tree* t, *and root-to-leaf path* $p = p_1 \ldots p_{max(p)}$ *in* t, *the set* $\{i|\ t, p_i \models \psi \wedge a(x)\}$ *can be partitioned into a set of at most* $|\psi|^2$ a-*intervals.*

From Lemma 3, we will show that $\text{FO}^2[\downarrow_+]$ sentences that are satisfiable over UAR trees always have polynomial-depth witnesses:

Lemma 4. *If an* $\text{FO}^2[\downarrow_+]$ *formula* φ *is satisfied over a UAR tree, then it is satisfied by a model of depth bounded by a polynomial in* $|\varphi|$.

Let us prove this fact. Suppose that φ is satisfied over a UAR tree t. On each path p, for each letter b, let a b, φ-interval be a maximal b-interval on which every one-variable subformula of φ has constant truth value. By the lemma above, the total number of such intervals is polynomially bounded. We let W contain the endpoints of each b, φ-interval for all symbols b. We note the following crucial property of W: for every node m in p which is not in W, there is a node in W with the same φ-type as m that is strictly above m, and also one strictly below m.

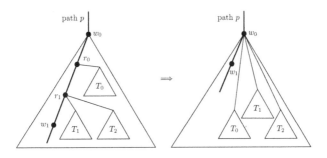

Fig. 1. Tree Promotion

The idea is now to remove all those points on path p that are not in W. This must be done in a slightly unusual way, by "promoting" subtrees that are off the path. For every child c of a removed node r that does not lie on path p we attach $\text{SubTree}(t, c)$ to the closest node of W above r (see Figure 1). Let t' denote the tree obtained as a result of this surgery.

Let f be the partial function taking a node in t that is not removed to its image in t'. We claim that t' still satisfies φ, and more generally that for any subformula $\rho(x)$ of φ and node m of t, we have $t, m \models \rho$ iff $t', f(m) \models \rho$. This is proved by induction on ρ, with the base cases and the cases for boolean operators being straightforward. For an existential formula $\exists y \beta(x, y)$, we give just the "only if" direction, which is via case analysis on the position of a witness node w such that $t, m, w \models \beta$.

If w is in t' then $t', m, w \models \beta$ by the induction hypothesis and the fact that w is an ancestor (or descendant) of m in t' if and only if it is an ancestor (or descendant) of m in t.

If w is not in t', then it must be that w lies on the path p and is not one the protected witnesses in W. But then w has both an ancestor w' and descendant w'' in W that satisfy all the same one-variable subformulas as w does in t, with both w' and w'' preserved in the tree t'. If m and w'' are distinct then $t', m, w'' \models \beta$ by the induction hypothesis and the fact that m and w'' have the same ancestor/descendant relationship in t' as do m and w in t. If m is identical to w'' then $t', m, w' \models \beta$ by similar reasoning. In any case we deduce that $t', m \models \exists y \beta$.

Since this process reduces both the length of the chosen path p and does not increase the length of any other path, it is clear that iterating it yields a tree of polynomial depth.

Note that we can guess a tree as above in NEXPTIME, and hence we have the following bound:

Theorem 5. *Satisfiability for* $FO^2[\downarrow_+]$ *formulas over UAR unranked trees is in* NEXPTIME, *and hence is* NEXPTIME-*complete.*

Bounds on Subtrees and Satisfiability of $FO^2[\downarrow_+]$ with Respect to a Ranked Schema. The collapse argument above relied heavily on the fact that trees were unranked, since over a fixed rank we could not apply "pathwise collapse". Indeed, we can show that over ranked trees, an $FO^2[\downarrow_+]$ formula satisfiable over UAR trees need *not* have a witness of polynomial depth:

Theorem 6. *There are* $FO^2[\downarrow_+]$ *formulas* φ_n *of size* $O(n)$ *that are satisfiable over UAR binary trees, where the minimum depth of satisfying UAR binary trees grows as* 2^n.

Nevertheless, we can still obtain an NEXPTIME bound for UAR trees of a given rank, and even for satisfiability with respect to a ranked schema.

Theorem 7. *The satisfiability problem for* $FO^2[\downarrow_+]$ *over ranked schemas is in* NEXPTIME, *and is thus* NEXPTIME-*complete.*

We give the argument only for satisfiability with respect to rank-k UAR trees, leaving the extension to schemas for the full paper. The idea will be to create a model with only an exponential number of distinct subtrees, which can be represented by an exponential-sized DAG. We do this by creating an equivalence relation that is globally φ-preserving (not just pathwise) and which has exponential index (not just path index). We will then collapse equivalent nodes, as in Lemma 2. There are several distinctions from that lemma: to identify nodes that are not necessarily comparable we can not afford to abstract a node by the set of *all* the types realized below it, since within the tree as a whole there can be doubly-exponentially many such sets. Instead we will make use of some "global information" about the tree, in the form of a set of "protected witnesses", which we denote W.

By Lemma 1 we know that a satisfiable $\text{FO}^2[\downarrow_+]$ formula φ has a model t of depth at most exponential in φ. Fix such a t. For each φ-type τ, let w_τ be a node of t with maximal depth satisfying τ. We include all w_τ and all of their ancestors in a set W, and call these *basic global witnesses*. For any m that is an ancestor or equal to a basic global witness w_τ, and any subformula $\rho(x) = \exists y \beta(x, y)$ of φ, if there is w' incomparable (by the descendant relation) to m such that $t, m, w' \models \beta$ we add one such w' to W, along with all its ancestors – these are the *incomparable global witnesses*.

We need one more definition. Given a node m in a tree, for every φ-type τ realized by some ancestor m' of m, for every subformula $\exists y \beta(x, y)$ of τ, if there is a descendant w of m such that $t, m', w \models \beta(x, y)$, choose one such witness w and let $\mathsf{SelectedDescTypes}(m)$ include the φ-type of that witness. Note that the same witness will suffice for every ancestor m' realizing τ, and since there are only polynomially many φ-types realized on the path, the collection $\mathsf{SelectedDescTypes}(m)$ will be of polynomial size.

Now we transform t to t' such that $t' \models \varphi$ and t' has only exponentially many different subtrees. We make use of a well-founded linear order $\prec$ on trees with a given rank and label alphabet, such that: 1. $\mathsf{SubTree}(t, n') \prec \mathsf{SubTree}(t, n)$ implies n' is not an ancestor of n; 2. for every tree C with a distinguished leaf, for trees t_1, t_2 with $t_1 \prec t_2$, we have $C[t_1] \prec C[t_2]$, where $C[t_i]$ is the tree obtained by replacing the distinguished leaf of C with t_i. There are many such orderings, e.g. using standard string encodings of a tree.

For any model t if there are two nodes n, n' in t such that 1. $n, n' \notin W$, 2. $\mathsf{Tp}_\varphi(n) = \mathsf{Tp}_\varphi(n')$, 3. $\mathsf{AncTypes}(n) = \mathsf{AncTypes}(n')$, 4. $\mathsf{SelectedDescTypes}(n) = \mathsf{SelectedDescTypes}(n')$, 5. $\mathsf{SubTree}(t, n') \prec \mathsf{SubTree}(t, n)$ (which implies that n' cannot be an ancestor of n), then let $t' = \mathsf{Update}(t)$ be obtained by choosing such n and n' and replacing the subtree rooted at n by the subtree rooted at n'.

Let T_1 be the nodes in t that were not in $\mathsf{SubTree}(t, n)$, and for any node $m \in T_1$ let $f(m)$ denote the same node considered within t'. Let T_2 denote the nodes in t' that are images of a node in $\mathsf{SubTree}(t, n')$. For each $m \in T_2$, let $f^{-1}(m)$ denote the node in $\mathsf{SubTree}(t, n')$ from which it derives. We claim the following:

Lemma 5. *For all $m \in T_1$ the φ-type of n in t is the same as the φ-type of $f(m)$ in t'. Moreover, for every node m' in T_2, the φ-type of m' in t' is the same as that of $f^{-1}(m)$ in t.*

Applying the lemma above to the root of t, which is necessarily in T_1, it follows that the truth of the sentence φ is preserved by this operation.

We now iterate the procedure $t_{i+1} := \mathsf{Update}(t_i)$, until no more updates are possible. This procedure terminates, because the tree decreases in the order $\prec$ every step. We can thus represent the tree as an exponential-sized DAG, with one node for each subtree.

Thus we have shown that any satisfiable formula has an exponential-size DAG that unfolds into a model of the formula. Given such a DAG, we can check whether an FO^2 formula holds in polynomial time in the size of the DAG. This gives a NEXPTIME algorithm for checking satisfiability.

5   Satisfiability without Descendant

Recall that even on words with only the successor relation, the satisfiability problem for two-variable logic is NEXPTIME-hard [EVW02]. From this it is easy to see that the satisfiability for $FO^2[\downarrow]$ is NEXPTIME-hard, on ranked and unranked trees.

Theorem 8. *The satisfiability problem for* $FO^2[\downarrow]$ *is* NEXPTIME-*hard, even with the UAR.*

We now present a matching upper bound, which holds even in the presence of sibling relations, i.e., for $FO^2[\downarrow, \rightarrow, \rightarrow^+]$. The result is surprising, in that it is easy to write satisfiable $FO^2[\downarrow]$ sentences φ_n of polynomial size whose smallest tree model is of depth exponential in n, and whose size is doubly exponential. Indeed, such formulas can be obtained as a variation of the proof of Theorem 8, by coding a complete binary tree whose nodes are associated with n-bit numbers, increasing the number by 1 as we move from parent to either child.

Theorem 9. *The satisfiability problem for* $FO^2[\downarrow, \rightarrow, \rightarrow^+]$, *and the satisfiability problem with respect to a ranked schema, are in* NEXPTIME, *and hence are* NEXPTIME-*complete.*

We sketch the idea for satisfiability, which iteratively quotients the structure by an equivalence relation, while preserving certain global witnesses, along the lines of Theorem 7. By Lemma 1 we know that a satisfiable $FO^2[\downarrow, \rightarrow, \rightarrow^+]$ formula φ has a model t of depth at most exponential in φ, where the outdegree of nodes is bounded by an exponential.

For each φ-type that is satisfied in t, choose a witness and include it along with all its ancestors in a set W – that is, we include the "basic witnesses" as in Theorem 7. We also include all children of each basic witness – call these "child witnesses".

Thus the size of the set of "protected witnesses" W is again at most exponential. Now we transform t to t' such that $t' \models \varphi$ and at the same time t' has only exponentially many different subtrees. Our update procedure looks for nodes n, n' in t such that 1. $n, n' \notin W$; 2. $\mathsf{SubTree}(t, n') \prec \mathsf{SubTree}(t, n)$, where $\prec$ is an appropriate ordering (as in Theorem 7); 3. $\mathsf{Tp}_\varphi(n) = \mathsf{Tp}_\varphi(n')$ and $\mathsf{Tp}_\varphi(\mathrm{parent}(n)) = \mathsf{Tp}_\varphi(\mathrm{parent}(n'))$. We then obtain $t' = \mathsf{Update}(t)$ by choosing such n and n' and replacing $\mathsf{SubTree}(t, n)$ by $\mathsf{SubTree}(t, n')$.

The theorem is proved by showing that this update operation preserves φ. Iterating it until no two nodes can be found produces a tree that can be represented as an exponential-size DAG.

6   Restricting the Logic

For $FO^2[\downarrow_+]$ over trees the complexity drop from EXPSPACE to NEXPTIME, resulting from restricting the class of models to those satisfying UAR, is slightly less spectacular than in the case of words, where an analogous restriction decreases the complexity from NEXPTIME to NP. However, to obtain NEXPTIME-lower bound, we need to speak about pairs x, y of elements in *free position*, i.e., such elements that y is neither an ascendant nor descendant of x. Thus it is natural to look at the situation where quantification is restricted to only pairs of elements that are connected by binary relations. To capture the former kind of scenario we consider the restriction of FO^2 to the two-variable guarded fragment, GF^2, in which all quantifiers have to be relativised by binary predicates. It is easy to see that GF^2 on trees still embeds NavXP, while still being exponentially more succinct. We are able to show a PSPACE bound on satisfiability of $GF^2[\downarrow_+]$ for UAR trees. The following observation is crucial.

Lemma 6. *Let φ be a $GF^2[\downarrow_+]$ formula and let t be a UAR tree satisfying φ. Then, there exists a tree t', obtained by removing some subtrees from t, still satisfying φ, such that the degree of nodes in t' is bounded polynomially in $|\varphi|$ and the depth of t.*

For the proof assume w.l.o.g. that φ is written in negation normal form, i.e, negations occur only in front of atomic formulas. For every subformula of φ of the form $\exists x \psi(x)$ which is satisfied in t choose a single node satisfying ψ and mark it together with all its ancestors. Analogously for formulas $\exists y \psi(y)$. For every formula $\exists y (x \downarrow_+ y \wedge \psi(x, y))$ belonging to the φ-type of the root of t choose a witness and mark it, together with all its ancestors. Then remove all subtrees rooted at unmarked successors of the root. Note that the obtained structure still satisfies φ. Analogously as with the root proceed with all marked elements, e.g., in a depth-first manner. Let t' be the tree obtained after the final step of the above process. Note that the number of descendants of a node in t' at depth l is bounded by $(l + 1) \cdot |\varphi|$. This justifies the bound from the statement of the lemma.

Theorem 10. *The satisfiability problem for $GF^2[\downarrow_+]$ over finite UAR trees is PSPACE-complete.*

We propose an alternating procedure solving the problem. Note that by combining Lemma 4 and Lemma 6 we may restrict our attention to trees whose depth and degree are polynomially bounded in the size of the input formula φ. First, our procedure guesses labels and φ-types of the root and its children, and checks if the guessed information is consistent. Then it universally chooses one of the children, guesses labels and φ-types of its children, and proceeds analogously. In this way, the procedure builds a single path of the tree, together with the immediate successors of all its nodes. This is sufficient to determine if a model satisfies φ, as φ is guarded and cannot speak about pairs of elements not belonging to a common path. Note that our procedure works in alternating polynomial time, and thus can be also implemented in PSPACE. The matching lower bound can be shown by reduction from the QBF problem. The crux is enforcing a full binary tree of depth n with internal nodes at depth i coding truth values of the i-th propositional variable from the QBF formula. In $GF^2[\downarrow_+]$ we can measure the depth of a node in a tree, and thus we can determine the identity of the variable encoded at a given node. Then evaluating a formula at a leaf node, we can reconstruct the valuation stored on the path leading to it from the root.

Augmenting $GF^2[\downarrow_+]$ with any of the remaining binary navigational predicates leads to an EXPSPACE lower bound over UAR trees.

Theorem 11. *The satisfiability problem over finite UAR trees for each of the logics* $GF^2[\downarrow, \downarrow_+]$, $GF^2[\downarrow_+, \rightarrow]$, $GF^2[\downarrow_+, \rightarrow^+]$ *is* EXPSPACE-*hard.*

In the proof we can follow the construction from [Kie02] showing EXPSPACE-hardness of $GF^2[\downarrow_+]$. Combinations of unary predicates holding in a single node in that proof can be now simulated by means of additional binary predicates. In the case of $GF^2[\downarrow, \downarrow_+]$ we code them using auxiliary children of a node. In the case of $GF^2[\downarrow_+, \rightarrow]$ and $GF^2[\downarrow_+, \rightarrow^+]$ we employ siblings for this task. Details will be given in the full version of the paper.

This completes the picture for the case of signatures containing $\downarrow_+$. (Recall Theorem 3, which shows that without the UAR restriction already $GF^2[\downarrow_+]$ is EXPSPACE-hard.) We now consider the case of signatures containing $\downarrow$ but not containing $\downarrow_+$.

Theorem 12. *The satisfiability problem for* $GF^2[\downarrow, \rightarrow]$ *over finite trees is in* EXPTIME. *The satisfiability problem for* $GF^2[\downarrow]$ *is* EXPTIME-*hard, even under* UAR *assumption.*

For the upper bound we propose an alternating procedure working in polynomial space. The procedure looks for a model of an input formula φ of depth and degree exponentially bounded in the size of φ, as guaranteed by Lemma 1. Again we assume that φ is in negation normal form. At each moment during its execution the procedure stores information about a node of the tree and polynomially many of its children. For each node this information consists of its label, φ-type, and, for every subformula of φ of the form $\exists x\psi(x)$ or $\exists y\psi(y)$, a note whether ψ is satisfied at some point in the subtree rooted at this node. The procedure first guesses the information about the root. For every formula $\exists y(x{\downarrow}y \wedge \psi(x,y))$ belonging to its φ-type it guesses a witness. Additionally, for every formula $\exists x\psi(x)$ or $\exists y\psi(y)$, for which it is declared the ψ will hold below, a witness satisfying ψ or declaring that ψ will hold below it is guessed. (Note that in total at most polynomially many witnesses in are required.) Also the order in which all those witnesses appear on the list of the children of the root is guessed. Now the procedure universally chooses a pair of consecutive witnesses n_1, n_2 and, starting from n_1, tries to reach n_2 by a $\rightarrow$-chain of elements. At each of the nodes it checks if the guessed information is consistent with the information about its neighbours, and additionally it makes a universal choice between continuing the horizontal path to n_2 or going down the tree (in a way similar to the one described for the root).

The lower bound in the above theorem can be shown by an encoding of an alternating Turing machine working in polynomial space.

Equipping the logic with $\rightarrow^+$ allows us to lift the lower bound, even assuming UAR:

Theorem 13. *The satisfiability problem for* $GF^2[\downarrow, \rightarrow^+]$ *over finite UAR trees is* NEXPTIME-*hard.*

The proof of this theorem relies on the fact that without the UAR assumption FO^2 is NEXPTIME-hard even if only unary relations are allowed in the signature [EVW02]. This can be simulated in our scenario: we use the children of the root to encode the elements in a model of such a unary formula. Then the relation $\rightarrow^+$ may be used as a guard, allowing to refer to any pair of these. The combination of unary predicates holding at a given position can be simulated by means of the $\downarrow$-successors.

Recall that an upper bound matching the lower bound from Theorem 13 holds for $FO^2[\downarrow, \rightarrow, \rightarrow^+]$ even without the UAR assumption (Theorem 9).

7 Conclusions and Acknowledgements

The main result of the paper is that the satisfiability problem for FO^2 over finite trees, with four navigational predicates: $\downarrow, \downarrow_+, \rightarrow, \rightarrow^+$, is EXPSPACE-complete. We also consider an additional semantic restriction that at a single node precisely one unary predicate holds (UAR). Under UAR the full logic remains EXPSPACE-complete, but for some of its weakened variants this assumption makes difference. Namely, $FO^2[\downarrow_+]$ becomes NEXPTIME-complete and $GF^2[\downarrow_+]$ is even PSPACE-complete under UAR, even though both logics are still EXPSPACE-complete without UAR.

We go on to establish the precise complexity bounds for all logics $GF^2[\tau_{bin}]$ and $FO^2[\tau_{bin}]$, with $\tau_{bin} \subseteq \{\downarrow, \downarrow_+, \rightarrow, \rightarrow^+\}$ containing at least $\downarrow$ or $\downarrow_+$, for arbitrary finite trees or under UAR. These bounds can be inferred from the results below: Each of 48 possible variants lies between two logics, one for which we have a lower bound and one for which we have the same upper bound.

Lower Bounds:

- PSPACE: $GF^2[\downarrow_+]$ with UAR (Thm. 10)
- EXPTIME: $GF^2[\downarrow]$ with UAR (Thm. 12)
- NEXPTIME: $FO^2[\downarrow_+]$ with UAR (Thm. 4), $FO^2[\downarrow]$ with UAR (Thm. 8), $GF^2[\downarrow, \rightarrow^+]$ with UAR (Thm. 13)
- EXPSPACE: $GF^2[\downarrow_+]$ (Thm. 3), $GF^2[\downarrow, \downarrow_+]$, $GF^2[\downarrow_+, \rightarrow]$, $GF^2[\downarrow_+, \rightarrow^+]$, all with UAR (Thm. 11)

Upper Bounds:

- PSPACE: $GF^2[\downarrow_+]$ with UAR (Thm. 10)
- EXPTIME: $GF^2[\downarrow, \rightarrow]$ (Thm. 12)
- NEXPTIME: $FO^2[\downarrow_+]$ with UAR (Thm. 5), $FO^2[\downarrow, \rightarrow, \rightarrow^+]$ (Thm. 9)
- EXPSPACE: $FO^2[\downarrow, \downarrow_+, \rightarrow, \rightarrow^+]$(Thm. 1)

We also obtain some results concerning satisfiability over ranked trees and satisfiability in the presence of schemas.

One direction of future research is to extend the analysis to infinite trees. It seems that the complexity results we have obtained here can be transferred to this case without major difficulties.

Acknowledgements. This paper is a merger of two independently-developed works, [BBLW13] and [CKM13]. We thank the anonymous reviewers of ICALP for many helpful remarks on both works.

References

[AvBN98] Andréka, H., van Benthem, J., Németi, I.: Modal languages and bounded fragments of predicate logic. J. Phil. Logic 27, 217–274 (1998)

[BBLW13] Benaim, S., Benedikt, M., Lenhardt, R., Worrell, J.: Controlling the depth, size, and number of subtrees in two variable logic over trees. CoRR abs/1304.6925 (2013)

[BFG08] Benedikt, M., Fan, W., Geerts, F.: XPath satisfiability in the presence of DTDs. J. ACM 55(2), 8:1–8:79 (2008)

[BK08] Benedikt, M., Koch, C.: XPath Leashed. ACM Comput. Surv. 41(1), 3:1–3:54 (2008)

[BLW12] Benedikt, M., Lenhardt, R., Worrell, J.: Verification of two-variable logic revisited. In: QEST, pp. 114–123. IEEE (2012)

[BMSS09] Bojańczyk, M., Muscholl, A., Schwentick, T., Segoufin, L.: Two-variable logic on data trees and XML reasoning. J. ACM 56(3) (2009)

[CKM13] Charatonik, W., Kieroński, E., Mazowiecki, F.: Satisfiability of the two-variable fragment of first-order logic over trees. CoRR abs/1304.7204 (2013)

[CW13] Charatonik, W., Witkowski, P.: Two-variable logic with counting and trees. In: LICS. IEEE (to appear, 2013)

[EVW02] Etessami, K., Vardi, M.Y., Wilke, T.: First-order logic with two variables and unary temporal logic. Inf. Comput. 179(2), 279–295 (2002)

[Fig12] Figueira, D.: Satisfiability for two-variable logic with two successor relations on finite linear orders. CoRR abs/1204.2495 (2012)

[GKV97] Grädel, E., Kolaitis, P.G., Vardi, M.Y.: On the decision problem for two-variable first-order logic. Bull. Symb. Logic 3(1), 53–69 (1997)

[Kie02] Kieroński, E.: EXPSPACE-complete variant of guarded fragment with transitivity. In: Alt, H., Ferreira, A. (eds.) STACS 2002. LNCS, vol. 2285, pp. 608–619. Springer, Heidelberg (2002)

[Kie05] Kieroński, E.: Results on the guarded fragment with equivalence or transitive relations. In: Ong, L. (ed.) CSL 2005. LNCS, vol. 3634, pp. 309–324. Springer, Heidelberg (2005)

[Mar04] Marx, M.: XPath with conditional axis relations. In: Bertino, E., Christodoulakis, S., Plexousakis, D., Christophides, V., Koubarakis, M., Böhm, K. (eds.) EDBT 2004. LNCS, vol. 2992, pp. 477–494. Springer, Heidelberg (2004)

[MdR04] Marxand, M., de Rijke, M.: Semantic characterization of navigational XPath. In: TDM. CTIT Workshop Proceedings Series, pp. 73–79 (2004)

[ST13] Szwast, W., Tendera, L.: FO² with one transitive relation is decidable. In: STACS. LIPIcs, vol. 20, pp. 317–328, Schloss Dagstuhl - Leibniz-Zentrum fuer Informatik (2013)

[Sto74] Stockmeyer, L.J.: The Complexity of Decision Problems in Automata Theory and Logic. PhD thesis, Massachusetts Institute of Technology (1974)

[Tho97] Thomas, W.: Languages, automata, and logic. In: Rozenberg, G., Salomaa, A. (eds.) Handbook of Formal Languages. Springer (1997)

[Wei11] Weis, P.: Expressiveness and Succinctness of First-Order Logic on Finite Words. PhD thesis, University of Massachusetts (2011)

Nondeterminism in the Presence
of a Diverse or Unknown Future*

Udi Boker[1], Denis Kuperberg[2], Orna Kupferman[2], and Michał Skrzypczak[3]

[1] IST Austria, Klosterneuburg, Austria
[2] The Hebrew University, Jerusalem, Israel
[3] University of Warsaw, Poland

Abstract. Choices made by nondeterministic word automata depend on both the past (the prefix of the word read so far) and the future (the suffix yet to be read). In several applications, most notably synthesis, the future is diverse or unknown, leading to algorithms that are based on deterministic automata. Hoping to retain some of the advantages of nondeterministic automata, researchers have studied restricted classes of nondeterministic automata. Three such classes are nondeterministic automata that are *good for trees* (GFT; i.e., ones that can be expanded to tree automata accepting the derived tree languages, thus whose choices should satisfy diverse futures), *good for games* (GFG; i.e., ones whose choices depend only on the past), and *determinizable by pruning* (DBP; i.e., ones that embody equivalent deterministic automata). The theoretical properties and relative merits of the different classes are still open, having vagueness on whether they really differ from deterministic automata. In particular, while DBP ⊆ GFG ⊆ GFT, it is not known whether every GFT automaton is GFG and whether every GFG automaton is DBP. Also open is the possible succinctness of GFG and GFT automata compared to deterministic automata. We study these problems for ω-regular automata with all common acceptance conditions. We show that GFT=GFG⊃DBP, and describe a determinization construction for GFG automata.

1 Introduction

Nondeterminism is very significant in word automata: it allows for exponential succinctness [14] and in some cases, such as Büchi automata, it also increases the expressive power [9]. In the automata-theoretic approach to formal verification, temporal logic formulas are translated to nondeterministic word automata [16]. In some applications, such as model checking, algorithms can proceed on the nondeterministic automaton, whereas in other applications, such as synthesis and control, they cannot. There, the advantages of nondeterminism are lost, and the algorithms involve a complicated determinization construction [15] or acrobatics for circumventing determinization [8].

To see the inherent difficulty of using nondeterminism in synthesis, let us review the current approach for solving the synthesis problem, going through games [4].

---

* This work was supported in part by the Polish Ministry of Science grant no. N206 567840, Poland's NCN grant no. DEC-2012/05/N/ST6/03254, Austrian Science Fund NFN RiSE (Rigorous Systems Engineering), ERC Advanced Grant QUAREM (Quantitative Reactive Modeling), and ERC Grant QUALITY. The full version is available at the authors' URLs.

F.V. Fomin et al. (Eds.): ICALP 2013, Part II, LNCS 7966, pp. 89–100, 2013.

Let L be a language of infinite words over an alphabet $2^{I \cup O}$, where I and O are sets of input and output signals, respectively. The synthesis problem for L is to build a reactive system that outputs signals from 2^O upon receiving input signals from 2^I, such that the generated sequence (an infinite word over the alphabet $2^{I \cup O}$) is in L [12]. The problem is solved by taking a deterministic automaton $\mathcal{D}$ for L and conducting a two-player game on top of it. The players, "system" and "environment", generate words over $2^{I \cup O}$, where in each turn the environment first chooses the 2^I component of the next letter, the system responds with the 2^O component, and $\mathcal{D}$ moves to the successor state. The goal of the system is to generate an accepting run of $\mathcal{D}$ no matter which sequence of input assignments is generated by the environment. The system has a winning strategy iff the language L can be synthesized.

Now, if one tries to replace $\mathcal{D}$ with a nondeterministic automaton $\mathcal{A}$ for L, the system should also choose a transition to proceed with. Then, it might be that L is synthesizable and still the system has no winning strategy, as each choice of Σ may "cover" a strict subset of the possible futures.

Some nondeterministic automata are, however, good for games: in these automata it is possible to resolve the nondeterminism in a way that only depends on the past and still accepts all the words in the language. This notion, of *good for games* (GFG) automata was first introduced in [5].[1] Formally, a nondeterministic automaton over the alphabet Σ is GFG if there is a strategy that maps each word $x \in \Sigma^*$ to the transition to be taken after x is read. Note that a state q of the automaton may be reachable via different words, and the strategy may suggest different transitions from q after different words are read. Still, the strategy depends only on the past, meaning on the word read so far. Obviously, there exist GFG automata: deterministic ones, or nondeterministic ones that are *determinizable by pruning* (DBP); that is, ones that just add transitions on top of a deterministic automaton. In fact, these are the only examples known so far of GFG automata. [2] A natural question is whether all GFG automata are DBP.

More generally, a central question is what role nondeterminism can play in automata used for games, or abstractly put, in cases that the future is unknown. Specifically, can such nondeterminism add expressive power? Can it contribute to succinctness? Is it "real" or must it embody a deterministic choice?

Before addressing these questions, one should consider their tight connection to nondeterminism in tree automata for derived languages [7]: A nondeterministic word automaton $\mathcal{A}$ with language L is *good for trees* (GFT) if, when expanding its transition function to get a symmetric tree automaton, it recognizes the *derived language*, denoted $\mathrm{der}(L)$, of L; that is, all trees all of whose branches are in L [7]. Tree automata for derived languages were used for solving the synthesis problem [12] and are used when translating branching temporal logics such as CTL* to tree automata [3]. Analogously to GFG automata, the problem in using nondeterminism in GFT automata stems from the need to satisfy different futures (the different branches in the tree). For example,

---

[1] GFGness is also used in [2] in the framework of cost functions under the name "history-determinism".

[2] As explained in [5], the fact the GFG automata constructed there are DBP does not contradict their usefulness in practice, as their transition relation is simpler than the one of the embodied deterministic automaton and it can be defined symbolically.

solving the synthesis problem, the branches of the tree correspond to the possible input sequences, and when the automaton makes a guess, the guess has to be successful for all input sequences. The main difference between GFG and GFT is that the former can only use the past, whereas the latter can possibly take advantage of the future, except that the future is diverse.

A principal question is whether GFG and GFT automata are the same, meaning whether nondeterminism can take some advantage of a diverse future, or is it the same as only considering the past.

It is not difficult to answer all the above questions for safety languages; that is, when the language $L = L(\mathcal{A}) \subseteq \Sigma^\omega$ is such that all the words in L can be arranged in one tree. Then, a memoryless accepting run of $\mathcal{A}$ (that is, its expansion to a symmetric tree automaton for $\mathrm{der}(L)$) on this tree induces a deterministic automaton embodied in $\mathcal{A}$, meaning that $\mathcal{A}$ is DBP. Moving to general ω-regular languages, the first question, concerning expressiveness of deterministic versus GFT automata, was answered in [7] with respect to Büchi automata, and in [11] with respect to all levels of the Mostowski hierarchy. It is shown in these works that if $\mathrm{der}(L)$ can be recognized by a nondeterministic Büchi tree automaton, then L can be recognized by a deterministic Büchi word automaton, and similarly for parity conditions of a particular index. Thus, nondeterminism in the presence of unknown or diverse future does not add expressive power. The other questions, however, are open since the 90s.

In this paper we examine these questions further for automata with all common acceptance conditions. We first show that a Muller automaton is GFG iff it is GFT. As the Muller condition can describe all the common acceptance conditions (Büchi, co-Büchi, parity, Streett, and Rabin), the result follows to all of them. Intuitively, a GFT automaton $\mathcal{A}$ (or, equivalently, a nondeterministic tree automaton for a derived language) is limited in using information about the future, as different branches of the tree challenge it with different futures. Formally, we prove that $\mathcal{A}$ is GFG by using determinacy of a well-chosen game. The same game allows us to show that there is a deterministic automaton for $L(\mathcal{A})$ with the same acceptance condition as $\mathcal{A}$. This also simplifies the result of [11] and generalizes it to Muller conditions. Indeed, the proof in [11] is based on intricate arguments that heavily rely on the structure of parity condition.

Can GFG automata take some advantage of nondeterminism or do they simply hide determinism? We show the existence of GFG Büchi and co-Büchi automata that use the past in order to make decisions, and thus cannot have a memoryless strategy. Note that we use the basic acceptance conditions for these counter examples, thus the result follows to all common acceptance conditions. This is different from known results on GFG automata over finite words or weak GFG automata, where GFG automata are DBP [7,10]. This result is quite surprising, as strategies in parity games are memoryless. We further build a GFG automaton that cannot be pruned into a deterministic automaton even with a finite unbounded look-ahead, meaning that even an unbounded yet finite view of the future cannot compensate on memorylessness.

Regarding succinctness, the currently known upper bound for the state blowup involved in determinizing a GFG parity automaton is exponential [7], with no nontrivial lower bound. We provide some insights on GFG automata, showing that in some cases its determinization is efficient. We show that if $\mathcal{A}$ and $\mathcal{B}$ are GFG Rabin automata that recognize a language L and its complement, then there is a deterministic Rabin automaton for L of size $|\mathcal{A} \times \mathcal{B}|$. Thus, in the context of GFG automata, determinization is essentially the same problem as complementation. Moreover, our construction shows that determinization cannot induce an exponential blowup both for an automaton and its complement. This is in contrast with standard nondeterminism, even over finite words. For example, both the language $L_k = (a+b)^*a(a+b)^k$ and its complement admit non-deterministic automata that are linear in k, while the deterministic ones are exponential in k.

Due to lack of space, some proofs are omitted, or shortened, and can be found in the full version.

2 Preliminaries

2.1 Trees and Labeled Trees

We consider trees over a set $\mathbb{D}$ of directions. A tree T is a prefix-closed subset of $\mathcal{T} = \mathbb{D}^*$. We refer to $\mathcal{T}$ as the *complete* $\mathbb{D}$-*tree*. The elements in T are called *nodes*, and ε is the *root* of T. For a node $u \in \mathbb{D}^*$ and $d \in \mathbb{D}$, the node ud is the *child* of u with *direction* d. A *path* of T is a set $\pi \subseteq T$, such that $\varepsilon \in \pi$ and for all $u \in \pi$, there is a unique $d \in \mathbb{D}$ with $ud \in \pi$. Note that each path π corresponds to an infinite word in $\mathbb{D}^\omega$.

For an alphabet Σ, a Σ-*labeled* $\mathbb{D}$-*tree* is a $\mathbb{D}$-tree in which each edge is labeled by a letter from Σ. We choose to label edges instead of nodes in order to be able to compose a set of words into a single tree, even when the set contains words that do not agree on their first letter. Formally, a Σ-labeled $\mathbb{D}$-tree is a pair $\langle T, t \rangle$ where $T \subseteq \mathcal{T}$ is a $\mathbb{D}$-tree and $t : T \setminus \{\varepsilon\} \to \Sigma$ labels each edge (or equivalently its target node) by a letter in Σ. Let $\mathcal{T}_{\mathbb{D}, \Sigma}$ be the set of Σ-labeled $\mathbb{D}$-trees (not necessarily complete). We say that a word $w \in \Sigma^\omega$ is a *branch* of a tree $\langle T, t \rangle \in \mathcal{T}_{\mathbb{D}, \Sigma}$ if there is a path $\pi = \{\varepsilon, u_1, u_2, \ldots\} \subseteq T$ such that $w = t(\pi) = t(u_1)t(u_2) \ldots$ We use $branches(\langle T, t \rangle)$ to denote the set of branches of $\langle T, t \rangle$. Note that $branches(\langle T, t \rangle)$ is a subset of Σ^ω.

2.2 Automata

Automata on words An automaton on infinite words is a tuple $\mathcal{A} = \langle \Sigma, Q, q_0, \Delta, \alpha \rangle$, where Σ is the input alphabet, Q is a finite set of states, $q_0 \in Q$ is an (for simplicity, single) initial state, $\Delta \subseteq Q \times \Sigma \times Q$ is a transition relation such that $\langle q, a, q' \rangle \in \Delta$ if the automaton in state q, reading a, can move to state q'. The state $q_0 \in Q$ is the initial state, and α is an acceptance condition. Here we will use *Büchi, co-Büchi, parity, Rabin, Streett* and *Muller* automata. In a Büchi (resp. co-Büchi) conditions, $\alpha \subseteq Q$ is a set of accepting (resp. rejecting) states. In a parity condition of index $[i, j]$, the acceptance condition $\alpha : Q \to [i, j]$ is a function mapping each state to its priority (we use $[i, j]$ to denote the set $\{i, i + 1, \ldots, j\}$). In a Rabin (resp. Streett) condition, $\alpha \subseteq 2^{2^Q \times 2^Q}$ is a set of pairs of sets of states, and in a Muller condition, $\alpha \subseteq 2^{2^Q}$ is a set of sets of states.

Since the transition relation may specify many possible transitions for each state and letter, the automaton $\mathcal{A}$ may be *nondeterministic*. If Δ is such that for every $q \in Q$ and $a \in \Sigma$, there is a single state $q' \in Q$ such that $\langle q, a, q' \rangle \in \Delta$, then $\mathcal{A}$ is a *deterministic* automaton.

Given an input word $w = a_0 \cdot a_1 \cdots$ in Σ^ω, a *run* of $\mathcal{A}$ on w is a function $r : \mathbb{N} \to Q$ where $r(0) = q_0$ and for every $i \geq 0$, we have $\langle r(i), a_i, r(i+1)\rangle \in \Delta$; i.e., the run starts in the initial state and obeys the transition function. For a run r, let $\inf(r)$ denote the set of states that r visits infinitely often. That is, $\inf(r) = \{q \in Q : \text{for infinitely}$ many $i \geq 0$, we have $r(i) = q\}$. The run r is *accepting* iff

- $\inf(r) \cap \alpha \neq \emptyset$, for a Büchi condition.
- $\inf(r) \cap \alpha = \emptyset$, for a co-Büchi condition.
- $\max\{\alpha(q) : q \in \inf(r)\}$ is even, for a parity condition.
- there exists $\langle E, F\rangle \in \alpha$, such that $\inf(r) \cap E = \emptyset$ and $\inf(r) \cap F \neq \emptyset$ for a Rabin condition.
- for all $\langle E, F\rangle \in \alpha$, we have $\inf(r) \cap E \neq \emptyset$ or $\inf(r) \cap F \neq \emptyset$ for a Streett condition.
- $\inf(r) \in \alpha$ for a Muller condition.

Note that Büchi and co-Büchi are dual, as well as Rabin and Streett. Parity and Muller are self-dual. Also note that Büchi and co-Büchi are a special case of parity, which is a special case of Rabin and Streett, which in turn are special cases of the Muller condition. An automaton $\mathcal{A}$ accepts an input word w iff there exists an accepting run of $\mathcal{A}$ on w. The *language* of $\mathcal{A}$, denoted $L(\mathcal{A})$, is the set of all words in Σ^ω that $\mathcal{A}$ accepts.

Automata on Trees. An automaton on Σ-labeled $\mathbb{D}$-trees is a tuple $\mathcal{A} = \langle \Sigma, \mathbb{D}, Q, q_0, \Delta, \alpha\rangle$, where Σ, Q, q_0, and α are as in automata on words, and $\Delta \subseteq Q \times (\Sigma \times Q)^\mathbb{D}$. Recall that we label the edges of the input trees. Accordingly, $\langle q, (a_d, q_d)_{d \in \mathbb{D}}\rangle \in \Delta$ if the automaton in state q, reading for each $d \in \mathbb{D}$ the letter a_d in direction d, can send a copy in q_d to the child in direction d. If for all $q \in Q$ and $(a_d)_{d \in \mathbb{D}} \in \Sigma^\mathbb{D}$, there is a single tuple $(q_d)_{d \in \mathbb{D}}$ such that $\langle q, (a_d, q_d)_{d \in \mathbb{D}}\rangle \in \Delta$, then $\mathcal{A}$ is deterministic.

A *run* of $\mathcal{A}$ on a Σ-labeled tree $\langle T, t\rangle$ is a function $r : T \to Q$ such that $r(\varepsilon) = q_0$ and for all $u \in T$, we have that $\langle r(u), (t(ud), r(ud))_{d \in \mathbb{D}}\rangle \in \Delta$. If for some directions d, the nodes ud are not in T, we assume that the requirement on them is satisfied. A run r on a tree T is accepting if the acceptance condition of the automaton is satisfied on all infinite paths of $\langle T, r\rangle$. For instance when $\mathcal{A}$ is a Büchi automaton, the run r is accepting if on all infinite paths in T it visits α infinitely often. As in automata on words, a tree $\langle T, t\rangle$ is accepted by $\mathcal{A}$ if there exists an accepting run of $\mathcal{A}$ on $\langle T, t\rangle$, and the language of $\mathcal{A}$, denoted $L(\mathcal{A})$, is the set of all trees in $\mathcal{T}_{\mathbb{D},\Sigma}$ that $\mathcal{A}$ accepts.

We use three letter acronyms in $\{D, N\} \times \{F, B, C, P, R, S, M\} \times \{W, T\}$ to denote classes of automata, with the first letter indicating whether this is a deterministic or nondeterministic automaton, the second whether it is an automaton on finite words or a Büchi / co-Büchi / parity / Rabin / Streett / Muller automaton, and the third whether it runs on words or trees. For example, a DBW is a deterministic Büchi automaton on infinite words.

2.3 Between Deterministic and Nondeterministic Automata

Let $L \subseteq \Sigma^\omega$ be a language of infinite words. We define the *derived language* of L, denoted $\operatorname{der}(L)$, as the set of Σ-labeled $\mathbb{D}$-trees all of whose branches are in L. Note that the definition has $\mathbb{D}$ as a parameter.

Since membership of a tree $\langle T, t \rangle$ in $\mathrm{der}(L)$ only depends on $branches(\langle T, t \rangle)$, we do not lose generality if we consider, in the context of derivable languages, trees in a *normal form* in which $\mathbb{D} = \Sigma$ and labels agree with directions. We note that examining trees for which $|\mathbb{D}| < |\Sigma|$ introduces an extra assumption on the set of possible futures, of which a nondeterministic automaton may take advantage.

Formally, we say that a Σ-labeled $\mathbb{D}$-tree $\langle T, t \rangle$ is in a normal form if $\Sigma = \mathbb{D}$, and for all $ua \in \Sigma^+$, we have $t(ua) = a$. Clearly, each Σ-labeled $\mathbb{D}$-tree $\langle T, t \rangle$ has a unique Σ-labeled Σ-tree $\langle T', t' \rangle$ in a normal form such that $branches(\langle T, t \rangle) = branches(\langle T', t' \rangle)$. Working with trees in a normal form enables us to identify the domain T with its labeling t. Thus, from now on we refer to a Σ-tree T, with the understanding that we talk about the unique Σ-labeled Σ-tree in normal form that has T as its underlying Σ-tree. For a Σ-tree T, the branch associated with a path $\{\epsilon, d_1, d_1 d_2, d_1 d_2 d_3, \ldots\}$ is the infinite word $d_1 d_2 d_3 \cdots$. The tree automata we consider also have $\mathbb{D} = \Sigma$ (and we omit $\mathbb{D}$ from the specification of the automaton).

Consider a nondeterministic word automaton $\mathcal{A} = \langle \Sigma, Q, q_0, \Delta, \alpha \rangle$. Let $\mathcal{A}_t$ be the *expansion* of $\mathcal{A}$ to a tree automaton. Recall that we restrict attention to automata with $\mathbb{D} = \Sigma$. That is, $\mathcal{A}_t = \langle \Sigma, Q, q_0, \Delta_t, \alpha \rangle$ is such that for every $\langle q, (a_d, q_d)_{d \in \Sigma} \rangle \in Q \times (\Sigma \times Q)^\Sigma$, we have that $\langle q, (a_d, q_d)_{d \in \Sigma} \rangle \in \Delta_t$ iff for all $d \in \Sigma$, there is a transition $\langle q, a_d, q_d \rangle$ is in Δ. We say that $\mathcal{A}$ is *good for trees* (GFT, for short), if $L(\mathcal{A}_t) = \mathrm{der}(L(\mathcal{A}))$.

It is easy to see that when $\mathcal{A}$ is deterministic, then $\mathcal{A}$ is GFT. Indeed, $\mathcal{A}_t$ only accepts trees in $\mathrm{der}(L(\mathcal{A}))$, so $L(\mathcal{A}_t) \subseteq \mathrm{der}(L(\mathcal{A}))$. Conversely, since each prefix of a word in Σ^ω corresponds to a single prefix of a run of $\mathcal{A}$, we can compose the accepting runs of $\mathcal{A}$ of the words in $L(\mathcal{A})$ to an accepting run on $\mathcal{A}_t$ on every tree in $\mathrm{der}(L(\mathcal{A}))$.

General nondeterministic automata are not GFT. For example, let $\mathcal{A} = \langle \{a, b\}, \{q_0, q_1\}, q_0, \{\langle q_0, a, q_0 \rangle, \langle q_0, b, q_0 \rangle, \langle q_0, a, q_1 \rangle, \langle q_1, a, q_1 \rangle\}, \{q_1\} \rangle$ be the canonical NBW recognizing $L = (a+b)^* a^\omega$. Then, $\mathcal{A}_t$ cannot accept the tree $T = a^* \cup a^* b a^*$. Indeed, $\mathcal{A}_t$ has to move to q_1 at some point on the a^ω branch, but it then fails to accept other branches from that point, as there is no transition leaving q_1 labeled with b. In fact no NBT can recognize $\mathrm{der}(L(\mathcal{A}))$ [13].

A nondeterministic word automaton $\mathcal{A} = \langle \Sigma, Q, q_0, \Delta, \alpha \rangle$ is *good for games* (GFG, for short) if there is a strategy $\sigma : \Sigma^* \to Q$ such that the following hold: (1) The strategy σ is compatible with Δ. That is, for all $\langle u, a \rangle \in \Sigma^* \times \Sigma$, we have $\langle \sigma(u), a, \sigma(ua) \rangle \in \Delta$. (2) The restriction imposed by σ does not exclude words from $L(\mathcal{A})$; that is, for all $u = u_0 \cdot u_1 \cdot u_2, \cdots \in L(\mathcal{A})$, the sequence $\sigma(\varepsilon), \sigma(u_0), \sigma(u_0 u_1), \sigma(u_0 u_1 u_2), \ldots$ satisfies the acceptance condition α.

Finally, Σ is *determinizable by pruning* (DBP, for short) if it can be determinized to an equivalent automaton by removing some of its transitions.

A DBP automaton is obviously also GFG, using the strategy that follows the unpruned transitions. A GFG automaton is also GFT, as the latter can resolve its nondeterminism using the strategy that witnesses the GFGness.

Proposition 1. *If an automaton $\mathcal{A}$ is DBP, then it is GFG. If $\mathcal{A}$ is GFG then $\mathcal{A}$ is GFT.*

Let $\mathcal{A} = \langle \Sigma, Q, q_0, \Delta, \alpha \rangle$ be a tree automaton. The word automaton associated with $\mathcal{A}$ is $\mathcal{A}_w = \langle \Sigma, Q, q_0, \Delta_w, \alpha \rangle$, where Δ_w is such that $\langle q, a, q' \rangle \in \Delta_w$ iff Δ has a transition from q in which q' is sent to some direction along an edge labeled a. Formally, there is a transition $\langle q, (a_d, q_d)_{d \in \Sigma} \rangle \in \Delta$ with $(a_d, q_d) = (a, q')$ for some $d \in \Sigma$. It is easy to see that $\mathcal{A}_w$ accepts exactly all infinite words that appear as a branch of some tree accepted by $\mathcal{A}$. Note that if $L(\mathcal{A}) = \text{der}(L)$, then $L(\mathcal{A}_w) = L$, and $L((\mathcal{A}_w)_t) = \text{der}(L)$, so $\mathcal{A}_w$ is GFT.

3 From GFT to GFG

In this section we prove that if an NMW is GFT then it is also GFG. In addition, we show that GFG automata admit finite memory strategies and we study connections with [11].

The crucial tool in the proof is the following infinite-duration perfect-information game between two players $\exists$ and $\forall$. Let $\mathcal{A} = \langle \mathbb{D}, Q^{\mathcal{A}}, q_I^{\mathcal{A}}, \Delta^{\mathcal{A}}, \alpha^{\mathcal{A}} \rangle$ be an arbitrary NMW. Let $\mathcal{D} = \langle \mathbb{D}, Q^{\mathcal{D}}, q_I^{\mathcal{D}}, \Delta^{\mathcal{D}}, \alpha^{\mathcal{D}} \rangle$ be a DSW recognizing $L(\mathcal{A})$. The arena of the game $\mathcal{G}(\mathcal{A})$ is $Q^{\mathcal{A}} \times Q^{\mathcal{D}}$ and its initial position $\langle q_0, p_0 \rangle$ is the pair of initial states $(q_I^{\mathcal{A}}, q_I^{\mathcal{D}})$. In the i-th round of a play, $\forall$ chooses a letter $d_i \in \mathbb{D}$ and $\exists$ chooses a state q_{i+1} such that $\langle q_i, d_i, q_{i+1} \rangle \in \Delta^{\mathcal{A}}$. The successive position is (q_{i+1}, p_{i+1}), where p_{i+1} is the unique state of $\mathcal{D}$ such that $\langle p_i, d_i, p_{i+1} \rangle \in \Delta^{\mathcal{D}}$.

An infinite play $\Pi = (\langle q_0, p_0 \rangle, d_0), (\langle q_1, p_1 \rangle, d_1), \ldots$ is won by $\exists$ if either the run $\Pi_{\mathcal{A}} := (q_i)_{i \in \mathbb{N}}$ is accepting or the run $\Pi_{\mathcal{D}} := (p_i)_{i \in \mathbb{N}}$ is rejecting. Note that since $\mathcal{D}$ recognizes $L(\mathcal{A})$, it follows that $\Pi_{\mathcal{D}}$ is rejecting iff $\Pi_{\mathbb{D}} := (d_i)_{i \in \mathbb{N}}$ does not belong to $L(\mathcal{A})$.

Since the game is ω-regular, it admits finite-memory winning strategies. The winning condition for $\exists$ in $\mathcal{G}(\mathcal{A})$ is the disjunction of $\alpha^{\mathcal{A}}$ with the Rabin condition that is dual to $\alpha^{\mathcal{D}}$. In particular, when $\mathcal{A}$ is an NRW, then the winning condition is a Rabin condition, thus if $\exists$ has a winning strategy in $\mathcal{G}(\mathcal{A})$, she also has a memoryless one.

Obviously, a strategy for $\exists$ is a strategy for resolving the nondeterminism in $\mathcal{A}$. Hence, we have the following.

Lemma 1. *If $\exists$ has a winning strategy in $\mathcal{G}(\mathcal{A})$ then $\mathcal{A}$ is GFG. Additionally, there exists a finite-memory strategy σ witnessing its GFGness. If $\mathcal{A}$ is an NRW (NMW), then σ is at most exponential (resp. doubly exponential) in the size of $\mathcal{A}$.*

Lemma 2. *If $\forall$ has a winning strategy in $\mathcal{G}(\mathcal{A})$, then $\mathcal{A}$ is not GFT.*

Proof. Let $\sigma_{\forall} : (Q^{\mathcal{A}})^+ \to \mathbb{D}$ be a winning strategy of $\forall$ in $\mathcal{G}(\mathcal{A})$. Thus, for a sequence $\mathbf{q}$ of states, namely the history of the game so far, the strategy $\sigma_{\forall}$ assigns the letter to be played by $\forall$. Note that for some sequences $\mathbf{q} \in (Q^{\mathcal{A}})^+$, the value $\sigma_{\forall}(\mathbf{q})$ is set arbitrarily, as there is no play corresponding to such a sequence (e.g., if $q_0 \neq q_I^{\mathcal{A}}$).

Let $u \in \mathbb{D}^*$ be a word and $\mathbf{q} = (q_0, \ldots, q_{j-1}) \in (Q^{\mathcal{A}})^+$ be a sequence of states. We say that $\mathbf{q}$ *forces* u if $(\mathbf{q}, u)$ is a prefix of a play in $\mathcal{G}(\mathcal{A})$ in which $\forall$ plays according to the strategy $\sigma_{\forall}$. Formally, $\mathbf{q}$ forces u if the following hold: (1) $|u| = |\mathbf{q}| = j > 0$, (2) $q_0 = q_I^{\mathcal{A}}$, (3) for every $i < j - 1$, the tuple $(q_i, u(i), q_{i+1})$ is a transition of $\mathcal{A}$, and (4) for every $i < j$, the letter $u(i)$ equals $\sigma_{\forall}(q_0 \ldots q_i)$.

Let $T \subseteq \mathbb{D}^*$ be the set of words $u \in \mathbb{D}^*$ such that there is a sequence $\mathbf{q} \in (Q^{\mathcal{A}})^*$ that forces u. Note that T is prefix-closed, so it is a $\mathbb{D}$-branching tree. We first show that $T \in \text{der}(L(\mathcal{A}))$. Consider an infinite path π of T. Let $\pi = \{\epsilon, u_1, u_2, \ldots\}$ and let

$(\boldsymbol{q}_i)_{i>0}$ be sequences $\boldsymbol{q}_i \in (Q^{\mathcal{A}})^*$ such that $\boldsymbol{q}_i$ forces u_i. Note that $|\boldsymbol{q}_i| = |u_i| = i$. Since there are finitely many states in $\mathcal{A}$, there exists a subsequence of $(\boldsymbol{q}_i)_{i>0}$ that is pointwise convergent to a limit $\rho \in (Q^{\mathcal{A}})^{\omega}$. For instance, this sequence can be built by iteratively choosing states that appear in infinitely many of the $\boldsymbol{q}_i$. For a finite or infinite sequence of states $\boldsymbol{q}$ and an index $j \in \mathbb{N}$, let $\boldsymbol{q}_{|j}$ be the prefix of $\boldsymbol{q}$ of length j. It follows that for every $j \in \mathbb{N}$ there exists $i > 0$ such that $\rho_{|j} = (\boldsymbol{q}_i)_{|j}$.

Let Π be the play that is the outcome of $\forall$ playing $\sigma_{\forall}$ and $\exists$ playing successive states of ρ. By the above, for every $j \in \mathbb{N}$, we have $\langle \rho(j), \pi(j), \rho(j+1) \rangle \in \Delta^{\mathcal{A}}$. Therefore, the play Π is well defined, $\Pi_{\mathbb{D}} = \pi$, and $\Pi_{\mathcal{A}} = \rho$. Since $\sigma_{\forall}$ is a winning strategy, $\Pi_{\mathcal{D}}$ is accepting, and therefore $\pi \in L(\mathcal{A})$. Since we showed the above for all paths π of T, we conclude that $T \in \mathrm{der}(L(\mathcal{A}))$.

Assume now, by way of contradiction, that $\mathcal{A}$ is GFT. Thus, $L(\mathcal{A}_t) = \mathrm{der}(L(\mathcal{A}))$. Since $T \in \mathrm{der}(L(\mathcal{A}))$, there is an accepting run ρ_t of $\mathcal{A}_t$ on T. Let Π be the infinite play of $\mathcal{G}(\mathcal{A})$ that is the outcome of $\forall$ playing $\sigma_{\forall}$ and $\exists$ playing transitions of ρ_t: if $\forall$ played $u \in \mathbb{D}^+$, then $\exists$ plays $\rho_t(u)$. Since ρ_t is accepting, $\Pi_{\mathcal{A}}$ is also accepting, and $\exists$ wins Π, contradicting the fact that $\forall$ plays his winning strategy. □

Observe that the arena of the game $\mathcal{G}(\mathcal{A})$ is finite and the winning condition for $\exists$ is ω-regular. Thus, the game is determined (see [1,4]) and one of the conditions in Lemma 1 or 2 holds. Hence $\mathcal{A}$ is either GFG or not GFT, and we can conclude with the following:

Theorem 1. *If an NMW is GFT then it is GFG. Moreover, there exists a finite-memory strategy σ witnessing its GFGness. If $\mathcal{A}$ is an NRW (NMW), then σ is at most exponential (resp. doubly exponential) in the size of $\mathcal{A}$.*

The following observation can be seen as an extension of [11] from parity condition to general Muller acceptance conditions. The only difference here is that we work with Σ-labelled $\mathbb{D}$-trees with $|\mathbb{D}| \geq |\Sigma|$, while [11] was working on binary trees with arbitrary alphabets. Again, we believe that these differences in the formalisms do not reflect essential behaviors of automata on infinite trees, since a simple encoding always allows to go from one formalism to another. Notice that the proof in [11] relies crucially on the structure of parity conditions and does not seem to generalize to arbitrary Muller conditions. In the following statement we use γ to denote an acceptance condition, e.g. a parity $[i, j]$ condition, a Rabin condition with k pairs, a Muller condition with k sets, etc.

Corollary 1. *Consider an ω-regular word language L. If $\mathrm{der}(L)$ can be recognized by a nondeterministic γ tree automaton, then L can be recognized by a deterministic γ word automaton.*

Proof. The word automaton $\mathcal{A}_w = \langle \mathbb{D}, Q, q_I, \Delta, \alpha \rangle$ associated with $\mathcal{A}$ is GFT, so by Theorem 1 it is GFG. By Theorem 1, the fact that $\mathcal{A}_w$ is GFG is witnessed by a strategy σ, using a finite memory structure M with an initial state $m_0 \in M$. That is to say, $\sigma : Q \times M \times \mathbb{D} \to Q \times M$ can be used to guide choices in $\mathcal{A}_w$, ensuring that all words in L are accepted. Therefore, we can build the required deterministic automaton $\mathcal{D}$ with states $Q \times M$, where the transition function maps a state $\langle q, m \rangle$ and letter $d \in \mathbb{D}$ to the state $\sigma(q, m, d)$. The acceptance condition of $\mathcal{D}$ is identical to α, and needs only to consider the Q-components of the states (the M component does not play a role for acceptance), and is thus of type γ. Since an accepting run of $\mathcal{D}$ induces an accepting run of $\mathcal{A}_w$, we have $L(\mathcal{D}) \subseteq L$. Conversely, if π is a word in L, the unique run of $\mathcal{D}$ on π corresponds to the execution of the GFG strategy σ in $\mathcal{A}_w$, and it thus accepting. Hence, $L(\mathcal{D}) = L$, for the deterministic γ automaton $\mathcal{D}$. $\square$

Observe that since deterministic automata are clearly GFT, the other direction of Corollary 1 is trivial.

4 From GFG to Deterministic Automata

In this section we study determinization of GFG automata. As discussed in Section 2, every DBP automaton (that is, a nondeterministic automaton that is determinizable by pruning) is GFG. The first question we consider is whether the converse is also valid, thus whether every GFG is DBP. We show that, surprisingly, not all GFG Büchi and co-Büchi automata are DBP. Note that since these counter examples are with basic acceptance conditions, the result follows for all common acceptance conditions. This gives rise to a second question, of the blowup involved in determinizing GFG NRWs. We describe a determinization construction that generates a DRW whose size is bounded by the product of the input GFG NRW and a GFG NRW for its complement.

4.1 GFG Büchi and Co-Büchi Automata Are not DBP

There is a strong intuition that a GFG NBW can be determinized by pruning: By definition, the choices of a GFG NBW are independent on the future. Accordingly, the question about GFG NBWs being DBP amounts to asking whether the choices are independent of the past. Since Büchi games are memoryless, it is tempting to believe that the answer is positive. A positive answer is also supported by the fact that GFG NBWs that recognize safety languages are DBP. In addition, all GFG NBWs studied so far, and in particular these constructed in [5] are DBP. Yet, as we show in this section, GFG NBWs are not DBP.

We start with a simple meta-NBW that operates over infinite words composed of two finite words (tokens), x and y. Afterwards, we formalize it to a GFG NBW. Moreover, we show that even more flexible versions of DBP, such as ones that allow the "deterministic" automata to have finite look-ahead are not sufficient.

A Meta Example. The meta-NBW $\mathcal{M}$, described in Figure 1, accepts exactly all words that contain infinitely many 'xx's or 'yy's. That is, $L(\mathcal{M}) = [(x + y)^*(xx + yy)]^\omega$. It is not hard to see that $\mathcal{M}$ is GFG by using the following strategy in its single nondeterministic state q_0: "if the last token was x then go to q_1 else go to q_2". On the other hand, determinizing q_0 to always choose q_1 loses y^ω and always choosing q_2 loses x^ω. Hence, $\mathcal{M}$ is not DBP.

A Concrete Example. Using the above meta-NBW with $x = aaa$ and $y = aba$ provides the NBW $\mathcal{A}$, described in Figure 2, whose language is $L = [(aaa + aba)^*(aaa\,aaa + aba\,aba)]^\omega$. Essentially, it follows from the simple observations that $\mathcal{A}$ has an infinite run on a word w iff $w \in (aaa + aba)^\omega$. Also, after a prefix whose length divides by 3, a run of $\mathcal{A}$ can only be in either q_0, p or g.

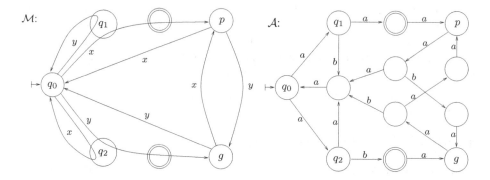

Fig. 1. A meta GFG NBW that is not DBP

Fig. 2. A GFG NBW that is not DBP

A Co-Büchi Example. In order to show that these counter examples are not specific to the Büchi condition, we give another example of GFG which is not DBP, using the co-Büchi condition. For simplicity, the acceptance is now specified via the transitions instead of the states. Dashed transitions are co-Büchi, i.e. accepting runs must take them only finitely often. (It is not hard to build a counter-example with co-Büchi condition on states from this automaton.)

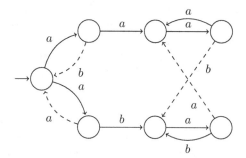

Fig. 3. A co-Büchi automaton that recognizes the language $(aa + ab)^*[a^\omega + (ab)^\omega]$. It is GFG but not DBP. Note that unlike the Büchi counter-example, one good choice is enough for getting an accepting run.

Theorem 2. *GFG NPWs are not DBP, even for Büchi and co-Büchi conditions.*

Proof. We prove that the NBW $\mathcal{A}$ from Figure 2 is GFG and is not DBP. First, the only nondeterminism of $\mathcal{A}$ is in q_0. The following strategy, applied in q_0, witnesses that $\mathcal{A}$ is GFG: "if the last three letters were 'aaa' then go to q_1 else go to q_2". Now, to see that $\mathcal{A}$ is not DBP, recall that the only nondeterminism of $\mathcal{A}$ is in q_0. Therefore, there are two possible prunings to consider: the DBW $\mathcal{A}'$ in which $\delta(q_0, a) = q_1$ and the DBW $\mathcal{A}''$ in which $\delta(q_0, a) = q_2$. With the former, $(aba)^\omega \in L(\mathcal{A}) \setminus L(\mathcal{A}')$ and with the latter $(aaa)^\omega \in L(\mathcal{A}) \setminus L(\mathcal{A}'')$.   □

While the GFG NBW $\mathcal{A}$ used in the proof of Theorem 2 is not DBP, it can be determinized by merging the states q_1 and q_2, to which q_0 goes nondeterministically, and then pruning. Furthermore, $\mathcal{A}$ is "almost deterministic", in the sense that a look-ahead of one letter into the future is sufficient for resolving its nondeterminism. One may wonder whether GFG NBWs are determinizable with more flexible definitions of pruning. We answer this to the negative, describing (in the full version) a GFG NBW in which merging the target states of the nondeterminism cannot help, and no finite look-ahead suffices for resolving the nondeterminism.

Theorem 3. *There are GFG NBWs that cannot be pruned into deterministic automata with unbounded yet finite look-ahead, or by merging concurrent target states.*

4.2 A Determinization Construction

In this section, we show that determinization in the context of GFG automata cannot induce an exponential blowup for both a language and its complement. This gives a serious hint towards the fact that determinization is simpler in the case of GFG. Indeed, for general nondeterministic automata, the blowup can occur on both sides. For example, consider the family of languages of finite words $L_k = (a + b)^* a (a + b)^k$. While for all $k \geq 1$, both L_k and its complement have nondeterministic automata with $O(k)$ states, a deterministic automaton for L_k must have at least 2^k states.

We now assume that we have a Rabin GFG (NRW-GFG) automaton for L, and an NRW-GFG for the complement $\text{comp}(L)$ of L. We show the following.

Theorem 4. *If $\mathcal{A}$ is an NRW-GFG for L with n states, and $\mathcal{B}$ is an NRW-GFG for $\text{comp}(L)$ with m states, then we can build a DRW for L with nm states.*

Proof. Let $\mathcal{A} = \langle \Sigma, Q, q_0, \Delta_{\mathcal{A}}, \alpha \rangle$ be an NRW-GFG for L, and $\mathcal{B} = \langle \Sigma, P, p_0, \Delta_{\mathcal{B}}, \beta \rangle$ be an NRW-GFG for $\text{comp}(L)$. We construct a Rabin game $\mathcal{G}$ between two players, $\exists$ and $\forall$, as follows. The arena of $\mathcal{G}$ is the product $\mathcal{A} \times \mathcal{B}$. Formally, the positions of the game are pairs of states $(q, p) \in Q \times P$, and there is an edge $(q, p) \xrightarrow{a} (q', p')$ in $\mathcal{G}$ if $(q, a, q') \in \Delta_{\mathcal{A}}$ and $(p, a, p') \in \Delta_{\mathcal{B}}$. The initial position of the game is (q_0, p_0).

A turn from position (p, q) is played as follows: First, $\forall$ chooses a letter a in Σ. Then, $\exists$ chooses an edge $(q, p) \xrightarrow{a} (q', p')$. The game then continues from (q', p'). Thus, the outcome of a play is an infinite sequence $\pi = (q_0, p_0), (q_1, p_1), (q_2, p_2), \ldots$ of positions. Note that π combines the run $\pi_{\mathcal{A}} = q_0, q_1, q_2, \ldots$ of $\mathcal{A}$ and the run $\pi_{\mathcal{B}} = p_0, p_1, p_2, \ldots$ of $\mathcal{B}$.

The winning condition for $\exists$ is that either $\pi_{\mathcal{A}}$ satisfies α or $\pi_{\mathcal{B}}$ satisfies β. These objectives can be easily specified by a Rabin winning condition. It is easy to see that $\exists$ has a winning strategy: it suffices to play in both automata according to their respecting GFG strategies. By definition of GFG automata, if $\forall$ generates a word u in L, the run $\pi_{\mathcal{A}}$ is accepting in $\mathcal{A}$ and thus satisfies α. Likewise, if $u \in \text{comp}(L)$, then the run $\pi_{\mathcal{B}}$ is accepting in $\mathcal{B}$ and thus satisfies β. Since every word is either in L or in $\text{comp}(L)$, the winning condition for $\exists$ is always satisfied.

It is known that Rabin games admit memoryless strategies [6]. Hence, $\exists$ actually has a memoryless winning strategy in $\mathcal{G}$. Such a strategy maps each position $(q, p) \in Q \times P$ and letter $a \in \Sigma$ to a destination (q', p'). Hence, by keeping only edges used by the memoryless strategy, we can prune the nondeterministic product automaton $\mathcal{A} \times \mathcal{B}$ into a deterministic automaton that accepts all words in Σ^ω. Moreover, by simply forgetting the acceptance condition of $\mathcal{B}$, and keeping only the one from $\mathcal{A}$, we get a DRW $\mathcal{D}$ recognizing L. Notice that if $\mathcal{A}$ was for instance Büchi, or parity with index $[i, j]$, the automaton $\mathcal{D}$ has the same acceptance condition. The number of states of $\mathcal{D}$ is $|P \times Q|$.

$\square$

References

1. Büchi, J.R., Landweber, L.H.: Solving Sequential Conditions by Finite State Strategies. CSD TR (1967)
2. Colcombet, T.: The theory of stabilisation monoids and regular cost functions. In: Albers, S., Marchetti-Spaccamela, A., Matias, Y., Nikoletseas, S., Thomas, W. (eds.) ICALP 2009, Part II. LNCS, vol. 5556, pp. 139–150. Springer, Heidelberg (2009)
3. Emerson, E.A., Sistla, A.P.: Deciding branching time logic. In: Proc. 16th ACM Symp. on Theory of Computing, pp. 14–24 (1984)
4. Grädel, E., Thomas, W., Wilke, T. (eds.): Automata, Logics, and Infinite Games. LNCS, vol. 2500. Springer, Heidelberg (2002)
5. Henzinger, T.A., Piterman, N.: Solving games without determinization. In: Ésik, Z. (ed.) CSL 2006. LNCS, vol. 4207, pp. 395–410. Springer, Heidelberg (2006)
6. Klarlund, N.: Progress measures, immediate determinacy, and a subset construction for tree automata. Ann. Pure Appl. Logic 69(2-3), 243–268 (1994)
7. Kupferman, O., Safra, S., Vardi, M.Y.: Relating word and tree automata. Ann. Pure Appl. Logic 138(1-3), 126–146 (2006)
8. Kupferman, O., Vardi, M.Y.: Safraless decision procedures. In: Proc. 46th IEEE Symp. on Foundations of Computer Science, pp. 531–540 (2005)
9. Landweber, L.H.: Decision problems for ω–automata. Mathematical Systems Theory 3, 376–384 (1969)
10. Morgenstern, G.: Expressiveness results at the bottom of the ω-regular hierarchy. M.Sc. Thesis, The Hebrew University (2003)
11. Niwinski, D., Walukiewicz, I.: Relating hierarchies of word and tree automata. In: Meinel, C., Morvan, M. (eds.) STACS 1998. LNCS, vol. 1373, Springer, Heidelberg (1998)
12. Pnueli, A., Rosner, R.: On the synthesis of a reactive module. In: Proc. 16th ACM Symp. on Principles of Programming Languages, pp. 179–190 (1989)
13. Rabin, M.O.: Weakly definable relations and special automata. In: Proc. Symp. Math. Logic and Foundations of Set Theory, pp. 1–23. North-Holland (1970)
14. Rabin, M.O., Scott, D.: Finite automata and their decision problems. IBM Journal of Research and Development 3, 115–125 (1959)
15. Safra, S.: On the complexity of ω-automata. In: Proc. 29th IEEE Symp. on Foundations of Computer Science, pp. 319–327 (1988)
16. Vardi, M.Y., Wolper, P.: Reasoning about infinite computations. Information and Computation 115(1), 1–37 (1994)

Coalgebraic Announcement Logics

Facundo Carreiro[1], Daniel Gorín[2], and Lutz Schröder[2]

[1] Institute for Logic, Language and Computation, Universiteit van Amsterdam
[2] Department of Computer Science, Universität Erlangen-Nürnberg

Abstract. In epistemic logic, dynamic operators describe the evolution of the knowledge of participating agents through communication, one of the most basic forms of communication being public announcement. Semantically, dynamic operators correspond to transformations of the underlying model. While metatheoretic results on dynamic epistemic logic so far are largely limited to the setting of Kripke models, there is evident interest in extending its scope to non-relational modalities capturing, e.g., uncertainty or collaboration. We develop a generic framework for non-relational dynamic logic by adding dynamic operators to coalgebraic logic. We discuss a range of examples and establish basic results including bisimulation invariance, complexity, and a small model property.

1 Introduction

Dynamic epistemic logics [5] are tools for reasoning about knowledge and belief of agents in a setting where interaction is of crucial interest. These logics extend epistemic logic (EL) [11] with dynamic operators, used to denote knowledge-changing actions. The most common of these is *public announcement*, first introduced in [17], which supports formulas of the form $\langle \phi \rangle \psi$ stating that after publicly (and faithfully) announcing that a certain fact ϕ holds (such as 'agent b does not know that agent a knows p'), ψ will hold (e.g. 'agent b knows p').

EL and its extension with public announcements (PAL) are typically interpreted on epistemic models, i.e. Kripke models where each accessibility relation is an equivalence; the points of the model represent *epistemic alternatives*. Evaluating a formula $\langle \phi \rangle \psi$ at a point c of an epistemic model I (notation $c \Vdash_I \langle \phi \rangle \psi$) amounts to verifying that the announcement is faithful (i.e., $c \Vdash_I \phi$) and that ψ holds at c after *removing* from I all epistemic alternatives where ϕ does *not* hold (notation $c \Vdash_{I \restriction \phi} \psi$). The term 'dynamic' refers precisely to the fact that models are changed during evaluation in this way.

Dynamic operators are of independent interest outside an epistemic setting. E.g., they occur as soon as one tries to express resiliency-related properties in verification (cf. van Benthem's *sabotage logic* [3] for an example); and they can turn a logic-based database query language into one supporting *hypothetical queries* (as in "return the aggregated sales we would have if we assumed that December sales corresponded to March").

Moreover, dynamic effects need not be restricted to a relational setting as found in Kripke models. E.g., the notion of announcing that a formula ψ holds

F.V. Fomin et al. (Eds.): ICALP 2013, Part II, LNCS 7966, pp. 101–112, 2013.

has a natural analogue in probabilistic modal logic [12,7,10] where announcements have the effect of conditioning the current distribution, as discussed along with other examples of dynamic actions in non-relational settings by Baltag [1].

Of course, dynamic operators are subject to the usual tension between expressive power and computational complexity. The extension of robustly decidable modal languages with dynamic operators can quickly lead to undecidability (see e.g. [13,18,9]; also, the $\downarrow$-binder of hybrid logic can be seen as an example of a dynamic operator, and in general leads to undecidability). On the other hand, PAL is well-behaved: it is as expressive as EL but exponentially more succinct with the same complexity (PSPACE-complete in the multi-agent case [14,8]).

In this paper we study announcement operators in a broad sense for modal logics beyond Kripke semantics. To deal with these at the appropriate level of generality, we work in the setting of coalgebraic modal logic [15], which uniformly covers a broad range of modal operators including, e.g., probabilistic, graded, and game-theoretic modalities (Sect. 2). A coalgebraic announcement can then be seen as the *global application* of a certain form of *local transformation* (contrasting with the global transformations considered in [1]).

A pervasive principle that transpires is that adding announcement operators preserves invariance under bisimulation and hence does not add fundamentally new expressive power; it may however be necessary to add new static modalities in order to eliminate announcements. We deal with generic announcement operators at increasing levels of generality, starting with a very well-behaved class of *strong announcements* that allow for a straightforward translation into the modal base language without requiring additional modalities. These constitute a particular type of *deterministic update* on models (i.e. certain transformation of the behaviour functor); which are also extended to account for announcements that are enriched with effects, such as non-determinism or uncertainty. As a unifying notion, we arrive at *backwards transformations* (Sect. 4) which act on predicates on the behaviour functor rather than the behaviour functor itself. We refer to the overall framework as *coalgebraic announcement logic* (CAL).

Besides bisimulation invariance, our technical results on these logics include: i) an equivalent translation of CAL into the modal base language, usually inducing an exponential blowup; ii) satisfiability preserving polynomial reductions to the base language for strong announcements, thus enabling the transfer of upper complexity bounds; and iii) a *constructive filtration* argument that yields a small model property independently of the presence of a master modality. The latter contribution appears to be a novel observation even for the static case, and substantially clarifies the original coalgebraic filtration construction [19].

2   Preliminaries

The framework of coalgebraic modal logics uniformly deals with a broad range of modal operators and a variety of different structures. This is achieved by recognizing the latter as instances of the concept of *coalgebra*. Given a functor $T : \mathsf{Set} \to \mathsf{Set}$, a *$T$-coalgebra* is a pair $\langle X, \gamma \rangle$ consisting of a non-empty set of *states* X and a *transition map* $\gamma : X \to TX$. We often identify $\langle X, \gamma \rangle$ with γ. For $x \in X$, we shall refer to $\gamma(x)$ as the T-*description of* x.

Example 1. Many structures that are well-known from theoretical computer science or from modal logic admit a natural presentation as coalgebras.

(i) Kripke frames are coalgebras for the covariant powerset functor $\mathcal{P}$. The map $\gamma_R : X \to \mathcal{P}X$ encodes a Kripke frame $\langle W, R \rangle$ with $\gamma_R(x) := \{w \mid xRw\}$.

(ii) A Kripke model $\langle W, R, V : \mathsf{P} \to \mathcal{P}W \rangle$ for a set P of propositions, corresponds to the K-coalgebra $\langle W, \gamma \rangle$ where $\mathsf{K}W := \mathcal{P}W \times \mathcal{P}\mathsf{P}$. The structure is recovered with $\gamma(x) := (\gamma_R(x), V^\flat(x))$ where $V^\flat(x) = \{p \in \mathsf{P} \mid x \in V(p)\}$.

(iii) The neighbourhood functor is $\mathcal{N} := \breve{\mathcal{P}} \circ \breve{\mathcal{P}}$, where $\breve{\mathcal{P}}$ is the contravariant powerset functor.[1] $\mathcal{N}$-coalgebras are the neighbourhood frames of modal logic [22], used in dynamic logics for reasoning about evidence and belief [6].

(iv) Let $\mathcal{M}$ be the subfunctor of $\mathcal{N}$ given by $\mathcal{M}X := \{S \in \mathcal{N}X \mid S$ is upwards closed$\}$. $\mathcal{M}$-coalgebras are monotone neighbourhood frames.

(v) The discrete distribution functor $\mathcal{D}_\omega$ maps X to the set of discrete probability distributions over X. $\mathcal{D}_\omega$-coalgebras are Markov chains. The subprobability functor $\mathcal{S}_\omega$ is similar but requires only that the measure of the whole space is at most 1 (instead of equal to 1).

(vi) Similarly, the finite multiset functor $\mathcal{B}_\omega$ maps X to the set of functions $\mu : X \to \mathbb{N}$ with finite support. Coalgebras for $\mathcal{B}_\omega$ are *multigraphs*, i.e. $\mathbb{N}$-weighted transition systems.

Coalgebras for a functor T form a category CoAlg_T where morphisms $f : \gamma \to \sigma$ between $\gamma : X \to TX$ and $\sigma : Y \to TY$ are maps $f : X \to Y$ with $\sigma \circ f = Tf \circ \gamma$. For $x \in X$ and $y \in Y$, we write $(x, \gamma) \sim (y, \sigma)$, read "$x$ and y are *behaviourally equivalent*", if there exists a coalgebra $\xi : Z \to TZ$ with morphisms $f : \gamma \to \xi$ and $g : \sigma \to \xi$ such that $f(x) = g(y)$. Functors are assumed wlog. to preserve injective maps [2] and to be non-trivial, in the sense that $TX = \emptyset$ implies $X = \emptyset$.

The syntax of coalgebraic modal logics is parametrized by a modal *similarity type* Λ. The language $\mathrm{CML}(\Lambda)$ is then given by the grammar

$$\phi ::= \bot \mid \phi \to \phi \mid \heartsuit_k(\phi_1, \dots \phi_k) \tag{1}$$

where $\heartsuit_k \in \Lambda$ is a modal operator of arity $k \geq 0$. We shall use the usual Boolean abbreviations $\wedge, \vee$, etc. when convenient. Each modality $\heartsuit_k \in \Lambda$ is interpreted as a k-ary *predicate lifting* $[\![\heartsuit_k]\!]$, i.e., a natural transformation $[\![\heartsuit_k]\!] : \breve{\mathcal{P}}^k \dot{\to} \breve{\mathcal{P}} \circ T^{op}$. The extension of ϕ in a coalgebra γ is given by $[\![\bot]\!]_\gamma = \emptyset$, $[\![\phi \to \psi]\!]_\gamma = (X \setminus [\![\phi]\!]_\gamma) \cup [\![\psi]\!]_\gamma$ and $[\![\heartsuit_k(\phi_1, \dots \phi_k)]\!]_\gamma = \{x \mid \gamma(x) \in [\![\heartsuit_k]\!]_X([\![\phi_1]\!]_\gamma, \dots, [\![\phi_k]\!]_\gamma)\}$. For the sake of readability, we sometimes pretend that all modal operators are unary.

Example 2. Some predicate liftings for the functors of Example 1 are

(i) For $\mathcal{P}$ we get the usual diamond with $[\![\Diamond]\!]_X(A) := \{t \in \mathcal{P}X \mid A \cap t \neq \emptyset\}$. The box is defined as $[\![\Box]\!]_X(A) := \{t \in \mathcal{P}X \mid t \subseteq A\}$.

---

[1] Formally, $\breve{\mathcal{P}} : \mathsf{Set}^{op} \to \mathsf{Set}$ with $\breve{\mathcal{P}}X = 2^X$ and, for $f : X \to Y$, $\breve{\mathcal{P}}f(A) = f^{-1}[A]$.

(ii) The diamond and box for K are obtained analogously. Propositions corre-
spond to *nullary* liftings $[\![p]\!]_X := \{(s, C) \in \mathsf{K}X \mid p \in C\}$ for every $p \in \mathsf{P}$.

(iii) For $\mathcal{N}$ and $\mathcal{M}$, we have $[\![\Box]\!]_X(A) := \{s \in \mathcal{N}X \mid A \in s\}$.

(iv) For $\mathcal{D}_\omega$ (and $\mathcal{S}_\omega$), the modalities L_p of probabilistic modal logic correspond
to the liftings $[\![L_p]\!]_X(A) := \{\mu \in \mathcal{D}_\omega X \mid \mu(A) \geq p\}$ for $p \in \mathbb{Q} \cap [0,1]$.

(v) The counting modalities $\Diamond_k$ of graded modal logic are given as predicate
liftings for $\mathcal{B}_\omega$ with $[\![\Diamond_k]\!]_X(A) := \{\mu \in \mathcal{B}_\omega X \mid \mu(A) \geq k\}$ for $k \in \mathbb{N}$.

It is well-known that $\mathrm{CML}(\Lambda)$ is invariant under behavioural equivalence; i.e., if
$(x, \gamma) \sim (y, \sigma)$ then $x \in [\![\phi]\!]_\gamma$ iff $y \in [\![\phi]\!]_\sigma$, for all $\phi \in \mathrm{CML}(\Lambda)$.

An operator $\heartsuit \in \Lambda$ is *monotone* if $A \subseteq B \subseteq X$ implies $[\![\heartsuit]\!]_X A \subseteq [\![\heartsuit]\!]_X B$. For
example, all operators of Example 2 are monotone except the one for $\mathcal{N}$. We
say that Λ is *separating* [16] if $t \in TX$ is uniquely determined by $\{([\![\heartsuit]\!], A) \in \Lambda \times \breve{\mathcal{P}}X \mid t \in [\![\heartsuit]\!]_X(A)\}$.

3   Strong Coalgebraic Announcements

Announcements (and, more generally, dynamic operators) are accounted for at
the syntactic level by extending $\mathrm{CML}(\Lambda)$ with a set Π of *dynamic modalities*
which we call the *dynamic similarity type* (as opposed to *static modalities* of the
static similarity type Λ). The syntax of $\mathrm{CAL}(\Pi, \Lambda)$ is obtained by extending the
grammar in (1) with the clause $\Delta_{\phi_1}\phi_2$ for $\Delta \in \Pi$.

At this point, one may informally read $\Delta_\psi \phi$ as "after announcing ψ, ϕ holds";
more generally, Δ_ψ will represent an *update* operation on the model that is
parameterized by a formula ψ. The update affects every state of the model but
does so in a *local* way. That is, for each state x of γ, Δ_ψ updates $\gamma(x)$ in a way
that depends only on $\gamma(x)$ and $[\![\psi]\!]_\gamma$.

Definition 3. An *update* is a natural transformation $\tau : T \xrightarrow{\cdot} (\breve{\mathcal{P}} \twoheadrightarrow T)$, where
$\breve{\mathcal{P}} \twoheadrightarrow T$ is the Set-functor defined by $(\breve{\mathcal{P}} \twoheadrightarrow T)X := (TX)^{\breve{\mathcal{P}}X}$ and, for $f : X \to Y$,
by $(\breve{\mathcal{P}} \twoheadrightarrow T)f := \lambda h : (TX)^{\breve{\mathcal{P}}X}.(Tf \circ h \circ \breve{\mathcal{P}}f)$.

Intuitively, every component τ_X takes as input the extension of a formula and the
T-description of an element and returns an updated T-description. Naturality
says that $Tf(\tau_X(t, \breve{\mathcal{P}}fA)) = \tau_Y(Tf(t), A)$, for $f : X \to Y$, $t \in TX$ and $A \subseteq Y$. We interpret $\Delta \in \Pi$ as an update $[\![\Delta]\!]$, and extend the semantics with
the clause $[\![\Delta_\psi\phi]\!]_\gamma = [\![\phi]\!]_{[\![\Delta]\!]_X([\![\psi]\!]_\gamma)\circ\gamma}$ — i.e., Δ_ψ applies local changes to the
entire coalgebra γ. We often identify Δ and $[\![\Delta]\!]$. The basic example is public
announcement logic over unrestricted frames (as considered in [14]; which we call
standard PAL although it is not interpreted over epistemic models), for which
we take $[\![\Delta]\!](S)(A) = A \cap S$ and then rewrite an announcement $\langle\psi\rangle\phi$ to $\psi \wedge \Delta_\psi\phi$
— this induces essentially the standard semantics, since restricting all successors
to satisfy ψ is modally indistinguishable from restricting the whole model to ψ.

Example 4. On relational models, the update $[\![!!]\!] : \mathcal{P} \xrightarrow{\cdot} (\breve{\mathcal{P}} \twoheadrightarrow \mathcal{P})$ defined as
$[\![!!]\!]_X(S)(A) := \texttt{if } A \neq \emptyset \texttt{ then } S \cap A \texttt{ else } S$ gives the *total announcements*

of [23]. That is, the announcement need not be truthful. If we think of A as the extension of a formula ϕ then this transformation removes the successors not satisfying ϕ. If an impossible formula is announced, it is ignored.

Example 5. For the functor $\mathcal{D}_\omega$ we can define an update $\tau : \mathcal{D}_\omega \dot{\to} (\breve{\mathcal{P}} \to \mathcal{D}_\omega)$ that has the effect of conditioning all probabilities to a given formula as $\tau_X(\mu)(A) := \lambda x.\text{if } \mu(A) > 0 \text{ then } \mu(x \mid A) \text{ else } \mu(x)$. Again, this update simply ignores the announcement of impossible events (i.e. those with probability 0). We also write this update as $\mu_{\restriction A} := \tau(\mu)(A)$.

It is clear that there is a resemblance between Examples 4 and 5: both updates give rise to dynamic operators that restrict the successors of a node to the points that satisfy certain formula. This connection can be made more precise, which will allow us to discuss this type of announcements in a uniform way.

Definition 6. An update τ is called a *strong announcement on Λ* if

(a) the partial application $\tau_X(-, A) : TX \to TX$ factors through the inclusion $i_A : TA \hookrightarrow TX$, for every $A \subseteq X$ (intuitively, $\tau_X(-, A) : TX \to TA$); and
(b) $\tau_X(s, A) \in \llbracket \heartsuit \rrbracket_X(C)$ iff $s \in \llbracket \heartsuit \rrbracket_X(C)$, for all $s \in TX$, $C \subseteq A \subseteq X$, $\heartsuit \in \Lambda$.

Condition (a) intuitively says that when ψ is announced, the resulting model should be based on the states satisfying ψ, while (b) ensures that all states satisfying ψ are retained. (Note that (b) is purely local and hence does not imply that whenever $\phi \to \psi$ is valid, then $\Delta_\psi \heartsuit \phi \leftrightarrow \heartsuit \phi$ is valid; this fails already in standard PAL.) In most cases, condition (b) is sufficient for naturality (so, for instance, we are exempt from proving it in the examples below).

Proposition 7. *A set-indexed family of maps $\tau_X : TX \to (2^X \to TX)$ satisfying condition (b) of Definition 6 for a separating set Λ of predicate liftings is a natural transformation $T \dot{\to} (\breve{\mathcal{P}} \to T)$; i.e. it is an update.*

Example 8.   (i) In slight modification of Example 4, putting $\llbracket ! \rrbracket(S)(A) := A \cap S$ defines a strong announcement on $\{\Diamond\}$ (but not on $\{\Box\}$); it induces standard PAL. For the differences between $\llbracket ! \rrbracket$ and $\llbracket !! \rrbracket$ see [14,23].
  (ii) Putting $\tau(\mu)(A) := \lambda x.\text{if } x \in A \text{ then } \mu(x) \text{ else } 0$ for $\mu \in \mathcal{B}_\omega X$ defines a strong announcement on $\Lambda = \{\Diamond_0, \Diamond_1, \dots\}$. The case for the subdistribution functor $\mathcal{S}_\omega$ is similar.
 (iii) For the neighbourhood functor $\mathcal{N}$, putting $\tau(t)(A) := t \cap \mathcal{P}A$ defines a strong announcement on $\Lambda = \{\Box\}$. The same definition (sic!) works for the monotone neighbourhood functor $\mathcal{M}$ and $\Lambda = \{\Diamond\}$.
 (iv) Probabilistic conditioning (cf. Example 5) is *not* a strong announcement.

These examples show that strong announcements occur in varying settings. For monotone logics, they are actually uniquely determined.

Theorem 9. *Let Λ consist of monotone operators. If τ is a strong announcement on Λ, then we have an adjunction $Ti_A \dashv \tau_X(-, A)$ where the ordering on TX is given by $s \leq t \iff \forall \heartsuit \in \Lambda, A \subseteq X. s \in \llbracket \heartsuit \rrbracket(A) \Rightarrow t \in \llbracket \heartsuit \rrbracket(A)$. In particular, τ is uniquely determined.*

This applies to all the updates of Example 8 except the one for $\mathcal{N}$ (since the predicate lifting involved is not monotone). In PAL, the announcement operator can be removed by means of well-known *reduction laws* [5,14], and hence does not add expressive power. This generalizes to strong announcements:

Theorem 10. *Let Δ be a strong announcement on Λ, and let $\heartsuit \in \Lambda$; then $\Delta_\psi \heartsuit \phi \equiv \heartsuit(\psi \wedge \Delta_\psi \phi)$.*

Remark 11. Theorem 10 can be used on duals of strong announcements, yielding, e.g., that in standard PAL, $!_\psi \Box \phi \equiv !_\psi \neg \Diamond \neg \phi \equiv \neg !_\psi \Diamond \neg \phi \equiv \Box(\psi \rightarrow !_\psi \phi)$.

Corollary 12. *Let Π be a set of strong announcements on Λ. For every $\phi \in \mathrm{CAL}(\Pi, \Lambda)$ there is $\phi^* \in \mathrm{CML}(\Lambda)$ such that $\phi \equiv \phi^*$. Hence, $\mathrm{CAL}(\Pi, \Lambda)$ is invariant under behavioural equivalence.*

In general, the translation induced by Theorem 10 (and commutation of announcements with Boolean operators) induces an exponential blowup. In Section 5 we will look at this in more detail, and show that one can obtain a satisfiability-preserving polynomial translation in many cases.

4   Imperfect Announcements and Other Effects

The updates introduced in the previous section are all of a *deterministic* nature, in the sense that points of a coalgebra are updated in a unique way. While this is a sensible condition for many applications, one can also think of updates where the outcome is, e.g. non-deterministic or governed by a probability distribution (what are usually called *effects* in the context of programming languages).

Let us consider non-determinism for a moment. It seems reasonable to extend the (deterministic) updates of Definition 3 to natural transformations of the form $T \dot{\rightarrow} (\breve{\mathcal{P}} \rightarrow \mathcal{P}T)$ which return a set of possible T-descriptions to choose from. The question is now how to interpret a formula such as $\Delta_\psi \phi$ in this setting. Notice that there are at least two sensible readings: we could declare $\Delta_\psi \phi$ true at x in γ if ϕ is true at x in *all* possible transformations of $\gamma(x)$ (demonic interpretation); or, alternatively, if ϕ is true at x in *at least one* of them (angelic interpretation). This example shows that such a notion of non-deterministic update by itself does not suffice; we will see that the missing behaviour can be specified by means of predicate liftings.

In order to suitably generalize the deterministic updates of the previous section we involve a different notion of transformation that acts directly on the involved predicates. As such, unsurprisingly, it is contravariant, generalizing as it does the preimage under an update.

Definition 13. *A regenerator is a natural transformation $\rho : \breve{\mathcal{P}} \times \breve{\mathcal{P}}T \dot{\rightarrow} \breve{\mathcal{P}}T$.*

The arguments of a regenerator should be thought of as the extension of the announced formula ψ and a predicate on TX of the form $[\![\heartsuit]\!][\![\phi]\!]_{\gamma_{\restriction \psi}}$, where $\gamma_{\restriction \psi}$ denotes an updated version of γ; the regenerator transforms this back into a predicate on TX as seen from the original γ. We can now define the *coalgebraic logic of announcements with effects* $\mathrm{CAL}^\circ(\Pi, \Lambda)$, which syntactically coincides with $\mathrm{CAL}(\Pi, \Lambda)$. In $\mathrm{CAL}^\circ(\Pi, \Lambda)$, each $\Delta \in \Pi$ is interpreted by a regenerator $[\![\Delta]\!]^\circ$. The semantics of formulas requires not only a T-coalgebra $\langle X, \gamma \rangle$ but also a map $\rho : 2^{TX} \to 2^{TX}$ (the *global regenerator*) that keeps track of the updates applied so far. The extension $[\![\cdot]\!]^\circ_{\rho, \gamma}$ of formulas of $\mathrm{CAL}^\circ(\Pi, \Lambda)$ is defined as usual for Boolean connectives and by

$$[\![\Delta_\psi \phi]\!]^\circ_{\rho, \gamma} = [\![\phi]\!]^\circ_{[\![\Delta]\!]^\circ_X ([\![\psi]\!]^\circ_{\rho, \gamma}, -) \circ \rho, \gamma} \qquad [\![\heartsuit \phi]\!]^\circ_{\rho, \gamma} = (\breve{\mathcal{P}}\gamma \circ \rho \circ [\![\heartsuit]\!]_X)[\![\phi]\!]^\circ_{\rho, \gamma} \ .$$

When no ambiguity arises, we may write $[\![\cdot]\!]$ instead of $[\![\cdot]\!]^\circ$. We will also use $[\![\phi]\!]_\gamma$ instead of $[\![\phi]\!]_{\iota_X, \gamma}$ where $\iota : \breve{\mathcal{P}}TX \to \breve{\mathcal{P}}T$ is the identity.

The connection between regenerators and "updates with effects" as in the non-deterministic update discussed above can now be made precise. The crucial observation is that any natural transformation $\tau : T \dashrightarrow (\breve{\mathcal{P}} \to FT)$ equipped with a predicate lifting (for F) $\lambda : \breve{\mathcal{P}} \to \breve{\mathcal{P}}F$ induces the regenerator (for T) $\rho^{\tau, \lambda} : \breve{\mathcal{P}} \times \breve{\mathcal{P}}T \dashrightarrow \breve{\mathcal{P}}T$ defined by $\rho^{\tau, \lambda}_X (A, S) := \breve{\mathcal{P}}(\tau_X(-)(A))[\lambda_{TX}(S)]$. In fact, $\mathrm{CAL}(\Lambda, \Pi)$ is just $\mathrm{CAL}^\circ(\Lambda, \Pi)$ with $F = \mathrm{Id}$ and $\lambda = id$.

Example 14. The non-deterministic announcements discussed above correspond to taking $F = \mathcal{P}$; the angelic interpretation is induced by $\lambda^a_X(t) := \{s \mid t \cap s \neq \emptyset\}$ and the demonic one by $\lambda^d_X(t) := \{s \mid t \subseteq s\}$ (i.e. $[\![\Diamond]\!]$ and $[\![\Box]\!]$ from Example 2.i). Examples of other updates for various choices of T and τ are:

(i) *Lossy announcements*: take $T = \mathcal{P}$ and $\tau_X(S, A) := \{S \cap A, S\}$; this models an announcement that can fail (leaving the set of successors unchanged). If $[\![\Delta]\!]^\circ$ is based on λ^d, then $\Delta_\psi \phi$ means that ϕ has to hold regardless of whether the announcement of ψ succeeds or not. The angelic case is dual.

(ii) *Controlled sabotage*: again for $T = \mathcal{P}$, but define $\tau_X(S, A) := \{S \setminus A, S\}$. If we think of A as a delicate area of a network, this transformation models links that may fail every time we want to go through them.

(iii) *Unstable (pseudo-)Markov chains*: let $T = \mathcal{S}_\omega$ and, for each $\varepsilon \in \mathbb{Q} \cap [0, 1]$, define a non-deterministic update $\tau^\varepsilon_X(\mu, A) = \{\tilde{\mu}_p \mid 0 \leq p \leq \varepsilon, \tilde{\mu}_p \in \mathcal{S}_\omega X\}$ where $\tilde{\mu}_p(x) := $ if $x \in A$ then $\mu(x) + p$ else $\mu(x)$. This update non-deterministically augments the probability of each $a \in A$ by at most ε.

Example 15. Taking $F = \mathcal{D}_\omega$ we get a probability distribution over the outcomes of an update. For $p \in \mathbb{Q} \cap [0, 1]$ we can define $\lambda^p_X(A) := \{\mu \mid \mu(A) \geq p\}$, obtaining dynamic operators Δ^p such that $\Delta^p_\psi \phi$ is true if the probability of the effect of announcing ψ (in some unspecified way) making ϕ true is greater than p. Note that the underlying coalgebra need not be probabilistic: in this example the coalgebra type T is arbitrary and F only plays a role in the liftings.

Remark 16. One is tempted to think of non-deterministic or probabilistic updates as changing the coalgebra γ, non-deterministically or randomly, to a fixed γ'. Although this is not accurate in that the choice is made again *every time* the evaluation encounters a static modality, it becomes formally correct by restricting to *tree-shaped* coalgebras, i.e. those where the underlying Kripke frame is a tree, which, in the light of Theorem 17 below, is without loss of generality since every coalgebra is behaviourally equivalent to a tree-shaped one [20]. One still needs to keep in mind, however, that the choice is made *per state*, e.g. a lossy announcement may succeed in some states and fail in others.

We now show that even in the presence of effects, dynamic modalities can be rewritten in terms of static modalities (albeit not necessarily of the base logic), and hence coalgebraic announcement logic in the more general sense remains invariant under behavioural equivalence. The crucial observation is that composing a predicate lifting and a regenerator yields a predicate lifting of a higher arity. That is, given $\lambda : \breve{\mathcal{P}}^n \dashrightarrow \breve{\mathcal{P}}T$ and $\rho : \breve{\mathcal{P}} \times \breve{\mathcal{P}}T \dashrightarrow \breve{\mathcal{P}}T$, we have that the composite $\lambda'_X(A, B_1 \ldots B_n) := \rho_X(A, \lambda_X(B_1 \ldots B_n))$ is a predicate lifting $\lambda' : \breve{\mathcal{P}}^{n+1} \dashrightarrow \breve{\mathcal{P}}T$. Given a static modality $\heartsuit$ and a dynamic modality Δ, we introduce a static modality $\boxtimes_{(\Delta.\heartsuit)}$ interpreted by the composite of $[\![\Delta]\!]$ and $[\![\heartsuit]\!]$ in this sense; one easily shows that

$$\Delta_\psi \heartsuit(\phi_1, \ldots, \phi_n) \equiv \boxtimes_{(\Delta.\heartsuit)}(\psi, \Delta_\psi \phi_1, \ldots, \Delta_\psi \phi_n).$$

Iterating this, we obtain static modalities $\boxtimes_m$ for all strings $m = \Delta^1 \cdots \Delta^n \heartsuit$; we denote the similarity type extending Λ by these modalities as $\mathsf{CL}_\Pi(\Lambda)$. We say that Λ is *closed for Π* if for every $\boxtimes_m \in \mathsf{CL}_\Pi(\Lambda)$, $\boxtimes_m(a_1, \ldots, a_n)$ (for propositional variables a_i) can be expressed as a polynomial-sized (in n) formula that is a propositional combination of formulae $\heartsuit\phi$ where ϕ is a propositional combination of the a_i. Note that when Π consists of strong announcements on Λ, then Λ is closed for Π.

Theorem 17. *For all $\phi \in \mathrm{CAL}^\circ(\Pi, \Lambda)$ there is $\phi^* \in \mathrm{CML}(\mathsf{CL}_\Pi(\Lambda))$ s.t. $\phi \equiv \phi^*$. Hence, $\mathrm{CAL}^\circ(\Pi, \Lambda)$ is invariant under behavioural equivalence.*

5   Decidability and Complexity

We have shown in the previous sections that coalgebraic announcement logic can be reduced to basic coalgebraic modal logic, albeit incurring an exponential blow-up. For PAL, it is known that this blow-up is unavoidable [14,8] and yet its computational complexity is the same as that of the base logic. We now show that under mild assumptions, the complexity result generalizes to the coalgebraic setting. We consider two standard decision problems: i) the *satisfiability problem* (SAT) "given a formula ϕ, decide if there is $\langle X, \gamma \rangle$ such that $[\![\phi]\!]_\gamma \neq \emptyset$"; and ii) the *constrained satisfiability problem* (CSAT) "given two formulas ψ and ϕ, decide if there is $\langle X, \gamma \rangle$ such that $[\![\phi]\!]_\gamma \neq \emptyset$ and $[\![\psi]\!]_\gamma = X$", in which case we say that ϕ is *satisfiable with respect to ψ*. In the terminology of description logic, CSAT corresponds to reasoning with a general TBox (given by ψ). SAT is a special case of CSAT but tends to have lower complexity.

Our results have different levels of generality. First we prove a small model property which holds unconditionally. For the case that the static similarity type Λ is closed for Π, we moreover provide a polynomial reduction of CSAT to the base logic, which allows inheriting the complexity of CSAT for the latter, typically EXPTIME. For a polynomial reduction of SAT to the base logic, we need to assume that Λ contains a *master modality*, which then again allows inheriting the complexity, typically PSPACE. We illustrate these methods for the logic of probabilistic conditioning.

5.1 Constructive Filtrations and the Small Model Property

For Σ a set of formulas, a Σ-filtered model is understood as one whose states are subsets of Σ and such that each state satisfies all the Σ-formulas it contains. Hence, a Σ-filtered model has at most $2^{|\Sigma|}$ states. We shall prove that every satisfiable formula ϕ of $\mathrm{CAL}^\circ(\Lambda, \Pi)$ is satisfied in some Σ-filtered model with $|\Sigma| = O(|\phi|)$. In fact, we show how to derive a Σ-filtered models from any model for ϕ, rather simplifying the construction given for $\mathrm{CML}(\Lambda)$ in [19].

In what follows, let Σ be a fixed set of $\mathrm{CAL}^\circ(\Lambda, \Pi)$-formulas, closed under subformulas and negation (identifying $\neg\neg\phi$ with ϕ as usual). Let $H_\Sigma \subseteq 2^\Sigma$ be the set of all maximal satisfiable subsets of Σ. For a given coalgebra $\gamma : X \to TX$, let $f_\Sigma : X \to H_\Sigma$ be the mapping $f_\Sigma(x) = \{\phi \in \Sigma \mid x \in \llbracket \phi \rrbracket_\gamma\}$. A coalgebra $\gamma_\Sigma : H_\Sigma \to TH_\Sigma$ is said to be a Σ-*filtration* of γ whenever for all $x \in X$, there exists $y \in X$ such that $f_\Sigma(x) = f_\Sigma(y)$ and $\gamma_\Sigma(f_\Sigma(x)) = Tf_\Sigma(\gamma(y))$.

Intuitively, we take the quotient of X by satisfaction of formulas in Σ and allow *any* of the members of the equivalence class to be the representative (each choice of representatives induces a potentially different filtration).[2]

To state the filtration theorem we need to relate global regenerators based on a coalgebra γ with global regenerators based on a Σ-filtration γ_Σ. So we say that two maps $\rho_X : 2^{TX} \to 2^{TX}$ and $\rho_{H_\Sigma} : 2^{TH_\Sigma} \to 2^{TH_\Sigma}$ are f_Σ-*synchronized* if $\rho_X \circ \breve{\mathcal{P}}Tf = \breve{\mathcal{P}}Tf \circ \rho_{H_\Sigma}$ (the naturality diagram for f_Σ if ρ was a natural transform). By induction over ϕ, one shows

Theorem 18. *Let γ_Σ be a Σ-filtration of $\langle X, \gamma\rangle$. For all $\phi \in \Sigma$, $x \in X$ and every pair of f_Σ-synchronized ρ_X and ρ_{H_Σ}, we have $x \in \llbracket\phi\rrbracket_{\rho_X,\gamma}$ iff $f_\Sigma(x) \in \llbracket\phi\rrbracket_{\rho_{H_\Sigma},\gamma_\Sigma}$.*

Observing that $id_{2^{TX}}$ and $id_{2^{TH_\Sigma}}$ are f_Σ-synchronized, we obtain

Corollary 19. $\mathrm{CAL}^\circ(\Lambda, \Pi)$ *has the small (exponential) model property.*

It is easy to exploit the small model property to give an upper bound NEXP-TIME for CSAT under mild additional conditions; in view of the results in the next section, we refrain from spelling out details.

5.2 Polynomial Satisfiability-Preserving Translations

From Theorem 17 we already know that when Λ is closed for Π, then every formula in $\mathrm{CAL}^\circ(\Lambda, \Pi)$ is equivalent to one of $\mathrm{CML}(\Lambda)$, perhaps exponentially

---

[2] The simpler definition $\gamma_\Sigma(f_\Sigma(x)) = Tf_\Sigma(\gamma(x))$ is not well-defined when f_Σ is not injective. Also, f_Σ may not be a coalgebra morphism, even with $\Sigma = \mathrm{CAL}^\circ(\Lambda, \Pi)$.

larger. This implies that the complexity of the decision problems for $\mathrm{CAL}^{\circ}(\Lambda, \Pi)$ is at most one exponential higher than for $\mathrm{CML}(\Lambda)$. But one can do better. The main observation is that, although the translated formula ϕ^{*} may be of size exponential in $|\phi|$, it contains only polynomially many *different* subformulas. Using essentially the same argument as in [14, Lemma 9], one can prove:

Theorem 20. *Let Λ be closed for Π. Then CSAT for $\mathrm{CAL}^{\circ}(\Lambda, \Pi)$ has the same complexity as for $\mathrm{CML}(\Lambda)$.*

The proof is by introducing propositional variables as abbreviations for subformulas, using the constraint. To deal with satisfiability in the absence of a constraint, we need a master modality to make abbreviations work up to the modal depth of the target formula. Coalgebraically, a *master modality* for Λ is a static modality $\boxdot$ such that $\boxdot\top$ and $\boxdot\phi \to (\heartsuit\psi \leftrightarrow \heartsuit(\phi \wedge \psi))$, for all $\heartsuit \in \Lambda$ and $\phi, \psi \in \mathrm{CML}(\Lambda)$, are valid. In the presence of a master modality one can give better bounds for SAT than those from Theorem 20.

Theorem 21. *Let Λ be closed for Π, and contain a master modality. Then the complexity of SAT for $\mathrm{CAL}^{\circ}(\Lambda, \Pi)$ is the same as for $\mathrm{CML}(\Lambda)$.*

Interestingly, master modalities abound: if T preserves inverse images then the predicate lifting $[\![\boxdot]\!]_{X}(A) := TA$ induces a master modality. Preserving inverse images is weaker than the frequent assumption of preservation of weak pullbacks. E.g., in graded modal logic $\Box_{1} := \neg\Diamond_{1}\neg$ is a master modality, and in probabilistic modal logic L_{1} is a master modality. Having observed that Λ is closed for strong announcements on Λ, we note explicitly

Theorem 22. *If Π consists of strong announcements on Λ, then CSAT for $\mathrm{CAL}(\Lambda, \Pi)$ has the same complexity as for $\mathrm{CML}(\Lambda)$; the same holds for SAT if Λ contains a master modality.*

In particular, we regain the known complexity of standard PAL, and we obtain, as new results, PSPACE and EXPTIME as the complexity of SAT and CSAT, respectively, for graded modal logic with the strong announcement operator (Example 8), as well as, e.g., NP as the complexity of neighbourhood logic and monotone modal logic with strong announcement.

5.3   Case Study: Conditionings in Probabilistic Logic

We now turn the attention to a logic where a master modality is available but the announcements that we are interested in are not strong: the logic of probabilistic conditioning (cf. Examples 5 and 8), the latter denoted Δ, with L_{1} being the master modality.

First, we observe that the static similarity type $\Lambda = \{L_{p} \mid p \in \mathbb{Q} \cap [0, 1]\}$ likely fails to be closed for $\{\Delta\}$. To see this, we first move to an extended modal language with linear inequalities over probabilities of formulas ϕ, the latter denoted $\ell(\phi)$ for 'likelihood' (i.e. essentially the probabilistic part of the logic introduced in [7]). In this notation, we have

$$\Delta_{\psi}L_{p}\phi \equiv (\ell(\psi) = 0 \to \ell(\Delta_{\psi}\phi) \geq 0) \wedge \ell(\psi \wedge \Delta_{\psi}\phi) \geq p \cdot \ell(\psi) \qquad (2)$$

where the first conjunct takes care of the exceptional case of impossible announcements. It seems unlikely that one could express the right-hand-side of (2) with a finite formula using only the operators L_p. However, we can extend Λ to a closed similarity type. A very conservative solution is to let $L_p(\phi \mid \psi)$ be a binary modal operator abbreviating $\ell(\phi \wedge \psi) \geq p \cdot \ell(\psi)$; then

$$\Delta_\psi L_p(\phi \mid \chi) \equiv (L_1(\neg\psi \mid \top) \to L_p(\Delta_\psi\phi \mid \chi)) \wedge L_p(\Delta_\psi\phi \mid \chi \wedge \psi),$$

i.e. the $L_p(-,-)$ are closed for $\{\Delta\}$. More generally, one may verify that the full language of linear inequalities (with n-ary modal operators $\sum_{i=1}^{n} a_i\ell(\_\_i) \geq b$ for all $n \geq 0$ and $a_1, \ldots, a_n, b \in \mathbb{Q}$) is closed. SAT for the modal logic of linear inequalities over probabilities is known to be in PSPACE [7], hence the complexity of SAT for the above logics of probabilistic conditioning is PSPACE.

6 Conclusions

We have introduced the framework of coalgebraic announcement logics and seen that it transfers to a setting of richer structures and general effects many nice properties enjoyed by (relational) public announcement logics. Our work fits in the spirit of [1], which also studies dynamic epistemic operators on a coalgebraic setting; although with a rather different perspective and a completely different technical machinery. That framework gains much generality from defining updates as natural transformations in CoAlg_T instead of Set. Giving up locality in this way (for now updates can look at the whole coalgebra structure) has its consequences: one loses small (or even finite) model properties, general decidability results, etc. The framework of [4] avoids explicit updates of models but otherwise has a comparable level of generality, with similar advantages and drawbacks. We expect that all coalgebraic announcement logics can be shown to be expressible in those frameworks.

It is only a slight simplification to claim that *all coalgebraic results are compositional*. One can reduce the study of composite functors to that of multi-sorted functors and almost all coalgebraic results extend straightforwardly from the single-sorted to the multi-sorted one at the expense of no more than additional indexes in the notation [21]. Applying this mechanism to the case of coalgebraic announcement logics requires some concentration (e.g. one needs to realize that regenerators apply to multi-sorted predicates) but does not pose any essential problems. Effectively, this means that all our results — invariance under behavioural equivalence, complexity analysis and the small model property — carry over to (complex) composite settings, such as a probabilistic logic about beliefs in a multi-agent system with group announcement operators that communicate facts only to selected agents by probabilistic conditioning.

References

1. Baltag, A.: A coalgebraic semantics for epistemic programs. In: Coalgebraic Methods in Computer Science. ENTCS, vol. 82, pp. 17–38. Elsevier (2003)
2. Barr, M.: Terminal coalgebras in well-founded set theory. Theoret. Comput. Sci. 114, 299–315 (1993)
3. van Benthem, J.: An essay on sabotage and obstruction. In: Hutter, D., Stephan, W. (eds.) Mechanizing Mathematical Reasoning. LNCS (LNAI), vol. 2605, pp. 268–276. Springer, Heidelberg (2005)
4. Cîrstea, C., Sadrzadeh, M.: Coalgebraic epistemic update without change of model. In: Mossakowski, T., Montanari, U., Haveraaen, M. (eds.) CALCO 2007. LNCS, vol. 4624, pp. 158–172. Springer, Heidelberg (2007)
5. van Ditmarsch, H., van der Hoek, W., Kooi, B.: Dynamic epistemic logics. Springer (2007)
6. Duque, D.F., van Benthem, J., Pacuit, E.: Evidence logic: a new look at neighborhood structures. In: Advances in Modal Logics. College Publications (2012)
7. Fagin, R., Halpern, J.Y.: Reasoning about knowledge and probability. J. ACM 41, 340–367 (1994)
8. French, T., van der Hoek, W., Iliev, P., Kooi, B.: Succinctness of epistemic languages. In: Int. Joint Conf. on Artif. Int., pp. 881–886 (2011)
9. French, T., van Ditmarsch, H.: Undecidability for arbitrary public announcement logic. In: Advances in Modal Logics, pp. 23–42. College Publications (2008)
10. Heifetz, A., Mongin, P.: Probabilistic logic for type spaces. Games and Economic Behavior 35, 31–53 (2001)
11. Hintikka, J.: Knowledge and belief. Cornell University Press (1962)
12. Larsen, K., Skou, A.: Bisimulation through probabilistic testing. Inf. Comput. 94, 1–28 (1991)
13. Löding, C., Rohde, P.: Model checking and satisfiability for sabotage modal logic. In: Pandya, P.K., Radhakrishnan, J. (eds.) FSTTCS 2003. LNCS, vol. 2914, pp. 302–313. Springer, Heidelberg (2003)
14. Lutz, C.: Complexity and succinctness of public announcement logic. In: Joint Conference on Autonomous Agents and Multi-Agent Systems, pp. 137–143 (2006)
15. Pattinson, D.: Coalgebraic modal logic: Soundness, completeness and decidability of local consequence. Theoret. Comput. Sci. 309, 177–193 (2003)
16. Pattinson, D.: Expressive logics for coalgebras via terminal sequence induction. Notre Dame J. Formal Logic 45, 2004 (2002)
17. Plaza, J.A.: Logics of public communications. In: International Symposium on Methodologies for Intelligent Systems, pp. 201–216 (1989)
18. Rohde, P.: Moving in a crumbling network: The balanced case. In: Marcinkowski, J., Tarlecki, A. (eds.) CSL 2004. LNCS, vol. 3210, pp. 310–324. Springer, Heidelberg (2004)
19. Schröder, L.: A finite model construction for coalgebraic modal logic. J. Log. Algebr. Prog. 73, 97–110 (2007)
20. Schröder, L., Pattinson, D.: Coalgebraic correspondence theory. In: Ong, L. (ed.) FOSSACS 2010. LNCS, vol. 6014, pp. 328–342. Springer, Heidelberg (2010)
21. Schröder, L., Pattinson, D.: Modular algorithms for heterogeneous modal logics via multi-sorted coalgebra. Math. Struct. Comput. Sci. 21(2), 235–266 (2011)
22. Segerberg, K.: An essay in classical modal logic. No. 1 in Filosofiska studier utgivna av Filosofiska föreningen och Filosofiska institutionen vid Uppsala univ. (1971)
23. Steiner, D., Studer, T.: Total public announcements. In: Artemov, S., Nerode, A. (eds.) LFCS 2007. LNCS, vol. 4514, pp. 498–511. Springer, Heidelberg (2007)

Self-shuffling Words[*]

Émilie Charlier[1], Teturo Kamae[2], Svetlana Puzynina[3,5],
and Luca Q. Zamboni[4,5]

[1] Département de Mathématique, Université de Liège, Belgium
echarlier@ulg.ac.be
[2] Advanced Mathematical Institute, Osaka City University, Japan
kamae@apost.plala.or.jp
[3] Sobolev Institute of Mathematics, Novosibirsk, Russia
svepuz@utu.fi
[4] Institut Camille Jordan, Université Lyon 1, France
lupastis@gmail.com
[5] FUNDIM, University of Turku, Finland

Abstract. In this paper we introduce and study a new property of infinite words which is invariant under the action of a morphism: We say an infinite word $x \in \mathbb{A}^{\mathbb{N}}$, defined over a finite alphabet $\mathbb{A}$, is self-shuffling if x admits factorizations: $x = \prod_{i=1}^{\infty} U_i V_i = \prod_{i=1}^{\infty} U_i = \prod_{i=1}^{\infty} V_i$ with $U_i, V_i \in \mathbb{A}^+$. In other words, there exists a shuffle of x with itself which reproduces x. The morphic image of any self-shuffling word is again self-shuffling. We prove that many important and well studied words are self-shuffling: This includes the Thue-Morse word and all Sturmian words (except those of the form aC where $a \in \{0, 1\}$ and C is a characteristic Sturmian word). We further establish a number of necessary conditions for a word to be self-shuffling, and show that certain other important words (including the paper-folding word and infinite Lyndon words) are not self-shuffling. In addition to its morphic invariance, which can be used to show that one word is not the morphic image of another, this new notion has other unexpected applications: For instance, as a consequence of our characterization of self-shuffling Sturmian words, we recover a number theoretic result, originally due to Yasutomi, which characterizes pure morphic Sturmian words in the orbit of the characteristic.

1 Introduction

Let $\mathbb{A}$ be a finite non-empty set. We denote by $\mathbb{A}^*$ the set of all finite words $u = x_1 x_2 \ldots x_n$ with $x_i \in \mathbb{A}$. The quantity n is called the length of u and is denoted $|u|$. For a letter $a \in \mathbb{A}$, by $|u|_a$ we denote the number of occurrences of

---

[*] The first and fourth authors are supported in part by FiDiPro grant of the Academy of Finland. The third author is supported in part by the Academy of Finland under grant 251371, by Russian Foundation of Basic Research (grant 12-01-00448), and by RF President grant MK-4075.2012.1. Preliminary version: http://arxiv.org/abs/1302.3844.

F.V. Fomin et al. (Eds.): ICALP 2013, Part II, LNCS 7966, pp. 113–124, 2013.
© Springer-Verlag Berlin Heidelberg 2013

a in u. The empty word, denoted ε, is the unique element in $\mathbb{A}^*$ with $|\varepsilon| = 0$. We set $\mathbb{A}^+ = \mathbb{A} - \{\varepsilon\}$. We denote by $\mathbb{A}^\mathbb{N}$ the set of all one-sided infinite words $x = x_0 x_1 x_2 \ldots$ with $x_i \in \mathbb{A}$.

Given k finite or infinite words $x^{(1)}, x^{(2)}, \ldots, x^{(k)} \in A^* \cup \mathbb{A}^\mathbb{N}$ we denote by

$$\mathscr{S}(x^{(1)}, x^{(2)}, \ldots, x^{(k)}) \subset \mathbb{A}^* \cup \mathbb{A}^\mathbb{N}$$

the collection of all words z for which there exists a factorization

$$z = \prod_{i=0}^{\infty} U_i^{(1)} U_i^{(2)} \cdots U_i^{(k)}$$

with each $U_i^{(j)} \in \mathbb{A}^*$ and with $x^{(j)} = \prod_{i=0}^{\infty} U_i^{(j)}$ for $1 \leq j \leq k$. Intuitively, z may be obtained as a *shuffle* of the words $x^{(1)}, x^{(2)}, \ldots, x^{(k)}$. In case $x^{(1)}, x^{(2)}, \ldots, x^{(k)} \in A^*$, each of the above products can be taken to be finite.

Finite word shuffles were extensively studied in [5]. Given $x \in \mathbb{A}^*$, it is generally a difficult problem to determine whether there exists $y \in \mathbb{A}^*$ such that $x \in \mathscr{S}(y, y)$ (see Open Problem 4 in [5]). However, in the context of infinite words, this question is essentially trivial: In fact, it is readily verified that if $x \in \mathbb{A}^\mathbb{N}$ is such that each $a \in \mathbb{A}$ occurring in x occurs an infinite number of times in x, then there exist infinitely many $y \in \mathbb{A}^\mathbb{N}$ with $x \in \mathscr{S}(y, y)$. Instead, in the framework of infinite words, a far more delicate question is the following:

Question 1. Given $x \in \mathbb{A}^\mathbb{N}$, does there exist an integer $k \geq 2$ such that $x \in \mathscr{S}(\underbrace{x, x, \ldots, x}_{k})$?

If such a k exists, we say x is *k-self-shuffling*.

Given $x = x_0 x_1 x_2 \ldots \in \mathbb{A}^\mathbb{N}$ and an infinite subset $N = \{N_0 < N_1 < N_2 < \ldots\} \subseteq \mathbb{N}$, we put $x[N] = x_{N_0} x_{N_1} x_{N_2} \ldots \in \mathbb{A}^\mathbb{N}$. Alternatively,

Definition 1. For $x \in \mathbb{A}^\mathbb{N}$ and $k = 2, 3, \ldots$, we say x is *k-self-shuffling* if there exists a k-element partition $\mathbb{N} = \bigcup_{i=1}^{k} N^i$ with $x[N^i] = x$ for each $i = 1, \ldots, k$.

In case $k = 2$, we say simply x is *self-shuffling*. We note that if x is k-self-shuffling, then x is ℓ-self-shuffling for each $\ell \geq k$ but not conversely (see §2), whence each self-shuffling word is k-self-shuffling for all $k \geq 2$. In this paper we are primarily interested in self-shuffling words, however, many of the results presented here extend to general k. Thus $x \in \mathbb{A}^\mathbb{N}$ is self-shuffling if and only if x admits factorizations

$$x = \prod_{i=1}^{\infty} U_i V_i = \prod_{i=1}^{\infty} U_i = \prod_{i=1}^{\infty} V_i$$

with $U_i, V_i \in \mathbb{A}^+$.

The property of being self-shuffling is an intrinsic property of the word (and not of the associated language) and seems largely independent of its complexity (examples exist from the lowest to the highest possible complexity). The simplest class of self-shuffling words consists of all (purely) periodic words $x = u^\omega$. It is clear that if x is self-shuffling, then every letter $a \in \mathbb{A}$ occurring in x must occur an infinite number of times. Thus for instance, the ultimately periodic word 01^ω is not self-shuffling. As we shall see, many well-known words which are of interest in both combinatorics on words and symbolic dynamics, are self-shuffling. This includes for instance the famous *Thue-Morse* word

$$\mathbf{T} = 0110100110010110100101100110100110010110\ldots$$

whose origins go back to the beginning of the last century with the works of the Norwegian mathematician Axel Thue [9]. The nth entry t_n of $\mathbf{T}$ is defined as the sum modulo 2 of the digits in the binary expansion of n. While the Thue-Morse word appears naturally in many different areas of mathematics (from discrete mathematics to number theory to differential geometry-see [1] or [2]), proving that Thue-Morse is self-shuffling is somewhat more involved than expected.

Sturmian words constitute another important class of aperiodic self-shuffling words. Sturmian words are infinite words over a binary alphabet having exactly $n+1$ factors of length n for each $n \geq 0$ [7]. Their origin can be traced back to the astronomer J. Bernoulli III in 1772. They arise naturally in many different areas of mathematics including combinatorics, algebra, number theory, ergodic theory, dynamical systems and differential equations. Sturmian words are also of great importance in theoretical physics and in theoretical computer science and are used in computer graphics as digital approximation of straight lines. We show that all Sturmian words are self-shuffling except those of the form aC where $a \in \{0, 1\}$ and C is a characteristic Sturmian word. Thus for every irrational number α, all (uncountably many) Sturmian words of slope α are self-shuffling except for two. Our proof relies on a geometric characterization of Sturmian words via irrational rotations on the circle.

So while there are many natural examples of aperiodic self-shuffling words, the property of being self-shuffling is nevertheless quite restrictive. We obtain a number of necessary (and in some cases sufficient) conditions for a word to be self-shuffling. For instance, if a word x is self-shuffling, then x begins in only finitely many Abelian border-free words. As an application of this we show that the well-known *paper folding word* is not self-shuffling. Infinite Lyndon words (i.e., infinite words which are lexicographically smaller than each of its suffixes) are also shown not to be self-shuffling.

One important feature of self-shuffling words stems from its invariance under the action of a morphism: The morphic image of a self-shuffling word is again self-shuffling. In some instances this provides a useful tool for showing that one word is not the morphic image of another. So for instance, the paper folding word is not the morphic image of any self-shuffling word. However this application requires knowing a priori whether a given word is or is not self-shuffling. In general, to show that a word is self-shuffling, one must actually exhibit a shuffle. Self-shuffling words have other unexpected applications particularly in the study of fixed points of substitutions. For instance, as an almost immediate consequence

of our characterization of self-shuffling Sturmian words, we recover a result, first
proved by Yasutomi via number theoretic methods, which characterizes pure
morphic Sturmian words in the orbit of the characteristic.

2   Examples and Non-examples

In this section we list some examples and non-examples of self-shuffling words.
As usual in combinatorics on words, we follow notation from [7].

Fibonacci Word: The Fibonacci infinite word

$$x = 0100101001001010010100\ldots$$

is defined as the fixed point of the morphism φ given by $0 \mapsto 01$, $1 \mapsto 0$. It
is readily verified that $\varphi^2(a) = \varphi(a)a$ for each $a \in \{0,1\}$. Whence, writing
$x = x_0 x_1 x_2 \ldots$ with each $x_i \in \{0,1\}$ we obtain

$$x = x_0 x_1 x_2 \ldots = \varphi(x_0)\varphi(x_1)\varphi(x_2)\ldots = \varphi^2(x_0)\varphi^2(x_1)\varphi^2(x_2)\ldots =$$
$$\varphi(x_0)x_0\varphi(x_1)x_1\varphi(x_2)x_2 \ldots$$

which shows that x is self-shuffling. In contrast, the word $y = 0x$ is not self-
shuffling. The word y starts with infinitely many prefixes of the form $0B1$ with
B a palindrome. It follows that $0B1$ is *Abelian border-free* (i.e., no proper suffix
of $0B1$ is Abelian equivalent to a proper prefix of $0B1$). By Proposition 3 the
word y is not self-shuffling.

Paper-folding Word: The paper-folding word

$$x = 00100110001101100010\ldots$$

is a Toeplitz word generated by the pattern $u = 0?1?$ (see, e.g., [4]). It is readily
verified that x begins in arbitrarily long Abelian border-free words and hence by
Proposition 3 is not self-shuffling. More precisely, the prefixes u_j of x of length
$n_j = 2^j - 1$ are Abelian border-free. Indeed, it is verified that for each $k < n_j$, we
have $|\mathrm{pref}_k(u_j)|_0 > k/2$ while $|\mathrm{suff}_k(u_j)|_0 \leq k/2$. Here $\mathrm{pref}_k(u)$ (resp., $\mathrm{suff}_k(u)$)
denotes the prefix (resp., suffix) of length k of a word u.

A 3-Self-shuffling Word Which Is Not Self-shuffling: Let y denote the
fixed point of the morphism $\sigma : 0 \mapsto 0001$ and $1 \mapsto 0101$, and put

$$x = 0^{-2}y = 010001000101010001000100010101000100010001010100010101010 0\ldots,$$

where the notation $w = v^{-k}u$ means that $u = v^k w$. Then for each prefix u_j of
x of length $4^j - 2$, the longest Abelian border of u_j of length less than or equal
to $(4^j - 2)/2$ has length 2. Hence x is not self-shuffling (see Proposition 3). The
3-shuffle is given by the following:

$U_0 = 0100, U_1 = 01,$ $\quad \ldots, \ U_{4i+2} = \varepsilon,$ $\qquad\qquad U_{4i+3} = \sigma^{i+1}(0100),$

$\qquad\qquad\qquad\qquad U_{4i+4} = \sigma(0),$ $\qquad\qquad U_{4i+5} = (\sigma(0))^{-1}\sigma^{i+1}(01),$

$V_0 = 0100, V_1 = 01,$ $\quad \ldots, \ V_{4i+2} = (\sigma(0))^{-1}\sigma^{i+1}(0),$ $\quad V_{4i+3} = \sigma(0),$

$\qquad\qquad\qquad\qquad V_{4i+4} = (\sigma(0))^{-1}\sigma^{i+1}(01)\sigma(0),$ $V_{4i+5} = \varepsilon,$

$W_0 = 01, \ \ W_1 = (\sigma(0))^2, \ldots, \ W_{4i+2} = \varepsilon,$ $\qquad\qquad W_{4i+3} = (\sigma(0))^{-1}\sigma^{i+1}(01),$

$\qquad\qquad\qquad\qquad W_{4i+4} = \varepsilon,$ $\qquad\qquad W_{4i+5} = \sigma^{i+2}(0)\sigma(0).$

It is then verified that

$$x = \prod_{i=0}^{\infty} U_i V_i W_i = \prod_{i=0}^{\infty} U_i = \prod_{i=0}^{\infty} V_i = \prod_{i=0}^{\infty} W_i,$$

from which it follows that x is 3-self-shuffling.

A Recurrent Binary Self-shuffling Word with Full Complexity: For each positive integer n, let z_n denote the concatenation of all words of length n in increasing lexicographic order. For example, $z_2 = 00011011$. For $i \geq 0$ put

$$v_i = \begin{cases} z_n, & \text{if } i = n2^{n-1} \text{ for some } n, \\ 0^i 1^i, & \text{otherwise,} \end{cases}$$

and define

$$x = \prod_{i=0}^{\infty} X_i = 0101001 10^3 011^3 0^4 010^2 1^2 011^4 \ldots,$$

where $X_0 = X_1 = 01$, $X_2 = 0011$, and for $i \geq 3$, $X_i = 0^i y_{i-2} 1^i$, where $y_{i-2} = y_{i-3} v_{i-2} y_{i-3}$, and $y_0 = \varepsilon$. We note that x is recurrent (i.e., each prefix occurs twice) and has full complexity (since it contains z_n as a factor for every n).

To show that the word x is self-shuffling, we first show that $X_{i+1} \in \mathscr{S}(X_i, X_i)$. Take $N_i = \{0, \ldots, i-1, i+1, \ldots, 2^i - i, 2^i - i + v_{i-1}|_1, 2^{i+1} - i - 1\}$, where $u|_1$ denotes the positions j of a word u in which the j-th letter u_j of u is equal to 1. Then it is straightforward to see that $X_i = X_{i+1}[N_i] = X_{i+1}[\{1, \ldots, 2^{i+1}\} \backslash N_i]$. The self-shuffle of x is built in a natural way concatenating shuffles of X_i starting with $U_0 = V_0 = 01$, so that $X_0 \ldots X_{i+1} \in \mathscr{S}(X_0 \ldots X_i, X_0 \ldots X_i)$.

3   General Properties

In this section we develop several fundamental properties of self-shuffling words. The next two propositions show the invariance of self-shuffling words with respect to the action of a morphism:

Proposition 1. *Let $\mathbb{A}$ and $\mathbb{B}$ be finite non-empty sets and $\tau : \mathbb{A} \to \mathbb{B}^*$ a morphism. If $x \in \mathbb{A}^{\mathbb{N}}$ is self-shuffling, then so is $\tau(x) \in \mathbb{B}^{\mathbb{N}}$.*

Proof. If $x \in \mathscr{S}(x, x)$, then we can write $x = \prod_{i=1}^{\infty} U_i V_i = \prod_{i=1}^{\infty} U_i = \prod_{i=1}^{\infty} V_i$. Whence $\tau(x) = \prod_{i=1}^{\infty} \tau(U_i V_i) = \prod_{i=1}^{\infty} \tau(U_i)\tau(V_i) = \prod_{i=1}^{\infty} \tau(U_i) = \prod_{i=1}^{\infty} \tau(V_i)$ as required.

Proposition 2. *Let $\tau : \mathbb{A} \to \mathbb{A}^*$ be a morphism, and $x \in \mathbb{A}^{\mathbb{N}}$ be a fixed point of τ.*

1. *Let u be a prefix of x and k be a positive integer such that $\tau^k(a)$ begins in u for each $a \in \mathbb{A}$. Then if x is self-shuffling, then so is $u^{-1}x$.*
2. *Let $u \in \mathbb{A}^*$, and let k be a positive integer such that $\tau^k(a)$ ends in u for each $a \in \mathbb{A}$. Then if x is self-shuffling, then so is ux.*

Proof. We prove only item (1) since the proof of (2) is essentially identical. Suppose $x = \prod_{i=1}^{\infty} U_i V_i = \prod_{i=1}^{\infty} U_i = \prod_{i=1}^{\infty} V_i$. Then by assumption, for each $i \geq 1$ we can write $\tau^k(U_i) = uU_i'$ and $\tau^k(V_i) = uV_i'$ for some $U_i', V_i' \in \mathbb{A}^*$. Put $X_i = U_i'u$ and $Y_i = V_i'u$. Then since

$$x = \tau^k(x) = \prod_{i=1}^{\infty} \tau^k(U_i V_i) = \prod_{i=1}^{\infty} \tau^k(U_i)\tau^k(V_i) = \prod_{i=1}^{\infty} \tau^k(U_i) = \prod_{i=1}^{\infty} \tau^k(V_i),$$

we deduce that

$$u^{-1}x = \prod_{i=1}^{\infty} X_i Y_i = \prod_{i=1}^{\infty} X_i = \prod_{i=1}^{\infty} Y_i.$$

Corollary 1. *Let $\tau : \mathbb{A} \to \mathbb{A}^*$ be a primitive morphism, and $a \in \mathbb{A}$. Suppose $\tau(b)$ begins (respectively ends) in a for each letter $b \in \mathbb{A}$. Suppose further that the fixed point $\tau^{\infty}(a)$ is self-shuffling. Then every right shift (respectively left shift) of $\tau^{\infty}(a)$ is self-shuffling.*

Remark 1. Since the Fibonacci word is self-shuffling and is fixed by the primitive morphism $0 \mapsto 01$, $1 \mapsto 0$, it follows from Corollary 1 that every tail of the Fibonacci word is self-shuffling.

There are a number of necessary conditions that a self-shuffling word must satisfy, which may be used to deduce that a given word is not self shuffling. For instance:

Proposition 3. *If $x \in \mathbb{A}^{\mathbb{N}}$ is self-shuffling, then for each positive integer N there exists a positive integer M such that every prefix u of x with $|u| \geq M$ has an Abelian border v with $|u|/2 \geq |v| \geq N$. In particular, x must begin in only a finite number of Abelian border-free words.*

Proof. Suppose to the contrary that there exist factorizations $x = \prod_{i=0}^{\infty} U_i V_i = \prod_{i=0}^{\infty} U_i = \prod_{i=0}^{\infty} V_i$ with $U_i, V_i \in \mathbb{A}^+$, and there exists N such that for every M there exists a prefix u of x with $|u| \geq M$ which has no Abelian borders of length between N and $|u|/2$. Take $M = |\prod_{i=0}^{N} U_i V_i|$ and a prefix u satisfying these conditions. Then there exist non-empty proper prefixes U' and V' of u such that $u \in \mathscr{S}(U', V')$ with $|U'|, |V'| > N$. Writing $u = U'U''$ it follows that U'' and V' are Abelian equivalent. This contradicts that u has no Abelian borders of length between N and $|u|/2$.

An extension of this argument gives both a necessary and sufficient condition for self-shuffling in terms of Abelian borders (which is however difficult to check in practice). For $u \in \mathbb{A}^*$ let $\Psi(u)$ denote the *Parikh vector* of u, i. e., $\Psi(u) = (|u|_a)_{a \in \mathbb{A}}$.

Definition 2. Given $x \in \mathbb{A}^{\mathbb{N}}$, we define a directed graph $G_x = (V_x, E_x)$ with vertex set

$$V_x = \{(n, m) \in \mathbb{N}^2 \mid \Psi(\mathrm{pref}_n x) + \Psi(\mathrm{pref}_m x) = \Psi(\mathrm{pref}_{n+m} x)\}$$

and the edge set

$$E_x = \{ ((n, m), (n', m')) \in V_x \times V_x \mid \\ n' = n + 1 \text{ and } m' = m \text{ or } m' = m + 1 \text{ and } n' = n\}.$$

We say that G_x *connects* $\mathbf{0}$ *to* ∞ if there exists an infinite path $\prod_{j=1}^{\infty}(n_j, m_j)$ in G_x such that $(n_0, m_0) = (0, 0)$ and $n_j, m_j \to \infty$ as $j \to \infty$.

Theorem 1. *A word $x \in \mathbb{A}^{\mathbb{N}}$ is self-shuffling if and only if the graph G_x connects $\mathbf{0}$ to ∞.*

The theorem gives a constructive necessary and sufficient condition for self-shuffling since a path to infinity defines a self-shuffle.

As we shall now see, lexicographically extremal words are never self-shuffling. Let $(\mathbb{A}, \leq)$ be a finite linearly ordered set. Then $\leq$ induces the lexicographic ordering $\leq_{\mathrm{lex}}$ on $\mathbb{A}^+$ and $\mathbb{A}^{\mathbb{N}}$ defined as follows: If $u, v \in \mathbb{A}^+$ (or $\mathbb{A}^{\mathbb{N}}$) we write $u \leq_{\mathrm{lex}} v$ if either $u = v$ or if u is lexicographically smaller than v. In the latter case we write $u <_{\mathrm{lex}} v$.

Let $x \in \mathbb{A}^{\mathbb{N}}$. A factor u of x is called *minimal* (in x) if $u \leq_{\mathrm{lex}} v$ for all factors v of x with $|v| = |u|$. An infinite word y in the shift orbit closure S_x of x is called *Lyndon* (in S_x) if every prefix of y is minimal in x. The proof of the following result is omitted for space considerations:

Theorem 2. *Let $(A, \leq)$ be a linearly ordered finite set and let $x \in \mathbb{A}^{\mathbb{N}}$. Let $y, z \in S_x$ with y Lyndon and aperiodic. Then for each $w \in \mathscr{S}(y, z)$, we have $w <_{\mathrm{lex}} z$. In particular, taking $z = y$ we deduce that y is not self-shuffling.*

Let $\mathbb{A}$ be a finite non-empty set. We say $x \in \mathbb{A}^{\mathbb{N}}$ is *extremal* if there exists a linear ordering $\leq$ on $\mathbb{A}$ with respect to which x is Lyndon. As an immediate consequence of Theorem 2 we obtain:

Corollary 2. *Let $\mathbb{A}$ be a finite non-empty set and $x \in \mathbb{A}^{\mathbb{N}}$ be an aperiodic extremal word. Then x is not self-shuffling.*

Remark 2. Let $x = 11010011001011010010110\ldots$ denote the first shift of the Thue-Morse infinite word. It is easily checked that x is extremal and hence is not self-shuffling; yet it can be verified that x begins in only a finite number of Abelian border-free words.

4   The Thue-Morse Word Is Self-shuffling

Theorem 3. *The Thue-Morse word* $\mathbf{T} = 011010011001\dots$ *fixed by the morphism* τ *mapping* $0 \mapsto 01$ *and* $1 \mapsto 10$ *is self-shuffling.*

Proof. For $u \in \{0,1\}^*$ we denote by $\bar{u}$ the word obtained from u by exchanging 0s and 1s. Let $\sigma : \{1,2,3,4\} \to \{1,2,3,4\}^*$ be the morphism defined by

$$\sigma(1) = 12, \quad \sigma(2) = 31, \quad \sigma(3) = 34, \quad \sigma(4) = 13.$$

Set $u = 01101$ and $v = 001$; note that uv is a prefix of $\mathbf{T}$. Also define morphisms $g, h : \{1,2,3,4\} \to \{0,1\}^*$ by

$$g(1) = v\bar{u}, \quad g(2) = \bar{v}\bar{u}, \quad g(3) = \bar{v}u, \quad g(4) = vu$$

and

$$h(1) = uv, \quad h(2) = \bar{u}\bar{v}, \quad h(3) = \bar{u}\bar{v}, \quad h(4) = uv$$

We will make use of the following lemmas:

Lemma 1. $g(\sigma(a)) \in \mathscr{S}(g(a), h(a))$ *for each* $a \in \{1,2,3,4\}$. *In particular* $ug(\sigma(1)) \in \mathscr{S}(ug(1), h(1))$.

Proof. For $a = 1$ we note that

$$g(\sigma(1)) = g(12) = v\bar{u}\bar{v}\bar{u} = 0011001011010010.$$

Factoring $0011001011010010 = 0 \cdot 011 \cdot 0 \cdot 010 \cdot 11 \cdot 01 \cdot 0010$ we obtain

$$g(\sigma(1)) \in \mathscr{S}(00110010, 01101001) = \mathscr{S}(v\bar{u}, uv) = \mathscr{S}(g(1), h(1)).$$

Similarly, for $a = 2$ we have

$$g(\sigma(2)) = g(31) = \bar{v}uv\bar{u} = 1100110100110010.$$

Factoring $1100110100110010 = 1 \cdot 100 \cdot 1 \cdot 1 \cdot 010 \cdot 0110 \cdot 010$ we obtain

$$g(\sigma(2)) \in \mathscr{S}(11010010, 10010110) = \mathscr{S}(\bar{v}\bar{u}, \bar{u}\bar{v}) = \mathscr{S}(g(2), h(2)).$$

Exchanging 0s and 1s in the previous two shuffles yields

$$g(\sigma(3)) = g(34) = \bar{v}uvu \in \mathscr{S}(\bar{v}u, \bar{u}\bar{v}) = \mathscr{S}(g(3), h(3))$$

and

$$g(\sigma(4)) = g(13) = v\bar{u}\bar{v}u \in \mathscr{S}(vu, uv) = \mathscr{S}(g(4), h(4)).$$

It is readily verified that

Lemma 2. $h(\sigma(a)) = \tau(h(a))$ *for each* $a \in \{1,2,3,4\}$.

Let $w = w_0 w_1 w_2 w_3 \dots$ with $w_i \in \{1,2,3,4\}$ denote the fixed point of σ beginning in 1. As a consequence of the previous lemma we deduce that

Lemma 3. T $= h(w)$.

Proof. In fact $\tau(h(w)) = h(\sigma(w)) = h(w)$ from which it follows that $h(w)$ is one of the two fixed points of τ. Since $h(w)$ begins in $h(1)$ which in turn begins in 0, it follows that $\mathbf{T} = h(w)$.

Lemma 4. T $= ug(w)$.

Proof. It is readily verified that:

$$ug(1) = h(1)\bar{u}$$

$$\bar{u}g(2) = h(2)\bar{u}$$

$$\bar{u}g(3) = h(3)u$$

$$ug(4) = h(4)u.$$

Moreover, each occurrence of $g(1)$ and $g(4)$ in $ug(w)$ is preceded by u while each occurrence of $g(2)$ and $g(3)$ in $ug(w)$ is preceded by $\bar{u}$. It follows that $ug(w) = h(w)$ which by the preceding lemma equals $\mathbf{T}$.

Set

$$A_0 = ug(\sigma(w_0)) \quad \text{and} \quad A_i = g(\sigma(w_i)), \quad \text{for} \quad i \geq 1$$

$$B_0 = ug(w_0)) \quad \text{and} \quad B_i = g(w_i)), \quad \text{for} \quad i \geq 1$$

and

$$C_i = h(w_i) \quad \text{for} \quad i \geq 0.$$

It follows from Lemma 3 and Lemma 4 that

$$\mathbf{T} = \prod_{i=0}^{\infty} A_i = \prod_{i=0}^{\infty} B_i = \prod_{i=0}^{\infty} C_i$$

and it follows from Lemma 1 that $A_i \in \mathscr{S}(B_i, C_i)$ for each $i \geq 0$. Hence $\mathbf{T} \in \mathscr{S}(\mathbf{T}, \mathbf{T})$ as required.

5   Self-shuffling Sturmian Words

In this section we characterize self-shuffling Sturmian words. Sturmian words admit various types of characterizations of geometric and combinatorial nature, e.g., they can be defined via balance, complexity, morphisms, etc. (see Chapter 2 in [7]). In [8], Morse and Hedlund showed that each Sturmian word may be realized geometrically by an irrational rotation on the circle. More precisely, every Sturmian word x is obtained by coding the symbolic orbit of a point $\rho(x)$ on the circle (of circumference one) under a rotation by an irrational angle α where the circle is partitioned into two complementary intervals, one of length α (labeled 1) and the other of length $1 - \alpha$ (labeled 0) (see Fig. 1). And conversely each such coding gives rise to a Sturmian word. The irrational α is called the *slope* and the point $\rho(x)$ is called the *intercept* of the Sturmian word x. A Sturmian word x of slope α with $\rho(x) = \alpha$ is called a *characteristic Sturmian word*. It is well known that every prefix u of a characteristic Sturmian word is *left special*, i.e., both $0u$ and $1u$ are factors of x [7]. Thus if x is a characteristic Sturmian word of slope α, then both $0x$ and $1x$ are Sturmian words of slope α and $\rho(0x) = \rho(1x) = 0$. The fact that ρ is not one-to-one stems from the ambiguity of the coding of the boundary points 0 and $1 - \alpha$.

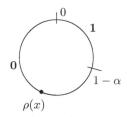

Fig. 1. Geometric picture of a Sturmian word of slope α

Theorem 4. *Let S, M and L be Sturmian words of the same slope α, $0 < \alpha < 1$, satisfying $S \leq_{\mathrm{lex}} M \leq_{\mathrm{lex}} L$. Then $M \in \mathscr{S}(S, L)$ if and only if the following conditions hold: If $\rho(M) = \rho(S)$ (respectively, $\rho(M) = \rho(L)$), then $\rho(L) \neq 0$ (respectively $\rho(S) \neq 0$).*

In particular (taking $S = M = L$), we obtain

Corollary 3. *A Sturmian word $x \in \{0, 1\}^{\mathbb{N}}$ is self-shuffling if and only if $\rho(x) \neq 0$, or equivalently, x is not of the form aC where $a \in \{0, 1\}$ and C is a characteristic Sturmian word.*

Our proof explicitly describes an algorithm for shuffling S and L so as to produce M. It is formulated in terms of the circle rotation description of Sturmian words. Geometrically speaking, points $\rho(S)$ and $\rho(L)$ will take turns following the trajectory of $\rho(M)$ so that the respective codings agree; as one follows the other waits its turn (remains neutral). The algorithm specifies this following rule depending on the relative positions of the trajectories of all three points and is broken down into several cases. The proof can be summarized by the directed graph in Fig. 2 in which each state n corresponds to "case n" in the proof.

We let s, m, and ℓ denote the current tail of the words S, M, and L. They are initialized as

$$s := S, \ \ell := L, \text{ and } m := M.$$

While m is always a tail of M, the letters s and ℓ may be tails of S or L, depending on which is the current lexicographically largest[1]. Each directed edge

---

[1] The choice of the letter s, m, and ℓ is intended to refer to *small, medium,* and *large* respectively.

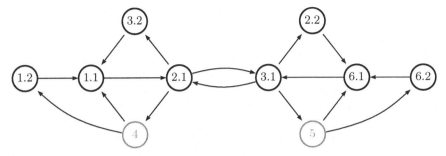

Fig. 2. Graphical depiction of the proof of Theorem 4

corresponds to a precise set of instructions which specify which of s or ℓ is neutral, which of s or ℓ follows m and for how long, and in the end a possible relabeling of the variables s and ℓ. In each case the outcome leads to a new case in which there is a switch in the follower. In other words, if there is an edge from case i to case j in the graph, then either the instructions for case i and case j specify different followers (as is the case for cases 1.1 and 2.1) in which case the passage from i to j leaves the labeling of s and ℓ unchanged, or the instructions for case i and case j specify the same follower (as is the case for cases 1.2 and 1.1) in which case the passage from i to j exchanges the labeling of s and ℓ.

The proof of Theorem 4 amounts to showing that for each state n in the graph, the specified instructions will take n to an adjacent state in the graph.

As an almost immediate application of Corollary 3 we recover the following result originally proved by Yasutomi in [10] and later reproved by Berthé, Ei, Ito and Rao in [3] and independently by Fagnot in [6]. We say an infinite word is *pure morphic* if it is a fixed point of some morphism different from the identity.

Theorem 5 (Yasutomi [10]). *Let $x \in \{0,1\}^{\mathbb{N}}$ be a characteristic Sturmian word. If y is a pure morphic word in the orbit of x, then $y \in \{x, 0x, 1x, 01x, 10x\}$.*

Proof. We begin with some preliminary observations. Let $\Omega(x)$ denote the set of all left and right infinite words y such that $\mathcal{F}(x) = \mathcal{F}(y)$ where $\mathcal{F}(x)$ and $\mathcal{F}(y)$ denote the set of all factors of x and y respectively. If $y \in \Omega(x)$ is a right infinite word, and $0y, 1y \in \Omega(x)$, then $y = x$. This is because every prefix of y is a left special factor and hence also a prefix of the characteristic word x. Similarly if y is a left infinite word and $y0, y1 \in \Omega(x)$, then y is equal to the reversal of x. If τ is a morphism fixing some point $y \in \Omega(x)$, then $\tau(z) \in \Omega(x)$ for all $z \in \Omega(x)$.

Suppose to the contrary that $\tau \neq id$ is a morphism fixing a proper tail y of x. Then y is self-shuffling by Corollary 3. Put $x = uy$ with $u \in \{0,1\}^{+}$. Using the characterization of Sturmian morphisms (see Theorem 2.3.7 & Lemma 2.3.13 in [7]) we deduce that τ must be primitive. Thus we can assume that $|\tau(a)| > 1$ for each $a \in \{0,1\}$. If $\tau(0)$ and $\tau(1)$ end in distinct letters, then as both $0\tau(x), 1\tau(x) \in \Omega(x)$, it follows that $\tau(x) = x$. Since also $\tau(y) = y$ and $|\tau(u)| > |u|$, it follows that y is a proper tail of itself, a contradiction since x is aperiodic. Thus $\tau(0)$ and $\tau(1)$ must end in the same letter. Whence by Corollary 1 it follows that every left extension of y is self-shuffling, which is again a contradiction since $0x$ and $1x$ are not self-shuffling.

Next suppose $\tau \neq id$ is a morphism fixing a point $y = uabx \in \Omega(x)$ where $u \in \{0,1\}^+$ and $\{a,b\} = \{0,1\}$. Again we can suppose τ is primitive and $|\tau(0)| > 1$ and $|\tau(1)| > 1$. If $\tau(0)$ and $\tau(1)$ begin in distinct letters, then $\tau(\tilde{x})0, \tau(\tilde{x})1 \in \Omega(x)$ where $\tilde{x}$ denotes the reverse of x. Thus $\tau(\tilde{x}) = \tilde{x}$. Thus for each prefix v of abx we have $\tau(\tilde{x}v) = \tilde{x}\tau(v)$ whence $\tau(v)$ is also a prefix of abx. Hence $\tau(abx) = abx$. As before this implies that abx is a proper tail of itself which is a contradiction. Thus $\tau(0)$ and $\tau(1)$ begin in the same letter. Whence by Corollary 1 it follows that every tail of y is self-shuffling, which is again a contradiction since $0x$ and $1x$ are not self-shuffling.

Remark 3. In the case of the Fibonacci infinite word x, each of $\{x, 0x, 1x, 01x, 10x\}$ is pure morphic. For a general characteristic word x, since every point in the orbit of x except for $0x$ and $1x$ is self-shuffling, it follows that if τ is a morphism fixing x (respectively $01x$ or $10x$), then $\tau(0)$ and $\tau(1)$ must end (respectively begin) in distinct letters.

References

1. Allouche, J.-P., Shallit, J.: The ubiquitous Prouhet-Thue-Morse sequence. In: Ding, C., Helleseth, T., Niederreiter, H. (eds.) Proceedings of Sequences and Their Applications, SETA 1998, pp. 1–16. Springer (1999)
2. Allouche, J.-P., Shallit, J.: Automatic sequences. In: Theory, Applications, Generalizations. Cambridge University Press (2003)
3. Berthé, V., Ei, H., Ito, S., Rao, H.: On substitution invariant Sturmian words: an application of Rauzy fractals. Theor. Inform. Appl. 41, 329–349 (2007)
4. Cassaigne, J., Karhumäki, J.: Toeplitz Words, Generalized Periodicity and Periodically Iterated Morphisms. European J. Combin. 18, 497–510 (1997)
5. Henshall, D., Rampersad, N., Shallit, J.: Shuffling and unshuffling. Bull. EATCS 107, 131–142 (2012)
6. Fagnot, I.: A little more about morphic Sturmian words. Theor. Inform. Appl. 40, 511–518 (2006)
7. Lothaire, M.: Algebraic Combinatorics on Words. Encyclopedia of Mathematics and its Applications, vol. 90. Cambridge University Press, U.K (2002)
8. Morse, M., Hedlund, G.A.: Symbolic dynamics II: Sturmian sequences. Amer. J. Math. 62, 1–42 (1940)
9. Thue, A.: Über unendliche Zeichenreihen. Norske Vid. Selsk. Skr. I Math-Nat. Kl. 7, 1–22 (1906)
10. Yasutomi, S.-I.: On sturmian sequences which are invariant under some substitutions. In: Kanemitsu, S., et al. (eds.) Number Theory and Its Applications, Proceedings of the Conference held at the RIMS, Kyoto, Japan, November 10-14, 1997, pp. 347–373. Kluwer Acad. Publ., Dordrecht (1999)

Block-Sorted Quantified Conjunctive Queries

Hubie Chen[1,*] and Dániel Marx[2,**]

[1] Universidad del País Vasco and IKERBASQUE, E-20018 San Sebastián, Spain
[2] Computer and Automation Research Institute, Hungarian Academy of Sciences
(MTA SZTAKI), Budapest, Hungary

Abstract. We study the complexity of model checking in quantified conjunctive logic, that is, the fragment of first-order logic where both quantifiers may be used, but conjunction is the only permitted connective. In particular, we study block-sorted queries, which we define to be prenex sentences in multi-sorted relational first-order logic where two variables having the same sort must appear in the same quantifier block. We establish a complexity classification theorem that describes precisely the sets of block-sorted queries of bounded arity on which model checking is fixed-parameter tractable. This theorem strictly generalizes, for the first time, the corresponding classification for existential conjunctive logic (which is known and due to Grohe) to a logic in which both quantifiers are present.

1 Introduction

Model checking, the problem of deciding if a logical sentence holds on a structure, is a fundamental computational task that appears in many guises throughout computer science. Witness its appearance in areas such as computational logic, verification, artificial intelligence, constraint satisfaction, and computational complexity. The case where one wishes to evaluate a first-order sentence on a finite structure is a problem of principal interest in database theory and is the topic of this article. This problem is well-known to be quite intractable in general: it is PSPACE-complete.

As has been articulated in the literature [7], the typical situation in the database setting is the posing of a relatively short query to relatively large database, or in logical parlance, the evaluation of a short formula on a large relational structure. It has consequently been argued that, in measuring the time complexity of this task, one could reasonably allow a slow (that is, possibly non-polynomial-time) computable preprocessing of the formula, so long as

* Research supported by the Spanish Project FORMALISM (TIN2007-66523), by the Basque Government Project S-PE12UN050(SAI12/219), and by the University of the Basque Country under grant UFI11/45.
** Research supported by the European Research Council (ERC) grant "PARAMTIGHT: Parameterized complexity and the search for tight complexity results," reference 280152.

F.V. Fomin et al. (Eds.): ICALP 2013, Part II, LNCS 7966, pp. 125–136, 2013.
© Springer-Verlag Berlin Heidelberg 2013

the desired evaluation can be performed in polynomial time following this pre-processing. Relaxing polynomial-time computation so that an arbitrary dependence in a *parameter* is tolerated yields, in essence, the notion of *fixed-parameter tractability*. This notion of tractability is the base of *parameterized complexity theory*, which provides a taxonomy for reasoning about and classifying problems where each instance has an associated parameter. We follow this paradigm, and focus the discussion on this form of tractability.

First-order model checking is intractable even if one restricts the connectives and quantifiers permitted; for instance, model checking of *existential conjunctive queries*, by which we mean sentences formed using atoms, conjunction ($\wedge$), and existential quantification ($\exists$), is well-known to be intractable (it is NP-complete). Thus, a typical way to gain insight into which sentences exhibit tractable behavior is to consider model checking relative to a set Φ of sentences. In the context of existential conjunctive logic, there is a mature understanding of sentence sets. It was proved by Grohe [6] that when Φ is a set of existential conjunctive queries having bounded arity, model checking on Φ is fixed-parameter tractable if there is a constant $k \geq 1$ such that each sentence in Φ is logically equivalent to one whose treewidth is bounded above by k, and is intractable otherwise (under a standard assumption from parameterized complexity). The treewidth of a conjunctive sentence (in prenex form) is measured here via the graph on the sentence's variables wherein two variables are adjacent if they co-occur in an atom.

An important precursor to Grohe's theorem was the complexity classification of graph sets for existential conjunctive logic. Grohe, Schwentick, and Segoufin [7] defined model checking relative to a graph set $\mathcal{G}$ as the problem of deciding, given a structure and an existential conjunctive query whose graph is in $\mathcal{G}$, whether or not the query is true on the structure; they showed that the problem is fixed-parameter tractable when $\mathcal{G}$ has bounded treewidth, and intractable otherwise. In this paper, we restrict our attention to queries of bounded arity (the case of unbounded arity leads to a different theory, where complexity may depend on the choice of representation of relations [3,8]). For bounded-arity structures, this result is *coarser* than Grohe's theorem, as it can be taken as a classification of sentence sets Φ that obey the closure property that if a sentence is in Φ, then all sentences having the same graph are also in Φ; in contrast, Grohe's theorem classifies arbitrary sentence sets.

This graph classification was recently generalized to quantified conjunctive logic, wherein both quantifiers ($\forall, \exists$) are permitted in addition to conjunction ($\wedge$). Define a *prefixed graph* to be a quantifier prefix $Q_1 v_1 \ldots Q_n v_n$ paired with a graph on the variables $\{v_1, \ldots, v_n\}$; each quantified conjunctive query in prenex form can naturally be mapped to a prefixed graph, by simply taking the quantifier prefix of the query along with the graph of the quantifier-free, conjunctive portion of the query. Chen and Dalmau [2] defined a width measure for prefixed graphs, which generalizes treewidth, and proved that model checking on a set of prefixed graphs is fixed-parameter tractable if the set has bounded width, and intractable otherwise. This result generalizes the graph classification by Grohe, Schwentick, and Segoufin, and provides a unified view of this classification as well as earlier complexity results [5] on quantified conjunctive logic. Note, however, that the present result is incomparable to Grohe's result: Grohe's result is on arbitrary sentence sets in a less expressive logic, while the result of Chen and Dalmau considers sentences in more expressive logic, but considers them from the coarser graph-based viewpoint, that is, it classifies sentence sets obeying the (analog of the) described closure property.

In this article, we present a veritable generalization of Grohe's theorem in quantified conjunctive logic. In the bounded-arity case, our theorem naturally unifies together both Grohe's theorem and the classification of prefixed graphs in quantified conjunctive logic. The sentences studied by our theorem are of the following type. Define a *block-sorted query* to be a quantified conjunctive sentence in multi-sorted, relational first-order logic where two variables having the same sort must occur in the same quantifier block. This class of sentences includes each sentence having a sort for each quantifier block. As an example, consider the sentence

$$\exists x_1, x_2 \forall y_1, y_2, y_3 \exists z_1, z_2$$
$$R(x_1, y_1) \land R(x_2, y_3) \land S(x_2, y_2, y_3, z_1) \land S(x_1, y_1, y_2, z_2) \land T(x_1, x_2, y_2),$$

where the variables x_i have the same sort e, the variables y_i have the same sort u, and the variables z_i have the same sort e'; the arities of the relation symbols R, S, and T are eu, $euue'$, and eeu, respectively. The definitions impose that a structure **B** on which such a sentence can be evaluated needs to provide a domain B_s (which is a set) for each sort; quantifying a variable of sort s is performed over the domain B_s. (See the next section for the precise formalization that is studied.)

Our main theorem is the classification of block-sorted queries. We show how to computably derive from each query a second logically equivalent query, and demonstrate that, for a bounded-arity set of block-sorted queries, model checking is fixed-parameter tractable if the width of the derived queries is bounded (with respect to the mentioned width measure [2]), and is intractable otherwise. This studied class of queries encompasses existential conjunctive queries, which can be viewed as block-sorted queries in which there is one existential quantifier block, and all variables have the same sort. Observe that, given any sentence in quantified conjunctive logic (either one-sorted or multi-sorted) and any structure on which the sentence is to be evaluated, one can view the sentence as a block-sorted query. (This is done as follows: for each sort s that appears in more than one quantifier block, introduce a new sort s^b for each block b where it appears; correspondingly, introduce new relation symbols.) Our theorem can thus be read as providing a general tractability result which is applicable to all of quantified conjunctive logic, and a matching intractability result that proves optimality of this tractability result for the class of block-sorted queries.

Our theorem is the first generalization of Grohe's theorem to a logic where both quantifiers are present. The previous work suggests that we should proceed the following way: take the width measure of Chen and Dalmau [2], and apply it to some analog of the logically equivalent core of Grohe [6]. However, the execution of these ideas are not at all obvious and we have to overcome a number of technical barriers. For instance, Grohe's theorem statement (in the formulation given here) makes reference to logical equivalence. While there is a classical and simple characterization of logical equivalence in existential conjunctive logic [1], logical equivalence for first-order logic is of course well-known to be an undecidable property; logical equivalence for quantified conjunctive logic is now known (in the one-sorted case) to be decidable [4], but is perhaps still not well-understood (for instance, its exact complexity is quite open). Despite this situation, we succeed in identifying, for each block-sorted sentence, a logically equivalent sentence whose width characterizes the original sentence's complexity, obtaining a statement parallel to that of Grohe's theorem; the definition of this equivalent sentence is a primary contribution of this article. In carrying out this identification, we present a notion of *core* for block-sorted sentences and develop its basic theory; the core of an existential conjunctive sentence (an established notion) is, intuitively, a minimal equivalent sentence, and Grohe's theorem can be stated in terms of the treewidth of the cores of a sentence set. Another technical contribution of the article is to develop a graph-theoretic understanding of variable interactions (see Section 4), which understanding is sufficiently strong so as to allow for the delicate embedding of hard sentences from the previous work [2] into the sentences under consideration, to obtain the intractability result. Overall, we believe that the notions, concepts, and techniques that we introduce in this article will play a basic role in the investigation of model checking in logics that are more expressive than the one considered here.

2   Preliminaries

2.1   Terminology and Setup

We will work with the following formalization of multi-sorted relational first-order logic. A *signature* is a pair $(\sigma, \mathcal{S})$ where $\mathcal{S}$ is a set of *sorts* and σ is a set of relation symbols; each relation symbol $R \in \sigma$ has associated with it an element of $\mathcal{S}^*$, called the arity of R and denoted $\mathsf{ar}(R)$. In formulas over signature $(\sigma, \mathcal{S})$, each variable v has associated with it a sort $s(v)$ from $\mathcal{S}$; we use *atom* to refer to an atomic formula $R(v_1, \ldots, v_k)$ where $R \in \sigma$ and $s(v_1) \ldots s(v_k) = \mathsf{ar}(R)$. A *structure* $\mathbf{B}$ on signature $(\sigma, \mathcal{S})$ consists of an $\mathcal{S}$-sorted family $\{B_s \mid s \in \mathcal{S}\}$ of sets called the *universe* of $\mathbf{B}$, and, for each symbol $R \in \sigma$, an interpretation $R^{\mathbf{B}} \subseteq B_{\mathsf{ar}(R)}$. Here, for a word $w = w_1 \ldots w_k \in \mathcal{S}^*$, we use B_w to denote the product $B_{w_1} \times \cdots \times B_{w_k}$. We say that two structures are *similar* if they are defined on the same signature. Let $\mathbf{B}$ and $\mathbf{C}$ be two similar structures defined on the same signature $(\sigma, \mathcal{S})$. We say that $\mathbf{B}$ is a *substructure* of $\mathbf{C}$ if for each $s \in \mathcal{S}$, it holds that $B_s \subseteq C_s$, and for each $R \in \sigma$, it holds that $R^{\mathbf{B}} \subseteq R^{\mathbf{C}}$. We say that $\mathbf{B}$ is an *induced substructure* of $\mathbf{C}$ if, in addition, for each $R \in \sigma$ one has that $R^{\mathbf{B}} = R^{\mathbf{C}} \cap B_{\mathsf{ar}(R)}$.

A *quantified conjunctive query* is a sentence built from atoms, conjunction, existential quantification, and universal quantification. It is well-known that such sentences can be efficiently translated into prenex normal form, that is, of the form $Q_1 v_1 \ldots Q_n v_n \phi$ where each Q_i is a quantifier and where ϕ is a conjunction of atoms. For such a sentence, it is well-known that the conjunction ϕ can be encoded as a structure $\mathbf{A}$ where A_s contains the variables of sort s that appear in ϕ and, for each relation symbol R, the relation $R^{\mathbf{A}}$ consists of all tuples $(v_1, \ldots, v_k)$ such that $R(v_1, \ldots, v_k)$ appears in ϕ. In the other direction, any structure $\mathbf{A}$ can be viewed as encoding the conjunction $\bigwedge_{(v_1, \ldots, v_k) \in R^{\mathbf{B}}} R(v_1, \ldots, v_k)$. We will typically denote a quantified conjunctive query $Q_1 v_1 \ldots Q_n v_n \phi$ as a pair $(P, \mathbf{A})$ consisting of the quantifier prefix $P = Q_1 v_1 \ldots Q_n v_n$ and a structure $\mathbf{A}$ that encodes the quantifier-free part ϕ. Note that when discussing the evaluation of a sentence $(P, \mathbf{A})$ on a structure, we can and often will assume that all variables appearing in P are elements of $\mathbf{A}$.

We define a *block-sorted query* to be a quantified conjunctive query in prenex normal form where for all variables v, v', if $s(v) = s(v')$ then v, v' occur in the same quantifier block. By a quantifier block, we mean a subsequence $Q_i v_i \ldots Q_j v_j$ of the quantifier prefix (with $i \leq j$) having maximal length such that $Q_i = \cdots = Q_j$. We number the quantifier blocks from left to right (that is, the outermost quantifier block is considered the first). For each sort s having a variable that appears in such a query, either all variables of sort s are universal, in which case we call s a *universal sort* or a $\forall$-sort, or all variables or sort s are existential, in which case we call s a *existential sort* or a $\exists$-sort.

2.2   Conventions

In general, when $\mathbf{A}$ is a structure with universe $\{A_s \mid s \in \mathcal{S}\}$, we assume that the sets A_s are pairwise disjoint, and use A to denote $\cup_{s \in \mathcal{S}} A_s$. Correspondingly, we assume that in forming formulas over signature $(\sigma, \mathcal{S})$, the sets of permitted variables for different sorts are pairwise disjoint. Relative to a quantified conjunctive query $(P, \mathbf{A})$, we use $A_{\exists}$ to denote the set $\{a \in A \mid s(a) \text{ is an } \exists\text{-sort}\}$; likewise, we use $A_{\forall}$ to denote the set $\{a \in A \mid s(a) \text{ is an } \forall\text{-sort}\}$. In dealing with sets such as these, for a variable v we use a subscript $< v$ to restrict to variables coming before v in the quantifier prefix P; for instance, we will use $A_{\forall, <v}$ to denote the set of all universally quantified variables that occur before v. When discussing a function from a set whose elements are sorted to another such set, we assume tacitly that the function preserves sort, that is, for each sort s, each elements of sort s in the first set is mapped to an element of sort s in the second set.

Let $(P, \mathbf{A})$ be a block-sorted query and let $\mathbf{B}$ be a structure similar to $\mathbf{A}$; we say that a homomorphism $\phi : \mathbf{A} \to \mathbf{B}$ is *universal-injective* if ϕ is injective on $A_{\forall}$.

2.3   Basic Facts

Intuitively, evaluating the query $(P, \mathbf{A})$ on the structure $\mathbf{B}$ can be interpreted as game with two players "universal" and "existential." In the order given by the prefix P, the two players assign values to the variables; existential and universal sets the values of the existential and the universal variables, respectively. The aim of existential is to ensure that the resulting assignment satisfies the formula, that is, gives a homomorphism from $\mathbf{A}$ to $\mathbf{B}$, while universal tries to prevent this. The query $(P, \mathbf{A})$ is true on $\mathbf{B}$ if existential has a winning strategy. We formalize this intuition by the following definition:

Definition 1. *Let $(P, \mathbf{A})$ be a quantified conjunctive query, and let $\mathbf{B}$ be a structure similar to $\mathbf{A}$. An existential strategy for $(P, \mathbf{A})$ on $\mathbf{B}$ is a set of mappings $(f_x : (A_{\forall, <x} \to B) \to B_{s(x)})_{x \in A_\exists}$ such that the following holds: for any $h : A_\forall \to B$, a homomorphism from $\mathbf{A}$ to $\mathbf{B}$ is given by the map $(f, h) : A \to B$ defined by $(f, h)(x) = f_x(h \upharpoonright A_{\forall, <x})$ for each existential variable x, and $(f, h)(y) = h(y)$ for each universal variable y.*

Proposition 2. *Let $(P, \mathbf{A})$ be a quantified conjunctive query, and let $\mathbf{B}$ be a structure similar to $\mathbf{A}$. Then $\mathbf{B} \models (P, \mathbf{A})$ if and only if there is an existential strategy.*

The transitivity of homomorphisms allows us to quickly deduce consequences of the existence of a homomorphism $\mathbf{A} \to \mathbf{B}$. For example, we know that there is also a homomorphism $\mathbf{A}' \to \mathbf{B}$ whenever there is a homomorphism $\mathbf{A}' \to \mathbf{A}$; and there is a also homomorphism $\mathbf{A} \to \mathbf{B}'$ whenever there is a homomorphism $\mathbf{B} \to \mathbf{B}'$. These quick observations are very useful in the study of the homomorphism problem, where they allow us to restrict our attention to specific type of structures. In our setting, however, the quantified nature of the problem makes such consequences less obvious. In the following, we find analogs of these observations in our setting, that is, assuming that $\mathbf{B} \models (P_A, \mathbf{A})$ holds, we explore under what conditions the structure $\mathbf{B}$ or the query $(P_A, \mathbf{A})$ can be replaced to obtain another true statement.

First we give a sufficient condition under which the query can be replaced. Let us say that two similar block-sorted queries $(P_A, \mathbf{A})$ and $(P_C, \mathbf{C})$ having the same number of quantifier blocks are *mutually respecting* if for each sort s and for each $i \geq 1$, it holds that s is used in the ith quantifier block of P_A if and only if it is used only in the ith quantifier block of P_C.

Proposition 3. *Let $(P_A, \mathbf{A})$ and $(P_C, \mathbf{C})$ be similar block-sorted queries that are mutually respecting. Suppose that $i : \mathbf{A} \to \mathbf{C}$ is a universal-injective homomorphism. Then it holds that $(P_C, \mathbf{C})$ entails $(P_A, \mathbf{A})$.*

The following proposition gives a sufficient condition for replacing the structure $\mathbf{B}$ on which the query is evaluated:

Proposition 4. *Let σ be a signature, let $(P, \mathbf{A})$ be a block-sorted query over σ, and let $\mathbf{B}, \mathbf{B}'$ be structures over σ. Suppose that $\mathbf{B} \models (P, \mathbf{A})$ and that there exists a homomorphism $g : \mathbf{B} \to \mathbf{B}'$ that is* universal-surjective *in the sense that $f(B_\forall) = B'_\forall$. Then, it holds that $\mathbf{B}' \models (P, \mathbf{A})$.*

Note that this proposition can be viewed as a variant of the known fact that, in standard (one-sorted) first-order logic, if a quantified conjunctive query Φ holds on a structure $\mathbf{B}$ and $\mathbf{B}$ admits a surjective homomorhpism to $\mathbf{B}'$, then Φ also holds on $\mathbf{B}'$ (see for example [4, Lemma 1]).

3   The Selfish Core

Let $(P, \mathbf{C})$ be a block-sorted query on signature $(\sigma, \mathcal{S})$. When $\mathbf{A}$ is similar to $\mathbf{C}$, we say that $\mathbf{A}$ is an $\exists$-*substructure* of $\mathbf{C}$ if $\mathbf{A}$ is a substructure of $\mathbf{C}$; for each $\forall$-sort u it holds that $A_u = C_u$; and, for each $\exists$-sort e it holds that $A_e \subseteq C_e$. We say that $\mathbf{A}$ is a *proper $\exists$-substructure* of $\mathbf{C}$ if, in addition, there exists an $\exists$-sort e such that the containment $A_e \subseteq C_e$ is proper.

We say that a block-sorted query $(P, \mathbf{C})$ is *selfish* if $\mathbf{C} \models (P, \mathbf{C})$. We say that a block-sorted query $(P, \mathbf{C})$ is a *selfish core* if it is selfish and for any proper $\exists$-substructure $\mathbf{A}$ of $\mathbf{C}$, either $(P, \mathbf{A})$ is not selfish or the queries $(P, \mathbf{A})$ and $(P, \mathbf{C})$ are not logically equivalent.

We give characterizations of the notion of selfish core in the following proposition; afterwards, we show that each block-sorted query has (in a sense made precise) a selfish core. Let us say that an endomorphism $h : \mathbf{C} \to \mathbf{C}$ of a structure $\mathbf{C}$ is *proper* if its image is proper, that is, if there exists a sort $s \in \mathcal{S}$ such that $h(C_s) \subsetneq C_s$.

Proposition 5. *Let $(P, \mathbf{C})$ be a selfish block-sorted query. The following are equivalent.*

1. *$(P, \mathbf{C})$ is a selfish core.*
2. *There does not exist a proper endomorphism of $\mathbf{C}$ that fixes each universal variable.*
3. *There does not exist a proper endomorphism of $\mathbf{C}$ that, for each universal sort u, is injective on C_u.*

Define a *selfish core of a block-sorted query* $(P, \mathbf{A})$ to be a block-sorted query that is a selfish core and that is logically equivalent to $(P, \mathbf{A})$. We now show that each block-sorted query has a selfish core which is computable (from the query).

Definition 6. *Let $(P, \mathbf{A})$ be a block-sorted query; we define the block-sorted query $(P^*, \mathbf{A}^*)$ the following way.*

- *For each $\forall$-sort u, define $A_u^* = A_u$.*
- *For each $\exists$-sort e, define $A_e^* = \{x^g \mid x \in A_e, g : A_{\forall, <x} \to A_{\forall, <x}\}$.*
- *P^* is obtained from P by replacing each quantification $\exists x$ with $\exists x^{g_1} \ldots \exists x^{g_m}$ where $g_1, \ldots, g_m$ is a list of all the mappings from $A_{\forall, <x}$ to $A_{\forall, <x}$.*
- *$R^{\mathbf{A}^*} = \{(g'(a_1), \ldots, g'(a_k)) \mid (a_1, \ldots, a_k) \in R^{\mathbf{A}}, g : A_\forall \to A_\forall\}$ where g' is the extension of g that maps a value $x \in A_\exists$ to $x^{g|A_{\forall, <x}}$.*

Example 7. Consider the query $(P, \mathbf{A}) = \forall y_1, y_2 \exists x : R_1(x, y_1) \wedge R_2(x, y_2)$. For $i, j \in \{1, 2\}$, let g_{ij} be the mapping defined by $g_{ij}(y_1) = y_i$ and $g_{ij}(y_2) = y_j$. Then the query $(P^*, \mathbf{A}^*)$ can be defined as

$$\forall y_1, y_2 \exists x^{g_{11}}, x^{g_{12}}, x^{g_{21}}, x^{g_{22}} :$$
$$[R_1(x^{g_{11}}, y_1) \wedge R_2(x^{g_{11}}, y_1) \wedge R_1(x^{g_{12}}, y_1) \wedge R_2(x^{g_{12}}, y_2)$$
$$\wedge R_1(x^{g_{21}}, y_2) \wedge R_2(x^{g_{21}}, y_1) \wedge R_1(x^{g_{22}}, y_2) \wedge R_2(x^{g_{22}}, y_2)].$$

Proposition 8. *Let* $(P, \mathbf{A})$ *be a block-sorted query. The following statements concerning* $(P^*, \mathbf{A}^*)$ *hold.*

1. $(P^*, \mathbf{A}^*)$ *and* $(P, \mathbf{A})$ *are logically equivalent.*
2. $(P^*, \mathbf{A}^*)$ *is selfish.*
3. *The structure* $\mathbf{A}^*$ *contains an induced substructure* $\mathbf{C}$ *such that* $(P^*, \mathbf{C})$ *is a selfish core of* $(P, \mathbf{A})$*; moreover,* $(P^*, \mathbf{C})$ *is computable from* $(P, \mathbf{A})$*.*

4   Strong and Weak Elements

Throughout this section, we assume that $(P, \mathbf{A})$ is a block-sorted query; the definitions and claims are all relative to this query. We use $G_\mathbf{A}$ to denote the Gaifman graph of the structure $\mathbf{A}$, that is, the graph with vertex set A and containing an edge $\{a, a'\}$ if and only if a and a' are distinct and co-occur in a tuple of a relation of $\mathbf{A}$. Relative to $(P, \mathbf{A})$, when i is the number of a quantifier block, we will use notation such as $A_{\geq i}$ to denote the set of variables occurring in block i or later, and define for example $A_{<i}$ analogously.

Definition 9. *A level* i *component is a maximal connected set of* $\exists$*-variables in* $G_\mathbf{A}[A_{\geq i}]$*.*

Definition 10. *Let* $x \in A_\exists$ *be an* $\exists$*-variable in the* i*th quantifier block.*

– *For* $j \leq i$*, use* $C_\mathbf{A}(x, j)$ *to denote the level* j *component containing* x*.*
– *Define* $N_\mathbf{A}(x, j)$*, the neighborhood of* $C_\mathbf{A}(x, j)$*, to be the set of all universal variables in* $A_{<j}$ *adjacent to* $C_\mathbf{A}(x, j)$ *in* $G_\mathbf{A}$*.*
– *Define* $U_\mathbf{A}(x)$ *to be* $\bigcup_{j \leq i} N_\mathbf{A}(x, j)$*.*

In other words, a universal variable y on level j is in $U_\mathbf{A}(x)$ if and only if y can be reached from x on a path in $G_\mathbf{A}$ such that all the vertices of the path other than y are existential variables on levels greater than j. We remark that the definition of $C_\mathbf{A}(x, j)$, as well as that of the other sets, depends on $\mathbf{A}$ as well as the quantifier prefix P; however, this prefix will be clear from context, and we omit it from the notation.

Definition 11. *We say that an* $\exists$*-variable* $x^g \in A_\exists^*$ *is degenerate if* g *is non-injective on* $U_\mathbf{A}(x)$*.*

Definition 12. *An* $\exists$*-variable* $x \in A_\exists$ *is weak if there exists a universal-injective homomorphism* $\psi : \mathbf{A} \to \mathbf{A}^*$ *where* $\psi(x)$ *is a degenerate element of* $\mathbf{A}^*$*; the* $\exists$*-variable* x *is strong otherwise.*

Example 13. Consider the folllowing query $(P, \mathbf{A})$:

$$\forall y_1, y_2, y_3 \exists x_1 x_2, x_3, x_4, x_5$$
$$R_1(x_1, y_1) \land R_2(x_2, y_2) \land R_3(x_1, x_2) \land R_1(x_3, y_3) \land R_1(x_5, y_3)$$
$$\land R_2(x_4, y_3) \land R_3(x_3, x_4) \land R_3(x_5, x_4).$$

If g is the mapping with $g(y_1) = g(y_2) = g(y_3) = y_3$, then there is a homomorphism ψ from $\mathbf{A}$ to $\mathbf{A}^*$ that is identity on y_1, y_2, y_3, x_1, x_2 and $\psi(x_3) = \psi(x_5) = x_1^g$ and $\psi(x_4) = x_2^g$. Hence x_3, x_4, x_5 are weak elements.

Definition 14. *Define the* strong substructure *of* **A** *to be the substructure of* **A** *induced by the union of* $A_\forall$ *with the strong elements of* **A**.

The main result of the section is showing that removing the weak elements does not change the sentence. In the proof of the classification theorem, this will allow us to consider the width of the strong substructure as the classification criteria.

Theorem 15. *Let* **S** *be the strong substructure of* **A**. *The queries* $(P, \mathbf{S})$ *and* $(P, \mathbf{A})$ *are logically equivalent.*

We conclude this section with a simple lemma that will be of help in establishing the complexity hardness result.

Lemma 16. *Suppose that* ϕ *is a universal-injective endomorphism of* **A** *and that* $x \in A_\exists$ *is a strong variable. Then* $\phi(x)$ *is strong as well.*

Proof. Assume for contradiction that $\phi(x)$ is not strong: there is a universal-injective homomorphism $\psi : \mathbf{A} \to \mathbf{A}^*$ where $\psi(\phi(x))$ is a degenerate element. Now $\psi(\phi)$ is a universal-injective homomorphism $\mathbf{A} \to \mathbf{A}^*$ that maps x to a degenerate element of $\mathbf{A}^*$, contradicting the assumption that x is strong.   □

5   Classification Theorem

When Φ is a set of (possibly multi-sorted) first-order sentences, define Φ-MC to be the model checking problem of deciding, given a sentence $\phi \in \Phi$ and a finite structure **B** over the same signature, whether or not $\mathbf{B} \models \phi$. We study this problem using parameterized complexity; we use the terminology and conventions for parameterized complexity defined in [2], and take ϕ to be the parameter of an instance $(\phi, \mathbf{B})$.

As defined in [2], a *prefixed graph* consists of a quantifier prefix P paired with an undirected graph whose vertices are the variables appearing in P. In [2], a width measure is defined that associates a natural number with each prefixed graph; we refer the reader to that article for the precise definition. As we use both the algorithmic and hardness results of [2] as black box, the exact definition does not matter for the purposes of this paper. Fix a computable mapping M that, given a block-sorted query ϕ, computes a selfish core $(P, \mathbf{A})$ of ϕ, and then computes the strong substructure **S** of $(P, \mathbf{A})$, and outputs $(P, \mathbf{S})$.

Theorem 17. *Let* Φ *be a set of block-sorted queries of bounded arity. If the set of prefixed graphs* $\{(P, G_{\mathbf{S}}) \mid (P, \mathbf{S}) \in M(\Phi)\}$ *has bounded width, then the problem* Φ-MC *is in FPT; otherwise, the problem* Φ-MC *is not in FPT, unless* $W[1] \subseteq nuFPT$.

The remainder of this section is devoted to the proof of Theorem 17.

The positive FPT result is obtained as follows. Given an instance $(\phi, \mathbf{B})$ of the problem Φ-MC, the algorithm is to evaluate $\mathbf{B} \models M(\phi)$ using the algorithm of [2]; this evaluation can be performed in polynomial time given $M(\phi)$, and since the computation of $M(\phi)$ depends only on the parameter of the instance $(\phi, \mathbf{B})$, the whole computation is in FPT.

We now give the hardness result. For a block-sorted query $(P, \mathbf{S})$ over signature $(\sigma, \mathcal{S})$, we define the relativization $(P, \mathbf{S})_{\mathsf{rel}}$ of $(P, \mathbf{S})$ in the following way. Denote P by $Q_1 v_1 \ldots Q_n v_n$, and let θ be the conjunction of atoms corresponding to $\mathbf{S}$. Define $(P, \mathbf{S})_{\mathsf{rel}}$ to be the one-sorted sentence $Q_1 v_1 \in W_{v_1} \ldots Q_n v_n \in W_{v_n} \theta$ over signature $\sigma \cup \{W_{v_1}, \ldots, W_{v_n}\}$ where each W_{v_i} is a fresh unary relation symbol and the arity of a symbol $R \in \sigma$ is the *length* of $\mathsf{ar}_{(\sigma, \mathcal{S})}(R)$. Here, $\exists v \in W \psi$ is syntactic shorthand for $\exists v (W(v) \wedge \psi)$; and, $\forall v \in W \psi$ is syntactic shorthand for $\forall v (W(v) \rightarrow \psi)$. Assuming that the set of prefixed graphs given in the theorem statement has unbounded width, the hardness result of [2, Section 6] implies that $\Phi_{\mathsf{rel}} = \{(M(\phi))_{\mathsf{rel}} \mid \phi \in \Phi\}$ is W[1]-hard or coW[1]-hard under nuFPT reductions. It thus suffices to give an nuFPT reduction from Φ_{rel}-MC to Φ-MC, which we now do. Let $((P, \mathbf{S})_{\mathsf{rel}}, \mathbf{B})$ be an instance of Φ_{rel}-MC, and let $\phi \in \Phi$ be such that $(P, \mathbf{S}) = M(\phi)$; let $(P, \mathbf{A})$ denote the selfish core of ϕ computed by $M(\phi)$ (note that $\mathbf{S}$ is a substructure of $\mathbf{A}$).

We will work with the structure $\mathbf{A}^*$. Let A^{id} denote the subuniverse of $\mathbf{A}^*$ containing all universal variables of A^* and each existential variable of A^* of the form a^{id}, where id is the identity mapping. (We use id generically to denote the identity mapping, but note that this is defined on $A_{\forall, <a}$ for an existential variable a.) Observe that A^{id} induces in $\mathbf{A}^*$ a copy of the structure $\mathbf{A}$. With this correspondence, let S^{id} denote the union of $A_\forall$ and the strong elements of A^{id}, and let $\mathbf{S}^{\mathsf{id}}$ denote the induced substructure of $\mathbf{A}^*$ on S^{id}. Let D denote the subuniverse of $\mathbf{A}^*$ containing all degenerate elements of A^*. We will sometimes drop the id superscript when it is clear from context.

Define a structure $\mathbf{B}'$ over signature $(\sigma, \mathcal{S})$ as follows. The universe is denoted by $\{B'_s \mid s \in \mathcal{S}\}$ and is defined by $B'_s = \{(a, b) \in (A^{\mathsf{id}}_s \cup D_s) \times (B \cup \{\bot\}) \mid (a \in S^{\mathsf{id}}_s \rightarrow b \in U^{\mathbf{B}}_s)$ and $(a \notin S^{\mathsf{id}}_s \rightarrow b = \bot)\}$. Here, B denotes the universe of the one-sorted structure $\mathbf{B}$. Now, for each $R \in \sigma$, define $R^{\mathbf{B}'}$ to be the relation $\{((a_1, b_1), \ldots, (a_k, b_k)) \in B'_{\mathsf{ar}(R)} \mid (a_1, \ldots, a_k) \in R^{\mathbf{A}^*}$ and $((a_1, \ldots, a_k) \in R^{\mathbf{S}^{\mathsf{id}}} \rightarrow (b_1, \ldots, b_k) \in R^{\mathbf{B}})\}$. We will use π_i to denote the mapping that projects a tuple onto the ith coordinate.

We claim that $\mathbf{B} \models (P, \mathbf{S})_{\mathsf{rel}}$ if and only if $\mathbf{B}' \models (P, \mathbf{A})$.

We first prove the backwards direction. We will use the following lemma.

Lemma 18. *Suppose that $h : A \rightarrow (A^{\mathsf{id}} \cup D)$ is a homomorphism from $\mathbf{A}$ to $\mathbf{A}^*$ that is identity on $A_\forall$. Then $h(S) = S^{\mathsf{id}}$.*

Proof. Let h_0 be the homomorphism $\mathbf{A} \rightarrow \mathbf{A}^*$ that is identitiy on universals and maps x to x^{id}. As $\mathbf{A}$ is selfish and $(P, \mathbf{A})$ and $(P, \mathbf{A}^*)$ are logically equivalent by Proposition 8(1), we have that $\mathbf{A} \models (P, \mathbf{A}^*)$, implying that there is a homomorphism h^* from $\mathbf{A}^*$ to $\mathbf{A}$ that is identity on the universals.

We claim that h^* is injective on A^{id}. Indeed, otherwise $h^*(h_0)$ is noninjective (as A^{id} is the image of h_0), hence it is a proper endomorphism of $\mathbf{A}$ that is identity on the universals. By Proposition 5, this contradicts the assumption that $\mathbf{A}$ is a selfish core.

Next we claim that h^* maps S^{id} to S. Otherwise, the endomorphism $h^*(h_0)$ of $\mathbf{A}$ maps a strong element to a weak element, contradicting Lemma 16. Together with the fact that h^* is injective on A^{id}, it follows that h^* maps $A^{\text{id}} \setminus S^{\text{id}}$ to $A \setminus S$.

The homomorphism h cannot map a strong element $x \in S$ to D by definition of strong elements. If h maps a strong element $x \in S$ to $A^{\text{id}} \setminus S^{\text{id}}$, then (as shown in the previous paragraph) endomorphism $h^*(h)$ of $\mathbf{A}$ maps x to $A \setminus S$, that is, to a weak element. As $h^*(h)$ is an endomorphism fixing the universals, this contradicts Lemma 16. Thus we have proved that h maps every strong element to S^{id}. □

Let $(f'_x)_{x \in A_\exists}$ be an existential strategy witnessing $\mathbf{B}' \models (P, \mathbf{A})$. Let $x \in S$ be an existential variable of $(P, \mathbf{S})_{\text{rel}}$, and let s be the sort of x. For any mapping $h : A_{\forall, <x} \to B$, define $h' : A_{\forall, <x} \to B'$ by $h'(y) = (y, h(y))$. Observe that under any such mapping h, we have $\pi_1(\{f'_x(h') \mid x \in S_s\}) = S_s$, since we can extend h' to a mapping $h'' : A_\forall \to B'$ and then the homomorphism (f', h'') given by Definition 1 is from A to $A^{\text{id}} \cup D$, and Lemma 18 can be applied.

We can thus define a strategy (f_x) for $(P, \mathbf{S})_{\text{rel}}$ on $\mathbf{B}$ as follows: for an existential variable $x \in S_\exists$ and a map $h : A_{\forall, <x} \to B$, define $f_x(h) = b$ if and only if (x, b) is in $\{f'_x(h') \mid x \in S_s\}$. This mapping is well-defined by the observation of the previous paragraph, and for any $h : A_\forall \to B$ obeying $h(a) \in U^{\mathbf{B}}_a$, we obtain from the definition of $\mathbf{B}'$ that the homomorphism (f, h) given by Definiton 1 is from $\mathbf{S}$ to $\mathbf{B}$ with $(f, h)(x) \in U^{\mathbf{B}}_x$ for each $\exists$-variable x.

We now prove the forwards direction. We will make use of the following lemma.

Lemma 19. *There is an existential strategy (f'_x) for $(P, \mathbf{A})$ on the substructure of $\mathbf{A}^*$ induced by $(A^{\text{id}} \cup D)$ where $f'(h)$ is a degenerate element in D whenever $h : A_{\forall, x} \to A_{\forall, x}$ is not injective on $U(x)$.*

Let (f_t) witness $\mathbf{B} \models (P, \mathbf{S})_{\text{rel}}$. We will define a strategy (F_x) to witness $\mathbf{B}' \models (P, \mathbf{A})$.

For each partial map $H : A_\forall \to B'$ and subset $Y \subseteq A_\forall$ containing the domain of H, fix $e(H, Y)$ to be an extension of H defined on Y such that

- if for some universal sort s it holds that $H_1|A_s = H_2|A_s$, then $e(H_1, Y)|A_s = e(H_2, Y)|A_s$; and,
- if for some universal sort s the map $\pi_1(H)$ is injective on A_s, then $\pi_1(e(H, Y))$ is as well.

It is straightforward to verify that such a mapping e exists; note that when using this mapping, Y will be of the form $A_{\forall, <x}$ for an existential variable x.

We define the strategy (F_x) as follows. Let x be an existential variable of $(P, \mathbf{A})$, and let $H : A_{\forall, <x} \to B'$ be a map. Define $H[x]$ as $e(H|U(x), A_{\forall, <x})$. Set $F_x(H)$ to be the pair (c_1, c_2) where

- $c_1 = f'_x(\pi_1(H[x]))$, where (f'_x) is the strategy from Lemma 19; and,
- c_2 is $\bot$ if c_1 is not in S, and otherwise is equal to $f_{c_1}((H[x])(A_{\forall, <x}))$. Note that $H[x])(A_{\forall, <x})$ is a set of pairs that should be viewed as a function, in passing it to f_{c_1}.

Observe that if c_1 is not degenerate, then by the just-given lemma, the mapping $\pi_1(H[x])$ is injective on $U(x)$; it follows that $\pi_1(H[x])$ is injective on $A_{\forall,<x}$ by the definition of $H[x]$ and the second condition in the definition of e. This implies that $(H[x])(A_{\forall,<x})$ is the graph of a mapping defined on $A_{\forall,<x}$, and c_2 as described above is well-defined.

By the definition of c_1, we have that (F_x) has the property that for any $H : A_\forall \to B'$, it holds that $\pi_1(F, H)$ is a homomorphism from $\mathbf{A}$ to $\mathbf{A}^*$. It remains to verify that if $(a_1, \ldots, a_k) \in R^{\mathbf{A}}$, then the image $((t_1, b_1), \ldots, (t_k, b_k))$ of $(a_1, \ldots, a_k)$ under (F, H) has the property that $(t_1, \ldots, t_k) \in R^{\mathbf{S}}$ implies $(b_1, \ldots, b_k) \in R^{\mathbf{B}}$. For each existential variable x occurring in $(a_1, \ldots, a_k)$, observe that for any universal variable y coming before it in the quantifier prefix, one has $y \in U(x)$ and thus $H(y) = (H[x])(y)$. It thus suffices to show that if x and x' are existential variables in this tuple where x occurs before x', $H : A_{\forall,<x} \to B'$ and $H' : A_{\forall,<x'} \to B'$ are mappings where H' extends H, then $(H'[x'])(A_{\forall,<x'})$ extends $(H[x])(A_{\forall,<x})$. It suffices to show that $H'[x']$ and $H[x]$ agree on $A_{\forall,<x}$. It follows by definition of U that $U(x)|A_{\forall,<x} = U(x')|A_{\forall,<x}$. Thus, for an $\forall$-sort s occurring before x, we have $(H'|U(x'))|A_s = (H|U(x))|A_s$. So thus by the first condition in the definition of e, it holds that $e(H'|U(x'), A_{\forall,<x'})|A_s = e(H|U(x), A_{\forall,<x})|A_s$ from which we obtain the desired agreement.

References

1. Chandra, A.K., Merlin, P.M.: Optimal implementation of conjunctive queries in relational data bases. In: Procceddings of STOC 1977, pp. 77–90 (1977)
2. Chen, H., Dalmau, V.: Decomposing quantified conjunctive (or disjunctive) formulas. In: LICS (2012)
3. Chen, H., Grohe, M.: Constraint satisfaction with succinctly specified relations. Journal of Computer and System Sciences 76(8), 847–860 (2010)
4. Chen, H., Madelaine, F., Martin, B.: Quantified constraints and containment problems. In: Twenty-Third Annual IEEE Symposium on Logic in Computer Science, LICS (2008)
5. Gottlob, G., Greco, G., Scarcello, F.: The complexity of quantified constraint satisfaction problems under structural restrictions. In: IJCAI 2005 (2005)
6. Grohe, M.: The complexity of homomorphism and constraint satisfaction problems seen from the other side. Journal of the ACM 54(1) (2007)
7. Grohe, M., Schwentick, T., Segoufin, L.: When is the evaluation of conjunctive queries tractable? In: STOC 2001 (2001)
8. Marx, D.: Tractable hypergraph properties for constraint satisfaction and conjunctive queries. In: Proceedings of the 42nd ACM Symposium on Theory of Computing, pp. 735–744 (2010)

From Security Protocols to Pushdown Automata[*]

Rémy Chrétien[1,2], Véronique Cortier[1], and Stéphanie Delaune[2]

[1] LORIA, CNRS, France
[2] LSV, ENS Cachan & CNRS & INRIA Saclay Île-de-France

Abstract. Formal methods have been very successful in analyzing security protocols for reachability properties such as secrecy or authentication. In contrast, there are very few results for equivalence-based properties, crucial for studying e.g. privacy-like properties such as anonymity or vote secrecy.
We study the problem of checking equivalence of security protocols for an unbounded number of sessions. Since replication leads very quickly to undecidability (even in the simple case of secrecy), we focus on a limited fragment of protocols (standard primitives but pairs, one variable per protocol's rules) for which the secrecy preservation problem is known to be decidable. Surprisingly, this fragment turns out to be undecidable for equivalence. Then, restricting our attention to deterministic protocols, we propose the first decidability result for checking equivalence of protocols for an unbounded number of sessions. This result is obtained through a characterization of equivalence of protocols in terms of equality of languages of (generalized, real-time) deterministic pushdown automata.

1 Introduction

Formal methods have been successfully applied for rigorously analyzing security protocols. In particular, many algorithms and tools (see [13,4,9,2,11] to cite a few) have been designed to automatically find flaws in protocols or prove security. Most of these results focus on reachability properties such as authentication or secrecy: for any execution of the protocol, an attacker should never learn a secret (secrecy property) or make Alice think she's talking to Bob while Bob did not engage a conversation with her (authentication property). However, privacy properties such as vote secrecy, anonymity, or untraceability cannot be expressed as such. They are instead defined as indistinguishability properties in [1,6]. For example, Alice's identity remains private if an attacker cannot distinguish a session where Alice is talking from a session where Bob is talking.

Studying indistinguishability properties for security protocols amounts into checking a behavioral equivalence between processes. Processes represent protocols and are specified in some process algebras such as CSP or the pi-calculus, except that messages are no longer atomic actions but terms, in order to faithfully represent cryptographic messages. Of course, considering terms instead of atomic actions considerably

---

[*] Full version available at http://hal.inria.fr/hal-00817230. The research leading to these results has received funding from the European Research Council under the European Union's Seventh Framework Programme (FP7/2007-2013) / ERC grant agreement n° 258865, project ProSecure, and the ANR project JCJC VIP n° 11 JS02 006 01.

F.V. Fomin et al. (Eds.): ICALP 2013, Part II, LNCS 7966, pp. 137–149, 2013.

increases the difficulty of checking equivalence. As a matter of fact, there are just a few results for checking equivalence of processes that manipulate terms.

- Based on a procedure developed by M. Baudet [3], it has been shown that trace equivalence is decidable for deterministic processes with no else branches, and for a family of equational theories that captures most standard primitives [10]. A simplified proof of [3] has been proposed by Y. Chevalier and M. Rusinowitch [8].
- A. Tiu and J. Dawson [17] have designed and implemented a procedure for open bisimulation, a notion of equivalence stronger than the standard notion of trace equivalence. This procedure only works for a limited class of processes.
- V. Cheval *et al.* [7] have proposed and implemented a procedure for trace equivalence, and for a quite general class of processes. They consider non deterministic processes that use standard primitives, and that may involve else branches.

However, these decidability results analyse equivalence for a *bounded number of sessions* only, that is assuming that protocols are executed a limited number of times. This is of course a strong limitation. Even if no flaw is found when a protocol is executed n times, there is absolutely no guarantee that the protocol remains secure when it is executed $n+1$ times. And actually, the existing tools for a bounded number of sessions can only analyse protocols for a very limited number of sessions, typically 2 or 3. Another approach consists in implementing a procedure that is not guaranteed to terminate. This is in particular the case of ProVerif [4], a well-established tool for checking security of protocols. ProVerif is able to check equivalence although it does not always succeed [5]. Of course, Proverif does not correspond to any decidability result.

Our Contribution. We study the decidability of equivalence of security protocols for an unbounded number of sessions. Even in the case of reachability properties such as secrecy, the problem is undecidable in general. We therefore focus on a class of protocols for which secrecy is decidable [9]. This class typically assumes that each protocol rule manipulates at most one variable. Surprisingly, even a fragment of this class (with only symmetric encryption) turns out to be undecidable for equivalence properties. We consequently further assume our protocols to be deterministic (that is, given an input, there is at most one possible output). We show that equivalence is decidable for an unbounded number of sessions and for protocols with standard primitives but pairs. Interestingly, we show that checking for equivalence of protocols actually amounts into checking equality of languages of deterministic pushdown automata. The decidability of equality of languages of deterministic pushdown automata is a difficult problem, shown to be decidable at Icalp in 1997 [14]. We actually characterize equivalence of protocols in terms of equivalence of deterministic generalized real-time pushdown automata, that is deterministic pushdown automata with no epsilon-transition but such that the automata may unstack several symbols at a time. More precisely, we show how to associate to a process P an automata $\mathcal{A}_P$ such that two processes are equivalent if, and only if, their corresponding automata yield the same language and, reciprocally, we show how to associate to an automata $\mathcal{A}$ a process P_A such that two automata yield the same language if, and only if, their corresponding processes are equivalent, that is:

$$P \approx Q \Leftrightarrow L(\mathcal{A}_P) = L(\mathcal{A}_Q), \text{ and } L(\mathcal{A}) = L(\mathcal{B}) \Leftrightarrow P_A \approx P_B.$$

Therefore, checking for equivalence of protocols is as difficult as checking equivalence of deterministic generalized real-time pushdown automata.

2 Model for Security Protocols

Security protocols are modeled through a process algebra that manipulates terms.

2.1 Syntax

Term algebra. As usual, messages are represented by terms. More specifically, we consider a *sorted signature* with six sorts rand, key, msg, SimKey, PrivKey and PubKey that represent respectively random numbers, keys, messages, symmetric keys, private keys and public keys. We assume that msg subsumes the five other sorts, key subsumes SimKey, PrivKey and PubKey. We consider six function symbols senc and sdec, aenc and adec, sign and check that represent symmetric, asymmetric encryption and decryption as well as signatures. Since we are interested in the analysis of indistinguishability properties, we consider randomized primitives:

$$
\begin{aligned}
\mathsf{senc} &: \mathsf{msg} \times \mathsf{SimKey} \times \mathsf{rand} \to \mathsf{msg} & \mathsf{sdec} &: \mathsf{msg} \times \mathsf{SimKey} \to \mathsf{msg} \\
\mathsf{aenc} &: \mathsf{msg} \times \mathsf{PubKey} \times \mathsf{rand} \to \mathsf{msg} & \mathsf{adec} &: \mathsf{msg} \times \mathsf{PrivKey} \to \mathsf{msg} \\
\mathsf{sign} &: \mathsf{msg} \times \mathsf{PrivKey} \times \mathsf{rand} \to \mathsf{msg} & \mathsf{check} &: \mathsf{msg} \times \mathsf{PubKey} \to \mathsf{msg}
\end{aligned}
$$

We further assume an infinite set Σ_0 of *constant symbols* of sort key or msg, an infinite set Ch of constant symbols of sort channel, two infinite sets of *variables* $\mathcal{X}, \mathcal{W}$, and an infinite set $\mathcal{N} = \mathcal{N}_{\mathsf{pub}} \uplus \mathcal{N}_{\mathsf{prv}}$ of *names* of sort rand: $\mathcal{N}_{\mathsf{pub}}$ represents the random numbers drawn by the attacker while $\mathcal{N}_{\mathsf{prv}}$ represents the random numbers drawn by the protocol's participants. As usual, *terms* are defined as names, variables, and function symbols applied to other terms. We denote by $\mathcal{T}(\mathcal{F}, \mathcal{N}, \mathcal{X})$ the set of terms built on function symbols in $\mathcal{F}$, names in $\mathcal{N}$, and variables in $\mathcal{X}$. We simply write $\mathcal{T}(\mathcal{F}, \mathcal{N})$ when $\mathcal{X} = \emptyset$. We consider three particular signatures:

$$
\begin{aligned}
\Sigma_{\mathsf{pub}} &= \{\mathsf{senc}, \mathsf{sdec}, \mathsf{aenc}, \mathsf{adec}, \mathsf{sign}, \mathsf{check}, \mathsf{start}\} \\
\Sigma^+ &= \Sigma_{\mathsf{pub}} \cup \Sigma_0 \qquad \Sigma = \{\mathsf{senc}, \mathsf{aenc}, \mathsf{sign}, \mathsf{start}\} \cup \Sigma_0
\end{aligned}
$$

where start $\notin \Sigma_0$ is a constant symbol of sort msg. Σ_{pub} represents the functions/data available to the attacker, Σ^+ is the most general signature, while Σ models actual messages (with no failed computation). We add a bijection between elements of sort PrivKey and PubKey. If k is a constant of sort PrivKey, k^{-1} will denotes its image by this function, called *inverse*. We will write the inverse function the same, so that $(\mathsf{k}^{-1})^{-1} = \mathsf{k}$. To keep homogeneous notations, we will extend this function to symmetric keys: if k is of sort SimKey, then $\mathsf{k}^{-1} = \mathsf{k}$. The relation between encryption and decryption is represented through the following rewriting rules, yielding a convergent rewrite system:

$$
\mathsf{sdec}(\mathsf{senc}(x, y, z), y) \to x \qquad \mathsf{adec}(\mathsf{aenc}(x, y, z), y^{-1}) \to x
$$
$$
\mathsf{check}(\mathsf{sign}(x, y, z), y^{-1}) \to x
$$

This rule models the fact that the decryption of a ciphertext will return the associated plaintext when the right key is used to perform decryption. We denote by $t\!\downarrow$ the *normal form* of a term $t \in \mathcal{T}(\Sigma^+, \mathcal{N}, \mathcal{X})$.

Example 1. The term $m = \mathsf{senc}(\mathsf{s}, \mathsf{k}, r)$ represents an encryption of the constant s with the key k using the random $r \in \mathcal{N}$, whereas $t = \mathsf{sdec}(m, \mathsf{k})$ models the application of the decryption algorithm on m using k. We have that $t{\downarrow} = \mathsf{s}$.

An attacker may build his own messages by applying functions to terms he already knows. Formally, a computation done by the attacker is modeled by a *recipe*. i.e. a term in $\mathcal{T}(\Sigma_{\mathsf{pub}}, \mathcal{N}_{\mathsf{pub}}, \mathcal{W})$. The variables in $\mathcal{W}$ intuitively refer to variables used to store messages learnt by the attacker.

Process algebra. The intended behavior of a protocol can be modelled by a *process* defined by the following grammar where $u \in \mathcal{T}(\Sigma, \mathcal{N}, \mathcal{X})$, $n \in \mathcal{N}$, and $c \in Ch$:

$$P, Q := 0 \mid \mathsf{in}(c, u).P \mid \mathsf{out}(c, u).P \mid (P \mid Q) \mid !P \mid \mathsf{new}\ n.P$$

The process "$\mathsf{in}(c, u).P$" expects a message m of the form u on channel c and then behaves like $P\theta$ where θ is a substitution such that $m = u\theta$. The process "$\mathsf{out}(c, u).P$" emits u on channel c, and then behaves like P. The variables that occur in u will be instantiated when the evaluation will take place. The process $P \mid Q$ runs P and Q in parallel. The process $!P$ executes P some arbitrary number of times. The process $\mathsf{new}\ n.P$ invents a new name n and continues as P.

Sometimes, we will omit the null process. We write $fv(P)$ for the set of *free variables* that occur in P, *i.e.* the set of variables that are not in the scope of an input. A *protocol* is a ground process, *i.e.* a process P such that $fv(P) = \emptyset$.

Example 2. For the sake of illustration, we consider a naive protocol, where A sends a value v (e.g. a vote) to B, encrypted by a short-term key exchanged through a server.

$$
\begin{aligned}
&1.\ A \rightarrow S : \mathsf{senc}(\mathsf{k}_{AB}, \mathsf{k}_{AS}, r_A) \\
&2.\ S \rightarrow B : \mathsf{senc}(\mathsf{k}_{AB}, \mathsf{k}_{BS}, r_S) \\
&3.\ A \rightarrow B : \mathsf{senc}(v, \mathsf{k}_{AB}, r)
\end{aligned}
$$

The agent A sends a symmetric key k_{AB} encrypted with the key k_{AS} (using a fresh random number r_A). The server answers to this request by decrypting this message and encrypting it with k_{BS}. The agent A can now send his vote v encrypted with k_{AB}.

The role of A is modeled by a process $P_A(v)$ while the role of S is modeled by P_S. The role of B (which does not output anything) is omitted for concision.

$$
\begin{aligned}
P_A(v) &\overset{\mathsf{def}}{=} \ !\mathsf{in}(c_A, \mathsf{start}).\mathsf{new}\ r_A.\mathsf{out}(c_A, \mathsf{senc}(\mathsf{k}_{AB}, \mathsf{k}_{AS}, r_A)) && (1) \\
&\mid \ !\mathsf{in}(c_A', \mathsf{start}).\mathsf{new}\ r.\mathsf{out}(c_A, \mathsf{senc}(v, \mathsf{k}_{AB}, r)) && (2) \\
P_S &\overset{\mathsf{def}}{=} \ !\mathsf{in}(c_S, \mathsf{senc}(x, \mathsf{k}_{AS}, z)).\mathsf{new}\ r_S.\mathsf{out}(c_S, \mathsf{senc}(x, \mathsf{k}_{BS}, r_S)) && (3) \\
&\mid \ !\mathsf{in}(c_S', \mathsf{senc}(x, \mathsf{k}_{AS}, z)).\mathsf{new}\ r_S.\mathsf{out}(c_S', \mathsf{senc}(x, \mathsf{k}_{CS}, r_S)) && (4)
\end{aligned}
$$

where c_A, c_A', c_S, c_S' are constants of sort channel, $\mathsf{k}_{AB}, \mathsf{k}_{AS}, \mathsf{k}_{BS}$, and k_{CS} are (private) constants in Σ_0 of sort SimKey, whereas r_A, r_S, r are names of sort rand, and x (resp. z) is a variable of sort msg (resp. rand).

Intuitively, $P_A(v)$ sends k_{AB} encrypted by k_{AS} to the server (branch 1), and then her vote encrypted by k_{AB} (branch 2). The process P_S models the server, answering both requests from A to B (branch 3), as well as requests from A to C (branch 4). More generally the server answers requests from any agent to any agent but only two cases are considered here, again for concision. The whole protocol is given by $P(v)$, where $P_A(v)$ and P_S evolve in parallel and additionally, the secret key k_{CS} is sent in clear, to model the fact that the attacker may learn keys of some corrupted agents:

$$P(v) \stackrel{\text{def}}{=} P_A(v) \mid P_S \mid !\,\text{in}(c, \text{start}).\text{out}(c, k_{CS})$$

2.2 Semantics

A *configuration* of a protocol is a pair $(\mathcal{P}; \sigma)$ where:

- $\mathcal{P}$ is a multiset of processes. We often write $P \cup \mathcal{P}$, or $P \mid \mathcal{P}$, instead of $\{P\} \cup \mathcal{P}$.
- $\sigma = \{w_1 \triangleright m_1, \ldots, w_n \triangleright m_n\}$ is a *frame*, i.e. a substitution where $w_1, \ldots, w_n$ are variables in $\mathcal{W}$, and $m_1, \ldots, m_n$ are terms in $\mathcal{T}(\Sigma, \mathcal{N})$. Those terms represent the messages that are known by the attacker.

The operational semantics of protocol is defined by the relation $\xrightarrow{\alpha}$ over configurations. For sake of simplicity, we often write P instead of $(P; \emptyset)$.

$(\text{in}(c, u).P \cup \mathcal{P}; \sigma) \xrightarrow{\text{in}(c,R)} (P\theta \cup \mathcal{P}; \sigma)$
    where R is a recipe such that $R\sigma{\downarrow} \in \mathcal{T}(\Sigma, \mathcal{N})$ and $R\sigma{\downarrow} = u\theta$ for some θ

$(\text{out}(c, u).P \cup \mathcal{P}; \sigma) \xrightarrow{\text{out}(c,w_{i+1})} (P \cup \mathcal{P}; \sigma \cup \{w_{i+1} \triangleright u\})$
    where i is the number of elements in σ

$(!P \cup \mathcal{P}; \sigma) \xrightarrow{\tau} (P \cup !P \cup \mathcal{P}; \sigma)$

$(\text{new } n.P \cup \mathcal{P}; \sigma) \xrightarrow{\tau} (P\{^{n'}/_n\} \cup \mathcal{P}; \sigma)$ where n' is a fresh name in $\mathcal{N}_{\text{prv}}$

A process may input any term that an attacker can build (rule IN). The process $\text{out}(c, u).P$ outputs u (which is stored in the attacker's knowledge) and then behaves like P. The two remaining rules are unobservable (τ action) from the point of view of the attacker. The relation $\xrightarrow{w}$ between configurations (where w is a sequence of actions) is defined in a usual way. Given a sequence of observable actions w, we write $K \xRightarrow{w} K'$ when there exists w' such that $K \xrightarrow{w'} K'$ and w is obtained from w' by erasing all occurrences of τ. For every configuration K, we define its *set of traces* as follows:

$$\text{trace}(K) = \{(\text{tr}, \sigma) \mid K \xRightarrow{\text{tr}} (\mathcal{P}; \sigma) \text{ for some configuration } (\mathcal{P}; \sigma)\}.$$

Example 3. Going back to the protocol introduced in Example 2, consider the following scenario: *(i)* the corrupted agent C discloses his secret key k_{CS}; *(ii)* the agent A initiates a session with B, and for this she sends a request to the server S; *(iii)* the attacker intercepts this message and sends it to S as a request coming from A to establish a key with C. Instead of answering to this request with $\text{senc}(k_{AB}, k_{BS}, r_S)$, the server sends $\text{senc}(k_{AB}, k_{CS}, r_S)$, and the attacker will learn k_{AB}. More formally, we have that:

$$K_0 \stackrel{\text{def}}{=} (P(v); \emptyset) \xrightarrow{\text{in}(c,\text{start}).\text{out}(c,w_1).\text{in}(c_A,\text{start}).\text{out}(c_A,w_2).\text{in}(c'_S,w_2).\text{out}(c'_S,w_3).} (P(v); \sigma)$$

where $\sigma = \{w_1 \triangleright k_{CS}, \ w_2 \triangleright \text{senc}(k_{AB}, k_{AS}, r_A), \ w_3 \triangleright \text{senc}(k_{AB}, k_{CS}, r_S)\}$, and r_A, r_S are (fresh) names in $\mathcal{N}_{\text{prv}}$. In this execution trace, first the key k_{CS} is sent after having called the corresponding process. Then, branches (1) and (4) of $P(v)$ are triggered.

2.3   Trace Equivalence

Intuitively, two processes are equivalent if they cannot be distinguished by any attacker. Trace equivalence can be used to formalise many interesting security properties, in particular privacy-type properties, such as those studied for instance in [1,6]. We first introduce a notion of intruder's knowledge well-suited to cryptographic primitives for which the success of decrypting or checking a signature is visible.

Definition 1. *Two frames σ_1 and σ_2 are statically equivalent, $\sigma_1 \sim \sigma_2$, when we have that $dom(\sigma_1) = dom(\sigma_2)$, and:*

- *for any recipe R, $R\sigma_1\!\downarrow \in \mathcal{T}(\Sigma, \mathcal{N})$ if, and only if, $R\sigma_2\!\downarrow \in \mathcal{T}(\Sigma, \mathcal{N})$; and*
- *for all recipes R_1 and R_2 such that $R_1\sigma_1\!\downarrow, R_2\sigma_1\!\downarrow \in \mathcal{T}(\Sigma, \mathcal{N})$, we have that $R_1\sigma_1\!\downarrow = R_2\sigma_1\!\downarrow$ if, and only if, $R_1\sigma_2\!\downarrow = R_2\sigma_2\!\downarrow$.*

Intuitively, two frames are equivalent if an attacker cannot see the difference between the two situations they represent: if some computation fails in σ_1 it should fail in σ_2 as well, and σ_1 and σ_2 should satisfy the same equalities.

Example 4. Assume some agent publishes her vote encrypted. The possible values for the votes are typically public. Therefore the question is not whether an attacker may know the value of the vote (that he knows anyway) but instead, whether he may distinguish between two executions where A votes differently. Consider the two frames:

$$\sigma_i \stackrel{\text{def}}{=} \{w_4 \triangleright v_0, \ w_5 \triangleright v_1, \ w_6 \triangleright \mathsf{senc}(v_i, k_{AB}, r)\} \text{ with } i \in \{0, 1\}$$

where $v_0, v_1 \in \Sigma_0$, and $r \in \mathcal{N}_{\mathsf{prv}}$. We have that $\sigma_0 \sim \sigma_1$. Intuitively, there is no test that allows the attacker to distinguish the two frames since the key k_{AB} is not available. In this scenario, the vote v_i remains private. Now, consider the frames $\sigma'_i = \sigma \cup \sigma_i$ with $i \in \{0, 1\}$ and σ as defined in Example 3. We have that $\sigma'_0 \not\sim \sigma'_1$. Indeed, consider the recipes $R_1 = \mathsf{sdec}(w_6, \mathsf{sdec}(w_3, w_1))$ and $R_2 = w_4$. We have that $R_1\sigma'_0\!\downarrow = R_2\sigma'_0\!\downarrow = v_0$, whereas $R_1\sigma'_1\!\downarrow = v_1$ and $R_2\sigma'_1\!\downarrow = v_0$. Intuitively, an attacker can learn k_{AB} and then compare the encrypted vote to the values v_0 and v_1.

Intuitively, two processes are *trace equivalent* if, however they behave, the resulting sequences of messages observed by the attacker are in static equivalence.

Definition 2. *Let P and Q be two protocols. We have that $P \sqsubseteq Q$ if for every $(\mathsf{tr}, \sigma) \in \mathsf{trace}(P)$, there exists $(\mathsf{tr}', \sigma') \in \mathsf{trace}(Q)$ such that $\mathsf{tr} = \mathsf{tr}'$ and $\sigma \sim \sigma'$. They are* trace equivalent*, written $P \approx Q$, if $P \sqsubseteq Q$ and $Q \sqsubseteq P$.*

Example 5. Continuing Example 2, our naive protocol is secure if the vote of A remains private. This is typically expressed by $P(v_0) \mid Q \approx P(v_1) \mid Q$. An attacker should not distinguish between two instances of the protocol where A votes two different values. The purpose of Q is to disclose the two values v_0 and v_1.

$$Q \stackrel{\text{def}}{=} !\,\mathsf{in}(c_0, \mathsf{start}).\mathsf{out}(c_0, v_0) \ \mid\ !\,\mathsf{in}(c_1, \mathsf{start}).\mathsf{out}(c_1, v_1)$$

However, our protocol is insecure. As seen in Example 3, an attacker may learn k_{AB}, and therefore distinguish between the two processes described above. Formally, we have that $P(v_0) \mid Q \not\approx P(v_1) \mid Q$. This is reflected by the trace tr' described below:

$$\mathsf{tr}' \stackrel{\text{def}}{=} \mathsf{tr}.\mathsf{in}(c_0, \mathsf{start}).\mathsf{out}(c_0, w_4).\mathsf{in}(c_1, \mathsf{start}).\mathsf{out}(c_1, w_5).\mathsf{in}(c'_A, \mathsf{start}).\mathsf{out}(c'_A, w_6).$$

We have that $(\mathrm{tr}', \sigma_0') \in \mathrm{trace}(K_0)$ with $K_0 = (P(\mathsf{v}_0) \mid Q; \emptyset)$ and σ_0' as defined in Example 4. Because of the existence of only one branch using each channel, there is only one possible execution of $P(\mathsf{v}_1) \mid Q$ (up to a bijective renaming of the private names of sort rand) matching the labels in tr', and the corresponding execution will allow us to reach the frame σ_1' as described in Example 4. We have already seen that static equivalence does not hold, i.e. $\sigma_0' \not\approx \sigma_1'$.

3 Ping-Pong Protocols

We aim at providing a decidability result for the problem of trace equivalence between protocols in presence of replication. However, it is well-known that replication leads to undecidability even for the simple case of reachability properties. Thus, we consider a class of protocols, called $\mathcal{C}_{\mathsf{pp}}$, for which (in a slightly different setting), reachability has already been proved decidable [9].

3.1 Class $\mathcal{C}_{\mathsf{pp}}$

We basically consider ping-pong protocols (an output is computed using only the message previously received in input), and we assume a kind of determinism. Moreover, we restrict the terms that are manipulated throughout the protocols: only one unknown message (modelled by the use of a variable of sort msg) can be received at each step.

We fix a variable $x \in \mathcal{X}$ of sort msg. An *input term* u (resp. *output term* v) is a term defined by the grammars given below:

$$u := x \mid \mathsf{s} \mid \mathsf{f}(u, \mathsf{k}, z) \qquad v := x \mid \mathsf{s} \mid \mathsf{f}(v, \mathsf{k}, r)$$

where $\mathsf{s}, \mathsf{k} \in \Sigma_0 \cup \{\mathsf{start}\}$, $z \in \mathcal{X}$, $\mathsf{f} \in \{\mathsf{senc}, \mathsf{aenc}, \mathsf{sign}\}$ and $r \in \mathcal{N}$. Moreover, we assume that each variable (resp. name) occurs at most once in u (resp. v).

Definition 3. $\mathcal{C}_{\mathsf{pp}}$ *is the class of protocol of the form:*

$$P = \prod_{i=1}^{n} \prod_{j=1}^{p_i} !\mathrm{in}(c_i, u_j^i).\mathrm{new}\ r_1. \ldots .\mathrm{new}\ r_{k_j^i}.\mathrm{out}(c_i, v_j^i) \qquad such\ that:$$

1. *for all $i \in \{1, \ldots, n\}$, and $j \in \{1, \ldots, p_i\}$, $k_j^i \in \mathbb{N}$, u_j^i is an input term, and v_j^i is an output term where names occurring in v_j^i are included in $\{r_1, \ldots, r_{k_j^i}\}$;*
2. *for all $i \in \{1, \ldots, n\}$, and $j_1, j_2 \in \{1, \ldots, p_i\}$, if $j_1 \neq j_2$ then for any renaming of variables, $u_{j_1}^i$ and $u_{j_2}^i$ are not unifiable[1].*

Note that the purpose of item 2 is to restrict the class of protocols to those that have a deterministic behavior (a particular input action can only be accepted by one branch of the protocol). This is a natural restriction since most of the protocols are indeed deterministic: an agent should usually know exactly what to do once he has received a message. Actually, the main limitations of the class $\mathcal{C}_{\mathsf{pp}}$ are stated in item 1: we consider a restricted signature (e.g. no pair, no hash function), and names can only be used to produce randomized ciphertexts/signatures.

---

[1] i.e. there does not exist θ such that $u_{j_1}^i \theta = u_{j_2}^i \theta$.

Example 6. The protocols described in Example 5 are in $\mathcal{C}_{\mathsf{pp}}$. For instance, we can check that $\mathsf{senc}(x, \mathsf{k}_{AS}, z)$ is an input term whereas $\mathsf{senc}(x, \mathsf{k}_{BS}, r_S)$ is an output term. Moreover, the determinism condition (item 2) is clearly satisfied: each branch of the protocol $P(\mathsf{v}_0) \mid Q$ (resp. $P(\mathsf{v}_0) \mid Q$) uses a different channel.

Our main contribution is a decision procedure for trace equivalence of processes in $\mathcal{C}_{\mathsf{pp}}$. Details of the procedure are provided in Section 4.

Theorem 1. *Let P and Q be two protocols in $\mathcal{C}_{\mathsf{pp}}$. The problem whether P and Q are trace equivalent, i.e. $P \approx Q$, is decidable.*

3.2 Undecidability Results

The class $\mathcal{C}_{\mathsf{pp}}$ is somewhat limited but surprisingly, extending $\mathcal{C}_{\mathsf{pp}}$ to non deterministic processes immediately yields undecidability of trace equivalence. More precisely, trace inclusion of processes in $\mathcal{C}_{\mathsf{pp}}$ is already undecidable.

Theorem 2. *Let P and Q be two protocols in $\mathcal{C}_{\mathsf{pp}}$. The problem whether P is trace included in Q, i.e. $P \sqsubseteq Q$, is undecidable.*

This result is shown by encoding the Post Correspondence Problem (PCP). Alternatively, it results from the reduction result established in Section 5 and the undecidability result established in [12]. Undecidability of trace inclusion actually implies undecidability of trace equivalence as soon as processes are non deterministic. Indeed consider the choice operator $+$ whose (standard) semantics is given by the following rules:

$$(\{P + Q\} \cup \mathcal{P}; \sigma) \xrightarrow{\tau} (P \cup \mathcal{P}; \sigma) \qquad (\{P + Q\} \cup \mathcal{P}; \sigma) \xrightarrow{\tau} (Q \cup \mathcal{P}; \sigma)$$

Corollary 1. *Let P, Q_1, and Q_2 be three protocols in $\mathcal{C}_{\mathsf{pp}}$. The problem whether P is equivalent to $Q_1 + Q_2$, i.e. $P \approx Q_1 + Q_2$, is undecidable.*

Indeed, consider P and Q_1, for which trace inclusion encodes PCP, and let $Q_2 = P$. Trivially, $P \sqsubseteq Q_1 + Q_2$. Thus $P \approx Q_1 + Q_2$ if, and only if, $Q_1 + Q_2 \sqsubseteq P$, i.e. if, and only if, $Q_1 \sqsubseteq P$, hence the undecidability result.

4 From Trace Equivalence to Language Equivalence

This section is devoted to a sketch of proof of Theorem 1. Deciding trace equivalence is done in two main steps. First, we show how to reduce the trace equivalence problem between protocols in $\mathcal{C}_{\mathsf{pp}}$, to the problem of deciding trace equivalence (still between protocols in $\mathcal{C}_{\mathsf{pp}}$) when the attacker acts as a *forwarder*.

Then, we encode the problem of deciding trace equivalence for forwarding attackers into the problem of language equivalence for real-time generalized pushdown deterministic automata (GPDA).

4.1 Generalized Pushdown Automata

GPDA differ from deterministic pushdown automata (DPA) as they can unstack several symbols at a time. We consider real-time GPDA with final-state acceptance.

Definition 4. *A real-time GPDA is a 7-tuple* $\mathcal{A} = (Q, \Pi, \Gamma, q_0, \omega, Q_f, \delta)$ *where Q is the finite set of states, $q_0 \in Q$ is the initial state, $Q_f \subseteq Q$ is the set of accepting states, Π is the finite input-alphabet, Γ is the finite stack-alphabet, ω is the initial stack symbol, and $\delta : (Q \times \Pi \times \Gamma_0) \to Q \times \Gamma_0$ is the partial transition function such that:*

- Γ_0 *is a finite subset of Γ^*; and*
- *for any $(q, a, x) \in dom(\delta)$ and y suffix strict of x, we have that $(q, a, y) \notin dom(\delta)$.*

Let $q, q' \in Q$, $w, w', \gamma \in \Gamma^*$, $m \in \Pi^*$, $a \in \Pi$; we note $(qw\gamma, am) \rightsquigarrow_{\mathcal{A}} (q'ww', m)$ if $(q', w') = \delta(q, a, \gamma)$. The relation $\rightsquigarrow_{\mathcal{A}}^*$ is the reflexive and transitive closure of $\rightsquigarrow_{\mathcal{A}}$. For every qw, $q'w'$ in $Q\Gamma^*$ and $m \in \Pi^*$, we note $qw \xrightarrow{m}_{\mathcal{A}} q'w'$ if, and only if, $(qw, m) \rightsquigarrow_{\mathcal{A}}^* (q'w', \epsilon)$. For sake of clarity, a transition from q to q' reading a, popping γ from the stack and pushing w' will be denoted by $q \xrightarrow{a;\gamma/w'} q'$.

Let $\mathcal{A}$ be a GPDA. The language recognized by $\mathcal{A}$ is defined by:

$$\mathcal{L}(\mathcal{A}) = \{m \in \Pi^* \mid q_0\omega \xrightarrow{m}_{\mathcal{A}} q_f w \text{ for some } q_f \in Q_f \text{ and } w \in \Gamma^*\}.$$

A real-time GPDA can easily be converted into a DPA by adding new states and ϵ-transitions. Thus, the problem of language equivalence for two real-time GPDA $\mathcal{A}_1$ and $\mathcal{A}_2$, i.e. deciding whether $\mathcal{L}(\mathcal{A}_1) = \mathcal{L}(\mathcal{A}_2)$ is decidable [15].

4.2 Getting Rid of the Attacker

We define the actions of a forwarder by modifying our semantics. We restrict the recipes R, R_1, and R_2 that are used in the IN rule and in static equivalence (Definition 1) to be either the public constant start or a variable in $\mathcal{W}$. This leads us to consider a new relation $\Rightarrow_{\mathsf{fwd}}$ between configurations, and a new notion of static equivalence $\sim_{\mathsf{fwd}}$. We denote by $\approx_{\mathsf{fwd}}$ the trace equivalence relation induced by this new semantics.

Example 7. The trace exhibited in Example 3 is still a valid one according to the forwarder semantics, and the frames σ_0' and σ_1' described in Example 4 are in equivalence according to $\sim_{\mathsf{fwd}}$. Actually, we have that $P(\mathsf{v}_0) \mid Q \approx_{\mathsf{fwd}} P(\mathsf{v}_1) \mid Q$. Indeed, the fact that a forwarder simply acts as a relay prevents him to mount the aforementioned attack.

As shown above, the forwarder semantics is very restrictive: a forwarder can not rely on his deduction capabilities to mount an attack. To counterbalance the effects of this semantics, the key idea consists in modifying the protocols under study by adding new rules that encrypt/sign and decrypt/check messages on demand for the forwarder.

Formally, we define a transformation $\mathcal{T}_{\mathsf{fwd}}$ that associates to a pair of protocols in $\mathcal{C}_{\mathsf{pp}}$ a finite set of pairs of protocols (still in $\mathcal{C}_{\mathsf{pp}}$), and we show the following result:

Proposition 1. *Let P and Q be two protocols in $\mathcal{C}_{\mathsf{pp}}$. We have that:*

$$P \approx Q \text{ if, and only if, } P' \approx_{\mathsf{fwd}} Q' \text{ for some } (P', Q') \in \mathcal{T}_{\mathsf{fwd}}(P, Q).$$

Roughly the transformation $\mathcal{T}_{\mathsf{fwd}}$ consists in first guessing among the keys of the protocols P and the keys of the protocols Q those that are deducible by the attacker, as well as a bijection α between these two sets. We can show that such a bijection necessarily exists when $P \approx Q$. Then, to compensate the fact that the attacker is a simple forwarder, we give him access to oracles for any deducible key k, adding the corresponding branches in the processes, i.e. in case k is of sort SimKey, we add

$$! \mathsf{in}(c_\mathsf{k}^{\mathsf{senc}}, x).\mathsf{new}\, r.\mathsf{out}(c_\mathsf{k}^{\mathsf{senc}}, \mathsf{senc}(x, \mathsf{k}, r)) \mid\, ! \mathsf{in}(c_\mathsf{k}^{\mathsf{sdec}}, \mathsf{senc}(x, \mathsf{k}, z)).\mathsf{out}(c_\mathsf{k}^{\mathsf{sdec}}, x)$$

To maintain the equivalence, we do a similar transformation in both P and Q relying on the bijection α. We ensure that the set of deducible keys has been correctly guessed by adding of some extra processes. Then the main step of the proof consists in showing that the forwarder has now the same power as a full attacker, although he cannot reuse the same randomness in two distinct encryptions/signatures, as a real attacker could.

4.3   Encoding a Protocol into a Real-Time GPDA

For any process $P \in \mathcal{C}_{\mathsf{pp}}$, we can show that it is possible to define a polynomial-sized real-time GPDA $\mathcal{A}_P$ such that trace equivalence against forwarder of two processes coincides with language equivalence of the two corresponding automata.

Theorem 3. *Let P and Q in $\mathcal{C}_{\mathsf{pp}}$, we have that: $P \approx_{\mathsf{fwd}} Q \Longleftrightarrow \mathcal{L}(\mathcal{A}_P) = \mathcal{L}(\mathcal{A}_Q)$.*

The idea is that the automaton $\mathcal{A}_P$ associated to a protocol P recognizes the words (a sequence of channels) that correspond to a possible execution in P. The stack of $\mathcal{A}_P$ is used to store a (partial) representation of the last outputted term. This requires to convert a term into a word, and we use the following representation:

$$\bar{\mathsf{s}} = \mathsf{s} \text{ for any constant } \mathsf{s} \in \Sigma_0 \cup \{\mathsf{start}\}; \text{ and } \overline{f(v, \mathsf{k}, r)} = \bar{v}.\mathsf{k} \text{ otherwise.}$$

Note that, even if our signature is infinite, we show that only a finite number of constants of sort msg and a finite number of constants of sort channel need to be considered (namely those that occur in the protocols under study). Thus, the stack-alphabet and the input-alphabet of the automaton are both finite.

To construct the automaton associated to a process $P \in \mathcal{C}_{\mathsf{pp}}$, we need to construct an automaton that recognizes any execution of P and the corresponding valid tests. For the sake of illustration, we present only the automaton (depicted below) that recognizes tests of the form w $=$ w$'$ such that the corresponding term is actually a constant.

Intuitively, the basic building blocks (e.g. q_0 with the transitions from q_0 to itself) mimic an execution of P where each input is fed with the last outputted term. Then, to recognize the tests of the form w $=$ w$'$ that are true in such an execution, it is sufficient to memorize the constant s_i that is associated to w (adding a new state q_i), and to see whether it is possible to reach a state where the stack contains s_i again.

Capturing tests that lead to non-constant symbols (i.e. terms of the form $\mathsf{senc}(u, \mathsf{k}, r)$) is more tricky for several reasons. First, it is not possible anymore to memorize the resulting term in a state of the automaton. Second, names of sort rand play a role in such a test, while they are forgotten in our encoding. We therefore have to, first, characterize more precisely trace equivalence and secondly, construct more complex automata that use some special track symbols to encode when randomized ciphertexts may be reused.

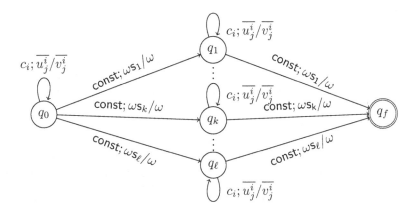

5   From Language Equivalence to Trace Equivalence

We have just seen how to encode equivalence of processes in C_{pp} into real-time GPDA. The equivalence of processes in C_{pp} is actually *equivalent* to language equivalence of real-time GPDA. Indeed, we can conversely encode any real-time GPDA into a process in C_{pp}, preserving equivalence. The transformation works as follows.

Given a word $u = \alpha_1 \ldots \ldots \alpha_p$, for sake of concision, the expression $x.u$ will denote either the term $\mathsf{senc}(\ldots \mathsf{senc}(x, \alpha_1, z_1), \ldots), \alpha_p, z_p)$ when it occurs as an input term; or $\mathsf{senc}(\ldots \mathsf{senc}(x, \alpha_1, r_1), \ldots), \alpha_p, r_p)$ when it occurs as an output term. Then given an automaton $\mathcal{A} = (Q, \Pi, \Gamma, q_0, \omega, Q_f, \delta)$, the corresponding process $P_{\mathcal{A}}$ is defined as follows:

$$
\begin{aligned}
P_{\mathcal{A}} \stackrel{\text{def}}{=} \; & !\, \mathsf{in}(c_0, \mathsf{start}).\mathsf{new}\; r.\mathsf{out}(c_0, \mathsf{senc}(\omega, q_0, r)) \\
& |\; !\, \mathsf{in}(c_a, \mathsf{senc}(x.u, q, z)).\mathsf{new}\; \tilde{r}.\mathsf{out}(c_a, \mathsf{senc}(x.v, q', r)) \\
& |\; !\, \mathsf{in}(c_f, \mathsf{senc}(x, q_f, z)).\mathsf{out}(c_f, \mathsf{start}) \\
& |\; P_{\mathcal{A}}'
\end{aligned}
$$

where a quantifies over Π, q over Q, u over words in Γ^* such that $(q, a, u) \in dom(\delta)$, q_f over Q_f, and $(q', v) = \delta(q, a, u)$.

Intuitively, the stack of the automata $\mathcal{A}$ is encoded as a pile of encryptions (where each key encodes a tile of the stack). Then, upon receiving a stack s encrypted by q on channel c_a, the process $P_{\mathcal{A}}$ mimics the transition of $\mathcal{A}$ at state q and stack s, upon reading a. The resulting stack is sent encrypted by the resulting state. This polynomial encoding (with some additional technical details hidden in $P_{\mathcal{A}}'$) preserves equivalence.

Proposition 2. *Let $\mathcal{A}$ and $\mathcal{B}$ be two real-time GPDA:* $\mathcal{L}(\mathcal{A}) \subseteq \mathcal{L}(\mathcal{B}) \iff P_{\mathcal{A}} \sqsubseteq P_{\mathcal{B}}$.

Therefore, checking for equivalence of protocols is as difficult as checking equivalence of real-time generalized pushdown deterministic automata. It follows that the exact complexity of checking equivalence of protocols is unknown. The only upper bound is that equivalence is at most primitive recursive. This bound comes from the algorithm proposed by C. Stirling for equivalence of DPA [16] (Icalp 2002). Whether equivalence of DPA (or even real-time GPDA) is e.g. at least NP-hard is unknown.

6   Conclusion

We have shown a first decidability result for equivalence of security protocols for an unbounded number of sessions by reducing it to the equality of languages of deterministic pushdown automata. We further show that deciding equivalence of security protocols is actually at least as hard as deciding equality of languages of deterministic, generalized, real-time pushdown automata.

Our class of security protocols handles only randomized primitives, namely symmetric/asymmetric encryptions and signatures. Our decidability result could be extended to handle deterministic primitives instead of the randomized one (the reverse encoding - from real-time GPDAs to processes with deterministic encryption - may not hold anymore). Due to the use of pushdown automata, extending our decidability result to protocols with pair is not straightforward. A direction is to use pushdown automata for which stacks are terms.

G. Sénizergues is currently implementing his procedure for pushdown automata [14]. As soon as the tool will be available, we plan to implement our translation, yielding a tool for automatically checking equivalence of security protocols, for an unbounded number of sessions.

References

1. Arapinis, M., Chothia, T., Ritter, E., Ryan, M.: Analysing unlinkability and anonymity using the applied pi calculus. In: 23rd Computer Security Foundations Symposium (CSF 2010), pp. 107–121. IEEE Computer Society Press (2010)
2. Basin, D., Mödersheim, S., Viganò, L.: A symbolic model checker for security protocols. Journal of Information Security 4(3), 181–208 (2005)
3. Baudet, M.: Deciding security of protocols against off-line guessing attacks. In: 12th ACM Conference on Computer and Communications Security (CCS 2005). ACM Press (2005)
4. Blanchet, B.: An efficient cryptographic protocol verifier based on prolog rules. In: 14th Computer Security Foundations Workshop (CSFW 2001). IEEE Computer Society Press (2001)
5. Blanchet, B., Abadi, M., Fournet, C.: Automated Verification of Selected Equivalences for Security Protocols. In: 20th Symposium on Logic in Computer Science (2005)
6. Bruso, M., Chatzikokolakis, K., den Hartog, J.: Formal verification of privacy for RFID systems. In: 23rd Computer Security Foundations Symposium, CSF 2010 (2010)
7. Cheval, V., Comon-Lundh, H., Delaune, S.: Trace equivalence decision: Negative tests and non-determinism. In: 18th ACM Conference on Computer and Communications Security (CCS 2011). ACM Press (2011)
8. Chevalier, Y., Rusinowitch, M.: Decidability of equivalence of symbolic derivations. J. Autom. Reasoning 48(2), 263–292 (2012)
9. Comon-Lundh, H., Cortier, V.: New decidability results for fragments of first-order logic and application to cryptographic protocols. In: Nieuwenhuis, R. (ed.) RTA 2003. LNCS, vol. 2706, pp. 148–164. Springer, Heidelberg (2003)
10. Cortier, V., Delaune, S.: A method for proving observational equivalence. In: 22nd IEEE Computer Security Foundations Symposium (CSF 2009). IEEE Computer Society Press (2009)
11. Cremers, C.: Unbounded verification, falsification, and characterization of security protocols by pattern refinement. In: 15th ACM Conference on Computer and Communications Security (CCS 2008). ACM (2008)

12. Friedman, E.P.: The inclusion problem for simple languages. Theor. Comput. Sci. 1(4), 297–316 (1976)

13. Rusinowitch, M., Turuani, M.: Protocol Insecurity with Finite Number of Sessions and Composed Keys is NP-complete. Theoretical Computer Science 299, 451–475 (2003)

14. Sénizergues, G.: The equivalence problem for deterministic pushdown automata is decidable. In: Degano, P., Gorrieri, R., Marchetti-Spaccamela, A. (eds.) ICALP 1997. LNCS, vol. 1256, pp. 671–681. Springer, Heidelberg (1997)

15. Sénizergues, G.: L(A)=L(B)? Decidability results from complete formal systems. Theor. Comput. Sci. 251(1-2), 1–166 (2001)

16. Stirling, C.: Deciding DPDA equivalence is primitive recursive. In: Widmayer, P., Triguero, F., Morales, R., Hennessy, M., Eidenbenz, S., Conejo, R. (eds.) ICALP 2002. LNCS, vol. 2380, pp. 821–832. Springer, Heidelberg (2002)

17. Tiu, A., Dawson, J.E.: Automating open bisimulation checking for the SPI calculus. In: 23rd IEEE Computer Security Foundations Symposium (CSF 2010), pp. 307–321 (2010)

Efficient Separability of Regular Languages by Subsequences and Suffixes

Wojciech Czerwiński, Wim Martens, and Tomáš Masopust

Institute for Computer Science, University of Bayreuth
wczerwin@mimuw.edu.pl, wim.martens@uni-bayreuth.de, masopust@math.cas.cz

Abstract. When can two regular word languages K and L be separated by a simple language? We investigate this question and consider separation by piecewise- and suffix-testable languages and variants thereof. We give characterizations of when two languages can be separated and present an overview of when these problems can be decided in polynomial time if K and L are given by nondeterministic automata.

1 Introduction

In this paper we are motivated by scenarios in which we want to describe something complex by means of a simple language. The technical core of our scenarios consists of *separation* problems, which are usually of the following form:

Given are two languages K and L. Does there exist a language S, coming from a family $\mathcal{F}$ of *simple* languages, such that S contains everything from K and nothing from L?

The family $\mathcal{F}$ of simple languages could be, for example, languages definable in FO, piecewise testable languages, or languages definable with small automata.

Our work is specifically motivated by two seemingly orthogonal problems coming from practice: (a) increasing the user-friendliness of XML Schema and (b) efficient approximate query answering. We explain these next.

Our first motivation comes from simplifying XML Schema. XML Schema is currently the only industrially accepted and widely supported schema language for XML. Historically, it is designed to alleviate the limited expressiveness of Document Type Definition (DTD) [6], thereby making DTDs obsolete. Unfortunately, XML Schema's extra expressiveness comes at the cost of simplicity. Its code is designed to be machine-readable rather than human-readable and its logical core, based on *complex types*, does not seem well-understood by users [16]. One reason may be that the specification of XML Schema's core [8] consists of over 100 pages of intricate text. The BonXai schema language [16,17] is an attempt to overcome these issues and to combine the simplicity of DTDs with the expressiveness of XML Schema. It has exactly the same expressive power as XML Schema, is designed to be human-readable, and avoids the use of complex types. Therefore, it aims at simplifying the development or analysis of XSDs. In its core, a BonXai schema is a set of rules $L_1 \rightarrow R_1, \ldots, L_n \rightarrow R_n$ in which all L_i and R_i are regular expressions. An unranked tree t (basically, an XML

F.V. Fomin et al. (Eds.): ICALP 2013, Part II, LNCS 7966, pp. 150–161, 2013.
© Springer-Verlag Berlin Heidelberg 2013

document) is in the language of the schema if, for every node u, the word formed by the labels of u's children is in the language R_k, where k is the largest number such that the word of ancestors of u is in L_k. This semantical definition is designed to ensure full back-and-forth compatibility with XML Schema [16].

When translating an XML Schema Definition (XSD) into an equivalent BonXai schema, the regular expressions L_i are obtained from a finite automaton that is embedded in the XSD. Since the current state-of-the-art in translating automata to expressions does not yet generate human-readable results, we are investigating simpler classes of expressions which we expect to suffice in practice. Practical and theoretical studies show evidence that regular expressions of the form $\Sigma^* w$ (with $w \in \Sigma^+$) and $\Sigma^* a_1 \Sigma^* \cdots \Sigma^* a_n$ (with $a_1, \ldots, a_n \in \Sigma$) and variations thereof seem to be quite well-suited [9,13,18]. We study these kinds of expressions in this paper.

Our second motivation comes from efficient approximate query answering. Efficiently evaluating regular expressions is relevant in a very wide array of fields. We choose one: in graph databases and in the context of the SPARQL language [5,10,14,19] for querying RDF data. Typically, regular expressions are used in this context to match paths between nodes in a huge graph. In fact, the data can be so huge that exact evaluation of a regular expression r over the graph (which can lead to a product construction between an automaton for the expression and the graph [14,19]) may not be feasible within reasonable time. Therefore, as a compromise to exact evaluation, one could imagine that we try to rewrite the regular expression r as an expression that we can evaluate much more efficiently and is close enough to r. Concretely, we could specify two expressions r_{pos} (resp., r_{neg}) that define the language we want to (resp., do not want to) match in our answer and ask whether there exists a simple query (e.g., defining a piecewise testable language) that satisfies these constraints. Notice that the scenario of approximating an expression r in this way is very general and not even limited to databases. (Also, we can take r_{neg} to be the complement of r_{pos}.)

At first sight, these two motivating scenarios may seem to be fundamentally different. In the first, we want to compute an *exact* simple description of a complex object and in the second one we want to compute an *approximate* simple query that can be evaluated more efficiently. However, both scenarios boil down to the same underlying question of language separation. Our contributions are:
(1) We formally define separation problems that closely correspond to the motivating scenarios. Query approximation will be abstracted as *separation* and schema simplification as *layer-separation* (Section 2.1).
(2) We prove the equivalece of separability of languages K and L by boolean combinations of simple languages, layer-separability, and the existence of an infinite sequence of words that goes back and forth between K and L. This characterization shows how the exact and approximate scenario are related and does not require K and L to be regular (Sec. 3). Our characterization generalizes a result by Stern [23] that says that a regular language L is piecewise testable iff every increasing infinite sequence of words (w.r.t. subsequence ordering) alternates finitely many times between L and its complement.

(3) In Section 4 we prove a decomposition characterization for separability of regular languages by piecewise testable languages and we give an algorithm that decides separability. The decomposition characterization is in the spirit of an algebraic result by Almeida [1]. It is possible to prove our characterization using Almeida's result but we provide a self-contained, elementary proof which can be understood without a background in algebra. We then use this characterization to distill a polynomial time decision procedure for separability of languages of NFAs (or regular expressions) by piecewise testable languages. The state-of-the-art algorithm for separability by piecewise testable languages ([2,4]) runs in time $O(\text{poly}(|Q|) \cdot 2^{|\Sigma|})$ when given DFAs for the regular languages, where $|Q|$ is the number of states in the DFAs and $|\Sigma|$ is the alphabet size. Our algorithm runs in time $O(\text{poly}(|Q| + |\Sigma|))$ even for NFAs. Notice that $|\Sigma|$ can be large (several hundreds and more) in the scenarios that motivate us, so we believe the improvement with respect to the alphabet to be relevant in practice.

(4) Whereas Section 4 focuses exclusively on separation by piecewise testable languages, we broaden our scope in Section 5. Let's say that a *subsequence language* is a language of the form $\Sigma^* a_1 \Sigma^* \cdots \Sigma^* a_n \Sigma^*$ (with all $a_i \in \Sigma$). Similarly, a *suffix language* is of the form $\Sigma^* a_1 \cdots a_n$. We present an overview of the complexities of deciding whether regular languages can be separated by subsequence languages, suffix languages, finite unions thereof, or boolean combinations thereof. We prove all cases to be in polynomial time, except separability by a single subsequence language which is NP-complete. By combining this with the results from Section 3 we also have that layer-separability is in polynomial time for all languages we consider.

We now discuss further related work. There is a large body of related work that has not been mentioned yet. Piecewise testable languages are defined and studied by Simon [20,21], who showed that a regular language is piecewise testable iff its syntactic monoid is J-trivial and iff both the minimal DFA for the language and the minimal DFA for the reversal are partially ordered. Stern [24] suggested an $O(n^5)$ algorithm in the size of a DFA to decide whether a regular language is piecewise testable. This was improved to quadratic time by Trahtman [25]. (Actually, from our proof, it now follows that this question can be decided in polynomial time if an NFA and its complement NFA are given.)

Almeida [2] established a connection between a number of separation problems and properties of families of monoids called pseudovarieties. Almeida shows, e.g., that deciding whether two given regular languages can be separated by a language with its syntactic monoid lying in pseudovariety V is algorithmically equivalent to computing two-pointlike sets for a monoid in pseudovariety V. It is then shown by Almeida et al. [3] how to compute these two-pointlike sets in the pseudovariety J corresponding to piecewise testable languages. Henckell et al. [11] and Steinberg [22] show that the two-pointlike sets can be computed for pseudovarieties corresponding to languages definable in first order logic and languages of dot depth at most one, respectively. By Almeida's result [2] this implies that the separation problem is also decidable for these classes.

2   Preliminaries and Definitions

For a finite set S, we denote its cardinality by $|S|$. By Σ we always denote an alphabet, that is, a finite set of symbols. A $(\Sigma\text{-})word$ w is a finite sequence of symbols $a_1 \cdots a_n$, where $n \geq 0$ and $a_i \in \Sigma$ for all $i = 1, \ldots, n$. The *length of* w, denoted by $|w|$, is n and the *alphabet of* w, denoted by $\mathtt{Alph}(w)$, is the set $\{a_1, \ldots, a_n\}$ of symbols occurring in w. The empty word is denoted by ε. The set of all Σ-words is denoted by Σ^*. A *language* is a set of words. For $v = a_1 \cdots a_n$ and $w \in \Sigma^* a_1 \Sigma^* \cdots \Sigma^* a_n \Sigma^*$, we say that v is a *subsequence* of w, denoted by $v \preceq w$.

A *(nondeterministic) finite automaton* or *NFA* $\mathcal{A}$ is a tuple $(Q, \Sigma, \delta, q_0, F)$, where Q is a finite set of states, $\delta : Q \times \Sigma \to 2^Q$ is the transition function, $q_0 \in Q$ is the initial state, and $F \subseteq Q$ is the set of accepting states. We sometimes denote that $q_2 \in \delta(q_1, a)$ as $q_1 \xrightarrow{a} q_2 \in \delta$ to emphasize that $\mathcal{A}$ being in state q_1 can go to state q_2 reading an $a \in \Sigma$. A *run of* $\mathcal{A}$ *on word* $w = a_1 \cdots a_n$ is a sequence of states $q_0 \cdots q_n$ where, for each $i = 1, \ldots, n$, we have $q_{i-1} \xrightarrow{a_i} q_i \in \delta$. The run is *accepting* if $q_n \in F$. Word w is *accepted* by $\mathcal{A}$ if there is an accepting run of $\mathcal{A}$ on w. The *language of* $\mathcal{A}$, denoted by $L(\mathcal{A})$, is the set of all words accepted by $\mathcal{A}$. By δ^* we denote the extension of δ to words, that is, $\delta^*(q, w)$ is the set of states that can be reached from q by reading w. The *size* $|\mathcal{A}| = |Q| + \sum_{q,a} |\delta(q, a)|$ of $\mathcal{A}$ is the total number of transitions and states. An NFA is *deterministic* (a *DFA*) when every $\delta(q, a)$ consists of at most one element.

The *regular expressions (RE)* over Σ are defined as follows: ε and every Σ-symbol is a regular expression; whenever r and s are regular expressions, then so are $(r \cdot s)$, $(r + s)$, and $(s)^*$. In addition, we allow $\emptyset$ as a regular expression, but we assume that $\emptyset$ does not occur in any other regular expression. For readability, we usually omit concatenation operators and parentheses in examples. We sometimes abbreviate an n-fold concatenation of r by r^n. The *language* defined by an RE r is denoted by $L(r)$ and is defined as usual. Often we simply write r instead of $L(r)$. Whenever we say that expressions or automata are *equivalent*, we mean that they define the same language. The *size* $|r|$ of r is the total number of occurrences of alphabet symbols, epsilons, and operators in r, i.e., the number of nodes in its parse tree. A regular expression is *union-free* if it does not contain the operator $+$. A language is *union-free* if it is defined by a union-free regular expression.

A quasi-order is a reflexive and transitive relation. For a quasi-order $\preccurlyeq$, the *(upward)* $\preccurlyeq$-*closure* of a language L is the set $\text{closure}^{\preccurlyeq}(L) = \{w \mid v \preccurlyeq w \text{ for some } v \in L\}$. We denote the $\preccurlyeq$-closure of a word w as $\text{closure}^{\preccurlyeq}(w)$ instead of $\text{closure}^{\preccurlyeq}(\{w\})$. Language L is *(upward)* $\preccurlyeq$-*closed* if $L = \text{closure}^{\preccurlyeq}(L)$.

A quasi-order $\preccurlyeq$ on a set X is a *well-quasi-ordering* (a *WQO*) if for every infinite sequence $(x_i)_{i=1}^{\infty}$ of elements of X there exist indices $i < j$ such that $x_i \preccurlyeq x_j$. It is known that every WQO is also *well-founded*, that is, there exist no infinite descending sequences $x_1 \succcurlyeq x_2 \succcurlyeq \cdots$ such that $x_i \not\preccurlyeq x_{i+1}$ for all i.

Higman's Lemma [12] (which we use multiple times) states that, for every alphabet Σ, the subsequence relation $\preceq$ is a WQO on Σ^*. Notice that, as a corollary to Higman's Lemma, every $\preceq$-closed language is a finite union of languages of the form $\Sigma^* a_1 \Sigma^* \ldots \Sigma^* a_n \Sigma^*$ which means that it is also regular, see also [7]. A language is *piecewise testable* if it is a finite boolean combination of $\preceq$-closed languages (or, finite boolean combination of languages $\Sigma^* a_1 \Sigma^* \cdots \Sigma^* a_n \Sigma^*$). In this paper, all boolean combinations are finite.

2.1   Separability of Languages

A language S *separates language K from L* if S contains K and does not intersect L. We say that S *separates K and L* if it either separates K from L or L from K. Let $\mathcal{F}$ be a family of languages. Languages K and L are *separable by $\mathcal{F}$* if there exists a language S in $\mathcal{F}$ that separates K and L. Languages K and L are *layer-separable by $\mathcal{F}$* if there exists a finite sequence of languages $S_1, \ldots, S_m$ in $\mathcal{F}$ such that

1. for all $1 \le i \le m$, language $S_i \setminus \bigcup_{j=1}^{i-1} S_j$ intersects at most one of K and L;
2. K or L (possibly both) is included in $\bigcup_{j=1}^{m} S_j$.

Notice that separability always implies layer-separability. However, the opposite implication does not hold, as we demonstrate next.

Example 1. *Let $\mathcal{F} = \{a^n a^* \mid n \ge 0\}$ be a family of $\preceq$-closed languages over $\Sigma = \{a\}$, $K = \{a, a^3\}$, and $L = \{a^2, a^4\}$. We first show that languages K and L are not separable by $\mathcal{F}$. Indeed, assume that $S \in \mathcal{F}$ separates K and L. If K is included in S, then $aa^* \subseteq S$, hence L and S are not disjoint. Conversely, if $L \subseteq S$, then $a^2 a^* \subseteq S$ and therefore S and K are not disjoint. This contradicts that S separates K and L. Now we show that the languages are layer-separable by $\mathcal{F}$. Consider languages $S_1 = a^4 a^*$, $S_2 = a^3 a^*$, $S_3 = a^2 a^*$, and $S_4 = aa^*$. Then both K and L are included in S_4, and S_1 intersects only L, $S_2 \setminus S_1 = a^3$ intersects only K, $S_3 \setminus (S_1 \cup S_2) = a^2$ intersects only L, and $S_4 \setminus (S_1 \cup S_2 \cup S_3) = a$ intersects only K; see Fig. 1.*

Example 1 illustrates some intuition behind layered separability. Our motivation for layered separability comes from the BonXai schema language which is discussed in the introduction. We need to solve layer-separability if we want to decide whether an XML Schema has an equivalent BonXai schema with simple regular expressions (defining lan-

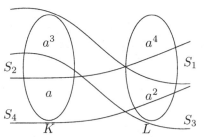

Fig. 1. An example of a layer-separation

guages in $\mathcal{F}$). Layered separability implies that languages are, in a sense, separable by languages from $\mathcal{F}$ in a priority-based system: If we consider the ordered sequence of languages S_1, S_2, S_3, S_4 then, in order to classify a word $w \in K \cup L$ in either K or L, we have to match it against the S_i in increasing order of the index i. If we know the lowest index j for which $w \in S_j$, we know whether $w \in K$ or $w \in L$.

We now define a tool (similar to and slightly more general than the *alternating towers* of Stern [23]) that allows us to determine when languages are *not* separable. For languages K and L and a quasi-order $\preccurlyeq$, we say that a sequence $(w_i)_{i=1}^k$ of words is a $\preccurlyeq$-*zigzag between* K *and* L if $w_1 \in K \cup L$ and, for all $i = 1, \ldots, k-1$:

(1) $w_i \preccurlyeq w_{i+1}$; (2) $w_i \in K$ implies $w_{i+1} \in L$; and (3) $w_i \in L$ implies $w_{i+1} \in K$.

We say that k is the *length* of the $\preccurlyeq$-zigzag. We similarly define an infinite sequence of words to be an *infinite $\preccurlyeq$-zigzag between* K *and* L. If the languages K and L are clear from the context then we sometimes omit them and refer to the sequence as a *(infinite) $\preccurlyeq$-zigzag*. If we consider the subsequence order $\preceq$, then we simply write a *zigzag* instead of a $\preceq$-zigzag. Notice that we do not require K and L to be disjoint. If there is a $w \in K \cap L$ then there clearly exists an infinite zigzag: $w, w, w, \ldots$

Example 2. *In order to illustrate infinite zigzags consider the languages* $K = \{a(ab)^{2k}c(ac)^{2\ell} \mid k, \ell \geq 0\}$ *and* $L = \{b(ab)^{2k+1}c(ac)^{2\ell+1} \mid k, \ell \geq 0\}$. *Then the following infinite sequence is an infinite zigzag between* K *and* L:

$$w_i = \begin{cases} b(ab)^i c(ac)^i & \text{if } i \text{ is odd} \\ a(ab)^i c(ac)^i & \text{if } i \text{ is even} \end{cases}$$

Indeed $w_1 \in L$, *words from the sequence alternately belong to* K *and* L, *and for all* $i \geq 1$ *we have* $w_i \preceq w_{i+1}$. □

3   A Characterization of Separability

The aim of this section is to prove the following theorem. It extends a result by Stern that characterizes piecewise testable languages [23]. In particular, it also applies to non-regular languages and does not require K to be the complement of L.

Theorem 3. *For languages* K *and* L *and a WQO* $\preccurlyeq$ *on words, the following are equivalent.*

(1) K and L are separable by a boolean combination of $\preccurlyeq$-closed languages.
(2) K and L are layer-separable by $\preccurlyeq$-closed languages.
(3) There does not exist an infinite $\preccurlyeq$-zigzag between K and L.

Some of the equivalences in the theorem still hold when the assumptions are weakened. For example the equivalence between (1) and (2) does not require $\preccurlyeq$ to be a WQO.

Since the subsequence order $\preceq$ is a WQO on words, we know from Theorem 3 that languages are separable by piecewise testable languages if and only if they are layer-separable by $\preceq$-closed languages. Actually, since $\preceq$ is a WQO (and therefore only has finitely many minimal elements within a language), the latter is equivalent to being layer-separable by languages of the form $\Sigma^* a_1 \Sigma^* \cdots \Sigma^* a_n \Sigma^*$.

In Example 1 we illustrated two languages K and L that are layer-separable by $\preceq$-closed languages. Notice that K and L can also be separated by a boolean combination of the languages a^*a^1, a^*a^2, a^*a^3, and a^*a^4 from $\mathcal{F}$, as $K \subseteq ((a^*a^1 \setminus a^*a^2) \cup (a^*a^3 \setminus a^*a^4))$ and $L \cap ((a^*a^1 \setminus a^*a^2) \cup (a^*a^3 \setminus a^*a^4)) = \emptyset$.

We now give an overview of the proof of Theorem 3. The next lemma proves the equivalence between (1) and (2), but is slightly more general. In particular, it does not rely on a WQO.

Lemma 4. *Let $\mathcal{F}$ be a family of languages closed under intersection and containing Σ^*. Then languages K and L are separable by a finite boolean combination of languages from $\mathcal{F}$ if and only if K and L are layer-separable by $\mathcal{F}$.*

The proof is constructive. The *only if* direction is the more complex one and shows how to exploit the implicit negation in the first condition in the definition of layer-separability in order to simulate separation by boolean combinations. Notice that the families of $\preceq$-closed languages in Theorem 3 always contain Σ^* and are closed under intersection.

The following lemma shows that the implication (2) $\Rightarrow$ (3) in Theorem 3 does not require well-quasi ordering.

Lemma 5. *Let $\preceq$ be a quasi order on words and assume that languages K and L are layer-separable by $\preceq$-closed languages. Then there is no infinite $\preceq$-zigzag between K and L.*

To prove that (3) implies (2), we need the following technical lemma in which we require $\preceq$ to be a WQO. In the proof of the lemma, we argue how we can see $\preceq$-zigzags in a tree each of whose nodes is labeled by a word. If a node labelled w_1 is the parent of a node labeled w_2, then we have $w_1 \preceq w_2$. Intuitively, every path in the tree structure corresponds to a $\preceq$-zigzag. We need the fact that $\preceq$ is a WQO in order to show that we can assume that every node in this tree structure has a finite number of children. We then apply König's lemma to show that arbitrarily long $\preceq$-zigzags imply the existence of an infinite $\preceq$-zigzag. The lemma then follows by contraposition.

Lemma 6. *Let $\preceq$ be a WQO on words. If there is no infinite $\preceq$-zigzag between languages K and L, then there exists a constant $k \in \mathbb{N}$ such that no $\preceq$-zigzag between K and L is longer than k.*

If there is no infinite $\preceq$-zigzag, then we can put a bound on the maximal length of zigzags by Lemma 6. This bound actually has a close correspondence to the number of "layers" we need to separate K and L.

Lemma 7. *Let $\preceq$ be a WQO on words and assume that there is no infinite $\preceq$-zigzag between languages K and L. Then the languages K and L are layer-separable by $\preceq$-closed languages.*

4   Testing Separability by Piecewise Testable Languages

Whereas Section 3 proves a result for general WQOs, we focus in this section exclusively on the ordering $\preceq$ of subsequences. Therefore, if we say *zigzag* in this section, we always mean $\preceq$-*zigzag*. We show here how to decide the existence of an infinite zigzag between two regular word languages, given by their regular expressions or NFAs, in polynomial time. According to Theorem 3, this is equivalent to deciding if the two languages can be separated by a piecewise testable language.

To this end, we first prove a decomposition result that is reminiscent of a result of Almeida ([1], Theorem 4.1 in [3]). We show that, if there is an infinite zigzag between regular languages, then there is an infinite zigzag of a special form in which every word can be decomposed in some synchronized manner. We can find these special forms of zigzags in polynomial time in the NFAs for the languages. The main features are that our algorithm runs exponentially faster in the alphabet size than the current state-of-the-art [4] and that our algorithm and its proof of correctness do not require knowledge of the algebraic perspective on regular languages.

A regular language is a *cycle language* if it is of the form $u(v)^*w$, where u, v, w are words and $(\mathtt{Alph}(u) \cup \mathtt{Alph}(w)) \subseteq \mathtt{Alph}(v)$. We say that v is a *cycle* of the language and that $\mathtt{Alph}(v)$ is its *cycle alphabet*. Regular languages $L^{\mathcal{A}}$ and $L^{\mathcal{B}}$ are *synchronized in one step* if they are of one of the following forms:

- $L^{\mathcal{A}} = L^{\mathcal{B}} = \{w\}$, that is, they are the same singleton word, or
- $L^{\mathcal{A}}$ and $L^{\mathcal{B}}$ are cycle languages with equal cycle alphabets.

We say that regular languages $L^{\mathcal{A}}$ and $L^{\mathcal{B}}$ are *synchronized* if they are of the form $L^{\mathcal{A}} = D_1^{\mathcal{A}} D_2^{\mathcal{A}} \dots D_k^{\mathcal{A}}$ and $L^{\mathcal{B}} = D_1^{\mathcal{B}} D_2^{\mathcal{B}} \dots D_k^{\mathcal{B}}$ where, for all $1 \leq i \leq k$, languages $D_i^{\mathcal{A}}$ and $D_i^{\mathcal{B}}$ are synchronized in one step. So, languages are synchronized if they can be decomposed into (equally many) components that can be synchronized in one step. Notice that synchronized languages are always non-empty.

Example 8. *Languages* $L^{\mathcal{A}} = a(ba)^* aab\, ca\, bb(bc)^*$ *and* $L^{\mathcal{B}} = b(aab)^* ba\, ca\, cc(cbc)^* b$ *are synchronized. Indeed,* $L^{\mathcal{A}} = D_1^{\mathcal{A}} D_2^{\mathcal{A}} D_3^{\mathcal{A}}$ *and* $L^{\mathcal{B}} = D_1^{\mathcal{B}} D_2^{\mathcal{B}} D_3^{\mathcal{B}}$ *for* $D_1^{\mathcal{A}} = a(ba)^* aab$, $D_2^{\mathcal{A}} = ca$, $D_3^{\mathcal{A}} = bb(cb)^*$ *and* $D_1^{\mathcal{B}} = b(aab)^* ba$, $D_2^{\mathcal{B}} = ca$, *and* $D_3^{\mathcal{B}} = cc(cbc)^* b$.

The next lemma shows that, in order to search for infinite zigzags, it suffices to search for synchronized sublanguages. The proof goes through a sequence of lemmas that gradually shows how the sublanguages of $L^{\mathcal{A}}$ and $L^{\mathcal{B}}$ can be made more and more specific.

Lemma 9 (Synchronization / Decomposition). *There is an infinite zigzag between regular languages* $L^{\mathcal{A}}$ *and* $L^{\mathcal{B}}$ *if and only if there exist synchronized languages* $K^{\mathcal{A}} \subseteq L^{\mathcal{A}}$ *and* $K^{\mathcal{B}} \subseteq L^{\mathcal{B}}$.

We now use this result to obtain a polynomial-time algorithm solving our problem. The first step is to define what it means for NFAs to contain synchronized sublanguages.

For an NFA $\mathcal{A}$ over an alphabet Σ, two states p, q, and a word $w \in \Sigma^*$, we write $p \xrightarrow{w} q$ if $q \in \delta^*(p, w)$ or, in other words, the automaton can go from state p to state q by reading w. For $\Sigma_0 \subseteq \Sigma$, states p and q are Σ_0-*connected* in $\mathcal{A}$ if there exists a word $uvw \in \Sigma_0^*$ such that:

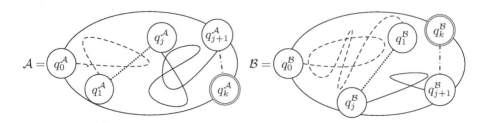

Fig. 2. Synchronization of automata $\mathcal{A}$ and $\mathcal{B}$

1. $\mathtt{Alph}(v) = \Sigma_0$ and
2. there is a state m such that $p \xrightarrow{u} m$, $m \xrightarrow{v} m$, and $m \xrightarrow{w} q$.

Consider two NFAs $\mathcal{A} = (Q^{\mathcal{A}}, \Sigma, \delta^{\mathcal{A}}, q_0^{\mathcal{A}}, F^{\mathcal{A}})$ and $\mathcal{B} = (Q^{\mathcal{B}}, \Sigma, \delta^{\mathcal{B}}, q_0^{\mathcal{B}}, F^{\mathcal{B}})$. Let $(q^{\mathcal{A}}, q^{\mathcal{B}})$ and $(\bar{q}^{\mathcal{A}}, \bar{q}^{\mathcal{B}})$ be in $Q^{\mathcal{A}} \times Q^{\mathcal{B}}$. We say that $(q^{\mathcal{A}}, q^{\mathcal{B}})$ and $(\bar{q}^{\mathcal{A}}, \bar{q}^{\mathcal{B}})$ are *synchronizable in one step* if one of the following situations occurs:

– there exists a symbol a in Σ such that $q^{\mathcal{A}} \xrightarrow{a} \bar{q}^{\mathcal{A}}$ and $q^{\mathcal{B}} \xrightarrow{a} \bar{q}^{\mathcal{B}}$,
– there exists an alphabet $\Sigma_0 \subseteq \Sigma$ such that $q^{\mathcal{A}}$ and $\bar{q}^{\mathcal{A}}$ are Σ_0-connected in $\mathcal{A}$ and $q^{\mathcal{B}}$ and $\bar{q}^{\mathcal{B}}$ are Σ_0-connected in $\mathcal{B}$.

We say that automata $\mathcal{A}$ and $\mathcal{B}$ are *synchronizable* if there exists a sequence of pairs $(q_0^{\mathcal{A}}, q_0^{\mathcal{B}}), \ldots, (q_k^{\mathcal{A}}, q_k^{\mathcal{B}}) \in Q^{\mathcal{A}} \times Q^{\mathcal{B}}$ such that:

1. for all $0 \le i < k$, $(q_i^{\mathcal{A}}, q_i^{\mathcal{B}})$ and $(q_{i+1}^{\mathcal{A}}, q_{i+1}^{\mathcal{B}})$ are synchronizable in one step;
2. states $q_0^{\mathcal{A}}$ and $q_0^{\mathcal{B}}$ are initial states of $\mathcal{A}$ and $\mathcal{B}$, respectively; and
3. states $q_k^{\mathcal{A}}$ and $q_k^{\mathcal{B}}$ are accepting states of $\mathcal{A}$ and $\mathcal{B}$, respectively.

Notice that if the automata $\mathcal{A}$ and $\mathcal{B}$ are synchronizable, then the languages $L(\mathcal{A})$ and $L(\mathcal{B})$ are not necessarily synchronized, only some of its sublanguages are necessarily synchronized.

Lemma 10 (Synchronizability of automata). *For two NFAs $\mathcal{A}$ and $\mathcal{B}$, the following conditions are equivalent.*

1. *Automata $\mathcal{A}$ and $\mathcal{B}$ are synchronizable.*
2. *There exist synchronized languages $K^{\mathcal{A}} \subseteq L(\mathcal{A})$ and $K^{\mathcal{B}} \subseteq L(\mathcal{B})$.*

The intuition behind Lemma 10 is depicted in Figure 2. The idea is that there is a sequence $(q_0^{\mathcal{A}}, q_0^{\mathcal{B}}), \ldots, (q_k^{\mathcal{A}}, q_k^{\mathcal{B}})$ that witnesses that $\mathcal{A}$ and $\mathcal{B}$ are synchronizable. The pairs of paths that have the same style of lines depict parts of the automaton that are synchronizable in one step. In particular, the dotted path from $q_1^{\mathcal{A}}$ to $q_j^{\mathcal{A}}$ has the same word as the one from $q_1^{\mathcal{B}}$ to $q_j^{\mathcal{B}}$. The other two paths contain at least one loop.

The following theorem states that synchronizability in automata captures exactly the existence of infinite zigzags between their languages. The theorem statement uses Theorem 3 for the connection between infinite zigzags and separability.

Theorem 11. *Let $\mathcal{A}$ and $\mathcal{B}$ be two NFAs. Then the languages $L(\mathcal{A})$ and $L(\mathcal{B})$ are separable by a piecewise testable language if and only if the automata $\mathcal{A}$ and $\mathcal{B}$ are not synchronizable.*

We can now show how the algorithm from [4] can be improved to test in polynomial time whether two given NFAs are synchronizable or not. Our algorithm computes quadruples of states that are synchronizable in one step and links such quadruples together so that they form a pair of paths as illustrated in Figure 2.

Theorem 12. *Given two NFAs $\mathcal{A}$ and $\mathcal{B}$, it is possible to test in polynomial time whether $L(\mathcal{A})$ and $L(\mathcal{B})$ can be separated by a piecewise testable language.*

5   Asymmetric Separation and Suffix Order

We present a bigger picture on efficient separations that are relevant to the scenarios that motivate us. For example, we consider what happens when we restrict the allowed boolean combinations of languages. Technically, this means that separation is no longer symmetric. Orthogonally, we also consider the suffix order $\preceq_s$ between words in which $v \preceq_s w$ if and only if v is a (not necessarily strict) suffix of w. An important technical difference with the rest of the paper is that the suffix order is not a WQO. Indeed, the suffix order $\preceq_s$ has an infinite antichain, e.g., $a, ab, abb, abbb, \ldots$ The results we present here for suffix order hold true for prefix order as well.

Let $\mathcal{F}$ be a family of languages. Language K is *separable from a language L by $\mathcal{F}$* if there exists a language S in $\mathcal{F}$ that separates K from L, i.e., contains K and does not intersect L. Thus, if $\mathcal{F}$ is closed under complement, then K is separable from L implies L is separable from K. The *separation problem by $\mathcal{F}$* asks, given an NFA for K and an NFA for L, whether K is separable from L by $\mathcal{F}$.

We consider separation by families of languages $\mathcal{F}(O, C)$, where O ("order") specifies the ordering relation and C ("combinations") specifies how we are allowed to combine (upward) O-closed languages. Concretely, O is either the subsequence order $\preceq$ or the suffix order $\preceq_s$. We allow C to be one of *single, unions*, or *bc* (boolean combinations), meaning that each language in $\mathcal{F}(O, C)$ is either the O-closure of a single word, a finite union of the O-closures of single words, or a finite boolean combination of the O-closures of single words. Thus, $\mathcal{F}(\preceq, bc)$ is the family of piecewise testable languages and $\mathcal{F}(\preceq_s, bc)$ is the family of suffix-testable languages. With this convention in mind, the main result of this section is to provide a complete complexity overview of the six possible cases of separation by $\mathcal{F}(O, C)$. The case $\mathcal{F}(\preceq, bc)$ has already been proved in Section 4.

Theorem 13. *For $O \in \{\preceq, \preceq_s\}$ and C being one of single, unions, or boolean combinations, we have that the complexity of the separation problem by $\mathcal{F}(O, C)$ is as indicated in Table 1.*

Since the separation problem for prefix order is basically the same as the separation for suffix order and has the same complexity we didn't list it separately in the table.

Table 1. The complexity of deciding separability for regular languages K and L

$\mathcal{F}(O, C)$	single	unions	bc (boolean combinations)
$\preceq$ (subsequence)	NP-complete	PTIME	PTIME
$\preceq_s$ (suffix)	PTIME	PTIME	PTIME

6   Conclusions and Further Questions

Subsequence and suffix languages seem to be promising for obtaining "simple" separations of regular languages, since we can often efficiently decide if two given regular languages are separable (Table 1). Layer-separability is even in PTIME in all cases (since it has the same complexity. Looking back at our motivating scenarios, the obvious next questions are: if a separation exists, can we efficiently compute one? How large is it?

If we look at the broader picture, we are interested in how characterization of separability can be used in a wider context than regular languages and subsequence ordering. Another concrete question is whether we can decide in polynomial time if a given NFA defines a piecewise-testable language. Furthermore, we are also interested in efficient separation results by combinations of languages of the form $\Sigma^* w_1 \Sigma^* \cdots \Sigma^* w_n$ or variants thereof.

We discovered that Theorem 12 and a characterization similar to Theorem 11 also have been obtained in [26], which was submitted to ArXiv 3 weeks after the ICALP deadline.

Acknowledgments. We thank Jean-Eric Pin and Marc Zeitoun for patiently answering our questions about the algebraic perspective on this problem. We are grateful to Mikołaj Bojańczyk, who pointed out the connection between layered separability and boolean combinations. We also thank Piotr Hofman for pleasant and insightful discussions about our proofs during his visit to Bayreuth. This work was supported by DFG grant MA 4938/2-1.

References

1. Almeida, J.: Implicit operations on finite J-trivial semigroups and a conjecture of I. Simon. Journal of Pure and Applied Algebra 69, 205–218 (1990)
2. Almeida, J.: Some algorithmic problems for pseudovarieties. Publicationes Mathematicae Debrecen 54, 531–552 (1999)
3. Almeida, J., Costa, J.C., Zeitoun, M.: Pointlike sets with respect to R and J. Journal of Pure and Applied Algebra 212(3), 486–499 (2008)
4. Almeida, J., Zeitoun, M.: The pseudovariety J is hyperdecidable. RAIRO Informatique Théorique et Applications 31(5), 457–482 (1997)
5. Arenas, M., Conca, S., Pérez, J.: Counting beyond a yottabyte, or how SPARQL 1.1 property paths will prevent the adoption of the standard. In: World Wide Web Conference, pp. 629–638 (2012)

6. Bray, T., Paoli, J., Sperberg-McQueen, C.M., Maler, E., Yergeau, F.: Extensible Markup Language XML 1.0, 5th edn. Tech. report, W3C Recommendation (November 2008), http://www.w3.org/TR/2008/REC-xml-20081126/

7. Ehrenfeucht, A., Haussler, D., Rozenberg, G.: On regularity of context-free languages. Theoretical Computer Science 27(3), 311–332 (1983)

8. Gao, S., Sperberg-McQueen, C.M., Thompson, H.S., Mendelsohn, N., Beech, D., Maloney, M.: W3C XML Schema Definition Language (XSD) 1.1 part 1. Tech. report, W3C (2009), http://www.w3.org/TR/2009/CR-xmlschema11-1-20090430/

9. Gelade, W., Neven, F.: Succinctness of pattern-based schema languages for XML. Journal of Computer and System Sciences 77(3), 505–519 (2011)

10. Harris, S., Seaborne, A.: SPARQL 1.1 query language. Tech. report, W3C (2010)

11. Henckell, K., Rhodes, J., Steinberg, B.: Aperiodic pointlikes and beyond. International Journal of Algebra and Computation 20(2), 287–305 (2010)

12. Higman, G.: Ordering by divisibility in abstract algebras. Proceedings of the London Mathematical Society s3-2(1), 326–336 (1952)

13. Kasneci, G., Schwentick, T.: The complexity of reasoning about pattern-based XML schemas. In: Principles of Database Systems, pp. 155–164 (2007)

14. Losemann, K., Martens, W.: The complexity of evaluating path expressions in SPARQL. In: Principles of Database Systems, pp. 101–112 (2012)

15. Maier, D.: The complexity of some problems on subsequences and supersequences. Journal of the ACM 25(2), 322–336 (1978)

16. Martens, W., Neven, F., Niewerth, M., Schwentick, T.: Developing and analyzing XSDs through BonXai. Proc. of the VLDB Endowment 5(12), 1994–1997 (2012)

17. Martens, W., Neven, F., Niewerth, M., Schwentick, T.: BonXai: Combining the simplicity of DTD with the expressiveness of XML Schema (manuscript 2013)

18. Martens, W., Neven, F., Schwentick, T., Bex, G.J.: Expressiveness and complexity of XML Schema. ACM Trans. on Database Systems 31(3), 770–813 (2006)

19. Pérez, J., Arenas, M., Gutierrez, C.: nSPARQL: A navigational language for RDF. Journal of Web Semantics 8(4), 255–270 (2010)

20. Simon, I.: Hierarchies of Events with Dot-Depth One. PhD thesis, Dep. of Applied Analysis and Computer Science, University of Waterloo, Canada (1972)

21. Simon, I.: Piecewise testable events. In: Brakhage, H. (ed.) GI Conference on Automata Theory and Formal Languages. LNCS, vol. 33, pp. 214–222. Springer, Heidelberg (1975)

22. Steinberg, B.: A delay theorem for pointlikes. Semigroup Forum 63, 281–304 (2001)

23. Stern, J.: Characterizations of some classes of regular events. Theoretical Computer Science 35(1985), 17–42 (1985)

24. Stern, J.: Complexity of some problems from the theory of automata. Information and Control 66(3), 163–176 (1985)

25. Trahtman, A.N.: Piecewise and local threshold testability of DFA. In: Freivalds, R. (ed.) FCT 2001. LNCS, vol. 2138, pp. 347–358. Springer, Heidelberg (2001)

26. van Rooijen, L., Zeitoun, M.: The separation problem for regular languages by piecewise testable languages (March 8, 2013), http://arxiv.org/abs/1303.2143

On the Complexity of Verifying Regular Properties on Flat Counter Systems[*],[**]

Stéphane Demri[2,3], Amit Kumar Dhar[1], and Arnaud Sangnier[1]

[1] LIAFA, Univ Paris Diderot, Sorbonne Paris Cité, CNRS, France
[2] New York University, USA
[3] LSV, CNRS, France

Abstract. Among the approximation methods for the verification of counter systems, one of them consists in model-checking their flat unfoldings. Unfortunately, the complexity characterization of model-checking problems for such operational models is not always well studied except for reachability queries or for Past LTL. In this paper, we characterize the complexity of model-checking problems on flat counter systems for the specification languages including first-order logic, linear mu-calculus, infinite automata, and related formalisms. Our results span different complexity classes (mainly from PTime to PSpace) and they apply to languages in which arithmetical constraints on counter values are systematically allowed. As far as the proof techniques are concerned, we provide a uniform approach that focuses on the main issues.

1 Introduction

Flat Counter Systems. Counter systems, finite-state automata equipped with program variables (counters) interpreted over non-negative integers, are known to be ubiquitous in formal verification. Since counter systems can actually simulate Turing machines [17], it is undecidable to check the existence of a run satisfying a given (reachability, temporal, etc.) property. However it is possible to approximate the behavior of counter systems by looking at a subclass of witness runs for which an analysis is feasible. A standard method consists in considering a finite union of path schemas for abstracting the whole bunch of runs, as done in [14]. More precisely, given a finite set of transitions Δ, a *path schema* is an ω-regular expression over Δ of the form $L = p_1(l_1)^* \cdots p_{k-1}(l_{k-1})^* p_k(l_k)^\omega$ where both p_i's and l_i's are paths in the control graph and moreover, the l_i's are loops. A path schema defines a set of infinite runs that respect a sequence of transitions that belongs to L. We write $\mathtt{Runs}(c_0, L)$ to denote such a set of runs starting at the initial configuration c_0 whereas $\mathtt{Reach}(c_0, L)$ denotes the set of configurations occurring in the runs of $\mathtt{Runs}(c_0, L)$. A counter system is *flattable* whenever the set of configurations reachable from c_0 is equal to $\mathtt{Reach}(c_0, L)$ for

---

[*] Work partially supported by the EU Seventh Framework Programme under grant agreement No. PIOF-GA-2011-301166 (DATAVERIF).

[**] A version with proofs is available as [5].

F.V. Fomin et al. (Eds.): ICALP 2013, Part II, LNCS 7966, pp. 162–173, 2013.
© Springer-Verlag Berlin Heidelberg 2013

some finite union of path schemas L. Similarly, a *flat counter system*, a system in which each control state belongs to at most one simple loop, verifies that the set of runs from c_0 is equal to $\mathtt{Runs}(c_0, L)$ for some finite union of path schemas L. Obviously, flat counter systems are flattable. Moreover, reachability sets of flattable counter systems are known to be Presburger-definable, see e.g. [1,3,7]. That is why, verification of flat counter systems belongs to the core of methods for model-checking arbitrary counter systems and it is desirable to character-ize the computational complexity of model checking problems on this kind of systems (see e.g. results about loops in [2]). Decidability results for verifying safety and reachability properties on flat counter systems have been obtained in [3,7,2]. For the verification of temporal properties, it is much more difficult to get sharp complexity characterization. For instance, it is known that verifying flat counter systems with CTL* enriched with arithmetical constraints is decid-able [6] whereas it is only NP-complete with Past LTL [4] (NP-completeness already holds with flat Kripke structures [10]).

Our Motivations. Our objectives are to provide a thorough classification of model-checking problems on flat counter systems when linear-time properties are considered. So far complexity is known with Past LTL [4] but even the de-cidability status with linear μ-calculus is unknown. Herein, we wish to consider several formalisms specifying linear-time properties (FO, linear μ-calculus, in-finite automata) and to determine the complexity of model-checking problems on flat counter systems. Note that FO is as expressive as Past LTL but much more concise whereas linear μ-calculus is strictly more expressive than Past LTL, which motivates the choice for these formalisms dealing with linear properties.

Our Contributions. We characterize the computational complexity of model-checking problems on flat counter systems for several prominent linear-time specification languages whose alphabets are related to atomic propositions but also to linear constraints on counter values. We obtain the following results:

- The problem of **model-checking first-order formulae on flat counter systems is PSPACE-complete** (Theorem 9). Note that model-checking classical first-order formulae over arbitrary Kripke structures is already known to be non-elementary. However the flatness assumption allows to drop the complexity to PSPACE even though linear constraints on counter values are used in the specification language.
- **Model-checking linear μ-calculus formulae on flat counter systems is PSPACE-complete** (Theorem 14). Not only linear μ-calculus is known to be more expressive than first-order logic (or than Past LTL) but also the decidability status of the problem on flat counter systems was open [6]. So, we establish decidability and we provide a complexity characterization.
- **Model-checking Büchi automata over flat counter systems is NP-complete** (Theorem 12).
- **Global model-checking is possible** for all the above mentioned for-malisms (Corollary 16).

2   Preliminaries

2.1   Counter Systems

Counter constraints are defined below as a subclass of Presburger formulae whose free variables are understood as counters. Such constraints are used to define guards in counter systems but also to define arithmetical constraints in temporal formulae. Let $C = \{x_1, x_2, \ldots\}$ be a countably infinite set of *counters* (variables interpreted over non-negative integers) and $AT = \{p_1, p_2, \ldots\}$ be a countable infinite set of propositional variables (abstract properties about program points). We write C_n to denote the restriction of C to $\{x_1, x_2, \ldots, x_n\}$. The set of *guards* g using the counters from C_n, written $G(C_n)$, is made of Boolean combinations of *atomic guards* of the form $\sum_{i=0}^{n} a_i \cdot x_i \sim b$ where the a_i's are in $\mathbb{Z}$, $b \in \mathbb{N}$ and $\sim \in \{=, \leq, \geq, <, >\}$. For $g \in G(C_n)$ and a vector $\mathbf{v} \in \mathbb{N}^n$, we say that $\mathbf{v}$ satisfies g, written $\mathbf{v} \models g$, if the formula obtained by replacing each x_i by $\mathbf{v}[i]$ holds. For $n \geq 1$, a *counter system* of dimension n (shortly a counter system) S is a tuple $\langle Q, C_n, \Delta, l \rangle$ where: Q is a finite set of *control states*, $l : Q \to 2^{AT}$ is a *labeling function*, $\Delta \subseteq Q \times G(C_n) \times \mathbb{Z}^n \times Q$ is a finite set of transitions labeled by guards and updates. As usual, to a counter system $S = \langle Q, C_n, \Delta, l \rangle$, we associate a labeled transition system $TS(S) = \langle C, \to \rangle$ where $C = Q \times \mathbb{N}^n$ is the set of *configurations* and $\to \subseteq C \times \Delta \times C$ is the *transition relation* defined by: $\langle \langle q, \mathbf{v} \rangle, \delta, \langle q', \mathbf{v}' \rangle \rangle \in \to$ (also written $\langle q, \mathbf{v} \rangle \xrightarrow{\delta} \langle q', \mathbf{v}' \rangle$) iff $\delta = \langle q, g, \mathbf{u}, q' \rangle \in \Delta$, $\mathbf{v} \models g$ and $\mathbf{v}' = \mathbf{v} + \mathbf{u}$. Note that in such a transition system, the counter values are non-negative since $C = Q \times \mathbb{N}^n$.

Given an initial configuration $c_0 \in Q \times \mathbb{N}^n$, a *run* ρ starting from c_0 in S is an infinite path in the associated transition system $TS(S)$ denoted as: $\rho := c_0 \xrightarrow{\delta_0} \cdots \xrightarrow{\delta_{m-1}} c_m \xrightarrow{\delta_m} \cdots$ where $c_i \in Q \times \mathbb{N}^n$ and $\delta_i \in \Delta$ for all $i \in \mathbb{N}$. We say that a counter system is *flat* if every node in the underlying graph belongs to at most one simple cycle (a cycle being simple if no edge is repeated twice in it) [3,14,4]. We denote by $\mathcal{CFS}$ the class of flat counter systems. A *Kripke structure* S can be seen as a counter system without counter and is denoted by $\langle Q, \Delta, l \rangle$ where $\Delta \subseteq Q \times Q$ and $l : Q \to 2^{AT}$. Standard notions on counter systems, as configuration, run or flatness, naturally apply to Kripke structures.

2.2   Model-Checking Problem

We define now our main model-checking problem on flat counter systems parameterized by a specification language $\mathcal{L}$. First, we need to introduce the notion of constrained alphabet whose letters should be understood as Boolean combinations of atomic formulae (details follow). A *constrained alphabet* is a triple of the form $\langle at, ag_n, \Sigma \rangle$ where at is a finite subset of AT, ag_n is a finite subset of atomic guards from $G(C_n)$ and Σ is a subset of $2^{at \cup ag_n}$. The size of a constrained alphabet is given by $size(\langle at, ag_n, \Sigma \rangle) = card(at) + card(ag_n) + card(\Sigma)$ where $card(X)$ denotes the cardinality of the set X. Of course, any standard alphabet (finite set of letters) can be easily viewed as a constrained alphabet (by ignoring the structure of letters). Given an infinite run $\rho := \langle q_0, \mathbf{v}_0 \rangle \to \langle q_1, \mathbf{v}_1 \rangle \cdots$ from a counter system with n counters and an ω-word over a constrained alphabet $w = a_0, a_1, \ldots \in \Sigma^\omega$, we say that ρ *satisfies* w, written $\rho \models w$, whenever for $i \geq 0$, we have $\mathbf{p} \in \mathrm{l}(q_i)$ [resp. $\mathbf{p} \notin \mathrm{l}(q_i)$] for every $\mathbf{p} \in (a_i \cap at)$ [resp. $\mathbf{p} \in (at \setminus a_i)$] and $\mathbf{v}_i \models \mathbf{g}$ [resp. $\mathbf{v}_i \not\models \mathbf{g}$] for every $\mathbf{g} \in (a_i \cap agn)$ [resp. $\mathbf{g} \in (agn \setminus a_i)$].

A *specification language* $\mathcal{L}$ over a constrained alphabet $\langle at, agn, \Sigma \rangle$ is a set of *specifications* A, each of it defining a set $\mathrm{L}(A)$ of ω-words over Σ. We will also sometimes consider specification languages over (unconstrained) standard finite alphabets (as usually defined). We now define the *model-checking problem* over flat counter systems with specification language $\mathcal{L}$ (written $\mathrm{MC}(\mathcal{L}, \mathcal{CFS})$): it takes as input a flat counter system S, a configuration c and a specification A from $\mathcal{L}$ and asks whether there is a run ρ starting at c and $w \in \Sigma^\omega$ in $\mathrm{L}(A)$ such that $\rho \models w$. We write $\rho \models A$ whenever there is $w \in \mathrm{L}(A)$ such that $\rho \models w$.

2.3   A Bunch of Specification Languages

Infinite Automata. Now let us define the specification languages BA and ABA, respectively with nondeterministic Büchi automata and with alternating Büchi automata. We consider here transitions labeled by Boolean combinations of atoms from $at \cup agn$. A specification A in ABA is a structure of the form $\langle Q, E, q_0, F \rangle$ where E is a finite subset of $Q \times \mathbb{B}(at \cup agn) \times \mathbb{B}^+(Q)$ and $\mathbb{B}^+(Q)$ denotes the set of positive Boolean combinations built over Q. Specification A is a concise representation for the alternating Büchi automaton $\mathcal{B}_A = \langle Q, \delta, q_0, F \rangle$ where $\delta : Q \times 2^{at \cup agn} \to \mathbb{B}^+(Q)$ and $\delta(q, a) \overset{\text{def}}{=} \bigvee_{\langle q, \psi, \psi' \rangle \in E, \, a \models \psi} \psi'$. We say that A is over the constrained alphabet $\langle at, agn, \Sigma \rangle$, whenever, for all edges $\langle q, \psi, \psi' \rangle \in E$, ψ holds at most for letters from Σ (i.e. the transition relation of $\mathcal{B}_A$ belongs to $Q \times \Sigma \to \mathbb{B}^+(Q)$). We have then $\mathrm{L}(A) = \mathrm{L}(\mathcal{B}_A)$ with the usual acceptance criterion for alternating Büchi automata. The specification language BA is defined in a similar way using Büchi automata. Hence the transition relation E of $A = \langle Q, E, q_0, F \rangle$ in BA is included in $Q \times \mathbb{B}(at \cup agn) \times Q$ and the transition relation of the Büchi automaton $\mathcal{B}_A$ is then included in $Q \times 2^{at \cup agn} \times Q$.

Linear-Time Temporal Logics. Below, we present briefly three logical languages that are tailored to specify runs of counter systems, namely ETL (see e.g.[25,19]), Past LTL (see e.g. [21]) and linear μ-calculus (or μTL), see e.g. [23]. A specification in one of these logical specification languages is just a formula. The differences with their standard versions in which models are ω-sequences of propositional valuations are listed below: models are infinite runs of counters systems; atomic formulae are either propositional variables in AT or atomic guards; given an infinite run $\rho := \langle q_0, \mathbf{v}_0 \rangle \to \langle q_1, \mathbf{v}_1 \rangle \cdots$, we will have $\rho, i \models \mathbf{p}$ $\overset{\text{def}}{\Leftrightarrow} \mathbf{p} \in \mathrm{l}(q_i)$ and $\rho, i \models \mathbf{g} \overset{\text{def}}{\Leftrightarrow} \mathbf{v}_i \models \mathbf{g}$. The temporal operators, fixed point operators and automata-based operators are interpreted then as usual. A formula ϕ built over the propositional variables in at and the atomic guards in agn defines a language $\mathrm{L}(\phi)$ over $\langle at, agn, \Sigma \rangle$ with $\Sigma = 2^{at \cup agn}$. There is no need to recall here the syntax and semantics of ETL, Past LTL and linear μ-calculus since with their standard definitions and with the above-mentioned differences, their variants for counter systems are defined unambiguously (see a lengthy presentation

of Past LTL for counter systems in [4]). However, we may recall a few definitions
on-the-fly if needed. Herein the size of formulae is understood as the number of
subformulae.

Example. In adjoining figure, we present a flat counter system with two counters
and with labeling function l such that $l(q_3) = \{p, q\}$ and $l(q_5) = \{p\}$. We would
like to characterize the set of configurations c with control state q_1 such that
there is some infinite run from c for which after some position i, all future even
positions j (i.e. $i \equiv_2 j$) satisfy that p holds and the first counter is equal to the
second counter.

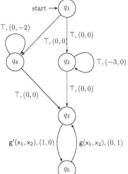

This can be specified in linear μ-calculus using as
atomic formulae either propositional variables or
atomic guards. The corresponding formula in linear
μ-calculus is: $\mu z_1.(X(\nu z_2.(p \wedge (x_1 - x_2 = 0) \wedge XXz_2) \vee$
$Xz_1)$. Clearly, such a position i occurs in any run
after reaching the control state q_3 with the same
value for both counters. Hence, the configurations
$\langle q_1, \mathbf{v} \rangle$ satisfying these properties have counter values $\mathbf{v} \in \mathbb{N}^2$ verifying the Presburger formula below:

$$\exists\, y\, (((x_1 = 3y + x_2) \wedge (\forall\, y'\; g(x_2 + y', x_2 + y') \wedge g'(x_2 + y', x_2 + y' + 1)))\vee$$

$$((x_2 = 2y + x_1) \wedge (\forall\, y'\; g(x_1 + y', x_1 + y') \wedge g'(x_1 + y', x_1 + y' + 1))))$$

In the paper, we shall establish how to compute systematically such formulae
(even without universal quantifications) for different specification languages.

3   Constrained Path Schemas

In [4] we introduced minimal path schemas for flat counter systems. Now, we
introduce *constrained path schemas* that are more abstract than path schemas.
A *constrained path schema* cps is a pair $\langle p_1(l_1)^* \cdots p_{k-1}(l_{k-1})^* p_k(l_k)^\omega, \phi(x_1,$
$\ldots, x_{k-1})\rangle$ where the first component is an ω-regular expression over a con-
strained alphabet $\langle at, ag_n, \Sigma \rangle$ with p_i, l_i's in Σ^*, and $\phi(x_1, \ldots, x_{k-1}) \in G(C_{k-1})$.
Each constrained path schema defines a language $L(\text{cps}) \subseteq \Sigma^\omega$ given by $L(\text{cps}) \overset{\text{def}}{=}$
$\{p_1(l_1)^{n_1} \cdots p_{k-1}(l_{k-1})^{n_{k-1}} p_k(l_k)^\omega \;:\; \phi(n_1, \ldots, n_{k-1})$ holds true$\}$. The size of
cps, written size(cps), is equal to $2k + \text{len}(p_1 l_1 \cdots p_{k-1} l_{k-1} p_k l_k) + \text{size}(\phi(x_1, \ldots,$
$x_{k-1}))$. Observe that in general constrained path schemas are defined under con-
strained alphabet and so will the associated specifications unless stated
otherwise.

Let us consider below the three decision problems on constrained path schemas
that are useful in the rest of the paper. *Consistency problem* checks whether

L(cps) is non-empty. It amounts to verify the satisfiability status of the second component. Let us recall the result below.

Theorem 1. [20] There are polynomials $\text{pol}_1(\cdot)$, $\text{pol}_2(\cdot)$ and $\text{pol}_3(\cdot)$ such that for every guard g, say in $G(C_n)$, of size N, we have (I) there exist $B \subseteq [0, 2^{\text{pol}_1(N)}]^n$ and $P_1, \ldots, P_\alpha \in [0, 2^{\text{pol}_1(N)}]^n$ with $\alpha \leq 2^{\text{pol}_2(N)}$ such that for every $y \in \mathbb{N}^n$, $y \models g$ iff there are $b \in B$ and $a \in \mathbb{N}^\alpha$ such that $y = b + a[1]P_1 + \cdots + a[\alpha]P_\alpha$; (II) if g is satisfiable, then there is $y \in [0, 2^{\text{pol}_3(N)}]^n$ s.t. $y \models g$.

Consequently, the consistency problem is NP-complete (the hardness being obtained by reducing SAT). The *intersection non-emptiness problem*, clearly related to model-checking problem, takes as input a constrained path schema cps and a specification $A \in \mathcal{L}$ and asks whether $L(\text{cps}) \cap L(A) \neq \emptyset$. Typically, for several specification languages $\mathcal{L}$, we establish the existence of a computable map $f_{\mathcal{L}}$ (at most exponential) such that whenever $L(\text{cps}) \cap L(A) \neq \emptyset$ there is $p_1(l_1)^{n_1} \cdots p_{k-1}(l_{k-1})^{n_{k-1}} p_k(l_k)^\omega$ belonging to the intersection and for which each n_i is bounded by $f_{\mathcal{L}}(A, \text{cps})$. This motivates the introduction of the *membership problem* for $\mathcal{L}$ that takes as input a constrained path schema cps, a specification $A \in \mathcal{L}$ and $n_1, \ldots, n_{k-1} \in \mathbb{N}$ and checks whether $p_1(l_1)^{n_1} \cdots p_{k-1}(l_{k-1})^{n_{k-1}} p_k(l_k)^\omega \in L(A)$. Here the n_i's are understood to be encoded in binary and we do not require them to satisfy the constraint of the path schema.

Since constrained path schemas are abstractions of path schemas used in [4], from this work we can show that runs from flat counter systems can be represented by a finite set of constrained path schemas as stated below.

Theorem 2. Let at be a finite set of atomic propositions, ag_n be a finite set of atomic guards from $G(C_n)$, S be a flat counter system whose atomic propositions and atomic guards are from $at \cup ag_n$ and $c_0 = \langle q_0, v_0 \rangle$ be an initial configuration. One can construct in exponential time a set X of constrained path schemas s.t.: (I) Each constrained path schema cps in X has an alphabet of the form $\langle at, ag_n, \Sigma \rangle$ (Σ may vary) and cps is of polynomial size. (II) Checking whether a constrained path schema belongs to X can be done in polynomial time. (III) For every run ρ from c_0, there is a constrained path schema cps in X and $w \in L(\text{cps})$ such that $\rho \models w$. (IV) For every constrained path schema cps in X and for every $w \in L(\text{cps})$, there is a run ρ from c_0 such that $\rho \models w$.

In order to take advantage of Theorem 2 for the verification of flat counter systems, we need to introduce an additional property: $\mathcal{L}$ has the *nice subalphabet property* iff for all specifications $A \in \mathcal{L}$ over $\langle at, ag_n, \Sigma \rangle$ and for all constrained alphabets $\langle at, ag_n, \Sigma' \rangle$, one can build a specification A' over $\langle at, ag_n, \Sigma' \rangle$ in polynomial time in the sizes of A and $\langle at, ag_n, \Sigma' \rangle$ such that $L(A) \cap (\Sigma')^\omega = L(A')$. We need this property to build from A and a constraint path schema over $\langle at, ag_n, \Sigma' \rangle$, the specification A'. This property will also be used to transform a specification over $\langle at, ag_n, \Sigma \rangle$ into a specification over the finite alphabet Σ'.

Lemma 3. *BA, ABA, μTL, ETL, Past LTL have the nice subalphabet property.*

The abstract Algorithm 1 which performs the following steps (1) to (3) takes as input S, a configuration c_0 and $A \in \mathcal{L}$ and solves $MC(\mathcal{L}, \mathcal{CFS})$: (1) Guess cps over $\langle at, ag_n, \Sigma' \rangle$ in X; (2) Build A' such that $L(A) \cap (\Sigma')^\omega = L(A')$; (3) Return $L(\mathsf{cps}) \cap L(A') \neq \emptyset$. Thanks to Theorem 2, the first guess can be performed in polynomial time and with the nice subalphabet property, we can build A' in polynomial time too. This allows us to conclude the following lemma which is a consequence of the correctness of the above algorithm (see [5]).

Lemma 4. *If* $\mathcal{L}$ *has the nice subalphabet property and its intersection non-emptiness problem is in* NP *[resp.* PSPACE*], then* $\mathrm{MC}(\mathcal{L}, \mathcal{CFS})$ *is in* NP *[resp.* PSPACE*]*

We know that the membership problem for Past LTL is in PTIME and the inter-section non-emptiness problem is in NP (as a consequence of [4, Theorem 3]). By Lemma 4, we are able to conclude the main result from [4]: $\mathrm{MC}(\mathrm{PastLTL}, \mathcal{CFS})$ is in NP. This is not surprising at all since in this paper we present a general method for different specification languages that rests on Theorem 2 (a conse-quence of technical developments from [4]).

4   Taming First-Order Logic and Flat Counter Systems

In this section, we consider first-order logic as a specification language. By Kamp's Theorem, first-order logic has the same expressive power as Past LTL and hence model-checking first-order logic over flat counter systems is decid-able too [4]. However this does not provide us an optimal upper bound for the model-checking problem. In fact, it is known that the satisfiability problem for first-order logic formulae is non-elementary and consequently the translation into Past LTL leads to a significant blow-up in the size of the formula.

4.1   First-Order Logic in a Nutshell

For defining first-order logic formulae, we consider a countably infinite set of variables Z and a finite (unconstrained) alphabet Σ. The syntax of *first-order logic* over atomic propositions FO_Σ is then given by the following grammar: $\phi ::= a(\mathsf{z}) \mid S(\mathsf{z}, \mathsf{z}') \mid \mathsf{z} < \mathsf{z}' \mid \mathsf{z} = \mathsf{z}' \mid \neg\phi \mid \phi \wedge \phi' \mid \exists\mathsf{z} \; \phi(\mathsf{z})$ where $a \in \Sigma$ and $\mathsf{z}, \mathsf{z}' \in Z$. For a formula ϕ, we will denote by $free(\phi)$ its set of free variables de-fined as usual. A formula with no free variable is called a *sentence*. As usual, we define the *quantifier height* $qh(\phi)$ of a formula ϕ as the maximum nesting depth of the operators $\exists$ in ϕ. Models for FO_Σ are ω-words over the alphabet Σ and variables are interpreted by positions in the word. A *position assignment* is a partial function $f : Z \to \mathbb{N}$. Given a model $w \in \Sigma^\omega$, a FO_Σ formula ϕ and a position assignment f such that $f(\mathsf{z}) \in \mathbb{N}$ for every variable $\mathsf{z} \in free(\phi)$, the satisfaction relation $\models_f$ is defined as usual. Given a FO_Σ sentence ϕ, we write $w \models \phi$ when $w \models_f \phi$ for an arbitrary position assignment f. The language of ω-words w over Σ associated to a sentence ϕ is then $\mathcal{L}(\phi) = \{w \in \Sigma^\omega \mid w \models \phi\}$. For $n \in \mathbb{N}$, we define the equivalence relation $\approx_n$ between ω-words over Σ as: $w \approx_n w'$ when for every sentence ϕ with $qh(\phi) \leq n$, $w \models \phi$ iff $w' \models \phi$.

FO on CS. FO formulae interpreted over infinite runs of counter systems are defined as FO formulae over a finite alphabet except that atomic formulae of the form $a(z)$ are replaced by atomic formulae of the form $p(z)$ or $g(z)$ where p is an atomic formula or g is an atomic guard from $G(C_n)$. Hence, a formula ϕ built over atomic formulae from a finite set at of atomic propositions and from a finite set ag_n of atomic guards from $G(C_n)$ defines a specification for the constrained alphabet $\langle at, at_n, 2^{at \cup ag_n} \rangle$. Note that the alphabet can be of exponential size in the size of ϕ and $p(z)$ actually corresponds to a disjunction $\bigvee_{p \in a} a(z)$.

Lemma 5. *FO has the nice subalphabet property.*

We have taken time to properly define first-order logic for counter systems (whose models are runs of counter systems, see also Section 2.2) but below, we will mainly operate with FO_Σ over a standard (unconstrained) alphabet. Let us state our first result about FO_Σ which allows us to bound the number of times each loop is taken in a constrained path schema in order to satisfy a formula. We provide a stuttering theorem equivalent for FO_Σ formulas as is done in [4] for PLTL and in [12] for LTL. The lengthy proof of Theorem 6 uses Ehrenfeucht-Fraïssé game (see [5]).

Theorem 6 (Stuttering Theorem). *Let $w = w_1 s^M w_2, w' = w_1 s^{M+1} w_2 \in \Sigma^\omega$ such that $N \geq 1$, $M > 2^{N+1}$ and $s \in \Sigma^+$. Then $w \approx_N w'$.*

4.2   Model-Checking Flat Counter Systems with FO

Let us characterize the complexity of $MC(FO, \mathcal{CFS})$. First, we will state the complexity of the intersection non-emptiness problem. Given a constrained path schema **cps** and a FO sentence ψ, Theorem 1 provides two polynomials $\mathtt{pol}_1$ and $\mathtt{pol}_2$ to represent succinctly the solutions of the guard in **cps**. Theorem 6 allows us to bound the number of times loops are visited. Consequently, we can compute a value $f_{FO}(\psi, \mathtt{cps})$ exponential in the size of ψ and **cps**, as explained earlier, which allows us to find a witness for the intersection non-emptiness problem where each loop is taken a number of times smaller than $f_{FO}(\psi, \mathtt{cps})$.

Lemma 7. *Let **cps** be a constrained path schema and ψ be a FO_Σ sentence. Then $L(\mathtt{cps}) \cap L(\psi)$ is non-empty iff there is an ω-word in $L(\mathtt{cps}) \cap L(\psi)$ in which each loop is taken at most $2^{(qh(\psi)+2)+\mathtt{pol}_1(\mathrm{size}(\mathtt{cps}))+\mathtt{pol}_2(\mathrm{size}(\mathtt{cps}))}$ times.*

Hence $f_{FO}(\psi, \mathtt{cps})$ has the value $2^{(qh(\psi)+2)+(\mathtt{pol}_1+\mathtt{pol}_2)(\mathrm{size}(\mathtt{cps}))}$. Furthermore checking whether $L(\mathtt{cps}) \cap L(\psi)$ is non-empty amounts to guess some $\mathbf{n} \in [0, 2^{(qh(\psi)+2)+\mathtt{pol}_1(\mathrm{size}(\mathtt{cps}))+\mathtt{pol}_2(\mathrm{size}(\mathtt{cps}))}]^{k-1}$ and verify whether $w = p_1(l_1)^{\mathbf{n}[1]} \cdots p_{k-1}(l_{k-1})^{\mathbf{n}[k-1]} p_k(l_k)^\omega \in L(\mathtt{cps}) \cap L(\psi)$. Checking if $w \in L(\mathtt{cps})$ can be done in polynomial time in $(qh(\psi)+2)+\mathtt{pol}_1(\mathrm{size}(\mathtt{cps}))+\mathtt{pol}_2(\mathrm{size}(\mathtt{cps}))$ (and therefore in polynomial time in $\mathrm{size}(\psi)+\mathrm{size}(\mathtt{cps})$) since this amounts to verify whether $\mathbf{n} \models \phi$. Checking whether $w \in L(\psi)$ can be done in exponential space in $\mathrm{size}(\psi)+\mathrm{size}(\mathtt{cps})$ by using [15, Proposition 4.2]. Hence, this leads to a nondeterministic exponential space decision procedure for the intersection non-emptiness problem but it is possible to get down to nondeterministic polynomial space using the succinct representation of constrained path schema as stated by Lemma 8 below for which the lower bound is deduced by the fact that model-checking ultimately periodic words with first-order logic is PSPACE-hard [15].

Lemma 8. *Membership problem with* FO_Σ *is* PSPACE-*complete.*

Note that the membership problem for FO is for unconstrained alphabet, but due to the nice subalphabet property of FO, the same holds for constrained alphabet since given a FO formula over $\langle at, agn, \Sigma \rangle$, we can build in polynomial time a FO formula over $\langle at, agn, \Sigma' \rangle$ from which we can build also in polynomial time a formula of $\mathrm{FO}_{\Sigma'}$ (where Σ' is for instance the alphabet labeling a constrained path schema). We can now state the main results concerning FO.

Theorem 9. *(I) The intersection non-emptiness problem with FO is* PSPACE-*complete. (II)* $\mathrm{MC}(\mathrm{FO}, \mathcal{CFS})$ *is* PSPACE-*complete. (III) Model-checking flat Kripke structures with FO is* PSPACE-*complete.*

Proof. (I) is a consequence of Lemma 7 and Lemma 8. We obtain (II) from (I) by applying Lemma 4 and Lemma 5. (III) is obtained by observing that flat Kripke structures form a subclass of flat counter systems. To obtain the lower bound, we use that model-checking ultimately periodic words with first-order logic is PSPACE-hard [15]. □

5 Taming Linear μ-calculus and Other Languages

We now consider several specification languages defining ω-regular properties on atomic propositions and arithmetical constraints. First, we deal with BA by establishing Theorem 10 and then deduce results for ABA, ETL and μTL.

Theorem 10. *Let* $\mathcal{B} = \langle Q, \Sigma, q_0, \Delta, F \rangle$ *be a Büchi automaton (with standard definition) and* $\mathrm{cps} = \langle p_1(l_1)^* \cdots p_{k-1}(l_{k-1})^* p_k(l_k)^\omega, \phi(\mathsf{x}_1, \ldots, \mathsf{x}_{k-1}) \rangle$ *be a constrained path schema over* Σ. *We have* $\mathrm{L}(\mathrm{cps}) \cap \mathrm{L}(\mathcal{B}) \neq \emptyset$ *iff there exists* $\boldsymbol{y} \in [0, 2^{\mathrm{pol}_1(\mathrm{size}(\mathrm{cps}))} + 2.\mathrm{card}(Q)^k \times 2^{\mathrm{pol}_1(\mathrm{size}(\mathrm{cps})) + \mathrm{pol}_2(\mathrm{size}(\mathrm{cps}))}]^{k-1}$ *such that* $p_1(l_1)^{\boldsymbol{y}[1]} \cdots p_{k-1}(l_{k-1})^{\boldsymbol{y}[k-1]} p_k l_k^\omega \in \mathrm{L}(\mathcal{B}) \cap \mathrm{L}(\mathrm{cps})$ *(pol_1 and pol_2 are from Theorem 1).*

Theorem 10 can be viewed as a pumping lemma involving an automaton and semilinear sets. Thanks to it we obtain an exponential bound for the map f_{BA} so that $f_{\mathrm{BA}}(\mathcal{B}, \mathrm{cps}) = 2^{\mathrm{pol}_1(\mathrm{size}(\mathrm{cps}))} + 2.\mathrm{card}(Q)^{\mathrm{size}(\mathrm{cps})} \times 2^{\mathrm{pol}_1(\mathrm{size}(\mathrm{cps})) + pol_2(\mathrm{size}(\mathrm{cps}))}$. So checking $\mathrm{L}(\mathrm{cps}) \cap \mathrm{L}(\mathcal{B}) \neq \emptyset$ amounts to guess some $\mathbf{n} \in [0, 2^{\mathrm{pol}_1(\mathrm{size}(\mathrm{cps}))} + 2.\mathrm{card}(Q)^{\mathrm{size}(\mathrm{cps})} \times 2^{\mathrm{pol}_1(\mathrm{size}(\mathrm{cps})) + pol_2(\mathrm{size}(\mathrm{cps}))}]^{k-1}$ and to verify whether the word $w = p_1(l_1)^{\mathbf{n}[1]} \cdots p_{k-1}(l_{k-1})^{\mathbf{n}[k-1]} p_k(l_k)^\omega \in \mathrm{L}(\mathrm{cps}) \cap \mathrm{L}(\mathcal{B})$. Checking whether $w \in \mathrm{L}(\mathrm{cps})$ can be done in polynomial time in $\mathrm{size}(\mathcal{B}) + \mathrm{size}(\mathrm{cps})$ since this amounts to check $\mathbf{n} \models \phi$. Checking whether $w \in \mathrm{L}(\mathcal{B})$ can be also done in polynomial time by using the results from [15]. Indeed, w can be encoded in polynomial time as a pair of straight-line programs and by [15, Corollary 5.4] this can be done in polynomial time. So, the membership problem for Büchi automata is in PTIME. By using that BA has the nice subalphabet property and that we can create a polynomial size Büchi automata from a given BA specification and cps, we get the following result.

Lemma 11. *The intersection non-emptiness problem with BA is* NP*-complete.*

Now, by Lemma 3, Lemma 4 and Lemma 11, we get the result below for which
the lower bound is obtained from an easy reduction of SAT.

Theorem 12. MC(BA, $\mathcal{CFS}$) *is* NP*-complete.*

We are now ready to deal with ABA, ETL and linear μ-calculus. A language
$\mathcal{L}$ has the *nice BA property* iff for every specification A from $\mathcal{L}$, we can build a
Büchi automaton $\mathcal{B}_A$ such that L(A) = L$(\mathcal{B}_A)$, each state of $\mathcal{B}_A$ is of polynomial
size, it can be checked if a state is initial [resp. accepting] in polynomial space
and the transition relation can be decided in polynomial space too. So, given a
language $\mathcal{L}$ having the nice BA property, a constrained path schema cps and
a specification in $A \in \mathcal{L}$, if L(cps) $\cap$ L(A) is non-empty, then there is an ω-
word in L(cps) $\cap$ L(A) such that each loop is taken at most a number of times
bounded by $f_{BA}(\mathcal{B}_A, \text{cps})$. So $f_{\mathcal{L}}(A, \text{cps})$ is obviously bounded by $f_{BA}(\mathcal{B}_A, \text{cps})$.
Hence, checking whether L(cps)$\cap$L(A) is non-empty amounts to guess some $\mathbf{n} \in$
$[0, f_{\mathcal{L}}(A, \text{cps})]^{k-1}$ and check whether $w = p_1(l_1)^{\mathbf{n}[1]} \cdots p_{k-1}(l_{k-1})^{\mathbf{n}[k-1]} p_k(l_k)^{\omega} \in$
L(cps) $\cap$ L(A). Checking whether $w \in$ L(cps) can be done in polynomial time
in size(A) + size(cps) since this amounts to check $\mathbf{n} \models \phi$. Checking whether
$w \in$ L(A) can be done in nondeterministic polynomial space by reading w while
guessing an accepting run for $\mathcal{B}_A$. Actually, one guesses a state q from $\mathcal{B}_A$ and
check whether the prefix $p_1(l_1)^{\mathbf{n}[1]} \cdots p_{k-1}(l_{k-1})^{\mathbf{n}[k-1]} p_k$ can reach it and then
nonemptiness between $(l_k)^{\omega}$ and the Büchi automaton $\mathcal{B}_A^q$ in which q is an initial
state is checked. Again, this can be done in nondeterministic polynomial space
thanks to the nice BA property. We obtain the lemma below.

Lemma 13. *Membership problem and intersection non-emptiness problem for*
$\mathcal{L}$ *having the nice BA property are in* PSPACE.

Let us recall consequences of results from the literature. ETL has the nice BA
property by [24], linear μ-calculus has the nice BA property by [23] and ABA
has the nice BA property by [18]. Note that the results for ETL and ABA can
be also obtained thanks to translations into linear μ-calculus. By Lemma 13,
Lemma 4 and the above-mentioned results, we obtain the following results.

Theorem 14. MC(ABA, $\mathcal{CFS}$), MC(ETL, $\mathcal{CFS}$) *and* MC(μTL, $\mathcal{CFS}$) *are in*
PSPACE.

Note that for obtaining the PSPACE upper bound, we use the same procedure for
all the logics. Using that the emptiness problem for finite alternating automata
over a single letter alphabet is PSPACE-hard [8], we are also able to get lower
bounds.

Theorem 15. *(I) The intersection non-emptiness problem for* ABA *[resp. μTL]*
is PSPACE*-hard. (II)* MC(ABA, $\mathcal{CFS}$) *and* MC(μTL, $\mathcal{CFS}$) *are* PSPACE*-hard.*

According to the proof of Theorem 15 (see [5]), PSPACE-hardness already holds
for a fixed Kripke structure, that is actually a simple path schema. Hence, for linear μ-caluclus, there is a complexity gap between model-checking unconstrained

path schemas with two loops (in UP∩co-UP [9]) and model-checking unconstrained path schemas (Kripke structures) made of a single loop, which is in contrast to Past LTL for which model-checking unconstrained path schemas with a bounded number of loops is in PTime [4, Theorem 9].

As an additional corollary, we can solve the global model-checking problem with existential Presburger formulae. The global model-checking consists in characterizing the set of initial configurations from which there exists a run satisfying a given specification. We knew that Presburger formulae exist for global model-checking [6] for Past LTL (and therefore for FO) but we can conclude that they are structurally simple and we provide an alternative proof. Moreover, the question has been open for μTL since the decidability status of MC(μTL, $\mathcal{CFS}$) has been only resolved in the present work.

Corollary 16. *Let $\mathcal{L}$ be a specification language among FO, BA, ABA, ETL or μTL. Given a flat counter system S, a control state q and a specification A in $\mathcal{L}$, one can effectively build an existential Presburger formula $\phi(z_1, \ldots, z_n)$ such that for all $v \in \mathbb{N}^n$. $v \models \phi$ iff there is a run ρ starting at $\langle q, v \rangle$ verifying $\rho \models A$.*

6   Conclusion

We characterized the complexity of MC($\mathcal{L}, \mathcal{CFS}$) for prominent linear-time specification languages $\mathcal{L}$ whose letters are made of atomic propositions and linear constraints. We proved the PSPACE-completeness of the problem with linear μ-calculus (decidability was open), for alternating Büchi automata and also for FO. When specifications are expressed with Büchi automata, the problem is shown NP-complete. Global model-checking is also possible on flat counter systems with such specification languages. Even though the core of our work relies on small solutions of quantifier-free Presburger formulae, stuttering properties, automata-based approach and on-the-fly algorithms, our approach is designed to be generic. Not only this witnesses the robustness of our method but our complexity characterization justifies further why verification of flat counter systems can be at the core of methods for model-checking counter systems. Our main results are in the table below with useful comparisons ('Ult. periodic KS' stands for ultimately periodic Kripke structures namely a path followed by a loop).

	Flat counter systems	Kripke struct.	Flat Kripke struct.	Ult. periodic KS
μTL	PSPACE-C (Thm. 14)	PSPACE-C [23]	PSPACE-C (Thm. 14)	in UP∩co-UP [16]
ABA	PSPACE-C (Thm. 14)	PSPACE-C	PSPACE-C (Thm. 14)	in PTime (see e.g. [11, p. 3])
ETL	in PSPACE (Thm. 14)	PSPACE-C [21]	in PSPACE [21]	in PTime (see e.g. [19,11])
BA	NP-C (Thm.12)	in PTime	in PTime	in PTime
FO	PSPACE-C (Thm. 9)	Non-el. [22]	PSPACE-C (Thm. 9)	PSPACE-C [15]
Past LTL	NP-C [4]	PSPACE-C [21]	NP-C [10,4]	PTime [13]

References

1. Boigelot, B.: Symbolic methods for exploring infinite state spaces. PhD thesis, Université de Liège (1998)
2. Bozga, M., Iosif, R., Konečný, F.: Fast acceleration of ultimately periodic relations. In: Touili, T., Cook, B., Jackson, P. (eds.) CAV 2010. LNCS, vol. 6174, pp. 227–242. Springer, Heidelberg (2010)
3. Comon, H., Jurski, Y.: Multiple counter automata, safety analysis and PA. In: Vardi, M.Y. (ed.) CAV 1998. LNCS, vol. 1427, pp. 268–279. Springer, Heidelberg (1998)
4. Demri, S., Dhar, A.K., Sangnier, A.: Taming Past LTL and Flat Counter Systems. In: Gramlich, B., Miller, D., Sattler, U. (eds.) IJCAR 2012. LNCS (LNAI), vol. 7364, pp. 179–193. Springer, Heidelberg (2012)
5. Demri, S., Dhar, A.K., Sangnier, A.: On the complexity of verifying regular properties on flat counter systems (2013), http://arxiv.org/abs/1304.6301
6. Demri, S., Finkel, A., Goranko, V., van Drimmelen, G.: Model-checking CTL* over flat Presburger counter systems. JANCL 20(4), 313–344 (2010)
7. Finkel, A., Leroux, J.: How to compose presburger-accelerations: Applications to broadcast protocols. In: Agrawal, M., Seth, A.K. (eds.) FSTTCS 2002. LNCS, vol. 2556, pp. 145–156. Springer, Heidelberg (2002)
8. Jančar, P., Sawa, Z.: A note on emptiness for alternating finite automata with a one-letter alphabet. IPL 104(5), 164–167 (2007)
9. Jurdziński, M.: Deciding the winner in parity games is in UP ∩ co-UP. IPL 68(3), 119–124 (1998)
10. Kuhtz, L., Finkbeiner, B.: Weak kripke structures and LTL. In: Katoen, J.-P., König, B. (eds.) CONCUR 2011. LNCS, vol. 6901, pp. 419–433. Springer, Heidelberg (2011)
11. Kupferman, O., Vardi, M.: Weak alternating automata are not that weak. ACM Transactions on Computational Logic 2(3), 408–429 (2001)
12. Kučera, A., Strejček, J.: The stuttering principle revisited. Acta Informatica 41(7-8), 415–434 (2005)
13. Laroussinie, F., Markey, N., Schnoebelen, P.: Temporal logic with forgettable past. In: LICS 2002, pp. 383–392. IEEE (2002)
14. Leroux, J., Sutre, G.: Flat counter systems are everywhere! In: Peled, D.A., Tsay, Y.-K. (eds.) ATVA 2005. LNCS, vol. 3707, pp. 489–503. Springer, Heidelberg (2005)
15. Markey, N., Schnoebelen, P.: Model checking a path. In: Amadio, R.M., Lugiez, D. (eds.) CONCUR 2003. LNCS, vol. 2761, pp. 251–265. Springer, Heidelberg (2003)
16. Markey, N., Schnoebelen, P.: Mu-calculus path checking. IPL 97(6) (2006)
17. Minsky, M.: Computation, Finite and Infinite Machines. Prentice Hall (1967)
18. Miyano, S., Hayashi, T.: Alternating finite automata on ω-words. Theor. Comput. Sci. 32, 321–330 (1984)
19. Piterman, N.: Extending temporal logic with ω-automata. Master's thesis, The Weizmann Institute of Science (2000)
20. Pottier, L.: Minimal Solutions of Linear Diophantine Systems: Bounds and Algorithms. In: Book, R.V. (ed.) RTA 1991. LNCS, vol. 488, pp. 162–173. Springer, Heidelberg (1991)
21. Sistla, A., Clarke, E.: The complexity of propositional linear temporal logic. JACM 32(3), 733–749 (1985)
22. Stockmeyer, L.J.: The complexity of decision problems in automata and logic. PhD thesis, MIT (1974)
23. Vardi, M.: A temporal fixpoint calculus. In: POPL 1988, pp. 250–259. ACM (1988)
24. Vardi, M., Wolper, P.: Reasoning about infinite computations. I&C 115 (1994)
25. Wolper, P.: Temporal logic can be more expressive. I&C 56, 72–99 (1983)

Multiparty Compatibility in Communicating Automata: Characterisation and Synthesis of Global Session Types

Pierre-Malo Deniélou and Nobuko Yoshida

[1] Royal Holloway, University of London
[2] Imperial College London

Abstract. Multiparty session types are a type system that can ensure the safety and liveness of distributed peers via the global specification of their interactions. To construct a global specification from a set of distributed uncontrolled behaviours, this paper explores the problem of fully characterising multiparty session types in terms of communicating automata. We equip global and local session types with labelled transition systems (LTSs) that faithfully represent asynchronous communications through unbounded buffered channels. Using the equivalence between the two LTSs, we identify a class of communicating automata that exactly correspond to the projected local types. We exhibit an algorithm to synthesise a global type from a collection of communicating automata. The key property of our findings is the notion of *multiparty compatibility* which non-trivially extends the duality condition for binary session types.

1 Introduction

Over the last decade, *session types* [12,18] have been studied as data types or functional types for communications and distributed systems. A recent discovery by [4,20], which establishes a Curry-Howard isomorphism between binary session types and linear logics, confirms that session types and the notion of duality between type constructs have canonical meanings. Multiparty session types [2,13] were proposed as a major generalisation of binary session types. They can enforce communication safety and deadlock-freedom for more than two peers thanks to a choreographic specification (called *global type*) of the interaction. Global types are projected to end-point types (*local types*), against which processes can be statically type-checked and verified to behave correctly.

The motivation of this paper comes from our practical experiences that, in many situations, even where we start from the end-point projections of a choreography, we need to reconstruct a global type from distributed specifications. End-point specifications are usually available, either through inference from the control flow, or through existing service interfaces, and always in forms akin to individual communicating finite state machines. If one knows the precise conditions under which a global type can be constructed (i.e. the conditions of *synthesis*), not only the global safety property which multiparty session types ensure is guaranteed, but also the generated global type can be used as a refinement and be integrated within the distributed system development life-cycle (see [17]). This paper attempts to give the synthesis condition as a sound and complete characterisation of multiparty session types with respect to Communicating Finite State Machines (CFSMs) [3]. CFSMs have been a well-studied formalism for analysing distributed safety properties and are widely present in industry tools.

F.V. Fomin et al. (Eds.): ICALP 2013, Part II, LNCS 7966, pp. 174–186, 2013.

They can been seen as generalised end-point specifications, therefore an excellent target for a common comparison ground and for synthesis. As explained below, to identify a complete set of CFSMs for synthesis, we first need to answer a question – *what is the canonical duality notion in multiparty session types?*

Characterisation of Binary Session Types as Communicating Automata. The subclass which fully characterises *binary session types* was actually proposed by Gouda, Manning and Yu in 1984 [11] in a pure communicating automata context. Consider a simple business protocol between a Buyer and a Seller from the Buyer's viewpoint: Buyer sends the title of a book, Seller answers with a quote. If Buyer is satisfied by the quote, then he sends his address and Seller sends back the delivery date; otherwise it retries the same conversation. This can be described by the following session type:

$$\mu t.\,!\,\text{title};\ ?\text{quote};\ !\{\ \text{ok}:!\,\text{addrs};\,?\text{date};\,\text{end},\quad \text{retry}:t\ \} \qquad (1.1)$$

where the operator $!\,\text{title}$ denotes an output of the title, whereas $?\text{quote}$ denotes an input of a quote. The output choice features the two options ok and retry and ; denotes sequencing. end represents the termination of the session, and μt is recursion.

The simplicity and tractability of binary sessions come from the notion of *duality* in interactions [10]. The interaction pattern of the Seller is fully given as the dual of the type in (1.1) (exchanging input ! and output ? in the original type). When composing two parties, we only have to check they have mutually dual types, and the resulting communication is guaranteed to be deadlock-free. Essentially the same characterisation is given in communicating automata. Buyer and Seller's session types are represented by the following two machines.

We can observe that these CFSMs satisfy three conditions. First, the communications are *deterministic*: messages that are part of the same choice, ok and retry here, are distinct. Secondly, there is no mixed state (each state has either only sending actions or only receiving actions). Third, these two machines have *compatible* traces (i.e. dual): the Seller machine can be defined by exchanging sending to receiving actions and vice versa. Breaking one of these conditions allows deadlock situations and breaking one of the first two conditions makes the compatibility checking undecidable [11, 19].

Multiparty Compatibility. This notion of duality is no longer effective in multiparty communications, where the whole conversation cannot be reconstructed from only a single behaviour. To bypass the gap between binary and multiparty, we take the *synthesis* approach, that is to find conditions which allow a global choreography to be built from the local machine behaviour. Instead of directly trying to decide whether the communications of a system will satisfy safety (which is undecidable in the general case), inferring a global type guarantees the safety as a direct consequence.

We give a simple example above to illustrate the problem. The Commit protocol involves three machines: Alice A, Bob B and Carol C. A orders B to act or quit. If act is sent, B sends a signal to C, and A sends a commitment to C and continues. Otherwise B informs C to save the data and A gives the final notification to C to terminate the protocol.

This paper presents a decidable notion of *multiparty compatibility* as a generalisation of duality of binary sessions, which in turns characterises a synthesis condition. The idea is to check the duality between each automaton and the rest, up to the internal communications (1-bounded executions in the terminology of CFSMs, see § 2) that the other machines will independently perform. For example, in the Commit example, to check the compatibility of trace $AB!$quit $AC!$finish in A, we observe the dual trace $AB?$quit $\cdot AC?$finish from B and C executing the internal communications between B and C such that $BC!$save $\cdot BC?$save. If this extended duality is valid for all the machines from any 1-bounded reachable state, then they satisfy multiparty compatibility and can build a well-formed global choreography.

Contributions and Outline. Section 3 defines new labelled transition systems for global and local types that represent the abstract observable behaviour of typed processes. We prove that a global type behaves exactly as its projected local types, and the same result between a single local type and its CFSMs interpretation. These correspondences are the key to prove the main theorems. Section 4 defines *multiparty compatibility*, studies its safety and liveness properties, gives an algorithm for the synthesis of global types from CFSMs, and proves the soundness and completeness results between global types and CFSMs. Section 5 discusses related work and concludes. The full proofs and applications of this work can be found in [17].

2   Communicating Finite State Machines

This section starts from some preliminary notations (following [6]). ε is the empty word. $\mathbb{A}$ is a finite alphabet and $\mathbb{A}^*$ is the set of all finite words over $\mathbb{A}$. $|x|$ is the length of a word x and $x.y$ or xy the concatenation of two words x and y. Let $\mathcal{P}$ be a set of *participants* fixed throughout the paper: $\mathcal{P} \subseteq \{A, B, C, \ldots, p, q, \ldots\}$.

Definition 2.1 (CFSM). A communicating finite state machine is a finite transition system given by a 5-tuple $M = (Q, C, q_0, \mathbb{A}, \delta)$ where (1) Q is a finite set of *states*; (2) $C = \{pq \in \mathcal{P}^2 \mid p \neq q\}$ is a set of channels; (3) $q_0 \in Q$ is an initial state; (4) $\mathbb{A}$ is a finite *alphabet* of messages, and (5) $\delta \subseteq Q \times (C \times \{!, ?\} \times \mathbb{A}) \times Q$ is a finite set of *transitions*.

In transitions, $pq!a$ denotes the *sending* action of a from process p to process q, and $pq?a$ denotes the *receiving* action of a from p by q. ℓ, ℓ' range over actions and we define the *subject* of an action ℓ as the principal in charge of it: $subj(pq!a) = subj(qp?a) = p$.

A state $q \in Q$ whose outgoing transitions are all labelled with sending (resp. receiving) actions is called a *sending* (resp. *receiving*) state. A state $q \in Q$ which does not have any outgoing transition is called *final*. If q has both sending and receiving outgoing transitions, q is called *mixed*. We say q is *directed* if it contains only sending (resp. receiving) actions to (resp. from) the same (identical) participant. A *path* in M is a finite sequence of $q_0, \ldots, q_n$ $(n \geq 1)$ such that $(q_i, \ell, q_{i+1}) \in \delta$ $(0 \leq i \leq n - 1)$, and we write

$q\xrightarrow{\ell}q'$ if $(q,\ell,q') \in \delta$. M is *connected* if for every state $q \neq q_0$, there is a path from q_0 to q. Hereafter we assume each CFSM is connected.

A CFSM $M = (Q,C,q_0,\mathbb{A},\delta)$ is *deterministic* if for all states $q \in Q$ and all actions ℓ, $(q,\ell,q'),(q,\ell,q'') \in \delta$ imply $q' = q''$.[1]

Definition 2.2 (CS). A (communicating) system S is a tuple $S = (M_{\mathsf{p}})_{\mathsf{p}\in\mathcal{P}}$ of CFSMs such that $M_{\mathsf{p}} = (Q_{\mathsf{p}},C,q_{0\mathsf{p}},\mathbb{A},\delta_{\mathsf{p}})$.

For $M_{\mathsf{p}} = (Q_{\mathsf{p}},C,q_{0\mathsf{p}},\mathbb{A},\delta_{\mathsf{p}})$, we define a *configuration* of $S = (M_{\mathsf{p}})_{\mathsf{p}\in\mathcal{P}}$ to be a tuple $s = (\vec{q};\vec{w})$ where $\vec{q} = (q_{\mathsf{p}})_{\mathsf{p}\in\mathcal{P}}$ with $q_{\mathsf{p}} \in Q_{\mathsf{p}}$ and where $\vec{w} = (w_{\mathsf{pq}})_{\mathsf{p}\neq\mathsf{q}\in\mathcal{P}}$ with $w_{\mathsf{pq}} \in \mathbb{A}^*$. The element $\vec{q}$ is called a *control state* and $q \in Q_{\mathsf{p}}$ is the *local state* of machine M_{p}.

Definition 2.3 (reachable state). Let S be a communicating system. A configuration $s' = (\vec{q}';\vec{w}')$ is *reachable* from another configuration $s = (\vec{q};\vec{w})$ by the *firing of the transition* t, written $s \rightarrow s'$ or $s\xrightarrow{t}s'$, if there exists $a \in \mathbb{A}$ such that either: (1) $t = (q_{\mathsf{p}},\mathsf{pq}!a,q'_{\mathsf{p}}) \in \delta_{\mathsf{p}}$ and (a) $q'_{\mathsf{p}'} = q_{\mathsf{p}'}$ for all $\mathsf{p}' \neq \mathsf{p}$; and (b) $w'_{\mathsf{pq}} = w_{\mathsf{pq}}.a$ and $w'_{\mathsf{p}'\mathsf{q}'} = w_{\mathsf{p}'\mathsf{q}'}$ for all $\mathsf{p}'\mathsf{q}' \neq \mathsf{pq}$; or (2) $t = (q_{\mathsf{q}},\mathsf{pq}?a,q'_{\mathsf{q}}) \in \delta_{\mathsf{q}}$ and (a) $q'_{\mathsf{p}'} = q_{\mathsf{p}'}$ for all $\mathsf{p}' \neq \mathsf{q}$; and (b) $w_{\mathsf{pq}} = a.w'_{\mathsf{pq}}$ and $w'_{\mathsf{p}'\mathsf{q}'} = w_{\mathsf{p}'\mathsf{q}'}$ for all $\mathsf{p}'\mathsf{q}' \neq \mathsf{pq}$.

The condition (1-b) puts the content a to a channel pq, while (2-b) gets the content a from a channel pq. The reflexive and transitive closure of $\rightarrow$ is $\rightarrow^*$. For a transition $t = (s,\ell,s')$, we refer to ℓ by $act(t)$. We write $s_1\xrightarrow{t_1\cdots t_m}s_{m+1}$ for $s_1\xrightarrow{t_1}s_2\cdots\xrightarrow{t_m}s_{m+1}$ and use φ to denote $t_1\cdots t_m$. We extend act to these sequences: $act(t_1\cdots t_n) = act(t_1)\cdots act(t_n)$.

The *initial configuration* of a system is $s_0 = (\vec{q}_0;\vec{\varepsilon})$ with $\vec{q}_0 = (q_{0\mathsf{p}})_{\mathsf{p}\in\mathcal{P}}$. A *final configuration* of the system is $s_f = (\vec{q};\vec{\varepsilon})$ with all $q_{\mathsf{p}} \in \vec{q}$ final. A configuration s is *reachable* if $s_0 \rightarrow^* s$ and we define the *reachable set* of S as $RS(S) = \{s \mid s_0 \rightarrow^* s\}$. We define the traces of a system S to be $Tr(S) = \{act(\varphi) \mid \exists s \in RS(S), s_0\xrightarrow{\varphi}s\}$.

We now define several properties about communicating systems and their configurations. These properties will be used in § 4 to characterise the systems that correspond to multiparty session types. Let S be a communicating system, t one of its transitions and $s = (\vec{q};\vec{w})$ one of its configurations. The following definitions of configuration properties follow [6, Definition 12].

1. s is *stable* if all its buffers are empty, i.e., $\vec{w} = \vec{\varepsilon}$.
2. s is a *deadlock configuration* if s is not final, and $\vec{w} = \vec{\varepsilon}$ and each q_{p} is a receiving state, i.e. all machines are blocked, waiting for messages.
3. s is an *orphan message configuration* if all $q_{\mathsf{p}} \in \vec{q}$ are final but $\vec{w} \neq \emptyset$, i.e. there is at least an orphan message in a buffer.
4. s is an *unspecified reception configuration* if there exists $\mathsf{q} \in \mathcal{P}$ such that q_{q} is a receiving state and $(q_{\mathsf{q}},\mathsf{pq}?a,q'_{\mathsf{q}}) \in \delta$ implies that $|w_{\mathsf{pq}}| > 0$ and $w_{\mathsf{pq}} \notin a\mathbb{A}^*$, i.e q_{q} is prevented from receiving any message from buffer pq.

A sequence of transitions is said to be *k-bounded* if no channel of any intermediate configuration s_i contains more than k messages. We define the *k-reachability set* of S to be the largest subset $RS_k(S)$ of $RS(S)$ within which each configuration s can be

---

[1] "Deterministic" often means the same channel should carry a unique value, i.e. if $(q,c!a,q') \in \delta$ and $(q,c!a',q'') \in \delta$ then $a = a'$ and $q' = q''$. Here we follow a different definition [6] in order to represent branching type constructs.

reached by a k-bounded execution from s_0. Note that, given a communicating system S, for every integer k, the set $RS_k(S)$ is finite and computable. We say that a trace φ is n-bound, written $bound(\varphi) = n$, if the number of send actions in φ never exceeds the number of receive actions by n. We then define the equivalences: (1) $S \approx S'$ is $\forall \varphi,\ \varphi \in Tr(S) \Leftrightarrow \varphi \in Tr(S')$; and (2) $S \approx_n S'$ is $\forall \varphi,\ bound(\varphi) \leq n \Rightarrow (\varphi \in Tr(S) \Leftrightarrow \varphi \in Tr(S'))$.

The following key properties will be examined throughout the paper as properties that multiparty session type can enforce. They are undecidable in general CFSMs.

Definition 2.4 (safety and liveness). (1) A communicating system S is *deadlock-free* (resp. *orphan message-free, reception error-free*) if for all $s \in RS(S)$, s is not a deadlock (resp. orphan message, unspecified reception) configuration. (2) S satisfies the *liveness property* if for all $s \in RS(S)$, there exists $s \longrightarrow^* s'$ such that s' is final.

3   Global and Local Types: The LTSs and Translations

This section presents multiparty session types, our main object of study. For the syntax of types, we follow [2] which is the most widely used syntax in the literature. We introduce two labelled transition systems, for local types and for global types, and show the equivalence between local types and communicating automata.

Syntax. A *global type*, written $G, G', ..$, describes the whole conversation scenario of a multiparty session as a type signature, and a *local type*, written by $T, T', ..$, type-abstract sessions from each end-point's view. $p, q, \cdots \in \mathcal{P}$ denote participants (see § 2 for conventions). The syntax of types is given as:

$$G \ ::= \ p \to p' : \{a_j.G_j\}_{j \in J} \ | \ \mu t.G \ | \ t \ | \ end$$
$$T \ ::= \ p?\{a_i.T_i\}_{i \in I} \ | \ p!\{a_i.T_i\}_{i \in I} \ | \ \mu t.T \ | \ t \ | \ end$$

$a_j \in \mathbb{A}$ corresponds to the usual message label in session type theory. We omit the mention of the carried types from the syntax in this paper, as we are not directly concerned by typing processes. Global branching type $p \to p' : \{a_j.G_j\}_{j \in J}$ states that participant p can send a message with one of the a_i labels to participant p' and that interactions described in G_j follow. We require $p \neq p'$ to prevent self-sent messages and $a_i \neq a_k$ for all $i \neq k \in J$. Recursive type $\mu t.G$ is for recursive protocols, assuming that type variables $(t, t', \dots)$ are guarded in the standard way, i.e. they only occur under branchings. Type end represents session termination (often omitted). $p \in G$ means that p appears in G.

Concerning local types, the *branching type* $p?\{a_i.T_i\}_{i \in I}$ specifies the reception of a message from p with a label among the a_i. The *selection type* $p!\{a_i.T_i\}_{i \in I}$ is its dual. The remaining type constructors are the same as global types. When branching is a singleton, we write $p \to p' : a.G'$ for global, and $p!a.T$ or $p?a.T$ for local.

Projection. The relation between global and local types is formalised by projection. Instead of the restricted original projection [2], we use the extension with the merging operator $\bowtie$ from [7]: it allows each branch of the global type to actually contain different interaction patterns. The *projection of G onto p* (written $G \upharpoonright p$) is defined as:

$$p \to p' : \{a_j.G_j\}_{j \in J} \upharpoonright q = \begin{cases} p!\{a_j.G_j \upharpoonright q\}_{j \in J} & q = p \\ p?\{a_j.G_j \upharpoonright q\}_{j \in J} & q = p' \\ \sqcup_{j \in J} G_j \upharpoonright q & \text{otherwise} \end{cases} \qquad (\mu t.G) \upharpoonright p = \begin{cases} \mu t.G \upharpoonright p & G \upharpoonright p \neq t \\ end & \text{otherwise} \end{cases}$$

$$t \upharpoonright p = t \qquad\qquad\qquad\qquad\qquad end \upharpoonright p = end$$

The mergeability relation $\bowtie$ is the smallest congruence relation over local types such that:

$$\frac{\forall i \in (K \cap J).T_i \bowtie T_i' \quad \forall k \in (K \setminus J), \forall j \in (J \setminus K).a_k \neq a_j}{p?\{a_k.T_k\}_{k \in K} \bowtie p?\{a_j.T_j'\}_{j \in J}}$$

When $T_1 \bowtie T_2$ holds, we define the operation $\sqcup$ as a partial commutative operator over two types such that $T \sqcup T = T$ for all types and that:

$$p?\{a_k.T_k\}_{k \in K} \sqcup p?\{a_j.T_j'\}_{j \in J} = p?(\{a_k.(T_k \sqcup T_k')\}_{k \in K \cap J} \cup \{a_k.T_k\}_{k \in K \setminus J} \cup \{a_j.T_j'\}_{j \in J \setminus K})$$

and homomorphic for other types (i.e. $\mathscr{C}[T_1] \sqcup \mathscr{C}[T_2] = \mathscr{C}[T_1 \sqcup T_2]$ where $\mathscr{C}$ is a context for local types). We say that G is *well-formed* if for all $p \in \mathcal{P}$, $G \restriction p$ is defined.

Example 3.1 (Commit). The global type for the commit protocol in § 1 is:

$\mu t.A \to B : \{act.B \to C : \{sig.A \to C : commit.t\}, quit.B \to C : \{save.A \to C : finish.end\}\}$

Then C's local type is: $\mu t.B?\{sig.A?\{commit.t\}, save.A?\{finish.end\}\}$.

We now present labelled transition relations (LTS) for global and local types and their sound and complete correspondence.

LTS over Global Types. We first designate the observables $(\ell, \ell', \ldots)$. We choose here to follow the definition of actions for CFSMs where a label ℓ denotes the sending or the reception of a message of label a from p to p': $\ell ::= pp'!a \mid pp'?a$

In order to define an LTS for global types, we need to represent intermediate states in the execution. For this reason, we introduce in the grammar of G the construct $p \rightsquigarrow p' : j \{a_i.G_i\}_{i \in I}$ to represent the fact that a_j has been sent but not yet received.

Definition 3.1 (LTS over global types.). The relation $G \xrightarrow{\ell} G'$ is defined as $(subj(\ell)$ is defined in § 2):

$$[GR1] \quad p \to p' : \{a_i.G_i\}_{i \in I} \xrightarrow{pp'!a_j} p \rightsquigarrow p' : j \{a_i.G_i\}_{i \in I} \quad (j \in I)$$

$$[GR2] \quad p \rightsquigarrow p' : j \{a_i.G_i\}_{i \in I} \xrightarrow{pp'?a_j} G_j \qquad [GR3] \frac{G[\mu t.G/t] \xrightarrow{\ell} G'}{\mu t.G \xrightarrow{\ell} G'}$$

$$[GR4] \frac{\forall j \in I \quad G_j \xrightarrow{\ell} G_j' \quad p,q \notin subj(\ell)}{p \to q : \{a_i.G_i\}_{i \in I} \xrightarrow{\ell} p \to q : \{a_i.G_i'\}_{i \in I}} [GR5] \frac{G_j \xrightarrow{\ell} G_j' \quad q \notin subj(\ell) \quad \forall i \in I \setminus j, G_i' = G_i}{p \rightsquigarrow q : j \{a_i.G_i\}_{i \in I} \xrightarrow{\ell} p \rightsquigarrow q : j \{a_i.G_i'\}_{i \in I}}$$

[GR1] represents the emission of a message while [GR2] describes the reception of a message. [GR3] governs recursive types. [GR4,5] define the asynchronous semantics of global types, where the syntactic order of messages is enforced only for the participants that are involved. For example, when the participants of two consecutive communications are disjoint, as in: $G_1 = A \to B : a.C \to D : b.$end, we can observe the emission (and possibly the reception) of b before the interactions of a (by [GR4]). A more interesting example is: $G_2 = A \to B : a.A \to C : b.$end. We write $\ell_1 = AB!a$, $\ell_2 = AB?a$, $\ell_3 = AC!b$ and $\ell_4 = AC?b$. The LTS allows the following three sequences:

$$G_2 \xrightarrow{\ell_1} A \rightsquigarrow B : a.A \to C : b.\text{end} \xrightarrow{\ell_2} A \to C : b.\text{end} \xrightarrow{\ell_3} A \rightsquigarrow C : b.\text{end} \xrightarrow{\ell_4} \text{end}$$

$$G_2 \xrightarrow{\ell_1} A \rightsquigarrow B : a.A \to C : b.\text{end} \xrightarrow{\ell_3} A \rightsquigarrow B : a.A \rightsquigarrow C : b.\text{end} \xrightarrow{\ell_2} A \rightsquigarrow C : b.\text{end} \xrightarrow{\ell_4} \text{end}$$

$$G_2 \xrightarrow{\ell_1} A \rightsquigarrow B : a.A \to C : b.\text{end} \xrightarrow{\ell_3} A \rightsquigarrow B : a.A \rightsquigarrow C : b.\text{end} \xrightarrow{\ell_4} A \rightsquigarrow B : a.\text{end} \xrightarrow{\ell_2} \text{end}$$

The last sequence is the most interesting: the sender A has to follow the syntactic order but the receiver C can get the message b before B receives a. The respect of these constraints is enforced by the conditions $p, q \notin subj(\ell)$ and $q \notin subj(\ell)$ in rules [GR4,5].

LTS over Local Types. We define the LTS over local types. This is done in two steps, following the model of CFSMs, where the semantics is given first for individual automata and then extended to communicating systems. We use the same labels $(\ell, \ell', ...)$ as the ones for CFSMs.

Definition 3.2 (LTS over local types). The relation $T \xrightarrow{\ell} T'$, for the local type of role p, is defined as:

$$[LR1]\ q!\{a_i.T_i\}_{i \in I} \xrightarrow{pq!a_i} T_i \quad [LR2]\ q?\{a_i.T_i\}_{i \in I} \xrightarrow{qp?a_j} T_j \quad [LR3]\ \frac{T[\mu t.T/t] \xrightarrow{\ell} T'}{\mu t.T \xrightarrow{\ell} T'}$$

The semantics of a local type follows the intuition that every action of the local type should obey the syntactic order. We define the LTS for collections of local types.

Definition 3.3 (LTS over collections of local types). A configuration $s = (\vec{T}; \vec{w})$ of a system of local types $\{T_p\}_{p \in \mathcal{P}}$ is a pair with $\vec{T} = (T_p)_{p \in \mathcal{P}}$ and $\vec{w} = (w_{pq})_{p \neq q \in \mathcal{P}}$ with $w_{pq} \in \mathbb{A}^*$. We then define the transition system for configurations. For a configuration $s_T = (\vec{T}; \vec{w})$, the visible transitions of $s_T \xrightarrow{\ell} s'_T = (\vec{T}'; \vec{w}')$ are defined as: (1) $T_p \xrightarrow{pq!a} T'_p$ and (a) $T'_{p'} = T_{p'}$ for all $p' \neq p$; and (b) $w'_{pq} = w_{pq} \cdot a$ and $w'_{p'q'} = w_{p'q'}$ for all $p'q' \neq pq$; or (2) $T_q \xrightarrow{pq?a} T'_q$ and (a) $T'_{p'} = T_{p'}$ for all $p' \neq q$; and (b) $w_{pq} = a \cdot w'_{pq}$ and $w'_{p'q'} = w_{p'q'}$ for all $p'q' \neq pq$.

The semantics of local types is therefore defined over configurations, following the definition of the semantics of CFSMs. w_{pq} represents the FIFO queue at channel pq. We write $Tr(G)$ to denote the set of the visible traces that can be obtained by reducing G. Similarly for $Tr(T)$ and $Tr(S)$. We extend the trace equivalences $\approx$ and $\approx_n$ in § 2 to global types and configurations of local types.

We now state the soundness and completeness of projection w.r.t. the LTSs.

Theorem 3.1 (soundness and completeness). [2] *Let G be a global type with participants $\mathcal{P}$ and let $\vec{T} = \{G \restriction p\}_{p \in \mathcal{P}}$ be the local types projected from G. Then $G \approx (\vec{T}; \vec{\varepsilon})$.*

Local types and CFSMs Next we show how to algorithmically go from local types to CFSMs and back while preserving the trace semantics. We start by translating local types into CFSMs.

Definition 3.4 (translation from local types to CFSMs). Write $T' \in T$ if T' occurs in T. Let T_0 be the local type of participant p projected from G. The automaton corresponding to T_0 is $\mathcal{A}(T_0) = (Q, C, q_0, \mathbb{A}, \delta)$ where: (1) $Q = \{T' \mid T' \in T_0, T' \neq t, T' \neq \mu t.T\}$; (2) $q_0 = T'_0$ with $T_0 = \mu \vec{t}.T'_0$ and $T'_0 \in Q$; (3) $C = \{pq \mid p, q \in G\}$; (4) $\mathbb{A}$ is the set of $\{a \in G\}$; and (5) δ is defined as:

---

[2] The local type abstracts the behaviour of multiparty typed processes as proved in the subject reduction theorem in [13]. Hence this theorem implies that processes typed by global type G by the typing system in [2, 13] follow the LTS of G.

If $T = \mathrm{p'}!\{a_j.T_j\}_{j \in J} \in Q$, then $\begin{cases} (T,(\mathrm{pp'}!a_j),T_j) \in \delta & T_j \neq \mathrm{t} \\ (T,(\mathrm{pp'}!a_j),T') \in \delta & T_j = \mathrm{t}, \ \mu\vec{\mathrm{t}}\mathrm{t}.T' \in T_0, T' \in Q \end{cases}$

If $T = \mathrm{p'}?\{a_j.T_j\}_{j \in J} \in Q$, then $\begin{cases} (T,(\mathrm{p'p}?a_j),T_j) \in \delta & T_j \neq \mathrm{t} \\ (T,(\mathrm{p'p}?a_j),T') \in \delta & T_j = \mathrm{t} \ \mu\vec{\mathrm{t}}\mathrm{t}.T' \in T_0, T' \in Q \end{cases}$

The definition says that the set of states Q are the suboccurrences of branching or selection or end in the local type; the initial state q_0 is the occurrence of (the recursion body of) T_0; the channels and alphabets correspond to those in T_0; and the transition is defined from the state T to its body T_j with the action $\mathrm{pp'}!a_j$ for the output and $\mathrm{pp'}?a_j$ for the input. If T_j is a recursive type variable t, it points the state of the body of the corresponding recursive type. As an example, see C's local type in Example 3.1 and its corresponding automaton in § 1.

Proposition 3.1 (local types to CFSMs). *Assume T_p is a local type. Then $\mathcal{A}(T_\mathrm{p})$ is deterministic, directed and has no mixed states.*

We say that a CFSM is *basic* if it is deterministic, directed and has no mixed states. Any basic CFSM can be translated into a local type.

Definition 3.5 (translation from a basic CFSM to a local type). From a basic $M_\mathrm{p} = (Q, C, q_0, \mathbb{A}, \delta)$, we define the translation $\mathcal{T}(M_\mathrm{p})$ such that $\mathcal{T}(M_\mathrm{p}) = \mathcal{T}_\varepsilon(q_0)$ where $\mathcal{T}_{\tilde{q}}(q)$ is defined as:

(1) $\mathcal{T}_{\tilde{q}}(q) = \mu\mathrm{t}_q.\mathrm{p'}!\{a_j.\mathcal{T}^\circ_{\tilde{q}\cdot q}(q_j)\}_{j \in J}$ if $(q, \mathrm{pp'}!a_j, q_j) \in \delta$;

(2) $\mathcal{T}_{\tilde{q}}(q) = \mu\mathrm{t}_q.\mathrm{p'}?\{a_j.\mathcal{T}^\circ_{\tilde{q}\cdot q}(q_j)\}_{j \in J}$ if $(q, \mathrm{p'p}?a_j, q_j) \in \delta$;

(3) $\mathcal{T}^\circ_{\tilde{q}}(q) = \mathcal{T}_\varepsilon(q) = \mathrm{end}$ if q is final; (4) $\mathcal{T}_{\tilde{q}}(q) = \mathrm{t}_{q_k}$ if $(q, \ell, q_k) \in \delta$ and $q_k \in \tilde{q}$; and

(5) $\mathcal{T}^\circ_{\tilde{q}}(q) = \mathcal{T}_{\tilde{q}}(q)$ otherwise.

Finally, we replace $\mu\mathrm{t}.T$ by T if t is not in T.

In $\mathcal{T}_{\tilde{q}}$, $\tilde{q}$ records visited states; (1,2) translate the receiving and sending states to branching and selection types, respectively; (3) translates the final state to end; and (4) is the case of a recursion: since q_k was visited, ℓ is dropped and replaced by the type variable.

The following proposition states that these translations preserve the semantics.

Proposition 3.2 (translations between CFSMs and local types). *If a CFSM M is basic, then $M \approx \mathcal{T}(M)$. If T is a local type, then $T \approx \mathcal{A}(T)$.*

4 Completeness and Synthesis

This section studies the synthesis and sound and complete characterisation of multiparty session types as communicating automata. A first idea would be to restrict basic CFSMs to the natural generalisation of half-duplex systems [6, § 4.1.1], in which each pair of machines linked by two channels, one in each direction, communicates in a half-duplex way. In this class, the safety properties of Definition 2.4 are however undecidable [6, Theorem 36]. We therefore need a stronger (and decidable) property to force basic CFSMs to behave as if they were the result of a projection from global types.

Multiparty compatibility In the two machines case, there exists a sound and complete condition called *compatible* [11]. Let us define the isomorphism $\Phi : (C \times \{!,?\} \times \mathbb{A})^* \longrightarrow (C \times \{!,?\} \times \mathbb{A})^*$ such that $\Phi(j?a) = j!a$, $\Phi(j!a) = j?a$, $\Phi(\varepsilon) = \varepsilon$, $\Phi(t_1 \cdots t_n) = \Phi(t_1) \cdots \Phi(t_n)$. Φ exchanges a sending action with the corresponding receiving one and vice versa. The compatibility of two machines can be immediately defined as $Tr(M_1) = \Phi(Tr(M_2))$ (i.e. the traces of M_1 are exactly the set of dual traces of M_2). The idea of the extension to the multiparty case comes from the observation that from the viewpoint of the participant p, the rest of all the machines $(M_q)_{q \in \mathcal{P} \setminus p}$ should behave as if they were one CFSM which offers compatible traces $\Phi(Tr(M_p))$, up to internal synchronisations (i.e. 1-bounded executions). Below we define a way to group CFSMs.

Definition 4.1 (Definition 37, [6]). Let $M_i = (Q_i, C_i, q_{0i}, \mathbb{A}_i, \delta_i)$. The *associated CFSM* of a system $S = (M_1, .., M_n)$ is $M = (Q, C, q_0, \Sigma, \delta)$ such that: $Q = Q_1 \times Q_2 \times \cdots \times Q_n$, $q_0 = (q_{01}, \ldots, q_{0n})$ and δ is the smallest relation for which: if $(q_i, \ell, q_i') \in \delta_i$ $(1 \leq i \leq n)$, then $((q_1, ..., q_i, ..., q_n), \ell, (q_1, ..., q_i', ..., q_n)) \in \delta$.

We now define a notion of compatibility extended to more than two CFSMs. We say that φ is an *alternation* if φ is an alternation of sending and corresponding receive actions (i.e. the action pq!a is immediately followed by pq?a).

Definition 4.2 (multiparty compatible system). A system $S = (M_1, .., M_n)$ $(n \geq 2)$ is *multiparty compatible* if for any 1-bounded reachable stable state $s \in RS_1(S)$, for any sending action ℓ and for at least one receiving action ℓ from s in M_i, there exists a sequence of transitions $\varphi \cdot t$ from s in a CFSM corresponding to $S^{-i} = (M_1, \ldots, M_{i-1}, M_{i+1}, \ldots, M_n)$ where φ is either empty or an alternation and $\ell = \Phi(act(t))$ and $i \notin act(\varphi)$ (i.e. φ does not contain actions to or from channel i).

The above definition states that for each M_i, the rest of machines S^{-i} can produce the compatible (dual) actions by executing alternations in S^{-i}. From M_i, these intermediate alternations can be seen as non-observable internal actions.

Example 4.1 (multiparty compatibility). As an example, we can test the multiparty compatibility property on the commit example in § 1. We only detail here how to check the compatibility from the point of view of A. To check the compatibility for the actions $act(t_1 \cdot t_2) = AB!$quit $\cdot AC!$finish, the only possible action is $\Phi(act(t_1)) = AB?$quit from B, then a 1-bounded excecution is $BC!$save $\cdot BC?$save, and $\Phi(act(t_2)) = AC?$finish from C. To check the compatibility for the actions $act(t_3 \cdot t_4) = AB!$act $\cdot AC!$commit, $\Phi(act(t_3)) = AB?$act from B, the 1-bound execution is $BC!$sig $\cdot BC?$sig, and $\Phi(act(t_4)) = AC?$commit from C.

Remark 4.1. In Definition 4.2, we check the compatibility from any 1-bounded reachable stable state in the case one branch is selected by different senders. Consider the following machines:

In A, B and C, each action in each machine has its dual but they do not satisfy multiparty compatibility. For example, if $BA!a \cdot BA?a$ is executed, $CA!d$ does not have a dual action (hence they do not satisfy the safety properties). On the other hand, the machines A', B and C satisfy the multiparty compatibility.

Theorem 4.1. *Assume $S = (M_p)_{p \in \mathcal{P}}$ is basic and multiparty compatible. Then S satisfies the three safety properties in Definition 2.4. Further, if there exists at least one M_q which includes a final state, then S satisfies the liveness property.*

Proposition 4.1. *If all the CFSMs M_p ($p \in \mathcal{P}$) are basic, there is an algorithm to check whether $(M_p)_{p \in \mathcal{P}}$ is multiparty compatible.*

The proof of Theorem 4.1 is non-trivial, using a detailed analysis of causal relations. The proof of Proposition 4.1 comes from the finiteness of $RS_1(S)$. See [17] for details.

Synthesis. Below we state the lemma which will be crucial for the proof of synthesis and completeness. The lemma comes from the intuition that the transitions of multiparty compatible systems are always permutations of one-bounded executions as it is the case in multiparty session types. See [17] for the proof.

Lemma 4.1 (1-buffer equivalence). *Suppose S_1 and S_2 are two basic and multiparty compatible communicating systems such that $S_1 \approx_1 S_2$, then $S_1 \approx S_2$.*

Theorem 4.2 (synthesis). *Suppose S is a basic system and multiparty compatible. Then there is an algorithm which successfully builds well-formed G such that $S \approx G$ if such G exists, and otherwise terminates.*

Proof. We assume $S = (M_p)_{p \in \mathcal{P}}$. The algorithm starts from the initial states of all machines $(q^{p_1}{}_0, ..., q^{p_n}{}_0)$. We take a pair of the initial states which is a sending state q_0^p and a receiving state q_0^q from p to q. We note that by directness, if there are more than two pairs, the participants in two pairs are disjoint, and by [G4] in Definition 3.1, the order does not matter. We apply the algorithm with the invariant that all buffers are empty and that we repeatedly pick up one pair such that q_p (sending state) and q_q (receiving state). We define $G(q_1, ..., q_n)$ where $(q_p, q_q \in \{q_1, ..., q_n\})$ as follows:

- if $(q_1, ..., q_n)$ has already been examined and if all participants have been involved since then (or the ones that have not are in their final state), we set $G(q_1, ..., q_n)$ to be $t_{q_1,...,q_n}$. Otherwise, we select a pair sender/receiver from two participants that have not been involved (and are not final) and go to the next step;
- otherwise, in q_p, from machine p, we know that all the transitions are sending actions towards p' (by directedness), i.e. of the form $(q_p, pq!a_i, q_i) \in \delta_p$ for $i \in I$.
  - we check that machine q is in a receiving state q_q such that $(q_q, pq?a_j, q'_j) \in \delta_{p'}$ with $j \in J$ and $I \subseteq J$.
  - we set $\mu t_{q_1,...,q_n}.p \to q: \{a_i.G(q_1, ..., q_p \leftarrow q_i, ..., q_q \leftarrow q'_i, ..., q_n)\}_{i \in I}$ (we replace q_p and q_q by q_i and q'_i, respectively) and continue by recursive calls.
  - if all sending states in $q_1, ..., q_n$ become final, then we set $G(q_1, ..., q_n) = \text{end}$.
- we erase unnecessary μt if $t \notin G$.

Since the algorithm only explores 1-bounded executions, the reconstructed G satisfies $G \approx_1 S$. By Theorem 3.1, we know that $G \approx (\{G \upharpoonright p\}_{p \in \mathcal{P}}; \vec{\varepsilon})$. Hence, by Proposition 3.2, we have $G \approx S'$ where S' is the communicating system translated from the projected local types $\{G \upharpoonright p\}_{p \in \mathcal{P}}$ of G. By Lemma 4.1, $S \approx S'$ and therefore $S \approx G$.   □

The algorithm can generate the global type in Example 3.1 from CFSMs in § 1and the global type $B \rightarrow A\{a : C \rightarrow A : \{c : \mathsf{end}, d : \mathsf{end}\}, b : C \rightarrow A : \{c : \mathsf{end}, d : \mathsf{end}\}\}$ from A', B and C in Remark 4.1. Note that $B \rightarrow A\{a : C \rightarrow A : \{c : \mathsf{end}\}, b : C \rightarrow A : \{d : \mathsf{end}\}\}$ generated by A, B and C in Remark 4.1 is not projectable, hence not well-formed.

By Theorems 3.1 and 4.1, and Proposition 3.2, we can now conclude:

Theorem 4.3 (soundness and completeness). *Suppose S is basic and multiparty compatible. Then there exists G such that $S \approx G$. Conversely, if G is well-formed, then there exists a basic and multiparty compatible system S such that $S \approx G$.*

5 Conclusion and Related Work

This paper investigated the sound and complete characterisation of multiparty session types into CFSMs and developed a decidable synthesis algorithm from basic CFSMs. The main tool we used is a new extension to multiparty interactions of the duality condition for binary session types, called *multiparty compatibility*. The basic condition (coming from binary session types) and the multiparty compatibility property are a *necessary and sufficient condition* to obtain safe global types. Our aim is to offer a duality notion which would be applicable to extend other theoretical foundations such as the Curry-Howard correspondence with linear logics [4,20] to multiparty communications. Basic multiparty compatible CFSMs also define one of the few non-trivial decidable subclass of CFSMs which satisfy deadlock-freedom. The methods proposed here are palatable to a wide range of applications based on choreography protocol models and more widely, finite state machines. Multiparty compatibility is applicable for extending the synthesis algorithm to build more expressive graph-based global types (*general global types* [8]) which feature fork and join primitives [9].

Our previous work [8] presented the first translation from global and local types into CFSMs. It only analysed the properties of the automata resulting from such a translation. The complete characterisation of global types independently from the projected local types was left open, as was synthesis. This present paper closes this open problem. There are a large number of paper that can be found in the literature about the synthesis of CFSMs. See [16] for a summary of recent results. The main distinction with CFSM synthesis is, apart from the formal setting (i.e. types), about the kind of the target specifications to be generated (global types in our case). Not only our synthesis is concerned about trace properties (languages) like the standard synthesis of CFSMs (the problem of the closed synthesis of CFSMs is usually defined as the construction from a regular language L of a machine satisfying certain conditions related to buffer boundedness, deadlock-freedom and words swapping), but we also generate concrete syntax or choreography descriptions as *types* of programs or software. Hence they are directly applicable to programming languages and can be straightforwardly integrated into the existing frameworks that are based on session types.

Within the context of multiparty session types, [15] first studied the reconstruction of a global type from its projected local types up to asynchronous subtyping and [14] recently offers a typing system to synthesise global types from local types. Our synthesis based on CFSMs is more general since CFSMs do not depend on the syntax. For example, [14, 15] cannot treat the synthesis for A', B and C in Remark 4.1. These works also do not study the completeness (i.e. they build a global type from a set of projected local types (up to subtyping), and do not investigate necessary and sufficient conditions to build a well-formed global type). A difficulty of the completeness result is that it is generally unknown if the global type constructed by the synthesis can simulate executions with arbitrary buffer bounds since the synthesis only directly looks at 1-bounded executions. In this paper, we proved Lemma 4.1 and bridged this gap towards the complete characterisation. Recent work by [1, 5] focus on proving the semantic correspondence between global and local descriptions (see [8] for more detailed comparison), but no synthesis algorithm is studied.

Acknowledgement. The work has been partially sponsored by the Ocean Observatories Initiative and EPSRC EP/K011715/1, EP/K034413/1 and EP/G015635/1.

References

1. Basu, S., Bultan, T., Ouederni, M.: Deciding choreography realizability. In: POPL 2012, pp. 191–202. ACM (2012)
2. Bettini, L., Coppo, M., D'Antoni, L., De Luca, M., Dezani-Ciancaglini, M., Yoshida, N.: Global progress in dynamically interleaved multiparty sessions. In: van Breugel, F., Chechik, M. (eds.) CONCUR 2008. LNCS, vol. 5201, pp. 418–433. Springer, Heidelberg (2008)
3. Brand, D., Zafiropulo, P.: On communicating finite-state machines. J. ACM 30, 323–342 (1983)
4. Caires, L., Pfenning, F.: Session types as intuitionistic linear propositions. In: Gastin, P., Laroussinie, F. (eds.) CONCUR 2010. LNCS, vol. 6269, pp. 222–236. Springer, Heidelberg (2010)
5. Castagna, G., Dezani-Ciancaglini, M., Padovani, L.: On global types and multi-party session. LMCS 8(1) (2012)
6. Cécé, G., Finkel, A.: Verification of programs with half-duplex communication. Inf. Comput. 202(2), 166–190 (2005)
7. Deniélou, P.-M., Yoshida, N.: Dynamic multirole session types. In: POPL, pp. 435–446. ACM, Full version, Prototype at http://www.doc.ic.ac.uk/~pmalo/dynamic
8. Deniélou, P.-M., Yoshida, N.: Multiparty session types meet communicating automata. In: Seidl, H. (ed.) ESOP 2012. LNCS, vol. 7211, pp. 194–213. Springer, Heidelberg (2012)
9. http://arxiv.org/abs/1304.1902
10. Girard, J.-Y.: Linear logic. TCS 50 (1987)
11. Gouda, M., Manning, E., Yu, Y.: On the progress of communication between two finite state machines. Information and Control 63, 200–216 (1984)
12. Honda, K., Vasconcelos, V.T., Kubo, M.: Language primitives and type discipline for structured communication-based programming. In: Hankin, C. (ed.) ESOP 1998. LNCS, vol. 1381, pp. 122–138. Springer, Heidelberg (1998)
13. Honda, K., Yoshida, N., Carbone, M.: Multiparty Asynchronous Session Types. In: POPL 2008, pp. 273–284. ACM (2008)

14. Lange, J., Tuosto, E.: Synthesising choreographies from local session types. In: Koutny, M., Ulidowski, I. (eds.) CONCUR 2012. LNCS, vol. 7454, pp. 225–239. Springer, Heidelberg (2012)
15. Mostrous, D., Yoshida, N., Honda, K.: Global principal typing in partially commutative asynchronous sessions. In: Castagna, G. (ed.) ESOP 2009. LNCS, vol. 5502, pp. 316–332. Springer, Heidelberg (2009)
16. Muscholl, A.: Analysis of communicating automata. In: Dediu, A.-H., Fernau, H., Martín-Vide, C. (eds.) LATA 2010. LNCS, vol. 6031, pp. 50–57. Springer, Heidelberg (2010)
17. DoC Technical Report, Imperial College London, Computing, DTR13-5 (2013)
18. Takeuchi, K., Honda, K., Kubo, M.: An interaction-based language and its typing system. In: Halatsis, C., Philokyprou, G., Maritsas, D., Theodoridis, S. (eds.) PARLE 1994. LNCS, vol. 817, pp. 398–413. Springer, Heidelberg (1994)
19. Villard, J.: Heaps and Hops. PhD thesis, ENS Cachan (2011)
20. Wadler, P.: Proposition as Sessions. In: ICFP 2012, pp. 273–286 (2012)

Component Reconfiguration in the Presence of Conflicts[*]

Roberto Di Cosmo[1], Jacopo Mauro[2], Stefano Zacchiroli[1], and Gianluigi Zavattaro[2]

[1] Univ Paris Diderot, Sorbonne Paris Cité, PPS, UMR 7126, CNRS, F-75205 Paris, France
roberto@dicosmo.org, zack@pps.univ-paris-diderot.fr
[2] Focus Team, Univ of Bologna/INRIA, Italy, Mura A. Zamboni, 7, Bologna
{jmauro,zavattar}@cs.unibo.it

Abstract. Components are traditionally modeled as black-boxes equipped with interfaces that indicate provided/required ports and, often, also conflicts with other components that cannot coexist with them. In modern tools for automatic system management, components become *grey*-boxes that show relevant internal states and the possible actions that can be acted on the components to change such state during the deployment and reconfiguration phases. However, state-of-the-art tools in this field do not support a systematic management of conflicts. In this paper we investigate the impact of conflicts by precisely characterizing the increment of complexity on the reconfiguration problem.

1 Introduction

Modern software systems are more and more based on interconnected software components (e.g. packages or services) deployed on clusters of heterogeneous machines that can be created, connected and reconfigured on-the-fly. Traditional component models represent components as black-boxes with interfaces indicating their *provide* and *require* ports. In many cases also *conflicts* are considered in order to deal with frequent situations in which components cannot be co-installed.

In software systems where components are frequently reconfigured (e.g. "cloud" based applications that elastically reacts to client demands) more expressive component models are considered: a component becomes a grey-box showing relevant internal states and the actions that can be acted on the component to change state during deployment and reconfiguration. For instance, in the popular system configuration tool Puppet [10] or the novel deployment management system Engage [8], components can be in the *absent, present, running* or *stopped* states, and the actions *install, uninstall, start, stop* and *restart* can be executed upon them. Rather expressive dependencies among components can be declared. The aim of these tools is to allow the system administrator to declaratively express the desired component configuration and automatically execute a correct sequence of low-level actions that bring the current configuration to a new one satisfying the administrator requests respecting dependencies. We call *reconfigurability* the problem of checking the existence of such sequence of low-level actions.

---

[*] Work partially supported by Aeolus project, ANR-2010-SEGI-013-01, and performed at IR-ILL, center for Free Software Research and Innovation in Paris, France, www.irill.org

Despite the importance of conflicts in many component models, see e.g. package-based software distributions used for Free and Open Source Software (FOSS) [5], the Eclipse plugin model [3], or the OSGi component framework [12], state-of-the-arts management systems like the above do not take conflicts into account. This is likely ascribable to the increased complexity of the reconfigurability problem in the presence of conflicts. In this paper we precisely characterize this increment of complexity.

In a related paper [6] we have proposed the Aeolus component model that, despite its simplicity, is expressive enough to capture the main features of tools like Puppet and Engage. We have proved that the reconfigurability problem is Polynomial-Time for Aeolus⁻, the fragment without numerical constraints. In this paper we consider Aeolus core, the extension of this fragment with conflicts, and we prove that even if the reconfigurability problem remains decidable, it turns out to be Exponential-Space hard. We consider this result a fundamental step towards the realization of tools that manage conflicts systematically. In fact, we shed some light on the specific sources of the increment of complexity of the reconfigurability problem.

The technical contribution of the paper and its structure is as follows. In Section 2 we formalize the reconfigurability problem in the presence of conflicts. In Section 3 we prove its decidability by resorting to the theory of Well-Structured Transition Systems [2,7]. We consider this decidability result interesting also from a foundational viewpoint: despite our component model has many commonalities with concurrent models like Petri nets, in our case the addition of conflicts (corresponding to inhibitor arcs in Petri nets) does not make the analysis of reachability problems undecidable. The closed relationship between our model and Petri nets is used in Section 4 where we prove the Exponential-Space hardness of the reconfigurability problem by reduction from the coverability problem in Petri nets. In Section 5 we discuss related work and report concluding remarks. Missing proofs are available in [4].

2 The Aeolus core Model

The Aeolus core model represents relevant internal states of components by means of a *finite state automaton* (see Fig. 1): depending on its state components activate *provide* and *require* functionalities (called *ports*), and get in *conflict* with ports provided by others (in Fig. 1 active ports are black while inactive ones are grey). Each port is identified by an interface name. Bindings can be established between provide and require ports with the same interface. Fig. 1 shows the graphical representation of a typical deployment of the popular WordPress blog. According to the Debian packages metadata, WordPress requires a Web server providing httpd in order to be installed, and an active MySQL database server in order to be in production. The chosen Web server is Apache2 which is broken into various packages (e.g. `apache2`, `apache2-bin`) that shall be simultaneously installed. Notice that Apache2 is not co-installable with other Web servers, such as lighttpd.

We now move to the formal definition of Aeolus core. We assume given a set $\mathscr{I}$ of interface names.

Definition 1 (Component type). *The set Γ of component types of the Aeolus core model, ranged over by $\mathscr{T}, \mathscr{T}_1, \mathscr{T}_2, \ldots$ contains 4-ples $\langle Q, q_0, T, D \rangle$ where:*

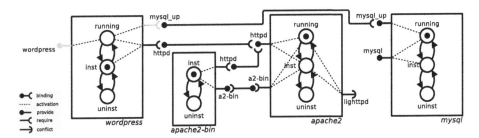

Fig. 1. Typical Wordpress/Apache/MySQL deployment, modeled in Aeolus core

- Q is a finite set of states containing the initial state q_0;
- $T \subseteq Q \times Q$ is the set of transitions;
- D is a function from Q to a 3-ple $\langle \mathbf{P}, \mathbf{R}, \mathbf{C} \rangle$ of interface names (i.e. $\mathbf{P}, \mathbf{R}, \mathbf{C} \subseteq \mathscr{I}$) indicating the provide, require, and conflict ports that each state activates. We assume that the initial state q_0 has no requirements and conflicts (i.e. $D(q_0) = \langle \mathbf{P}, \emptyset, \emptyset \rangle$).

We now define configurations that describe systems composed by components and their bindings. Each component has a unique identifier taken from the set $\mathscr{Z}$. A configuration, ranged over by $\mathscr{C}_1, \mathscr{C}_2, \ldots$, is given by a set of component types, a set of components in some state, and a set of bindings.

Definition 2 (Configuration). *A configuration $\mathscr{C}$ is a 4-ple $\langle U, Z, S, B \rangle$ where:*

- $U \subseteq \Gamma$ is the finite universe of the available component types;
- $Z \subseteq \mathscr{Z}$ is the set of the currently deployed components;
- S is the component state description, i.e. a function that associates to components in Z a pair $\langle \mathscr{T}, q \rangle$ where $\mathscr{T} \in U$ is a component type $\langle Q, q_0, T, D \rangle$, and $q \in Q$ is the current component state;
- $B \subseteq \mathscr{I} \times Z \times Z$ is the set of bindings, namely 3-ple composed by an interface, the component that requires that interface, and the component that provides it; we assume that the two components are different.

Configuration are equivalent if they have the same instances up to instance renaming.

Definition 3 (Configuration equivalence). *Two configurations $\langle U, Z, S, B \rangle$ and $\langle U, Z', S', B' \rangle$ are equivalent ($\langle U, Z, S, B \rangle \equiv \langle U, Z', S', B' \rangle$) iff there exists a bijective function ρ from Z to Z' s.t.*

- $S(z) = S'(\rho(z))$ for every $z \in Z$;
- $\langle r, z_1, z_2 \rangle \in B$ iff $\langle r, \rho(z_1), \rho(z_2) \rangle \in B'$.

Notation. We write $\mathscr{C}[z]$ as a lookup operation that retrieves the pair $\langle \mathscr{T}, q \rangle = S(z)$, where $\mathscr{C} = \langle U, Z, S, B \rangle$. On such a pair we then use the postfix projection operators `.type` and `.state` to retrieve $\mathscr{T}$ and q, respectively. Similarly, given a component type $\langle Q, q_0, T, D \rangle$, we use projections to decompose it: `.states`, `.init`, and `.trans` return the first three elements; $.\mathbf{P}(q)$, $.\mathbf{R}(q)$, and $.\mathbf{C}(q)$ return the three elements of the $D(q)$ tuple. Moreover, we use `.prov` (resp. `.req`) to denote the union of all the provide ports (resp. require ports) of the states in Q. When there is no ambiguity we take the liberty to apply the component type projections to $\langle \mathcal{T}, q \rangle$ pairs. *Example:* $\mathcal{C}[z].\mathbf{R}(q)$ stands for the require ports of component z in configuration $\mathcal{C}$ when it is in state q.

We can now formalize the notion of configuration correctness.

Definition 4 (Correctness). *Let us consider the configuration* $\mathcal{C} = \langle U, Z, S, B \rangle$.

We write $\mathcal{C} \models_{req} (z, r)$ *to indicate that the require port of component z, with interface r, is bound to an active port providing r, i.e. there exists a component $z' \in Z \setminus \{z\}$ such that $\langle r, z, z' \rangle \in B$, $\mathcal{C}[z'] = \langle \mathcal{T}', q' \rangle$ and r is in $\mathcal{T}'.\mathbf{P}(q')$. Similarly, for conflicts, we write* $\mathcal{C} \models_{cnf} (z, c)$ *to indicate that the conflict port c of component z is satisfied because no other component has an active port providing c, i.e. for every $z' \in Z \setminus \{z\}$ with $\mathcal{C}[z'] = \langle \mathcal{T}', q' \rangle$ we have that $c \notin \mathcal{T}'.\mathbf{P}(q')$.*

The configuration $\mathcal{C}$ is correct *if for every component $z \in Z$ with $S(z) = \langle \mathcal{T}, q \rangle$ we have that $\mathcal{C} \models_{req} (z, r)$ for every $r \in \mathcal{T}.\mathbf{R}(q)$ and $\mathcal{C} \models_{cnf} (z, c)$ for every $c \in \mathcal{T}.\mathbf{C}(q)$.*

Configurations evolve at the granularity of actions.

Definition 5 (Actions). *The set $\mathscr{A}$ contains the following actions:*

- $stateChange(\langle z_1, q_1, q_1' \rangle, \ldots, \langle z_n, q_n, q_n' \rangle)$ *where $z_i \in \mathscr{Z}$ and $\forall i \neq j \, . \, z_i \neq z_j$;*
- $bind(r, z_1, z_2)$ *where $z_1, z_2 \in \mathscr{Z}$ and $r \in \mathscr{I}$;*
- $unbind(r, z_1, z_2)$ *where $z_1, z_2 \in \mathscr{Z}$ and $r \in \mathscr{I}$;*
- $newRsrc(z : \mathscr{T})$ *where $z \in \mathscr{Z}$ and $\mathscr{T} \in U$ is the component type of z;*
- $delRsrc(z)$ *where $z \in \mathscr{Z}$.*

Notice that we consider a set of state changes in order to deal with simultaneous installations like the one needed for Apache2 and Apache2-bin in Fig. 1. The execution of actions is formalized as configuration transitions.

Definition 6 (Reconfigurations). *Reconfigurations are denoted by transitions $\mathcal{C} \xrightarrow{\alpha} \mathcal{C}'$ meaning that the execution of $\alpha \in \mathscr{A}$ on the configuration $\mathcal{C}$ produces a new configuration $\mathcal{C}'$. The transitions from a configuration $\mathcal{C} = \langle U, Z, S, B \rangle$ are defined as follows:*

$$\mathcal{C} \xrightarrow{stateChange(\langle z_1, q_1, q_1' \rangle, \ldots, \langle z_n, q_n, q_n' \rangle)} \langle U, Z, S', B \rangle$$
$$\text{if } \forall i \, . \, \mathcal{C}[z_i].\mathtt{state} = q_i$$
$$\text{and } \forall i \, . \, (q_i, q_i') \in \mathcal{C}[z_i].\mathtt{trans}$$
$$\text{and } S'(z') = \begin{cases} \langle \mathcal{C}[z_i].\mathtt{type}, q_i' \rangle & \text{if } \exists i \, . \, z' = z_i \\ \mathcal{C}[z'] & \text{otherwise} \end{cases}$$

$$\mathcal{C} \xrightarrow{bind(r, z_1, z_2)} \langle U, Z, S, B \cup \langle r, z_1, z_2 \rangle \rangle$$
$$\text{if } \langle r, z_1, z_2 \rangle \notin B$$
$$\text{and } r \in \mathcal{C}[z_1].\mathtt{req} \cap \mathcal{C}[z_2].\mathtt{prov}$$

$$\mathcal{C} \xrightarrow{unbind(r, z_1, z_2)} \langle U, Z, S, B \setminus \langle r, z_1, z_2 \rangle \rangle \quad \text{if } \langle r, z_1, z_2 \rangle \in B$$

$$\mathcal{C} \xrightarrow{newRsrc(z : \mathscr{T})} \langle U, Z \cup \{z\}, S', B \rangle$$
$$\text{if } z \notin Z, \, \mathscr{T} \in U$$
$$\text{and } S'(z') = \begin{cases} \langle \mathscr{T}, \mathscr{T}.\mathtt{init} \rangle & \text{if } z' = z \\ \mathcal{C}[z'] & \text{otherwise} \end{cases}$$

$$\mathcal{C} \xrightarrow{delRsrc(z)} \langle U, Z \setminus \{z\}, S', B' \rangle$$
$$\text{if } S'(z') = \begin{cases} \bot & \text{if } z' = z \\ \mathcal{C}[z'] & \text{otherwise} \end{cases}$$
$$\text{and } B' = \{ \langle r, z_1, z_2 \rangle \in B \mid z \notin \{z_1, z_2\} \}$$

We can now define a *reconfiguration run* as the effect of the execution of a sequence of actions (atomic or multiple state changes).

Definition 7 (Reconfiguration Run). *A reconfiguration run is a sequence of reconfigurations* $\mathscr{C}_0 \xrightarrow{\alpha_1} \mathscr{C}_1 \xrightarrow{\alpha_2} \cdots \xrightarrow{\alpha_m} \mathscr{C}_m$ *such that* $\mathscr{C}_i$ *is correct, for every* $0 \leq i \leq m$.

As an example, a reconfiguration run to reach the scenario depicted in Fig. 1 starting from a configuration where only `apache2` and `mysql` are running and `apache2-bin` is installed is the one involving in sequence the creation of `wordpress`, the bindings of `wordpress` with `mysql` and `apache2`, and finally the installation of `wordpress`.

We now have all the ingredients to define the *reconfigurability* problem: given a universe of component types and an initial configuration, we want to know whether there exists a reconfiguration run leading to a configuration that includes at least one component of a given type $\mathscr{T}$ in a given state q.

Definition 8 (Reconfigurability Problem). *The* reconfigurability problem *has as input a universe U of component types, an initial configuration $\mathscr{C}$, a component type $\mathscr{T}$, and a state q. It returns as output* ***true*** *if there exists a reconfiguration run* $\mathscr{C} \xrightarrow{\alpha_1} \mathscr{C}_1 \xrightarrow{\alpha_2} \cdots \xrightarrow{\alpha_m} \mathscr{C}_m$ *and* $\mathscr{C}_m[z] = \langle \mathscr{T}, q \rangle$, *for some component* $z \in \mathscr{C}_m$. *Otherwise, it returns* ***false***.

The restriction to only one component in a given state is not limiting: we can encode any given combination of component types and states by adding dummy provide ports enabled only by the final states of interest, and a target dummy component with requirements on all such provide ports.

3   Reconfigurability is Decidable in Aeolus core

We demonstrate decidability of the reconfigurability problem by resorting to the theory of Well-Structured Transition Systems (WSTS) [2,7].

A reflexive and transitive relation is called *quasi-ordering*. A *well-quasi-ordering* (wqo) is a quasi-ordering $(X, \leq)$ such that, for every infinite sequence $x_1, x_2, x_3, \cdots$, there exist $i < j$ with $x_i \leq x_j$. Given a quasi-order $\leq$ over X, an *upward-closed set* is a subset $I \subseteq X$ such that the following holds: $\forall x, y \in X : (x \in I \wedge x \leq y) \Rightarrow y \in I$. Given $x \in X$, its upward closure is $\uparrow x = \{y \in X \mid x \leq y\}$. This notion can be extended to sets in the obvious way: given a set $Y \subseteq X$ we define its upward closure as $\uparrow Y = \bigcup_{y \in Y} \uparrow y$. A *finite basis* of an upward-closed set I is a finite set B such that $I = \bigcup_{x \in B} \uparrow x$.

Definition 9. *A WSTS is a transition system* $(\mathscr{S}, \rightarrow, \preceq)$ *where* $\preceq$ *is a wqo on $\mathscr{S}$ which is* compatible *with* $\rightarrow$*, i.e., for every* $s_1 \preceq s_1'$ *such that* $s_1 \rightarrow s_2$*, there exists* $s_1' \rightarrow^* s_2'$ *such that* $s_2 \preceq s_2'$ *($\rightarrow^*$ is the reflexive and transitive closure of $\rightarrow$). Given a state $s \in \mathscr{S}$, $Pred(s)$ is the set* $\{s' \in \mathscr{S} \mid s' \rightarrow s\}$ *of immediate predecessors of s. Pred is extended to sets in the obvious way:* $Pred(S) = \bigcup_{s \in S} Pred(s)$*. A WSTS has* effective pred-basis *if there exists an algorithm that, given $s \in \mathscr{S}$, returns a finite basis of $\uparrow Pred(\uparrow s)$.*

The following proposition is a special case of Proposition 3.5 in [7].

Proposition 1. *Let* $(\mathscr{S}, \rightarrow, \preceq)$ *be a finitely branching WSTS with decidable $\preceq$ and effective pred-basis. Let I be any upward-closed subset of $\mathscr{S}$ and let $Pred^*(I)$ be the set* $\{s' \in S \mid s' \rightarrow^* s\}$ *of predecessors of states in I. A finite basis of $Pred^*(I)$ is computable.*

In the remainder of the section, we assume a given universe U of component types; so we can consider that the sets of possible component types $\mathscr{T}$ and of possible internal states q are both finite. We will resort to the theory of WSTS by considering an abstract model of configurations in which bindings are not taken into account.

Definition 10 (Abstract Configuration). *An abstract configuration $\mathscr{B}$ is a finite multiset of pairs $\langle \mathscr{T}, q \rangle$ where $\mathscr{T}$ is a component type and q is a corresponding state. We use Conf to denote the set of abstract configurations.*

A concretization of an abstract configuration is simply a correct configuration that for every component-type and state pair $\langle \mathscr{T}, q \rangle$ has as many instances of component $\mathscr{T}$ in state q as pairs $\langle \mathscr{T}, q \rangle$ in the abstract configuration.

Definition 11 (Concretization). *Given an abstract configuration $\mathscr{B}$ we say that a correct configuration $\mathscr{C} = \langle U, Z, S, B \rangle$ is one concretization of $\mathscr{B}$ if there exists a bijection f from the multiset $\mathscr{B}$ to Z s.t. $\forall \langle \mathscr{T}, q \rangle \in \mathscr{B}$ we have that $S(f(\langle \mathscr{T}, q \rangle)) = \langle \mathscr{T}, q \rangle$. We denote with $\gamma(\mathscr{B})$ the set of concretizations of $\mathscr{B}$. We say that an abstract configuration $\mathscr{B}$ is correct if it has at least one concretization (formally $\gamma(\mathscr{B}) \neq \emptyset$).*

An interesting property of an abstract configuration is that from one of its concretizations it is possible to reach via bind and unbind actions all the other concretizations up to instance renaming. This is because it is always possible to switch one binding from one provide port to another one by adding a binding to the new port and then removing the old binding.

Property 1. Given an abstract configuration $\mathscr{B}$ and configurations $\mathscr{C}_1, \mathscr{C}_2 \in \gamma(\mathscr{B})$ there exists $\alpha_1, \ldots, \alpha_n$ sequence of binding and unbinding actions s.t. $\mathscr{C}_1 \xrightarrow{\alpha_1} \ldots \xrightarrow{\alpha_n} \mathscr{C} \equiv \mathscr{C}_2$.

We now move to the definition of our quasi-ordering on abstract configurations. In order to be compatible with the notion of correctness we cannot adopt the usual multiset inclusion ordering. In fact, the addition of one component to a correct configuration could introduce a conflict. If the type-state pair of the added component was absent in the configuration, the conflict might be with a component of a different type-state. If the type-state pair was present in a single copy, the conflict might be with that component if the considered type-state pair activates one provide and one conflict port on the same interface. This sort of self-conflict is revealed when there are at least two instances, as one component cannot be in conflict with itself. If the type-state pair was already present in at least two copies, no new conflicts can be added otherwise such conflicts were already present in the configuration (thus contradicting its correctness).

In the light of the above observation, we define an ordering on configurations that corresponds to the product of three orderings: the identity on the set of type-state pairs that are absent, the identity on the pairs that occurs in one instance, and the multiset inclusion for the projections on the remaining type-state pairs.

Definition 12 ($\leq$). *Given a pair $\langle \mathscr{T}, q \rangle$ and an abstract configuration $\mathscr{B}$, let $\#_{\mathscr{B}}(\langle \mathscr{T}, q \rangle)$ be the number of occurrences in $\mathscr{B}$ of the pair $\langle \mathscr{T}, q \rangle$. Given two abstract configurations $\mathscr{B}_1, \mathscr{B}_2$ we write $\mathscr{B}_1 \leq \mathscr{B}_2$ if for every component type $\mathscr{T}$ and state q we have that $\#_{\mathscr{B}_1}(\langle \mathscr{T}, q \rangle) = \#_{\mathscr{B}_2}(\langle \mathscr{T}, q \rangle)$ when $\#_{\mathscr{B}_1}(\langle \mathscr{T}, q \rangle) \in \{0, 1\}$ or $\#_{\mathscr{B}_2}(\langle \mathscr{T}, q \rangle) \in \{0, 1\}$, and $\#_{\mathscr{B}_1}(\langle \mathscr{T}, q \rangle) \leq \#_{\mathscr{B}_2}(\langle \mathscr{T}, q \rangle)$ otherwise.*

As discussed above, this ordering is compatible with correctness.

Property 2. If an abstract configuration $\mathscr{B}$ is correct than all the configurations $\mathscr{B}'$ such that $\mathscr{B} \leq \mathscr{B}'$ are also correct.

Another interesting property of the $\leq$ quasi-ordering is that from one concretization of an abstract configuration, it is always possible to reconfigure it to reach a concretization of a smaller abstract configuration. In this case it is possible to first add from the starting configuration the bindings that are present in the final configuration. Then the extra components present in the starting configuration can be deleted because not needed to guarantee correctness (they are instances of components that remain available in at least two copies). Finally the remaining extra bindings can be removed.

Property 3. Given two abstract configurations $\mathscr{B}_1, \mathscr{B}_2$ s.t. $\mathscr{B}_1 \leq \mathscr{B}_2$, $\mathscr{C}_1 \in \gamma(\mathscr{B}_1)$, and $\mathscr{C}_2 \in \gamma(\mathscr{B}_2)$ we have that there exists a reconfiguration run $\mathscr{C}_2 \xrightarrow{\alpha_1} \ldots \xrightarrow{\alpha_n} \mathscr{C} \equiv \mathscr{C}_1$.

We have that $\leq$ is a wqo on *Conf* because, as we consider finitely many component type-state pairs, the three distinct orderings that compose $\leq$ are themselves wqo.

Lemma 1. $\leq$ *is a wqo over Conf.*

We now define a transition system on abstract reconfigurations and prove it is a WSTS with respect to the ordering defined above.

Definition 13 (Abstract reconfigurations). *We write* $\mathscr{B} \to \mathscr{B}'$ *if there exists* $\mathscr{C} \xrightarrow{\alpha} \mathscr{C}'$ *for some* $\mathscr{C} \in \gamma(\mathscr{B})$ *and* $\mathscr{C}' \in \gamma(\mathscr{B}')$.

By Property 3 and Lemma 1 we have the following.

Lemma 2. *The transition system* $(Conf, \to, \leq)$ *is a WSTS.*

The following lemma is rather technical and it will be used to prove that $(Conf, \to, \leq)$ has effective pred-basis. Intuitively it will allow us to consider, in the computation of the predecessors, only finitely many different state change actions.

Lemma 3. *Let k be the number of distinct component type-state pairs. If $\mathscr{B}_1 \to \mathscr{B}_2$ then there exists $\mathscr{B}'_1 \to \mathscr{B}'_2$ such that $\mathscr{B}'_1 \leq \mathscr{B}_1$, $\mathscr{B}'_2 \leq \mathscr{B}_2$ and $|\mathscr{B}'_2| \leq 3k + 2k^2$.*

Proof. If $|\mathscr{B}_2| \leq 3k + 2k^2$ the thesis trivially holds. Consider now $|\mathscr{B}_2| > 3k + 2k^2$ and a transition $\mathscr{C}_1 \xrightarrow{\alpha} \mathscr{C}_2$ such that $\mathscr{C}_1 \in \gamma(\mathscr{B}_1)$ and $\mathscr{C}_2 \in \gamma(\mathscr{B}_2)$. Since $|\mathscr{B}_2| > 3k$ there are three components z_1, z_2 and z_3 having the same component type and internal state. We consider two subcases.

Case 1. z_1, z_2 and z_3 do not perform a state change in the action α. W.l.o.g we can assume that z_3 does not appear in α (this is not restrictive because at most two components that do not perform a state change can occur in an action). We can now consider the configuration $\mathscr{C}'_1$ obtained by $\mathscr{C}_1$ after removing z_3 (if there are bindings connected to provide ports of z_3, these can be rebound to ports of z_1 or z_2). Consider now $\mathscr{C}'_1 \xrightarrow{\alpha} \mathscr{C}'_2$ and the corresponding abstract configurations $\mathscr{B}'_1$ and $\mathscr{B}'_2$. It is easy to see that $\mathscr{B}'_1 \to \mathscr{B}'_2$,

$\mathscr{B}'_1 \leq \mathscr{B}_1, \mathscr{B}'_2 \leq \mathscr{B}_2$ and $|\mathscr{B}'_2| < |\mathscr{B}_2|$. If $|\mathscr{B}'_2| \leq 3k + 2k^2$ the thesis is proved, otherwise we repeat this deletion of components.

Case 2. There are no three components of the same type-state that do not perform a state change. Since $|\mathscr{B}_2| > 2k^2 + 2$ we have that α is a state change involving strictly more than $2k^2$ components. This ensures the existence of three components z'_1, z'_2 and z'_3 of the same type that perform the same state change from q to q'. As in the previous case we consider the configuration $\mathscr{C}'_1$ obtained by $\mathscr{C}_1$ after removing z'_3 and α' the state change similar to α but without the state change of z'_3. Consider now $\mathscr{C}'_1 \xrightarrow{\alpha'} \mathscr{C}'_2$ and the corresponding abstract configurations $\mathscr{B}'_1$ and $\mathscr{B}'_2$. As above, $\mathscr{B}'_1 \leq \mathscr{B}_1$, $\mathscr{B}'_2 \leq \mathscr{B}_2$ and $|\mathscr{B}'_2| < |\mathscr{B}_2|$. If $|\mathscr{B}'_2| \leq 3k + 2k^2$ the thesis is proved, otherwise we repeat the deletion of components. □

We are now in place to prove that $(Conf, \rightarrow, \leq)$ has effective pred-basis.

Lemma 4. *The transition system* $(Conf, \rightarrow, \leq)$ *has effective pred-basis.*

Proof. We first observe that given an abstract configuration the set of its concretizations up to configuration equivalence is finite, and that given a configuration $\mathscr{C}$ the set of preceding configurations $\mathscr{C}'$ such that $\mathscr{C}' \xrightarrow{\alpha} \mathscr{C}$ is also finite (and effectively computable). Consider now an abstract configuration $\mathscr{B}$. We now show how to compute a finite basis for $\uparrow Pred(\uparrow \mathscr{B})$. First of all we consider the configuration $\mathscr{B}$ if $|\mathscr{B}| > 3k + 2k^2$, the (finite) set of configurations $\mathscr{B}'$ such that $\mathscr{B} \leq \mathscr{B}'$ and $|\mathscr{B}'| \leq 3k + 2k^2$ otherwise. Then we consider the (finite) set of concretizations of all such abstract configurations. And finally we compute the (finite) set of the preceding configurations of all such concretizations. The set of abstract configuration corresponding to the latter is a finite basis for $\uparrow Pred(\uparrow \mathscr{B})$ as a consequence of Lemma 3. □

We are finally ready to prove our decidability result.

Theorem 1. *The reconfigurability problem in* Aeolus core *is decidable.*

Proof. Let k be the number of distinct component type-state pairs according to the considered universe of component types. We first observe that if there exists a correct configuration containing a component of type $\mathscr{T}$ in state q then it is possible to obtain via some binding, unbinding, and delete actions another correct configuration with k or less components. Hence, given a component type $\mathscr{T}$ and a state q, the number of target configurations that need to be considered is finite. Moreover, given a configuration $\mathscr{C}' \in \gamma(\mathscr{B}')$ there exists a reconfiguration run from $\mathscr{C} \in \gamma(\mathscr{B})$ to $\mathscr{C}'$ iff $\mathscr{B} \in Pred^*(\uparrow \mathscr{B}')$.

To solve the reconfigurability problem it is therefore possible to consider only the (finite set of) abstractions of the target configurations. For each of them, say $\mathscr{B}'$, by Proposition 1, Lemma 2 and Lemma 4 we know that a finite basis for $Pred^*(\uparrow \mathscr{B}')$ can be computed. It is sufficient to check whether at least one of the abstract configurations in such basis is $\leq$ w.r.t. the abstraction of the initial configuration. □

4   Reconfigurability is ExpSpace-hard in Aeolus core

We prove that the reconfigurability problem in Aeolus core is ExpSpace-hard by reduction from the coverability problem in Petri nets, a problem which is indeed known to be ExpSpace-complete [11,13]. We start with some background on Petri nets.

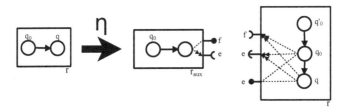

Fig. 2. Example of a component type transformation $\eta(\ )$

A *Petri net* is a tuple $N = (P, T, m_0)$, where P and T are finite sets of *places* and *transitions*, respectively. A finite multiset over the set P of places is called a *marking*, and m_0 is the initial marking. Given a marking m and a place p, we say that the place p contains a number of *tokens* equal to the number of instances of p in m. A transition $t \in T$ is a pair of markings denoted with $\bullet t$ and $t \bullet$. A transition t can fire in the marking m if $\bullet t \subseteq m$ (where $\subseteq$ is multiset inclusion); upon transition firing the new marking of the net becomes $n = (m \setminus m') \uplus m''$ (where $\setminus$ and $\uplus$ are the difference and union operators for multisets, respectively). This is written as $m \Rightarrow n$. We use $\Rightarrow^*$ to denote the reflexive and transitive closure of $\Rightarrow$. We say that m' is *reachable from m* if $m \Rightarrow^* m'$. The *coverability* problem for marking m consists of checking whether $m_0 \Rightarrow^* m'$ for some $m \subseteq m'$.

We now discuss how to encode Petri nets in Aeolus core component types. Before entering into the details we observe that given a component type $\mathcal{T}$ it is always possible to modify it in such a way that its instances are persistent and unique. The uniqueness constraint can be enforced by allowing all the states of the component type to provide a new port with which they are in conflict. To avoid the component deletion it is sufficient to impose its reciprocal dependence with a new type of component. When this dependence is established the components be deleted without violating it. In Fig. 2 we show an example of how a component type having two states can be modified in order to reach our goal. A new auxiliary initial state q_0' is created. The new port e ensures that the instances of type $\mathcal{T}$ in a state different from q_0' are unique. The require port f provided by a new component type $\mathcal{T}_{aux}$ forbids the deletion of the instances of type $\mathcal{T}$, if they are not in state q_0'. We assume that the ports e and f are fresh. We can therefore consider w.l.o.g. components that, when deployed, are unique and persistent. Given a component type $\mathcal{T}$ we denote this component type transformation with $\eta(\mathcal{T})$.

We now describe how to encode a Petri net in the Aeolus core model. We will use three types of components: one modeling the tokens, one for transitions and one for defining a counter. The components for transitions and the counter are unique and persistent, while those for the tokens cannot be unique because the number of tokens in a Petri net can be unbounded. The simplest component is the one used to model a token in a given place. Intuitively one token in a place is encoded as one instance of a corresponding component type in an *on* state. There could be more than one of these components deployed simultaneously representing multiple tokens in a place. In Fig. 3a we represent the component type for the tokens in the place p of the Petri net. The initial state is the *off* state. The token could be created following a protocol consisting of requiring the port a_p and then providing the port b_p to signal the change of status. Similarly a token can be deleted requiring the port c_p and then providing the port d_p.

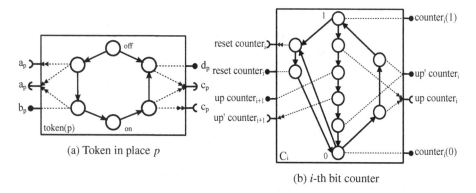

(a) Token in place p

(b) i-th bit counter

Fig. 3. Token and counter component types

Even if multiple instances of the token component can be deployed simultaneously, the conflict ports a_p and c_p guarantee that only one at a time can initiate the protocol to change its state. We denote with *token(p)* the component type representing the tokens in the place p.

In order to model the transitions with component types without having an exponential blow up of the size of the encoding we need a mechanism to count up to a fixed number. Indeed a transition can consume and produce up to a given number of tokens. To count a number up to n we will use $C_1, \ldots, C_{\lceil \log(n) \rceil}$ components; every C_i will represent the i-th less significant bit of the binary representation of the counter that, for our purposes, needs just to support the increment and reset operations. In Fig. 3b we represent one of the bits implementing the counter. The initial state is 0. To reset the bit it is possible to provide the *reset counter$_i$* port while to increment it the *up counter$_i$* should be provided. If the bit is in state 1 the increment will trigger the increment of the next bit except for the component representing the most significant bit that will never need to do that. We transform all the component types representing the counter using the η transformation to ensure uniqueness and persistence of its instances. The instance of $\eta(C_i)$ can be used to count how many tokens are consumed or produced checking if the right number is reached via the ports *counter$_i$(1)* and *counter$_i$(0)*.

A transition can be represented with a single component interacting with token and counter components. The state changes of the transition component can be intuitively divided in phases. In each of those phases a fixed number of tokens from a given place is consumed or produced. The counter is first reset providing the *reset counter$_i$* and requiring the *reset' counter$_i$* ports for all the counter bits. Then a cycle starts incrementing the counter providing and requiring the ports *up counter$_1$* and *up' counter$_1$* and consuming or producing a token. The production of a token in place p is obtained providing and requiring ports a_p and b_p while the consumption providing and requiring the ports c_p and d_p. The phase ends when all the bits of the counter represent in binary the right number of tokens that need to be consumed or produced. If instead at least one bit is wrong the cycle restarts. In Fig. 4 we depict the phase of a consumption of n tokens.

Starting from the initial state of the component representing the transition, the consumption phases need to be performed first. When the final token has been produced

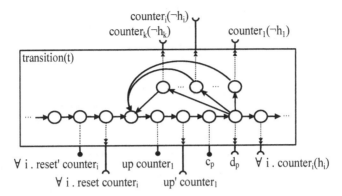

Fig. 4. Consumption phase of n tokens from place p for a transition t ($k = \lceil \log(n) \rceil$ and h_i is the i-th least significant bit of the binary representation of n)

the transition component can restart from the initial state. Given a transition t we will denote with *transition(t)* the component type explained above.

Definition 14 (Petri net encoding in Aeolus core). *Given a Petri net* $N = (P, T, m_0)$ *if* n *is the largest number of tokens that can be consumed or produced by a transition in* T, *the encoding of* N *in Aeolus core is the set of component types* $\Gamma_N = \{token(p) \mid p \in P\} \cup \{\eta(C_i) \mid i \in [1..\lceil \log(n) \rceil]\} \cup \{\eta(transition(t)) \mid t \in T\}$.

An important property of the previous encoding is that it is polynomial w.r.t. the size of the Petri net. This is due to the fact that the counter and place components have a constant amount of states and ports while the transition components have a number of states that grows linearly w.r.t. the number of places involved in a transition.

The proof that the reconfiguration problem for Aeolus core is ExpSpace-hard thus follows from the following correspondence between a Petri net N and its set of component types Γ_N: every computation in N can be faithfully reproduced by a corresponding reconfiguration run on the components types Γ_N; every reconfiguration run on Γ_N corresponds to a computation in N excluding the possibility for components of kind $token(p)$ to be deleted (because η is not applied to those components) and of components $transition(t)$ to execute only partially the consumption of the tokens (because e.g. some token needed by the transition is absent). In both cases, the effect is to reach a configuration in which some of the token was lost during the reconfiguration run, but this is not problematic as we deal with coverability. In fact, if a configuration is reached with at least some tokens, then also the corresponding Petri nets will be able to reach a marking with at least those tokens (possibly more).

Theorem 2. *The reconfiguration problem for Aeolus core is ExpSpace-hard.*

5   Related Work and Conclusions

Engage [8] is very close to Aeolus purposes: it provides a declarative language to define resource configurations and a deployment engine. However, it lacks conflicts.

This might make a huge computational differences, as it is precisely the introduction of conflicts that makes reconfigurability ExpSpace-hard in Aeolus core (the problem is polynomial in Aeolus⁻ [6]). ConfSolve [9] is a DSL used to specify system configurations with constraints suitable for modern Constraint Satisfaction Problems solvers. ConfSolve allocates virtual machines to physical ones considering constraints like CPU, RAM, ... This differs from reconfigurability in Aeolus. Package-based software management [1,5] is a degenerate case of Aeolus reconfigurability. Package managers are used to compute a new configuration, but they use simple heuristics to reach it, ignoring transitive incoherences met during deployment.

In this work we have studied the impact of adding conflicts to a realistic component model, onto the complexity of reconfigurability: the problem remains decidable—while in other models, like Petri nets, the addition of tests-for-absence makes the model Turing powerful—but becomes ExpSpace-hard.

We consider our decidability and hardness proofs useful for at least two future intertwined research directions. On the one hand, we plan to extend existing tools with techniques inspired by our decidability proof in order to also deal with conflicts and produce a reconfiguration run. On the other hand, the hardness proof sheds some light on the specific combination of component model features that make the reconfigurability problem ExpSpace-hard. We plan to investigate realistic restrictions on the Aeolus component model for which efficient reconfigurability algorithms could be devised.

References

1. Abate, P., Di Cosmo, R., Treinen, R., Zacchiroli, S.: Dependency solving: a separate concern in component evolution management. J. Syst. Software 85, 2228–2240 (2012)
2. Abdulla, P.A., Cerans, K., Jonsson, B., Tsay, Y.K.: General decidability theorems for infinite-state systems. In: LICS, pp. 313–321. IEEE (1996)
3. Clayberg, E., Rubel, D.: Eclipse Plug-ins, 3rd edn. Addison-Wesley (2008)
4. Di Cosmo, R., Mauro, J., Zacchiroli, S., Zavattaro, G.: Component reconfiguration in the presence of conflicts. Tech. rep. Aeolus Project (2013),
   http://hal.archives-ouvertes.fr/hal-00816468
5. Di Cosmo, R., Trezentos, P., Zacchiroli, S.: Package upgrades in FOSS distributions: Details and challenges. In: HotSWup 2008 (2008)
6. Di Cosmo, R., Zacchiroli, S., Zavattaro, G.: Towards a formal component model for the cloud. In: Eleftherakis, G., Hinchey, M., Holcombe, M. (eds.) SEFM 2012. LNCS, vol. 7504, pp. 156–171. Springer, Heidelberg (2012)
7. Finkel, A., Schnoebelen, P.: Well-structured transition systems everywhere! Theoretical Computer Science 256, 63–92 (2001)
8. Fischer, J., Majumdar, R., Esmaeilsabzali, S.: Engage: a deployment management system. In: PLDI 2012: Programming Language Design and Implementation, pp. 263–274. ACM (2012)
9. Hewson, J.A., Anderson, P., Gordon, A.D.: A declarative approach to automated configuration. In: LISA 2012: Large Installation System Administration Conference, pp. 51–66 (2012)
10. Kanies, L.: Puppet: Next-generation configuration management. The USENIX Magazine 31(1), 19–25 (2006)
11. Lipton, R.J.: The Reachability Problem Requires Exponential Space. Research report 62, Department of Computer Science, Yale University (1976)
12. OSGi Alliance: OSGi Service Platform, Release 3. IOS Press, Inc. (2003)
13. Rackoff, C.: The covering and boundedness problems for vector addition systems. Theoret. Comp. Sci. 6, 223–231 (1978)

Stochastic Context-Free Grammars, Regular Languages, and Newton's Method[*]

Kousha Etessami[1], Alistair Stewart[1], and Mihalis Yannakakis[2]

[1] School of Informatics, University of Edinburgh
kousha@inf.ed.ac.uk, stewart.al@gmail.com
[2] Department of Computer Science, Columbia University
mihalis@cs.columbia.edu

Abstract. We study the problem of computing the probability that a given stochastic context-free grammar (SCFG), G, generates a string in a given regular language $L(D)$ (given by a DFA, D). This basic problem has a number of applications in statistical natural language processing, and it is also a key necessary step towards quantitative ω-regular model checking of stochastic context-free processes (equivalently, 1-exit recursive Markov chains, or stateless probabilistic pushdown processes).

We show that the probability that G generates a string in $L(D)$ can be computed to within arbitrary desired precision in polynomial time (in the standard Turing model of computation), under a rather mild assumption about the SCFG, G, and with no extra assumption about D. We show that this assumption is satisfied for SCFG's whose rule probabilities are learned via the well-known inside-outside (EM) algorithm for maximum-likelihood estimation (a standard method for constructing SCFGs in statistical NLP and biological sequence analysis). Thus, for these SCFGs the algorithm always runs in P-time.

1 Introduction

Stochastic (or *Probabilistic*) *Context-Free Grammars* (SCFG) are context-free grammars where the rules (productions) have associated probabilities. They are a central stochastic model, widely used in natural language processing [14], with applications also in biology (e.g. [2, 13]). A SCFG G generates a language $L(G)$ (like an ordinary CFG) and assigns a probability to every string in the language. SCFGs have been extensively studied since the 1970's. A number of important problems on SCFGs can be viewed as instances of the following *regular pattern matching problem* for different regular languages:

Given a SCFG G and a regular language L, given e.g., by a deterministic finite automaton (DFA) D, compute the probability $\mathbb{P}_G(L)$ that G generates a string in L, i.e. compute the sum of the probabilities of all the strings in L.

A simple example is when $L = \Sigma^*$, the set of all strings over the terminal alphabet Σ of the SCFG G. Then this problem simply asks to compute the

---

[*] The full version of this paper is available at arxiv.org/abs/1302.6411. Research partially supported by the Royal Society and by NSF Grant CCF-1017955.

F.V. Fomin et al. (Eds.): ICALP 2013, Part II, LNCS 7966, pp. 199–211, 2013.
© Springer-Verlag Berlin Heidelberg 2013

probability $\mathbb{P}_G(L(G))$ of the language $L(G)$ generated by the grammar G. Alternatively, if we view the SCFG as a stochastic process that starts from the start nonterminal, repeatedly applies the probabilistic rules to replace (say, leftmost) nonterminals, and terminates when a string of terminals is reached, then $\mathbb{P}_G(L(G))$ is simply the probability that this process terminates. Another simple example is when L is a singleton, $L = \{w\}$, for some string w; in this case the problem corresponds to the basic parsing question of computing the probability that a given string w is generated by the SCFG G. Another basic well-studied problem is the computation of *prefix probabilities*: given a SCFG G and a string w, compute the probability that G generates a string with prefix w [12, 21]. This is useful in online processing in speech recognition [12] and corresponds to the case $L = w\Sigma^*$. A more complex problem is the computation of *infix probabilities* [1, 18], where we wish to compute the probability that G generates a string that contains a given string w as a substring, which corresponds to the language $L = \Sigma^* w \Sigma^*$. In general, even when rule probabilities of the SCFG G are rational, the probabilities we wish to compute can be irrational. Thus the typical aim for "computing" them is to approximate them to desired precision.

Stochastic context-free grammars are closely related to *1-exit recursive Markov chains* (1-RMC) [9], and to *stateless probabilistic pushdown automata* (also called pBPA) [5]; these are two equivalent models for a subclass of probabilistic programs with recursive procedures. The above regular pattern matching problem for SCFGs is equivalent to the problem of computing the probability that a computation of a given 1-RMC (or pBPA) terminates and satisfies a given regular property. In other words, it corresponds to the quantitative model checking problem for 1-RMCs with respect to regular *finite string* properties.

We first review some prior related work, and then describe our results.

Previous Work. As mentioned above, there has been, on the one hand, substantial work in the NLP literature on different cases of the problem for various regular languages L, and on the other hand, there has been work in the verification and algorithms literature on the analysis and model checking of recursive Markov chains and probabilistic pushdown automata. Nevertheless, even the simple special case of $L = \Sigma^*$, the question of whether it is possible to compute (approximately) in polynomial time the desired probability for a given SCFG G (i.e. the probability $\mathbb{P}_G(L(G))$ of $L(G)$) was open until very recently. In [7] we showed that $\mathbb{P}_G(L(G))$ can be computed to arbitrary precision in polynomial time in the size of the input SCFG G and the number of bits of precision. From a SCFG G, one can construct a multivariate system of equations $x = P_G(x)$, where x is a vector of variables and P_G is a vector of polynomials with positive coefficients which sum to (at most) 1. Such a system is called a *probabilistic polynomial system* (PPS), and it always has a non-negative solution that is smallest in every coordinate, called the *least fixed point* (LFP). A particular coordinate of the LFP of the system $x = P_G(x)$ is the desired probability $\mathbb{P}_G(L(G))$. To compute $\mathbb{P}_G(L(G))$, we used a variant of Newton's method on $x = P_G(x)$, with suitable rounding after each step to control the bit-size of numbers, and showed that it converges in P-time to the LFP [7]. Building on this, we also showed that the probability $\mathbb{P}_G(\{w\})$ of string w under SCFG G can also be computed to any precision in P-time in the size of G, w and the number of bits of precision.

The use of Newton's method was proposed originally in [9] for computing termination probabilities for (multi-exit) RMC's, which requires the solution of equations from a more general class of polynomial systems $x = P(x)$, called *monotone polynomial systems* (MPS), where the polynomials of P have positive coefficients, but their sum is not restricted to ≤ 1. An arbitrary MPS may not have any non-negative solution, but if it does then it has a LFP, and a version of Newton provably converges to the LFP [9]. There are now implementations of variants of Newton's method in several tools [22, 16] and experiments show that they perform well on many instances. The rate of convergence of Newton for general MPSs was studied in detail in [4], and was further studied most recently in [20] (see below). In certain cases, Newton converges fast, but in general there are exponential bad examples. Furthermore, there are negative results indicating it is very unlikely that any non-trivial approximation of termination probabilities of multi-exit RMCs, and the LFP of MPSs, can be done in P-time (see [9]).

The model checking problem for RMCs (equivalently pPDAs) and ω-regular properties was studied in [5, 10]. This is of course a more general problem than the problem for SCFGs (which correspond to 1-RMCs) and regular languages (the finite string case of ω-regular languages). It was shown in [10] that in the case of 1-RMCs, the qualitative problem of determining whether the probability that a run satisfies the property is 0 or 1 can be solved in P-time in the size of the 1-RMC, but for the quantitative problem of approximating the probability, the algorithm runs in PSPACE, and no better complexity bound was known.

The particular cases of computing prefix and infix probabilities for a SCFG have been studied in the NLP literature, but no polynomial time algorithm for general SCFGs is known. Jelinek and Lafferty gave an algorithm for grammars in Chomsky Normal Form (CNF) [12]. Note that a general SCFG G may not have any equivalent CNF grammar with rational rule probabilities, thus one can only hope for an "approximately equivalent" CNF grammar; constructing such a grammar in the case of stochastic grammars G is non-trivial, at least as difficult as computing the probability of $L(G)$, and the first P-time algorithm was given in [7]. Another algorithm for prefix probabilities by Stolcke [21] applies to general SCFGs, but in the presence of unary and ϵ-rules, the algorithm does not run in polynomial time. The problem of computing infix probabilities was studied in [1, 16, 18], and in particular [16, 18] cast it in the general regular language framework, and studied the general problem of computing the probability $\mathbb{P}_G(L(D))$ of the language $L(D)$ of a DFA D under a SCFG G. From G and D they construct a product *weighted context-free grammar* (WCFG) G': a CFG with (positive) weights on the rules, which may not be probabilities, in particular the weights on the rules of a nonterminal may sum to more than 1. The desired probability $\mathbb{P}_G(L(D))$ is the weight of $L(G')$. As in the case of SCFGs, this weight is given by the LFP of a monotone system of equations $y = P_{G'}(y)$, however, unlike the case of SCFGs the system now is not a probabilistic system (thus our result of [7] does not apply). Nederhof and Satta then solve the system using the decomposed Newton method from [9] and Broyden's (quasi-Newton) method, and present experimental results for infix probability computations.

Most recently, in [20], we have obtained worst-case upper bounds on (rounded and exact) Newton's method applied to arbitrary MPSs, $x = P(x)$, as a function of the input encoding size $|P|$ and $\log(1/\epsilon)$, to converge to within additive error $\epsilon > 0$ of the LFP solution q^*. However, our bounds in [20], even when $0 < q^* \leq 1$, are exponential in the depth of (not necessarily critical) strongly connected components of $x = P(x)$, and furthermore they also depend linearly on $\log(\frac{1}{q^*_{\min}})$, where $q^*_{\min} = \min_i q^*_i$, which can be $\approx \frac{1}{2^{2^{|P|}}}$. As we describe next, we do far better in this paper for the MPSs that arise from the "product" of a SCFG and a DFA.

Our Results. We study the general problem of computing the probability $\mathbb{P}_G(L(D))$ that a given SCFG G generates a string in the language $L(D)$ of a given DFA D. We show that, under a certain mild assumption on G, this probability can be computed to any desired precision in time polynomial in the encoding sizes of G & D and the number of bits of precision.

We now sketch briefly the approach and state the assumption on G. First we construct from G and D the product weighted CFG $G' = G \otimes D$ as in [16] and construct the corresponding MPS $y = P_{G'}(y)$, whose LFP contains the desired probability $\mathbb{P}_G(L(D))$ as one of its components. The system is monotone but not probabilistic. We eliminate (in P-time) those variables that have value 0 in the LFP, and apply Newton, with suitable rounding in every step. The heart of the analysis shows there is a tight algebraic correspondence between the behavior of Newton's method on this MPS and its behavior on the probabilistic polynomial system (PPS) $x = P_G(x)$ of G. In particular, this correspondence shows that, with exact arithmetic, the two computations converge at the same rate. By exploiting this, and by extending recent results we established for PPSs, we obtain the conditional polynomial upper bound. Specifically, call a PPS $x = P(x)$ *critical* if the spectral radius of the Jacobian of $P(x)$, evaluated at the LFP q^* is equal to 1 (it is always ≤ 1). We can form a dependency graph between the variables of a PPS, and decompose the variables and the system into strongly connected components (SCCs); an SCC is called critical if the induced subsystem on that SCC is critical. The *critical depth* of a PPS is the maximum number of critical SCCs on any path of the DAG of SCCs (i.e. the max nesting depth of critical SCCs). We show that if the PPS of the given SCFG G has bounded (or even logarithmic) critical depth, then we can compute $\mathbb{P}_G(L(D))$ (for any DFA D) in polynomial time in the size of G, D and the number of bits of precision.

Furthermore, we show this condition is satisfied by a broad class of SCFGs used in applications. Specifically, a standard way the probabilities of rules of a SCFG are set is by using the EM (inside-outside) algorithm. We show that the SCFGs constructed in this way are guaranteed to be noncritical (i.e., have critical depth 0). So for these SCFGs, and any DFA, the algorithm runs in P-time.

Proofs are in the full version [8].

2   Definitions and Background

A *weighted context-free grammar* (WCFG), $G = (V, \Sigma, R, p)$, has a finite set V of *nonterminals*, a finite set Σ of *terminals* (alphabet symbols), and a finite list of *rules*, $R \subset V \times (V \cup \Sigma)^*$, where each rule $r \in R$ is a pair (A, γ), which we usually denote by $A \to \gamma$, where $A \in V$ and $\gamma \in (V \cup \Sigma)^*$. Finally $p : R \to \mathbb{R}^+$ maps each rule $r \in R$ to a positive *weight*, $p(r) > 0$. We often denote a rule $r = (A \to \gamma)$ together with its weight by writing $A \xrightarrow{p(r)} \gamma$. We will sometimes also specify a specific non-terminal $S \in V$ as the starting symbol.

Note that we allow $\gamma \in (V \cup \Sigma)^*$ to possibly be the empty string, denoted by ϵ. A rule of the form $A \to \epsilon$ is called an *ϵ-rule*. For a rule $r = (A \to \gamma)$, we let $\text{left}(r) := A$ and $\text{right}(r) := \gamma$. We let $R_A = \{r \in R \mid \text{left}(r) = A\}$. For $A \in V$, let $p(A) = \sum_{r \in R_A} p(r)$. A WCFG, G, is called a *stochastic* or *probabilistic context-free grammar* (SCFG or PCFG; we shall use SCFG), if for $\forall A \in V$, $p(A) \leq 1$. An SCFG is called *proper* if $\forall A \in V$, $p(A) = 1$.

We will say that a WCFG, $G = (V, \Sigma, R, p)$ is in *Simple Normal Form* (SNF) if every nonterminal $A \in V$ belongs to one of the following three types:

1. type L: every rule $r \in R_A$, has the form $A \xrightarrow{p(r)} B$.

2. type Q: there is a single rule in R_A: $A \xrightarrow{1} BC$, for some $B, C \in V$.

3. type T: there is a single rule in R_A: either $A \xrightarrow{1} \epsilon$, or $A \xrightarrow{1} a$ for some $a \in \Sigma$.

For a WCFG, G, strings $\alpha, \beta \in (V \cup \Sigma)^*$, and $\pi = r_1 \ldots r_k \in R^*$, we write $\alpha \xRightarrow{\pi} \beta$ if the leftmost derivation starting from α, and applying the sequence π of rules, derives β. We let $p(\alpha \xRightarrow{\pi} \beta) = \prod_{i=1}^k p(r_k)$ if $\alpha \xRightarrow{\pi} \beta$, and $p(\alpha \xRightarrow{\pi} \beta) = 0$ otherwise. If $A \xRightarrow{\pi} w$ for $A \in V$ and $w \in \Sigma^*$, we say that π is a *complete* derivation from A and its *yield* is $y(\pi) = w$. There is a natural one-to-one correspondence between the complete derivations of w starting at A and the *parse trees* of w rooted at A, and this correspondence preserves weights.

For a WCFG, $G = (V, \Sigma, R, p)$, nonterminal $A \in V$, and terminal string $w \in \Sigma^*$, we let $p_A^{G,w} = \sum_{\{\pi \mid y(\pi) = w\}} p(A \xRightarrow{\pi} w)$. For a general WCFG, $p_A^{G,w}$ need not be a finite value (it may be $+\infty$, since the sum may not converge). Note however that if G is an SCFG, then $p_A^{G,w}$ defines the probability that, starting at nonterminal A, G generates w, and thus it is clearly finite.

The *termination probability* (*termination weight*) of an SCFG (WCFG), G, starting at nonterminal A, denoted q_A^G, is defined by $q_A^G = \sum_{w \in \Sigma^*} p_A^{G,w}$. Again, for an arbitrary WCFG q_A^G need not be a finite number. A WCFG G is called *convergent* if q_A^G is finite for all $A \in V$. We will only encounter convergent WCFGs in this paper, so when we say WCFG we mean convergent WCFG, unless otherwise specified. In G is an SCFG, then q_A^G is just the total probability with which the derivation process starting at A eventually generates a finite string and (thus) stops, so SCFGs are clearly convergent.

An SCFG, G, is called *consistent starting at* A if $q_A^G = 1$, and G is called *consistent* if it is consistent starting at every nonterminal. Note that even if a SCFG, G, is proper this does not necessarily imply that G is consistent. For an

SCFG, G, we can decide whether $q_A^G = 1$ in P-time ([9]). The same decision problem is PosSLP-hard for convergent WCFGs ([9]).

For any WCFG, $G = (V, \Sigma, R, p)$, with $n = |V|$, assume the nonterminals in V are indexed as $A_1, \ldots, A_n$. We define the following **monotone polynomial system of equations** (MPS) associated with G, denoted $x = P_G(x)$. Here $x = (x_1, \ldots, x_n)$ denotes an n-vector of variables. Likewise $P_G(x) = (P_G(x)_1, \ldots, P_G(x)_n)$ denotes an n-vector of multivariate polynomials over the variables $x = (x_1, \ldots, x_n)$. For a vector $\kappa = (\kappa_1, \kappa_2, \ldots, \kappa_n) \in \mathbb{N}^n$, we use the notation x^κ to denote the monomial $x_1^{\kappa_1} x_2^{\kappa_2} \ldots x_n^{\kappa_n}$. For a non-terminal $A_i \in V$, and a string $\alpha \in (V \cup \Sigma)^*$, let $\kappa_i(\alpha) \in \mathbb{N}$ denote the number of occurrences of A_i in the string α. We define $\kappa(\alpha) \in \mathbb{N}^n$ to be $\kappa(\alpha) = (\kappa_1(\alpha), \kappa_2(\alpha), \ldots, \kappa_n(\alpha))$.

In the MPS $x = P_G(x)$, corresponding to each nonterminal $A_i \in V$, there will be one variable x_i and one equation, namely $x_i = P_G(x)_i$, where: $P_G(x)_i \equiv \sum_{r=(A \to \alpha) \in R_{A_i}} p(r) x^{\kappa(\alpha)}$. If there are no rules associated with A_i, i.e., if $R_{A_i} = \emptyset$, then by default we define $P_G(x)_i \equiv 0$. Note that if $r \in R_{A_i}$ is a terminal rule, i.e., $\kappa(r) = (0, \ldots, 0)$, then $p(r)$ is one of the constant terms of $P_G(x)_i$.

Note: *Throughout this paper, for any n-vector z, whose i'th coordinate z_i "corresponds" to nonterminal A_i, we often find it convenient to use z_{A_i} to refer to z_i. So, e.g., we alternatively use x_{A_i} and $P_G(x)_{A_i}$, instead of x_i and $P_G(x)_i$.*

Note that if G is a SCFG, then in $x = P_G(x)$, by definition, the sum of the monomial coefficients and constant terms of each polynomial $P_G(x)_i$ is at most 1, because $\sum_{r \in R_{A_i}} p(r) \leq 1$ for every $A_i \in V$. An MPS that satisfies this extra condition is called a **probabilistic polynomial system of equations** (PPS).

Consider any MPS, $x = P(x)$, with n variables, $x = (x_1, \ldots, x_n)$. Let $\mathbb{R}_{\geq 0}$ denote the non-negative real numbers. Then $P(x)$ defines a monotone operator on the non-negative orthant $\mathbb{R}_{\geq 0}^n$. In general, an MPS need not have any real-valued solution: consider $x = x + 1$. However, by monotonicity of $P(x)$, if there exists $a \in \mathbb{R}_{\geq 0}^n$ such that $a = P(a)$, then there is a *least fixed point* (LFP) solution $q^* \in \mathbb{R}_{\geq 0}^n$ such that $q^* = P(q^*)$, and such that $q^* \leq a$ for all solutions $a \in \mathbb{R}_{\geq 0}^n$.

Proposition 1. *(cf. [9] or see [17]) For any SCFG (or convergent WCFG), G, with n nonterminals $A_1, \ldots, A_n$, the LFP solution of $x = P_G(x)$ is the n-vector $q^G = (q_{A_1}^G, \ldots, q_{A_n}^G)$ of termination probabilities (termination weights) of G.*

For computation purposes, we assume that the input probabilities (weights) associated with rules of input SCFGs or WCFGs are positive rationals encoded by giving their numerator and denominator in binary. We use $|G|$ to denote the encoding size (i.e., number of bits) of an input WCFG G.

Given any WCFG (SCFG) $G = (V, \Sigma, R, p)$ we can compute in linear time an SNF form WCFG (resp. SCFG) $G' = (V', \Sigma, R', p')$ of size $|G'| = O(|G|)$ with $V' \supseteq V$ such that $q_A^{G,w} = q_A^{G',w}$ for all $A \in V$, $w \in \Sigma^*$ (cf. [9] and Proposition 2.1 of [7]). Thus, for the problems studied in this paper, we may assume wlog that a given input WCFG or SCFG is in SNF form.

A DFA, $D = (Q, \Sigma, \Delta, s_0, F)$, has states Q, alphabet Σ, transition function $\Delta : Q \times \Sigma \to Q$, start state $s_0 \in Q$ and final states $F \subseteq Q$. We extend Δ to strings: $\Delta^* : Q \times \Sigma^* \to Q$ is defined by induction on the length $|w| \geq 0$ of

$w \in \Sigma^*$: for $s \in Q$, $\Delta^*(s, \epsilon) := s$. Inductively, if $w = aw'$, with $a \in \Sigma$, then $\Delta^*(s, w) := \Delta^*(\Delta(s, a), w')$. We define $L(D) = \{w \in \Sigma^* \mid \Delta^*(s_0, w) \in F\}$.

Given a WCFG G and a DFA D over the same terminal alphabet, for any nonterminal A of G, we define $q_A^{G,D} = \sum_{w \in L(D)} q_A^{G,w}$. If G is a SCFG, $q_A^{G,D}$ simply denotes the probability that G, starting at A, generates a string in $L(D)$. Our goal is to compute $q_A^{G,D}$, given SCFG G and DFA D. In general, $q_A^{G,D}$ may be irrational, even when all rule probabilities of G are rational values. So one natural goal is to approximate $q_A^{G,D}$ with desired precision. More precisely, the approximation problem is: given an SCFG, G, a nonterminal A, a DFA, D, over the same terminal alphabet Σ, and a rational error threshold $\delta > 0$, output a rational value $v \in [0, 1]$ such that $|v - q_A^{G,D}| < \delta$. We would like to do this as efficiently as possible as a function of the input size: $|G|$, $|D|$, and $\log(1/\delta)$.

To compute $q_A^{G,D}$, it will be useful to define a WCFG obtained as the *product* of a SCFG and a DFA. We assume, wlog, that the input SCFG is in SNF form. The **product** (or **intersection**) of a SCFG $G = (V, \Sigma, R, p)$ in SNF form, and DFA, $D = (Q, \Sigma, \Delta, s_0, F)$, is defined to be a new WCFG, $G \otimes D = (V', \Sigma, R', p')$, where the set of nonterminals is $V' = Q \times V \times Q$. Assuming $n = |V|$ and $d = |Q|$, then $|V'| = d^2 n$. The rules R' and rule probabilities p' of the product $G \otimes D$ are defined as follows (recall G is assumed to be in SNF):

- Rules of form L: For every rule of the form $(A \xrightarrow{p} B) \in R$, and every pair of states $s, t \in Q$, there is a rule $(sAt) \xrightarrow{p} (sBt)$ in R'.
- Rules of form Q: for every rule $(A \xrightarrow{1} BC) \in R$, and for all states $s, t, u \in Q$, there is a rule $(sAu) \xrightarrow{1} (sBt)(tCu)$ in R'.
- Rules of form T: for every rule $(A \xrightarrow{1} a) \in R$, where $a \in \Sigma$, and for every state $s \in Q$, if $\Delta(s, a) = t$, then there is a rule $(sAt) \xrightarrow{1} a$ in R'.

  For every rule $(A \xrightarrow{1} \epsilon) \in R$, and every $s \in Q$, there is a rule $(sAs) \xrightarrow{1} \epsilon$

Associated with the WCFG, $G \otimes D$, is the MPS $y = P_{G \otimes D}(y)$, where y is now a $d^2 n$-vector of variables, where $n = |V|$ and $d = |Q|$. The LFP solution of this MPS captures the probabilities $q_A^{G,D}$ in the following sense:

Proposition 2. *(cf. [18], or [10] for a variant of this) For any SCFG, $G = (V, \Sigma, R, p)$, and DFA, $D = (Q, \Sigma, \Delta, s_0, F)$, the LFP solution $q^{G \otimes D}$ of the MPS $x = P_{G \otimes D}(x)$, satisfies $0 \le q^{G \otimes D} \le 1$. Furthermore, for any $A \in V$ and $s, t \in Q$, $q_{(sAt)}^{G \otimes D} = \sum_{\{w \mid \Delta^*(s, w) = t\}} q_A^{G,w}$. Thus, for every $A \in V$, $q_A^{G,D} = \sum_{t \in F} q_{(s_0 At)}^{G \otimes D}$.*

Newton's Method (NM). For an MPS (or PPS), $x = P(x)$, in n variables, let $B(x) := P'(x)$ denote the Jacobian matrix of $P(x)$. In other words, $B(x)$ is an $n \times n$ matrix such that $B(x)_{i,j} = \frac{\partial P_i(x)}{\partial x_j}$. For a vector $z \in \mathbb{R}^n$, assuming that matrix $(I - B(z))$ is non-singular, we define a single iteration of Newton's method (NM) for $x = P(x)$ on z via the following operator:

$$\mathcal{N}(z) := z + (I - B(z))^{-1}(P(z) - z) \tag{1}$$

Using Newton iteration, starting at n-vector $x^{(0)} := \mathbf{0}$, yields the following iteration: $x^{(k+1)} := \mathcal{N}(x^{(k)})$, for $k = 0, 1, 2, \ldots$.

For every MPS, we can detect in P-time all the variables x_j such that $q_j^* = 0$ [9]. We can then remove these variables and their corresponding equation $x_j = P(x)_j$, and substitute their values on the right hand sides of remaining equations. This yields a new MPS, with LFP $q' > 0$, which corresponds to the non-zero coordinates of q^*. It was shown in [9] that one can always apply a decomposed Newton's method to this MPS, to converge monotonically to the LFP solution.

Proposition 3. *(cf. Theorem 6.1 of [9] and Theorem 4.1 of [4]) Let $x = P(x)$ be a MPS, with LFP $q^* > 0$. Then starting at $x^{(0)} := 0$, the Newton iterations $x^{(k+1)} := \mathcal{N}(x^{(k)})$ are well defined and monotonically converge to q^*, i.e. $\lim_{k\to\infty} x^{(k)} = q^*$, and $x^{(k+1)} \geq x^{(k)} \geq 0$ for all $k \geq 0$.*

Unfortunately, it was shown in [9] that obtaining any non-trivial additive approximation to the LFP solution of a general MPS, even one whose LFP is $0 < q^* \leq 1$, is **PosSLP**-hard, so we can not compute the termination weights of general WCFGs in P-time (nor even in NP), without a major breakthrough in the complexity of numerical computation. (See [9] for more information.)

Fortunately, for the class of PPSs, we can do a lot better. First we can identify in P-time also all the variables x_j such that $q_j^* = 1$ [9] and remove them from the system. We showed recently in [7] that by then applying a suitably *rounded down* variant of Newton's method to the resulting PPS, we can approximate q^* within additive error 2^{-j} in time polynomial in the size of the PPS and j.

3   Balance, Collapse, and Newton's Method

For an SCFG, $G = (V, \Sigma, R, p)$, and a DFA, $D = (Q, \Sigma, \Delta, s_0, F)$, we want to relate the behavior of Newton's method on the MPS associated with the WCFG, $G \otimes D$, to that of the PPS associated with the SCFG G. We shall show that there is indeed a tight correspondence, regardless of what the DFA D is. This holds even when G itself is a convergent WCFG, and thus $x = P_G(x)$ is an MPS. We need an abstract algebraic way to express this correspondence. A key notion will be *balance*, and the *collapse* operator defined on balanced vectors and matrices.

Consider the LFP q^G of $x = P_G(x)$, and LFP $q^{G \otimes D}$ of $y = P_{G \otimes D}(y)$. By Propos. 1 and 2, for any $A \in V$, $q_A^G = \sum_{w \in \Sigma^*} q_A^{G,w}$ is the probability (weight) that G, starting at A, generates any finite string. Likewise $q_{(sAt)}^{G \otimes D} = \sum_{\{w | \Delta^*(s,w)=t\}} q_A^{G,w}$ is the probability (weight) that, starting at A, G generates a finite string w such that $\Delta^*(s, w) = t$. Thus, for any $A \in V$ and $s \in Q$, $q_A^G = \sum_{t \in Q} q_{(sAt)}^{G \otimes D}$.

It turns out that analogous relationships hold between many other vectors associated with G and $G \otimes D$, including between the Newton iterates obtained by applying Newton's method to their respective PPS (or MPS) and the product MPS. Furthermore, associated relationships also hold between the Jacobian matrices $B_G(x)$ and $B_{G \otimes D}(y)$ of $P_G(x)$ and $P_{G \otimes D}(y)$, respectively.

Let $n = |V|$ and let $d = |Q|$. A vector $y \in \mathbb{R}^{d^2 n}$, whose coordinates are indexed by triples $(sAt) \in Q \times V \times Q$, is called **balanced** if for any non-terminal A, and any pair of states $s, s' \in Q$, $\sum_{t \in Q} y_{(sAt)} = \sum_{t \in Q} y_{(s'At)}$. In other words, y is balanced if the value of the sum $\sum_{t \in Q} y_{(sAt)}$ is independent of the state s. As already observed, $q^{G \otimes D} \in \mathbb{R}_{\geq 0}^{d^2 n}$ is balanced. Let $\mathfrak{B} \subseteq \mathbb{R}^{d^2 n}$ denote the set of balanced vectors. Let us define the **collapse** mapping $\mathfrak{C} : \mathfrak{B} \to \mathbb{R}^n$. For any $A \in V$, $\mathfrak{C}(y)_A := \sum_t y_{(sAt)}$. Note: $\mathfrak{C}(y)$ is well-defined, because for $y \in \mathfrak{B}$, and any $A \in V$, the sum $\sum_t y_{(sAt)}$ is by definition independent of the state s.

We next extend the definition of balance to matrices. A matrix $M \in \mathbb{R}^{d^2 n \times d^2 n}$ is called **balanced** if, for any non-terminals $B, C \in V$ and states $s, u \in Q$, and for any pair of states $v, v' \in Q$, $\sum_t M_{(sBt),(uCv)} = \sum_t M_{(sBt),(uCv')}$, and for any $s, v \in Q$ and $s', v' \in Q$, $\sum_{t,u} M_{(sBt),(uCv)} = \sum_{t,u} M_{(s'Bt),(uCv')}$. Let $\mathfrak{B}^\times \subseteq \mathbb{R}^{d^2 n \times d^2 n}$ denote the set of balanced matrices. We extend the **collapse** map $\mathfrak{C}$ to matrices. $\mathfrak{C} : \mathfrak{B}^\times \to \mathbb{R}^{n \times n}$ is defined as follows. For any $M \in \mathfrak{B}^\times$, and any $B, C \in V$, $\mathfrak{C}(M)_{BC} := \sum_{t,u} M_{(sBt),(uCv)}$. Note, again, $\mathfrak{C}(M)$ is well-defined.

We denote the Newton operator, $\mathcal{N}$, applied to a vector $x' \in \mathbb{R}^n$ for the PPS $x = P_G(x)$ associated with G by $\mathcal{N}_G(x')$. Likewise, we denote the Newton operator applied to a vector $y' \in \mathbb{R}^{d^2 n}$ for the MPS $y = P_{G \otimes D}(y)$ associated with $G \otimes D$ by $\mathcal{N}_{G \otimes D}(y')$. For a real square matrix M, let $\rho(M)$ denote the spectral radius of M. The main result of this section is the following:

Theorem 1. *Let $x = P_G(x)$ be any PPS (or MPS), with n variables, associated with a SCFG (or WCFG) G, and let $y = P_{G \otimes D}(y)$ be the corresponding product MPS, for any DFA D, with d states. For any balanced vector $y \in \mathfrak{B} \subseteq \mathbb{R}^{d^2 n}$, with $y \geq 0$, $\rho(B_{G \otimes D}(y)) = \rho(B_G(\mathfrak{C}(y)))$. Furthermore, if $\rho(B_{G \otimes D}(y)) < 1$, then $\mathcal{N}_{G \otimes D}(y)$ is defined and balanced, $\mathcal{N}_G(\mathfrak{C}(y))$ is defined, and $\mathfrak{C}(\mathcal{N}_{G \otimes D}(y)) = \mathcal{N}_G(\mathfrak{C}(y))$. Thus, $\mathcal{N}_{G \otimes D}$ preserves balance, and the collapse map $\mathfrak{C}$ "commutes" with $\mathcal{N}$ over non-negative balanced vectors, irrespective of what the DFA D is.*

We prove this in [8] via a series of lemmas that reveal many algebraic/analytic properties of balance, collapse, and Newton's method. Key is:

Lemma 1. *Let $\mathfrak{B}_{\geq 0} = \mathfrak{B} \cap \mathbb{R}_{\geq 0}^{d^2 n}$ and $\mathfrak{B}_{\geq 0}^\times = \mathfrak{B} \cap \mathbb{R}_{\geq 0}^{d^2 n \times d^2 n}$. We have $q^{G \otimes D} \in \mathfrak{B}_{\geq 0}$ and $\mathfrak{C}(q^{G \otimes D}) = q^G$, and:*

(i) If $y \in \mathfrak{B}_{\geq 0} \subseteq \mathbb{R}_{\geq 0}^{d^2 n}$ then $B_{G \otimes D}(y) \in \mathfrak{B}_{\geq 0}^\times$, and $\mathfrak{C}(B_{G \otimes D}(y)) = B_G(\mathfrak{C}(y))$.

(ii) If $y \in \mathfrak{B}_{\geq 0}$, then $P_{G \otimes D}(y) \in \mathfrak{B}_{\geq 0}$, and $\mathfrak{C}(P_{G \otimes D}(y)) = P_G(\mathfrak{C}(y))$.

(iii) If $y \in \mathfrak{B}_{\geq 0}$ and $\rho(B_G(\mathfrak{C}(y))) < 1$, then $I - B_{G \otimes D}(y)$ is non-singular, $(I - B_{G \otimes D}(y))^{-1} \in \mathfrak{B}_{\geq 0}^\times$, and $\mathfrak{C}((I - B_{G \otimes D}(y))^{-1}) = (I - B_G(\mathfrak{C}(y)))^{-1}$.

(iv) If $y \in \mathfrak{B}_{\geq 0}$ and $\rho(B_G(\mathfrak{C}(y))) < 1$, then $\mathcal{N}_{G \otimes D}(y) \in \mathfrak{B}^\times$ and $\mathfrak{C}(\mathcal{N}_{G \otimes D}(y)) = \mathcal{N}_G(\mathfrak{C}(y))$.

An easy consequence of Thm. 1 (and Prop. 3) is that if we use NM with exact arithmetic on the PPS or MPS, $x = P_G(x)$, and on the product MPS, $y = P_{G \otimes D}(y)$, they converge at the same rate:

Corollary 1. *For any PPS or MPS, $x = P_G(x)$, with LFP $q^G > 0$, and corresponding product MPS, $y = P_{G \otimes D}(y)$, if we use Newton's method with exact arithmetic, starting at $x^{(0)} := 0$, and $y^{(0)} := 0$, then all the Newton iterates $x^{(k)}$ and $y^{(k)}$ are well-defined, and for all k: $x^{(k)} = \mathfrak{C}(y^{(k)})$.*

4 Rounded Newton on PPSs and Product MPSs

To work in the Turing model of computation (as opposed to the unit-cost RAM model) we have to consider *rounding* between iterations of NM, as in [7].

Definition 1. *(Rounded-down Newton's method (R-NM), with parameter h.) Given an MPS, $x = P(x)$, with LFP q^*, where $q^* > \mathbf{0}$, in R-NM with integer rounding parameter $h > 0$, we compute a sequence of iteration vectors $x^{[k]}$. Starting with $x^{[0]} := \mathbf{0}$, $\forall k \geq 0$ we compute $x^{[k+1]}$ as follows:*

1. *Compute $x^{\{k+1\}} := \mathcal{N}_P(x^{[k]})$, where $\mathcal{N}_P(x)$ is the Newton op. defined in (1).*
2. *For each coordinate $i = 1, \ldots, n$, set $x_i^{[k+1]}$ to be equal to the maximum multiple of 2^{-h} which is $\leq \max(x_i^{\{k+1\}}, 0)$. (In other words, round down $x^{\{k+1\}}$ to the nearest multiple of 2^{-h}, while ensuring the result is non-negative.)*

Rounding can cause iterates $x^{[k]}$ to become unbalanced, but we can handle this. For any PPS, $x = P(x)$, with Jacobian matrix $B(x)$, and LFP q^*, $\rho(B(q^*)) \leq 1$ ([9, 7]). If $\rho(B(q^*)) < 1$, we call the PPS **non-critical**. Otherwise, if $\rho(B(q^*)) = 1$, we call the PPS **critical**. For SCFGs whose PPS $x = P_G(x)$ is non-critical, we get good bounds, even though R-NM iterates can become unbalanced:

Theorem 2. *For any $\epsilon > 0$, and for an SCFG, G, if the PPS $x = P_G(x)$ has LFP $0 < q^G \leq 1$ and $\rho(B_G(q^G)) < 1$, then if we use R-NM with parameter $h + 2$ to approximate the LFP solution of the MPS $y = P_{G \otimes D}(y)$, then $\|q^{G \otimes D} - y^{[h+1]}\|_\infty \leq \epsilon$ where $h := 14|G| + 3 + \lceil \log(1/\epsilon) + \log d \rceil$.*

Thus we can compute the probability $q_A^{G,D} = \sum_{t \in F} q_{s_0 At}^{G \otimes D}$ within additive error $\delta > 0$ in time polynomial in the input size: $|G|$, $|D|$ and $\log(1/\delta)$, in the standard Turing model of computation.

We in fact obtain a much more general result. For any SCFG, G, and corresponding PPS, $x = P_G(x)$, with LFP $q^* > 0$, the *dependency graph*, $H_G = (V, E)$, has the variables (or the nonterminals of G) as nodes and has the following edges: $(x_i, x_j) \in E$ iff x_j appears in some monomial in $P_G(x)_i$ with a positive coefficient. We can decompose the dependency graph H_G into its SCCs, and form the DAG of SCCs, H'_G. For each SCC, $\mathcal{S}$, suppose its corresponding equations are $x_\mathcal{S} = P_G(x_\mathcal{S}, x_{D(\mathcal{S})})_\mathcal{S}$, where $D(\mathcal{S})$ is the set of variables $x_j \notin \mathcal{S}$ such that there is a path in H_G from some variable $x_i \in \mathcal{S}$ to x_j. We call a SCC, $\mathcal{S}$, of H_G, a *critical SCC* if the PPS $x_\mathcal{S} = P_G(x_\mathcal{S}, q_{D(\mathcal{S})}^G)_\mathcal{S}$ is critical. In other words, the SCC $\mathcal{S}$ is critical if we plug in the LFP values q^G into variables that are in lower SCCs, $D(\mathcal{S})$, then the resulting PPS is critical. We note that an arbitrary PPS, $x = P_G(x)$ is non-critical if and only if it has no critical SCC. We define the *critical depth*, $\mathfrak{c}(G)$, of $x = P_G(x)$ as follows: it is the maximum length, k, of any sequence $\mathcal{S}_1, \mathcal{S}_2, \ldots, \mathcal{S}_k$, of SCCs of H_G, such that for all $i \in \{1, \ldots, k-1\}$, $\mathcal{S}_{i+1} \subseteq D(\mathcal{S}_i)$, and furthermore, such that for all $j \in \{1, \ldots, k\}$, $\mathcal{S}_j$ is critical. Let us call a critical SCC, $\mathcal{S}$, of H_G a *bottom-critical SCC*, if $D(\mathcal{S})$ does not contain any critical SCCs. By using earlier results ([9, 3]) we can compute in P-time the critical SCCs of a PPS, and its critical depth (see [8]).

PPSs with nested critical SCCs are hard to analyze directly. It turns out we can circumvent this by "tweaking" the probabilities in the SCFG G to obtain an SCFG G' with no critical SCCs, and showing that the "tweaks" are small enough so that they do not change the probabilities of interest by much. Concretely:

Theorem 3. *For any $\epsilon > 0$, and for any SCFG, G, in SNF form, with $q^G > 0$, with critical depth $\mathfrak{c}(G)$, consider the new SCFG, G', obtained from G by the following process: for each bottom-critical SCC, $\mathcal{S}$, of $x = P_G(x)$, find any rule $r = A \xrightarrow{p} B$ of G, such that A and B are both in $\mathcal{S}$ (since G is in SNF, such a rule must exist in every critical SCC). Reduce the probability p, by setting it to $p' = p(1 - 2^{-(14|G|+3)2^{\mathfrak{c}(G)}}\epsilon^{2^{\mathfrak{c}(G)}})$. Do this for all bottom-critical SCCs. This defines G', which is non-critical. Using G' instead of G, if we apply R-NM, with parameter $h + 2$ to approximate the LFP $q^{G' \otimes D}$ of MPS $y = P_{G' \otimes D}(y)$, then $\|q^{G \otimes D} - y^{[h+1]}\|_\infty \le \epsilon$ where $h := \lceil \log d + (3 \cdot 2^{\mathfrak{c}(G)} + 1)(\log(1/\epsilon) + 14|G| + 3)\rceil$. Thus we can compute $q_A^{G,D} = \sum_{t \in F} q_{s_0 A t}^{G \otimes D}$ within additive error $\delta > 0$ in time polynomial in: $|G|$, $|D|$, $\log(1/\delta)$, and $2^{\mathfrak{c}(G)}$, in the Turing model of computation.*

The proof is very involved, and is in [8]. There, we also give a family of SCFGs, and a 3-state DFA that checks the infix probability of string aa, and we explain why these examples indicate it will likely be difficult to overcome the exponential dependence on the critical-depth $\mathfrak{c}(G)$ in the above bounds.

5   Non-criticality of SCFGs Obtained by EM

In doing parameter estimation for SCFGs, in either the supervised or unsupervised (EM) settings (see, e.g., [17]), we are given a CFG, $\mathcal{H}$, with start nonterminal S, and we wish to extend it to an SCFG, G, by giving probabilities to the rules of $\mathcal{H}$. We also have some probability distribution, $\mathcal{P}(\pi)$, over the complete derivations, π, of $\mathcal{H}$ that start at start non-terminal S. (In the unsupervised case, we begin with an SCFG, and the distribution $\mathcal{P}$ arises from the prior rule probabilities, and from the training corpus of strings.) We then assign each rule of $\mathcal{H}$ a (new) probability as follows to obtain (or update) G:

$$p(A \to \gamma) := \frac{\sum_\pi \mathcal{P}(\pi)C(A \to \gamma, \pi)}{\sum_\pi \mathcal{P}(\pi)C(A, \pi)} \qquad (2)$$

where $C(r, \pi)$ is the number of times the rule r is used in the complete derivation π, and $C(A, \pi) = \sum_{r \in R_A} C(r, \pi)$. Equation (2) only makes sense when the sums $\sum_\pi \mathcal{P}(\pi)C(A, \pi)$ are finite and nonzero, which we assume; we also assume every non-terminal and rule of $\mathcal{H}$ appears in some complete derivation π with $\mathcal{P}(\pi) > 0$.

Proposition 4. *If we use parameter estimation to obtain SCFG G using equation (2), under the stated assumptions, then G is consistent[1], i.e. $q^G = 1$, and furthermore the PPS $x = P_G(x)$ is non-critical, i.e., $\rho(B_G(1)) < 1$.*

---

[1] Consistency of the obtained SCFGs is well-known; see, e.g., [15, 17] & references therein; also [19] has results related to Prop. 4 for restricted grammars.

It follows from Prop. 4 and Thm. 2, that for SCFGs obtained by parameter estimation and EM, we can compute the probability $q_A^{G,D}$ of generating a string in $L(D)$ to within any desired precision in P-time, for any DFA D.

References

[1] Corazza, A., De Mori, R., Gretter, D., Satta, G.: Computation of probabilities for an island-driven parser. IEEE Trans. PAMI 13(9), 936–950 (1991)

[2] Durbin, R., Eddy, S.R., Krogh, A., Mitchison, G.: Biological Sequence Analysis: Probabilistic models of Proteins and Nucleic Acids. Cambridge U. Press (1999)

[3] Esparza, J., Gaiser, A., Kiefer, S.: Computing least fixed points of probabilistic systems of polynomials. In: Proc. 27th STACS, pp. 359–370 (2010)

[4] Esparza, J., Kiefer, S., Luttenberger, M.: Computing the least fixed point of positive polynomial systems. SIAM J. on Computing 39(6), 2282–2355 (2010)

[5] Esparza, J., Kučera, A., Mayr, R.: Model checking probabilistic pushdown automata. Logical Methods in Computer Science 2(1), 1–31 (2006)

[6] Etessami, K., Stewart, A., Yannakakis, M.: Polynomial time algorithms for branching Markov decision processes and probabilistic min(max) polynomial Bellman equations. In: Czumaj, A., Mehlhorn, K., Pitts, A., Wattenhofer, R. (eds.) ICALP 2012, Part I. LNCS, vol. 7391, pp. 314–326. Springer, Heidelberg (2012); See full version at ArXiv:1202.4798

[7] Etessami, K., Stewart, A., Yannakakis, M.: Polynomial-time algorithms for multi-type branching processes and stochastic context-free grammars. In: Proc. 44th ACM STOC, Full version is available at ArXiv:1201.2374 (2012)

[8] Etessami, K., Stewart, A., Yannakakis, M.: Stochastic Context-Free Grammars, Regular Languages, and Newton's method, Full preprint of this paper: ArXiv:1302.6411 (2013)

[9] Etessami, K., Yannakakis, M.: Recursive Markov chains, stochastic grammars, and monotone systems of nonlinear equations. Journal of the ACM 56(1) (2009)

[10] Etessami, K., Yannakakis, M.: Model checking of recursive probabilistic systems. ACM Trans. Comput. Log. 13(2), 12 (2012)

[11] Horn, R.A., Johnson, C.R.: Matrix Analysis. Cambridge U. Press (1985)

[12] Jelinek, F., Lafferty, J.D.: Computation of the probability of initial substring generation by stochastic context-free grammars. Computational Linguistics 17(3), 315–323 (1991)

[13] Knudsen, B., Hein, J.: Pfold: RNA secondary structure prediction using stochastic context-free grammars. Nucleic Acids Res 31, 3423–3428 (2003)

[14] Manning, C., Schütze, H.: Foundations of Statistical Natural Language Processing. MIT Press (1999)

[15] Nederhof, M.-J., Satta, G.: Estimation of consistent probabilistic context-free grammars. In: HLT-NAACL (2006)

[16] Nederhof, M.-J., Satta, G.: Computing partition functions of PCFGs. Research on Language and Computation 6(2), 139–162 (2008)

[17] Nederhof, M.-J., Satta, G.: Probabilistic parsing. New Developments in Formal Languages and Applications 113, 229–258 (2008)

[18] Nederhof, M.-J., Satta, G.: Computation of infix probabilities for probabilistic context-free grammars. In: EMNLP, pp. 1213–1221 (2011)

[19] Sánchez, J., Benedí, J.-M.: Consistency of stochastic context-free grammars from probabilistic estimation based on growth transformations. IEEE Trans. Pattern Anal. Mach. Intell. 19(9), 1052–1055 (1997)

[20] Stewart, A., Etessami, K., Yannakakis, M.: Upper bounds for Newton's method on monotone polynomial systems, and P-time model checking of probabilistic one-counter automata, Arxiv:1302.3741 (2013) (conference version to appear in CAV 2013)

[21] Stolcke, A.: An efficient probabilistic context-free parsing algorithm that computes prefix probabilities. Computational Linguistics 21(2), 167–201 (1995)

[22] Wojtczak, D., Etessami, K.: Premo: an analyzer for probabilistic recursive models. In: Grumberg, O., Huth, M. (eds.) TACAS 2007. LNCS, vol. 4424, pp. 66–71. Springer, Heidelberg (2007)

Reachability in Two-Clock Timed Automata Is PSPACE-Complete[*]

John Fearnley[1] and Marcin Jurdziński[2]

[1] Department of Computer Science, University of Liverpool, UK
[2] Department of Computer Science, University of Warwick, UK

Abstract. Haase, Ouaknine, and Worrell have shown that reachability in two-clock timed automata is log-space equivalent to reachability in bounded one-counter automata. We show that reachability in bounded one-counter automata is PSPACE-complete.

1 Introduction

Timed automata [1] are a successful and widely used formalism, which are used in the analysis and verification of real time systems. A timed automaton is a non-deterministic finite automaton that is equipped with a number of real-valued *clocks*, which allow the automaton to measure the passage of time.

Perhaps the most fundamental problem for timed automata is the *reachability* problem: given an initial state, can we perform a sequence of transitions in order to reach a specified target state? In their foundational paper on timed automata [1], Alur and Dill showed that this problem is PSPACE-complete. To show hardness for PSPACE, their proof starts with a linear bounded automaton (LBA), which is a non-deterministic Turing machine with a fixed tape length n. They produced a timed automaton with $2n + 1$ clocks, and showed that the timed automaton can reach a specified state if and only if the LBA halts.

However, the work of Alur and Dill did not address the case where the number of clocks is small. This was rectified by Courcoubetis and Yannakakis [3], who showed that reachability in timed automata with only three clocks is still PSPACE-complete. Their proof cleverly encodes the tape of an LBA in a single clock, and then uses the two additional clocks to perform all necessary operations on the encoded tape. In contrast to this, Laroussinie et al. have shown that reachability in one-clock timed automata is complete for NLOGSPACE, and therefore no more difficult than computing reachability in directed graphs [6].

The complexity of reachability in two-clock timed automata has been left open. So far, the best lower bound was given by Laroussinie et al., who gave a proof that the problem is NP-hard via a very natural reduction from subset-sum [6]. Moreover, the problem lies in PSPACE, because reachability in two-clock

---

[*] A full version of this paper is available at http://arxiv.org/abs/1302.3109. This work was supported by EPSRC grants EP/H046623/1 *Synthesis and Verification in Markov Game Structures* and EP/D063191/1 *The Centre for Discrete Mathematics and its Applications (DIMAP)*.

F.V. Fomin et al. (Eds.): ICALP 2013, Part II, LNCS 7966, pp. 212–223, 2013.
© Springer-Verlag Berlin Heidelberg 2013

timed automata is no harder than reachability in three-clock timed automata. However, the PSPACE-hardness proof of Courcoubetis and Yannakakis seems to fundamentally require three clocks, and does not naturally extend to the two-clock case. Naves [7] has shown that several extensions to two-clock timed automata lead to PSPACE-completeness, but his work does not advance upon the NP-hard lower bound for unextended two-clock timed automata.

In a recent paper, Haase et al. have shown a link between reachability in timed automata and reachability in *bounded counter automata* [5]. A bounded counter automaton is a non-deterministic finite automaton equipped with a set of counters, and the transitions of the automaton may add or subtract arbitrary integer constants to the counters. The state space of each counter is bounded by some natural number b, so the counter may only take values in the range $[0, b]$. Moreover, transitions may only be taken if they do not increase or decrease a counter beyond the allowable bounds. This gives these seemingly simple automata a surprising amount of power, because the bounds can be used to implement inequality tests against the counters.

Haase et al. show that reachability in two-clock timed automata is log-space equivalent to reachability in bounded *one*-counter automata. Reachability in bounded one-counter automata has also been studied in the context of one-clock timed automata with energy constraints [2], where it was shown that the problem lies in PSPACE and is NP-hard. It has also been shown that the reachability problem for *unbounded* one-counter automata is NP-complete [4], but the NP membership proof does not seem to generalise to bounded one-counter automata.

Our Contribution. We show that satisfiability for quantified boolean formulas can be reduced, in polynomial time, to reachability in bounded one-counter automata. Hence, we show that reachability in bounded one-counter automata is PSPACE-complete, and therefore we resolve the complexity of reachability in two-clock timed automata. Our reduction uses two intermediate steps: *subset-sum games* and *safe counter-stack automata*.

Counter automata are naturally suited for solving subset-sum problems, so our reduction starts with a quantified version of subset-sum, which we call subset-sum games. One interpretation of satisfiability for quantified boolean formulas is to view the problem as a game between an *existential* player and a *universal* player. The players take turns to set their propositions to true or false, and the existential player wins if and only if the boolean formula is satisfied. Subset-sum games follow the same pattern, but apply it to subset-sum: the two players alternate in choosing numbers from sets, and the existential player wins if and only if the chosen numbers sum to a given target. Previous work by Travers can be applied to show that subset-sum games are PSPACE-complete [8].

We reduce subset-sum games to reachability in bounded one-counter automata. However, we will not do this directly. Instead, we introduce safe counter-stack automata, which are able to store multiple counters, but have a stack-like restriction on how these counters may be accessed. These automata are a convenient intermediate step, because having access to multiple counters makes it easier for us to implement subset-sum games. Moreover, the stack based restrictions mean that it is relatively straightforward to show that reachability in safe counter-stack automata is reducible, in polynomial time, to reachability in bounded one-counter automata, which completes our result.

2   Subset-Sum Games

A subset-sum game is played between an *existential* player and a *universal* player. The game is specified by a pair (ψ, T), where $T \in \mathbb{N}$, and ψ is a list:

$$\forall \{A_1, B_1\} \exists \{E_1, F_1\} \ \ldots \ \forall \{A_n, B_n\} \exists \{E_n, F_n\},$$

where A_i, B_i, E_i, and F_i, are all natural numbers.

The game is played in rounds. In the first round, the universal player chooses an element from $\{A_1, B_1\}$, and the existential player responds by choosing an element from $\{E_1, F_1\}$. In the second round, the universal player chooses an element from $\{A_2, B_2\}$, and the existential player responds by choosing an element from $\{E_2, F_2\}$. This pattern repeats for rounds 3 through n. Thus, at the end of the game, the players will have constructed a sequence of numbers, and the existential player wins if and only if the sum of these numbers is T.

Formally, the set of *plays* of the game is the set:

$$\mathcal{P} = \prod_{1 \le j \le n} \{A_j, B_j\} \times \{E_j, F_j\}.$$

A play $P \in \mathcal{P}$ is winning for the existential player if and only if $\sum P = T$.

A strategy for the existential player is a list of functions $\mathsf{s} = (s_1, s_2, \ldots, s_n)$, where each function s_i dictates how the existential player should play in the ith round. Thus, each function s_i is of the form:

$$s_i : \prod_{1 \le j \le i} \{A_j, B_j\} \to \{E_i, F_i\}.$$

This means that the function s_i maps the first i moves of the universal player to a decision for the existential player in the ith round.

A play P conforms to a strategy s if the decisions made by the existential player in P always agree with s. More formally, if $P = p_1 p_2 \ldots p_{2n}$ is a play, and $\mathsf{s} = (s_1, s_2, \ldots, s_n)$ is a strategy, then P conforms to s if and only if we have $s_i(p_1, p_3, \ldots, p_{2i-1}) = p_{2i}$ for all i. A strategy s is *winning* if every play $P \in \text{Plays}(\mathsf{s})$ is winning for the existential player. The *subset-sum game problem* is to decide, for a given SSG instance (ψ, T), whether the existential player has a winning strategy for (ψ, T).

The SSG problem clearly lies in PSPACE, because it can be solved on a polynomial time alternating Turing machine. A quantified version of subset-sum has been shown to be PSPACE-hard, via a reduction from quantified boolean formulas [8]. Since SSGs are essentially a quantified version of subset-sum, the proof of PSPACE-hardness easily carries over.

Lemma 1. *The subset-sum game problem is PSPACE-complete.*

3   Bounded One-Counter Automata

A bounded one-counter automaton has a single counter that can store values between 0 and some bound $b \in \mathbb{N}$. The automaton may add or subtract values from the counter, so long as the bounds of 0 and b are not overstepped. This can be used to test inequalities against the counter. For example, let $n \in \mathbb{N}$ be a number, and suppose that we want to test whether the counter is smaller-than or equal to n. We first attempt to add $b - n$ to the counter, then, if that works, we subtract $b - n$ from the counter. This creates a sequence of two transitions which can be taken if and only if the counter is smaller-than or equal to n. A similar construction can be given for greater-than tests. For the sake of convenience, we will include explicit inequality testing in our formal definition, with the understanding that this is not actually necessary.

We now give a formal definition. For two integers $a, b \in \mathbb{Z}$ we define $[a, b] = \{n \in \mathbb{Z} : a \leq n \leq b\}$ to be the subset of integers between a and b. A bounded one-counter automaton is defined by a tuple (L, b, Δ, l_0), where L is a finite set of locations, $b \in \mathbb{N}$ is a global counter bound, Δ specifies the set of transitions, and $l_0 \in L$ is the initial location. Each transition in Δ has the form (l, p, g_1, g_2, l'), where l and l' are locations, $p \in [-b, b]$ specifies how the counter should be modified, and $g_1, g_2 \in [0, b]$ give lower and upper guards for the counter.

Each state of the automaton consists of a location $l \in L$ along with a counter value c. Thus, we define the set of states S to be $L \times [0, b]$. A transition exists between a state $(l, c) \in S$, and a state $(l', c') \in S$ if there is a transition $(l, p, g_1, g_2, l') \in \Delta$, where $g_1 \leq c \leq g_2$, and $c' = c + p$.

The reachability problem for bounded one-counter automata is: starting at the state $(l_0, 0)$, can the automaton reach a specified target location l_t? It has been shown that the reachability problem for bounded one-counter automata is equivalent to the reachability problem for two-clock timed automata.

Theorem 2 ([5]). *Reachability in bounded one-counter automata is log-space equivalent to reachability in two-clock timed automata.*

4   Counter-Stack Automata

Outline. In this section we ask: can we use a bounded one-counter automaton to store multiple counters? The answer is yes, but doing so forces some interesting restrictions on the way in which the counters are accessed. By the end of this section, we will have formalised these restrictions as *counter-stack* automata.

Suppose that we have a bounded one-counter automaton with counter c and bound $b = 15$. Hence, the width of the counter is 4 bits. Now suppose that we wish to store two 2-bit counters c_1 and c_2 in c. We can do this as follows:

$$c = \boxed{1 \quad 0} \; \boxed{0 \quad 1}$$
$$\underset{c_2}{} \quad \underset{c_1}{}$$

We allocate the top two bits of c to store c_2, and the bottom two bits to store c_1. We can easily write to both counters: if we want to increment c_2 then we add 4 to c, and if we want to increment c_1 then we add 1 to c.

However, if we want to test equality, then things become more interesting. It is easy to test equality against c_2: if we want to test whether $c_2 = 2$, then we test whether $8 \leq c \leq 11$ holds. But, we cannot easily test whether $c_1 = 2$ because we would have to test whether c is 2, 6, 10, or 14, and this list grows exponentially as the counters get wider. However, if we know that $c_2 = 1$, then we only need to test whether $c = 6$. Thus, we arrive at the following guiding principle: if you want to test equality against c_i, then you must know the values of c_j for all $j > i$. Counter-stack automata are a formalisation of this principle.

Counter-Stack Automata. A counter-stack automaton has a set of k distinct counters, which are referred to as c_1 through c_k. For our initial definitions, we will allow the counters to take all values from $\mathbb{N}$, but we will later refine this by defining *safe* counter-stack automata. The defining feature of a counter-stack automaton is that the counters are arranged in a stack-like fashion:

- All counters may be increased at any time.
- c_i may only be tested for equality if the values of c_{i+1} through c_k are known.
- c_i may only be reset if the values of c_i through c_k are known.

When the automaton increases a counter, it adds a specified number $n \in \mathbb{N}$ to that counter. The automaton has the ability to perform equality tests against a counter, but the stack-based restrictions must be respected. An example of a valid equality test is $c_k = 3 \wedge c_{k-1} = 10$, because $c_{k-1} = 10$ only needs to be tested in the case where $c_k = 3$ is known to hold. Conversely, the test $c_{k-1} = 10$ by itself is invalid, because it places no restrictions on the value of c_k.

The automaton may also reset a counter, but the stack-based restrictions apply. Counter c_i may only be reset by a transition, if that transition tests equality against the values of c_i through c_k. For example, c_{k-1} may only be reset if the transition is guarded by a test of the form $c_{k-1} = n_1 \wedge c_{k-2} = n_2$.

Formal Definition. A counter-stack automaton is a tuple (L, C, Δ, l_0), where L is a finite set of locations, $C = [1, k]$ is a set of counter indexes, $l_0 \in L$ is an initial location, and Δ specifies the transition relation. Each transition in Δ has the form (l, E, I, R, l') where:

- $l, l' \in L$ is a pair of locations.
- E is a partial function from C to $\mathbb{N}$ which specifies the equality tests. If $E(i)$ is defined for some i, then $E(j)$ must be defined for all $j \in C$ with $j > i$.
- $I \in \mathbb{N}^k$ specifies how the counters must be increased.
- $R \subseteq C$ specifies the set of counters that must be reset. It is required that $E(r)$ is defined for every $r \in R$.

Each state of the automaton is a location annotated with values for each of the k counters. That is, the state space of the automaton is $L \times \mathbb{N}^k$. A state $(l, c_1, c_2, \ldots, c_k)$ can transition to a state $(l', c'_1, c'_2, \ldots, c'_k)$ if and only if there exists a transition $(l, E, I, R, l') \in \Delta$, where the following conditions hold:

- For every i for which $E(i)$ is defined, we must have $c_i = E(i)$.
- For every $i \in R$, we must have $c'_i = 0$.
- For every $i \notin R$, we must have $c'_i = c_i + I_i$.

A *run* is a sequence of states $s_0, s_1, \ldots, s_n$, where each s_i can transition to s_{i+1}. A counter-stack automaton is *b-safe*, for some $b \in \mathbb{N}$, if it is impossible for the automaton to increase a counter beyond b. Formally, this condition requires that, for every state $(l, c_1, c_2, \ldots, c_k)$ that can be reached by a run from $(l_0, 0, 0, \ldots, 0)$, we have $c_i \le b$ for all i. We say that a counter-stack automaton is safe, if it is b-safe for some $b \in \mathbb{N}$.

The reachability problem for safe counter-stack automata is a *promise problem*. The input to the problem is a triple $(\mathcal{S}, b, t)$, where $\mathcal{S}$ is a counter-stack automaton, $b \in \mathbb{N}$ is a bound, and t is a target location in $\mathcal{S}$. If $\mathcal{S}$ is b-safe, then the algorithm must decide whether there is a run from $(l_0, 0, 0, \ldots, 0)$ to $(t, 0, 0, \ldots, 0)$. If $\mathcal{S}$ is not b-safe, then the algorithm can output anything.

Simulation by a Bounded One-Counter Automaton. A safe counter-stack automaton is designed to be simulated by a bounded one-counter automaton. To do this, we follow the construction outlined at the start of this section: we split the bits of the counter c into k chunks, where each chunk represents one of the counters c_i. Note that the safety assumption is crucial, because otherwise incrementing c_i may overflow the allotted space, and inadvertently modify the value of c_{i+1}.

Lemma 3. *Reachability in safe counter-stack automata is polynomial-time reducible to reachability in bounded one-counter automata.*

5 Outline of the Construction

Our goal is to show that reachability in safe counter-stack automata is PSPACE-hard. To do this, we will show that subset-sum games can be solved by safe counter-stack automata. In this section, we give an overview of our construction using the following two-round subset-sum game.

$$\big(\forall \{A_1, B_1\} \; \exists \{E_1, F_1\} \; \forall \{A_2, B_2\} \; \exists \{E_2, F_2\}, \; T \big).$$

For brevity, we will refer to this instance as (ψ, T) for the rest of this section. The construction is split into two parts: the *play* gadget and the *reset* gadget.

The Play Gadget. The play gadget is shown in Figure 1. The construction uses nine counters. The locations are represented by circles and the transitions are represented by edges. The annotations on the transitions describe the increments, resets, and equality tests: the notation $c_i + n$ indicates that n is added to counter i, the notation $R(c_i)$ indicates that counter i is reset to 0, and the notation $c_i = n$ indicates that the transition may only be taken when $c_i = n$ is satisfied.

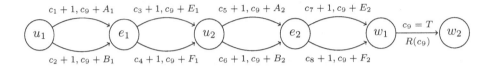

Fig. 1. The play gadget

This gadget allows the automaton to implement a play of the SSG. The locations u_1 and u_2 allow the automaton to choose the first and second moves of the universal player, while the locations e_1 and e_2 allow the automaton to choose the first and second moves for the existential player. As the play is constructed, a running total is stored in c_9, which is the top counter on the stack. The final transition between w_1 and w_2 checks whether the existential player wins the play, and then resets c_9. Thus, the set of runs between u_1 and w_2 corresponds precisely to the set of plays won by the existential player in the SSG.

In addition to this, each outgoing transition from u_i or e_i comes equipped with its own counter. This counter is incremented if and only if the corresponding edge is used during the play, and this allows us to check precisely which play was chosen. These counters will be used by the reset gadget. The idea behind our construction is to force the automaton to pass through the play gadget multiple times. Each time we pass through the play gadget, we will check a different play, and our goal is to check a set of plays that verify whether the existential player has a winning strategy for the SSG.

Which Plays Should Be Checked?. In our example, we must check four plays. The format of these plays is shown in Table 1.

Table 1. The set of plays that the automaton will check

Play	u_1	e_1	u_2	e_2
1	A_1	E_1 or F_1	A_2	E_2 or F_2
2	A_1	Unchanged	B_2	E_2 or F_2
3	B_1	E_1 or F_1	A_2	E_2 or F_2
4	B_1	Unchanged	B_2	E_2 or F_2

The table shows four different plays, which cover every possible strategy choice of the universal player. Clearly, if the existential player does have a winning strategy, then that strategy should be able to win against all strategy choices of the universal player. The plays are given in a very particular order: the first two plays contain A_1, while the second two plays contain B_1. Moreover, we always check A_2, before moving on to B_2.

We want to force the decisions made at e_1 and e_2 to form a coherent strategy for the existential player. In this game, a strategy for the existential player is a pair $\mathsf{s} = (s_1, s_2)$, where s_i describes the move that should be made at e_i. It is critical to note that s_1 only knows whether A_1 or B_1 was chosen at u_1. This

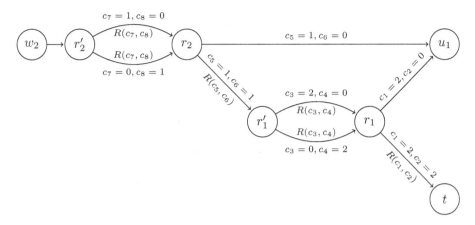

Fig. 2. The reset gadget

restriction is shown in the table: the automaton may choose freely between E_1 and F_1 in the first play. However, in the second play, the automaton must make the same choice as it did in the first play. The same relationship holds between the third and fourth plays. These restrictions ensure that the plays shown in Table 1 are a description of a strategy for the existential player.

The Reset Gadget. The reset gadget, shown in Figure 2, enforces the constraints shown in Table 1. The locations w_2 and u_1 represent the same locations as they did in Figure 1. To simplify the diagram, we have only included non-trivial equality tests. Whenever we omit a required equality test, it should be assumed that the counter is 0. For example, the outgoing transitions from r_2 implicitly include the requirement that c_7, c_8, and c_9 are all 0.

We consider the following reachability problem: can $(t, 0, 0, \ldots, 0)$ be reached from $(u_1, 0, 0, \ldots, 0)$? The structure of the reset gadget places restrictions on the runs that reach t. All such runs pass through the reset gadget exactly four times, and the following table describes each pass:

Pass	Path
1	$w_2 \to r_2' \to r_2 \to u_1$
2	$w_2 \to r_2' \to r_2 \to r_1' \to r_1 \to u_1$
3	$w_2 \to r_2' \to r_2 \to u_1$
4	$w_2 \to r_2' \to r_2 \to r_1' \to r_1 \to t$

To see why these paths must be taken, observe that, for every $i \in \{1, 3, 5, 7\}$, each pass through the play gadget increments either c_i or c_{i+1}, but not both. So, the first time that we arrive at r_2, we must take the transition directly to u_1, because the guard on the transition to r_1' cannot possibly be satisfied after a single pass through the play gadget. When we arrive at r_2 on the second pass, we are forced to take the transition to r_1', because we cannot have $c_5 = 1$ and $c_6 = 0$ after two passes through the play gadget. This transition resets both c_5

and c_6, so the pattern can repeat again on the third and fourth visits to r_2. The location r_1 behaves in the same way as r_2, but the equality tests are scaled up, because r_1 is only visited on every second pass through the reset gadget.

We can now see that all strategies of the universal player must be considered. The transition between r_2 and u_1 forces the play gadget to increment c_5, and therefore the first and third plays must include A_2. Similarly, the transition between r_2 and r_1' forces the second and fourth plays to include B_2. Meanwhile, the transition between r_1 and u_1 forces the first and second plays to include A_1, and the transition between r_1 and t forces the third and fourth plays to include B_1. Thus, we select the universal player strategies exactly as Table 1 prescribes.

The transitions between r_1' and r_1 check that the existential player is playing a coherent strategy. When the automaton arrives at r_1' during the second pass, it verifies that either E_1 was included in the first and second plays, or that F_1 was included in the first and second plays. If this is not the case, then the automaton gets stuck. The counters c_3 and c_4 are reset when moving to r_1, which allows the same check to occur during the fourth pass. For the sake of completeness, we have included the transitions between r_2' and r_2, which perform the same check for E_2 and F_2. However, since the existential player is allowed to change this decision on every pass, the automaton can never get stuck at r_2'.

The end result is that location t can be reached if and only if the existential player has a winning strategy for (ψ, T). As we will show in the next section, the construction extends to arbitrarily large SSGs, which then leads to a proof that reachability in counter-stack automata is PSPACE-hard. Note that this construction is safe: c_9 is clearly bounded by the maximum value that can be achieved by a play of the SSG, and reset gadget ensures that no other counter may exceed 4. Thus, we will have completed our proof of PSPACE-hardness for bounded one-counter automata and two-clock timed automata.

6   Formal Definition and Proof

Sequential Strategies for SSGs. We start by formalising the ideas behind Table 1. Recall that the table gives a strategy for the existential player in the form of a list of plays. Moreover, the table gave a very specific ordering in which these plays must appear. We now formalise this ordering.

We start by dividing the integers in the interval $[1, 2^n]$ into i-blocks. The 1-blocks partition the interval into two equally sized blocks. The first 1-block consists of the range $[1, 2^{n-1}]$, and the second 1-block consists of the range $[2^{n-1} + 1, 2^n]$. There are four 2-blocks, which partition the 1-blocks into two equally sized sub-ranges. This pattern continues until we reach the n-blocks.

Formally, for each $i \in \{1, 2, \ldots, n\}$, there are 2^i distinct i-blocks. The set of i-blocks can be generated by considering the intervals $[k + 1, k + 2^{n-i}]$ for the first 2^i numbers $k \geq 0$ that satisfy $k \bmod 2^{n-i} = 0$. An i-block is *even* if k is an even multiple of 2^{n-i}, and it is *odd* if k is an odd multiple of 2^{n-i}.

The ordering of the plays in Table 1 can be described using blocks. There are four 2-blocks, and A_2 appears only in even 2-blocks, while B_2 only appears in odd 2-blocks. Similarly, A_1 only appears in the even 1-block, while B_1 only appears in the odd 1-block. The restrictions on the existential player can also be described using blocks: the existential player's strategy may not change between E_i and F_i during an i-block. We generalise this idea in the following definition.

Definition 4 (Sequential strategy). *A sequential strategy for the existential player in (ψ, T) is a list of 2^n plays $\mathcal{S} = P_1, P_2, \ldots, P_{2^n}$, where for every i-block L we have:*

- *If L is an even i-block, then P_j must contain A_i for all $j \in L$.*
- *If L is an odd i-block, then P_j must contain B_i for all $j \in L$.*
- *We either have $E_i \in P_j$ for all $j \in L$, or we have $F_i \in P_j$ for all $j \in L$.*

We say that $\mathcal{S}$ is winning for the existential player if $\sum P_j = T$ for every $P_j \in \mathcal{S}$. Since a sequential strategy is simply a list of all plays that conform to a strategy, we have the following lemma.

Lemma 5. *The existential player has a winning strategy if and only if the existential player has a sequential winning strategy.*

The Base Automaton. We describe the construction in two steps. Recall, from Figures 1 and 2, that the top counter is used by the play gadget to store the value of the play, and to test whether the play is winning. We begin by constructing a version of the automaton that omits the top counter. That is, if c_k is the top counter, we modify the play gadget by removing all increases to c_k, and the equality test for c_k between w_1 and w_2. We call this the *base* automaton. Later, we will add the constraints for c_k back in, to construct the *full* automaton.

We now give a formal definition of the base automaton. The location and counter names are consistent with Figures 1 and 2. For each natural number n, we define a counter-stack automaton $\mathcal{A}_n$. The automaton has the following set of locations:

- locations u_i and e_i for each $i \in [1, n]$,
- locations w_1 and w_2,
- reset locations r_i and r_i' for each $i \in [1, n]$, and
- the goal location t.

The automaton uses $k = 2n + 1$ counters. The top counter c_k is reserved for the full automaton, and will not be used in this construction. We will identify counters 1 to $2n$ using the following shorthands. For each integer i we define $a_i = c_{4(i-1)+1}$, $b_i = c_{4(i-1)+2}$, $e_i = c_{4(i-1)+3}$, and $f_i = c_{4(i-1)+4}$. For example, in Figure 1, we have $a_1 = c_1$ and $a_2 = c_5$, and these are precisely the counters associated with A_1 and A_2, respectively. The same relationship holds between b_1 and B_1, between b_2 and B_2, and so on.

The transitions of the automaton are defined as follows. Whenever we omit a required equality test against a counter c_i, it should be assumed that the transition includes the test $c_i = 0$.

- Each location u_i has two transitions to e_i: a transition that adds 1 to a_i, and a transition that adds 1 to b_i.
- We define u_{n+1} to be a shorthand for w_1. Each location e_i has two transitions to u_{i+1}: a transition that adds 1 to e_i, and transition that adds 1 to f_i.
- Location w_1 has a transition to w_2, and w_2 has a transition to r'_n. These transitions do not increase any counter, and do not test any equalities.
- Each location r'_i has two outgoing transitions to r_i. Firstly, there is a transition that tests $e_i = 2^{n-i}$ and $f_i = 0$, and then resets e_i and f_i. Secondly, there is a transition that tests $e_i = 0$ and $f_i = 2^{n-i}$, and then resets both e_i and f_i.
- We define r'_0 to be shorthand for location t. Each location r_i has two outgoing transitions. Firstly, there is a transition to u_1 that tests $a_i = 2^{n-i}$ and $b_i = 0$. Secondly, there is a transition to r'_{i-1} that tests $a_i = 2^{n-i}$ and $b_i = 2^{n-i}$ and then resets both a_i and b_i.

Runs in the Base Automaton. We now describe the set of runs that are possible in the base automaton. We decompose every run of the automaton into segments, such that each segment contains a single pass through the play gadget. More formally, we decompose R into segments $R_1, R_2, \ldots$, where each segment R_i starts at u_1, and ends at the next visit to u_1. We say that a run gets *stuck* if the run does not end at $(t, 0, 0, \ldots, 0)$, and if the final state of the run has no outgoing transitions. We say that a run R gets stuck during an i-block L if there exists a $j \in L$ such that R_j gets stuck. The following lemma gives a characterisation of the runs in $\mathcal{A}_n$.

Lemma 6. *A run R in $\mathcal{A}_n$ does not get stuck if and only if, for every i-block L, all of the following hold.*

- *If L is an even i-block, then R_j must increment a_i for every $j \in L$.*
- *If L is an odd i-block, then R_j must increment b_i for every $j \in L$.*
- *Either R_j increments e_i for every $j \in L$, or R_j increments f_i for every $j \in L$.*

We say that a run is *successful* if it eventually reaches $(t, 0, 0, \ldots, 0)$. By definition, a run is successful if and only if it never gets stuck. Also, the transition from r_1 to t ensures that every successful run must have exactly 2^n segments. With these facts in mind, if we compare Lemma 6 with Definition 4, then we can see that the set of successful runs in $\mathcal{A}_n$ corresponds exactly to the set of sequential strategies for the existential player in the SSG.

Since we eventually want to implement $\mathcal{A}_n$ as a safe one-counter automaton, it is important to prove that $\mathcal{A}_n$ is safe. We do this in the following Lemma.

Lemma 7. *Along every run of $\mathcal{A}_n$ we have that counters a_i and b_i never exceed 2^{n-i+1}, and counters e_i and f_i never exceed 2^{n-i}.*

The Full Automaton. Let (ψ, T) be an SSG instance, where ψ is:

$$\forall \{A_1, B_1\} \exists \{E_1, F_1\} \ldots \forall \{A_n, B_n\} \exists \{E_n, F_n\}.$$

We will construct a counter-stack automaton $\mathcal{A}_\psi$ from $\mathcal{A}_n$. Recall that the top counter c_k is unused in $\mathcal{A}_n$. We modify the transitions of $\mathcal{A}_n$ as follows. Let δ be a transition. If δ increments a_i then it also adds A_i to c_k, if δ increments b_i then it also adds B_i to c_k, if δ increments e_i then it also adds E_i to c_k, and if δ increments f_i then it also adds F_i to c_k. We also modify the transition between w_1 and w_2, so that it checks whether $c_k = T$, and resets c_k.

Since we only add extra constraints to $\mathcal{A}_n$, the set of successful runs in $\mathcal{A}_\psi$ is contained in the set of successful runs of $\mathcal{A}_n$. Recall that the set of successful runs in $\mathcal{A}_n$ encodes the set of sequential strategies for the existential player in (ψ, T). In $\mathcal{A}_\psi$, we simply check whether each play in the sequential strategy is winning for the existential player. Thus, we have shown the following lemma.

Lemma 8. *The set of successful runs in $\mathcal{A}_\psi$ corresponds precisely to the set of winning sequential strategies for the existential player in (ψ, T).*

We also have that $\mathcal{A}_\psi$ is safe. Bounds for counters c_1 through c_{k-1} are shown in Lemma 7, and counter c_k may never exceed $\sum\{A_i, B_i, E_i, F_i : 1 \le i \le n\}$. This completes the reduction from subset-sum games to safe counter-stack automata, and gives us our main result.

Theorem 9. *Reachability in safe counter-stack automata is PSPACE-complete.*

Corollary 10.

- *Reachability in bounded one-counter automata is PSPACE-complete.*
- *Reachability in two-clock timed automata is PSPACE-complete.*

References

1. Alur, R., Dill, D.L.: A theory of timed automata. Theoretical Computer Science 126(2), 183–235 (1994)
2. Bouyer, P., Fahrenberg, U., Larsen, K.G., Markey, N., Srba, J.: Infinite runs in weighted timed automata with energy constraints. In: Cassez, F., Jard, C. (eds.) FORMATS 2008. LNCS, vol. 5215, pp. 33–47. Springer, Heidelberg (2008)
3. Courcoubetis, C., Yannakakis, M.: Minimum and maximum delay problems in real-time systems. Formal Methods in System Design 1(4), 385–415 (1992)
4. Haase, C., Kreutzer, S., Ouaknine, J., Worrell, J.: Reachability in succinct and parametric one-counter automata. In: Bravetti, M., Zavattaro, G. (eds.) CONCUR 2009. LNCS, vol. 5710, pp. 369–383. Springer, Heidelberg (2009)
5. Haase, C., Ouaknine, J., Worrell, J.: On the relationship between reachability problems in timed and counter automata. In: Finkel, A., Leroux, J., Potapov, I. (eds.) RP 2012. LNCS, vol. 7550, pp. 54–65. Springer, Heidelberg (2012)
6. Laroussinie, F., Markey, N., Schnoebelen, P.: Model checking timed automata with one or two clocks. In: Gardner, P., Yoshida, N. (eds.) CONCUR 2004. LNCS, vol. 3170, pp. 387–401. Springer, Heidelberg (2004)
7. Naves, G.: Accessibilité dans les automates temporisé à deux horloges. Rapport de Master, MPRI, Paris, France (2006)
8. Travers, S.: The complexity of membership problems for circuits over sets of integers. Theoretical Computer Science 369(13), 211–229 (2006)

Ramsey Goes Visibly Pushdown[*]

Oliver Friedmann[1], Felix Klaedtke[2], and Martin Lange[3]

[1] LMU Munich
[2] ETH Zurich
[3] University of Kassel

Abstract. Checking whether one formal language is included in another is vital to many verification tasks. In this paper, we provide solutions for checking the inclusion of the languages given by visibly pushdown automata over both finite and infinite words. Visibly pushdown automata are a richer automaton model than the classical finite-state automata, which allows one, e.g., to reason about the nesting of procedure calls in the executions of recursive imperative programs. The highlight of our solutions is that they do not comprise automata constructions for determinization and complementation. Instead, our solutions are more direct and generalize the so-called Ramsey-based inclusion-checking algorithms, which apply to classical finite-state automata and proved effective there, to visibly pushdown automata. We also experimentally evaluate our algorithms thereby demonstrating the virtues of avoiding determinization and complementation constructions.

1 Introduction

Various verification tasks can be stated more or less directly as inclusion problems of formal languages or comprise inclusion problems as subtasks. For example, the model-checking problem of non-terminating finite-state systems with respect to trace properties boils down to the question whether the inclusion $L(\mathcal{A}) \subseteq L(\mathcal{B})$ for two Büchi automata $\mathcal{A}$ and $\mathcal{B}$ holds, where $\mathcal{A}$ describes the traces of the system and $\mathcal{B}$ the property [22]. Another application of checking language inclusion for Büchi automata appears in size-change termination analysis [13,19]. Inclusion problems are in general difficult. For Büchi automata it is PSPACE-complete.

From the closure properties of the class of ω-regular languages, i.e., those languages that are recognizable by Büchi automata it is obvious that questions like the one above for model checking non-terminating finite-state systems can be effectively reduced to an emptiness question, namely, $L(\mathcal{A}) \cap L(\mathcal{C}) = \emptyset$, where $\mathcal{C}$ is a Büchi automaton that accepts the complement of $\mathcal{B}$. Building a Büchi automaton for the intersection of the languages and checking its emptiness is fairly easy: the automaton accepting the intersection can be quadratically bigger, the emptiness problem is NLOGSPACE-complete, and it admits efficient implementations, e.g., by a nested depth-first search. However, complementing Büchi automata is

---

[*] Extended abstract. Omitted details can be found in the full version [15], which is available from the authors' web pages.

F.V. Fomin et al. (Eds.): ICALP 2013, Part II, LNCS 7966, pp. 224–237, 2013.
© Springer-Verlag Berlin Heidelberg 2013

challenging. One intuitive reason for this is that not every Büchi automaton has an equivalent deterministic counterpart. Switching to a richer acceptance condition like the parity condition so that determinization would be possible is currently not an option in practice. The known determinization constructions for richer acceptance conditions are intricate, although complementation would then be easy by dualizing the acceptance condition. A lower bound on the complementation problem with respect to the automaton size is $2^{\Omega(n \log n)}$. Known constructions for complementing Büchi automata that match this lower bound are also intricate. As a matter of fact, all attempts so far that explicitly construct the automaton $\mathcal{C}$ from $\mathcal{B}$ scale poorly. Often, the implementations produce automata for the complement language that are huge, or they even fail to produce an output at all in reasonable time and space if the input automaton has more than 20 states, see, e.g., [5, 21].

Other approaches for checking the inclusion of the languages given by Büchi automata or solving the closely related but simpler universality problem for Büchi automata have recently gained considerable attention [1,2,8–10,13,14,19]. In the worst case, these algorithms have exponential running times, which are often worse than the $2^{\Omega(n \log n)}$ lower bound on complementing Büchi automata. However, experimental results, in particular, the ones for the so-called Ramsey-based algorithms show that the performance of these algorithms is superior. The name *Ramsey-based* stems from the fact that their correctness is established by relying on Ramsey's theorem [20].[1]

The Ramsey-based algorithms for checking universality $L(\mathcal{B}) = \Sigma^\omega$ iteratively build a set of finite graphs starting from a finite base set and closing it off under a composition operation. These graphs capture $\mathcal{B}$'s essential behavior on finite words. The language of $\mathcal{B}$ is not universal iff this set contains graphs with certain properties that witness the existence of an infinite word that is not accepted by $\mathcal{B}$. First, there must be a graph that is idempotent with respect to the composition operation. This corresponds to the fact that all the runs of $\mathcal{B}$ on the finite words described by the graph loop. We must also require that no accepting state occurs on these loops. Second, there must be another graph for the runs on a finite word that reach that loop. To check the inclusion $L(\mathcal{A}) \subseteq L(\mathcal{B})$ the graphs are annotated with additional information about runs of $\mathcal{A}$ on finite words. Here, in case of $L(\mathcal{A}) \nsubseteq L(\mathcal{B})$, the constructed set of graphs contains graphs that witness the existence of at least one infinite word that is accepted by $\mathcal{A}$ but all runs of $\mathcal{B}$ on that word are rejecting. The Ramsey-based approach generalizes to parity automata [16]. The parity condition is useful in modeling reactive systems in which certain modules are supposed to terminate and others are not supposed to terminate. Also, certain Boolean combinations of Büchi (non-termination) and co-Büchi (termination) conditions can easily be expressed as a parity condition. Although parity automata can be translated into Büchi automata, it algorithmically pays off to handle parity automata directly [16].

---

[1] Büchi's original complementation construction, which also relies on Ramsey's theorem, shares similarities with these algorithms. However, there is significantly less overhead when checking universality and inclusion directly and additional heuristics and optimizations are applicable [1,5].

In this paper, we extend the Ramsey-based analysis to visibly pushdown automata (VPAs) [4]. This automaton model restricts nondeterministic pushdown automata in the way that the input symbols determine when the pushdown automaton pushes or pops symbols from its stack. In particular, the stack heights are identical at the same positions in every run of any VPA on a given input. It is because of this syntactic restriction that the class of visibly pushdown languages retains many closure properties like intersection and complementation. VPAs allow one to describe program behavior in more detail than finite-state automata. They can account for the nesting of procedures in executions of recursive imperative programs. Non-regular properties like "an acquired lock must be released within the same procedure" are expressible by VPAs. Model checking of recursive state machines [3] and Boolean programs, which are widely used as abstractions in software model checking, can be carried out in this refined setting by using VPAs for representing the behavior of the programs and the properties. Similar to the automata-theoretic approach to model checking finite-state systems, checking the inclusion of the languages of VPAs is vital here. This time, the respective decision problem is even EXPTIME-complete. Other applications for checking language inclusion of VPAs when reasoning about recursive imperative programs also appear in conformance checking [11] and in the counterexample-guided-abstraction-refinement loop [17].

A generalization of the Ramsey-based approach to VPAs is not straightforward since the graphs that capture the essential behavior of an automaton must also account for the stack content in the runs. Moreover, to guarantee termination of the process that generates these graphs, an automaton's behavior of all runs must be captured within finitely many such graphs. In fact, when considering pushdown automata in general such a generalization is not possible since the universality problem for pushdown automata is undecidable. We circumvent this problem by only considering graphs that differ in their stack height by at most one, and by refining the composition of such graphs in comparison to the unrestricted way that graphs can be composed in the Ramsey-based approach to finite automata. Then the composition operation only needs to account for the top stack symbols in all the runs described by the graphs, which yields a finite set of graphs in the end.

The main contribution of this paper is the generalization of the Ramsey-based approach for checking universality and language inclusion for VPAs over infinite inputs, where the automata's acceptance condition is stated as a parity condition. This approach avoids determinization and complementation constructions. The respective problems where the VPAs operate over finite inputs are special cases thereof. We also experimentally evaluate the performance of our algorithms showing that the Ramsey-based inclusion checking is more efficient than methods that are based on determinization and complementation.

The remainder of this paper is organized as follows. In Sect. 2, we recall the framework of VPAs. In Sect. 3, we provide a Ramsey-based universality check for VPAs. Note that universality is a special case of language inclusion. We treat universality in detail to convey the fundamental ideas first. In Sect. 4, we extend this to a Ramsey-based inclusion check for VPAs. In Sect. 5, we report on the experimental evaluation of our algorithms. In Sect. 6, we draw conclusions.

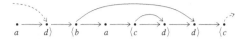

Fig. 1. Nested word $w = adbacddbc$ with $\Sigma_{\text{int}} = \{a\}$, $\Sigma_{\text{call}} = \{b, c\}$, and $\Sigma_{\text{ret}} = \{d\}$. Its pending positions are 1 and 7 with $w_1 = d$ and $w_7 = c$. The call position 2 with $w_2 = b$ matches with the return position 6 with $w_6 = d$. The positions 4 and 5 also match.

2   Preliminaries

Words. The set of finite words over the alphabet Σ is Σ^* and the set of infinite words over Σ is Σ^ω. Let $\Sigma^+ := \Sigma^* \setminus \{\varepsilon\}$, where ε is the empty word. The length of a word w is written as $|w|$, where $|w| = \omega$ when w is an infinite word. For a word w, w_i denotes the letter at position $i < |w|$ in w. That is, $w = w_0 w_1 \ldots$ if w is infinite and $w = w_0 w_1 \ldots w_{n-1}$ if w is finite and $|w| = n$. With $\inf(w)$ we denote the set of letters of Σ that occur infinitely often in $w \in \Sigma^\omega$.

Nested words [4] are linear sequences equipped with a hierarchical structure, which is imposed by partitioning an alphabet Σ into the pairwise disjoint sets Σ_{int}, Σ_{call}, and Σ_{ret}. For a finite or infinite word w over Σ, we say that the position $i \in \mathbb{N}$ with $i < |w|$ is an *internal position* if $w_i \in \Sigma_{\text{int}}$. It is a *call position* if $w_i \in \Sigma_{\text{call}}$ and it is a *return position* if $w_i \in \Sigma_{\text{ret}}$. When attaching an opening bracket $\langle$ to every call position and closing brackets $\rangle$ to the return positions in a word w, we group the word w into subwords. This grouping can be nested. However, not every bracket at a position in w needs to have a matching bracket. The call and return positions in a nested word without matching brackets are called *pending*. To emphasize this hierarchical structure imposed by the brackets $\langle$ and $\rangle$, we also refer to the words in $\Sigma^* \cup \Sigma^\omega$ as *nested words*. See Fig. 1 for illustration.

To ease the exposition, we restrict ourselves in the following to nested words without pending positions. Our results extend to nested words with pending positions; see [15]. For $\sharp \in \{*, \omega\}$, $NW^\sharp(\Sigma)$ denotes the set of words in $\Sigma^\sharp$ with no pending positions. These words are also called *well-matched*.

Automata. A *visibly pushdown automaton* [4], VPA for short, is a tuple $\mathcal{A} = (Q, \Gamma, \Sigma, \delta, q_I, \Omega)$, where Q is a finite set of states, Γ is a finite set of stack symbols, $\Sigma = \Sigma_{\text{int}} \cup \Sigma_{\text{call}} \cup \Sigma_{\text{ret}}$ is the input alphabet, δ consists of three transition functions $\delta_{\text{int}} : Q \times \Sigma_{\text{int}} \to 2^Q$, $\delta_{\text{call}} : Q \times \Sigma_{\text{call}} \to 2^{Q \times \Gamma}$, and $\delta_{\text{ret}} : Q \times \Gamma \times \Sigma_{\text{ret}} \to 2^Q$, $q_I \in Q$ is the initial state, and $\Omega : Q \to \mathbb{N}$ is the priority function. Since we restrict ourselves here to well-matched words, we do not need to consider a bottom stack symbol $\bot$. We write $\Omega(Q)$ to denote the set of all priorities used in $\mathcal{A}$, i.e. $\Omega(Q) := \{\Omega(q) \mid q \in Q\}$. The *size* of $\mathcal{A}$ is $|Q|$ and its *index* is $|\Omega(Q)|$.

A *run* of $\mathcal{A}$ on $w \in \Sigma^\omega$ is a word $(q_0, \gamma_0)(q_1, \gamma_1) \ldots \in (Q \times \Gamma^*)^\omega$ with $(q_0, \gamma_0) = (q_I, \varepsilon)$ and for each $i \in \mathbb{N}$, the following conditions hold:

1. If $w_i \in \Sigma_{\text{int}}$ then $q_{i+1} \in \delta_{\text{int}}(q_i, w_i)$ and $\gamma_{i+1} = \gamma_i$.
2. If $w_i \in \Sigma_{\text{call}}$ then $(q_{i+1}, B) \in \delta_{\text{call}}(q_i, w_i)$ and $\gamma_{i+1} = B\gamma_i$, for some $B \in \Gamma$.
3. If $w_i \in \Sigma_{\text{ret}}$ and $\gamma_i = Bu$ with $B \in \Gamma$ and $u \in \Gamma^*$ then $q_{i+1} \in \delta_{\text{ret}}(q_i, B, w_i)$ and $\gamma_{i+1} = u$.

The run is *accepting* if $\max\{\Omega(q) \mid q \in \inf(q_0q_1\dots)\}$ is even. Runs of $\mathcal{A}$ on finite words are defined as expected. In particular, a run on a finite word is accepting if the last state in the run has an even priority. For $\sharp \in \{*, \omega\}$, we define

$$L^{\sharp}(\mathcal{A}) := \left\{ w \in NW^{\sharp}(\Sigma) \mid \text{there is an accepting run of } \mathcal{A} \text{ on } w \right\}.$$

Priority and Reward Ordering. For an arbitrary set S, we always assume that $\dagger$ is a distinct element not occurring in S. We write $S_{\dagger}$ for $S \cup \{\dagger\}$. We use $\dagger$ to explicitly speak about partial functions into S, i.e., $\dagger$ denotes undefinedness.

We define the following two orders on $\mathbb{N}_{\dagger}$. The *priority ordering* is denoted $\sqsubseteq$ and is the standard order of type $\omega + 1$. Thus, we have $0 \sqsubset 1 \sqsubset 2 \sqsubset \cdots \sqsubset \dagger$. The *reward ordering* $\preceq$ is defined by $\dagger \prec \cdots \prec 5 \prec 3 \prec 1 \prec 0 \prec 2 \prec 4 \prec \cdots$. Note that $\dagger$ is maximal for $\sqsubseteq$ but minimal for $\preceq$. For a finite nonempty set $S \subseteq \mathbb{N}_{\dagger}$, $\bigsqcup S$ and $\curlyvee S$ denote the maxima with respect to the priority ordering $\sqsubseteq$ and the reward ordering $\preceq$, respectively. Furthermore, we write $c \sqcup c'$ for $\bigsqcup\{c, c'\}$.

The reward ordering reflects the intuition of how valuable a priority of a VPA's state is for acceptance: even priorities are better than odd ones, and the bigger an even one is the better, while small odd priorities are better than bigger ones because it is easier to subsume them in a run with an even priority elsewhere. The element $\dagger$ stands for the non-existence of a run.

3   Universality Checking

Throughout this section, we fix a VPA $\mathcal{A} = (Q, \Gamma, \Sigma, \delta, q_I, \Omega)$. We describe an algorithm that determines whether $L^{\omega}(\mathcal{A}) = NW^{\omega}(\Sigma)$, i.e., whether $\mathcal{A}$ accepts all well-matched infinite nested words over Σ. An extension of the algorithm to account for non-well-matched nested words and a universality check for VPAs over finite nested words is given in [15]. Moreover, in [15], we present a complementation construction for VPAs based on determinization and compare it to the presented algorithm.

Central to the algorithm are so-called transition profiles. They capture $\mathcal{A}$'s essential behavior on finite words.

Definition 1. There are three kinds of *transition profiles*, TP for short. The first one is an int-TP, which is a function of type $Q \times Q \to \Omega(Q)_{\dagger}$. We associate with a symbol $a \in \Sigma_{\text{int}}$ the int-TP f_a. It is defined as

$$f_a(q, q') := \begin{cases} \Omega(q') & \text{if } q' \in \delta_{\text{int}}(q, a) \text{ and} \\ \dagger & \text{otherwise.} \end{cases}$$

A call-TP is a function of type $Q \times \Gamma \times Q \to \Omega(Q)_\dagger$. With a symbol $a \in \Sigma_{\mathsf{call}}$ we associate the call-TP f_a. It is defined as

$$f_a(q, B, q') := \begin{cases} \Omega(q') & \text{if } (q', B) \in \delta_{\mathsf{call}}(q, a) \text{ and} \\ \dagger & \text{otherwise.} \end{cases}$$

Finally, a ret-TP is a function of type $Q \times \Gamma \times Q \to \Omega(Q)_\dagger$. With a symbol $a \in \Sigma_{\mathsf{ret}}$ we associate the ret-TP f_a. It is defined as

$$f_a(q, B, q') := \begin{cases} \Omega(q') & \text{if } q' \in \delta_{\mathsf{ret}}(q, B, a) \text{ and} \\ \dagger & \text{otherwise.} \end{cases}$$

A TP of the form f_a for an $a \in \Sigma$ is also called *atomic*. For $\tau \in \{\mathsf{int}, \mathsf{call}, \mathsf{ret}\}$, we define the set of atomic TPs as $T_\tau := \{f_a \mid a \in \Sigma_\tau\}$.

The above TPs describe $\mathcal{A}$'s behavior when $\mathcal{A}$ reads a single letter. In the following, we define how TPs can be composed to describe $\mathcal{A}$'s behavior on words of finite length. The composition, written $f \circ g$, can only be applied to TPs of certain kinds. This ensures that the resulting TP describes the behavior on a word w such that, after reading w, $\mathcal{A}$'s stack height has changed by at most one.

Definition 2. Let f and g be TPs. There are six different kinds of compositions, depending on the TPs' kind of f and g, which we define in the following. If f and g are both int-TPs, we define

$$(f \circ g)(q, q') := \curlyvee \{f(q, q'') \sqcup g(q'', q') \mid q'' \in Q\}.$$

If f is an int-TP and g is either a call-TP or a ret-TP, we define

$$(f \circ g)(q, B, q') := \curlyvee \{f(q, q'') \sqcup g(q'', B, q') \mid q'' \in Q\} \qquad \text{and}$$
$$(g \circ f)(q, B, q') := \curlyvee \{g(q, B, q'') \sqcup f(q'', q') \mid q'' \in Q\}.$$

If f is a call-TP and g a ret-TP, we define

$$(f \circ g)(q, q') := \curlyvee \{f(q, B, q'') \sqcup g(q'', B, q') \mid q'' \in Q \text{ and } B \in \Gamma\}.$$

Intuitively, the composition of two TPs f and g is obtained by following any edge through f from some state q to another state q'', then following any edge through g to some other state q'. The value of this path is the maximum of the two values encountered in f and g with respect to the priority ordering $\sqsubseteq$. Then one takes the maximum over all such possible values with respect to the reward ordering $\preceq$ and obtains a weighted path from q to q' in the composition.

We associate finite words with TPs as follows. With a letter $a \in \Sigma$ we associate the TP f_a as done in Def. 1. If the words $u, v \in \Sigma^+$ are associated with the TPs f and g, respectively, we associate the word uv with the TP $f \circ g$, provided that $f \circ g$ is defined. A word cannot be associated with two distinct TPs. This follows from the following lemma, which is easy to prove.

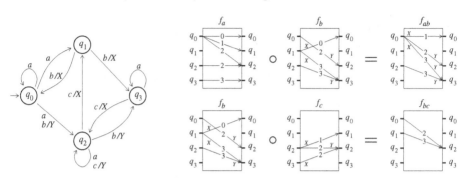

Fig. 2. VPA (left) and the TPs (right) from Example 4

Lemma 3. *Let f, g, h, and k be TPs. If $(h \circ f) \circ (g \circ k)$ and $h \circ ((f \circ g) \circ k)$ are both defined then $(h \circ f) \circ (g \circ k) = h \circ ((f \circ g) \circ k)$.*

If the word $u \in \Sigma^+$ is associated with the TP f, we write f_u for f. Note that two distinct words can be associated with the same TP, i.e., it can be the case that $f_u = f_v$, for $u, v \in \Sigma^+$ with $u \neq v$. Intuitively, if this is the case then $\mathcal{A}$'s behavior on u is identical to $\mathcal{A}$'s behavior on v.

The following example illustrates TPs and their composition.

Example 4. Consider the VPA on the left in Fig. 2 with the states q_0, q_1, q_2, and q_3. The states' priorities are the same as their indices. We assume that $\Sigma_{\text{int}} = \{a\}$, $\Sigma_{\text{call}} = \{b\}$, and $\Sigma_{\text{ret}} = \{c\}$. The stack alphabet is $\Gamma = \{X, Y\}$.

Fig. 2 also depicts the TPs f_a, f_b, f_c and their compositions $f_a \circ f_b = f_{ab}$ and $f_b \circ f_c = f_{bc}$. The VPA's states are in-ports and out-ports of a TP. Assume that f is a call-TP. An in-port q is connected with an out-port q' if $f(q, B, q') \neq \dagger$, for some $B \in \Gamma$. Moreover, this connection of the two ports is labeled with the stack symbol B and the priority. The number of a connection between an in-port and an out-port specifies its priority. For example, the connection in the TP f_a from the in-port q_0 to the out-port q_0 has priority 0 since $f_a(q_0, q_0) = 0$. Since f_a is an int-TP, connections are not labeled with stack symbols.

In a composition $f \circ g$, we plug f's out-ports with g's in-ports together. The priority from an in-port of $f \circ g$ to an out-port of $f \circ g$ is the maximum with respect to the priority ordering $\sqsubseteq$ of the priorities of the two connections in f and g. However, if f is a call-TP and g a ret-TP, we are only allowed to connect the ports in $f \circ g$, if the stack symbols of the connections in f and g match. Finally, since there can be more than one connection between ports in $f \circ g$, we take the maximum with respect to reward ordering $\preceq$.

We extend the composition operation $\circ$ to sets of TPs in the natural way, i.e., we define $F \circ G := \{f \circ g \mid f \in F \text{ and } g \in G \text{ for which } f \circ g \text{ is defined}\}$.

Definition 5. Define $\mathfrak{T}$ as the least solution to the equation

$$\mathfrak{T} = T_{\text{int}} \cup T_{\text{call}} \circ T_{\text{ret}} \cup T_{\text{call}} \circ \mathfrak{T} \circ T_{\text{ret}} \cup \mathfrak{T} \circ \mathfrak{T}.$$

Note that the operations $\circ$ and $\cup$ are monotonic, and the underlying lattice of the powerset of all int-TPs is finite. Thus, the least solution always exists and can be found using fixpoint iteration in a finite number of steps.

The following lemma is helpful in proving that the elements of $\mathfrak{T}$ can be used to characterize (non-)universality of $\mathcal{A}$.

Lemma 6. *For every TP f, we have $f \in \mathfrak{T}$ only if there is a well-matched $w \in \Sigma^+$ with $f = f_w$.*

We need the following notions to characterize universality in terms of the existence of TPs with certain properties.

Definition 7. Let f be an int-TP.
   (i) f is *idempotent* if $f \circ f = f$. Note that only an int-TP can be idempotent.
   (ii) For $q \in Q$, we write $f(q)$ for the set of all $q' \in Q$ that are connected to q in this TP, i.e., $f(q) := \{q' \in Q \mid f(q,q') \neq \dagger\}$. Moreover, for $Q' \subseteq Q$, we define $f(Q') := \bigcup_{q \in Q'} f(q)$.
   (iii) f is *bad* for the set $Q' \subseteq Q$ if $f(q,q)$ is either $\dagger$ or odd, for every $q \in f(Q')$. A *good* TP is a TP that is not bad. Note that any TP is bad for $\emptyset$. In the following, we consider bad TPs only in the context of idempotent TPs.

Example 8. Reconsider the VPA from Example 4 and its TPs. It is easy to see that TP $g := f_a \circ f_a$ is idempotent. Since $g(q_2, q_2) = 2$, g is good for any $Q' \subseteq \{q_0, q_1, q_2, q_3\}$ with $q_2 \in Q'$. The intuition is that there is at least one run on $(aa)^\omega$ that starts in q_2 and loops infinitely often through q_2. Moreover, on this run 2 is the highest priority that occurs infinitely often. So, if there is a prefix $v \in \Sigma^+$ with a run that starts in the initial state and ends in q_2, we have that $v(aa)^\omega$ is accepted by the VPA. The TP g is bad for $\{q_1, q_3\}$, since $g(q_1, q_1) = \dagger$ and $g(q_3, q_3) = 3$. So, if there is prefix $v \in \Sigma^+$ for which all runs that start in the initial state and end in q_1 or q_3 then $v(aa)^\omega$ is not accepted by the VPA. Another TP that is idempodent is the TP $g' := f_b \circ ((f_b \circ f_c) \circ f_c)$. Here, we have that $g'(q_1, q_1) = 2$ and $g'(q, q') = \dagger$, for all $q, q' \in \{q_0, q_1, q_2, q_3\}$ with not $q = q' = q_1$. Thus, g' is bad for every $Q' \subseteq Q$ with $q_1 \notin Q'$.

The following theorem characterizes universality of the VPA $\mathcal{A}$ in terms of the TPs that are contained in the least solution of the equation from Def. 5.

Theorem 9. $L^\omega(\mathcal{A}) \neq NW^\omega(\Sigma)$ *iff there are TPs $f, g \in \mathfrak{T}$ such that g is idempotent and bad for $f(q_I)$.*

Thm. 9 can be used to decide universality for VPAs with respect to the set of well-matched infinite words. The resulting algorithm, which we name UNIV, is depicted in Fig. 3. It computes $\mathfrak{T}$ by least-fixpoint iteration and checks at each stage whether two TPs exist that witness non-universality according to Thm. 9. The variable T stores the generated TPs and the variable N stores the newly generated TPs in an iteration. UNIV terminates if no new TPs are generated in an iteration. Termination is guaranteed since there are only finitely many TPs. For returning a witness of the VPA's non-universality, we assume that we have a word associated with a TP at hand. UNIV's asymptotic time complexity is as follows, where we assume that we use hash tables to represent T and N.

```
1  N ← T_int ∪ T_call ∘ T_ret
2  T ← N
3  while N ≠ ∅ do
4      forall (f_u, f_v) ∈ N × T ∪ T × N do
5          if f_v idempotent and f_v bad for f_u(q_I) then
6              return universality does not hold, witnessed by uv^ω
7      N ← (N ∘ T ∪ T ∘ N ∪ T_call ∘ N ∘ T_ret) \ T
8      T ← T ∪ N
9  return universality holds
```

Fig. 3. Universality check UNIV for VPAs with respect to well-matched words

Theorem 10. *Assume that the given VPA $\mathcal{A}$ has $n \geq 1$ states, index $k \geq 2$, and $m = \max\{1, |\Sigma|, |\Gamma|\}$, where Σ is the VPA's input alphabet and Γ its stack alphabet. The running time of the algorithm UNIV is in $m^3 \cdot 2^{\mathcal{O}(n^2 \cdot \log k)}$.*

There are various ways to tune UNIV. For instance, we can store the TPs in a single hash table and store pointers to the newly generated TPs. Furthermore, we can store pointers to idempotent TPs. Another optimization also concerns the badness check in the line 4 to 6. Observe that it is sufficient to know the sets $f_u(q_I)$, for $f_u \in T$, i.e, the sets $Q' \subseteq Q$ for which all runs for some well-matched word end in a state in Q'. We can maintain a set R to store this information. We initialize R with the singleton set $\{(\varepsilon, \{q_I\})\}$. We update it after line 8 in each iteration by assigning the set $R \cup \{(uv, f_v(Q')) \mid (u, Q') \in R \text{ and } f_v \in T\}$ to it. After this update, we can optimize R by removing an element (u, Q') from it if there is another element (u', Q'') in R with $Q'' \subseteq Q'$. These optimizations do not improve UNIV's worst-case complexity but they are of great practical value.

4   Inclusion Checking

In this section, we describe how to check language inclusion for VPAs. For the sake of simplicity, we assume a single VPA and check for inclusion of the languages that are defined by two states q_I^1 and q_I^2. It should be clear that it is always possible to reduce the case for two VPAs to this one by forming the disjoint union of the two VPAs. Thus, for $i \in \{1, 2\}$, let $\mathcal{A}_i = (Q, \Gamma, \Sigma, \delta, q_I^i, \Omega)$ be the respective VPA. We describe how to check whether $L^\omega(\mathcal{A}_1) \subseteq L^\omega(\mathcal{A}_2)$ holds.

Transition profiles for inclusion checking extend those for universality checking. A *tagged transition profile* (TTP) of the int-type is an element of

$$(Q \times \Omega(Q) \times Q) \times (Q \times Q \to \Omega(Q)_\dagger).$$

We write it as $f^{\langle p, c, p' \rangle}$ instead of (p, c, p', f) in order to emphasize the fact that the TP f is extended with a tuple of states and priorities. A call-TTP is of type

$$(Q \times \Gamma \times \Omega(Q) \times Q) \times (Q \times \Gamma \times Q \to \Omega(Q)_\dagger)$$

and a ret-TTP is of type

$$\left(Q \times \Omega(Q) \times \Gamma \times Q\right) \times \left(Q \times \Gamma \times Q \rightarrow \Omega(Q)_\dagger\right).$$

Accordingly, they are written $f^{\langle p,B,c,p' \rangle}$ and $f^{\langle p,c,B,p' \rangle}$, respectively.

The intuition of an int-TTP $f^{\langle p,c,p' \rangle}$ is as follows. The TP f describes the essential information of *all* runs of the VPA $\mathcal{A}_2$ on a well-matched word $u \in \Sigma^+$. The attached information $\langle p,c,p' \rangle$ describes the existence of *some* run of the VPA $\mathcal{A}_1$ on u. This run starts in state p, ends in state p', and the maximal occurring priority on it is c. The intuition behind a call-TTP or a ret-TTP is similar. The symbol B in the annotation is the topmost stack symbol that is pushed or popped in the run of $\mathcal{A}_2$ for the pending position in the word u.

For $a \in \Sigma$, we now associate a set F_a of TTPs with the appropriate type. Recall that f_a stands for the TP associated to the letter a as defined in Def. 1.

- If $a \in \Sigma_{\text{int}}$, let $F_a := \{f_a^{\langle p, \Omega(p'), p' \rangle} \mid p, p' \in Q \text{ and } p' \in \delta_{\text{int}}(p,a)\}$.
- If $a \in \Sigma_{\text{call}}$, let $F_a := \{f_a^{\langle p, B, \Omega(p'), p' \rangle} \mid p, p' \in Q, \ B \in \Gamma, \text{ and } (p', B) \in \delta_{\text{call}}(p,a)\}$.
- If $a \in \Sigma_{\text{ret}}$, let $F_a := \{f_a^{\langle p, \Omega(p'), B, p' \rangle} \mid p, p' \in Q, \ B \in \Gamma, \text{ and } p' \in \delta_{\text{ret}}(p,B,a)\}$.

As with TPs, the composition of TTPs is only allowed in certain cases. They are the same as for TPs, e.g., the composition of a call-TTP with an int-TTP results in a call-TTP, and with a ret-TTP it results in an int-TTP. However, the composition of TTPs is not a monoid operation but behaves like the composition of morphisms in a category in which the states in Q, respectively pairs of states and stack symbols in Γ, act as objects. A TTP $f^{\langle p,c,p' \rangle}$ for instance can be seen as a morphism from p to p', and it can therefore only be composed with a morphism from p' to anything else.

The composition of two TTPs extends the composition of the underlying TPs by explaining how the tag of the resulting TTP is obtained. For int-TTPs $f^{\langle p,c,p' \rangle}$ and $g^{\langle p',c',p'' \rangle}$, we define

$$f^{\langle p,c,p' \rangle} \circ g^{\langle p',c',p'' \rangle} \ := \ (f \circ g)^{\langle p,c \sqcup c',p'' \rangle}.$$

Composing an int-TTP $f^{\langle p,c,p' \rangle}$ and a call-TTP $g^{\langle q,B,c',q' \rangle}$ yields call-TTPs:

$$f^{\langle p,c,p' \rangle} \circ g^{\langle q,B,c',q' \rangle} \ := \ (f \circ g)^{\langle p,B,c \sqcup c',q' \rangle} \quad \text{if } p' = q$$

$$g^{\langle q,B,c',q' \rangle} \circ f^{\langle p,c,p' \rangle} \ := \ (g \circ f)^{\langle q,B,c \sqcup c',p' \rangle} \quad \text{if } q' = p.$$

The two possible compositions of an int-TTP with a ret-TTP are defined in exactly the same way. Finally, the composition of a call-TTP $f^{\langle p,B,c,p' \rangle}$ and a ret-TTP $g^{\langle p',c',B,p'' \rangle}$ is defined as

$$f^{\langle p,B,c,p' \rangle} \circ g^{\langle p',c',B,p'' \rangle} \ := \ (f \circ g)^{\langle p,c \sqcup c',p'' \rangle}.$$

Note that the stack symbol B is the same in both annotations. As for sets of TPs, we extend the composition of TTPs to sets.

Similar to Def. 5, we define a set $\hat{\mathfrak{T}}$ to be the least solution to the equation

$$\hat{\mathfrak{T}} \ = \ \hat{T}_{\text{int}} \ \cup \ \hat{T}_{\text{call}} \circ \hat{T}_{\text{ret}} \ \cup \ \hat{T}_{\text{call}} \circ \hat{\mathfrak{T}} \circ \hat{T}_{\text{ret}} \ \cup \ \hat{\mathfrak{T}} \circ \hat{\mathfrak{T}},$$

where $\hat{T}_\tau := \bigcup \{F_a \mid a \in \Sigma_\tau\}$, for $\tau \in \{\mathsf{int}, \mathsf{call}, \mathsf{ret}\}$. This allows us to characterize language inclusion between two VPAs in terms of the existence of certain TTPs.

Theorem 11. $L^\omega(\mathcal{A}_1) \not\subseteq L^\omega(\mathcal{A}_2)$ *iff there are TTPs* $f^{\langle q_I^1, c, p\rangle}$ *and* $g^{\langle p, d, p\rangle}$ *in* $\hat{\mathfrak{T}}$ *fulfilling the following properties:*
(1) The priority d is even.
(2) The TP g is idempotent and bad for $f(q_I^2)$.

Thm. 11 yields an algorithm INCL to check $L^\omega(\mathcal{A}_1) \not\subseteq L^\omega(\mathcal{A}_2)$, for given VPAs $\mathcal{A}_1$ and $\mathcal{A}_2$. It is along the same lines as the algorithm UNIV and we omit it. The essential difference lies in the sets $\hat{T}_{\mathsf{int}}$, $\hat{T}_{\mathsf{call}}$, and $\hat{T}_{\mathsf{ret}}$, which contain TTPs instead of TPs, and the refined way in which they are being composed. Each iteration now searches for two TTPs that witness the existence of some word of the form uv^ω that is accepted by $\mathcal{A}_1$ but not accepted by $\mathcal{A}_2$. Similar optimizations that we sketch for UNIV at the end of Sect. 3 also apply to INCL.

For the complexity analysis of the algorithm INCL below, we do not assume that the VPAs $\mathcal{A}_1$ and $\mathcal{A}_2$ necessarily share the state set, the priority function, the stack alphabet, and the transition functions as assumed at the beginning of this subsection. Only the input alphabet Σ is the same for $\mathcal{A}_1$ and $\mathcal{A}_2$.

Theorem 12. *Assume that for* $i \in \{1,2\}$, *the number of states of the VPA* $\mathcal{A}_i$ *is* $n_i \geq 1$, $k_i \geq 2$ *its index, and* $m_i = \max\{1, |\Sigma|, |\Gamma_i|\}$, *where* Σ *is the VPA's input alphabet and* Γ_i *its stack alphabet. The running time of the algorithm* INCL *is in* $n_1^4 \cdot k_1^2 \cdot m_1 \cdot m_2^3 \cdot 2^{\mathcal{O}(n_2^2 \cdot \log k_2)}$.

5   Evaluation

Our prototype tool FADecider implements the presented algorithms in the programming language OCaml.[2] To evaluate the tool's performance we carried out the following experiments for which we used a 64-bit Linux machine with 4 GB of main memory and two dual-core Xeon 5110 CPUs, each with 1.6 GHz. Our benchmark suite consists of VPAs from [11], which are extracted from real-world recursive imperative programs. Tab. 1 describes the instances, each consisting of two VPAs $\mathcal{A}$ and $\mathcal{B}$, in more detail. Tab. 2 shows FADecider's running times for the inclusion checks $L^*(\mathcal{A}) \subseteq L^*(\mathcal{B})$ and $L^\omega(\mathcal{A}) \subseteq L^\omega(\mathcal{B})$. For comparison, we used the OpenNWA library [12]. The inclusion check there is implemented by a reduction to an emptiness check via a complementation construction. Note that OpenNWA does not support infinite nested words at all and has no direct support for only considering well-matched nested words. We used therefore OpenNWA to perform the language-inclusion checks with respect to all finite nested words.

FADecider outperforms OpenNWA on these examples. Profiling the inclusion check based on the OpenNWA library yields that complementation requires about 90% of the overall running time. FADecider spends about 90% of its time

---

[2] The tool (version 0.4) is publicly available at www2.tcs.ifi.lmu.de/fadecider.

Table 1. Statistics on the input instances. The first row lists the number of states of the VPAs from an input instance and their alphabet sizes. The number of stack symbols of a VPA and its index are not listed, since in these examples the VPA's stack symbol set equals its state set and states are either accepting or non-accepting. The second row lists whether the inclusions $L^*(\mathcal{A}) \subseteq L^*(\mathcal{B})$ and $L^\omega(\mathcal{A}) \subseteq L^\omega(\mathcal{B})$ of the respective VPAs hold.

	ex	ex-§2.5	gzip	gzip-fix	png2ico
size $\mathcal{A}$ / size $\mathcal{B}$ / alphabet size	9 / 5 / 4	10 / 5 / 5	51 / 71 / 4	51 / 73 / 4	22 / 26 / 5
language relation	$\subseteq$ / $\subseteq$	$\not\subseteq$ / $\subseteq$	$\not\subseteq$ / ?	$\subseteq$ / $\subseteq$	$\subseteq$ / $\subseteq$

Table 2. Experimental results for the language-inclusion checks. The row "FADecider" lists the running times for the tool FADecider for checking $L^*(\mathcal{A}) \subseteq L^*(\mathcal{B})$ and $L^\omega(\mathcal{A}) \subseteq L^\omega(\mathcal{B})$. The row "#TTPs" lists the number of encountered TTPs. The symbol ‡ indicates that FADecider ran out of time (2 hours). The row "OpenNWA" lists the running times for the implementation based on the OpenNWA library for checking inclusion on finite words and the VPA's size obtained by complementing $\mathcal{B}$.

	ex	ex-§2.5	gzip	gzip-fix	png2ico
FADecider	0.00s / 0.00s	0.00s / 0.00s	36s / ‡	42s / 294s	0.10s / 0.11s
#TTPs	6 / 6	18 / 19	694 / ‡	518 / 1,117	586 / 609
OpenNWA	0.16s / 27	0.04s / 11	49s / 27	1,104s / 176	74.70s / 543

on composing TPs and about 5% on checking equality of TPs. The experiments also show that FADecider's performance on inclusion checks for infinite words can be worse than for finite words. Note that checking inclusion for infinite-word languages is more expensive than for finite-word languages, since, in addition to reachability, one needs to account for loops.

6   Conclusion

Checking universality and language inclusion for automata by avoiding determinization and complementation has recently attracted a lot of attention, see, e.g., [1, 9, 10, 13, 16]. We have shown that Ramsey-based methods for Büchi automata generalize to the richer automaton model of VPAs with a parity acceptance condition. Another competitive approach based on antichains has recently also been extended to VPAs, however, only to VPAs over finite words [6]. It remains to be seen if optimizations for the Ramsey-based algorithms for Büchi automata [1] extend, with similar speed-ups, to this richer setting. Another direction of future work is to investigate Ramsey-based approaches for automaton models that extend VPAs like multi-stack VPAs [18].

Acknowledgments. We are grateful to Evan Driscoll for providing us with VPAs.

References

1. Abdulla, P.A., Chen, Y.-F., Clemente, L., Holík, L., Hong, C.-D., Mayr, R., Vojnar, T.: Advanced Ramsey-based Büchi automata inclusion testing. In: Katoen,
   J.-P., König, B. (eds.) CONCUR 2011. LNCS, vol. 6901, pp. 187–202. Springer,
   Heidelberg (2011)
2. Abdulla, P.A., Chen, Y.-F., Holík, L., Mayr, R., Vojnar, T.: When simulation meets
   antichains. In: Esparza, J., Majumdar, R. (eds.) TACAS 2010. LNCS, vol. 6015,
   pp. 158–174. Springer, Heidelberg (2010)
3. Alur, R., Benedikt, M., Etessami, K., Godefroid, P., Reps, T.W., Yannakakis, M.:
   Analysis of recursive state machines. ACM Trans. Progr. Lang. Syst. 27(4), 786–818
   (2005)
4. Alur, R., Madhusudan, P.: Adding nesting structure to words. J. ACM 56(3), 1–43
   (2009)
5. Breuers, S., Löding, C., Olschewski, J.: Improved Ramsey-based Büchi complementation. In: Birkedal, L. (ed.) FOSSACS 2012. LNCS, vol. 7213, pp. 150–164.
   Springer, Heidelberg (2012)
6. Bruyère, V., Ducobu, M., Gauwin, O.: Visibly pushdown automata: Universality
   and inclusion via antichains. In: Dediu, A.-H., Martín-Vide, C., Truthe, B. (eds.)
   LATA 2013. LNCS, vol. 7810, pp. 190–201. Springer, Heidelberg (2013)
7. Büchi, J.R.: On a decision method in restricted second order arithmetic. In: Proc.
   of the 1960 Internat. Congr. on Logic, Method, and Philosophy of Science, pp. 1–11
   (1960)
8. Dax, C., Hofmann, M., Lange, M.: A proof system for the linear time μ-calculus.
   In: Arun-Kumar, S., Garg, N. (eds.) FSTTCS 2006. LNCS, vol. 4337, pp. 273–284.
   Springer, Heidelberg (2006)
9. De Wulf, M., Doyen, L., Henzinger, T.A., Raskin, J.-F.: Antichains: A new algorithm for checking universality of finite automata. In: Ball, T., Jones, R.B. (eds.)
   CAV 2006. LNCS, vol. 4144, pp. 17–30. Springer, Heidelberg (2006)
10. Doyen, L., Raskin, J.-F.: Antichains for the automata-based approach to model-
    checking. Log. Methods Comput. Sci. 5(1) (2009)
11. Driscoll, E., Burton, A., Reps, T.: Checking conformance of a producer and a
    consumer. In: ESEC/FSE 2011, pp. 113–123.
12. Driscoll, E., Thakur, A., Reps, T.: OpenNWA: A nested-word automaton library.
    In: Madhusudan, P., Seshia, S.A. (eds.) CAV 2012. LNCS, vol. 7358, pp. 665–671.
    Springer, Heidelberg (2012)
13. Fogarty, S., Vardi, M.Y.: Büchi complementation and size-change termination. In:
    Kowalewski, S., Philippou, A. (eds.) TACAS 2009. LNCS, vol. 5505, pp. 16–30.
    Springer, Heidelberg (2009)
14. Fogarty, S., Vardi, M.Y.: Efficient Büchi universality checking. In: Esparza, J., Ma-
    jumdar, R. (eds.) TACAS 2010. LNCS, vol. 6015, pp. 205–220. Springer, Heidelberg
    (2010)
15. Friedmann, O., Klaedtke, F., Lange, M.: Ramsey goes visibly pushdown (2012)
    (Manuscript); Available at authors' web pages
16. Friedmann, O., Lange, M.: Ramsey-based analysis of parity automata. In: Flana-
    gan, C., König, B. (eds.) TACAS 2012. LNCS, vol. 7214, pp. 64–78. Springer,
    Heidelberg (2012)
17. Heizmann, M., Hoenicke, J., Podelski, A.: Nested interpolants. In: POPL 2010, pp.
    471–482 (2010)

18. La Torre, S., Madhusudan, P., Parlato, G.: A robust class of context-sensitive languages. In: LICS 2007, pp. 161–170 (2007)
19. Lee, C.S., Jones, N.D., Ben-Amram, A.M.: The size-change principle for program termination. In: POPL 2001, pp. 81–92 (2001)
20. Ramsey, F.P.: On a problem of formal logic. Proc. London Math. Soc. 30, 264–286 (1928)
21. Tsai, M.-H., Fogarty, S., Vardi, M.Y., Tsay, Y.-K.: State of büchi complementation. In: Domaratzki, M., Salomaa, K. (eds.) CIAA 2010. LNCS, vol. 6482, pp. 261–271. Springer, Heidelberg (2011)
22. Vardi, M.Y., Wolper, P.: An automata-theoretic approach to automatic program verification (preliminary report). In: LICS 1986, pp. 332–344 (1986)

Checking Equality and Regularity
for Normed BPA with Silent Moves*

Yuxi Fu

BASICS, Department of Computer Science, Shanghai Jiao Tong University
MOE-MS Key Laboratory for Intelligent Computing and Intelligent Systems

Abstract. The decidability of weak bisimilarity on normed BPA is a
long standing open problem. It is proved in this paper that branching
bisimilarity, a standard refinement of weak bisimilarity, is decidable for
normed BPA and that the associated regularity problem is also decidable.

1 Introduction

In [BBK87] Baeten, Bergstra and Klop proved a surprising result that strong
bisimilarity between context free grammars without empty production is decidable. The decidability is in sharp contrast to the well known fact that language
equivalence between these grammars is undecidable. After [BBK87] decidability
and complexity issues of equivalence checking of infinite systems *à la* process
algebra have been intensively investigated. As regards BPA, Hüttel and Stirling [HS91] improved Baeten, Bergstra and Klop's proof by a more straightforward one using tableau system. Hüttel [Hüt92] then repeated the tableau
construction for branching bisimilarity on totally normed BPA processes. Later
Hirshfeld [Hir96] applied the tableau method to the weak bisimilarity on the
totally normed BPA. An affirmative answer to the decidability of the strong
bisimilarity on general BPA is given by Christensen, Hüttel and Stirling by applying the technique of bisimulation base [CHS92].

The complexity aspect of BPA has also been investigated over the years. Balcazar, Gabarro and Santha [BGS92] pointed out that strong bisimilarity is P-
hard. Huynh and Tian [HT94] showed that the problem is in Σ_2^p, the second level
of the polynomial hierarchy. Hirshfeld, Jerrum and Moller [HJM96] completed
the picture by offering a remarkable polynomial algorithm for the strong bisimilarity of normed BPA. For the general BPA, Burkart, Caucal and Steffen [BCS95]
showed that the strong bisimilarity problem is elementary. They claimed that
their algorithm can be optimized to get a 2-EXPTIME upper bound. A further
elaboration of the 2-EXPTIME upper bound is given in [Jan12] with the introduction of infinite regular words. The current known best lower bound of the
problem, EXPTIME, is obtained by Kiefer [Kie13], improving both the PSPACE
lower bound result and its proof of Srba [Srb02]. Much less is known about the
weak bisimilarity on BPA. Stříbrná's PSPACE lower bound [Stř98] is subsumed

---

* The full paper can be found at http://basics.sjtu.edu.cn/~yuxi/.

F.V. Fomin et al. (Eds.): ICALP 2013, Part II, LNCS 7966, pp. 238–249, 2013.
© Springer-Verlag Berlin Heidelberg 2013

by both the result of Srba [Srb02] and that of Mayr [May03], all of which are subsumed by Kiefer's recent result. A slight modification of Mayr's proof shows that the EXPTIME lower bound holds for the branching bisimilarity as well.

It is generally believed that weak bisimilarity, as well as branching bisimilarity, on BPA is decidable. There has been however a lack of technique to resolve the difficulties caused by silent transitions. This paper aims to advance our understanding of the decidability problems of BPA in the presence of silent transitions. The main contributions of the paper are as follows:

– We introduce branching norm, which is the least number of nontrivial actions a process has to do to become an empty process. With the help of this concept one can carry out a much finer analysis on silent actions than one would have using weak norm. Branching norm turns out to be crucial in our approach.
– We reveal that in normed BPA the length of a state preserving silent transition sequence can be effectively bounded. As a consequence we show that branching bisimilarity on normed BPA processes can be approximated by a sequence of finite branching bisimulations.
– We establish the decidability of branching bisimilarity on normed BPA by constructing a sound and complete tableau system for the equivalence.
– We demonstrate how to derive the decidability of the associated regularity problem from the decidability of the branching bisimilarity of normed BPA.

The result of this paper is significantly stronger than previous decidability results on the branching bisimilarity of totally normed BPA [Hüt92, CHT95]. It is easy to derive effective size bound for totally normed BPA since a totally normed BPA process with k variable occurrences has a norm at least k. For the same reason right cancellation property holds. Hence the decidability. The totality condition makes the branching bisimilarity a lot more like strong bisimilarity.

2   Branching Bisimilarity for BPA

A *basic process algebra* (BPA for short) Γ is a triple $(\mathcal{V}, \mathcal{A}, \Delta)$ where $\mathcal{V} = \{X_1, \ldots, X_n\}$ is a finite set of *variables*, $\mathcal{A} = \{a_1, \ldots, a_m\} \cup \{\tau\}$ is a finite set of *actions* ranged over by ℓ, and Δ is a finite set of *transition rules*. The special symbol τ denotes a *silent* action. A *BPA process* defined in Γ is an element of the set $\mathcal{V}^*$ of finite string of element of $\mathcal{V}$. The set $\mathcal{V}$ will be ranged over by capital letters and $\mathcal{V}^*$ by lower case Greek letters. The empty string is denoted by ϵ. We will use = for the grammar equality on $\mathcal{V}^*$. A transition rule is of the form $X \xrightarrow{\ell} \alpha$, where ℓ ranges over $\mathcal{A}$. The transitional semantics is closed under *composition* in the sense that $X\gamma \xrightarrow{\ell} \alpha\gamma$ for all γ whenever $X \xrightarrow{\ell} \alpha$. We shall assume that every variable of a BPA is defined by at least one transition rule and every action in $\mathcal{A}$ appears in some transition rule. Accordingly we sometimes refer to a BPA by its set of transition rules. We write $\longrightarrow$ for $\xrightarrow{\tau}$ and $\Longrightarrow$ for the reflexive transitive closure of $\xrightarrow{\tau}$. The set $\mathcal{A}^*$ will be ranged over by ℓ^*. If $\ell^* = \ell_1 \ldots \ell_k$ for some $k \geq 0$, then $\alpha \xrightarrow{\ell^*} \alpha'$ stands for $\alpha \xrightarrow{\ell_1} \alpha_1 \ldots \xrightarrow{\ell_{k-1}} \alpha_{k-1} \xrightarrow{\ell_k} \alpha'$ for some $\alpha_1, \ldots, \alpha_{k-1}$. We say that α' is a *descendant* of α if $\alpha \xrightarrow{\ell^*} \alpha'$ for some ℓ^*.

A BPA process α is *normed* if there are some actions $\ell_1, \ldots \ell_j$ such that $\alpha \xrightarrow{\ell_1} \ldots \xrightarrow{\ell_j} \epsilon$. A process is *unnormed* if it is not normed. The *norm* of a BPA process α, denoted by $\|\alpha\|$, is the least k such that $\alpha \xrightarrow{\ell_1} \ldots \xrightarrow{\ell_k} \epsilon$ for some $\ell_1, \ldots \ell_k$. A *normed BPA*, or *nBPA*, is one in which every variable is normed.

For each given BPA Δ, we introduce the following notations:

- m_Δ is the number of transition rules; and n_Δ is the number of variables.
- r_Δ is $\max \left\{ |\gamma| \;\middle|\; X \xrightarrow{\lambda} \gamma \in \Delta \right\}$, where $|\gamma|$ denotes the length of γ.
- $\|\Delta\|$ is $\max \{ \|X_i\| \mid 1 \le i \le n_\Delta \text{ and } X_i \text{ is normed} \}$.

Each of m_Δ, n_Δ, r_Δ and $\|\Delta\|$ can be effectively calculated from Δ.

2.1 Branching Bisimilarity

The idea of the branching bisimilarity of van Glabbeek and Weijland [vGW89] is that not all silent actions can be ignored. What can be ignored are those that do not change system states irreversibly. For BPA we need to impose additional condition to guarantee congruence. In what follows $x \mathcal{R} y$ stands for $(x, y) \in \mathcal{R}$.

Definition 1. *A symmetric relation $\mathcal{R}$ on BPA processes is a branching bisimulation if the following statements are valid whenever $\alpha \mathcal{R} \beta$:*

1. *If $\beta \mathcal{R} \alpha \xrightarrow{\ell} \alpha'$ then one of the following statements is valid:*
   (i) $\ell = \tau$ and $\alpha' \mathcal{R} \beta$.
   (ii) $\beta \Longrightarrow \beta'' \mathcal{R} \alpha$ for some β'' such that $\beta'' \xrightarrow{\ell} \beta' \mathcal{R} \alpha'$ for some β'.
2. *If $\alpha = \epsilon$ then $\beta \Longrightarrow \epsilon$.*

The branching bisimilarity $\simeq$ is the largest branching bisimulation.

The branching bisimilarity $\simeq$ satisfies the standard property of observational equivalence stated in the next lemma [vGW89].

Lemma 1. *Suppose $\alpha_0 \xrightarrow{\tau} \alpha_1 \xrightarrow{\tau} \ldots \xrightarrow{\tau} \alpha_k \simeq \alpha_0$. Then $\alpha_0 \simeq \alpha_1 \simeq \ldots \simeq \alpha_k$.*

Using Lemma 1 it is easy to show that $\simeq$ is a congruence and that whenever $\beta \simeq \alpha \xrightarrow{\ell} \alpha'$ is simulated by $\beta \xrightarrow{\tau} \beta_1 \xrightarrow{\tau} \beta_2 \ldots \xrightarrow{\tau} \beta_k \xrightarrow{\ell} \beta'$ such that $\beta_k \simeq \alpha$ and $\beta' \simeq \alpha'$ then $\beta \simeq \beta_1 \simeq \ldots \simeq \beta_k$.

Having defined an equality for BPA, we can formally draw a line between the silent actions that change the capacity of systems and those that do not. We say that a silent action $\alpha \xrightarrow{\tau} \alpha'$ is *state preserving* if $\alpha \simeq \alpha'$; it is a *change-of-state* if $\alpha \not\simeq \alpha'$. We will write $\alpha \to \alpha'$ if $\alpha \xrightarrow{\tau} \alpha'$ is state preserving and $\alpha \xrightarrow{\iota} \alpha'$ if it is a change-of-state. The reflexive and transitive closure of $\to$ is denoted by $\to^*$. Since both external actions and change-of-state silent actions must be explicitly bisimulated, we let $\jmath$ range over the set $(\mathcal{A} \setminus \{\tau\}) \cup \{\iota\}$. So $\alpha \xrightarrow{\jmath} \alpha'$ means either $\alpha \xrightarrow{a} \alpha'$ for some $a \ne \tau$ or $\alpha \xrightarrow{\iota} \alpha'$.

Let's see an example.

Example 1. The BPA Γ_1 is defined by the following transition rules:

$$A \xrightarrow{a} A, \; A \xrightarrow{\tau} \epsilon, \; B \xrightarrow{b} B, \; B \xrightarrow{\tau} \epsilon, \; C \xrightarrow{a} C, \; C \xrightarrow{b} C, \; C \xrightarrow{\tau} \epsilon.$$

Clearly $AC \simeq BC$, although $A \not\simeq B$. In this example all variables are normed.

2.2 Bisimulation Base

An *axiom system* $\mathcal{B}$ is a finite set of equalities on nBPA processes. An element $\alpha = \beta$ of $\mathcal{B}$ is called an *axiom*. Write $\mathcal{B} \vdash \alpha = \beta$ if the equality $\alpha = \beta$ can be derived from the axioms of $\mathcal{B}$ by repetitive use of any of the three equivalence rules and two congruence rules. For our purpose the most useful axiom systems are those that generate branching bisimulations. These are bisimulation bases originally due to Caucal. The following definition is Hüttel's adaptation to the branching scenario [Hüt92].

Definition 2. *A finite axiom system $\mathcal{B}$ is a* bisimulation base *if the following* bisimulation base property *hold for every axiom* (α_0, β_0) *of $\mathcal{B}$:*

1. *If $\beta_0 \longrightarrow \beta_1 \longrightarrow \ldots \longrightarrow \beta_n \overset{\ell}{\longrightarrow} \beta'$ then there are $\alpha_1, \ldots, \alpha_n, \alpha'$ such that $\mathcal{B} \vdash \beta_1 = \alpha_1, \ldots, \mathcal{B} \vdash \beta_n = \alpha_n, \mathcal{B} \vdash \beta' = \alpha'$ and the following hold:*
   (i) *For each i with $0 \leq i < n$, either $\alpha_i = \alpha_{i+1}$, or $\alpha_i \longrightarrow \alpha_{i+1}$, or there are $\alpha_i^1, \ldots, \alpha_i^{k_i}$ such that $\alpha_i \longrightarrow \alpha_i^1 \longrightarrow \ldots \longrightarrow \alpha_i^{k_i} \longrightarrow \alpha_{i+1}$ and $\mathcal{B} \vdash \beta_i = \alpha_i^1, \ldots, \mathcal{B} \vdash \beta_i = \alpha_i^{k_i}$.*
   (ii) *Either $\ell = \tau$ and $\alpha_n = \alpha'$, or $\alpha_n \overset{\ell}{\longrightarrow} \alpha'$, or there are $\alpha_n^1, \ldots, \alpha_n^{k_n}$ such that $\alpha_n \longrightarrow \alpha_n^1 \longrightarrow \ldots \longrightarrow \alpha_n^{k_n} \overset{\ell}{\longrightarrow} \alpha'$ and $\mathcal{B} \vdash \beta_n = \alpha_n^1, \ldots, \mathcal{B} \vdash \beta_n = \alpha_n^{k_n}$.*
2. *If $\beta_0 = \epsilon$ then either $\alpha_0 = \epsilon$ or $\alpha_0 \longrightarrow \alpha_1 \longrightarrow \ldots \longrightarrow \alpha_k \longrightarrow \epsilon$ for some $\alpha_1, \ldots, \alpha_k$ with $k \geq 0$ such that $\mathcal{B} \vdash \alpha_1 = \epsilon, \ldots, \mathcal{B} \vdash \alpha_k = \epsilon$.*
3. *The conditions symmetric to 1 and 2.*

The next lemma justifies the above definition [Hüt92].

Lemma 2. *If $\mathcal{B}$ is a bisimulation base then $\mathcal{B}^\vdash = \{(\alpha, \beta) \mid \mathcal{B} \vdash \alpha = \beta\} \subseteq \simeq$.*

Proof. If $\mathcal{B} \vdash \alpha = \beta$, then an inductive argument shows that there exist $\gamma_1\delta_1\lambda_1$, $\gamma_2\delta_2\lambda_2, \gamma_3\delta_3\lambda_3, \ldots, \gamma_{k-1}\delta_{k-1}\lambda_{k-1}, \gamma_k\delta_k\lambda_k$ and $\delta_1', \ldots, \delta_k'$ for $k \geq 1$ such that $\alpha = \gamma_1\delta_1\lambda_1, \gamma_k\delta_k'\lambda_k = \beta$ and $\gamma_1\delta_1\lambda_1 \ \mathcal{B} \ \gamma_1\delta_1'\lambda_1 = \gamma_2\delta_2\lambda_2 \ \mathcal{B} \ \gamma_2\delta_2'\lambda_2 = \ldots \gamma_{k-1}\delta_{k-1}'\lambda_{k-1} = \gamma_k\delta_k\lambda_k \ \mathcal{B} \ \gamma_k\delta_k'\lambda_k$. The transitive closure makes it easy to see that $\mathcal{B}^\vdash$ satisfies the bisimulation base property. Consequently it is a branching bisimulation. □

3 Approximation of Branching Bisimilarity

To look at the algebraic property of the branching bisimilarity $\simeq$ more closely, we introduce a notion of normedness appropriate for the equivalence.

Definition 3. *The* branching norm *of an nBPA process α is the least number k such that $\exists j_1 \ldots j_k.\exists \alpha_1 \ldots \alpha_k.\alpha \rightarrow^* \overset{j_1}{\longrightarrow} \alpha_1 \rightarrow^* \overset{j_2}{\longrightarrow} \ldots \alpha_{k-1} \rightarrow^* \overset{j_k}{\longrightarrow} \alpha_k \rightarrow^* \epsilon$. The branching norm of α is denoted by $\|\alpha\|_b$.*

For example the branching norm of B defined by $\{B \overset{a}{\longrightarrow} B, B \overset{\tau}{\longrightarrow} \epsilon\}$ is 1. It is easy to prove that if $\alpha \simeq \beta$ then $\|\alpha\|_b = \|\beta\|_b$ and that if $\|\alpha\|_b = 0$ then $\alpha \simeq \epsilon$. It follows that $\|\alpha'\|_b = \|\alpha\|_b$ whenever $\alpha \rightarrow^* \alpha'$. Also notice that $\|\alpha\|_b \leq \|\alpha\|$.

An important property of branching norm is stated next.

Lemma 3. *Suppose α is normed. Then $\alpha \simeq \delta\alpha$ if and only if $\|\alpha\|_b = \|\delta\alpha\|_b$.*

Proof. If $\|\alpha\|_b = \|\delta\alpha\|_b$ then every silent action sequence from $\delta\alpha$ to α must contain only state preserving silent transitions according to Lemma 1. Moreover there must exist such a silent action path for otherwise $\|\alpha\|_b < \|\delta\alpha\|_b$.   □

It does not follow from $\alpha \simeq \delta\alpha$ that $\delta \simeq \epsilon$. A counter example is given by the BPA defined in Example 1. One has $AC \simeq C \simeq BC$. But clearly $\epsilon \not\simeq A \not\simeq B \not\simeq \epsilon$. To deal with situations like this we need the notion of relative norm.

Definition 4. *The relative norm $\|\alpha\|_b^\sigma$ of α with respect to σ is the least k such that $\alpha\sigma \to^* \xrightarrow{\jmath_1} \alpha_1\sigma \ldots \alpha_{k-1}\sigma \to^* \xrightarrow{\jmath_k} \alpha_k\sigma \to^* \sigma$ for some $\jmath_1, \ldots, \jmath_k, \alpha_1, \ldots, \alpha_k$.*

Obviously $0 \le \|\alpha\|_b^\sigma \le \|\alpha\|_b$. Returning to the BPA Γ_1 defined in Example 1, we see that $\|A\|_b^B = 1$ and $\|A\|_b^C = 0$. Using the notion of relative norm we may introduce the following terminologies:

- A transition $X\sigma \xrightarrow{\ell} \eta\sigma$ is *norm consistent* if either $\|\eta\|_b^\sigma = \|X\|_b^\sigma$ and $\ell = \tau$ or $\|\eta\|_b^\sigma = \|X\|_b^\sigma - 1$ and $\ell \ne \tau \lor \ell = \iota$.
- If $X\sigma \longrightarrow \eta\sigma$ is norm consistent with $\|X\|_b^\sigma > 0$, then it is *norm splitting* if at least two variables in η have (smaller) nonzero relative norms in $\eta\sigma$.

For an nBPA Δ no silent transition sequence contains more than $\|\Delta\|_b$ norm splitting transitions, where $\|\Delta\|_b$ is $\max\{\|X_i\|_b \mid 1 \le i \le n_\Delta \text{ and } X_i \text{ is normed}\}$.
    The crucial property about relative norm is described in the following lemma.

Lemma 4. *Let $\alpha, \beta, \delta, \gamma$ be normed with $\|\alpha\|_b^\gamma = \|\beta\|_b^\delta$. If $\alpha\gamma \simeq \beta\delta$ then $\gamma \simeq \delta$.*

Proof. Suppose $\|\alpha\|_b^\gamma = \|\beta\|_b^\delta$. Now $\|\alpha\|_b^\gamma + \|\gamma\|_b = \|\alpha\gamma\|_b = \|\beta\delta\|_b = \|\beta\|_b^\delta + \|\delta\|_b$. Therefore $\|\gamma\|_b = \|\delta\|_b$. A norm consistent action sequence $\alpha\gamma \to^* \xrightarrow{\jmath_1} \ldots \to^* \xrightarrow{\jmath_k} \to^* \gamma$ must be matched up by $\beta\delta \to^* \xrightarrow{\jmath_1} \ldots \to^* \xrightarrow{\jmath_k} \beta'\delta$ for some β'. Clearly $\|\beta'\delta\|_b = \|\gamma\|_b = \|\delta\|_b$. It follows from Lemma 3 that $\delta \simeq \beta'\delta \simeq \gamma$.   □

Lemma 4 describes a weak form of left cancelation property. The general left cancelation property fails. Fortunately there is a nice property of nBPA that allows us to control the size of common suffix of a pair of bisimilar processes.

Definition 5. *A process α is irredundant over γ if $\|\alpha\|_b^\gamma > 0$. It is redundant over γ if $\|\alpha\|_b^\gamma = 0$. A process α is head irredundant if either $\alpha = \epsilon$ or $\alpha = X\alpha'$ for some X, α' such that $\alpha \not\simeq \alpha'$. It is head redundant otherwise. We write $Hirred(\alpha)$ to indicate that α is head irredundant. A process α is completely irredundant if every suffix of α is head irredundant. We write $Cirred(\alpha)$ to mean that α is completely irredundant.*

If α is normed, then α is irredundant over γ if and only if $\alpha\gamma \not\simeq \gamma$. In nBPA a redundant process consists solely of redundant variables.

Lemma 5. *Suppose $X_1, \ldots, X_k, \sigma$ are normed. Then $X_1 \ldots X_k$ is redundant over σ if and only if X_i is redundant over σ for every $X_i \in \{X_1, \ldots, X_k\}$.*

Proof. Suppose $X_1, \ldots, X_k, \sigma$ are normed and $X_1 \ldots X_k$ is redundant over σ. Then $X_1 \ldots X_k \sigma \implies X_2 \ldots X_k \sigma \implies \ldots \implies X_k \sigma \implies \sigma \simeq X_1 \ldots X_k \sigma$. It follows from Lemma 1 that $X_1 \ldots X_k \sigma \simeq X_2 \ldots X_k \sigma \simeq \ldots \simeq X_k \sigma \simeq \sigma$. We are done by using the congruence property.   □

For each σ, let the *redundant set* $\mathcal{R}_\sigma$ of σ be $\{X \mid X\sigma \simeq \sigma\}$. Let $\mathcal{V}(\alpha)$ be the set of variables appearing in α. We have two useful corollaries.

Corollary 1. *Suppose* α, σ *are normed. Then* $\alpha\sigma \simeq \sigma$ *if and only if* $\mathcal{V}(\alpha) \subseteq \mathcal{R}_\sigma$.

Corollary 2. *Suppose* $\alpha, \beta, \sigma_0, \sigma_1$ *are defined in an nBPA and* $\mathcal{R}_{\sigma_0} = \mathcal{R}_{\sigma_1}$. *Then* $\alpha\sigma_0 \simeq \beta\sigma_0$ *if and only if* $\alpha\sigma_1 \simeq \beta\sigma_1$.

Proof. Suppose $\mathcal{R}_{\sigma_0} = \mathcal{R}_{\sigma_1}$. Let $\mathcal{S}$ be $\{(\alpha\sigma_0, \beta\sigma_0) \mid \alpha\sigma_1 \simeq \beta\sigma_1\}$. It is not difficult to see that $\mathcal{S} \cup \simeq$ is a branching bisimulation.   □

We now take a look at the state preserving transitions of nBPA processes. We are particularly interested in knowing if the quotient set $\{\theta \mid \alpha \to^* V\theta\}/\simeq$ of the equivalence classes is finite for every nBPA process α and every variable V. It turns out that all such sets are finite with effective size bound.

Lemma 6. *For each nBPA process* $\alpha = X\omega$, *there is an effective bound* H_α, *uniformly computable from* α, *satisfying the following: If* $\alpha \to^* V\theta$ *then* $\alpha \to^* V\eta$ *for some* η *such that* $\theta \simeq \eta$ *and the length of* $\alpha \to^* V\eta$ *is no more than* H_α.

Proof. The basic idea is to show that in an effectively bounded number of steps α can reach, via norm consistent and norm splitting silent transitions, terms $V\theta$ with all possible variable V and all possible relative norm of V. We then apply Lemma 4. The bound H_α is computed from $|\alpha|$ and the transition system.   □

Under the assumption $\gamma \not\simeq \beta\gamma$ we can repeat the proof of Lemma 6 for $\beta\gamma$ in a way that γ is not affected. Hence the next corollary.

Corollary 3. *Suppose* $\alpha, \beta\gamma$ *are nBPA processes and* $\gamma \not\simeq \beta\gamma$. *If* $\beta\gamma \simeq \alpha \xrightarrow{\jmath} \alpha'$, *then there is a transition sequence* $\beta\gamma \to^* \beta''\gamma \xrightarrow{\jmath} \beta'\gamma$ *with its length bounded by* H_β *such that* $\beta''\gamma \simeq \alpha$ *and* $\beta'\gamma \simeq \alpha'$.

We are now in a position to prove the following.

Proposition 1. *The relation* $\not\simeq$ *on nBPA processes is semi-decidable.*

Proof. We define $\simeq_k$, the branching bisimilarity up to depth k, by exploiting Corollary 3. The inductive definition is as follows:

- $\alpha \simeq_0 \beta$ for all α, β.
- $\alpha \simeq_{i+1} \beta$ if the following condition and its symmetric version hold: If $\alpha \simeq_i$ $\beta \xrightarrow{\ell} \beta'$ then one of the following statements is valid:
  (i) $\ell = \tau$ and $\alpha \simeq_i \beta'$.
  (ii) $\alpha \implies \alpha'' \simeq_i \beta$ for some α'' such that $\alpha'' \xrightarrow{\ell} \alpha' \simeq_i \beta'$ for some α' and the length of $\alpha \implies \alpha''$ is bounded by H_α.

Each $\simeq_k$ is decidable. Using Corollary 3 one easily sees that $\simeq \subseteq \bigcap_{k \in \omega} \simeq_k$. The proof of the converse inclusion is standard.   □

4   Equality Checking

A straightforward approach to proving an equality between two processes is to construct a finite bisimulation tree for the equality. A tree of this kind has been called a tableau system [HS91, Hüt92]. To apply this approach we need to make sure that the following properties are satisfied: (i) Every tableau for an equality $\alpha = \beta$ is finite. (ii) The set of tableaux for an equality $\alpha = \beta$ is finite. We can achieve (i) by using Corollary 2 and Corollary 3. This is because if σ is long enough then according to Corollary 2 it can be decomposed into some $\sigma_0\sigma_1\sigma_2$ such that $\mathcal{R}_{\sigma_1\sigma_2} = \mathcal{R}_{\sigma_2}$. Then $\lambda\sigma_0\sigma_1\sigma_2 \simeq \gamma\sigma_0\sigma_1\sigma_2$ can be simplified to $\lambda\sigma_0\sigma_2 \simeq \gamma\sigma_0\sigma_2$. The equivalence provides a method to control the size of labels of a tableau. Now (ii) is a consequence of (i), Corollary 3 and König lemma.

The building blocks for tableaux are matches. Suppose $\alpha_0\alpha \not\simeq \alpha$ and $\beta_0\beta \not\simeq \beta$. A *match* for the equality $\alpha_0\alpha = \beta_0\beta$ over (α, β) is a finite symmetric relation $\{\gamma_i\alpha = \lambda_i\beta\}_{i=1}^k$ containing only those equalities accounted for in the following condition: For each transition $\alpha_0\alpha \xrightarrow{\ell} \alpha'\alpha$, one of the following holds:

- $\ell = \tau$ and $\alpha'\alpha = \beta_0\beta \in \{\gamma_i\alpha = \lambda_i\beta\}_{i=1}^k$;
- there is a sequence $\beta_0\beta \xrightarrow{\tau} \beta_1\beta \xrightarrow{\tau} \ldots \xrightarrow{\tau} \beta_n\beta \xrightarrow{\ell} \beta'\beta$, for $n < H_{\beta_0}$, such that $\{\alpha_0\alpha = \beta_1\beta, \ldots, \alpha_0\alpha = \beta_n\beta, \alpha'\alpha = \beta'\beta\} \subseteq \{\gamma_i\alpha = \lambda_i\beta\}_{i=1}^k$.

If $\alpha_0\sigma \not\simeq \sigma \not\simeq \beta_0\sigma$, a match for $\alpha_0\sigma = \beta_0\sigma$ over (σ, σ) is said to be a match for $\alpha_0\sigma = \beta_0\sigma$ over σ. The computable bound H_{β_0}, given by Corollary 3, guarantees that the number of matches for $\alpha_0\alpha = \beta_0\beta$ is effectively bounded.

Suppose α_0, β_0 are nBPA processes. A *tableau* for $\alpha_0 = \beta_0$ is a tree with each of its nodes labeled by an equality between nBPA processes. The root is labeled by $\alpha_0 = \beta_0$. We shall distinguish between *global tableau* and *local tableau*. The global tableau is the overall tableau whose root is labeled by the goal $\alpha_0 = \beta_0$. It is constructed from the rules given in Fig. 1. Decmp rule decomposes a goal into several subgoals. We shall find it useful to use SDecmp, which is a stronger version of Decmp. The side condition of SDecmp ensures that it is unnecessary to apply it consecutively. When applying Decmp rule we assume that an equality $\gamma\sigma = \sigma$, respectively $\sigma = \gamma\sigma$, is always decomposed in the following manner

$$\frac{\gamma\sigma = \sigma}{\sigma = \sigma \quad \{V\sigma = \sigma\}_{V\in\mathcal{V}(\gamma)}} \text{ respectively } \frac{\sigma = \gamma\sigma}{\sigma = \sigma \quad \{V\sigma = \sigma\}_{V\in\mathcal{V}(\gamma)}}.$$

Accordingly $\gamma = \epsilon$, respectively $\epsilon = \gamma$, is decomposed in the following fashion

$$\frac{\gamma = \epsilon}{\epsilon = \epsilon \quad \{V = \epsilon\}_{V\in\mathcal{V}(\gamma)}} \text{ respectively } \frac{\epsilon = \gamma}{\epsilon = \epsilon \quad \{V = \epsilon\}_{V\in\mathcal{V}(\gamma)}}.$$

SubstL and SubstR allow one to create common suffix for the two processes in an equality. ContrL and ContrR are used to remove a redundant variable inside a process. In the side conditions of these two rules, α_0, β_0 are the processes appearing in the root of the global tableau. ContrC deletes redundant variables from the common suffix of a node label whenever the size of the common suffix

Decmp	$\dfrac{\gamma\alpha = \lambda\beta}{\alpha = \beta \quad \{U\alpha = \alpha\}_{U\in\mathcal{V}(\gamma)} \quad \{V\beta = \beta\}_{V\in\mathcal{V}(\lambda)}}$	$\begin{array}{l}|\gamma| + |\lambda| > 0,\\ \forall U \in \mathcal{V}(\gamma).U \Longrightarrow \epsilon,\\ \forall V \in \mathcal{V}(\lambda).V \Longrightarrow \epsilon.\end{array}$
SDecmp	$\dfrac{\gamma\alpha = \lambda\beta}{\alpha = \beta \quad \{U\alpha = \alpha\}_{U\in\mathcal{V}(\gamma)} \quad \{V\beta = \beta\}_{V\in\mathcal{V}(\lambda)}}$	$\begin{array}{l}|\gamma| + |\lambda| > 0,\\ Hirred(\alpha),\ Hirred(\beta),\\ \forall U \in \mathcal{V}(\gamma).U \Longrightarrow \epsilon,\\ \forall V \in \mathcal{V}(\lambda).V \Longrightarrow \epsilon.\end{array}$
Match	$\dfrac{\gamma\alpha = \lambda\beta}{\alpha_1\alpha = \beta_1\beta \ \ldots \ \alpha_k\alpha = \beta_k\beta}$	$\gamma\alpha \not\simeq \alpha,\ \lambda\beta \not\simeq \beta,$ and $\{\alpha_i\alpha = \beta_i\beta\}_{i=1}^{k}$ is a match for $\gamma\alpha = \lambda\beta$ over (α,β).
SubstL	$\dfrac{\gamma\alpha = \lambda\beta}{\gamma\delta\beta = \lambda\beta}$	$\alpha = \delta\beta$ is the residual.
SubstR	$\dfrac{\gamma\alpha = \lambda\beta}{\gamma\alpha = \lambda\delta\alpha}$	$\delta\alpha = \beta$ is the residual.
ContrL	$\dfrac{\gamma Z\delta = \lambda}{\gamma\delta = \lambda \quad Z\delta = \delta}$	$Hirred(\delta),\ Z \Longrightarrow \epsilon$ and $|\gamma Z\delta| > \max\{|\alpha_0|,|\beta_0|\}\|\Delta\|.$
ContrR	$\dfrac{\gamma = \lambda Z\delta}{\gamma = \lambda\delta \quad Z\delta = \delta}$	$Hirred(\delta),\ Z \Longrightarrow \epsilon$ and $|\lambda Z\delta| > \max\{|\alpha_0|,|\beta_0|\}\|\Delta\|.$
ContrC	$\dfrac{\gamma\sigma'\sigma_0\sigma_1 = \lambda\sigma'\sigma_0\sigma_1}{\gamma\sigma'\sigma_1 = \lambda\sigma'\sigma_1 \quad \{V\sigma_1 = \sigma_1\}_{V\in\mathcal{V}(\sigma_0)}}$	$\begin{array}{l}|\sigma'\sigma_0\sigma_1| > 2^{n_\Delta},\ |\sigma_0| > 0,\\ Hirred(\sigma_1),\\ \forall V \in \mathcal{V}(\sigma_0).V \Longrightarrow \epsilon.\end{array}$

Fig. 1. Rules for Global Tableaux

is over limit. Notice that all the side conditions on the rules are semi-decidable due to the semi-decidability of $\not\simeq$. So we can effectively enumerate tableaux.

In what follows a node $Z\eta = W\kappa$ to which Match rule is applied with the condition $Z\eta \not\simeq \eta \wedge W\kappa \not\simeq \kappa$ is called an *M-node*. A node of the form $Z\sigma = \sigma$ with σ being head irredundant is called a *V-node*. We now describe how a global tableau for $\alpha_0 = \beta_0$ is constructed. Assuming $\alpha_0 = \gamma X\alpha_1$ and $\beta_0 = \lambda Y\beta_1$ such that $X\alpha_1 \not\simeq \alpha_1$ and $Y\beta_1 \not\simeq \beta_1$, we apply the following instance of SDecmp rule:

$$\frac{\gamma X\alpha_1 = \lambda Y\beta_1}{X\alpha_1 = Y\beta_1 \quad \{UX\alpha_1 = X\alpha_1\}_{U\in\mathcal{V}(\gamma)} \quad \{VY\beta_1 = Y\beta_1\}_{V\in\mathcal{V}(\lambda)}}.$$

By definition $X\alpha_1 = Y\beta_1$ is an M-node and $\{UX\alpha_1 = X\alpha_1\}_{U\in\mathcal{V}(\gamma)} \cup \{VY\beta_1 = Y\beta_1\}_{V\in\mathcal{V}(\lambda)}$ is a set of V-nodes. These nodes are the roots of new subtableaux. Starting from $X\alpha_1 = Y\beta_1$ we apply Match rule under the condition that neither α_1 nor β_1 is affected. The application of Match rule is repeated to grow the subtableau rooted at $X\alpha_1 = Y\beta_1$. The construction of the tree is done in a breadth first fashion. So the tree grows level by level. At some stage we apply Decmp rule to all the current leaves. This particular application of Decmp must meet the following conditions: (i) Both α_1 and β_1 must be kept intact in all the current leaves; (ii) Either α_1 or β_1 is exposed in at least one current leaf. Choose a leaf labeled by either $\alpha_1 = \delta_1\beta_1$ for some δ_1 or by $\delta_1'\alpha_1 = \beta_1$ for some δ_1' and call it the *residual node* or *R-node*. Suppose the residual node is $\alpha_1 = \delta_1\beta_1$. All the other current leaves, the *non-residual nodes*, must be labeled by an equality of the form $\gamma_1\alpha_1 = \lambda_1\beta_1$. A non-residual node with label $\gamma_1\alpha_1 = \lambda_1\beta_1$ is then attached with a single child labeled by $\gamma_1\delta_1\beta_1 = \lambda_1\beta_1$. This is an application of

		$\lvert \gamma \rvert > 0$ and $\lvert \lambda \rvert > 0$; $\lvert \sigma' \sigma_0 \sigma_1 \rvert > 2^{n_\Delta}$,
Localization	$\dfrac{\gamma \sigma' \sigma_0 \sigma_1 = \lambda \sigma' \sigma_0 \sigma_1}{\gamma \sigma' \sigma_1 = \lambda \sigma' \sigma_1}$ $\{X_i \sigma_1 = \sigma_1\}_{i \in I}$ $\{X_i \sigma_0 \sigma_1 = \sigma_0 \sigma_1\}_{i \in I}$	$2^{n_\Delta} \geq \lvert \sigma_1 \rvert > 0$ and $\lvert \sigma_0 \rvert > 0$; $Cirred(\sigma' \sigma_0 \sigma_1)$ and $Cirred(\sigma' \sigma_1)$; $\gamma \sigma' \sigma_0 \sigma_1 \not\simeq \sigma' \sigma_0 \sigma_1,\ \gamma \sigma' \sigma_1 \not\simeq \sigma' \sigma_1$; $\lambda \sigma' \sigma_0 \sigma_1 \not\simeq \sigma' \sigma_0 \sigma_1,\ \lambda \sigma' \sigma_1 \not\simeq \sigma' \sigma_1$; $I \cap J = \emptyset,\ I \cup J = \{1, \dots, n_\Delta\}$; $\forall j \in J.\ X_j \sigma_0 \sigma_1 \not\simeq \sigma_0 \sigma_1$ and $X_j \sigma_1 \not\simeq \sigma_1$; $X_i \Longrightarrow \epsilon$ for all $i \in I$.

Fig. 2. Rule for Local Tableaux

SubstL rule. Now we can recursively apply the global tableau construction to $\gamma_1 \delta_1 \beta_1 = \lambda_1 \beta_1$ to produce a new subtableau. The treatment of a V-node child, say $UX\alpha_1 = X\alpha_1$, is similar. We keep applying Match rule over α_1 as long as the side condition is met. At certain stage we apply Decmp rule to all the leaves. The application should meet the following conditions: (i) No occurrence of α_1 is affected; (ii) There is an application of Decmp that takes the following shape

$$\frac{\gamma_1 \alpha_1 = \lambda_1 \alpha_1}{\alpha_1 = \alpha_1 \quad \{V\alpha_1 = \alpha_1\}_{V \in \mathcal{V}(\gamma_1)} \quad \{V\alpha_1 = \alpha_1\}_{V \in \mathcal{V}(\lambda_1)}}.$$

We then recursively apply the tableau construction to create new subtableaux.

In the above construction the R-node $\alpha_1 = \delta_1 \beta_1$ can be the root of a new subtableau, which might contain another R-node. In fact a chain of R-nodes is possible. ContrL/ContrR is used to control the size of R-nodes.

After an application of SubstL/SubstR rule we may get a *C-node* $\alpha' \sigma' \sigma_0 \sigma_1 = \beta' \sigma' \sigma_0 \sigma_1$ if ContrC rule is applicable. Once a C-node appears, we immediately apply ContrC rule to reduce the size of its common suffix. Intuitively we should apply ContrC rule sufficiently often so that the common suffix becomes completely irredundant. Eventually either the length of the common suffix has become no more than 2^{n_Δ}, in which case we continue to build up the global tableau, or Localization rule as defined in Fig. 2 is applicable, in which case we get an *L-node*. The soundness of Localization rule is guaranteed by Corollary 2.

Suppose Localization rule is applied to an L-node $\alpha' \sigma' \sigma_0 \sigma_1 = \beta' \sigma' \sigma_0 \sigma_1$:

$$\frac{\alpha' \sigma' \sigma_0 \sigma_1 = \beta' \sigma' \sigma_0 \sigma_1}{\{X_i \sigma_1 = \sigma_1\}_{i \in I} \quad \alpha' \sigma' \sigma_1 = \beta' \sigma' \sigma_1 \quad \{X_i \sigma_0 \sigma_1 = \sigma_0 \sigma_1\}_{i \in I}}.$$

The node $\alpha' \sigma' \sigma_1 = \beta' \sigma' \sigma_1$ is a new L-node. We call $\{X_i \mid i \in I\}$ the *R-set* of the new L-node. If the size of the common suffix of $\alpha' \sigma' \sigma_1 = \beta' \sigma' \sigma_1$ is still larger than 2^{n_Δ}, we continue to apply Localization rule. Otherwise we get an *L-root*, which is the root of a local tableau. Now suppose $\alpha' \sigma' \sigma_1 = \beta' \sigma' \sigma_1$ is an L-root. The construction of the local tableau should stick to two principles described as follows: (I) *Locality*. No application of Decmp, SDecmp, SubstL, SubsR and ContrC should ever affect $\sigma' \sigma_1$ or any suffix of $\sigma' \sigma_1$. Notice that applications of SubstL or SubstR can never affect $\sigma' \sigma_1$ or any suffix of $\sigma' \sigma_1$. (II) *Consistency*.

Suppose $\gamma\alpha = \lambda\beta$ is a node to which Match rule is applied using a match over (α, β). Then either $\sigma'\sigma_1$ is a suffix of both α and β, or $\alpha = \beta = \sigma''\sigma_1$ for some σ'' satisfying the following: (i) σ'' is a proper suffix of σ'; (ii) $\gamma = UZ$ and $\lambda = Z$ such that $Z\sigma''$ is a suffix of σ'; and (iii) the match is over $\sigma''\sigma_1$. The locality and consistency conditions basically say that choices made in the construction of the local tableau should not contradict to the fact that $\sigma'\sigma_1$ is completely irredundant.

The construction of a path in a tableau ends with a leaf. A *successful leaf* is either a node labeled by $\varsigma = \varsigma$ for some ς, or a node labeled by $\epsilon = V$ ($V = \epsilon$) with $V \simeq \epsilon$, or a node that has the same label as one of its ancestors. An *unsuccessful leaf* is produced if the node is either labeled by $\epsilon = V$ ($V = \epsilon$) with $V \not\simeq \epsilon$, or labeled by some $\varsigma = \varsigma'$ with distinct ς, ς' such that no rule is applicable to $\varsigma = \varsigma'$. A local tableau has additionally two new kind of successful/unsuccessful leaves: (i) An L-root is a successful leaf if it shares the same label with one of its ancestors that is also an L-root. (ii) Suppose $\alpha'\sigma'\sigma_0\sigma_1 = \beta'\sigma'\sigma_0\sigma_1$ is an L-node and its child $\alpha'\sigma'\sigma_1 = \beta'\sigma'\sigma_1$ is an L-root. In the local tableau rooted at $\alpha'\sigma'\sigma_1 = \beta'\sigma'\sigma_1$, a node of the form $Z\sigma_1 = \sigma_1$ is deemed as a leaf. It is a successful leaf if Z is in the R-set of the L-root; it is an unsuccessful leaf otherwise.

Tableau constructions always terminate. In fact we have the following.

Lemma 7. *The size of every tableau for an equality is effectively bounded. The number of tableaux for an equality is effectively bounded.*

A tableau is *successful* if all of its leaves are successful. Successful tableaux generate bisimulation bases.

Proposition 2. *Suppose $X\alpha, Y\beta$ are nBPA processes. Then $X\alpha \simeq Y\beta$ if and only if there is a successful tableau for $X\alpha = Y\beta$.*

Proof. If $X\alpha \simeq Y\beta$ we can easily construct a tableau using the bisimulation property, Corollary 2 and Corollary 3. Conversely suppose there is a successful tableau $\mathfrak{T}$ for $X\alpha = Y\beta$. Let $\mathcal{A} = \mathcal{A}_b \cup \mathcal{A}_z \cup \mathcal{A}_l$. The set $\mathcal{A}_b$ of basic axioms is given by $\{\gamma = \lambda \mid \gamma = \lambda$ is a label of a node in $\mathfrak{T}\}$. The set $\mathcal{A}_z$ is defined by

$$\mathcal{A}_z = \left\{ V\sigma = \theta\sigma, \theta\sigma = \sigma \,\middle|\, \begin{array}{l} V\sigma = \sigma \text{ is in } \mathcal{A}_b, \text{ and } V \stackrel{\tau}{\Longrightarrow} \theta \stackrel{\tau}{\Longrightarrow} \epsilon \\ \text{is a chosen shortest path from } V \text{ to } \epsilon. \end{array} \right\}.$$

Suppose $\gamma\sigma'\sigma_1 = \lambda\sigma'\sigma_1$ is an L-root and $\gamma\sigma'\sigma_0\sigma_1 = \lambda\sigma'\sigma_0\sigma_1$ is its parent. A node $\eta\sigma'\sigma_1 = \kappa\sigma'\sigma_1$ in the local tableau rooted at $\gamma\sigma'\sigma_1 = \lambda\sigma'\sigma_1$ must be lifted to $\eta\sigma'\sigma_0\sigma_1 = \kappa\sigma'\sigma_0\sigma_1$ in order to show that $\gamma\sigma'\sigma_0\sigma_1 = \lambda\sigma'\sigma_0\sigma_1$ satisfies the bisimulation base property. Since local tableaux may be nested, the node might have several lifted versions. The set $\mathcal{A}_l$ is defined to be the collection of all such lifted pairs. We can prove by induction on the nodes of the tableau, starting with the leaves, that $\mathcal{A}$ is a bisimulation base. Hence $X\alpha \simeq Y\beta$ by Lemma 2.   □

Our main result follows from Proposition 1, Lemma 7 and Proposition 2.

Theorem 1. *The branching bisimilarity on nBPA processes is decidable.*

5   Regularity Checking

Regularity problem asks if a process is bisimilar to a finite state process. For strong regularity problem of nBPA, Kučera [Kuč96] showed that it is decidable in polynomial time. Srba [Srb02] observed that it is actually NL-complete. The decidability of strong regularity problem for the general BPA was proved by Burkart, Caucal and Steffen [BCS95, BCS96]. It was shown to be PSPACE-hard by Srba [Srb02]. The decidability of almost all weak regularity problems of process rewriting systems [May00] are unknown. The only exception is Jancar and Esparza's undecidability result of weak regularity problem of Petri Net and its extension [JE96]. Srba [Srb03] proved that weak regularity is both NP-hard and co-NP-hard for nBPA. Using a result by Srba [Srb03], Mayr proved that weak regularity problem of nBPA is EXPTIME-hard [May03].

The present paper improves our understanding of the issue by the following.

Theorem 2. *The regularity problem of $\simeq$ on nBPA is decidable.*

Proof. One proves by a combinatorial argument that, in the transition tree of an infinite state BPA process, (i) a path $V_0\sigma_0 \xrightarrow{\ell_1^*} V_1\sigma_1 \xrightarrow{\ell_2^*} V_2\sigma_2 \ldots \xrightarrow{\ell_m^*} V_m\sigma_m$ exists such that (ii) $|\sigma_0| < |\sigma_1| < |\sigma_2| < \ldots < |\sigma_m|$ and (iii) $\|V_0\sigma_0\|_b < \|V_1\sigma_1\|_b < \|V_2\sigma_2\|_b < \ldots < \|V_m\sigma_m\|_b$. We can choose m large enough such that $0 \leq i < j \leq m$ for some i, j satisfying $V_i = V_j$ and $\mathcal{R}_{\sigma_i} = \mathcal{R}_{\sigma_j}$. Let $\sigma_j = \sigma\sigma_i$ for some σ. Clearly $\|\sigma_i\|_b < \|\sigma_j\|_b$. Using Corollary 2 one can prove by induction that $\sigma^i\sigma_i \not\simeq \sigma^j\sigma_i$ whenever $i \neq j$. It is semi-decidable to find (i) with properties (ii,iii). The converse implication is proved by a tree construction using Theorem 1.   □

6   Remark

For parallel processes (BPP/PN) with silent actions, the only known decidability result on equivalence checking is due to Czerwiński, Hofman and Lasota [CHL11]. This paper provides the analogous decidability result for the sequential processes (BPA/PDA) with silent actions. For further research one could try to apply the technique developed in this paper to general BPA and normed PDA.

Acknowledgement. I am indebted to He, Huang, Long, Shen, Tao, Yang, Yin and the anonymous referees. The support from NSFC (60873034, 61033002, ANR 61261130589) and STCSM (11XD1402800) are gratefully acknowledged.

References

[BBK87] Baeten, J., Bergstra, J., Klop, J.: Decidability of bisimulation equivalence for processes generating context-free languages. In: de Bakker, J.W., Nijman, A.J., Treleaven, P.C. (eds.) PARLE 1987. LNCS, vol. 259, pp. 94–113. Springer, Heidelberg (1987)

[BCS95] Burkart, O., Caucal, D., Steffen, B.: An elementary bisimulation decision procedure for arbitrary context free processes. In: Hájek, P., Wiedermann, J. (eds.) MFCS 1995. LNCS, vol. 969, pp. 423–433. Springer, Heidelberg (1995)

[BCS96]   Burkart, O., Caucal, D., Steffen, B.: Bisimulation collapse and the process taxonomy. In: Sassone, V., Montanari, U. (eds.) CONCUR 1996. LNCS, vol. 1119, pp. 247–262. Springer, Heidelberg (1996)

[BGS92]   Balcazar, J., Gabarro, J., Santha, M.: Deciding bisimilarity is p-complete. Formal Aspects of Computing 4, 638–648 (1992)

[CHL11]   Czerwiński, W., Hofman, P., Lasota, S.: Decidability of branching bisimulation on normed commutative context-free processes. In: Katoen, J.-P., König, B. (eds.) CONCUR 2011. LNCS, vol. 6901, pp. 528–542. Springer, Heidelberg (2011)

[CHS92]   Christensen, S., Hüttel, H., Stirling, C.: Bisimulation equivalence is decidable for all context-free processes. In: Cleaveland, W.R. (ed.) CONCUR 1992. LNCS, vol. 630, pp. 138–147. Springer, Heidelberg (1992)

[CHT95]   Caucal, D., Huynh, D., Tian, L.: Deciding branching bisimilarity of normed context-free processes is in σ_2^p. Information and Computation 118, 306–315 (1995)

[Hir96]   Hirshfeld, Y.: Bisimulation trees and the decidability of weak bisimulations. Electronic Notes in Theoretical Computer Science 5, 2–13 (1996)

[HJM96]   Hirshfeld, Y., Jerrum, M., Moller, F.: A polynomial algorithm for deciding bisimilarity of normed context free processes. Theoretical Computer Science 158(1-2), 143–159 (1996)

[HS91]   Hüttel, H., Stirling, C.: Actions speak louder than words: Proving bisimilarity for context-free processes. In: LICS 1991, pp. 376–386 (1991)

[HT94]   Huynh, T., Tian, L.: Deciding bisimilarity of normed context free processes is in σ_2^p. Theoretical Computer Science 123, 83–197 (1994)

[Hüt92]   Hüttel, H.: Silence is golden: Branching bisimilarity is decidable for context free processes. In: Larsen, K.G., Skou, A. (eds.) CAV 1991. LNCS, vol. 575, pp. 2–12. Springer, Heidelberg (1992)

[Jan12]   Jančar, P.: Bisimilarity on basic process algebra is in 2-exptime (2012)

[JE96]   Jančar, P., Esparza, J.: Deciding finiteness of petri nets up to bisimulation. In: Meyer auf der Heide, F., Monien, B. (eds.) ICALP 1996. LNCS, vol. 1099, pp. 478–489. Springer, Heidelberg (1996)

[Kie13]   Kiefer, S.: BPA bisimilarity is exptime-hard. Information Processing Letters 113, 101–106 (2013)

[Kuč96]   Kučera, A.: Regularity is decidable for normed BPA and normed BPP processes in polynomial time. In: Král, J., Bartosek, M., Jeffery, K. (eds.) SOFSEM 1996. LNCS, vol. 1175, pp. 377–384. Springer, Heidelberg (1996)

[May00]   Mayr, R.: Process rewrite systems. Information and Computation 156, 264–286 (2000)

[May03]   Mayr, R.: Weak bisimilarity and regularity of BPA is exptime-hard. In: EXPRESS 2003 (2003)

[Srb02]   Srba, J.: Strong bisimilarity and regularity of basic process algebra is pspace-hard. In: Widmayer, P., Triguero, F., Morales, R., Hennessy, M., Eidenbenz, S., Conejo, R. (eds.) ICALP 2002. LNCS, vol. 2380, pp. 716–727. Springer, Heidelberg (2002)

[Srb03]   Srba, J.: Complexity of weak bisimilarity and regularity for BPA and BPP. Mathematical Structures in Computer Science 13, 567–587 (2003)

[Stř98]   Stříbrná, J.: Hardness results for weak bisimilarity of simple process algebras. Electronic Notes in Theoretical Computer Science 18, 179–190 (1998)

[vGW89]   van Glabbeek, R., Weijland, W.: Branching time and abstraction in bisimulation semantics. In: Information Processing 1989, pp. 613–618. North-Holland (1989)

FO Model Checking of Interval Graphs[*]

Robert Ganian[1], Petr Hliněný[2], Daniel Král'[3], Jan Obdržálek[2],
Jarett Schwartz[4], and Jakub Teska[5]

[1] Vienna University of Technology, Austria
rganian@gmail.com
[2] Masaryk University, Brno, Czech Republic
{hlineny,obdrzalek}@fi.muni.cz
[3] University of Warwick, Coventry, United Kingdom
D.Kral@warwick.ac.uk
[4] UC Berkeley, Berkeley, United States
jarett@cs.berkeley.edu
[5] University of West Bohemia, Pilsen, Czech Republic
teska@kma.zcu.cz

Abstract. We study the computational complexity of the FO model
checking problem on interval graphs, i.e., intersection graphs of intervals
on the real line. The main positive result is that this problem can be
solved in time $O(n \log n)$ for n-vertex interval graphs with representations containing only intervals with lengths from a prescribed finite set.
We complement this result by showing that the same is not true if the
lengths are restricted to any set that is dense in some open subset, e.g.,
in the set $(1, 1 + \varepsilon)$.

Keywords: FO model checking, parameterized complexity, interval
graph, clique-width.

1 Introduction

Results on the existence of an efficient algorithm for a class of problems have recently attracted a significant amount of attention. Such results are now referred
to as algorithmic meta-theorems, see a recent survey [15]. The most prominent
example is a theorem of Courcelle [1] asserting that every MSO property can be
model checked in linear time on the class of graphs with bounded tree-width.
Another example is a theorem of Courcelle, Makowski and Rotics [2] asserting that the same conclusion holds for graphs with bounded clique-width when
quantification is restricted to vertices and their subsets.

---

[*] A full version of this contribution, which contains all proofs, can be downloaded
from http://arxiv.org/abs/1302.6043. All the authors except for Jarett Schwartz
acknowledge support of the Czech Science Foundation under grant P202/11/0196.
Robert Ganian also acknowledges support by the ERC grant (COMPLEX REA-
SON 239962) held by Stefan Szeider. Jarett Schwartz acknowledges support of the
Fulbright and NSF Fellowships.

F.V. Fomin et al. (Eds.): ICALP 2013, Part II, LNCS 7966, pp. 250–262, 2013.
© Springer-Verlag Berlin Heidelberg 2013

In this paper, we focus on more restricted graph properties, specifically those expressible in the first order logic. Clearly, every such property can be tested in polynomial time if we allow the degree of the polynomial to depend on the property of interest. But can these properties be tested in so-called *fixed parameter tractable* (FPT [6]) time, i.e., in polynomial time where the degree of the polynomial does not depend on the considered property? The first result in this direction could be that of Seese [22]: Every FO property can be tested in linear time on graphs with bounded maximum degree. A breakthrough result of Frick and Grohe [11] asserts that every FO property can be tested in almost linear time on classes of graphs with locally bounded tree-width. Here, an almost linear algorithm stands for an algorithm running in time $O(n^{1+\varepsilon})$ for every $\varepsilon > 0$. A generalization to graph classes locally excluding a minor (with worse running time) was later obtained by Dawar, Grohe and Kreutzer [4].

Research in this direction so far culminated in establishing that every FO property can be tested in almost linear time on classes of graphs with locally bounded expansion, as shown (independently) by Dawar and Kreutzer [5] (also see [13] for the complete proof), and by Dvořák, Král' and Thomas [7]. The concept of graph classes with bounded expansion has recently been introduced by Nešetril and Ossona de Mendéz [18,19,20]; examples of such graph classes include classes of graphs with bounded maximum degree or proper minor-closed classes of graphs. A holy grail of this area is establishing the fixed parameter tractability of testing FO properties on nowhere-dense classes of graphs.

In this work, we investigate whether structural properties which do not yield (locally) bounded width parameters could lead to similar results. Specifically, we study the intersection graphs of intervals on the real line, which are also called interval graphs. When we restrict to unit interval graphs, i.e., intersection graphs of intervals with unit lengths, one can easily deduce the existence of a linear time algorithm for testing FO properties from Gaifman's theorem, using the result of Courcelle et al [2] and that of Lozin [17] asserting that every proper hereditary subclass of unit interval graphs, in particular, the class of unit interval graphs with bounded radius, has bounded clique-width. This observation is a starting point for our research presented in this paper.

Let us now give a definition. For a set L of reals, an interval graph is called an L-interval graph if it is an intersection graph of intervals with lengths from L. For example, unit interval graphs are $\{1\}$-interval graphs. If L is a finite set of rationals, then any L-interval graph with bounded radius has bounded clique-width (see Section 4 for further details). So, FO properties of such graphs can be tested in the fixed parameter way. However, if L is not a set of rationals, there exist L-interval graphs with bounded radius and unbounded clique-width, and so the easy argument above does not apply.

Our main algorithmic result says that every FO property can be tested in time $O(n \log n)$ for L-interval graphs when L is any finite set of reals. To prove this result, we employ a well-known relation of FO properties to Ehrenfeucht-Fraïssé games. Specifically, we show using the notion of game trees, which we introduce, that there exists an algorithm transforming an input L-interval graph to another

L-interval graph that has bounded maximum degree and that satisfies the same properties expressible by FO sentences with bounded quantifier rank. We remark that encoding a game associated with a model checking problem by a tree, which describes the course of the game, was also applied in designing fast algorithms for MSO model checking [14,16].

On the negative side, we show that if L is an infinite set that is dense in some open set, then L-interval graphs can be used to model arbitrary graphs. Specifically, we show that L-interval graphs for such sets L allow polynomially bounded FO interpretations of all graphs. Consequently, testing FO properties for L-intervals graphs for such sets L is W[2]-hard (see Corollary 2). In addition, we show that unit interval graphs allow polynomially bounded MSO interpretations of all graphs.

The property of being *W[2]-hard* comes from the theory of parameterized complexity [6], and it is equivalent to saying that the considered problem is at least as hard as the *d-dominating* set problem, asking for an existence of a dominating set of fixed parameter size d in a graph. It is known that, unless the Exponential time hypothesis fails, W[2]-hard problems cannot have polynomial algorithms with the degree a constant independent of the parameter (of the considered FO property in our case).

In Section 2, we introduce the notation and the computational model used in the paper. In the following section, we present an $O(n \log n)$ algorithm for deciding FO properties of L-interval graphs for finite sets L. In Section 4, we present proofs of the facts mentioned above on the clique-width of L-interval graphs with bounded radius. Finally, we establish FO interpretability of graphs in L-interval graphs for sets L which are dense in an open set in Section 5.

2   Notation

An *interval graph* is a graph G such that every vertex v of G can be associated with an interval $J(v) = [\ell(v), r(v))$ such that two vertices v and v' of G are adjacent if and only if $J(v)$ and $J(v')$ intersect (it can be easily shown that the considered class of graphs remains the same regardless of whether we consider open, half-open or closed intervals in the definition). We refer to such an assignment of intervals to the vertices of G as a *representation* of G. The point $\ell(v)$ is the *left end point* of the interval $J(v)$ and $r(v)$ is its *right end point*.

If L is a set of reals and $r(v) - \ell(v) \in L$ for every vertex v, we say that G is an *L-interval* graph and we say that the representation is an *L-representation* of G. For example, if $L = \{1\}$, we speak about unit interval graphs. Finally, if $r(v) - \ell(v) \in L$ and $0 \le \ell(v) \le r(v) \le d$ for some real d, i.e., all intervals are subintervals of $[0, d)$, we speak about *(L, d)-interval graphs*. Note that if G is an interval graph of radius k, then G is also an $(L, (2k + 1) \max L)$-interval graph (we use $\max L$ and $\min L$ to denote the maximum and the minimum elements, respectively, of the set L).

We now introduce two technical definitions related to manipulating intervals and their lengths. These definitions are needed in the next section. If L is a set of reals, then $L^{(k)}$ is the set of all integer linear combinations of numbers from L with the sum of the absolute values of their coefficients bounded by k. For instance, $L^{(0)} = \{0\}$ and $L^{(1)} = L \cup \{0\}$. An L-*distance* of two intervals $[a, b)$ and $[c, d)$ is the smallest k such that $c - a \in L^{(k)}$. If no such k exists, then the L-distance of two intervals is defined to be ∞.

Since we do not restrict our attention to L-interval graphs where L is a set of rationals, we should specify the computational model considered. We use the standard RAM model with infinite arithmetic precision and unit cost of all arithmetic operation, but we refrain from trying to exploit the power of this computational model by encoding any data in the numbers we store. In particular, we only store the end points of the intervals in the considered representations of graphs in numerical variables with infinite precision.

2.1 Clique-Width

We now briefly present the notion of clique-width introduced in [3]. Our results on interval graphs related to this notion are given in Section 4. A k-labeled graph is a graph with vertices that are assigned integers (called labels) from 1 to k. The *clique-width* of a graph G equals the minimum k such that G can be obtained from single vertex graphs with label 1 using the following four operations: relabeling all vertices with label i to j, adding all edges between the vertices with label i and the vertices with label j (i and j can be the same), creating a vertex labeled with 1, and taking a disjoint union of graphs obtained using these operations.

2.2 First Order Properties

In this subsection, we introduce concepts from logic and model theory which we use. A *first order (FO) sentence* is a formula with no free variables with the usual logical connectives and quantification allowed only over variables for elements (vertices in the case of graphs). A *monadic second order (MSO) sentence* is a formula with no free variables with the usual logical connectives where, unlike in FO sentences, quantification over subsets of elements is allowed. An FO *property* is a property expressible by an FO sentence; similarly, an *MSO property* is a property expressible by an MSO sentence. Finally, the *quantifier rank* of a formula is the maximum number of nested quantifiers in it.

FO sentences are closely related to the so-called Ehrenfeucht-Fraïssé games. The *d-round Ehrenfeucht-Fraïssé game* is played on two relational structures R_1 and R_2 by two players referred to as the *spoiler* and the *duplicator*. In each of the d rounds, the spoiler chooses an element in one of the structures and the duplicator chooses an element in the other. Let x_i be the element of R_1 chosen in the i-th round and y_i be the element of R_2. We say that the duplicator *wins* the game if there is a strategy for the duplicator such that the substructure of R_1 induced by the elements $x_1, \ldots, x_d$ is always isomorphic to the substructure of R_2 induced by the elements $y_1, \ldots, y_d$, with the isomorphism mapping each x_i to y_i.

The following theorem [8,10] relates this notion to FO sentences of quantifier rank at most d.

Theorem 1. *Let d be an integer. The following statements are equivalent for any two structures R and R':*

- *The structures R and R' satisfy the same FO sentences of quantifier rank at most d.*
- *The duplicator wins the d-round Ehrenfeucht-Fraïssé game for R and R'.*

We describe possible courses of the d-round Ehrenfeucht-Fraïssé game on a single relational structure by a rooted tree, which we call a *d-EF-tree*. All the leaves of an d-EF-tree are at depth d and each of them is associated with a relational structure with elements labeled from 1 to d. The d-EF-tree $\mathcal{T}$ for the game played on a relational structure R is obtained as follows. The number of children of every internal node of $\mathcal{T}$ is the number of elements of R and the edges leaving the internal node to its children are associated in the one-to-one way with the elements of R. So, every path from the root of $\mathcal{T}$ to a leaf u of $\mathcal{T}$ yields a sequence $x_1, \ldots, x_d$ of the elements of R (those associated with the edges of that path) and the substructure of R induced by $x_1, \ldots, x_d$ with x_i labeled with i is the one associated with the leaf u.

A mapping f from a d-EF-tree $\mathcal{T}$ to another d-EF-tree $\mathcal{T}'$ is an EF-homomorphism if the following three conditions hold:

1. if u is a parent of v in $\mathcal{T}$, then $f(u)$ is a parent of $f(v)$ in $\mathcal{T}'$, and
2. if u is a leaf of $\mathcal{T}$, then $f(u)$ is a leaf of $\mathcal{T}'$, and
3. the structures associated with u and $f(u)$ are isomorphic through the bijection given by the labelings.

Two trees $\mathcal{T}$ and $\mathcal{T}'$ are *EF-equivalent* if there exist an EF-homomorphism from $\mathcal{T}$ to $\mathcal{T}'$ and an EF-homomorphism from $\mathcal{T}'$ to $\mathcal{T}$.

Let $\mathcal{T}$ and $\mathcal{T}'$ be the d-EF-trees for the game played on relational structures R and R', respectively. Suppose that $\mathcal{T}$ and $\mathcal{T}'$ are EF-equivalent and let f_1 and f_2 be the EF-homomorphism from $\mathcal{T}$ to $\mathcal{T}'$ and $\mathcal{T}'$ to $\mathcal{T}$, respectively. We claim that the duplicator wins the d-round Ehrenfeucht-Fraïssé game for R and R'. Let us describe a possible strategy for the duplicator. We restrict to the case when the spoiler chooses an element x_i of R in the i-th step, assuming elements $x_1, \ldots, x_{i-1}$ in R and elements $y_1, \ldots, y_{i-1}$ in R' have been chosen in the previous rounds. Let $u_0, \ldots, u_i$ be the path in $\mathcal{T}$ corresponding to $x_1, \ldots, x_i$. The duplicator chooses the element y_i in R' that corresponds to the edge $f_1(u_{i-1})f_1(u_i)$ in $\mathcal{T}'$. It can be verified that if the duplicator follows this strategy, then the substructures of R and R' induced by $x_1, \ldots, x_d$ and $y_1, \ldots, y_d$, respectively, are isomorphic.

Let us summarize our findings from the previous paragraph.

Theorem 2. *Let d be an integer. If the d-EF-trees for the game played on two relational structure R and R' are EF-equivalent, then the duplicator wins the d-round Ehrenfeucht-Fraïssé game for R and R'.*

The converse implication, i.e., that if the duplicator can win the d-round Ehrenfeucht-Fraïssé game for R and R', then the d-EF-trees for the game played on relational structures R and R' are EF-equivalent, is also true, but we omit further details as we only need the implication given in Theorem 2 in our considerations.

We finish this section with some observations on minimal d-EF-trees in EF-equivalence classes. Let $\mathcal{T}$ be a d-EF-tree. Suppose that an internal node at level $d-1$ has two children, which are leaves, associated with the same labeled structure. Observe that deleting one of them yields an EF-equivalent d-EF-tree. Suppose that we have deleted all such leaves and an internal node at level $d-2$ has two children with their subtrees isomorphic (in the usual sense). Again, deleting one of them (together with its subtree) yields an EF-equivalent d-EF-tree. So, if $\mathcal{T}'$ is a minimal subtree of $\mathcal{T}$ that is EF-equivalent to $\mathcal{T}$ and K is the number of non-isomorphic d-labeled structures, then the degree of nodes at depth $d-1$ does not exceed K, those at depth $d-2$ does not exceed 2^K, those at depth $d-3$ does not exceed 2^{2^K}, etc. We conclude that the size of a minimal subtree of $\mathcal{T}$ that is EF-equivalent to $\mathcal{T}$ is bounded by a function of d and the type of relational structures considered only.

3   FO Model Checking

Using Theorems 1 and 2, we prove the following kernelization result for L-interval graphs.

Theorem 3. *For every finite subset L of reals and every d, there exists an integer K_0 and an algorithm $\mathcal{A}$ with the following properties. The input of $\mathcal{A}$ is an L-representation of an n-vertex L-interval graph G and $\mathcal{A}$ outputs in time $O(n \log n)$ an L-representation of an induced subgraph G' of G such that*

- *every unit interval contains at most K_0 left end points of the intervals corresponding to vertices of G', and*
- *G and G' satisfy the same FO sentences with quantifier rank at most d.*

Proof. We first focus on proving the existence of the number K_0 and the subgraph G' and we postpone the algorithmic considerations to the end of the proof.

As the first step, we show that we can assume that all the left end points are distinct. Choose δ to be the minimum distance between distinct end points of intervals in the representation. Suppose that the intervals are sorted by their left end points (resolving ties arbitrarily). Shifting the i-th interval by $i\delta/2n$, for $i = 1, \ldots, n$, to the right does not change the graph represented by the intervals and all the end points become distinct.

Choose ε to be the minimum positive element of $L^{(2^{d+1})}$. Fix any real a and let $\mathcal{I}$ be the set of all intervals $[x, x+\varepsilon)$ such that $x - a \in L^{(2^{d+1})}$. By the choice of ε, the intervals of $\mathcal{I}$ are disjoint. In addition, the set $\mathcal{I}$ is finite (since L is finite). Let W be the set of vertices w of G such that $\ell(w)$ lies in an interval from $\mathcal{I}$, and for such a vertex w, let $i(w)$ be the left end point of that interval from $\mathcal{I}$. Define a linear order on W such that $w \leq w'$ for w and w' from W iff $\ell(w) - i(w) \leq \ell(w') - i(w')$ and resolve the cases of equality for distinct vertices w and w' arbitrarily.

We view W as a linearly ordered set with elements associated with intervals from $\mathcal{I}$ (specifically the interval with the left end point at $i(w)$) as well as associated with the lengths of their corresponding intervals in the representation of G. Let us establish the following claim.

Claim. There exists a number K depending on only $|\mathcal{I}|$ and d such that if W contains more than K elements associated with the interval $[a, a + \varepsilon)$, then there exists an element $w \in W$ associated with $[a, a + \varepsilon)$ such that the d-EF-trees for the game played on W and $W \setminus \{w\}$ are EF-equivalent.

Indeed, let $\mathcal{T}$ be the d-EF-tree for the game played on W and let $\mathcal{T}'$ a minimal subtree of $\mathcal{T}$ that is EF-equivalent to $\mathcal{T}$. Recall that the size of $\mathcal{T}'$ does not exceed a number K depending on only $|\mathcal{I}|$ and d. If W contains more than K elements associated with $[a, a + \varepsilon)$, then one of them is not associated with edges that are present in $\mathcal{T}'$. We set w to be this element. This finishes the proof of the claim.

Since the d-EF-trees for the game played on W and $W \setminus \{w\}$ are EF-equivalent, the duplicator wins the d-round Ehrenfeucht-Fraïssé game for W and $W \setminus \{w\}$ by Theorems 1 and 2.

We describe a strategy for the duplicator to win the d-round Ehrenfeucht-Fraïssé game for the graphs G and $G \setminus w$. During the game, some intervals from $\mathcal{I}$ will be marked as *altered*. At the beginning, the only altered interval is the interval $[a, a + \varepsilon)$.

The duplicator strategy in the i-th round of the game is the following.

- If the spoiler chooses a vertex w' with $\ell(w')$ in an interval of $\mathcal{I}$ at L-distance at most 2^{d+1-i} from an altered interval, then the duplicator follows its winning strategy for the d-round Ehrenfeucht-Fraïssé game for W and $W \setminus \{w\}$ which gives the vertex to choose in the other graph. In addition, the duplicator marks the interval of $\mathcal{I}$ that contains $\ell(w')$ as altered. Note that the interval of the chosen vertex has its left end point in the same interval of $\mathcal{I}$ as w'.
- Otherwise, the duplicator chooses the same vertex in the other graph. No new intervals are marked as altered.

It remains to argue that the subgraphs of G and $G \setminus w$ obtained in this way are isomorphic. Let $w_1, \ldots, w_d$ be the chosen vertices of G and $w'_1, \ldots, w'_d$ the chosen vertices of G'. For brevity, let us refer to vertices corresponding to the intervals with left end points in the altered intervals as to *altered vertices*. If w_i is not altered, then $w_i = w'_i$. If w_i is altered, then $\ell(w_i)$ and $\ell(w'_i)$ are in the same interval $J \in \mathcal{I}$ and the only intervals that might intersect the intervals corresponding to w_i and w'_i differently are those with left end points in the intervals of $\mathcal{I}$ at L-distance at most two from J. However, if some chosen vertices have their left end points in such intervals, then these intervals must also be altered and these chosen vertices are altered. Since we have followed a winning strategy for the duplicator for W and $W \setminus \{w\}$ when choosing altered vertices, the subgraphs of G and $G \setminus w$ induced by the altered vertices are isomorphic. We conclude that the subgraphs of G and $G \setminus w$ induced by the vertices $w_1, \ldots, w_d$ and $w'_1, \ldots, w'_d$, respectively, are isomorphic. So, the duplicator wins the game.

Let us summarize our findings. If an interval of length ε contains more than K left end points of intervals in the given L-representation of G, then one of the vertices corresponding to these intervals can be removed from G without changing the set of FO sentences with rank at most d that are satisfied by G. So, the statement of the theorem is true with K_0 set to $K \lceil \varepsilon^{-1} \rceil$.

It remains to consider the algorithmic aspects of the theorem. The values of ε and K_0 are determined by L and d. The algorithm sorts the left end points of all the intervals (this requires $O(n \log n)$ time) and for each of these points computes the distance to the left end of the interval that is K positions later in the obtained order. If all these distances are at least ε, then every interval of length at most ε contains at most K_0 left end points of the intervals and the representation is of the desired form.

Otherwise, we choose the smallest of these distances and consider the corresponding interval $[a, b)$, $b - a < \varepsilon$, containing at least K_0 left end points of the intervals from the representation. By the choice of this interval, any interval of length $b - a$ at L-distance at most 2^{d+1} from $[a, b]$ contains at most $K_0 + 1$ left end points of the intervals from the representation. So, the size of the d-EF-tree for the game played on the vertices v with $\ell(v)$ in such intervals is bounded by a function of K_0, d and $|L|$. Since this quantity is independent of the input graph, we can identify in constant time a vertex w with $\ell(w) \in [a, b)$ whose removal from G does not change the set of FO sentences with quantifier rank d satisfied by G.

We then update the order of the left end points and the at most K_0 computed distances affected by removing w, and iterate the whole process. Since at each step we alter at most K_0 distances, using a heap to store the computed distances and choose the smallest of them requires $O(\log n)$ time per vertex removal. So, the running time of the algorithm is bounded by $O(n \log n)$. □

It is possible to think of several strategies to efficiently decide FO properties of L-interval graphs given Theorem 3. We present one of them. Fix an FO sentence Φ with quantifier rank d and apply the algorithm from Theorem 3 to get an L-interval graph and a representation of this graph such that every unit interval contains at most K left end points of the intervals of the representation. After this preprocessing step, every vertex of the new graph has at most $K(|L| + 1) \lceil \max L \rceil$ neighbors. In particular, the maximum degree of the new graph is bounded. The result of Seese [22] asserts that every FO property can be decided in linear time for graphs with bounded maximum degree, and so we conclude:

Theorem 4. *For every finite subset L of reals and every FO sentence Φ, there exists an algorithm running in time $O(n \log n)$ that decides whether an input n-vertex L-interval graph G given by its L-representation satisfies Φ.*

4   Clique-Width of Interval Graphs

Unit interval graphs can have unbounded clique-width [12], but Lozin [17] noted that every proper hereditary subclass of unit interval graphs has bounded clique-width. In particular, the class of $(\{1\}, d)$-interval graphs has bounded clique-width for every $d > 0$. Using Gaifman's theorem, it follows that testing FO properties of unit interval graphs can be performed in linear time if the input graph is given by its $\{1\}$-representation with the left end points of the intervals sorted. We provide an easy extension of this, and outline how it can be used to prove the special case of our main result for FO model checking when L is a finite set of rational numbers (the proof of the lemma is omitted due to space constraints).

Lemma 1. *Let L be a finite set of positive rational numbers. For any $d > 0$, the class of (L, d)-interval graphs has bounded clique-width.*

From Lemma 1 and Gaifman's theorem, one can approach the FO model checking problem on L-interval graphs with L containing rational numbers only as follows. L-interval graphs with radius d are $(L, (2d + 1) \max L)$-interval graphs. By Gaifman's theorem, every FO model checking instance can be reduced to model checking of basic local FO sentences, i.e., to FO model checking on L-interval graphs with bounded radius. Since such graphs have bounded clique-width, the latter can be solved in linear time by [2]. Combining this with the covering technique from [11], which can be adapted to run in linear time in the case of L-interval graphs, we obtain the following.

Corollary 1. *Let L be a finite set of positive rational numbers. The FO model checking problem can be solved in linear time on the class of L-interval graphs if the input graph is given by its L-representation with the left end points of the intervals sorted.*

However, Corollary 1 is just a fortunate special case, since aside of rational lengths one can prove the following.

Lemma 2. *For any irrational $q > 0$ there is d such that the class of $(\{1, q\}, d)$-interval graphs has unbounded clique-width.*

Proof. This proof is in a sense complementary to that of Lemma 1. We may assume $q > 1$ (otherwise, we rescale and consider the set $\{1, 1/q\}$). So, fix $L = \{1, q\}$, $d = q + 3$ and an integer n (to be specified later).

Our task is to construct a $(\{1, q\}, d)$-interval representation of a graph G with large clique-width. Since q is irrational, for every ℓ we can find n such that $L^{(n)} \cap [0, d - q)$ contains more than ℓ points. We actually construct an arbitrarily long sequence $P = (a_1, a_2, \ldots, a_n)$ of such points as follows: $a_1 = 0, a_2 = 1$, and for $i > 2$ set

- $a_i = a_{i-1} + 1$, provided that $|a_{i-2} - a_{i-1}| = 1$ and $a_{i-1} < d - 2$,
- $a_i = a_{i-1} - 1$, provided that $a_{i-2} - a_{i-1} = q$, and
- $a_i = a_{i-1} - q$ otherwise (we call this a_i a q-*element* of P).

Informally, we are "folding" a long sequence with differences from $L^{(n)}$ into a bounded length interval, avoiding as much collisions of points as possible.

Let $\delta > 0$ be such that $n\delta$ is smaller than the smallest number in $L^{(n)} \cap (0, d - q)$. Let us introduce the following shorthand notation: if J is an interval and r a real, then $J + r$ is the interval J shifted by r to the right. Similarly, if $\mathcal{I}$ is a set of intervals, then $\mathcal{I} + r$ is the set of the intervals from $\mathcal{I}$ shifted by r to the right. We define sets of intervals $\mathcal{U}_1 := \{[i\delta, 1 + i\delta) : i = 0, \ldots, n - 1\}$ and $\mathcal{U}_q := \{[i\delta, q + i\delta) : i = 0, \ldots, n - 1\}$. For further reference we say that intervals $[i\delta, 1 + i\delta)$ or $[i\delta, q + i\delta)$ are *at level* i.

For $i = 1, \ldots, n$, we set $\mathcal{W}_i = \mathcal{U}_q + a_i$ if a_i is a q-element of P, and $\mathcal{W}_i = \mathcal{U}_1 + a_i$ otherwise. Then every interval of $\mathcal{W}_i$ is a subinterval of $[0, d)$. Let G be a graph on n^2 vertices represented by the union of the interval sets $\mathcal{W}_1 \cup \mathcal{W}_2 \cup \cdots \cup \mathcal{W}_n$. Let W_i, $i = 1, \ldots, n$, be the vertices represented by $\mathcal{W}_i$. We claim that the clique-width of G exceeds any fixed number $k \in \mathbb{N}$ when n sufficiently large.

Assume, for a contradiction, that the clique-width of G is at most k. We can view the construction of G as a binary tree and conclude a k-labelled subgraph G_1 of G with $\frac{1}{3}n^2 \leq |V(G_1)| \leq \frac{2}{3}n^2$ appeared during the construction of G. However, this implies that vertices of G_1 have at most k different neighborhoods in $G \setminus V(G_1)$. We will show that this is not possible (assuming that n is large).

For $2 \leq i \leq n$, vertices $x \in W_{i-1}$ and $y \in W_i$ are *mates* if they are represented by copies of the same-level intervals from $\mathcal{U}_1$ or $\mathcal{U}_q$ above. Our first observation is that, up to symmetry between $i-1$ and i, $0 \leq |W_{i-1} \cap V(G_1)| - |W_i \cap V(G_1)| \leq k$. Suppose not. Then there exist $k + 1$ vertices in $W_{i-1} \cap V(G_1)$ whose mates are in $W_i \setminus V(G_1)$, and thus certify pairwise distinct neighborhoods of the former ones in $G \setminus V(G_1)$.

A set W_i is *crossing* G_1 if $\emptyset \neq W_i \cap V(G_1) \neq W_i$. The arguments given in the previous paragraph and $\frac{1}{3}n^2 \leq |V(G_1)| \leq \frac{2}{3}n^2$ imply that for any m, if n is large, there exist sets $W_{i_0}, W_{i_0+1}, \ldots, W_{i_0+m}$ in G all crossing G_1. So, we can select an arbitrarily large index set $I \subseteq \{i_0, \ldots, i_0 + m - 1\}$, $|I| = \ell$, such that for each $i \in I$ the element a_{i+1} is to the right of a_i, and that all intervals in $\bigcup_{i \in I} \mathcal{W}_i$ share a common point. In particular, a_i is not a q-element and so both $\mathcal{W}_i$ and $\mathcal{W}_{i+1}$ are shifted copies of $\mathcal{U}_1$. Let $i_1, \ldots, i_\ell$ be the elements of $\mathcal{I}$ ordered according to the (strictly) increasing values of a_i, i.e., $a_{i_1} < \cdots < a_{i_\ell}$.

Finally, for any $j, j' \in \{1, \ldots, \ell\}$ such that $j' > j+1$, we see that each vertex of $W_{i_j} \cap V(G_1)$ cannot have the same neighborhood as any vertex of $W_{i_{j'}} \cap V(G_1)$: this is witnessed by the non-empty set $W_{i_{j+1}+1} \setminus V(G_1)$ (represented to the right of the intervals from $\mathcal{W}_{i_j}$ while intersecting every interval from $\mathcal{W}_{i_{j'}}$). Therefore, the vertices of G_1 have at least $\ell/2 > k$ distinct neighborhoods in $G \setminus V(G_1)$, which contradicts the fact that the clique-width of G is at most k. $\qquad\square$

5   Graph Interpretation in Interval Graphs

A useful tool when solving the model checking problem on a class of structures is the ability to "efficiently translate" an instance of the problem to a different class of structures, for which we already may have an efficient model checking algorithm. To this end we introduce simple FO graph interpretation, which is an instance of the general concept of interpretability of logic theories [21] restricted to simple graphs with vertices represented by singletons.

An FO *graph interpretation* is a pair $\mathcal{I} = (\nu, \mu)$ of FO formulae (with 1 and 2 free variables respectively) where μ is symmetric, i.e., $G \models \mu(x, y) \leftrightarrow \mu(y, x)$ in every graph G. If G is a graph, then $\mathcal{I}(G)$ is the graph defined as follows:

- The vertex set of $\mathcal{I}(G)$ is the set of all vertices v of G such that $G \models \nu(v)$, and
- the edge set of $\mathcal{I}(G)$ is the set of all the pairs $\{u, v\}$ of vertices of G such that $G \models \nu(u) \wedge \nu(v) \wedge \mu(u, v)$.

We say that a class $\mathcal{C}_1$ of graphs has an FO interpretation in a class $\mathcal{C}_2$ if there exists an FO graph interpretation $\mathcal{I}$ such that every graph from $\mathcal{C}_1$ is isomorphic to $\mathcal{I}(G)$ for some $G \in \mathcal{C}_2$.

A proof of the next lemma is omitted due to space constraints.

Lemma 3. *If L is a subset of non-negative reals that is dense in some non-empty open set, then there exists a polynomially bounded simple FO interpretation of the class of all graphs in the class of L-interval graphs.*

Since many FO properties are W[2]-hard for general graphs, we can immediately conclude the following.

Corollary 2. *If L is a subset of non-negative reals that is dense in some non-empty open set, then FO model checking is W[2]-hard on L-interval graphs.*

We now turn our attention to interpretation in unit interval graphs. The price we pay for restricting to a smaller class of interval graphs is the strength of the interpretation language used, namely that of MSO logic. At this point we remark that there exist two commonly used MSO frameworks for graphs; the MSO_1 language which is allowed to quantify over vertices and vertex sets only, and MSO_2 which is in addition allowed to quantify over edges and edge sets. We stay with the former weaker one in this paper.

An MSO_1 graph interpretation is defined in the analogous way to former FO interpretation with the formulas μ and ν being MSO_1 formulas (we omit a proof due to space limitations).

Lemma 4. *There is a polynomially bounded simple MSO_1 interpretation of the class of all graphs into the class of unit interval graphs.*

Again, we can immediately conclude the following.

Corollary 3. MSO_1 *model checking is W[2]-hard on unit interval graphs.*

This corollary is rather tight since the aforementioned result of Lozin [17] claims that every proper hereditary subclass of unit interval graphs has bounded clique-width, and hence MSO_1 model checking on this class is in linear time [2].

Lastly, we remark that Fellows et al [9] have shown that testing FO properties on unit two-interval graphs (i.e., such that each vertex corresponds to a pair of intervals, each on a distinct line) is W[1]-hard.

References

1. Courcelle, B.: The monadic second order logic of graphs I: Recognizable sets of finite graphs. Inform. and Comput. 85, 12–75 (1990)
2. Courcelle, B., Makowsky, J.A., Rotics, U.: Linear time solvable optimization problems on graphs of bounded clique-width. Theory Comput. Syst. 33, 125–150 (2000)
3. Courcelle, B., Olariu, S.: Upper bounds to the clique width of graphs. Discrete Appl. Math. 101, 77–114 (2000)
4. Dawar, A., Grohe, M., Kreutzer, S.: Locally excluding a minor. In: LICS 2007, pp. 270–279. IEEE Computer Society (2007)
5. Dawar, A., Kreutzer, S.: Parameterized complexity of first-order logic. ECCC TR09-131 (2009)
6. Downey, R., Fellows, M.: Parameterized complexity. Monographs in Computer Science. Springer (1999)
7. Dvořák, Z., Král', D., Thomas, R.: Deciding first-order properties for sparse graphs. In: FOCS 2010, pp. 133–142. IEEE Computer Society (2010)
8. Ehrenfeucht, A.: An application of games to the completeness problem for formalized theories. Fund. Math. 49, 129–141 (1961)
9. Fellows, M., Hermelin, D., Rosamond, F., Vialette, S.: On the parameterized complexity of multiple-interval graph problems. Theoret. Comput. Sci. 410, 53–61 (2009)
10. Fraïssé, R.: Sur quelques classifications des systèmes de relations. Université d'Alger, Publications Scientifiques, Série A 1, 35–182 (1954)
11. Frick, M., Grohe, M.: Deciding first-order properties of locally tree-decomposable structures. J. ACM 48, 1184–1206 (2001)
12. Golumbic, M., Rotics, U.: On the clique-width of some perfect graph classes. Int. J. Found. Comput. Sci. 11, 423–443 (2000)
13. Grohe, M., Kreutzer, S.: Methods for algorithmic meta theorems. In: Model Theoretic Methods in Finite Combinatorics Contemporary Mathematics, pp. 181–206. AMS (2011)
14. Kneis, J., Langer, A., Rossmanith, P.: Courcelle's theorem — a game-theoretic approach. Discrete Optimization 8(4), 568–594 (2011)
15. Kreutzer, S.: Algorithmic meta-theorems. ECCC TR09-147 (2009)
16. Langer, A., Reidl, F., Rossmanith, P., Sikdar, S.: Evaluation of an mso-solver. In: ALENEX 2012, pp. 55–63. SIAM / Omnipress (2012)
17. Lozin, V.: From tree-width to clique-width: Excluding a unit interval graph. In: Hong, S.-H., Nagamochi, H., Fukunaga, T. (eds.) ISAAC 2008. LNCS, vol. 5369, pp. 871–882. Springer, Heidelberg (2008)
18. Nešetřil, J., Ossona de Mendez, P.: Grad and classes with bounded expansion I. Decompositions. European J. Combin. 29, 760–776 (2008)

19. Nešetřil, J., Ossona de Mendez, P.: Grad and classes with bounded expansion II. Algorithmic aspects. European J. Combin. 29, 777–791 (2008)
20. Nešetřil, J., Ossona de Mendez, P.: Grad and classes with bounded expansion III. Restricted graph homomorphism dualities. European J. Combin. 29, 1012–1024 (2008)
21. Rabin, M.O.: A simple method for undecidability proofs and some applications. In: Logic, Methodology and Philosophy of Sciences, vol. 1, pp. 58–68. North-Holland (1964)
22. Seese, D.: Linear time computable problems and first-order descriptions. Math. Structures Comput. Sci. 6, 505–526 (1996)

Strategy Composition in Compositional Games

Marcus Gelderie[*]

RWTH Aachen, Lehrstuhl für Informatik 7,
Logic and Theory of Discrete Systems,
D-52056 Aachen
gelderie@automata.rwth-aachen.de

Abstract. When studying games played on finite arenas, the arena is given explicitly, hiding the underlying structure of the arena. We study games where the global arena is a product of several smaller, constituent arenas. We investigate how these "global games" can be solved by playing "component games" on the constituent arenas. To this end, we introduce two kinds of products of arenas. Moreover, we define a suitable notion of strategy composition and show how, for the first notion of product, winning strategies in reachability games can be composed from winning strategies in games on the constituent arenas. For the second kind of product, the complexity of solving the global game shows that a general composition theorem is equivalent to proving PSPACE = EXPTIME.

1 Introduction

Infinite games with ω-regular winning conditions have been studied extensively over the past decades [1–5]. This research has been most successful in establishing results about solving ω-regular games on an "abstract" arena. A fundamental open problem, which is of intrinsic interest in the area of automated synthesis, is to exploit the compositional structure of an arena to derive a compositional representation of a winning strategy. For instance, if an arena is viewed as a product of several smaller transition systems, is it possible to lift this structure to strategies in games on this arena?

The classical results on ω-regular games depend on the representation of a winning strategy by an automaton. None of these results allows to transfer a given composition of an arena into a composition of automata in such a way that a winning strategy is implemented. Since there is no lack of methods for composing automata (for example, the cascade product), it rather seems that automata are too "coarse" a tool to capture this compositional structure.

We study the compositional nature of winning strategies in games played on products of arenas. Products of arenas can be defined in a variety of ways (see e.g. [6]). As a first step towards a compositional approach to synthesis, we restrict ourselves to two notions, *parallel* and *synchronized* product. Our notion of strategy composition relies on a Turing machine based model for strategy

---

[*] Supported by DFG research training group 1298, "Algorithmic Synthesis of Reactive and Discrete-Continuous Systems" (AlgoSyn).

F.V. Fomin et al. (Eds.): ICALP 2013, Part II, LNCS 7966, pp. 263–274, 2013.

representation, called a *strategy machine*. Using this model, we show how winning strategies in reachability games can be composed from winning strategies in games over the constituent factors of the overall product arena. We study the complexity of such a composition: its size, its runtime and the computational complexity of finding it. This entails a study of the complexity of deciding who wins the game.

Compositionality in an arena is closely linked to a succinct representation of that arena. Likewise, composing a winning strategy from smaller winning strategies may yield a much smaller representation for that strategy. Finding succinct transition systems from specifications was studied in [7]. The authors consider the problem of finding a succinct representation of a model for a given CTL formula. They show that such succinct models are unlikely to exist in general.

Transition systems which are obtained by "multiplying" smaller transition systems have also been studied in [8]. The authors consider the problem of model checking such systems. They show the model checking problem for such systems to be of high complexity for various notions of behavioral specification and model checking problem.

Strategy machines were introduced in [9] (the model has been studied in a different setting in [10]). They allow for a broader range of criteria by which to compare strategies. Being based on Turing machines, strategy machines allow, for instance, to investigate the "runtime" of a strategy and to quantify and compare "dynamic" memory (the tape content) and "static memory" (the control states).

The complexity of deciding the winner of a game has been subject to extensive research in the case of games on an abstract arena [11–13]. These complexity results depend on the size of the abstract arena. We investigate the complexity of deciding the winner based on a composite representation of the arena.

Our paper is structured as follows: We first define two notions of product of arenas, the *parallel product* and the *synchronized product*. The games we study are played on arenas that are composed from smaller arenas using these two operators. Having defined the notion of arena composition, we define strategy machines and use them to introduce our notion of strategy composition. Subsequently, we study reachability games. We do this separately for the parallel product and the synchronized product. To this end, we first introduce two natural ways of defining a reachability condition on a composite arena, *local* and *synchronized reachability*. For the parallel product we obtain a compositionality theorem for both local and synchronized reachability. For the synchronized product we show that deciding the game is EXPTIME complete. From this we deduce that finding a general composition theorem is equivalent to showing EXPTIME = PSPACE.

2   Games on Composite Arenas

An *arena* is a directed, bipartite graph $\mathcal{A} = (V, E)$. The partition of $\mathcal{A}$ is $V = V^{(0)} \uplus V^{(1)}$. Given $v \in V$ let $vE = \{v' \in V \mid (v, v') \in E\}$. We define two operators on arenas: the *parallel product* and the *synchronized product*. Let $\mathbb{B} = \{0, 1\}$.

Definition 1 (Parallel Product). *Consider arenas* $\mathcal{A}_1, \ldots, \mathcal{A}_k$ *with* $\mathcal{A}_i = (V_i, E_i)$. *The* parallel product $\mathcal{A}_1 \parallel \cdots \parallel \mathcal{A}_k = (V, E)$ *is the arena given by*

- $V = \mathbb{B} \times \prod_{i=1}^{k} V_i$
- $E = \{((\sigma, v), (1 - \sigma, v')) \mid \exists i \colon v_i \in V_i^{(\sigma)} \wedge (v_i, v_i') \in E_i \wedge \forall j \neq i \colon v_j = v_j'\}$
- $V^{(\sigma)} = \{(b, v) \in V \mid b = \sigma\}$

Note that the parallel product again gives a bipartite arena. Note furthermore that, given a vertex $(\sigma, v) \in \mathbb{B} \times \prod_i V_i$, the number of vertices $v_i \in V_i^{(0)}$ alternatingly increases and decreases by one along all paths starting in (σ, v). The number of components player 0 controls in each of his moves is given by:

$$\mathrm{rank}_0(\sigma, v) = |\{i \mid v_i \in V_i^{(0)}\}| + \sigma$$

Player 1 controls $\mathrm{rank}_1(\sigma, v) = k - \mathrm{rank}_0(\sigma, v) + 1$ components during his moves. For every $p \in \mathbb{B}$ we have $\mathrm{rank}_p(\sigma', v') = \mathrm{rank}_p(\sigma, v)$ for all (σ', v') reachable from (σ, v). If (σ, v) is clear from context, we thus simply write rank_p.

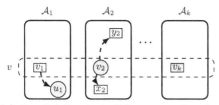 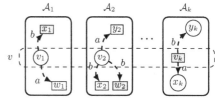

(a) Parallel Product: Edges are taken locally. The square player may move in, e.g., $\mathcal{A}_1$ but not in $\mathcal{A}_2$.

(b) Synchronized Product: Where transitions permit it, edges are taken globally. The circle player may choose transitions in $\mathcal{A}_1$ and $\mathcal{A}_2$. $\mathcal{A}_k$ is not affected.

Fig. 1. A vertex v in the parallel and synchronized product of (labeled) arenas $\mathcal{A}_1, \ldots, \mathcal{A}_k$. The shaded vertices define a possible successor state of v.

To define the synchronized product we use *labeled arenas*. A labeled arena is a triple $\mathcal{A} = (V, \Delta, \Sigma)$ with $V = V^{(0)} \uplus V^{(1)}$ and $\Delta \subseteq \bigcup_{\sigma \in \mathbb{B}} V^{(\sigma)} \times \Sigma \times V^{(1-\sigma)}$ for some finite set Σ of *letters*.

Definition 2 (Synchronized Product). *Consider labeled arenas* $\mathcal{A}_1, \ldots, \mathcal{A}_k$ *with* $\mathcal{A}_i = (V_i, \Delta_i, \Sigma_i)$. *The* synchronized product $\mathcal{A}_1 \otimes \cdots \otimes \mathcal{A}_k = (V, \Sigma, \Delta)$ *is given by*

- $V = \mathbb{B} \times \prod_{i=1}^{k} V_i$
- $\Sigma = \bigcup_{i=1}^{k} \Sigma_i$
- $((\sigma, v), a, (1 - \sigma, v')) \in \Delta$ *iff for all* i, *whenever* $a \in \Sigma_i$ *and* $v_i \in V_i^{(\sigma)}$, *then also* $(v_i, a, v_i') \in \Delta_i$ *and* $v_i = v_i'$ *otherwise*.

Remark 1. Neither the parallel product nor the synchronized product are associative in general. This is due to the fact that we absorb the information about whose turn it is into the arena. We do so for technical reasons. It is nonessential for the results.

In this paper we study ω-regular games. We assume the reader is familiar with the elementary theory of ω-regular games. For an introduction see [4, 5]. In the following, we recall some terminology. A *game* is a tuple $\mathbf{G} = (\mathcal{A}, W, v_0) = (\mathcal{A}, W)$ consisting of an arena $\mathcal{A} = (V, E)$ and a *winning condition* $W \subseteq V^\omega$ and an *initial vertex* v_0. We always assume that there is a designated initial vertex, even if we do not always list it explicitly. $\mathbf{G}$ is ω-regular if W is ω-regular. We denote the players by player 0 and player 1. A *play* in $\mathbf{G}$ is an infinite path $\pi = v_0 v_1 v_2 \cdots$ through $\mathcal{A}$, starting from v_0. On nodes in $V^{(0)}$ player 0 chooses the next vertex. Otherwise, player 1 chooses. The play is won by player 0 if $\pi \in W$. We denote the *winning set* of player σ by $\mathcal{W}^{(\sigma)} = \mathcal{W}^{(\sigma)}(\mathbf{G})$. The *attractor* for player p on a set F is denoted by $\mathrm{Attr}_p^{\mathcal{A}}(F)$ and defined as usual. It is the set of vertices from which p can enforce a visit to F.

To study games on composite arenas, we require some additional notation. Consider a game $\mathbf{G} = (\mathcal{A}, W)$ on a *composite arena* $\mathcal{A} = \mathcal{A}_1 * \cdots * \mathcal{A}_k$, with $* \in \{\|, \otimes\}$. We call $\mathcal{A}_1, \ldots, \mathcal{A}_k$ the *constituent* arenas of $\mathcal{A}$. A game $\mathbf{G}_i = (\mathcal{A}_i, W_i)$ for some $W_i \subseteq V_i^\omega$ is called a *component game*.

The winning condition W is necessarily given by means of some finite representation. In this paper we consider mainly reachability conditions, which are determined by a set $F \subseteq V$. A play π satisfies the reachability condition F if $\pi(i) \in F$ for some $i \in \mathbb{N} = \{0, 1, 2, \ldots\}$.

It is sometimes convenient to specify properties on a path in some logic. In this paper we use LTL to express temporal properties on paths. We again assume the reader is familiar with LTL (see [4, 5] for an introduction). We write $\psi \cup \phi$ for the strict until (ϕ is true eventually, and, until then, ψ holds).

3   Strategy Machines and Strategy Composition

Classically, strategies are represented using (usually finite) automata. Automata are a state space view on a computational system. They abstract away from implementation details. This comes at the price of loosing information about important implementation aspects, such as runtime and space usage.

The abstract view of automata is sometimes coarser than required. To augment this view, in [9] *strategy machines* were introduced. Strategy machines are Turing machines with three designated tapes, the IO-tape, the computation tape and the memory tape. The semantics of a strategy machine can intuitively be described as follows. A vertex (encoded in binary) appears on the IO-tape. The strategy machine inspects the content of its memory tape and computes a new vertex (using all three tapes). The content of the memory tape is updated and the computation tape is cleared. The new vertex is written on the output tape and the process repeats. We now recall the definition from [9]. Let $\hat{\mathbb{B}} = \mathbb{B} \cup \{\#\}$.

Definition 3 (Strategy Machine). *A strategy machine is a deterministic 3-tape Turing machine* $\mathcal{M} = (Q, \mathbb{B}, \hat{\mathbb{B}}, q_I, q_O, \delta)$ *with*

- *a finite set Q of states*
- *tape alphabet $\hat{\mathbb{B}}$ and input alphabet $\mathbb{B}$*
- *two designated states, q_I, the input state and q_O, the output state*
- *a designated IO-tape t_{IO}*
- *a designated memory tape t_M*
- *a designated computation tape t_C*

The partial transition function $\delta\colon Q \times \hat{\mathbb{B}}^3 \dashrightarrow Q \times \hat{\mathbb{B}}^3 \times \{-1, 0, 1\}^3$ *satisfies*

- $\delta(q, b) \neq (q_I, b', d)$ *for all $q \in Q$, $b, b' \in \hat{\mathbb{B}}^3$ and $d \in \{-1, 0, 1\}^3$.*
- $\delta(q_O, b)$ *is undefined for all $b \in \hat{\mathbb{B}}^3$.*

Since the tape and input alphabets are always the same, we usually omit them in the list of components of a strategy machine.

We sketch the semantics of a strategy machine (a formal definition can be found in [9]). Configurations are defined as usual. An *iteration of $\mathcal{M}$* is a sequence of configurations beginning with an *initial configuration* (with state q_I) and a *terminal configuration* (with state q_O). By definition of δ, q_I and q_O appear exactly at the beginning and at the end of an iteration. The iteration beginning in configuration c is unique (if it exists) and depends only on the input $t_{IO}(c)$ on the IO-tape and on the content $t_M(c)$ of the memory tape. We write c' for the unique terminal configuration reachable from c and we write (c, c') for the entire iteration. Its *length*, the number of computation steps, is denoted by $L(c, c')$.

Strategy machines are intended to implement functions on sequences of input. Let $\pi \in (\mathbb{B}^*)^\infty = (\mathbb{B}^*)^* \cup (\mathbb{B}^*)^\omega$, i.e. $\pi(i) \in \mathbb{B}^*$ for all $i \in \mathrm{dom}(\pi)$. Let $c_{\pi,0} = (q_I, \pi(0), \varepsilon, \varepsilon, 0, 0, 0)$. We define $c_{\pi,i+1}$ to be the configuration which inherits the memory tape content from the terminal configuration of the i-th iteration and has $\pi(i+1)$ as input on the IO-tape. Formally, $c_{\pi,i+1} = (q_I, \pi(i+1), t_M(c'_{\pi,i}), \varepsilon, 0, 0, 0)$. We also say that $(c_{\pi,i}, c'_{\pi,i})$ and $(c_{\pi,i+1}, c'_{\pi,i+1})$ are *compatible*. An iteration is *admissible* if it is of the form $(c_{\pi,i}, c'_{\pi,i})$ for some $\pi \in (\mathbb{B}^*)^\omega$. A sequence of iterations $(c_0, c'_0)(c_1, c'_1) \cdots$ which are compatible is called a *computation of $\mathcal{M}$*. Define $f_{\mathcal{M}}(\pi) = t_{IO}(c'_{\pi,0}) \cdot t_{IO}(c'_{\pi,1}) \cdots$. $f_{\mathcal{M}}$ maps strings from $(\mathbb{B}^*)^*$ to $(\mathbb{B}^*)^*$ (where, in our setting, elements from $\mathbb{B}^*$ are encoded vertices). We say $\mathcal{M}$ *implements* $f_{\mathcal{M}}$. We sometimes identify $\mathcal{M}$ with $f_{\mathcal{M}}$ and say, for instance, that $\mathcal{M}$ is a winning strategy.

One of the benefits of strategy machines is the complexity measures they offer to evaluate a strategy. We define the *latency* $T(\mathcal{M})$ and the *space requirement* $S(\mathcal{M})$ of a strategy machine $\mathcal{M}$ as follows:

$$T(\mathcal{M}) = \sup_{\pi \in (\mathbb{B}^*)^\omega} \sup_{n \in \mathbb{N}} L(c_{\pi,n}, c'_{\pi,n})$$

$$S(\mathcal{M}) = \sup_{\pi \in (\mathbb{B}^*)^\omega} \sup_{n \in \mathbb{N}} |t_M(c_{\pi,n})|$$

Finally, the *size* of $\mathcal{M}$ is the number $\|\mathcal{M}\| = |Q|$ of its control states.

In modular programming subroutines are a central concept. Let us formalize this notion. An n-*template* is a strategy machine $\mathcal{M} = (Q_M, q_{M,I}, q_{M,O}, \delta_M)$ with $2n$ distinguished states $\text{sub}_1, \text{ret}_1, \ldots, \text{sub}_n, \text{ret}_n$ such that $\delta_M(\text{sub}_i, b)$ is undefined for all $1 \leq i \leq n$ and all $b \in \hat{\mathbb{B}}^3$. Let $\mathcal{M}$ be an n-template and let $\mathcal{S}_i = (Q_i, q_{i,I}, q_{i,O}, \delta_i)$, $1 \leq i \leq n$, be strategy machines. Define the strategy machine $\mathcal{M}[\mathcal{S}_1, \ldots, \mathcal{S}_n] = (Q_M \uplus \biguplus_i Q_i, q_{M,I}, q_{M,O}, \delta)$ by $\delta(\text{sub}_i, b) = \delta_i(q_{i,I}, b)$ and $\delta(q_{i,O}, b) = (\text{ret}_i, b, 0, 0, 0)$. In all other cases δ coincides with δ_M or δ_i, whenever this makes sense. Such a machine is called a *composition* of $\mathcal{S}_1, \ldots, \mathcal{S}_k$.

Definition 4. *Let $* \in \{\|, \otimes\}$ and let $\mathbf{G} = (\mathcal{A}_1 * \cdots * \mathcal{A}_k, W)$.*

1. *Let $\mathbf{G}_1, \ldots, \mathbf{G}_k$ be component games with winning strategies $\mathcal{S}_1, \ldots, \mathcal{S}_k$ for player 0. A k-template $\mathcal{M}$ such that $\mathcal{M}[\mathcal{S}_1, \ldots, \mathcal{S}_k]$ is a winning strategy in $\mathbf{G}$ is called a* winning composition *of $\mathcal{S}_1, \ldots, \mathcal{S}_k$. If $\mathcal{M}$ is a winning composition of any choice $\mathcal{S}_1, \ldots, \mathcal{S}_k$ of winning strategies for player 0 in $\mathbf{G}_1, \ldots, \mathbf{G}_k$, it is called a* winning composition *of $\mathbf{G}_1, \ldots, \mathbf{G}_k$.*

2. *A class Λ of games on composite arenas is said to* admit polynomial compositions, *if for every $\mathbf{G} = (\mathcal{A}_1 * \cdots * \mathcal{A}_k, W) \in \Lambda$ there exist component games $\mathbf{G}_1, \ldots, \mathbf{G}_k$ and a polynomial sized winning composition $\mathcal{M}$ of $\mathbf{G}_1, \ldots, \mathbf{G}_k$ such that for some choice $\mathcal{S}_1, \ldots, \mathcal{S}_k$ of component winning strategies $T(\mathcal{M}[\mathcal{S}_1, \ldots, \mathcal{S}_k]) \in \text{poly}(\|\mathbf{G}\|)$. Such a tuple $\mathcal{M}, \mathcal{S}_1, \ldots, \mathcal{S}_k$ is called a* polynomial composition.

Part of the appeal of polynomial compositions is their efficiency: By definition, a class of games admitting polynomial compositions enables us to find strategies with a polynomial latency (and thus a polynomial space requirement) which, depending on $\mathbf{G}_1, \ldots, \mathbf{G}_k$, have polynomial size. Note that for positionally determined component games this is always the case. The converse holds for trivial reasons. Any strategy machine $\mathcal{M}$ of polynomial size implementing a winning strategy with latency bounded polynomially can trivially be seen as a polynomial composition of any choice of machines $\mathcal{S}_1, \ldots, \mathcal{S}_k$.

We elaborate a bit on the restrictions we impose in the above definition. The requirement that all strategies in component games are interchangeable is to ensure that we compose general component games, not specific choices of strategies. In the definition of polynomial composition, the restriction on the size is to avoid templates which never call their subroutines and instead implement the entire global winning strategy on their own. Likewise, the restriction on the latency is to avoid enabling too powerful computations during the course of a single iteration (such as, for instance, solving the entire game every turn).

4   Games on Parallel Products

In this section we study reachability games over parallel products of arenas. If the arena is given as a composition of smaller arenas, it is natural to also study several compositional ways of specifying the reachability condition. Let $F_i \subseteq V_i$. We study the following formalisms:

1. *local reachability*, where $F = \mathcal{F}_{\mathrm{loc}}(F_1, \ldots, F_k)$ is the set of all $v \in \prod_i V_i$ with $v_i \in F_i$ for some i
2. *synchronized reachability*, where $F = \mathcal{F}_{\mathrm{sync}}(F_1, \ldots, F_k)$ is the set of all $v \in \prod_i V_i$ with $v_i \in F_i$ for every i

If $F = \mathcal{F}(F_1, \ldots, F_k)$, $\mathcal{F} \in \{\mathcal{F}_{\mathrm{loc}}, \mathcal{F}_{\mathrm{sync}}\}$, the set player 0 has to reach is $\mathbb{B} \times F$, i.e. the $\mathbb{B}$-component does not influence the outcome of the game.

Remark 2. One might consider *asynchronous* reachability, where all components F_i must be reached, but not necessarily at the same time. We omit this condition here, because it is not expressible as a reachability condition on the composite arena.

Theorem 1. *Games of the form* $\mathbf{G} = (\mathcal{A}_1 \parallel \cdots \parallel \mathcal{A}_k, \mathcal{F}_{\mathrm{loc}}(F_1, \ldots, F_k))$ *admit polynomial compositions where the component games in def. 4 can be chosen as reachability games. In particular, component positional strategies suffice. Moreover, a polynomial composition can be computed in polynomial time.*

We omit the full proof due to space constraints. However, the idea is to show that deciding the winning set can be done by deciding conditions on the components: Player 0 wins from (σ_0, v_0) iff one of the following two applies

1. There exists i with $v_{i,0} \in V_i^{(0)}$ and $v_{i,0} E_i \cap F_i \neq \emptyset$.
2. $|\{i \mid v_{i,0} \in \mathrm{Attr}_0^{\mathcal{A}_i}(F_i)\}| \geq \mathrm{rank}_1$

The proof of this characterization gives component strategies for both players, which can be composed to a winning strategy for the respective player in $\mathbf{G}$.

Next, we consider synchronized reachability. We have:

Theorem 2. *Games of the form* $\mathbf{G} = (\mathcal{A}_1 \parallel \cdots \parallel \mathcal{A}_k, \mathcal{F}_{\mathrm{sync}}(F_1, \ldots, F_k))$ *admit polynomial compositions. The component games in def. 4 can be chosen as positionally determined games. A polynomial composition can be computed in polynomial time.*

The proof idea is to characterize the winning set by considering component games. The proof of the characterization gives polynomial compositions for both players. The characterization is more involved than in thm. 1. We split the problem into subcases. For space reasons, we only state the characterization for the case $\mathrm{rank}_1 = 1$ and the case where both $\mathrm{rank}_1 > 1$ and $\mathrm{rank}_0 > 1$ and $\sigma_0 = 1$.

Lemma 1. *In thm. 2, let* (σ_0, v_0) *be such that* $\mathrm{rank}_1 = 1$. *Then player 0 wins from* (σ_0, v_0) *iff all of the following hold:*

1. *for all i we have* $v_{i,0} \in \mathrm{Attr}_0^{\mathcal{A}_i}(F_i)$
2. $|\{i \mid v_{i,0} \in \mathrm{Attr}_0^{\mathcal{A}_i}(F_i \cap V_i^{(0)})\}| \geq k - 1$
3. *if $\sigma_0 = 1$ and $v_{j,0} \in F_j \cap V_j^{(1)}$ for some j, then $v_{i,0} \in F_i$ for all $i \neq j$ or* $v_{j,0} E_j \subseteq \mathrm{Attr}_0^{\mathcal{A}_j}(F_j)$

A polynomial winning composition for player 0 (resp. player 1) with respect to positional strategies in component reachability (resp. safety) games can be computed in polynomial time.

In lem. 1 the component games are also reachability games (just like in thm. 1). In the next lemma, where $\text{rank}_0 > 1$ and $\text{rank}_1 > 1$, this is no longer the case. Instead, we use games with a temporal winning condition of low complexity. We call a game $\mathbf{G} = (\mathcal{A}, W)$ a *reachability game with safety constraint* if W is given by an LTL-formula $\varphi = S \, \mathsf{U} \, F$ with sets $S, F \subseteq V_{\mathcal{A}}$. We have:

Proposition 1. *Every reachability game with safety constraint is determined with positional strategies. Moreover, a winning strategy for both players can be computed in time $\mathcal{O}(|V_{\mathcal{A}}| + |E_{\mathcal{A}}|)$, if it exists.*

For simplicity, we exclude the trivial case where the initial position is already in $\mathcal{F}_{\text{sync}}(F_1, \ldots, F_k)$. We consider the case where player 1 moves first.

Lemma 2. *In thm. 2, let $\text{rank}_0 > 1$ and $\text{rank}_1 > 1$. Then player 0 wins from $(1, v_0) \notin \mathcal{F}_{\text{sync}}(F_1, \ldots, F_k)$ iff all of the following constraints are met:*

1. $v_{i,0} \in V_i^{(1)} \implies (v_{i,0} \in F_i \wedge \forall v_i' \in v_{i,0}E_i : v_i' \in F_i \vee v_i'E_i \cap F_i \neq \emptyset)$
2. $v_{i,0} \in V_i^{(0)} \setminus F_i \implies v_{i,0}E_i \cap F_i \neq \emptyset$
3. $|\{i \mid v_{i,0} \in \mathcal{W}^{(0)}(\mathcal{A}_i, (V_i^{(0)} \cup F_i) \, \mathsf{U} (V_i^{(0)} \cap F_i))\}| \geq k - 1$

A polynomial winning composition for player 0 (resp. player 1) with respect to positional strategies in component reachability games with safety constraint can be computed in polynomial time.

We see that in the case of synchronized reachability we have to use a stronger notion of winning condition in the component games than we did in the global game. What kind of component games we need depends on the initial position.

The polynomial composition in thm. 2 relies on positional winning strategies in component games. This is closely tied to complexity in the following way:

Decision Problem (PARALLEL-REACH[$\mathcal{F}$]).
Input: Arenas $\mathcal{A}_1, \ldots, \mathcal{A}_k$, a vertex $s = (\sigma_0, v) \in \mathbb{B} \times \prod_{i=1}^{k} V_i$ and sets $F_i \subseteq V_i$
Decide: $s \in \mathcal{W}^{(0)}$ in the game $\mathbf{G} = (\mathcal{A}_1 \parallel \cdots \parallel \mathcal{A}_k, \mathcal{F}(F_1, \ldots, F_k))$?

Corollary 1. *Let $\mathcal{F} \in \{\mathcal{F}_{\text{loc}}, \mathcal{F}_{\text{sync}}\}$. PARALLEL-REACH[$\mathcal{F}$] is PTIME-complete.*

We also have the following corollary of thm. 1 and thm. 2:

Corollary 2. *Let $\mathbf{G} = (\mathcal{A}_1 \parallel \cdots \parallel \mathcal{A}_k, \mathcal{F}(F_1, \ldots, F_k))$, where $\mathcal{F} \in \{\mathcal{F}_{\text{loc}}, \mathcal{F}_{\text{sync}}\}$, be a reachability game. There exists a winning strategy machine $\mathcal{M}$ for player 0 of size $\|\mathcal{M}\| \in \text{poly}\left(\sum_{i=1}^{k} \|\mathcal{A}_i\|\right)$ and latency $T(\mathcal{M}) \in \text{poly}\left(\sum_{i=1}^{k}(\|\mathcal{A}_i\|)\right)$.*

5   Games on Synchronized Products

In this section we investigate arenas obtained via the synchronized product. Here the situation is quite different. We first consider the complexity of deciding the winning region in those games:

Decision Problem (SYNC-REACH[$\mathcal{F}$]).

Input: Arenas $\mathcal{A}_1, \ldots, \mathcal{A}_k$, a vertex $s = (\sigma_0, v) \in \mathbb{B} \times \prod_{i=1}^{k} V_i$ and sets $F_i \subseteq V_i$
Decide: $s \in \mathcal{W}^{(0)}$ in the game $\mathbf{G} = (\mathcal{A}_1 \otimes \cdots \otimes \mathcal{A}_k, \mathcal{F}(F_1, \ldots, F_k))$?

Theorem 3. *Let* $\mathcal{F} \in \{\mathcal{F}_{\mathrm{loc}}, \mathcal{F}_{\mathrm{sync}}\}$. SYNC-REACH[$\mathcal{F}$] *is* EXPTIME-*complete.*

Proof. We only show the claim for $\mathcal{F} = \mathcal{F}_{\mathrm{loc}}$. The proof for $\mathcal{F} = \mathcal{F}_{\mathrm{sync}}$ is an adaption of this proof.

Membership in EXPTIME is trivial (for instance, using a classical attractor on an in-memory explicit graph). We therefore focus on hardness.

The main idea is to reduce the acceptance problem of an APSPACE-Turing-machine. Since it is APSPACE, we have only polynomially many tape cells in use. We introduce a labeled arena $\mathcal{A}_i$ for each cell plus one additional labeled arena $\mathcal{A}_H$ storing both the state of the machine and the head position. Letters in $\mathcal{A}_i$ are indexed by i so that transitions may target a specific tape cell. The tape cell is updated by having players choose a transition label aligning the head position in $\mathcal{A}_H$ with the index of of the cell to be updated. Since the transition in $\mathcal{A}_H$ cannot observe the state of $\mathcal{A}_i$, the players may cheat with respect to the content of the tape cells. The construction below introduces a mechanism to enable players to challenge an opponent's cheating moves and win.

Let $\mathcal{M} = (Q_\exists, Q_\forall, \hat{\Gamma}, \Gamma, q_0, \Delta_{\mathcal{M}}, \{q_F\})$ be a bipartite APSPACE-machine. We suppose $\hat{\Gamma}$ is the input alphabet and $\Gamma \supseteq \hat{\Gamma}$ is the tape alphabet. Let $Q = Q_\exists \uplus Q_\forall$. Suppose $w \in \hat{\Gamma}^*$ is an input to $\mathcal{M}$. We assume that $\mathcal{M}$ accepts with exactly one final state $q_F \in Q_\exists$ and that q_F is never visited before termination. We may also assume that every configuration has at least one outgoing transition. Suppose p is a polynomial bounding the space of $\mathcal{M}$.

We define $p(|w|) = n$ automata $\mathcal{A}_1, \ldots, \mathcal{A}_n$ as follows. For every $i = 1, \ldots, n$, let $\Gamma_i = \Gamma \times \{i\}$. Write γ_i for $(\gamma, i) \in \Gamma_i$. All n automata have the same alphabet $\Sigma = \bigcup_{i=1}^{n} \{\mathrm{veto}_i\} \cup \bigcup_{i=1}^{n} \Gamma_i^2$. We define $\mathcal{A}_i = (A_i, \Sigma, \delta_i)$ with $A_i = (\Gamma \cup \{\bot_0, \bot_1\}) \times \mathbb{B}$ where $\delta_i \colon A_i \times \Sigma \dashrightarrow A_i$ is as follows:

$$\delta_i((\hat{\gamma}, \sigma), (\gamma_j, \gamma_j')) = \begin{cases} (\gamma', 1 - \sigma) & \text{if } i = j \text{ and } \hat{\gamma} = \gamma \\ (\bot_\sigma, 1 - \sigma) & \text{if } i = j \text{ but } \hat{\gamma} \neq \gamma \\ (\hat{\gamma}, 1 - \sigma) & \text{if } i \neq j \end{cases}$$

for all $j \in \{1, \ldots, k\}$, all $\hat{\gamma} \in \Gamma$, $\gamma_j, \gamma_j' \in \Gamma_j$ and all $\sigma \in \mathbb{B}$. We also define

$$\delta_i((\bot_p, \sigma), (\gamma_j, \gamma_j')) = (\bot_p, 1 - \sigma)$$

for all $j \in \{1, \ldots, k\}$, all $p, \sigma \in \mathbb{B}$ and all $\gamma_j, \gamma_j' \in \Gamma_j$.

Furthermore, a transition labeled with veto_i is defined on all player 1 states:

$$\delta_i((s, 1), \mathrm{veto}_j) = \begin{cases} (\bot_1, 0) & \text{if } i = j \text{ and } s = \gamma_i \in \Gamma_i \\ (s, 0) & \text{if } i \neq j \text{ or } s = \bot_p, \ p \in \mathbb{B} \end{cases}$$

In particular, player 0 can never play veto_i on his components. The partition of A_i into player 0 and player 1 states is given by $A_i^{(\sigma)} = \{(s, \sigma) \mid s \in \Gamma \cup \{\bot_0, \bot_1\}\}$.

Next, we define $\mathcal{A}_H$ with states $A_H = (Q \times \{1, \ldots, n\}) \uplus \{C, (\top, 0), (\bot, 0)\}$, where $A_H^{(0)} = \{(q, h) \mid q \in Q_\exists\} \cup \{(\top, 0), (\bot, 0)\}$ and $A_H^{(1)} = \{(q, h) \mid q \in Q_\forall\} \cup \{C\}$. The alphabet of this automaton is again Σ (as defined above). Its transition relation Δ_H is defined by

$$((q, h), (\gamma_j, \gamma_j'), (q', h')) \in \Delta_H \iff h = j \wedge (q, \gamma, q', \gamma', d) \in \Delta_\mathcal{M} \wedge h' = h + d$$

Note that "illegal" transitions are impossible. The players can only *cheat* with respect to the content of the h-th tape cell. Also, no transition labeled with veto_i for any i is possible from a state (q, h). In addition, we now have the following transitions:

$$\begin{aligned} ((q_F, h), (\gamma_i, \gamma_i'), C) &\in \Delta_H &&\text{for all } i, h \in \{1, \ldots, k\},\ \gamma, \gamma' \in \Gamma \\ (C, \text{veto}_i, (\bot, 0)) &\in \Delta_H &&\text{for all } i \in \{1, \ldots, k\} \\ (C, (\gamma_i, \gamma_i'), (\top, 0)) &\in \Delta_H &&\text{for all } i \in \{1, \ldots, k\},\ \gamma, \gamma' \in \Gamma \end{aligned}$$

Suppose $q_0 \in Q_\exists$. The play begins in position $((q_0, 1, 0), (\#_1, 0), \ldots, (\#_n, 0))$, where $\#_i \in \Gamma_i$ is the blank symbol of $\mathcal{M}$. Player 0 moves (i.e. picks a letter) at all states in which $\mathcal{A}_H$ is in a state from $Q_\exists$, player 1 if it is in a state from $Q_\forall$. This is ensured by the definition of the transition function δ_i, which guarantees that each component changes from a σ-state to a $(1 - \sigma)$-state in every round. The states $(\bot, 0)$ and $(\top, 0)$ in $\mathcal{A}_H$ are 0-states without outgoing transitions. The set player 0 tries to reach is $\mathcal{F}_{\text{loc}}(\{(\top, 0)\}, \{(\bot_1, 0)\}, \ldots, \{(\bot_1, 0)\})$.

We now show the correctness of the above construction. If $\mathcal{M}$ accepts w, then player 0 has a winning strategy in the reachability game on the configuration graph of $\mathcal{M}$ on w. If player 1 does not cheat and player 0 plays according to his strategy, the play will finally reach a state $((q_F, h), x_1, \ldots, x_n)$ with $x_i \neq (\bot_p, 0)$ for all $p \in \mathbb{B}$ and $i \in \{1, \ldots, n\}$. Recall that $q_F \in Q_\exists$. Now player 0 must move to C. Unless player 0 cheats (which is clearly a suboptimal choice at this point), this implies that every component i moves from $x_i = (\gamma_i, 0)$ to $(\gamma_i, 1)$ by the definition of δ_i. Player 1 can play veto_i for some i. However, since the i-th component is in state $(\gamma_i, 1)$ for some $\gamma_i \in \Gamma_i$, we have that this results in the i-th component making a transition to state $(\bot_1, 0)$. Thus player 0 wins. If player 1 plays (γ_i, γ_i') for some i, the play reaches $(\top, 0)$ and thus player 0 wins. If player 1 made an illegal transition at some point in the play, then for some i, the state of $\mathcal{A}_i$ loops between $(\bot_1, 0)$ and $(\bot_1, 1)$ from that point onwards and, again, player 0 wins.

Conversely, if $\mathcal{M}$ rejects w, then player 1 has a winning strategy in the safety game on the configuration graph of $\mathcal{M}$ on w. This implies that, unless player 0 uses an illegal transition, the play never reaches state q_F. On the other hand, if player 0 does make an illegal transition, one component, say i, changes to state $(\bot_0, 1)$ and remains in $\{(\bot_0, \sigma) \mid \sigma \in \mathbb{B}\}$ from this point onwards. If the play ever reaches q_F after that, and thereafter reaches C, player 1 can play veto_i moving $\mathcal{A}_H$ into state $(\bot, 0)$. Component i is in state $(\bot_0, 1)$ when $\mathcal{A}_H$ is in C whereby $\mathcal{A}_i$ never reaches state $(\bot_1, 0)$. Since player 1 never has to make an illegal transition, no component j is in a state $(\bot_1, 0)$. Hence player 0 loses. $\square$

The high complexity of SYNC-REACH[$\mathcal{F}$] for $\mathcal{F} \in \{\mathcal{F}_{loc}, \mathcal{F}_{sync}\}$ prohibits polynomially computable polynomial compositions for reachability games on synchronous arenas (with local or synchronized reachability conditions). Indeed, finding such compositions would amount to showing EXPTIME = PTIME. What can we do differently in order to succeed?

In order to find a winning composition of polynomial size one might suspect that more complex component games are necessary. In this event, finding strategies in the the component games would be more difficult, sparing us the complexity dilemma. Unfortunately, this turns out to be false.

Remark 3. In the reduction above, the constituent arenas are of size $\leq c$ for some constant $c \in \mathbb{N}$ (essentially the size of some APSPACE Turing machine $\mathcal{M}$ deciding an EXPTIME-complete problem). Thus, more complex winning conditions on those arenas will not increase the complexity of finding winning strategies.

Another possibility is to loosen the notion of a polynomial composition. Recall that a winning composition $\mathcal{M}$ of strategies $\mathcal{S}_1, \ldots, \mathcal{S}_k$ is a polynomial winning composition if $\mathcal{M}[\mathcal{S}_1, \ldots, \mathcal{S}_k]$ has polynomial latency and $\mathcal{M}$ has polynomial size. We now loosen this requirement as follows: A winning composition $\mathcal{M}$ is a *poly-space composition* if $\mathcal{M}$ is of polynomial size and the space requirement of $\mathcal{M}[\mathcal{S}_1, \ldots, \mathcal{S}_k]$ is polynomial.

Lemma 3. *Let* $\mathbf{G} = (\mathcal{A}_1 \otimes \cdots \otimes \mathcal{A}_k, \mathcal{F}(F_1, \ldots, F_k))$, *where* $\mathcal{F} \in \{\mathcal{F}_{loc}, \mathcal{F}_{sync}\}$. *Given* $\mathbf{G}$, *a strategy machine* $\mathcal{M}$ *with polynomial space requirement implementing a strategy for player 0 in* $\mathbf{G}$ *from position* $p_0 = (\sigma_0, v_0)$, *and sets* $F_1, \ldots, F_k$ *it is decidable in* PSPACE *whether or not* $\mathcal{M}$ *implements a winning strategy from* p_0.

The proof uses APTIME = PSPACE to verify that there is an $\mathcal{M}$-consistent loop which does not visit $F = \mathcal{F}(F_1, \ldots, F_k)$ and can be reached without visiting F.

Theorem 4. *The class of reachability games over synchronized products admits poly-space compositions iff* PSPACE = EXPTIME.

Proof. Clearly PSPACE = EXPTIME implies that a strategy machine with a polynomial space requirement can compute the next move in some attractor strategy in the course of a single iteration.

Conversely, if the class admits poly-space compositions, there always exists a polynomial sized winning strategy with a polynomial space requirement (assuming the component games are bounded by some constant, cf. Rem. 3) for player 0 (if he wins). Hence, the following NPSPACE procedure is correct: We guess a strategy machine of polynomial size and verify in PSPACE if it implements a winning strategy. By Savitch's theorem, PSPACE = NPSPACE. □

6 Conclusion

We studied the relation between the compositional nature of an arena and the structure of a winning strategy. To this end we introduced two kinds of products on arenas, the parallel and the synchronized product. We defined a notion of strategy composition which relies on strategy machines. This notion of composition allows to translate winning strategies in component games to winning strategies in the global game. We proved such a composition theorem for the class of reachability games on parallel products. We also showed why a similar result holds on synchronized products iff EXPTIME = PSPACE.

The results of this paper carry through to Büchi games with only minor modifications. We also have results on the case where the reachability condition is given explicitly (instead of as a sequence of k sets). For future research we want to consider more complex winning conditions, such as parity and weak parity. Also, we want to treat different ways of modeling the composite game from constituent arenas, addressing notions of composition from the field of process algebra and formal verification.

Acknowledgments. I would like to thank the anonymous reviewers for many helpful suggestions, both for the presentation and for future research.

References

1. Büchi, J.R., Landweber, L.H.: Solving Sequential Conditions by Finite-State Strategies. Trans. of the AMS 138, 295–311 (1969)
2. McNaughton, R.: Infinite games played on finite graphs. Annals of Pure and Applied Logic 65(2), 149–184 (1993)
3. Zielonka, W.: Infinite games on finitely coloured graphs with applications to automata on infinite trees. Theor. Comput. Sci. 200, 135–183 (1998)
4. Grädel, E., Thomas, W., Wilke, T. (eds.): Automata logics, and infinite games: a guide to current research. Springer, New York (2002)
5. Löding, C.: Infinite games and automata theory. In: Apt, K.R., Grädel, E. (eds.) Lectures in Game Theory for Computer Scientists. Cambridge U. P. (2011)
6. Baier, C., Katoen, J.: Principles of Model Checking. MIT Press (2008)
7. Fearnley, J., Peled, D., Schewe, S.: Synthesis of succinct systems. In: Chakraborty, S., Mukund, M. (eds.) ATVA 2012. LNCS, vol. 7561, pp. 208–222. Springer, Heidelberg (2012)
8. Harel, D., Kupferman, O., Vardi, M.Y.: On the complexity of verifying concurrent transition systems. Inf. Comput. 173(2), 143–161 (2002)
9. Gelderie, M.: Strategy machines and their complexity. In: Rovan, B., Sassone, V., Widmayer, P. (eds.) MFCS 2012. LNCS, vol. 7464, pp. 431–442. Springer, Heidelberg (2012)
10. Goldin, D.Q., Smolka, S.A., Wegner, P.: Turing machines, transition systems, and interaction. Electr. Notes Theor. Comput. Sci. 52(1), 120–136 (2001)
11. Hunter, P., Dawar, A.: Complexity bounds for regular games (extended abstract). In: Jedrzejowicz, J., Szepietowski, A. (eds.) MFCS 2005. LNCS, vol. 3618, pp. 495–506. Springer, Heidelberg (2005)
12. Dawar, A., Horn, F., Hunter, P.: Complexity Bounds for Muller Games. Theoretical Computer Science (2011) (submitted)
13. Horn, F.: Explicit Muller Games are PTIME. In: FSTTCS, pp. 235–243 (2008)

Asynchronous Games over Tree Architectures

Blaise Genest[1], Hugo Gimbert[2], Anca Muscholl[2], and Igor Walukiewicz[2]

[1] IRISA, CNRS, Rennes, France
[2] LaBRI, CNRS/Université Bordeaux, France

Abstract. We consider the distributed control problem in the setting of Zielonka asynchronous automata. Such automata are compositions of finite processes communicating via shared actions and evolving asynchronously. Most importantly, processes participating in a shared action can exchange complete information about their causal past. This gives more power to controllers, and avoids simple pathological undecidable cases as in the setting of Pnueli and Rosner. We show the decidability of the control problem for Zielonka automata over acyclic communication architectures. We provide also a matching lower bound, which is l-fold exponential, l being the height of the architecture tree.

1 Introduction

Synthesis is by now well understood in the case of sequential systems. It is useful for constructing small, yet safe, critical modules. Initially, the *synthesis problem* was stated by Church, who asked for an algorithm to construct devices transforming sequences of input bits into sequences of output bits in a way required by a specification [2]. Later Ramadge and Wonham proposed the *supervisory control* formulation, where a plant and a specification are given, and a controller should be designed such that its product with the plant satisfies the specification [18]. So control means restricting the behavior of the plant. Synthesis is the particular case of control where the plant allows for every possible behavior.

For synthesis of *distributed* systems, a common belief is that the problem is in general undecidable, referring to work by Pnueli and Rosner [17]. They extended Church's formulation to an architecture of *synchronously* communicating processes, that exchange messages through one slot communication channels. Undecidability in this setting comes mainly from *partial information*: specifications permit to control the flow of information about the global state of the system. The only decidable type of architectures is that of pipelines.

The setting we consider here is based on a by now well-established model of distributed computation using shared actions: *Zielonka's asynchronous automata* [20]. Such a device is an asynchronous product of finite-state processes synchronizing on common actions. Asynchronicity means that processes can progress at different speed. Similarly to [6,12] we consider the control problem for such automata. Given a Zielonka automaton (plant), find another Zielonka automaton (controller) such that the product of the two satisfies a given specification. In particular, the controller does not restrict the parallelism of the

F.V. Fomin et al. (Eds.): ICALP 2013, Part II, LNCS 7966, pp. 275–286, 2013.

system. Moreover, during synchronization the individual processes of the controller can exchange all their information about the global state of the system. This gives more power to the controller than in the Pnueli and Rosner model, thus avoiding simple pathological scenarios leading to undecidability. It is still open whether the control problem for Zielonka automata is decidable.

In this paper we prove decidability of the control problem for reachability objectives on tree architectures. In such architectures every process can communicate with its parent, its children, and with the environment. If a controller exists, our algorithm yields a controller that is a finite state Zielonka automaton exchanging information of *bounded* size. We also provide the first non-trivial lower bound for asynchronous distributed control. It matches the l-fold exponential complexity of our algorithm (l being the height of the architecture tree).

As an example, our decidability result covers client-server architectures where a server communicates with clients, and server and clients have their own interactions with the environment (cf. Figure 1). Our algorithm providing a controller for this architecture runs in exponential time. Moreover, each controller adds polynomially many bits to the state space of the process. Note also that this architecture is undecidable for [17] (each process has inputs), and is neither covered by [6] (the action alphabet is not a co-graph), nor by [12] (there is no bound on the number of actions performed concurrently).

Related work. The setting proposed by Pnueli and Rosner [17] has been thoroughly investigated in past years. By now we understand that, suitably using the interplay between specifications and an architecture, one can get undecidability results for most architectures rather easily. While specifications leading to undecidability are very artificial, no elegant solution to eliminate them exists at present.

The paper [10] gives an automata-theoretic approach to solving pipeline architectures and at the same time extends the decidability results to CTL* specifications and variations of the pipeline architecture, like one-way ring architectures. The synthesis setting is investigated in [11] for local specifications, meaning that each process has its own, linear-time specification. For such specifications, it is shown that an architecture has a decidable synthesis problem if and only if it is a sub-architecture of a pipeline with inputs at both endpoints. The paper [5] proposes information forks as an uniform notion explaining the (un)decidability results in distributed synthesis. In [15] the authors consider distributed synthesis for knowledge-based specifications. The paper [7] studies an interesting case of external specifications and well-connected architectures.

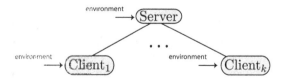

Fig. 1. Server/client architecture

Synthesis for asynchronous systems has been strongly advocated by Pnueli and Rosner in [16]. Their notion of asynchronicity is not exactly the same as ours: it means roughly that system/environment interaction is not turn-based, and processes observe the system only when scheduled. This notion of asynchronicity appears in several subsequent works, such as [19,9] for distributed synthesis.

As mentioned above, we do not know whether the control problem in our setting is decidable in general. Two related decidability results are known, both of different flavor than ours. The first one [6] restricts the alphabet of actions: control with reachability condition is decidable for co-graph alphabets. This restriction excludes among others client-server architectures. The second result [12] shows decidability by restricting the plant: roughly speaking, the restriction says that every process can have only bounded missing knowledge about the other processes (unless they diverge). The proof of [12] goes beyond the controller synthesis problem, by coding it into monadic second-order theory of event structures and showing that this theory is decidable when the criterion on the plant holds. Unfortunately, very simple plants have a decidable control problem but undecidable MSO-theory of the associated event structure. Melliès [14] relates game semantics and asynchronous games, played on event structures. More recent work [3] considers finite games on event structures and shows a determinacy result for such games under some restrictions.

Organization of the Paper. The next section presents basic definitions. The two consecutive sections present the algorithm and the matching lower bound. The full version of the paper is available at `http://hal.archives-ouvertes.fr/hal-00684223`.

2   Basic Definitions and Observations

We start by introducing Zielonka automata and state the control problem for such automata. We also give a game-based formulation of the problem.

2.1   Zielonka Automata

Zielonka automata are simple parallel finite-state devices. Such an automaton is a parallel composition of several finite automata, called *processes*, synchronizing on shared actions. There is no global clock, so between two synchronizations, two processes can do a different number of actions. Because of this Zielonka automata are also called asynchronous automata.

A *distributed action alphabet* on a finite set $\mathbb{P}$ of processes is a pair (Σ, dom), where Σ is a finite set of *actions* and $dom : \Sigma \to (2^{\mathbb{P}} \setminus \emptyset)$ is a *location function*. The location $dom(a)$ of action $a \in \Sigma$ comprises all processes that need to synchronize in order to perform this action. A (deterministic) *Zielonka automaton* $\mathcal{A} = \langle \{S_p\}_{p \in \mathbb{P}}, s_{in}, \{\delta_a\}_{a \in \Sigma} \rangle$ is given by:

- for every process p a finite set S_p of (local) states,
- the initial state $s_{in} \in \prod_{p \in \mathbb{P}} S_p$,

- for every action $a \in \Sigma$ a partial transition function $\delta_a : \prod_{p \in dom(a)} S_p \dashrightarrow \prod_{p \in dom(a)} S_p$ on tuples of states of processes in $dom(a)$.

For convenience, we abbreviate a tuple $(s_p)_{p \in P}$ of local states by s_P, where $P \subseteq \mathbb{P}$. We also talk about S_p as the set of p-states and of $\prod_{p \in \mathbb{P}} S_p$ as global states. Actions from $\Sigma_p = \{a \in \Sigma \mid p \in dom(a)\}$ are called p-actions. For $p, q \in \mathbb{P}$, let $\Sigma_{p,q} = \{a \in \Sigma \mid dom(a) = \{p, q\}\}$ be the set of synchronization actions between p and q. We write Σ_p^{loc} instead of $\Sigma_{p,p}$ for the set of local actions of p, and $\Sigma_p^{com} = \Sigma_p \setminus \Sigma_p^{loc}$ for the synchronization actions of p.

A Zielonka automaton can be seen as a sequential automaton with the state set $S = \prod_{p \in \mathbb{P}} S_p$ and transitions $s \xrightarrow{a} s'$ if $(s_{dom(a)}, s'_{dom(a)}) \in \delta_a$, and $s_{\mathbb{P} \setminus dom(a)} = s'_{\mathbb{P} \setminus dom(a)}$. By $L(\mathcal{A})$ we denote the set of words labeling runs of this sequential automaton that start from the initial state. Notice that $L(\mathcal{A})$ is closed under the congruence $\sim$ generated by $\{ab = ba \mid dom(a) \cap dom(b) = \emptyset\}$, in other words, it is a trace-closed language. A (Mazurkiewicz) trace is an equivalence class $[w]_\sim$ for some $w \in \Sigma^*$. The notion of trace and the idea of describing concurrency by a fixed independence relation on actions goes back to the late seventies, to Mazurkiewicz [13] (see also [4]).

Consider a Zielonka automaton $\mathcal{A}$ with two processes: $P, \overline{P}$. The local actions of process P are $\{u_0, u_1, c_0, c_1\}$ and those of process $\overline{P}$ are $\{\overline{u_0}, \overline{u_1}, \overline{c_0}, \overline{c_1}\}$. In addition, there is a shared action \$ with $dom(\$) = \{P, \overline{P}\}$. From the initial state, P can reach state (i, j) by executing $u_i c_j$, same for $\overline{P}$ with $\overline{u_k}\,\overline{c_l}$ and state $\overline{(k, l)}$. Action \$ is enabled in every pair $((i, j), \overline{(k, l)})$ satisfying $i = l$ or $k = j$, it leads to the final state. So $L(\mathcal{A}) = \{[u_i c_j \overline{u_k}\,\overline{c_l}\$] \mid i = l \text{ or } k = j\}$.

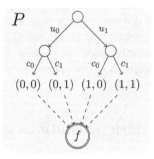

As the notion of a trace can be formulated without a reference to an accepting device, one can ask if the model of Zielonka automata is powerful enough. Zielonka's theorem says that this is indeed the case, hence these automata are a right model for the simple view of concurrency captured by Mazurkiewicz traces.

Theorem 1. *[20] Let $dom : \Sigma \to (2^{\mathcal{P}} \setminus \{\emptyset\})$ be a distribution of letters. If a language $L \subseteq \Sigma^*$ is regular and trace-closed then there is a deterministic Zielonka automaton accepting L (of size exponential in the number of processes and polynomial in the size of the minimal automaton for L, see [8]).*

2.2 The Control Problem

In Ramadge and Wonham's control setting [18] we are given an alphabet Σ of actions partitioned into system and environment actions: $\Sigma^{sys} \cup \Sigma^{env} = \Sigma$. Given a plant P we are asked to find a controller C over Σ such that the product $P \times C$ satisfies a given specification (the product being the standard product of the two automata). Both the plant and the controller are finite, deterministic automata over the same alphabet Σ. Additionally, the controller is required not to block environment actions, which in technical terms means that from every state of the controller there should be a transition on every action from Σ^{env}.

The definition of our problem is the same with the difference that we take Zielonka automata instead of finite automata. Given a distributed alphabet (Σ, dom) as above, and a Zielonka automaton P, find a Zielonka automaton C over the same distributed alphabet such that $P \times C$ satisfies a given specification. Additionally it is required that from every state of C there is a transition for every action from Σ^{env}. The important point here is that the controller has the same distributed structure as the plant. Hence concurrency in the controlled system is the same as in the plant. Observe that in the controlled system $P \times C$ the states carry the additional information computed by the controller.

Example: Reconsider the automaton on page 278, and assume that $u_i, \overline{u_k} \in \Sigma^{env}$ are the uncontrollable actions $(i, k \in \{0, 1\})$. So the controller needs to propose controllable actions c_j and $\overline{c_k}$, resp., in such a way that both P and $\overline{P}$ reach their final states $f, \overline{f}$ by executing the shared action \$. At first sight this may seem impossible to guarantee, as it looks like process P needs to know what $\overline{u_k}$ process $\overline{P}$ has received, or vice-versa. Nevertheless, such a controller exists. It consists of P allowing after u_i only action c_i, and $\overline{P}$ allowing after $\overline{u_k}$ only action $\overline{c_{1-k}}$. Regardless if the environment chooses $i = j$ or $i \neq j$, the action \$ is enabled in state $((i, i), (j, 1 - j))$, so both $P, \overline{P}$ can reach their final states.

It will be more convenient to work with a game formulation of this problem, as in [6,12]. Instead of talking about controllers we will talk about distributed strategies in a game between *system* and *environment*. A plant defines a game arena, with plays corresponding to initial runs of $\mathcal{A}$. Since $\mathcal{A}$ is deterministic, we can view a play as a word from $L(\mathcal{A})$ – or a trace, since $L(\mathcal{A})$ is trace-closed. Let $Plays(\mathcal{A})$ denote the set of traces associated with words from $L(\mathcal{A})$.

A strategy for the system will be a collection of individual strategies for each process. The important notion here is the view each process has about the global state of the system. Intuitively this is the part of the current play that the process could see or learn about from other processes during a communication with them. Formally, the *p-view* of a play u, denoted $view_p(u)$, is the smallest trace $[v]$ such that $u \sim vy$ and y contains no action from Σ_p. We write $Plays_p(\mathcal{A})$ for the set of plays that are *p*-views: $Plays_p(\mathcal{A}) = \{view_p(u) \mid u \in Plays(\mathcal{A})\}$.

A *strategy for a process* p is a function $\sigma_p : Plays_p(\mathcal{A}) \to 2^{\Sigma_p^{sys}}$, where $\Sigma_p^{sys} = \{a \in \Sigma^{sys} \mid p \in dom(a)\}$. We require in addition, for every $u \in Plays_p(\mathcal{A})$, that $\sigma_p(u)$ is a subset of the actions that are possible in the *p*-state reached on u. A *strategy* is a family of strategies $\{\sigma_p\}_{p \in \mathbb{P}}$, one for each process.

The set of plays respecting a strategy $\sigma = \{\sigma_p\}_{p \in \mathbb{P}}$, denoted $Plays(\mathcal{A}, \sigma)$, is the smallest set containing the empty play ε, and such that for every $u \in Plays(\mathcal{A}, \sigma)$:

1. if $a \in \Sigma^{env}$ and $ua \in Plays(\mathcal{A})$ then ua is in $Plays(\mathcal{A}, \sigma)$;
2. if $a \in \Sigma^{sys}$ and $ua \in Plays(\mathcal{A})$ then $ua \in Plays(\mathcal{A}, \sigma)$ provided that $a \in \sigma_p(view_p(u))$ for all $p \in dom(a)$.

Plays from $Plays(\mathcal{A}, \sigma)$ are called σ-plays and we write $Plays_p(\mathcal{A}, \sigma)$ for the set $Plays(\mathcal{A}, \sigma) \cap Plays_p(\mathcal{A})$. The above definition says that actions of the environment are always possible, whereas actions of the system are possible only if they are allowed by the strategies of all involved processes.

Our winning conditions in this paper are *local reachability* conditions: every process has a set of target states $F_p \subseteq S_p$. We also assume that states in F_p are *blocking*, that is, they have no outgoing transitions. This means that if $(s_{dom(a)}, s'_{dom(a)}) \in \delta_a$ then $s_p \notin F_p$ for all $p \in dom(a)$. For defining winning strategies, we need to consider also infinite σ-plays. By $Plays^\infty(\mathcal{A}, \sigma)$ we denote the set of finite or infinite σ-plays in $\mathcal{A}$. Such plays are defined as finite ones, replacing u in the definition of $Plays(\mathcal{A}, \sigma)$ by a possibly infinite, initial run of $\mathcal{A}$. A play $u \in Plays^\infty(\mathcal{A}, \sigma)$ is *maximal*, if there is no action c such that the trace uc is a σ-play (note that uc is defined only if no process in $dom(c)$ is scheduled infinitely often in u).

Definition 1. *The* control problem *for a plant $\mathcal{A}$ and a local reachability condition $(F_p)_{p \in \mathbb{P}}$ is to determine if there is a strategy $\sigma = (\sigma_p)_{p \in \mathbb{P}}$ such that every maximal trace $u \in Plays^\infty(\mathcal{A}, \sigma)$ is finite and ends in $\prod_{p \in \mathbb{P}} F_p$. Such traces and strategies are called* winning.

3   The Upper Bound for Acyclic Communication Graphs

We impose two simplifying assumptions on the distributed alphabet (Σ, dom). The first one is that all actions are at most binary: $|dom(a)| \leq 2$, for every $a \in \Sigma$. The second requires that all uncontrollable actions are local: $|dom(a)| = 1$, for every $a \in \Sigma^{env}$. The first restriction makes the technical reasoning much simpler. The second restriction reflects the fact that each process is modeled with its own, local environment.

Since actions are at most binary, we can define an undirected graph $\mathcal{CG}$ with node set $\mathbb{P}$ and edges $\{p, q\}$ if there exists $a \in \Sigma$ with $dom(a) = \{p, q\}$, $p \neq q$. Such a graph is called *communication graph*. We assume throughout this section that $\mathcal{CG}$ is acyclic and has at least one edge. This allows us to choose a leaf $\ell \in \mathbb{P}$ in $\mathcal{CG}$, with $\{r, \ell\}$ an edge in $\mathcal{CG}$. So, in this section ℓ denotes this fixed leaf process and r its parent process. Starting from a control problem with input $\mathcal{A}$, $(F_p)_{p \in \mathbb{P}}$ we will reduce it to a control problem over the smaller (acyclic) graph $\mathcal{CG}' = \mathcal{CG}_{\mathbb{P} \setminus \{\ell\}}$. The reduction will work in exponential-time. If we represent $\mathcal{CG}$ as a tree of depth l then applying this construction iteratively we will get an l-fold exponential algorithm to solve the control problem for the $\mathcal{CG}$ architecture.

The main idea is that process r can simulate the behavior of process ℓ. Indeed, after each synchronization between r and ℓ, the views of both processes are identical, and until the next synchronization (or termination) ℓ evolves locally. The way r simulates ℓ is by "guessing" the future local evolution of ℓ until the next synchronizations (or termination) in a summarized form. Correctness is ensured by letting the environment challenge the guesses.

In order to proceed this way, we first show that winning control strategies can be assumed to satisfy a "separation" property concerning the synchronizations of process r (cf. 2nd item of Lemma 1):

Lemma 1. *If there exists a winning strategy for controlling $\mathcal{A}$, then there is one, say σ, such that for every $u \in Plays(\mathcal{A}, \sigma)$ the following holds:*

1. *For every process p and $A = \sigma_p(view_p(u))$, we have either $A \subseteq \Sigma_p^{com}$ or $A = \{a\}$ for some $a \in \Sigma_p^{loc}$.*
2. *Let $A = \sigma_r(view_r(u))$ with $A \subseteq \Sigma_r^{com}$. Then either $A \subseteq \Sigma_{r,\ell}$ or $A \subseteq \Sigma_r^{com} \setminus \Sigma_{r,\ell}$ holds.*

It is important to note that the 2nd item of Lemma 1 only holds when final states are blocking. To see this, consider the client-server example in Fig. 1 and assume that environment can either put a client directly to a final state, or oblige him to synchronize with the server before going to the final state. Suppose that all states of the server are final. In this case, the server's strategy must propose synchronization with *all* clients at the same time in order to guarantee that all clients can reach their final states.

Lemma 1 implies that the behavior of process ℓ can be divided in phases consisting of a local game ending in states where the strategy proposes communications with r and no local actions. This allows to define summaries of results of local plays of the leaf process ℓ. We denote by $state_\ell(v)$ the ℓ-component of the state reached on $v \in \Sigma^*$ from the initial state. Given a strategy $\sigma = (\sigma_p)_{p \in \mathbb{P}}$ and a play $u \in Plays_\ell(\mathcal{A}, \sigma)$, we define:

$$Sync_\ell^\sigma(u) = \{(t_\ell, A) \mid \exists x \in (\Sigma_\ell^{loc})^* . \ ux \text{ is a } \sigma\text{-play}, state_\ell(ux) = t_\ell,$$
$$\sigma_\ell(ux) = A \subseteq \Sigma_{r,\ell}, \text{ and } A = \emptyset \text{ iff } t_\ell \text{ is final}\}.$$

Since our winning conditions are local reachability conditions, we can show that it suffices to consider memoryless local strategies for process ℓ until the next synchronization with r (or until termination). Moreover, since final states are blocking, either all possible local plays from a given ℓ-state ultimately require synchronization with r, or they all terminate in a final state of ℓ (mixing the two situations would result in a process blocked on communication).

Lemma 2. *If there exists a winning strategy for controlling $\mathcal{A}$, then there is one, say $\sigma = (\sigma_p)_{p \in \mathbb{P}}$, such that for all plays $u \in Plays_\ell(\mathcal{A}, \sigma)$ the following hold:*

1. *Either $Sync_\ell^\sigma(u) \subseteq (S_\ell \setminus F_\ell) \times (2^{\Sigma_{r,\ell}} \setminus \{\emptyset\})$ or $Sync_\ell^\sigma(u) \subseteq F_\ell \times \{\emptyset\}$.*
2. *If uy is a σ-play with $y \in (\Sigma \setminus \Sigma_\ell)^*$, $\sigma_r(view_r(uy)) = B \subseteq \Sigma_{r,\ell}$ and $B \neq \emptyset$, then for every $(t_\ell, A) \in Sync_\ell^\sigma(u)$ some action from $A \cap B$ is enabled in $(state_r(uy), t_\ell)$.*
3. *There is a memoryless local strategy $\tau : S_\ell \to (\Sigma_\ell^{sys} \cap \Sigma_\ell^{loc})$ to reach from $state_\ell(u)$ the set of local states $\{t_\ell \mid (t_\ell, A) \in Sync_\ell^\sigma(u) \text{ for some } A\}$.*

The second item of the lemma says that every evolution of r should be compatible with every evolution of ℓ. The memoryless strategy from the third item proposes local actions of ℓ based only on the current state of ℓ and not on the history of the play. This strategy is used in a game on the transition graph of process ℓ. The third item of the lemma follows from the fact that 2-player games with reachability objectives admit memoryless winning strategies.

Definition 2. *An admissible plan* T *from* $s_\ell \in S_\ell$ *for process* ℓ *is either a subset of* $(S_\ell \setminus F_\ell) \times (2^{\Sigma_{r,\ell}} \setminus \{\emptyset\})$ *or a subset of* $F_\ell \times \{\emptyset\}$, *such that there exists a memoryless local strategy* $\tau : S_\ell \to (\Sigma_\ell^{sys} \cap \Sigma_\ell^{loc})$ *to reach from* s_ℓ *the set* $\{t_\ell \mid (t_\ell, A) \in T \text{ for some } A\}$. *An admissible plan* T *is* final *if* $T \subseteq F_\ell \times \{\emptyset\}$.

So Lemma 2 states that if there exists a winning strategy then there is one, say σ such that $Sync_\ell^\sigma(u)$ is an admissible plan for every σ-play u. Note also that we can check in polynomial time whether T as above is an admissible plan.

We are now ready to define informally the reduced plant $\mathcal{A}'$ on the process set $\mathbb{P}' = \mathbb{P} \setminus \{\ell\}$, that is the result of eliminating process ℓ. The only part that changes in $\mathcal{A}$ concerns process r, who now simulates former processes r and ℓ. The new process r starts in state $\langle s_{in,r}, s_{in,\ell} \rangle$. It will get into a state from $S_r \times S_\ell$ every time it simulates a synchronization between the former r and ℓ. Between these synchronizations its behaviour is as follows.

- From a state of the form $\langle s_r, s_\ell \rangle$, process r can do a controllable action $ch(T)$, for every admissible plan T from s_ℓ, and go to state $\langle s_r, T \rangle$.
- From a state of the form $\langle s_r, T \rangle$ process r can behave as the former r: it can either do a local action (controllable or not) or a shared action with some $p \neq \ell$, that updates the S_r-component to some $\langle s_r', T \rangle$.
- From a state $\langle s_r, T \rangle$ process r can also do a controllable action $ch(B)$ for some $B \subseteq \Sigma_{r,\ell}$ and go to state $\langle s_r, T, B \rangle$; from $\langle s_r, T, B \rangle$ there are new, uncontrollable actions of the form (a, t_ℓ) where $(t_\ell, A) \in T$ and $a \in A \cap B$ such that $(s_r, t_\ell) \xrightarrow{a} (s_r', t_\ell')$ in $\mathcal{A}$. This case simulates r choosing a set of synchronization actions with ℓ, and the synchronization itself. For correctness of this step it is important that B is chosen such that for every $(t_\ell, A) \in T$ there is some $a \in A \cap B$ enabled in (s_r, t_ℓ).

Finally, accepting states of r in $\mathcal{A}'$ are $F_r \times F_\ell$, and $\langle s_r, T \rangle$ for $s_r \in F_r$ and T a final plan. The proof showing that this construction is correct provides a reduction from the control game on $\mathcal{A}$ to the control game on $\mathcal{A}'$.

Theorem 2. *Let* ℓ *be the fixed leaf process with* $\mathbb{P}' = \mathbb{P} \setminus \{\ell\}$ *and* r *its parent. Then the system has a winning strategy for* $\mathcal{A}, (F_p)_{p\in\mathbb{P}}$ *iff it has one for* $\mathcal{A}', (F_p')_{p\in\mathbb{P}'}$. *The size of* $\mathcal{A}'$ *is* $|\mathcal{A}| + \mathcal{O}(M_r 2^{M_\ell |\Sigma_{r\ell}|})$, *where* M_r *and* M_ℓ *are the sizes of processes* r *and* ℓ *in* $\mathcal{A}$, *respectively.*

Remark 1. Note that the bound on $|\mathcal{A}'|$ is better than $|\mathcal{A}| + \mathcal{O}(M_r 2^{M_\ell 2^{|\Sigma_{r\ell}|}})$ obtained by simply counting all possible states in the description above. The reason is that we can restrict admissible plans to be (partial) functions from S_ℓ into $2^{\Sigma_{r,\ell}}$. That is, we do not need to consider different sets of communication actions for the same state in S_ℓ.

Let us reconsider the example from Figure 1 of a server with k clients. Applying our reduction k times we reduce out all the clients and obtain the single process plant whose size is $M_s 2^{(M_1 + \cdots + M_k)c}$ where M_s is the size of the server, M_i is the size of client i, and c is the maximal number of communication actions between a client and a server. Our first main result also follows by applying the above reduction iteratively.

Theorem 3. *The control problem for distributed plants with acyclic communication graph is decidable. There is an algorithm for solving the problem (and computing a finite-state controller, if it exists) whose running time is bounded by a tower of exponentials of height equal to half of the diameter of the graph.*

4   The Lower Bound

Our main objective now is to show how using a communication architecture of diameter l one can code a counter able to represent numbers of size $Tower(2, l)$ (with $Tower(n, l) = 2^{Tower(n, l-1)}$ and $Tower(n, 1) = n$). Then an easy adaptation of the construction will allow to encode computations of Turing machines with the same space bound as the capabilities of the counters.

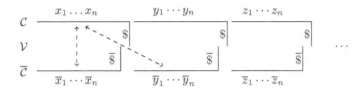

Fig. 2. Shape of a trace with 3 processes. Dashed lines show two types of tests.

Let us first explain the mechanism we will use. Consider a trace of the shape presented in Figure 2. There are three processes $\mathcal{C}, \overline{\mathcal{C}}$ and $\mathcal{V}$. Process $\mathcal{C}$ repeatedly generates a sequence of n local actions and then synchronizes on action $\$$ with the verifier process $\mathcal{V}$. Process $\overline{\mathcal{C}}$ does the same. The alphabets of $\mathcal{C}$ and $\overline{\mathcal{C}}$ are of course disjoint. The verifier process $\mathcal{V}$ always synchronizes first with $\overline{\mathcal{C}}$ and then with $\mathcal{C}$. Observe that the actions $\overline{y}_1 \cdots \overline{y}_n$ are concurrent to both $x_1 \cdots x_n$ and $y_1 \cdots y_n$, but they are before z_1. Suppose that we allow the environment to stop this generation process at any moment. Say it stops $\mathcal{C}$ at some x_i, and $\overline{\mathcal{C}}$ at $\overline{x}_i$. We can then set the processes in such a way that they are forced to communicate x_i and $\overline{x}_i$ to $\mathcal{V}$; who can verify if they are correct. The other possibility is that the environment stops $\mathcal{C}$ at x_i and $\overline{\mathcal{C}}$ at $\overline{y}_i$ forcing the comparison of x_i with $\overline{y}_i$. This way we obtain a mechanism allowing to compare position by position the sequence $x_1 \cdots x_n$ both with $\overline{x}_1 \cdots \overline{x}_n$ and with $\overline{y}_1 \cdots \overline{y}_n$. Observe that $\mathcal{V}$ knows which of the two cases he deals with, since the comparison with the latter sequence happens after some $\overline{\$}$ and before the next $\$$. Now, we can use sequences of n letters to encode numbers from 0 to $2^n - 1$. Then this mechanism permits us to verify if $x_1 \cdots x_n$ represents the same number as $\overline{x}_1 \cdots \overline{x}_n$ and the predecessor of $\overline{y}_1 \cdots \overline{y}_n$. Applying the same reasoning to $y_1 \cdots y_n$ we can test that it represents the same number as $\overline{y}_1 \cdots \overline{y}_n$ and the predecessor of $\overline{z}_1 \cdots \overline{z}_n$. If some test fails, the environment wins. If the environment does not stop $\mathcal{C}$ and $\overline{\mathcal{C}}$ at the same position, or stops only one of them, the system wins. So this way we force the processes $\mathcal{C}$ and $\overline{\mathcal{C}}$ to cycle through representations of numbers from 0 to

$2^n - 1$. Building on this idea we can encode alternating polynomial space Turing machines, and show that the control problem for this three process architecture (with diameter 2) is EXPTIME-hard. The algorithm from the previous section provides the matching upper bound.

After this explanation let us introduce general counters. We start with their alphabets. Let $\Sigma_i = \{a_i, b_i\}$ for $i = 1, \ldots, n$. We will think of a_i as 0 and b_i as 1, mnemonically: 0 is round and 1 is tall. Let $\Sigma_i^\# = \Sigma_i \cup \{\#_i\}$ be the alphabet extended with an end marker.

A 1-counter is just a letter from Σ_1 followed by $\#_1$. The value of a_1 is 0, and the one of b_1 is 1. An $(l+1)$-*counter* is a word $x_0 u_0 x_1 u_1 \cdots x_{k-1} u_{k-1} \#_{l+1}$ where $k = Tower(2, l)$, and for every $i < k$ we have: $x_i \in \Sigma_{l+1}$ and u_i is an l-counter with value i. The value of the above $(l+1)$-counter is $\sum_{i=0,\ldots,k} x_i 2^i$. The end marker $\#_{l+1}$ is there for convenience. An *iterated* $(l+1)$-*counter* is a nonempty sequence of $(l+1)$-counters (we do not require that the values of consecutive $(l+1)$-counters are consecutive).

Suppose that we have already constructed a plant $\mathcal{C}^l$ with root process r_l, such that every winning strategy in $\mathcal{C}^l$ needs to produce an iterated l-counter on r_l. We now define $\mathcal{C}^{l+1}$, a plant where every winning strategy needs to produce an iterated $(l + 1)$-counter on its root process r_{l+1}. Recall that such a counter is a sequence of l-counters with values $0, 1, \ldots, (Tower(2, l) - 1), 0, 1, \ldots$

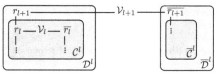

The plant $\mathcal{C}^{l+1}$ is made of two copies of $\mathcal{C}^l$, that we name $\mathcal{D}^l$ and $\overline{\mathcal{D}^l}$. We add three processes: $r_{l+1}, \overline{r_{l+1}}, \mathcal{V}_{l+1}$. The root r_{l+1} of $\mathcal{C}^{l+1}$ communicates with $\mathcal{V}_{l+1}$ and with the root r_l of $\mathcal{D}^l$, while $\overline{r_{l+1}}$ communicates with $\mathcal{V}_{l+1}$ and with the root of $\overline{\mathcal{D}^l}$.

In order to force $\mathcal{C}^{l+1}$ to generate an $(l+1)$-counter, we allow the environment to compare using $\mathcal{V}_{l+1}$ the sequence generated by r_{l+1} and the sequence generated by $\overline{r_{l+1}}$. The mechanism is similar to the example above. After each letter of Σ_l, we add an uncontrollable action that triggers the comparison between the current letters of Σ_l on r_{l+1} and on $\overline{r_{l+1}}$. This may correspond to two types of tests: equality or successor. For equality, $\mathcal{V}_{l+1}$ enters a losing state if (1) the symbols from r_{l+1} and from $\overline{r_{l+1}}$ are different; and (2) the number of remaining letters of Σ_l before $\#_l$ is the same on both r_{l+1} and $\overline{r_{l+1}}$. The latter test ensures that the environment has put the challenge at the same positions of the two counters. The case for successor is similar, accounting for a possible carry. In any other case (for instance, if the test was issued on one process only instead of both r_{l+1} and $\overline{r_{l+1}}$), the test leads to a winning configuration.

The challenge with this schema is to keep r_{l+1} and $\overline{r_{l+1}}$ synchronized in a sense that either (i) the two should be generating the same l-counter, or (ii) r_{l+1} should be generating the consecutive counter with respect to the one generated by $\overline{r_{l+1}}$. For this, a similar communication mechanism based on \$ symbols as in the example above is used. An action $\$_l$ shared by r_{l+1} and $\mathcal{V}_{l+1}$ is executed after each l-counter, that is after each $\#_l$ shared between r_{l+1} and r_l. Similarly with action $\overline{\$_l}$ shared by $\overline{r_{l+1}}$ and $\mathcal{V}_{l+1}$. Process $\mathcal{V}_{l+1}$ switches between state eq and state $succ$ when receiving $\overline{\$_l}$, and back when receiving $\$_l$ so it knows whether $\overline{r_{l+1}}$ is generating the same l-counter as r_{l+1}, or the next one. As $\overline{r_{l+1}}$ does not synchronize (unless there is a challenge) with $\mathcal{V}_{l+1}$ between two $\overline{\$_l}$, it does not know whether r_{l+1} has already started producing the same l-counter or whether it is still producing the previous one. Another important point about the flow of knowledge is that while r_l is informed when r_{l+1} is being challenged (as it synchronizes frequently with r_{l+1}, and could thus be willing to cheat to produce a different l-counter), $\overline{r_l}$ does not know that r_{l+1} is being challenged, and thus cheating on r_l would be caught by verifier $\mathcal{V}_l$.

Proposition 1. *For every l, the system has a winning strategy in $\mathcal{C}^l$. For every such winning strategy σ, if we consider the unique σ-play without challenges then its projection on $\bigcup_{i=1,\ldots,l} \Sigma_i^\#$ is an iterated l-counter.*

Proposition 1 is the basis for encoding Turing machines, with $\mathcal{C}^l$ ensuring that the space bound is equal to $Tower(n,l)$.

Theorem 4. *Let $l > 0$. There is an acyclic architecture of diameter $(4l-2)$ and with $3(2^l - 1)$ processes such that the space complexity of the control problem for it is $\Omega(Tower(n,l))$-complete.*

5   Conclusions

Distributed synthesis is a difficult and at the same time promising problem, since distributed systems are intrinsically complex to construct. We have considered here an asynchronous, shared-memory model. Already Pnueli and Rosner in [16] strongly argue in favour of asynchronous distributed synthesis. The choice of transmitting additional information while synchronizing is a consequence of the model we have adopted. We think that it is interesting from a practical point of view, since it is already used in multithreaded computing (e.g., CAS primitive) and it offers more decidable settings (e.g., client-server architecture).

Under some restrictions we have shown that the resulting control problem is decidable. The assumption about uncontrollable actions being local represents the most common situation where each process comes with its own environment (e.g., a client). The assumption on binary synchronizations simplifies the definition of architecture graph and is common in distributed algorithms. The most important restriction is that on architectures being a tree. Tree architectures are quite rich and allow to model hierarchical situations, like server/clients (recall that such cases are undecidable in the setting of Pnueli and Rosner). Nevertheless, it would be very interesting to know whether the problem is still decidable e.g. for ring architectures. Such an extension would require new proof ideas. A more immediate task is to consider more general winning conditions. A further interesting research direction is the synthesis of open, concurrent recursive programs, as considered e.g. in [1].

Our non-elementary lower bound result is somehow surprising. Since we have full information sharing, all the complexity is hidden in the uncertainty about actions performed in parallel by other processes.

References

1. Bollig, B., Grindei, M.-L., Habermehl, P.: Realizability of concurrent recursive programs. In: de Alfaro, L. (ed.) FOSSACS 2009. LNCS, vol. 5504, pp. 410–424. Springer, Heidelberg (2009)
2. Church, A.: Logic, arithmetics, and automata. In: Proceedings of the International Congress of Mathematicians, pp. 23–35 (1962)
3. Clairambault, P., Gutierrez, J., Winskel, G.: The winning ways of concurrent games. In: LICS, pp. 235–244. IEEE (2012)
4. Diekert, V., Rozenberg, G. (eds.): The Book of Traces. World Scientific (1995)
5. Finkbeiner, B., Schewe, S.: Uniform distributed synthesis. In: LICS, pp. 321–330. IEEE (2005)
6. Gastin, P., Lerman, B., Zeitoun, M.: Distributed games with causal memory are decidable for series-parallel systems. In: Lodaya, K., Mahajan, M. (eds.) FSTTCS 2004. LNCS, vol. 3328, pp. 275–286. Springer, Heidelberg (2004)
7. Gastin, P., Sznajder, N., Zeitoun, M.: Distributed synthesis for well-connected architectures. Formal Methods in System Design 34(3), 215–237 (2009)
8. Genest, B., Gimbert, H., Muscholl, A., Walukiewicz, I.: Optimal Zielonka-type construction of deterministic asynchronous automata. In: Abramsky, S., Gavoille, C., Kirchner, C., Meyer auf der Heide, F., Spirakis, P.G. (eds.) ICALP 2010. LNCS, vol. 6199, pp. 52–63. Springer, Heidelberg (2010)
9. Katz, G., Peled, D., Schewe, S.: Synthesis of distributed control through knowledge accumulation. In: Gopalakrishnan, G., Qadeer, S. (eds.) CAV 2011. LNCS, vol. 6806, pp. 510–525. Springer, Heidelberg (2011)
10. Kupferman, O., Vardi, M.: Synthesizing distributed systems. In: LICS (2001)
11. Madhusudan, P., Thiagarajan, P.S.: Distributed controller synthesis for local specifications. In: Orejas, F., Spirakis, P.G., van Leeuwen, J. (eds.) ICALP 2001. LNCS, vol. 2076, p. 396. Springer, Heidelberg (2001)
12. Madhusudan, P., Thiagarajan, P.S., Yang, S.: The MSO theory of connectedly communicating processes. In: Sarukkai, S., Sen, S. (eds.) FSTTCS 2005. LNCS, vol. 3821, pp. 201–212. Springer, Heidelberg (2005)
13. Mazurkiewicz, A.: Concurrent program schemes and their interpretations. DAIMI Rep. PB 78, Aarhus University, Aarhus (1977)
14. Melliès, P.-A.: Asynchronous games 2: The true concurrency of innocence. TCS 358(2-3), 200–228 (2006)
15. van der Meyden, R., Wilke, T.: Synthesis of distributed systems from knowledge-based specifications. In: Abadi, M., de Alfaro, L. (eds.) CONCUR 2005. LNCS, vol. 3653, pp. 562–576. Springer, Heidelberg (2005)
16. Pnueli, A., Rosner, R.: On the synthesis of an asynchronous reactive module. In: Ronchi Della Rocca, S., Ausiello, G., Dezani-Ciancaglini, M. (eds.) ICALP 1989. LNCS, vol. 372, pp. 652–671. Springer, Heidelberg (1989)
17. Pnueli, A., Rosner, R.: Distributed reactive systems are hard to synthesize. In: FOCS, pp. 746–757 (1990)
18. Ramadge, P.J.G., Wonham, W.M.: The control of discrete event systems. Proceedings of the IEEE 77(2), 81–98 (1989)
19. Schewe, S., Finkbeiner, B.: Synthesis of asynchronous systems. In: Puebla, G. (ed.) LOPSTR 2006. LNCS, vol. 4407, pp. 127–142. Springer, Heidelberg (2007)
20. Zielonka, W.: Notes on finite asynchronous automata. RAIRO–Theoretical Informatics and Applications 21, 99–135 (1987)

Querying the Guarded Fragment with Transitivity

Georg Gottlob[1], Andreas Pieris[1], and Lidia Tendera[2]

[1] Department of Computer Science, University of Oxford, UK
`firstname.lastname@cs.ox.ac.uk`
[2] Institute of Mathematics and Informatics, Opole University, Poland
`tendera@math.uni.opole.pl`

Abstract. We study the problem of answering a union of Boolean conjunctive queries q against a database Δ, and a logical theory φ which falls in the guarded fragment with transitive guards (GF+TG). We trace the frontier between decidability and undecidability of the problem under consideration. Surprisingly, we show that query answering under GF^2+TG, i.e., the two-variable fragment of GF+TG, is already undecidable (even without equality), whereas its monadic fragment is decidable; in fact, it is 2EXPTIME-complete in combined complexity and coNP-complete in data complexity. We also show that for a restricted class of queries, query answering under GF+TG is decidable.

1 Introduction

The Guarded Fragment. The *guarded fragment* of first-order logic (GF) was introduced by Andréka et al. [1] with the aim of explaining and generalizing the good properties of modal logic. Guarded formulas are constructed as usual first-order formulas with the exception that all quantification must be bounded, i.e., of the form $\forall \bar{x}(\alpha \rightarrow \varphi)$ or $\exists \bar{x}(\alpha \wedge \varphi)$, where α is an atomic formula which guards φ in the sense that it contains all the free variables of φ. Andréka et al. showed that modal logic can be embedded in GF, and they argued in a convincing way that GF inherits the good properties of modal logic. In [2], Grädel has established that GF enjoys several nice model-theoretic properties, and he also proved that satisfiability of GF-sentences is 2EXPTIME-complete, and EXPTIME-complete for sentences with relations of bounded arity.

The guarded fragment has since been intensively studied and extended in various ways. An interesting extension is the *guarded fragment with transitivity*, a natural representative language for multi-modal logics that are used to formalize epistemic logics. The obvious formalization of the transitivity of a binary relation R, namely $\forall x \forall y \forall z(R(x, y) \wedge R(y, z) \rightarrow R(x, z))$, is not guarded and there is no way to express it in GF [2]. As shown by Ganzinger et al. [3], the *two-variable guarded fragment* (GF^2) with transitivity is already undecidable, improving an analogous result for the three-variable guarded fragment proved by Grädel [2]. In [3], a logic which restricts the guarded fragment with transitivity by allowing transitive relations to appear only in guards has been proposed. This formalism, which was dubbed the *guarded fragment with transitive guards* (GF+TG) [4,5], is indeed expressive enough to be able to capture multi-modal logics of type K4, S4 or S5. The decidability of the *monadic* fragment of GF^2+TG (MGF^2+TG), where all non-unary relations may appear in guards only,

F.V. Fomin et al. (Eds.): ICALP 2013, Part II, LNCS 7966, pp. 287–298, 2013.
© Springer-Verlag Berlin Heidelberg 2013

was established in [3]; its exact complexity, as well as the decidability of GF + TG, was left as an open problem. Satisfiability of GF + TG-sentences is 2EXPTIME-complete [5], while the 2EXPTIME-hardness holds also for $MGF^2 + TG$ [6].

Querying Guarded-Based Fragments. It is evident that a large corpus of works on GF (and extensions of it) has focused on satisfiability. More recently, the attention has shifted on the problem of *query answering*, a central reasoning task in database theory [7] and description logics [8]. An extensional database Δ, which is actually a conjunction of ground atoms, is combined with a first-order sentence φ describing constraints which derive new knowledge. The database does not necessarily satisfy φ, and may thus be incomplete. A query is not just answered against Δ, as in the classical setting, but against the logical theory $(\Delta \wedge \varphi)$. Here we focus on *union of Boolean conjunctive queries* (UCQ). A *Boolean conjunctive query* (BCQ) q consists of an existentially closed conjunction of atoms, while a UCQ is a disjunction of a finite number of BCQ. Thus, given a UCQ q, one checks whether $(\Delta \wedge \varphi) \models q$, written $(\Delta, \varphi) \models q$.

Several fragments of GF have been considered for query answering in the context of database theory. A notable example is the class of *guarded tuple-generating dependencies* (or guarded TGDs) [7], that is, sentences of the form $\forall \bar{x}(\varphi(\bar{x}) \rightarrow \exists \bar{y} \, \psi(\bar{x}, \bar{y}))$, where φ and ψ are conjunctions of atoms, and in φ an atom exists which contains all the variables of $\bar{x}$. Although guarded TGDs are, strictly speaking, not GF-sentences, since their heads may be unguarded, they can be rewritten as guarded sentences [9]. Several extensions of guarded TGDs have been investigated, see, e.g., [10,7,11]. Fragments of GF have been also considered in the context of description logics. A prominent language is DL-Lite$_\mathcal{R}$ [8], which forms the OWL 2 QL profile of W3Cs standard ontology language for modeling Semantic Web ontologies. In fact, each DL-Lite$_\mathcal{R}$ axiom can be written as a GF-sentence of the form $\forall \bar{x}(\alpha(\bar{x}) \rightarrow \exists \bar{y} \, \beta(\bar{x}, \bar{y}))$. Following a more general approach, Bárány et al. studied the problem of query answering for the whole guarded fragment [9]. Query answering under GF is coNP-complete in *data complexity*, i.e., when only the database is part of the input, and 2EXPTIME-complete in *combined complexity*, i.e., when also the theory and the query are part of the input. Notice that the data complexity is widely regarded as more meaningful in practice, since the theory and the query are typically of a size that can be productively assumed to be fixed.

Research Challenges. While the decidability and complexity landscape of query answering under GF (and fragments of it) is clearing up, the picture for extensions of GF is still foggy. Notable exceptions are the *two-variable guarded fragment with counting quantifiers*, under which query answering is coNP-complete in data complexity [12], and the *guarded negation* fragment, where query answering is 2EXPTIME-complete in combined complexity [13]. In this paper we focus on GF + TG. Our goal is to better understand the problem of query answering, and give answers to the following questions:

- Is query answering under GF + TG decidable, and if so, what is the exact data and combined complexity?
- In case the previous question is answered negatively: *(i)* What is the frontier between decidability and undecidability, and what is the exact data and combined complexity of query answering under the decidable fragment of GF + TG? *(ii)* Can we gain decidability for GF + TG by restricting the syntax of the query?

We provide answers to all these questions. Notice that query answering under GF+TG is at least as hard as (un)satisfiability of GF+TG-sentences; in fact, $(\Delta, \varphi) \models q$ iff $(\Delta \wedge \varphi \wedge \neg q)$ is unsatisfiable. However, previous results on GF+TG are not immediately applicable for the following two reasons: Δ contains constants which are forbidden in the original definition of GF+TG, and $\neg q$ may not be a GF+TG-sentence. Therefore, we had either to come up with novel techniques beyond the state of the art, or significantly extend existing procedures.

Contribution. Our contributions can be summarized as follows:

1. We show that query answering under GF+TG is undecidable even without equality. This is done by forcing an infinite grid to appear in every model of a GF^2+TG-sentence, and then, by a further conjunction of formulas, we simulate a deterministic Turing machine. The same proof shows undecidability of guarded disjunctive TGDs (i.e., guarded TGDs extended with disjunction in rule-heads) with transitive guards. Although the question whether the same undecidability result holds also for non-disjunctive guarded TGDs remains open, we establish that transitivity without the restriction to guards cannot be safely combined with guarded TGDs.
2. We trace the frontier between decidability and undecidability of query answering by establishing that for the monadic fragment of GF^2+TG (MGF^2+TG) it is decidable; in fact, it is 2EXPTIME-complete in combined complexity and coNP-complete in data complexity. The proof of this result is constituted by two steps. First, we show that satisfiability of an MGF^2+TG-sentence combined with a database Δ, is 2EXPTIME-complete, and NP-complete if we consider only the database as part of the input. Then, given $q \in UCQ$, we construct a sentence $\Phi_{\Delta,q}$ such that for every MGF^2+TG-sentence φ, $(\Delta, \varphi) \models q$ iff $(\Delta^\star \wedge \varphi \wedge \neg\Phi_{\Delta,q})$ is unsatisfiable, where $\Delta^\star$ is obtained from Δ by adding some auxiliary atoms, and $(\varphi \wedge \neg\Phi_{\Delta,q})$ is an MGF^2+TG-sentence.
3. We show decidability of query answering under GF+TG if we consider unions of single-transitive-acyclic BCQs, that is, a restricted class of queries; it is 2EXPTIME-complete in combined complexity, and coNP-complete in data complexity.

2 Preliminaries

We work with finite relational signatures. Let us fix such a signature τ, and let $width(\tau)$ be the maximal arity of any of the predicate symbols in τ. The *guarded fragment* of first-order logic (GF), introduced by Andréka et al. [1], is the collection of first-order formulas with some syntactic restrictions in the quantification pattern, which is analogous to the relativised nature of modal logic. The set GF of formulas over τ is the smallest set *(i)* containing all atomic τ-formulas and equalities, *(ii)* closed under logical connectives $\neg, \wedge, \vee, \rightarrow$, and *(iii)* if $\bar{x}$ and $\bar{y}$ are tuples of variables, α is a τ-atom or an equality atom containing all the variables of $\{\bar{x}, \bar{y}\}$, and $\varphi \in$ GF with free variables contained in $\{\bar{x}, \bar{y}\}$, then $\forall \bar{x}(\alpha \rightarrow \varphi)$ and $\exists \bar{x}(\alpha \wedge \varphi)$ belong to GF as well. Equality atoms are allowed to occur anywhere including as guards. To define the *guarded fragment with transitive guards* (GF+TG), we additionally fix a subset $\tau_0 \subseteq \tau$ of *transitive* predicates, and consider only those constant-free GF-formulas where the transitive predicates do not appear outside guards. For a τ-structure $\mathfrak{A}$ to be a model of a GF+TG-sentence φ, we require all predicates of τ_0 to be interpreted as transitive relations in $\mathfrak{A}$. The *two-variable guarded fragment with transitive guards* is denoted GF^2+TG. In the *monadic* fragment of GF^2+TG (MGF^2+TG) all non-unary relations may appear in guards only.

If $\mathfrak{A}$ is a τ-structure with the universe A and $B \subseteq A$, then $\mathfrak{A}{\upharpoonright}B$ denotes the substructure of $\mathfrak{A}$ induced on B. An *atomic τ-type* $t(x_1, \ldots, x_n)$ is a maximal consistent set of τ-literals (atoms or negated atoms) whose constituent terms are among the variables $x_1, \ldots, x_n$. In a τ-structure $\mathfrak{A}$ the atomic type $\mathrm{tp}^{\mathfrak{A}}(\bar{a})$ of a tuple $\bar{a}$ is the unique atomic type $t(\bar{x})$ such that $\mathfrak{A} \models t(\bar{a})$. A *conjunctive query* (CQ) is a sentence of the form $\exists \bar{x}\, \varphi(\bar{x})$, where φ is a conjunction of atomic formulas possibly with constants. *Boolean conjunctive queries* (BCQ) are conjunctive queries without free variables. A *union of (Boolean) conjunctive queries* (UCQ) is a disjunction of a finite number of (Boolean) conjunctive queries. By abuse of notation, sometimes we consider a query $q \in$ UCQ as a set of conjunctive queries. To every query $q \in$ BCQ of the form $\exists \bar{x}\, \varphi(\bar{x})$ over τ, one can associate the τ-structure $\mathfrak{Q}$ having as universe the set of variables in $\bar{x}$ and atoms as prescribed by φ. Then, $\mathfrak{A} \models q$ iff there exists a homomorphism $h : \mathfrak{Q} \to \mathfrak{A}$. We say that q is *acyclic* if the associated structure $\mathfrak{Q}$ is acyclic. By $var(q)$ (resp., $const(q)$) we denote the set of variables (resp., constants) occurring in q. The main problem tackled in this paper, called *UCQ answering*, is defined as follows: given a database Δ, which is actually a conjunction of ground atoms, a first-order theory φ, and a Boolean query $q \in$ UCQ, decide whether $(\Delta \wedge \varphi) \models q$, written $(\Delta, \varphi) \models q$. The decision version of the problem for non-Boolean queries, which asks whether a tuple of constants belongs to the answer of the query, can be reduced to the same problem for Boolean queries.

3 Undecidability of Querying the GF^2+TG

In contrast to decidability of satisfiability for GF+TG, we show undecidability of UCQ answering under GF^2+TG.

Theorem 1. *UCQ answering for* GF^2+TG *is undecidable, even if we consider only one transitive relation, equality-free sentences, and an empty database.*

Proof (sketch). We first construct a GF^2+TG-sentence φ_{grid} such that every infinite model of φ_{grid} is grid-like. The signature of φ_{grid} is constituted by two binary relations H and V (grid relations), a binary relation $\bar{H}$ (an auxiliary relation which will allow us to encode a crucial non-guarded sentence as part of the query), a transitive relation T, and unary relations $c_{i,j}$, where $0 \leq i \leq 3$ and $0 \leq j \leq 1$ ($c_{i,j}$ describes elements of the grid whose column number modulo 4 is i, and whose row number modulo 2 is j).

Let φ_0 be the conjunction of the initial formulas which fixes the starting point, and constructs the horizontal and vertical edges of the grid. Let φ_1 be the additional conjunction of formulas which asserts that certain elements connected by T are also connected by the horizontal grid relation:

$$\bigwedge_{i=0,2}\bigwedge_{j=0,1} \Big(\forall x \forall y \big(T(x,y) \wedge c_{i,j}(x) \wedge c_{i+1,j}(y) \to (H(x,y) \leftrightarrow \neg \bar{H}(x,y)) \big) \Big)$$

$$\bigwedge_{i=1,3}\bigwedge_{j=0,1} \Big(\forall x \forall y \big(T(y,x) \wedge c_{i,j}(x) \wedge c_{i+1,j}(y) \to (H(x,y) \leftrightarrow \neg \bar{H}(x,y)) \big) \Big).$$

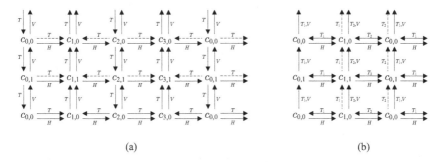

(a)   (b)

Fig. 1. Grid structure for (a) $\mathsf{GF}^2 + \mathsf{TG}$ and (b) GTGD^2 + transitivity

We need to guarantee that H is complete over V. Denote $\gamma_{i,j} := c_{i,j}(x) \wedge c_{i+1,j}(y) \wedge c_{i,j+1}(x') \wedge c_{i+1,j+1}(y')$, $\psi_0 = \psi_2 := T(x',x) \wedge T(x,y) \wedge T(y,y') \wedge T(x',y')$, and $\psi_1 = \psi_3 := T(x,x') \wedge T(y,x) \wedge T(y',y) \wedge T(y',x')$. The completeness of H over V is achieved by the conjunction of formulas φ_2:

$$\bigwedge_{i=0,1,2,3} \bigwedge_{j=0,1} \Big(\forall x \forall y \forall x' \forall y' \big(\gamma_{i,j} \wedge \psi_i \wedge H(x,y) \wedge V(x,x') \wedge V(y,y') \rightarrow H(x',y') \big) \Big).$$

Let $\varphi_{grid} = \varphi_0 \wedge \varphi_1 \wedge \varphi_2$. It can be shown that a grid structure as the one in Figure 1(a), where dashed arrows represent induced edges due to transitivity, appears in every infinite model of φ_{grid}. By using the infinite grid, where its i-th horizontal line represents the i-th configuration of a deterministic Turing machine M over an empty input tape, we can now simulate M by constructing a GF^2-sentence φ_M such that M halts iff $\varphi_{grid} \wedge \varphi_M \wedge \neg \exists x\, halt(x)$ is unsatisfiable; thus, the latter is an undecidable problem. We now show that this undecidable problem can be reduced to UCQ answering for $\mathsf{GF}^2 + \mathsf{TG}$. The non-guarded sentence φ_2 is equivalent to $\neg \hat{\varphi}_2$, where $\hat{\varphi}_2$ is:

$$\bigvee_{i=0,1,2,3} \bigvee_{j=0,1} \Big(\exists x \exists y \exists x' \exists y' \big(\gamma_{i,j} \wedge \psi_i \wedge H(x,y) \wedge V(x,x') \wedge V(y,y') \wedge \bar{H}(x',y') \big) \Big).$$

Thus, by letting $\hat{\varphi}_{grid} = \varphi_0 \wedge \varphi_1$, we get that $\varphi_{grid} \wedge \varphi_M \wedge \neg \exists x\, halt(x)$ is equivalent to $\hat{\varphi}_{grid} \wedge \varphi_M \wedge \neg(\hat{\varphi}_2 \vee \exists x\, halt(x))$. Hence, $\varphi_{grid} \wedge \varphi_M \wedge \neg \exists x\, halt(x)$ is unsatisfiable iff $\hat{\varphi}_{grid} \wedge \varphi_M \models \hat{\varphi}_2 \vee \exists x\, halt(x)$. The claim follows by observing that $\hat{\varphi}_{grid} \wedge \varphi_M$ is a $\mathsf{GF}^2 + \mathsf{TG}$-sentence, while $(\hat{\varphi}_2 \vee \exists x\, halt(x)) \in \mathsf{UCQ}$. $\qquad\square$

Interestingly, the above proof shows that query answering for two-variable guarded disjunctive TGDs (i.e., guarded TGDs extended with disjunction in rule-heads [14]) with transitive guards ($\mathsf{GDTGD}^2 + \mathsf{TG}$) is undecidable. In fact, the sentence φ_1 of the form $\forall \bar{x}(\varphi \rightarrow (\alpha \leftrightarrow \neg \beta))$, which is the only part of the above construction that is not constituted by guarded TGDs, is equivalent to the following conjunction of formulas: $\neg \exists \bar{x}(\varphi \wedge \alpha \wedge \beta) \wedge \forall \bar{x}(\varphi \rightarrow \alpha \vee \beta)$. Notice that the first conjunct forms a negated query, while the second one is a guarded disjunctive TGD. The next result follows.

Corollary 1. *UCQ answering for* $\mathsf{GDTGD}^2 + \mathsf{TG}$ *is undecidable, even if we consider only one transitive relation, and an empty database.*

Querying the GTGD^2 + **Transitivity.** A challenging issue which remains open is whether the above undecidability result holds also for guarded TGDs with transitive guards. However, transitivity without the restriction to guards cannot be safely combined with guarded TGDs, which extends the known undecidability result for GF with two transitive relations [15,16]. This result is of high relevance in the context of classical databases, where TGDs form a central class of integrity constraints [17].

Theorem 2. *UCQ answering for* GTGD^2 + *transitivity is undecidable, even if we consider only two transitive relations, an empty database, and an atomic CQ.*

Proof (sketch). It is possible to construct a GTGD^2 + transitivity-sentence φ_{grid} such that a grid structure as the one depicted in Figure 1(b) appears in every infinite model of φ_{grid}. Having the grid relations H and V in place, we can exploit the sentence φ_M, used also in the proof of Theorem 1, in order to simulate the behavior of a deterministic Turing machine M over an empty input tape. □

4   Querying the Monadic Fragment of $\mathsf{GF}^2 + \mathsf{TG}$

We show that UCQ answering under $\mathsf{MGF}^2 + \mathsf{TG}$ is decidable. If a definition or a result is related to a two-variable logic, since there is little to be gained by allowing predicates of arity bigger than two, we concentrate on unary and binary predicates only.

Ramified Models and Δ**-satisfiability.** Query answering is at least as hard as (un)satisfiability: if P is a predicate not occurring in Δ or φ, then $(\Delta, \varphi) \models \exists x\, P(x)$ iff $(\Delta \wedge \varphi)$ is unsatisfiable. Hence, we first study the related problem of Δ-satisfiability: given a conjunction of ground atoms Δ and a formula φ, decide whether $(\Delta \wedge \varphi)$ is satisfiable. We establish decidability of the problem and exact complexity bounds. Notice that in the presence of Δ, existing algorithms deciding satisfiability of fragments of GF with transitivity cannot be applied directly since constants are not allowed there.

Recall that one of the main properties of GF is the tree-model property saying that every satisfiable guarded formula has a model $\mathfrak{A}$ whose treewidth is bounded by the number of variables in the formula [2]. Also, it is known that there exists a tree decomposition of $\mathfrak{A}$ such that each of its bags is guarded in $\mathfrak{A}$ [9]. It is easy to see that these properties are not preserved if we consider GF + TG-formulas. However, it was shown that any satisfiable GF + TG-formula has a special *ramified* model [5]. We show that models with similar properties can be also found in the presence of Δ. Ramified models will be useful for both Δ-satisfiability and UCQ answering. Grädel's analysis for GF [2] uses the so-called Scott normal form corresponding to a relational Skolemisation. For GF + TG the following variant of the normal form turned out to be useful.

Definition 1. *A* GF + TG-*sentence in* normal form *is a conjunction of sentences of the form: (1)* $\exists x\, (\alpha(x) \wedge \vartheta(x))$*, (2)* $\forall \bar{x}\, (\alpha(\bar{x}) \rightarrow \exists y\, (\beta(\bar{x}, y) \wedge \vartheta(\bar{x}, y)))$*, or (3)* $\forall \bar{x}\, (\alpha(\bar{x}) \rightarrow \vartheta(\bar{x}))$*, where* $y \notin \bar{x}$*,* α*,* β *are atomic, and* ϑ *is quantifier-free without a transitive letter.*

Lemma 1 ([5]). *With every* GF + TG-*sentence* φ *of length* n *over* τ *one can associate a set* Φ *of* GF + TG-*sentences in normal form over an extended signature* σ *such that:*

(1) φ is satisfiable iff $\bigvee_{\psi \in \Phi} \psi$ is satisfiable, (2) $|\Phi| \leq \mathcal{O}(2^n)$, $|\sigma| \leq n$, $width(\sigma) = width(\tau)$, and for every $\psi \in \Phi$, $|\psi| = \mathcal{O}(n \log n)$, and (3) Φ can be computed in 2EXPTIME, *and every sentence $\psi \in \Phi$ can be computed in* PTIME *w.r.t. n.*

Also in our case we might restrict attention to formulas in normal form.

Lemma 2. *Let Δ be a conjunction of ground atoms, and $\varphi \in$ GF+TG. Then, $(\Delta \wedge \varphi)$ has a (finite) model iff $(\Delta \wedge \bigvee_{\psi \in \Phi} \psi)$ has a (finite) model, where Φ is as in Lemma 1.*

Intuitively, the key property of the *ramified* models for GF+TG-sentences can be described as follows: if we eliminate the atoms induced due to transitivity during the construction of a ramified model, then the obtained structure $\mathfrak{A}$ has bounded treewidth, and there exists a tree decomposition of $\mathfrak{A}$ such that each of its bags is guarded and single-transitive. For the monadic case, the graph of a ramified model after removing atoms induced due to transitivity can be seen as a forest with roots arbitrarily connected through Δ, and where every edge is labeled with only one binary relation.

Definition 2. *Let Δ be a conjunction of ground atoms with $D = dom(\Delta)$, φ be an* MGF^2+TG*-sentence in normal form, and $\mathfrak{R} \models \Delta \wedge \varphi$ with $R \supseteq D$. We say that $\mathfrak{R}$ is Δ-ramified if there exists a set S of root choices, where $D \subseteq S \subseteq R$, and a finite set of trees $\{\mathcal{T}_s\}_{s \in S}$ rooted at elements from S such that the following conditions hold:*

1. *every $a \in R$ is a node in one of the trees of $\{\mathcal{T}_s\}_{s \in S}$ and for every $s, s' \in S$ such that $s \neq s'$, $\mathcal{T}_s$ and $\mathcal{T}_{s'}$ are disjoint;*
2. *for every conjunct γ of φ of form (1), there is $a \in S$ such that a is a witness of γ;*
3. *for every $a \in R$, for every conjunct γ of φ of form (2), if a is not a self-witness of γ, then one of the successors of a in its tree is a witness of γ for a;*
4. *for every pair of distinct elements $a, b \in R \setminus D$, for every $P, P' \in \tau$, if $\mathfrak{R} \models P(a, b)$ and $\mathfrak{R} \models P'(a, b)$, then $P = P'$;*
5. *for every $a \in R \setminus D$, for every $T, T' \in \tau_0$, if $\mathfrak{R} \models T(a, a)$ and $\mathfrak{R} \models T'(a, a)$, then $T = T'$ and $a \in S$;*
6. *for every pair of distinct elements $a, b \in R \setminus D$, for every $P \in \tau \setminus \tau_0$, if $\mathfrak{R} \models P(a, b)$, then a and b are neighbors in one of the trees of $\{\mathcal{T}_s\}_{s \in S}$;*
7. *for every pair of distinct elements $a, b \in R \setminus D$, for every $T \in \tau_0$, if $\mathfrak{R} \models T(a, b)$ then either a and b are in the same subtree $\mathcal{T}_s$ and there is a T-path in $\mathcal{T}_s$ from a to b, or there are $c, c' \in D$ such that there is a T-path from a to c in $\mathcal{T}_c$, there is a T-path from c' to b in $\mathcal{T}_{c'}$ and, if $c \neq c'$, then $\mathfrak{R} \models T(c, c')$.*

Theorem 3. *Let Δ be a conjunction of ground atoms, φ an* MGF^2+TG*-sentence in normal form, and $\mathfrak{A} \models \Delta \wedge \varphi$. Then, there is a Δ-ramified model $\mathfrak{R}$ for $(\Delta \wedge \varphi)$ and a homomorphism h that maps $\mathfrak{R}$ to $\mathfrak{A}$.*

Proof (sketch). Let $\mathfrak{A}$ be a model of $(\Delta \wedge \varphi)$ and $D = dom(\Delta)$. For every conjunct γ of φ of form (1) pick a witness $a_\gamma \in A$, and let S be the union of D with the set of the elements a_γ. The structure $\mathfrak{R}$ is defined as an unraveling of $\mathfrak{A}$ from elements of S. More precisely, we start defining $\mathfrak{R}_0 = \mathfrak{A} \upharpoonright S$ and taking $h : \mathfrak{R}_0 \to \mathfrak{A}$ to be the identity on $\mathfrak{R}_0$. Then, we proceed inductively. For every $i \geq 0$, we extend the structure $\mathfrak{R}_i$ to $\mathfrak{R}_{i+1}$ by adding, for every element added to $\mathfrak{R}_i$ in step i, witnesses for conjuncts of φ of the form (2). In a single step of this stage, we have $a \in R_i$, $\gamma = \forall x(\alpha(x) \to \exists y \beta(x, y) \wedge \vartheta(x, y))$ and $h(a) \in A$. As $\mathfrak{A} \models \gamma$ we can find a witness b' of γ for $h(a)$ in $\mathfrak{A}$. We add a new element b to R_{i+1}, define $h(b) = b'$ and define $\text{tp}^{\mathfrak{R}_{i+1}}(a, b)$ using the relevant part (identified by the guard β) of $\text{tp}^{\mathfrak{A}}(h(a), b')$. After adding b as a witness of a conjunct γ with β, where β is a T-atom, we ensure that b is connected by the transitive relation T to all elements c connected to a via a T-path in $\mathfrak{R}_i$, defining $\text{tp}^{\mathfrak{R}_{i+1}}(b, c)$ from the corresponding 2-type $\text{tp}^{\mathfrak{A}}(h(b), h(c))$. Other pairs of distinct elements in $\mathfrak{R}_{i+1}$ are connected using only negative 2-types, i.e., they are not in the interpretation of any binary relation. One can show that $\mathfrak{R}$ is a ramified model as defined in Definition 2, and that h is a homomorphism from $\mathfrak{R}$ to $\mathfrak{A}$. $\qquad \square$

We can now show that Δ-satisfiability of $\text{MGF}^2 + \text{TG}$-sentences is decidable.

Theorem 4. *Δ-satisfiability of $\text{MGF}^2 + \text{TG}$ is* 2EXPTIME-*complete in combined complexity, and* NP-*complete in data complexity.*

Proof (sketch). We compute (in exponential space w.r.t. $|\varphi|$) the set Φ of Lemma 1. Then we guess a sentence $\varphi \in \Phi$ and check whether φ has a Δ-ramified model. To check the latter we first guess a structure $\mathfrak{D}$ of size at most $|dom(\Delta)| + |\varphi|$ interpreting Δ and containing witnesses for conjuncts of φ of the form (1). Then we universally choose an element $d \in D$ and check whether $\mathfrak{D}$ can be extended to a Δ-ramified model with D being the set of root choices. Observe that in the next steps of the procedure, it suffices to keep for each element a in the model a description of 1-types occurring on transitive paths from and to a. This information can be stored using exponential size with respect to $|\varphi|$. An alternating procedure working in exponential space can be naturally derived from the construction. If only Δ is considered as part of the input, then the above procedure works in nondeterministic polynomial time. For the lower bounds, it is known that satisfiability of $\text{MGF}^2 + \text{TG}$ is already 2EXPTIME-hard [6]. For data complexity the corresponding NP-hard lower bound follows from Theorem 2 in [18]. $\square$

It is important to say that by adapting the notion of ramified models introduced in [5] for $\text{GF} + \text{TG}$, one can show that Δ-satisfiability for $\text{GF} + \text{TG}$-sentences is decidable, and of the same complexity. Details are omitted due to space limits.

Query Answering via Unsatisfiability. We now investigate query answering under $\text{MGF}^2 + \text{TG}$. Given a database Δ and a query $q \in \text{UCQ}$, our goal is to construct a sentence $\Phi_{\Delta,q}$ which enjoys the following properties: *(i)* for each $\varphi \in \text{MGF}^2 + \text{TG}$, $(\Delta, \varphi) \models q$ iff $(\Delta^\star, \varphi) \models \Phi_{\Delta,q}$, where $\Delta^\star \supseteq \Delta$, and *(ii)* $\Phi_{\Delta,q}$ is equivalent to an $\text{MGF}^2 + \text{TG}$-sentence. Since $(\Delta^\star, \varphi) \models \Phi_{\Delta,q}$ iff $(\Delta^\star \wedge \varphi \wedge \neg \Phi_{\Delta,q})$ is unsatisfiable, we can then rely on the results regarding Δ-satisfiability of $\text{MGF}^2 + \text{TG}$-sentences. Let us first introduce the class of *single-acyclic* queries.

Definition 3. *A query $q \in \text{BCQ}$ is* single-acyclic *if the following conditions hold: (i) q is acyclic, (ii) for every pair of distinct variables $x, y \in var(q)$, q contains at most one atom containing both x and y, and (iii) there exists at most one transitive predicate T such that a reflexive atom of the form $T(x, x)$, where $x \in var(q)$, occurs in q.*

As shown in [19], a query $q \in$ BCQ is *acyclic* iff there exists a query $p \in$ BCQ equivalent to q which is also in GF. By exploiting this result, one can easily show that a single-acyclic query is equivalent to a sentence of $\mathsf{MGF}^2 + \mathsf{TG}$.

Lemma 3. *For each single-acyclic query $q \in$ BCQ over τ, there is an $\mathsf{MGF}^2 + \mathsf{TG}$-sentence χ_q of size linear in $|q|$ such that, for every τ-structure $\mathfrak{A}$, $\mathfrak{A} \models q$ iff $\mathfrak{A} \models \chi_q$.*

Fix a database Δ and a query $q \in$ UCQ. Having the notion of single-acyclic queries in place, we are now ready to construct the sentence $\Phi_{\Delta,q} = \bigvee_{p \in q} \phi_{\Delta,p}$, where each disjunct $\phi_{\Delta,p}$ is a union of single-acyclic BCQs constructed as described below. Clearly, if $(\Delta, \varphi) \models p$, then there exists a homomorphism h that maps p to each model of $(\Delta \wedge \varphi)$, and thus to each Δ-ramified model of $(\Delta \wedge \varphi)$. The key idea underlying our construction is, for each such mapping h, to describe the image $h(p)$ of p in each Δ-ramified model of $(\Delta \wedge \varphi)$ by a union of single-acyclic BCQs. As we shall see below, for query answering purposes, it suffices to focus on the Δ-ramified models. The formal construction of $\phi_{\Delta,p}$ is as follows. If p is single-acyclic, then $\phi_{\Delta,p}$ coincides with p; otherwise, we apply the following steps:

1. **Enumerate all the possible mappings.** Let $H = \{h \mid h$ is a mapping $var(p) \to (var(p) \cup dom(\Delta))\}$. For $h \in H$, we denote $h(p)_{\blacktriangledown}$ the maximal subset of $h(p)$ which contains only constants of $dom(\Delta)$, while $h(p)_{\blacktriangle} = h(p) \setminus h(p)_{\blacktriangledown}$.
2. **Partition $h(p)$ into different subtrees of a Δ-ramified model.** For each $h \in H$, let P_h be the set of all possible n-tuples $\langle \langle S_1, c_1 \rangle, \dots, \langle S_n, c_n \rangle \rangle$, where $1 \le n \le |var(p)|$, such that: (i) $\{S_1, \dots, S_n\}$ is a partition of $var(p)$, and (ii) $\{c_1, \dots, c_n\}$ is a subset of $(dom(\Delta) \cup \{\star_1, \dots, \star_n\})$, where $\star_i$'s are auxiliary variables.
3. **Focus on a subtree by eliminating crossing transitive edges.** For each $h \in H$, let ϕ_h be the set of all possible BCQs that can be constructed as follows: for each $\langle \langle S_1, c_1 \rangle, \dots, \langle S_n, c_n \rangle \rangle \in P_h$, replace each atom $T(x, y)$ of $h(p)_{\blacktriangle}$, where $T \in \tau_0$, with $T(x, c_i) \wedge T(c_i, c_j) \wedge T(c_j, y)$ (or with $T(x, c) \wedge T(c, y)$ if $c_i = c_j = c$) if there exist $1 \le i \ne j \le n$ such that $x \in S_i$ and $y \in S_j$.
4. **Eliminate constants.** For each $h \in H$, let ϕ_h^- be the set of BCQs defined as follows: for each $p' \in \phi_h$, in ϕ_h^- there exists a BCQ obtained from $p' \wedge h(p)_{\blacktriangledown}$ by replacing, for each constant c in $p' \wedge h(p)_{\blacktriangledown}$, each occurrence of c with the variable $\Diamond_c$, and adding the conjunct $R_c(\Diamond_c)$, where R_c is an auxiliary unary predicate.
5. **Describe the image.** For each $h \in H$, and for each BCQ $p' \in \phi_h^-$:
   (a) Let p'_Δ be the maximal subset of p' such that $dom(p'_\Delta) \subseteq \{\Diamond_c\}_{c \in dom(\Delta)}$, and let $p'_1, \dots, p'_m$ be the maximal connected components of $p' \setminus p'_\Delta$.
   (b) For each $i \in \{1, \dots, m\}$, if p'_i is single-acyclic, then $Q_{p'_i} = \{p'_i\}$; otherwise, let $Q_{p'_i}$ be the set of all single-acyclic BCQs which *entail* p'_i that can be constructed by eliminating *induced*[1] transitive atoms from p'_i; let $Q_{p'_{m+1}} = \{p'_\Delta\}$.
   (c) If there exists $i \in \{1, \dots, m\}$ such that $Q_{p'_i} = \varnothing$, then $\overline{\phi_h^-} = \bot$; otherwise, $\overline{\phi_h^-}$ is defined as $\bigvee_{p' \in \phi_h^-} (\times_{1 \le i \le m+1} Q_{p'_i})$, where $\times_{1 \le i \le m+1} Q_{p'_i}$ are all the BCQs that can be constructed by keeping exactly one BCQ from each $Q_{p'_i}$.
6. **Finalization.** Let $\phi_{\Delta,p} = \bigvee_{h \in H} \overline{\phi_h^-}$.

---

[1] An atom $T(x, z)$ is induced if atoms of the form $T(x, y)$ and $T(y, z)$ already occur in p'_i.

Before showing soundness and completeness of our construction, let us first establish two auxiliary results. The first one states that, if we focus our attention on Δ-ramified models, then our construction is complete. In the sequel, let $\Delta^\star = \Delta \wedge \bigwedge_{c \in dom(\Delta)} R_c(c)$.

Lemma 4. *Consider a conjunction of ground atoms Δ, and a query $q \in$ UCQ. Given a sentence $\varphi \in$ MGF^2+TG, if $(\Delta \wedge \varphi)$ is consistent and $(\Delta, \varphi) \models q$, then $\mathfrak{R} \models \Phi_{\Delta,q}$, for each Δ-ramified model $\mathfrak{R}$ of $(\Delta^\star \wedge \varphi)$, and $\Phi_{\Delta,q} \not\equiv \bot$.*

Proof (sketch). By hypothesis, there exists $p \in$ BCQ in q such that $(\Delta, \varphi) \models p$. Fix a Δ-ramified model $\mathfrak{R}$ for $(\Delta \wedge \varphi)$, which exists by Theorem 3, and extend it to a ramified model $\mathfrak{R}^\star$ interpreting the auxiliary symbols from $\Delta^\star$. As $\mathfrak{R}^\star \models (\Delta^\star \wedge \varphi)$, we have $\mathfrak{R}^\star \models p$. Let h be the homomorphism that maps p into $\mathfrak{R}^\star$. Obviously, $h \in H$. Let S be the set of root choices in $\mathfrak{R}^\star$. Using S we define a partition of $var(p)$ into subsets mapped into the same subtree $\mathcal{T}_s$ of $\mathfrak{R}^\star$. By construction, ϕ_h is nonempty and one can show that there exists at least one disjunct γ in $\phi_{\Delta,p}$ such that $\mathfrak{R} \models \gamma$. $\qquad\square$

For query answering we can consider only the Δ-ramified models of a theory.

Lemma 5. *Consider a conjunction of ground atoms Δ, a sentence $\varphi \in$ MGF^2+TG, and a query $q \in$ UCQ. $(\Delta, \varphi) \models q$ iff $\mathfrak{R} \models q$, for each Δ-ramified model $\mathfrak{R}$ of $(\Delta \wedge \varphi)$.*

Proof. $(\Rightarrow)$ By hypothesis, each model of $(\Delta \wedge \varphi)$ entails q, and the claim follows. $(\Leftarrow)$ Towards a contradiction, assume that each Δ-ramified model entails q, but $(\Delta, \varphi) \not\models q$. The latter implies that there exists a model $\mathfrak{A}$ of $(\Delta \wedge \varphi)$ such that $\mathfrak{A} \not\models q$. By Theorem 3, there exists a ramified model $\mathfrak{R}$ of $(\Delta \wedge \varphi)$, and a homomorphism h that maps $\mathfrak{R}$ into $\mathfrak{A}$. Since $\mathfrak{R} \models q$, there exists a homomorphism μ that maps q into $\mathfrak{R}$. Therefore, $h \circ \mu$ maps q into $\mathfrak{A}$, and thus $\mathfrak{A} \models q$ which is a contradiction. The claim follows. $\qquad\square$

We are now ready to establish soundness and completeness of the construction.

Theorem 5. *Consider a conjunction of ground atoms Δ, and a query $q \in$ UCQ. For each $\varphi \in$ MGF^2+TG, $(\Delta, \varphi) \models q$ iff $(\Delta^\star, \varphi) \models \Phi_{\Delta,q}$.*

Proof. $(\Rightarrow)$ If $(\Delta \wedge \varphi)$ is not consistent, then also $(\Delta^\star \wedge \varphi)$ is not consistent and the claim follows. In case that $(\Delta \wedge \varphi)$ is consistent, the claim follows immediately from Lemmas 4 and 5. $(\Leftarrow)$ By hypothesis, there exists a BCQ $p \in \Phi_{\Delta,q}$ such that $(\Delta^\star, \varphi) \models p$. By construction, p entails q, and thus $(\Delta^\star, \varphi) \models q$. The auxiliary predicates of the form R_c, where $c \in dom(\Delta)$, being introduced only during the construction of $\Phi_{\Delta,q}$, do not match any predicate in q, and hence $(\Delta, \varphi) \models q$. $\qquad\square$

Let us now investigate the complexity of the obtained formula. For brevity, let $r = |\tau|$. Also, given a query $q \in$ UCQ, let $H_q = \max_{p \in q} |p|$ and $V_q = \max_{p \in q} |var(p)|$.

Lemma 6. *Consider a conjunction of ground atoms Δ, and a query $q \in$ BCQ. It holds that (1) $|\Phi_{\Delta,q}|$ is at most $|q| \cdot (V_q + |dom(\Delta)|)^{V_q} \cdot r^{\mathcal{O}(H_q)} \cdot H_q^{\mathcal{O}(H_q)}$, (2) $H_{\Phi_{\Delta,q}} = \mathcal{O}(H_q)$, and (3) $\Phi_{\Delta,q}$ can be constructed in EXPTIME w.r.t. q, and in PTIME w.r.t. Δ.*

From Theorem 5 we get that $(\Delta, \varphi) \models q$ iff $(\Delta^\star \wedge \varphi \wedge \neg\Phi_{\Delta,q})$ is unsatisfiable. By Lemma 6, we can construct $\Phi_{\Delta,q}$ in EXPTIME w.r.t. q. Then, for each of its disjuncts p, which are exponentially many, we call the 2EXPTIME algorithm for Δ-satisfiability of MGF^2+TG-sentences, provided by Theorem 4, in order to check whether $(\Delta^\star \wedge \varphi \wedge \neg p)$ is unsatisfiable. Since $|p|$ is linear w.r.t. q, the above describes a 2EXPTIME procedure for query answering. Now, in case that both φ and q are fixed, it is easy to see that Theorem 4 and Lemma 6 give us a coNP procedure for query answering. The double-exponential time lower bound follows from the fact that satisfiability of MGF^2+TG-sentences is 2EXPTIME-hard [6]. The coNP-hardness is inherited immediately from [20], where it was shown that query answering under a single sentence of the form $\forall x(R_1(x) \to R_2(x) \vee R_3(x))$ is coNP-hard in data complexity.

Theorem 6. *UCQ answering for* MGF^2+TG *is* 2EXPTIME-*complete in combined complexity, and* coNP-*complete in data complexity.*

Restricting the Query. We conclude this section by briefly discussing how we can gain decidability of query answering under GF+TG. Although single-acyclic queries were tailored for MGF^2+TG, it turned out that they can be naturally extended to *single-transitive-acyclic* queries, which are suitable for querying arbitrary GF+TG-sentences.

Definition 4. *A query $q \in$ BCQ is single-transitive-acyclic if the following holds: (i) q is acyclic, and (ii) for each hyperedge e in the hypergraph of q, there exists at most one pair of distinct variables $x, y \in e$, and at most one $T \in \tau_0$ such that $T(x, y)$ is in q.*

Since a BCQ q is acyclic iff there exists a guarded BCQ equivalent to q [19], one can show that a single-transitive-acyclic query is equivalent to a GF+TG-sentence.

Lemma 7. *For each single-transitive-acyclic query $q \in$ BCQ over τ, there is a GF+TG-sentence χ_q of size linear in $|q|$ such that, for every τ-structure $\mathfrak{A}$, $\mathfrak{A} \models q$ iff $\mathfrak{A} \models \chi_q$.*

Given a database Δ, a sentence $\varphi \in$ GF+TG, and a single-transitive-acyclic BCQ q, the above lemma implies that one can decide whether $(\Delta, \varphi) \models q$ just by checking if the sentence $(\Delta \wedge \varphi \wedge \neg\chi_q)$, where $(\varphi \wedge \neg\chi_q) \in$ GF+TG, is unsatisfiable. Since the results on Δ-satisfiability established above hold, not only for MGF^2+TG, but for the whole fragment under consideration, we get the following complexity result.

Theorem 7. *UCQ answering under* GF+TG *with single-transitive-acyclic BCQs is* 2EXPTIME-*complete in combined complexity, and* coNP-*complete in data complexity.*

5   Future Work

We state three open problems for query answering. The first one concerns the decidability of guarded TGDs with transitive guards. The second one is whether MGF^2+TG can be safely combined with counting quantifiers, an important feature for many computational logics. Finally, the third one is to pinpoint the complexity of MGF^2+TG under finite models; recall that in this work we considered arbitrary (finite or infinite) models. For the latter, since MGF^2+TG does not enjoy the finite model property, completely new techniques are needed.

Acknowledgements. Georg Gottlob and Lidia Tendera acknowledge the EPSRC Grant EP/H051511/1 "ExODA". Lidia Tendera also gratefully acknowledges her association with St. John's College during her visit to Oxford in 2012, and the support of Polish Ministry of Science and Higher Education Grant N N206 37133. Andreas Pieris acknowledges the ERC Grant 246858 "DIADEM" and the EPSRC Grant EP/G055114/1 "Constraint Satisfaction for Configuration: Logical Fundamentals, Algorithms and Complexity".

References

1. Andréka, H., van Benthem, J., Németi, I.: Modal languages and bounded fragments of predicate logic. J. Philosophical Logic 27, 217–274 (1998)
2. Grädel, E.: On the restraining power of guards. J. Symb. Log. 64(4), 1719–1742 (1999)
3. Ganzinger, H., Meyer, C., Veanes, M.: The two-variable guarded fragment with transitive relations. In: Proc. of LICS, pp. 24–34 (1999)
4. Szwast, W., Tendera, L.: On the decision problem for the guarded fragment with transitivity. In: Proc. of LICS, pp. 147–156 (2001)
5. Szwast, W., Tendera, L.: The guarded fragment with transitive guards. Ann. Pure Appl. Logic 128(1-3), 227–276 (2004)
6. Kieroński, E.: The two-variable guarded fragment with transitive guards is 2EXPTIME-hard. In: Gordon, A.D. (ed.) FOSSACS 2003. LNCS, vol. 2620, pp. 299–312. Springer, Heidelberg (2003)
7. Calì, A., Gottlob, G., Kifer, M.: Taming the infinite chase: Query answering under expressive relational constraints. In: Proc. of KR, pp. 70–80 (2008)
8. Calvanese, D., De Giacomo, G., Lembo, D., Lenzerini, M., Rosati, R.: Tractable reasoning and efficient query answering in description logics: The DL-Lite family. J. Autom. Reasoning 39(3), 385–429 (2007)
9. Bárány, V., Gottlob, G., Otto, M.: Querying the guarded fragment. In: Proc. of LICS, pp. 1–10 (2010)
10. Baget, J.F., Mugnier, M.L., Rudolph, S., Thomazo, M.: Walking the complexity lines for generalized guarded existential rules. In: Proc. of IJCAI, pp. 712–717 (2011)
11. Krötzsch, M., Rudolph, S.: Extending decidable existential rules by joining acyclicity and guardedness. In: Proc. of IJCAI, pp. 963–968 (2011)
12. Pratt-Hartmann, I.: Data-complexity of the two-variable fragment with counting quantifiers. Inf. Comput. 207(8), 867–888 (2009)
13. Bárány, V., ten Cate, B., Segoufin, L.: Guarded negation. In: Aceto, L., Henzinger, M., Sgall, J. (eds.) ICALP 2011, Part II. LNCS, vol. 6756, pp. 356–367. Springer, Heidelberg (2011)
14. Gottlob, G., Manna, M., Morak, M., Pieris, A.: On the complexity of ontological reasoning under disjunctive existential rules. In: Rovan, B., Sassone, V., Widmayer, P. (eds.) MFCS 2012. LNCS, vol. 7464, pp. 1–18. Springer, Heidelberg (2012)
15. Kazakov, Y.: Saturation-based decision procedures for extensions of the guarded fragment. PhD thesis, Universität des Saarlandes (2005)
16. Kieroński, E.: Results on the guarded fragment with equivalence or transitive relations. In: Ong, L. (ed.) CSL 2005. LNCS, vol. 3634, pp. 309–324. Springer, Heidelberg (2005)
17. Beeri, C., Vardi, M.Y.: A proof procedure for data dependencies. J. ACM 31(4), 718–741 (1984)
18. Pratt-Hartmann, I.: Complexity of the two-variable fragment with counting quantifiers. Journal of Logic, Language and Information 14(3), 369–395 (2005)
19. Gottlob, G., Leone, N., Scarcello, F.: Robbers, marshals, and guards: Game theoretic and logical characterizations of hypertree width. J. Comput. Syst. Sci. 66(4), 775–808 (2003)
20. Calvanese, D., Giacomo, G.D., Lembo, D., Lenzerini, M., Rosati, R.: Data complexity of query answering in description logics. Artif. Intell. 195, 335–360 (2013)

Contractive Signatures with Recursive Types, Type Parameters, and Abstract Types*

Hyeonseung Im[1], Keiko Nakata[2], and Sungwoo Park[3]

[1] LRI - Université Paris-Sud 11, Orsay, France
[2] Institute of Cybernetics at Tallinn University of Technology, Estonia
[3] Pohang University of Science and Technology, Republic of Korea

Abstract. Although theories of equivalence or subtyping for recursive types have been extensively investigated, sophisticated interaction between recursive types and abstract types has gained little attention. The key idea behind type theories for recursive types is to use syntactic contractiveness, meaning every μ-bound variable occurs only under a type constructor such as $\rightarrow$ or $*$. This syntactic contractiveness guarantees the existence of the unique solution of recursive equations and thus has been considered necessary for designing a sound theory for recursive types. However, in an advanced type system, such as OCaml, with recursive types, type parameters, and abstract types, we cannot easily define the syntactic contractiveness of types. In this paper, we investigate a sound type system for recursive types, type parameters, and abstract types. In particular, we develop a new semantic notion of contractiveness for types and signatures using mixed induction and coinduction, and show that our type system is sound with respect to the standard call-by-value operational semantics, which eliminates signature sealings. Moreover we show that while non-contractive types in signatures lead to unsoundness of the type system, they may be allowed in modules. We have also formalized the whole system and its type soundness proof in Coq.

1 Introduction

Recursive types are widely used features in most programming languages and the key constructs to exploit recursively defined data structures such as lists and trees. In type theory, there are two ways to exploit recursive types, namely by using the *iso-recursive* or *equi-recursive* formulation.

In the iso-recursive formulation, a recursive type $\mu X.\tau$ is considered isomorphic but not equal to its one-step unfolding $\{X \mapsto \mu X.\tau\}\tau$. Correspondingly the term language provides built-in coercion functions called fold and unfold, witnessing this isomorphism.

$$\begin{aligned} \mathsf{fold} \quad &: \{X \mapsto \mu X.\tau\}\tau \rightarrow \mu X.\tau \\ \mathsf{unfold} &: \mu X.\tau \rightarrow \{X \mapsto \mu X.\tau\}\tau \end{aligned}$$

---

* An expanded version of this paper, containing detailed proofs and omitted definitions, and the Coq development are available at `http://toccata.lri.fr/~im`.

F.V. Fomin et al. (Eds.): ICALP 2013, Part II, LNCS 7966, pp. 299–311, 2013.
© Springer-Verlag Berlin Heidelberg 2013

Although in the iso-recursive formulation programs have to be annotated with fold and unfold coercion functions, this formulation usually simplifies typecheck-ing in the presence of recursive types. For example, datatypes in SML [11] are a special form of iso-recursive types.

In contrast, the equi-recursive formulation defines a recursive type $\mu X.\tau$ to be equal to its one-step unfolding $\{X \mapsto \mu X.\tau\}\tau$ and does not require any coercion functions as in the iso-recursive formulation. For example, polymorphic variant and object types as implemented in OCaml [1] require structural recursive types and thus equi-recursive types. To use equi-recursive types, it suffices to add either of the two rules below into the type system:

$$\frac{\Gamma \vdash e : \tau \quad \tau \equiv \sigma}{\Gamma \vdash e : \sigma} \text{ typ-eq} \qquad \frac{\Gamma \vdash e : \tau \quad \tau \leq \sigma}{\Gamma \vdash e : \sigma} \text{ typ-sub}$$

Here the rule typ-eq exploits the equivalence relation $\equiv$ between recursive types, and the rule typ-sub the subtyping relation $\leq$. The equi-recursive formulation makes it easier to write programs using recursive types, but it raises a tricky problem in typechecking: we need to decide when two recursive types are in the equivalence or subtyping relation. In response, several authors have investigated theories of equivalence or subtyping for equi-recursive types [2,3,6,18].

The key idea behind type theories for equi-recursive types is to use syntactic contractiveness [9], meaning that given a recursive type $\mu X.\tau$, the use of the recursion variable X in τ must occur under a type constructor such as $\rightarrow$ or $*$. In other words, non-contractive types such as $\mu X.X$ and $\text{unit} * \mu X.X$ are rejected at the syntax level. For example, most previous work considers variants of the following type language:

$$\tau, \sigma \quad ::= \quad X \mid \tau \rightarrow \sigma \mid \mu X.(\tau \rightarrow \sigma)$$

The main reason for employing this syntactic contractiveness is to guarantee the existence of the unique solution of recursive equations introduced by equi-recursive types and obtain a sound type theory.

However, in an advanced type system, such as OCaml, with recursive types, type parameters, abstract types, and modules, we cannot easily define the syn-tactic contractiveness of types. To illustrate, consider the following code fragment which is allowed in OCaml using the "-rectypes" option. The "-rectypes" option allows arbitrary equi-recursive types.

```
module M : T = struct          module type T = sig
  type 'a t = 'a                 type 'a t
  type s = int and u = bool      type s = s t and u = u t
  let f x = x                    val f : int -> s
  let g x = x                    val g : u -> bool
end                            end
let h x = M.g (M.f x)
let y = h 3                    (* run-time error *)
```

Under the usual interpretation of ML signature sealings, the module M correctly implements, or satisfies, the signature T. Moreover the types s and u in T are considered contractive in OCaml since the type cycles are guarded by the parameterized abstract type t in T; hence the signature T is well-formed. Furthermore the types s and u in T are structurally equivalent, and thus the values h and y are well-formed with types int -> bool and bool, respectively. At run-time, however, the evaluation of y, *i.e.*, h 3, leads to an unknown constructor 3 of type bool, breaking type soundness.[1]

In this paper, we investigate a type system for equi-recursive types, type parameters, and abstract types. In our system, recursive types may be declared by using type definitions of the form type α $t = \tau$ where both the type parameter α and the recursive type t may appear in the type τ.[2] Abstract types may be declared by using the usual ML-style signature sealing operations (Section 2.1). For this system, we develop a new notion of semantic contractiveness for types and signatures using mixed induction and coinduction (Section 2.3). Our semantic contractiveness determines the types s and u in the signature T above to be non-contractive, and our type system rejects T. We then show that our type system with semantic contractiveness is sound with respect to the standard call-by-value operational semantics, which eliminates signature sealings (Section 2.4).

Another notable result is that even in the presence of non-contractive types in modules, we can develop a sound type system where well-typed programs cannot go wrong. This is particularly important since our type soundness result may give a strong hint about the soundness of OCaml, which allows us to define non-contractive types using recursive modules and signature sealings.

Our contributions are summarized as follows.

– To our knowledge, we are the first to consider a type system for equi-recursive types, type parameters, and abstract types, and define a type sound semantic notion of contractiveness.
– Since the OCaml type system allows both recursive types and abstract types, and non-contractive types in modules, our type soundness result gives a strong hint about how to establish the soundness of OCaml.
– We have formalized the whole system and its type soundness proof in Coq version 8.4. Our formalization extensively uses mixed induction and coinduction, so it may act as a good reference for using mixed induction and coinduction in Coq.

The remainder of the paper is organized as follows. Section 2 presents a type system for recursive types, type parameters, and abstract types. In particular, we consider a simple module system with a signature sealing operation and define a structural type equivalence and semantic contractiveness using mixed induction and coinduction. Section 3 discusses Coq mechanization and an algorithmic type equivalence and contractiveness. Section 4 discusses related work and Section 5 concludes.

---

[1] We discovered the above bug together with Jacques Garrigue and it has been fixed in the development version of OCaml (available in the OCaml svn repository).

[2] We do not use the usual μ-notation because encoding mutually recursive type definitions into μ-types requires type-level pairs and projection, complicating the theory. Moreover the use of type definitions better reflects the OCaml implementation.

Syntax

types	$\tau, \sigma ::= \mathsf{unit} \mid \alpha \mid \tau \to \sigma \mid \tau_1 * \tau_2 \mid \tau \, t$
expressions	$e ::= () \mid x \mid \lambda x : \tau.\, e \mid e_1 \, e_2 \mid (e_1, e_2) \mid \pi_i(e) \mid \mathsf{fix} \, x : \tau.\, e \mid l$
specifications	$D ::= \mathsf{type} \, \alpha \, t \mid \mathsf{type} \, \alpha \, t = \tau \mid \mathsf{val} \, l : \tau$
definitions	$d_\tau ::= \mathsf{type} \, \alpha \, t = \tau$ (type definitions)
	$d_e ::= \mathsf{let} \, l = e$ (value definitions)
signatures	$S ::= \cdot \mid D, S$
modules	$M ::= (\overline{d_\tau}, \overline{d_e})$
programs	$P ::= (M, S, e) \mid (M, e)$

Well-Formed types $\boxed{S; \Sigma \vdash \tau \, \mathsf{type}}$

$$\text{type variable sets } \Sigma ::= \cdot \mid \{\alpha\}$$
Standard well-formedness rules for types are omitted.

$$\frac{S \ni \mathsf{type} \, \alpha \, t \quad S; \Sigma \vdash \tau \, \mathsf{type}}{S; \Sigma \vdash \tau \, t \, \mathsf{type}} \; \text{wft-abs} \qquad \frac{S \ni \mathsf{type} \, \alpha \, t = \sigma \quad S; \Sigma \vdash \tau \, \mathsf{type}}{S; \Sigma \vdash \tau \, t \, \mathsf{type}} \; \text{wft-app}$$

Well-formed specifications and signatures $\boxed{S \vdash D \, \mathsf{ok}} \; \boxed{S \, \mathsf{ok}}$

$$\frac{}{S \vdash \mathsf{type} \, \alpha \, t \, \mathsf{ok}} \; \text{wfs-abs} \qquad \frac{S; \{\alpha\} \vdash \tau \, \mathsf{type}}{S \vdash \mathsf{type} \, \alpha \, t = \tau \, \mathsf{ok}} \; \text{wfs-type} \qquad \frac{S; \cdot \vdash \tau \, \mathsf{type}}{S \vdash \mathsf{val} \, l : \tau \, \mathsf{ok}} \; \text{wfs-val}$$

$$\frac{\mathrm{BN}(S) \, \text{distinct} \quad \forall D \in S, \; S \vdash D \, \mathsf{ok}}{S \, \mathsf{ok}} \; \text{wf-sig}$$

Fig. 1. Syntax and well-formedness

2   A Type System $\lambda_{\mathsf{abs}}^{\mathsf{rec}}$

This section presents a type system $\lambda_{\mathsf{abs}}^{\mathsf{rec}}$ for recursive types, type parameters, and abstract types, permitting non-contractive types in modules. Section 2.1 presents the syntax and inference rules for well-formedness. Section 2.2 defines a structural type equivalence and Section 2.3 defines a semantic contractiveness using mixed induction and coinduction. Finally Section 2.4 presents a type soundness result with respect to the standard call-by-value operational semantics.

2.1   Syntax and Well-Formedness

Figure 1 shows the syntax for $\lambda_{\mathsf{abs}}^{\mathsf{rec}}$ and inference rules for well-formedness. We use meta-variables α, β for type variables, s, t, u for type names, x, y, z for value variables, and l for value names. Both types and expressions are defined in the standard way as in the simply-typed λ-calculus except that they may refer to type definitions and value definitions in modules. For simplicity, we include simple signatures and modules only, which suffice to introduce abstract types, and exclude nested modules and functors. Abstract types are then introduced by a program of the form (M, S, e), called a signature sealing, which hides the implementation details of the module M behind the signature S.

Recursive types are introduced by type definitions of the form type $\alpha\ t = \tau$ where τ may refer to the name t with no restriction. For example, we may define recursive types such as type $\alpha\ t = \alpha\ t \to \alpha\ t$ and type $\alpha\ t = \alpha\ t \to (\alpha * \alpha)\ t$. We also permit a non-contractive type definition such as type $\alpha\ t = \alpha\ t$ in a module but reject a non-contractive type specification in a signature, since the latter breaks type soundness. Any sequence $\overline{d_\tau}$ of type definitions (or type specifications) in modules (or signatures) may be mutually recursive, whereas no sequence $\overline{d_e}$ of value definitions are mutually recursive. The main reason for this design choice is that our focus in this paper is to investigate the interaction between non-contractive recursive types and abstract types. Moreover, to simplify the discussion, we consider only those type constructors with a single parameter. We can easily add into the system nullary or multi-parameter type constructors.

As for well-formedness of types, we use a judgment $S; \Sigma \vdash \tau$ type to mean that type τ is well-formed under context (S, Σ). Here we use a type variable set Σ to denote either an empty set or a singleton set. We also use judgments $S \vdash D$ ok and S ok to mean that specification D and signature S are well-formed, respectively. Most of the rules are standard, and we only remark that a signature S is well-formed only if all bound type names in S are distinct from each other and each type definition is well-formed under S (rule wf-sig). These well-formedness conditions for signatures allow us to define arbitrarily mutually recursive type definitions. In the remainder of the paper, we assume that every type and signature that we mention is well-formed, without explicitly saying so.

2.2 Type Equivalence

In this section, we define a type equivalence relation in terms of unfolding type definitions. The usual β-equivalence is then embedded into our type equivalence relation by means of unfolding. We use a judgment $S \vdash \tau \rightharpoonup \sigma$ to mean that type τ unfolds into σ by expanding a type name in τ into its definition under S. The rule unfold below, which is the only rule for unfolding type definitions, implements this idea and given a type application $\tau\ t$, it replaces type variable α with argument τ in definition σ of t.

$$\frac{S \ni \text{type } \alpha\ t = \sigma}{S \vdash \tau\ t \rightharpoonup \{\alpha \mapsto \tau\}\sigma} \ \text{unfold}$$

We write $S \vdash \tau \rightharpoonup^* \sigma$ for the reflexive and transitive closure of unfolding. We say that a type τ is *vacuous* under signature S, if $\forall \tau'$, $S \vdash \tau \rightharpoonup^* \tau'$ implies $\exists \tau''$, $S \vdash \tau' \rightharpoonup \tau''$. In other words, vacuous types are those types that allow us to unfold type definitions infinitely using the rule unfold.

The type equivalence relation is defined by nesting induction into coinduction. In order to structurally compare types for their equivalence, we should be able to check equivalence for vacuous types. While using induction to compare structures of types, we use coinduction to compare infinite unfoldings of types and thus to

Coinductive type equivalence $\boxed{S; \Sigma \vdash \tau_1 \equiv \tau_2}$

$$\frac{S; \Sigma \vdash \tau \overset{\equiv}{=} \sigma}{S; \Sigma \vdash \tau \equiv \sigma} \text{ eq-ind} \qquad \frac{S \vdash \tau \rightharpoonup \tau' \quad S \vdash \sigma \rightharpoonup \sigma' \quad S; \Sigma \vdash \tau' \equiv \sigma'}{S; \Sigma \vdash \tau \equiv \sigma} \text{ eq-coind}$$

Inductive type equivalence $\boxed{S; \Sigma \vdash \tau_1 \overset{R}{=} \tau_2}$

$$\frac{}{S; \Sigma \vdash \text{unit} \overset{R}{=} \text{unit}} \text{ eq-unit} \qquad \frac{}{S; \{\alpha\} \vdash \alpha \overset{R}{=} \alpha} \text{ eq-var} \qquad \frac{S; \Sigma \vdash \tau_i \, R \, \sigma_i \quad (i = 1, 2)}{S; \Sigma \vdash \tau_1 \rightarrow \tau_2 \overset{R}{=} \sigma_1 \rightarrow \sigma_2} \text{ eq-fun}$$

$$\frac{S; \Sigma \vdash \tau_i \, R \, \sigma_i \quad (i = 1, 2)}{S; \Sigma \vdash \tau_1 * \tau_2 \overset{R}{=} \sigma_1 * \sigma_2} \text{ eq-prod} \qquad \frac{S \ni \text{type } \alpha \, t \quad S; \Sigma \vdash \tau \, R \, \sigma}{S; \Sigma \vdash \tau \, t \overset{R}{=} \sigma \, t} \text{ eq-abs}$$

$$\frac{S \vdash \tau \rightharpoonup \tau' \quad S; \Sigma \vdash \tau' \overset{R}{=} \sigma}{S; \Sigma \vdash \tau \overset{R}{=} \sigma} \text{ eq-lunfold} \qquad \frac{S \vdash \sigma \rightharpoonup \sigma' \quad S; \Sigma \vdash \tau \overset{R}{=} \sigma'}{S; \Sigma \vdash \tau \overset{R}{=} \sigma} \text{ eq-runfold}$$

Fig. 2. Type equivalence

check equivalence for vacuous types. Figure 2 shows inference rules for type equivalence, defined using the rule unfold. We use a judgment $S; \Sigma \vdash \tau_1 \equiv \tau_2$ to mean that τ_1 and τ_2 are coinductively equivalent under context (S, Σ) and a judgment $S; \Sigma \vdash \tau_1 \overset{R}{=} \tau_2$ to mean that τ_1 and τ_2 are inductively equivalent. Note that the inductive equivalence relation $\overset{R}{=}$ is parameterized over a relation R, which is instantiated with the coinductive equivalence relation $\equiv$ in the rule eq-ind. This way, we nest the inductive equivalence relation into the coinductive equivalence relation[3]. We use a double horizontal line for a coinductive rule and a single horizontal line for an inductive rule.

The rule eq-coind is a coinductive rule for checking equivalence between vacuous types. To show that two vacuous types τ and σ are equivalent, that is, $S; \Sigma \vdash \tau \equiv \sigma$, we repeatedly apply the rule eq-coind. When we get the very same proposition to be proved in the premise, the proof is completed by coinduction. Notably vacuous types are only equivalent to vacuous types. As for equivalence for types other than vacuous types, we use the rule eq-ind, which nests induction into coinduction, to compare their structures.

The inductive type equivalence compares structures of types. Given a pair of types, we apply the rule eq-lunfold or eq-runfold a finite number of times, unfolding type definitions, until we get a pair of the unit type, type variables, function types, or product types. Then we structurally compare them. Note that the rules eq-lunfold and eq-runfold are the only rules where induction plays a role. It is crucial that these rules are defined inductively; if we allow them to be used coinductively, a vacuous type becomes equivalent to any type. The rules eq-unit for the unit type and eq-var for type variables are standard. The rules eq-fun

---

[3] A definition of the form $\nu X.F(X, \mu Y.G(X, Y))$.

Contractive types and signatures $\boxed{S \Downarrow \tau}\ \boxed{S \downarrow_C \tau}\ \boxed{S \Downarrow}$

$$\frac{S \downarrow_\Downarrow \tau}{S \Downarrow \tau}\ \text{c-coind} \qquad \frac{}{S \downarrow_C \text{unit}}\ \text{c-unit} \qquad \frac{}{S \downarrow_C \alpha}\ \text{c-var} \qquad \frac{(S, \tau_i) \in C \quad (i = 1, 2)}{S \downarrow_C \tau_1 \to \tau_2}\ \text{c-fun}$$

$$\frac{(S, \tau_i) \in C \quad (i = 1, 2)}{S \downarrow_C \tau_1 * \tau_2}\ \text{c-prod} \qquad \frac{S \ni \text{type } \alpha\ t \quad S \downarrow_C \tau}{S \downarrow_C \tau\ t}\ \text{c-abs} \qquad \frac{S \vdash \tau \to \sigma \quad S \downarrow_C \sigma}{S \downarrow_C \tau}\ \text{c-type}$$

$$\frac{\forall (\text{type } \alpha\ t = \tau) \in S,\ S \Downarrow \tau}{S \Downarrow}\ \text{c-sig}$$

Fig. 3. Contractive types and signatures

for function types, eq-prod for product types, and eq-abs for abstract types are where the inductive equivalence goes back to the coinductive equivalence.

With this definition of type equivalence, for example, now we prove that the types s and u in the signature T in the introduction are equivalent as follows:

$$\frac{T \ni \text{type 'a } t \quad \overline{T; \cdot \vdash s \equiv u}\ \text{coinduction hypothesis}}{\frac{T; \cdot \vdash s\ t \stackrel{\equiv}{=} u\ t}{T; \cdot \vdash s \equiv u}\ \text{eq-ind, eq-lunfold, eq-runfold}}\ \text{eq-abs}$$

Our type equivalence is indeed an equivalence relation, *i.e.*, reflexive, symmetric, and transitive (see the expanded version for the proof).

2.3   Contractive Types and Signatures

Given a program (M, S, e), we restrict every type τ in a type specification type $\alpha\ t = \tau$ in S to be contractive to obtain type soundness as illustrated in the introduction. Intuitively a type is contractive if any sequence of its unfolding eventually produces a type constructor such as unit, α, $\to$, or $*$. A subtle case is a type application $\tau\ t$ where t is an abstract type: we require that τ be contractive. For instance, assuming t is an abstract type, type $\alpha\ s = (\alpha\ s * \alpha\ s)\ t$ and type $\alpha\ s = ((\alpha * \alpha)\ t)\ t$ are contractive, but type $\alpha\ s = (\alpha\ s)\ t$ is not. This way, we avoid the possibility of a type specification type $\alpha\ s = \tau$ in a signature to degenerate into type $\alpha\ s = \alpha\ s$ during subtyping.

Figure 3 shows inference rules for contractive types and signatures. We use two judgments $S \Downarrow \tau$ and $S \downarrow_C \tau$ to define contractive types: the former to define coinductive contractiveness and the latter inductive contractiveness. The basic idea of using nested induction into coinduction is the same as for type equivalence. Note that the rule c-coind is the only coinductive rule for checking contractiveness of a type, which nests induction into coinduction.

The inductive contractiveness is defined using six rules: two axioms, two rules going back to the coinductive contractiveness, and two inductive rules for type applications. The unit type and type variables are by definition inductively contractive (rules c-unit and c-var). In the rules c-fun and c-prod, a function type

$\tau_1 \to \tau_2$ and a product type $\tau_1 * \tau_2$ are inductively contractive under signature S if each component τ_i and S are related by C, which is instantiated with the coinductive contractiveness $\Downarrow$ in the rule c-coind. The rules c-abs and c-type are where induction plays a role. An abstract type $\tau\ t$ is inductively contractive if so is its argument τ (rule c-abs). Finally a type τ is inductively contractive if so is its unfolding σ (rule c-type).

A signature S is contractive, denoted by $S \Downarrow$, if for every type specification in S, its right hand side is contractive under S (rule c-sig). With this definition of contractiveness, for example, now the signature T in the introduction is not contractive because the type s t in the type specification type s = s t (or u t in type u = u t) cannot be proved to be contractive:

$$
\cfrac{
\text{T} \ni \text{type 'a t} \quad
\cfrac{
\text{T} \vdash \text{s} \to \text{s t} \quad
\cfrac{
\text{T} \ni \text{type 'a t} \quad
\cfrac{\quad\quad\quad \text{an infinite derivation}\quad\quad}{\text{T} \Downarrow_\Downarrow \text{s}}\;\text{c-abs}
}{\text{T} \Downarrow_\Downarrow \text{s t}}\;\text{c-type}
}{\text{T} \Downarrow_\Downarrow \text{s}}
}{\text{T} \Downarrow \text{s t}}\;\text{c-coind, c-abs}
$$

2.4   Type Soundness

In this section, we prove type soundness of $\lambda_{\text{abs}}^{\text{rec}}$ by the usual progress and preservation properties (Theorems 1 and 4). First we give an overview of typing rules and reduction rules in Figure 4 (where standard typing rules and reduction rules for expressions are omitted). Most of the typing rules are standard, and we only remark that a sealed program (M, S, e) is well-typed if the module M has some signature S', S is contractive, S' is a sub-signature of S, and e is well-typed under the sealed signature S (rule typ-prog-seal). Here we use a subtyping judgment $S' \leq S$, which is defined as for subtyping for record types, where width subtyping is allowed: S' may include more specifications than S. Furthermore any type specification is considered to be a sub-specification of an abstract type specification with the same bound name.

Reduction rules are also standard. Given a sealed program (M, S, e), we first remove the sealed signature. Then we evaluate the module M to a module value V, by sequentially evaluating the value definitions $\overline{d_e}$ in M to definition values $\overline{d_v}$. Finally using the module value V we evaluate the expression e to a value v, obtaining a program value (V, v).

Theorems 1 and 4 below now prove type soundness of $\lambda_{\text{abs}}^{\text{rec}}$. For the progress theorem, we also prove classification, inversion, and canonical forms lemmas as usual. In particular, the classification lemma ensures that we cannot prove equivalence of types with different constructors even in the presence of non-contractive types in modules. For example, $S; \Sigma \vdash \tau_1 * \tau_2 \not\equiv \sigma_1 \to \sigma_2$ for any types $\tau_1, \tau_2, \sigma_1, \sigma_2$.

Well-typed definitions and modules $\boxed{S \vdash \overline{d_e} : S_e}$ $\boxed{\vdash M : S}$

$$\frac{}{S \vdash \cdot : \cdot} \text{ typ-emp} \qquad \frac{S; \cdot \vdash e : \tau \quad S, \text{val } l : \tau \vdash \overline{d_e} : S_e}{S \vdash (\text{let } l = e, \overline{d_e}) : (\text{val } l : \tau, S_e)} \text{ typ-val}$$

$$\frac{\overline{d_\tau} \text{ ok} \quad \overline{d_\tau} \vdash \overline{d_e} : S_e \quad \text{BN}(\overline{d_e}) \text{ distinct}}{\vdash (\overline{d_\tau}, \overline{d_e}) : (\overline{d_\tau}, S_e)} \text{ typ-mod}$$

Well-typed programs $\boxed{\vdash P : (S, \tau)}$

$$\frac{\vdash M : S' \quad S \Downarrow \quad S' \leq S \quad S; \cdot \vdash e : \tau}{\vdash (M, S, e) : (S, \tau)} \text{ typ-prog-seal} \qquad \frac{\vdash M : S \quad S; \cdot \vdash e : \tau}{\vdash (M, e) : (S, \tau)} \text{ typ-prog}$$

Reduction rules

$$
\begin{array}{ll}
\text{values} & v ::= () \mid \lambda x : \tau.\, e \mid (v_1, v_2) \\
\text{definition values} & d_v ::= \text{let } l = v \\
\text{module values} & V ::= (\overline{d_\tau}, \overline{d_v}) \\
\text{program values} & P_v ::= (V, v)
\end{array}
$$

$$\frac{}{(M, S, e) \longmapsto (M, e)} \text{ red-p-seal}$$

$$\frac{M \longmapsto M'}{(M, e) \longmapsto (M', e)} \text{ red-p-mod} \qquad \frac{\overline{d_v} \vdash e \longmapsto e'}{(\overline{d_\tau}, \overline{d_v}, e) \longmapsto (\overline{d_\tau}, \overline{d_v}, e')} \text{ red-p-exp}$$

$$\frac{\overline{d_v} \vdash e \longmapsto e'}{(\overline{d_\tau}, \overline{d_v}, \text{let } l = e, \overline{d_e}) \longmapsto (\overline{d_\tau}, \overline{d_v}, \text{let } l = e', \overline{d_e})} \text{ red-mod} \qquad \frac{\overline{d_v} \ni \text{let } l = v}{\overline{d_v} \vdash l \longmapsto v} \text{ red-name}$$

Fig. 4. Typing rules and reduction rules

Theorem 1 (Progress).

(1) If $\vdash (\overline{d_\tau}, \overline{d_v}) : S$ and $S; \cdot \vdash e : \tau$, then either e is a value or $\exists e', \overline{d_v} \vdash e \longmapsto e'$.
(2) If $\vdash M : S$, then either M is a module value or $\exists M', M \longmapsto M'$.
(3) If $\vdash P : (S, \tau)$, then either P is a program value or $\exists P', P \longmapsto P'$.

The key lemma for the preservation theorem is that type equivalence is preserved by subtyping. In the lemma below, the signature S_2 being contractive is crucial. For example, assuming S is the inferred signature of the module M in the introduction, although $S \leq T$ and $T; \cdot \vdash s \equiv t$, we have $S; \cdot \vdash s \not\equiv t$.

Lemma 2. *If $S_1 \leq S_2$, $S_2 \Downarrow$, and $S_2; \Sigma \vdash \tau \equiv \sigma$, then $S_1; \Sigma \vdash \tau \equiv \sigma$.*

Now using Lemma 2, we show that if a sealed program (M, S, e) is well-typed, the program (M, e) where the sealed signature S is eliminated is also well-typed (Lemma 3), which proves the most difficult case (4) of Theorem 4. We then prove other cases of Theorem 4 as usual using induction and case analysis.

Lemma 3 (Contractive signature elimination). *If $\vdash (M, S, e) : (S, \tau)$, then there exists S' such that $\vdash (M, e) : (S', \tau)$ and $S' \leq S$.*

Theorem 4 (Preservation).

(1) If $\vdash (\overline{d_\tau}, \overline{d_v}) : S$, $S; \cdot \vdash e : \tau$, and $\overline{d_v} \vdash e \longmapsto e'$, then $S; \cdot \vdash e' : \tau$.
(2) If $\vdash M : S$ and $M \longmapsto M'$, then $\vdash M' : S$.
(3) If $P = (M, e)$, $\vdash P : (S, \tau)$ and $P \longmapsto P'$, then $\vdash P' : (S, \tau)$.
(4) If $\vdash (M, S, e) : (S, \tau)$, then $\exists S'$ such that $\vdash M : S'$, $S' \leq S$, and $S'; \cdot \vdash e : \tau$.

3   Discussion

3.1   Coq Mechanization

For the Coq mechanization, we use Mendler-style [10] coinductive rules for type equivalence and contractiveness in the style of Nakata and Uustalu [14], instead of the Park-style rules in Figures 2 and 3. The reason is that Coq's syntactic guardedness condition for induction nested into coinduction is too weak to work with the Park-style rules. We cannot construct corecursive functions (coinductive proofs) that we need. For example, to enable Coq's guarded corecursion, we use the following Mendler-style coinductive rule instead of the Park-style rule eq-ind:

$$\frac{R \subseteq \equiv \quad S; \Sigma \vdash \tau \overset{R}{\equiv} \sigma}{S; \Sigma \vdash \tau \equiv \sigma} \text{ eq-ind}'$$

The main difference is that we use in the rule eq-ind' a relation R that is stronger than the coinductive equivalence relation $\equiv$. Hence, to build a coinductive proof, we need to find such a relation R, and in many cases we cannot just use $\equiv$ for R. With this definition, the Park-style rules are derivable.

3.2   Algorithmic Type Equivalence and Contractiveness

Strictly speaking, equality of equi-recursive types with type parameters is decidable [4]. Solomon [17] has shown it to be equivalent to the equivalence problem for deterministic pushdown automata (DPDA), which has been shown decidable by Sénizergues [16]. There is, however, no known practical algorithm for DPDA-equivalence, and it is not known whether there exists any algorithm for unification either, which is required for type inference in the core language.

One possible approach to practical type equivalence would be to reject non-regular recursive types as in OCaml. We can then also algorithmically decide contractiveness of every type in a program by enumerating all the distinct type structures that can be obtained by unfolding each type used in the program and its subterms. Still we need a sound metatheory of non-regular recursive types to prove soundness of an OCaml-style recursive module system, because such types may be hidden behind signature sealings in the presence of recursive modules.

4 Related Work

The literature on subtyping for μ-types (hence without type definitions, type parameters, and abstract types) is abundant. In this setting, contractiveness can be checked syntactically: every μ-bound variable occurs under $\rightarrow$ or $*$. We mention three landmark papers. Amadio and Cardelli [2] were the first to give a subtyping algorithm. They define subtyping in three ways, which are proved equivalent: an inclusion between unfoldings of μ-types into infinite trees, a subtyping algorithm, and an inductive axiomatization. Brandt and Henglein [3] give a new inductive axiomatization in which the underlying coinductive nature of Amadio and Cardelli's system is internalized by allowing, informally speaking, construction of circular proofs. Gapeyev et al. [6] is a good self-contained introduction to subtyping for recursive types, including historical notes on theories of recursive types. They define a subtyping relation on contractive μ-types as the greatest fixed point of a suitable generating function.

Danielsson and Altenkirch [5] present an axiomatization of subtyping for μ-types using induction nested into coinduction. They formalized the development in Agda, which supports induction nested into coinduction as a basic form. Komendantsky [8] conducted a similar project in Coq using the Mendler-style coinduction.

Recursive types are indispensable in theories of recursive modules since recursive modules allow us to indirectly introduce recursion in types that span across module boundaries. In this setting, one has to deal with a more expressive language for recursive types, which may include, for instance, higher-order type constructors, type definitions, and abstract types. Montagu and Rémy [12,13] investigate existential types to model modular type abstraction in the context of a structural type system. They consider its extensions with recursion (i.e., equi-recursive types without type parameters) and higher-order type constructors separately but do not investigate a combination of the two extensions. Crary et al. [4] first propose a type system for recursive modules using an inductive axiomatization of (coinductive) type equivalence for equi-recursive types with higher-order type constructors, type definitions, and abstract types. However, the metatheory of their axiomatization such as type soundness is not investigated. Rossberg and Dreyer [15] use equi-recursive types with inductive type equivalence (i.e., they do not have a rule equivalent to contract in [2] to enable coinductive reasoning) to prove soundness of their mixin-style recursive module system. They do not intend to use equi-recursive types for the surface language. Our earlier work [7] on recursive modules considers equi-recursive types with type definitions and abstract types, but without type parameters. There we define a type equivalence relation using weak bisimilarity.

5 Conclusion and Future Work

This paper studies a type system for recursive types, type parameters, and abstract types. In particular, we investigate the interaction between non-contractive types and abstract types, and show that while non-contractive types in signatures lead to unsoundness of the type system, they may be allowed in modules. Our study is mainly motivated by OCaml, which allows us to define both abstract types and equi-recursive types with type parameters (with the "-rectypes" option). To obtain a sound type system, we develop a new notion of semantic contractiveness using mixed induction and coinduction and reject non-contractive types in signatures. We show that our type system is sound with respect to the standard call-by-value operational semantics, which eliminates signature sealings. We have also formalized the whole system and its soundness proof in Coq. Future work includes extending our type system to the full-scale module system including recursive modules, nested modules, and higher-order functors.

Acknowledgments. This work was supported by Mid-career Researcher Program through NRF funded by the MEST (2010-0022061). Hyeonseung Im was partially supported by the ANR TYPEX project n. ANR-11-BS02-007_02. Keiko Nakata was supported by the ERDF funded EXCS project, the Estonian Ministry of Education and Research research theme no. 0140007s12, and the Estonian Science Foundation grant no. 9398.

References

1. OCaml, `http://caml.inria.fr/ocaml/`
2. Amadio, R.M., Cardelli, L.: Subtyping recursive types. ACM Transactions on Programming Languages and Systems 15(4), 575–631 (1993)
3. Brandt, M., Henglein, F.: Coinductive axiomatization of recursive type equality and subtyping. In: de Groote, P., Hindley, J.R. (eds.) TLCA 1997. LNCS, vol. 1210, pp. 63–81. Springer, Heidelberg (1997)
4. Crary, K., Harper, R., Puri, S.: What is a recursive module? In: PLDI 1999 (1999)
5. Danielsson, N.A., Altenkirch, T.: Subtyping, declaratively: an exercise in mixed induction and coinduction. In: Bolduc, C., Desharnais, J., Ktari, B. (eds.) MPC 2010. LNCS, vol. 6120, pp. 100–118. Springer, Heidelberg (2010)
6. Gapeyev, V., Levin, M.Y., Pierce, B.C.: Recursive subtyping revealed. Journal of Functional Programming 12(6), 511–548 (2002)
7. Im, H., Nakata, K., Garrigue, J., Park, S.: A syntactic type system for recursive modules. In: OOPSLA 2011 (2011)
8. Komendantsky, V.: Subtyping by folding an inductive relation into a coinductive one. In: Peña, R., Page, R. (eds.) TFP 2011. LNCS, vol. 7193, pp. 17–32. Springer, Heidelberg (2012)
9. MacQueen, D., Plotkin, G., Sethi, R.: An ideal model for recursive polymorphic types. In: POPL 1984 (1984)
10. Mendler, N.P.: Inductive types and type constraints in the second-order lambda calculus. Annals of Pure and Applied Logic 51(1-2), 159–172 (1991)
11. Milner, R., Tofte, M., Harper, R., MacQueen, D.: The Definition of Standard ML (Revised). The MIT Press (1997)
12. Montagu, B.: Programming with first-class modules in a core language with subtyping, singleton kinds and open existential types. PhD thesis, École Polytechnique, Palaiseau, France (December 2010)

13. Montagu, B., Rémy, D.: Modeling abstract types in modules with open existential types. In: POPL 2009 (2009)
14. Nakata, K., Uustalu, T.: Resumptions, weak bisimilarity and big-step semantics for While with interactive I/O: An exercise in mixed induction-coinduction. In: SOS 2010, pp. 57–75 (2010)
15. Rossberg, A., Dreyer, D.: Mixin' up the ML module system. ACM Transactions on Programming Languages and Systems 35(1), 2:1–2:84 (2013)
16. Sénizergues, G.: The equivalence problem for deterministic pushdown automata is decidable. In: Degano, P., Gorrieri, R., Marchetti-Spaccamela, A. (eds.) ICALP 1997. LNCS, vol. 1256, pp. 671–681. Springer, Heidelberg (1997)
17. Solomon, M.: Type definitions with parameters (extended abstract). In: POPL 1978 (1978)
18. Stone, C.A., Schoonmaker, A.P.: Equational theories with recursive types (2005), http://www.cs.hmc.edu/~stone/publications.html

Algebras, Automata and Logic for Languages of Labeled Birooted Trees

David Janin*

Université de Bordeaux, LaBRI UMR 5800,
351, cours de la Libération,
F-33405 Talence, France
`janin@labri.fr`

Abstract. In this paper, we study the languages of labeled finite birooted trees: Munn's birooted trees extended with vertex labeling. We define a notion of finite state birooted tree automata that is shown to capture the class of languages that are upward closed w.r.t. the natural order and definable in Monadic Second Order Logic. Then, relying on the inverse monoid structure of labeled birooted trees, we derive a notion of recognizable languages by means of (adequate) premorphisms into finite (adequately) ordered monoids. This notion is shown to capture finite boolean combinations of languages as above. We also provide a simple encoding of finite (mono-rooted) labeled trees in an antichain of labeled birooted trees that shows that classical regular languages of finite (mono-rooted) trees are also recognized by such premorphisms and finite ordered monoids.

Introduction

Motivations and background. Semigroup theory has amply demonstrated its considerable efficiency over the years for the study and fine grain analysis of languages of finite words, that is subsets of the free monoid A^*. This can be illustrated most simply by the fact that a language $L \subseteq A^*$ is regular if and only if there is a finite monoid S and a monoid morphism $\theta : A^* \to S$ such that $L = \theta^{-1}(\theta(L))$. In this case, we say that the language L is *recognized* by the finite monoid S (and the morphism θ).

Even more effectively, for every language $L \subseteq A^*$, the notion of recognizability induces a notion of syntactic congruence $\simeq_L$ for the language L in such a way that the monoid $M(L) = A^*/\simeq^L$ is the smallest monoid that recognizes L. Then, many structural properties of the language L can be decided by analyzing the properties of its syntactic monoid $M(L)$, e.g. regularity, star freeness, etc (see [14] for more examples of such properties).

These results triggered the development of entire algebraic theories of languages of various structures elaborated on the basis of richer algebraic frameworks such as, among others, ω-semigroups for languages of infinite words [19,12],

---

* Partially funded by project INEDIT, ANR-12-CORD-0009

Complete version available at `http://hal.archives-ouvertes.fr/hal-00784898`

F.V. Fomin et al. (Eds.): ICALP 2013, Part II, LNCS 7966, pp. 312–323, 2013.

preclones or forest algebra for languages of trees [5,3,2], or indeed ω-hyperclones for languages of infinite trees [1]. With an aim to describing the more subtle properties of languages, several extensions of the notion of recognizability by monoids and morphisms were also taken into consideration, e.g. recognizability by monoids and relational morphisms [13] or recognizability by ordered monoids and monotonic morphisms [15].

A recent study of the languages of overlapping tiles [6,9] or, equivalently, subsets of the (inverse) monoid of McAlister [11,8], has led to the definition of quasi-recognizability: recognizability by means of *(adequate) premorphisms* into *(adequately ordered) ordered monoids*.

As (monotonic) morphisms are particular cases of premorphisms, this notion can be seen as a generalization of recognizability by (ordered) monoids and (monotonic) morphisms [15]. To some extent, quasi-recognizability can also be seen as a notion of co-algebraic recognizability in the sense that it is dual to the standard notion. Indeed, (adequate) premorphisms preserve *some* (and sufficiently many) decompositions while morphisms preserve *all* compositions.

However, this notion of quasi-recognizability has not yet been settled for we need to restrict both the class of allowed premorphisms and/or the class of finite ordered monoids for that notion to be effective. Without any restrictions, the inverse image by a premorphism of a finite subset of a finite ordered monoid may not even be computable [8]. Further still, there are several incomparable candidates for defining such an effective restriction as illustrated by a recent and complementary study of walking automata on birooted trees [7].

In this paper, we aim to stabilize the notion of recognizability by adequate premorphisms by applying it to the study of languages of labeled birooted trees. In doing so, it appears that this notion admits both simple automata theoretic characterization and robust logical characterization.

Outline. Birooted labeled trees, called birooted F-trees, are presented in Section 1. Equipped with an extension of Scheiblich's product of (unlabeled) birooted trees [16], the resulting algebraic structures are inverse monoids that are quite similar to discrete instances of Kellendonk's tiling semigroups [10]. Then, birooted F-trees can be ordered by the (inverse semigroup) *natural order* relation that is stable under product: the inverse monoid $\mathcal{B}^1(F)$ of labeled birooted F-trees is also a partially ordered monoid.

Birooted tree automata are defined and studied in Section 2. By construction, languages recognized by these finite automata are upward closed in the natural order. It follows that they fail to capture all languages definable by means of Monadic Second Order (MSO) formulae. However, this loss of expressive power is shown to be limited to the property of upward closure. Indeed, we prove (Theorem 2) that every upward closed language of birooted trees which is MSO definable is recognized by a finite state birooted tree automata.

As a case in point, when F is seen as a functional signature, by embedding the classical F-terms (see [18]) into birooted F-trees, we show (Theorem 3) that the birooted tree image of every regular language L of F-terms is of the form

$U_L \cap D_L$ for some MSO definable and upward closed (resp. downward closed) language U_L (resp. language D_L).

The algebraic counterpart of birooted tree automata is presented in Section 3 where the notions of adequately ordered monoids and adequate premorphisms are defined. The induced notion of quasi-recognizable languages of birooted F-trees is shown to be effective (Theorem 4).

As for expressive power, it is shown that every birooted tree automaton simply induces an adequate premorphism that recognizes the same language (Theorem 5) and that every quasi-recognizable language is MSO definable (Theorem 6). The picture is made complete by proving (Theorem 7) that quasi-recognizable languages of birooted trees correspond exactly to finite boolean combinations of upward closed MSO definable languages.

Together with Theorem 3, this result demonstrates that our proposal can also be seen as yet another algebraic characterization of regular languages of trees that complete that previously obtained by means of preclones [5], forest algebras [3] or ordered monoids and admissible premorphisms [7].

Related Works. We should also mention that the notion of birooted F-tree automata defined above is an extension of that previously defined [9] for languages of one-dimensional overlapping tiles: subsets of McAlister monoids [11].

Although closely related, we can observe that an extension of this type is by no means straightforward. Of course going from the linear structure of overlapping tiles to the tree shaped structure of birooted F-trees already induces a tangibly increased level of complexity. However, the main difference comes from edge directions. In overlapping tiles, all edges go in the same direction while, in birooted F-trees, edges can go back and forth (almost) arbitrarily. Proving Theorem 2 is thus much more complex than proving an analogous result for overlapping tiles.

Comparing our proposal with other known algebraic characterizations of languages of (mono-rooted) F-trees [5,3] is not easy. Of course, our proposal induces a larger class of definable languages since we are dealing with birooted F-trees and not just F-trees. However, a more relevant comparison would be to compare the classification of languages through a full series of approaches, by restricting even further the allowed recognizers: be them preclones as in [5], forest algebras [2] or adequately ordered monoids as proposed here.

With quasi-recognizability, recognizers are monoids (and premorphisms). It follows that the known restrictions applicable to the study of languages of words, e.g. aperiodic monoids [14], can simply be extended to adequately ordered monoids. Yet, the relevance of such restrictions for languages of mono-rooted or birooted F-trees still needs to be evaluated.

Another source of difficulty comes from the fact that adequate premorphisms are *not* morphisms : only *disjoint products* are preserved. To some extent, the notion of quasi-recognizability by premorphisms presented here is analogous, compared with classical recognizability by morphisms, to what unambiguous non deterministic automata are in comparison with deterministic automata. On the negative side, this means that the notion of quasi-recognizability has not yet been completely understood. On the positive side, this means that it may lead to radically new outcomes.

1   Semigroups and Monoids of Birooted F-trees

Simply said, a labeled birooted tree is a (non empty) finite connected subgraph of the Cayley graph of the free group $FG(A)$ with labeled vertices on some finite alphabet F and two distinguished vertices respectively called the input root and the output root. This definition and some of the associated properties are detailled in this section.

Formally, let A be a finite (edge) alphabet and let $\bar{A}$ be a disjoint copy of A with, for every letter $a \in A$, its copy $\bar{a} \in \bar{A}$. Let $u \mapsto \bar{u}$ be the mapping from $(A + \bar{A})^*$ to itself inductively defined by $\bar{1} = 1$ and $\overline{ua} = \bar{a}\,\bar{u}$ and $\overline{u\bar{a}} = a\,\bar{u}$, for every $u \in (A + \bar{A})^*$, every $a \in A$. This mapping is involutive, i.e. $\bar{\bar{u}} = u$ for every $u \in (A + \bar{A})^*$, and it is an anti-morphism, i.e. $\overline{uv} = \bar{v}\,\bar{u}$ for every word u and $v \in (A + \bar{A})^*$.

The *free group* $FG(A)$ generated by A is the quotient of $(A + \bar{A})^*$ by the least congruence $\simeq$ such that, for every letter $a \in A$, $a\bar{a} \simeq 1$ and $\bar{a}a \simeq 1$. This is indeed a group since, for every $u \in (A + \bar{A})^*$, we have $[u][\bar{u}] = [1]$ hence $[\bar{u}]$ is the group inverse of $[u]$.

It is known that every class $[u] \in FG(A)$ contains a unique element $red(u)$ (the *reduced form* of u) that contains no factors of the form $a\bar{a}$ nor $\bar{a}a$ for $a \in A$. In the sequel, every such class $[u] \in FG(A)$ is thus represented by its reduced form $red(u)$. Doing so, the product $u \cdot v$ of every two reduced words u and $v \in FG(A)$ is directly defined by $u \cdot v = red(uv)$.

Elements of $FG(A)$, when seen as reduced words, can then be ordered by the *prefix order relation* $\leq_p$ defined, for every (reduced word) u and $v \in FG(A)$ by $u \leq_p v$ when there exists (a reduced word) $w \in FG(A)$ such that $red(uw) = uw = v$. The associated *predecessor relation* $\prec_p$ is defined, for every v and $w \in FG(A)$, by $v \prec_p w$ when $v <_p w$ and $w = vx$ for some $x \in A + \bar{A}$.

A *labeled birooted tree* on the edge alphabet A and the vertex alphabet F is a pair $B = \langle t, u \rangle$ where $t : FG(A) \to F$ is a partial maps which domain $dom(t)$ is a prefix closed subset of $FG(A)$ with $u \in dom(t)$. In such a presentation, $1 \in dom(t)$ is the input root vertex and $u \in dom(t)$ is the output root vertex. Assuming the edge alphabet A is implicit, these labeled birooted trees are called *birooted F-trees* or, when F is also implicit, simply *birooted trees*.

For every birooted tree $B = \langle t, u \rangle$, for every $v \in dom(t)$, let $t_v : FG(A) \to F$ be the partial function defined by $dom(t_v) = \bar{v} \cdot dom(t)$ and $t_v(w) = t(vw)$ for every $w \in dom(t_v)$. Accordingly, let $B_v = \langle t_v, \bar{v}u \rangle$ be the v *translation* of the birooted tree B.

Observe that such a translation slightly differs from the classical notion of subtrees since $dom(t_v) = \bar{v} \cdot dom(t)$ contains as many vertices as $dom(t)$. A notion of sub-birooted tree B_v^p, with fewer vertices and thus more closely related with the classical notion of subtree, is defined below when proving the decomposition property (Lemma 1).

The partial product $\langle r, u \rangle \cdot \langle s, v \rangle$ of two birooted F-tree $\langle r, u \rangle$ and $\langle s, v \rangle$ is defined, when it exists, as the birooted F-tree $\langle t, w \rangle$ defined by $w = u \cdot v$, $dom(t) = dom(r) \cup u \cdot dom(s)$, $t(u') = r(u')$ for every $u' \in dom(r)$ and $t_u(v') = s(v')$ for every $v' \in dom(s)$.

Observe that such a product exists if and only if the tree r_u and the tree s agree on $dom(r_u) \cap dom(s)$, i.e. for every $v' \in dom(r_u) \cap dom(s)$, we have $r_u(v') = r(uv') = s(v')$. It follows that undefined products may arise when F is not a singleton.

Two examples of birooted F-trees B_1 and B_2 are depicted below, with a dangling input edge marking the input root and a dangling output edge marking the output root.

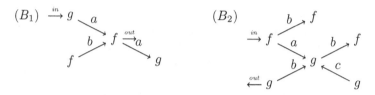

The (defined) product of the birooted F-trees B_1 and B_2 is then depicted below.

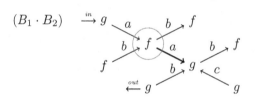

In that picture, the cercle marks the synchronization vertex that results from the merging of the output root of B_1 and the input root of B_2. The a-labeled edge $f \xrightarrow{a} g$ emanating from that vertex is the common edge resulting from the fusion of the two (synchronized) birooted F-trees.

The product is completed by adding a zero element for the undefined case with $0 \cdot \langle t, v \rangle = \langle t, v \rangle \cdot 0 = 0 \cdot 0 = 0$ for every (defined) birooted tree $\langle t, v \rangle$.

One can easily check that the completed product is associative. The resulting structure is thus a semigroup denoted by $\mathcal{B}(F)$: the *semigroup of birooted F-trees*. When F is a singleton, every birooted F-tree can just be seen as a pair (P, u) with an non empty prefix closed domain $P \subseteq FG(A)$ and an output root $u \in P$. Then, following Scheiblich presentation [16], the semigroup $\mathcal{B}(F)$ is the free monoid $FIM(A)$ generated by A with unit $1 = (\{1\}, 1)$. When F is not a singleton, we extend the set $\mathcal{B}(F)$ with a unit denoted by 1. The resulting structure is a monoid denoted by $\mathcal{B}^1(F)$: the *monoid of birooted F-trees*.

The monoid of birooted F-trees is an *inverse monoid*, i.e. for every $B \in \mathcal{B}^1(F)$ there is a unique $B^{-1} \in \mathcal{B}^1(F)$ such that $BB^{-1}B = B$ and $B^{-1}BB^{-1} = B^{-1}$. Indeed, we necessarily have $0^{-1} = 0$, $1^{-1} = 1$ and, for every non trivial birooted F-tree $\langle t, u \rangle$ one can check that $\langle t, u \rangle^{-1} = \langle t_u, \bar{u} \rangle$.

As an inverse monoid, elements of $\mathcal{B}^1(F)$ can be ordered by the *natural order* defined, for every B and $C \in \mathcal{B}^1(F)$ by $B \leq C$ when $B = BB^{-1}C$ (equivalently

$B = CB^{-1}B$). One can check that 0 is the least element and, for every defined birooted F-trees $\langle r, u \rangle$ and $\langle s, v \rangle$ we have $\langle r, u \rangle \leq \langle s, v \rangle$ if and only if $u = v$, $dom(r) \supseteq dom(s)$ and, for every $w \in dom(s)$, $t(w) = s(w)$.

Observe that, as far as trees only are concerned, the natural order is the reverse of the (often called) prefix order on trees. In particular, the bigger is the size of a birooted tree, the smaller is the birooted tree in the natural order.

One can easily check that the monoid of birooted F-trees is finitely generated. We prove here a stronger statement that will be extensively used in the remainder of the text.

A birooted tree is said *elementary* when it is either 0 or 1, or of the from $B_f = \langle \{1 \mapsto f\}, 1 \rangle$ for some $f \in F$ or of the form $B_{fxg} = \langle \{1 \mapsto f, x \mapsto g\}, x \rangle$ for some vertex label f and $g \in F$ and some letter $x \in A + \bar{A}$.

$$(B_{fag}) \xrightarrow{in} f \xrightarrow{a} g \xrightarrow{out} \qquad (B_f) \xrightarrow{in} f \xrightarrow{out} \qquad (B_{f\bar{a}g}) \xrightarrow{in} f \xleftarrow{a} g \xrightarrow{out}$$

The *right projection* B^R (resp. the *left projection* B^L) of a birooted tree $B = \langle t, u \rangle$ is defined to be $B^R = BB^{-1}$ (resp. $B^L = B^{-1}B$) or, equivalently, $B^R = \langle t, 1 \rangle$ (resp. $B^L = \langle t_u, 1 \rangle$). The right projection B^R of B is also called the *reset* of B.

The product $B_1 \cdot B_2$ of two birooted trees B_1 and B_2 is a *disjoint product* when $B_1 \cdot B_2 \neq 0$ and there is a unique elementary birooted tree B_f such that $B_1^L \leq B_f$ and $B_2^R \leq 1$, i.e. $B_1^L \vee B_2^R = 1$. This restricted product is called a disjoint product because, when $B_1 = \langle t_1, u_1 \rangle$ and $B_2 = \langle t_2, u_2 \rangle$, the product $B_1 \cdot B_2$ is disjoint if and only if $t(u_1) = t_2(1) = f$ and $dom(t_1) \cap u_1 \cdot dom(t_2) = \{u_1\}$, i.e. the set of edges in $B_1 \cdot B_2$ is the *disjoint union* of the set of edges of B_1 and the set of (translated) edges of B_2.

Lemma 1 (Strong Decomposition). *For every $B \in \mathcal{B}(F)$, the birooted F-tree B can be decomposed into a finite combination of elementary birooted trees by disjoint products and (right) resets.*

Proof. Let $B = \langle t, u \rangle$ be a birooted F-tree. We aim at proving it can be decomposed as stated above. We first define some specific sub-birooted trees of B that will be used for such a decomposition.

For every vertex v and $w \in dom(t)$ such that $v \prec_p w$, let $B^p_{v,w}$ be the two vertices birooted F-tree defined by $B^p_{v,w} = B_{fxg}$ where $f = t(v)$, $g = t(w)$ and $vx = w$.

Let $U = \{v \in dom(t) : 1 \leq_p v \leq_p u\}$ be the set of vertices that appears on the path from the input root 1 to the output root u.

For every $v \in dom(t)$, let $D^p(v)$ be the greatest prefix closed subset of the set $\{w \in dom(t_v) : v \leq_p vw, vw \in U \Rightarrow w = 1\}$ and let $B^p_v = \langle t_v | D^p(v), 1 \rangle$ be the idempotent birooted tree obtained from B by restricting the subtree t_v rooted at the vertex v to the domain $D^p(v)$.

Then, given $u_0 = 1 <_p u_1 <_p u_2 <_p \cdots <_p u_{n-1} <_p u_n = u$ the increasing sequence (under the prefix order) of all the prefixes of the output root u, we observe that $B = B^p_{u_0} B^p_{u_0,u_1} B^p_{u_1} \cdots B^p_{u_{n-1}} B^p_{u_{n-1},u_n} B^p_{u_n}$ with only disjoint products.

It remains thus to prove that every idempotent sub-birooted tree of the form B_v^p for some $v \in dom(t)$ can also be decomposed into an expression of the desired form. But this is easily done by induction on the size of the birooted trees B_c^p.

Indeed, Let $v \in dom(t)$. In the case v is a leaf (w.r.t. the prefix order) then $B_v^p = B_{t(v)}$ and we are done. Otherwise, we have $B_v^p = \langle r, 1 \rangle$ for some F-tree r and we observe that $B_v^p = \prod \{ (B_{v,w}^p \cdot B_w^p)^R : w \in dom(r), v \prec_p w \}$ with only disjoint products and resets. This concludes the proof.   □

The above decomposition of B as a combination of elementary birooted trees by disjoint products and right projections is called a *strong decomposition* of the birooted F-tree B.

2   Birooted F-tree Automata

In this section, we define the notion of birooted F-tree automata that is shown to capture the class of languages of birooted F-trees that are upward closed w.r.t. the natural order and definable in Monadic Second Order Logic (MSO).

A *birooted F-tree (finite) automaton* is a quintuple $\mathcal{A} = \langle Q, \delta, \Delta, W \rangle$ defined by a (finite) set of states Q, a (non deterministic) state table $\delta : F \to \mathcal{P}(Q)$, a (non deterministic) transition table $\Delta : A \to \mathcal{P}(Q \times Q)$ and an acceptance condition $W \subseteq Q \times Q$.

A *run* of the automaton $\mathcal{A}$ on a non trivial birooted F-tree $B = \langle t, u \rangle$ is a mapping $\rho : dom(t) \to Q$ such that for every $v \in dom(t)$:

▷ State coherence: $\rho(v) \in \delta(t(v))$,
▷ Transition coherence: for every $a \in A$, if $va \in dom(t)$ then $(\rho(v), \rho(va)) \in \Delta(a)$ and if $v\bar{a} \in dom(t)$ then $(\rho(v\bar{a}), \rho(v)) \in \Delta(a)$.

The run ρ is an *accepting run* when $(\rho(1), \rho(u)) \in W$. The set $L(\mathcal{A}) \subseteq \mathcal{B}(F)$ of birooted F-tree B such that there is an accepting run of $\mathcal{A}$ on B is the language recognized by the automaton $\mathcal{A}$.

Every non trivial birooted F-tree $B = \langle t, u \rangle$ can be seen as a (tree-shaped) FO-structure $\mathcal{M}_B$ with domain $dom(\mathcal{M}_B) = dom(t)$, constant $in_B = 1$ and constant $out_B = u$, unary relation $S_f = t^{-1}(f)$ for every $f \in F$ and binary relation $R_a = \{(v, w) \in dom(t) \times dom(t) : va = w\}$ for every $a \in A$.

We say that a language $L \subseteq \mathcal{B}(F)$ is definable in monadic second order logic (MSO) when there exists a closed MSO formula φ on the FO-signature $\{in, out\} \cup \{S_f\}_{f \in F} \cup \{R_a\}_{a \in A}$ such that $L = \{B \in \mathcal{B}(F) : \mathcal{M}_B \models \varphi\}$.

The following theorem gives a rather strong characterization of the languages recognized by finite state birooted F-tree automata.

Theorem 2. *Let $L \subseteq \mathcal{B}(F)$ be a language of birooted F-trees. The language is recognized by a finite birooted F-tree automaton if and only if L is upward closed (in the natural order) and MSO definable.*

Proof. Let $L \subseteq \mathcal{B}(F)$ be a language of birooted F-trees. We first prove the easiest direction, from birooted tree automata to MSO. Then, we prove the slightly more difficult direction from MSO to birooted tree automata.

From birooted tree automata to MSO. Assume that L is recognizable by a finite state birooted tree automaton $\mathcal{A}$. Without loss of generality, since $\mathcal{A}$ is finite, we assume that the set Q of states of $\mathcal{A}$ is such that $Q \subseteq \mathcal{P}([1, n])$ for some $n \geq \log_2 |Q|$.

Then, checking that a birooted tree $\langle t, u \rangle$ belongs to $L(\mathcal{A})$ just amounts to checking that there exists an accepting run. This can easily be described by an existential formula of monadic second order logic of the form $\exists X_1 X_2 \cdots X_n \varphi(in, out)$ with n set variables X_1, X_2, ..., X_n and a first order formula $\varphi(in, out)$.

Indeed, every mapping $\rho : dom(v) \to Q$ is encoded by saying, for every vertex $v \in dom(t)$, that $\rho(v) = \{k \in [1, n] : v \in X_k\}$. Then, checking that the mapping ρ encoded in such a way is indeed an accepting run amounts to checking that it satisfies state and transition coherence conditions and acceptance condition. This is easily encoded in the FO-formula $\varphi(x, y)$.

From MSO to Birooted Tree Automata. Conversely, assume that L is upward closed for the natural order and that L is definable in MSO. Observe that every $B = \langle t, u \rangle$ can just be seen as a (deterministic) tree rooted in the input root vertex 1 with edges labeled on the alphabet $A + \bar{A}$ (with edge "direction" being induced by the prefix order on $FG(A)$), vertices labeled on the alphabet $F \times \{0, 1\}$ (with 1 used to distinguish the output root u from the other vertices). An example of such an encoding of birooted trees into trees is depicted below.

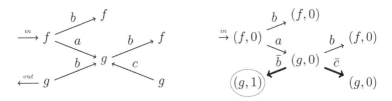

Since L is definable in MSO, applying (an adapted version of) the theorem of Doner, Thatcher and Wright (see for instance [18]), there exists a finite state tree automaton $\mathcal{A}$ that recognizes L. We conclude our proof by defining from the (finite) tree automaton $\mathcal{A}$ a (finite) birooted tree automaton $\mathcal{A}'$ such that $L(\mathcal{A}) = L(\mathcal{A}')$.

The major difficulty in defining $\mathcal{A}'$ is that the (one root) tree automaton $\mathcal{A}$ reads a tree from the (input) root to the leaves hence following the prefix relation order $\leq_p$. Moreover, birooted trees, such a prefix order is *not* encoded in the direction of edges. It follows that, when translating the tree automaton $\mathcal{A}$ into an equivalent birooted tree automaton $\mathcal{A}'$, we need to encode (and propagate) that direction information into states.

But this can be achieved by observing that for every vertex v and w such that $v \prec_p w$, the edge from v to w is uniquely defined by the letter $x \in (A + \bar{A})$ such that $vx = w$. It follows that every such a vertex w (distinct from the input root 1) will be marked in automaton $\mathcal{A}'$ by a state that will encode that letter

x; distinguishing thus the unique predecessor vertex v of w from all successor vertices w' such that $w \prec_p w'$.

□

From now on, a language of birooted F-trees that is definable by a finite birooted F-tree automaton is called a *regular language of birooted F-trees*.

We aim now at relating languages of birooted F-trees and languages of F-trees. Assume till the end of that section that the set F is now a finite functional signatures that is a finite set of symbols equipped with some arity mapping $\rho : F \rightarrow \mathcal{P}(A)$ that maps every function symbol f the set of its arguments' names $\rho(f) \subseteq A$.

A F-tree (also called F-term) is a function $t : A^* \rightarrow F$ with prefix closed finite domain $dom(t)$ such that for every $u \in dom(t)$, every $a \in A$, if $ua \in dom(t)$ then $a \in \rho(t(u))$. Such a finite tree t is said to be complete when, moreover, for every $u \in dom(t)$, for every $a \in A$, if $a \in \rho(t(u))$ then $ua \in dom(t)$.

Every F-tree t is encoded into a birooted F-tree $\langle t, 1 \rangle$ called the birooted image of tree t. By extension, for every set X of F-trees, the language $L_X = \{\langle t, 1 \rangle \in \mathcal{B}(F) : t \in X\}$ of birooted tree images of trees of X is called the birooted tree image of the language X.

Theorem 3. *For every regular language X of complete finite F-trees, we have $L_X = U_X \cap D_X$ for some regular language U_X of birooted F-trees and the complement D_X of some regular language $\mathcal{B}(F) - D_X$ of birooted F-trees.*

Proof. This essentially follows from Theorem 2.

□

3   Quasi-Recognizable Languages of Birooted F-trees

Intimately related to the theory of non-regular semigroups initiated by Fountain in the 70s (see e.g [4]), the notion of recognizability by premorphisms is proposed in [6] (generalized in [9]) to define languages of positive (resp. arbitrary) overlapping tiles. This notion is extended here to languages of birooted F-trees.

Let S be a monoid partially ordered by a relation $\leq_S$ (or just $\leq$ when there is no ambiguity). We always assume that the order relation $\leq$ is stable under product, i.e. if $x \leq y$ then $xz \leq yz$ and $zx \leq zy$ for every x, y and $z \in S$. The set $U(S)$ of *subunits* of the partially ordered monoid S is defined by $U(S) = \{y \in S : y \leq 1\}$.

A partially ordered monoid S is an *adequately ordered monoid* when all subunits of S are idempotents, and for every $x \in S$, both the minimum of right local units $x^L = \min\{y \in U(S) : xy = x\}$ and the minimum of left local units $x^R = \min\{y \in U(S) : yx = x\}$ exist and belong to $U(S)$. The subunits x^L and x^R are respectively called the *left projection* and the *right projection* of x.

Examples. Every inverse monoid S ordered by the natural order is an adequately ordered monoid with $x^L = x^{-1}x$ and $x^R = xx^{-1}$ for every $x \in S$.

As a particular case, the monoid $\mathcal{B}^1(F)$ ordered by the natural order is also an adequately ordered monoid. The subunits of $\mathcal{B}^1(F)$ are, when distinct from 0 or 1, the birooted F-trees of the form $\langle t, 1 \rangle$ and, indeed, for every birooted F-tree $B = \langle t, u \rangle$ we have $B^R = \langle t, 1 \rangle$ and $B^L = \langle t_u, 1 \rangle$.

For every set Q, the relation monoid $\mathcal{P}(Q \times Q)$ ordered by inclusion is also an adequately ordered monoid with, for every $X \subseteq Q \times Q$, $X^L = \{(q, q) \in Q \times Q : (p, q) \in X\}$ and $X^R = \{(p, p) \in Q \times Q : (p, q) \in X\}$.

A mapping $\theta : S \to T$ between two adequately ordered monoids is a *premorphism* when $\theta(1) = 1$ and, for every x and $y \in S$, we have $\theta(xy) \leq_T \theta(x)\theta(y)$ and if $x \leq_S y$ then $\theta(x) \leq_T \theta(y)$. A premorphism $\theta : S \to T$ is an *adequate premorphism* when for every x and $y \in S$ we have $\theta(x^L) = (\theta(x))^L$, $\theta(y^R) = (\theta(y))^R$ and, if $xy \neq 0$ with $x^L \vee y^R = z \prec 1$, i.e. the product xy is a disjoint product, then $\theta(xy) = \theta(x)\theta(y)$.

A language $L \subseteq \mathcal{B}(F)$ of birooted tree is a *quasi-recognizable language* when there exists a finite adequately ordered monoid S and an adequate premorphism $\theta : \mathcal{B}(F) \to S$ such that $L = \theta^{-1}(\theta(L))$.

Theorem 4. *Let $\theta : FIM(A) \to S$ be an adequate premorphism with finite S. For every $B \in \mathcal{B}(F)$ the image $\theta(B)$ of the birooted F-tree B by the adequate premorphism θ is uniquely determined by the structure of B, the structure of S and the image by θ of elementary birooted F-trees.*

Proof. This essentially follows from the adequacy assumption and the strong decomposition property (Lemma 1). □

Now we want to show that every finite state birooted automaton induces an adequate premorphism that recognizes the same language.

Theorem 5. *Let $L \subseteq \mathcal{B}(F)$ be a language of birooted F-trees. If L is recognizable by a finite state birooted tree automaton then it is recognizable by an adequate premorphism into a finite adequately ordered monoid.*

Proof. Let $L \subseteq \mathcal{B}(F)$ and let $\mathcal{A} = \langle Q, \delta, \Delta, T \rangle$ be a finite birooted tree automaton such that $L = L(\mathcal{A})$.

We define the mapping $\varphi_\mathcal{A} : \mathcal{B}(F) \to \mathcal{P}(Q \times Q)$ by saying that $\varphi_\mathcal{A}(B)$ is, for every birooted F-tree $B = \langle t, u \rangle \in \mathcal{B}(F)$, the set of all pairs of state $(p, q) \in Q \times Q$ such that there exists a run $\rho : dom(t) \to Q$ such that $p = \rho(1)$ and $q = \rho(u)$. The mapping $\varphi_\mathcal{A}$ is extended to 0 by taking $\varphi_\mathcal{A}(0) = \emptyset$ and, to 1 by taking $\varphi(1) = I_Q = \{(q, q) \in Q \times Q : q \in Q\}$.

The fact $\mathcal{P}(Q \times Q)$ is an adequately ordered monoid have already been detailled in the examples above. By definition we have $L = \varphi^{-1}(\mathcal{X})$ with $\mathcal{X} = \{X \subseteq Q \times Q : X \cap T \neq 0\}$. Then, we prove that $\varphi_\mathcal{A}$ is indeed an adequate premorphism. □

The following theorem tells how quasi-recognizability and MSO definability are related.

Theorem 6. *Let $\theta : FIM(A) \to S$ be an adequate premorphism with finite S. For every $X \subseteq S$, the language $\theta^{-1}(X)$ is definable in Monadic Second Order Logic.*

Proof. Let $\theta : FIM(A) \to S$ as above and let $X \subseteq S$. Uniformly computing the value of θ on every birooted tree by means of an MSO formula is done by adapting Shelah's decomposition techniques [17]. More precisely, we show that the strong decomposition provided by Lemma 1 is indeed definable in MSO. Then, the computation of the value of θ on every birooted rooted B can be done from the value of θ on the elementary birooted trees and the sub-birooted F-trees that occur in such a decomposition. $\square$

For the picture to be complete, it remains to characterize the class of quasi-recognizable languages w.r.t. the class of languages definable in Monadic Second Order Logic.

Theorem 7. *Let $L \subseteq \mathcal{B}(F)$ be a language of birooted F-trees. The following properties are equivalent:*

(1) the language L is quasi-recognizable,
(2) the language L is a finite boolean combination of upward closed MSO definable languages,
(3) the language L is a finite boolean combination of languages recognized by finite state birooted tree automata.

Proof. The fact that (1) implies (2) essentially follows from Theorem 6. The fact (2) implies (3) immediately follows from Theorem 2. Last, we prove, by classical argument (e.g. cartesian product of monoids) that the class of quasi-recognizable languages is closed under boolean operations. Then, by applying Theorem 5 this proves that (3) implies (1). $\square$

Corollary 8. *The birooted image of every regular languages of F-tree is recognizable by an adequate premorphism in a finite adequately ordered monoid.*

Proof. This follows from Theorem 3 and Theorem 7. $\square$

4   Conclusion

Studying languages of birooted F-trees, structures that generalize F-terms, we have thus defined a notion of automata, a related notion of quasi-recognizability and we have characterized quite in depth their expressive power in relationship with language definability in Monadic Second Order Logic.

As a particular case, our results provide a new algebraic characterization of the regular languages of finite F-trees. Potential links with the preclones approach [5] or the forest algebra approach [3,2] need to be investigated further.

References

1. Blumensath, A.: Recognisability for algebras of infinite trees. Theor. Comput. Sci. 412(29), 3463–3486 (2011)
2. Bojanczyk, M., Straubing, H., Walukiewicz, I.: Wreath products of forest algebras, with applications to tree logics. Logical Methods in Computer Science 8(3) (2012)
3. Bojańczyk, M., Walukiewicz, I.: Forest algebras. In: Logic and Automata, pp. 107–132 (2008)
4. Cornock, C., Gould, V.: Proper two-sided restriction semigroups and partial actions. Journal of Pure and Applied Algebra 216, 935–949 (2012)
5. Ésik, Z., Weil, P.: On logically defined recognizable tree languages. In: Pandya, P.K., Radhakrishnan, J. (eds.) FSTTCS 2003. LNCS, vol. 2914, pp. 195–207. Springer, Heidelberg (2003)
6. Janin, D.: Quasi-recognizable vs MSO definable languages of one-dimensional overlapping tiles. In: Rovan, B., Sassone, V., Widmayer, P. (eds.) MFCS 2012. LNCS, vol. 7464, pp. 516–528. Springer, Heidelberg (2012)
7. Janin, D.: Walking automata in the free inverse monoid. Technical Report RR-1464-12 (revised April 2013), LaBRI, Université de Bordeaux (2012)
8. Janin, D.: On languages of one-dimensional overlapping tiles. In: van Emde Boas, P., Groen, F.C.A., Italiano, G.F., Nawrocki, J., Sack, H. (eds.) SOFSEM 2013. LNCS, vol. 7741, pp. 244–256. Springer, Heidelberg (2013)
9. Janin, D.: Overlapping tile automata. In: Bulatov, A. (ed.) CSR 2013. LNCS, vol. 7913, pp. 431–443. Springer, Heidelberg (2013)
10. Kellendonk, J., Lawson, M.V.: Tiling semigroups. Journal of Algebra 224(1), 140–150 (2000)
11. Lawson, M.V.: McAlister semigroups. Journal of Algebra 202(1), 276–294 (1998)
12. Perrin, D., Pin, J.-E.: Semigroups and automata on infinite words. In: Fountain, J. (ed.) Semigroups, Formal Languages and Groups. NATO Advanced Study Institute, pp. 49–72. Kluwer Academic (1995)
13. Pin, J.-E.: Relational morphisms, transductions and operations on languages. In: Pin, J.E. (ed.) LITP 1988. LNCS, vol. 386, pp. 34–55. Springer, Heidelberg (1989)
14. Pin, J.-E.: Finite semigroups and recognizable languages: an introduction. In: Fountain, J. (ed.) Semigroups, Formal Languages and Groups. NATO Advanced Study Institute, pp. 1–32. Kluwer Academic (1995)
15. Pin, J.-.E.: Syntactic semigroups. In: Handbook of Formal Languages, ch. 10, vol. I, pp. 679–746. Springer (1997)
16. Scheiblich, H.E.: Free inverse semigroups. Semigroup Forum 4, 351–359 (1972)
17. Shelah, S.: The monadic theory of order. Annals of Mathematics 102, 379–419 (1975)
18. Thomas, W.: Languages, automata, and logic. In: Handbook of Formal Languages, ch. 7, vol. III, pp. 389–455. Springer (1997)
19. Wilke, T.: An algebraic theory for regular languages of finite and infinite words. Int. J. Alg. Comput. 3, 447–489 (1993)

One-Variable Word Equations in Linear Time*

Artur Jeż[1,2,**]

[1] Max Planck Institute für Informatik,
Campus E1 4, DE-66123 Saarbrücken, Germany
[2] Institute of Computer Science, University of Wrocław,
ul. Joliot-Curie 15, PL-50383 Wrocław, Poland
aje@cs.uni.wroc.pl

Abstract. In this paper we consider word equations with one variable (and arbitrary many appearances of it). A recent technique of recompression, which is applicable to general word equations, is shown to be suitable also in this case. While in general case it is non-deterministic, it determinises in case of one variable and the obtained running time is $\mathcal{O}(n)$ (in RAM model).

Keywords: Word equations, string unification, one variable equations.

1 Introduction

Word Equations. The problem of satisfiability of word equations was considered as one of the most intriguing in computer science. The first algorithm for it was given by Makanin [11] and his algorithm was improved several times, however, no essentially different approach was proposed for over two decades.

An alternative algorithm was proposed by Plandowski and Rytter [16], who presented a very simple algorithm with a (nondeterministic) running time polynomial in n and $\log N$, where N is the length of the length-minimal solution. However, at that time the only bound on such length followed from Makanin's work and it was triply exponential in n.

Soon after Plandowski showed, using novel factorisations, that N is at most doubly exponential [14], proving that satisfiability of word equations is in NEXPTIME. Exploiting the interplay between factorisations and compression he improved the algorithm so that it worked in PSPACE [15]. On the other hand, it is only known that the satisfiability of word equations is NP-hard.

One Variable. Constructing a cubic algorithm for the word equations with only one variable (and arbitrarily many appearances of it) is trivial. First non-trivial bound was given by Obono, Goralcik and Maksimenko, who devised an $\mathcal{O}(n \log n)$ algorithm [13]. This was improved by Dąbrowski and Plandowski [2] to $\mathcal{O}(n + \#_X \log n)$, where $\#_X$ is the number of appearances of the variable

---

* The full version of this paper is available at http://arxiv.org/abs/1302.3481
** This work was supported by Alexander von Humboldt Foundation.

in the equation. The latter work assumed that alphabet Σ is finite or that it can be identified with numbers. A general solution was presented by Laine and Plandowski [9], who gave an $\mathcal{O}(n \log \#_X)$ algorithm in a simpler model, in which the only operation on letters is their comparison.

Recompression. Recently, the author proposed a technique of *recompression* based on previous techniques of Mehlhorn et. al [12], Lohrey and Mathissen [10] and Sakamoto [17]. This method was successfully applied to various problems related to grammar-compressed strings [5,3,4]. Unexpectedly, this approach was also applicable to word equations, in which case alternative proofs of many known results were obtained [6].

The technique is based on iterative application of two replacement schemes performed on the text t:

pair compression of ab For two different letters a, b such that substring ab appears in t replace each of ab in t by a fresh letter c.

a**'s block compression** For each maximal block a^ℓ, where a is a letter and $\ell > 1$, that appears in t, replace all a^ℓs in t by a fresh letter a_ℓ.

In one phase, pair compression (block compression) is applied to all pairs (blocks, respectively) that appeared at the beginning of this phase. Ideally, each letter is compressed and so the length of t halves, in a worst-case scenario during one phase t is still shortened by a constant factor.

The surprising property is that such a schema can be efficiently applied even to grammar-compressed data [5,3] or to text given in an implicit way, i.e. as a solution of a word equation [6]. In order to do so, local changes of the variables (or nonterminals) are needed: X is replaced with $a^\ell X$ (or Xa^ℓ), where a^ℓ is a prefix (suffix, respectively) of substitution for X. In this way the solution that substitutes $a^\ell w$ for X is implicitly replaced with one that substitutes w.

Recompression and One-Variable Equations. As the recompression works for general word equations, it can be applied also to restricted subclasses. In the general case it relies on the nondeterminism, however, when restricted to one-variable equations it determinises. A simple implementation has $\mathcal{O}(n + \#_X \log n)$ running time, see Section 3. Adding a few heuristics, data structures and applying a more sophisticated analysis yields an $\mathcal{O}(n)$ running time, see Section 4.

Outline of the Algorithm. We present an algorithm for one-variable equation based on the recompression; it also provides a compact description of all solutions of such an equation. Intuitively: when pair compression is applied, say ab is replaced by c (assuming it *can* be applied), then there is a one-to-one correspondence of the solutions before and after the compression, this correspondence is simply exchange of all abs by cs and vice-versa. The same applies to the block compression. On the other hand, the modification of X can lead to loss of solutions (for technical reasons we do not consider the solution ϵ): when X is to be replaced with $a^\ell X$ then each solution of the form $a^\ell w$ has a corresponding solution w, but solution a^ℓ is lost in the process. So before the replacement, it is tested whether a^ℓ is a solution and if so, it is reported. The testing is performed by on-the-fly evaluation of both sides under substitution $X = a^\ell$ and comparing the obtained strings letter by letter until a mismatch is found or both strings end.

It is easy to implement the recompression so that one phase takes linear time. The cost is distributed to explicit words between the variables, each such w is charged $\mathcal{O}(|w|)$. If such w is long enough, its length decreases by a constant factor in one phase, see Lemma 8. Thus, such cost is charged to the lost length and sums to $\mathcal{O}(n)$ in total. However, this is not true when w is short (in particular, of constant length). In this case we use the fact that there are $\mathcal{O}(\log n)$ phases and in each phase such cost is at most $\mathcal{O}(\#_X)$ (i.e. proportional to the number of explicit words in total).

Using the following heuristics as well as more involved analysis the running time can be lowered to $\mathcal{O}(n)$ (see Section 4 for some details):

- We save space used for problematic 'short' words between the variables (and thus time needed to compress them in a phase): instead of storing multiple copies of the same short string we store it once and have pointers to it in the equation. Additionally we prove that those short words are substrings of 'long' words, which allows a bound on the sum of their lengths.
- when we compare $X w_1 X w_2 \ldots w_m X$ from one side of the equation with its copy appearing on the other side, we make such a comparison in $\mathcal{O}(1)$ time (using suffix arrays);
- the $(Xu)^m$ and $(Xu')^{m'}$ (under substitution for X) are compared in $\mathcal{O}(|u| + |u'|)$ time instead of naive $\mathcal{O}(m \cdot |u| + m' \cdot |u'|)$, using simple facts from combinatorics on words.

Furthermore a more insightful analysis shows that problematic 'short' words in the equation invalidate several candidate solutions. This allows a tighter estimation of the time spent on testing the solutions.

Model. To perform the recompression efficiently, an algorithm for grouping pairs is needed. When we identify the symbols in Σ with consecutive numbers, this is done using RadixSort in linear time[1]. Thus, all (efficient) applications of recompression technique make such an assumption. On the other hand, the second of the mentioned heuristics craves checking substring equality in $\mathcal{O}(1)$, to this end a suffix array [7] with a structure for answering *longest common prefix query* (lcp) [8] is employed on which we use range minimum queries [1]. The last structure needs the flexibility of the RAM model to run in $\mathcal{O}(1)$ time per query.

---

[1] The RadixSort runs in time linear in number of numbers plus the universe size. Since we introduce numbers in each phase, it might be that the latter is much larger than the equation length. However, after each phase in linear time we can replace the letters appearing in the equation so that they constitute an interval of numbers, which yields that the RadixSort has indeed linear running time.

2   Preliminaries

One-Variable Equations. Consider a word equation $\mathcal{A} = \mathcal{B}$ over one variable X, by $|\mathcal{A}| + |\mathcal{B}|$ we denote its length and n is the initial length of the equation. Without loss of generality one of $\mathcal{A}$ and $\mathcal{B}$ begins with a variable and the other with a letter [2]: If they both begin with the same symbol (be it letter or nonterminal), we can remove this symbol from them, without affecting the set of solutions; if they begin with different letters, this equation clearly has no solution. The same applies to the last symbols of $\mathcal{A}$ and $\mathcal{B}$. Thus, in the following we assume that the equation is of the form

$$A_0 X A_1 \ldots A_{n_{\mathcal{A}}-1} X A_{n_{\mathcal{A}}} = X B_1 \ldots B_{n_{\mathcal{B}}-1} X B_{n_{\mathcal{B}}} , \tag{1}$$

where $A_i, B_j \in \Sigma^*$ are called (explicit) words, $n_{\mathcal{A}}$ ($n_{\mathcal{B}}$) denotes the number of appearances of X in $\mathcal{A}$ ($\mathcal{B}$, respectively). A_0 (first word) is nonempty and exactly one of $A_{n_{\mathcal{A}}}$, $B_{n_{\mathcal{B}}}$ (last word) is nonempty. If this condition is violated for any reason, we greedily repair by cutting letters from appropriate strings.

A *substitution* S assigns a string to X, we extend S to $(X \cup \Sigma)^*$ with an obvious meaning. A *solution* is a substitution such that $S(\mathcal{A}) = S(\mathcal{B})$. For an equation $\mathcal{A} = \mathcal{B}$ we are looking for a description of all its solutions. We disregard the empty solution $S(X) = \epsilon$ and always assume that $S(X) \neq \epsilon$. In such a case by (1) we can determine the first (last) letter of $S(X)$ in $\mathcal{O}(1)$ time.

Lemma 1. *Let a be the first letter of A_0. If $A_0 \in a^+$ then $S(X) \in a^*$ for each solution S of $\mathcal{A} = \mathcal{B}$, all such solutions can be calculated and reported in $\mathcal{O}(|\mathcal{A}| + |\mathcal{B}|)$ time. If $A_0 \notin a^*$ then there is at most one solution $S(X) \in a^+$, the length of such a solution can be returned in $\mathcal{O}(|\mathcal{A}| + |\mathcal{B}|)$ time. For $S(X) \notin a^+$ the lengths of the a-prefixes of $S(X)$ and A_0 are the same.*

A symmetric version of Lemma 1 holds for the suffix of $S(X)$. By SimpleSolution(a) we denote a procedure that for $A_0 \notin a^*$ returns the unique ℓ such that $S(X) = a^\ell$ is a solution (or nothing, if there is no such solution).

Representation of Solutions. Consider any solution S of $\mathcal{A} = \mathcal{B}$. If $|S(X)| \leq |A_0|$ then $S(X)$ is a prefix of A_0. When $|S(X)| > |A_0|$ then $S(\mathcal{A})$ begins with $A_0 S(X)$ while $S(\mathcal{B})$ begins with $S(X)$ and thus $S(X)$ has a period A_0. Hence $S(X) = A_0^k A$, where A is a prefix of A_0 and $k > 0$. In both cases $S(X)$ is uniquely determined by $|S(X)|$, so it is enough to describe such lengths.

Each letter in the current instance of our algorithm represents some (compressed) string of the input equation, we store its *weight* which is the length of such a string. Furthermore, when we replace X with $a^\ell X$ (or $X a^\ell$) we keep track of the weight of a^ℓ. In this way, for each solution of the current equation we know what is the length of the corresponding solution of the original equation and this identifies it uniquely.

Preserving Solutions. All subprocedures of the algorithm should preserve solutions, i.e. there should be a one-to-one correspondence between solutions before and after the application of the subprocedure. However, as they replace X with $a^\ell X$ (or Xb^r), some solutions are lost in the process and so they should be reported. We formalise these notions.

We say that a subprocedure *preserves solutions* when given an equation $\mathcal{A} = \mathcal{B}$ it returns $\mathcal{A}' = \mathcal{B}'$ such that for some strings u and v

- some solutions of $\mathcal{A} = \mathcal{B}$ are reported by the subprocedure,
- S is an unreported solution of $\mathcal{A} = \mathcal{B}$ if and only if there is a solution S' of $\mathcal{A}' = \mathcal{B}'$ such that $S(X) = uS'(X)v \neq uv$.

By $\mathrm{PC}_{ab \to c}(w)$ we denote the string obtained from w by replacing each ab by c (we assume that $a \neq b$, so this is well-defined), this corresponds to pair compression. We say that a subprocedure *properly implements pair compression* for ab, if it satisfies the conditions for preserving solutions above, but with $\mathrm{PC}_{ab \to c}(S(X)) = uS'(X)v$ replacing $S(X) = uS'(X)v$. Similarly, by $\mathrm{BC}_a(w)$ we denote a string with maximal blocks a^ℓ replaced by a_ℓ (for each $\ell > 1$) and we say that a subprocedure *properly implements blocks compression* for a letter a.

Given an equation $\mathcal{A} = \mathcal{B}$, its solution S and a pair $ab \in \Sigma^2$ appearing in $S(\mathcal{A})$ (or $S(\mathcal{B})$) we say that this appearance is *explicit*, if it comes from substring ab of $\mathcal{A}$ (or $\mathcal{B}$, respectively); *implicit*, if it comes (wholly) from $S(X)$; *crossing* otherwise. A pair is *crossing* if it has a crossing appearance and *noncrossing* otherwise. A similar notion applies to maximal blocks of as, in which case we say that a *has a crossing block* or it *has no crossing blocks*. Alternatively, a pair ab is crossing if b is the first letter of $S(X)$ and aX appears in the equation or a is the last letter of $S(X)$ and Xb appears in the equation or a is the last and b the first letter of $S(X)$ and XX appears in the equation.

Unless explicitly stated, we consider crossing/noncrossing pairs ab in which $a \neq b$. As the first (last) letter of $S(X)$ is the same for each S, the definition of the crossing pair *does not depend on the solution*; the same applies to crossing blocks.

When a pair ab is noncrossing, its compression is easy, as it is enough to replace each explicit ab with a fresh letter c, we refer to this procedure as $\mathsf{PairCompNCr}(a, b)$. Similarly, when no block of a has a crossing appearance, the a's blocks compression consists simply of replacing explicit a blocks, we call this procedure $\mathsf{BlockCompNCr}(a)$.

Lemma 2. *If ab is a noncrossing pair then $\mathsf{PairCompNCr}(a, b)$ properly implements pair compression for ab. If a has no crossing blocks, then $\mathsf{BlockCompNCr}(a)$ properly implements the block compression for a.*

The main idea of the recompression method is the way it deals with the crossing pairs: imagine that ab is a crossing pair, this is because $S(X) = bw$ and aX appears in $\mathcal{A} = \mathcal{B}$ or $S(X) = wa$ and Xb appears in it (the remaining case, in which $S(X) = awb$ and XX appears in the equation is treated in the same way). The cases are symmetric, so we deal only with the first one. To 'uncross' ab in this case it is enough to 'left-pop' b from X: replace each X in the equation with bX and implicitly change the solution to $S(X) = w$.

Algorithm 1. Pop(a, b)

1: **if** b is the first letter of $S(X)$ **then**
2:     **if** SimpleSolution(b) returns 1 **then**   ▷ $S(X) = b$ is a solution
3:         report solution $S(X) = b$
4:     replace each X in $\mathcal{A} = \mathcal{B}$ by bX ▷ Implicitly change $S(X) = bw$ to $S(X) = w$
5:   ▷ perform symmetric actions for a

Lemma 3. Pop(a, b) *preserves solutions. After it the pair ab is noncrossing.*

The presented procedures are merged into PairComp(a, b) that turns crossing pairs into noncrossing ones and then compresses them.

Lemma 4. PairComp(a, b) *properly implements the pair compression of ab.*

The number of noncrossing pairs can be large, however, applying Pop(a, b), where b, a are the first and last letters of the $S(X)$ reduces their number to 2.

Lemma 5. *After* Pop(a, b), *where b, a are the first and last letters of the $S(X)$, the solutions are preserved and there are at most two crossing pairs.*

The problems with crossing blocks are solved in a similar fashion: a has a crossing block, if and only if aa is a crossing pair. So we 'left-pop' a from X until the first letter of $S(X)$ is different than a, we do the same with the ending letter b. This effectively removes the whole a-prefix (b-suffix, respectively) from X: suppose that $S(X) = a^\ell w b^r$, where w does not start with a nor end with b. Then we replace each X by $a^\ell X b^r$, implicitly changing the solution to $S(X) = w$. The corresponding procedure is called CutPrefSuff.

Lemma 6. CutPrefSuff *preserves solutions and after its application there are no crossing blocks of letters.*

BlockComp(a) compresses all blocks of a, regardless of whether it is crossing or not, by first applying CutPrefSuff and then BlockCompNCr(a).

Lemma 7. BlockComp(a) *properly implements the block compression for a before its application.*

3   Main Algorithm

The following algorithm OneVar is basically a simplification of the general algorithm for testing the satisfiability of word equations [6].

Algorithm 2. OneVar reports all solutions of a given word equation

---

1: **while** $|A_0| > 1$ **do**
2:     $Letters \leftarrow$ letters in $\mathcal{A} = \mathcal{B}$
3:     run CutPrefSuff   ▷ There are now crossing blocks
4:     **for** $a \in Letters$ **do**   ▷ Compressing blocks, time $\mathcal{O}(|\mathcal{A}| + |\mathcal{B}|)$ in total
5:         run BlockComp(a)
6:     Pop(a, b), where a is the first and b the last letter of $S(X)$
7:   ▷ Now there are only two crossing pairs
8:     $Crossing \leftarrow$ list of crossing pairs, $Non\text{-}Crossing \leftarrow$ list of noncrossing pairs
9:     **for** each $ab \in Non\text{-}Crossing$ **do**   ▷ Compress noncrossing pairs, $\mathcal{O}(|\mathcal{A}| + |\mathcal{B}|)$
10:         PairCompNCr(a, b)
11:     **for** $ab \in Crossing$ **do**   ▷ Compress the 2 crossing pairs, $\mathcal{O}(|\mathcal{A}| + |\mathcal{B}|)$
12:         PairComp(a, b)
13: TestSolution   ▷ Test solutions from a^*

We call one iteration of the main loop a *phase*.

Theorem 1. OneVar *runs in time* $\mathcal{O}(|\mathcal{A}| + |\mathcal{B}| + (n_\mathcal{A} + n_\mathcal{B}) \log(|\mathcal{A}| + |\mathcal{B}|))$ *and correctly reports all solutions of a word equation* $\mathcal{A} = \mathcal{B}$.

The most important property of OneVar is that the explicit strings between the variables shorten (assuming they are long enough): We say that a word A_i (B_j) is *short* if it consists of at most $C = 100$ letters and *long* otherwise.

Lemma 8. *If* A_i (B_j) *is long then its length is reduced by* $1/4$ *in this phase; if it is short then after the phase it still is.*
If the first word is short then its length is shortened by at least 1 in a phase.

It is relatively easy to estimate the running time of one phase.

Lemma 9. *One phase of* OneVar *can be performed in* $\mathcal{O}(|\mathcal{A}| + |\mathcal{B}|)$ *time.*

The cost of one phase is charged towards the words $A_0, \ldots, A_{n_\mathcal{A}}, B_1, \ldots, B_{n_\mathcal{B}}$ proportionally to their lengths. Since the lengths of the long words drop by a constant factor in each phase, in total such cost is $\mathcal{O}(n)$. For short words the cost is $\mathcal{O}(1)$ per phase and there are $\mathcal{O}(\log n)$ phases by Lemma 8.

4   Heuristics and Better Analysis

The main obstacle in the linear running time is the necessity of dealing with short words, as the time spent on processing them is difficult to charge. The improvement to linear running time is done by four major modifications:

several equations We store a system of several equations and look for a solution of such a system. This allows removal of some words from the equations.

small solutions We identify a class of particularly simple solutions, called *small*, and show that a solution is reported within $\mathcal{O}(1)$ phases from the moment when it became small. In several cases of the analysis we show that the solutions involved are small and so it is easier to charge the time spent on testing them.

storage All words are represented by a structure of size proportional to the size of the long words. In this way the storage space (and so also time used for compression) decreases by a constant factor in each phase.

testing The testing procedure is modified, so that the time it spends on the short words is reduced. We also improve the rough estimate that SimpleSolution takes time proportional to $|\mathcal{A}| + |\mathcal{B}|$ to an estimation that counts for each word whether it was included in the test or not.

Several Equations. We store several equations and look for substitutions that simultaneously satisfy all of them. Hence we have a collection $\mathcal{A}_i = \mathcal{B}_i$ of equations, for $i = 1, \ldots, m$, each of them is of the form (1). This system is obtained by replacing one equation $\mathcal{A}'_i \mathcal{A}''_i = \mathcal{B}'_i \mathcal{B}''_i$ with equivalent two equations $\mathcal{A}'_i = \mathcal{B}'_i$ and $\mathcal{A}''_i = \mathcal{B}''_i$.

Each of the equations $\mathcal{A}_i = \mathcal{B}_i$ in the system specifies the first and last letter of the solution, length of the a-prefix and suffix etc., exactly in the same way as it does for a single equation. However, it is enough to use only one of them, say $\mathcal{A}_1 = \mathcal{B}_1$, as if there is any conflict then there is no solution at all. The consistency is not checked, simply when we find out about inconsistency, we terminate immediately. We say that A_i (B_j) is first or last if it is in any of the stored equations.

All operations on a single equation from previous sections (popping letters, cutting prefixes/suffixes, pair/block compression, etc.) generalise to a system of equations and they preserve their properties and running times, with the length of a single equation $|\mathcal{A}| + |\mathcal{B}|$ replaced by a sum of lengths of all equations $\sum_{i=1}^{m} |\mathcal{A}_i| + |\mathcal{B}_i|$.

Small Solutions. We say that a word w represented as $w = w_1 w_2^\ell w_3$ (where ℓ is arbitrary) is *almost periodic*, with *period size* $|w_2|$ and *side size* $|w_1 w_3|$ (note that several such representations may exist, we use this notion for a particular representation that is clear from the context). A substitution S is *small*, if $S(X) = (w)^k v$, where w, v are almost periodic, with period size at most C and side size at most $6C$.

Lemma 10. *Suppose that S is a small solution. There is a constant c such that within c phases the corresponding solution is reported by* OneVar.

Storing. While the long words are stored exactly as they used to, the short words are stored more efficiently: we keep a table of short words and equations point to the table of short words instead of storing them. We say that such a representation is *succinct* and its size is the sum of lengths of words stored in it. Note that we do *not* include the size of the equation.

The correctness of such an approach is guaranteed by the fact that equality of two explicit words is not changed by OneVar, which is shown by a simple induction.

Lemma 11. *Consider any words A and B in the input equation. Suppose that during OneVar they were transformed to $A' = B'$, none of which is a first nor last word. Then $A = B$ if and only if $A' = B'$.*

Hence, to perform the compression it is enough to read the succinct representation without looking at the whole equation. In particular, the compression (both pair and block) can be performed in time proportional to the size of the succinct representation.

Lemma 12. *The compression in one phase of OneVar can be performed in time linear in size of the succinct representation.*

Ideally, we want to show that the succinct representation has size proportional to the length of long words. In this way its size would decrease by a constant factor in each phase and thus be $\mathcal{O}(n)$ in total. In reality, we are quite close to this: the words stored in the tables are of two types: normal and overdue. The *normal* words are substrings of the long words or A_0^2 and consequently the sum of their sizes is proportional to the size of the long words. A word becomes *overdue* if at the beginning of the phase it is not a substring of a long word or A_0^2. It might be that it becomes a substring of such a word later, it does not stop to be an overdue word in such a case. The new overdue words can be identified in linear time using standard operations on a suffix array for a concatenation of long and short strings appearing in the equations.

Lemma 13. *In time proportional to the sum of sizes of the long words plus the number of overdue words we can identify the new overdue words.*

The overdue words can be removed from the equations in $\mathcal{O}(1)$ phases after becoming overdue. This is shown by a serious of lemmata.

We say that for a substitution S the word A is *arranged against itself* if each A in $S(\mathcal{A})$ coming from explicit $A_i = A$ corresponds to $B_j = A$ at the same positions in $S(\mathcal{B})$ (and symmetrically, for the sides of the equation exchanged).

Lemma 14. *Consider a word A in a phase in which it becomes overdue and a solution S. Then either S is small or A is arranged against itself.*

The proof is rather easy: we consider the $A_i = A$ that is not arranged against some $B_j = A$ in $S(\mathcal{A}) = S(\mathcal{B})$. Since by definition it also cannot be arranged against a subword of a long word, case inspection gives that one of the $S(X)$ preceding or succeeding A_i overlaps with some other $S(X)$, yielding that $S(X)$ is periodical. Furthermore, this period has length at most $|A_i| \leq C$, hence $S(X)$ is small.

Due to Lemmata 10 and 14 the overdue words can be removed in $\mathcal{O}(1)$ phases after their introduction: suppose that A becomes an overdue word in phase ℓ. Any solution, in which an overdue word A is not arranged against another copy of A is small and so it is reported after $\mathcal{O}(1)$ phases. Then an equation $\mathcal{A}'_i X A X \mathcal{A}''_i = \mathcal{B}'_i X A X \mathcal{B}''_i$, where $\mathcal{A}'_i$ and $\mathcal{B}'_i$ do not have A as a word, is equivalent to two equations $\mathcal{A}'_i = \mathcal{B}'_i$ and $\mathcal{A}''_i = \mathcal{B}''_i$ and this procedure can be applied recursively to $\mathcal{A}''_i = \mathcal{B}''_i$. This removes all copies of A from the system.

Lemma 15. *Consider the set of overdue words introduced in phase ℓ. Then in phase $\ell + c$ (for some constant c) we can remove all words A from equations. The obtained set of equations has the same set of solutions. The time spend on removal of overdue words, over the whole run of* OneVar, *is $\mathcal{O}(n)$.*

This allows to bound the time spent on compression.

Lemma 16. *The running time of* OneVar, *except for time used to test the solutions, is $\mathcal{O}(n)$.*

Testing. SimpleSolution checks whether S is a solution by comparing $S(\mathcal{A}_i)$ and $S(\mathcal{B}_i)$ letter by letter, replacing X with a^ℓ on the fly. We say that in such a case a letter b in $S(\mathcal{A}_i)$ is *tested against* the corresponding letter in $S(\mathcal{B}_i)$.

Suppose that for a substitution S a letter from A_i is tested against a letter from $S(XB_j)$ (there is some asymmetry regarding A_is and B_js in the definition, this is a technical detail without an importance). We say that this test is:

protected if at least one of A_i, A_{i+1}, B_j, B_{j+1} is long
failed if A_i, A_{i+1}, B_j and B_{j+1} are short and a mismathch for S is found till the end of A_{i+1} or B_{j+1};
aligned if $A_i = B_j$ and $A_{i+1} = B_{j+1}$, all of them are short and the first letter of A_i is tested against the first letter of B_j;
misaligned if all of A_i, A_{i+1}, B_j, B_{j+1} are short, $A_{i+1} \neq A_i$ or $B_{j+1} \neq B_j$ and this is not an aligned test;
periodical if $A_{i+1} = A_i$, $B_{j+1} = B_j$, all of them are short and this is not an aligned test.

It is easy to show by case inspection that each test is of one of those type. We calculate the cost of each type of tests separately. For failed tests note that there are constantly many of them in each of the $\mathcal{O}(\log n)$ phases.

Lemma 17. *The number of all failed tests is $\mathcal{O}(\log n)$.*

For protected tests, we charge the cost of the protected test to the long word and only $\mathcal{O}(|A|)$ such tests can be charged to one long word A in a phase. On the other hand, each long word is shortened by a constant factor in a phase and so this cost can be charged to those removed letters and thus the total cost of those tests (over the whole run of OneVar) is $\mathcal{O}(n)$.

Lemma 18. *In one phase the number of protected tests is proportional to the length of long words. Thus there are $\mathcal{O}(n)$ such tests in total.*

In case of the misaligned tests, consider the phase in which the last of A_{i+1}, A_i, B_{j+1}, B_j becomes short. We show that the corresponding solution S' is small in this phase and so by Lemma 10 it is reported within $\mathcal{O}(1)$ following phases. The proof is quite technical, it follows a general idea of Lemma 14: we show that $S(X)$ overlaps with itself and so it has a period. A closer inspection proves that this period is almost periodical.

The cost of the misaligned test is charged to the last word among A_i, A_{i+1}, B_j, B_{j+1} that became short, say, B_j and only $\mathcal{O}(1)$ such tests are charged to this B_j (over the whole run of OneVar). Hence there are $\mathcal{O}(n)$ misaligned tests.

Lemma 19. *There are $\mathcal{O}(n)$ misaligned tests during the whole run of OneVar.*

Consider the maximal set of consecutive aligned tests, they correspond to comparison of $A_i X A_{i+1} \ldots A_{i+k} X$ and $B_j X B_{j+1} \ldots B_{j+k} X$, where $A_{i+\ell} = B_{j+\ell}$ for $\ell = 0, \ldots, k$. Then the next test is either misaligned, protected or failed, so if the cost of all those aligned tests can be bounded by $\mathcal{O}(1)$, they can be associated with the succeeding test. Note that instead of performing the aligned tests (by comparing letters), it is enough to identify the maximal (syntactically) equal substrings of the equation. From Lemma 11 it follows that this corresponds to the (syntactical) equality of substrings in the original equation. We identify such substrings in $\mathcal{O}(1)$ per substring using a suffix array constructed for the input equation.

Lemma 20. *The total cost of aligned tests is $\mathcal{O}(n)$.*

For the periodical tests we apply a similar charging strategy. Suppose that we are to test the equality of (suffix of) $S((A_i X)^\ell)$ and (prefix of) $S(X(B_j X)^k)$. Firstly, it is easy to show that the next test is either misaligned, protected or failed. Secondly, if $|A_i| = |B_j|$ then the test for $A_{i+\ell'}$ and $B_{j+\ell'}$ for $0 < \ell' \leq \ell$ is the same as for A_i and B_j and so they can be all skipped. If $|A_i| > |B_j|$ then the common part of $S((A_i X)^\ell)$ and $S(X(B_j X)^k)$ have periods $|S(A_i X)|$ and $|S(B_j X)|$ and consequently has a period $|A_i| - |B_j| \leq C$. So to test the equality of $S((A_i X)^\ell)$ and (prefix of) $S(X(B_j X)^k)$ it is enough to test first common $|A_i| - |B_j|$ letters and check whether both $S(A_i X)$ and $S(B_j X)$ have period $|A_i| - |B_j|$.

Lemma 21. *Performing all periodical tests takes in total $\mathcal{O}(n)$ time*

This yields that the total time of testing is linear.

Lemma 22. *The time spent on testing solutions during OneVar is $\mathcal{O}(n)$.*

Acknowledgements I would like to thank P. Gawrychowski for initiating my interest in compressed membership problems and compressed pattern matching, exploring which led to this work and for pointing to relevant literature [10,12]; J. Karhumäki, for his explicit question, whether the techniques of local recompression can be applied to the word equations; W. Plandowski for his numerous comments and suggestions on the recompression applied to word equations.

References

1. Berkman, O., Vishkin, U.: Recursive star-tree parallel data structure. SIAM J. Comput. 22(2), 221–242 (1993)
2. Dąbrowski, R., Plandowski, W.: On word equations in one variable. Algorithmica 60(4), 819–828 (2011)
3. Jeż, A.: Faster fully compressed pattern matching by recompression. In: Czumaj, A., Mehlhorn, K., Pitts, A., Wattenhofer, R. (eds.) ICALP 2012, Part I. LNCS, vol. 7391, pp. 533–544. Springer, Heidelberg (2012)
4. Jeż, A.: Approximation of grammar-based compression via recompression. In: Fischer, J., Sanders, P. (eds.) CPM 2013. LNCS, vol. 7922, pp. 165–176. Springer, Heidelberg (2013)
5. Jeż, A.: The complexity of compressed membership problems for finite automata. Theory of Computing Systems, 1–34 (2013),
   http://dx.doi.org/10.1007/s00224-013-9443-6
6. Jeż, A.: Recompression: a simple and powerful technique for word equations. In: Portier, N., Wilke, T. (eds.) STACS. LIPIcs, vol. 20, pp. 233–244. Schloss Dagstuhl–Leibniz-Zentrum fuer Informatik, Dagstuhl (2013),
   http://drops.dagstuhl.de/opus/volltexte/2013/3937
7. Kärkkäinen, J., Sanders, P., Burkhardt, S.: Linear work suffix array construction. J. ACM 53(6), 918–936 (2006)
8. Kasai, T., Lee, G., Arimura, H., Arikawa, S., Park, K.: Linear-time longest-common-prefix computation in suffix arrays and its applications. In: Amir, A., Landau, G.M. (eds.) CPM 2001. LNCS, vol. 2089, pp. 181–192. Springer, Heidelberg (2001)
9. Laine, M., Plandowski, W.: Word equations with one unknown. Int. J. Found. Comput. Sci. 22(2), 345–375 (2011)
10. Lohrey, M., Mathissen, C.: Compressed membership in automata with compressed labels. In: Kulikov, A., Vereshchagin, N. (eds.) CSR 2011. LNCS, vol. 6651, pp. 275–288. Springer, Heidelberg (2011)
11. Makanin, G.S.: The problem of solvability of equations in a free semigroup. Matematicheskii Sbornik 2(103), 147–236 (1977) (in Russian)
12. Mehlhorn, K., Sundar, R., Uhrig, C.: Maintaining dynamic sequences under equality tests in polylogarithmic time. Algorithmica 17(2), 183–198 (1997)
13. Obono, S.E., Goralcik, P., Maksimenko, M.N.: Efficient solving of the word equations in one variable. In: Privara, I., Ružička, P., Rovàn, B. (eds.) MFCS 1994. LNCS, vol. 841, pp. 336–341. Springer, Heidelberg (1994)
14. Plandowski, W.: Satisfiability of word equations with constants is in NEXPTIME. In: STOC, pp. 721–725 (1999)
15. Plandowski, W.: Satisfiability of word equations with constants is in PSPACE. J. ACM 51(3), 483–496 (2004)
16. Plandowski, W., Rytter, W.: Application of Lempel-Ziv encodings to the solution of word equations. In: Larsen, K.G., Skyum, S., Winskel, G. (eds.) ICALP 1998. LNCS, vol. 1443, pp. 731–742. Springer, Heidelberg (1998)
17. Sakamoto, H.: A fully linear-time approximation algorithm for grammar-based compression. J. Discrete Algorithms 3(2-4), 416–430 (2005)

The IO and OI Hierarchies Revisited[*]

Gregory M. Kobele[1],[**] and Sylvain Salvati[2],[***]

[1] University of Chicago
kobele@uchicago.edu
[2] INRIA, LaBRI, Université de Bordeaux
sylvain.salvati@labri.fr

Abstract. We study languages of λ-terms generated by IO and OI unsafe grammars. These languages can be used to model meaning representations in the formal semantics of natural languages following the tradition of Montague [19]. Using techniques pertaining to the denotational semantics of the simply typed λ-calculus, we show that the emptiness and membership problems for both types of grammars are decidable. In the course of the proof of the decidability results for OI, we identify a decidable variant of the λ-definability problem, and prove a stronger form of Statman's finite completeness Theorem [28].

1 Introduction

In the end of the sixties, similar but independent lines of research were pursued in formal language theory and in the formal semantics of natural language. Formal language theory was refining the Chomsky hierarchy so as to find an adequate syntactic model of programming languages lying in between the context-free and context-sensitive languages. Among others, this period resulted in the definition of *IO and OI macro languages* by Fischer [12] and the notion of *indexed languages* (which coincide with OI macro languages) by Aho [2]. At the same time, Richard Montague [19] was proposing a systematic way of mapping natural language sentences to logical formulae representing their meanings, providing thereby a solid foundation for the field of formal semantics. The main idea behind these two lines of research can be summed up in the phrase '*going higher-order.*' For macro and indexed grammars, this consisted in parameterizing non-terminals with strings and indices (stacks) respectively, and in Montague's work it consisted in using the simply typed λ-calculus to map syntactic structures to their meanings. Montague was ahead of the formal language theory community which took another decade to go higher-order with the work of Damm [7]. However, the way Damm defined higher-order grammars used (implicitly) a restricted version of the λ-calculus that is now known as the *safe λ-calculus*. This restriction was made explicit by Knapik *et al.* [16] and further studied by Blum and Ong [4].

---

[*] Long version: http://hal.inria.fr/hal-00818069
[**] The first author was funded by LaBRI while working on this research.
[***] This work has been supported by ANR-12-CORD-0004 POLYMNIE.

F.V. Fomin et al. (Eds.): ICALP 2013, Part II, LNCS 7966, pp. 336–348, 2013.
© Springer-Verlag Berlin Heidelberg 2013

For formal grammars this restriction has first been lifted by de Groote [8] and Muskens [21] in the context of computational linguistics and as a way of applying Montague's techniques to syntactic modeling.

In the context of higher-order recursive schemes, Ong showed that safety was not a necessary condition for the decidability of the MSO model checking problem. Moreover, the safety restriction has been shown to be a real restriction by Parys [23]. Nevertheless, concerning the IO and OI hierarchies, the question as to whether safety is a genuine restriction in terms of the definable languages is still an open problem. Aehlig *et al.* [1] showed that for second order OI grammars safety was in fact *not* a restriction. It is nevertheless generally conjectured that for higher-order grammars safety is in fact a restriction.

As we wish to extend Montague's technique with the OI hierarchy so as to enrich it with fixed-point computation as proposed by Moschovakis [20], or as in proposals to handle presuppositions in natural languages by Lebedeva and de Groote [10,9,17], we work with languages of λ-terms rather than with just languages of strings or trees. In the context of languages of λ-terms, safety clearly appears to be a restriction since, as shown by Blum and Ong [4], not every λ-term is safe. Moreover the terms generated by Montague's technique appear to be unsafe in general.

This paper is thus studying the formal properties of the unsafe IO and OI languages of λ-terms. A first property that the use of unsafe grammars brings into the picture is that the unsafe IO hierarchy is strictly included within the unsafe OI hierarchy. The inclusion can be easily shown using a standard CPS transform on the grammars and its strictness is implied by decidability results. Nevertheless, it is worth noting that such a transform cannot be performed on safe grammars, so that it is unclear whether safe IO languages are safe OI languages. This paper focuses primarily on the emptiness and the membership problems for unsafe IO and OI languages, by using simple techniques related to the denotational semantics of the λ-calculus. For the IO case, we are going to recast some known results from Salvati [25,24], so as to emphasize that they derive from the fact that for an IO language and a finite model of the λ-calculus, one can effectively compute the elements of the model which are the interpretations of terms in the language. This allows us to show that the emptiness problem is decidable, and also, using Statman's finite completeness theorem [28], to show that the membership problem is decidable. In contrast to the case for IO languages, we show that this property does not hold for OI languages. Indeed, we prove that the set of closed terms of a given type is an OI language, and thus, since λ-definability is undecidable [18], the set of elements in a finite model that are the interpretation of terms in an OI language cannot be effectively computed. To show the decidability of emptiness and of the membership problems for OI, we prove a theorem that we call the *Observability Theorem*; it characterizes some semantic properties of the elements of an OI language in monotonic models, and leads directly to the decidability of the emptiness problem. For the membership problem we prove a generalization of Statman's finite completeness theorem which, in combination with the Observability Theorem, entails the decidability of the membership problem of OI languages.

This work is closely related to the research that is being carried out on higher-order recursive schemes. It differs from it in one important respect: the main objects of study in the research on higher-order recursive schemes are the infinite trees generated by schemes, while our work is related to the study of the Böhm trees of λY-terms which may contain λ-binders. Such Böhm trees are closer to the configuration graphs of Higher-order Collapsible Pushdown Automata whose first-order theory has been shown undecidable [6]. If we were only interested in grammars generating trees or strings, the decidability of MSO for higher-order recursion schemes [22] would yield the decidability of both the emptiness and the membership problems of unsafe OI grammars, but this is no longer the case when we turn to languages of λ-terms.

Organization of the Paper. We start by giving the definitions related to the λ-calculus, its finitary semantics, and how to define higher-order grammars in section 2. We then present the decidability results concerning higher-order IO languages and explain why the techniques used there cannot be extended to OI languages in section 3. Section 4 contains the main contributions of the paper: the notion of hereditary prime elements of monotone models together with the Observability Theorem, and a strong form of Statman's finite completeness Theorem. Finally we present conclusions and a broader perspective on our results in section 5.

2   Preliminaries

In this section, we introduce the various calculi we are going to use in the course of the article. Then we show how those calculi may be used to define IO and OI grammars. We give two presentations of those grammars, one using traditional rewriting systems incorporating non-terminals, and the other as terms in one of the calculi; these two perspectives are equivalent. In the remainder of the paper we will switch between these two formats as is most convenient. Finally we introduce the usual notions of full and monotone models for the calculi we work with.

2.1 λ-Calculi

We introduce here various extensions of the simply typed λ-calculus. Given an atomic type 0 (our results extend with no difficulty to arbitrarily many atomic types), the set *type* of types is built inductively using the binary right-associative infix operator $\rightarrow$. We write $\alpha_1 \rightarrow \cdots \rightarrow \alpha_n \rightarrow \alpha_0$ for $(\alpha_1 \rightarrow (\cdots (\alpha_n \rightarrow \alpha_0)))$. As in [14], the order of a type is: $\text{order}(0) = 1$, $\text{order}(\alpha \rightarrow \beta) = \max(\text{order}(\alpha) + 1, \text{order}(\beta))$. Constants are declared in higher-order signatures Σ which are finite sets of typed constants $\{A_1^{\alpha_1}, \ldots, A_n^{\alpha_n}\}$. We use constants to represent non-terminal symbols.

We assume that we are given a countably infinite set of typed λ-variables $(x^\alpha, y^\beta, \ldots)$. The families of typed $\lambda Y + \Omega$-terms $(\Lambda^\alpha)_{\alpha \in type}$ built on a signature Σ are inductively constructed according to the following rules: x^α, c^α and Ω^α are in Λ^α; Y^α is in $\Lambda^{(\alpha \to \alpha) \to \alpha}$; if M is in $\Lambda^{\alpha \to \beta}$ and N is in Λ^α, then (MN) is in Λ^α; if M is in Λ^β then $(\lambda x^\alpha.M)$ is in $\Lambda^{\alpha \to \beta}$; if M and N are in Λ^0 then $M + N$ is in Λ^0. When M is in Λ^α we say that it has type α, we write M^α to indicate that M has type α; the order of a term M, is the order of its type. As it is customary, we omit type annotations when they can be easily inferred or when they are irrelevant. We adopt the usual conventions about dropping parentheses in the syntax of terms. We write $M^{\alpha \to \beta} + N^{\alpha \to \beta}$ as an abbreviation for $\lambda x^\alpha.Mx + Nx$. The set of free variables of the term M is written $FV(M)$; a term M is closed when $FV(M) = \emptyset$. Finally we write $M[x_1 \leftarrow N_1, \ldots, x_n \leftarrow N_n]$ for the simultaneous capture-avoiding substitutions of the terms $N_1, \ldots, N_n$ for the free occurrences of the variables $x_1, \ldots, x_n$ in M.

The set of λ-terms is the set of terms that do not contain occurrences of Y, $+$ or Ω, and for any $S \subseteq \{Y, +, \Omega\}$, the λS-terms are the λ-terms that may contain only constants that are in S. For example, $\lambda + \Omega$-terms are the terms that do not contain occurrences of Y.

We assume the reader is familiar with the notions of β-contractions and η-contraction and η-long forms (see [14]). The constant Ω^α stands for the undefined term of type α, Y^α is the fixpoint combinator of type $(\alpha \to \alpha) \to \alpha$, and $+$ is the non-deterministic choice operator. The families of terms that may contain occurrences of Ω are naturally ordered with the least compatible relation $\sqsubseteq$ such that $\Omega^\alpha \sqsubseteq M$ for every term M of type α; δ-contraction provides the operational semantics of the fixpoint combinator: $YM \to_\delta M(YM)$, and $+$-contraction gives the operational semantics of the non-deterministic choice operator: $M + N \to_+ M$ and $M + N \to_+ N$. Given a set $\mathcal{R}$ of symbols denoting compatible relations, for $S \subseteq \mathcal{R}$, S-contraction is the union of the contraction relations denoted by the symbols in S; it will generally be written as $\to_S$. For example, $\to_{\beta\eta+}$ denotes the $\beta\eta+$-contraction. S-reduction, written $\overset{*}{\to}_S$, is the reflexive transitive closure of S-contraction and S-conversion, $=_S$, is the smallest equivalence relation containing $\to_S$. The notion of S-normal form is defined as usual, we simply say *normal form* when S is obvious from the context. We recall (see [14]) that when $M \overset{*}{\to}_{\beta\eta\delta+} N$ and M' and N' are respectively the η-long forms of M and N, then $M' \overset{*}{\to}_{\beta\delta+} N'$. In the remainder of the paper we assume that we are working with terms in η-long form and forget about η-reduction.

2.2 IO/OI Grammars and $\lambda Y +$-Calculus

We define a higher-order macro grammar $\mathcal{G}$ as a triple (Σ, R, S) where Σ is a higher-order signature of non-terminals, R is a finite set of rules $A \to M$ where A is a non-terminal of Σ and M is a λ-term built on Σ that has the same type as A, and where S is a distinguished non-terminal of Σ, the *start symbol*. We do not require M to be a closed term, the free variables in the right hand side of grammatical rules play the role of terminal symbols. We also do not require

S to be of a particular type, this permits (higher-order) macro grammars to define languages of λ-terms of arbitrary types. As noted in [8], languages of strings and trees are particular cases of the languages we study. A grammar has order n when the highest order of its non-terminals is n. Our higher-order macro grammars generalize those of Damm [7] in two ways: first, they do not necessarily verify the safety condition that Damm's grammar implicitly verify; second, instead of only defining languages of strings or trees, they can define languages of λ-terms, following Montague's tradition in the formal semantics of natural languages. According to [4], a term M is safe when no subterm N of M contains free variables (excluding the free variables of M which play the role of constants) of order lower than that of N, unless N occurs as part of a subterm NP or $\lambda x.N$. A grammar is safe when the right hand sides of its rules are all safe terms. Safe terms can be *safely* reduced using substitution in place of capture-avoiding substitutions.

The rules of a grammar $\mathcal{G} = (\Sigma, R, S)$ define a natural relation $\to_{\mathcal{G}}$ on terms built on Σ. We write $M \to_{\mathcal{G}} N$ when N is obtained from M by replacing (without capturing free variables) an occurrence of a non-terminal A in M by a term P, such that $A \to P$ is a rule of $\mathcal{G}$. The grammar $\mathcal{G}$ defines two languages: $\mathcal{L}_{OI}(\mathcal{G}) = \{M \text{ in normal form} \mid S \xrightarrow{*}_{\beta\mathcal{G}} M\}$ and $\mathcal{L}_{IO}(\mathcal{G}) = \{M \text{ in normal form} \mid \exists P.S \xrightarrow{*}_{\mathcal{G}} P \wedge P \xrightarrow{*}_{\beta} M\}$. These two languages can be defined in a different manner, in particular M is in $\mathcal{L}_{OI}(\mathcal{G})$ iff S can be reduced to M with the head reduction strategy that consists in always contracting top-most redices of the relation $\to_{\beta\mathcal{G}}$. For a given grammar $\mathcal{G}$, we always have that $\mathcal{L}_{IO}(\mathcal{G}) \subseteq \mathcal{L}_{OI}(\mathcal{G})$, but, in general, $\mathcal{L}_{IO}(\mathcal{G}) \neq \mathcal{L}_{OI}(\mathcal{G})$. Here follows an example of a second order macro grammar $\mathcal{G}$ whose free variables (or terminals) are $\text{EX}^{(0\to0)\to0}$, $\text{AND}^{0\to0\to0}$, $\text{NOT}^{0\to0}$ and $\text{P}^{0\to0}$ and whose non-terminals are S^0 (the start symbol), $S'^{0\to0}$ and $\mathbf{cons}^{0\to0\to0}$ (extending BNF notation to macro grammars).

$$S \to \text{EX}(\lambda x.(S'x)) \mid \text{AND}\, S\, S \mid \text{NOT} S$$

$$S' \to \lambda y.\text{EX}(\lambda x.S'(\mathbf{cons}\, y\, x)) \mid \lambda y.\text{AND}(S'y)(S'y) \mid \lambda y.\text{NOT}(S'y) \mid \lambda y.Py$$

$$\mathbf{cons} \to \lambda xy.x \mid \lambda xy.y$$

The language $\mathcal{L}_{OI}(\mathcal{G})$ represents the set of formulae of first-order logic built with one predicate P (we use a similar construction later on to prove Theorem 8). The language $\mathcal{L}_{IO}(\mathcal{G})$ represents the formulae of first-order logic that can be built with only one variable (that is each subformula of a formula represented in $\mathcal{L}_{IO}(\mathcal{G})$ contains at most one free variable). Given the definition of safety given in [4], it is easily verified that the terms of these languages are not safe; this illustrates that unsafe IO and OI languages of λ-terms are more general than their safe counterparts. Moreover, when seen as graphs, the terms of $\mathcal{L}_{OI}(\mathcal{G})$ form a class of graphs which has an unbounded treewidth; the MSO theory of these terms is undecidable. This explains why the decidability results we obtain later on cannot be seen as corollaries of Ong's Theorem [22].

We extend the notions of IO and OI languages to $\lambda Y + \Omega$-terms. Given a $\lambda Y + \Omega$-term M, its IO language $\mathcal{L}_{IO}(M)$ is the set of λ-terms N in normal form such that there is P such that $M \xrightarrow{*}_{\delta+} P \xrightarrow{*}_{\beta} N$. Its OI language $\mathcal{L}_{OI}(M)$ is the set of λ-terms N in normal form such that $M \xrightarrow{*}_{\beta\delta+} N$ (we can also restrict our attention to head-reduction). An alternative characterization of $\mathcal{L}_{OI}(M)$ is the following. Given a term M we write $\omega(M)$ for the *immediate approximation of M*, that is the term obtained from M as follows: $\omega(\lambda x^\alpha.M) = \Omega^{\alpha\to\beta}$ if $\omega(M) = \Omega^\beta$, and $\lambda x^\alpha.\omega(M)$ otherwise; $\omega(MN) = \Omega^\beta$ if $\omega(M) = \Omega^{\alpha\to\beta}$ or $M = \lambda x.P$, and $\omega(MN) = \omega(M)\omega(N)$ otherwise; $\omega(Y^\alpha) = \Omega^{(\alpha\to\alpha)\to\alpha}$, $\omega(x^\alpha) = x^\alpha$, $\omega(\Omega^\alpha) = \Omega^\alpha$, and $\omega(N_1 + N_2) = \omega(N_1) + \omega(N_2)$. Note that $\omega(M)$ is a $\lambda+\Omega$-term that contains no β-redices. A $\lambda+\Omega$-term Q is a *finite approximation* of M if there is a P such that $M \xrightarrow{*}_{\beta\delta} P$ and $Q = \omega(P)$. The language $\mathcal{L}_{OI}(M)$ is the union of the languages $\mathcal{L}_{OI}(Q)$ so that Q is a finite approximation of M.

In both the IO and OI mode of evaluation, $\lambda+\Omega$-terms define finite languages, and $\lambda Y+\Omega$-calculus defines exactly the same classes of languages as higher-order macro grammars.

Theorem 1. *Given a higher-order macro grammar $\mathcal{G}$, there is a $\lambda Y+\Omega$-term M so that $\mathcal{L}_{OI}(\mathcal{G}) = \mathcal{L}_{OI}(M)$ and $\mathcal{L}_{IO}(\mathcal{G}) = \mathcal{L}_{IO}(M)$.*

Given a $\lambda Y+\Omega$-term M there is a higher-order macro grammar $\mathcal{G}$ so that $\mathcal{L}_{OI}(\mathcal{G}) = \mathcal{L}_{OI}(M)$ and $\mathcal{L}_{IO}(\mathcal{G}) = \mathcal{L}_{IO}(M)$.

The proof of this theorem is based on the correspondence between higher-order schemes and λY-calculus that is given in [27]. Going from a $\lambda Y+\Omega$-term to a grammar is simply a direct transposition of the procedure described in [27] with the obvious treatment for $+$. For the other direction, it suffices to see the grammar as a non-deterministic scheme, which is done by viewing all the rules $A \to M_1, \ldots, A \to M_n$, of a non-terminal A as a unique rule of a scheme $A \to M_1 + \cdots + M_n$; and then to transform the scheme into a $\lambda Y+$-term using the transformation given in [27]. There is a minor technicality concerning the IO languages; one needs to start with a grammar where every non-terminal can be rewritten into a $\mathcal{G}$-normal form using $\xrightarrow{*}_{\mathcal{G}}$ only.

2.3   Models of the λ-Calculi

Full models of the λ-calculus We start by giving the simplest notion of models of λ-calculus, that of full models. A full model $\mathcal{F}$ is a collection of sets indexed by types $(\mathcal{F}_\alpha)_{\alpha \in type}$ so that $\mathcal{F}_{\alpha\to\beta}$ is $\mathcal{F}_\beta^{\mathcal{F}_\alpha}$, the set of functions from $\mathcal{F}_\alpha$ to $\mathcal{F}_\beta$. Note that $\mathcal{F}$ is completely determined by $\mathcal{F}_0$. A full model is said to be finite when $\mathcal{F}_0$ is a finite set; in that case $\mathcal{F}_\alpha$ is finite for every $\alpha \in type$. A valuation ν is a function that maps variables to elements of $\mathcal{F}$ respecting typing, meaning that, for every x^α, $\nu(x^\alpha)$ is in $\mathcal{F}_\alpha$. Given a valuation ν and a in $\mathcal{F}_\alpha$, we write $\nu[x^\alpha \leftarrow a]$ for the valuation which maps the variable x^α to a but is otherwise equal to ν. We can now interpret λ-terms in $\mathcal{F}$, using the following interpretation scheme: $[\![x^\alpha]\!]_\mathcal{F}^\nu = \nu(x^\alpha)$, $[\![MN]\!]_\mathcal{F}^\nu = [\![M]\!]_\mathcal{F}^\nu([\![N]\!]_\mathcal{F}^\nu)$ and for a in $\mathcal{F}_\alpha$, $[\![\lambda x^\alpha.M]\!]_\mathcal{F}^\nu(a) = [\![M]\!]_\mathcal{F}^{\nu[x^\alpha \leftarrow a]}$. For a closed term M, $[\![M]\!]_\mathcal{F}^\nu$ does not depend on ν, and thus we simply write $[\![M]\!]_\mathcal{F}$. The following facts are known about full models:

Theorem 2. *If $M =_{\beta\eta} N$ then for every full model $\mathcal{F}$, $[\![M]\!]_\mathcal{F}^\nu = [\![N]\!]_\mathcal{F}^\nu$.*

Theorem 3 (Finite Completeness [28]). *Given a λ-term M, there is a finite full model $\mathcal{F}_M$ and a valuation ν so that, for every N, $[\![N]\!]^\nu_{\mathcal{F}_M} = [\![M]\!]^\nu_{\mathcal{F}_M}$ iff $N =_{\beta\eta} M$.*

In this theorem, the construction of $\mathcal{F}_M$ and ν is effective.

For a full model $\mathcal{F}$, an element f of $\mathcal{F}_\alpha$ is said to be λ-definable when there is a closed M such that $[\![M]\!]_{\mathcal{F}} = f$. The problem of λ-definability is the problem whose input is a finite full model $\mathcal{F}$ and an element f of $\mathcal{F}_\alpha$, and whose answer is whether f is λ-definable.

Theorem 4 (Loader [18]). *The λ-definability problem is undecidable.*

Given a language of λ-terms of type α, L, and a full model $\mathcal{F}$, we write $[\![L]\!]^\nu_{\mathcal{F}}$ for the set $\{[\![M]\!]^\nu_{\mathcal{F}} \mid M \in L\}$. So in particular, for a $\lambda Y + \Omega$-term M, we may write $[\![\mathcal{L}_{IO}(M)]\!]^\nu_{\mathcal{F}}$ or $[\![\mathcal{L}_{OI}(M)]\!]^\nu_{\mathcal{F}}$.

Monotone models of $\lambda Y + \Omega$-calculus Given two complete lattices L_1 and L_2, we write $L_3 = \mathbf{Mon}[L_1 \to L_2]$ for the lattice of monotonous functions from L_1 to L_2 ordered pointwise; f is monotonous if $a \leq_1 b$ implies $f(a) \leq_2 f(b)$, and given f and g in L_3, $f \leq_3 g$ whenever for every a in L_1, $f(a) \leq_2 g(a)$. Among the functions in $\mathbf{Mon}[L_1 \to L_2]$, of special interest are the step functions which are functions $a \mapsto f$ determined from elements a in L_1 and f in L_2, and are defined such that $(a \mapsto f)(b)$ is equal to f when $a \leq_1 b$ and to $\perp_2$ otherwise. A monotone model $\mathcal{M}$ is a collection of finite lattices indexed by types, $(\mathcal{M}_\alpha)_{\alpha \in type}$ where $\mathcal{M}_{\alpha \to \beta} = \mathbf{Mon}[\mathcal{M}_\alpha \to \mathcal{M}_\beta]$ (we write $\perp_\alpha$ and $\top_\alpha$ respectively for the least and greatest elements of $\mathcal{M}_\alpha$). The notion of valuation on monotone models is similar to the one on full models and we use the same notation. Terms are interpreted in monotone models according to the following scheme: $[\![x^\alpha]\!]^\nu_\mathcal{M} = \nu(x^\alpha)$, $[\![MN]\!]^\nu_\mathcal{M} = [\![M]\!]^\nu_\mathcal{M}([\![N]\!]^\nu_\mathcal{M})$ and for a in $\mathcal{M}_\alpha$, $[\![\lambda x^\alpha.M]\!]^\nu_\mathcal{M}(a) = [\![M]\!]^{\nu[x^\alpha \leftarrow a]}_\mathcal{M}$, $[\![\Omega^\alpha]\!]^\nu_\mathcal{M} = \perp_\alpha$, $[\![M + N]\!]^\nu_\mathcal{M} = [\![M]\!]^\nu_\mathcal{M} \vee [\![N]\!]^\nu_\mathcal{M}$ and for every a in $\mathcal{M}_{\alpha \to \alpha}$, $[\![Y]\!]^\nu_\mathcal{M}(a) = \bigvee\{a^n(\perp_\alpha) \mid n \in \mathbb{N}\}$.

The following Theorem gives well known results on monotone models (see [3]):

Theorem 5. *Given two $\lambda Y + \Omega$ terms of type α, M and N:*

1. *if $M =_{\beta\delta} N$ then for every monotone model, $\mathcal{M}$, $[\![M]\!]^\nu_\mathcal{M} = [\![N]\!]^\nu_\mathcal{M}$,*
2. *$[\![M]\!]^\nu_\mathcal{M} = \bigvee\{[\![Q]\!]^\nu_\mathcal{M} \mid Q$ is a finite approximation of $M\}$,*
3. *if $M \xrightarrow{*}_{\beta\delta+} N$ then $[\![N]\!]^\nu_\mathcal{M} \leq [\![M]\!]^\nu_\mathcal{M}$,*
4. *if $N \sqsubseteq M$ then $[\![N]\!]^\nu_\mathcal{M} \leq [\![M]\!]^\nu_\mathcal{M}$.*

3   Relations between IO, OI and Full Models

In this section, we investigate some basic properties of IO and OI languages. We will see that the class of higher-order OI languages strictly subsumes that of higher-order IO languages. We will then see that the emptiness and membership problems for higher-order IO languages are decidable, by showing that for a higher-order grammar $\mathcal{G}$, a finite full model $\mathcal{F}$, and a valuation ν, the set $[\![\mathcal{L}_{IO}(\mathcal{G})]\!]_{\mathcal{F}}^{\nu}$ is effectively computable. On the other hand, we show that $[\![L]\!]_{\mathcal{F}}^{\nu}$ is *not* in general effectively computable when L is an OI language.

A simple continuation passing style (CPS) transform witnesses that:

Theorem 6 (OI subsumes IO). *Given a higher-order grammar $\mathcal{G}$ there is a higher-order grammar $\mathcal{G}'$ so that $\mathcal{L}_{IO}(\mathcal{G}) = \mathcal{L}_{OI}(\mathcal{G}')$.*

The CPS transform naturally makes the order of $\mathcal{G}'$ be the order of $\mathcal{G}$ plus 2.

We now show that for a full model $\mathcal{F}$, a valuation ν and a given grammar $\mathcal{G}$, the set $[\![\mathcal{L}_{IO}(\mathcal{G})]\!]_{\mathcal{F}}^{\nu}$ can be effectively computed. A natural consequence of this is that the emptiness and the membership problems for higher-order IO languages are decidable. These results are known in the literature [24,25,26], nevertheless, we include them here so as to emphasize that they are related to the effectivity of the set $[\![\mathcal{L}_{IO}(\mathcal{G})]\!]_{\mathcal{F}}^{\nu}$, a property that, as we will see later, does not hold in the case of OI languages.

Theorem 7 (Effective finite interpretation of IO). *Given a higher-order macro grammar $\mathcal{G}$, a full model $\mathcal{F}$ and a valuation ν, one can effectively construct the set $[\![\mathcal{L}_{IO}(\mathcal{G})]\!]_{\mathcal{F}}^{\nu}$.*

Corollary 1. *Given a higher-order macro grammar $\mathcal{G}$, the problem of deciding whether $\mathcal{L}_{IO}(\mathcal{G}) = \emptyset$ is P-complete.*

Corollary 2. *Given a higher-order macro grammar $\mathcal{G}$ and a term M, it is decidable whether $M \in \mathcal{L}_{IO}(\mathcal{G})$.*

We are now going to see that the set of closed λ-terms of a given type α is an OI language. Combined with Theorem 4, we obtain that the set $[\![\mathcal{L}_{OI}(\mathcal{G})]\!]_{\mathcal{F}}$ cannot be effectively computed. Moreover, Theorems 6 and 7 imply that the IO hierarchy is strictly included in the OI hierarchy.

Theorem 8. *For every type α, there is a closed $\lambda Y+$-term M of type α such that $\mathcal{L}_{OI}(M)$ is the set of all closed normal λ-terms of type α.*

Theorem 9 (Undecidable finite interpretation of OI). *Given a higher-order macro grammar $\mathcal{G}$, a finite full model $\mathcal{F}$, and f an element of $\mathcal{F}$, it is undecidable whether $f \in [\![\mathcal{L}_{OI}(\mathcal{G})]\!]$.*

Proof. Direct consequence of Theorems 8 and 4.   □

Theorem 10. *The class of higher-order IO language is strictly included in the class of higher-order OI languages.*

Proof. If there were an IO grammar that could define the set of closed terms of type α, Theorem 7 would contradict Theorem 4.   □

This last theorem should be contrasted with the result of Haddad [13] which shows that OI and IO coincide for schemes. The two results do not contradict each other as IO is not defined in the same way on schemes and on grammars.

4   Emptiness and Membership for the OI Hierarchy

In this section we prove the decidability of the emptiness and membership problems for higher-order OI languages. For this we use monotone models as approximations of sets of elements of full models.

4.1   Hereditary Primality and the Observability Theorem

Theorem 9 implies that the decision techniques we used for the emptiness and the membership problems for IO do not extend to OI. So as to show that those problems are nevertheless decidable, we are going to prove a theorem that we call the *Observability Theorem*, which allows us to *observe* certain semantic properties of λ-terms in the OI language of a $\lambda Y + \Omega$-term M by means of the semantic values of M in monotone models. For this we introduce the notion of *hereditary prime elements* of a monotone model.

Definition 1. *In a lattice $\mathcal{L}$, an element f is* prime *(or $\vee$-prime) when for every g_1 and g_2 in $\mathcal{L}$, $f \leq g_1 \vee g_2$ implies that $f \leq g_1$ or $f \leq g_2$.*

Given a monotone model $\mathcal{M} = (\mathcal{M}_\alpha)_{\alpha \in type}$, for every type α we define the sets $\mathcal{M}_\alpha^+$ and $\mathcal{M}_\alpha^-$ by:

1. *$\mathcal{M}_0^+$ and $\mathcal{M}_0^-$ contain the prime elements of $\mathcal{M}_0$ that are different from $\perp_0$,*
2. *$\mathcal{M}_{\alpha \to \beta}^+ = \{(\bigvee F) \mapsto g \mid F \subseteq \mathcal{M}_\alpha^- \wedge g \in \mathcal{M}_\beta^+\}$,*
3. *$\mathcal{M}_{\alpha \to \beta}^- = \{f \mapsto g \mid f \in \mathcal{M}_\alpha^+ \wedge g \in \mathcal{M}_\beta^-\}$.*

A valuation ν on $\mathcal{M}$ is said hereditary prime *when, for every variable x^α, $\nu(x^\alpha) = \bigvee F$ for some $F \subseteq \mathcal{M}_\alpha^-$. The elements of $\mathcal{M}_\alpha^+$ are called the* hereditary prime elements *of $\mathcal{M}_\alpha$.*

The main interest of primality lies in that, if f is prime and $f \leq [\![M + N]\!]_{\mathcal{M}}^\nu$, then either $f \leq [\![M]\!]_{\mathcal{M}}^\nu$ or $f \leq [\![N]\!]_{\mathcal{M}}^\nu$. The notion of hereditary primality is simply a way of making primality compatible with all the constructs of $\lambda Y + \Omega$-terms. The proof of the following technical Lemma from which we derive the Observability Theorem, is mainly based on this idea.

Lemma 1. *Given a $\lambda + \Omega$-term M^α, a monotone model $\mathcal{M} = (\mathcal{M}_\alpha)_{\alpha \in type}$, a hereditary prime valuation ν and a hereditary prime element f of $\mathcal{M}_\alpha$, we have the equivalence:*

$$f \leq [\![M^\alpha]\!]_{\mathcal{M}}^\nu \Leftrightarrow \exists N \in \mathcal{L}_{OI}(M).f \leq [\![N]\!]_{\mathcal{M}}^\nu$$

Theorem 5 allows to extend Lemma 1 to $\lambda Y + \Omega$-terms.

Theorem 11 (Observability). *Given a $\lambda Y + \Omega$-term M, a monotone model $\mathcal{M} = (\mathcal{M}_\alpha)_{\alpha \in type}$, a hereditary prime valuation ν and a hereditary prime element f of $\mathcal{M}_\alpha$, we have the equivalence:*

$$f \leq [\![M^\alpha]\!]_{\mathcal{M}}^\nu \Leftrightarrow \exists N \in \mathcal{L}_{OI}(M).f \leq [\![N]\!]_{\mathcal{M}}^\nu$$

Proof. Since for every α, $\mathcal{M}_\alpha$ is finite, according to Theorem 5.2 (and the fact that the set of finite approximations of M is directed for the partial order $\sqsubseteq$), there is a finite approximation Q of M such that $[\![Q]\!]^\nu_\mathcal{M} = [\![M]\!]^\nu_\mathcal{M}$ and thus $f \leq [\![Q^\alpha]\!]^\nu_\mathcal{M}$. But then Q is a $\lambda+\Omega$-term and by the previous Lemma this is equivalent to there being some N in $\mathcal{L}_{OI}(Q)$ such that $f \leq [\![N]\!]^\nu_\mathcal{M}$. The conclusion follows from the fact that obviously $\mathcal{L}_{OI}(Q) \subseteq \mathcal{L}_{OI}(M)$. The other direction follows from Theorem 5.3.   □

4.2   Decidability Results

We are now going to use the Observability Theorem so as to prove the decidability of both the emptiness and the membership problems for OI languages.

Decidability of emptiness We consider the monotone model $\mathcal{E} = (\mathcal{E}_\alpha)_{\alpha \in type}$ so that $\mathcal{E}_0$ is the lattice with two elements $\{\top, \bot\}$ so that $\bot \leq \top$. We then define for every α, the element $\mathbf{e}_\alpha$ of $\mathcal{E}^+_\alpha \cap \mathcal{E}^-_\alpha$ so that: $\mathbf{e}_0 = \top$, and $\mathbf{e}_{\alpha \to \beta} = \mathbf{e}_\alpha \mapsto \mathbf{e}_\beta$. We let ξ be the valuation so that for each variable x^α, $\xi(x^\alpha) = \mathbf{e}_\alpha$. A simple induction gives the following Lemma which, combined with Theorem 11 gives proposition 1 and finally Theorem 12:

Lemma 2. *For every λ-term M of type γ, $\mathbf{e}_\gamma \leq [\![M]\!]^\xi_\mathcal{E}$.*

Proposition 1. *Given a $\lambda Y + \Omega$-term M of type α, we have that*

$$\mathcal{L}_{OI}(M) \neq \emptyset \Leftrightarrow \mathbf{e}_\alpha \leq [\![M]\!]^\xi_\mathcal{E}$$

Proof. If $\mathcal{L}_{OI}(M) \neq \emptyset$, then there is N in normal form so that $M \xrightarrow{*}_{\beta\delta+} N$. Lemma 2 implies that $\mathbf{e}_\alpha \leq [\![N]\!]^\xi_\mathcal{E}$ and thus, using Theorem 5, $\mathbf{e}_\alpha \leq [\![M]\!]^\xi_\mathcal{E}$. If $\mathbf{e}_\alpha \leq [\![M]\!]^\xi_\mathcal{E}$, since $\mathbf{e}_\alpha$ is in $\mathcal{E}^+_\alpha$, from Theorem 11, there is N in $\mathcal{L}_{OI}(M)$ so that $\mathbf{e}_\alpha \leq [\![N]\!]^\xi_\mathcal{E}$; so in particular that $\mathcal{L}_{OI}(M) \neq \emptyset$.   □

Theorem 12. *The emptiness problem for OI languages is decidable.*

Decidability of membership For the decidability of the membership problem, we are going to prove a stronger version of Statman's finite completeness Theorem. The proofs based mostly on logical relations (see [3]) can be found in the appendix.

Given a finite set A, we write $\mathcal{M}(A) = (\mathcal{M}_\alpha(A))_{\alpha \in type}$ for the monotone model so that $\mathcal{M}_0(A)$ is the lattice of subsets of A ordered by inclusion. We let $\bot_{A,\alpha}$ be the least element of $\mathcal{M}_\alpha(A)$.

Definition 2. *Given a λ-term M of type α, a triple $T = (A, \nu, f)$, where A is a finite set, ν is a valuation on $\mathcal{M}(A)$ and f is an element of $\mathcal{M}_\alpha(A)$, is characteristic of M when:*

1. *for every λ-term N of type α, $M =_\beta N$ iff $f \leq [\![N]\!]^\nu_{\mathcal{M}(A)}$,*
2. *f is a hereditary prime element of $\mathcal{M}_\alpha(A)$ and ν is a hereditary prime valuation.*

The stronger form of Statman finite complete is formulated as:

Theorem 13 (Monotone finite completeness). *For every type α and every pure term M, one can effectively construct a triple T that is characteristic of M.*

Using the Observability Theorem as in the proof of Proposition 1 we obtain:

Theorem 14. *Given a λ-term M in normal form of type α and a $\lambda Y + \Omega$-term N of type α, if $T = (A, \nu, f)$ is a characteristic triple of M then $f \leq [\![N]\!]^{\nu}_{\mathcal{M}(A)}$ iff $M \in \mathcal{L}_{OI}(N)$.*

Theorem 15. *Given M a λ-term of type α and N a $\lambda Y + \Omega$-term of type α, it is decidable whether $M \in \mathcal{L}_{OI}(N)$.*

5   Conclusion

We have seen how to use models of λ-calculus so as to solve algorithmic questions, namely the emptiness and membership problems, related to the classes of higher-order IO and OI languages of λ-terms. In so doing, we have revisited various questions related to finite models of the λ-calculus. In particular, we have seen that hereditary prime elements, via the Observability Theorem, play a key role in finding effective solutions for higher-order OI languages. In combination with Theorem 8, we obtain that it is decidable whether there is a term M whose interpretation in a monotone model is greater than a given hereditary prime element of that model, which gives a decidability result for a restricted notion of λ-definability. This raises at least two questions: (i) what kind of properties of λ-terms can be captured with hereditary prime elements, (ii) is there a natural extension of this notion that still defines some decidable variant of λ-definability.

On the complexity side, we expect that, using similar techniques as in [29], it might be possible to prove that verifying whether the value of a $\lambda Y + \Omega$-term is greater than a hereditary prime element of a monotone model is of the same complexity as the emptiness and membership problems for the safe OI hierarchy which is $(n-2)$-EXPTIME-complete for order n-grammars (see [11], with Huet's convention, the order of grammars is one plus the order of their corresponding higher-order pushdown automaton). Of course, such a high complexity makes the decidability results we obtained of little interest for practical applications in natural language processing. It does however underscore the need to identify linguistically motivated generalizations which point to tractable subclasses of OI grammars [30]. Some restricted classes of IO grammars are known to have low complexity [15,5]. A natural move is to see whether in the OI mode of derivation those grammars still have reasonable complexity for the emptiness and membership problems.

References

1. Aehlig, K., de Miranda, J.G., Ong, C.-H.L.: Safety is not a restriction at level 2 for string languages. In: Sassone, V. (ed.) FOSSACS 2005. LNCS, vol. 3441, pp. 490–504. Springer, Heidelberg (2005)
2. Aho, A.V.: Indexed grammars - an extension of context-free grammars. J. ACM 15(4), 647–671 (1968)
3. Amadio, R.M., Curien, P.-L.: Domains and Lambda-Calculi. Cambridge Tracts in Theoretical Computer Science. Cambridge University Press (1998)
4. Blum, W., Ong, C.-H.L.: The safe lambda calculus. Logical Methods in Computer Science 5(1:3), 1–38 (2009)
5. Bourreau, P., Salvati, S.: A datalog recognizer for almost affine λ-cfgs. In: Kanazawa, M., Kornai, A., Kracht, M., Seki, H. (eds.) MOL 12. LNCS, vol. 6878, pp. 21–38. Springer, Heidelberg (2011)
6. Broadbent, C.H.: The limits of decidability for first order logic on cpda graphs. In: STACS, pp. 589–600 (2012)
7. Damm, W.: The IO- and OI-hierarchies. Theor. Comput. Sci. 20, 95–207 (1982)
8. de Groote, P.: Towards abstract categorial grammars. In: ACL (ed.) Proceedings 39th Annual Meeting of ACL, pp. 148–155 (2001)
9. de Groote, P., Lebedeva, E.: On the dynamics of proper names. Technical report, INRIA (2010)
10. de Groote, P., Lebedeva, E.: Presupposition accommodation as exception handling. In: SIGDIAL, pp. 71–74. ACL (2010)
11. Engelfriet, J.: Iterated stack automata and complexity classes. Inf. Comput. 95(1), 21–75 (1991)
12. Fischer, M.J.: Grammars with macro-like productions. PhD thesis, Harvard University (1968)
13. Haddad, A.: IO vs OI in higher-order recursion schemes. In: FICS. EPTCS, vol. 77, pp. 23–30 (2012)
14. Huet, G.: Résolution d'équations dans des langages d'ordre 1,2,...,ω. Thèse de doctorat en sciences mathématiques, Université Paris VII (1976)
15. Kanazawa, M.: Parsing and generation as datalog queries. In: Proceedings of the 45th Annual Meeting of ACL, pp. 176–183. ACL (2007)
16. Knapik, T., Niwiński, D., Urzyczyn, P.: Higher-order pushdown trees are easy. In: Nielsen, M., Engberg, U. (eds.) FOSSACS 2002. LNCS, vol. 2303, pp. 205–222. Springer, Heidelberg (2002)
17. Lebedeva, E.: Expressing Discourse Dynamics Through Continuations. PhD thesis, Université de Lorraine (2012)
18. Loader, R.: The undecidability of λ-definability. In: Logic, Meaning and Computation: Essays in Memory of Alonzo Church, pp. 331–342. Kluwer (2001)
19. Montague, R.: Formal Philosophy: Selected Papers of Richard Montague. Yale University Press, New Haven (1974)
20. Moschovakis, Y.: Sense and denotation as algorithm and value. In: Logic Colloquium 1990: ASL Summer Meeting in Helsinki, vol. 2, p. 210. Springer (1993)
21. Muskens, R.: Lambda Grammars and the Syntax-Semantics Interface. In: Proceedings of the Thirteenth Amsterdam Colloquium, pp. 150–155 (2001)
22. Ong, C.-H.L.: On model-checking trees generated by higher-order recursion schemes. In: LICS, pp. 81–90 (2006)
23. Parys, P.: On the significance of the collapse operation. In: LICS, pp. 521–530 (2012)

24. Salvati, S.: Recognizability in the Simply Typed Lambda-Calculus. In: Ono, H., Kanazawa, M., de Queiroz, R. (eds.) WoLLIC 2009. LNCS, vol. 5514, pp. 48–60. Springer, Heidelberg (2009)
25. Salvati, S.: On the membership problem for non-linear acgs. Journal of Logic Language and Information 19(2), 163–183 (2010)
26. Salvati, S., Manzonetto, G., Gehrke, M., Barendregt, H.: Loader and Urzyczyn are logically related. In: Czumaj, A., Mehlhorn, K., Pitts, A., Wattenhofer, R. (eds.) ICALP 2012, Part II. LNCS, vol. 7392, pp. 364–376. Springer, Heidelberg (2012)
27. Salvati, S., Walukiewicz, I.: Recursive schemes, Krivine machines, and collapsible pushdown automata. In: Finkel, A., Leroux, J., Potapov, I. (eds.) RP 2012. LNCS, vol. 7550, pp. 6–20. Springer, Heidelberg (2012)
28. Statman, R.: Completeness, invariance and λ-definability. Journal of Symbolic Logic 47(1), 17–26 (1982)
29. Terui, K.: Semantic evaluation, intersection types and complexity of simply typed lambda calculus. In: RTA, pp. 323–338 (2012)
30. van Rooij, I.: The tractable cognition thesis. Cognitive Science 32, 939–984 (2008)

Evolving Graph-Structures and Their Implicit Computational Complexity

Daniel Leivant[1] and Jean-Yves Marion[2]

[1] Indiana University, USA
leivant@indiana.edu
[2] Université de Lorraine, LORIA, France
Jean-Yves.Marion@loria.fr

Abstract. Dynamic data-structures are ubiquitous in programming, and they use extensively underlying directed multi-graph structures, such as labeled trees, DAGs, and objects. This paper adapts well-established static analysis methods, namely data ramification and language-based information flow security, to programs over such graph structures. Our programs support the creation, deletion, and updates of both vertices and edges, and are related to pointer machines. The main result states that a function over graph-structures is computable in polynomial time iff it is computed by a terminating program whose graph manipulation is ramified, provided all edges that are both created and read in a loop have the same label.

1 Introduction

The interplay of algorithms and data-structures has been central to both theoretical and practical facets of programming. A core method of this relation is the organization of data-structures by underlying directed multi-graphs, such as trees, DAGs, and objects, where each vertex points to a record. Such data structures are often thought of as "dynamic", because they are manipulated by algorithms that modify the underlying graph, namely by creating, updating and removing vertices and edges. Our imperative language is inspired by pointer machines [6,10] and by abstract state machines [2].

In this work we propose a simple and effective static analysis method for guaranteeing the feasible time-complexity of programs over many dynamic data-structures. Most static analysis efforts have focused in recent years on program termination and on safety and security. Our work is thus a contribution to another strand of static analysis, namely computational complexity.

Static analysis of computational complexity is based on several methods, classified broadly into descriptive ones (i.e. related to Finite Model Theory), and applicative (i.e. identifying restrictions of programs and proof methods that guarantee upper bounds on the complexity of computation). One of the most fruitful applicative methods has been ramification, also referred to as tiering. Initially this method was used for inductive data, such as words and natural numbers, but lately the method has been applied to more general forms of data.

F.V. Fomin et al. (Eds.): ICALP 2013, Part II, LNCS 7966, pp. 349–360, 2013.

The intuition behind ramification is that programs' execution time depends on the nature of information flow during execution, and that such flow can be constrained naturally and effectively by imposing a precedence relation regulating the information flow, e.g. from higher tier to lower tier. Here we refer to a ramification that pertains to commands of our imperative language as well as to expressions denoting graph elements.

Our main result states that a function over graph-structures is computable in polynomial time iff it is computed by a terminating program whose graph manipulation is ramified, provided all edges that are both created and read in a loop have the same label. This result considerably extends previous uses of ramification in implicit computational complexity, and this extension touches on some of the most important aspects of programming, which have been disregarded in previous research. A simple modification of the proof gives an analogous characterization of Logspace computation over graph structures.[1] We thus believe that this work is of both theoretical and practical significance. Our results raise interesting questions about relations between data ramification and typed systems for program security [12,9], where the concept of information flow is explicit.

2   Evolving Graph-Structures

2.1   Sorted Partial Structures

The framework of sorted structures is natural for the graph-structures we wish to consider, with vertices and data treated as distinct sorts. Data might itself be sorted, but that does not concern us here. Recall that in a sorted structure, if $\mathcal{V}$ and $\mathcal{D}$ are sorts, then a function f is of type $\mathcal{V} \to \mathcal{D}$ if its domain consists of the structure elements of sort $\mathcal{V}$, and its range of elements of sort $\mathcal{D}$.

The graphs we consider are essentially deterministic transition graphs: edges are labeled (to which we refer as "actions"), and every vertex has at most one out-edge with a given label. Such graphs are conveniently represented by partial functions, corresponding to actions. An edge labeled $\mathbf{f}$ from vertex u to vertex v is represented by the equality $\mathbf{f}(u) = v$. When u has no $\mathbf{f}$-labeled out-edges, we leave $\mathbf{f}(u)$ undefined. To represent function partiality in the context of structures, we post a special constant $\mathbf{nil}$, assumed to lie outside the sorts (or, equivalently, in a special singleton sort)[2]. We write $f : \mathcal{V} \rightharpoonup \mathcal{D}$, and say that f is a *partial function from sort $\mathcal{V}$ to sort $\mathcal{D}$*, if $f : \mathcal{V} \to (\mathcal{D} \cup \{\mathbf{nil}\})$. We write $\mathcal{V} \rightharpoonup \mathcal{D}$ for the type of such partial functions.

2.2   Graph Structures

We consider sorted partial structures with a distinguished sort $\mathcal{V}$ of *vertices*. To account for the creation of new vertices, we also include a sort $\mathcal{R}$ of *reserved-vertices*. For simplicity, and without loss of generality, we assume only one additional sort $\mathcal{D}$, which we dub *data*. A *graph vocabulary* is a sorted vocabulary for

---

[1] We also present extensions of our language to incorporate recursion.

[2] This is a minor, but deliberate, departure from the usual ontological (i.e. Church-style) typing of sorted structure.

these sorts, where we have just five sets of identifiers: $\mathbb{V}$ (the vertex constants), $\mathbb{D}$ (the data constants), $\mathbb{F}$ (function-identifiers for labeled edges, of type $\mathcal{V} \rightharpoonup \mathcal{V}$), $\mathbb{G}$ (function-identifiers for data, of type $\mathcal{V} \to \mathcal{D}$), and $\mathbb{R}$ (the relation-identifiers), each of some type $\tau \times \cdots \times \tau$, where each τ is $\mathcal{V}$ or $\mathcal{D}$. As syntactic parameters we use $\mathbf{v} \in \mathbb{V}$, $\mathbf{d} \in \mathbb{D}$, $\mathbf{f} \in \mathbb{F}$, $\mathbf{g} \in \mathbb{G}$, and $\mathbf{R} \in \mathbb{R}$.

Given a sorted vocabulary Σ as above, a Σ-*structure* $\mathcal{S}$ consists of a vertex-universe $\mathcal{V}_\mathcal{S}$ which is finite, a potentially infinite reserve-universe $\mathcal{R}_\mathcal{S}$, a data-universe $\mathcal{D}_\mathcal{S}$, a distinct object $\bot$ to interpret **nil**, and a sort-correct interpretation $\mathbf{A}_\mathcal{S}$ for each Σ-identifier $\mathbf{A}$: $\mathbf{v}_\mathcal{S} \in \mathcal{V}_\mathcal{S}$; $\mathbf{d}_\mathcal{S} \in \mathcal{D}_\mathcal{S}$; $\mathbf{g}_\mathcal{S} \in [\mathcal{V}_\mathcal{S} \to \mathcal{D}_\mathcal{S}]$ (a *data-function*); $\mathbf{f}_\mathcal{S} \in [\mathcal{V}_\mathcal{S} \rightharpoonup \mathcal{V}_\mathcal{S}]$ (a *partial* function), and for a relation-identifier $\mathbf{R}$, a relation $\mathbf{R}_\mathcal{S} \subseteq \tau_\mathcal{S} \times \cdots \tau_\mathcal{S}$. Note that we do not refer to functions over data, nor to functions of arity > 1. Also, the fact that our graphs are edge-deterministic is reflected in our reprensetation of edges by functions.

Our graph structures bear similarity to the `struct` construct in the programming language C, and to objects (without behaviors or methods): a vertex identifies an object, and the state of that object is given by fields that are specified by the unary function identifiers. This is why Tarjan, in defining similar structures [11], talks about *records* and *items* rather than vertices and edges. The restriction of a graph-structure $\mathcal{S}$ to the sort $\mathcal{V}$ of vertices can be construed as a labeled directed multi-graph, in which there is an edge labeled by $\mathbf{f}$ from vertex u to vertex v exactly when $\mathbf{f}_\mathcal{S}(u) = v$. Thus the fan-out of each graph is bounded by the number of edge-labels in Σ. Examples of graph structures abound, see examples in Section 5. Linked-lists of data is an obvious one, of which words (represented as linked lists of alphabet-symbols) form a special case.

2.3   Expressions

Expressions are generated from a set $\mathbb{X}$ of vertex-variables, a set $\mathbb{Y}$ of data-variables, and the vocabulary identifiers, as follows. Equality here does not conform strictly to the sort discipline, in that we allow equations of the form $V = \mathbf{nil}^3$.

$V \in \text{VExpr} ::= X \mid \mathbf{nil} \mid \mathbf{v} \mid \mathbf{f}(V) \quad \text{where } X \in \mathbb{X}$

$D \in \text{DExpr} ::= Y \mid \mathbf{d} \mid \mathbf{g}(V) \quad \text{where } Y \in \mathbb{Y}$

$B \in \text{BExpr} ::= V = V \mid D = D \mid \neg(B) \mid \mathbf{R}(E_1 \dots E_n) \quad \text{where } \mathbf{R} : \tau^n, E_i : \tau$

3   Imperative Programs over Graph-Structures

3.1   Programs

We refer to a skeletal imperative language, which supports pointers:

$P \in \text{Prg} ::= X{:=}V \mid Y{:=}D \mid \mathbf{f}(X){:=}V \mid \mathbf{g}(X){:=}D \mid \mathbf{New}(X)$
$\qquad\qquad \mid \mathbf{skip} \mid P; P \mid \mathbf{if}\,(B)\,\{P\}\,\{P\} \mid \mathbf{while}\,(B)\,\{P\}$

---

[3] Negation is useful, however, in defining commands' semantics and dispensing with truth constants.

The boolean expression B of conditional and iterative commands is said to be their *guard*. We posit that each program is given with a finite set $\mathbb{X}_0 \subset \mathbb{X}$ of *input variables*.   •

3.2   Example: Tarjan's Union Algorithm

The following graph-algorithm is due to Tarjan [11, p.21]. It refers to a representation of sets by linked-lists, whose initial vertex also serves as a name for the set. The linked-list is represented by a partial-function **next**, and the function **parent** maps each node to the head of its linked-list. The algorithm generates, given as input two lists r, q representing disjoint sets, a list representing their union. It successively inserts right after r's head the entries of q; thus r is maintained as the name of the union.

```
while  (q ≠ nil)  {save  :=  next(q);
                   parent(q)  :=  r;   % parent and next are modified
                   next(q)  :=  next(r);
                   next(r)  :=  q;
                   q  :=  save}
```

We shall see that our tiering method admits the program above.

3.3   Evolving Structures

In defining the semantics of an "uninterpreted" imperative program one refers to structures for that program's vocabulary, augmented with a store (i.e. environment, valuation). For programs over a Σ-structure $\mathcal{S}$, a *store* consists then of a function $\mu = \mu_X \cup \mu_Y$, where $\mu_X : \mathbb{X} \to \mathcal{V}_{\mathcal{S}} \cup \{\bot\}$, and $\mu_Y : \mathbb{Y} \to \mathcal{D}_{\mathcal{S}}$. A Σ-*configuration* is a pair consisting of a Σ-structure $\mathcal{S}$ and a store μ. We chose the phrase *configuration* to stress their dynamic nature, as computation stages of an evolving structure.

Commands of imperative languages are interpreted semantically as partial-functions that map an initial configuration to a final one. Here we have two commands whose intended semantic interpretation is to modify the structure itself. A command $\textbf{New}(X)$ modifies the sorts, by moving an element of $\mathcal{R}_{\mathcal{S}}$ to $\mathcal{V}_{\mathcal{S}}$, and updating the store to have X point to the new element. We write $(\mathcal{S}, \mu)[\nu X]$ for the resulting configuration. Thus X points to a fresh vertex. More importantly, a command of the form $\mathbf{f}(X):=V$ modifies the semantics of the partial-function $\mathbf{f}_{\mathcal{S}}$, in that for $a = \mu(X)$ and w the value of V in $(\mathcal{S}, \mu)$ (defined formally below), an $\mathbf{f}$-edge $v \rightsquigarrow u$ is replaced by $v \rightsquigarrow w$. We write $(\mathcal{S}, \mu)[\mathbf{f}(X) \leftarrow w]$ for the resulting interpretation.

Our structure updates are obviously related to Gurevich's abstract sequential machines (ASM) [2]. Gurevich divides identifiers into two classes: static identifiers, whose interpretation remains constant, and dynamic identifiers, whose interpretation may evolve during computation. Here the only dynamic identifiers are the edge functions. An ASM computation progresses through *"states"*, where every state is a structure. In contrast, we refer to configurations, because program variables play a central role in our imperative programs. Thus, the execution of our programs progresses through configuration.

Our programming language is related, more broadly, to pointer machines. Tarjan [11] defined a pure reference machine, consisting of an expandable collection of records and a finite number of registers. Pure reference machines are easily simulated by our programs, and vice versa.

Our programs are also related to Schönhage machines [10]. Each such machine consists of a finite control program combined with a dynamic structure (which is essentially the same as our graph-structures). Schönhage machines are an extension of Kolmogorov-Uspensky machines [6]. Another source of inspiration is the work of Jones & als. on blob model of computations [3].

3.4 Semantics of Expressions

We give next the evaluation rules for Σ-expressions E in a Σ-configuration $(\mathcal{S}, \mu)$, writing $\mathcal{S}, \mu \models E \overset{e}{\Rightarrow} a$ to indicate that E evaluates to element a of $\mathcal{S}$.[4]

$$\frac{\mathbf{b} \in \mathbb{V} \cup \mathbb{D}}{\mathcal{S}, \mu \models \mathbf{b} \overset{e}{\Rightarrow} \mathbf{b}_{\mathcal{S}}} \qquad \frac{Z \in \mathbb{X} \cup \mathbb{Y}}{\mathcal{S}, \mu \models Z \overset{e}{\Rightarrow} \mu(Z)} \qquad \frac{\mathcal{S}, \mu \models E \overset{e}{\Rightarrow} a \qquad \mathbf{h} \in \mathbb{F} \cup \mathbb{G}}{\mathcal{S}, \mu \models \mathbf{h}(a) \overset{e}{\Rightarrow} \mathbf{h}_{\mathcal{S}}(a)}$$

$$\frac{\mathcal{S}, \mu \models E_i \overset{e}{\Rightarrow} a_i \quad \langle a_1..a_n \rangle \in \mathbf{R}_{\mathcal{S}}}{\mathcal{S}, \mu \models \mathbf{R}(E_1, \dots, E_n)} \qquad \frac{\mathcal{S}, \mu \models E_i \overset{e}{\Rightarrow} a_i \quad \langle a_1..a_n \rangle \notin \mathbf{R}_{\mathcal{S}}}{\mathcal{S}, \mu \models \neg \mathbf{R}(E_1, \dots, E_n)}$$

3.5 Semantics of Programs

The semantics of programs is defined below:

$$\frac{\mathcal{S}, \mu \models E \overset{e}{\Rightarrow} a}{\mathcal{S}, \mu \models Z := E \overset{s}{\Rightarrow} \mathcal{S}, \mu[Z \leftarrow a] \models \mathbf{skip}} \qquad \frac{}{\mathcal{S}, \mu \models \mathbf{New}(X) \overset{s}{\Rightarrow} (\mathcal{S}, \mu)[\nu X] \models \mathbf{skip}}$$

$$\frac{\mathcal{S}, \mu \models X \overset{e}{\Rightarrow} a \qquad \mathcal{S}, \mu \models V \overset{e}{\Rightarrow} b}{\mathcal{S}, \mu \models \mathbf{f}(X) := V \overset{s}{\Rightarrow} \mathcal{S}, \mu[\mathbf{f}(a) := b] \models \mathbf{skip}}$$

$$\frac{\mathcal{S}, \mu \models P_1 \overset{s}{\Rightarrow} \mathcal{S}', \mu' \models P_1'}{\mathcal{S}, \mu \models P_1; P_2 \overset{s}{\Rightarrow} \mathcal{S}', \mu' \models P_1'; P_2} \qquad \frac{\mathcal{S}, \mu \models P_1 \overset{s}{\Rightarrow} \mathcal{S}', \mu' \models \mathbf{skip}}{\mathcal{S}, \mu \models P_1; P_2 \overset{s}{\Rightarrow} \mathcal{S}', \mu' \models P_2}$$

$$\frac{\mathcal{S}, \mu \models B}{\mathcal{S}, \mu \models \mathbf{if} \ (B)\{P_0\}\{P_1\} \overset{s}{\Rightarrow} \mathcal{S}, \mu \models P_0} \qquad \frac{\mathcal{S}, \mu \models \neg B}{\mathcal{S}, \mu \models \mathbf{if} \ (B)\{P_0\}\{P_1\} \overset{s}{\Rightarrow} \mathcal{S}, \mu \models P_1}$$

$$\frac{\mathcal{S}, \mu \models B}{\mathcal{S}, \mu \models \mathbf{while}(B)\{P\} \overset{s}{\Rightarrow} \mathcal{S}, \mu \models P; \mathbf{while}(B)\{P\}} \qquad \frac{\mathcal{S}, \mu \models \neg B}{\mathcal{S}, \mu \models \mathbf{while}(B)\{P\} \overset{s}{\Rightarrow} \mathcal{S}, \mu \models \mathbf{skip}}$$

The phrase $\mathcal{S}, \mu \models P \overset{s}{\Rightarrow} \mathcal{S}', \mu' \models P'$ conveys that evaluating a program P starting with configuration $(\mathcal{S}, \mu)$ is reduced to evaluating P' in configuration $(\mathcal{S}', \mu')$; i.e., P reduces to P' while updating $(\mathcal{S}, \mu)$ to $(\mathcal{S}', \mu')$.

An *initial configuration* is a configuration $(\mathcal{S}, \mu)$ where $\mu(X) = \mathbf{nil}$ for every non-input variable X. A program P *computes* the partial function $[\![P]\!]$ with initial configurations as input, defined by: $[\![P]\!](\mathcal{S}, \mu) = (\mathcal{S}', \xi)$ iff $\mathcal{S}, \mu \models P \ (\overset{s}{\Rightarrow})^* \ \mathcal{T}, \xi \models \mathbf{skip}$.

---

[4] Here we consider equality as just another relation.

3.6   Run-Time

We say that a program P *runs in time t* on input $(\mathcal{S}, \mu)$, and write $Time_P(\mathcal{S}, \mu) = t$, when $\mathcal{S}, \mu \vDash P \ (\stackrel{\mathrm{s}}{\Longrightarrow})^t \ \mathcal{T}, \xi \vDash \textbf{skip}$ for some $(\mathcal{T}, \xi)$. We take the size $|\mathcal{S}, \mu|$ of a configuration $(\mathcal{S}, \mu)$ to be the number n of elements in the vertex-universe V. Since the number of edges is bounded by n^2, we disregard them here. We also disregard the size of the data-universe, because our programs do not modify the data present in records. A program P is *running in polynomial time* if there is a $k > 0$ such that $Time_P(\mathcal{S}, \mu) \leqslant k \cdot |\mathcal{S}, \mu|^k$ for all configurations $(\mathcal{S}, \mu)$,

4   Ramifiable Programs

4.1   Tiering

Program tiering, also referred to as *ramification*, has been introduced in [7] and used in restricted form already in [1]. It serves to syntactically control the run-time of programs. Here we adapt tiering to graph-structures. The main challenge here is the evolution of structures in course of computation. To address it, we consider a finite lattice $\mathbb{T} = (T, \preceq, \mathbf{0}, \vee, \wedge)$, and refer to the elements of T as *tiers*. However, in order to simplify soundness proofs, and without loss of generality, we will focus on the boolean lattice $\mathbb{T} = (\{\mathbf{0}, \mathbf{1}\}, \leqslant, \mathbf{0}, \vee, \wedge)$. We use lower case Greek letters α, β as discourse parameters for tiers.

Given $\mathbb{T}$, we consider $\mathbb{T}$-*environments* (Γ, Δ). Here Γ assigns a tier to each variable in $\mathbb{V}$, whereas Δ assigns to each function identifier $\mathbf{f} : \mathcal{V} \rightharpoonup \mathcal{V}$ one or several expressions of the form $\alpha \rightarrow \beta$, so that either

1. all types in $\Delta(\mathbf{f})$ are of the form $\alpha \rightarrow \alpha$, in which case we say that $\mathbf{f}$ is *stable* in the environment; or
2. all types in $\Delta(\mathbf{f})$ are of the form $\alpha \rightarrow \beta$, with $\beta \prec \alpha$, and we say that $\mathbf{f}$ is *reducing* in the environment.

A *tiering assertion* is a phrase of the form $\Gamma, \Delta \vdash V : \alpha$, where V is a vertex-expression and (Γ, Δ) a $\mathbb{T}$-environment. The correct tiering assertions are generated by the tiering system in Figure 1.

4.2   Ramifiable Programs

Given a lattice $\mathbb{T}$, a program P is $\mathbb{T}$-*ramifiable* if there is a $\mathbb{T}$-environment (Γ, Δ) for which $\Gamma, \Delta \vdash P : \alpha$ for some α, and such that $\Gamma(X) = \mathbf{1}$ for every input variable $X \in \mathbb{X}_0$.[5] Thus, ramifiable programs can be construed as programs decorated with tiering information.

Lemma 1 (Subject Reduction)
If $\mathcal{S}, \mu \vDash P \stackrel{\mathrm{s}}{\Longrightarrow} \mathcal{S}', \mu' \vDash P'$ and $\Gamma, \Delta \vdash P : \alpha$ then $\Gamma, \Delta \vdash P' : \alpha$.

Lemma 2 (Type Inference). *The problem of deciding, given a program P and a lattice $\mathbb{T}$, whether P is $\mathbb{T}$-ramifiable, is decidable in polynomial time.*

---

[5] Recall that each program is assumed given with a set $\mathbb{X}_0$ of input variables.

$$\frac{}{\Gamma, \Delta \vdash \mathbf{c} : \alpha} \qquad \frac{\Gamma(X) = \alpha}{\Gamma, \Delta \vdash X : \alpha} \qquad \frac{\alpha \rightarrow \beta \in \Delta(\mathbf{f}) \qquad \Gamma, \Delta \vdash V : \alpha}{\Gamma, \Delta \vdash \mathbf{f}(V) : \beta}$$

$$\frac{\Gamma, \Delta \vdash V_i : \alpha}{\Gamma, \Delta \vdash \mathbf{R}(V_1, \ldots, V_n) : \alpha} \qquad \frac{\Gamma, \Delta \vdash V_i : \alpha}{\Gamma, \Delta \vdash V_0 = V_1 : \alpha}$$

Fig. 1. Tiering rules for vertex and boolean expressions

$$\frac{\Gamma, \Delta \vdash X : \alpha \qquad \Gamma, \Delta \vdash V : \alpha}{\Gamma, \Delta \vdash X := V : \alpha} \qquad \frac{\Gamma, \Delta \vdash \mathbf{f}(X) : \alpha \qquad \Gamma, \Delta \vdash V : \alpha}{\Gamma, \Delta \vdash \mathbf{f}(X) := V : \alpha}$$

$$\frac{\Gamma, \Delta \vdash X : \mathbf{0}}{\Gamma, \Delta \vdash \mathbf{New}(X) : \mathbf{0}} \qquad \frac{\Gamma, \Delta \vdash B : \alpha \qquad \Gamma, \Delta \vdash P : \alpha}{\Gamma, \Delta \vdash \mathbf{while}(B)\{P\} : \alpha} \mathbf{0} \prec \alpha$$

$$\frac{}{\Gamma, \Delta \vdash \mathbf{skip} : \mathbf{0}} \qquad \frac{\Gamma, \Delta \vdash P : \alpha \qquad \Gamma, \Delta \vdash P' : \beta}{\Gamma, \Delta \vdash P' ; P' : \alpha \vee \beta}$$

$$\frac{\Gamma, \Delta \vdash B : \alpha \quad \Gamma, \Delta \vdash P_i : \alpha}{\Gamma, \Delta \vdash \mathtt{if}\ (B)\{P_0\}\{P_1\} : \alpha} \qquad \frac{\Gamma, \Delta \vdash P : \beta}{\Gamma, \Delta \vdash P : \alpha} (\beta \preceq \alpha)$$

Fig. 2. Tiering rules for programs

Proof. We associate with each vertex-variable X a "tier-variable" α_X, and with each function $\mathbf{f} \in \mathbb{F}$ two variables α_f and β_f, with the intent that $\alpha_f \rightarrow \beta_f$ is a possible tiering of $\mathbf{f}$. The typing rules for tiers give rise to a set of linear constraints on these tier-variables, a problem which is poly-time decidable.

4.3   Stationary and Tightly-Modifying Loops

We say that a function identifier $\mathbf{f}$ is *probed* in P if it occurs in P either in some assignment $X := V$ or in the guard of a loop or a branching command. For example, $\mathbf{f}$ is probed in $X := \mathbf{f}(V)$, as well as in $\mathtt{if}\ (\mathbf{f}(X) \neq \mathtt{nil})\{P\}\{P'\}$. The identifier $\mathbf{f}$ is *modified* in P if it occurs in an assignment $\mathbf{f}(X) := V$ in P.

Fix a lattice $\mathbb{T}$, and a $\mathbb{T}$-environment (Γ, Δ). By the tiering rules, if a loop $\mathbf{while}(B)\{P\}$ is of tier α then $\Gamma, \Delta \vdash B : \alpha$. We say that the loop is *stationary* if no $\mathbf{f} \in \mathbb{F}$ of type $\alpha \rightarrow \alpha$ is modified therein. The loop above is *tightly-modifying* if it has modified function-identifiers of type $\alpha \rightarrow \alpha$, but at most one of those is also probed. In other words, all edges that are both created and read in a loop have the same label. For instance, in Example 3.2 above, **next** is both modified and probed, but **parent** is modified without being probed. Thus the loop, with its obvious tiering environment, is tightly-modifying.[6]

---

[6] Note that **next** and **parent** are of type $\mathbf{1} \rightarrow \mathbf{1}$, and all variables are of tier $\mathbf{1}$. Set union can be iterated because the result r is of tier $\mathbf{1}$, unlike in most other works.

4.4 Main Characterization Theorem

Given a lattice $\mathbb{T}$ and $\Gamma, \Delta \vdash P : \alpha$, we say that (Γ, Δ) is a *tight* ramification of P if Γ is an initial tiering, and each loop of P is stationary or tightly-modifying. We say that P is *tightly-ramifiable* if it has a tight $\mathbb{T}$-ramification (Γ, Δ), with Γ initial, for some non-trivial $\mathbb{T}$.

Theorem 1. *A function over graph-structures is computable in polynomial time iff it is computed by a terminating and tightly-ramifiable program.*

The Theorem follows from the Soundness Lemma 6 and the Completeness Proposition 1 below.

5 Examples of Ramified Programs

Tree insertion. The program below inserts the tree T into the binary search tree whose root is pointed-to by x. The input variables are x and T.

```
if  (x¹  =  nil )
   {x¹:=T:1;} % then clause
   { % else clause
      while ((x¹ ≠ nil ) and (key(T¹) ≠ key(x¹)))
         {if (key(T¹) < key(x¹)) {p¹:=x¹; x¹ :=  left(x¹)¹}
                                 {p¹:=x¹; x¹:=right(x¹)¹} }:1;
if (key(T¹) < key(p¹)) {left(p¹) := T¹:1}
                       {right(p¹):= T¹:1}
```

Note that neither **left** nor **right** is modified in the loop, so the loop is stationary.

Copying lists. Here we use **New** to copy a list, where the copy is in reverse order. Note that the source list is of tier **1** while the copy is of tier **0**.

```
y⁰ = nil  :  0;
while  (x¹ ≠ nil )
       { z⁰:=y⁰:0;  New(y⁰);  suc(y⁰):=z⁰:0;
         x¹ := suc(x¹):1 }:1
```

The loop is stationary, because the updated occurrence of **suc** is of type $0 \to 0$.

6 Soundness of Programs for Feasibility: Run-Time Analysis

We show next that every tightly-ramified program computes a PTime function over configurations. The proof is based on the following observations, which we articulate more precisely below. First, if we start with a configuration where no vertex is assigned to variables of different tiers, then all configurations obtained in the course of computation have that property. We are thus assured that vertices can be ramified unambiguously.

The tiering rules imply that a program P of tier $\mathbf{0}$ cannot have loops, and is therefore evaluated in $\leqslant |P|$ steps. At the same time, the value of a variable of tier $\mathbf{1}$ depends only on vertices of tier $\mathbf{1}$. This implies, as we shall see, that the number of iterations of a given loop must be bounded by the number of possible configurations that may be generated by its body. Our restriction to tightly-modifying ramification guarantees that the number is polynomial.

6.1   Non-interference

Lemma 3 (Confinement). *Let (Γ, Δ) be an environment. If $\Gamma, \Delta \vdash P : \mathbf{0}$, then $\Gamma(X) = \mathbf{0}$ for every variable X assigned-to in P.*

The proof is a straightforward structural induction. Note also that a program P of tier $\mathbf{0}$ cannot have a loop, and is thus evaluated within $|P|$ steps.

We say that a vertex-tiering Γ is *compatible* with a store μ if $\Gamma(X) \neq \Gamma(X')$ implies $\mu(X) \neq \mu(X')$ for all $X, X' \in \mathbb{X}$. We say that Γ is an *initial tiering* if $\Gamma(X)$ is $\mathbf{1}$ for X initial (i.e. $X \in \mathbb{X}_0$), and $\mathbf{0}$ otherwise. Thus an initial tiering is always compatible with an initial store.

Lemma 4. *Suppose that* $\Gamma, \Delta \vdash P : \alpha$ *and* $S, \mu \models P \overset{s}{\Longrightarrow} S', \mu' \models P'$. *If μ is compatible with Γ then so is μ'.*

The proof is straightforward by structural induction on P.

We show next that tiering, when compatible with the initial configuration, guarantees the non-interference of lower-tiered values in the run-time of higher-tiered programs. A similar effect of tiering, albeit simpler, was observed already in [7]. This is also similar to the security-related properties considered in [12]. Non-interference can also be rendered algebraically, as in [8].

The (Γ, Δ)-*collapse* of a configuration (S, μ) is the configuration (S_Δ, μ_Γ), where $\mu_\Gamma(X) = \mu(X)$ if $\Gamma(X) = \mathbf{1}$, and $\mu_\Gamma(X)$ is undefined otherwise; whereas S_Δ is the structure identical to S except that each $\mathbf{f}$ for which $(\mathbf{1} \to \mathbf{1}) \notin \Delta(\mathbf{f})$ is interpreted as $\emptyset$. Thus (S_Δ, μ_Γ) disregards vertices that are not not reachable from some variable of tier $\mathbf{1}$ using edges of type $(\mathbf{1} \to \mathbf{1})$.

The next lemma states that a program's output vertices in tier $\mathbf{1}$ do not depend on vertices in tier $\mathbf{0}$, nor on edges that do not have tier $\mathbf{1} \to \mathbf{1}$.

Lemma 5. *Suppose $\Gamma, \Delta \vdash P : \alpha$, and $S, \mu \models P \overset{s}{\Longrightarrow} S', \mu' \models P'$. There is a configuration (S'', μ'') such that $S_\Delta, \mu_\Gamma \models P \overset{s}{\Longrightarrow} S'', \mu'' \models P'$, and $(S''_\Delta, \mu''_\Gamma) = (S'_\Delta, \mu'_\Gamma)$.*

The proof is straightforward by structural induction on programs.

6.2   Polynomial Bounds

(Soundness)

Lemma 6. *Assume that $\Gamma, \Delta \vdash P : \alpha$, where P is tightly-modifying. There is a $k > 0$ such that for every graph-structure S and every store μ compatible with Γ, if $S, \mu \models P \overset{s}{\Longrightarrow} S', \mu' \models P'$, then $S, \mu \models P (\overset{s}{\Longrightarrow})^t S', \mu' \models P'$ for some $t < k + |S|^k$.*

In proving Lemma 6 we will use the following combinatorial observation. Let $\mathcal{G}$ be a digraph of out-degree 1. We say that a set of vertices C *generates* $\mathcal{G}$ if every vertex in $\mathcal{G}$ is reachable by a path starting at C. The following lemma provides a polynomial, albeit crude, upper bound on the number of digraphs with k generators.

Lemma 7. *The number (up to isomorphism) of digraphs with n vertices, and a generator of size k, is $\leqslant n^{2k^2}$.*

Proof. A *connected* digraph of out-degree 1 must consist of a loop of vertices, with incoming linear spikes. If there are just k generators, then there are at most k such spikes. There are at most k entry points on the loop to choose for these spikes, and each spike is of size $\leqslant n$. So there are at most $n^k \times n^k$ non-isomorphic connected graphs with a generator of size k.

Also, with only k generating vertices we can have at most k connected components. In sum there are at most $(n^{2k})^k = n^{2k^2}$ non-isomorphic graphs of size n with k generators.

Proof of Lemma 6. We proceed by structural induction on P. The only non-trivial observations are as follows. For program composition, we use the Compatibility Lemma 4.

The crucial case is, of course, where P is of the form **while**$(B)Q$. Say $X_1, \ldots, X_m$ are the vertex-variables in B. The tiering rules require that B, and therefore $X_1, \ldots, X_m$, are all of tier **1**.

If Q updates only edges that are not probed in P (including the guard B), then neither the execution of Q nor the evaluation of B is affected, that is all configurations in the computation have the same vertices of tier **1**, with no change in edges that affect the execution of P, by Lemma 5. Thus the truth of B in each invocation of Q is determined by the combinations of values assigned to $X_1, \ldots, X_m$, while the structural changes caused by Q do not affect the execution of Q in subsequent invocations. Since we assume that P terminates, it follows that the combinations of values for $X_1, \ldots, X_m$ must be all different. If n is the number of vertices of tier **1**, then $n \leqslant |\mathcal{S}|$, and there are n^m such combinations. By IH Q terminates in PTime, and therefore so does P.

Suppose Q does update edges that are probed in P. Since P is assumed tightly-modifying, all such updates are for the same $\mathbf{f} \in \mathbb{F}$. Let C be the set of initial values of variables occurring in P (including $X_1 \ldots X_m$ and possibly others). Let U be the set of vertices reachable from C by some path of tier **1** (i.e. using edges labels by $\mathbf{h} \in \mathbb{F}$ assigned $\mathbf{1} \to \mathbf{1}$ by Δ from C vertices). By Lemma 5, the execution of P, including all iterated invocations of B and Q, has only vertices in U as value of tier **1**. Moreover, U is generated by C, whose size is fixed by the syntax of P. It follows, by Lemma 7, that the number of such configurations is polynomial in the size of U, which in turns in bounded by the size $|\mathcal{S}|$ of the vertex-universe of the structure $\mathcal{S}$.   □

7   Completeness of Tightly-Ramifiable Programs for Feasibility

Proposition 1. *Every polynomial-time function on graph-structures is computable by a terminating and tightly-ramifiable program.*

Proof. Suppose f is a unary function on graph-structures, which is computed by a Turing machine M over alphabet Σ, modulo some canonical coding of graph structures by strings. Assume that M operates in time $k \cdot n^k$. We posit that M uses a read-only input tape, and a work-tape. We simulate M by a tightly-ramifiable program $\Gamma, \Delta \vdash P : \mathbf{1}$ over graph-structures, as follows. The structures considered simulate each of the two tapes by a double-linked list of records, with the three fields **left**, **right**, and **val**, returning two pointers and an alphabet letter, respectively. We take our data constants to include each $\sigma \in \Sigma$. The input-tape is assigned tier **1**, and the work-tape tier **0**. Initially the work-tape is empty, and new cells are progressively created by using the command **New** at tier **0**. The machine's yield-relation between configurations is simulated at tier **0** by nested conditionals. Finally, we include in our simulation a clock, consisting of k nested loops, as in the following nesting of two loops:

```
u¹ :=  head¹ : 1  // head is a pointer to the input tape;
while  (u¹ ≠ nil )
              { v¹:=head¹ : 1 ;
                  while  (v¹ ≠ nil )
                  { v¹ :=  suc(v¹):1 ;
                     Transition function is here: 0 }: 1
                  u¹ :=  suc(u¹):1 } : 1
```

8   Characterization of Log-Space Languages

Hofmann and Schöpp [4] introduced pure pointer programs based on a uniform iterator **forall**, and related them to computation in logarithmic space. Our programs differ from these in supporting modification of the structure, in the guises of vertex creation and edge displacement and deletion. Moreover, our programs are based on a looping construct that treats individually each vertex of the structure.

These differences notwithstanding, our characterization of PTime can be modified to a characterization of log-space, by restricting the syntax of ramifiable-programs. Say that a program P over graph-structures is a *jumping-program* if it uses no edge update. Jones [5] showed that a simple cons-free imperative programming language $\text{WHILE}^{\backslash Ro}$ accepts precisely the languages decidable by Turing machines in logarithmic space. Since our jumping-programs can be rephrased in $\text{WHILE}^{\backslash Ro}$ they too accept only log-space languages (where input strings are represented as linked lists). Conversely, the store used by a jumping-program is essentially of size $k \cdot log(n) + log(Q)$, where n is the size of the vertex-universe, Q the size of the data-universe, and k the number of variables. Consequently, we have:

Theorem 2. *A language is accepted by a jumping-program iff it is decidable in* LOGSPACE.

9   Adding Recursion

It is not hard to augment our programing language with recursion. Here is a procedure that recursively searches for a path from vertex v to w.[7]

```
Proc search(v¹,w¹))
        {if (v=w)¹ return true:1;
         visited(v) = true:1;
         forall t¹ in AdjList(v) % List of adjacency nodes of t
         {if (visited(t)¹=false)
             if (search(t,w)¹=true) return true:1;}
         return false:1;}
```

A restricted form of recursion is *linear recursion,* where at most one recursive call is allowed in the definition of a recursive procedure. Moreover, we suppose that each function body is stationary or tightly-modifying.

Theorem 3. *On its domain of computation, a tightly-ramifiable program with linear recursive calls is computable in polynomial time.*

References

1. Bellantoni, S., Cook, S.A.: A new recursion-theoretic characterization of the poly-time functions. Computational Complexity 2, 97–110 (1992)
2. Gurevich, Y.: Sequential abstract state machines capture sequential algorithms. ACM Transactions on Computational Logic 1(1), 77–111 (2000)
3. Hartmann, L., Jones, N.D., Simonsen, J.G., Vrist, S.B.: Programming in biomolecular computation: Programs, self-interpretation and visualisation. Sci. Ann. Comp. Sci. 21(1), 73–106 (2011)
4. Hofmann, M., Schöpp, U.: Pure pointer programs with iteration. ACM Trans. Comput. Log. 11(4) (2010)
5. Jones, N.D.: Logspace and ptime characterized by programming languages. Theor. Comput. Sci. 228(1-2), 151–174 (1999)
6. Kolmogorov, A.N., Uspensky, V.: On the definition of an algorithm. Uspekhi Mat. Naut. 13(4) (1958)
7. Leivant, D.: Predicative recurrence and computational complexity I: Word recurrence and poly-time. In: Feasible Mathematics II. Birkhauser-Boston (1994)
8. Marion, J.-Y.: A type system for complexity flow analysis. In: LICS (2011)
9. Sabelfeld, A., Sands, D.: Declassification: dimensions and principles. J. Comput. Secur. 17, 517–548 (2009)
10. Schönhage, A.: Storage modification machines. SIAM J. Comp. 9(3), 490–508 (1980)
11. Tarjan, R.E.: Reference machines require non-linear time to maintain disjoint sets. In: STOC 1977, pp. 18–29. ACM (1977)
12. Volpano, D., Irvine, C., Smith, G.: A sound type system for secure flow analysis. Journal of Computer Security 4(2/3), 167–188 (1996)

---

[7] The construct `forall` X `in` $R(u)$, which is "blind," in the sense that it does not depend on node ordering. As a result, no function identifier is probed except **visited**.

Rational Subsets and Submonoids of Wreath Products[*]

Markus Lohrey[1], Benjamin Steinberg[2], and Georg Zetzsche[3]

[1] Universität Leipzig, Institut für Informatik
[2] City College of New York, Department of Mathematics
[3] Technische Universität Kaiserslautern, Fachbereich Informatik

Abstract. It is shown that membership in rational subsets of wreath products $H \wr V$ with H a finite group and V a virtually free group is decidable. On the other hand, it is shown that there exists a fixed finitely generated submonoid in the wreath product $\mathbb{Z} \wr \mathbb{Z}$ with an undecidable membership problem.

1 Introduction

The study of algorithmic problems in group theory has a long tradition. Dehn, in his seminal paper from 1911, introduced the word problem (Does a given word over the generators represent the identity?), the conjugacy problem (Are two given group elements conjugate?) and the isomorphism problem (Are two given finitely presented groups isomorphic?), see [25] for general references in combinatorial group theory. Starting with the work of Novikov and Boone from the 1950's, all three problems were shown to be undecidable for finitely presented groups in general. A generalization of the word problem is the *subgroup membership problem* (also known as the *generalized word problem*) for finitely generated groups: Given group elements $g, g_1, \ldots, g_n$, does g belong to the subgroup generated by $g_1, \ldots, g_n$? Explicitly, this problem was introduced by Mihailova in 1958, although Nielsen had already presented in 1921 an algorithm for the subgroup membership problem for free groups.

Motivated partly by automata theory, the subgroup membership problem was further generalized to the *rational subset membership problem*. Assume that the group G is finitely generated by the set X (where $a \in X$ if and only if $a^{-1} \in X$). A finite automaton A with transitions labeled by elements of X defines a subset $L(A) \subseteq G$ in the natural way; such subsets are the rational subsets of G. The rational subset membership problem asks whether a given group element belongs to $L(A)$ for a given finite automaton (in fact, this problem makes sense for any finitely generated monoid). The notion of a rational subset of a monoid can be traced back to the work of Eilenberg and Schützenberger from 1969 [8]. Other early references are [1,11]. Rational subsets of groups also found applications for the solution of word equations (here, quite often the term rational constraint is used) [6,20]. In automata theory, rational subsets are tightly related to valence automata (see [9,16,17] for details): For any group G, the emptiness problem for valence automata over G (which are also known as G-automata) is decidable if and only if G has a decidable rational subset membership problem.

---

[*] This work was supported by the DAAD research project RatGroup. The second author was partially supported by a grant from the Simons Foundation (#245268 to Benjamin Steinberg). Omitted proofs can be found in the long version [24] of this paper.

F.V. Fomin et al. (Eds.): ICALP 2013, Part II, LNCS 7966, pp. 361–372, 2013.

For free groups, Benois [2] proved that the rational subset membership problem is decidable using a classical automaton saturation procedure (which yields a polynomial time algorithm). For commutative groups, the rational subset membership can be solved using integer programming. Further (un)decidability results on the rational subset membership problem can be found in [21] for right-angled Artin groups, in [28] for nilpotent groups, and in [23] for metabelian groups. In general, groups with a decidable rational subset membership problem seem to be rare. In [22] it was shown that if the group G has at least two ends, then the rational subset membership problem for G is decidable if and only if the submonoid membership problem for G (Does a given element of G belong to a given finitely generated submonoid of G?) is decidable.

In this paper, we investigate the rational subset membership problem for wreath products. The wreath product is a fundamental operation in group theory. To define the wreath product $H \wr G$ of two groups G and H, one first takes the direct sum $K = \bigoplus_{g \in G} H$ of copies of H, one for each element of G. An element $g \in G$ acts on K by permuting the copies of H according to the left action of g on G. The corresponding semidirect product $K \rtimes G$ is the wreath product $H \wr G$.

In contrast to the word problem, decidability of the rational subset membership problem is not preserved under wreath products. For instance, in [23] it was shown that for every non-trivial group H, the rational subset membership problem for $H \wr (\mathbb{Z} \times \mathbb{Z})$ is undecidable. The proof uses an encoding of a tiling problem, which uses the grid structure of the Cayley graph of $\mathbb{Z} \times \mathbb{Z}$.

In this paper, we prove the following two new results concerning the rational subset membership problem and the submonoid membership problem for wreath products:

(i) The submonoid membership problem is undecidable for $\mathbb{Z} \wr \mathbb{Z}$. The wreath product $\mathbb{Z} \wr \mathbb{Z}$ is one of the simplest examples of a finitely generated group that is not finitely presented, see [4,5] for further results showing the importance of $\mathbb{Z} \wr \mathbb{Z}$.

(ii) For every finite group H and every virtually free group[1] V, the group $H \wr V$ has a decidable rational subset membership problem; this includes for instance the famous lamplighter group $\mathbb{Z}_2 \wr \mathbb{Z}$.

For the proof of (i) we encode the acceptance problem for a 2-counter machine (Minsky machine [26]) into the submonoid membership problem for $\mathbb{Z} \wr \mathbb{Z}$. One should remark that $\mathbb{Z} \wr \mathbb{Z}$ is a finitely generated metabelian group and hence has a decidable subgroup membership problem [29,30]. For the proof of (ii), an automaton saturation procedure is used. The termination of the process is guaranteed by a well-quasi-order (wqo) that refines the classical subsequence wqo considered by Higman [14].

Wqo theory has also been applied successfully for the verification of infinite state systems. This research led to the notion of well-structured transition systems [10]. Applications in formal language theory are the decidability of the membership problem for leftist grammars [27] and Kunc's proof of the regularity of the solutions of certain language equations [18]. A disadvantage of using wqo theory is that the algorithms it yields are not accompanied by complexity bounds. The membership problem for leftist grammars [15] and, in the context of well-structured transition systems, several natural reachability problems [3,32] (e.g. for lossy channel systems) have even been shown

---

[1] Recall that a group is virtually free if it has a free subgroup of finite index.

not to be primitive recursive. The complexity status for the rational subset membership problem for wreath products $H \wr V$ (H finite, V virtually free) thus remains open. Actually, we do not even know whether the rational subset membership problem for the lamplighter group $\mathbb{Z}_2 \wr \mathbb{Z}$ is primitive recursive.

2 Rational Subsets of Groups

Let G be a finitely generated group and X a finite symmetric generating set for G (symmetric means that $x \in X \Leftrightarrow x^{-1} \in X$). For a subset $B \subseteq G$ we denote with B^* (resp. $\langle B \rangle$) the *submonoid* (resp. subgroup) of G generated by B. The set of *rational subsets* of G is the smallest set that contains all finite subsets of G and that is closed under union, product, and $*$. Alternatively, rational subsets can be represented by finite automata. Let $A = (Q, G, E, q_0, Q_F)$ be a finite automaton, where transitions are labeled with elements of G: Q is the finite set of states, $q_0 \in Q$ is the initial state, $Q_F \subseteq Q$ is the set of final states, and $E \subseteq Q \times G \times Q$ is a finite set of transitions. Every transition label $g \in G$ can be represented by a finite word over the generating set X. The subset $L(A) \subseteq G$ accepted by A consists of all group elements $g_1 g_2 g_3 \cdots g_n$ such that there exists a sequence of transitions $(q_0, g_1, q_1), (q_1, g_2, q_2), (q_2, g_3, q_3), \ldots, (q_{n-1}, g_n, q_n) \in E$ with $q_n \in Q_F$. The *rational subset membership problem* for G is the following decision problem: Given a finite automaton A as above and an element $g \in G$, does $g \in L(A)$ hold? Since $g \in L(A)$ if and only if $1_G \in L(A)g^{-1}$, and $L(A)g^{-1}$ is rational, too, the rational subset membership problem for G is equivalent to the question whether a given automaton accepts the group identity.

The *submonoid membership problem* for G is the following decision problem: Given elements $g, g_1, \ldots, g_n \in G$, does $g \in \{g_1, \ldots, g_n\}^*$ hold? Clearly, decidability of the rational subset membership problem for G implies decidability of the submonoid membership problem for G. Moreover, the latter generalizes the classical subgroup membership problem for G (also known as the generalized word problem), where the input is the same as for the submonoid membership problem for G but it is asked whether $g \in \langle g_1, \ldots, g_n \rangle$ holds.

In our undecidability results in Sec. 5, we will actually consider the non-uniform variant of the submonoid membership problem, where the submonoid is fixed, i.e., not part of the input.

3 Wreath Products

Let G and H be groups. Consider the direct sum $K = \bigoplus_{g \in G} H_g$, where H_g is a copy of H. We view K as the set $H^{(G)} = \{f \in H^G \mid f^{-1}(H \setminus \{1_H\}) \text{ is finite}\}$ of all mappings from G to H with finite support together with pointwise multiplication as the group operation. The group G has a natural left action on $H^{(G)}$ given by $gf(a) = f(g^{-1}a)$, where $f \in H^{(G)}$ and $g, a \in G$. The corresponding semidirect product $H^{(G)} \rtimes G$ is the wreath product $H \wr G$. In other words:

- Elements of $H \wr G$ are pairs (f, g), where $f \in H^{(G)}$ and $g \in G$.
- The multiplication in $H \wr G$ is defined as follows: Let $(f_1, g_1), (f_2, g_2) \in H \wr G$. Then $(f_1, g_1)(f_2, g_2) = (f, g_1 g_2)$, where $f(a) = f_1(a) f_2(g_1^{-1} a)$.

The following intuition might be helpful: An element $(f, g) \in H \wr G$ can be thought of as a finite multiset of elements of $H \setminus \{1_H\}$ that are sitting at certain elements of G (the mapping f) together with the distinguished element $g \in G$, which can be thought of as a cursor moving in G. If we want to compute the product $(f_1, g_1)(f_2, g_2)$, we do this as follows: First, we shift the finite collection of H-elements that corresponds to the mapping f_2 by g_1: If the element $h \in H \setminus \{1_H\}$ is sitting at $a \in G$ (i.e., $f_2(a) = h$), then we remove h from a and put it to the new location $g_1 a \in H$. This new collection corresponds to the mapping $f_2' \colon a \mapsto f_2(g_1^{-1} a)$. After this shift, we multiply the two collections of H-elements pointwise: If in $a \in G$ the elements h_1 and h_2 are sitting (i.e., $f_1(a) = h_1$ and $f_2'(a) = h_2$), then we put the product $h_1 h_2$ into the location a. Finally, the new distinguished G-element (the new cursor position) becomes $g_1 g_2$.

If H (resp. G) is generated by the set A (resp. B) with $A \cap B = \emptyset$, then $H \wr G$ is generated by $A \cup B$.

Proposition 1. *Let K be a subgroup of G of finite index m and let H be a group. Then $H^m \wr K$ is isomorphic to a subgroup of index m in $H \wr G$.*

4   Decidability

We show that the rational subset membership problem is decidable for groups $G = H \wr V$, where H is finite and V is virtually free. First, we will show that the rational subset membership problem for $G = H \wr F_2$, where F_2 is the free group generated by a and b, is decidable. For this we make use of a particular well-quasi-order.

A Well-quasi-order. Recall that a *well-quasi-order (wqo)* on a set A is a reflexive and transitive relation $\preceq$ such that for every infinite sequence $a_1, a_2, a_3, \ldots$ with $a_i \in A$ there exist $i < j$ such that $a_i \preceq a_j$. In this paper, $\preceq$ will always be antisymmetric as well; so $\preceq$ will be a well partial order.

For a finite alphabet X and two words $u, v \in X^*$, we write $u \preceq v$ if there exist $v_0, \ldots, v_n \in X^*$, $u_1, \ldots, u_n \in X$ such that $v = v_0 u_1 v_1 \cdots u_n v_n$ and $u = u_1 \cdots u_n$. The following theorem was shown by Higman [14] (and independently Haines [13]).

Theorem 1 (Higman's Lemma). *The order $\preceq$ on X^* is a wqo.*

Let H be a group. For a monoid morphism $\alpha \colon X^* \to H$ and $u, v \in X^*$ let $u \preceq_\alpha v$ if there is a factorization $v = v_0 u_1 v_1 \cdots u_n v_n$ with $v_0, \ldots, v_n \in X^*$, $u_1, \ldots, u_n \in X$, $u = u_1 \cdots u_n$, and $\alpha(v_i) = 1$ for $0 \le i \le n$. It is easy to see that $\preceq_\alpha$ is indeed a partial order on X^*. Furthermore, let $\preceq_H$ be the partial order on X^* with $u \preceq_H v$ if $v = v_0 u_1 v_1 \cdots u_n v_n$ for some $v_0, \ldots, v_n \in X^*$, $u_1, \ldots, u_n \in X$, and $u = u_1 \cdots u_n$ such that $\alpha(v_i) = 1$ for every morphism $\alpha \colon X^* \to H$ and $0 \le i \le n$. Note that if H is finite, there are only finitely many morphisms $\alpha \colon X^* \to H$. The upward closure $U \subseteq X^*$ of $\{\varepsilon\}$ with respect to $\preceq_H$ is the intersection of all preimages $\alpha^{-1}(1)$ for all morphisms $\alpha \colon X^* \to H$, which is therefore regular if H is finite (and a finite automaton for this upward closure can be constructed from X and H). Since for $w = w_1 \cdots w_n$, $w_1, \ldots, w_n \in X$, the upward closure of $\{w\}$ equals $Uw_1 \cdots Uw_nU$, we can also construct a finite automaton for the upward closure of any given singleton provided that H is finite. In the latter case, we can also show that $\preceq_H$ is a wqo:

Lemma 1. *For every finite group H and finite alphabet X, $(X^*, \preceq_H)$ is a wqo.*[2]

Proof. There are only finitely many morphisms $\alpha \colon X^* \to H$, say $\alpha_1, \ldots, \alpha_\ell$. If $\beta \colon X^* \to H^\ell$ is the morphism with $\beta(w) = (\alpha_1(w), \ldots, \alpha_\ell(w))$, then for all words $w \in X^*$: $\beta(x) = 1$ if and only if $\alpha(x) = 1$ for all morphisms $\alpha \colon X^* \to H$. Thus, $\preceq_H$ coincides with $\preceq_\beta$, and it suffices to show that $\preceq_\beta$ is a wqo.

Let $w_1, w_2, \ldots \in X^*$ be an infinite sequence of words. Since H^ℓ is finite, we can assume that all the w_i have the same image under β; otherwise, choose an infinite subsequence on which β is constant. Consider the alphabet $Y = X \times H^\ell$. For every $w \in X^*$, $w = a_1 \cdots a_r$, let $\bar{w} \in Y^*$ be the word

$$\bar{w} = (a_1, \beta(a_1))(a_2, \beta(a_1a_2)) \cdots (a_r, \beta(a_1 \cdots a_r)). \tag{1}$$

Applying Thm. 1 to the sequence $\bar{w}_1, \bar{w}_2, \ldots$ yields $i < j$ with $\bar{w}_i \preceq \bar{w}_j$. This means $\bar{w}_i = u'_1 \cdots u'_r$, $\bar{w}_j = v'_0 u'_1 v'_1 \cdots u'_r v'_r$ for some $u'_1, \ldots, u'_r \in Y$, $v'_0, \ldots, v'_r \in Y^*$. By definition of $\bar{w}_i$ we have $u'_s = (u_s, h_s)$ for $1 \le s \le r$, where $h_s = \beta(u_1 \cdots u_s)$ and $w_i = u_1 \cdots u_r$. Let $\pi_1 \colon Y^* \to X^*$ be the morphism extending the projection onto the first component, and let $v_s = \pi_1(v'_s)$ for $0 \le s \le r$. Then clearly $w_j = v_0 u_1 v_1 \cdots u_r v_r$. We claim that $\beta(v_s) = 1$ for $0 \le s \le r$, from which $w_i \preceq_\beta w_j$ and hence the lemma follows. Since $\bar{w}_j$ is also obtained according to (1), we have $\beta(u_1 \cdots u_{s+1}) = h_{s+1} = \beta(v_0 u_1 v_1 \cdots u_s v_s u_{s+1})$ for $0 \le s \le r - 1$. By induction on s, this implies $\beta(v_s) = 1$ for $0 \le s \le r - 1$. Finally, $\beta(v_r) = 1$ follows from $\beta(u_1 \cdots u_r) = \beta(w_i) = \beta(w_j) = \beta(v_0 u_1 v_1 \cdots u_r v_r) = \beta(u_1 \cdots u_r v_r)$. □

Loops. Let $G = H \wr F_2$ and fix free generators $a, b \in F_2$. Recall that every element of F_2 can be represented by a unique word over $\{a, a^{-1}, b, b^{-1}\}$ that does not contain a factor of the form aa^{-1}, $a^{-1}a$, bb^{-1}, or $b^{-1}b$; such words are called *reduced*. For $f \in F_2$, let $|f|$ be the length of the reduced word representing f. Also recall that elements of G are pairs (k, f), where $k \in K = \bigoplus_{g \in F_2} H$ and $f \in F_2$. In the following, we simply write kf for the pair (k, f). Fix an automaton $A = (Q, G, E, q_0, Q_F)$ with labels from G for the rest of Sec. 4. We want to check whether $1 \in L(A)$. Since G is generated as a monoid by $H \cup \{a, a^{-1}, b, b^{-1}\}$, we can assume that $E \subseteq Q \times (H \cup \{a, a^{-1}, b, b^{-1}\}) \times Q$.

A *configuration* is an element of $Q \times G$. For configurations (p, g_1), (q, g_2), we write $(p, g_1) \to_A (q, g_2)$ if there is a $(p, g, q) \in E$ such that $g_2 = g_1g$. For elements $f, g \in F_2$, we write $f \le g$ ($f < g$) if the reduced word representing f is a (proper) prefix of the reduced word representing g. We say that an element $f \in F_2 \setminus \{1\}$ is *of type* $x \in \{a, a^{-1}, b, b^{-1}\}$ if the reduced word representing f ends with x. Furthermore, $1 \in F_2$ is *of type* 1. Hence, the set of *types* is $T = \{1, a, a^{-1}, b, b^{-1}\}$. When regarding

---

[2] One can actually show for any group H: $(X^*, \preceq_H)$ is a wqo if and only if for every $n \in \mathbb{N}$, there is $k \in \mathbb{N}$ with $|\langle g_1, \ldots, g_n \rangle| \le k$ for all $g_1, \ldots, g_n \in H$. See the full version [24].

the Cayley graph of F_2 as a tree with root 1, the children of a node of type t are of the types $C(t) = \{a, a^{-1}, b, b^{-1}\} \setminus \{t^{-1}\}$. Clearly, two nodes have the same type if and only if their induced subtrees of the Cayley graph are isomorphic. The elements of $D = \{a, a^{-1}, b, b^{-1}\}$ will also be called *directions*.

Let $p, q \in Q$ and $t \in T$. A sequence of configurations

$$(q_1, k_1 f_1) \to_A (q_2, k_2 f_2) \to_A \cdots \to_A (q_n, k_n f_n) \tag{2}$$

(recall that $k_i f_i$ denotes the pair $(k_i, f_i) \in G$) is called a *well-nested (p, q)-computation for t* if (i) $q_1 = p$ and $q_n = q$, (ii) $f_1 = f_n$ is of type t, and (iii) $f_i \geq f_1$ for $1 < i < n$ (this last condition is satisfied automatically if $f_1 = f_n = 1$). We define the *effect* of the computation to be $f_1^{-1} k_1^{-1} k_n f_n \in K$. Hence, the effect describes the change imposed by applying the corresponding sequence of transitions, independently of the configuration in which it starts. The *depth* of the computation (2) is the maximum value of $|f_1^{-1} f_i|$ for $1 \leq i \leq n$. We have $1 \in L(A)$ if and only if for some $q \in Q_F$, there is a well-nested (q_0, q)-computation for 1 with effect 1.

For $d \in C(t)$, a well-nested (p, q)-computation (2) for t is called a *(p, d, q)-loop for t* if in addition $f_1 d \leq f_i$ for $1 < i < n$. Note that there is a (p, d, q)-loop for t that starts in (p, kf) (where f is of type t) with effect e and depth m if and only if there exists a (p, d, q)-loop for t with effect e and depth m that starts in (p, t).

Given $p, q \in Q$, $t \in T$, $d \in C(t)$, it is decidable whether there is a (p, d, q)-loop for t: This amounts to checking whether a given automaton with input alphabet $\{a, a^{-1}, b, b^{-1}\}$ accepts a word representing the identity of F_2 such that no proper prefix represents the identity of F_2. Since this can be accomplished using pushdown automata, we can compute the set

$$X_t = \{(p, d, q) \in Q \times C(t) \times Q \mid \text{there is a } (p, d, q)\text{-loop for } t\}.$$

Loop Patterns. Given a word $w = (p_1, d_1, q_1) \cdots (p_n, d_n, q_n) \in X_t^*$, a *loop assignment for w* is a choice of a (p_i, d_i, q_i)-loop for t for each position i, $1 \leq i \leq n$. The *effect* of a loop assignment is $e_1 \cdots e_n \in K$, where $e_i \in K$ is the effect of the loop assigned to position i. The *depth* of a loop assignment is the maximum depth of an appearing loop. A *loop pattern for t* is a word $w \in X_t^*$ that has a loop assignment with effect 1. The *depth* of the loop pattern is the minimum depth of a loop assignment with effect 1. Note that applying the loops for the symbols in a loop pattern $(p_1, d_1, q_1) \cdots (p_n, d_n, q_n)$ does not have to be a computation: We do not require $q_i = p_{i+1}$. Instead, the loop patterns describe the possible ways in which a well-nested computation can enter (and leave) subtrees of the Cayley graph of F_2 in order to have effect 1. The sets

$$P_t = \{w \in X_t^* \mid w \text{ is a loop pattern for } t\}$$

for $t \in T$ will therefore play a central role in the decision procedure.

Recall the definition of the partial order $\preceq_H$ from Sec. 4. We have shown that $\preceq_H$ is a wqo (Lemma 1). The second important result on $\preceq_H$ is:

Lemma 2. *For each $t \in T$, P_t is an upward closed subset of X_t^* with respect to $\preceq_H$.*

Lemma 1 and 2 already imply that each P_t is a regular language, since the upward closure of each singleton is regular. This can also be deduced by observing that $\preceq_H$ is a monotone order in the sense of [7]. Therein, Ehrenfeucht et al. show that languages that are upward closed with respect to monotone well-quasi-orders are regular. Our next step is a characterization of the P_t that allows us to compute finite automata for them. In order to state this characterization, we need the following definitions.

Let X, Y be alphabets. A *regular substitution* is a map $\sigma \colon X \to 2^{Y^*}$ such that $\sigma(x) \subseteq Y^*$ is regular for every $x \in X$. For $w \in X^*$, $w = w_1 \cdots w_n$, $w_i \in X$, let $\sigma(w) = R_1 \cdots R_n$, where $\sigma(w_i) = R_i$ for $1 \le i \le n$. Given $R \subseteq Y^*$ and a regular substitution $\sigma \colon X \to 2^{Y^*}$, let $\sigma^{-1}(R) = \{w \in X^* \mid \sigma(w) \cap R \ne \emptyset\}$. If R is regular, then $\sigma^{-1}(R)$ is regular as well [31, Prop. 2.16], and an automaton for $\sigma^{-1}(R)$ can be obtained effectively from automata for R and the $\sigma(x)$. The alphabet Y_t is given by

$$Y_t = X_t \cup ((Q \times H \times Q) \cap E).$$

We will interpret a word in Y_t^* as that part of a computation that happens in a node of type t: A symbol in $Y_t \setminus X_t$ stands for a transition that stays in the current node and only changes the local H-value and the state. A symbol $(p, d, q) \in X_t$ represents the execution of a (p, d, q)-loop in a subtree of the current node. The morphism $\pi_t \colon Y_t^* \to X_t^*$ is the projection onto X_t^*, meaning $\pi_t(y) = y$ for $y \in X_t$ and $\pi_t(y) = \varepsilon$ for $y \in Y_t \setminus X_t$. The morphism $\nu_t \colon Y_t^* \to H$ is defined by

$$\nu_t((p, d, q)) = 1 \text{ for } (p, d, q) \in X_t$$
$$\nu_t((p, h, q)) = h \text{ for } (p, h, q) \in Y_t \setminus X_t.$$

Hence, when $w \in Y_t^*$ describes part of a computation, $\nu_t(w)$ is the change it imposes on the current node. For $p, q \in Q$ and $t \in T$, define the regular set

$$R_{p,q}^t = \{(p_0, g_1, p_1)(p_1, g_2, p_2) \cdots (p_{n-1}, g_n, p_n) \in Y_t^* \mid p_0 = p, p_n = q\}.$$

Then $\pi_t^{-1}(P_t) \cap \nu_t^{-1}(1) \cap R_{p,q}^t$ consists of those words over Y_t that admit an assignment of loops to occurrences of symbols in X_t so as to obtain a well-nested (p, q)-computation for t with effect 1. Given $d \in C(t)$, $t \in T$, the regular substitution $\sigma_{t,d} \colon X_t \to 2^{Y_d^*}$ is defined by

$$\sigma_{t,d}((p, d, q)) = \bigcup\{R_{p',q'}^d \mid (p, d, p'), (q', d^{-1}, q) \in E\}$$
$$\sigma_{t,d}((p, u, q)) = \{\varepsilon\} \text{ for } u \in C(t) \setminus \{d\}.$$

For tuples $(U_t)_{t \in T}$ and $(V_t)_{t \in T}$ with $U_t, V_t \subseteq X_t^*$, we write $(U_t)_{t \in T} \le (V_t)_{t \in T}$ if $U_t \subseteq V_t$ for each $t \in T$. We can now state the following fixpoint characterization:

Lemma 3. $(P_t)_{t \in T}$ is the smallest tuple such that for every $t \in T$ we have $\varepsilon \in P_t$ and

$$\bigcap_{d \in C(t)} \sigma_{t,d}^{-1}\left(\pi_d^{-1}(P_d) \cap \nu_d^{-1}(1)\right) \subseteq P_t.$$

Given a language $L \subseteq X_t^*$, let $L{\uparrow}_t = \{v \in X_t^* \mid u \preceq_H v \text{ for some } u \in L\}$.

Theorem 2. *The rational subset membership problem is decidable for every group $G = H \wr F$, where H is finite and F is a finitely generated free group.*

Proof. Since $H \wr F$ is a subgroup of $H \wr F_2$ (since F is a subgroup of F_2), it suffices to show decidability for $G = H \wr F_2$. First, we compute finite automata for the languages P_t. We do this by initializing $U_t^{(0)} := \{\varepsilon\}\!\uparrow_t$ for each $t \in T$ and then successively extending the sets $U_t^{(i)}$, which are represented by finite automata, until they equal P_t: If there is a $t \in T$ and a word

$$w \in \bigcap_{d \in C(t)} \sigma_{t,d}^{-1} \left(\pi_d^{-1}(U_d^{(i)}) \cap \nu_d^{-1}(1) \right) \setminus U_t^{(i)},$$

we set $U_t^{(i+1)} := U_t^{(i)} \cup \{w\}\!\uparrow_t$ and $U_u^{(i+1)} := U_u^{(i)}$ for $u \in T \setminus \{t\}$. Otherwise we stop. By induction on i, it follows from Lemma 2 and Lemma 3 that $U_t^{(i)} \subseteq P_t$.

In each step, we obtain $U_t^{(i+1)}$ by adding new words to $U_t^{(i)}$. Since the sets $U_t^{(i)}$ are upward closed by construction and there is no infinite (strictly) ascending chain of upward closed sets in a wqo, the algorithm above has to terminate with some tuple $(U_t^{(k)})_{t \in T}$. This, however, means that for every $t \in T$

$$\bigcap_{d \in C(t)} \sigma_{t,d}^{-1} \left(\pi_d^{-1}(U_d^{(k)}) \cap \nu_d^{-1}(1) \right) \subseteq U_t^{(k)}.$$

Since on the other hand $\varepsilon \in U_t^{(k)}$ and $U_t^{(k)} \subseteq P_t$, Lemma 3 yields $U_t^{(k)} = P_t$.

Now we have $1 \in L(A)$ if and only if $\pi_1^{-1}(P_1) \cap \nu_1^{-1}(1) \cap R_{q_0,q}^1 \neq \emptyset$ for some $q \in Q_F$, which can be reduced to non-emptiness for finite automata.   □

Theorem 3. *The rational subset membership problem is decidable for every group $H \wr V$ with H finite and V virtually free.*

Proof. This is immediate from Thm. 2 and Prop. 1: If F is a free subgroup of index m in V, then $H^m \wr F$ is isomorphic to a subgroup of index m in $H \wr V$ and decidability of rational subset membership is preserved by finite extensions [12,17].   □

5   Undecidability

In this section, we will prove the second main result of this paper: The wreath product $\mathbb{Z} \wr \mathbb{Z}$ contains a fixed submonoid with an undecidable membership problem. Our proof is based on the undecidability of the halting problem for 2-counter machines.

2-Counter Machines A *2-counter machine* (also known as Minsky machine) is a tuple $C = (Q, q_0, q_f, \delta)$, where Q is a finite set of *states*, $q_0 \in Q$ is the *initial state*, $q_f \in Q$ is the *final state*, and $\delta \subseteq (Q \setminus \{q_f\}) \times \{c_0, c_1\} \times \{+1, -1, = 0\} \times Q$ is the set of *transitions*. The set of *configurations* is $Q \times \mathbb{N} \times \mathbb{N}$, on which we define a binary relation $\rightarrow_C$ as follows: $(p, m_0, m_1) \rightarrow_C (q, n_0, n_1)$ if and only if one of the following holds:

– There is $(p, c_i, b, q) \in \delta$ such that $b \in \{-1, 1\}$, $n_i = m_i + b$, and $n_{1-i} = m_{1-i}$.
– There is $(p, c_i, = 0, q) \in \delta$ such that $n_i = m_i = 0$ and $n_{1-i} = m_{1-i}$.

It is well known that every Turing-machine can be simulated by a 2-counter machine (see e.g. [26]). In particular, we have:

Theorem 4. *There is a fixed 2-counter machine $C = (Q, q_0, q_f, \delta)$ such that the following problem is undecidable: Given $m, n \in \mathbb{N}$, does $(q_0, m, n) \to_C^* (q_f, 0, 0)$ hold?*

Submonoids of $\mathbb{Z} \wr \mathbb{Z}$. In this section, we only consider wreath products of the form $H \wr \mathbb{Z}$. An element $(f, m) \in H \wr \mathbb{Z}$ such that the support of f is contained in the interval $[a, b]$ (with $a, b \in \mathbb{Z}$) and $0, m \in [a, b]$ will also be written as a list $[f(a), \ldots, f(b)]$, where in addition the element $f(0)$ is labeled by an incoming (downward) arrow and the element $f(m)$ is labeled by an outgoing (upward) arrow.

We will construct a fixed finitely generated submonoid of the wreath product $\mathbb{Z} \wr \mathbb{Z}$ with an undecidable membership problem. For this, let $C = (Q, q_0, q_f, \delta)$ be the 2-counter machine from Thm. 4. W.l.o.g. we can assume that there exists a partition $Q = Q_0 \cup Q_1$ such that $q_0 \in Q_0$ and

$$\delta \subseteq (Q_0 \times \{c_0\} \times \{+1, -1, = 0\} \times Q_1) \cup (Q_1 \times \{c_1\} \times \{+1, -1, = 0\} \times Q_0).$$

In other words, C alternates between the two counters. Hence, a transition (q, c_i, x, p) can be just written as (q, x, p).

Let $\Sigma = Q \uplus \{c, \#\}$ and let $\mathbb{Z}^\Sigma$ be the free abelian group generated by Σ. First, we prove that there is a fixed finitely generated submonoid M of $\mathbb{Z}^\Sigma \wr \mathbb{Z}$ with an undecidable membership problem. Let $a \notin \Sigma$ be a generator for the right $\mathbb{Z}$-factor; hence $\mathbb{Z}^\Sigma \wr \mathbb{Z}$ is generated by $\Sigma \cup \{a\}$. Let $K = \bigoplus_{m \in \mathbb{Z}} \mathbb{Z}^\Sigma$. In the following, we will freely switch between the description of elements of $\mathbb{Z}^\Sigma \wr \mathbb{Z}$ by words over $(\Sigma \cup \{a\})^{\pm 1}$ and by pairs from $K \rtimes \mathbb{Z}$.

Our finitely generated submonoid M of $\mathbb{Z}^\Sigma \wr \mathbb{Z}$ is generated by the following elements. The right column shows the generators in list notation (elements of $\mathbb{Z}^\Sigma$ are written additively, i.e., as $\mathbb{Z}$-linear combinations of elements of Σ):

$$p^{-1}a\#a^2\#aq \text{ for } (p, = 0, q) \in \delta \qquad [\overset{\downarrow}{-p}, \#, 0, \#, \overset{\uparrow}{q}] \tag{3}$$

$$p^{-1}a\#aca^2qa^{-2} \text{ for } (p, +1, q) \in \delta \qquad [\overset{\downarrow}{-p}, \#, \overset{\uparrow}{c}, 0, q] \tag{4}$$

$$p^{-1}a\#a^3qa^6c^{-1}a^{-8} \text{ for } (p, -1, q) \in \delta \qquad [\overset{\downarrow}{-p}, \#, \overset{\uparrow}{0}, 0, q, 0, 0, 0, 0, 0, -c] \tag{5}$$

$$c^{-1}a^8ca^{-8} \qquad [\overset{\downarrow\uparrow}{-c}, 0, 0, 0, 0, 0, 0, 0, 0, c] \tag{6}$$

$$c^{-1}a\#a^7ca^{-6} \qquad [\overset{\downarrow}{-c}, \#, \overset{\uparrow}{0}, 0, 0, 0, 0, 0, 0, c] \tag{7}$$

$$q_f^{-1}a^{-1} \qquad [\overset{\uparrow}{0}, \overset{\downarrow}{-q_f}] \tag{8}$$

$$\#^{-1}a^{-2} \qquad [\overset{\uparrow}{0}, 0, \overset{\downarrow}{-\#}] \tag{9}$$

For initial counter values $m, n \in \mathbb{N}$ let $I(m, n) = aq_0a^2c^ma^4c^na^{-6}$; its list notation is

$$[\overset{\downarrow}{0}, \overset{\uparrow}{q_0}, 0, m \cdot c, 0, 0, 0, n \cdot c]. \tag{10}$$

Here is some intuition: The group element $I(m,n)$ represents the initial configuration (q_0, m, n) of the 2-counter machine C. Lemma 4 below states that $(q_0, m, n) \to_C^*$ $(q_f, 0, 0)$ is equivalent to the existence of $Y \in M$ with $I(m, n)Y = 1$, i.e., $I(m, n)^{-1} \in M$. Generators of type (3)–(7) simulate the 2-counter machine C. States of C will be stored at cursor positions $4k + 1$. The values of the first (resp., second) counter will be stored at cursor positions $8k + 3$ (resp., $8k + 7$). Note that $I(m, n)$ puts a single copy of the symbol $q_0 \in \Sigma$ at position 1, m copies of symbol c (which represents counter values) at position 3, and n copies of symbol c at position 7. Hence, indeed, $I(m, n)$ sets up the initial configuration (q_0, m, n) for C. Even cursor positions will carry the special symbol $\#$. Note that generator (8) is the only generator which changes the cursor position from even to odd or vice versa. It will turn out that if $I(m, n)Y = 1$ $(Y \in M)$, then generator (8) has to occur exactly once in Y; it terminates the simulation of the 2-counter machine C. Hence, Y can be written as $Y = U(q_f^{-1}a^{-1})V$ with $U, V \in M$. Moreover, it turns out that $U \in M$ is a product of generators (3)–(7), which simulate C. Thereby, even cursor positions will be marked with a single occurrence of the special symbol $\#$. In a second phase, which corresponds to $V \in M$, these special symbols $\#$ will be removed again and the cursor will be moved left to position 0. This is accomplished with generator (9). In fact, our construction enforces that V is a power of (9).

During the simulation phase (corresponding to $U \in M$), generators of type (3) implement zero tests, whereas generators of type (4) (resp., (5)) increment (resp., decrement) a counter. Finally, (6) and (7) copy the counter value to the next cursor position that is reserved for the counter (that is copied). During such a copy phase, (6) is first applied ≥ 0 many times. Finally, (7) is applied exactly once.

Lemma 4. *For all $m, n \in \mathbb{N}$ we have: $(q_0, m, n) \to_C^* (q_f, 0, 0)$ if and only if there exists $Y \in M$ such that $I(m, n)Y = 1$.*

The following result is an immediate consequence of Thm. 4 and Lemma 4.

Theorem 5. *There is a fixed finitely generated submonoid M of the wreath product $\mathbb{Z}^\Sigma \wr \mathbb{Z}$ with an undecidable membership problem.*

Finally, we can establish the main result of this section.

Theorem 6. *There is a fixed finitely generated submonoid M of the wreath product $\mathbb{Z} \wr \mathbb{Z}$ with an undecidable membership problem.*

Proof. By Thm. 5 it suffices to reduce the submonoid membership problem of $\mathbb{Z}^\Sigma \wr \mathbb{Z}$ to the submonoid membership problem of $\mathbb{Z} \wr \mathbb{Z}$. If $m = |\Sigma|$, then Prop. 1 shows that $\mathbb{Z}^\Sigma \wr \mathbb{Z} \cong \mathbb{Z}^m \wr m\mathbb{Z}$ is isomorphic to a subgroup of index m in $\mathbb{Z} \wr \mathbb{Z}$. So if $\mathbb{Z} \wr \mathbb{Z}$ had a decidable submonoid membership problem for each finitely generated submonoid, then the same would be true of $\mathbb{Z}^\Sigma \wr \mathbb{Z}$. □

Theorem 6 together with the undecidability of the rational subset membership problem for groups $H \wr (\mathbb{Z} \times \mathbb{Z})$ for non-trivial H [23] implies the following: For finitely generated non-trivial abelian groups G and H, $H \wr G$ has a decidable rational subset membership problem if and only if (i) G is finite[3] or (ii) G has rank 1 and H is finite.

---

[3] If G has size m, then by Prop. 1, $H^m \cong H^m \wr 1$ is isomorphic to a subgroup of index m in $H \wr G$. Since H^m is finitely generated abelian and decidability of the rational subset membership is preserved by finite extensions [12,17], decidability for $H \wr G$ follows.

By [4], $\mathbb{Z} \wr \mathbb{Z}$ is a subgroup of Thompson's group F as well as of Baumslag's finitely presented metabelian group $\langle a, s, t \mid [s, t] = [a^t, a] = 1, a^s = aa^t \rangle$. Hence, we get:

Corollary 1. *Thompson's group F and Baumslag's finitely presented metabelian group both contain finitely generated submonoids with an undecidable membership problem.*

6   Open Problems

As mentioned in the introduction, the rational subset membership problem is undecidable for every wreath product $H \wr (\mathbb{Z} \times \mathbb{Z})$, where H is a non-trivial group [23]. We conjecture that for every non-trivial group H and every non-virtually free group G, the rational subset membership problem for $H \wr G$ is undecidable. The reason is that the undecidability proof for $H \wr (\mathbb{Z} \times \mathbb{Z})$ [23] only uses the grid-like structure of the Cayley graph of $\mathbb{Z} \times \mathbb{Z}$. In [19] it was shown that the Cayley graph of a group G has bounded tree width if and only if the group is virtually free. Hence, if G is not virtually free, then the Cayley-graph of G has unbounded tree width, which means that finite grids of arbitrary size appear as minors in the Cayley-graph of G. One might therefore hope to again reduce a tiling problem to the rational subset membership problem for $H \wr G$ (for H non-trivial and G not virtually free).

Another interesting case, which is not resolved by our results, concerns the rational subset membership problem for wreath products $G \wr V$ with V virtually free and G a finitely generated infinite torsion group. Finally, all these questions can also be asked for the submonoid membership problem. We do not know any example of a group with decidable submonoid membership problem but undecidable rational subset membership problem. If such a group exists, it must be one-ended [22].

References

1. Anisimov, A.V.: Group languages. Kibernetika 4, 18–24 (1971) (in Russian); English translation. Cybernetics 4, 594–601 (1973)
2. Benois, M.: Parties rationnelles du groupe libre. C. R. Acad. Sci. Paris, Sér. A 269, 1188–1190 (1969)
3. Chambart, P., Schnoebelen, P.: Post embedding problem is not primitive recursive, with applications to channel systems. In: Arvind, V., Prasad, S. (eds.) FSTTCS 2007. LNCS, vol. 4855, pp. 265–276. Springer, Heidelberg (2007)
4. Cleary, S.: Distortion of wreath products in some finitely-presented groups. Pacific Journal of Mathematics 228(1), 53–61 (2006)
5. Davis, T.C., Olshanskii, A.Y.: Subgroup distortion in wreath products of cyclic groups. Journal of Pure and Applied Algebra 215(12), 2987–3004 (2011)
6. Diekert, V., Muscholl, A.: Solvability of equations in free partially commutative groups is decidable. International Journal of Algebra and Computation 16(6), 1047–1069 (2006)
7. Ehrenfeucht, A., Haussler, D., Rozenberg, G.: On regularity of context-free languages. Theor. Comput. Sci. 27, 311–332 (1983)
8. Eilenberg, S., Schützenberger, M.P.: Rational sets in commutative monoids. Journal of Algebra 13, 173–191 (1969)
9. Fernau, H., Stiebe, R.: Sequential grammars and automata with valences. Theor. Comput. Sci. 276(1-2), 377–405 (2002)

10. Finkel, A., Schnoebelen, P.: Well-structured transition systems everywhere! Theor. Comput. Sci. 256(1-2), 63–92 (2001)
11. Gilman, R.H.: Formal languages and infinite groups. In: Geometric and Computational Perspectives on Infinite Groups DIMACS Ser. Discrete Math. Theoret. Comput. Sci, vol. 25, pp. 27–51. AMS (1996)
12. Grunschlag, Z.: Algorithms in Geometric Group Theory. PhD thesis, University of California at Berkley (1999)
13. Haines, L.H.: On free monoids partially ordered by embedding. Journal of Combinatorial Theory 6, 94–98 (1969)
14. Higman, G.: Ordering by divisibility in abstract algebras. Proceedings of the London Mathematical Society. Third Series 2, 326–336 (1952)
15. Jurdziński, T.: Leftist grammars are non-primitive recursive. In: Aceto, L., Damgård, I., Goldberg, L.A., Halldórsson, M.M., Ingólfsdóttir, A., Walukiewicz, I. (eds.) ICALP 2008, Part II. LNCS, vol. 5126, pp. 51–62. Springer, Heidelberg (2008)
16. Kambites, M.: Formal languages and groups as memory. Communications in Algebra 37(1), 193–208 (2009)
17. Kambites, M., Silva, P.V., Steinberg, B.: On the rational subset problem for groups. Journal of Algebra 309(2), 622–639 (2007)
18. Kunc, M.: Regular solutions of language inequalities and well quasi-orders. Theor. Comput. Sci. 348(2–3), 277–293 (2005)
19. Kuske, D., Lohrey, M.: Logical aspects of Cayley-graphs: the group case. Annals of Pure and Applied Logic 131(1–3), 263–286 (2005)
20. Lohrey, M., Sénizergues, G.: Theories of HNN-extensions and amalgamated products. In: Bugliesi, M., Preneel, B., Sassone, V., Wegener, I. (eds.) ICALP 2006. LNCS, vol. 4052, pp. 504–515. Springer, Heidelberg (2006)
21. Lohrey, M., Steinberg, B.: The submonoid and rational subset membership problems for graph groups. Journal of Algebra 320(2), 728–755 (2008)
22. Lohrey, M., Steinberg, B.: Submonoids and rational subsets of groups with infinitely many ends. Journal of Algebra 324(4), 970–983 (2010)
23. Lohrey, M., Steinberg, B.: Tilings and submonoids of metabelian groups. Theory Comput. Syst. 48(2), 411–427 (2011)
24. Lohrey, M., Steinberg, B., Zetzsche, G.: Rational subsets and submonoids of wreath products. arXiv.org (2013), http://arxiv.org/abs/1302.2455
25. Lyndon, R.C., Schupp, P.E.: Combinatorial Group Theory. Springer (1977)
26. Minsky, M.L.: Computation: Finite and Infinite Machines. Prentice-Hall International (1967)
27. Motwani, R., Panigrahy, R., Saraswat, V.A., Venkatasubramanian, S.: On the decidability of accessibility problems (extended abstract). In: Proc. STOC 2000, pp. 306–315. ACM (2000)
28. Roman'kov, V.: On the occurence problem for rational subsets of a group. In: International Conference on Combinatorial and Computational Methods in Mathematics, pp. 76–81 (1999)
29. Romanovskii, N.S.: Some algorithmic problems for solvable groups. Algebra i Logika 13(1), 26–34 (1974)
30. Romanovskii, N.S.: The occurrence problem for extensions of abelian groups by nilpotent groups. Sibirsk. Mat. Zh. 21, 170–174 (1980)
31. Sakarovitch, J.: Elements of Automata Theory. Cambridge University Press (2009)
32. Schnoebelen, P.: Verifying lossy channel systems has nonprimitive recursive complexity. Inf. Process. Lett. 83(5), 251–261 (2002)

Fair Subtyping for Open Session Types[*]

Luca Padovani

Dipartimento di Informatica, Università di Torino, Italy
luca.padovani@unito.it

Abstract. Fair subtyping is a liveness-preserving refinement relation for session types akin to (but coarser than) the well-known should-testing precongruence. The behavioral characterization of fair subtyping is challenging essentially because fair subtyping is context-sensitive: two session types may or may not be related depending on the context in which they occur, hence the traditional coinductive argument for dealing with recursive types is unsound in general. In this paper we develop complete behavioral and axiomatic characterizations of fair subtyping and we give a polynomial algorithm to decide it.

1 Introduction

Session types [7,8] describe the type, order, and direction of messages that can be sent over channels. In essence, session types are simple CCS-like processes using a reduced set of operators [3,1]: termination, external and internal choices respectively guarded by input and output actions, and recursion. For example, the session type $T = \mu x.(!\mathsf{buy}.x \oplus !\mathsf{pay})$ denotes a channel for sending an arbitrary number of buy messages followed by a single pay message. The session type $S = \mu x.(?\mathsf{buy}.x + ?\mathsf{pay}.?\mathsf{vouch})$ denotes a channel for receiving an arbitrary number of buy messages, or a single pay message followed by a vouch message. We can describe a whole session in abstract terms as the parallel composition of the types of its endpoint channels. For instance, $T \mid S \mid !\mathsf{vouch}$ describes a session with a client that buys an arbitrary number of items and then pays, a shop that serves the client, and a bank that vouches for the client. Session type systems check that processes use session channels according to a session type. As an example, the typing derivation below proves that the process $\mathtt{rec}\ \mathscr{X}.k!\langle m \rangle.\mathscr{X}$ sending the message m on channel k is well typed in the channel environment $k : T$ provided that "m is a message of type buy" (the exact interpretation of this property is irrelevant):

$$
\frac{
\displaystyle \vdash m : \mathsf{buy} \quad
\frac{}{\mathscr{X} \mapsto \{k : x\}; k : x \vdash \mathscr{X}} \ [\text{VAR}]
}{
\frac{
\mathscr{X} \mapsto \{k : x\}; k : !\mathsf{buy}.x \oplus !\mathsf{pay} \vdash k!\langle m \rangle.\mathscr{X}
}{
k : \mu x.(!\mathsf{buy}.x \oplus !\mathsf{pay}) \vdash \mathtt{rec}\ \mathscr{X}.k!\langle m \rangle.\mathscr{X}
} \ [\text{REC}]
} \ [\text{OUTPUT}]
$$

Rule [REC] opens the recursive session type T in correspondence with recursion in the process and augments the process environment with the association $\mathscr{X} \mapsto \{k : x\}$. In this way, an occurrence of the process variable $\mathscr{X}$ in a channel environment where the

---

[*] Full version http://www.di.unito.it/~padovani/Papers/OpenFairSubtyping.pdf.

F.V. Fomin et al. (Eds.): ICALP 2013, Part II, LNCS 7966, pp. 373–384, 2013.
© Springer-Verlag Berlin Heidelberg 2013

channel k has type x can be declared well typed. Rule [OUTPUT] checks that the output performed by the process on channel k is allowed by the type of k. Finally, the residual process after $k!\langle m \rangle$ is checked against the residual type of k after !buy.

It is worth noting that *all* session type systems admit a derivation like the one above and that such derivation implicitly and crucially relies on the subtyping relation $!buy.x \oplus !pay \leqslant !buy.x$ saying that a channel of type $!buy.x \oplus !pay$ can be safely used where a channel of type $!buy.x$ is expected. Conventional works on subtyping for session types [6,3,1] establish that $\leqslant$ is contravariant for outputs, thereby making the theory of session types a conservative extension of the theory of channel types [11,4]. Nonetheless, it can be argued that this subtyping relation is inadequate. For instance, consider the session described earlier as $T \mid S \mid !vouch$ and observe that all non-terminated participants retain the potential to make progress: it is always *possible* for the client to pay. If that happens, the bank can send the vouch message to the shop, at which point all participants terminate. Accepting the typing derivation above means allowing a client that behaves according to $\mu x.!buy.x$ to interact with a shop and a bank that behave as $S \mid !vouch$. In the session $\mu x.!buy.x \mid S \mid !vouch$, however, the client never pays and the liveness of the session with respect to the bank is compromised. This example proves that the original subtyping relation for session types, for which $!buy.x \oplus !pay \leqslant !buy.x$ holds, is not liveness preserving in general: there exist a context $\mathscr{C} = \mu x.[\ ]$ and two behaviors $S \mid !vouch$ such that the session described by $\mathscr{C}[!buy.x \oplus !pay] \mid S \mid !vouch$ does have the liveness property while $\mathscr{C}[!buy.x] \mid S \mid !vouch$ does not.

The contribution of this work is the definition of a new subtyping relation, which we dub *fair subtyping*, as the coarsest liveness-preserving refinement for possibly *open* session types (like $!buy.x \oplus !pay$ and $!buy.x$ above) that is a *pre-congruence* for all the operators of the type language. With this definition in place, we can reject a derivation like the one above because it is based on the law $!buy.x \oplus !pay \leqslant !buy.x$ which is *invalid* for fair subtyping. It may be questioned whether dealing with open session types is really necessary, given that the above derivation can also be reformulated as follows

$$
\cfrac{
\cfrac{\vdash m : buy \qquad \mathscr{X} \mapsto \{k : T\}; k : T \vdash \mathscr{X}}{\mathscr{X} \mapsto \{k : T\}; k : !buy.T \oplus !pay \vdash k!\langle m \rangle.\mathscr{X}}\text{ [OUTPUT]}
}{
k : T \vdash \texttt{rec } \mathscr{X}.k!\langle m \rangle.\mathscr{X}
}\text{ [REC]}
$$

where the top right leaf is labeled [VAR].

where T is *unfolded* (instead of being opened) by rule [REC] and associated with the process variable $\mathscr{X}$ (this corresponds to using *equi-recursive* types, whereby a recursive type and its unfolding are deemed equal). Now, this derivation should be rejected just as the first one, with the difference that the second derivation relies on the law $!buy.T \oplus !pay \leqslant !buy.T$. It turns out that this law holds even for fair subtyping, intuitively because there are only finitely many differences between (the infinite unfoldings of) $!buy.T$ and $!pay \leqslant !buy.T$. In conclusion, we are not aware of alternative ways of detecting such invalid derivations other than forbidding subsumption within recursions, or using a theory of *open* session types like the one developed in the present paper.

A behavioral refinement called *should-testing* enjoying all the properties that we seek for in fair subtyping has been extensively studied in [12]. There, should-testing is shown to be the coarsest liveness-preserving pre-congruence of a process algebra considerably richer than session types. Therefore, given the correspondence between session types and processes, we could just take should-testing as the defining notion for fair subtyping. We find this shortcut unsatisfactory for several reasons: first, should-testing implies trace equivalence between related processes. In our context, this would amount to requiring *invariance* of outputs, essentially collapsing subtyping to type equality. Second, no complete axiomatization is known for should-testing and its alternative characterization is based on a complex denotational model. As a consequence, it is difficult to understand the basic laws that underlie should-testing. Third, the decision algorithm for should-testing is linear exponential, that is remarkably more expensive compared to the quadratic algorithm for the original subtyping [6]. Instead, by restricting the language of processes to that of session types, we are able to show that:

- Fair subtyping is coarser than should-testing and *does not* imply trace equivalence.
- Fair subtyping admits a complete axiomatization obtained from that of the original subtyping by plugging in a simple auxiliary relation in just two strategic places.
- Fair subtyping can be decided in $O(n^4)$ time.

In the rest of the paper we formalize session types as an appropriate subset of CCS (Section 2) and define fair subtyping as the relation that preserves session liveness in every context (Definition 2.2). Then, we provide a coinductive characterization of fair subtyping that unveils its properties (Section 3). The pre-congruence property is subtle to characterize because fair subtyping is *context sensitive* (two session types may or may not be related depending on the context in which they occur). For example, we have seen that $!\mathsf{buy}.x \oplus !\mathsf{pay} \nleqslant !\mathsf{buy}.x$ and yet $!\mathsf{buy}.(!\mathsf{buy}.x \oplus !\mathsf{pay}) \oplus !\mathsf{pay} \leqslant !\mathsf{buy}.!\mathsf{buy}.x \oplus !\mathsf{pay}$ despite the unrelated terms $!\mathsf{buy}.x \oplus !\mathsf{pay}$ and $!\mathsf{buy}.x$ occur in corresponding positions in the latter pair of related session types. The coinductive characterization also paves the way to the complete axiomatization of fair subtyping and to its decision algorithm (Section 4). In turn, the axiomatization shows how to incrementally patch the original subtyping for session types to ensure liveness preservation. A more detailed comparison with related work is given in the conclusions (Section 5). Because of space constraints, proofs of the results can only be found in the long version of the paper.

2 Syntax and Semantics of Session Types

We assume an infinite set $\mathbf{V}$ of variables x, y, ... and an infinite set of messages a, b, ...; we let $X, Y, \ldots$ range over subsets of $\mathbf{V}$. Session types are defined by the grammar

$$T \ ::= \ \mathsf{end} \ \mid \ x \ \mid \ \textstyle\sum_{i \in I} ?\mathsf{a}_i.T_i \ \mid \ \bigoplus_{i \in I} !\mathsf{a}_i.T_i \ \mid \ \mu x.T$$

where the set I is always finite and non-empty and choices are deterministic, in the sense that $\mathsf{a}_i = \mathsf{a}_j$ implies $i = j$ for every $i, j \in I$.

The term end denotes the type of channels on which no further operations are possible. We will often omit trailing occurrences of end. A term $\sum_{i \in I} ?\mathsf{a}_i.T_i$ is the type of a channel for receiving a message in the set $\{\mathsf{a}_i\}_{i \in I}$. According to the received message a_i, the channel must be used according to T_i afterwards. Terms $\bigoplus_{i \in I} !\mathsf{a}_i.T_i$ are analogous, but they denote the type of channels that can be used for sending messages. Note that output session types represent *internal choices* (the process using a channel with output type can choose any message in the set $\{\mathsf{a}_i\}_{i \in I}$) while input session

Table 1. Transition system of sessions

[T-OUTPUT]
$!a.T \xrightarrow{!a} T$

[T-CHOICE]	[T-INPUT]	[T-PAR]	[T-COMM]
$k \in I$	$k \in I$	$M \xrightarrow{\ell} M'$	$M \xrightarrow{\overline{\alpha}} M' \quad N \xrightarrow{\alpha} N'$
$\bigoplus_{i \in I} !a_i.T_i \xrightarrow{\tau} !a_k.T_k$	$\sum_{i \in I} ?a_i.T_i \xrightarrow{?a_k} T_k$	$M \mid N \xrightarrow{\ell} M' \mid N$	$M \mid N \xrightarrow{\tau} M' \mid N'$

types are *external choices* (the process using a channel with input type must be ready to deal with any message in the set $\{a_i\}_{i \in I}$). We will sometimes use an infix notation for choices writing $?a_1.T_1 + \cdots + ?a_n.T_n$ and $!a_1.T_1 \oplus \cdots \oplus !a_n.T_n$ instead of $\sum_{1 \le i \le n} ?a_i.T_i$ and $\bigoplus_{1 \le i \le n} !a_i.T_i$ respectively. Terms $\mu x.T$ and x are used for building recursive session types, as usual. We assume that session types are contractive, namely that they do not contain subterms of the form $\mu x_1 \cdots \mu x_n.x_1$. The notions of free and bound variables are standard and so are the definitions of open and closed session types. We take an equire-cursive point of view and identify session types modulo renaming of bound variables and folding/unfolding of recursions. That is, $\mu x.T = T\{\mu x.T/x\}$ where $T\{S/x\}$ is the capture-avoiding substitution of every free occurrence of x in T with S. We say that T and S are *strongly equivalent*, notation $T \approx S$, if their infinite unfoldings are the same regular tree [5].

Sessions $M, N, \ldots$ are abstracted as parallel compositions of session types $T, S, \ldots$, their grammar is:

$$M \quad ::= \quad T \quad \mid \quad (M \mid M)$$

We define the operational semantics of sessions by means of a labeled transition system mimicking the actions performed by processes that behave according to session types (in fact, we are abstracting processes into types). The transition system makes use of *actions* α of the form $?a$ and $!a$ describing the input/output of a messages and labels ℓ that are either actions or the invisible move τ. The transition system is defined in Table 1. Rules [T-OUTPUT], [T-CHOICE], and [T-INPUT] deal with prefixed terms. The first and last ones are standard. Rule [T-CHOICE] states that a process behaving according to the type $\bigoplus_{i \in I} !a_i.T_i$ may internally choose, through an invisible move τ, to send any message from the set $\{a_i\}_{i \in I}$. Rule [T-PAR] (and its symmetric, omitted) propagates labels across compositions while [T-COMM] is the synchronization rule between complementary actions resulting into an invisible move (we let $\overline{?a} = !a$ and $\overline{!a} = ?a$).

We use $\varphi, \psi, \ldots$ to range over strings of actions, ε to denote the empty string, and $\le$ to denote the usual prefix order between strings. We write $\xRightarrow{\tau}$ for the reflexive, transitive closure of $\xrightarrow{\tau}$ and $\xRightarrow{\alpha}$ for the composition $\xRightarrow{\tau}\xrightarrow{\alpha}\xRightarrow{\tau}$. We extend this notation to strings of actions so that $\xRightarrow{\alpha_1 \cdots \alpha_n}$ stands for the composition $\xRightarrow{\alpha_1} \cdots \xRightarrow{\alpha_n}$. We write $T \xRightarrow{\alpha}$ (respectively $T \xRightarrow{\varphi}$) if there exists S such that $T \xRightarrow{\alpha} S$ (respectively $T \xRightarrow{\varphi} S$). We write $T \xnRightarrow{\varphi}$ if not $T \xRightarrow{\varphi}$. We let $\text{tr}(T)$ denote the set of *traces* of T, namely $\text{tr}(T) \overset{\text{def}}{=} \{\varphi \mid T \xRightarrow{\varphi}\}$.

We say that a session M is successful if every computation starting from M can be extended to a state that emits the action $!\mathsf{OK}$, where OK is a special message that we assume is not used for synchronization purposes. Formally:

Definition 2.1 (Success). *We say that M is* successful *if $M \overset{\tau}{\Longrightarrow} N$ implies $N \overset{!\mathsf{OK}}{\Longrightarrow}$.*

Example 2.1. Consider $T = \mu x.(!\mathsf{a}.x \oplus !\mathsf{b})$ and $S = \mu x.(?\mathsf{a}.x + ?\mathsf{b}.?\mathsf{c})$ from the introduction, where for brevity we respectively use a, b, and c in place of buy, pay, and vouch. Then $T \mid S \mid !\mathsf{c}.!\mathsf{OK}$ is a successful session because, no matter how many a messages the first participant sends, it is always possible that a b message and a c message are sent. At that point, the third participant emits $!\mathsf{OK}$. The session $!\mathsf{b} \mid S \mid !\mathsf{c}.!\mathsf{OK}$ is successful as well. By contrast $\mu x.!\mathsf{a}.x \mid S \mid !\mathsf{c}.!\mathsf{OK}$ is unsuccessful even if the first two participants keep interacting with each other, because none of them is willing to receive the c message sent by the third participant. ∎

We can now define fair subtyping as the relation that preserves session success. To make sure that fair subtyping is a pre-congruence, we quantify over all possible contexts that apply to the two session types being compared. *Contexts*, ranged over by $\mathscr{C}$, are just session types with exactly one hole $[\,]$ in place of some subterm. We write $\mathscr{C}[T]$ for the session type obtained by filling the hole in $\mathscr{C}$ with T. This operation differs from variable substitution in that $\mathscr{C}$ may capture variables occurring free in T.

Fair subtyping is defined thus:

Definition 2.2 (Fair Subtyping). *We say that T is a* fair subtype *of S, written $T \leqslant S$, if $M \mid \mathscr{C}[T]$ successful implies $M \mid \mathscr{C}[S]$ successful for every M and $\mathscr{C}$.*

To provide the intuition underlying Definition 2.2, suppose that a (well-typed) process accesses a session using a channel k of type S. By replacing k with another channel k' of type $T \leqslant S$ we are in fact changing the session that the process accesses. However, the process is oblivious to the replacement, in the sense that it keeps behaving on k' as if it was a channel of type S even if in reality k' has type T. Now, suppose that the session accessed through k' is described by $M \mid T$ and that it is successful. The replacement changes the session to $M \mid S$. By the hypothesis $T \leqslant S$, we have the assurance that $M \mid S$ is also successful, therefore the substitution is safe.

Showing that $T \leqslant S$ is difficult in general, because of the universal quantification over sessions M and contexts $\mathscr{C}$ in Definition 2.2. Until we characterize precisely the properties of $\leqslant$ in Section 3, it is easier to find session types that are *not* related by fair subtyping.

Example 2.2. Consider $T = \mu x.(!\mathsf{a}.x \oplus !\mathsf{b})$ and $S = \mu x.!\mathsf{a}.x$ and take $M = \mu x.(?\mathsf{a}.x + ?\mathsf{b}.!\mathsf{OK})$. Note that M relies on receiving a b message to emit $!\mathsf{OK}$. Then we have $T \not\leqslant S$ because $M \mid T$ is successful but $M \mid S$ is not (no b message is ever sent). Now, for $\mathscr{C} = \mu x.[\,]$ we have $T = \mathscr{C}[!\mathsf{a}.x \oplus !\mathsf{b}]$ and $S = \mathscr{C}[!\mathsf{a}.x]$, therefore $!\mathsf{a}.x \oplus !\mathsf{b} \not\leqslant !\mathsf{a}.x$. ∎

3 Fair Subtyping

We begin our study of fair subtyping by recalling the traditional subtyping relation for session types, which we dub "unfair subtyping".

Definition 3.1 (Unfair Subtyping). *We say that $\mathscr{U}$ is a coinductive subtyping if $T \mathrel{\mathscr{U}} S$ implies either* (1) $T = S = x$ *or* (2) $T = S = \mathsf{end}$ *or* (3) $T = \sum_{i \in I} ?a_i.T_i$ *and* $S = \sum_{i \in I} ?a_i.S_i$ *and $T_i \mathrel{\mathscr{U}} S_i$ for every $i \in I$ or* (4) $T = \bigoplus_{i \in I \cup J} !a_i.T_i$ *and* $S = \bigoplus_{i \in I} !a_i.S_i$ *and $T_i \mathrel{\mathscr{U}} S_i$ for every $i \in I$.* Unfair subtyping, *denoted by* $\leqslant_\mathsf{U}$, *is the largest coinductive subtyping.*

Clauses (1–2) state the reflexivity of $\leqslant_\mathsf{U}$ for end and type variables, while clauses (3–4) respectively state invariance and contravariance of $\leqslant_\mathsf{U}$ with respect to external and internal choices. There is no need for a clause dealing with recursive types, because type equality already accounts for their unfolding and types are contractive. Unfair subtyping is essentially the standard subtyping relation for session types presented in [6].[1] The appeal for unfair subtyping comes from its simplicity and intuitive rationale. The key clause (4) states that the larger session type allows in general for fewer kinds of messages to be sent: when $T \leqslant_\mathsf{U} S$, a process behaving as S can be safely placed in a context where a process behaving as T is expected because S expresses a more deterministic behavior compared to T. Reducing non-determinism is generally perceived as harmless, but sometimes it may compromise liveness.

Example 3.1. Consider the session types $T = !a.x \oplus !b$ and $S = !a.x$ and the context $\mathscr{C} = \mu x.[\ ]$. Then both $\{(T, S), (x, x)\}$ and $\{(\mathscr{C}[T], \mathscr{C}[S])\}$ are coinductive subtyping relations, from which we deduce $T \leqslant_\mathsf{U} S$ and $\mathscr{C}[T] \leqslant_\mathsf{U} \mathscr{C}[S]$. Yet Example 2.2 shows that neither $\mathscr{C}[T] \leqslant \mathscr{C}[S]$ nor $T \leqslant S$ do hold.   ∎

Unfair subtyping is a necessary but not sufficient condition for fair subtyping.

Theorem 3.1. $\leqslant \subsetneq \leqslant_\mathsf{U}$.

Note that $T \leqslant_\mathsf{U} S$ implies $\mathtt{tr}(T) \supseteq \mathtt{tr}(S)$ and that $\leqslant_\mathsf{U}$ may compromise session success by letting S have "too few" traces compared to T (Example 3.1). Therefore, Theorem 3.1 suggests that fair subtyping should be characterized as a restriction of Definition 3.1 where we impose additional conditions to clause (4). The condition $\mathtt{tr}(T) \subseteq \mathtt{tr}(S)$ is clearly sufficient but too strong: it imposes invariance of fair subtyping with respect to outputs, collapsing fair subtyping to equality. Nonetheless, we will show that, when $T \leqslant S$, there must be an "inevitable" pair of corresponding states at some "finite distance" from T and S for which trace inclusion holds. We formalize this property saying that T *converges into* S. The precise definition of convergence is subtle because T and S may be *open*: we must be able to reason on the property of trace inclusion between corresponding states of T and S by considering the possibility that T and S occur in a context that binds (some of) their free variables.

We begin by introducing a notation for referring to the residual state of a session type after some sequence of actions.

Definition 3.2 (Continuation). *Let $\alpha \in \mathtt{tr}(T)$. The* continuation *of T after α is the session type S such that $T \stackrel{\varepsilon}{\Longrightarrow} \stackrel{\alpha}{\longrightarrow} S$ (note that S is uniquely determined because branches in session types are guarded by distinct actions). We extend the notion of continuation to sequences of actions so that $T(\varepsilon) = T$ and $T(\alpha\varphi) = T(\alpha)(\varphi)$ when $\alpha\varphi \in \mathtt{tr}(T)$.*

---

[1] In practice, subtyping can be relaxed so that it is covariant with respect to external choices [6]. This difference between unfair and standard subtyping does not affect our results.

Next, we define the traces of a session type leading to a free variable:

Definition 3.3 (*X*-**traces**). *The X-traces of T, denoted by X-$\mathtt{tr}(T)$, are the traces of T leading to a variable in X. That is, X-$\mathtt{tr}(T) = \{\varphi \mid \exists x \in X : T \stackrel{\varphi}{\Longrightarrow} x\}$.*

Convergence is the relation $\sqsubseteq_{X;Y}$ inductively defined by the rule:

$$\frac{\forall \varphi \in (\mathtt{tr}(T) \setminus \mathtt{tr}(S)) \cup Y\text{-}\mathtt{tr}(S) : \exists \psi \leq \varphi, \mathsf{a} : T(\psi!\mathsf{a}) \sqsubseteq_{\emptyset;X \cup Y} S(\psi!\mathsf{a})}{T \sqsubseteq_{X;Y} S} \tag{1}$$

where X and Y are two sets of so-called *safe* and *dangerous* variables. Because of its subtle definition, we explain convergence incrementally and, for the time being, we consider the particular instance when $X = Y = \emptyset$. In this case rule (1) reduces to

$$\frac{\forall \varphi \in \mathtt{tr}(T) \setminus \mathtt{tr}(S) : \exists \psi \leq \varphi, \mathsf{a} : T(\psi!\mathsf{a}) \sqsubseteq S(\psi!\mathsf{a})}{T \sqsubseteq S} \tag{2}$$

and it is easy to observe that its base case corresponds to the condition $\mathtt{tr}(T) \setminus \mathtt{tr}(S) = \emptyset$, that is $\mathtt{tr}(T) \subseteq \mathtt{tr}(S)$. Now, suppose $T \sqsubseteq S$ and imagine some session M composed with either T or S whose aim is to tell T and S apart in the sense that M succeeds (emitting !OK) as soon as it enters some trace of T that is not present in S. In order to achieve its goal, M will try to drive the interaction with T along some path $\varphi \in \mathtt{tr}(T) \setminus \mathtt{tr}(S)$. Rule (2) says that after following some prefix ψ of φ that is shared by both T and S, M encounters an internal choice having a branch (corresponding to some action !a) that may divert the interaction to a new stage where the residual behaviors of T and S (respectively $T(\psi!\mathsf{a})$ and $S(\psi!\mathsf{a})$) have sets of traces that are slightly less different. We say "slightly less different" because $T(\psi!\mathsf{a})$ and $S(\psi!\mathsf{a})$ are one step closer to the top of the derivation of $T \sqsubseteq S$, whose leaves imply trace inclusion. Since convergence is defined inductively, this means that T and S are a *finite number* of steps away from the point where trace inclusion holds. In conclusion, when $T \sqsubseteq S$ holds, it is impossible for M to solely rely on the traces in $\mathtt{tr}(T) \setminus \mathtt{tr}(S)$ in order to succeed; M can always be veered into a stage of the interaction where (some corresponding states of) T and S are no longer distinguishable as the traces of (such corresponding states of) T and S are the same.

Example 3.2. Take $T = \mu x.(!\mathsf{a}.x \oplus !\mathsf{b})$ and $S = \mu x.(!\mathsf{a}.!\mathsf{a}.x \oplus !\mathsf{b})$ and observe that $\mathtt{tr}(T) \setminus \mathtt{tr}(S)$ is the language of strings generated by the regular expression $!\mathsf{a}(!\mathsf{a}!\mathsf{a})^*!\mathsf{b}$. Given an arbitrary string in $\mathtt{tr}(T) \setminus \mathtt{tr}(S)$ we can take $\psi = \varepsilon$ and we have $T(!\mathsf{b}) = S(!\mathsf{b}) = \text{end}$ where $\text{end} \sqsubseteq \text{end}$, so we can conclude $T \sqsubseteq S$. ∎

Example 3.3. Consider again the session types $T = \mu x.(!\mathsf{a}.x \oplus !\mathsf{b})$ and $S = \mu x.!\mathsf{a}.!\mathsf{a}.x$ and recall that in Example 3.1 we showed $T \leqslant_\mathsf{U} S$. Let us try to build a derivation for $T \sqsubseteq S$. Note that $\mathtt{tr}(T) \setminus \mathtt{tr}(S)$ is the language of strings generated by the regular expression $(!\mathsf{a})^*!\mathsf{b}$. Taken $\varphi \in \mathtt{tr}(T) \setminus \mathtt{tr}(S)$ we have that any prefix ψ of φ that is in $\mathtt{tr}(T) \cap \mathtt{tr}(S)$ has the form $!\mathsf{a} \cdots !\mathsf{a}$ and now $T(\psi!\mathsf{a}) = T$ and $S(\psi!\mathsf{a}) = S$. Therefore, in order to prove $T \sqsubseteq S$, we need a derivation for $T \sqsubseteq S$. Since convergence is an inductive relation, $T \sqsubseteq S$ is not derivable which agrees with the fact that these two session types are not related by fair subtyping. ∎

We can now turn our attention to the general definition of $\sqsubseteq_{X;Y}$, whose base case adds the condition $Y\text{-}\mathrm{tr}(S) = \emptyset$ to trace inclusion that we have discussed earlier. First of all, observe that the naive extension of $\sqsubseteq$ with the axiom $x \sqsubseteq x$, whereby every variable x converges into itself, fails to yield a pre-congruence for recursion. For example, according to this extension we would have $!a.x \oplus !b \sqsubseteq !a.x$ and yet Example 3.3 shows that $\mu x.(!a.x \oplus !b) \not\sqsubseteq \mu x.!a.x$. It is the *context* in which a type variable x occurs that determines whether or not x converges into itself:

- If x only lies along traces that do *not* distinguish T from S then it is *safe*, in the sense that cycles created by contexts binding x do not allow sensing any difference between T and S.
- If x lies along a path that distinguishes T from S then it is *dangerous*, because cycles created by contexts binding x may enable such difference to be sensed.

In the general definition of convergence, the two sets X and Y respectively contain the variables that are *assumed* to be safe (but that may be found to be dangerous at some later stage while proving convergence) and the variables that are *known* to be dangerous. Whenever a trace that distinguishes T from S is discovered ($\varphi \in (\mathrm{tr}(T) \setminus \mathrm{tr}(S))$), the safe variables become dangerous ones. The condition $Y\text{-}\mathrm{tr}(S) = \emptyset$ then restricts the application of the axiom $x \sqsubseteq x$ to safe variables.

Example 3.4. Take $T = !a.(!a.x \oplus !b.\mathsf{end}) \oplus !b.\mathsf{end}$ and $S = !a.!a.x \oplus !b.\mathsf{end}$ which are obtained from the session types in Example 3.2 by possibly unfolding and then opening recursions. Note that the variable x is dangerous because it lies along the trace $!a!a$ which goes through corresponding states of T and S which differ, indeed $\mathrm{tr}(T) \setminus \mathrm{tr}(S) = \{!a!b\}$. However, we can divert from the trace $!a!b$ by taking $\psi = \varepsilon$ and now we have $T(!b) = S(!b) = \mathsf{end}$ where $\mathsf{end} \sqsubseteq_{\emptyset;\{x\}} \mathsf{end}$, so $T \sqsubseteq_{\{x\};\emptyset} S$.   ∎

Example 3.5. Let $T = !a.(!a.x \oplus !b.\mathsf{end}) \oplus !b.\mathsf{end}$ and $S = !a.!a.x$ and let us try to build a derivation for $T \sqsubseteq_{\{x\};\emptyset} S$. Note that $\mathrm{tr}(T) \setminus \mathrm{tr}(S) = \{!b, !a!b\}$. The prefixes of any $\varphi \in \mathrm{tr}(T) \setminus \mathrm{tr}(S)$ that are in $\mathrm{tr}(T) \cap \mathrm{tr}(S)$ are either ε or $!a$. If we take the prefix $\psi = !a$, we have $T(!a!a) = S(!a!a) = x$ and now $x \not\sqsubseteq_{\emptyset;\{x\}} x$. If we take the prefix $\psi = \varepsilon$, we have $T(!a) = !a.x \oplus !b.\mathsf{end}$ and $S(!a) = !a.x$ and also in this case we deduce $T(!a) \not\sqsubseteq_{\emptyset;\{x\}} S(!a)$ by iterating a similar argument. Therefore we conclude $T \not\sqsubseteq_{\{x\};\emptyset} S$ which was expected since from Example 3.3 we knew that $\mu x.T \not\sqsubseteq \mu x.S$.   ∎

From now on, we will often write $\sqsubseteq_X$ as an abbreviation for $\sqsubseteq_{X;\emptyset}$. The key property of the X set of safe variables is formalized thus:

Lemma 3.1. $T \sqsubseteq_{X \cup \{x\}} S$ *implies* $\mu x.T \sqsubseteq_X \mu x.S$.

In words, the variables in X can be safely bound by a recursive context without compromising convergence. We now show the characterization of fair subtyping:

Definition 3.4 (Coinductive Fair Subtyping). *We say that* $\mathscr{F}$ *is a* coinductive fair subtyping relation *if* $(X, T, S) \in \mathscr{F}$ *implies* $T \sqsubseteq_X S$ *and either:*

(1) $T = S = x$, *or*
(2) $T = S = \mathsf{end}$, *or*

Table 2. Axiomatization of convergence and fair subtyping

$$
\begin{array}{ccc}
& [\text{C-END}] & [\text{C-VAR}] \quad \dfrac{x \notin Y}{x \trianglelefteq_{X;Y} x} & [\text{C-REC}] \quad \dfrac{T \trianglelefteq_{X \cup \{x\};Y \setminus \{x\}} S}{\mu x.T \trianglelefteq_{X;Y} \mu x.S}
\end{array}
$$

$$
\text{end} \trianglelefteq_{X;Y} \text{end}
$$

[C-INPUT]
$$
\dfrac{\forall i \in I : T_i \trianglelefteq_{X;Y} S_i}{\sum_{i \in I} ?a_i.T_i \trianglelefteq_{X;Y} \sum_{i \in I} ?a_i.S_i}
$$

[C-OUTPUT 1]
$$
\dfrac{\forall i \in I : T_i \trianglelefteq_{X;Y} S_i}{\bigoplus_{i \in I} !a_i.T_i \trianglelefteq_{X;Y} \bigoplus_{i \in I} !a_i.S_i}
$$

[C-OUTPUT 2]
$$
\dfrac{\exists k \in I : T_k \trianglelefteq_{\emptyset;X \cup Y} S_k}{\bigoplus_{i \in I \cup J} !a_i.T_i \trianglelefteq_{X;Y} \bigoplus_{i \in I} !a_i.S_i}
$$

[F-END]
$$
\text{end} \leqslant_F \text{end}
$$

[F-VAR]
$$
x \leqslant_F x
$$

[F-REC]
$$
\dfrac{T \leqslant_F S \qquad T \trianglelefteq_{\{x\};\emptyset} S}{\mu x.T \leqslant_F \mu x.S}
$$

[F-INPUT]
$$
\dfrac{\forall i \in I : T_i \leqslant_F S_i}{\sum_{i \in I} ?a_i.T_i \leqslant_F \sum_{i \in I} ?a_i.S_i}
$$

[F-OUTPUT]
$$
\dfrac{\forall i \in I : T_i \leqslant_F S_i}{\bigoplus_{i \in I \cup J} !a_i.T_i \leqslant_F \bigoplus_{i \in I} !a_i.S_i}
$$

[A-SUBT]
$$
\dfrac{T \leqslant_F S \qquad T \trianglelefteq_{V;\emptyset} S}{T \leqslant_A S}
$$

(3) $T = \sum_{i \in I} ?a_i.T_i$ and $S = \sum_{i \in I} ?a_i.S_i$ and $(\emptyset, T_i, S_i) \in \mathscr{F}$ for every $i \in I$, or

(4) $T = \bigoplus_{i \in I \cup J} !a_i.T_i$ and $S = \bigoplus_{i \in I} !a_i.S_i$ and $(\emptyset, T_i, S_i) \in \mathscr{F}$ for every $i \in I$.

We write $T \preccurlyeq_X S$ if $(X,T,S) \in \mathscr{F}$ for some coinductive fair subtyping $\mathscr{F}$.

Theorem 3.2. $\leqslant \, = \, \preccurlyeq_V$

Structurally, Definition 3.4 and Definition 3.1 are very similar. The key difference between $T \leqslant_U S$ and $T \preccurlyeq_X S$ is that in the latter $T \sqsubseteq_X S$ must also hold. Note that, when checking that the continuations T_i and S_i are related, the set of safe variables is emptied. This twist is motivated by the fact that applying a context $\mu x.[\,]$ around T creates a cycle that necessarily goes through the initial state of T while no context applied to T can create loops "within" T. This property makes $\leqslant$ context sensitive: consider for example the session types $T = !a.x \oplus !b$ and $S = !a.x$ and the context $\mathscr{C} = !a.[\,] \oplus !b$ and note that $\mathscr{C}$ does not bind the variable x. Now we have $\mathscr{C}[T] \sqsubseteq_{\{x\}} \mathscr{C}[S]$ while $T \not\sqsubseteq_{\{x\}} S$. Therefore $\mathscr{C}[T] \leqslant \mathscr{C}[S]$ even if $T \not\leqslant S$.

4 Axioms and Algorithms

The characterization developed in Section 3 allows us to axiomatize fair subtyping. The axiomatization, besides being the first for a liveness-preserving refinement precongruence, shows how to guarantee liveness preservation by patching unfair subtyping and, more generally, the traditional subtyping for session types [6] with appropriate conditions in just two strategic places.

Table 2 defines three inductive relations $\trianglelefteq_{X;Y}$ (rules [C-*]), $\leqslant_F$ (rules [F-*]), and $\leqslant_A$ (rule [A-SUBT]) which will be shown to coincide with $\sqsubseteq_{X;Y}$, $\preccurlyeq_\emptyset$, and $\preccurlyeq_V$. The axiomatization of $\leqslant_F$ largely follows from clauses (1–4) of Definition 3.4, except that recursive session types are treated explicitly by rule [F-REC], which requires the condition

$T \unlhd_{\{x\};\emptyset} S$ that verifies whether it is safe to close T and S with the context $\mu x.[\ ]$ (see Lemma 3.1). Fair subtyping is defined by [A-SUBT], which is basically Theorem 3.2 in the form of inference rule. The axiomatization of convergence includes a core set of rules where [C-END], [C-INPUT], and [C-OUTPUT 1] enforce trace inclusion (condition $\text{tr}(T) \subseteq \text{tr}(S)$ in (1)) and rule [C-VAR] checks that x is not a dangerous variable (condition $Y\text{-tr}(S) = \emptyset$ in (1)). Rule [C-OUTPUT 2] deals with the case in which the larger session type provides strictly fewer choices with respect to the smaller one and corresponds to the "existential part" of the rule (1). In this case, there must be a common branch ($k \in I$) such that the corresponding continuations are in the convergence relation where all the safe variables have become dangerous ones ($T_k \unlhd_{\emptyset;X \cup Y} S_k$). Finally, rule [C-REC] deals with recursive contexts $\mu x.[\ ]$ by recording x as a safe variable.

Theorem 4.1 (Correctness). $\leqslant_A \subseteq \leqslant$.

The presented axiomatization is complete when session types have recursive terms binding the same variable in corresponding positions. We do not regard this as a limitation, though, because when $T \leqslant S$ it is always possible to find T' and S' that are strongly equivalent to T and S for which this property holds. For example, it is not possible to derive $\mu x.!a.x \leqslant_A \mu x.!a.!a.x$ using the rules in Table 2, but $\mu x.!a.!a.x \leqslant_A \mu x.!a.x \approx \mu x.!a.!a.x$. On the contrary, making this assumption allows us to focus on the interesting aspects of the axiomatization by leaving out some well-understood technicalities [2].

Theorem 4.2 (Completeness). *Let* $T \leqslant S$. *Then* $T \approx T' \leqslant_A S' \approx S$ *for some* T' *and* S'.

We briefly discuss an algorithm for deciding fair subtyping based on its axiomatization. The only two rules in Table 2 that are not syntax directed are [C-OUTPUT 1] and [C-OUTPUT 2] when $J \setminus I = \emptyset$ because the sets of variables may or may not change when going from the conclusion to the premise of these rules. A naive algorithm would have to backtrack in case the wrong choice is made, leading to exponential complexity. Table 3 presents an alternative set of syntax-directed rules for convergence. Space constraints prevent us from describing them in detail, but the guiding principle of these rules is simple: a judgment $T \#_X S \blacktriangleright Y$ *synthesizes*, whenever possible, the *smallest* subset Y of X such that $T \unlhd_{Y;X \setminus Y} S$ holds. This way, in [AC-OUTPUT 1] and [AC-OUTPUT 2] the index set X does not change from the conclusion to the premises, so the algorithm can just recur and then verify whether $T_k \#_X S_k \blacktriangleright \emptyset$ for some branch $k \in I$: if this is the case, then [AC-OUTPUT 2] applies; if not and $J \setminus I = \emptyset$, then [AC-OUTPUT 1] applies; otherwise, the algorithm fails. This new set of rules is sound and complete:

Theorem 4.3. *The following properties hold:*

1. $T \#_Y S \blacktriangleright X$ *implies* $T \unlhd_{X;Y \setminus X} S$;
2. $T \unlhd_{X;Y} S$ *implies* $T \#_{X \cup Y} S \blacktriangleright Z$ *and* $Z \subseteq X$.

Regarding the complexity of the proposed algorithm, observe that convergence can be decided in linear time using the rules in Table 3 and that, in Table 2, only [F-REC] and [A-SUBT] duplicate work. Moreover, rule [A-SUBT] is needed only once for each derivation. Therefore, the algorithm for fair subtyping is quadratic in the size of the proof tree for

Table 3. Algorithmic rules for convergence

$$
\begin{array}{ccc}
 & & \text{[AC-REC]} \\
\text{[AC-END]} & \text{[AC-VAR]} & \dfrac{T \,\#_{X \cup \{x\}}\, S \,\blacktriangleright\, Y}{\mu x.T \,\#_X\, \mu x.S \,\blacktriangleright\, Y \setminus \{x\}} \\
\text{end} \,\#_X\, \text{end} \,\blacktriangleright\, \emptyset & x \,\#_X\, x \,\blacktriangleright\, \{x\} \cap X &
\end{array}
$$

$$
\begin{array}{ccc}
\text{[AC-INPUT]} & \text{[AC-OUTPUT 1]} & \text{[AC-OUTPUT 2]} \\[4pt]
\dfrac{\forall i \in I : T_i \,\#_X\, S_i \,\blacktriangleright\, X_i}{\sum\limits_{i \in I} ?\mathsf{a}_i.T_i \,\#_X\, \sum\limits_{i \in I} ?\mathsf{a}_i.S_i \,\blacktriangleright\, \bigcup\limits_{i \in I} X_i} & \dfrac{\forall i \in I : (T_i \,\#_X\, S_i \,\blacktriangleright\, X_i \wedge X_i \neq \emptyset)}{\bigoplus\limits_{i \in I} !\mathsf{a}_i.T_i \,\#_X\, \bigoplus\limits_{i \in I} !\mathsf{a}_i.S_i \,\blacktriangleright\, \bigcup\limits_{i \in I} X_i} & \dfrac{\exists k \in I : T_k \,\#_X\, S_k \,\blacktriangleright\, \emptyset}{\bigoplus\limits_{i \in I \cup J} !\mathsf{a}_i.T_i \,\#_X\, \bigoplus\limits_{i \in I} !\mathsf{a}_i.S_i \,\blacktriangleright\, \emptyset}
\end{array}
$$

$T' \leqslant_{\mathsf{A}} S'$, which is the same as $\|S'\|$ (the number of distinct subtrees in S'). Since $\|S'\|$ in the (constructive) proof of Theorem 4.2 is bound by $\|T\| \cdot \|S\|$, the overall complexity for deciding $T \leqslant S$ is $O(n^4)$ where $n = \max\{\|T\|, \|S\|\}$.

5   Conclusion and Related Work

Subtyping is ubiquitous in session type systems, even when it is not explicitly mentioned in the type rules, and if the liveness of sessions is a major concern, the conventional subtyping relation for session types is inadequate. In this paper we have defined and characterized a liveness-preserving subtyping relation for possibly open session types that is a pre-congruence for all the operators of the type language.

Fair subtyping shares with the should-testing pre-congruence [12] its semantic definition (Definition 2.2), but differs in that should-testing is defined for a richer language of processes. In particular, our type language disallows parallel compositions inside T terms, while in [12] parallel composition can occur anywhere. This feature increases the discriminating power of tests in [12] and ultimately implies that $\mathsf{tr}(T) = \mathsf{tr}(S)$ when $T \leqslant S$. To see why, consider $T = !\mathsf{a} \oplus !\mathsf{b}$ and $S = !\mathsf{a}$ and the context $\mathscr{C} = \mu x.((?\mathsf{a}.x + ?\mathsf{b}.!\mathsf{done}) \mid [\,])$. The intuition is that the term $\mathscr{C}[T]$ "restarts" T if T chooses to emit a and terminates if T chooses to emit b. So, $?\mathsf{done}.!\mathsf{OK} \mid \mathscr{C}[T]$ is successful while $?\mathsf{done}.!\mathsf{OK} \mid \mathscr{C}[S]$ is not, determining $T \nleqslant S$. In our context, the ability to restart a process an arbitrary number of times is too powerful: session types are associated with *linear channels* which have exactly one owner at any given point in time. Also, the type of the channel reduces as the channel is used and the type system forbids "jumps" to previous stages of the protocol described by the session type, unless the session type itself allows to do so by means of an explicit recursion. If the condition $\mathsf{tr}(T) \subseteq \mathsf{tr}(S)$ were to hold also in our context, fair subtyping would collapse to equality (modulo folding/unfolding of recursions) and would therefore lack any interest whatsoever. We conjecture that the restriction of should-testing to finite-state processes (where parallel composition is forbidden underneath recursions) shares the same trace-related properties (and possibly the behavioral characterization) of fair subtyping.

Fair subtyping was introduced for the first time in [10], compared to which this work presents major differences. First of all, the subtyping relation in [10] is defined for closed session types only and essentially coincides with the should-testing relation in [12]. Previous works [9,12] have shown that the extension of liveness-preserving refinements to open terms is a challenging task with substantial impact on the properties of such refinements (this extension was left as an open problem in [9] and it was discovered to induce trace equivalence in [12]). By contrast, fair subtyping for open session types *does not* induce trace equivalence and turns out to be an original liveness-preserving pre-congruence that is not investigated elsewhere. Moreover, in the present work we purposefully adopt a notion of "session correctness" (Definition 2.1) that is weaker (i.e. more general) than the analogous notion in [10]. Since fair subtyping is defined as the relation that preserves success (Definition 2.2), the net effect is that the results presented here apply to all session type theories based on stronger notions of session correctness. Technically, the consequence is that all session types are *inhabited* and therefore the technique described in [10] based on session type difference is no longer applicable. By contrast, here we are able to give a direct definition of convergence with no need for auxiliary operators or notions of type emptiness.

Acknowledgments. This work was partially supported by a visiting professor position of the Université Paris Diderot, by MIUR PRIN 2010-2011 CINA, and by Ateneo/CSP Project SALT. The author is grateful to the anonymous referees for their comments and to Daniele Varacca and Viviana Bono for discussions on the topics of the paper.

References

1. Barbanera, F., de'Liguoro, U.: Two notions of sub-behaviour for session-based client/server systems. In: Kutsia, T., Schreiner, W., Fernández, M. (eds.) PPDP 2010, pp. 155–164. ACM (2010)
2. Brandt, M., Henglein, F.: Coinductive axiomatization of recursive type equality and subtyping. Fundamenta Informaticae 33(4), 309–338 (1998)
3. Castagna, G., Dezani-Ciancaglini, M., Giachino, E., Padovani, L.: Foundations of session types. In: Porto, A., López-Fraguas, F.J. (eds.) PPDP 2009, pp. 219–230. ACM (2009)
4. Castagna, G., De Nicola, R., Varacca, D.: Semantic subtyping for the pi-calculus. Theoretical Computer Science 398(1-3), 217–242 (2008)
5. Courcelle, B.: Fundamental properties of infinite trees. Theoretical Computer Science 25, 95–169 (1983)
6. Gay, S., Hole, M.: Subtyping for session types in the π-calculus. Acta Informatica 42(2-3), 191–225 (2005)
7. Honda, K.: Types for dyadic interaction. In: Best, E. (ed.) CONCUR 1993. LNCS, vol. 715, pp. 509–523. Springer, Heidelberg (1993)
8. Honda, K., Vasconcelos, V.T., Kubo, M.: Language primitives and type disciplines for structured communication-based programming. In: Hankin, C. (ed.) ESOP 1998. LNCS, vol. 1381, pp. 122–138. Springer, Heidelberg (1998)
9. Natarajan, V., Cleaveland, R.: Divergence and fair testing. In: Fülöp, Z. (ed.) ICALP 1995. LNCS, vol. 944, pp. 648–659. Springer, Heidelberg (1995)
10. Padovani, L.: Fair Subtyping for Multi-Party Session Types. In: De Meuter, W., Roman, G.-C. (eds.) COORDINATION 2011. LNCS, vol. 6721, pp. 127–141. Springer, Heidelberg (2011)
11. Pierce, B., Sangiorgi, D.: Typing and subtyping for mobile processes. Mathematical Structures in Computer Science 6(5), 409–453 (1996)
12. Rensink, A., Vogler, W.: Fair testing. Information and Computation 205(2), 125–198 (2007)

Coeffects: Unified Static Analysis of Context-Dependence

Tomas Petricek, Dominic Orchard, and Alan Mycroft

University of Cambridge, UK
{tp322,dao29,am}@cl.cam.ac.uk

Abstract. Monadic effect systems provide a unified way of tracking effects of computations, but there is no unified mechanism for tracking how computations rely on the environment in which they are executed. This is becoming an important problem for modern software – we need to track where distributed computations run, which resources a program uses and how they use other capabilities of the environment.

We consider three examples of context-dependence analysis: *liveness* analysis, tracking the use of *implicit parameters* (similar to tracking of *resource usage* in distributed computation), and calculating caching requirements for *dataflow* programs. Informed by these cases, we present a unified calculus for tracking context dependence in functional languages together with a categorical semantics based on *indexed comonads*. We believe that indexed comonads are the right foundation for constructing context-aware languages and type systems and that following an approach akin to monads can lead to a widespread use of the concept.

Modern applications run in diverse environments – such as mobile phones or the cloud – that provide additional resources and meta-data about provenance and security. For correct execution of such programs, it is often more important to understand how they *depend* on the environment than how they *affect* it.

Understanding how programs affect their environment is a well studied area: *effect systems* [13] provide a static analysis of effects and *monads* [8] provide a unified semantics to different notions of effect. Wadler and Thiemann unify the two approaches [17], *indexing* a monad with effect information, and showing that the propagation of effects in an effect system matches the semantic propagation of effects in the monadic approach.

No such unified mechanism exists for tracking the context requirements. We use the term *coeffect* for such contextual program properties. Notions of context have been previously captured using *comonads* [14] (the dual of monads) and by languages derived from *modal logic* [12,9], but these approaches do not capture many useful examples which motivate our work. We build mainly on the former comonadic direction (§3) and discuss the modal logic approach later (§5).

We extend a simply-typed lambda calculus with a *coeffect system* based on comonads, replicating the successful approach of effect systems and monads.

Examples of Coeffects. We present three examples that do not fit the traditional approach of effect systems and have not been considered using the modal

F.V. Fomin et al. (Eds.): ICALP 2013, Part II, LNCS 7966, pp. 385–397, 2013.

logic perspective, but can be captured as coeffect systems (§1) – the tracking of implicit dynamically-scoped parameters (or resources), analysis of variable liveness, and tracking the number of required past values in dataflow computations.

Coeffect Calculus. Informed by the examples, we identify a general algebraic structure for coeffects. From this, we define a general *coeffect calculus* that unifies the motivating examples (§2) and discuss its syntactic properties (§4).

Indexed Comonads. Our categorical semantics (§3) extends the work of Uustalu and Vene [14]. By adding annotations, we generalize comonads to *indexed comonads*, which capture notions of computation not captured by ordinary comonads.

1 Motivation

Effect systems, introduced by Gifford and Lucassen [5], track *effects* of computations, such as memory access or message-based communication [6]. Their approach augments typing judgments with effect information: $\Gamma \vdash e : \tau, F$. In Moggi's semantics, well-typed terms $\Gamma \vdash e : \tau$ are mapped to morphisms $[\![\Gamma]\!] \to M[\![\tau]\!]$ where M encodes effects and has the structure of a monad [8]. Wadler and Thiemann annotate monads with effect information, written M^F [17].

In contrast to the analysis of effects, our analysis of *context-dependence* differs in the treatment of lambda abstraction. Wadler and Thiemann explain that *"in the rule for abstraction, the effect is empty because evaluation immediately returns the function, with no side effects. The effect on the function arrow is the same as the effect for the function body, because applying the function will have the same side effects as evaluating the body"* [17]. We instead consider systems where λ-abstraction places requirements on both the *call-site* (latent requirements) and *declaration-site* (immediate requirements), resulting in different program properties. We informally discuss three examples that demonstrate how contextual requirements propagate. Section 2 unifies these in a single calculus.

We write coeffect judgements $C^s\Gamma \vdash e : \tau$ where the coeffect annotation s associates context requirements with the free-variable context Γ. Function types have the form $C^s\tau_1 \to \tau_2$ associating *latent* coeffects s with the parameter. The $C^s\Gamma$ syntax and $C^s\tau$ types are a result of the indexed comonadic semantics (§3).

Implicit Parameters and Resources. Implicit parameters [7] are *dynamically-scoped* variables. They can be used to parameterize a computation without propagating arguments explicitly through a chain of calls and are part of the context in which expressions evaluate. As correctly expected [7], they can be modelled by comonads. Rebindable resources in distributed computations (*e.g.*, a local clock) follow a similar pattern, but we discuss implicit parameters for simplicity.

The following function prints a number using implicit parameters ?culture (determining the decimal mark) and ?format (the number of decimal places):

$$\lambda n.\text{printNumber } n \text{ ?culture ?format}$$

$$(var) \quad \frac{x : \tau \in \Gamma}{C^\emptyset \Gamma \vdash x : \tau} \qquad (app) \quad \frac{C^r \Gamma \vdash e_1 : C^t \tau_1 \to \tau_2 \qquad C^s \Gamma \vdash e_2 : \tau_1}{C^{r \cup s \cup t} \Gamma \vdash e_1 \, e_2 : \tau_2}$$

$$(access) \quad \frac{}{C^{\{?a\}} \Gamma \vdash ?a : \rho} \qquad (abs) \quad \frac{C^{r \cup s} (\Gamma, x : \tau_1) \vdash e : \tau_2}{C^r \Gamma \vdash \lambda x.e : C^s \tau_1 \to \tau_2}$$

Fig. 1. Selected coeffect rules for implicit parameters

Figure 1 shows a type-and-coeffect system tracking the set of an expression's implicit parameters. For simplicity here, all implicit parameters have type ρ.

Context requirements are created in $(access)$, while (var) requires no implicit parameters; (app) combines requirements of both sub-expressions as well as the latent requirements of the function.

The (abs) rule is where the example differs from effect systems. Function bodies can access the union of the parameters (or resources) available at the declaration-site $(C^r \Gamma)$ and at the call-site $(C^s \tau_1)$. Two of the nine permissible judgements for the above example are:

$$C^\emptyset \Gamma \vdash (\ldots) : C^{\{?\text{culture}, ?\text{format}\}} \text{int} \to \text{string}$$
$$C^{\{?\text{culture}, ?\text{format}\}} \Gamma \vdash (\ldots) : C^{\{?\text{format}\}} \text{int} \to \text{string}$$

The coeffect system infers multiple, *i.e.* non-principal, coeffects for functions. Different judgments are desirable depending on how a function is used. In the first case, both parameters have to be provided by the caller. In the second, both are available at declaration-site, but ?format may be rebound (the precise meaning is provided by the semantics, discussed in §3).

Implicit parameters can be captured by the *reader* monad, where parameters are associated with the function codomain $M^\emptyset(\text{int} \to M^{\{?\text{culture}, ?\text{format}\}} \text{string})$, modelling only the first case. Whilst the reader monad can be extended to model rebinding, the next example cannot be structured by *any* monad.

Liveness Analysis. Liveness analysis detects whether a free variable of an expression may be used (*live*) or whether it is definitely not used (*dead*). A compiler can remove bindings to dead variables as the result is never used.

We start with a restricted analysis and briefly mention how to make it practical later (§5). The restricted form is interesting theoretically as it gives rise to the *indexed partiality comonad* (§3), which is a basic but instructive example.

The coeffect system in Fig. 2 detects whether all free variables are dead $(C^\mathsf{D} \Gamma)$ or whether at least one variable is live $(C^\mathsf{L} \Gamma)$. Variable use (var) is annotated with L and constants with D, *i.e.*, if $c \in \mathbb{N}$ then $C^\mathsf{D} \Gamma \vdash c : \text{int}$. A dead context may be marked as live by letting $\mathsf{D} \sqsubseteq \mathsf{L}$ and adding sub-coeffecting (§2).

$$(var) \quad \frac{x : \tau \in \Gamma}{C^\mathsf{L} \Gamma \vdash x : \tau} \qquad (app) \quad \frac{C^r \Gamma \vdash e_1 : C^t \tau_1 \to \tau_2 \qquad C^s \Gamma \vdash e_2 : \tau_1}{C^{r \sqcup (s \sqcap t)} \Gamma \vdash e_1 \, e_2 : \tau_2}$$

Fig. 2. Selected coeffect rules for liveness analysis

$$(var) \; \frac{x : \tau \in \Gamma}{C^0 \Gamma \vdash x : \tau} \qquad (app) \; \frac{C^m \Gamma \vdash e_1 : C^p \tau_1 \to \tau_2 \qquad C^n \Gamma \vdash e_2 : \tau_1}{C^{max(m, n+p)} \Gamma \vdash e_1 \; e_2 : \tau_2}$$

$$(prev) \; \frac{C^n \Gamma \vdash e : \tau}{C^{n+1} \Gamma \vdash \mathbf{prev} \; e : \tau} \qquad (abs) \; \frac{C^{min(m,n)} (\Gamma, x : \tau_1) \vdash e : \tau_2}{C^m \Gamma \vdash \lambda x.e : C^n \tau_1 \to \tau_2}$$

Fig. 3. Selected coeffect rules for causal data flow

The (app) rule can be understood by discussing its semantics. Consider semantic functions f, g, h annotated by r, s, t respectively. The *sequential composition* $g \circ f$ is live in its parameter only when both f and g are live. In the coeffect semantics, f is not evaluated if g ignores its parameter (regardless of evaluation order). Thus, $g \circ f$ is annotated by conjunction $r \sqcap s$ (where $\mathsf{L} \sqcap \mathsf{L} = \mathsf{L}$). A *pointwise composition* of g and h, passing the same parameter to both, is live in its parameter if either g or h is live (*i.e.*, disjunction $s \sqcup t$). Application uses both compositions, thus Γ is live if it is needed by e_1 *or* by the function *and* by e_2.

An (abs) rule (not shown) compatible with the structure in Fig. 1 combines the context annotations using $\sqcap$. Thus, if the body uses some variables, both the function argument and the context of the declaration-site are marked as live.

The coeffect system thus provides a call-by-name-style semantics, where redundant computations are omitted. Liveness cannot be modelled using monads with denotations $\tau_1 \to M^r \tau_2$. In call-by-value languages, the argument τ_1 is always evaluated. Using indexed comonads (§3), we model liveness as a morphism $C^r \tau_1 \to \tau_2$ where C^r is the parametric type $\mathsf{Maybe} \; \tau = \tau + 1$ (which contains a value τ when $r = \mathsf{L}$ and does not contain value when $r = \mathsf{D}$).

Efficient Dataflow. Dataflow languages (*e.g.*, Lucid [16]) declaratively describe computations over streams. In *causal* data flow, programs may access past values. In this setting, a function $\tau_1 \to \tau_2$ becomes a function from a list of historical values $[\tau_1] \to \tau_2$. A coeffect system here tracks how many past values to cache.

Figure 3 annotates contexts with an integer specifying the maximum number of required past values. The current value is always present, so (var) is annotated with 0. The expression $\mathbf{prev} \; e$ gets the previous value of stream e and requires one additional past value $(prev)$; *e.g.* $\mathbf{prev} \; (\mathbf{prev} \; e)$ requires 2 past values.

The (app) rule follows the same intuition as for liveness. Sequential composition adds the tags (the first function needs $n + p$ past values to produce p past inputs for the second function); passing the context to two subcomputations requires the maximum number of the elements required by the two subcomputations. The (abs) rule for data-flow needs a distinct operator – min – therefore, the declaration-site and call-site must each provide at least the number of past values required by the function body (as the body may use variables coming from the declaration-site as well as the argument).

Soundness follows from our categorical model (§3). Uustalu and Vene model causal dataflow by a non-empty list comonad $\mathsf{NeList} \; \tau = \tau \times (\mathsf{NeList} \; \tau + 1)$ [14]. However, this model leads to (inefficient) unbounded lists of past elements.

The coeffect system above infers a (sound) over-approximation of the number of required past elements and so fixed-length lists may be used instead.

2   Generalized Coeffect Calculus

The previous three examples exhibit a number of commonalities. We capture these in the *coeffect calculus*. We do not overly restrict the calculus to allow for notions of context-dependent computations not discussed above.

The syntax of our calculus is that of the simply-typed lambda calculus (where v ranges over variables, T over base types, and r over coeffect annotations):

$$e ::= v \mid \lambda v.e \mid e_1\ e_2 \qquad\qquad \tau ::= \mathsf{T} \mid \tau_1 \to \tau_2 \mid C^r \tau$$

The type $C^r \tau$ captures values of type τ in a context specified by the annotation r. This type appears only on the left-hand side of a function arrow $C^r \tau_1 \to \tau_2$. In the semantics, C^r corresponds to some data type (*e.g.*, List or Maybe). Extensions such as explicit *let*-binding are discussed later (§4).

The coeffect tags r, that were demonstrated in the previous section, can be generalized to a structure with three binary operators and a particular element.

Definition 1. *A* coeffect algebra $(S, \oplus, \vee, \wedge, \mathsf{e})$ *is a set* S *with an element* $\mathsf{e} \in S$, *a semi-lattice* $(S, \vee)$, *a monoid* $(S, \oplus, \mathsf{e})$, *and a binary* $\wedge$. *That is,* $\forall r, s, t \in S$:

$$r \oplus (s \oplus t) = (r \oplus s) \oplus t \qquad \mathsf{e} \oplus r = r = r \oplus \mathsf{e} \qquad\qquad \text{(monoid)}$$

$$r \vee s = s \vee r \qquad r \vee (s \vee t) = (r \vee s) \vee t \qquad r \vee r = r \qquad \text{(semi-lattice)}$$

The generalized coeffect calculus captures the three motivating examples (§1), where some operators of the coeffect algebra may coincide.

The $\oplus$ operator represents *sequential* composition; guided by the categorical model (§3), we require it to form a monoid with e. The operator $\vee$ corresponds to merging of context requirements in *pointwise composition* and the semi-lattice $(S, \vee)$ defines a partial order: $r \leq s$ when $r \vee s = s$. This ordering implies a sub-coeffecting rule. The coeffect e is often the top or bottom of the lattice.

The $\wedge$ operator corresponds to splitting requirements of a function body between the call- and definition-site. This operator is unrestricted in the general system, though it has additional properties in *some* coeffects systems, *e.g.*, semi-lattice structure on $\wedge$. Possibly these laws should hold for all coeffect systems, but we start with as few laws as possible to avoid limiting possible uses of the calculus. We consider constrained variants with useful properties later (§4).

Implicit parameters use sets of names $S = \mathcal{P}(\mathsf{Id})$ as tags with union $\cup$ for all three operators. Variable use is annotated with $\mathsf{e} = \emptyset$ and $\leq$ is subset ordering.

Liveness uses a two point lattice $S = \{\mathsf{D}, \mathsf{L}\}$ where $\mathsf{D} \sqsubseteq \mathsf{L}$. Variables are annotated with the top element $\mathsf{e} = \mathsf{L}$ and constants with bottom D. The $\vee$ operation is $\sqcup$ (join) and $\wedge$ and $\oplus$ are both $\sqcap$ (meet).

$$(var) \ \frac{x : \tau \in \Gamma}{C^e \Gamma \vdash x : \tau} \qquad (app) \ \frac{C^r \Gamma \vdash e_1 : C^t \tau_1 \to \tau_2 \quad C^s \Gamma \vdash e_2 : \tau_1}{C^{r \vee (s \oplus t)} \Gamma \vdash e_1 \ e_2 : \tau_2}$$

$$(sub) \ \frac{C^s \Gamma \vdash e : \tau}{C^r \Gamma \vdash e : \tau} \ (s \le r) \qquad (abs) \ \frac{C^{r \wedge s}(\Gamma, x : \tau_1) \vdash e : \tau_2}{C^r \Gamma \vdash \lambda x.e : C^s \tau_1 \to \tau_2}$$

Fig. 4. Type and coeffect system for the coeffect calculus

Dataflow tags are natural numbers $S = \mathbb{N}$ and operations $\vee, \wedge$ and $\oplus$ correspond to *max*, *min* and $+$, respectively. Variable use is annotated with $e = 0$ and the order $\le$ is the standard ordering of natural numbers.

Coeffect Typing Rules. Figure 4 shows the rules of the coeffect calculus, given some coeffect algebra $(S, \oplus, \vee, \wedge, e)$. The context required by a variable (*var*) is annotated with e. The sub-coeffecting rule (*sub*) allows the contextual requirements of an expression to be generalized.

The (*abs*) rule checks the body of the function in a context $r \wedge s$, which is a combination of the coeffects available in the context r where the function is defined and in a context s provided by the caller of the function. Note that none of the judgements create a *value* of type $C^r \tau$. This type appears only immediately to the left of an arrow $C^r \tau_1 \to \tau_2$.

In function application (*app*), context requirements of both expressions and the function are combined as previously: the pointwise composition $\vee$ is used to combine the coeffect r of the expression representing a function and the coeffects of the argument, sequentially composed with the coeffects of the function: $s \oplus t$.

For space reasons, we omit recursion. We note that this would require adding coeffect variables and extending the coeffect algebra with a fixed-point operation.

3 Coeffect Semantics Using Indexed Comonads

The approach of *categorical semantics* interprets terms as morphisms in some category. For typed calculi, typing judgments $x_1 : \tau_1 \ldots x_n : \tau_n \vdash e : \tau$ are usually mapped to morphisms $[\![\tau_1]\!] \times \ldots \times [\![\tau_n]\!] \to [\![\tau]\!]$. Moggi showed the semantics of various effectful computations can be captured generally using the (*strong*) *monad* structure [8]. Dually, Uustalu and Vene showed that (*monoidal*) *comonads* capture various kinds of context-dependent computation [14].

We extend Uustalu and Vene's approach to give a semantics for the coeffect calculus by generalising comonads to *indexed comonads*. We emphasise semantic intuition and abbreviate the categorical foundations for space reasons.

Indexed Comonads. Uustalu and Vene's approach interprets well-typed terms as morphisms $C(\tau_1 \times \ldots \times \tau_n) \to \tau$, where C encodes contexts and has a comonad structure [14]. Indexed comonads comprise a *family* of object mappings C^r indexed by a coeffect r describing the contextual requirements satisfied by the encoded context. We interpret judgments $C^r(x_1 : \tau_1, \ldots, x_n : \tau_n) \vdash e : \tau$ as morphisms $C^r([\![\tau_1]\!] \times \ldots \times [\![\tau_n]\!]) \to [\![\tau]\!]$.

The indexed comonad structure provides a notion of composition for computations with different contextual requirements.

Definition 2. *Given a monoid* $(S, \oplus, \mathsf{e})$ *with binary operator* $\oplus$ *and unit* e, *an* indexed comonad *over a category* $\mathcal{C}$ *comprises a family of object mappings* C^r *where for all* $r \in S$ *and* $A \in \mathrm{obj}(\mathcal{C})$ *then* $C^r A \in \mathrm{obj}(\mathcal{C})$ *and:*

- *a natural transformation* $\varepsilon_A : C^{\mathsf{e}} A \to A$, *called the* counit;
- *a family of mappings* $(-)^{\dagger}_{r,s}$ *from morphisms* $C^r A \to B$ *to* *morphisms* $C^{r \oplus s} A \to C^s B$ *in* $\mathcal{C}$, *natural in* A, B, *called* coextend;

such that for all $f : C^r \tau_1 \to \tau_2$ *and* $g : C^s \tau_2 \to \tau_3$ *the following equations hold:*

$$\varepsilon \circ f^{\dagger}_{r,\mathsf{e}} = f \qquad (\varepsilon)^{\dagger}_{\mathsf{e},r} = \mathsf{id} \qquad (g \circ f^{\dagger}_{r,s})^{\dagger}_{(r \oplus s),t} = g^{\dagger}_{s,t} \circ f^{\dagger}_{r,(s \oplus t)}$$

The *coextend* operation gives rise to an associative composition operation for computations with contextual requirements (with *counit* as the identity):

$$\hat{\circ} : (C^r \tau_1 \to \tau_2) \to (C^s \tau_2 \to \tau_3) \to (C^{r \oplus s} \tau_1 \to \tau_3) \qquad g \,\hat{\circ}\, f = g \circ f^{\dagger}_{r,s}$$

The composition $\hat{\circ}$ best expresses the intention of indexed comonads: contextual requirements of the composed functions are combined. The properties of the composition follow from the indexed comonad laws and the monoid $(S, \oplus, \mathsf{e})$.

Example 1. Indexed comonads are analogous to comonads (in coKleisli form), but with the additional monoidal structure on indices. Indeed, comonads are a special case of indexed comonads with a trivial singleton monoid, *e.g.*, $(\{1\}, *, 1)$ with $1 * 1 = 1$ where C^1 is the underlying functor of the comonad and ε and $(-)^{\dagger}_{1,1}$ are the usual comonad operations. However, as demonstrated next, not all indexed comonads are derived from ordinary comonads.

Example 2. The *indexed partiality comonad* encodes free-variable contexts of a computation which are either *live* or *dead* (*i.e.*, have *liveness* coeffects) with the monoid $(\{\mathsf{D}, \mathsf{L}\}, \sqcap, \mathsf{L})$, where $C^{\mathsf{L}} A = A$ encodes live contexts and $C^{\mathsf{D}} A = 1$ encodes dead contexts, where 1 is the unit type inhabited by a single value (). The *counit* operation $\varepsilon : C^{\mathsf{L}} A \to A$ and *coextend* operations $f^{\dagger}_{r,s} : C^{r \sqcap s} A \to C^s B$ (for all $f : C^r A \to B$), are defined:

$$\varepsilon\, x = x \qquad f^{\dagger}_{\mathsf{D},\mathsf{D}} x = () \qquad f^{\dagger}_{\mathsf{D},\mathsf{L}} x = f() \qquad f^{\dagger}_{\mathsf{L},\mathsf{D}} x = () \qquad f^{\dagger}_{\mathsf{L},\mathsf{L}} x = f\, x$$

The indexed family C^r here is analogous to the non-indexed Maybe (or *option*) data type Maybe $A = A + 1$. This type does not permit a comonad structure since $\varepsilon : $ Maybe $A \to A$ is undefined at $(\mathsf{inr}\, ())$. For the indexed comonad, ε need only be defined for $C^{\mathsf{L}} A = A$. Thus, indexed comonads capture a broader range of contextual notions of computation than comonads.

Moreover, indexed comonads are not restricted by the *shape preservation* property of comonads [11]: that a coextended function cannot change the *shape* of the context. For example, in the second case above $f^{\dagger}_{\mathsf{D},\mathsf{L}} : C^{\mathsf{D}} A \to C^{\mathsf{L}} B$ where the shape changes from 1 (empty context) to B (available context).

$$\llbracket C^r \Gamma \vdash \lambda x.e : C^s \tau_1 \to \tau_2 \rrbracket = curry\,(\llbracket C^{r \wedge s}(\Gamma, x : \tau_1) \vdash e : \tau_2 \rrbracket \circ m_{r,s})$$
$$\llbracket C^{r \vee (s \oplus t)} \Gamma \vdash e_1\ e_2 : \tau \rrbracket = (uncurry\,\llbracket C^r \Gamma \vdash e_1 : C^t \tau_1 \to \tau_2 \rrbracket) \circ$$
$$(id \times \llbracket C^s \Gamma \vdash e_2 : \tau_1 \rrbracket_{s,t}^\dagger) \circ n_{r,s \oplus t} \circ C^{r \vee (s \oplus t)} \Delta$$
$$\llbracket C^e \Gamma \vdash x_i : \tau_i \rrbracket = \pi_i \circ \varepsilon$$

Fig. 5. Categorical semantics for the coeffect calculus

Monoidal Indexed Comonads. Indexed comonads provide a semantics to sequential composition, but additional structure is needed for the semantics of the full coeffect calculus. Uustalu and Vene [14] additionally require a (*lax semi-*)*monoidal comonad* structure, which provides a monoidal operation m : $CA \times CB \to C(A \times B)$ for merging contexts (used in the semantics of abstraction).

The semantics of the coeffect calculus requires an indexed lax semi-monoidal structure for combining contexts *as well as* an indexed *colax* monoidal structure for *splitting* contexts. These are provided by two families of morphisms (given a coeffect algebra with $\vee$ and $\wedge$):

- $m_{r,s} : C^r A \times C^s B \to C^{(r \wedge s)}(A \times B)$ natural in A, B;
- $n_{r,s} : C^{(r \vee s)}(A \times B) \to C^r A \times C^s B$ natural in A, B;

The $m_{r,s}$ operation merges contextual computations with tags combined by $\wedge$ (greatest lower-bound), elucidating the behaviour of $m_{r,s}$: that merging may result in the loss of some parts of the contexts r and s.

The $n_{r,s}$ operation splits context-dependent computations and thus the contextual requirements. To obtain coeffects r and s, the input needs to provide *at least* r and s, so the tags are combined using the $\vee$ operator (least upper-bound).

For the sake of brevity, we elide the indexed versions of the laws required by Uustalu and Vene (*e.g.*, most importantly, merging and splitting are associative).

Example 3. For the indexed partiality comonad, given the liveness coeffect algebra $(\{D, L\}, \sqcap, \sqcup, \sqcap, L)$, the additional lax/colax monoidal operations are:

$$m_{L,L}(x, y) = (x, y) \quad n_{D,D}\ () \quad = ((), ()) \quad n_{D,L}(x, y) = ((), y)$$
$$m_{r,s}\ (x, y) = () \qquad n_{L,D}(x, y) = (x, ()) \quad n_{L,L}(x, y) = (x, y)$$

Example 4. Uustalu and Vene model causal dataflow computations using the non-empty list comonad NEList $A = A \times (1 + \text{NEList } A)$ [14]. Whilst this comonad implies a trivial indexed comonad, we define an indexed comonad with integer indices for the number of past values demanded of the context.

We define $C^n A = A \times (A \times \ldots \times A)$ where the first A is the current (always available) value, followed by a finite product of n past values. The definition of the operations is a straightforward extension of the work of Uustalu and Vene.

Categorical Semantics. Figure 5 shows the categorical semantics of the coeffect calculus using additional operations π_i for projection of the ith element of a product, usual *curry* and *uncurry* operations, and $\Delta : A \to A \times A$ duplicating a value. While C^r is a family of object mappings, it is promoted to a family of functors with the derived morphism mapping $C^r(f) = (f \circ \varepsilon)_{e,r}^\dagger$.

The semantics of variable use and abstraction are the same as in Uustalu and Vene's semantics, modulo coeffects. Abstraction uses $\mathsf{m}_{r,s}$ to merge the outer context with the argument context for the context of the function body. The indices of e for ε and r, s for $\mathsf{m}_{r,s}$ match the coeffects of the terms. The semantics of application is more complex. It first duplicates the free-variable values inside the context and then splits this context using $\mathsf{n}_{r,s\oplus t}$. The two contexts (with different coeffects) are passed to the two sub-expressions, where the argument subexpression, passed a context $(s \oplus t)$, is coextended to produce a context t which is passed into the parameter of the function subexpression ($cf.$ given $f : A \to (B \to C)$, $g : A \to B$, then $uncurry\, f \circ (id \times g) \circ \Delta : A \to C$).

A semantics for sub-coeffecting is omitted, but may be provided by an operation $\iota_{r,s} : C^r A \to C^s A$ natural in A, for all $r, s \in S$ where $s \leq r$, which transforms a value $C^r A$ to $C^s A$ by ignoring some of the encoded context.

4   Syntax-Based Equational Theory

The operational semantics of every context-dependent language here differs as the notion of context is always different. However, for coeffect calculi satisfying certain conditions we can define a universal equational theory. This suggests a pathway to an operational semantics for two out of our three examples (the notion of context for data-flow is more complex).

In a pure λ-calculus, β- and η-equality for functions (also called *local soundness* and *completeness* respectively [12]) describe how pairs of abstraction and application can be eliminated: $(\lambda x.e_2)e_1 \equiv_\beta e_1[x \leftarrow e_2]$ and $(\lambda x.e\,x) \equiv_\eta e$. The β-equality rule, using the usual Barendregt convention of syntactic substitution, implies a *reduction*, giving part of an operational semantics for the calculus.

The call-by-name evaluation strategy modelled by β-reduction is not suitable for impure calculi therefore a restricted β rule, corresponding to call-by-value, is used, *i.e.* $(\lambda x.e_2)v \equiv e_2[x \leftarrow v]$. Such reduction can be encoded by a *let*-binding term, **let** $x = e_1$ **in** e_2, which corresponds to sequential composition of two computations, where the resulting pure value of e_1 is substituted into e_2 [4,8].

For an equational theory of coeffects, consider first a notion of *let*-binding equivalent to $(\lambda x.e_2)\, e_1$, which has the following type and coeffect rule:

$$\frac{C^s \Gamma \vdash e_1 : \tau_1 \qquad C^{r_1 \wedge r_2}(\Gamma, x : \tau_1) \vdash e_2 : \tau_2}{C^{r_1 \vee (r_2 \oplus s)} \Gamma \vdash \mathbf{let}\ x = e_1\ \mathbf{in}\ e_2 : \tau_2} \tag{1}$$

For our examples, $\wedge$ is idempotent (*i.e.*, $r \wedge r = r$) implying a simpler rule:

$$\frac{C^s \Gamma \vdash e_1 : \tau_1 \qquad C^r(\Gamma, x : \tau_1) \vdash e_2 : \tau_2}{C^{r \vee (r \oplus s)} \Gamma \vdash \mathbf{let}\ x = e_1\ \mathbf{in}\ e_2 : \tau_2} \tag{2}$$

For our examples (but not necessarily *all* coeffect systems), this defines a more "precise" coeffect with respect to $\leq$ where $r \vee (r \oplus s) \leq r_1 \vee (r_2 \oplus s)$.

This rule removes the non-principality of the first rule (*i.e.*, multiple possible typings). However, using idempotency to split coeffects in abstraction would remove additional flexibility needed by the implicit parameters example.

The coeffect $r \vee (r \oplus s)$ can also be simplified for all our examples, leading to more intuitive rules – for implicit parameters $r \cup (r \cup s) = r \cup s$; for liveness we get that $r \sqcup (r \sqcap s) = r$ and for dataflow we obtain $max(r, r + s) = r + s$.

Our calculus can be extended with *let*-binding and (2). However, we also consider the cases when a syntactic substitution $e_2[x \leftarrow e_1]$ has the coeffects specified by the above rule (2) and prove *subject reduction* theorem for certain coeffect calculi. We consider two common special cases when the coeffect of variables e is the greatest ($\top$) or least ($\bot$) element of the semi-lattice $(S, \vee)$ and derive additional properties that hold about the coeffect algebra:

Lemma 1 (Substitution). *Given $C^r(\Gamma, x : \tau_2) \vdash e_1 : \tau_1$ and $C^s\Gamma \vdash e_2 : \tau_2$ then $C^{r \vee (r \oplus s)}\Gamma \vdash e_2[x \leftarrow e_1] : \tau_1$ if the coeffect algebra satisfies the conditions that e is either the greatest or least element of the semi-lattice, $\oplus = \wedge$, and $\oplus$ distributes over $\vee$, i.e., $X \oplus (Y \vee Z) = (X \oplus Y) \vee (X \oplus Z)$.*

Proof. By induction over $\vdash$, using the laws (§2) and additional assumptions. ☐

Assuming $\rightarrow_\beta$ is the usual call-by-name reduction, the following theorem models the evaluation of coeffect calculi with coeffect algebra that satisfies the above requirements. We do not consider *call-by-value*, because our calculus does not have a notion of *value*, unless explicitly provided by *let*-binding (even a function "value" $\lambda x.e$ may have immediate contextual requirements).

Theorem 1 (Subject Reduction). *For a coeffect calculus, satisfying the conditions of Lemma 1, if $C^r\Gamma \vdash e : \tau$ and $e \rightarrow_\beta e'$ then $C^r\Gamma \vdash e' : \tau$.*

Proof. A direct consequence of Lemma 1.   ☐

The above theorem holds for both the liveness and resources examples, but not for dataflow. In the case of liveness, e is the greatest element ($r \vee e = e$); in the case of resources, e is the *least* element ($r \vee e = r$) and the proof relies on the fact that additional context requirements can be placed at the context $C^r\Gamma$ (without affecting the type of function when substituted under λ-abstraction).

However, the coeffect calculus also captures context-dependence in languages with more complex evaluation strategies than *call-by-name* reduction based on syntactic substitution. In particular, syntactic substitution does not provide a suitable evaluation for dataflow (because a substituted expression needs to capture the context of the original scope).

Nevertheless, the above results show that – unlike effects – context-dependent properties can be integrated with *call-by-name* languages. Our work also provides a model of existing work, namely Haskell implicit parameters [7].

5   Related and Further Work

This paper follows the approaches of effect systems [5,13,17] and categorical semantics based on monads and comonads [8,14]. Syntactically, *coeffects* differ from *effects* in that they model systems where λ-abstraction may split contextual requirements between the declaration-site and call-site.

Our *indexed (monoidal) comonads* (§3) fill the gap between (non-indexed) *(monoidal) comonads* of Uustalu and Vene [14] and indexed monads of Atkey [2], Wadler and Thiemann [17]. Interestingly, *indexed* comonads are *more general* than comonads, capturing more notions of context-dependence (§1).

Comonads and Modal Logics. Bierman and de Paiva [3] model the □ modality of an intuitionistic S4 modal logic using monoidal comonads, which links our calculus to modal logics. This link can be materialized in two ways.

Pfenning et al. and Nanevski et al. derive term languages using the Curry-Howard correspondence [12,3,9], building a *metalanguage* (akin to Moggi's monadic metalanguage [8]) that includes □ as a type constructor. For example, in [12], the modal type □τ represents closed terms. In contrast, the *semantic* approach uses monads or comonads *only* in the semantics. This has been employed by Uustalu and Vene and (again) Moggi [8,14]. We follow the semantic approach.

Nanevski et al. extend an S4 term language to a *contextual* modal type theory (CMTT) [9]. The *context* is a set of variables required by a computation, which makes CMTT useful for meta-programming and staged computations. Our contextual types are indexed by a coeffect algebra, which is more general and can capture variable contexts, but also integers, two-point lattices, *etc.*.

The work on CMTT suggests two extensions to coeffects. The first is developing the logical foundations. We briefly considered special cases of our system that permits local soundness in §4; local completeness can be treated similarly. The second is developing a coeffect *metalanguage*. The use of coeffect algebras provides an additional flexibility over CMTT, allowing a wider range of applications via a richer metalanguage.

Relating Effects and Coeffects. The difference between effects and coeffects is mainly in the (*abs*) rule. While the semantic models (monads vs. comonads) are different, they can be extended to obtain equivalent syntactic rules. To allow splitting of implicit parameters in lambda abstraction, the reader monad needs an operation that eagerly performs some effects of a function: $(\tau_1 \to M^{r \oplus s}\tau_2) \to M^r(\tau_1 \to M^s\tau_2)$. To obtain a pure lambda abstraction for coeffects, we need to restrict the $\mathsf{m}_{r,s}$ operation of indexed comonads, so that the first parameter is annotated with e (meaning no effects): $C^{\mathsf{e}}A \times C^r B \to C^r(A \times B)$.

Structural Coeffects. To make the liveness analysis practical, we need to associate information with individual variables (rather than the entire context). We can generalize the calculus from this paper by adding a product operation × to the coeffect algebra. A variable context $x : \tau_1, y : \tau_2, z : \tau_3$ is then annotated with $r \times s \times t$ where each component of the tag corresponds to a single variable. The system is then extended with structural rules such as:

$$(abs)\frac{C^{r \times s}(\Gamma, x : \tau_1) \vdash e : \tau_2}{C^r \Gamma \vdash \lambda x.e : C^s \tau_1 \to \tau_2} \qquad (contr)\frac{C^{r \times s}(x : \tau_1, y : \tau_1) \vdash e : \tau_2}{C^{r \vee s}(z : \tau_1) \vdash e[x \leftarrow z][y \leftarrow z] : \tau_2}$$

The context requirements associated with function are exactly those linked to the specific variable of the lambda abstraction. Rules such as contraction manipulate variables and perform a corresponding operation on the indices.

The structural coeffect system is related to bunched typing [10] (but generalizes it by adding indices). We are currently investigating how to use structural coeffects to capture fine-grained context-dependence properties such as secure information flow [15] or, more generally, those captured by the dependency core calculus [1].

6   Conclusions

We examined three simple calculi with associated coeffect systems (liveness analysis, implicit parameters, and dataflow analysis). These were unified in the *coeffect calculus*, providing a general coeffect system parameterised by an algebraic structure describing propagation of context requirements throughout a program.

We model the semantics of the coeffect calculus using the *indexed (monoidal) comonad* structure – a novel structure, which is more powerful than (monoidal) comonads. Indices of the indexed comonad operations manifest the semantic propagation of context so that the propagation of information in the general coeffect type system corresponds exactly to the semantic propagation of context in our categorical model.

We consider the analysis of context to be essential, not least for the examples here but also given increasingly rich and diverse distributed systems.

Acknowledgements. We thank Gavin Bierman, Tarmo Uustalu, Varmo Vene and reviewers of earlier drafts. This work was supported by the EPSRC and CHESS.

References

1. Abadi, M., Banerjee, A., Heintze, N., Riecke, J.G.: A core calculus of dependency. In: Proceedings of POPL (1999)
2. Atkey, R.: Parameterised notions of computation. J. Funct. Program. 19 (2009)
3. Bierman, G.M., de Paiva, V.C.V.: On an intuitionistic modal logic. Studia Logica 65, 2000 (2001)
4. Filinski, A.: Monads in action. In: Proceedings of POPL (2010)
5. Gifford, D.K., Lucassen, J.M.: Integrating functional and imperative programming. In: Proceedings of Conference on LISP and func. prog., LFP 1986 (1986)
6. Jouvelot, P., Gifford, D.K.: Communication Effects for Message-Based Concurrency. Technical report, Massachusetts Institute of Technology (1989)
7. Lewis, J.R., Shields, M.B., Meijert, E., Launchbury, J.: Implicit parameters: dynamic scoping with static types. In: Proceedings of POPL, POPL 2000 (2000)
8. Moggi, E.: Notions of computation and monads. Inf. Comput. 93, 55–92 (1991)
9. Nanevski, A., Pfenning, F., Pientka, B.: Contextual modal type theory. ACM Trans. Comput. Logic 9(3), 23:1–23:49 (2008)

10. O'Hearn, P.: On bunched typing. J. Funct. Program. 13(4), 747–796 (2003)
11. Orchard, D., Mycroft, A.: A Notation for Comonads. In: Post-Proceedings of IFL 2012. LNCS. Springer, Heidelberg (2012) (to appear)
12. Pfenning, F., Davies, R.: A judgmental reconstruction of modal logic. Mathematical. Structures in Comp. Sci. 11(4), 511–540 (2001)
13. Talpin, J., Jouvelot, P.: The type and effect discipline. In: Logic in Computer Science, 1992. LICS, pp. 162–173 (1994)
14. Uustalu, T., Vene, V.: Comonadic Notions of Computation. Electron. Notes Theor. Comput. Sci. 203, 263–284 (2008)
15. Volpano, D., Irvine, C., Smith, G.: A sound type system for secure flow analysis. J. Comput. Secur. 4, 167–187 (1996)
16. Wadge, W.W., Ashcroft, E.A.: LUCID, the dataflow programming language. Academic Press Professional, Inc., San Diego (1985)
17. Wadler, P., Thiemann, P.: The marriage of effects and monads. ACM Trans. Comput. Logic 4, 1–32 (2003)

Proof Systems for Retracts in Simply Typed Lambda Calculus

Colin Stirling

School of Informatics
University of Edinburgh
cps@inf.ed.ac.uk

Abstract. This paper[1] concerns retracts in simply typed lambda calculus assuming $\beta\eta$-equality. We provide a simple tableau proof system which characterises when a type is a retract of another type and which leads to an exponential decision procedure.

1 Introduction

Type ρ is a retract of type τ if there are functions $C : \rho \to \tau$ and $D : \tau \to \rho$ with $D \circ C = \lambda x.x$. This paper concerns retracts in the case of simply typed lambda calculus [1]. Various questions can be asked. The decision problem is: given ρ and τ, is ρ a retract of τ? Is there an independent characterisation of when ρ is a retract of τ? Is there an inductive method, such as a proof system, for deriving assertions of the form "ρ is a retract of τ"? If so, can one also construct (inductively) the witness functions C and D?

Bruce and Longo [2] provide a simple proof system that solves when there are retracts in the case that $D \circ C =_\beta \lambda x.x$. The problem is considerably more difficult if β-equality is replaced with $\beta\eta$-equality. De Liguoro, Piperno and Statman [3] show that the retract relation with respect to $\beta\eta$-equality coincides with the surjection relation: ρ is a retract of τ iff for any model there is a surjection from τ to ρ. They also provide a proof system for the affine case (when each variable in C and D occurs at most once) assuming a single ground type. Regnier and Urzyczyn [9] extend this proof system to cover multiple ground types. The proof systems yield simple inductive nondeterministic algorithms belonging to NP for deciding whether ρ is an affine retract of τ. Schubert [10] shows that the problem of affine retraction is NP-complete and how to derive witnesses C and D from the proof system in [9]. Under the assumption of a single ground type, decidability of when ρ is a retract of τ is shown by Padovani [8] by explicit witness construction (rather than by a proof system) of a special form.

More generally, decidability of the retract problem follows from decidability of higher-order matching in simply typed lambda calculus [13]: ρ is a retract of τ iff the equation $\lambda z^\rho.x_1^{\tau \to \rho}(x_2^{\rho \to \tau} z) =_{\beta\eta} \lambda z^\rho.z$ has a solution (the witnesses D and C for x_1, x_2). Since the complexity of matching is non-elementary [15] this decidability result leaves open whether there is a better algorithm, or even a proof

---

[1] For a full version see http://www.homepages.inf.ed.ac.uk/cps/ret.pdf

F.V. Fomin et al. (Eds.): ICALP 2013, Part II, LNCS 7966, pp. 398–409, 2013.

system, for the problem. In the case of β-equality matching is no guide to solvability: the retract problem is simply solvable whereas β-matching is undecidable [4].

In this paper we provide an independent solution to the retract problem. We show it is decidable by exhibiting sound and complete tableau proof systems. We develop two proof systems for retracts, one for the (slightly easier) case when there is a single ground type and the other for when there are multiple ground types. Both proof systems appeal to paths in terms. Their correctness depend on properties of such paths. We appeal to a dialogue game between witnesses of a retract to prove such properties: a similar game-theoretic characterisation of β-reduction underlies decidability of matching.

In Section 2 we introduce retracts in simply typed lambda calculus and fix some notation for terms as trees and for their paths. The two tableau proof systems for retracts are presented in Section 3 where we also briefly examine how they generate a decision procedure for the retract problem. In Section 4 we sketch the proof of soundness of the tableau proof systems (and completeness and further details are provided in the full version).

2   Preliminaries

Simple types are generated from ground types using the binary function operator $\rightarrow$. We let $a, b, o, \ldots$ range over ground types and $\rho, \sigma, \tau, \ldots$ range over simple types. Assuming $\rightarrow$ associates to the right, so $\rho \rightarrow \sigma \rightarrow \tau$ is $\rho \rightarrow (\sigma \rightarrow \tau)$, if a type ρ is not a ground type then it has the form $\rho_1 \rightarrow \ldots \rightarrow \rho_n \rightarrow a$. We say that a is the *target* type of a and of any type $\rho_1 \rightarrow \ldots \rightarrow \rho_n \rightarrow a$.

Simply typed terms in Church style are generated from a countable set of typed variables x^σ using lambda abstraction and function application [1]. We write S^σ, or sometimes $S : \sigma$, to mean term S has type σ. The usual typing rules hold: if S^τ then $\lambda x^\sigma . S^\tau : \sigma \rightarrow \tau$; if $S^{\sigma \rightarrow \tau}$ and U^σ then $(S^{\sigma \rightarrow \tau} U^\sigma) : \tau$. In a sequence of unparenthesised applications we assume that application associates to the left, so $S U_1 \ldots U_k$ is $((\ldots (S U_1) \ldots) U_k)$. Another abbreviation is $\lambda z_1 \ldots z_m$ for $\lambda z_1 \ldots \lambda z_m$. Usual definitions of when a variable occurrence is free or bound and when a term is closed are assumed.

We also assume the usual dynamics of β and η-reductions and the consequent $\beta\eta$-equivalence between terms (as well as α-equivalence). Confluence and strong normalisation ensure that terms reduce to (unique) normal forms. Moreover, we assume the standard notion of η-long β-normal form (a term in normal form which is not an η-reduct of some other term) which we abbreviate to lnf. The syntax of such terms reflects their type: a lnf of type a is a variable x^a, or $x U_1 \ldots U_k$ where $x^{\rho_1 \rightarrow \ldots \rightarrow \rho_k \rightarrow a}$ and each $U_i^{\rho_i}$ is a lnf; a lnf of type $\rho_1 \rightarrow \ldots \rightarrow \rho_n \rightarrow a$ has the form $\lambda x_1^{\rho_1} \ldots x_n^{\rho_n} . S$, where S^a is a lnf.

The following definition introduces retracts between types [2,3].

Definition 1. *Type ρ is a* retract *of type τ, written $\models \rho \trianglelefteq \tau$, if there are terms $C : \rho \rightarrow \tau$ and $D : \tau \rightarrow \rho$ such that $D \circ C =_{\beta\eta} \lambda x^\rho . x$.*

The witnesses C and D to a retract can always be presented as lnfs. We can think of C as a "coder" and D as a "decoder" [9]. Assume $\rho = \rho_1 \to \ldots \to \rho_l \to a$ and $\tau = \tau_1 \to \ldots \to \tau_n \to a$: in a retract the types must share target type [9]. We instantiate the bound ρ_i variables in a decoder D to $D(z_1^{\rho_1}, \ldots, z_l^{\rho_l})$, often abbreviated to $D(\overline{z})$, and the bound variable of type ρ in C to $C(x^\rho)$: so, $\models \rho \trianglelefteq \tau$ if $D(z_1^{\rho_1}, \ldots, z_l^{\rho_l})(C(x^\rho)) =_{\beta\eta} xz_1 \ldots z_l$. From [9], we can restrict a decoder to be of the form $\lambda f^\tau. f S_1^{\tau_1} \ldots S_n^{\tau_n}$ with f as head variable and a coder $C(x)$ has the form $\lambda y_1^{\tau_1} \ldots y_n^{\tau_n}. H(x T_1^{\rho_1} \ldots T_l^{\rho_l})$.

Definition 2. *We say that the decoder* $D(z_1, \ldots, z_l) = \lambda f^\tau. f S_1^{\tau_1} \ldots S_n^{\tau_n}$ *and the coder* $C(x) = \lambda y_1^{\tau_1} \ldots y_n^{\tau_n}. H(x T_1^{\rho_1} \ldots T_l^{\rho_l})$ *are* canonical *witnesses for* $\rho \trianglelefteq \tau$ *if* $D(\overline{z})(C(x)) =_{\beta\eta} xz_1 \ldots z_l$ *and they obey the following properties:*

1. *variables* $f, z_1, \ldots, z_l$ *occur only once in* $D(\overline{z})$,
2. x *occurs only once in* $C(x)$,
3. H *is* ε *if* ρ *and* τ *are constructed from a single ground type*,
4. *if* $T_i^{\rho_i}$ *contains an occurrence of* y_j *then it is the head variable of* $T_i^{\rho_i}$, z_i *occurs in* $S_j^{\tau_j}$ *and* $T_i^{\rho_i}$ *contains no other occurrences of any* y_k, $1 \leq k \leq n$.

The next result follows from observations in [3,9].

Proposition 1. $\models \rho \trianglelefteq \tau$ *iff there exist canonical witnesses for* $\rho \trianglelefteq \tau$.

So, if there is only a single ground type then $C(x)$ can be restricted to have the form $\lambda y_1^{\tau_1} \ldots y_n^{\tau_n}. x T_1^{\rho_1} \ldots T_l^{\rho_l}$ with x as head variable [3].

Example 1. From [3]. Let $\rho = \rho_1 \to \rho_2 \to o$ where $\rho_1 = \rho_2 = \sigma \to o$ and let $\tau = \tau_1 \to o$ where $\tau_1 = \sigma \to (o \to o \to o) \to o$ and σ is arbitrary. It follows that $\models \rho \trianglelefteq \tau$. A decoder $D(z_1^{\rho_1}, z_2^{\rho_2})$ is $\lambda f^\tau. f(\lambda u^\sigma v^{o \to o \to o}. v(z_1 u)(z_2 u))$ and a coder $C(x^\rho)$ is $\lambda y^{\tau_1}. x(\lambda w^\sigma. yw(\lambda s^o t^o. s))(\lambda w^\sigma. yw(\lambda s^o t^o. t))$; so, $(D(z_1, z_2))C(x) \to_\beta^* x(\lambda w^\sigma. z_1 w)(\lambda w^\sigma. z_2 w) =_{\beta\eta} xz_1 z_2$.   □

Example 2. From [9] with multiple ground types. Let $\rho = \rho_1 \to \rho_2 \to a$ where $\rho_1 = b \to a$, $\rho_2 = a$ and let $\tau = \tau_1 \to a$ where $\tau_1 = b \to (a \to o \to a) \to a$. A decoder is $D(z_1^{\rho_1}, z_2^{\rho_2})$ is $\lambda f^\tau. f(\lambda u_1^b u_2^{a \to o \to a}. u_2(z_1 u_1) z_2)$ and a coder $C(x^\rho)$ is $\lambda y^{\tau_1}. y s^b (\lambda w_1^a w_2^o. x(\lambda v^b. yv(\lambda w_1^a w_2^o. w_1)) w_2)$; so, $(D(z_1, z_2))C(x) \to_\beta^* x(\lambda v^b. z_1 v) z_2 =_{\beta\eta} xz_1 z_2$.   □

Terms are represented as special kinds of tree (that we call *binding* trees in [12,14]) with dummy lambdas and an explicit binding relation. A term of the form y^a is represented as a tree with a single node labelled y^a. In the case of $y U_1 \ldots U_k$, when $y^{\rho_1 \to \cdots \to \rho_k \to a}$, we assume that a dummy lambda with the empty sequence of variables is placed directly above any subterm U_i in its tree representation if ρ_i is a ground type. With this understanding, the tree for $y U_1 \ldots U_k$ consists of a root node labelled $y^{\rho_1 \to \cdots \to \rho_k \to a}$ and k-successor trees representing $U_1, \ldots, U_k$. We also use the abbreviation $\lambda \overline{y}$ for $\lambda y_1 \ldots y_m$ for $m \geq 0$, so $\overline{y}$ is possibly the empty sequence of variables in the case of a dummy lambda. The tree representation of $\lambda \overline{y}. S : \rho_1 \to \ldots \to \rho_k \to a$ consists of a root node

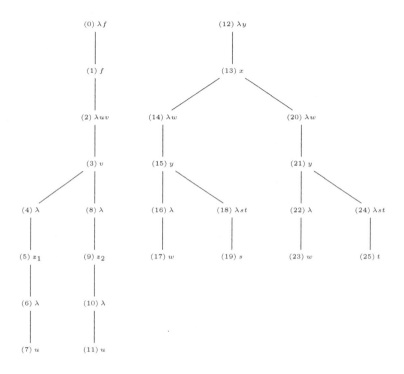

Fig. 1. $D(z_1, z_2)$ and $C(x)$ of Example 1

labelled $\lambda \overline{y}$ and a single successor tree for S^a. The trees for $C(x)$ and $D(z_1, z_2)$ of Example 1, where we have omitted the types, are in Figure 1.

We say that a node is a *lambda* (*variable*) node if it is labelled with a lambda abstraction (variable). The *type* (*target type*) of a variable node is the type (target type) of the variable at that node and the *type* (*target type*) of a lambda node is the type (target type) of the subterm rooted at that node.

The other elaboration is that we assume an extra binary relation $\downarrow$ between nodes in a tree that represents *binding*; that is, between a node labelled $\lambda y_1 \ldots y_n$ and a node below it labelled y_j (that it binds). A binder $\lambda \overline{y}$ is such that either $\overline{y}$ is empty and therefore is a dummy lambda and cannot bind a variable occurrence or $\overline{y} = y_1 \ldots y_k$ and $\lambda \overline{y}$ can only then bind variable occurrences of the form y_i, $1 \le i \le k$. Consequently, we also employ the following abbreviation $n \downarrow_i m$ if $n \downarrow m$ and n is labelled $\lambda y_1 \ldots y_k$ and m is labelled y_i. In Figure 1 we have not included the binding relation; however, for instance, $(2) \downarrow_1 (7)$.

Definition 3. *Lambda node n is a* descendant *(k-descendant) of m if either $m \downarrow m'$ ($m \downarrow_k m'$), n is a successor of m' for some m' and the target types of m, m' and n are the same, or n' is a descendant (k-descendant) of m and n is a descendant of n' for some n'.*

We assume a standard presentation of nodes of a tree as sequences of integers: an initial sequence, typically ε, is the root node; if n is a node and m is the ith successor of n then $m = ni$. For the sake of brevity we have not followed this approach in Figure 1 where we have presented each node as a unique integer (i).

Definition 4. *A* path *of the tree of a term of type σ is a sequence of nodes $\bar{n} = n_1, \ldots, n_k$ where n_1 is the root of the tree, each n_{i+1} is a successor of n_i and if n_j is a variable node then for some $i < j$, $n_i \downarrow n_j$ (hence is a closed path).*

For paths $\bar{m} = m_1, \ldots, m_l$ and $\bar{n} = n_1, \ldots, n_k$ of type σ we write $\bar{m} \sqsubset \bar{n}$ if for some $i > 0$, for all $h \le 2i$, $m_h = n_h$, $m_{2i+1} = m_{2i}p$, $n_{2i+1} = n_{2i}q$ and $p < q$.

A (closed) subtree *of a tree of a term of type σ is a set of paths P of type σ such that if $\bar{m}, \bar{n}$ are distinct paths in P then $\bar{m} \sqsubset \bar{n}$ or $\bar{n} \sqsubset \bar{m}$.*

A path $\bar{n} = n_1, \ldots, n_k$ is a contiguous sequence of nodes in a tree of a term starting at the root; for $i \ge 1$, each n_{2i-1} is a lambda node and each n_{2i} is a variable node (whose binder occurs earlier in the path). Path $\bar{m}$ is before $\bar{n}$, $\bar{m} \sqsubset \bar{n}$, if they have a common even length prefix and then differ as to their successors (the one in $\bar{m}$ before that in $\bar{n}$). These paths could, therefore, be in the same term: therefore, a closed subtree is a set of such paths.

Definition 5. *A path $\bar{n} = n_1, \ldots, n_l$ is k-minimal provided that for each binding node n_i there are at most k distinct nodes n_j, $i < j \le l$, such that $n_i \downarrow n_j$. A subtree P is k-minimal if each path in P is k-minimal.*

Not every path or subtree is useful in a term. So, we define when a path or subtree is realisable meaning that their nodes are "accessible" [7] or "reachable" [6] in an applicative context.

Definition 6. *Assume $\bar{n} = n_1, \ldots, n_l$ is a path of odd length of a closed term T of type σ, m is the node below n_l in T and T' is the term T when the variable u^τ at node m is replaced with a fresh free variable z^τ. We say that $\bar{n}$ is realisable if there is a closed term $U = \lambda y^\sigma . y S_1 \ldots S_k$ such that $UT' =_{\beta\eta} \lambda \bar{x}.z W_1 \ldots W_q$ for some $q \ge 0$.*

Definition 7. *Assume P is a subtree of closed term T of type σ where each path has even length, $m_1, \ldots, m_q$ are the leaves of P and T_i, $1 \le i \le q$, is the term T when the variable $u_i^{\tau_i}$ at m_i is replaced with a fresh free variable $z_i^{\tau_i}$. We say that P is realisable if there is a closed term $U = \lambda y^\sigma . y S_1 \ldots S_k$ such that for each i, $UT_i =_{\beta\eta} \lambda \bar{x}.z_i W_1 \ldots W_{q_i}$ for $q_i \ge 0$.*

Next we define two useful operations on paths, restriction relative to a suffix and the subtype after a prefix.

Definition 8. *Assume that $\bar{n} = n_1, \ldots, n_p$ is a path, $\sigma = \sigma_1 \to \ldots \to \sigma_k \to a$, n_i is a lambda node of type σ and $w = n_i, \ldots, n_p$ is a suffix of $\bar{n}$.*

1. *The suffix w* admits *σ_j, $1 \le j \le k$, if either there is no n_q, $i \le q \le p$, such that $n_i \downarrow_j n_q$ or there is a j-descendant n_q of n_i whose type is $\tau_1 \to \ldots \to \tau_l \to a$ and for some r there is not a $t : q < t \le p$ such that $n_q \downarrow_r n_t$ and a is the target type of τ_r.*

2. *The restriction of σ to w, $\sigma \restriction w$, is defined as σ_w where*

   - $a_w = a$,
   - $(\sigma_j \to \ldots \to \sigma_k \to a)_w =$ *if w admits σ_j then* $\sigma_j \to (\sigma_{j+1} \to \ldots \to \sigma_k \to a)_w$ *else* $(\sigma_{j+1} \to \ldots \to \sigma_k \to a)_w$.

Definition 9. *Assume that $\bar{n} = n_1, \ldots, n_p$ is a path of type σ. For a prefix w of $\bar{n}$ we define the* subtype *of σ after w, $w(\sigma)$:*

   - *if $w = \varepsilon$ (the empty prefix) then σ,*
   - *if $w = n_1, \ldots, n_q$, $q \leq p$, then the type of node n_q.*

We also define a canonical presentation of a (prefix or suffix of a) path $\bar{n} = n_1, \ldots, n_k$ of type σ as a *word* w. If w is the empty prefix we write $w = \varepsilon$. Otherwise, $w = (w_1, \ldots, w_j)$, $j \leq k$, where for each $i \geq 0$, $w_{2i+1} = n_{2i+1}$ and if $n_h \downarrow_m n_{2i}$ then $w_{2i} = n_h m$. Thus, we distinguish between $w = \varepsilon$ (the empty word) and $w = (\varepsilon)$ the prefix of length 1 consisting of the root node. Also, we can present a subtree as a set of words. Words will occur in our proof systems as presentations of paths. For example, $w = (\varepsilon, 1, 11, 112, 1112)$ of type τ as in Example 1 represents the path labelled $\lambda f, f, \lambda uv, v, \lambda$ of $D(z_1, z_2)$ in Figure 1 when its root is ε. To illustrate Definitions 8 and 9, for the prefix $w' = (\varepsilon, 1, 11)$ and the suffix $w'' = (11, 112, 1122)$ of w we have $w'(\tau) = \tau_1$ where $\tau_1 = \sigma \to (o \to o \to o) \to o$ as in Example 1 and $\tau_1 \restriction w'' = \sigma \to o$: word w'' of type τ_1 has labelling $\lambda u^\sigma v^{o \to o \to o}, v, \lambda$; so, w'' admits the first component σ of τ_1 but not the second $(o \to o \to o)$. The final element of w' is the same as the first element of w''; in such a case we define their concatenation to be w.

Definition 10. *The* concatenation *of (a prefix) v and (a suffix) w, $v^\wedge w$, is:* $\varepsilon^\wedge w = w$; *if $v_k = w_1$ then $v_1, \ldots, v_k^\wedge w_1, \ldots, w_n = v_1, \ldots, v_k, w_2, \ldots, w_n$.*

3 Proof Systems for Retracts

We now develop goal directed tableau proof systems for showing retracts. By inverting the rules one has more classical axiomatic systems: we do it this way because it thereby provides an immediate nondeterministic decision procedure for deciding retracts. We present two such proof systems: a slightly simpler system for the restricted case when there is a single ground type and one for the general case.

3.1 Single Ground Type

Assertions in our proof system are of two kinds. First is $\rho \trianglelefteq \tau$ with meaning ρ is a retract of τ. The second has the form $[\rho_1, \ldots, \rho_k] \trianglelefteq \tau$ which is based on the "product" as defined in [3]. We follow [9] in allowing reordering of components of types since $\rho \to \sigma \to \tau$ is isomorphic to $\sigma \to \rho \to \tau$. Instead we could include explict rules for reordering (as with the axiom in [3]). Moreover, we assume that $[\rho_1, \ldots, \rho_k]$ is a multi-set and so elements can be in any order.

$$I \quad \rho \trianglelefteq \rho$$

$$W \ \dfrac{\rho \trianglelefteq \sigma \to \tau}{\rho \trianglelefteq \tau}$$

$$C \ \dfrac{\delta \to \rho \trianglelefteq \sigma \to \tau}{\delta \trianglelefteq \sigma \qquad \rho \trianglelefteq \tau}$$

$$P_1 \ \dfrac{\rho_1 \to \ldots \to \rho_k \to \rho \trianglelefteq \sigma \to \tau}{[\rho_1, \ldots, \rho_k] \trianglelefteq \sigma \qquad \rho \trianglelefteq \tau}$$

$$P_2 \ \dfrac{[\rho_1, \ldots, \rho_k] \trianglelefteq \sigma}{\rho_1 \trianglelefteq \sigma \upharpoonright w_1 \quad \ldots \quad \rho_k \trianglelefteq \sigma \upharpoonright w_k} \quad \text{where}$$

– $w_1 \sqsubset \ldots \sqsubset w_k$ are k-minimal realisable paths of odd length of type σ

Fig. 2. Goal directed proof rules

The proof rules are given in Figure 2. There is a single axiom I, identity, a weakening rule W, a covariance rule C, and two product rules P_1 and P_2. The rules are goal directed: for instance, C allows one to decompose the goal $\delta \to \rho \trianglelefteq \sigma \to \tau$ into the two subgoals $\delta \trianglelefteq \sigma$ and $\rho \trianglelefteq \tau$. I, W and C (or their variants) occur in the proof systems for affine retracts (when variables in witnesses can only occur at most once) [3,9]. The new rules are the product rules: P_2 appeals to k-minimal realisable paths (presented as words), and the restriction operator of Definition 8. The proof system does not require the axiom A4 of [3], $\sigma \trianglelefteq (\sigma \to a) \to a$: all instances are provable using W and C.

Definition 11. *A successful* proof tree *for $\rho \trianglelefteq \tau$ is a finite tree whose root is labelled with the goal $\rho \trianglelefteq \tau$, the successor nodes of a node are the result of an application of one of the rules to it, and each leaf is labelled with an axiom. We write $\vdash \rho \trianglelefteq \tau$ if there is a successful proof tree for $\rho \trianglelefteq \tau$.*

For some intuition about the product rules assume $\rho = \rho_1 \to \ldots \to \rho_l \to a$ and $\tau = \tau_1 \to \ldots \to \tau_n \to a$. Now, $\models \rho \trianglelefteq \tau$ iff there are canonical, Definition 2, witnesses $D(z_1^{\rho_1}, \ldots, z_l^{\rho_l}) = \lambda f^\tau . f S_1^{\tau_1} \ldots S_n^{\tau_n}$. Since we can reorder components of ρ and τ we can assume that z_1 is in $S_1^{\tau_1}$. Suppose $z_1, \ldots, z_k$, where $k > 1$, are in $S_1^{\tau_1}$ and so y_1 must occur in $T_1^{\rho_1}, \ldots, T_k^{\rho_k}$. Therefore, there is a common coder $S_1^{\tau_1}(x_1/z_1, \ldots, x_k/z_k)$ and k decoders $T_i(\bar{z}_i)$ where $\bar{z}_i = z_{i1}^{\rho_{i1}}, \ldots, z_{il_i}^{\rho_{il_i}}$ and $\rho_{i1}, \ldots, \rho_{il_i}$ are the components of ρ_i such that $T_i(\bar{z}_i)(S_1^{\tau_1}(x_1, \ldots, x_k)) =_{\beta\eta} x_i \bar{z}_i$ (which is similar to the product in [3]). In $S_1^{\tau_1}(x_1/z_1, \ldots, x_k/z_k)$ there are distinct odd length paths $w_1, \ldots, w_k$ of type τ_1 to the lambda nodes above $x_1, \ldots, x_k$. These paths are realisable, Definition 6, because each x_i belongs to the normal form of $T_i(\bar{z}_i)(S_1^{\tau_1}(x_1, \ldots, x_k))$. Using a combinatorial argument, see the full version, $S_1^{\tau_1}$ can be chosen so that these words are k-minimal and by reordering ρ's components $w_1 \sqsubset \ldots \sqsubset w_k$. We may not be able to reduce

$$\frac{\dfrac{(\sigma \to o) \to (\sigma \to o) \to o \trianglelefteq (\sigma \to (o \to o \to o) \to o) \to o}{[\sigma \to o, \sigma \to o] \trianglelefteq \sigma \to (o \to o \to o) \to o \quad o \trianglelefteq o}}{\sigma \to o \trianglelefteq \sigma \to o \quad \sigma \to o \trianglelefteq \sigma \to o}$$

Fig. 3. A proof tree for Example 1

to the subgoals $\rho_1 \trianglelefteq \tau_1, \dots, \rho_k \trianglelefteq \tau_1$ as w_i may prescribe the form of $T_i(\bar{z}_i)$: if $T_i(\bar{z}_i) = \lambda f^{\tau_1}.f S_1^i \dots S_m^i$ then path w_i may prevent S_j^i containing elements of $\bar{z}_i$; so, this may restrict the possible distribution of $\bar{z}_i$ within the subterms $S_1^i, \dots, S_m^i$ which is captured using $\tau_1 \upharpoonright w_i$.

An example proof tree is in Figure 3 for the retract of Example 1 (which is not affine). Rule P_1 is applied to the root and then P_2 to the first subgoal where $w_1 = (\varepsilon, 2, 21)$ and $w_2 = (\varepsilon, 2, 22)$. Let $\sigma' = \sigma \to (o \to o \to o) \to o$. Now, $\sigma' \upharpoonright w_1 = \sigma \to o = \sigma' \upharpoonright w_2$; in both cases only the first component of σ' is admitted.

3.2 Multiple Ground Types

We extend the proof system to include multiple ground types. Again, assertions are of the two kinds $\rho \trianglelefteq \tau$ and $[\rho_1, \dots, \rho_k] \trianglelefteq \tau$. However, we now assume that to be a well-formed assertion $\rho \trianglelefteq \tau$ both ρ and τ must share the same target type (which is guaranteed when there is a single ground type). The rules for this assertion are as before the axiom I, weakening W, covariance C and the product rule P_1 in Figure 2: however, C carries the requirement that the target types of δ and σ coincide. The other product rule P_2', presented in Figure 4, is different: the *arity* of $\rho_1 \to \dots \to \rho_n \to a$ is the maximum of n and the arities of each ρ_i where a gound type a has arity 0.

In $[\rho_1, \dots, \rho_k] \trianglelefteq \sigma$ it is not required that ρ_j and σ share the same target type. Instead rule P_2' requires that ρ_i and $v_i(\sigma)$, see Definition 9, do share target types:

$$P_2' \quad \frac{[\rho_1, \dots, \rho_k] \trianglelefteq \sigma}{\rho_1 \trianglelefteq v_1(\sigma) \upharpoonright w_1 \quad \dots \quad \rho_k \trianglelefteq v_k(\sigma) \upharpoonright w_k} \quad \text{where}$$

- k' is the maximum of k and h^2 where h is the arity of σ
- there is a k'-minimal realisable subtree U of type σ where each path has even length (which can be $\emptyset$),
- each v_i is ε, a prefix of a path in U of odd length or the extension of a path in U with a single node,
- $v_1^\wedge w_1 \sqsubset \dots \sqsubset v_k^\wedge w_k$ and each $v_i^\wedge w_i$ is a k'-minimal realisable path of type σ of odd length and if $U \neq \emptyset$, $v_i^\wedge w_i$ extends some path in U.

Fig. 4. Product proof rule for multiple gound types

$$\frac{\dfrac{(b \to a) \to o \to a \trianglelefteq (b \to (a \to o \to a) \to a) \to a}{[b \to a, o] \trianglelefteq b \to (a \to o \to a) \to a \quad a \trianglelefteq a}}{b \to a \trianglelefteq b \to a \quad o \trianglelefteq o}$$

Fig. 5. A proof tree for Example 2

for the concatenation $v_i^\wedge w_i$ see Definition 10. The specialisation to the case of the single ground type is when $U = \emptyset$ and $v = \varepsilon$.

Let $\rho = \rho_1 \to \ldots \to \rho_l \to a$ and $\tau = \tau_1 \to \ldots \to \tau_n \to a$. So, $\models \rho \trianglelefteq \tau$ iff there are canonical witnesses $D(z_1^{\rho_1}, \ldots, z_l^{\rho_l}) = \lambda f^\tau . f S_1^{\tau_1} \ldots S_n^{\tau_n}$ and $C(x) = \lambda y_1^{\tau_1} \ldots y_n^{\tau_n} . H(x T_1^{\rho_1} \ldots T_l^{\rho_l})$. Assume $z_1, \ldots, z_k$, where $k \geq 1$, occur in $S_1^{\tau_1}$. There is a path v in $C(x)$ to the node above x which determines a subtree U of $S_1^{\tau_1}$. The head variable in $T_i^{\rho_i}$ bound in v has the same target type as ρ_i. There are distinct paths $v_1^\wedge w_1, \ldots, v_k^\wedge w_k$ of odd length to the lambda nodes above $z_1, \ldots, z_k$ in $S_1^{\tau_1}$: v_i is decided by the meaning of the head variable in $T_i^{\rho_i}$; so, $v_i(\tau_1)$ has the same target type as ρ_i. The rest of the path is the tail of w_i: so we need to consider whether $\models \rho_i \trianglelefteq v_i(\tau_1) \restriction w_i$.

Figure 5 is the proof tree for the retract in Example 2. There is an application of P_1 followed by P_2'. In the application of P_2' the subtree $U = \{(\varepsilon, 2)\}$, $v_1 = \varepsilon$, $w_1 = (\varepsilon, 2, 21) = v_1^\wedge w_1$, $v_2 = (\varepsilon, 2, 22) = v_2^\wedge w_2$ when $w_2 = (22)$. So, $v_1(b \to (a \to o \to a) \to a) \restriction w_1 = b \to a$ as the first component is admitted (unlike the second); and $v_2(b \to (a \to o \to a) \to a) = o = o \restriction w_2$.

3.3   Complexity

The proof systems provide nondeterministic decision procedures for checking retracts. Each subgoal of a proof rule has smaller size than the goal. Hence, by focussing on one subgoal at a time a proof witness can be presented in PSPACE. However, this does not take into account checking that a subgoal obeys the side conditions in the case of the product rules. Given any type σ, there are boundedly many realisable k-minimal paths (with an upper bound of k^n where n is size of σ). So, this means that overall the decision procedure requires at most EXPSPACE.

4   Soundness and Completeness

To show soundness and completeness of our proof systems, we define a dialogue game $\mathsf{G}(D(\overline{z}), C(x))$ played by a single player $\forall$ on the trees of potential witnesses for a retract that characterises when $(D(\overline{z}))C(x) =_{\beta\eta} x\overline{z}$, similar to game semantics [5]. The game is defined in the full version of the paper.

To provide intuition for the reader we briefly describe $\mathsf{G}(D(z_1, z_2), C(x))$ where these terms are from Figure 1. Play starts at node (0), the binder λf at that node is associated with $C(x)$ rooted at (12); so, the next position is at node (1) and therefore jumps to (12); the binder at (12) λy is associated with node (2) (the successor of (1)). Play proceeds to (13) and $\forall$ chooses to go left or right; suppose it is left, so play is then at (14); nodes (13) and (14) are part of the normal form. Play descends to (15) and, therefore, jumps to (2); so, with the binder at (2), u is associated with the the the subtree at (16) and v with the subtree at (18). Play proceeds to (3) and so jumps to (18); now, s is associated with (4) and t with (8). Play proceeds to (19) and so jumps to (4), descends to (5) and then to (6) and then to (7) and jumps to (16) before finishing at (17). This play captures the path $x\lambda w.z_1 w$ of the normal form.

Some of the key properties, defined in the full version, we appeal to in the correctness proofs below *associate* subtrees with realisable paths and vice versa. For instance, as illustrated in the play above the path rooted at (0) downto (7) is associated with the subtree rooted at (12) and with leaves (17) and (19). Let $\rho = \rho_1 \to \ldots \to \rho_l \to a$ and $\tau = \tau_1 \to \ldots \to \tau_n \to a$ and let $\tau_1 = \sigma = \sigma_1 \to \ldots \to \sigma_m \to b$.

Theorem 1. *(Soundness) If* $\vdash \rho \trianglelefteq \tau$ *then* $\models \rho \trianglelefteq \tau$.

Proof. By induction on the depth of a proof. For the base case, the result is clear for a proof that uses the axiom I. So, assume the result for all proofs of depth $< d$. Consider now a proof of depth d. We proceed by examining the first rule that is applied to show $\vdash \rho \trianglelefteq \tau$. If it is W or C the result follows using the same arguments as in [3]. Assume the rule is W and suppose $\models \rho \trianglelefteq \tau$. Therefore there are terms D_1 and C_1 such that $D_1^{\tau \to \rho}(C_1^{\rho \to \tau x}) =_{\beta\eta} x$. Now $D^{(\sigma \to \tau) \to \rho} = \lambda f^{\sigma \to \tau} y^\sigma. D_1(fy)$ and $C^{\rho \to (\sigma \to \tau)} x = \lambda s^\sigma. C_1(x)$ are witnesses for $\models \rho \trianglelefteq \sigma \to \tau$. Assume that the rule is C, so $\models \delta \trianglelefteq \sigma$ and $\models \rho \trianglelefteq \tau$. So there are terms D_1, C_1, D_2, C_2 such that $D_1^{\sigma \to \delta}(C_1^{\delta \to \sigma} x) =_{\beta\eta} x$ and $D_2^{\tau \to \rho}(C_2^{\rho \to \tau} x) =_{\beta\eta} x$. Now $D^{(\sigma \to \tau) \to (\delta \to \rho)} = \lambda x y. C_2(x(D_1 y))$ and $C^{(\delta \to \rho) \to (\sigma \to \tau)} = \lambda u z. D_2(u(C_1 z))$ are witnesses for $\models \delta \to \rho \trianglelefteq \sigma \to \tau$.

Consider next that the first rule is P_1. So after P_1 there is either an application of P_2 or P_2': in the former case, there are k-minimal realisable paths $w_1 \sqsubset \ldots \sqsubset w_k$ of odd length of type σ such that $\vdash \rho_i \trianglelefteq \sigma \restriction w_i$; in the latter case, there is a k'-minimal realisable subtree U of type σ where each path has even length; and there are paths $v_1^\wedge w_1 \sqsubset \ldots \sqsubset v_k^\wedge w_k$ where each element is a k'-minimal realisable path of type σ of odd length and if $U \neq \emptyset$, it extends some path in U and where each v_i is ε, a prefix of a path in U of odd length path or an extension of a path in U with a single node and $\vdash \rho_i \trianglelefteq v_i(\sigma) \restriction w_i$; where k' is the maximum of k and the square of the arity of σ. So, by the induction hypothesis there are terms $D_i(\bar{z}_i)$ and $C_i(x_i)$ such that $D_i(\bar{z}_i)(C_i(x_i)) =_{\beta\eta} x_i\bar{z}_i$, witnesses for $\rho_i \trianglelefteq \sigma \restriction w_i$ or $\rho_i \trianglelefteq v_i(\sigma) \restriction w_i$, and terms $D'(z_{k+1}, \ldots, z_l)$ and $C'(x')$ such that $D'(z_{k+1}, \ldots, z_l)(C'(x')) =_{\beta\eta} x' z_{k+1} \ldots z_l$, witnesses for $\rho_{k+1} \to \ldots \to \rho_l \to a \trianglelefteq \tau'$ where $\tau' = \tau_2 \to \ldots \to \tau_n \to a$. We assume that all these terms are canonical witnesses. The term $D'(z_{k+1}, \ldots, z_l)$ is $\lambda f^{\tau'}.fS_2^{\tau_2} \ldots S_n^{\tau_n}$ and $C'(x')$ is $\lambda y_2^{\tau_2} \ldots y_n^{\tau_n}.H'(x'T_{k+1}^{\rho_{k+1}} \ldots T_l^{\rho_l})$ where $H' = \varepsilon$ if the rule applied was P_2.

We need to show that there are terms $D(z_1, \ldots, z_l)$ and $C(x)$ that are witnesses for $\models \rho \trianglelefteq \tau$. $D(\bar{z})$ will have the form $\lambda f^\tau.fS_1^{\tau_1} \ldots S_n^{\tau_n}$ and $C(x)$ the form

$\lambda y_1^{\tau_1} \dots y_n^{\tau_n}.H(xT_1^{\rho_1} \dots T_l^{\rho_l})$ where $H = \varepsilon$ in the case of a single ground type. All that remains is to define $S_1^{\tau_1}$ so it contains $z_1, \dots, z_k$, $T_1^{\rho_1}, \dots, T_k^{\rho_k}$ and H (as an extension of H'). If $U = \emptyset$ then $H = H'$. Otherwise, let u be an odd length path such that U is associated with (so, its head variable is $y_1^{\tau_1}$). H consists of the suffix of u followed by the subtree H'. The head variable of each $T_i^{\rho_i}$ is y_1 in the case of the single ground type and $g_i^{v_i(\sigma)}$ in the general case (which is either y_1 or bound in u). We assume that S_i' is the subterm of S_1^σ that is rooted at the initial vertex of the path w_i: which is S_1^σ itself in the single ground type. To complete these terms we require that $T_i^{\rho_i}(S_1^\sigma(z_1, \dots, z_k)) =_{\beta\eta} z_i$. Therefore, removing lambda abstraction over variables z_{ij} and changing z_i to x_i, we require that $T_i(\bar{z}_i)(S_i'(x_1, \dots, x_k)) =_{\beta\eta} x_i\bar{z}_i$. We construct a term $C''(x_i)$ that occurs after the path w_i in S_i' (and which has root x_i when there is a single ground type). We also complete $T_i(\bar{z}_i)$ whose initial part is the tree U_i associated with the path w_i.

First, we examine the single ground type case. So, S_1^σ will have the form $\lambda u_1 \dots u_m.S_1'$, $C''(x_i)$ the form $x_iC_{i1}'' \dots C_{ip}''$ and $T_i(\bar{z}_i)$ the form $\lambda f_i^\sigma.f_iV_1^i \dots V_m^i$. Assume $D_i(\bar{z}_i)$ is $\lambda g_i^{\sigma\upharpoonright w_i}.g_iW_{i_1}^i \dots W_{i_l}^i$ and $C_i(x_i)$ is $\lambda u_{i_1} \dots u_{i_l}.x_iC_1^i \dots C_p^i$. Assume w_i admits σ_{i_j}: therefore, for some $r : 1 \leq r \leq m$, $i_j = r$ (so, W_r^i may contain occurrences of variables in $\bar{z}_i$). If u_r does not occur in the path w_i then we set $V_r^i = W_r^i$. Otherwise, there is a non-empty subpath w_{ir} of w_i generated by u_r, and a subtree U_r^i of V_r^i associated with w_{ir}. Each C_j^i contains a single u_{i_k} (as head variable). Assume C_s^i contains u_r. Assume that the path in W_r^i to the lambda node above z_{is} is w_s'. If we can build the same path in V_r^i (by copying nodes of C_s^i to C_{is}'') then we are done (letting V_r^i include this path followed by the subterm of W_r^i rooted at z_{is}). Otherwise, we initially include w_{ir} in C_{is}'' and then try to build w_s' in V_r^i by copying nodes of C_s^i to C_{is}'': in V_r^i and, therefore in U_r^i, there is a path whose prefix except for its final variable vertex is the same as a prefix of w_s' and then differ. In the game $G(C_{is}'', V_r^i)$, play jumps from that variable in V_r^i to a lambda node in w_{ir}. By definition of admits, there is a binder n' labelled $\lambda\bar{v}$ in w_{ir} such that for some q not$(n' \downarrow_q n_i')$ for all nodes n_i' after n' in w_i (and in w_{ir}). Therefore, we add a variable node labelled v_q to the end of w_{ir} in C_{is}''; so play jumps to a lambda node in V_r^i which is a successor of a leaf of U_r^i; below this node, we build the path w_s' except for its root node (by adding further nodes to C_{is}'' and add the subtree rooted at z_{is} in W_r^i to V_r^i).

For the general case, assume $v_i(\sigma) = \sigma_1' \to \dots \to \sigma_m' \to b$. So, $S_i^{v_i(\sigma)}$ will have the form $\lambda u_1 \dots u_m.S_1'$, $C''(x_i)$ the form $H_i(x_iC_{i1}'' \dots C_{ip}'')$ and $T_i(\bar{z}_i)$ the form $\lambda f_i^\sigma.f_iV_1^i \dots V_m^i$. Assume $D_i(\bar{z}_i)$ is $\lambda g_i^{v_i(\sigma)\upharpoonright w_i}.g_iW_{i_1}^i \dots W_{i_l}^i$ and $C_i(x_i)$ is $\lambda u_{i_1}' \dots u_{i_l}'.H_i'(x_iC_1^i \dots C_p^i)$. We set $H_i = H_i'$. Then we proceed in a similar fashion to the single base type case. If some u_r' does not occur in the path w_i then $V_r^i = W_r^i$; otherwise we need to build similar paths to z_{is} in W_r^i in V_r^i (by copying vertices from C_s^i to C_{is}'' and using that w_i admits $(v_i(\sigma))_r$). $\quad\square$

The proof of completeness (by induction on the size of ρ) is easier.

Theorem 2. *(Completeness) If* $\models \rho \trianglelefteq \tau$ *then* $\vdash \rho \trianglelefteq \tau$.

5   Conclusion

We have provided tableau proof systems that characterise when a type is a retract of another type in simply typed lambda calculus (with respect to $\beta\eta$-equality). They offer a a nondeterministic decision procedure for the retract problem in EXPSPACE: it may be possible to improve on the rather crude k-minimality bounds used on paths within the proof systems. Given the constructive proof of correctness, we also expect to be able to extract witnesses for a retract from a successful tableau proof tree (similar in spirit to [10]).

References

1. Barendregt, H.: Lambda calculi with types. In: Abramsky, S., Gabbay, D., Maibaum, T. (eds.) Handbook of Logic in Computer Science, vol. 2, pp. 118–309. Oxford University Press (1992)
2. Bruce, K., Longo, G.: Provable isomorphisms and domain equations in models of typed languages. In: Proc. 17th Symposium on Theory of Computing, pp. 263–272. ACM (1985)
3. de 'Liguoro, U., Piperno, A., Statman, R.: Retracts in simply typed $\lambda\beta\eta$-calculus. In: Procs. LICS 1992, pp. 461–469 (1992)
4. Loader, R.: Higher-order β-matching is undecidable. Logic Journal of the IGPL 11(1), 51–68 (2003)
5. Ong, C.-H.L.: On model-checking trees generated by higher-order recursion schemes. In: Procs. LICS 2006, pp. 81–90 (2006)
6. Ong, C.-H.L., Tzevelekos, N.: Functional Reachability. In: Procs. LICS 2009, pp. 286–295 (2009)
7. Padovani, V.: Decidability of fourth-order matching. Mathematical Structures in Computer Science 10(3), 361–372 (2000)
8. Padovani, V.: Retracts in simple types. In: Abramsky, S. (ed.) TLCA 2001. LNCS, vol. 2044, pp. 376–384. Springer, Heidelberg (2001)
9. Regnier, L., Urzyczyn, P.: Retractions of types with many atoms, pp. 1–16 (2005), http://arxiv.org/abs/cs/0212005
10. Schubert, A.: On the building of affine retractions. Math. Struct. in Comp. Science 18, 753–793 (2008)
11. Stirling, C.: Higher-order matching, games and automata. In: Procs. LICS 2007, pp. 326–335 (2007)
12. Stirling, C.: Dependency tree automata. In: de Alfaro, L. (ed.) FOSSACS 2009. LNCS, vol. 5504, pp. 92–106. Springer, Heidelberg (2009)
13. Stirling, C.: Decidability of higher-order matching. Logical Methods in Computer Science 5(3:2), 1–52 (2009)
14. Stirling, C.: An introduction to decidability of higher-order matching (2012) (Submitted for Publication), Availble at author's website
15. Vorobyov, S.: The "hardest" natural decidable theory. In: Procs. LICS 1997, pp. 294–305 (1997)

Presburger Arithmetic, Rational Generating Functions, and Quasi-Polynomials*

Kevin Woods

Oberlin College, Oberlin, Ohio, USA
Kevin.Woods@oberlin.edu

Abstract. A Presburger formula is a Boolean formula with variables in $\mathbb{N}$ that can be written using addition, comparison ($\leq$, $=$, etc.), Boolean operations (and, or, not), and quantifiers ($\forall$ and $\exists$). We characterize sets that can be defined by a Presburger formula as exactly the sets whose characteristic functions can be represented by rational generating functions; a geometric characterization of such sets is also given. In addition, if $\mathbf{p} = (p_1, \ldots, p_n)$ are a subset of the free variables in a Presburger formula, we can define a counting function $g(\mathbf{p})$ to be the number of solutions to the formula, for a given $\mathbf{p}$. We show that every counting function obtained in this way may be represented as, equivalently, either a piecewise quasi-polynomial or a rational generating function. In the full version of this paper, we also translate known computational complexity results into this setting and discuss open directions.

1 Introduction

A broad and interesting class of sets are those that can be defined over $\mathbb{N} = \{0, 1, 2, \ldots\}$ with first order logic and addition.

Definition 1. *A* Presburger formula *is a Boolean formula with variables in $\mathbb{N}$ that can be written using addition, comparison ($\leq$, $=$, etc.), Boolean operations (and, or, not), and quantifiers ($\forall$ and $\exists$). We will denote a generic Presburger formula as $F(\mathbf{u})$, where $\mathbf{u}$ are the free variables (those not associated with a quantifier); we use bold notation like $\mathbf{u}$ to indicate vectors of variables.*

We say that a set $S \subseteq \mathbb{N}^d$ is a Presburger set *if there exists a Presburger formula $F(\mathbf{u})$ such that $S = \{\mathbf{u} \in \mathbb{N}^d : F(\mathbf{u})\}$.*

Example 1. The Presburger formula

$$F(u) = \big(u > 1 \text{ and } \exists b \in \mathbb{N} : \ b + b + 1 = u\big)$$

defines the Presburger set $\{3, 5, 7, \ldots\}$. Since multiplication by an integer is the same as repeated addition, we can conceive of a Presburger formula as a Boolean combination of integral linear (in)equalities, appropriately quantified: $\exists b \, (u > 1$ and $2b + 1 = u)$.

---

* Full version available at http://www.oberlin.edu/faculty/kwoods/papers.html

F.V. Fomin et al. (Eds.): ICALP 2013, Part II, LNCS 7966, pp. 410–421, 2013.
© Springer-Verlag Berlin Heidelberg 2013

Presburger proved [35] that the truth of a Presburger *sentence* (a formula with no free variables) is decidable. In contrast, a broader class of sentences, where multiplication of variables is allowed, is undecidable; this is a consequence of the negative solution to Hilbert's 10th problem, given by Davis, Putnam, Robinson, and Matiyasevich (see, for example, [19]).

We would like to understand more clearly the *structure* of a given Presburger set. One way to attempt to do this is to encode the elements of the set into a generating function.

Definition 2. *Given a set $S \subseteq \mathbb{N}^d$, its associated generating function is*

$$f(S; \mathbf{x}) = \sum_{\mathbf{s} \in S} \mathbf{x}^{\mathbf{s}} = \sum_{(s_1, \ldots, s_d) \in S} x_1^{s_1} x_2^{s_2} \cdots x_d^{s_d}.$$

For example, if S is the set defined by Example 1, then

$$f(S; x) = x^3 + x^5 + x^7 + \cdots = \frac{x^3}{1 - x^2}.$$

We see that, in this instance, the generating function has a nice form; this is not a coincidence.

Definition 3. *A rational generating function is a function that can be written in the form*

$$\frac{q(\mathbf{x})}{(1 - \mathbf{x}^{\mathbf{b}_1}) \cdots (1 - \mathbf{x}^{\mathbf{b}_k})},$$

where $q(\mathbf{x})$ is a polynomial in $\mathbb{Q}[\mathbf{x}]$ and $\mathbf{b}_i \in \mathbb{N}^d \setminus \{\mathbf{0}\}$.

We will prove that $S \subseteq \mathbb{N}^d$ is a Presburger set if and only if $f(S; \mathbf{x})$ is a rational generating function. These are Properties 1 and 3 in the following theorem:

Theorem 1. *Given a set $S \subseteq \mathbb{N}^d$, the following are equivalent:*
1. *S is a Presburger set,*
2. *S is a finite union of sets of the form $P \cap (\lambda + \Lambda)$, where P is a polyhedron, $\lambda \in \mathbb{Z}^d$, and $\Lambda \subseteq \mathbb{Z}^d$ is a lattice.*
3. *$f(S; \mathbf{x})$ is a rational generating function.*

Property 2 gives a nice geometric characterization of Presburger sets; the set in Example 1 can be written as $[3, \infty) \cap (1 + 2\mathbb{Z})$.

We are particularly interested in generating functions because of their powerful flexibility: we can use algebraic manipulations to answer questions about the set. For example, $f(S; 1, 1, \ldots, 1)$ is exactly the cardinality of S (if finite). More generally, we may want to count solutions to a Presburger formula as a function of several parameter variables:

Definition 4. *The Presburger counting function for a given Presburger formula $F(\mathbf{c}, \mathbf{p})$ is*

$$g_F(\mathbf{p}) = \#\{\mathbf{c} \in \mathbb{N}^d : F(\mathbf{c}, \mathbf{p})\}.$$

Note that $\mathbf{c}$ (the *counted* variables) and $\mathbf{p}$ (the *parameter* variables) are free variables. We will restrict ourselves to counting functions such that $g_F(\mathbf{p})$ is finite for all $\mathbf{p} \in \mathbb{N}^n$. One could instead either include ∞ in the codomain of g_F or restrict the domain of g_F to where $g_F(\mathbf{p})$ is finite (this domain would itself be a Presburger set).

A classic example is to take $F(\mathbf{c}, p)$ to be the conjunction of linear inequalities of the form $a_1 c_1 + \cdots + a_d c_d \leq a_0 p$, where $a_i \in \mathbb{Z}$. Then $g_F(p)$ counts the number of integer points in the pth dilate of a polyhedron.

Example 2. If $F(c_1, c_2, p)$ is $2c_1 + 2c_2 \leq p$, then the set of solutions $(c_1, c_2) \in \mathbb{N}^2$ lies in the triangle with vertices $(0,0)$, $(0, p/2)$, $(p/2, 0)$, and

$$g_F(p) = \frac{1}{2}\left(\left\lfloor \frac{p}{2} \right\rfloor + 1\right)\left(\left\lfloor \frac{p}{2} \right\rfloor + 2\right)$$

$$= \begin{cases} \frac{1}{8}p^2 + \frac{3}{4}p + 1 & \text{if } p \text{ is even,} \\ \frac{1}{8}p^2 + \frac{1}{2}p + \frac{3}{8} & \text{if } p \text{ is odd.} \end{cases}$$

The nice form of this function is also not a coincidence. For this particular type of Presburger formula (dilates of a polyhedron), Ehrhart proved [21] that the counting functions are *quasi-polynomials*:

Definition 5. *A* quasi-polynomial *(over $\mathbb{Q}$) is a function $g : \mathbb{N}^n \to \mathbb{Q}$ such that there exists an n-dimensional lattice $\Lambda \subseteq \mathbb{Z}^n$ together with polynomials $q_{\bar{\lambda}}(\mathbf{p}) \in \mathbb{Q}[\mathbf{p}]$, one for each $\bar{\lambda} \in \mathbb{Z}^n / \Lambda$, such that*

$$g(\mathbf{p}) = q_{\bar{\lambda}}(\mathbf{p}), \text{ for } \mathbf{p} \in \bar{\lambda}.$$

In Example 2, we can take the lattice $\Lambda = 2\mathbb{Z}$ and each coset (the evens and the odds) has its associated polynomial. We need something slightly more general to account for all Presburger counting functions:

Definition 6. *A* piecewise quasi-polynomial *is a function $g : \mathbb{N}^n \to \mathbb{Q}$ such that there exists a finite partition $\bigcup_i (P_i \cap \mathbb{N}^n)$ of $\mathbb{N}^n$ with P_i polyhedra (which may not all be full-dimensional) and there exist quasi-polynomials g_i such that*

$$g(\mathbf{p}) = g_i(\mathbf{p}) \text{ for } \mathbf{p} \in P_i \cap \mathbb{N}^n.$$

One last thing that is not a coincidence: For the triangle in Example 2, we can compute

$$\sum_{p \in \mathbb{N}} g_F(p) x^p = 1 + x + 3x^2 + 3x^3 + 6x^4 + \cdots$$

$$= \frac{1}{(1-x)(1-x^2)^2},$$

a rational generating function! The following theorem says that these ideas are – almost – equivalent.

Theorem 2. *Given a function $g : \mathbb{N}^n \to \mathbb{Q}$ and the following three possible properties:*

A. g is a Presburger counting function,
B. g is a piecewise quasi-polynomial, and
C. $\sum_{\boldsymbol{p} \in \mathbb{N}^n} g(\boldsymbol{p}) x^{\boldsymbol{p}}$ is a rational generating function,

we have the implications

$$A \Rightarrow B \Leftrightarrow C.$$

Remark 1. Proving Theorem 2 will give us much of Theorem 1, using the following idea. A set $S \subseteq \mathbb{Z}^d$ corresponds exactly to its characteristic function

$$\chi_S(\mathbf{u}) = \begin{cases} 1 & \text{if } \mathbf{u} \in S, \\ 0 & \text{if } \mathbf{u} \notin S. \end{cases}$$

If S is a Presburger set defined by $F(\mathbf{u})$, then

$$\chi_S(\mathbf{u}) = \#\{c \in \mathbb{N} : \ F(\mathbf{u}) \text{ and } c = 0\}$$

is a Presburger counting function.

In light of Theorem 1, we might wonder if there is a sense in which $B \Rightarrow A$. Of course we would have to restrict g, for example requiring that its range be in $\mathbb{N}$ (Theorem 1 essentially restricts the range of g to $\{0, 1\}$, as it must be a characteristic function). The implication still does not hold, however. For example, suppose the polynomial

$$g(s, t) = (t - s^2)^2$$

were a Presburger counting function given by a Presburger formula $F(\boldsymbol{c}, s, t)$, that is,

$$g(s, t) = \#\{\boldsymbol{c} \in \mathbb{N}^d : \ F(\boldsymbol{c}, s, t)\}.$$

Then the set

$$\begin{aligned} \{(s, t) \in \mathbb{N}^2 : \ \nexists \boldsymbol{c} \ F(\boldsymbol{c}, s, t)\} &= \{(s, t) \in \mathbb{N}^2 : \ g(s, t) = 0\} \\ &= \{(s, s^2) : \ s \in \mathbb{N}\} \end{aligned}$$

would be a Presburger set. This is not the case, however, as it does not satisfy Property 2 in Theorem 1. If the parameter is univariate, however, the following proposition shows that we do have the implication $B \Rightarrow A$.

Proposition 1. *Given a function $g : \mathbb{N} \to \mathbb{Q}$, if g is a piecewise quasi-polynomial whose range is in $\mathbb{N}$, then g is a Presburger counting function.*

In Section 4, we prove Theorems 1 and 2 (the proof of Proposition 1 appears in the full version of this paper). In Section 2, we survey related work. In Section 3, we present the primary tools we need for the proofs. In the full version of this paper, we also turn to computational questions; we survey known results, but restate them in terms of Presburger arithmetic.

2   Related Work

Presburger arithmetic is a classical first order theory of logic, proven decidable by Presburger [35]. Various upper and lower bounds on the complexity of decision algorithms for the general theory have occupied the theoretical computer science community, see [8,17,22,24,26,33].

A finite automata approach to Presburger arithmetic was pioneered in [12,15], and continues to be an active area of research (see, for example, [10,16,30,47]). This approach is quite different from the present paper's, but it can attack similar questions: for example, see [34] for results on counting solutions to Presburger formulas (non-parametrically).

The importance of understanding Presburger Arithmetic is highlighted by the fact that many problems in computer science and mathematics can be phrased in this language: for example, integer programming [31,40], geometry of numbers [13,29], Gröbner bases and algebraic integer programming [43,45], neighborhood complexes and test sets [38,44], the Frobenius problem [37], Ehrhart theory [7,21], monomial ideals [32], and toric varieties [23]. Several of the above references analyze the computational complexity of their specific problem. In most of the above references, the connection to Presburger arithmetic is only implicit.

The algorithmic complexity of specific rational generating function problems has been addressed in, for example, [1,5,9,20,27,28]. Several of these results are summarized in the full version of this current paper.

Connections between subclasses of Presburger arithmetic and generating functions are made explicit in [3,4,5]. Connections between rational generating functions and quasi-polynomials have been made in [21,41,42], and the algorithmic complexity of their relationship was examined in [46]. Counting solutions to Presburger formulas has been examined in [36], though the exact scope of the results is not made explicit, and rational generating functions are not used. Similar counting algorithms appear in [14], and [18] proves that the counting functions for a special class of Presburger formuals (those whose parameters p only appear in terms $c_i \leq p_i$) are piecewise quasi-polynomials. This current paper is the first to state and prove a general connection between Presburger arithmetic, quasi-polynomials, and rational generating functions.

Theorem 1 was originally proved in the author's thesis [48]; in this paper, it is put into context as a consequence of the more general Theorem 2. A simpler geometric characterization of Presburger sets (equivalent to Property 2 of Theorem 1) was given in [25]: they are the *semi-linear* sets, those sets that can be written as a finite union of sets of the form $S = \{a_0 + \sum_{i=1}^{k} n_i a_i : n_i \in \mathbb{N}\}$, where $a_i \in \mathbb{N}^d$. Furthermore, if one takes these S to be disjoint and requires the $a_1, \ldots, a_k$ to be linearly independent, for each S (as [25] implicitly prove can be done, made explicit in [18] as *semi-simple* sets), then each S can be encoded with the rational generating function

$$\frac{x^{a_0}}{(1 - x_1^{a_1}) \cdots (1 - x_k^{a_k})}$$

and we obtain a slightly different version of $2 \Rightarrow 3$ in Theorem 1. There seems to be no previous result analogous to $3 \Rightarrow 2$.

3   Primary Background Theorems

Here we detail several tools we will use. The first tool we need is a way to simplify Presburger formulas. As originally proved [35] by Presburger (see [33] for a nice exposition), we can completely eliminate the quantifiers if we are allowed to also use modular arithmetic.

Definition 7. *An* extended Presburger formula *is a Boolean formula with variables in $\mathbb{N}$ expressible in the elementary language of Presburger Arithmetic extended by the* mod k *operations, for constants $k > 1$.*

Theorem 3. *Given a formula $F(\mathbf{u})$ in extended Presburger arithmetic (and hence any formula in Presburger arithmetic), there exists an equivalent* quantifier free *formula $G(\mathbf{u})$ such that*

$$\{\mathbf{u} \in \mathbb{N}^d : \ F(\mathbf{u})\} = \{\mathbf{u} \in \mathbb{N}^d : \ G(\mathbf{u})\}.$$

For instance, the set from Example 1 can be written as $(u > 1$ and $u \bmod 2 = 1)$.

Next, we give two theorems that tie in generating functions. The first gives us a way to convert from a specific type of Presburger set to a generating function.

Theorem 4. *Given a point $\lambda \in \mathbb{Z}^d$, a lattice $\Lambda \subseteq \mathbb{Z}^d$, and a rational polyhedron $P \subseteq \mathbb{R}_{\geq 0}^d$, $f(P \cap (\lambda + \Lambda); \mathbf{x})$ (as given in Definition 2) is a rational generating function.*

The first step to proving this is to use Brion's Theorem [11], which says that the generating function can be decomposed into functions of the form $f(K \cap (\lambda + \Lambda); \mathbf{x})$, where K is a cone. Then, one can notice that integer points in cones have a natural structure that can be encoded as geometric series.

Example 3. Let $K \subseteq \mathbb{R}^2$ be the cone with vertex at the origin and extreme rays $\mathbf{u} = (1, 0)$ and $\mathbf{v} = (1, 2)$. Using the fact that the lattice $(u\mathbb{Z} + v\mathbb{Z})$ has index 2 in $\mathbb{Z}^2$, with coset representatives $(0, 0)$ and $(1, 1)$, every integer point in K can be written as either $(0, 0) + \lambda_1 \mathbf{u} + \lambda_2 \mathbf{v}$ or $(1, 1) + \lambda_1 \mathbf{u} + \lambda_2 \mathbf{v}$, where $\lambda_1, \lambda_2 \in \mathbb{N}$. Therefore

$$f(K \cap \mathbb{Z}^2; \mathbf{x}) = (\mathbf{x}^{(0,0)} + \mathbf{x}^{(1,1)})(1 + \mathbf{x}^{\mathbf{u}} + \mathbf{x}^{2\mathbf{u}} + \cdots)(1 + \mathbf{x}^{\mathbf{v}} + \mathbf{x}^{2\mathbf{v}} + \cdots)$$
$$= \frac{\mathbf{x}^{(0,0)} + \mathbf{x}^{(1,1)}}{(1 - \mathbf{x}^{\mathbf{u}})(1 - \mathbf{x}^{\mathbf{v}})}.$$

See [2, Chapter VIII], for example, for more details.

Next, we would like to be able to perform substitutions on the variables in a rational generating function and still retain a rational generating function; particularly, we would like to substitute in 1's for several of the variables.

Theorem 5. *Given a rational generating function $f(x)$, then*

$$g(z) = f(z^{l_1}, z^{l_2}, \ldots, z^{l_d}),$$

with $l_i \in \mathbb{N}^k$, is also a rational generating function, assuming the substituted values do not lie entirely in the poles of f. In particular, substituting in $x_i = z^0 = 1$ yields a rational function, if 1 is not a pole of f.

The proof is immediate: if substituting in $x_i = z^{l_i}$ would make any of the binomials in the denominator of f zero (when f is written in the form from Definition 3), then that binomial must be a factor of the numerator (or else such z^{l_i} would lie entirely in the poles of f); therefore, substituting in $x_i = z^{l_i}$ yields a new rational generating function.

Finally, we need a connection between Presburger formulas and quasi-polynomials. This is given by Sturmfels [42]:

Definition 8. *Given $a_1, \ldots, a_d \in \mathbb{N}^n$, the* vector partition function $g : \mathbb{N}^n \to \mathbb{N}$ *is defined by*

$$g(p) = \#\{(\lambda_1, \ldots, \lambda_d) \in \mathbb{N}^d : p = \lambda_1 a_1 + \cdots + \lambda_d a_d\},$$

that is, the number of ways to partition the vector p into parts taken from $\{a_i\}$.

Theorem 6. *Any vector partition function is a piecewise quasi-polynomial.*

See [6] for a self-contained explanation utilizing the partial fraction expansion of the generating function

$$\sum_{\mathbf{p} \in \mathbb{N}^n} g(\mathbf{p}) \mathbf{x}^{\mathbf{p}} = \frac{1}{(1 - \mathbf{x}^{a_1}) \cdots (1 - \mathbf{x}^{a_d})};$$

this equality can be obtained by rewriting the rational function as a product of infinite geometric series:

$$(1 + \mathbf{x}^{a_1} + \mathbf{x}^{2a_1} + \cdots) \cdots (1 + \mathbf{x}^{a_d} + \mathbf{x}^{2a_d} + \cdots).$$

4   Proofs

4.1   Proof of Theorem 2

A $\Rightarrow$ C.

Given a Presburger counting function, $g(p) = \#\{c \in \mathbb{N}^d : F(c, p)\}$, we first apply Presburger Elimination (Theorem 3) to F to obtain a quantifier free formula, $G(c, p)$, in extended Presburger arithmetic such that $g(p) = \#\{c \in \mathbb{N}^d : G(c, p)\}$. Integers which satisfy a statement of the form

$$a_1 p_1 + \cdots + a_n p_n + a_{n+1} c_1 + \cdots + a_{n+d} c_d \equiv a_0 \pmod{m}$$

are exactly sets $\lambda + \Lambda$, where $\lambda \in \mathbb{Z}^{n+d}$ and Λ is a lattice in $\mathbb{Z}^{n+d}$. Since $G(\boldsymbol{c}, \boldsymbol{p})$ is a Boolean combination of linear inequalities and these linear congruences, we may write the set, S, of points $(\boldsymbol{c}, \boldsymbol{p})$ which satisfy $G(\boldsymbol{c}, \boldsymbol{p})$ as a *disjoint* union

$$S = \bigcup_{i=1}^{k} P_i \cap (\lambda_i + \Lambda_i),$$

where, for $1 \leq i \leq k$, $P_i \subseteq \mathbb{R}_{\geq 0}^{n+d}$ is a polyhedron, Λ_i is a sublattice of $\mathbb{Z}^{n+d}$, and λ_i is in $\mathbb{Z}^{n+d}$. (To see this, convert the formula into disjunctive normal form; each conjunction will be of this form $P_i \cap (\lambda_i + \Lambda_i)$; these sets may overlap, but their overlap will also be of this form.)

Let $S_i = P_i \cap (\lambda_i + \Lambda_i)$. By Theorem 4, we know we can write $f(S_i; \boldsymbol{y}, \boldsymbol{x})$ as a rational generating function, and so

$$f(S; \boldsymbol{y}, \boldsymbol{x}) = \sum_i f(S_i; \boldsymbol{y}, \boldsymbol{x}) = \sum_{(\boldsymbol{c},\boldsymbol{p}):\; G(\boldsymbol{c},\boldsymbol{p})} \boldsymbol{y}^{\boldsymbol{c}} \boldsymbol{x}^{\boldsymbol{p}}$$

can be written as a rational generating function. Finally, we substitute $\boldsymbol{y} = (1, 1, \ldots, 1)$, using Theorem 5, to obtain the rational generating function

$$\sum_{\boldsymbol{p}} \#\{\boldsymbol{c} \in \mathbb{N}^d :\; G(\boldsymbol{c}, \boldsymbol{p})\} \boldsymbol{x}^{\boldsymbol{p}} = \sum_{\boldsymbol{p}} g(\boldsymbol{p}) \boldsymbol{x}^{\boldsymbol{p}}.$$

C $\Rightarrow$ B.

It suffices to prove this for functions g such that $\sum_{\boldsymbol{p}} g(\boldsymbol{p}) \boldsymbol{x}^{\boldsymbol{p}}$ is a rational generating function of the form

$$\frac{\boldsymbol{x}^{\boldsymbol{q}}}{(1 - \boldsymbol{x}^{\boldsymbol{a}_1})(1 - \boldsymbol{x}^{\boldsymbol{a}_2}) \cdots (1 - \boldsymbol{x}^{\boldsymbol{a}_k})},$$

where $\boldsymbol{q} \in \mathbb{N}^n, \boldsymbol{a}_i \in \mathbb{N}^n \setminus \{0\}$, because the property of being a piecewise quasi-polynomial is preserved under linear combinations. Furthermore, we may take $\boldsymbol{q} = (0, 0, \ldots, 0)$, because multiplying by $\boldsymbol{x}^{\boldsymbol{q}}$ only shifts the domain of the function g. Expanding this rational generating function as a product of infinite geometric series,

$$\sum_{\boldsymbol{p}} g(\boldsymbol{p}) \boldsymbol{x}^{\boldsymbol{p}} = (1 + \boldsymbol{x}^{\boldsymbol{a}_1} + \boldsymbol{x}^{2\boldsymbol{a}_1} + \cdots) \cdots (1 + \boldsymbol{x}^{\boldsymbol{a}_k} + \boldsymbol{x}^{2\boldsymbol{a}_k} + \cdots),$$

and we see that

$$g(\boldsymbol{p}) = \#\{(\lambda_1, \ldots, \lambda_k) \in \mathbb{N}^k :\; \boldsymbol{p} = \lambda_1 \boldsymbol{a}_1 + \cdots + \lambda_k \boldsymbol{a}_k\}.$$

This is exactly a vector partition function, which Theorem 6 tells us is a piecewise quasi-polynomial.

B ⇒ C.

Any piecewise quasi-polynomial can be written as a linear combination of functions of the form

$$g(\boldsymbol{p}) = \begin{cases} \boldsymbol{p}^{\boldsymbol{a}} & \text{if } \boldsymbol{p} \in P \cap (\lambda + \Lambda), \\ 0 & \text{otherwise,} \end{cases}$$

where $\boldsymbol{a} \in \mathbb{N}^n$, $P \subseteq \mathbb{R}^n_{\geq 0}$ is a polyhedron, $\lambda \in \mathbb{Z}^n$, and Λ is a sublattice of $\mathbb{Z}^n$. Since linear combinations of rational generating functions are rational generating functions, it suffices to prove it for such a g. Let c_{ij}, for $1 \leq i \leq n$ and $1 \leq j \leq a_i$, be variables, and define the polyhedron

$$Q = \{(\boldsymbol{p}, \boldsymbol{c}) \in \mathbb{N}^{n+a_1+\cdots+a_n} :$$

$$\boldsymbol{p} \in P \text{ and } 1 \leq c_{ij} \leq p_i \text{ for all } c_{ij}\}.$$

This Q is defined so that $\#\{\boldsymbol{c} : (\boldsymbol{p}, \boldsymbol{c}) \in Q\}$ is $p_1^{a_1} \cdots p_n^{a_n} = \boldsymbol{p}^{\boldsymbol{a}}$ for $\boldsymbol{p} \in P$ (and 0 otherwise). Using Theorem 4, we can find the generating function for the set $Q \cap (\lambda + \Lambda)$ as a rational generating function. Substituting $\boldsymbol{c} = (1, 1, \ldots, 1)$, using Theorem 5, gives us $\sum_{\boldsymbol{p}} g(\boldsymbol{p}) \mathbf{x}^{\boldsymbol{p}}$ as a rational generating function.

4.2   Proof of Theorem 1

Given a set $S \subseteq \mathbb{Z}^d$, define the characteristic function, $\chi_S : \mathbb{N}^d \to \{0, 1\}$, as in Remark 1. Define a new property:

2′. χ_S is a piecewise quasi-polynomial.

Translating Theorem 2 into properties of S and χ_S, we have

$$1 \Rightarrow (2' \Leftrightarrow 3).$$

So we need to prove $2 \Rightarrow 1$ and $2' \Rightarrow 2$.

2 ⇒ 1.

This is straightforward: the property of being an element of $\lambda + \Lambda$ can be written using linear congruences and existential quantifiers, and the property of being an element of P can be written as a set of linear inequalities.

2′ ⇒ 2.

Since χ_S is a piecewise quasi-polynomial, it is constituted from associated polynomials. Let us examine such a polynomial $q(\mathbf{p})$ that agrees with χ_S on some $P \cap (\lambda + \Lambda)$, where $P \subseteq \mathbb{R}^n_{\geq 0}$ is a polyhedron, $\lambda \in \mathbb{Z}^n$, and Λ a sublattice of $\mathbb{Z}^n$. It suffices to prove that 2 holds for $S \cap P \cap (\lambda + \Lambda)$, since S is the disjoint union of such pieces.

Ideally, we would like to argue that, since q only takes on the values 0 and 1, the polynomial q must be constant on $P \cap (\lambda + \Lambda)$, at least if P is unbounded. This is not quite true; for example, if

$$P = \{(x, y) \in \mathbb{R}^2 : x \geq 0 \text{ and } 0 \leq y \leq 1\},$$

then the polynomial $q(x, y) = y$ is 1 for $y = 1$ and 0 for $y = 0$.

What we can say is that q must be constant on any infinite ray contained in $P \cap (\lambda + \Lambda)$: if we parametrize the ray by $\mathbf{x}(t) = (x_1(t), \cdots, x_n(t))$, then $q(\mathbf{x}(t))$ is a *univariate* polynomial that is either 0 or 1 at an infinite number of points, and so must be constant. Inductively, we can similarly show that q must be constant on any cone contained in P.

Let K be the cone with vertex at the origin

$$K = \{\mathbf{y} \in \mathbb{R}^n : \mathbf{y} + P \subseteq P\}.$$

Then K is the largest cone such that the cones $\mathbf{x} + K$ are contained in P, for all $\mathbf{x} \in P$; K is often called the *recession cone* or *characteristic cone* of P (see Section 8.2 of [39]), and the polyhedron P can be decomposed into a Minkowski sum $K + Q$, where Q is a *bounded* polyhedron. We can write $P \cap (\lambda + \Lambda)$ as a finite union (possibly with overlap) of sets of the form

$$Q_j = (v_j + K) \cap (\lambda + \Lambda),$$

for some v_j, and on each of these pieces q must be constant. If q is the constant 1 on Q_j, then Q_j is contained in S, and if q is the constant 0, then none of Q_j is in S. Since S is a finite union of the appropriate Q_j, S has the form needed for Property 2.

Acknowledgements. My thanks to the referees, particularly for pointers to relevant related works.

References

1. Barvinok, A.: A polynomial time algorithm for counting integral points in polyhedra when the dimension is fixed. Math. Oper. Res. 19(4), 769–779 (1994)
2. Barvinok, A.: A Course in Convexity. Graduate Studies in Mathematics, vol. 54. American Mathematical Society, Providence (2002)
3. Barvinok, A.: The complexity of generating functions for integer points in polyhedra and beyond. In: International Congress of Mathematicians, vol. III, pp. 763–787. Eur. Math. Soc., Zürich (2006)
4. Barvinok, A., Pommersheim, J.: An algorithmic theory of lattice points in polyhedra. In: New Perspectives in Algebraic Combinatorics (Berkeley, CA, 1996–1997). Math. Sci. Res. Inst. Publ., vol. 38, pp. 91–147. Cambridge Univ. Press, Cambridge (1999)
5. Barvinok, A., Woods, K.: Short rational generating functions for lattice point problems. J. Amer. Math. Soc. 16(4), 957–979 (2003) (electronic)
6. Beck, M.: The partial-fractions method for counting solutions to integral linear systems. Discrete Comput. Geom. 32(4), 437–446 (2004)
7. Beck, M., Robins, S.: Computing the continuous discretely. Undergraduate Texts in Mathematics. Springer, New York (2007); Integer-point enumeration in polyhedra
8. Berman, L.: The complexity of logical theories. Theoret. Comput. Sci. 11(1), 57, 71–77 (1980); With an introduction "On space, time and alternation"

9. Blanco, V., García-Sánchez, P.A., Puerto, J.: Counting numerical semigroups with short generating functions. Internat. J. Algebra Comput. 21(7), 1217–1235 (2011)

10. Boudet, A., Comon, H.: Diophantine equations, Presburger arithmetic and finite automata. In: Kirchner, H. (ed.) CAAP 1996. LNCS, vol. 1059, pp. 30–43. Springer, Heidelberg (1996)

11. Brion, M.: Points entiers dans les polyèdres convexes. Ann. Sci. École Norm. Sup. 4, 653–663 (1988)

12. Büchi, J.R.: Weak second-order arithmetic and finite automata. Z. Math. Logik Grundlagen Math. 6, 66–92 (1960)

13. Cassels, J.W.S.: An introduction to the geometry of numbers. Classics in Mathematics. Springer, Berlin (1997); Corrected reprint of the 1971 edition

14. Clauss, P., Loechner, V.: Parametric analysis of polyhedral iteration spaces. Journal of VLSI Signal Processing 19(2), 179–194 (1998)

15. Cobham, A.: On the base-dependence of sets of numbers recognizable by finite automata. Math. Systems Theory 3, 186–192 (1969)

16. Comon, H., Jurski, Y.: Multiple counters automata, safety analysis and Presburger arithmetic. In: Vardi, M.Y. (ed.) CAV 1998. LNCS, vol. 1427, pp. 268–279. Springer, Heidelberg (1998)

17. Cooper, D.: Theorem proving in arithmetic without multiplication. Machine Intelligence 7, 91–99 (1972)

18. D'Alessandro, F., Intrigila, B., Varricchio, S.: On some counting problems for semilinear sets. CoRR abs/0907.3005 (2009)

19. Davis, M.: Hilbert's tenth problem is unsolvable. Amer. Math. Monthly 80, 233–269 (1973)

20. De Loera, J., Haws, D., Hemmecke, R., Huggins, P., Sturmfels, B., Yoshida, R.: Short rational functions for toric algebra. To appear in Journal of Symbolic Computation (2004)

21. Ehrhart, E.: Sur les polyèdres rationnels homothétiques à n dimensions. C. R. Acad. Sci. Paris 254, 616–618 (1962)

22. Fischer, M., Rabin, M.: Super-exponential complexity of Presburger arithmetic. In: Complexity of Computation. SIAM–AMS Proc., vol. VII, pp. 27–41. Amer. Math. Soc., Providence (1974); Proc. SIAM-AMS Sympos., New York (1973)

23. Fulton, W.: Introduction to Toric Varieties. Annals of Mathematics Studies, vol. 131. Princeton University Press, Princeton (1993)

24. Fürer, M.: The complexity of Presburger arithmetic with bounded quantifier alternation depth. Theoret. Comput. Sci. 18(1), 105–111 (1982)

25. Ginsburg, S., Spanier, E.: Semigroups, Presburger formulas and languages. Pacific Journal of Mathematics 16(2), 285–296 (1966)

26. Grädel, E.: Subclasses of Presburger arithmetic and the polynomial-time hierarchy. Theoret. Comput. Sci. 56(3), 289–301 (1988)

27. Guo, A., Miller, E.: Lattice point methods for combinatorial games. Adv. in Appl. Math. 46(1-4), 363–378 (2011)

28. Hoşten, S., Sturmfels, B.: Computing the integer programming gap. To appear in Combinatorics (2004)

29. Kannan, R.: Test sets for integer programs, ∀∃ sentences. In: Polyhedral Combinatorics. DIMACS Ser. Discrete Math. Theoret. Comput. Sci, vol. 1, pp. 39–47. Amer. Math. Soc., Providence (1990); Morristown, NJ (1989)

30. Klaedtke, F.: Bounds on the automata size for Presburger arithmetic. ACM Trans. Comput. Log. 9(2), 34 (2008)

31. Lenstra Jr., H.: Integer programming with a fixed number of variables. Math. Oper. Res. 8(4), 538–548 (1983)

32. Miller, E., Sturmfels, B.: Combinatorial commutative algebra. Graduate Texts in Mathematics, vol. 227. Springer, New York (2005)
33. Oppen, D.: A superexponential upper bound on the complexity of Presburger arithmetic. J. Comput. System Sci. 16(3), 323–332 (1978)
34. Parker, E., Chatterjee, S.: An automata-theoretic algorithm for counting solutions to presburger formulas. In: Duesterwald, E. (ed.) CC 2004. LNCS, vol. 2985, pp. 104–119. Springer, Heidelberg (2004)
35. Presburger, M.: On the completeness of a certain system of arithmetic of whole numbers in which addition occurs as the only operation. Hist. Philos. Logic 12(2), 225–233 (1991); Translated from the German and with commentaries by Dale Jacquette
36. Pugh, W.: Counting solutions to presburger formulas: how and why. SIGPLAN Not. 29(6), 121–134 (1994)
37. Ramírez Alfonsín, J.L.: The Diophantine Frobenius problem. Oxford Lecture Series in Mathematics and its Applications, vol. 30. Oxford University Press, Oxford (2005)
38. Scarf, H.: Test sets for integer programs. Math. Programming Ser. B 79(1-3), 355–368 (1997)
39. Schrijver, A.: Theory of Linear and Integer Programming. Interscience Series in Discrete Mathematics. John Wiley & Sons Ltd., Chichester (1986)
40. Schrijver, A.: Combinatorial optimization. Polyhedra and efficiency. Algorithms and Combinatorics, vol. 24. Springer, Berlin (2003)
41. Stanley, R.P.: Decompositions of rational convex polytopes. Ann. Discrete Math. 6, 333–342 (1980); Combinatorial mathematics, optimal designs and their applications. In: Proc. Sympos. Combin. Math. and Optimal Design, Colorado State Univ., Fort Collins, Colo. (1978)
42. Sturmfels, B.: On vector partition functions. J. Combin. Theory Ser. A 72(2), 302–309 (1995)
43. Sturmfels, B.: Gröbner Bases and Convex Polytopes. University Lecture Series, vol. 8. American Mathematical Society, Providence (1996)
44. Thomas, R.: A geometric Buchberger algorithm for integer programming. Math. Oper. Res. 20(4), 864–884 (1995)
45. Thomas, R.: The structure of group relaxations. In: Aardal, K., Nemhauser, G., Weismantel, R. (eds.) Handbook of Discrete Optimization (2003)
46. Verdoolaege, S., Woods, K.: Counting with rational generating functions. J. Symbolic Comput. 43(2), 75–91 (2008)
47. Wolper, P., Boigelot, B.: An automata-theoretic approach to Presburger arithmetic constraints. In: Mycroft, A. (ed.) SAS 1995. LNCS, vol. 983, pp. 21–32. Springer, Heidelberg (1995)
48. Woods, K.: Rational Generating Functions and Lattice Point Sets. PhD thesis, University of Michigan (2004)

Revisiting the Equivalence Problem
for Finite Multitape Automata

James Worrell*

Department of Computer Science, University of Oxford, UK

Abstract. The decidability of determining equivalence of deterministic multitape automata (or transducers) was a longstanding open problem until it was resolved by Harju and Karhumäki in the early 1990s. Their proof of decidability yields a **co-NP** upper bound, but apparently not much more is known about the complexity of the problem. In this paper we give an alternative proof of decidability, which follows the basic strategy of Harju and Karhumäki but replaces their use of group theory with results on matrix algebras. From our proof we obtain a simple randomised algorithm for deciding equivalence of deterministic multitape automata, as well as automata with transition weights in the field of rational numbers. The algorithm involves only matrix exponentiation and runs in polynomial time for each fixed number of tapes. If the two input automata are inequivalent then the algorithm outputs a word on which they differ.

1 Introduction

One-way multitape finite automata were introduced in the seminal 1959 paper of Rabin and Scott [15]. Such automata (under various restrictions) are also commonly known as transducers—see Elgot and Mezei [6] for an early reference. A multitape automaton with k tapes accepts a k-ary relation on words. The class of relations recognised by deterministic automata coincides with the class of k-ary rational relations [6].

Two multitape automata are said to be equivalent if they accept the same relation. Undecidability of equivalence of non-deterministic automata is relatively straightforward [8]. However the deterministic case remained open for many years, until it was shown decidable by Harju and Karhumäki [9]. Their solution made crucial use of results about ordered groups—specifically that a free group can be endowed with a compatible order [13] and that the ring of formal power series over an ordered group with coefficients in a division ring and with well-ordered support is itself a division ring (due independently to Malcev [11] and Neumann [14]). Using these results [9] established the decidability of *multiplicity equivalence* of non-deterministic multitape automata, i.e., whether two non-deterministic multitape automata have the same number of accepting computations on each input. Decidability in the deterministic case (and, more

---

* Supported by EPSRC grant EP/G069727/1.

generally, the unambiguous case) follows immediately. We refer the reader to [16] for a self-contained account of the proof, including the underlying group theory.

Harju and Karhumäki did not address questions of complexity in [9]. However the existence of a **co-NP** *guess-and-check* procedure for deciding equivalence of deterministic multitape automata follows directly from [9, Theorem 8]. This theorem states that two inequivalent automata are guaranteed to differ on a tuple of words whose total length is at most the total number of states of the two automata. Such a tuple can be guessed, and it can be checked in polynomial time whether the tuple is accepted by one automaton and rejected by the other. In the special case of two-tape deterministic automata, a polynomial-time algorithm was given in [7], before decidability was shown in the general case.

A **co-NP** upper bound also holds for multiplicity equivalence of k-tape automata for each fixed k. However, as we observe below, if the number of tapes is not fixed, computing the number of accepting computations of a given nondeterministic multitape automata on a tuple of input words is #**P**-hard. Thus the guess-and-check method does not yield a **co-NP** procedure for multiplicity equivalence in general.

It is well-known that the equivalence problem for single-tape weighted automata with rational transition weights is solvable in polynomial time [18,19]. Now the decision procedure in [9] reduces multiplicity equivalence of multitape automata to equivalence of single-tape automata with transition weights in a division ring of power series over an ordered group. However the complexity of arithmetic in this ring seems to preclude an application of the polynomial-time procedures of [18,19]. Leaving aside issues of representing infinite power series, even the operation of multiplying a family of polynomials in two non-commuting variables yields a result with exponentially many monomials in the length of its input.

In this paper we give an alternative proof that multiplicity equivalence of multitape automata is decidable, which also yields new complexity bounds on the problem. We use the same basic idea as [9]—reduce to the single-tape case by enriching the set of transition weights. However we replace their use of power series on ordered groups with results about matrix algebras and Polynomial Identity rings (see Remark 1 for a more technical comparison). In particular, we use the Amitsur-Levitzki theorem concerning polynomial identities in matrix algebras. Our use of the latter is inspired by the work of [3] on non-commutative polynomial identity testing, and our starting point is a simple generalisation of the approach of [3] to what we call *partially commutative* polynomial identity testing.

Our construction for establishing decidability immediately yields a simple randomised algorithm for checking multiplicity equivalence of multitape automata (and hence also equivalence of deterministic automata). The algorithm involves only matrix exponentiation, and runs in polynomial time for each fixed number of tapes.

2   Partially Commutative Polynomial Identities

2.1   Matrix Algebras and Polynomial Identities

Let F be an infinite field. Recall that an F-algebra is a vector space over F equipped with an associative bilinear product that has an identity . Write $F\langle X \rangle$ for the free F-algebra over a set X. The elements of $F\langle X \rangle$ can be viewed as polynomials over a set of non-commuting variables X with coefficients in F. Each such polynomial is an F-linear combination of monomials, where each monomial is an element of X^*. The degree of a polynomial is the maximum of the lengths of its monomials.

Let A be an F-algebra and $f \in F\langle X \rangle$. If f evaluates to 0 for all valuations of its variables in A then we say that A satisfies the *polynomial identity* $f = 0$. For example, an algebra satisfies the polynomial identity $xy - yx = 0$ if and only if it is commutative. Note that since the variables x and y do not commute, the polynomial $xy - yx$ is not identically zero.

We denote by $M_n(F)$ the F-algebra of $n \times n$ matrices with coefficients in F. The Amitsur-Levitzki theorem [1,4] is a fundamental result about polynomial identities in matrix algebras.

Theorem 1 (Amitsur-Levitzki). *The algebra $M_n(F)$ of $n \times n$ matrices over a commutative ring F satisfies the polynomial identity*

$$\sum_{\sigma \in S_{2n}} x_{\sigma(1)} \dots x_{\sigma(2n)} = 0 \,,$$

where the sum is over the $(2n)!$ elements of the symmetric group S_{2n}. Moreover $M_n(F)$ satisfies no identity of degree less than $2n$.

Given a finite set X of non-commuting variables, the *generic $n \times n$ matrix algebra* $F_n\langle X \rangle$ is defined as follows. For each variable $x \in X$ we introduce a family of commuting indeterminates $\{t_{ij}^{(x)} : 1 \leq i, j \leq n\}$ and define $F_n\langle X \rangle$ to be the F-algebra of $n \times n$ matrices generated by the matrices $(t_{ij}^{(x)})$ for each $x \in X$. Then $F_n\langle X \rangle$ has the following universal property: any homomorphism from $F\langle X \rangle$ to a matrix algebra $M_n(R)$, with R an F-algebra, factors uniquely through the map $\Phi_n^X : F\langle X \rangle \to F_n\langle X \rangle$ given by $\Phi_n^X(x) = (t_{ij}^{(x)})$.

Related to the map Φ_n^X we also define an F-algebra homomorphism

$$\Psi_n^X : F\langle X \rangle \to M_n(F\langle t_{ij}^{(x)} \mid x \in X, 1 \leq i, j \leq n \rangle)$$

by

$$\Psi_n^X(x) = \begin{pmatrix} 0 & t_{12}^{(x)} & & \\ & \ddots & \ddots & \\ & & & t_{n-1,n}^{(x)} \\ & & & 0 \end{pmatrix}$$

where the matrix on the right has zero entries everywhere but along the super-diagonal.

2.2 Partially Commutative Polynomial Identities

In this section we introduce a notion of *partially commutative polynomial identity*. We first establish notation and recall some relevant facts about tensor products of algebras.

Write $A \otimes B$ for the tensor product of F-algebras A and B, and write $A^{\otimes k}$ for the k-fold tensor power of A. If A is an algebra of $a \times a$ matrices and B an algebra of $b \times b$ matrices, then we identify the tensor product $A \otimes B$ with the algebra of $ab \times ab$ matrices spanned by the matrices $M \otimes N$, $M = (m_{ij}) \in A$ and $N = (n_{ij}) \in B$, where

$$M \otimes N = \begin{pmatrix} m_{11}N & \cdots & m_{1a}N \\ \vdots & & \vdots \\ m_{a1}N & \cdots & m_{aa}N \end{pmatrix}$$

In particular we have $F^{\otimes k} = F$.

A *partially commuting set of variables* is a tuple $\boldsymbol{X} = (X_1, \ldots, X_k)$, where the X_i are disjoint sets. Write $F\langle \boldsymbol{X} \rangle$ for the tensor product $F\langle X_1 \rangle \otimes \cdots \otimes F\langle X_k \rangle$. We think of $F\langle \boldsymbol{X} \rangle$ as a set of polynomials in partially commuting variables. Intuitively two variables $x, y \in X_i$ do not commute, whereas $x \in X_i$ commutes with $y \in X_j$ if $i \neq j$. Note that if each X_i is a singleton $\{x_i\}$ then $F\langle \boldsymbol{X} \rangle$ is the familiar ring of polynomials in commuting variables $x_1, \ldots, x_k$. At the other extreme, if $k = 1$ then we recover the non-commutative case.

An arbitrary element $f \in F\langle \boldsymbol{X} \rangle = F\langle X_1 \rangle \otimes \cdots \otimes F\langle X_k \rangle$ can be written uniquely as a finite sum of distinct *monomials*, where each monomial is a tensor product of elements of X_1^*, $X_2^*, \ldots$, and X_k^*. Formally, we can write

$$f = \sum_{i \in I} \alpha_i (m_{i,1} \otimes \cdots \otimes m_{i,k}), \tag{1}$$

where $\alpha_i \in F$ and $m_{i,j} \in X_j^*$ for each $i \in I$ and $1 \leq j \leq k$. Thus we can identify $F\langle \boldsymbol{X} \rangle$ with the free F-algebra over the product monoid $X_1^* \times \ldots \times X_k^*$.

Define the *degree* of a monomial $m_1 \otimes \ldots \otimes m_k$ to be the total length $|m_1| + \ldots + |m_k|$ of its constituent words. The degree of a polynomial is the maximum of the degrees of its constituent monomials.

Let $\boldsymbol{A} = (A_1, \ldots, A_k)$ be a k-tuple of F-algebras. A *valuation* of $F\langle \boldsymbol{X} \rangle$ in $\boldsymbol{A}$ is a tuple of functions $\boldsymbol{v} = (v_1, \ldots, v_k)$, where $v_i : X_i \to A_i$. Each v_i extends uniquely to an F-algebra homomorphism $\widetilde{v}_i : F\langle X_i \rangle \to A_i$, and we define the map $\widetilde{\boldsymbol{v}} : F\langle \boldsymbol{X} \rangle \to A_1 \otimes \ldots \otimes A_k$ by $\widetilde{\boldsymbol{v}} = \widetilde{v}_1 \otimes \ldots \otimes \widetilde{v}_k$. Often we will abuse terminology slightly and speak of a valuation of $F\langle \boldsymbol{X} \rangle$ in $A_1 \otimes \cdots \otimes A_k$. Given $f \in F\langle \boldsymbol{X} \rangle$, we say that $\boldsymbol{A}$ satisfies the partially commutative identity $f = 0$ if $\widetilde{\boldsymbol{v}}(f) = 0$ for all valuations $\boldsymbol{v}$.

Next we introduce two valuations that will play an important role in the subsequent development. Recall that given a set of non-commuting variables X, we have a map $\Phi_n^X : F\langle X \rangle \to F_n\langle X \rangle$ from the free F-algebra to the generic n-dimensional matrix algebra. We now define a valuation

$$\Phi_n^{\boldsymbol{X}} : F\langle \boldsymbol{X} \rangle \longrightarrow F_n\langle X_1 \rangle \otimes \cdots \otimes F_n\langle X_k \rangle \tag{2}$$

by $\Phi_n^{\boldsymbol{X}} = \Phi_n^{X_1} \otimes \cdots \otimes \Phi_n^{X_k}$. Likewise we define

$$\Psi_n^{\boldsymbol{X}} : F\langle \boldsymbol{X} \rangle \longrightarrow M_n(F\langle t_{ij}^{(x)} \mid x \in X_1, 1 \le i, j \le n \rangle) \otimes \cdots$$
$$\otimes M_n(F\langle t_{ij}^{(x)} \mid x \in X_k, 1 \le i, j \le n \rangle)$$

by $\Psi_n^{\boldsymbol{X}} = \Psi_n^{X_1} \otimes \cdots \otimes \Psi_n^{X_k}$. We will usually elide the superscript $\boldsymbol{X}$ from $\Phi_n^{\boldsymbol{X}}$ and $\Psi_n^{\boldsymbol{X}}$ when it is clear from the context.

The following result generalises (part of) the Amitsur-Levitzki theorem, by giving a lower bound on the degrees of partially polynomial identities holding in tensor products of matrix algebras.

Proposition 1. *Let* $f \in F\langle \boldsymbol{X} \rangle$ *and let* L *be a field extending* F. *Then the following are equivalent: (i) The partially commutative identity* $f = 0$ *holds in* $M_n(L) \otimes_F \cdots_F \otimes M_n(L)$; *(ii)* $\Phi_n(f) = 0$. *Moreover, if* f *has degree strictly less than* n *then (i) and (ii) are both equivalent to (iii)* $\Psi_n(f) = 0$; *and (iv)* f *is identically* 0 *in* $F\langle \boldsymbol{X} \rangle$.

Proof. The implication (ii) $\Rightarrow$ (i) follows from the fact that any valuation from $F\langle \boldsymbol{X} \rangle$ to $M_n(L) \otimes_F \cdots_F \otimes M_n(L)$ factors through Φ_n. To see that (i) $\Rightarrow$ (ii), observe that $\Phi_n(f)$ is an $n^k \times n^k$ matrix in which each entry is a polynomial in the commuting variables $t_{ij}^{(x)}$. Condition (i) implies in particular that each such polynomial evaluates to 0 for all valuations of its variables in F. Since F is an infinite field, it must be that each such polynomial is identically zero, i.e., (ii) holds.

The implications (ii) $\Rightarrow$ (iii) and (iv) $\Rightarrow$ (i) are both straightforward, even without the degree restriction on f.

Finally we show that (iii) $\Rightarrow$ (iv). Let $m_1 \otimes \ldots \otimes m_k$ be a monomial in $F\langle \boldsymbol{X} \rangle$, where $m_i = m_{i,1} \ldots m_{i,l_i} \in X_i^*$ has length $l_i < n$. Then $\Psi_n(m_1 \otimes \cdots \otimes m_k)$ is an $n^k \times n^k$ matrix whose first row has a single non-zero entry, which is the monomial

$$t_{12}^{(m_{1,1})} \ldots t_{l_1, l_1+1}^{(m_{1,l_1})} \cdots t_{12}^{(m_{k,1})} \ldots t_{l_k, l_k+1}^{(m_{k,l_k})} \tag{3}$$

at index $(1, \ldots, 1), (l_1 + 1, \ldots, l_k + 1)$.

It follows that Ψ_n maps the set of monomials in $F\langle \boldsymbol{X} \rangle$ of degree less than n injectively into a linearly independent set of matrices, each with a single monomial entry over the commuting indeterminates $t_{ij}^{(x)}$. Condition (iv) immediately follows. $\square$

The hypothesis that f have degree less than n in Proposition 1 can be weakened somewhat, but is sufficient for our purposes.

2.3   Division Rings and Ore Domains

A ring R with no zero divisors is a *domain*. If moreover each non-zero element of R has a two-sided multiplicative inverse, then we say that R is a *division ring* (also called a *skew field*). A domain R is a *(right) Ore domain* if for all $a, b \in R \setminus \{0\}$, $aR \cap bR \neq 0$. The significance of this notion is that an Ore domain can be embedded in a division ring of fractions [4, Corollary 7.1.6], something that need not hold for an arbitrary domain. If the Ore condition fails then it can easily be shown that the subalgebra of R generated by a and b is free on a and b. It follows that a domain R that satisfies some polynomial identity is an Ore domain [4, Corollary 7.5.2].

Proposition 2. *The tensor product of generic matrix algebras* $F_n\langle X_1 \rangle \otimes \cdots \otimes F_n\langle X_k \rangle$ *is an Ore domain for each* $n \in \mathbb{N}$.

Proof (sketch). We give a proof sketch here, deferring the details to Appendix A.

By the Amitsur-Levitzki theorem, $F_n\langle X_1 \rangle \otimes \cdots \otimes F_n\langle X_n \rangle$ satisfies a polynomial identity. Thus it suffices to show that $F_n\langle X_1 \rangle \otimes \cdots \otimes F_n\langle X_n \rangle$ is a domain for each n. Now it is shown in [4, Proposition 7.7.2] that $F_n\langle X \rangle$ is a domain for each n and set of variables X. While the tensor product of domains need not be a domain (e.g., $\mathbb{C} \otimes_{\mathbb{R}} \mathbb{C} \cong \mathbb{C} \times \mathbb{C}$), the proof in [4] can be adapted *mutatis mutandis* to show that $F\langle X_1 \rangle \otimes \cdots \otimes F\langle X_k \rangle$ is also a domain.

To prove the latter, it suffices to find central simple F-algebras $D_1, \ldots, D_k$, each of degree n, such that the k-fold tensor product $D \otimes_F \cdots \otimes_F D$ is a domain. Such an example can be found, e.g.,in [17, Proposition 1.1]. Then, using the fact that $D \otimes_F L \cong M_n(L)$ for any algebraically closed extension field of F, one can infer that $F_n\langle X_1 \rangle \otimes \cdots \otimes F_n\langle X_k \rangle$ is also a domain.   □

3   Multitape Automata

Let $\Sigma = (\Sigma_1, \ldots, \Sigma_k)$ be a tuple of finite alphabets. We denote by S the product monoid $\Sigma_1^* \times \cdots \times \Sigma_k^*$. Define the length of $s = (w_1, \ldots, w_k) \in S$ to be $|s| = |w_1| + \ldots + |w_k|$ and write $S^{(l)}$ for the set of elements of S of length l. A *multitape automaton* is a tuple $A = (\Sigma, Q, E, Q_0, F)$, where Q is a set of *states*, $E \subseteq Q \times S^{(1)} \times Q$ is a set of *edges*, $Q_0 \subseteq Q$ is a set of *initial states*, and $Q_f \subseteq Q$ is a set of *final states*. A *run* of A from state q_0 to state q_m is a finite sequence of edges $\rho = e_1 e_2 \ldots e_m$ such that $e_i = (q_{i-1}, s_i, q_i)$. The *label* of ρ is the product $s_1 s_2 \ldots s_m \in S$. Define the *multiplicity* $A(s)$ of an input $s \in S$ to be the number of runs with label s such that $q_0 \in Q_0$ and $q_m \in Q_f$. An automaton is *deterministic* if each state reads letters from a single tape and has a single transition for every input letter. Thus a deterministic automaton has a single run on each input $s \in S$.

3.1   Multiplicity Equivalence

We say that two automata A and B over the same alphabet are *multiplicity equivalent* if $A(s) = B(s)$ for all $s \in S$. The following result implies that multiplicity equivalence of multitape automata is decidable.

Theorem 2 (Harju and Karhumäki). *Given automata* A *and* B *with* n *states in total,* A *and* B *are equivalent if and only if* $A(s) = B(s)$ *for all* $s \in S$ *of length at most* $n - 1$.

Theorem 2 immediately yields a **co-NP** bound for checking language equivalence of deterministic multitape automata. Given two inequivalent automata A and B, a distinguishing input s can be guessed, and it can be verified in polynomial time that only one of A and B accepts s. A similar idea also gives a **co-NP** bound for multiplicity equivalence in case the number of tapes is fixed. In general we note however that the *evaluation problem*—given an automaton A and input s, compute $A(s)$—is #**P**-complete. Thus it is not clear that the **co-NP** upper bound applies to the multiplicity equivalence problem without bounding the number of tapes.

Proposition 3. *The evaluation problem for multitape automata is #**P**-complete.*

Proof. Membership in #**P** follows from the observation that a non-deterministic polynomial-time algorithm can enumerate all possible runs of an automaton A on an input $s \in S$.

The proof of #**P**-hardness is by reduction from #SAT, the problem of counting the number of satisfying assignments of a propositional formula. Consider such a formula φ with k variables, each with fewer than n occurrences. We define a k-tape automaton A, with each tape having alphabet $\{0,1\}$, and consider as input the k-tuple $s = ((01)^n, \ldots, (01)^n)$. The automaton A is constructed such that its runs on input s are in one-to-one correspondence with satisfying assignments of φ. Each run starts with the automaton reading the symbol 0 from a non-deterministically chosen subset of its tapes, corresponding to the set of false variables. Thereafter it evaluates the formula φ by repeatedly guessing truth values of the propositional variables. If the i-th variable is guessed to be true then the automaton reads 01 from the i-th tape; otherwise it reads 10 from the i-th tape. The final step is to read the symbol 1 from a non-deterministically chosen subset of the input tapes—again corresponding to the set of false variables. The consistency of the guesses is ensured by the requirement that the automaton have read s by the end of the computation. $\qquad\square$

3.2   Decidability

We start by recalling from [9] an equivalence-respecting transformation from multitape automata to single-tape weighted automata.

Recall that a single-tape automaton on a unary alphabet with transition weights in a ring R consists of a set of *states* $Q = \{q_1, \ldots, q_n\}$, *initial states* $Q_0 \subseteq Q$, *final states* $Q_f \subseteq Q$, and *transition matrix* $M \in M_n(R)$. Given such an automaton, define the *initial-state vector* $\boldsymbol{\alpha} \in R^{1 \times n}$ and *final-state vector* $\boldsymbol{\eta} \in R^{n \times 1}$ respectively by

$$\alpha_i = \begin{cases} 1 & \text{if } q_i \in Q_0 \\ 0 & \text{otherwise} \end{cases} \quad \text{and} \quad \eta_i = \begin{cases} 1 & \text{if } q_i \in Q_f \\ 0 & \text{otherwise} \end{cases}$$

Then $\boldsymbol{\alpha} M^l \boldsymbol{\eta}$ is the weight of the (unique) input word of length l.

Consider a k-tape automaton $A = (\boldsymbol{\Sigma}, Q, E, Q_0, Q_f)$, where $\boldsymbol{\Sigma} = (\Sigma_1, \ldots, \Sigma_k)$, and write $S = \Sigma_1^* \times \cdots \times \Sigma_k^*$. Recall the ring of polynomials

$$F\langle \boldsymbol{\Sigma} \rangle = F\langle \Sigma_1 \rangle \otimes \cdots \otimes F\langle \Sigma_n \rangle,$$

as defined in Section 2. Recall also that we can identify the monoid S with the set of monomials in $F\langle\Sigma\rangle$, where $(w_1, \ldots, w_k) \in S$ corresponds to $w_1 \otimes \cdots \otimes w_k$—indeed $F\langle\Sigma\rangle$ is the free F-algebra on S.

We derive from A an $F\langle\Sigma\rangle$-weighted automaton $\widetilde{A}$ (with a single tape and unary input alphabet) that has the same sets of states, initial states, and final states as A. We define the transition matrix M of $\widetilde{A}$ by combining the different transitions of A into a single matrix with entries in $F\langle\Sigma\rangle$. To this end, suppose that the set of states of A is $Q = \{q_1, \ldots, q_n\}$. Define the matrix $M \in M_n(F\langle\Sigma\rangle)$ by $M_{ij} = \sum_{(q_i,s,q_j)\in E} s$ for $1 \le i, j \le n$.

Let $\boldsymbol{\alpha}$ and $\boldsymbol{\eta}$ be the respective initial- and final-state vectors of $\widetilde{A}$. Then the following proposition is straightforward. Intuitively it says that the weight of the unary word of length l in $\widetilde{A}$ represents the language of all length-l tuples accepted by A.

Proposition 4. *For all $l \in \mathbb{N}$ we have $\boldsymbol{\alpha} M^l \boldsymbol{\eta} = \sum_{s\in S^{(l)}} A(s) \cdot s$.*

Now consider two k-tape automata A and B. Let the weighted single-tape automata derived from A and B have respective transition matrices M_A and M_B, initial-state vectors $\boldsymbol{\alpha}_A$ and $\boldsymbol{\alpha}_B$, and final-state vectors $\boldsymbol{\eta}_A$ and $\boldsymbol{\eta}_B$. We combine the latter into a single weighted automaton with transition matrix M, initial-state vector $\boldsymbol{\alpha}$, and final-state vector $\boldsymbol{\eta}$, respectively defined by:

$$\boldsymbol{\alpha} = (\boldsymbol{\alpha}_A \ \boldsymbol{\alpha}_B) \qquad M = \begin{pmatrix} M_A & 0 \\ 0 & M_B \end{pmatrix} \qquad \boldsymbol{\eta} = \begin{pmatrix} \boldsymbol{\eta}_A \\ -\boldsymbol{\eta}_B \end{pmatrix}$$

Proposition 5. *Automata A and B are multiplicity equivalent if and only if $\boldsymbol{\alpha} M^l \boldsymbol{\eta} = 0$ for $l = 0, 1, \ldots, n - 1$, where n is the total number of states of the two automata.*

Proof. Since S is a linearly independent subset of $F\langle\Sigma\rangle$, from Proposition 4 it follows that A and B are multiplicity equivalent just in case $\boldsymbol{\alpha}_A(M_A)^l\boldsymbol{\eta}_A = \boldsymbol{\alpha}_B(M_B)^l\boldsymbol{\eta}_B$ for all $l \in \mathbb{N}$. The latter is clearly equivalent to $\boldsymbol{\alpha} M^l \boldsymbol{\eta} = 0$ for all $l \in \mathbb{N}$. It remains to show that we can check equivalence by looking only at exponents l in the range $0, 1, \ldots, n - 1$.

Suppose that $\boldsymbol{\alpha} M^i \boldsymbol{\eta} = 0$ for $i = 0, \ldots, n - 1$. We show that $\boldsymbol{\alpha} M^l \boldsymbol{\eta} = 0$ for an arbitrary $l \ge n$.

Consider the map $\Phi_l : F\langle\Sigma\rangle \to F_l\langle\Sigma_1\rangle \otimes \cdots \otimes F_l\langle\Sigma_k\rangle$, as defined in (2). Observe that $\boldsymbol{\alpha} M^l \boldsymbol{\eta}$ is a polynomial expression in $F\langle\Sigma\rangle$ of degree at most l. Therefore by Proposition 1 ((ii) $\Leftrightarrow$ (iv)), to show that $\boldsymbol{\alpha} M^l \boldsymbol{\eta} = 0$ it suffices to show that

$$\Phi_l(\boldsymbol{\alpha} M^l \boldsymbol{\eta}) = 0. \tag{4}$$

Let us write $\Phi_l(M)$ for the pointwise application of Φ_l to the matrix M, so that $\Phi_l(M)$ is an $n \times n$ matrix, each of whose entries is an $n^k \times n^k$ matrix belonging to $F_l\langle\Sigma_1\rangle \otimes \cdots \otimes F_l\langle\Sigma_k\rangle$. Since Φ_l is a homomorphism and $\boldsymbol{\alpha}$ and $\boldsymbol{\eta}$ are integer vectors, (4) is equivalent to:

$$\alpha \, \Phi_l(M)^l \, \eta \, = \, 0 \, . \tag{5}$$

Recall from Proposition 2 that the tensor product of generic matrix algebras $F_l\langle \Sigma_1 \rangle \otimes \cdots \otimes F_l\langle \Sigma_k \rangle$ is an Ore domain and hence can be embedded in a division ring. Now a standard result about single-tape weighted automata with transition weights in a division ring is that such an automaton with n states is equivalent to the zero automaton if and only if it assigns zero weight to all words of length n (see [5, pp143–145] and [18]). Applying this result to the unary weighted automaton defined by α, M, and η, we see that (5) is implied by

$$\alpha \, \Phi_l(M)^i \, \eta \, = \, 0 \qquad i = 0, 1, \ldots, n-1 \, . \tag{6}$$

But, since Φ_l is a homomorphism, (6) is implied by

$$\alpha M^i \eta = 0 \qquad i = 0, 1, \ldots, n-1 \, . \tag{7}$$

This concludes the proof.   □

Theorem 2 immediately follows from Proposition 5.

Remark 1. The difference between our proof of Theorem 2 and the proof in [9] is that we consider a family of homomorphisms of $F\langle \Sigma \rangle$ into Ore domains of matrices—the maps Φ_l—rather than a single "global" embedding of $F\langle \Sigma \rangle$ into a division ring of power series over a product of free groups. None of the maps Φ_l is an embedding, but it suffices to use the lower bound on the degrees of polynomial identities in Proposition 1 in lieu of injectivity. On the other hand, the fact that $F_l\langle \Sigma_1 \rangle \otimes \cdots \otimes F_l\langle \Sigma_k \rangle$ satisfies a polynomial identity makes it relatively straightforward to exhibit an embedding of the latter into a division ring. As we now show, this approach leads directly to a very simple randomised polynomial-time algorithm for solving the equivalence problem.

3.3   Randomised Algorithm

Proposition 5 reduces the problem of checking multiplicity equivalence of multi-tape automata A and B to checking the partially commutative identities $\alpha M^l \eta = 0, l = 0, 1, \ldots, n-1$ in $F\langle \Sigma \rangle$. Since each identity has degree less than n, applying Proposition 1 ((iii) $\Leftrightarrow$ (iv)) we see that A and B are equivalent if and only if

$$\alpha \, \Psi_n(M)^l \, \eta \, = \, 0 \qquad l = 0, 1, \ldots, n-1 \, . \tag{8}$$

Each equation $\alpha \Psi_n(M)^l \eta = 0$ in (8) asserts the zeroness of an $n^k \times n^k$ matrix of polynomials in the commuting variables $t_{ij}^{(x)}$, with each polynomial having degree less than n. Suppose that $\alpha \Psi_n(M)^l \eta \neq 0$ for some l—say the matrix entry with index $((1, \ldots, 1), (l_1 + 1, \ldots, l_k + 1))$ contains a monomial with non-zero coefficient. By (3) such a monomial determines a term $s \in \Sigma_1^{l_1} \times \cdots \times \Sigma_k^{l_k}$ with non-zero coefficient in $\alpha M^l \eta$, and by Proposition 4 we have $A(s) \neq B(s)$.

We can verify each polynomial identity in (8), outputting a monomial of any non-zero polynomial, using a classical identity testing procedure based on the *isolation lemma* of [12].

Lemma 1 ([12]). *There is a randomised polynomial-time algorithm that inputs a multilinear polynomial $f(x_1, \ldots, x_m)$, represented as an algebraic circuit, and either outputs a monomial of f or that f is zero. Moreover the algorithm is always correct if f is the zero polynomial and is correct with probability at least $1/2$ if f is non-zero.*

The idea behind the algorithm described in Lemma 1 is to choose a weight $w_i \in \{1, \ldots, 2m\}$ for each variable x_i of f independently and uniformly at random. Defining the weight of a monomial $x_{i_1} \ldots x_{i_t}$ to be $w_{i_1} + \ldots + w_{i_t}$, then with probability at least $1/2$ there is a unique minimum-weight monomial. The existence of a minimum-weight monomial can be detected by computing the polynomial $g(y) = f(y^{w_1}, \ldots, y^{w_k})$, since a monomial with weight w in f yields a monomial of degree w in g. Using similar ideas one can moreover determine the composition of a minimum-weight monomial in f.

Applying Lemma 1 we obtain our main result:

Theorem 3. *Let k be fixed. Then multiplicity equivalence of k-tape automata can be decided in randomised polynomial time. Moreover there is a randomised polynomial algorithm for the function problem of computing a distinguishing input given two inequivalent automata.*

The reason for the requirement that k be fixed is because the dimension of the entries of the transition matrix M, and thus the number of polynomials to be checked for equality, depends exponentially on k.

The above use of the isolation technique generalises [10], where it is used to generate counterexample words of weighted single-tape automata. A very similar application in [2] occurs in the context of identity testing for non-commutative algebraic branching programs.

4   Conclusion

We have given a simple randomised algorithm for deciding language equivalence of deterministic multitape automata and multiplicity equivalence of nondeterministic automata. The algorithm arises directly from algebraic constructions used to establish decidability of the problem, and runs in polynomial time for each fixed number of tapes. We leave open the question of whether there is a deterministic polynomial-time algorithm for deciding the equivalence of deterministic and weighted multitape automata with a fixed number of tapes. (Recall that the 2-tape case is already known to be in polynomial time [7].) We also leave open whether there is a deterministic or randomised polynomial time algorithm for solving the problem in case the number of tapes is not fixed.

A   Proof of Proposition 2

We first recall a construction of a *crossed product division algebra* from [17, Proposition 1.1]. Let $z_1, \ldots, z_k$ be commuting indeterminates and write $F =$

$\mathbb{Q}(z_1^n, \ldots, z_k^n)$ for the field of rational functions obtained by adjoining $z_1^n, \ldots, z_k^n$ to $\mathbb{Q}$. Furthermore, let K/F be a field extension whose Galois group is generated by commuting automorphisms $\sigma_1, \ldots, \sigma_k$, each of order n, which has fixed field F. (Such an extension can easily be constructed by adjoining extra indeterminates to F, and having the σ_i be suitable permutations of the new indeterminates.) For each i, $1 \leq i \leq k$, write K_i for the subfield of K that is fixed by each σ_j for $j \neq i$; then define D_i to be the F-algebra generated by K_i and z_i such that $az_i = z_i\sigma_i(a)$ for all $a \in K_i$. Then each D_i is a simple algebra of dimension n^2 over its centre F. It is shown in [17, Proposition 1.1] that the tensor product $D_1 \otimes_F \cdots \otimes_F D_k$ can be characterised as the localisation of an iterated skew polynomial ring—and is therefore a domain.

The following two propositions are straightforward adaptations of [4, Proposition 7.5.5.] and [4, Proposition 7.7.2] to partially commutative identities.

Proposition 6. *Let $f \in F\langle X_1 \rangle \otimes \cdots \otimes F\langle X_k \rangle$. If the partially commutative identity $f = 0$ holds in $D_1 \otimes_F \cdots \otimes_F D_k$ then it also holds in $(D_1 \otimes_F L) \otimes_F \cdots \otimes_F (D_k \otimes_F L)$ for any extension field L of F.* •

Proof. Noting that the D_i are all isomorphic as F-algebras, let $\{e_1, \ldots, e_{n^2}\}$ be a basis of each D_i over its centre F. For each variable x appearing in f, introduce commuting indeterminates t_{xj}, $1 \leq j \leq n^2$, and write $x = \sum_{j=1}^{n^2} t_{xj}e_j$. Then we can express f in the form

$$f = \sum_{\nu \in \{1, \ldots, n^2\}^k} f_\nu \cdot (e_{\nu(1)} \otimes \cdots \otimes e_{\nu(k)}), \tag{9}$$

where $f_\nu \in F\langle t_{xj} : x \in X_1, 1 \leq j \leq n^2 \rangle \otimes_F \cdots \otimes_F F\langle t_{xj} : x \in X_k, 1 \leq j \leq n^2 \rangle$.

By assumption, each f_ν evaluates to 0 for all values of the t_{xj} in F. Since F is an infinite field it follows that each f_ν must be identically zero. Now we can also regard $\{e_1, \ldots, e_{n^2}\}$ as a basis for $D_i \otimes_F L$ over L. Then by (9), $f = 0$ also on $(D_1 \otimes_F L) \otimes_F \cdots \otimes_F (D_k \otimes_F L)$. □

Proposition 7. $F_n\langle X_1 \rangle \otimes \cdots \otimes F_n\langle X_k \rangle$ *is a domain.*

Proof. Recall that if L is an algebraically closed field extension of F, then we have $D_i \otimes_F L \cong M_n(L)$ for each i. By Proposition 6 it follows that an identity $f = 0$ holds in $D_1 \otimes_F \cdots \otimes_F D_k$ if and only if it holds in $M_n(L) \otimes_F \cdots \otimes_F M_n(L)$. But by Proposition 1 the latter holds if and only if $\Phi_n(f)$ is identically zero.

To prove the proposition it will suffice to show that the image of Φ_n contains no zero divisors, since the latter is a surjective map. Now given $f, g \in F\langle X_1 \rangle \otimes \cdots \otimes F\langle X_k \rangle$ with $\Phi_n(fg) = 0$, we have that $D_1 \otimes_F \cdots \otimes_F D_k$ satisfies the identity $fg = 0$. Since $D_1 \otimes_F \cdots \otimes_F D_k$ is a domain, it follows that it satisfies the identity $fhg = 0$ for any h in $F\langle X_1 \rangle \otimes \cdots \otimes F\langle X_k \rangle$. But now $M_n(L) \otimes_F \cdots \otimes_F M_n(L)$ satisfies the identity $fhg = 0$ for any h. Since h can take the value of an arbitrary matrix (in particular, any matrix unit) it follows that $M_n(L) \otimes_F \cdots \otimes_F M_n(L)$ satisfies either the identity $f = 0$ or $g = 0$, and so, by Proposition 1 again, either $\Phi_n(f) = 0$ or $\Phi_n(g) = 0$. □

Acknowledgments. The author is grateful to Louis Rowen for helpful pointers in the proof of Proposition 2.

References

1. Amitsur, S.A., Levitzki, J.: Minimal identities for algebras. Proceedings of the American Mathematical Society 1, 449–463 (1950)
2. Arvind, V., Mukhopadhyay, P.: Derandomizing the isolation lemma and lower bounds for circuit size. In: Goel, A., Jansen, K., Rolim, J.D.P., Rubinfeld, R. (eds.) APPROX and RANDOM 2008. LNCS, vol. 5171, pp. 276–289. Springer, Heidelberg (2008)
3. Bogdanov, A., Wee, H.: More on noncommutative polynomial identity testing. In: IEEE Conference on Computational Complexity, pp. 92–99. IEEE Computer Society (2005)
4. Cohn, P.M.: Further Algebra and Applications. Springer (2003)
5. Eilenberg, S.: Automata, Languages, and Machines, vol. A. Academic Press (1974)
6. Elgot, C.C., Mezei, J.E.: Two-sided finite-state transductions (abbreviated version). In: SWCT (FOCS), pp. 17–22. IEEE Computer Society (1963)
7. Friedman, E.P., Greibach, S.A.: A polynomial time algorithm for deciding the equivalence problem for 2-tape deterministic finite state acceptors. SIAM J. Comput. 11(1), 166–183 (1982)
8. Griffiths, T.V.: The unsolvability of the equivalence problem for ϵ-free nondeterministic generalized machines. J. ACM 15(3), 409–413 (1968)
9. Harju, T., Karhumäki, J.: The equivalence problem of multitape finite automata. Theor. Comput. Sci. 78(2), 347–355 (1991)
10. Kiefer, S., Murawski, A., Ouaknine, J., Wachter, B., Worrell, J.: On the complexity of equivalence and minimisation for Q-weighted automata. Logical Methods in Computer Science 9 (2013)
11. Malcev, A.I.: On the embedding of group algebras in division algebras. Dokl. Akad. Nauk 60, 1409–1501 (1948)
12. Mulmuley, K., Vazirani, U.V., Vazirani, V.V.: Matching is as easy as matrix inversion. In: STOC, pp. 345–354 (1987)
13. Neumann, B.H.: On ordered groups. Amer. J. Math. 71, 1–18 (1949)
14. Neumann, B.H.: On ordered division rings. Trans. Amer. Math. Soc. 66, 202–252 (1949)
15. Rabin, M., Scott, D.: Finite automata and their decision problems. IBM Journal of Research and Development 3(2), 114–125 (1959)
16. Sakarovich, J.: Elements of Automata Theory. Cambridge University Press (2003)
17. Saltman, D.: Lectures on Division Algebras. American Math. Soc. (1999)
18. Schützenberger, M.-P.: On the definition of a family of automata. Inf. and Control 4, 245–270 (1961)
19. Tzeng, W.: A polynomial-time algorithm for the equivalence of probabilistic automata. SIAM Journal on Computing 21(2), 216–227 (1992)

Silent Transitions in Automata with Storage[*]

Georg Zetzsche

Fachbereich Informatik, Technische Universität Kaiserslautern,
Postfach 3049, 67653 Kaiserslautern, Germany
zetzsche@cs.uni-kl.de

Abstract. We consider the computational power of silent transitions in one-way automata with storage. Specifically, we ask which storage mechanisms admit a transformation of a given automaton into one that accepts the same language and reads at least one input symbol in each step. We study this question using the model of valence automata. Here, a finite automaton is equipped with a storage mechanism that is given by a monoid. This work presents generalizations of known results on silent transitions. For two classes of monoids, it provides characterizations of those monoids that allow the removal of silent transitions. Both classes are defined by graph products of copies of the bicyclic monoid and the group of integers. The first class contains pushdown storages as well as the blind counters while the second class contains the blind and the partially blind counters.

1 Introduction

In a one-way automaton, a transition is called *silent* if it reads no input symbol. If it has no silent transitions, such an automaton is called *real-time*. We consider the problem of removing silent transitions from one-way automata with various kinds of storage. Specifically, we ask for which kinds of storage the real-time and the general version have equal computational power. This is an interesting problem for two reasons. First, it has consequences for the time and space complexity of the membership problem for these automata. For automata with silent transitions, it is not even clear whether the membership problem is decidable. If, however, an automaton has no silent transitions, we only have to consider paths that are at most as long as the word at hand. In particular, if we can decide whether a sequence of storage operations is valid using linear space, we can also solve the membership problem (nondeterministically) with a linear space bound. Similarly, if we can decide validity of such a sequence in polynomial time, we can solve the membership problem in (nondeterministic) polynomial time. Second, we can interpret the problem as a question on resource consumption of restricted machine models: we ask for which storage mechanisms we can process every input word by executing only a bounded number of operations per symbol.

---

[*] This is an extended abstract. The full version of this work is available under
http://arxiv.org/abs/1302.3798.

F.V. Fomin et al. (Eds.): ICALP 2013, Part II, LNCS 7966, pp. 434–445, 2013.
© Springer-Verlag Berlin Heidelberg 2013

There is a wide variety of machine models that consist of a finite state control with a one-way input and some mechanism to store data, for example (higher order) pushdown automata, various kinds of counter automata [8], or off-line Turing machines that can only move right on the input tape.

For some of these models, it is known whether silent transitions can be eliminated. For example, the Greibach normal form allows their removal from *pushdown automata*. Furthermore, for *blind counter automata* (i.e., the counters can go below zero and a zero-test is only performed in the end), Greibach also has also shown that silent transitions can be avoided [8]. However, for *partially blind counter automata* (i.e., the counters cannot go below zero and are only zero-tested in the end) or, equivalently, Petri nets, there are languages for which silent transitions are indeed necessary [8, 11].

The aim of this work is to generalize these results and obtain insights into how the structure of the storage mechanism influences the computational power of the real-time variant. In order to study the expressive power of real-time computations in greater generality, we use the model of *valence automata*. For our purposes, a storage mechanism consists of a (possibly infinite) set of states and partial transformations operating on them. Such a mechanism often works in a way such that a computation is considered valid if the composition of the applied transformations is the identity. For example, in a pushdown storage, the operations *push* and *pop* (for each participating stack symbol) and compositions thereof are partial transformations on the set of words over some alphabet. In this case, a computation is valid if, in the end, the stack is brought back to the initial state, i.e., the identity transformation has been applied. Furthermore, in a partially blind counter automaton, a computation is valid if it leaves the counters with value zero, i.e., the composition of the applied operations *increase* and *decrease* is the identity. Therefore, the set of all compositions of the partial transformations forms a monoid such that, for many storage mechanisms, a computation is valid if the composition of the transformations is the identity.

A valence automaton is a finite automaton in which each edge carries, in addition to an input word, an element of a monoid. A word is then accepted if there is a computation that spells the word and for which the product of the monoid elements is the identity. Valence automata have been studied throughout the last decades [4, 5, 10, 12, 16, 17].

The contribution of this work is threefold. On the one hand, we introduce a class of monoids that accommodates, among others, all storage mechanisms for which we mentioned previous results on silent transitions. The monoids in this class are graph products of copies of the bicyclic monoid and the integers. On the other hand, we present two generalizations of those established facts. Our first main result is a characterization of those monoids in a certain subclass for which silent transitions can be eliminated. This subclass contains, among others, both the monoids corresponding to pushdown storages as well as those corresponding to blind multicounter storages. Thus, we obtain a generalization and unification of two of the three λ-removal results above. For those storage mechanisms in this subclass for which we can remove silent transitions, there is a simple intuitive description. As the simplest example of storages covered by our result beyond pushdowns and blind multicounters, *Parikh pushdown automata* [13] are also provided with a λ-removal procedure.

The second main result is a characterization of the previous kind for the class of those storage mechanisms that consist of a number of blind counters and a number of partially blind counters. Specifically, we show that we can remove silent transitions if and only if there is at most one partially blind counter. Again, this generalizes and unifies two of the three results above. It should be noted that *all our results are effective*.

In Section 2, we will fix notation and define some basic concepts. In Section 3, we state the main results, describe how they relate to what is known, and explain key ideas. Sections 4, 5, and 6 contain auxiliary results needed in Section 7, which presents an outline of the proofs of the main results.

2   Basic Notions

A *monoid* is a set M together with an associative operation and a neutral element. Unless defined otherwise, we will denote the neutral element of a monoid by 1 and its operation by juxtaposition. That is, for a monoid M and $a, b \in M$, $ab \in M$ is their product. For $a, b \in M$, we write $a \sqsubseteq b$ if there is a $c \in M$ such that $b = ac$. By **1**, we denote the trivial monoid that consists of just one element.

We call a monoid *commutative* if $ab = ba$ for any $a, b \in M$. A subset $N \subseteq M$ is said to be a *submonoid of M* if $1 \in N$ and $a, b \in N$ implies $ab \in N$. In each monoid M, we have the submonoids $\mathsf{H}(M) = \{a \in M \mid \exists b \in M : ab = ba = 1\}$, $\mathsf{R}(M) = \{a \in M \mid \exists b \in M : ab = 1\}$, and $\mathsf{L}(M) = \{a \in M \mid \exists b \in M : ba = 1\}$. When using a monoid M as part of a control mechanism, the subset $\mathsf{J}(M) = \{a \in M \mid \exists b, c \in M : bac = 1\}$ will play an important role. By M^n, we denote the n-fold direct product of M, i.e. $M^n = M \times \cdots \times M$ with n factors.

Let $S \subseteq M$ be a subset. If there is no danger of confusion with the n-fold direct product, we write S^n for the set of all elements of M that can be written as a product of n factors from S.

Let Σ be a fixed countable set of abstract symbols, the finite subsets of which are called *alphabets*. For an alphabet X, we will write X^* for the set of words over X. The empty word is denoted by $\lambda \in X^*$. Together with concatenation as its operation, X^* is a monoid. For a symbol $x \in X$ and a word $w \in X^*$, let $|w|_x$ be the number of occurrences of x in w. Given an alphabet X and a monoid M, subsets of X^* and $X^* \times M$ are called *languages* and *transductions*, respectively. A *family* is a set of languages that is closed under isomorphism and contains at least one non-trivial member.

Given an alphabet X, we write $X^\oplus$ for the set of maps $\alpha : X \to \mathbb{N}$. Elements of $X^\oplus$ are called *multisets*. By way of pointwise addition, written $\alpha + \beta$, $X^\oplus$ is a commutative monoid. We write 0 for the empty multiset, i.e. the one that maps every $x \in X$ to $0 \in \mathbb{N}$. For $\alpha \in X^\oplus$, let $|\alpha| = \sum_{x \in X} \alpha(x)$. The *Parikh mapping* is the mapping $\Psi : \Sigma^* \to \Sigma^\oplus$ with $\Psi(w)(x) = |w|_x$ for $w \in \Sigma^*$ and $x \in \Sigma$.

Let A be a (not necessarily finite) set of symbols and $R \subseteq A^* \times A^*$. The pair (A, R) is called a *(monoid) presentation*. The smallest congruence of A^*

containing R is denoted by $\equiv_R$ and we will write $[w]_R$ for the congruence class of $w \in A^*$. The *monoid presented by* (A, R) is defined as $A^*/\equiv_R$. Note that since we did not impose a finiteness restriction on A, every monoid has a presentation. Furthermore, for monoids M_1, M_2 we can find presentations (A_1, R_1) and (A_2, R_2) such that $A_1 \cap A_2 = \emptyset$. We define the *free product* $M_1 * M_2$ to be presented by $(A_1 \cup A_2, R_1 \cup R_2)$. Note that $M_1 * M_2$ is well-defined up to isomorphism. By way of the injective morphisms $[w]_{R_i} \mapsto [w]_{R_1 \cup R_2}$, $w \in A_i^*$ for $i = 1, 2$, we will regard M_1 and M_2 as subsets of $M_1 * M_2$. In analogy to the n-fold direct product, we write $M^{(n)}$ for the n-fold free product of M.

Rational Sets. Let M be a monoid. An *automaton over* M is a tuple $A = (Q, M, E, q_0, F)$, in which Q is a finite set of *states*, E is a finite subset of $Q \times M \times Q$ called the set of *edges*, $q_0 \in Q$ is the *initial state*, and $F \subseteq Q$ is the set of *final states*. The *step relation* $\Rightarrow_A$ of A is a binary relation on $Q \times M$, for which $(p, a) \Rightarrow_A (q, b)$ iff there is an edge (p, c, q) such that $b = ac$. The set generated by A is then $S(A) = \{a \in M \mid \exists q \in F : (q_0, 1) \Rightarrow_A^* (q, a)\}$.

A set $R \subseteq M$ is called *rational* if it can be written as $R = S(A)$ for some automaton A over M. The set of rational subsets of M is denoted by $\mathsf{RAT}(M)$. Given two subsets $S, T \subseteq M$, we define $ST = \{st \mid s \in S, t \in T\}$. Since $\{1\} \in \mathsf{RAT}(M)$ and $ST \in \mathsf{RAT}(M)$ whenever $S, T \in \mathsf{RAT}(M)$, this operation makes $\mathsf{RAT}(M)$ a monoid itself.

Let C be a commutative monoid for which we write the composition additively. For $n \in \mathbb{N}$ and $c \in C$, we use nc to denote $c + \cdots + c$ (n summands). A subset $S \subseteq C$ is *linear* if there are elements $s_0, \ldots, s_n$ such that $S = \{s_0 + \sum_{i=1}^n a_i s_i \mid a_i \in \mathbb{N}, 1 \leq i \leq n\}$. A set $S \subseteq C$ is called *semilinear* if it is a finite union of linear sets. By $\mathsf{SL}(C)$, we denote the set of semilinear subsets of C. It is well-known that $\mathsf{RAT}(C) = \mathsf{SL}(C)$ for commutative C (we will, however, sometimes still use $\mathsf{SL}(C)$ to make explicit that the sets at hand are semilinear). Moreover, $\mathsf{SL}(C)$ is a commutative monoid by way of the product $(S, T) \mapsto S + T = \{s + t \mid s \in S, t \in T\}$. It is well-known that the class of semilinear subsets of a free commutative monoid is closed under intersection [6].

In slight abuse of terminology, we will sometimes call a language L semilinear if the set $\Psi(L)$ is semilinear. If there is no danger of confusion, we will write $S^\oplus$ instead of $\langle S \rangle$ if S is a subset of a commutative monoid C. Note that if X is regarded as a subset of $X^\oplus$, the two meanings of $X^\oplus$ coincide.

Valence Automata. A *valence automaton over* M is an automaton A over the monoid $X^* \times M$, where X is an alphabet. An edge (p, w, m, q) in A is called a λ-*transition* if $w = \lambda$. A is called λ-*free* if it has no λ-transitions. The *language accepted by* A is defined as $L(A) = \{w \in X^* \mid (w, 1) \in S(A)\}$. The class of languages accepted by valence automata and λ-free valence automata over M is denoted by $\mathsf{VA}(M)$ and $\mathsf{VA}^+(M)$, respectively.

A *finite automaton* is a valence automaton over the trivial monoid $\mathbf{1}$. For a finite automaton $A = (Q, X^* \times \mathbf{1}, E, q_0, F)$, we also write $A = (Q, X, E, q_0, F)$. Languages accepted by finite automata are called *regular languages*. The finite automaton A is *spelling*, if $E \subseteq Q \times X \times Q$, i.e. every edges carries exactly one letter. Let M and C be monoids. A *valence transducer over M with output in C* is an automaton A over $X^* \times M \times C$, where X is an alphabet. The *transduction performed by A* is $T(A) = \{(x,c) \in X^* \times C \mid (x,1,c) \in S(A)\}$. A valence transducer is called λ-*free* if it is λ-free as a valence automaton. We denote the class of transductions performed by (λ-free) valence transducers over M with output in C by $\mathsf{VT}(M,C)$ ($\mathsf{VT}^+(M,C)$).

Graphs. A *graph* is a pair $\Gamma = (V,E)$ where V is a finite set and $E \subseteq \{S \subseteq V \mid 1 \le |S| \le 2\}$. The elements of V are called *vertices* and those of E are called *edges*. Vertices $v,w \in V$ are *adjacent* if $\{v,w\} \in E$. If $\{v\} \in E$ for some $v \in V$, then v is called a *looped* vertex, otherwise it is *unlooped*. A *subgraph* of Γ is a graph (V',E') with $V' \subseteq V$ and $E' \subseteq E$. Such a subgraph is called *induced (by V')* if $E' = \{S \in E \mid S \subseteq V'\}$, i.e. E' contains all edges from E incident to vertices in V'. By $\Gamma \setminus \{v\}$, for $v \in V$, we denote the subgraph of Γ induced by $V \setminus \{v\}$. Given a graph $\Gamma = (V,E)$, its *underlying loop-free graph* is $\Gamma' = (V,E')$ with $E' = E \cap \{S \subseteq V \mid |S| = 2\}$. For a vertex $v \in V$, the elements of $N(v) = \{w \in V \mid \{v,w\} \in E\}$ are called *neighbors* of v. A *looped clique* is a graph in which $E = \{S \subseteq V \mid 1 \le |S| \le 2\}$. Moreover, a *clique* is a loop-free graph in which any two distinct vertices are adjacent. Finally, an *anti-clique* is a graph with $E = \emptyset$.

A presentation (A,R) in which A is a finite alphabet is a *Thue system*. To each graph $\Gamma = (V,E)$, we associate the Thue system $T_\Gamma = (X_\Gamma, R_\Gamma)$ over the alphabet $X_\Gamma = \{a_v, \bar{a}_v \mid v \in V\}$. R_Γ is defined as

$$R_\Gamma = \{(a_v \bar{a}_v, \lambda) \mid v \in V\} \cup \{(xy, yx) \mid x \in \{a_v, \bar{a}_v\},\ y \in \{a_w, \bar{a}_w\},\ \{v,w\} \in E\}.$$

In particular, we have $(a_v \bar{a}_v, \bar{a}_v a_v) \in R_\Gamma$ whenever $\{v\} \in E$. To simplify notation, the congruence $\equiv_{T_\Gamma}$ is then also denoted by $\equiv_\Gamma$ and $[w]_{T_\Gamma}$ is also denoted $[w]_\Gamma$. In order to describe the monoids we use to model storage mechanisms, we define monoids using graphs. To each graph Γ, we associate the monoid

$$\mathbb{M}\Gamma = X_\Gamma^* / \equiv_\Gamma.$$

If Γ consists of one vertex and has no edges, $\mathbb{M}\Gamma$ is also denoted as $\mathbb{B}$ and we will refer to it as the *bicyclic monoid*. The generators a_v and $\bar{a}_v$ are then also written a and $\bar{a}$, respectively.

3   Overview of Results

Storage Mechanisms as Monoids. First, we will see how pushdown storages and (partially) blind counters can be regarded as monoids of the form $\mathbb{M}\Gamma$. See Table 1 for examples. Clearly, in the bicyclic monoid $\mathbb{B}$, a word over the generators a and $\bar{a}$ is the identity if and only if in every prefix of the word, there are at least as many a's as there are $\bar{a}$'s and in the whole word, there are as many a's as there are $\bar{a}$'s. Thus, a valence automaton over $\mathbb{B}$ is an automaton with one counter that cannot go below zero and is zero in the end. Here, the increment operation corresponds to a and the decrement corresponds to $\bar{a}$.

Table 1. Examples of storage mechanisms

Graph Γ	Monoid $M\Gamma$	Storage mechanism
	$\mathbb{B}^{(3)}$	Pushdown (with three symbols)
	$\mathbb{B}^3$	Three partially blind counters
	$\mathbb{Z}^3$	Three blind counters
	$\mathbb{B}^{(2)} \times \mathbb{Z}^2$	Pushdown (with two symbols) and two blind counters

Observe that building the direct product means that both storage mechanisms (described by the factors) are available and can be used simultaneously. Thus, valence automata over $\mathbb{B}^n$ are automata with n partially blind counters. Therefore, if Γ is a clique, then $M\Gamma \cong \mathbb{B}^n$ corresponds to a partially blind multicounter storage.

Furthermore, the free product of a monoid M with $\mathbb{B}$ yields what can be seen as a stack of elements of M: a valence automaton over $M * \mathbb{B}$ can store a sequence of elements of M (separated by a) such that it can only remove the topmost element if it is the identity element. The available operations are those available for M (which then operate on the topmost entry) and in addition *push* (represented by a) and *pop* (represented by $\bar{a}$). Thus, $\mathbb{B} * \mathbb{B}$ corresponds to a stack over two symbols. In particular, if Γ is an anti-clique (with at least two vertices), then $M\Gamma \cong \mathbb{B}^{(n)}$ represents a pushdown storage.

Finally, valence automata over $\mathbb{Z}^n$ (regarded as a monoid by way of addition) correspond to automata with n blind counters. Hence, if Γ is a looped clique, then $M\Gamma \cong \mathbb{Z}^n$ corresponds to a blind multicounter storage.

Main Results. Our class of monoids that generalizes pushdown and blind multicounter storages is the class of $M\Gamma$ where in Γ, any two looped vertices are adjacent and no two unlooped vertices are adjacent. Our first main result is the following.

Theorem 1. *Let Γ be a graph such that any two looped vertices are adjacent and no two unlooped vertices are adjacent. Then, the following are equivalent:*

(1) $\mathsf{VA}^+(M\Gamma) = \mathsf{VA}(M\Gamma)$.
(2) Every language in $\mathsf{VA}(M\Gamma)$ *is context-sensitive.*
(3) The membership problem of each language in $\mathsf{VA}(M\Gamma)$ *is in* NP.
(4) Every language in $\mathsf{VA}(M\Gamma)$ *is decidable.*
(5) Γ *does not contain* *as an induced subgraph.*

Note that this generalizes the facts that in pushdown automata and in blind counter automata, λ-transitions can be avoided. Furthermore, while Greibach's construction triples the number of counters, we do not need any additional ones.

It turns out that the storages that satisfy the equivalent conditions of Theorem 1 (and the hypothesis), are exactly those in the following class.

Definition 1. *Let $\mathcal{C}$ be the smallest class of monoids such that $1 \in \mathcal{C}$ and whenever $M \in \mathcal{C}$, we also have $M \times \mathbb{Z} \in \mathcal{C}$ and $M * \mathbb{B} \in \mathcal{C}$.*

Thus, $\mathcal{C}$ contains those storage types obtained by successively *adding blind counters* and *building a stack of elements*. For example, we could have a stack each of whose entries contains n blind counters. Or we could have an ordinary pushdown and a number of blind counters. Or a stack of elements, each of which is a pushdown storage and a blind counter, etc. The simplest example of a storage mechanism in $\mathcal{C}$ beyond blind multicounters and pushdowns is given by the monoids $(\mathbb{B} * \mathbb{B}) \times \mathbb{Z}^n$ for $n \in \mathbb{N}$. It is not hard to see that these yield the same languages as *Parikh pushdown automata* [13]. Hence, our result implies that the latter also permit the removal of λ-transitions.

Our second main result concerns storages consisting of a number of blind counters and a number of partially blind counters.

Theorem 2. *Let Γ be a graph such that any two distinct vertices are adjacent. Then, $\mathsf{VA}^+(\mathbb{M}\Gamma) = \mathsf{VA}(\mathbb{M}\Gamma)$ if and only if $r \leq 1$, where r is the number of unlooped vertices in Γ.*

In other words, when one has r partially blind counters and s blind counters, λ-transitions can be eliminated if and only if $r \leq 1$. Note that this generalizes Greibach's result that in partially blind multicounter automata, λ-transitions are indispensable.

Key Technical Ingredients. As a first step, we show that for $M \in \mathcal{C}$, all languages in $\mathsf{VA}(M)$ are semilinear. This is needed in various situations throughout the proof. We prove this using an old result by van Leeuwen [14], which states that languages that are algebraic over a class of semilinear languages are semilinear themselves. Thereby, the corresponding Lemma 2 slightly generalizes one of the central components in a decidability result by Lohrey and Steinberg on the rational subset membership problem for graph groups [15] and provides a simpler proof (relying, however, on van Leeuwen's result).

Second, we use an undecidability result by Lohrey and Steinberg [15] concerning the rational subset membership problem for certain graph groups. We deduce that for monoids M outside of $\mathcal{C}$ (and satisfying the hypothesis of Theorem 1), $\mathsf{VA}(M)$ contains an undecidable language.

Third, in order to prove our claim by induction on the construction of $M \in \mathcal{C}$, we use a significantly stronger induction hypothesis: We show that it is not only possible to remove λ-transitions from valence automata, but also from valence transducers with output in a commutative monoid. Here, however, the constructed valence transducer is allowed to output a semilinear set in each step. Monoids that admit such a transformation will be called *strongly λ-independent*.

Fourth, we develop a normal form result for rational subsets of monoids in $\mathcal{C}$ (see Section 6). Such normal form results have been available for monoids described by monadic rewriting systems (see, for example, [1]), which was applied by Render and Kambites to monoids representing pushdown storages [17]. Under different terms, this normal form trick has been used by Bouajjani, Esparza, and Maler [2] and by Caucal [3] to describe rational sets of pushdown operations. However, since the monoids in $\mathcal{C}$ allow commutation of certain non-trivial elements, a technique more general than those was necessary here. In the case of monadic rewriting systems, one transforms a finite automaton according to rewriting rules by gluing in new edges. Here, we glue in automata accepting sets that are semilinear by earlier steps in the proof. See Lemma 6 for details.

Fifth, we have three new techniques to eliminate λ-transitions from valence transducers while retaining the output in a commutative monoid. Here, we need one technique to show that if M is strongly λ-independent, then $M \times \mathbb{Z}$ is as well. This technique again uses the semilinearity of certain sets and a result that provides small preimages for morphisms from multisets to the integers.

The second technique is to show that $\mathbb{B}$ is strongly λ-independent. We use a construction that allows the postponement of increment operations and the early execution of decrement operations. This is used to show that one can restrict to computations in which a sequence of increments, followed by a sequence of decrements, will in the end change the counter only by a bounded amount.

The third technique is to show that if M is strongly λ-independent, where M is non-trivial, then $M * \mathbb{B}$ is as well. Here, the storage consists of a stack of elements of M. The construction works by encoding rational sets over $M * \mathbb{B}$ as elements on the stack. We have to use the semilinearity results again in order to be able to compute the set of all possible outputs when elements from two given rational sets cancel each other out (in the sense that push operations are followed by pop operations).

4 Semilinear Languages

This section contains semilinearity results that will be needed in later sections. The first lemma guarantees small preimages of morphisms from multisets to the integers. This will be used to bound the number of necessary operations on a blind counter in order to obtain a given counter value.

Lemma 1. *Let $\varphi : X^{\oplus} \to \mathbb{Z}$ be a morphism. Then for any $n \in \mathbb{Z}$, the set $\varphi^{-1}(n)$ is semilinear. In particular, $\ker \varphi$ is finitely generated. Furthermore, there is a constant $k \in \mathbb{N}$ such that for any $\mu \in X^{\oplus}$, there is a $\nu \sqsubseteq \mu$ with $\mu \in \nu + \ker \varphi$ and $|\nu| \leq k \cdot |\varphi(\mu)|$.*

Another fact used in later sections is that languages in $\mathsf{VA}(M)$ are semilinear if $M \in \mathcal{C}$. This will be employed in various constructions, for instance when the effect of computations (that make use of M as storage) on the output in a commutative monoid is to be realized by a finite automaton. We prove this using a result of van Leeuwen [14], which states that semilinearity of all languages in a family is inherited by languages that are algebraic over this family. A language is called algebraic over a family of languages if it is generated by a grammar in which each production allows a non-terminal to be replaced by any word from a language in this family.

Note that in [15], a group G is called *SLI-group* if every language in $\mathsf{VA}(G)$ is semilinear (in different terms, however). Thus, the following recovers the result from [15] that the class of SLI-groups is closed under taking the free product.

Lemma 2. *Every $L \in \mathsf{VA}(M_0 * M_1)$ is algebraic over $\mathsf{VA}(M_0) \cup \mathsf{VA}(M_1)$.*

Combining the latter lemma with van Leeuwen's result and a standard argument for the preservation of semilinearity when builing the direct product with $\mathbb{Z}$ yields the following.

Lemma 3. *Let $M \in \mathcal{C}$. Then, every language in $\mathsf{VA}(M)$ is semilinear.*

5   Membership Problems

In this section, we study decidability and complexity of the membership problem for valence automata over $M\Gamma$. Specifically, we show in this section that for certain graphs Γ, the class $\mathsf{VA}(M\Gamma)$ contains undecidable languages (Lemma 5), while for every Γ, membership for languages in $\mathsf{VA}^+(M\Gamma)$ is (uniformly) decidable. We present two nondeterministic algorithms, one of them uses linear space and one runs in polynomial time (Lemma 4).

These results serve two purposes. First, for those graphs Γ for which there are undecidable languages in $\mathsf{VA}(M\Gamma)$, it follows that silent transitions are indispensable. Second, if we can show that silent transitions can be removed from valence automata over $M\Gamma$, the algorithms also apply to languages in $\mathsf{VA}(M\Gamma)$.

Our algorithms rely on the convergence property of certain reduction systems. For more information on reduction systems, see [1, 9]. The following lemma makes use of two algorithms to decide, given a word $w \in X_\Gamma^*$, whether $[w]_\Gamma = [\lambda]_\Gamma$. Specifically, we have a deterministic polynomial-time algorithm that employs a convergent trace rewriting system to successively reduce a dependence graph, which is then checked for emptiness. On the other hand, the convergence of the same rewriting system is used in a (nondeterministic) linear space algorithm to decide the equality above. These two algorithms are then used to verify the validity of a guessed run to decide the membership problem for languages in $\mathsf{VA}^+(M\Gamma)$.

Lemma 4. *For each $L \in \mathsf{VA}^+(M\Gamma)$, the membership problem can be decided by a nondeterministic polynomial-time algorithm as well as a nondeterministic linear-space algorithm. Hence, the languages in $\mathsf{VA}^+(M\Gamma)$ are context-sensitive.*

The undecidability result is shown by reducing the rational subset membership problem of the graph group corresponding to a path on four vertices, which was proven undecidable by Lohrey and Steinberg [15], to the membership problem of languages $L \in \mathsf{VA}(M\Gamma)$.

Lemma 5. *Let Γ be a graph whose underlying loop-free graph is a path on four vertices. Then, $\mathsf{VA}(M\Gamma)$ contains an undecidable language.*

6   Rational Sets

When removing silent transitions, we will regard an automaton with silent transition as an automaton that is λ-free but is allowed to multiply a rational subset (of the storage monoid) for each input symbol. In order to restrict the ways in which elements can cancel out, these rational sets are first brought into a normal form. Our normal form result essentially states that there is an automaton that reads the generators in an order such that certain cancellations do not occur on any path. Note that in a valence automaton over M, we can remove all edges labeled with elements outside of $J(M)$. This is due to the fact that they cannot be part of a valid computation. In a valence transducer over M with output in C, the edges carry elements from $X^* \times M \times C$. Therefore, in the situation outlined above, a rational set $S \subseteq M \times C$ will be replaced by $S \cap (J(M) \times C)$.

Lemma 6. *Let $M \in \mathcal{C}$ and C be a commutative monoid and $S \subseteq M \times C$ a rational set. Then, we have $S \cap (J(M) \times C) = \bigcup_{i=1}^n L_i U_i R_i$, in which*

$$L_i \in \mathsf{RAT}(\mathsf{L}(M) \times C), \quad U_i \in \mathsf{RAT}(\mathsf{H}(M) \times C), \quad R_i \in \mathsf{RAT}(\mathsf{R}(M) \times C)$$

for $1 \leq i \leq n$. Moreover,

$$S \cap (\mathsf{L}(M) \times C) = \bigcup_{\substack{1 \leq i \leq n \\ 1 \in R_i}} L_i U_i, \quad S \cap (\mathsf{R}(M) \times C) = \bigcup_{\substack{1 \leq i \leq n \\ 1 \in L_i}} U_i R_i.$$

7   Silent Transitions

The first lemma in this section can be shown using a simple combinatorial argument.

Lemma 7. *Let Γ be a graph such that any two looped vertices are adjacent, no two unlooped vertices are adjacent, and Γ does not contain* ●———○——○———● *as an induced subgraph. Then, $M\Gamma$ is in $\mathcal{C}$.*

We prove Theorem 1 by showing that $\mathsf{VA}^+(M) = \mathsf{VA}(M)$ for every $M \in \mathcal{C}$. This will be done using an induction with respect to the definition of $\mathcal{C}$. In order for this induction to work, we need to strengthen the induction hypothesis. The latter will state that for any $M \in \mathcal{C}$ and any commutative monoid C, we can transform a valence transducer over M with output in C into another one that has no λ-transitions but is allowed to output a semilinear set of elements in each step. Formally, we will show that each $M \in \mathcal{C}$ is *strongly λ-independent*: Let C be a commutative monoid and $T \subseteq X^* \times \mathsf{SL}(C)$ be a transduction. Then $\Phi(T) \subseteq X^* \times C$ is defined as $\Phi(T) = \{(w,c) \in X^* \times C \mid \exists (w,S) \in T : c \in S\}$. For a class $\mathcal{F}$ of transductions, $\Phi(\mathcal{F})$ is the class of all $\Phi(T)$ with $T \in \mathcal{F}$.

A monoid M is called *strongly λ-independent* if for any commutative monoid C, we have $\mathsf{VT}(M,C) = \Phi(\mathsf{VT}^+(M,\mathsf{SL}(C)))$. Note that $\Phi(\mathsf{VT}^+(M,\mathsf{SL}(C))) \subseteq \mathsf{VT}(M,C)$ holds for any M and C. In order to have equality, it is necessary to grant the λ-free transducer the ability to output semilinear sets, since valence transducers without λ-transitions and with output in C can only output finitely many elements per input word. With λ-transitions, however, a valence transducer can output an infinite set for one input word.

By choosing the trivial monoid for C, we can see that for every strongly λ-independent monoid M, we have $\mathsf{VA}^+(M) = \mathsf{VA}(M)$. Indeed, given a valence automaton A over M, add an output of 1 to each edge and transform the resulting valence transducer into a λ-free one with output in $\mathsf{SL}(1)$. The latter can then clearly be turned into a valence automaton for the language accepted by A.

The following three lemmas each employ a different technique to eliminate silent transitions. Together with Lemma 7 and the results in Section 5, they yield the main result.

Lemma 8. $\mathbb{B}$ *is strongly λ-independent.*

Lemma 9. *If $M \in \mathcal{C}$ is strongly λ-independent, then $M \times \mathbb{Z}$ is as well.*

Lemma 10. *Suppose $M \in \mathcal{C}$ is non-trivial and strongly λ-independent. Then, $M * \mathbb{B}$ is strongly λ-independent as well.*

We will now outline the proof of Theorem 2. By Theorem 1, we already know that when $r \leq 1$, we have $\mathsf{VA}^+(M\Gamma) = \mathsf{VA}(M\Gamma)$. Hence, it suffices to show that $\mathsf{VA}^+(M\Gamma) \subsetneq \mathsf{VA}(M\Gamma)$ if $r \geq 2$. Greibach [8] and, independently, Jantzen [11] have shown that the language $L_1 = \{wc^n \mid w \in \{0,1\}^*,\ n \leq \mathrm{bin}(w)\}$ can be accepted by a partially blind counter machine with two counters, but not without λ-transitions. Here, $\mathrm{bin}(w)$ denotes the number obtained by interpreting w as a base 2 representation: $\mathrm{bin}(w1) = 2 \cdot \mathrm{bin}(w) + 1$, $\mathrm{bin}(w0) = 2 \cdot \mathrm{bin}(w)$, $\mathrm{bin}(\lambda) = 0$. Since we have to show $\mathsf{VA}^+(\mathbb{B}^r \times \mathbb{Z}^s) \subsetneq \mathsf{VA}(\mathbb{B}^r \times \mathbb{Z}^s)$ and we know $L_1 \in \mathsf{VA}(\mathbb{B}^r \times \mathbb{Z}^s)$, it suffices to prove $L_1 \notin \mathsf{VA}^+(\mathbb{B}^r \times \mathbb{Z}^s)$. We do this by transforming Greibach's and Jantzen's proof into a general property of languages accepted by valence automata without λ-transitions. We will then apply this to show that $L_1 \notin \mathsf{VA}^+(\mathbb{B}^r \times \mathbb{Z}^s)$.

Let M be a monoid. For $x, y \in M$, write $x \equiv y$ iff x and y have the same set of right inverses. For a finite subset $S \subseteq M$ and $n \in \mathbb{N}$, let $f_{M,S}(n)$ be the number of equivalence classes of $\equiv$ in $S^n \cap R(M)$. The following notion is also used as a tool to prove lower bounds in state complexity of finite automata [7]. Here, we use it to prove lower bounds on the number of configurations that an automaton must be able to reach in order to accept a language L. Let $n \in \mathbb{N}$. An n-*fooling set for a language* $L \subseteq \Theta^*$ is a set $F \subseteq \Theta^n \times \Theta^*$ such that (i) for each $(u, v) \in F$, we have $uv \in L$, and (ii) for $(u_1, v_1), (u_2, v_2) \in F$ such that $u_1 \neq u_2$, we have $u_1 v_2 \notin L$ or $u_2 v_1 \notin L$. Let $g_L : \mathbb{N} \to \mathbb{N}$ be defined as $g_L(n) = \max\{|F| \mid F \text{ is an } n\text{-fooling set for } L\}$.

The following three lemmas imply that $L_1 \notin \mathsf{VA}^+(\mathbb{B}^r \times \mathbb{Z}^s)$ for any $r, s \in \mathbb{N}$.

Lemma 11. *Let M be a monoid and $L \in \mathsf{VA}^+(M)$. Then, there is a constant $k \in \mathbb{N}$ and a finite set $S \subseteq M$ such that $g_L(n) \leq k \cdot f_{M,S}(n)$ for all $n \in \mathbb{N}$.*

Lemma 12. *For $L = L_1$, we have $g_L(n) \geq 2^n$ for every $n \in \mathbb{N}$.*

Lemma 13. *Let $M = \mathbb{B}^r \times \mathbb{Z}^s$ for $r, s \in \mathbb{N}$ and $S \subseteq M$ a finite set. Then, $f_{M,S}$ is bounded by a polynomial.*

Acknowledgements. The author would like to thank Nils Erik Flick, Reiner Hüchting, Matthias Jantzen, and Klaus Madlener for comments that improved the presentation of the paper.

References

[1] Book, R.V., Otto, F.: String-Rewriting Systems. Springer, New York (1993)

[2] Bouajjani, A., Esparza, J., Maler, O.: Reachability Analysis of Pushdown Automata: Application to Model-Checking. In: Mazurkiewicz, A., Winkowski, J. (eds.) CONCUR 1997. LNCS, vol. 1243, pp. 135–150. Springer, Heidelberg (1997)

[3] Caucal, D.: On infinite transition graphs having a decidable monadic theory. Theor. Comput. Sci. 290(1), 79–115 (2003)

[4] Elder, M., Kambites, M., Ostheimer, G.: On Groups and Counter Automata. Internat. J. Algebra Comput. 18(8), 1345–1364 (2008)

[5] Gilman, R.H.: Formal Languages and Infinite Groups. In: Geometric and Computational Perspectives on Infinite Groups. DIMACS Series in Discrete Mathematics and Theoretical Computer Science, vol. 25 (1996)

[6] Ginsburg, S., Spanier, E.H.: Bounded Algol-Like Languages. Trans. Amer. Math. Soc. 113(2), 333–368 (1964)

[7] Glaister, I., Shallit, J.: A lower bound technique for the size of nondeterministic finite automata. Inf. Process. Lett. 59(2), 75–77 (1996)

[8] Greibach, S.A.: Remarks on blind and partially blind one-way multicounter machines. Theor. Comput. Sci. 7(3), 311–324 (1978)

[9] Huet, G.: Confluent Reductions: Abstract Properties and Applications to Term Rewriting Systems. J. ACM 27(4), 797–821 (1980)

[10] Ibarra, O.H., Sahni, S.K., Kim, C.E.: Finite automata with multiplication. Theor. Comput. Sci. 2(3), 271–294 (1976)

[11] Jantzen, M.: Eigenschaften von Petrinetzsprachen. German. PhD thesis. Universität Hamburg (1979)

[12] Kambites, M.: Formal Languages and Groups as Memory. Communications in Algebra 37(1), 193–208 (2009)

[13] Karianto, W.: Adding Monotonic Counters to Automata and Transition Graphs. In: De Felice, C., Restivo, A. (eds.) DLT 2005. LNCS, vol. 3572, pp. 308–319. Springer, Heidelberg (2005)

[14] van Leeuwen, J.: A generalisation of Parikh's theorem in formal language theory. In: Loeckx, J. (ed.) ICALP 1974. LNCS, vol. 14, pp. 17–26. Springer, Heidelberg (1974)

[15] Lohrey, M., Steinberg, B.: The submonoid and rational subset membership problems for graph groups. J. Algebra 320(2), 728–755 (2008)

[16] Mitrana, V., Stiebe, R.: Extended finite automata over groups. Discrete Applied Mathematics 108(3), 287–300 (2001)

[17] Render, E., Kambites, M.: Rational subsets of polycyclic monoids and valence automata. Inform. and Comput. 207(11), 1329–1339 (2009)

New Online Algorithms for Story Scheduling in Web Advertising

Susanne Albers and Achim Passen

Department of Computer Science, Humboldt-Universität zu Berlin
{albers,passen}@informatik.hu-berlin.de

Abstract. We study *storyboarding* where advertisers wish to present sequences of ads (stories) uninterruptedly on a major ad position of a web page. These jobs/stories arrive online and are triggered by the browsing history of a user who at any time continues surfing with probability β. The goal of an ad server is to construct a schedule maximizing the expected reward. The problem was introduced by Dasgupta, Ghosh, Nazerzadeh and Raghavan (SODA'09) who presented a 7-competitive online algorithm. They also showed that no deterministic online strategy can achieve a competitiveness smaller than 2, for general β.

We present improved algorithms for storyboarding. First we give a simple online strategy that achieves a competitive ratio of $4/(2-\beta)$, which is upper bounded by 4 for any β. The algorithm is also $1/(1-\beta)$-competitive, which gives better bounds for small β. As the main result of this paper we devise a refined algorithm that attains a competitive ratio of $c = 1+\phi$, where $\phi = (1+\sqrt{5})/2$ is the Golden Ratio. This performance guarantee of $c \approx 2.618$ is close to the lower bound of 2. Additionally, we study for the first time a problem extension where stories may be presented simultaneously on several ad positions of a web page. For this parallel setting we provide an algorithm whose competitive ratio is upper bounded by $1/(3-2\sqrt{2}) \approx 5.828$, for any β. All our algorithms work in phases and have to make scheduling decisions only every once in a while.

1 Introduction

Online advertising has grown steadily over the last years. The worldwide online ad spending reached \$100 billion in 2012 and is expected to surpass the print ad spending during the next few years [4,9]. In this paper we study an algorithmic problem in advertising introduced by Dasgupta, Ghosh, Nazerzadeh and Raghavan [3]. An advanced online ad format is *storyboarding*, which was first launched by New York Times Digital and is also referred to as surround sessions [10]. In storyboarding, while a user surfs the web and visits a particular website, a single advertiser controls a major ad position for a certain continuous period of time. The advertiser can use these time slots to showcase a range of products and build a linear story line. Typically several advertisers compete for the ad position, depending on the user's browsing history and current actions. The goal of an ad server is to allocate advertisers to the time slots of a user's browsing session so as to maximize the total revenue.

F.V. Fomin et al. (Eds.): ICALP 2013, Part II, LNCS 7966, pp. 446–458, 2013.
© Springer-Verlag Berlin Heidelberg 2013

Dasgupta, Ghosh, Nazerzadeh and Raghavan [3] formulated storyboarding as an online job scheduling problem. Consider a user that starts a web session at time $t = 0$. Time is slotted. At any time t the user continues surfing with probability β, where $0 < \beta \leq 1$, and stops surfing with probability $1 - \beta$. Hence the surfing time is a geometrically distributed random variable. Over time jobs (advertisers) arrive online. These jobs arise based on the user's browsing history and accesses to web content. Each job i is specified by an arrival time a_i, a length l_i and a per-unit value v_i. Here l_i is the length of the ad sequence the advertiser would like to present and v_i is the reward obtained by the server in showing one unit of job i. This reward has to be discounted by the time when the job unit is shown. Considering all incoming jobs, we obtain a problem instance $\mathcal{I} = (a_i, v_i, l_i)_{i=1}^N$, where $N \in \mathbb{N} \cup \{\infty\}$. We allow $N = \infty$ to model potentially infinitely long browsing sessions and associated job arrivals.

A schedule S for $\mathcal{I}$ specifies which job to process at any time $t \geq 0$. The schedule does not have to contain all jobs; it is allowed to leave out (unattractive) jobs. Schedule S is feasible if every scheduled job i is processed at times $t \geq a_i$ for up to l_i time units. Moreover, it is required that each scheduled job is processed *continuously without* interruption so that an advertiser can build a story. *Preemption* of jobs is allowed, i.e. a job i may be processed for less than l_i time units. In this case no value can be attained for the preempted unscheduled portion of a job. Given a schedule S, its *value* is defined as the expected value $\sum_{t=0}^{\infty} \beta^t v(t)$, where $v(t)$ is the per-unit value of the job scheduled at time t. The goal is to maximize this reward. Let ALG be an online algorithm that, given any input $\mathcal{I}$, constructs a schedule of value $ALG(\mathcal{I})$. Let $OPT(\mathcal{I})$ be the value of an optimal offline schedule for $\mathcal{I}$. Algorithm ALG is *c-competitive* if there exists a constant α such that $c \cdot ALG(\mathcal{I}) + \alpha \geq OPT(\mathcal{I})$ holds for all $\mathcal{I}$, cf. [12].

Previous Work: Algorithmic problems in online advertising have received considerable research interest lately, see e.g. [1,2,5,6,7,8,11] and references therein. To the best of our knowledge storyboarding, from an algorithmic perspective, has only been studied so far by Dasgupta et al. [3]. A first observation is that if $\beta = 1$, then the scheduling problem is trivial. Every schedule that never preempts jobs and sequences them in an arbitrary order, subject to arrival constraints, achieves an optimal value. Therefore we concentrate on the case that the discount factor β satisfies $0 < \beta < 1$.

Dasgupta et al. [3] showed that no deterministic online algorithm can achieve a competitive ratio smaller than $\beta + \beta^2$. This ratio can be arbitrarily close to 2 as $\beta \to 1$. Hence, for general β, no deterministic online strategy can achieve a competitiveness smaller than 2. As a main result Dasgupta et al. devised a greedy algorithm that is 7-competitive. At any time the algorithm checks if it is worthwhile to preempt the job i currently being executed. To this end the strategy compares the reward obtained in scheduling another unit of job i to the loss incurred in delaying jobs of per-unit value higher than v_i for one time unit.

Furthermore, Dasgupta et al. addressed a problem extension where jobs have increasing rather than constant per-unit values. They focused on the case that value is obtained only when a job is completely finished. The authors showed that no algorithm can achieve a constant competitive ratio and gave a strategy with a logarithmic competitiveness. Finally Dasgupta et al. studied an extension where a job must be scheduled immediately upon arrival; otherwise it is lost. Here they proved a logarithmic lower bound on the performance of any randomized online strategy.

Our Contribution: We present new and improved online algorithms for story-boarding. All strategies follow the paradigm of processing a given job sequence $\mathcal{I}$ in phases, where a phase consists of k consecutive time steps in the scheduling horizon, for some $k \in \mathbb{N}$. At the beginning of each phase an algorithm computes a schedule for the phase, ignoring jobs that may arrive during the phase. Hence the strategies have to make scheduling decisions only every once in a while.

First in Section 2 we give a simple algorithm that computes an optimal schedule for each phase and preempts jobs that are not finished at the end of the respective phase. We prove that the competitive ratio of this strategy is exactly $1/(\beta^{k-1}(1 - \beta^k))$, for all $k \in \mathbb{N}$ and all β. The best choice of k gives a competitiveness of $4/(2 - \beta)$, which is upper bounded by 4 for any β. If k is set to 1, the resulting algorithm is $1/(1 - \beta)$-competitive. This gives further improved bounds for small β, i.e. when $\beta < 2/3$.

In Section 3, as our main contribution, we devise a refined algorithm that prefers not to preempt jobs sequenced last in a phase but rather tries to continue them in the following phase. The competitive ratio of this strategy is upper bounded by $1/\beta^{k-1} \cdot \max\{1/\beta^{k-1}, 1/(1 - \beta^{2k}), \beta^{3k}/(1 - \beta^k)\}$. Using the best choice of k, we obtain a competitive factor of $c = 1 + \phi$, where $\phi = (1 + \sqrt{5})/2$ is the Golden Ratio. Hence $c \approx 2.618$ and this performance guarantee is close to the lower bound of 2 presented by Dasgupta et al. [3] for general β.

In Section 4 we consider for the first time a problem extension where a web page features not only one but several ad positions where stories can be presented simultaneously. This is a natural extension because many web pages do contain a (small) number of ad positions. Again a job sequence $\mathcal{I} = (a_i, v_i, l_i)_{i=1}^{N}$ is triggered by the browsing history of a user. We assume that an ad server may assign these jobs to a general number m of ad positions. Following the scheduling terminology we refer to these ad positions as machines. In a feasible schedule each job must be processed continuously without interruption on one machine. A migration of jobs among machines is not allowed. The value of a schedule is $\sum_{t=0}^{\infty} \sum_{j=1}^{m} \beta^t v(t, j)$, where $v(t, j)$ is the per-unit value of the job scheduled on machine j at time t. We extend our first algorithm to this parallel setting and derive a strategy that achieves a competitive ratio of $(1 + 1/(1 - \beta(2 - \sqrt{2})))/(2 - \sqrt{2})$. For small β, this ratio can be as low as $2/(2 - \sqrt{2}) \approx 3.414$. For any β, the ratio is upper bounded by $1/(3 - 2\sqrt{2}) \approx 5.828$.

In the analyses of the algorithms we consider quantized inputs in which job arrival times are integer multiples of k. For the setting where one ad position is available (Sections 2 and 3), we are able to prove an interesting property given any quantized input: In an online schedule or a slight modification thereof, no job starts later than in an optimal offline schedule. This property has the important consequence that, for its scheduled job portions, an online algorithm achieves a total value that is at least as high as that of an optimal schedule. Hence the competitive analyses essentially reduce to bounding the loss incurred by an online strategy in preempting jobs. For the refined algorithm this loss analysis is quite involved and in order to prove a small competitive ratio we have to amortize the loss of a preempted job over several phases. In the setting were multiple ad positions are available (Section 4), such a property on job starting times unfortunately does not hold. Therefore we construct a specific optimal schedule S^* that allows us to match job units sequenced in S^* to job units sequenced online. Using this matching we can upper bound the additional value achieved by an optimal solution.

Remark: Due to space limitations the proofs of many lemmas, theorems and corollaries are omitted in this extended abstract. They are presented in the full version of the paper.

2   A 4-competitive Algorithm

As mentioned before, all algorithms we present in this paper process a job sequence in phases. Let $k \geq 1$ be an integer. A *k-phase* consists of k consecutive time steps in the scheduling horizon. More specifically, the n-th k-phase is the subsequence of time steps $P_n = (n-1)k, \ldots, nk - 1$, for any $n \geq 1$. Our first algorithm, called $ALG1_k$, computes an optimal schedule for any phase, given the jobs that are available at the beginning of the phase. Such an optimal schedule is obtained by simply sequencing the available jobs in order of non-increasing per-unit value. Jobs that arrive during the phase are deferred until the beginning of the next phase.

Formally, $ALG1_k$ works as follow. We say that a job i is *available at time t* if the job has arrived by time t, i.e. $a_i \leq t$, and has not been scheduled so far at any time $t' < t$. Consider an arbitrary phase P_n and let Q_n be the set of jobs that are available at the beginning of P_n. We note that Q_n includes the jobs that arrive at time $(n-1)k$. $ALG1_k$ constructs a schedule for P_n by first sorting the jobs of Q_n in order of non-increasing per-unit value. Jobs having the same per-unit value are sorted in order of increasing arrival times; ties may be broken arbitrarily. Given this sorted sequence, $ALG1_k$ then assigns the jobs one by one to P_n until the k time steps are scheduled or the job sequence ends. In the former case, the last job assigned to P_n is preempted at the end of the phase unless the job completes by the end of P_n. $ALG1_k$ executes this schedule for P_n, ignoring jobs that may arrive during the phase at times $t = (n-1)k + 1, \ldots, nk - 1$.

We first evaluate the performance of $ALG1_k$, for general k. Then we will determine the best choice of k.

Theorem 1. *For all $k \in \mathbb{N}$ and all probabilities β, $ALG1_k$ is $1/(\beta^{k-1}(1 - \beta^k))$-competitive.*

In the following we prove the above theorem. Let $\mathcal{I} = (a_i, v_i, l_i)_{i=1}^N$ be an arbitrary input. In processing $\mathcal{I}$, $ALG1_k$ defers jobs arriving after the beginning of a phase until the start of the next phase. Consider a *k-quantized input* $\mathcal{I}_k$

in which the arrival time of any job is set to the next integer multiple of k, i.e. $\mathcal{I}_k = (a_i', v_i, l_i)_{i=1}^N$, where $a_i' = k\lceil a_i/k \rceil$. If a_i is a multiple of k and hence coincides with the beginning of a k-phase, the job is not delayed. Otherwise the job is delayed until the beginning of the next phase. The schedule generated by $ALG1_k$ for $\mathcal{I}_k$ is identical to that computed by $ALG1_k$ for $\mathcal{I}$. Thus $ALG1_k(\mathcal{I}_k) = ALG1_k(\mathcal{I})$. In order to prove Theorem 1 it will be convenient to compare $ALG1_k(\mathcal{I}_k)$ to $OPT(\mathcal{I}_k)$. The next lemma ensures that $OPT(\mathcal{I}_k)$ and the true optimum $OPT(\mathcal{I})$ differ by a factor of at most $1/\beta^{k-1}$.

Lemma 1. *For all $k \in \mathbb{N}$ and all probabilities β, $1/\beta^{k-1} \cdot OPT(\mathcal{I}_k) \geq OPT(\mathcal{I})$.*

In order to estimate $OPT(\mathcal{I}_k)$ we consider a stronger optimal offline algorithm that was also proposed by Dasgupta et al. [3]. This algorithm is allowed to resume interrupted jobs at a later point in time. We call this offline strategy $CHOP$. For any input, at any time t $CHOP$ schedules a job having the highest per-unit value among the unfinished jobs that have arrived until time t. Obviously, $CHOP(\mathcal{I}_k) \geq OPT(\mathcal{I}_k)$. Let S be the schedule computed by $ALG1_k$ for $\mathcal{I}_k$ and let $S^\star$ be the schedule generated by $CHOP$ for $\mathcal{I}_k$. We assume w.l.o.g. that in $S^\star$ all jobs having a certain per-unit value v are processed in the same order as in S. More specifically, all jobs having per-unit value v are processed in order of increasing arrival times. Jobs of per-unit value v arriving at the same time are processed in the same order as in S. Schedule $S^\star$ can be easily modified so that this property is satisfied. For any job i, let $t_S(i)$ denote its starting time in S and let $t_{S^\star}(i)$ be its starting time in $S^\star$. If job i is never processed in S (or $S^\star$), then we set $t_S(i) = \infty$ (or $t_{S^\star}(i) = \infty$). The following lemma states that $ALG1_k$ starts each job at least as early as $CHOP$.

Lemma 2. *For any job i, $t_S(i) \leq t_{S^\star}(i)$.*

Lemma 3. *For all $k \in \mathbb{N}$ and all probabilities β, $1/(1-\beta^k) \cdot ALG1_k(\mathcal{I}_k) \geq OPT(\mathcal{I}_k)$.*

Proof. For any $n \geq 1$, let $I_n = \{i \mid (n-1)k \leq t_S(i) \leq nk - 1\}$ be the set of jobs scheduled by $ALG1_k$ in phase P_n. Let $ALG1_k(P_n)$ be the value achieved by $ALG1_k$ in scheduling the jobs of I_n, and let $CHOP(P_n)$ be the value achieved by $CHOP$ in processing these jobs. There holds $ALG1_k(\mathcal{I}_k) = \sum_n ALG1_k(P_n)$. A consequence of Lemma 2 is that all jobs ever scheduled by $CHOP$ also occur in $ALG1_k$'s schedule. Thus $CHOP(\mathcal{I}_k) = \sum_n CHOP(P_n)$. We will show that $CHOP(P_n)/ALG1_k(P_n) \leq 1/(1-\beta^k)$ holds for every $n \geq 1$. This implies $CHOP(\mathcal{I}_k)/ALG1_k(\mathcal{I}_k) \leq 1/(1-\beta^k)$. The lemma then follows because $CHOP(\mathcal{I}_k) \leq OPT(\mathcal{I}_k)$.

Consider any k-phase P_n. In the schedule S let j be the last job started in P_n and let λ_j be the number of time units for which j is sequenced in P_n and thus in the entire schedule S. By Lemma 2, for any job i, there holds $t_S(i) \leq t_{S^\star}(i)$. Hence the total value achieved by $CHOP$ in scheduling the jobs $i \in I_n$ with $i \neq j$ as well as the first λ_j time units of job j cannot be higher than $ALG1_k(P_n)$. If job j is preempted in S at the end of P_n, then $CHOP$ can achieve an additional value in scheduling units $\lambda_j + 1, \ldots, l_j$ of job j in $S^\star$. Again, since $t_S(j) \leq t_{S^\star}(j)$, these units cannot be sequenced before the beginning of phase P_{n+1}, i.e. at time nk. Thus the additional value achievable for units $\lambda_j + 1, \ldots, l_j$ is upper bounded by $\beta^{nk}/(1 - \beta) \cdot v_j$, which is obtained if a job of per-unit value v_j and infinite length is sequenced starting at time nk.

Thus $CHOP(P_n) \leq ALG1_k(P_n) + \beta^{nk}/(1 - \beta) \cdot v_j$. In each phase $ALG1_k$ sequences jobs in order of non-increasing per-unit value. Hence each job of I_n has a per-unit value of at least v_j. We conclude $ALG1_k(P_n) \geq (\beta^{(n-1)k} - \beta^{nk})/(1 - \beta) \cdot v_j$ and $CHOP(P_n)/ALG1_k(P_n) \leq 1 + \beta^{nk}/(\beta^{(n-1)k} - \beta^{nk}) = 1/(1 - \beta^k)$. □

Combining Lemmas 1 and 3 together with the fact $ALG1_k(\mathcal{I}) = ALG1_k(\mathcal{I}_k)$, we obtain Theorem 1. We determine the best value of k.

Corollary 1. *For $k = \lceil -\log_\beta 2 \rceil$, the resulting $ALG1_k$ is $4/(2 - \beta)$-competitive.*

Theorem 2. *For all $k \in \mathbb{N}$ and all probabilities β, the competitive ratio of $ALG1_k$ is not smaller than $1/(\beta^{k-1}(1 - \beta^k))$.*

The above theorem shows our analysis of $ALG1_k$ is tight. Finally we consider the algorithm $ALG1_1$ in which the phase length k is set to 1.

Corollary 2. *For all probabilities β, the competitive ratio of $ALG1_1$ is exactly $1/(1 - \beta)$.*

Corollary 3. *Set $k = 1$ if $\beta \leq 2/3$ and $k = \lceil -\log_\beta 2 \rceil$ otherwise. Then $ALG1_k$ that achieves a competitive ratio of $\min\{1/(1 - \beta), 4/(2 - \beta)\}$.*

3   A Refined Algorithm

We present a second algorithm that, compared to $ALG1_k$, reduces loss incurred in preempting jobs. The algorithm also operates in k-phases. Its crucial property is that it continues processing a job scheduled last in a phase if this job is among the highest-valued jobs available at the beginning of the next phase.

The refined algorithm, called $ALG2_k$, works in two steps. Again, let P_n be any k-phase. Step (1) is defined as follows. If $n > 1$, then let i_n be the job that was scheduled last in P_{n-1} and can potentially be continued in P_n. If this job has been scheduled for less than l_{i_n} time units in the prior schedule, until the end of P_{n-1}, then define a residual job i_n^r by $(a_{i_n}, v_{i_n}, l_{i_n}^r)$. Here $l_{i_n}^r$ is the remaining length of job i_n, i.e. $l_{i_n}^r$ further units have to be processed to complete the job. Let Q_n be the set consisting of job i_n^r and the jobs available at the beginning of P_n. $ALG2_k$ schedules the jobs of Q_n in order of non-increasing per-unit values in P_n. Again, jobs having the same per-unit value are scheduled in order of increasing arrival times, where ties may be broken arbitrarily. Among jobs having a per-unit value of $v = v_{i_n}$, job i_n^r is scheduled first. Let $S'(P_n)$ denote the schedule obtained for P_n at this point.

We next describe Step (2). If $S'(P_n)$ does not contain job i_n^r, then $S'(P_n)$ is equal to the final schedule $S(P_n)$ for the phase. If $S'(P_n)$ contains job i_n^r and this job is scheduled for s_n^r time units starting at time t_n^r in P_n, then $ALG2_k$

modifies $S'(P_n)$ so as to obtain a feasible schedule. Loosely speaking, job i_n^r is shifted to the beginning of P_n. More precisely, the original job i_n is scheduled for s_n^r time units at the beginning of P_n. The start of all jobs scheduled from time $(n-1)k$ to time $t_n^r - 1$ in $S'(P_n)$ is delayed by s_n^r time units. Between time $t_n^r + s_n^r$ and the end of P_n, no modification is required. The resulting schedule is the final output $S(P_n)$. While this schedule is executed, newly arriving jobs are deferred until the beginning of the next phase.

A pseudo-code description of $ALG2_k$ is given below. We remark that a long job i may be executed over several phases, provided that its per-unit value is sufficiently high.

> **Algorithm** $ALG2_k$: Each phase P_n is handled as follows. (1) If $n > 1$, let i_n be the job scheduled last in P_{n-1}. If job i_n has been scheduled for less than l_{i_n} time units so far, define job i_n^r by $(a_{i_n}, v_{i_n}, l_{i_n}^r)$ and add it to Q_n. Let $S'(P_n)$ be the schedule obtained by sequencing the jobs of Q_n in order of non-increasing per-unit value in P_n. (2) If $S'(P_n)$ processes job i_n^r for s_n^r time units staring at time t_n^r, then schedule job i_n for s_n^r time units at the beginning of P_n. Jobs originally processed from time $(n-1)k$ to $t_n^r - 1$ are delayed by s_n^r time units. Execute this schedule $S(P_n)$ for P_n, ignoring jobs that arrive during the phase.

Theorem 3. *For all $k \in \mathbb{N}$ and all probabilities β, algorithm $ALG2_k$ achieves a competitive ratio of $1/\beta^{k-1} \cdot \max\{1/\beta^{k-1}, 1/(1 - \beta^{2k}), \beta^{3k}/(1 - \beta^k)\}$.*

We proceed to prove the above theorem. Compared to the proof of Theorem 1 the analysis is more involved because we have to take care of the delays incurred by $ALG2_k$ in Step (2) when scheduling a portion of job i_n at the beginning of phase P_n and thereby postponing the start of jobs with higher per-unit values. Furthermore, in order to achieve a small competitive ratio we have to charge the loss of a job preempted in a phase to several adjacent phases. To this end we have to classify phases and form schedule segments of up to three phases.

Again, for any input $\mathcal{I} = (a_i, v_i, l_i)_{i=1}^N$, we consider the k-quantized input $\mathcal{I}_k = (a_i', v_i, l_i)_{i=1}^N$, where the arrival time of any job i is set to $a_i' = k\lceil a_i/k \rceil$. There holds $ALG2_k(\mathcal{I}_k) = ALG2_k(\mathcal{I})$ and, as shown in Lemma 1, $1/\beta^{k-1} OPT(\mathcal{I}_k) \geq OPT(\mathcal{I})$. We will compare $ALG2_k(\mathcal{I}_k)$ to $CHOP(\mathcal{I}_k)$, where $CHOP$ is the stronger optimal offline algorithm described in Section 2. Again let S denote the schedule computed by $ALG2_k$ for $\mathcal{I}_k$ and let $S^\star$ be $CHOP$'s schedule for $\mathcal{I}_k$. In $S^\star$ jobs having a certain per-unit value v are processed in the same order as in S.

In order to evaluate $ALG2_k(\mathcal{I}_k)$, we define a schedule S' that allows us to prove a statement analogous to Lemma 2 and, moreover, to compare the per-unit values of jobs scheduled in S' and $S^\star$. For any phase P_n, consider the schedule $S'(P_n)$ computed in Step (1) of $ALG2_k$. If $n > 1$ and the residual job i_n^r is scheduled for s_n^r time units starting at time t_n^r in P_n, then modify $S'(P_n)$ by scheduling the *original* job i_n for s_n^r time units starting at time t_n^r. By slightly overloading notation, we refer to this modified schedule as $S'(P_n)$. Schedule S' is the concatenation of the $S'(P_n)$, for all $n \geq 1$.

In $S'(P_n)$ jobs are sequenced in order of non-increasing per-unit value. Among jobs of per-unit value $v = v_{i_n}$, job i_n is processed first. Schedule $S'(P_n)$ differs from $S(P_n)$ only in that job i_n is sequenced after the jobs having a strictly higher per-unit value than v_{i_n}. Each such job starts and finishes in P_n. The shift of the job portion of i_n does not affect the relative order of jobs having the same per-unit value. Hence in S' and S, and thus in S' and $S^\star$, jobs of a certain per-unit value v occur in the same relative order. We note that schedule S' is infeasible in that a job i_n may be interrupted at the end of P_{n-1} and resumed later in P_n.

For any job i, let $t_{S'}(i)$ be its starting time in S', i.e. the earliest time when a portion of job i is processed. As usual $t_S(i)$ and $t_{S^\star}(i)$ denote the starting time of job i in S and $S^\star$, respectively.

Lemma 4. *For any job i, $t_{S'}(i) \leq t_{S^\star}(i)$.*

A main goal of the subsequent analysis is to bound the loss incurred by $ALG2_k$ in preempting jobs. The following lemma will be crucial as it specifies the earliest time when a job preempted in S can occur again in S^*.

Lemma 5. *If job i is preempted in $S(P_n)$ and the following phase schedules $S(P_{n+1}), \ldots, S(P_{n'})$ only process jobs of per-unit value higher than v_i, then $S^\star$ does not schedule job i in phases $P_{n+1}, \ldots, P_{n'}$.*

The proof of Lemma 5 relies on another lemma that compares per-unit values of jobs scheduled in S and S'. At any time t, let $v_{S^\star}(t)$ be the per-unit value of the job scheduled in $S^\star$ and let $v_{S'}(t)$ be the per-unit value of the job scheduled in S'. Then $v_{S^\star}(t) \geq v_{S'}(t)$.

Phase Classification: We classify phases, considering the original schedule S. A phase P_n is called *preempted* if a job is preempted in $S(P_n)$. Phase P_n is called *continued* if the job scheduled last in $S(P_n)$ is also scheduled at the beginning of $S(P_{n+1})$. Phase P_n is *complete* if all jobs scheduled in $S(P_n)$ are finished by the end of P_n.

We mention some properties of these phases in the schedule S. (a) In each phase P_n at most one job is preempted in $S(P_n)$. (b) If P_n is a continued or complete phase, no job is preempted in $S(P_n)$. (c) If P_n is a preempted phase, then the job preempted is one having the smallest per-unit value among jobs scheduled in $S(P_n)$. These properties can be verified as follows. Let P_n be an arbitrary phase. When $ALG2_k$ constructs a schedule for P_n, it firsts sorts the jobs of Q_n in order of non-increasing per-unit value. In this sorted sequence only the last job, say job i, assigned to P_n might not be scheduled completely in the phase and hence is a candidate for preemption. Job i is one having the smallest per-unit value among jobs scheduled in the phase. This shows properties (a) and (c). If job i is not moved to the beginning of the phase in Step (2) of $ALG2_k$ and continued at the beginning of the next phase, then P_n is a continued phase and no job is preempted in $S(P_n)$. By definition, no job is preempted in a complete phase. This shows property (b). We observe that in the schedule S each phase is either preempted, continued or complete.

Schedule Segments: For the further analysis we partition the schedule S into *segments* where a segment consists of up to three consecutive phases. The purpose of these segments is to combine "expensive" preempted phases with other phases so as to amortize preemption loss. First we build segments consisting of three phases. Phases P_n, P_{n+1}, P_{n+2} form a segment if P_n is a preempted phase that is not preceded by a continued phase, P_{n+1} is a continued phase and P_{n+2} is a preempted phase. Among the remaining phases we build segments consisting of two phases. Phases P_n, P_{n+1} form a segment if (a) P_n is a preempted phase that is not preceded by a continued phase and P_{n+1} is a continued or complete phase or (b) P_n is a continued phase followed by a preempted phase P_{n+1}. Each remaining phase forms a separate segment. We observe that a preempted phase that forms a separate one-phase segment is not preceded by a continued phase and is followed by a preempted phase.

For a segment σ, let $ALG2_k(\sigma)$ be the value obtained by $ALG2_k$ on σ. More specifically, let I be the set of jobs scheduled by $ALG2_k$ in the phases of σ. Set I also includes those jobs that are only partially processed in σ and might also be scheduled in phases before or after σ. Suppose that job $i \in I$ is processed for δ_i time units starting at time t_i is σ. Then

$$ALG2_k(\sigma) = \sum_{i \in I} \beta^{t_i}(1 - \beta^{\delta_i})/(1 - \beta) \cdot v_i.$$

Let $CHOP(\sigma)$ denote the value achieved by $CHOP$ in processing the jobs and job portions scheduled by $ALG2_k$ in σ. More specifically, suppose that in S job $i \in I$ has been processed for λ_i time units before the beginning of σ. Then $CHOP(\sigma)$ represents the value achieved by $CHOP$ in processing the units $\lambda_i + 1, \dots, \lambda_i + \delta_i$ of job i in $S^\star$. If job i is preempted in the segment σ of S, then $CHOP(\sigma)$ additionally represents the value achieved by processing units $u > \lambda_i + \delta_i$ in $S^\star$. There holds

$$CHOP(\sigma) \leq \sum_{i \in I} \beta^{t_{S^\star}(i) + \lambda_i}(1 - \beta^{\delta_i})/(1 - \beta) \cdot v_i + v_p(\sigma),$$

where $v_p(\sigma)$ denotes the additional value achieved by $CHOP$ for jobs preempted by $ALG2_k$ in σ. We have $ALG2_k(\mathcal{I}_k) = \sum_\sigma ALG2_k(\sigma)$ and $CHOP(\mathcal{I}_k) = \sum_\sigma CHOP(\sigma)$ because, by Lemma 4, every job scheduled by $CHOP$ is also scheduled by $ALG2_k$.

Segment Analysis: We develop three lemmas that upper bound the ratio $CHOP(\sigma)/ALG2_k(\sigma)$, for the various segments. In the proofs we use the following notation. For any phase P_n let I_n denote the set of jobs that are partially or completely processed in $S(P_n)$. For any $i \in I_n$, let $\delta_{i,n}$ be the number of time units for which job i is processed in $S(P_n)$. If job i is only scheduled in phase P_n of S, then we simply set $\delta_i = \delta_{i,n}$. Furthermore, let i_n^1 be the first job scheduled in $S(P_n)$. Suppose that P_n is preceded by a continued phase. When constructing $S(P_n)$, $ALG2_k$ might have delayed the starting times of some jobs of I_n in Step (2) in order to move job i_n^1 to the beginning of the phase. Let $I_n' \subset I_n$ be the set of these delayed jobs. If P_n is not preceded by a continued phase, there are no delayed jobs and we set $I'_n = \emptyset$. We observe that $t_S(i) = t_{S'}(i)$, for all $i \in I_n \setminus I'_n \cup \{i_n^1\}$, and $t_S(i) = t_{S'}(i) + \delta_{i_n^1,n}$, for all $i \in I'_n$.

For ease of exposition, let $w(t, \delta, v) = \beta^t(1 - \beta^\delta)/(1 - \beta) \cdot v$ be the value achieved in processing a job of per-unit value v for δ time units starting at time t. We allow $\delta = \infty$, i.e. a job of infinite length is scheduled starting at time t.

Lemma 6. *For any σ consisting of one phase, $CHOP(\sigma)/ALG2_k(\sigma) \leq \max\{1/\beta^{k-1}, 1 + \beta^{3k}/(1 - \beta^k)\}$.*

Proof. We first study the case that the phase P_n of σ is a continued or complete phase, i.e. no job is preempted in $S(P_n)$. Let $i_1 = i_n^1$ be the first job scheduled in $S(P_n)$. There holds

$$ALG2_k(\sigma) = w((n - 1)k, \delta_{i_1,n}, v_{i_1}) + \sum_{i \in I'_n} w(t_{S'}(i) + \delta_{i_1,n}, \delta_{i,n}, v_i)$$
$$+ \sum_{i \in I_n \setminus (I'_n \cup \{i_1\})} w(t_{S'}(i), \delta_{i,n}, v_i)$$
$$\geq \beta^{\delta_{i_1,n}}(w((n - 1)k, \delta_{i_1,n}, v_{i_1}) + \sum_{i \in I_n \setminus \{i_1\}} w(t_{S'}(i), \delta_{i,n}, v_i)).$$

The last inequality holds because $w(t_{S'}(i) + \delta_{i_1,n}, \delta_i, v_i) = \beta^{\delta_{i_1,n}} w(t_{S'}(i), \delta_{i,n}, v_i)$. Every job $i \in I_n$, except for possibly i_1, is started in $S(P_n)$. If job i_1 is started in $S(P_{n'})$, where $n' < n$, then the job is scheduled at the end of $S(P_{n'})$. Hence when $ALG2_k$ constructed $S(P_{n'})$, the job was not delayed in Step (2) of the algorithm in order to move another job to the beginning of the phase. Hence $t_S(i_1) = t_{S'}(i_1) \leq t_{S^\star}(i_1)$. Suppose that before P_n job i_1 was processed for λ_{i_1} time units in S. Since $t_S(i_1) \leq t_{S^\star}(i_1)$, the units $\lambda_{i_1} + 1, \ldots, \lambda_{i_1} + \delta_{i_1,n}$ of job i_1 cannot be started before the beginning of P_n in $S^\star$. For all jobs $i \in I_n \setminus \{i_1\}$, there holds $t_{S'}(i) \leq t_{S^\star}(i)$. Hence

$$CHOP(\sigma) \leq w((n - 1)k, \delta_{i_1,n}, v_{i_1}) + \sum_{j \in I_n \setminus \{i_1\}} w(t_{S'}(i), \delta_{i,n}, v_i).$$

We obtain $CHOP(\sigma)/ALG2_k(\sigma) \leq 1/\beta^{\delta_{i_1,n}} \leq 1/\beta^{k-1}$ because a job $i \in I'_n$ can be delayed by at most $k - 1$ time units.

We next study the case that P_n is a preempted phase. The preceding phase P_{n-1} is not a continued phase while the following phase P_{n+1} is also a preempted phase. Since P_n is not preceded by a continued phase all jobs of I_n are started in $S(P_n)$ and $t_S(i) = t_{S'}(i)$, for all $i \in I_n$. We obtain

$$ALG2_k(\sigma) = \sum_{i \in I_n} w(t_{S'}(i), \delta_i, v_i).$$

Let $i_p \in I_n$ be the job preempted in $S(P_n)$. The job is preempted at the end of $S(P_n)$. Moreover, its per-unit value is strictly smaller than the per-unit value of any job scheduled in $S(P_{n+1})$, the schedule of the following phase, since otherwise $ALG2_k$ would have scheduled job i_p in $S(P_{n+1})$. Phase P_{n+1} is also a preempted phase and the job preempted in $S(P_{n+1})$ is scheduled at the end of $S(P_{n+1})$. Thus the job preempted in $S(P_{n+1})$ has a strictly smaller per-unit value than any job scheduled in $S(P_{n+2})$. It follows that job i_p has a strictly smaller per-unit value than any job scheduled in $S(P_{n+1})$ and $S(P_{n+2})$. Lemma 5 ensures that $CHOP$ does not schedule i_p in phases P_{n+1} and P_{n+2}. Thus the value achieved by $CHOP$ for the preempted portion of i_p is upper bounded by $w((n+2)k, \infty, v_{i_p})$ and

$$CHOP(\sigma) \le \sum_{i \in I_n} w(t_{S'}(i), \delta_i, v_i) + w((n+2)k, \infty, v_{i_p})$$
$$= ALG2_k(\sigma) + w((n+2)k, \infty, v_{i_p}).$$

Job i_p has the smallest per-unit value among jobs scheduled in $S(P_n)$. Thus $ALG2_k(\sigma) \ge w((n-1)k, k, v_{i_p}) = \beta^{(n-1)k}(1 - \beta^k)/(1 - \beta) \cdot v_{i_p}$. Also $w((n+2)k, \infty, v_{i_p}) = \beta^{(n+2)k}/(1 - \beta)v_{i_p}$. We conclude $CHOP(\sigma)/ALG2_k(\sigma) \le 1 + \beta^{3k}/(1 - \beta^k)$. $\qquad\square$

The next two lemmas address segments consisting of at least two phases, the analysis of which is more involved.

Lemma 7. *Let σ be a segment consisting of at least two phases. If σ consists of two phases, assume that the first one is a preempted phase. If σ consists of three phases, assume that the per-unit value of the job preempted in the first phase is at least as high as that of the job preempted in the third phase. There holds $CHOP(\sigma)/ALG2_k(\sigma) \le \max\{1/\beta^{k-1}, 1/(1 - \beta^{2k})\}$.*

Lemma 8. *Let σ be a segment consisting of at least two phases. If σ consists of two phases, assume that the first one is a continued phase. If σ consists of three phases, assume that the per-unit value of the job preempted in the first phase is smaller than that of the job preempted in the third phase. There holds $CHOP(\sigma)/ALG2_k(\sigma) \le \max\{1/\beta^{k-1}, 1/(1 - \beta^{2k})\}$.*

Theorem 3 now follows from the three above lemmas, taking into account that Lemmas 7 and 8 cover all possible cases of 2-phase and 3-phase segments.

Corollary 4. *Let $c = 1 + \phi$, where $\phi = (1 + \sqrt{5})/2$ is the Golden Ratio. For $k = \lfloor -\frac{1}{2} \log_\beta c \rfloor + 1$, $ALG2_k$ achieves a competitive ratio of $c \approx 2.618$.*

4   An Algorithm for Multiple ad Positions

We study the setting where m parallel machines (ad positions) are available. Let $\mathcal{I} = (a_i, v_i, l_i)_{i=1}^N$ be any input. Each job may be processed on one machine only, i.e. an interruption and migration of jobs is not allowed.

We define an algorithm $ALG(m)_k$ that generalizes $ALG1_k$ of Section 2. Let $P_n = (n-1)k, \dots, nk - 1$ be any phase. Again let Q_n be the set of jobs i that have arrived until the beginning of P_n, i.e. $a_i \le (n-1)k$, and have not been processed in the past phases $P_1, \dots, P_{n-1}$. $ALG(m)_k$ constructs a schedule for P_n by considering the k time steps of the phase. For $t = (n-1)k, \dots, nk - 1$, $ALG(m)_k$ determines the m jobs having the highest per-unit values among the jobs of Q_n that are unfinished at time t. Each of these jobs is scheduled for one time unit. If a job was also scheduled at time $t-1$, then it is assigned to the same machine at time t. We specify a tie breaking rule if, among the unfinished jobs in Q_n, the m-th largest per-unit value is v and there exist several jobs having this value. In this case preference is given to those jobs that have already been started at times t', with $t' < t$. Jobs that have not been started yet are considered in increasing order of arrival time, where ties may be broken arbitrarily. Of course, if at time t set Q_n contains at most m unfinished jobs, then each of them is scheduled at that time. We observe that a feasible phase schedule, in which each job is processed without interruption on one machine, can be constructed easily: If a job of Q_n is among those having the m highest per-unit values, then the job will remain in this subset until it is finished. Hence the job can be sequenced continuously on the same machine. We also observe that on each machine jobs are sequenced in order of non-increasing per-unit value. Let $S(P_n)$ denote the schedule constructed for phase P_n. While $S(P_n)$ is executed, newly arriving jobs are deferred until the beginning of the next phase. At the end of $S(P_n)$ unfinished jobs are preempted.

Theorem 4. *For all $k \in \mathbb{N}$ and all probabilities β, $ALG(m)_k$ achieves a competitive ratio of $1/\beta^{k-1} \cdot (1 + 1/(1 - \beta^k))$.*

We finally determine the best choice of k. For any β, the competitive ratio of Corollary 5 is upper bounded by $(1+1/(\sqrt{2}-1))/(2-\sqrt{2}) = 1/(3-2\sqrt{2}) \approx 5.828$.

Corollary 5. *For $k = \lceil \log_\beta(2 - \sqrt{2}) \rceil$, the resulting algorithm $ALG(m)_k$ achieves a competitive ratio of $(1 + 1/(1 - \beta(2 - \sqrt{2})))/(2 - \sqrt{2})$.*

References

1. Buchbinder, N., Feldman, M., Ghosh, A., Naor, J(S.): Frequency capping in online advertising. In: Dehne, F., Iacono, J., Sack, J.-R. (eds.) WADS 2011. LNCS, vol. 6844, pp. 147–158. Springer, Heidelberg (2011)
2. Buchbinder, N., Jain, K., Naor, J(S.): Online primal-dual algorithms for maximizing ad-auctions revenue. In: Arge, L., Hoffmann, M., Welzl, E. (eds.) ESA 2007. LNCS, vol. 4698, pp. 253–264. Springer, Heidelberg (2007)
3. Dasgupta, A., Ghosh, A., Nazerzadeh, H., Raghavan, P.: Online story scheduling in web adverstising. In: Proc. 20th Annual ACM-SIAM Symposium on Discrete Algorithms, pp. 1275–1284 (2009)
4. http://www.emarketer.com/Article/Digital-Account-One-Five-Ad-Dollars/1009592
5. Feldman, J., Korula, N., Mirrokni, V., Muthukrishnan, S., Pál, M.: Online ad assignment with free disposal. In: Leonardi, S. (ed.) WINE 2009. LNCS, vol. 5929, pp. 374–385. Springer, Heidelberg (2009)
6. Feige, U., Immorlica, N., Mirrokni, V.S., Nazerzadeh, H.: A combinatorial allocation mechanism with penalties for banner advertising. In: Proc. 17th International Conferene on World Wide Web, pp. 169–178 (2008)
7. Feldman, J., Mehta, A., Mirrokni, V.S., Muthukrishnan, S.: Online stochastic matching: Beating 1-1/e. In: Proc. 50th Annual IEEE Symposium on Foundations of Computer Science, pp. 117–126 (2009)

8. Ghosh, A., Sayedi, A.: Expressive auctions for externalities in online advertising. In: Proc. 19th International Conferene on World Wide Web, pp. 371–380 (2010)

9. http://www.marketingcharts.com/wp/television/global-online-ad-spend-forecast-to-exceed-print-in-2015-25105/

10. marketingterms.com. Surround session, http://www.marketingterms.com/dictionary/surround_session/

11. Mehta, A., Saberi, A., Vazirani, U.V., Vazirani, V.V.: AdWords and generalized online matching. Journal of the ACM 54(5) (2007)

12. Sleator, D.D., Tarjan, R.E.: Amortized efficiency of list update and paging rules. Communications of the ACM 28, 202–208 (1985)

Sketching for Big Data Recommender Systems Using Fast Pseudo-random Fingerprints

Yoram Bachrach[1] and Ely Porat[2]

[1] Microsoft Research, Cambridge, UK
[2] Bar-Ilan University, Ramat-Gan, Israel

Abstract. A key building block for collaborative filtering recommender systems is finding users with similar consumption patterns. Given access to the full data regarding the items consumed by each user, one can directly compute the similarity between any two users. However, for massive recommender systems such a naive approach requires a high running time and may be intractable in terms of the space required to store the full data. One way to overcome this is using sketching, a technique that represents massive datasets concisely, while still allowing calculating properties of these datasets. Sketching methods maintain very short fingerprints of the item sets of users, which allow approximately computing the similarity between sets of different users.

The state of the art sketch [22] has a very low space complexity, and a recent technique [14] shows how to exponentially speed up the computation time involved in building the fingerprints. Unfortunately, these methods are incompatible, forcing a choice between low running time or a small sketch size. We propose an alternative sketching approach, which achieves both a low space complexity similar to that of [22] and a low time complexity similar to [14]. We empirically evaluate our algorithm using the Netflix dataset. We analyze the running time and the sketch size of our approach and compare them to alternatives. Further, we show that in practice the accuracy achieved by our approach is even better than the accuracy guaranteed by the theoretical bounds, so it suffices to use even shorter fingerprints to obtain high quality results.

1 Introduction

The amount of data generated and processed by computers has shown consistent exponential growth. There are currently over 20 billion webpages on the internet, and major phone companies process tens of gigabytes of call data each day. Analyzing such vast amounts of data requires extremely efficient algorithms, both in terms of running time and storage. This has given rise to the field of massive datasets processing. We focus on massive recommender systems, which provide users with recommendations for items that they are likely to find interesting, such as music, videos, or web pages. These systems keep a profile of each user and compare it to reference characteristics. One approach is *collaborative filtering* (CF), where the stored information is the items consumed or rated by the user in the past. CF systems predict whether an item is likely to interest the target user by seeking users who share similar rating patterns with the target user and then using the ratings from those like-minded users to generate a prediction for the target user. Various user similarity measures have been proposed, the

F.V. Fomin et al. (Eds.): ICALP 2013, Part II, LNCS 7966, pp. 459–471, 2013.

most prominent one being the Jaccard similarity [8,4]. A naive approach, which maintains the entire dataset of the items examined by each user and their ratings and directly computes the similarity between any two users, may not be tractable for big data applications, both in terms of space and time complexity. A recommender system may have tens of millions of users[1], and may need to handle a possible set of billions of items[2]. A tractable alternative requires representing knowledge about users concisely while still allowing inference on relations between users (e.g. user similarity). One method for concisely representing knowledge in such settings is sketching (also known as fingerprinting) [2,20,9,1]. Such methods store *fingerprints*, which are concise descriptions of the dataset, and can be though of as an extreme lossy compression method. These fingerprints are extremely short, far more concise than traditional compression techniques would achieve. On the other hand, as opposed to compression techniques, they do not allow a full reconstruction of the original data (even approximately), but rather only allow inferring very specific properties of the original dataset. Sketching allows keeping short fingerprints of the item sets of users in recommender systems, that can still be used to approximately compute similarity measures between any two users [5,4,3]. Most such sketches use random hashes [2,20,9,1,5].

Our Contribution. The state of the art sketch in terms of space complexity applies many random hashes to build a fingerprint [22], and stores only a single bit per hash. A drawback of this approach is its high running time, caused by the many applications of hashes to elements in the dataset. The state of the art method in terms of running time is [14], which exponentially speeds up the computation time for building the fingerprints. Unfortunately, the currently best sketching techniques in terms of space complexity [22] and time complexity [14] are mutually incompatible, forcing a choice between reducing space or reducing runtime. Also, the low time complexity method [14] is tailored for computing Jaccard similarity between users, but is unsuitable for other similarity measures, such as the more sensitive rank correlation similarity measures [4]. We propose an alternative general sketching approach, which achieves both a low space complexity similar to that of [22] and a low time complexity similar to [14]. Our sketch uses random hashing and has a similar space complexity to [22], storing a single bit per hash, thus outperforming previous approaches such as [2,13,5] in *space* complexity. Similarly to [14], we get an *exponential* speedup in one factor of the *computation time*. Our discussion focuses on Jaccard similarity [5], but our approach is more general than [14], capturing other fingerprints, such as frequency moments [2], L_p sketches [17], rarity [13] and rank correlations [4]. Our sketch "ties" hashes in a novel way, allowing an *exponential* runtime speedup while storing only a single bit per hash. We also make an **empirical contribution** and evaluate our method using the Netflix [6] dataset. We analyze the running time and space complexity of our sketch, comparing it to the above state of the

---

[1] For example, Netflix is a famous provider of on-demand internet streaming media and employs a recommender system with over 20 million users. Our empirical evaluation is based on the dataset released by Netflix in a challenge to improve their algorithms [6].

[2] An example is a recommender system for webpages. Each such a webpage is a potential information item, so there are billions of such items that can potentially be recommended.

art methods. We show that it in practice the accuracy of our sketch is higher than the theoretical bounds, so even shorter sketches achieve high quality results.

Related Work. Recent work [5,4,3] already suggests sketching for recommender systems. Rather than storing the full lists of items consumed by each user and their ratings, they only store fingerprints of the information for each user, designed so that given the fingerprints of any two users one can accurately estimate the similarity between them. These fingerprints are constructed using min-wise independent families of hash functions, *MWIFs* for short. MWIFs were introduced in [23,8], and are useful in many applications as they resemble random permutations. Much of the research on sketching focused on reducing *space complexity* while accurately computing data stream properties, but much less attention was given to *time complexity*. Recent work that *does* focus on running time as well as space is [18,19] and [12] which propose low runtime sketches for locally sensitive hashing under the l_2 norm. Many streaming algorithms apply *many* hashes to each element in a very long stream of elements, leading to a high and sometimes intractable computation time. Similarly to [14], our method achieves an *exponential* speedup in one factor of the *computation time* for constructing fingerprints of massive data streams. The heart of the method lies in using a specific family of pseudo-random hashes shown to be approximately-MWIF [16], and for which we can quickly locate the hashes resulting in a small value of an element under the hash. Similarly to [24] we use the fact that family members are pairwise independent between themselves. Whereas previous models examine only one hash at a time, we read and process "chunks" of hashes to find important elements in the chunk, exploiting the chunk's structure to significantly speed up computation. We show that our technique is compatible with storing a *single* bit rather than the full element IDs, improving the fingerprint size, similarly to [22].

Improving Time and Space Complexity. Rather than storing the full stream of b items of a universe $[u]$, sketching methods only store a fingerprint of the stream. Any sketch achieves an estimate that is "probably approximately correct", i.e. with high probability the estimation error is small. Thus the size of the fingerprint and the time required to compute it depend on the accuracy of the method ϵ and its confidence δ. The accuracy ϵ is the allowed error (the difference between the estimated similarity and the true similarity), and the confidence δ is the maximal allowed probability of obtaining an estimate with a "large error", higher than the allowed error ϵ. Similarly to [22] we store a single bit per hash function, which results in a fingerprint of length $O(\frac{\ln \frac{1}{\delta}}{\epsilon^2})$ bits, rather than $O(\log u \cdot \frac{\ln \frac{1}{\delta}}{\epsilon^2})$ bits required by previous approaches such as [14]. On the other hand, similarly to [14], rather than computing the fingerprint of a stream of b items in time of $O(\frac{b \ln \frac{1}{\delta}}{\epsilon^2})$ as required by [22], we can compute it in time $O(b \cdot \log \frac{1}{\delta} \cdot \log \frac{1}{\epsilon})$, achieving an exponential speedup for the fingerprint construction. In addition to the theoretical guarantees, in Section 3 we evaluate our approach on the Netflix dataset, contrasting it with previous approaches in terms of time and space complexity. We also show that a high accuracy can be obtained even for very small fingerprints. Our theoretical results relate the required storage to the accuracy of the Jaccard similarity estimates, but only provide an upper bound regarding the storage required; we show that in practice the storage required to achieve a good accuracy can be much lower.

Preliminaries. Let H be a family of hashes over source X and target Y, so $h \in H$ is a function $h : X \to Y$, where Y is ordered. We say H is min-wise independent if when randomly choosing $h \in H$, for any subset $C \subseteq X$, any $x \in C$ has an equal probability to be minimal after applying h.

Definition 1. *H is min-wise independent (MWIF), if for all $C \subseteq X$, for any $x \in C$, $Pr_{h \in H}[h(x) = min_{a \in C} h(a)] = \frac{1}{|C|}$.*

Definition 2. *H is a γ-approximately min-wise independent (γ-MWIF), if for all $C \subseteq X$, for any $x \in C$: $\left| Pr_{h \in H}[h(x) = min_{a \in C} h(a)] - \frac{1}{|C|} \right| \leq \frac{\gamma}{|C|}$*

Definition 3. *H is k-wise independent, if for all $x_1, x_2, \ldots, x_k \subseteq X, y_1, y_2, \ldots, y_k \subseteq Y$, $Pr_{h \in H}[(h(x_1) = y_1) \wedge \ldots \wedge (h(x_k) = y_k)] = \frac{1}{|Y|^k}$*

Pseudo-random Family of Hashes. We describe our hashes. Given a universe of item IDs $[u]$, consider a prime p, such that $p > u$. Consider taking random coefficients for a d-degree polynomial in $\mathbb{Z}_p$. Let $a_0, a_1, \ldots, a_d \in [p]$ be chosen uniformly at random from $[p]$, and the polynomial in $\mathbb{Z}_p$: $f(x) = a_0 + a_1 x + a_2 x^2 + \ldots + a_d x^d$. Denote by F_d all d-degree polynomials in $\mathbb{Z}_p$ with coefficients in $\mathbb{Z}_p$. Our method chooses members of this family uniformly at random. Indyk [16] shows that choosing a function f from F_d uniformly at random results in F_d being a γ-MWIF for $d = O(\log \frac{1}{\gamma})$. Randomly choosing $a_0, \ldots, a_d$ is equivalent to choosing a member of F_d uniformly at random, so $f(x) = a_0 + a_1 x + a_2 x^2 + \ldots + a_d x^d$ is a hash chosen at random from the γ-MWIF F_d. Similarly, let $b_0, b_1, \ldots, b_d \in [p]$ be chosen uniformly at random from $[p]$, and $g(x) = b_0 + b_1 x + b_2 x^2 + \ldots + b_d x^d$, also a hash chosen at random from the γ-MWIF F_d. Consider the hashes $h_0(x) = f(x), h_1(x) = f(x) + g(x), h_2(x) = f(x) + 2g(x), \ldots, h_i(x) = f(x) + ig(x), \ldots, h_{k-1}(x) = f(x) + (k-1)g(x)$. We call the construction procedure for $f(x), g(x)$ the *base random construction*, and the construction of h_i the *composition construction*. We prove properties of such hashes. We denote the probability of an event E when the hash h is constructed by choosing f, g using the base random construction and composing $h(x) = f(x) + i \cdot g(x)$ (for some $i \in [p]$) as $Pr_h(E)$. These constructions are similar to the constructions used in [21], however our hashes are min-wise independent and pair-wise independent between themselves.

Lemma 1 (Uniform Minimal Values). *Let f, g be constructed using the base construction, using $d = O(\log \frac{1}{\gamma})$. For any $z \in [u]$, any $X \subseteq [u]$ and any value i used to compose $h(x) = f(x) + i \cdot g(x)$: $Pr_h[h(z) \leq min_{y \in X}(h(y))] = (1 \pm \gamma) \frac{1}{|X|}$. Proof in full version.*

Lemma 2 (Pairwise Interaction). *Let f, g be constructed using the base construction for $d = O(\log \frac{1}{\gamma})$. For all $x_1, x_2 \in [u]$ and $X_1, X_2 \subseteq [u]$, and all $i \neq j$ used to compose $h_i(x) = f(x) + i \cdot g(x)$ and $h_j(x) = f(x) + j \cdot g(x)$: $Pr_{f,g \in F_d}[(h_i(x_1) \leq min_{y \in X_1} h_i(y)) \wedge (h_j(x_2) \leq min_{y \in X_2} h_i(y))] = (1 \pm \gamma)^2 \frac{1}{|X_1| \cdot |X_2|}$. Proof in full version.*

2 Collaborative Filtering Using Pseudo-random Fingerprints

Collaborative filtering systems provide a target user recommendations for items based on the consumption patterns of other users. Many such systems rank users by their *similarity* to the target user in their past consumption, then find items many such users have consumed by the target user has not yet consumed and recommend them to the target user [26,27,11]. We focus on estimating the similarity between users (see [28] for a survey of methods for generating recommendations based on similarity information). A common measure of similarity between two users is Jaccard similarity. Our item universe consists of the IDs of all items users may consume, $[u] = \{1, 2, 3, \ldots, u\}$. There are u different items in the universe, but each user only examined some of these. Consider one user who has consumed the items $C_1 \subseteq [u]$ in the past, and another user who has consumed the items $C_2 \subseteq [u]$. The Jaccard similarity between the two users is $J_{1,2} = \frac{|C_1 \cap C_2|}{|C_1 \cup C_2|}$. Several fingerprinting methods were proposed for approximating relations between massive datasets, including the Jaccard similarity [8]. We use hashes defined earlier to exponentially speed up such computations. We use pseudo-random effects, so we must relax the MWIF requirement to a pairwise independence requirement (2-wise independence). We briefly review earlier sketches for estimating Jaccard similarity. Consider a hash $h \in H$ chosen from a MWIF H (or a γ-MWIF). We apply h on all elements C_1 and examine the minimal integer we get, $m_1^h = \arg\min_{x \in C_1} h(x)$. We do the same to C_2 and examine $m_2^h = \arg\min_{x \in C_2} h(x)$. Jaccard similarity sketches are based on computing the probability that $m_1 = m_2$: $Pr_{h \in H}[m_1^h = m_2^h] = Pr_{h \in H}[\arg\min_{x \in C_1} h(x) = \arg\min_{x \in C_2} h(x)]$. Theorems 1 and 2 are proven in [7,8] regarding a hash h chosen uniformly at random from a MWIF H and γ-MWIF, correspondingly.[3]

Theorem 1 (Jaccard / Collision Probability (MWIF)). $Pr_{h \in H}[m_i^h = m_j^h] = J_{i,j}$.

Theorem 2 (Jaccard / Collision Probability (γ-MWIF)). $|Pr_{h \in H}[m_i^h = m_j^h] - J_{i,j}| \leq \gamma$.

Rather than storing the full C_i's, previous approaches [7,8] store their fingerprints. Given k hashes $h_1, \ldots, h_k$ randomly chosen from an γ-MWIF, we can store $m_i^{h_1}, \ldots, m_i^{h_k}$. Given C_i, C_j, for any $x \in [k]$, the probability that $m_i^{h_x} = m_j^{h_x}$ is $J_{i,j} \pm \gamma$. A hash h_x where we have $m_i^{h_x} = m_j^{h_x}$ is a *hash collision*. We can estimate $J_{i,j}$ by counting the proportion of collision hashes out of all the chosen hashes. In this approach, the fingerprint contains k item identities in $[u]$, since for any x, $m_i^{h_x}$ is in $[u]$. Thus, the fingerprint requires $k \log u$ bits. To achieve an accuracy ϵ and confidence δ, such approaches require $k = O(\frac{\ln \frac{1}{\delta}}{\epsilon^2})$.

Our General Approach. We use a "block fingerprint" that estimates $J = J_{i,j}$ with accuracy ϵ and confidence of $\frac{7}{8}$. It stores a *single bit* per hash (where many previous approaches store $\log u$ bits per hash). Later we show how to get a given accuracy ϵ and confidence δ, by combining several such blocks. To get a single bit per hash, we use a hash mapping elements in $[u]$ to a single bit — $\phi : [u] \rightarrow \{0, 1\}$, taken from

---

[3] The full version includes a proof for Theorem 1.

a pairwise independent family (PWIF for short) of hashes. Rather than using $m_i^h = \arg\min_{x \in C_1} h(x)$ we use $m_i^{\phi,h} = \phi(\arg\min_{x \in C_1} h(x))$. Storing $m_i^{\phi,h}$ rather than m_i^{ϕ} shortens the sketch by a $\log u$ factor.

Theorem 3. $Pr_{h \in H}[m_i^{\phi,h} = m_j^{\phi,h}] = \frac{J_{i,j}}{2} + \frac{1}{2} \pm \frac{\gamma}{2}$.

Proof. $Pr_{h \in H, \phi \in H'}[m_i^{\phi,h} = m_j^{\phi,h}] = Pr[m_i^{\phi,h} = m_j^{\phi,h} | m_i^h = m_j^h] \cdot Pr_{h \in H}[m_i^h = m_j^h] + Pr[m_i^{\phi,h} = m_j^{\phi,h} | m_i^h \neq m_j^h] \cdot Pr_{h \in H}[m_i^h \neq m_j^h] = 1 \cdot Pr_{h \in H}[m_i^h = m_j^h] + \frac{1}{2} \cdot (1 - Pr_{h \in H}[m_i^h = m_j^h]) = \frac{1 + J_{i,j} \pm \gamma}{2}$

The purpose of the fingerprint block is to estimate of $J_{i,j}$ with accuracy ϵ. We use $k = \frac{8.02}{\epsilon^2}$ hashes. Denote $\alpha = \frac{2^{10} - 1}{2^{10}}$, and let $\gamma = (1 - \alpha) \cdot \epsilon = \frac{1}{2^{10}}\epsilon$. We construct a γ-MWIF[4]. To construct the family, consider choosing $a_0, \ldots, a_d$ and $b_0, b_1, \ldots, b_d$ uniformly at random from $[p]$, constructing the polynomials $f(x) = a_0 + a_1 x + a_2 x^2 + \ldots + a_d x^d$, $g(x) = b_0 + b_1 x + b_2 x^2 + \ldots + b_d x^d$, and using the k hashes $h_i(x) = f(x) + i g(x)$, where $i \in \{0, 1, \ldots, k-1\}$.[5] We also use a hash $\phi : [u] \to \{0, 1\}$ chosen from the PWIF of such hashes. We say there is a collision on h_l if $m_i^{\phi,h_l} = m_j^{\phi,h_l}$, and denote the random variable Z_l where $Z_l = 1$ if there is a collision on h_l for users i, j and $Z_l = 0$ if there is no such collision. $Z_l = 1$ with probability $\frac{1}{2} + \frac{J}{2} \pm \frac{\gamma}{2}$ and $Z_l = 0$ with probability $\frac{1}{2} - \frac{J}{2} \pm \frac{\gamma}{2}$. Thus $E(Z_l) = \frac{1}{2} + \frac{J}{2} \pm \frac{\gamma}{2}$. Denote $X_l = 2Z_l - 1$. $E(X_l) = 2E(Z_l) - 1 = J \pm \gamma$. X_l can take two values, -1 when $Z_l = 0$, and 1 when $Z_l = 1$. Thus X_l^2 always takes the value of 1, so $E(X_l^2) = 1$. Consider $X = \sum_{l=1}^{k} X_l$, and take $Y = \hat{J} = \frac{X}{k}$ as an estimator for J. We show that for the above k, Y is accurate up to ϵ with probability at least $\frac{7}{8}$.

Theorem 4 (Simple Estimator). $Pr(|Y - J| \leq \epsilon) \geq \frac{7}{8}$. *Proof in full version.*

Due to Theorem 4, we approximate J with accuracy ϵ and confidence $\frac{7}{8}$ using a "block fingerprint" for C_i, composed of $m_i^{h_1, \phi_1}, \ldots, m_i^{h_k, \phi_k}$, where $h_1, \ldots, h_k$ are random members of a γ-MWIF and $\phi_1, \ldots, \phi_k$ are chosen from the PWIF of hashes $\phi : [u] \to \{0, 1\}$. It suffices to take $k = O(\frac{1}{\epsilon^2})$ to achieve this. Constructing each h_i can be done by choosing f, g using the base random construction and composing $h_i(x) = f(x) + i \cdot g(x)$. The base random construction chooses f, g uniformly at random from F_d, the family of d-degree polynomials in $\mathbb{Z}_p$, where $d = O(\log \frac{1}{\epsilon})$. This achieves a γ-MWIF where $\gamma = (1 - \alpha) \cdot \epsilon = \frac{1}{2^{10}}\epsilon$.

Achieving a Desired Confidence. We combine several *independent* fingerprints to increase the confidence to a level δ. Earlier this section we proposed a fingerprint of length k to get a confidence of $\frac{7}{8}$. Consider taking m fingerprints for each stream, each of length k. Given two streams, i, j, we have m pairs of fingerprints, each approximating J with accuracy ϵ, and confidence $\frac{7}{8}$. Denote the obtained estimators by $\hat{J}_1, \hat{J}_2, \ldots, \hat{J}_m$, and the *median* of these values by $\hat{J}$. Consider using $m > \frac{32}{9} \ln \frac{1}{\delta}$ "blocks".

---

[4] The accuracy γ is stronger than the ϵ required of the full fingerprint, for reasons examined later.

[5] This is similar to [21], but here the hashes are MWIF and pair-wise independent between themselves.

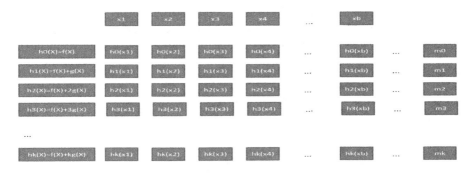

Fig. 1. A fingerprint "chunk" for a stream

Theorem 5 (Median Estimator). $Pr(|\hat{J} - J| \le \epsilon) \ge 1 - \delta$. *Proof in full version.*

By Theorem 5, to get $|\hat{J} - J| \le \epsilon$ it suffices to take $m > \frac{32}{9} \ln \frac{1}{\delta}$ blocks, each with $k = \frac{8.02}{\epsilon^2}$ hashes, or $\frac{32}{9} \ln \frac{1}{\delta} \cdot \frac{8.02}{\epsilon^2} \le \frac{28.45 \ln \frac{1}{\delta}}{\epsilon^2}$ hashes in total. Thus, we use $O(\frac{\ln \frac{1}{\delta}}{\epsilon^2})$ hashes.

2.1 Fast Fingerprint Computation

We discuss speeding up the fingerprint computation for a set of b items $X = \{x_1, \ldots, x_b\}$ where $x_i \in [u]$. The fingerprint has m "block fingerprints", with block r constructed using k hashes $h_1^r, \ldots, h_k^r$, built using $2 \cdot d$ random coefficients in $\mathbb{Z}_p$. The i'th location in the block is the minimal item in X under h_i: $m_i = \arg \min_{x \in X} h_i(x)$, which is then hashed through a hash ϕ mapping elements in $[u]$ to a single bit. We show how to quickly compute the block fingerprint $(m_1, \ldots, m_k)$. A naive way to do this is applying $k \cdot b$ hashes to compute $h_i(x_j)$ for $i \in [k], j \in [b]$. The values $h_i(x_i)$ where $i \in [k], j \in [b]$ form a matrix, where row i has the values $(h_i(x_1), \ldots, h_i(x_b))$, shown in Figure 1.

Once all $h_i(x_j)$ values are computed for $i \in [k], j \in [b]$, for each row i we check for which column j the row's minimal value occurs, and store $m_i = x_j$. Computing the fingerprint requires finding the minimal value across the rows (and the value x_j for the column j where this minimal value occurs). To speed up the process, we use a method similar to [25] as a building block. Recall the hashes h_i were defined as $h_i(x) = f(x) + ig(x)$ where $f(x), g(x)$ are d-degree polynomials with random coefficients in $\mathbb{Z}_p$. Our algorithm is based on a procedure that gets a value $x \in [u]$ and a threshold t, and returns all elements in $(h_0(x), h_1(x), \ldots, h_{k-1}(x))$ which are smaller than t, as well as their locations. Formally, the method returns the index list $I_t = \{i | h_i(x) \le t\}$ and the value list $V_t = \{h_i(x) | i \in I_t\}$ (note these are lists, so the j'th location in V_t, $V_t[j]$, contains $h_{I_t[j]}(x)$). We call this the *column procedure*, and denote by $pr - small - loc(f(x), g(x), k, x, t)$ the function that returns I_t, and by $pr - small - val(f(x), g(x), k, x, t)$ the function that returns V_t. We describe an implementation of these operations later in this section, with a running time of $O(\log k + |I_t|)$, rather than the naive algorithm which evaluates $O(k)$ hashes. Thus, this procedure quickly finds small elements across columns (by "small" we mean smaller than t). Our algorithm keeps a bound for the minimal value for each row. It goes through the columns, finding the small values in each, and updates the row bounds where these occur.

block-update $((x_1, \ldots, x_b), f(x), g(x), k, t)$:

1. Let $m_i = \infty$ for $i \in [k]$ and let $p_i = 0$ for $i \in [k]$
2. For $j = 1$ to b:
   (a) Let $I_t = pr - small - val(f(x), g(x), k, x_j, t)$
   (b) Let $V_t = pr - small - loc(f(x), g(x), k, x_j, t)$
   (c) For $y \in I_t$: // Indices of the small elements
       i. If $m_{I_t[y]} > V_t[y]$ // Update to row x required
          A. $m_{I_t[y]} = V_t[y], p_{I_t[y]} = x_j$

If our method updates m_i, p_i for row i, m_i indeed contains the minimal value in that row, and p_i the column where this minimal value occurs, since if even a single update occurred then the row indeed contains an item that is smaller than t, so the minimal item in that row is smaller than t and an update would occur for that item. On the other hand, if all the items in a row are bigger than t, an update would not occur for that row. The running time of the column procedure is $O(\log k + |I_t|)$, which is a random variable, that depends on the number of elements returned for that column, $|I_t|$. Denote by L_j the number of elements returned for column j (i.e. $|I_t|$ for column j). Since we have b columns, the running time of the block update is $O(b \log k) + O(\sum_{j=1}^b L_j)$. The total number of returned elements is $\sum_{j=1}^b L_j$, which is the total number of elements that are smaller than t. We denote by $Y_t = \sum_{j=1}^b L_j$ the random variable which is the number of all elements in the block that are smaller than t. The running time of our block update is thus $O(b \log k + Y_t)$. The random variable Y_t depends on t, since the smaller t is the less elements are returned and the faster the column procedure runs. On the other hand, we only update rows whose minimal value is below t, so if t is too low we have a high probability of having rows which are not updated correctly. A certain compromise t value combines a good running time of the block update with a good probability of correctly computing the values for all the rows.

Theorem 6. *Given the threshold $t = \frac{12 \cdot p \cdot l'}{b}$, where $l' = 80 + 2 \log \frac{1}{\epsilon}$ (so $l' = O(\log \frac{1}{\epsilon})$), the runtime of the block $-$ update procedure is $O(b \log \frac{1}{\epsilon} + \frac{1}{\epsilon^2} \log \frac{1}{\epsilon})$. Proof in full version.*

Computing Minimal Elements in the Series. We recursively implement $pr - small - loc(f(x), g(x), k, x, t)$ and $pr - small - val(f(x), g(x), k, x, t)$, the procedures for computing V_t and I_t. The hashes h_i were defined as $h_i(x) = f(x) + ig(x)$ where $f(x), g(x)$ are d-degree polynomials with random coefficients in $\mathbb{Z}_p$. Consider $x \in \mathbb{Z}_p$ for which we seek all the values (and indices) in $(h_0(x), h_2(x), \ldots, h_{k-1}(x))$ smaller than t. Given x, we evaluate $f(x), g(x)$ in time $O(d) = O(\log \frac{1}{\gamma})^6$, and denote $a =$

---

[6] Using multipoint evaluation we can calculate it in amortized time $O(\log^2 \log \frac{1}{\gamma})$. We can use other constructions for d-wise independent which can be evaluated in $O(1)$ time but use more space.

$f(x) \in \mathbb{Z}_p$ and $b = g(x) \in \mathbb{Z}_p$. Thus, we seek all values in $\{a \mod p, (a + b) \mod p, (a + 2b) \mod p, \ldots, (a + (k-1)b) \mod p\}$ smaller than t, and the indices i where they occur. Consider the series $S = (s_1, \ldots, s_k)$ where $s_i = (a + ib) \mod p$ and $i = \{0, 1, \ldots, k - 1\}$. We denote the arithmetic series $a + bi \mod p$ for $i \in \{0, 1, \ldots, k - 1\}$ as $S(a, b, k, p)$, so under this notation $S = S(a, b, k, p)$. Given a value we can find the index where it occurs, and vice versa. To get the value for index i, we compute $(a + ib) \mod p$. To get the index i where value v occurs, we solve $v = a + ib$ in $\mathbb{Z}_p$ (i.e. $i = \frac{v-a}{b} \mod p$). This can be done in $O(\log p)$ time using Euclid's algorithm. We compute b^{-1} in $\mathbb{Z}_p$ once to transform all values to generating indices. We call a location i where $s_i < s_{i-1}$ a *flip location*. The first index is a flip location if $a - b \mod p > a$. First, consider $b < \frac{p}{2}$. If s_i is a flip location, we have $s_{i-1} < p$ but $s_{i-1} + b > p$, so $s_i < b$. As $b < \frac{p}{2}$ there's at least one location that is *not* a flip location between any two flip locations. Given $S = S(a, b, k, p)$, denote by $f(S)$ the flip locations in S. The flip locations of S are $f(S)$. Denote $f_0(S) = f(S)$, and by $f_i(S)$ elements occurring i places after the closest flip location. Lemmas 3 and 4 are proven in the full version.

Lemma 3 (Flip Locations Are Small). *If $b < \frac{p}{2}$, at most $\frac{k}{2}$ elements are flip locations, and all elements that are smaller than b are flip locations.*

Lemma 4 (Element Comparison). *If $b < \frac{p}{2}$, $x \in f_i(S), y \in f_j(S)$ for $i > j$, then $x > y$.*

The first flip location is $\lceil \frac{p-a}{b} \rceil$, as to exceed p we add $b \lceil \frac{p-a}{b} \rceil$ times. There are $\lfloor \frac{a+bk}{p} \rfloor$ flip locations. Denote the first flip location as $j = \lceil \frac{p-a}{b} \rceil$, with value $a' = (a + jb) \mod p$. Denote $b' = (b - p) \mod b$ and the number of flip locations as $k' = \lfloor \frac{(a+bk)}{p} \rfloor$. The flip locations have an arithmetic progression [25].

Lemma 5 (Flip Locations Arithmetic Progression). *The flip locations of $S = S(a, b, k, p)$ are also an arithmetic progression $S' = (a', b', k', b)$.*

Using these lemmas, we search for the elements smaller than t by examining the flip locations series in recursion. If case $b < t$, given $q = \lceil t \rceil b$, due to Lemma 4 $f(S), f_1(S), \ldots f_{q-1}(S)$ are smaller then t, and all of their elements must be returned. We must also scan $f_q(S)$ and also return all the elements of $f_q(S)$ which are smaller then t. This additional scan requires $O(|f_q(S)|)$ time $|f_q(S)| \leq |f(S)|$. Thus the case of $b < t$ examines $O(|I_t|)$ elements. By Lemma 3, if $b > t$, all non-flip locations are bigger than b and thus bigger than t, so we need only consider the flip-locations as candidates. Using Lemma 5 we scan the flip locations recursively, examining the arithmetic series of the flip locations. If at most half of the elements in each recursion are flip locations, this gives a logarithmic running time, but if b is high, more than half the elements are flip locations. When $b > \frac{p}{2}$ we examine the same flip-location series S', in reverse order. The first element in the reversed series is the last element of the current series, and rather than progressing in steps of b, we progress in steps of $p - b$. Thus we obtain the same elements, but in reverse order. In this reversed series, at most half the elements are flip locations. The procedure below implements our method. It finds elements smaller then t in time $O(\log k) = O(\log \frac{1}{\epsilon} + |I_t|)$ where $|I_t|$ is the number of such values. Given the returned indices, we get the values in them. We use the same b for all $|I_t|$, so this can be done in time $O(c \log c + |I_t|)$ (usually c is a constant).

ps-min(a, b, p, k, t) :

1. if $b < t$:
   (a) $V_t = []$; if $a < t$ then $V_t = V_t + [a + ib$ for i in range $(\lceil \frac{t-a}{b} \rceil)]$
   (b) $j = \lceil \frac{p-a}{b} \rceil$ // First flip (excluding first location)
   (c) while $j < k$:
       i. $v = (a + jb) \mod p$
       ii. while $j < k$ and $v < t$:
           A. V_t.append(v); $j = j + 1$; $v = v + b$
       iii. $j = j + \lceil \frac{p-v}{b} \rceil$ //next flip location
       iv. return list1
   (d) if $b > \frac{p}{2}$ then return $f((a + (k-1) \cdot b) \mod p, p - b, p, k, t)$
   (e) $j = \lceil \frac{p-a}{b} \rceil$; $new_k = \lfloor \frac{a+bk}{p} \rfloor$
   (f) if $a < b$ then $j = 0$ and $new_k = new_k + 1$// get first flip location
   (g) return $f((a + jb) \mod p, -p \mod b, b, new_k, t)$

3   Empirical Analysis

We empirically evaluated our sketch using the Netflix dataset [6]. This is a movie ratings dataset, with 100 million movie ratings, provided by roughly half a million users on a collection of 17,000 movies. As there are 100 million ratings, even this meduim-sized dataset is difficult to fit in memory[7], so a massive recommender systems dataset certainly cannot fit in the main memory, making sketching necessary to handle such datasets [5]. The state of the art space complexity is achieved using the sketching technique of [22]. Consider using it to estimate Jaccard similarity, with a reasonable accuracy of $\epsilon = 0.01$ and confidence level of $\delta = 0.001$. The approach of [22] applies roughly 100,000 hash functions for each entry. Each hash computation requires 20 multiplication operations, and as there are 100 million entries in the dataset, sketching the entire dataset requires over than $2 \cdot 10^{14}$ multiplications. This takes more than a day to run on a powerful machine. On the other hand, although the approach of [14] allows a much shorter running time (less than an hour), it requires sacrificing the low space achieved by the method of [22]. We first compare our approach with [22,14] in terms of the running time. Figure 2 shows the running time for generating a fingerprint for a target with 1,000 items, both under our method (FPRF - Fast Pseudo Random Fingerprints), and under the sketch of [22] (appearing under "1-bit", as it maintains a single bit per hash used) and the sketch of [14] ("FPS", after the names of the authors). The Figure indicates the massive saving in computation time our approach offers over the approach of [22], and shows that the running time of our approach and that of [14] is very similar.

We now examine the accuracy achieved by our approach, which depends on the sketch size. To analyze empirical accuracy, we isolated users who provided ratings for

---

[7] The Netflix data can easily be stored on disk. It is even possible to store it in memory on machine with a large RAM, by compressing it or using a sparse matrix representation.

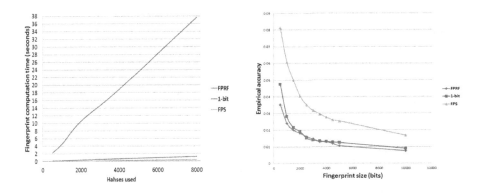

Fig. 2. Left: running time of fingerprint computation. Right: accuracy depending on size.

over 1,000 movies. There are over 10,000 such users in the dataset, and as these users have rated many movies, the Jaccard similarity between two such users is very fine-grained. We tested the fingerprint size required to achieve a target accuracy level for the Jaccard similarity. Consider a fingerprint size of k bits. Given two users, denote the true Jaccard similarity between their lists of rated movies as J. J can be easily computed using the entire dataset. Alternatively, we can use a fingerprint of size k, resulting in an estimate $\hat{J}$ that has a certain error. The error for a pair of users is $e = |J - \hat{J}|$. We can sample many such user pairs, and examine the average error obtained using a fingerprint of size k, which we call the *empirical inaccuracy*. We wish to minimize the error in our estimates, but to reduce the inaccuracy we must use larger fingerprints. As each user in our sample rated at least 1,000 movies, storing the full list of rated movies for a user takes 1,000 integers. The Netflix dataset only has 17,000 movies, so we require at least 15 bits to store the ID of each movie. Thus the full data for a user takes *at least* 15,000 bits. The space required for this data grows linearly with the numbers of movies a user has rated. Increasing the size of the universe of movies also increases the storage requirements, as more bits would be required to represent the ID of each movie.

Using our sketch, the required space does not depend on the number of ratings per user, or on the number of movies in the system, but rather on the target accuracy for the similarity estimate. Earlier sketches [5,3] eliminated the dependency on the number of ratings, but not on the number of movies in the system. Also, our fingerprints are faster to compute.

We tested how the average accuracy of our Jaccard similarity estimates changes as we chage the fingerprint size. We have tried fingerprints of different sizes, ranging from 500 bits to 10,000 bits. For each such size we sampled many pairs of users, and computed the average inaccuracy of the Jaccard similarity estimates. The results are given in Figure 2, for both our approach and for the sketch of [14] ("FPS"), as well as the "1-bit" sketch of [22]. Lower numbers indicate better empirical accuracy. Figure 2 shows that our sketch achieves a very high accuracy in estimating Jaccard similarity, even for small fingerprints. Even for a fingerprint size of 2500 bits per user, the Jaccard similarity can be estimated with an error smaller than 1.5%. Thus using fingerprints reduces the required storage to roughly 10% of that of the full dataset, without sacrifising much accuracy in estimating user similarity. The figure also indicates that for any sketch size, the accuracy achieved by our approach is superior to that of the FPS sketch [14]. This is predictable since the theoretical accuracy guarantee for our approach is better than that for the sketch of [14]. The figure shows no significant difference in accuracy between our sketch and the 1-bit sketch [22].

Figure 2 shows that on the Netflix dataset, our sketch has the good properties of the mutually exclusive sketches of [22,14], and outperforms each of these state of the art methods in either running time or accuracy. The Netflix dataset is a small dataset, and the saving in space is much greater for larger datasets. A recommender system for web pages is likely to have several orders of magnitude more users and information items. While the storage requirements for such a massive recommender system grow by several orders of magnitude when storing the full data, the required space remains almost the same using our sketch. Previous approaches [5,3] compute the sketch in time *quadratic* in the required accuracy. Using our approach, computing the sketch only requires time *logarithmic* in the accuracy, which makes it tractable even when the required accuracy is very high.

4 Conclusions

We presented a fast method for sketching massive datasets, based on pseudo-random hashes. Though we focused on collaborative filtering and examined the Jaccard similarity in detail, the same technique can be used for any fingerprint based on minimal elements under several hashes. Our approach is thus a general technique for exponentially speeding up computation of various fingerprints, while maintaining a single bit per hash. We showed that even for these small fingerprints which can be quickly computed, the required number of hashes is asymptotically similar to previously known methods, and is logarithmic in the required confidence and polynomial in the required accuracy. Our empirical analysis shows that for the Netflix dataset the required storage is even smaller than the theoretical bounds.

Several questions remain open. Can we speed up the sketch computation further? Can similar methods be used that are not based on minimal elements under hashes?

References

1. Aggarwal, C.C.: Data streams: models and algorithms. Springer-Verlag New York Inc. (2007)
2. Alon, N., Matias, Y., Szegedy, M.: The Space Complexity of Approximating the Frequency Moments. J. Computer and System Sciences 58(1), 137–147 (1999)
3. Bachrach, Y., Herbrich, R.: Fingerprinting Ratings for Collaborative Filtering — Theoretical and Empirical Analysis. In: Chavez, E., Lonardi, S. (eds.) SPIRE 2010. LNCS, vol. 6393, pp. 25–36. Springer, Heidelberg (2010)
4. Bachrach, Y., Herbrich, R., Porat, E.: Sketching algorithms for approximating rank correlations in collaborative filtering systems. In: Karlgren, J., Tarhio, J., Hyyrö, H. (eds.) SPIRE 2009. LNCS, vol. 5721, pp. 344–352. Springer, Heidelberg (2009)
5. Bachrach, Y., Porat, E., Rosenschein, J.S.: Sketching techniques for collaborative filtering. In: IJCAI, Pasadena, California (July 2009)

6. Bennett, J., Lanning, S.: The netflix prize. In: KDD Cup and Workshop (2007)
7. Broder, A.Z.: On the resemblance and containment of documents. Sequences (1998)
8. Broder, A.Z., Charikar, M., Frieze, A.M., Mitzenmacher, M.: Min-wise independent permutations. Journal of Computer and System Sciences 60(3), 630–659 (2000)
9. Cormode, G., Muthukrishnan, S.: An improved data stream summary: the count-min sketch and its applications. Journal of Algorithms 55(1), 58–75 (2005)
10. Cormode, G., Muthukrishnan, S., Rozenbaum, I.: Summarizing and mining inverse distributions on data streams via dynamic inverse sampling. In: VLDB (2005)
11. Das, A.S., Datar, M., Garg, A., Rajaram, S.: Google news personalization: scalable online collaborative filtering. In: WWW. ACM (2007)
12. Dasgupta, A., Kumar, R., Sarlos, T.: Fast locality-sensitive hashing. In: SIGKDD (2011)
13. Datar, M., Muthukrishnan, S.: Estimating rarity and similarity over data stream windows. In: Möhring, R., Raman, R. (eds.) ESA 2002. LNCS, vol. 2461, pp. 323–335. Springer, Heidelberg (2002)
14. Feigenblat, G., Shiftan, A., Porat, E.: Exponential time improvement for min-wise based algorithms. In: SODA (2011)
15. Hoeffding, W.: Probability inequalities for sums of bounded random variables. Journal of the American Statistical Association 58(301), 13–30 (1963)
16. Indyk, P.: A Small Approximately Min-Wise Independent Family of Hash Functions. Journal of Algorithms 38(1), 84–90 (2001)
17. Indyk, P.: Stable distributions, pseudorandom generators, embeddings, and data stream computation. Journal of the ACM (JACM) 53(3), 323 (2006)
18. Kane, D.M., Nelson, J., Porat, E., Woodruff, D.P.: Fast moment estimation in data streams in optimal space. In: STOC (2011)
19. Kane, D.M., Nelson, J., Woodruff, D.P.: An optimal algorithm for the distinct elements problem. In: PODS, pp. 41–52. ACM (2010)
20. Karp, R.M., Shenker, S., Papadimitriou, C.H.: A simple algorithm for finding frequent elements in streams and bags. ACM Transactions on Database Systems (TODS) 28(1), 51–55 (2003)
21. Kirsch, A., Mitzenmacher, M.: Less hashing, same performance: a better Bloom filter. In: Azar, Y., Erlebach, T. (eds.) ESA 2006. LNCS, vol. 4168, pp. 456–467. Springer, Heidelberg (2006)
22. Li, P., Koenig, C.: b-Bit minwise hashing. In: WWW (2010)
23. Mulmuley, K.: Randomized geometric algorithms and pseudorandom generators. Algorithmica (1996)
24. Pătraşcu, M., Thorup, M.: On the k-Independence Required by Linear Probing and Minwise Independence. In: Abramsky, S., Gavoille, C., Kirchner, C., Meyer auf der Heide, F., Spirakis, P.G. (eds.) ICALP 2010. LNCS, vol. 6198, pp. 715–726. Springer, Heidelberg (2010)
25. Pavan, A., Tirthapura, S.: Range-efficient counting of distinct elements in a massive data stream. SIAM Journal on Computing 37(2), 359–379 (2008)
26. Resnick, P., Iacovou, N., Suchak, M., Bergstrom, P., Riedl, J.: Grouplens: an open architecture for collaborative filtering of netnews. In: Computer Supported Cooperative Work (1994)
27. Sarwar, B., Karypis, G., Konstan, J., Reidl, J.: Item-based collaborative filtering recommendation algorithms. In: WWW (2001)
28. Su, X., Khoshgoftaar, T.M.: A survey of collaborative filtering techniques. Advances in Artificial Intelligence 2009, 4 (2009)

Physarum Can Compute Shortest Paths: Convergence Proofs and Complexity Bounds

Luca Becchetti[1], Vincenzo Bonifaci[2], Michael Dirnberger[3],
Andreas Karrenbauer[3], and Kurt Mehlhorn[3]

[1] Dipartimento di Informatica e Sistemistica, Sapienza Università di Roma, Italy
[2] Istituto di Analisi dei Sistemi ed Informatica "Antonio Ruberti",
Consiglio Nazionale delle Ricerche, Rome, Italy
[3] Max Planck Institute for Informatics, Saarbrücken, Germany

Abstract. *Physarum polycephalum* is a slime mold that is apparently able to solve shortest path problems. A mathematical model for the slime's behavior in the form of a coupled system of differential equations was proposed by Tero, Kobayashi and Nakagaki [TKN07]. We prove that a discretization of the model (Euler integration) computes a $(1 + \epsilon)$-approximation of the shortest path in $O(mL(\log n + \log L)/\epsilon^3)$ iterations, with arithmetic on numbers of $O(\log(nL/\epsilon))$ bits; here, n and m are the number of nodes and edges of the graph, respectively, and L is the largest length of an edge. We also obtain two results for a directed Physarum model proposed by Ito et al. [IJNT11]: convergence in the general, nonuniform case and convergence and complexity bounds for the discretization of the uniform case.

1 Introduction

Physarum polycephalum is a slime mold [BD97] that is apparently able to solve shortest path problems. In [NYT00], Nakagaki, Yamada, and Tóth report on the following experiment (see Figure 1): They built a maze, that was later covered with pieces of Physarum (the slime can be cut into pieces that will merge if brought into each other's vicinity), and then fed the slime with oatmeal at two locations. After a few hours, the slime retracted to the shortest path connecting the food sources in the maze. The experiment was repeated with different mazes; in all experiments, Physarum retracted to the shortest path. Tero, Kobayashi and Nakagaki [TKN07] propose a mathematical model for the behavior of the mold. Physarum is modeled as a tube network traversed by liquid flow, with the flow satisfying the standard Poiseuille assumption from fluid mechanics. In the following, we use terminology from the theory of electrical networks, relying on the fact that equations for electrical flow and Poiseuille flow are the same [Kir10].

In particular, let G be an undirected graph[1] with node set N, edge set E, length labels $l \in \mathbb{R}_{++}^E$ [2] and two distinguished nodes $s_0, s_1 \in N$. In our discussion,

---

[1] One can easily generalize the model and extend our results to multigraphs at the expense of heavier notation. Details will appear in the full version of the paper.

[2] We let $\mathbb{R}^A$, $\mathbb{R}_+^A$ and $\mathbb{R}_{++}^A$ denote the set of real, nonnegative real, and positive real vectors (respectively) whose components are indexed by A.

F.V. Fomin et al. (Eds.): ICALP 2013, Part II, LNCS 7966, pp. 472–483, 2013.

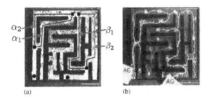  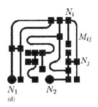

α_2—
α_1—
—β_1
—β_2

(a)   (b)   (c)

N_i
M_{ij}
N_j
N_1 N_2
(d)

Fig. 1. The experiment in [NYT00] (reprinted from there): (a) shows the maze uniformly covered by Physarum; the yellow color indicates the presence of Physarum. Food (oatmeal) is provided at the locations labelled AG. After a while, the mold retracts to the shortest path connecting the food sources as shown in (b) and (c). (d) shows the underlying abstract graph. The video [You] shows the experiment.

$x \in \mathbb{R}_+^E$ will be a state vector representing the diameters of the tubular channels of the Physarum (edges of the graph). The value x_e is called the *capacity* of edge e. The nodes s_0 and s_1 represent the location of two food sources. Physarum's dynamical system is described by the system of differential equations [TKN07]

$$\dot{x} = |q(x,l)| - x. \qquad (1)$$

Equation (1) is called the *evolution equation*, as it determines the dynamics of the system over time. It is a compact representation of a system of ordinary differential equations, one for every edge of the graph; the absolute value operator $|\cdot|$ is applied componentwise. The vector $q \in \mathbb{R}^E$, known as the *current flow*, is determined by the capacities and lengths of the edges, as follows (see Section 2 for the precise definitions). Force one unit of current from the source to the sink in an electrical network, where the resistance r_e of edge e is given by $r_e \overset{\text{def}}{=} l_e/x_e$, and call q_e the resulting current across edge e. In [BMV12, Bon13], it was shown that the dynamics (1) converges to the shortest source-sink path in the following sense: the potential difference between source and sink converges to the length of the shortest source-sink path, the capacities of the edges on the shortest source-sink path[3] converge to one, and the capacities of all other edges converge to zero.

Our first contribution relies on a numerical approximation of (1), as given by Euler's method [SM03],

$$\Delta x = h \cdot (|q(x,l)| - x), \qquad (2)$$

or, making the dependency on time explicit,

$$x(t+1) - x(t) = h \cdot (|q(x(t),l)| - x(t)), \qquad (3)$$

where $h \in (0,1)$ is the step size of the discretization. We prove that the dynamics (3) converges to the shortest source-sink path. More precisely, let opt be the length of the shortest path, n and m be the number of nodes and edges of the

---

[3] We assume uniqueness of the shortest path for simplicity of exposition.

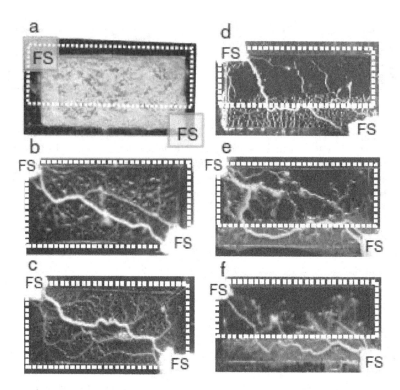

Fig. 2. Photographs of the connecting paths between two food sources (FS). (a) The rectangular sheet-like morphology of the organism immediately before the presentation of two FS and illumination of the region indicated by the dashed white lines. (b),(c) Examples of connecting paths in the control experiment in which the field was uniformly illuminated. A thick tube was formed in a straight line (with some deviations) between the FS. (d)-(f) Typical connecting paths in a nonuniformly illuminated field (95 K lx). Path length was reduced in the illuminated field, although the total path length increased. Note that fluctuations in the path are exhibited from experiment to experiment. (Figure and caption reprinted from [NIU+07, Figure 2].)

graph, and L be the largest length of an edge. We show that, *for $\epsilon \in (0, 1/300)$ and for $h = \epsilon/mL$, the discretized model yields a solution of value at most $(1 + O(\epsilon))$opt in $O(mL(\log n + \log L)/\epsilon^3)$ steps, even when $O(\log(nL/\epsilon))$-bit number arithmetic is used.* For bounded L, the time bound is therefore polynomial in the size of the input data.

Our second contribution was inspired by the following experiment of Nakagaki et al., reported in [NIU+07] (see also Figure 2). They cover a rectangular plate with Physarum and feed it at opposite corners of the plate. Two-thirds of the plate are put under a bright light, and one-third is kept in the dark. Under uniform lighting conditions, Physarum would retract to a straight-line path connecting the food sources [NYT00]. However, Physarum does not like light and therefore forms a path with one kink connecting the food sources. The path is such that the part under light is shorter than in a straight-line connection. In the theory section of [NYT00], a reactivity parameter $a_e > 0$ is introduced into (1):

$$\dot{x}_e(t) = |q_e(x,l)| - a_e x_e(t). \tag{4}$$

Note that if, for example, $q_e(x,l) = 0$, the capacity of edge e decreases with a rate that depends on a_e. To model the experiment, $a_e = 1$ for edges in the dark part of the plate, and $a_e = C > 1$ for the edges in the lighted area, where C is a constant. The authors of [NIU$^+$07] report that in computer simulations, the dynamics (4) converges to the shortest source-sink path with respect to the modified length function $a_e l_e$. A proof of convergence is currently only available for the uniform case $a_e = 1$ for all e, see [BMV12, Bon13].

A directed version of model (4) was proposed in [IJNT11]. The graph $G = (N, E)$ is now a directed graph. For a state vector $x(t)$, the flows are defined as above. A flow $q_e(x,l)$ is positive if it flows in the direction of e and is negative otherwise. The dynamics becomes

$$\dot{x}_e(t) = q_e(x,l) - a_e x_e(t). \tag{5}$$

Although this model apparently has no physical counterpart, it has the advantage of allowing one to treat directed graphs. Ito et al. [IJNT11] prove convergence to the shortest source-sink path in the uniform case ($a_e = 1$ for all e). In fact, they show convergence for a somewhat more general problem, the transportation problem, as does [BMV12] for the undirected model.

We show that the dynamics (5) converges to the shortest directed source-sink path under the modified length function $a_e l_e$. This generalizes the convergence result of [IJNT11] from the uniform ($a_e = 1$ for all e) to the nonuniform case, albeit only for the shortest path problem. Our proof combines arguments from [MO07, MO08, IJNT11, BMV12, Bon13] and we believe it is simpler than the one in [IJNT11]. Moreover, for the uniform case (that is, $a_e = 1$ for all e), we can prove convergence for the discretized model

$$x_e(t+1) = x_e(t) + h(q_e(x,l) - x_e(t)), \tag{6}$$

where $h \leq 1/(n(4nm^2 L X_0^2)^2)$ is the step size; here, X_0 is the maximum between the largest capacity and the inverse of the smallest capacity at time zero. In particular, let P^* be the shortest directed source-sink path and let $\epsilon \in (0,1)$ be arbitrary: *we show $x_e(t) \geq 1 - 2\epsilon$ for $e \in P^*$ and $x_e(t) \leq \epsilon$ for $e \notin P^*$, whenever* $t \geq \frac{4nL}{h}\left(3\ln X_0 + 2\ln \frac{2m}{\epsilon}\right)$.

Outline of the paper. The remainder of the paper is structured as follows. In Section 2 we give basic definitions and properties. In Section 3 we study the discrete dynamics (3). Section 4 concerns the directed models (5) and (6). We close with some concluding remarks in Section 5.

2   Electrical Networks

Without loss of generality, assume that $N = \{1, 2, \ldots, n\}$, $E = \{1, 2, \ldots, m\}$ and assume an arbitrary orientation of the edges.[4] Let $A = (a_{ve})_{v \in N, e \in E}$ be the incidence matrix of G under this orientation, that is, $a_{ve} = +1$ if v is the tail of e, $a_{ve} = -1$ if v is the head of e, and $a_{ve} = 0$, otherwise. Then q is defined as the unit-value flow from s_0 to s_1 of minimum energy, that is, as the unique optimal solution to the following continuous quadratic optimization problem, related to *Thomson's principle* from physics [Bol98, Theorem IX.2]:

$$\min\ q^T R q \quad \text{such that} \quad A q = b. \tag{7}$$

Here, $R \overset{\text{def}}{=} \operatorname{diag}(l/x) \in \mathbb{R}^{E \times E}$ is the diagonal matrix with value $r_e \overset{\text{def}}{=} l_e/x_e$ for the e-th element of the main diagonal, and $b \in \mathbb{R}^N$ is the vector defined by $b_v = +1$ if $v = s_0$, $b_v = -1$ if $v = s_1$, and $b_v = 0$, otherwise. The value r_e is called the *resistance* of edge e. Node s_0 is called the *source*, node s_1 the *sink*. The quantity $\eta \overset{\text{def}}{=} q^T R q$ is the *energy*; the quantity $b_{s_0} = 1$ is the *value* of the flow q. The optimality conditions for (7) imply that there exist values $p_1, \ldots, p_n \in \mathbb{R}$ (*potentials*) that satisfy *Ohm's law* [Bol98, Section II.1]:

$$q_e = (p_u - p_v)/r_e, \qquad \text{whenever edge } e \text{ is oriented from } u \text{ to } v. \tag{8}$$

By the *conservation of energy* principle, the total energy equals the difference between the source and sink potentials, times the value of the flow [Bol98, Corollary IX.4]:

$$\eta = (p_{s_0} - p_{s_1})b_{s_0} = p_{s_0} - p_{s_1}. \tag{9}$$

3   Convergence of the Undirected Physarum Model

In this section we characterize Physarum's convergence properties in the undirected model, as given by equation (3):

$$x(t + 1) = x(t) + h \cdot (|q(x(t), l)| - x(t)).$$

Assumptions on the input data: We assume that the length labels l and the initial conditions $x(0)$ satisfy the following:

a. each s_0-s_1 path in G has a distinct overall length; in particular, there is a unique shortest s_0-s_1 path;
b. all capacities are initialized to one:

$$x(0) = 1; \tag{10}$$

---

[4] In the directed model discussed in Section 4, this orientation is simply the one given by the directed graph.

c. the initially minimum capacity cut is the source cut, and it has unit capacity:

$$1_S^T \cdot x(0) \geq 1_0^T \cdot x(0) = 1, \qquad \text{for any } s_0\text{-}s_1 \text{ cut } S, \tag{11}$$

where 1_S is the characteristic vector of the set of edges in the cut S, and 1_0 is the characteristic vector of the set of edges incident to the source. Notice that this can be achieved even when s_0 has not degree 1, by connecting a new source s_0' to s_0 via a length 1, capacity 1 edge.

d. every edge has length at least 1.

Basic properties: The first property we show is that the set of fractional $s_0\text{-}s_1$ paths is an invariant for the dynamics.

Lemma 1. *Let* $x = x(t)$ *be the solution of* (3) *under the initial conditions* $x(0) = 1$. *The following properties hold at any time* $t \geq 0$: *(a)* $x > 0$, *(b)* $1_S^T \cdot x \geq 1_0^T \cdot x = 1$, *and (c)* $x \leq 1$.

Proof. (a.) Let $e \in E$ be any edge. Since $|q_e| \geq 0$, by the evolution equation (3) we have $\Delta x_e(t) = h(|q_e| - x_e(t)) \geq -hx_e(t)$. Therefore, by induction, $x_e(t+1) \geq x_e(t) - hx_e(t) = (1 - h)x_e(t) > 0$ as long as $h < 1$.

(b.) We use induction. The property is true for $x(0)$ by the assumptions on the input data. Then, using (3), induction, and the fact that $1_S^T \cdot |q| \geq 1$ for any cut S,

$$1_S^T \cdot x(t+1) = 1_S^T \cdot (x(t) + h(|q| - x(t))) = (1-h)1_S^T \cdot x(t) + h1_S^T \cdot |q| \geq 1 - h + h = 1.$$

The fact that $1_0^T \cdot x = 1$ can be shown similarly.

(c.) Easy induction, along the same lines as the proof of (a.).   □

An *equilibrium point* of (3) is a vector $x \in \mathbb{R}_+^E$ such that $\Delta x = 0$. Our assumptions imply that there are a finite number of equilibrium points: each equilibrium corresponds to an $s_0\text{-}s_1$ path of the network, and vice versa.

Lemma 2. *If* $x = 1_P$ *for some* $s_0\text{-}s_1$ *path* P, *then* x *is an equilibrium point. Conversely, if* x *is an equilibrium point, then* $x = 1_P$ *for some* $s_0\text{-}s_1$ *path* P.

Proof. The proof proceeds along the same lines as for the continuous case, see [Bon13, Lemma 2.3].   □

Convergence: Recall that, by (9),

$$\eta = \sum_{e \in E} r_e q_e^2 = q^T R q = p_{s_0} - p_{s_1}, \tag{12}$$

and let

$$V \stackrel{\text{def}}{=} l^T x = \sum_{e \in E} l_e x_e = \sum_{e \in E} r_e x_e^2 = x^T R x. \tag{13}$$

Here η is the energy dissipated by the system, as well as the potential difference between source and sink. Notice that the quantity V can be interpreted as the "infrastructural cost" of the system; in other terms, it is the cost that would be incurred if every link were traversed by a flow equal to its current capacity. While η may decrease or increase during the evolution of the system, we will show that $\eta \leq V$ and that V is always decreasing, except on equilibrium points.

Lemma 3. $\eta \leq V$.

Proof. To see the inequality, consider any flow f of maximum value subject to the constraint that $0 \leq f \leq x$. The minimum capacity of a source-sink cut is 1 at any time, by Lemma 1(b). Therefore, by the Max Flow-Min Cut Theorem, the value of the flow f must be 1. Then by (7),

$$\eta = q^T R q \leq f^T R f \leq x^T R x = V. \qquad \square$$

Lemma 4. V *is a Lyapunov function for* (3); *in other words, it is continuous and satisfies* (i) $V \geq 0$ *and* (ii) $\Delta V \leq 0$. *Moreover,* $\Delta V = 0$ *if and only if* $\Delta x = 0$.

Proof. V is continuous and nonnegative by construction. Moreover,

$$
\begin{aligned}
\Delta V/h = l^T \Delta x/h = l^T(|q| - x) & \qquad \text{by (3),} \\
= x^T R |q| - x^T R x & \qquad \text{by } l = Rx, \\
= (x^T R^{1/2}) \cdot (R^{1/2} |q|) - x^T R x & \\
\leq (x^T R x)^{1/2} \cdot (q^T R q)^{1/2} - x^T R x & \qquad \text{by Cauchy-Schwarz [Ste04],} \\
= (\eta V)^{1/2} - V, & \\
\leq V - V & \qquad \text{by Lemma 3.} \\
= 0. &
\end{aligned}
$$

Observe that $\Delta V = 0$ is possible only when equality holds in the Cauchy-Schwarz inequality. This, in turn, implies that the two vectors $R^{1/2} x$ and $R^{1/2} |q|$ are parallel, that is, $|q| = \lambda x$ for some $\lambda \in \mathbb{R}$. However, by Lemma 1(b), the capacity of the source cut is 1 and, by (7), the sum of the currents across the source cut is 1. Therefore, $\lambda = 1$ and $\Delta x = h(|q| - x) = 0$. $\qquad \square$

Corollary 1. *As* $t \to \infty$, $x(t)$ *approaches an equilibrium point of* (3), *and* $\eta(t)$ *approaches the length of the corresponding* s_0-s_1 *path.*

Proof. The existence of a Lyapunov function V implies [LaS76, Theorem 6.3] that $x(t)$ approaches the set $\{x \in \mathbb{R}_+^E : \Delta V = 0\}$, which by Lemma 4 is the same as the set $\{x \in \mathbb{R}_+^E : \Delta x = 0\}$. Since this set consists of isolated points (Lemma 2), $x(t)$ must approach one of those points, say the point $\mathbf{1}_P$ for some s_0-s_1 path P. When $x = \mathbf{1}_P$, one has $\eta = V = \mathbf{1}_P^T \cdot l$. $\qquad \square$

Convergence to an approximate shortest path and convergence time: We will track the convergence process via three main quantities: two of these, η and V, have already been introduced. The third one is defined as

$$W \stackrel{\text{def}}{=} \sum_{e \in P^*} l_e \ln x_e,$$

where P^* is the shortest path. Recall that opt denotes the length of P^*. Observe that $W(t) \leq 0$ for all t (due to Lemma 1(c)) and $W(0) = 0$ due to the choice of initial conditions. Also observe that $V(0) = l^T \cdot x(0) = \sum_{e \in E} l_e \leq mL$, where m is the number of edges of the graph and L is the length of the longest edge.

For a fixed $\epsilon \in (0, 1/300)$, we set $h = \epsilon/mL$. We will bound the number of steps before V falls below $(1 + 3\epsilon)^3 \text{opt} < (1 + 10\epsilon)\text{opt}$.

Definition 1. *We call a V-step any time step t such that $\eta(t) \leq (1+3\epsilon)\text{opt}$ and $V(t) > (1+3\epsilon)^3\text{opt}$. We call a W-step any time step t such that $\eta(t) > (1+3\epsilon)\text{opt}$ and $V(t) > (1 + 3\epsilon)^3\text{opt}$.*

Lemma 5. *The number k_V of V-steps is at most $O((\log n + \log L)/(h\epsilon))$.*

Proof. For any V-step t we have, by the proof of Lemma 4 and the assumptions on η and V,

$$\Delta V \leq h((\eta V)^{1/2} - V) = hV((\eta/V)^{1/2} - 1)$$
$$\leq hV(1/(1 + 3\epsilon) - 1) \leq -hV(3\epsilon/(1 + \epsilon)) \leq -h\epsilon V$$

so that $V(t + 1) \leq (1 - h\epsilon)V(t)$. In other words, V decreases by at least an $h\epsilon$ factor at each V-step. Moreover, V is nonincreasing at every step of the whole process, and after it gets below $(1+3\epsilon)^3\text{opt}$ there are no more V-steps. Therefore, the number of V-steps, k_V, is at most $\log_{1/(1-h\epsilon)}(V(0)/\text{opt}) \leq (\ln V(0))/(h\epsilon) = O(\log(mL)/(h\epsilon))$ (we used the assumption that $\text{opt} \geq 1$). □

Lemma 6. *At every W-step, W increases by at least $\text{opt} \cdot h\epsilon/2$.*

Proof. Let P^* be the shortest path, so that $1_{P^*}^T \cdot l = \text{opt}$. For a W-step t, we have

$$W(t + 1) - W(t) = \sum_{e \in P^*} l_e \ln \frac{x_e(t + 1)}{x_e(t)} = \sum_{e \in P^*} l_e \ln \left(1 + h\left(\frac{|p_u - p_v|}{l_e} - 1\right)\right),$$

where u, v are the endpoints of edge e. Using the bound $\ln(1 + z) \geq z/(1 + z)$, which is valid for any $z > -1$ (recall that $h < 1$), we obtain

$$W(t + 1) - W(t) \geq \sum_{e \in P^*} l_e \frac{h\left(\frac{|p_u - p_v|}{l_e} - 1\right)}{1 + h\left(\frac{|p_u - p_v|}{l_e} - 1\right)} = \sum_{e \in P^*} \frac{h(|p_u - p_v| - l_e)}{1 + h\left(\frac{|p_u - p_v|}{l_e} - 1\right)}$$

$$= h \cdot \left(\sum_{e \in P^*} \frac{|p_u - p_v|}{1 + h\left(\frac{|p_u - p_v|}{l_e} - 1\right)} - \sum_{e \in P^*} \frac{l_e}{1 + h\left(\frac{|p_u - p_v|}{l_e} - 1\right)}\right)$$

$$\geq h \cdot \left(\sum_{e \in P^*} \frac{|p_u - p_v|}{1 + h\eta} - \sum_{e \in P^*} \frac{l_e}{1 - h}\right),$$

where we used $\frac{|p_u - p_v|}{l_e} - 1 < \eta$ (we are using the assumption $l_e \geq 1$ for all e). Since $\sum_{e \in P*} |p_u - p_v| \geq \eta$ and $\eta \leq V \leq mL$, we obtain further

$$W(t+1) - W(t) \geq h\left(\frac{\eta}{1 + hmL} - \frac{\text{opt}}{1 - h}\right) = h\left(\frac{(1-h)\eta - (1 + hmL)\text{opt}}{(1-h)(1 + hmL)}\right)$$

$$\geq \text{opt} \cdot h\left(\frac{(1 - \epsilon)(1 + 3\epsilon) - (1 + \epsilon)}{(1 - h)(1 + \epsilon)}\right) > \text{opt} \cdot h\frac{\epsilon - 3\epsilon^2}{1 + \epsilon} \geq \frac{\text{opt} \cdot h\epsilon}{2},$$

where the third inequality follows since $\epsilon = hmL$ by definition of h and since $h = \epsilon/(mL) \leq \epsilon$ (note that $mL \geq 1$ from the definition of L). The fourth inequality follows from simple calculus, while the fifth follows since $(1 - 3\epsilon^2)/(1 + \epsilon) \geq 1/2$, whenever $\epsilon \leq 1/3$.   □

Lemma 7. *At every V-step, W decreases by at most $2\text{opt} \cdot h$.*

Proof. Trivially, $x_e(t+1) \geq (1-h)x_e(t)$, hence $\ln x_e(t+1) \geq \ln x_e(t) - \ln(1/(1 - h)) \geq \ln x_e(t) - 2h$ (since $h < 1/2$). The claim follows from the definition of W.   □

Lemma 8. *The number k_W of W-steps is at most $4k_V/\epsilon = O(mL(\log n + \log L)/\epsilon^3)$.*

Proof. At every W-step, W increases by at least $\text{opt} \cdot h\epsilon/2$. But W is always bounded above by 0, is decreased by at most $2\text{opt} \cdot h \cdot k_V$, and starts with $W(0) = 0$. The claim follows.   □

Theorem 1. *After at most $O(mL(\log n + \log L)/\epsilon^3)$ steps, V decreases below $(1 + 10\epsilon)\text{opt}$.*

Proof. Until the time that V gets below $(1 + 3\epsilon)^3\text{opt} \leq (1 + 10\epsilon)\text{opt}$, every step is either a V-step or a W-step, of which there can be at most $k_V + k_W = O(mL(\log n + \log L)/\epsilon^3)$ in total.   □

Approximate Computation. Real arithmetic is not needed for the results of the preceding section; in fact, arithmetic with $O(\log(nL/\epsilon))$ bits suffices. The proof that approximate arithmetic suffices mimics the proof in the preceding section; details are deferred to a full version of the paper.

4   Convergence of the Directed Physarum Model

We characterize Physarum's convergence properties in the directed model. We assume (A1) $x_e(0) > 0$ for all e, (A2) There is a directed path from the source to the sink, (A3) Edge lengths are integral, and (A4) The shortest source-sink path is unique. It is convenient to study the dynamics

$$\dot{x}_e(t) = a_e(q_e(t) - x_e(t)) \tag{14}$$

instead of (5). This is simply a change of variables and a rescaling of the edge lengths. We define several constants: $a_{\min} = \min(1, \min_e a_e)$, $x_{\max}(0) = \max(1, \max_e x_e(0))$, $x_{\min}(0) = \min(1, \min_e x_e(0))$, $X_0 = \max\left(x_{\max}(0), \frac{1}{x_{\min}(0)}\right)$, and $L = \max_e l_e$. P^* denotes the shortest directed source-sink path. We prove:

Theorem 2. *Assume (A1)–(A4) and let $\epsilon \in (0,1)$ be arbitrary. If $t \geq \frac{nL}{a_{\min}}$. $\left(3 \ln X_0 + 2 \ln \frac{2m}{\epsilon}\right)$, then $x_e(t) \geq 1 - 2\epsilon$ for $e \in P^*$ and $x_e(t) \leq \epsilon$ for $e \notin P^*$.*

Electrical flows are uniquely determined by Kirchhoff's and Ohm's laws. In our setting, the electrical flow $q(t)$ and the vertex potentials $p(t)$ are functions of time. For an edge $e = (u, v)$, let $\eta_e(t) = p_u(t) - p_v(t)$, and let $\eta(t) = p_{s_0}(t) - p_{s_1}(t)$. We have the following facts: (1) For any directed source-sink path P, $\sum_{e \in P} \eta_e(t) = \eta(t)$. (2) $x_e(t) \leq \max(1, x_e(0)) \leq x_{\max}(0)$ for all t. (3) $x_e(t) > 0$ for all $e \in E$ and all t (the existence of a directed source-sink path is crucial here). (4) $\ln x_e(t) = \ln x_e(0) + a_e \left(\frac{\hat{\eta}_e(t)}{l_e} - 1\right) \cdot t$, where $\hat{\eta}_e(t) = (1/t)\int_0^t \eta_e(s)ds$ is the average potential drop on edge e up to time t. For a directed source-sink path P, let

$$l_P = \sum_{e \in P} l_e \quad \text{and} \quad w_P(t) = \sum_{e \in P} \frac{l_e}{a_e} \ln x_e(t).$$

be its length and its weighted sum of log capacities, respectively. The quantity w_P was introduced in [MO07, MO08], and the following property (15) was derived in these papers.

Lemma 9. *Assume (A1), (A2) and let P be any directed source-sink path. Then*

$$\dot{w}_P(t) = \eta(t) - l_P \quad \text{and} \quad \frac{d}{dt}(w_P(t) - w_{P^*}(t)) = l_{P^*} - l_P. \tag{15}$$

Moreover, $w_P(t) \leq (3nL \ln X_0)/a_{\min} - t$, if P is a non-shortest source-sink path and (A3) holds: For $\epsilon \in (0,1)$, let $t_1 = nL(3 \ln X_0 + \ln(1/\epsilon))/a_{\min}$. Then $\min_{e \in P} x_e(t) \leq \epsilon$ for $t \geq t_1$.

The last claim states that for any non-shortest path P, $\min_{e \in P} x_e(t)$ goes to zero. This is not the same as stating that there is an edge in P whose capacity converges to zero. Such a stronger property will be shown in the proof of the main theorem.

The Convergence Proof: The proof proceeds in two steps. We first show that the vector of edge capacities becomes arbitrarily close to a nonnegative non-circulatory flow and then prove the main theorem. A flow is *nonnegative* if $f_e \geq 0$ for all e, and it is *non-circulatory* if $f_e \leq 0$ for at least one edge e on every directed cycle.

Lemma 10. *Assume (A1) and (A2): For $t > t_0 \stackrel{\text{def}}{=} (1/a_{\min}) \ln(3mX_0)$, there is a nonnegative non-circulatory flow $f(t)$ with*

$$|f_e(t) - x_e(t)| \leq 5mX_0 \cdot e^{-a_{\min}t}. \tag{16}$$

Proof. We follow the analysis in [IJNT11], taking reactivities into account. $\square$

We are now ready for the proof of the main theorem.

Proof (of Theorem 2). Let $\mathcal{P}$ be the set of non-shortest simple source-sink paths, and let $t > t_0$, where t_0 is defined as in Lemma 10. The nonnegative non-circulatory flow $f(t)$ can be written as a sum of flows along simple directed source-sink paths, i.e.,

$$f(t) = \alpha_{P^*}(t)\mathbf{1}_{P^*} + \sum_{P \in \mathcal{P}} \alpha_P(t)\mathbf{1}_P$$

with nonnegative coefficients α_P. This representation is not unique. However, there is always a representation with at most m nonzero coefficients.[5] For any edge e and any path P with $e \in P$, the flow $f_e(t)$ is at least $\alpha_P(t)$.

Let $\epsilon \in (0,1)$ be arbitrary. For

$$t \geq \frac{1}{a_{\min}} \max\left(\ln \frac{10m^2 X_0}{\epsilon}, nL\left(3\ln X_0 + \ln \frac{2m}{\epsilon}\right)\right),$$

we have $|f_e(t) - x_e(t)| \leq \epsilon/(2m)$ for all e (Lemma 10) and $\min_{e \in P} x_e(t) \leq \epsilon/(2m)$ for every non-shortest path P (Lemma 9). Thus, every non-shortest path contains an edge e with $f_e(t) \leq \epsilon/m$. Thus, $\alpha_P(t) \leq \epsilon/m$ for all non-shortest paths P, and hence,

$$x_e(t) \leq m\epsilon/m \leq \epsilon \quad \text{for all } e \notin P^*.$$

The value of the flow f is one. The total flow along the non-shortest paths is at most ϵ. Thus the flow along P^* is at least $1 - \epsilon$. Hence $x_e(t) \geq 1 - \epsilon - \epsilon/(2m) \geq 1 - 2\epsilon$ for all $e \in P^*$. Finally, $\ln \frac{10m^2 X_0}{\epsilon} \leq nL(3\ln X_0 + 2\ln \frac{2m}{\epsilon})$.   □

Discretization. We study the discretization of the system of differential equations (14). We proceed in discrete time steps $t = 0, 1, 2, \ldots$ and define the dynamics

$$x_e(t+1) = x_e(t) + ha_e(q_e(t) - x_e(t)), \tag{17}$$

where h is the step size. We will need the following additional assumptions: (A5) $a_e = 1$ for all e, and (A6) there is an edge $e_0 = (s_0, s_1)$ of length nL and initial capacity 0. Observe that the existence of this edge does not change the shortest directed source-sink path. Our main theorem becomes the following; the proof structure for the discrete case is similar to the one for the continuous case.

Theorem 3. *Assume (A1)–(A6) and $h \leq \frac{1}{24 \cdot n(4nm^2 L(X_0)^2)^2}$. Let $\epsilon \in (0,1)$ be arbitrary. For*

$$t \geq \frac{4nL}{h}\left(3\ln X_0 + 2\ln \frac{2m}{\epsilon}\right),$$

$x_e(t) \geq 1 - 2\epsilon$ *for* $e \in P^*$ *and* $x_e(t) \leq \epsilon$ *for* $e \notin P^*$.

---

[5] Let $\alpha_{P^*}(t)$ be the minimum value of $f_e(t)$ for $e \in P^*$. Subtract $\alpha_{P^*}(t)\mathbf{1}_{P^*}$ from $f(t)$. As long as $f(t)$ is not the zero flow, determine a source-sink path P carrying nonzero flow and set $\alpha_P(t)$ to the minimum value of $f_e(t)$ for $e \in P$. Subtract $\alpha_P(t)\mathbf{1}_P$ from $f(t)$.

5   Conclusions and Future Work

We summarize our three main results: the discretization (3) of the undirected Physarum model computes an $(1 + \epsilon)$-approximation of the shortest source-sink path in $O(mL(\log n + \log L)/\epsilon^3)$ iterations with arithmetic on numbers of $O(\log(nL/\epsilon))$ bits. The dynamics (5) of the nonuniform directed Physarum model converges to the shortest directed source-sink path under the modified length function $a_e l_e$. Within time $nLa_{\min}^{-1} \cdot \left(3 \ln X_0 + 2 \ln \frac{2m}{\epsilon}\right)$, an ϵ-approximation is reached. For the uniform model ($a_e = 1$), we also prove convergence of the discretization.

There are many open questions: (i) Convergence of the nonuniform undirected model; (ii) Convergence of the discretized nonuniform directed model; (iii) Are our bounds best possible? In particular, can the dependency on L be replaced by a dependency on $\log L$?

References

[BD97]   Baldauf, S.L., Doolittle, W.F.: Origin and evolution of the slime molds (Mycetozoa). Proc. Natl. Acad. Sci. USA 94, 12007–12012 (1997)

[BMV12]   Bonifaci, V., Mehlhorn, K., Varma, G.: Physarum can compute shortest paths. Journal of Theoretical Biology 309, 121–133 (2012); A preliminary version of this paper appeared at SODA 2012, pp. 233–240

[Bol98]   Bollobás, B.: Modern Graph Theory. Springer, New York (1998)

[Bon13]   Bonifaci, V.: Physarum can compute shortest paths: A short proof. Information Processing Letters 113(1-2), 4–7 (2013)

[IJNT11]   Ito, K., Johansson, A., Nakagaki, T., Tero, A.: Convergence properties for the Physarum solver. arXiv:1101.5249v1 (January 2011)

[Kir10]   Kirby, B.J.: Micro- and Nanoscale Fluid Mechanics: Transport in Microfluidic Devices. Cambridge University Press, Cambridge (2010)

[LaS76]   LaSalle, J.B.: The Stability of Dynamical Systems. SIAM (1976)

[MO07]   Miyaji, T., Ohnishi, I.: Mathematical analysis to an adaptive network of the Plasmodium system. Hokkaido Mathematical Journal 36(2), 445–465 (2007)

[MO08]   Miyaji, T., Ohnishi, I.: Physarum can solve the shortest path problem on Riemannian surface mathematically rigorously. International Journal of Pure and Applied Mathematics 47(3), 353–369 (2008)

[NIU+07]   Nakagaki, T., Iima, M., Ueda, T., Nishiura, Y., Saigusa, T., Tero, A., Kobayashi, R., Showalter, K.: Minimum-risk path finding by an adaptive amoebal network. Physical Review Letters 99(068104), 1–4 (2007)

[NYT00]   Nakagaki, T., Yamada, H., Tóth, Á.: Maze-solving by an amoeboid organism. Nature 407, 470 (2000)

[SM03]   Süli, E., Mayers, D.: Introduction to Numerical Analysis. Cambridge University Press (2003)

[Ste04]   Steele, J.: The Cauchy-Schwarz Master Class: An Introduction to the Art of Mathematical Inequalities. Cambridge University Press (2004)

[TKN07]   Tero, A., Kobayashi, R., Nakagaki, T.: A mathematical model for adaptive transport network in path finding by true slime mold. Journal of Theoretical Biology 244, 553–564 (2007)

[You]   http://www.youtube.com/watch?v=czk4xgdhdY4

On Revenue Maximization for Agents with Costly Information Acquisition
Extended Abstract

L. Elisa Celis[1], Dimitrios C. Gklezakos[2], and Anna R. Karlin[2]

[1] Xerox Research Centre India
elisa.celis@xerox.com
[2] University of Washington
{gklezd,karlin}@cs.washington.edu

Abstract. A prevalent assumption in traditional mechanism design is that buyers know their precise value for an item; however, this assumption is rarely true in practice. In most settings, buyers can "deliberate", i.e., spend money or time, in order improve their estimate of an item's value. It is known that the deliberative setting is fundamentally different than the classical one, and desirable properties of a mechanism such as equilibria, revenue maximization, or truthfulness, may no longer hold.

In this paper we introduce a new general deliberative model in which users have independent private values that are a-priori unknown, but can be learned. We consider the design of dominant-strategy revenue-optimal auctions in this setting. Surprisingly, for a wide class of environments, we show the optimal revenue is attained with a sequential posted price mechanism (SPP). While this result is not constructive, we show how to construct approximately optimal SPPs in polynomial time. We also consider the design of Bayes-Nash incentive compatible auctions for a simple deliberative model.

1 Introduction

In many real-world scenarios, people rarely know precisely how they value an item, but can pay some cost (e.g., money, time or effort) to attain more certainty. This not only occurs in online ad markets (where advertisers can buy information about users), but also in everyday life. Suppose you want to buy a house. You would research the area, school district, commute, and possibly pay experts such as a real estate agent or an inspection company. Each such action has some cost, but also helps better evaluate the worth of the house. In some cases, e.g., if you find out your commute would be more than an hour, you may simply walk away. However, if the commute is reasonable, you may choose to proceed further and take more actions (at more cost) in order to gain even more information. This continues until you take a final decision. A *deliberative agent* as defined in this paper has this kind of multiple-round information-buying capability.

Previous work shows that mechanism design for deliberative agents is fundamentally different than classical mechanism design due to the greater flexibility

F.V. Fomin et al. (Eds.): ICALP 2013, Part II, LNCS 7966, pp. 484–495, 2013.
© Springer-Verlag Berlin Heidelberg 2013

in the agents' strategies. A classical agent has to decide how much information to reveal; a deliberative agent has to additionally decide how much information to acquire. This affects equilibrium behavior. For example, in second-price auctions, deliberative agents do not have dominant strategies [20], and standard mechanisms and techniques do not apply. Revenue-optimal mechanisms have remained elusive, and the majority of positive results are for simple models where agents can determine their value exactly in one round, often restricting further to binary values or single-item auctions [2,5,3,24,7]. Positive results for more general deliberative settings restrict agents in other ways, such as forcing agents to acquire information before the mechanism begins [22,8], to commit to participate before deliberating [13], or to deliberate in order to be served [4]. (Also see [1,19,14].) Furthermore, impossibility results exist for certain types of dominant strategy[1] deliberative mechanisms [20,21]. This result, however, relies crucially on the fact that agents are assumed to have the ability to deliberate about other agent's values. In an independent value model, this is not a natural assumption.

In this paper we continue a line of research begun by Thompson and Leyton-Brown [24,7] related to dominant strategy mechanism design in the independent value model. Specifically, we extend results to a general deliberative model in which agents can repeatedly refine their information. Our first main result is that the profit maximizing mechanism (a.k.a. optimal mechanism) is, without loss of generality, a sequential posted price mechanism (SPP).[2] Our second main result is that, via a suitable reduction, we can leverage classical results (see [10,11,18]) that show revenue-optimal mechanisms can be approximated with SPPs, in order to construct construct approximately optimal SPPs in our setting. These are first results in an interesting model that raises many more questions than it answers. In the final section, we take first steps towards understanding Bayes-Nash incentive-compatible mechanisms in a simpler deliberative setting.

2   Deliberative Model

In our model, each agent has a set of "deliberative possibilities" that describe the ways in which they can acquire information about their value.

Definition 1 (Deliberative Possibilities). *An agent's deliberative possibilities are represented by tuples $(\mathcal{F}, \mathcal{D}, c)$ where*

- *$\mathcal{F}$ is a probability distribution over the possible values the agent can have;*
- *$\mathcal{D}$ is a set of* deliberations *the agent can perform; and*
- *$c : \mathcal{D} \to \mathbf{R}_+$, where $c(d) > 0$ represents the cost to perform deliberation d.*

In any given state $(\mathcal{F}, \mathcal{D}, c)$, the agent may choose one of the deliberations $d \in \mathcal{D}$ to perform. A deliberation is a random function that maps the agent to a new

---

[1] Dominant strategy equilibria occur when each agent has a strategy that is optimal against any (potentially suboptimal) strategies the other agents play.

[2] In a sequential posted price mechanism (SPP), agents are offered take-it-or-leave-it prices in sequence; the mechanism is committed to sell to an agent at the offered price if she accepts, and will not serve the agent at any point point if she rejects.

state $(\mathcal{F}', \mathcal{D}', c')$, where $\mathcal{F}'$ is the new prior the agent has over his value, $\mathcal{D}'$ is the new set of deliberations the agent can perform, and $c'(\cdot)$ the corresponding costs. The distribution over new priors is such that the marginals agree with $\mathcal{F}$; i.e., while $v \sim \mathcal{F}'$ is drawn from a new distribution (the updated prior), $v \sim d(\mathcal{F})$ is identically distributed as $v \sim \mathcal{F}$. [3]

We focus on the design of mechanisms in *single-parameter* environments [17,16], where each agent has a single private (in our case, unknown) value for "service", and there is a combinatorial feasibility constraint on the set of agents that can be served simultaneously. [4]

A **mechanism** in the deliberative setting is a (potentially) multi-stage process in which the mechanism designer interacts with the agents. It concludes with an allocation and payment rule $(\mathbf{x}, \mathbf{p})$ where $x_i = 1$ if agent i is served and is 0 otherwise [5] and agent i is charged p_i. [6] The mechanism designer knows the agents' deliberative possibilities and initial priors. At any point during the execution of the mechanism, an agent is free to perform any of her deliberation possibilities according to her current state $(\mathcal{F}, \mathcal{D}, c)$. Indeed, it may be in the mechanism designer's best interest to incentivize her to do certain deliberations.

We focus in this paper on a **public communication** model; i.e., every agent observes the interaction between any other agent and the mechanism. Versions of our results also extend to the private communication model. We also make the standard assumption that agents have full knowledge of the mechanism to be executed. Crucially however, if and when an agent deliberates, there is no way for other agents or the mechanism to certify that the deliberation occurred. Moreover, the outcome of a deliberation is *always private*. Hence, the mechanism designer must incentivize the agent appropriately in order to extract this information.

If, over the course of the execution, an agent performs deliberations $d_1, \ldots, d_k$, then her expected utility is

$$x_i \cdot \mathbf{E}[\mathcal{F}^k | \text{result of } \{d_1, \ldots, d_k\}] - p_i - \sum_{1 \leq j \leq k} c(d_j).$$

We assume that agents choose their actions so as to maximize their expected utility. However, note that it is possible that an agent's realized utility turns out to be negative.

In classical settings, the revelation principle [15,23] is a crucial step in narrowing the search for good mechanisms. In our deliberative setting, we cannot simply

---

[3] For example, the initial prior might be $U[0, 1]$, and the agent might be able to repeatedly determine which of two quantiles her value is in, at some cost. Note that in general we do not impose any restriction on the type of distributions or deliberations except, in some cases, that they be finite (see, e.g., Definition 4).

[4] Examples include single-item auctions, k-item auctions, and matroid environments.

[5] Clearly, this allocation must satisfy the feasibility constraints of the given single-parameter environment; e.g. in a single item auction, then, $\sum_i x_i \leq 1$.

[6] As is standard, the mechanism does not charge agents that are not served. Hence, $p_i = 0$ when $x_i = 0$.

apply the revelation principle to "flatten" mechanisms to a single stage, since the mechanism cannot simulate information gathering on behalf of the agent. However, a minor variant of the revelation principle developed by Thompson and Leyton-Brown [24] can be shown to also apply to our general setting. To state it formally, we consider the following type of mechanism:

Definition 2 (Simple Deliberative Mechanism[7]). *A simple deliberative mechanism (SDM) is a multi-stage mechanism where at each stage either*

- *a single agent (or set of agents) is asked to perform and report the result of a specified deliberation[8],*
- *the mechanism outputs an allocation and payment rule* $(\mathbf{x}, \mathbf{p})$. *(Note that allocation can be made to agents that were never asked to deliberate.)*

Note that for deterministic dominant strategy mechanisms, SDMs can interact with a single agent at a time and restrict interactions to soliciting and receiving the results of their deliberations without loss of generality.

A **strategy** of a deliberative agent against an SDM consists of either

1. not deliberating and reporting a result $\widehat{r}$,
2. performing each requested deliberation and reporting a result $\widehat{r}$ (which may depend on the real result r of the deliberation), or
3. performing other or additional deliberation(s) and reporting a result $\widehat{r}$ (which may depend on the real result(s) r of the deliberation(s)).

Definition 3. *A **truthful strategy** always takes option (2) and reports* $\widehat{r} = r$, *the true result of the deliberation. A **truthful SDM** is one in which truthtelling is a dominant strategy for every agent. That is, no matter how other agents behave, it is in her best interest to execute the truthful strategy.*

A crucial step in narrowing the search for good mechanisms is the Revelation Principle. A version for simple deliberative agents generalizes to our setting without complication.

Lemma 1 (Revelation Principle [24]). *For any deliberative mechanism* $\mathcal{M}$ *and equilibrium* σ *of* $\mathcal{M}$, *there exists a truthful SDM* $\mathcal{N}$ *which implements the same outcome as* $\mathcal{M}$ *in equilibrium* σ.

3   Revenue Domination by SPPs

Theorem 1 (SPPs are Optimal) *Any deterministic truthful mechanism* $\mathcal{M}$ *in a single-parameter deliberative environment is revenue-dominated by a sequential posted price mechanism* $\mathcal{M}'$ *under the assumption that* $\mathcal{M}$ *does not exploit indifference points.[9]*

---

[7] This definition was given in [24] under the name Dynamically Direct Mechanism.

[8] The deliberation the agent is asked to perform must be one of her current deliberative possibilities, assuming she did as she was told up to that point.

[9] I.e., if the agent is indifferent between receiving and not receiving the item, the mechanism commits to either serving or not serving this agent without probing other agents.

In other words, any optimal mechanism can be transformed into a revenue-equivalent SPP.

We use the following lemma which extends results from [7,24] to our more general setting. We omit the proof, which is a straightforward extension.

Lemma 2 (Generalized Influence Lemma).

Consider a truthful SDM $\mathcal{M}$, and a bounded agent who has performed t deliberations resulting in a prior distribution $\mathcal{F}^{(t)}$. Let $S = support(\mathcal{F}^{(t)})$. If the agent is asked to perform deliberation d, then there exist thresholds $L_d, H_d \in S$ such that

- *If an agent deliberates and reports value $\leq L_d$ she will not be served.*
- *If an agent deliberates and reports value $\geq H_d$ she will be served.*
- *It is possible that the deliberation results in a value $\leq L_d$ or $\geq H_d$.*

Note that "value" above refers to an agent's effective value $\mathbf{E}[\mathcal{F}^{(t+1)}]$ where $\mathcal{F}^{(t+1)} \sim d(\mathcal{F}^{(t)})$.

We now show that at the time the agent is first approached, the price she is charged does not depend on the future potential actions of any agent.

Lemma 3. *Let $\mathcal{M}$ be a dominant-strategy truthful SDM for deliberative agents. If $\mathcal{M}$ asks an agent to deliberate, we can modify $\mathcal{M}$ so that there is a single price p determined by the history before $\mathcal{M}$'s first interaction with this agent. Moreover, the agent will win if her effective value is above p, and lose if it is below p. The modification preserves truthfulness and can only improve revenue.*

Proof. Let $\mathcal{M}$ be an SDM as above. Note that $\mathcal{M}$ can be thought of as a tree where at each node the mechanism asks an agent to perform a deliberation (say d). Each child corresponds to a potential result reported by the agent, at which point, the mechanism either probes an (potentially the same) agent or terminates at a leaf with an allocation and payment rule. With some abuse of notation, we say an agent reports a value in H if her value is above H_d, reports a value in L if her value is below L_d, and reports a value in M otherwise.

Consider an execution of $\mathcal{M}$ in which agent i is asked to deliberate. Consider the *path* of execution. We show that we can modify $\mathcal{M}$ without any loss to revenue so that after i is *first* asked to deliberate, the price at which she gets the item (assuming she is served) is effectively fixed and does not depend on the rest of the path.

Firstly, note that if an agent i reports a value in H, from Lemma 2, then she is served. Assume that i determines her value is in H after deliberation. Consider the case where there is only one possible value i can report in H and she is not asked to deliberate again after this point. If this can result in multiple possible prices, the difference must depend only on the behavior of other agents, and is not intrinsic to i. In fact, all such prices must be acceptable to her due to truthfulness, since, for a fixed set of strategies the other agents take, it should not be the case that she prefer to lie and say her value is in L and hence avoid service at a price that is too high. Therefore, any such mechanism can be modified by replacing all these prices with the maximum such price.

Otherwise, if there are two different prices, p and p', that are be reached depending on i's report(s), then this again contradicts dominant-strategy truthfulness. This is due to the fact that an agent's behavior must be truthful against *any set of fixed strategies* for the other agents. Thus, whenever i reports a value in H, she must be charged the same price p.

If agent i is never served when i reports $v \in M$, the proof is complete. Assume otherwise. Let p_m be some price that she is served at along a path in M and let p_h be the price she is charged if she reports a value in H_d. Clearly, if $p_m > p_h$ then if i's value is in M she would have incentive to lie. Additionally, if $p_m < p_h$, then there is a set of strategies we can fix for the other agents for which i would again have incentive to lie. Hence, by dominant strategy truthfulness, $p_m = p_h$.

We conclude by observing that, by truthfulness, it is straightforward to see that an agent must be served at the above price p whenever her effective value is above p (and not served otherwise). □

Lemma 4. *Let $\mathcal{M}$ be a truthful SDM such that the price it charges i, assuming i is served, only depends on the history before $\mathcal{M}$'s first interaction with i. Then, $\mathcal{M}$ is revenue-equivalent to an SPP $\mathcal{N}$.*

Proof. Given $\mathcal{M}$, we construct $\mathcal{N}$. Consider any node in the SDM $\mathcal{M}$ where some agent, say i, is asked to deliberate. Let h_i^{b} be the history leading up to this node. Lemma 3 implies that from this point forward, whenever agent i is served, she is charged p_i. Let $\mathcal{N}$, under the same h_i^{b}, offer price p_i to agent i.

Since $\mathcal{M}$ is truthful, each deliberation she asks i to perform is in her best interest. Additionally, there is no deliberation i would like to perform which the mechanism did not ask her to perform. Let this final effective value be $\widehat{v}_i$. Clearly, by truthfulness, i gets the item if and only if $p_i \leq \widehat{v}_i$. Similarly, when $\mathcal{N}$ offers her price p_i it will be in her interest to take the same sequence of deliberations, otherwise $\mathcal{M}$ was not truthful to begin with. She accepts if and only if $p_i \leq \widehat{v}_i$, matching the scenario under which she accepts $\mathcal{M}$'s offer.

The above holds for any deliberative node. Hence, $\mathcal{M}$ and $\mathcal{N}$ have the same expected revenue, completing the proof. □

Combining Lemmas 4 and 1 concludes the proof of Theorem 1.

4   Approximating Optimal Revenue

While our result above is nonconstructive, it turns out to be easy to construct approximately-optimal SPPs. The basic approach is simple:

1. For any price $p \in [0, \infty)$, determine the utility-maximizing set of deliberations the agent would perform, and the probability $\alpha(p)$ that the agent accepts this price when she deliberates optimally. We denote the agent's optimal expected utility when offered a price of p by $u(p)$.
2. Note that $f(p) = 1 - \alpha(p)$ defines a cumulative distribution function on $[0, \infty]$.

3. Observe that this implies any SPP has the same expected revenue in the deliberative setting as it does in the classical setting when agents' values are drawn from the distribution $v \sim 1 - \alpha(\cdot)$.[10]
4. Use known approximation results [9,10,11,18] that show how to derive approximately optimal SPPs in the classical setting to derive an approximately optimal SPP in the deliberative setting.

We apply this recipe to bounded deliberative agents for which we can efficiently compute the distribution $f(p) = 1 - \alpha(p)$.

Definition 4 (Bounded Deliberative Agent). *An deliberative agent is bounded if the following holds:*

1. *Every prior has bounded expectation, i.e,* $\mathbf{E}[\mathcal{F}^{(t)}] < \infty$ *for all* i, t.
2. *Every set of deliberative actions is finite, i.e,* $\left|D^{(t)}\right| < \infty$ *for all* i, t.
3. *Every deliberative action results in one of finitely-many potential priors* $\mathcal{F}^{(t)}$.
4. *There is some finite* T *such that* $D^{(t)} = \emptyset$ *for all* $t \geq T$, *i.e., no further deliberation is possible.*

We now define a lemma that contains the key insight for this result.

Lemma 5. *The probability and utility functions* $\alpha(p)$ *and* $u(p)$ *have the following properties:*

1. u *is piecewise linear and convex.*
2. α *is a step function and decreasing.*
3. *If the agents are bounded,* α *and* u *can be constructed in polynomial time in the size of the deliberation tree.*

Proof. Consider the deliberative decision-making an agent faces when offered service at a particular price p. We think of her as alternating between decisions (deliberate, accept the price or reject it), and receiving the random results of a deliberation. This process defines a deliberation tree (which is finite, under the boundedness assumption above).

We call the decision nodes *choice nodes*, and each has a corresponding deliberation possibilities $(\mathcal{F}, \mathcal{D}, c)$. From each choice node, the agent can proceed to an accept or reject leaf. When such a node is selected, the deliberation process terminates with the agent having effective value equal to the expected value of his updated prior. Alternately, the agent can proceed to a *deliberation node*. One such node exists for each potential deliberation $d \in \mathcal{D}$, and proceeding to such a node comes at cost $c(d)$. The strategy for the agent at a choice node consists of deciding which child to select. Obviously, under optimal play, the child that yields the maximum expected utility is chosen.

At a deliberation node, the chosen deliberation d is performed. The children of a deliberation node d are the set of possible $(\mathcal{F}', \mathcal{D}', c')$ that can be returned

---

[10] Clearly, the expected revenue of this SPP in the classical setting is generally less than the expected revenue of the optimal mechanism in the classical setting for agents with values are drawn from these distributions.

when d is applied to $\mathcal{F}$. Note that the agent's expected utility at a deliberation node is simply a convex combination of the expected utilities of its children.

Consider a specific deliberation subtree T. Let $u_T(p)$ be the agent's optimal expected utility conditioned on reaching the root of this subtree. Also let $\alpha_T(p)$ be the probability the agent accepts an offer of p conditioned on reaching the root of this subtree when she uses an optimal deliberation strategy (from this point onwards). The lemma states that for any T, $u_T(p)$ is a piecewise linear and convex function of p, $\alpha_T(p)$ is a decreasing step function in p, and that both functions can be constructed in polynomial time in the size of the deliberation tree. The proof is by induction on the height of the deliberation subtree.

Base Case: The base case is a single node, which is either an accept or reject node. For an accept node A reached via a sequence of deliberations $d_1, \ldots, d_t$ with final prior $F^{(t)}$, the probability of acceptance is 1 and $u_A = \mathbf{E}[F^{(t)}] - p - \sum_{i=1}^{t} c(d_i)$. For a reject node, the probability of acceptance is 0 and $u_R = -\sum_{i=1}^{t} c(d_i)$. In this case, all three propositions of the theorem hold trivially.

Inductive Step: Let us first consider the claims about $u(p)$. By the inductive hypothesis, for any tree of height h, the utility function is convex and piecewise linear. A tree of height $h + 1$ is constructed by conjoining a set of m trees $\{T_1, \ldots, T_m\}$, the maximum height of which is h, via a single root. The convexity of the utility function is immediate from the fact that it is is either a convex combination of convex functions or the maximum of a set of convex functions. When taking the convex combination of the children, the utility function of the root has a breakpoint (a price where the utility function changes slope) for each breakpoint a child node has, and thus if $b(T_i)$ is the number of breakpoints in the utility curve for the child T_i, then the total number of breakpoints at the root is $|b(T)| \leq \sum_{i=1}^{m} |b(T_i)|$.

When the utility function at the root is the pointwise maximum of the utility functions associated with the children, order the set of all breakpoints $|b(T)| \leq \sum_{i=1}^{m} |b(T_i)|$ associated with any of the children of the root by their p value. Within any of these intervals, the utility function is the maximum of m lines, which can generate at most $m - 1$ new breakpoints. Thus, the total number of breakpoints is at most $|b(T)| \leq m \sum_{i=1}^{m} |b(T_i)|$. Inductively, this implies that the total time to compute the utility function is $O(|T|^2)$.

Now consider $\alpha(p)$. Let $\alpha_{T_i}(p)$ be the probability that the agent will accept an offer at price p conditioned on reaching the root of T_i. The inductive hypothesis is that $\alpha_{T_i}(p) = -u'_{T_i}(p)$. where we extend the derivative $(-u'_{T_i}(p))$ to all breakpoints by right continuity.

If the root of T is a deliberation node, then in each interval in the partition defined by the breakpoints the slope of the linear function representing $u_T(p)$ is the convex combination of the slopes of the individual utilities $u_{T_i}(p)$. Therefore $u'_T(p) = \sum_{i=1}^{m} q(T_i) u'_{T_i}(p)$, and $\alpha_T(p) = \sum_{i=1}^{m} q(T_i) \alpha_{T_i}(p)$. where $q(T_i)$ is the probability of the deliberation outcome associated with T_i. It then follows from the IH that $\alpha_T(p) = -\sum_{i=1}^{m} q(T_i) u'_{T_i}(p) = -u'_T(p)$.

When T is a choice node, the acceptance probability for each p is precisely that associated with the T_i that has maximum utility for that p, and we obtain $\alpha_T(p) = \alpha_{T_i}(p) = -u'_{T_i}(p) = -u'_T(p)$

Finally, by convexity, for $p < p'$, we have $u'_T(p') \geq u'_T(p)$ which, by the previous equalities implies that $-\alpha_T(p') \geq -\alpha_T(p)$ or equivalently $\alpha_T(p') \leq \alpha_T(p)$, concluding the proof.   □

The following lemma is then immediate.

Lemma 6. *Let $\mathcal{M}$ be any SPP in a single-parameter deliberative setting. Then the expected revenue of $\mathcal{M}$ in this deliberative setting is equal to the expected revenue of $\mathcal{M}$ in the classical setting with agents whose values v are drawn from the distribution $\mathcal{F}(v) = 1 - \alpha(v)$.*

This gives us a direct connection between the deliberative and classical settings, and, crucially, allows us to apply results from the classical setting. Specifically, using the fact that optimal mechanisms in the classical settings are well-approximated by SPPs [9,10,11,18] and that SPPs are optimal for deliberative settings, we obtain the following theorem.

Theorem 2 (Constructive SPP 2-Approximation) *Consider a collection of bounded deliberative agents. An SPP that 2-approximates the revenue of optimal deterministic deliberative mechanisms for public communication matroid environments can be constructed in polynomial time.*

5   Bayes-Nash Incentive Compatible Mechanisms

In this section, we consider the design of Bayes-Nash incentive compatible (BIC) mechanisms in the following simplified setting.

Definition 5. *A public communication, single deliberation environment is a setting where an agent's value is drawn from a distribution with continuous and atomless density function $f(\cdot)$, and each agent has a single deliberative possibility after which they learn their exact value v.*[11]

Suppose that agent is asked to deliberate at some point during the execution of the mechanism. Denote by $a(v)$ the probability that an agent receives allocation when her value is v.[12] Recall we assume a public communication model, hence the agent knows the reports of all agents that deliberated before her. Her expected payment $p(v)$ can be characterized as in the classical setting [23] and her utility is $u(v) = a(v)v - p(v)$.

Note that an SDM is Bayes-Nash incentive compatible (BIC) if and only if each agent's expected utility is maximized by complying with the requests of the mechanism whenever other agents are also compliant. We provide a characterization for BIC mechanisms.

---

[11] Note that a mechanism will ask an agent to deliberate at most once.

[12] This probability is taken over the values of all agents asked to deliberate at the same time or later than this agent.

Proposition 1. *Consider a set of single-parameter agents in a single deliberation enviornment. A simple deliberative mechanism is BIC if and only if for each agent i:*

1. *If i is asked to deliberate then*
   (a) $a_i'(v) \geq 0$, *with payment rule as in the classical setting.*
   (b) $\int_0^\infty \left(\int_0^v a_i(x)dx \right) f_i(v_i)dv_i \geq \int_0^{\mu_i} a_i(x)dx + c_i$.
2. *If i is offered the item at price e without being asked to deliberate, then*
   $\int_0^e F_i(v)dv \leq c_i$.

Proof. For notational simplicity, we present the proof when $F_i = F$ and $c_i = c$ for all i.

Condition 1a: Follows as in the classical setting [23] after deliberation.

Condition 1b: In the last case, we have to consider what the agent could gain from this action. The utility of an agent in BNE with allocation probability $a(x)$ that performs a deliberation is $u(v) = \int_0^v a(x)dx - c$. Given that $v \sim F$, the expected utility $u_d(v)$ of an agent that performs a deliberation with cost c is:

$$u_d = \int_0^\infty \left(\int_0^v a(x)dx \right) f(v)dv - c.$$

The utility of a player that does not deliberate and reports valuation w is:

$$u_E(w) = \mu a(w) - p(w) = \mu a(w) - wa(w) + \int_0^w a(x)dx$$

where $\mu = E[F]$. Observing that u_E is maximized at $w = \mu$, and simplifying the constraint that $u_d \geq u_E(w)$, we obtain

$$\int_0^\infty \left(\int_0^v a(x)dx \right) f(v)dv \geq \int_0^\mu a(x)dx + c$$

Condition 2: If the mechanism offers the item to the agent and expects her to take it without deliberation at price e, it must be that her utility $u_E = \mu - e$ is greater than the utility she could obtain from deliberating, that is:

$$\mu - e \geq \int_e^\infty (v - e)f(v)dv - c \text{ which after simplification is } \int_0^e F(v)dv \leq c.$$

$\square$

To give an example, consider a single item auction in the classical setting, with two agents whose values are drawn uniformly on $[0, 1]$. Hence, the revenue-optimal mechanism is a Vickrey auction with reserve price $1/2$, which achieves an expected revenue of $5/12$. Note that this acution is dominant strategy truthful. In the deliberative setting this is no longer the case, since that would require that, ex-post, an agent has "no regrets". However, a deliberative agent will regret having paid a deliberation cost c if she ends up losing. It follows though from Proposition 1 that VCG is BIC for for $c < \frac{1}{12}$. It also follows from condition 2 that an agent will take the item without deliberating at a price up to $\sqrt{2c}$. Thus, when $c = 1/12 - \epsilon$ the following mechanism raises more revenue than VCG:

- Approach agent 1, ask her to deliberate and report her value v_1. If it is above $p^* = 1 - \sqrt{2c}$, sell her the item at price p.
- Otherwise, approach agent 2, and offer her the item at price $\sqrt{2c}$ without deliberation.

Surprisingly, the problem of designing an optimal mechanisms, even for 2 agents and a single item, seems to be difficult.

6   Future Work

We view this as very preliminary work in the setting of deliberative environments; numerous open problems remain. In the specific model studied, directions for future research include understanding other communication models, and the power of randomization. Beyond dominant strategies, revenue maximization using other solution concepts is wide open. It would also be interesting to study objectives other than revenue maximization. Finally, it would be interesting to derive "price of anarchy" style results that compare optimal revenue in deliberative and non-deliberative settings.

Acknowledgements. We would like to thank Kevin Leyton-Brown and Dave Thompson for introducing us to the fascinating setting of deliberative agents.

References

1. Babaioff, M., Kleinberg, R., Leme, R.P.: Optimal mechanisms for selling information. In: 12th International World Wide Web Conference (2012)
2. Bergemann, D., Valimaki, J.: Information acquisition and efficient mechanism design. Econometrica 70(3) (2002)
3. Bergemann, D., Valimaki, J.: Information acquisition and efficient mechanism design. Econometrica 70(3) (2002)
4. Bikhchandani, S.: Information acquisition and full surplus extraction. Journal of Economic Theory (2009)
5. Cavallo, R., Parkes, D.C.: Efficient metadeliberation auctions. In: AAAI, pp. 50–56 (2008)
6. Celis, L.E., Gklezakos, D.C., Karlin, A.R.: On revenue maximization for agents with costly information acquisition (2013), Full version
   http://homes.cs.washington.edu/~gklezd/publications/deliberative.pdf
7. Celis, L.E., Karlin, A., Leyton-Brown, K., Nguyen, T., Thompson, D.: Approximately revenue-maximizing mechanisms for deliberative agents. In: Association for the Advancement of Artificial Intelligence (2011)
8. Chakraborty, I., Kosmopoulou, G.: Auctions with edogenous entry. Economic Letters 72(2) (2001)
9. Chawla, S., Hartline, J., Kleinberg, R.: Algorithmic pricing via virtual valuations. In: Proc. 9th ACM Conf. on Electronic Commerce (2007)
10. Chawla, S., Hartline, J., Malec, D., Sivan, B.: Sequential posted pricing and multi-parameter mechanism design. In: Proc. 41st ACM Symp. on Theory of Computing (2010)

11. Chawla, S., Malec, D., Sivan, B.: The power of randomness in bayesian optimal mechanism design. In: ACM Conference on Electronic Commerce, pp. 149–158 (2010)
12. Compte, O., Jehiel, P.: Auctions and information acquisition: Sealed-bid or dynamic formats? Levine's Bibliography 784828000000000495, UCLA Department of Economics (October 2005)
13. Cramer, J., Spiegel, Y., Zheng, C.: Optimal selling mechanisms wth costly information acquisition. Technical report (2003)
14. Cremer, J., McLean, R.P.: Full extraction of surplus in bayesian and dominant strategy auctions. Econometrica 56(6) (1988)
15. Gibbard, A.: Manipulation of voting schemes: a general result. Econometrica 41, 211–215 (1973)
16. Hartline, J.: Lectures on approximation and mechanism design. Lecture notes (2012)
17. Hartline, J., Karlin, A.: Profit maximization in mechanism design. In: Nisan, N., Roughgarden, T., Tardos, É., Vazirani, V. (eds.) Algorithmic Game Theory, ch. 13, pp. 331–362. Cambridge University Press (2007)
18. Kleinberg, R., Weinberg, S.M.: Matroid prophet inequalities. In: Symposium on Theoretical Computer Science (2012)
19. Larson, K.: Reducing costly information acquisition in auctions. In: AAMAS, pp. 1167–1174 (2006)
20. Larson, K., Sandholm, T.: Strategic deliberation and truthful revelation: an impossibility result. In: ACM Conference on Electronic Commerce, pp. 264–265 (2004)
21. Lavi, R., Swamy, C.: Truthful and near-optimal mechanism design via linear programming. In: Proc. 46th IEEE Symp. on Foundations of Computer Science (2005)
22. Levin, D., Smith, J.L.: Equilibrium in auctions with entry. American Economic Review 84, 585–599 (1994)
23. Myerson, R.: Optimal auction design. Mathematics of Operations Research 6, 58–73 (1981)
24. Thompson, D.R., Leyton-Brown, K.: Dominant-strategy auction design for agents with uncertain, private values. In: Twenty-Fifth Conference of the Association for the Advancement of Artificial Intelligence, AAAI 2011 (2011)

Price of Stability in Polynomial Congestion Games[*]

George Christodoulou and Martin Gairing

Department of Computer Science, University of Liverpool, U.K.

Abstract. The Price of Anarchy in congestion games has attracted a lot of research over the last decade. This resulted in a thorough understanding of this concept. In contrast the Price of Stability, which is an equally interesting concept, is much less understood.

In this paper, we consider congestion games with polynomial cost functions with nonnegative coefficients and maximum degree d. We give matching bounds for the Price of Stability in such games, i.e., our technique provides the exact value for any degree d.

For linear congestion games, tight bounds were previously known. Those bounds hold even for the more restricted case of dominant equilibria, which may not exist. We give a separation result showing that already for congestion games with quadratic cost functions this is not possible; that is, the Price of Anarchy for the subclass of games that admit a dominant strategy equilibrium is strictly smaller than the Price of Stability for the general class.

1 Introduction

During the last decade, the quantification of the inefficiency of game-theoretic equilibria has been a popular and successful line of research. The two most widely adopted measures for this inefficiency are the Price of Anarchy (PoA) [17] and the Price of Stability (PoS) [3].

Both concepts compare the social cost in a Nash equilibrium to the optimum social cost that could be achieved via central control. The PoA is pessimistic and considers the worst-case such Nash equilibrium, while the PoS is optimistic and considers the best-case Nash equilibrium. Therefore, the PoA can be used as an absolute worst-case guarantee in a scenario where we have no control over equilibrium selection. On the other hand, the PoS gives an estimate of what is the best we can hope for in a Nash equilibrium; for example, if the players collaborate to find the optimal Nash equilibrium, or if a trusted mediator suggest this solution to them. Moreover, it is a much more accurate measure for those instances that possess unique Nash equilibria.

Congestion games [20] have been a driving force in recent research on these inefficiency concepts. In a congestion game, we are given a set of resources and each player selects a subset of them (e.g. a path in a network). Each resource

---

[*] This work was supported by EPSRC grants EP/K01000X/1 and EP/J019399/1.

F.V. Fomin et al. (Eds.): ICALP 2013, Part II, LNCS 7966, pp. 496–507, 2013.

has a cost function that only depends on the number of players that use it. Each player aspires to minimise the sum of the resources' costs in its strategy given the strategies chosen by the other players. Congestion games always admit a *pure* Nash equilibrium [20], where players pick a single strategy and do not randomize. Rosenthal [20] showed this by means of a *potential function* having the following property: if a single player deviates to a different strategy then the value of the potential changes by the same amount as the cost of the deviating player. Pure Nash equilibria correspond to local optima of the potential function. Games admitting such a potential function are called potential games and every potential game is isomorphic to a congestion game [19].

Today we have a strong theory which provides a thorough understanding of the PoA in congestion games [1,4,5,11,21]. This theory includes the knowledge of the exact value of the PoA for games with linear [4,11] and polynomial [1] cost functions, a recipe for computing the PoA for general classes of cost functions [21], and an understanding of the "complexity" of the strategy space required to achieve the worst case PoA [5].

In contrast, we still only have a very limited understanding of the Price of Stability (PoS) in congestion games. Exact values for the PoS are only known for congestion games with linear cost functions [9,13] and certain network cost sharing games [3]. The reason for this is that there are more considerations when bounding the PoS as compared to bounding the PoA. For example, for linear congestion games, the techniques used to bound the PoS are considerably more involved than those used to bound the PoA.

A fundamental concept in the design of games is the notion of a *dominant-strategy equilibrium*. In such an equilibrium each player chooses a strategy which is better than any other strategy no matter what the other players do. It is well-known that such equilibria do not always exist, as the requirements imposed are too strong. However, it is appealing for a game designer, as it makes outcome prediction easy. It also simplifies the strategic reasoning of the players and is therefore an important concept in mechanism design. If we restrict to instances where such equilibria exist, it is natural to ask how inefficient those equilibria can be. Interestingly, for linear congestion games, they can be as inefficient as the PoS [9,13,14].

1.1   Contribution and High-Level Idea

Results. In this paper we study the fundamental class of congestion games with polynomial cost functions of maximum degree d and nonnegative coefficients. Our main result reduces the problem of finding the value of the Price of Stability to a single-parameter optimization problem. It can be summarized in the following theorem (which combines Theorem 2 and Theorem 3).

Theorem 1. *For congestion games with polynomial cost functions with maximum degree d and nonnegative coefficients, the Price of Stability is given by*

$$\mathsf{PoS} = \max_{r>1} \frac{(2^d d + 2^d - 1) \cdot r^{d+1} - (d+1) \cdot r^d + 1}{(2^d + d - 1) \cdot r^{d+1} - (d+1) \cdot r^d + 2^d d - d + 1}.$$

For any degree d, this gives the exact value of the Price of Stability. For example, for $d = 1$ and $d = 2$, we get

$$\max_r \frac{3\,r^2 - 2\,r + 1}{2\,r^2 - 2\,r + 2} = 1 + \frac{\sqrt{3}}{3} \approx 1.577 \quad \text{and} \quad \max_r \frac{11\,r^3 - 3\,r^2 + 1}{5\,r^3 - 3\,r^2 + 7} \approx 2.36,$$

respectively. The PoS converges to $d + 1$ for large d.

We further show that in contrast to linear congestion games [13,14], already for $d = 2$, there is no instance which admits a dominant strategy equilibrium and achieves this value. More precisely, we show in Theorem 4 that for the subclass of games that admit a dominant strategy equilibrium the Price of Anarchy is strictly smaller than the Price of Stability for the general class.

Upper Bound Techniques. Both finding upper and lower bounds for the PoS, seem to be a much more complicated task than bounding the PoA. For the PoA of a class of games, one needs to capture the worst-case example of *any* Nash equilibrium, and the PoA methodology has been heavily based on this fact. On the other hand, for the PoS of the same class one needs to capture the worst-case instance of the *best* Nash equilibrium. So far, we do not know a useful characterization of the set of best-case Nash equilibria. It is not straightforward to transfer the techniques for the PoA to solve the respective PoS problem.

A standard approach that has been followed for upper bounding the PoS can be summarised as follows:

1. Define a restricted subset $\mathcal{R}$ of Nash equilibria.
2. Find the Price of Anarchy with respect to Nash equilibria that belong in $\mathcal{R}$.

The above recipe introduces new challenges: What is a good choice for $\mathcal{R}$, and more importantly, how can we incorporate the description of $\mathcal{R}$ in the Price of Anarchy methodology? For example, if $\mathcal{R}$ is chosen to be the set of *all* Nash equilibria, then one obtains the PoA bound. Finding an appropriate restriction is a non-trivial task and might depend on the nature of the game, so attempts vary in the description level of $\mathcal{R}$ from natural, "as the set of equilibria with optimum potential", to the rather more technical definitions like "the equilibria that can be reached from a best-response path starting from an optimal setup".

Like previous work (see for example [3,6,9,13,14]) we consider the PoA of Nash equilibria with minimum potential (or in fact with potential smaller than the one achieved in the optimum).

Then we use a linear combination of two inequalities, which are derived from the potential and the Nash equilibrium conditions, respectively. Using only the Nash inequality gives the PoA value [1]. Using only the potential inequality gives an upper bound of $d + 1$. The question is what is the best way to combine these inequalities to obtain the minimum possible upper bound? Caragiannis et al. [9] showed how to do this for linear congestion games. Our analysis shows how to combine them optimally for all polynomials (cf. parameter $\widehat{\nu}$ in Definition 3).

The main technical challenge is to extend the techniques used for proving upper bounds for the PoA [1,11,21]. In general those techniques involve optimizing over two parameters λ, μ such that the resulting upper bound on the PoA is minimized and certain technical conditions are satisfied – Roughgarden [21] refers to those conditions as (λ, μ)-smoothness. The linear combination of the two inequalities mentioned above adds a third parameter ν, which makes the analysis much more involved.

Lower Bound Techniques. Proving lower bounds for the PoA and PoS is usually done by constructing specific classes of instances. However, there is a conceptual difference: Every Nash equilibrium provides a lower bound on the PoA, while for the PoS we need to give a Nash equilibrium and prove that this is the best Nash equilibrium. To guarantee optimality, the main approach is based on constructing games with *unique* equilibria. One way to guarantee this is to define a game with a dominant-strategy equilibrium. This approach gives tight lower bounds in congestion games with linear cost functions [13,14]. Recall, that our separation result (Theorem 4) shows that, already for $d = 2$, dominant-strategy equilibria will not give us a tight lower bound. Thus, we use a different approach. We construct an instance with a unique Nash equilibrium and show this by using an inductive argument (Lemma 1).

The construction of our lower bound was governed by the inequalities used in the proof of the upper bound. At an abstract level, we have to construct an instance that uses the cost functions and loads on the resource that make all used inequalities tight. This is not an easy task as there are many inequalities: most prominently, one derived from the Nash equilibrium condition, one from the potential, and a third one that upper bounds a linear combination of them (see Proposition 1). To achieve this we had to come up with a completely novel construction.

1.2 Related Work

The term Price of Stability was introduced by Anshelevich et al. [3] for a network design game, which is a congestion game with special decreasing cost functions. For such games with n players, they showed that the Price of Stability is exactly H_n, i.e., the n'th harmonic number. For the special case of *undirected* networks, the PoS is known to be strictly smaller than H_n [15,7,12,3], but while the best general upper bound [15] is close to H_n, the best current lower bound is a constant [8]. For special cases better upper bound can be achieved. Li [18] showed an upper bound of $O(\log n / \log \log n)$ when the players share a common sink, while Fiat et al. [16] showed a better upper bound of $O(\log \log n)$ when in addition there is a player in every vertex of the network. Chen and Roughgarden [10] studied the PoS for the *weighted* variant of this game, where each player pays for a share of each edge cost proportional to her weight, and Albers [2] showed that the PoS is $\Omega(\log W / \log \log W)$, where W is the sum of the players' weights.

The PoS has also been studied in congestion games with increasing cost functions. For linear congestion games, the PoS is equal to $1 + \sqrt{3}/3 \approx 1.577$ where the lower bound was shown in [13] and the upper bound in [9]. Bilo[6] showed upper bounds on the PoS of 2.362 and 3.322 for congestion games with quadratic and cubic functions respectively. He also gives non-matching lower bounds, which are derived from the lower bound for linear cost functions in [14].

Awerbuch et al. [4] and Christodoulou and Koutsoupias [11] showed that the PoA of congestion games with linear cost functions is $\frac{5}{2}$. Aland et. al. [1] obtained the exact value on the PoA for polynomial cost functions. Roughgarden's [21] smoothness framework determines the PoA with respect to any set of allowable cost functions. These results have been extended to the more general class of *weighted* congestion games [1,4,5,11].

Roadmap. The rest of the paper is organized as follows. In Section 2 we introduce polynomial congestion games. In Section 3 and 4, we present our matching lower and upper bounds on the PoS. We present a separation result in Section 5. Due to space constraints, some of the proofs are deferred to the full version of this paper.

2   Definitions

For any positive integer $k \in \mathbb{N}$, denote $[k] = \{1, \ldots, k\}$. A *congestion game* [20] is a tuple $(N, E, (\mathcal{S}_i)_{i \in N}, (c_e)_{e \in E})$, where $N = [n]$ is a set of n players and E is a set of resources. Each player chooses as her pure *strategy* a set $s_i \subseteq E$ from a given *set of available strategies* $S_i \subseteq 2^E$. Associated with each resource $e \in E$ is a nonnegative *cost function* $c_e : \mathbb{N} \to \mathbb{R}^+$. In this paper we consider polynomial cost functions with maximum degree d and nonnegative coefficients; that is every cost function is of the form $c_e(x) = \sum_{j=0}^{d} a_{e,j} \cdot x^j$ with $a_{e,j} \geq 0$ for all j.

A *pure strategy profile* is a choice of strategies $\mathbf{s} = (s_1, s_2, \ldots s_n) \in \mathcal{S} = \mathcal{S}_1 \times \cdots \times \mathcal{S}_n$ by players. We use the standard notation $\mathbf{s}_{-i} = (s_1, \ldots, s_{i-1}, s_{i+1}, \ldots s_n)$, $\mathcal{S}_{-i} = \mathcal{S}_1 \times \cdots \times S_{i-1} \times S_{i+1} \times \cdots \times \mathcal{S}_n$, and $\mathbf{s} = (s_i, \mathbf{s}_{-i})$. For a pure strategy profile $\mathbf{s}$ define the *load* $n_e(\mathbf{s}) = |i \in N : e \in s_i|$ as the number of players that use resource e. The *cost* for player i is defined by $C_i(\mathbf{s}) = \sum_{e \in s_i} c_e(n_e(\mathbf{s}))$.

Definition 1. *A pure strategy profile s is a pure Nash equilibrium if and only if for every player $i \in N$ and for all $s_i' \in \mathcal{S}_i$, we have $C_i(\mathbf{s}) \leq C_i(s_i', \mathbf{s}_{-i})$.*

Definition 2. *A pure strategy profile s is a (weakly) dominant strategy equilibrium if and only if for every player $i \in N$ and for all $s_i' \in \mathcal{S}_i$ and $\mathbf{s}_{-i} \in \mathcal{S}_{-i}$, we have $C_i(\mathbf{s}) \leq C_i(s_i', \mathbf{s}_{-i})$.*

The *social cost* of a pure strategy profile s is the sum of the players costs

$$SC(\mathbf{s}) = \sum_{i \in N} C_i(\mathbf{s}) = \sum_{e \in E} n_e(\mathbf{s}) \cdot c_e(n_e(\mathbf{s})).$$

Denote $\mathrm{OPT} = \min_\mathbf{s} SC(\mathbf{s})$ as the *optimum social cost* over all strategy profiles $\mathbf{s} \in \mathcal{S}$. The *Price of Stability* of a congestion game is the social cost of the best-case Nash equilibrium over the optimum social cost

$$\mathsf{PoS} = \min_{\mathbf{s} \text{ is a Nash Equilibrium}} \frac{SC(\mathbf{s})}{\mathrm{OPT}}.$$

The PoS for a class of games is the largest PoS among all games in the class.

For a class of games that admit dominant strategy equilibria, the *Price of Anarchy of dominant strategies*, dPoA, is the worst case ratio (over all games) between the social cost of the dominant strategies equilibrium and the optimum social cost.

Congestion games admit a potential function $\Phi(\mathsf{s}) = \sum_{e \in E} \sum_{j=1}^{n_e(\mathsf{s})} c_e(j)$ which was introduced by Rosenthal [20] and has the following property: for any two strategy profiles s and (s_i', s_{-i}) that differ only in the strategy of player $i \in N$, we have $\Phi(\mathsf{s}) - \Phi(s_i', \mathsf{s}_{-i}) = C_i(\mathsf{s}) - C_i(s_i', \mathsf{s}_{-i})$. Thus, the set of pure Nash equilibria correspond to local optima of the potential function. More importantly, there exists a pure Nash eqilibrium s, s.t.

$$\Phi(\mathsf{s}) \leq \Phi(\mathsf{s}') \quad \text{for all } \mathsf{s}' \in S. \tag{1}$$

3   Lower Bound

In this section we use the following instance to show a lower bound on PoS.

Example 1. Given nonnegative integers n, k and d, define a congestion game as follows:

- The set of resources E is partitioned into $E = \mathcal{A} \cup \mathcal{B} \cup \{\Gamma\}$ where $\mathcal{A}$ consists of n resources $\mathcal{A} = \{A_i | i \in [n]\}$, $\mathcal{B}$ consists of $n(n-1)$ resources $\mathcal{B} = \{B_{ij} | i, j \in [n], i \neq j\}$, and Γ is a single resource.
- All cost functions are monomials of degree d given as follows:
  - For $i \in [n]$ the cost of resource A_i is given by $c_{A_i}(x) = \alpha_i \cdot x^d$, where

$$\alpha_i = (k+i)^d + \varepsilon \qquad \text{for sufficiently small } \varepsilon > 0.$$

  - Denote $T_i = \frac{(k+i)^d - (k+i-1)^d}{(2^{2d}-1)}$. Resource B_{ij} with $i, j \in [n], i \neq j$ has cost

$$c_{B_{ij}}(x) = \beta_{ij} \cdot x^d \qquad \text{where } \beta_{ij} = \begin{cases} T_j & \text{, if } i < j, \\ 2^d T_i & \text{, if } i > j. \end{cases}$$

  - For resource Γ we have $c_\Gamma(x) = x^d$.
- There are $n + k$ players. Each player $i \in [n]$ has two strategies s_i, s_i^* where

$$s_i = \Gamma \cup \{B_{ij} | j \in [n], j \neq i\} \text{ , and}$$
$$s_i^* = A_i \cup \{B_{ji} | j \in [n], j \neq i\}.$$

The remaining players $i \in [n+1, n+k]$ are fixed to choose the single resource Γ. To simplify notation denote by $\mathsf{s} = (s_1, \ldots, s_n)$ and $\mathsf{s}^* = (s_1^*, \ldots, s_n^*)$ the corresponding strategy profiles. Those profiles correspond to the unique Nash equilibrium and to the optimal allocation respectively.

In the following lemma we show that s is the unique Nash equilibrium for the game in Example 1. To do so, we show that s_1 is a dominant strategy for player 1 and that given that the first $i - 1$ players play $s_1, \ldots, s_{i-1}$, then s_i is a dominant strategy for player $i \in [n]$.

Lemma 1. *In the congestion game from Example 1, s is the unique Nash equilibrium.*

We use the instance from Example 1 to show the lower bound in the following theorem. We define $\rho = \frac{k}{n}$ and $r = \frac{k+n}{k} = 1 + \frac{1}{\rho} > 1$. We let $n \to \infty$ and determine the $r > 1$ which maximises the resulting lower bound[1]. Note that $r > 1$ is the ratio of the loads on resource Γ in Example 1.

Theorem 2. *For congestion games with polynomial cost functions with maximum degree d and nonnegative coefficients, we have*

$$\mathsf{PoS} \geq \max_{r>1} \frac{(2^d d + 2^d - 1) \cdot r^{d+1} - (d+1) \cdot r^d + 1}{(2^d + d - 1) \cdot r^{d+1} - (d+1) \cdot r^d + 2^d d - d + 1}. \tag{2}$$

4   Upper Bound

In this section we show an upper bound on the PoS for polynomial congestion games. We start with two technical lemmas and a definition, all of which will be used in the proof of Proposition 1. This proposition is the most technical part of the paper. It shows an upper bound on a linear combination of two expressions; one is derived from the Nash equilibrium condition and the other one from the potential. Equipped with this, we prove our upper bound in Theorem 3.

Lemma 2. *Let f be a nonnegative and convex function, then for all nonnegative integers x, y with $x \geq y$, $\sum_{i=y+1}^{x} f(i) \geq \int_y^x f(t)dt + \frac{1}{2}(f(x) - f(y))$.*

Definition 3. *Define $\hat{\nu}$ as the minimum ν such that*

$$f(\nu) := \left(2^d + (d-1)\left(1 - \frac{1}{r^{d+1}}\right) - \frac{1}{r}\right) \cdot \nu - d\left(1 - \frac{1}{r^{d+1}}\right) \geq 0$$

for all $r > 1$.

Observe that for all $d \geq 1$ and $r > 1$, $f(\nu)$ is a monotone increasing function in ν. Thus $\hat{\nu} \in (0, 1]$ is well defined since $f(0) < 0$ and $f(1) > 0$ for all $r > 1$. Moreover, $f(\nu) \geq 0$ for all $\nu \geq \hat{\nu}$. We will make use of the following bounds on $\hat{\nu}$.

Lemma 3. *Define $\hat{\nu}$ as in Definition 3. Then $\frac{d}{2^d + d - 1} \leq \hat{\nu} < \frac{d+1}{2^d + d - 1}$.*

Proposition 1. *Let $1 \geq \nu \geq \hat{\nu}$ and define $\lambda = d + 1 - d\nu$ and $\mu = (2^d + d - 1)\nu - d$. Then for all polynomial cost functions c with maximum degree d and nonnegative coefficients and for all nonnegative integers x, y we have*

$$\nu \cdot y \cdot c(x+1) + (1-\nu)(d+1)\left(\sum_{i=1}^{y} c(i) - \sum_{i=1}^{x} c(i)\right) \leq (\mu + \nu - 1) \cdot x \cdot c(x) + \lambda \cdot y \cdot c(y).$$

---

[1] Notice that the value r that optimizes the right hand side expression of (2) might not be rational. The lower bound is still valid as we can approximate an irrational r arbitrarily close by a rational.

Proof. Since c is a polynomial cost function with maximum degree d and non-negative coefficients it is sufficient to show the claim for all monomials of degree t where $0 \leq t \leq d$. Thus, we will show that

$$(\mu + \nu - 1) \cdot x^{t+1} + \lambda \cdot y^{t+1} - \nu \cdot y(x+1)^t + (1 - \nu)(d+1) \left(\sum_{i=1}^{x} i^t - \sum_{i=1}^{y} i^t \right) \geq 0 \quad (3)$$

for all nonnegative integers x, y and degrees $0 \leq t \leq d$.

Fix some $0 \leq t \leq d$. First observe that (3) is trivially fulfilled for $y = 0$, as all the negative terms disappear. So in the following we assume $y \geq 1$.

Elementary calculations show that (3) holds when $0 \leq x \leq y$. So in the following we assume $x > y \geq 1$. By Lemma 2, we have

$$\sum_{i=1}^{x} i^t - \sum_{i=1}^{y} i^t = \sum_{i=y+1}^{x} i^t \geq \frac{1}{t+1}(x^{t+1} - y^{t+1}) + \frac{1}{2}(x^t - y^t)$$

$$\geq \frac{1}{d+1}(x^{t+1} - y^{t+1}) + \frac{1}{2}(x^t - y^t). \quad (4)$$

Moreover, since $x \geq 2$ we can bound

$$(x+1)^t = \sum_{i=0}^{t} \binom{t}{i} x^{t-i} \leq x^t + x^{t-1} \cdot \sum_{i=1}^{t} \binom{t}{i} \left(\frac{1}{2} \right)^{i-1}$$

$$= x^t + x^{t-1} \cdot 2 \left(\left(\frac{3}{2} \right)^t - 1 \right). \quad (5)$$

Using (4) and (5) and by defining $r = \frac{x}{y} > 1$, we can lower bound the left-hand-side of (3) by

$$\mu \cdot x^{t+1} + (\lambda + \nu - 1) \cdot y^{t+1} - \nu \cdot y(x+1)^t + \frac{1}{2}(1 - \nu)(d+1)(x^t - y^t)$$

$$\geq \left(\mu + \frac{\lambda + \nu - 1}{r^{t+1}} \right) \cdot x^{t+1} - \frac{\nu}{r} \cdot \left(x^{t+1} + x^t \cdot 2 \left(\left(\frac{3}{2} \right)^t - 1 \right) \right)$$

$$+ \frac{1}{2}(1 - \nu)(d+1) \left(1 - \frac{1}{r^t} \right) \cdot x^t$$

$$= \underbrace{\left(\mu + \frac{\lambda + \nu - 1}{r^{t+1}} - \frac{\nu}{r} \right)}_{:=A(\nu)} \cdot x^{t+1} + \underbrace{\left(\frac{1}{2}(1 - \nu)(d+1) \left(1 - \frac{1}{r^t} \right) - \frac{2\nu}{r} \left(\left(\frac{3}{2} \right)^t - 1 \right) \right)}_{:=B(\nu)} \cdot x^t.$$

First observe that by using the definitions of λ, μ, we get

$$A(\nu) = \left(2^d + (d-1) \left(1 - \frac{1}{r^{t+1}} \right) - \frac{1}{r} \right) \cdot \nu - d \left(1 - \frac{1}{r^{t+1}} \right)$$

$$\geq \left(2^d + (d-1) \left(1 - \frac{1}{r^{d+1}} \right) - \frac{1}{r} \right) \cdot \nu - d \left(1 - \frac{1}{r^{d+1}} \right)$$

$$\geq 0,$$

where the first inequality holds since $\nu \leq 1$ and the second inequality is by Definition 3 and $\nu \geq \hat{\nu}$. Since $x \geq 2$, we get

$$A(\nu) \cdot x^{t+1} + B(\nu) \cdot x^t \geq (2A(\nu) + B(\nu)) \cdot x^t.$$

To complete the proof we show that $2A(\nu) + B(\nu) \geq 0$ for $\nu \geq \hat{\nu}$.

$2A(\nu) + B(\nu)$

$$= \left(2^{d+1} + 2(d-1)\left(1 - \frac{1}{r^{t+1}}\right) - \frac{2}{r} - \frac{1}{2}(d+1)\left(1 - \frac{1}{r^t}\right) - \frac{2}{r}\left(\left(\frac{3}{2}\right)^t - 1\right) \right) \cdot \nu$$

$$- 2d\left(1 - \frac{1}{r^{t+1}}\right) + \frac{1}{2}(d+1)\left(1 - \frac{1}{r^t}\right)$$

$$= \left(2^{d+1} + 2(d-1)\left(1 - \frac{1}{r^{t+1}}\right) - \frac{1}{2}(d+1)\left(1 - \frac{1}{r^t}\right) - \frac{2}{r}\left(\frac{3}{2}\right)^t \right) \cdot \nu$$

$$- 2d\left(1 - \frac{1}{r^{t+1}}\right) + \frac{1}{2}(d+1)\left(1 - \frac{1}{r^t}\right),$$

which again is a monotone increasing function in ν. Since $\hat{\nu} \geq \frac{d}{2^d + d - 1}$ by Lemma 3 and $\nu \geq \hat{\nu}$, we get

$2A(\nu) + B(\nu)$

$$\geq \left(2^{d+1} + 2(d-1)\left(1 - \frac{1}{r^{t+1}}\right) - \frac{1}{2}(d+1)\left(1 - \frac{1}{r^t}\right) - \frac{2}{r}\left(\frac{3}{2}\right)^t \right) \cdot \frac{d}{2^d + d - 1}$$

$$- 2d\left(1 - \frac{1}{r^{t+1}}\right) + \frac{1}{2}(d+1)\left(1 - \frac{1}{r^t}\right)$$

$$= \frac{(d+1)(2^d - 1) \cdot r^{t+1} - 4d\left(\frac{3}{2}\right)^t \cdot r^t - (d+1)(2^d - 1) \cdot r + 4d2^d}{2(2^d + d - 1)r^{t+1}}.$$

Define $D(d, t)$ as the numerator of this term. Thus,

$$D(d, t) = (d+1)(2^d - 1) \cdot r^{t+1} - 4d\left(\frac{3}{2}\right)^t \cdot r^t - (d+1)(2^d - 1) \cdot r + 4d2^d.$$

If $r \leq \frac{4}{3}$ then $\qquad D(d, t) \geq (d+1)(2^d - 1) \cdot r(r^t - 1) \geq 0,$

for all integer $d \geq 1$ and $0 \leq t \leq d$. If $r \geq \frac{4}{3}$ then

$$D(d, t) \geq (d+1)(2^d - 1) \cdot r(r^t - 1) - 4d\left(\frac{3}{2}\right)^t \cdot (r^t - 1)$$

$$\geq \left((d+1)(2^d - 1) \cdot \frac{4}{3} - 4d\left(\frac{3}{2}\right)^d \right) \cdot (r^t - 1),$$

which is nonnegative for all integer $d \geq 4$ and $0 \leq t \leq d$. For $t \leq d \leq 3$, $D(d, t) \geq 0$ can be checked using elementary calculus. $\qquad \square$

We are now ready to prove the upper bound of our main result.

Theorem 3. *For congestion games with polynomial cost functions with maximum degree d and nonnegative coefficients, we have*

$$PoS \leq \max_{r>1} \frac{(2^d d + 2^d - 1) \cdot r^{d+1} - (d+1) \cdot r^d + 1}{(2^d + d - 1) \cdot r^{d+1} - (d+1) \cdot r^d + 2^d d - d + 1}.$$

Proof. Let s^* be an optimum assignment and let s be a pure Nash equilibrium with $\Phi(\mathsf{s}) \leq \Phi(\mathsf{s}^*)$. Such a Nash equilibrium exists by (1). Define $x_e = n_e(\mathsf{s})$ and $y_e = n_e(\mathsf{s}^*)$. Then

$$SC(\mathsf{s}) \leq SC(\mathsf{s}) + (d+1)(\Phi(\mathsf{s}^*) - \Phi(\mathsf{s}))$$

$$= \sum_{e \in E} n_e(\mathsf{s}) \cdot c_e(n_e(\mathsf{s})) + (d+1) \sum_{e \in E} \left(\sum_{i=1}^{n_e(\mathsf{s}^*)} c_e(i) - \sum_{i=1}^{n_e(\mathsf{s})} c_e(i) \right)$$

$$= \sum_{e \in E} x_e \cdot c_e(x_e) + (d+1) \sum_{e \in E} \left(\sum_{i=1}^{y_e} c_e(i) - \sum_{i=1}^{x_e} c_e(i) \right). \qquad (6)$$

Moreover, since s is a pure Nash equilibrium, we have

$$SC(\mathsf{s}) = \sum_{i=1}^{n} C_i(\mathsf{s}) \leq \sum_{i=1}^{n} C_i(s_i^*, \mathsf{s}_{-i})$$

$$\leq \sum_{i=1}^{n} \sum_{e \in s_i^*} c_e(n_e(\mathsf{s}) + 1) = \sum_{e \in E} y_e \cdot c_e(x_e + 1). \qquad (7)$$

Let $\widehat{\nu}$ as defined in Definition 3. Taking the convex combination $\widehat{\nu} \cdot (7) + (1 - \widehat{\nu}) \cdot (6)$ of those inequalities gives

$$SC(\mathsf{s}) \leq \sum_{e \in E} \left[\widehat{\nu} \cdot y_e \cdot c_e(x_e + 1) + (1 - \widehat{\nu}) x_e \cdot c_e(x_e) + (1 - \widehat{\nu})(d+1) \left(\sum_{i=1}^{y_e} c_e(i) - \sum_{i=1}^{x_e} c_e(i) \right) \right]$$

With $\lambda = d + 1 - d\widehat{\nu}$ and $\mu = (2^d + d - 1)\widehat{\nu} - d$, applying Proposition 1 gives

$$SC(\mathsf{s}) \leq \sum_{e \in E} [\mu \cdot x_e \cdot c_e(x_e) + \lambda \cdot y_e \cdot c_e(y_e)] = \mu \cdot SC(\mathsf{s}) + \lambda SC(\mathsf{s}^*).$$

Thus,

$$\frac{SC(\mathsf{s})}{SC(\mathsf{s}^*)} \leq \frac{\lambda}{1 - \mu} = \frac{d + 1 - d\widehat{\nu}}{d + 1 - (2^d + d - 1)\widehat{\nu}},$$

By Definition 3, for all real numbers $r > 1$, we have

$$\widehat{\nu} \geq \frac{d(1 - \frac{1}{r^{d+1}})}{2^d + (d-1)(1 - \frac{1}{r^{d+1}}) - \frac{1}{r}}. \qquad (8)$$

Denote $\hat{r}$ as the value for $r > 1$ which makes inequality (8) tight. Such a value $\hat{r}$ must exist since $\hat{\nu}$ is the minimum value satisfying this inequality. So,

$$\hat{\nu} = \frac{d(\hat{r}^{d+1} - 1)}{2^d \hat{r}^{d+1} + (d-1)(\hat{r}^{d+1} - 1) - \hat{r}^d}.$$

Substituting this in the bound from Theorem 3 gives

$$
\begin{aligned}
PoS &\leq \frac{d + 1 - d\hat{\nu}}{d + 1 - (2^d + d - 1)\hat{\nu}} \\
&= \frac{(d+1)2^d \hat{r}^{d+1} + (d^2-1)(\hat{r}^{d+1}-1) - (d+1)\hat{r}^d - d^2(\hat{r}^{d+1}-1)}{2^d(d+1)\hat{r}^{d+1} + (d^2-1)(\hat{r}^{d+1}-1) - (d+1)\hat{r}^d - 2^d d(\hat{r}^{d+1}-1) - d(d-1)(\hat{r}^{d+1}-1)} \\
&= \frac{(2^d d + 2^d - 1)\hat{r}^{d+1} - (d+1)\hat{r}^d + 1}{(2^d + d - 1)\hat{r}^{d+1} - (d+1)\hat{r}^d + 2^d d - d + 1} \\
&\leq \max_{r>1} \frac{(2^d d + 2^d - 1) \cdot r^{d+1} - (d+1) \cdot r^d + 1}{(2^d + d - 1) \cdot r^{d+1} - (d+1) \cdot r^d + 2^d d - d + 1},
\end{aligned}
$$

which proves the upper bound.   □

5   Separation

For the linear case, the Price of Stability was equal to the Price of Anarchy of dominant strategies, as the matching lower bound instance would hold for dominant strategies. Here, we show that linear functions was a degenerate case, and that this is not true for higher order polynomials. We show that for games that possess dominant equilibria, the Price of Anarchy for them is strictly smaller[2]. Our separation leaves as an open question what is the exact value of the Price of Anarchy of dominant strategies for these games.

Theorem 4. *Consider a congestion game with quadratic cost functions which admits a dominant strategy equilibrium s. Then $\frac{SC(s)}{\text{OPT}} \leq \frac{7}{3}$.*

Observe that this upper bound is strictly smaller than the exact value of the PoS for *general* congestion games with quadratic cost functions from Theorem 1, which was ≈ 2.36.

References

1. Aland, S., Dumrauf, D., Gairing, M., Monien, B., Schoppmann, F.: Exact price of anarchy for polynomial congestion games. SIAM Journal on Computing 40(5), 1211–1233 (2011)
2. Albers, S.: On the value of coordination in network design. SIAM Journal on Computing 38(6), 2273–2302 (2009)

---

[2] By a more elaborate analysis one can come up with an upper bound of ≈ 2.242 Here we just wanted to demonstrate the separation of the two measures.

3. Anshelevich, E., Dasgupta, A., Kleinberg, J.M., Tardos, É., Wexler, T., Roughgarden, T.: The price of stability for network design with fair cost allocation. SIAM Journal on Computing 38(4), 1602–1623 (2008)
4. Awerbuch, B., Azar, Y., Epstein, A.: Large the price of routing unsplittable flow. In: Proceedings of STOC, pp. 57–66 (2005)
5. Bhawalkar, K., Gairing, M., Roughgarden, T.: Weighted congestion games: Price of anarchy, universal worst-case examples, and tightness. In: de Berg, M., Meyer, U. (eds.) ESA 2010, Part II. LNCS, vol. 6347, pp. 17–28. Springer, Heidelberg (2010)
6. Bilò, V.: A unifying tool for bounding the quality of non-cooperative solutions in weighted congestion games. In: Erlebach, T., Persiano, G. (eds.) WAOA 2012. LNCS, vol. 7846, pp. 215–228. Springer, Heidelberg (2013)
7. Bilò, V., Bove, R.: Bounds on the price of stability of undirected network design games with three players. Journal of Interconnection Networks 12(1-2), 1–17 (2011)
8. Bilò, V., Caragiannis, I., Fanelli, A., Monaco, G.: Improved lower bounds on the price of stability of undirected network design games. In: Kontogiannis, S., Koutsoupias, E., Spirakis, P.G. (eds.) SAGT 2010. LNCS, vol. 6386, pp. 90–101. Springer, Heidelberg (2010)
9. Caragiannis, I., Flammini, M., Kaklamanis, C., Kanellopoulos, P., Moscardelli, L.: Tight bounds for selfish and greedy load balancing. Algorithmica 61, 606–637 (2011)
10. Chen, H.-L., Roughgarden, T.: Network design with weighted players. Theory of Computing Systems 45, 302–324 (2009)
11. Christodoulou, G., Koutsoupias, E.: The price of anarchy of finite congestion games. In: Proceedings of STOC, pp. 67–73 (2005)
12. Christodoulou, G., Chung, C., Ligett, K., Pyrga, E., van Stee, R.: On the price of stability for undirected network design. In: Bampis, E., Jansen, K. (eds.) WAOA 2009. LNCS, vol. 5893, pp. 86–97. Springer, Heidelberg (2010)
13. Christodoulou, G., Koutsoupias, E.: On the price of anarchy and stability of correlated equilibria of linear congestion games. In: Brodal, G.S., Leonardi, S. (eds.) ESA 2005. LNCS, vol. 3669, pp. 59–70. Springer, Heidelberg (2005)
14. Christodoulou, G., Koutsoupias, E., Spirakis, P.G.: On the performance of approximate equilibria in congestion games. Algorithmica 61(1), 116–140 (2011)
15. Disser, Y., Feldmann, A.E., Klimm, M., Mihalák, M.: Improving the h_k-bound on the price of stability in undirected shapley network design games. CoRR, abs/1211.2090 (2012); To appear in CIAC 2013
16. Fiat, A., Kaplan, H., Levy, M., Olonetsky, S., Shabo, R.: On the price of stability for designing undirected networks with fair cost allocations. In: Bugliesi, M., Preneel, B., Sassone, V., Wegener, I. (eds.) ICALP 2006. LNCS, vol. 4051, pp. 608–618. Springer, Heidelberg (2006)
17. Koutsoupias, E., Papadimitriou, C.: Worst-case equilibria. In: Meinel, C., Tison, S. (eds.) STACS 1999. LNCS, vol. 1563, pp. 404–413. Springer, Heidelberg (1999)
18. Li, J.: An $O(\frac{\log n}{\log \log n})$ upper bound on the price of stability for undirected Shapley network design games. Information Processing Letters 109(15), 876–878 (2009)
19. Monderer, D., Shapley, L.: Potential games. Games and Economics Behavior 14, 124–143 (1996)
20. Rosenthal, R.W.: A class of games possessing pure-strategy Nash equilibria. International Journal of Game Theory 2, 65–67 (1973)
21. Roughgarden, T.: Intrinsic robustness of the price of anarchy. Communications of the ACM 55(7), 116–123 (2012)

Localization for a System of Colliding Robots

Jurek Czyzowicz[1], Evangelos Kranakis[2], and Eduardo Pacheco[2]

[1] Université du Québec en Outaouais, Gatineau, Québec J8X 3X7, Canada
[2] Carleton University, Ottawa, Ontario K1S 5B6, Canada

Abstract. We study the *localization problem* in the ring: a collection of n anonymous mobile robots are deployed in a continuous ring of perimeter one. All robots start moving at the same time along the ring with arbitrary velocity, starting in clockwise or counterclockwise direction around the ring. The robots bounce against each other when they meet. The task of each robot is to find out, in finite time, the initial position and the initial velocity of every deployed robot. The only way that robots perceive the information about the environment is by colliding with their neighbors; any type of communication among robots is not possible.

We assume the principle of momentum conservation as well as the conservation of energy, so robots exchange velocities when they collide. The capabilities of each robot are limited to: observing the times of its collisions, being aware of its velocity at any time, and processing the collected information. Robots have no control of their walks or velocities. Robots' walks depend on their initial positions, velocities, and the sequence of collisions. They are not equipped with any visibility mechanism.

The localization problem for bouncing robots has been studied previously in [1,2] in which robots are assumed to move at the same velocity. The configuration of initial positions of robots and their speeds is considered *feasible*, if there is a finite time, after which every robot starting at this configuration knows initial positions and velocities of all other robots. Authors of [1] conjectured that if robots have arbitrary velocities, the problem might be solvable, if the momentum conservation and energy preservation principles are assumed.

In this paper we prove that the conjecture in [1] is false. We show that the feasibility of any configuration and the required time for solving it under such stronger constraints depend only on the collection of velocities of the robots. More specifically, if $v_0, v_1, \ldots, v_{n-1}$ are the velocities of a given robot configuration $\mathcal{S}$, we prove that $\mathcal{S}$ is feasible if and only if $v_i \neq \bar{v}$ for all $0 \leq i \leq n-1$, where $\bar{v} = \frac{v_0 + \ldots + v_{n-1}}{n}$. To figure out the initial positions of all robots no more than $\frac{2}{min_{0 \leq i \leq n-1}|v_i - \bar{v}|}$ time is required.

1 Introduction

Due to their simplicity, efficiency, and flexibility mobile agents or robots have been widely used in diverse areas namely artificial intelligence, computational economics, and robotics [3]. Mobile robots are autonomous entities that possess

F.V. Fomin et al. (Eds.): ICALP 2013, Part II, LNCS 7966, pp. 508–519, 2013.

the ability to move within their environment, to interact with other robots, to perceive the information of the environment and to process this information. Some examples of tasks carried out by mobile robots are environment exploration, perimeter patrolling, mapping, pattern formation and localization.

In order to reduce power consumption and to prevent scalability problems, minimum communication among robots and robots with limited capabilities are frequently sought. With this in mind, in this paper, we study distributed systems of mobile robots that allow no communication whatsoever and have limited capabilities. Part of our motivation is to understand the algorithmic limitations of what a set of such robots can compute.

The task of our interest is that each robot localizes the initial position of every other robot also deployed on the ring with respect to its own position. Our model assumes robot anonymity, collisions as the only way of interaction between robots, and robots' movements completely out of their control. The abilities of each robot are limited to observing the time of any of its collisions, awareness of its velocity at any time and the capacity to process this information.

Distributed applications often concern mobile robots of very limited communication and sensing capabilities, mainly due to the limited production cost, size and battery power. Such collections of mobile robots, called *swarms*, often perform exploration or monitoring tasks in hazardous or hard to access environments. The usual swarm robot attributes assumed for distributed models include anonymity, negligible dimensions, no explicit communication, no common coordinate system (see [4]). In most situations involving such weak robots the fundamental research question concerns the feasibility of solving the given task (cf. [5,4]).

In our paper, besides the limited sensing and communication capabilities, a robot has absolutely no control on its movement, which is determined by the bumps against its neighbors. In [6,7] the authors introduced *population protocols*, modeling wireless sensor networks by limited finite-state computational devices. The agents of population protocols also follow mobility patterns totally out of their control. This is called *passive mobility*, intended to model, e.g., some unstable environment, like a flow of water, chemical solution, human blood, wind or unpredictable mobility of agents' carriers (e.g. vehicles or flocks of birds).

Pattern formation is sometimes considered as one of the steps of more complex distributed tasks. Our interest in the problem of this paper was fueled by the patrolling problem [8]. Patrolling is usually more efficient if the robots are evenly distributed around the environment. Clearly, location discovery is helpful in uniform spreading of the collection. [9] investigated a related problem, where uniform spreading in one-dimensional environment has been studied.

The dynamics of the robots in our model is similar to the one observed in some systems of gas particles which have motivated some applications for mobile robots. The study of the dynamics of particles sliding in a surface that collide among themselves has been of great interest in physics for a long time. Much of such work has been motivated for the sake of understanding the dynamic properties of gas particles [10,11,12,13]. The simplest models of such particle systems assume either a line or a ring as the environment in which particles move. For instance, Jepsen [14], similarly to our paper, considers particles of equal mass and arbitrary velocity moving in a ring. He assumes the conservation of momentum and conservation of energy principles, such that when two particles collide they exchange velocities. Jepsen studies the probabilistic movement of particles because of its importance for understanding some gas equilibrium properties. Other works have found applications of particle systems in different fields. For example, Cooley and Newton [15,16] described a method to generate pseudo random numbers efficiently by using the dynamics of particle systems.

The distributed computing community has exploited the simple dynamics of some particle systems to design algorithms for mobile robots. [17] consider a system of mobile robots that imitate the impact behavior of n elastic particles moving in a ring. They considered a set of n synchronous robots that collide elastically moving in a frictionless ring in order to carry out perimeter surveillance. Sporadic communication at the times of collision is assumed. Another example of a system of robots that mimics particle's dynamics is found in [18], where the problem of motion synchronization in the line is studied.

In our paper we also assume a system wherein robots imitate the dynamics of gas particles moving in a ring, with the restriction that robots can not communicate. The task we consider is localization of the initial position of every robot in the ring. We call this problem the *localization problem*. The localization problem has been previously studied in [2] from a randomized approach and robots operating in synchronous rounds. Czyzowicz et.al [1], considered a simplified version of the localization task from a deterministic point of view. They assumed that all robots have equal speed and that when two of them collide they *bounce back*. Czyzowicz et.al show that all robot configurations in which not all the robots have the same initial direction are feasible and provided a position detection algorithm for all feasible configurations.

When robots' velocities are arbitrary and each robot is aware only of its current velocity, as we assume in this paper, the characterization of all feasible robot configurations becomes much more complex. In [1] infeasible configurations can be detected by robots in finite time, while we show here, that without the knowledge of initial velocities, there always exists some robot for which it is impossible to decide whether the given configuration is feasible, even if the total number of robots is known in advance.

We provide a complete characterization of all feasible configurations. If $v_0, v_1, \ldots, v_{n-1}$ is the collection of velocities of a given robot configuration $\mathcal{S}$, we prove that $\mathcal{S}$ is feasible if and only if for all i, $0 \leq i \leq n-1$ we have $v_i \neq \frac{v_0 + \ldots + v_{n-1}}{n}$. Moreover, we provide an upper bound of $\frac{2}{min_{0 \leq i \leq n-1}|v_i - \bar{v}|}$ on the necessary time to solve the localization problem. Hence the feasibility of any robot configuration is independent of the starting positions of the robots.

2   The Model

We consider a set of n synchronous and anonymous robots $r_0, r_1, \ldots, r_{n-1}$ deployed on a continuous, one-dimensional ring of perimeter one, represented by the real segment $[0,1)$ wherein 0 and 1 are identified as the same point. Each robot r_i starts moving at time $t = 0$ with velocity v_i either in counterclockwise or in clockwise direction. By $|v_i|$ we denote the speed of velocity v_i.

By $r_i(t) \in [0,1)$ we denote the position in the ring of robot r_i at time t. Let $\mathcal{P}$ denote the sequence $p_0, p_1, \ldots, p_{n-1}$ of initial positions of the robots, meaning $p_i = r_i(0), 0 \le i \le n - 1$. W.l.o.g, we assume $p_0 = 0$ as well as a non-decreasing order in the counterclockwise direction of the initial positions of the robots. If robot r_i moves freely along the ring with velocity v_i during the time interval $[t_1, t_2]$, then its position at time $t \in [t_1, t_2]$ is given by $r_i(t) = r_i(t_1) + v_i \cdot (t - t_1)$. When two robots collide they exchange velocities following the principle of momentum conservation and conservation of energy in classical mechanics for objects of equal mass [19]. We assume that in any collision no more than two robots are involved.

Regarding the capabilities of the robots, we assume that each of them has a clock which can measure time in a continuous way. Each robot is always aware of its clock, current velocity and the time of any of its collisions. The movement of a robot is beyond its control in that it depends solely on its initial position and velocity, as well as the collisions with other robots along the way. At the time of deployment, no robot is aware of the initial position and the velocity of any other robot nor of the total number of robots deployed in the ring. Moreover, robots do not have a common *sense of direction*.

Let $\mathcal{S} = (\mathcal{P}, V)$ be a system of n mobile robots $r_0, r_1, \ldots, r_{n-1}$ with initial positions $\mathcal{P} = (p_0, p_1, \ldots, p_{n-1})$ and velocities $V = (v_0, v_1, \ldots, v_{n-1})$ respectively; we denote by $\bar{v}$ the average of the velocities in V. We say that the *localization problem* for $\mathcal{S}$ is feasible if there exists a finite time T, such that each robot can determine the initial positions, and the initial velocities of all robots in the system with respect to its own starting position and its own orientation of the ring. This should be accomplished by each robot by observing the times of a sequence of collisions taking place within some time interval $[0, T]$. Note that each collision is accompanied by the measurement of collision time and a corresponding exchange of velocities.

At issue here is not only to determine the feasibility of the localization problem for the given system $\mathcal{S}$, but also to characterize all such feasible system instances.

3 Feasibility of the Localization Problem

For two points p, q in the ring, by $d^{(+)}(p,q)$ we denote the counterclockwise distance from p to q in the ring, i.e. the distance which needs to be travelled in the counterclockwise direction in order to arrive at q starting from p. Note that for $p \ne q$ we have $0 < d^{(+)}(p,q) < 1$, and $d^{(+)}(p,q) = 1 - d^{(+)}(q,p)$.

In order to visualize the dynamics of the robots in the ring, we consider an infinite line $L = (-\infty, \infty)$ and for each robot r_i we create an infinite number of its copies $r_i^{(j)}$, all having the same initial velocity, such that their initial positions in L are $r_i^{(j)}(0) = j + r_i(0)$ for all integer values of $j \in \mathbb{Z}$.

We use the idea of *baton*, applied previously in [1,2], in order to simplify our arguments and to gain intuition of the dynamics of the robots. Assume that each robot holds a *virtual object*, called baton, and when two robots collide they exchange their batons. By $b_i^{(j)}$ we denote the baton originally held by robot $r_i^{(j)}$ and by $b_i^{(j)}(t)$ we denote the position of this baton on L at time t. Notice that the velocity of baton $b_i^{(j)}$ is constant so its trajectory corresponds to the line of slope $1/v_i$.

By putting together the infinite line and the trajectories of batons, we can depict the walk of the robots up to any given time. For instance, in Fig. 1, the dynamics of a system of three mobile robots is depicted. The walk of robot r_0 along the ring corresponds to the thick polyline.

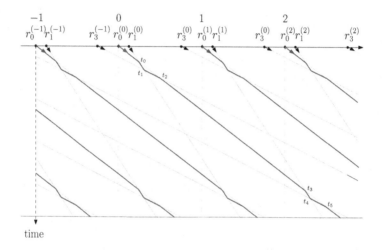

Fig. 1. Trajectory of robot r_0 corresponds to the thick polyline. The times of its first six collisions are also shown.

When a robot moves from any given position p on line L to the position $p+1$ (or $p-1$) such a robot has completed a tour along the ring in counterclockwise (or resp. clockwise) direction. For example r_0 in Fig. 1 has completed two counterclockwise tours along the ring between time t_0 and t_3. We show first, that the feasibility of the localization problem does not change when the initial speeds of all robots are increased, or decreased by the same value.

Definition 1 *A translation of a system* $\mathcal{S} = (\mathcal{P}, (v_0, \ldots, v_{n-1}))$ *is a system* $\mathcal{S}_c = (\mathcal{P}, (v_0', \ldots, v_{n-1}'))$, *where* $v_i' = v_i - c, 0 \le i \le n-1$, *for* $c \in \mathbb{R}$.

Lemma 1. *Let* $\mathcal{S}$ *be a system of robots and let* $\mathcal{S}_c$ *be any of its translations. For every time* t, *velocities* v_i *and* v_j *are exchanged in* $\mathcal{S}$ *at time* t *if, and only if at time* t *velocities* $v_i - c$ *and* $v_j - c$ *are exchanged in* $\mathcal{S}_c$.

Proof. Since $\mathcal{S}$ is also a translation of $\mathcal{S}_c$, it is enough to prove the lemma in one direction. Consider any translation $\mathcal{S}_c$ of system $\mathcal{S}$ and let v'_i and v'_j be such that $v'_i = v_i - c$ and $v'_j = v_j - c$ for some $c \in \mathbb{R}$ and let b_i, b_j, b'_i, and b'_j be the corresponding batons of velocities v_i, v_j, v'_i, and v'_j respectively.

Since the times of exchange of velocities v'_i and v'_j coincide with the times of exchange of batons b'_i and b'_j, it is enough to prove that for every time t in which batons b_i and b_j are exchanged in $\mathcal{S}$, so are batons b'_i and b'_j in $\mathcal{S}_c$.

We prove first that the time of the first meeting of b_i and b_j in $\mathcal{S}$ is the same as the time of the first meeting of b'_i and b'_j in $\mathcal{S}_c$. The statement is clearly true when $v_i = v_j$, since in both systems $\mathcal{S}$ and $\mathcal{S}_c$ batons b_i and b_j stay forever at the same distance on the ring and no meeting ever occurs. Suppose then, by symmetry, that $v_i \geq 0$ and $v_i > v_j$. Let d be the initial counterclockwise distance (i.e. at time $t = 0$) from b_i to b_j, i.e. $d = d^{(+)}(b_i(0), b_j(0))$. Observe that the batons approach at the speed equal to $|v_i - v_j|$. (Note that this holds as well when robots have different directions, i.e. $v_j < 0$: then $|v_i - v_j| = |v_i| + |v_j|$). Hence the first meeting of b_i and b_j occurs at time $t^* = \frac{d}{|v_i - v_j|}$. However in $\mathcal{S}_c$ we have $v'_i = v_i - c > v_j - c = v'_j$ and again the two batons approach reducing their original distance d with speed $|v'_i - v'_j| = |v_i - v_j|$ meeting eventually after the same time t^*.

A careful reader may observe that the above argument holds independently of the directions that the batons b_j, b'_i and b'_j may have.

Observe that the same analysis holds by induction for the k-th meeting of the robots, for $k = 2, 3, \ldots$ Indeed, if $i < j$, i.e. $p_i < p_j$, and $d = p_j - p_i$, then the k-th meeting of b_i and b_j corresponds to the intersection of the trajectory of the copy $b_i^{(0)}$ of baton b_i with the copy $b_j^{(k-1)}$ of baton b_j. As their initial distance on L equals $d + k - 1$, this meeting occurs at time $t^* = \frac{d+k-1}{|v_i-v_j|}$ in both systems $\mathcal{S}$ and $\mathcal{S}_c$. If $i > j$ we have $d = 1 - (p_j - p_i)$ and the k-th meeting of b_i and b_j corresponds to the intersection of the trajectories of $b_i^{(0)}$ and $b_j^{(k)}$, which happens at time $t^* = \frac{d+k}{|v_i-v_j|}$ in both systems $\mathcal{S}$ and $\mathcal{S}_c$. ∎

Fig. 2 illustrates Lemma 1. The walk of robot r_1 is represented by a thick polyline to illustrate how the walk of a robot is affected in a translation of a system.

Lemma 2. *Let $\mathcal{S}$ be a system of robots and $\mathcal{S}_c$ any of its translations. If t_i is the time of the i-th collision of robot r_q in $\mathcal{S}$, and t'_i the time of the i-th collision of robot r_q in $\mathcal{S}_c$, then $t_i = t'_i$ for $i \geq 1$, where $t_0 = t'_0 = 0$. Moreover, if $v(t_i)$ is the velocity of robot r_q at time t_i, then the velocity of r_q at time t'_i is $v(t_i) - c$.*

Proof. Assume the lemma holds for i, so at time t_i robot r_q obtains some baton b_j in $\mathcal{S}$, while at the same time r_q obtains the corresponding baton b'_j in $\mathcal{S}_c$. Let t_{i+1} denote the first time moment after t_i when baton b_j meets another baton in $\mathcal{S}$, say b_k. By Lemma 1 at the same time t_{i+1} baton b'_j meets b'_k in $\mathcal{S}_c$. As $v'_k = v_k - c$ the claim follows. ∎

We show below that every robot, by each of its collisions, acquires information about the initial position (relative to its own initial position) and initial velocity

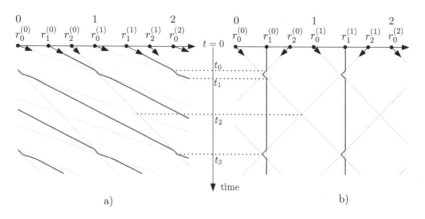

Fig. 2. $a)$ depicts a system of three robots r_0, r_1, r_2 whose velocities are $3, 2, 1$ respectively, $b)$ depicts a translation of $a)$ with new velocities $1, 0, -1$ (after subtracting 2 to each original velocity). Notice that the time at which each collision takes place does not get affected.

of some other robot of the system. We show later that if S is feasible, at some time moment the collision revealing the position and velocity of any other robot will eventually arise. However it is worth noting that up to that time moment, some collisions revealing the positions of the same robot may arise several times. We assume that, at time $t = 0$ each robot learns about its initial velocity.

Lemma 3. *Consider the collisions obtained by robot r_q deployed at its initial position p_q in the ring. Suppose that the last collision of robot r_q, at time t_i revealed some robot r_s, of initial velocity v_s and initial position at counterclockwise distance d_s from p_q, i.e. $d_s = d^{(+)}(p_q, p_s)$. Further assume that, at time t_{i+1}, robot r_q collides obtaining velocity v_t. Then there exists in S a robot r_t with initial velocity v_t such that $d^{(+)}(p_q, p_t) = ((v_s - v_t)t_{i+1} + d_s) \mod 1$.*

Proof. Between time t_i and t_{i+1} robot r_q moves with velocity v_s so we may assume that it holds baton b_s. The time of collision t_{i+1} corresponds to the time of intersection of the trajectory of some copy $b_s^{(j)}$ of this baton with the trajectory of some copy $b_t^{(k)}$ of the baton moving with velocity v_t. The absolute distance in L between the starting positions $b_s^{(j)}(0)$ and $b_t^{(k)}(0)$ equals $|v_s - v_t|t_{i+1}$. Therefore $d^{(+)}(p_s, p_t) = (v_s - v_t)t_{i+1} \mod 1$. Since by the assumption of the Lemma $d^{(+)}(p_q, p_s) = d_s$, we have $d^{(+)}(p_q, p_t) = (d^{(+)}(p_q, p_s) + d^{(+)}(p_s, p_t)) \mod 1 = ((v_s - v_t)t_{i+1} + d_s) \mod 1$. ∎

It follows from Lemma 3 that for a robot to figure out the starting position of every other robot it should acquire every velocity of the system in a finite amount of time. Lemma 3 provides the core of an algorithm for robots to report the starting position of every robot. We describe such an algorithm later on. The next lemma is an immediate consequence of Lemma 2 and Lemma 3.

Lemma 4. *For any system $\mathcal{S}$ and its translation $\mathcal{S}_c$, the position discovery problem is solvable for $\mathcal{S}$, if and only if it is solvable for $\mathcal{S}_c$.*

Given a fixed point ρ in the ring, which we call the *reference point* and $\mathcal{S}$ a system of robots, we associate with each robot r_i an integer counter c_i that we call *cycle counter*. A cycle counter c_i increases its value by one each time robot r_i traverses the reference point ρ in the counterclockwise direction and decreases by one when traversing ρ in clockwise direction. We denote by $c_i(t)$ the value of cycle counter c_i at time t. The initial value of c_i is set to 0, meaning $c_i(0) = 0$.

Let $D_i^{(+)}(t)$ denote the total distance that robot r_i travelled until time t in the counterclockwise direction, and $D_i^{(-)}(t)$ - the total distance travelled by r_i in the clockwise direction. Denote $D_i(t) = D_i^{(+)}(t) - D_i^{(-)}(t)$. The following observation is the immediate consequence of $\sum_{i=0}^{n-1} v_i = 0$ for system $\mathcal{S}_{\bar{v}}$:

Observation 1 *For any system $\mathcal{S}_{\bar{v}}$ at any time moment t we have $\sum_{i=0}^{n-1} D_i(t) = 0$.*

Lemma 5. *Consider the translation $\mathcal{S}_{\bar{v}}$ of any system $\mathcal{S}$. At any time t, no two cycle counters differ by more than 1, i.e $|c_i(t) - c_j(t)| \leq 1, 0 \leq i, j \leq n - 1$. Moreover, there should be a cycle counter $c_{k(t)}$ such that $c_{k(t)}(t) = 0$ for some $0 \leq k(t) \leq n - 1$.*

Proof. Let us observe that since robots can not overpass each other they always keep their initial cyclic order. Therefore, we can simulate the traversals on ρ by the robots by assuming that robots remain static while ρ is moving in one of the two directions along the ring; when ρ traverses a robot r_i in clockwise direction, counter c_i increases by one and decreases by one if ρ traverses r_i in counterclockwise direction.

We prove first that $|c_i(t)| \leq 1$, for each $0 \leq i \leq n - 1$. Indeed, suppose to the contrary, that $|c_i(t)| \geq 2$. Consider first the case when $c_i(t) \geq 2$. In such a case, r_i must have traversed point ρ at least two more times in the counterclockwise direction than in the clockwise one. Since the robots do not change their relative order around the ring, each other robot r_j must have traversed ρ at least once more in the counterclockwise direction than in the clockwise one. Hence $D_i^{(+)}(t) > D_i^{(-)}(t)$ for each $i = 0, \ldots, n - 1$. This contradicts $\sum_{i=0}^{n-1} D_i(t) = 0$. The argument for $c_i(t) \leq -2$ is symmetric.

It is easy to see that there are no two robots r_i, r_j, such that $c_i(t) = 1$ and $c_j(t) = -1$. Indeed in such a case these robots must have traversed point ρ in opposite directions which would have forced them to overpass - a contradiction.

Hence the values of all cycle counters at time t belong to the set $\{0, 1\}$ or to the set $\{0, -1\}$. However $c_i(t) \neq 0$, for all $i = 0, \ldots, n - 1$ would imply $D_i(t)$ be all positive or all negative, contradicting $\sum_{i=0}^{n-1} D_i(t) = 0$, which concludes the proof. ∎

We can conclude with the following Corollary:

Corollary 1. *For each robot r_i of $\mathcal{S}_{\bar{v}}$ and any time t we have $|D_i(t)| < 1$.*

Proof. Suppose to the contrary that $|D_i(t)| \geq 1$ or, by symmetry, that $D_i^{(+)}(t) - D_i^{(-)}(t) \geq 1$. In such a case, r_i at time t made a full counterclockwise tour around the ring. By putting the reference point $\rho = r_i(0)$, we notice that this forces each other robot r_j to have $c_j(t) \geq 1$, which contradicts Observation 1. ∎

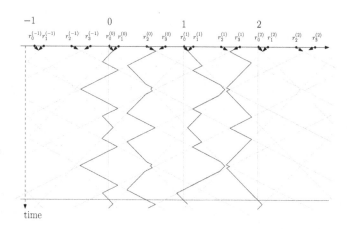

Fig. 3. An example of a system of robots where the average of the velocities is equal to 0. Notice that no robot completes more than one round in any direction.

Fig. 3 depicts a system of mobile robots, where the average of the velocities is equal to 0. Notice that every robot in the picture never completes more than one round along the ring in any direction. In the picture the movements of r_0 are shown with a thick polyline to illustrate this.

Theorem 1. *For any system of n mobile robots $\mathcal{S} = (\mathcal{P}, V)$ the localization problem is feasible if, and only if $v_i \neq \bar{v}$, for every $v_i \in V$. Moreover, if the problem is feasible, then each robot knows the positions and the velocities of other robots before time $T = \dfrac{2}{min_{0 \leq i \leq n-1}|v_i - \bar{v}|}$.*

Proof. By Lemma 4, it is sufficient to prove the theorem for $\mathcal{S}_{\bar{v}} = (\mathcal{P}, V_{\bar{v}})$.

We prove first, that if some robot r_i has the initial velocity $v_i = \bar{v} = 0$, then the system is not feasible. For the localization problem to be feasible in $\mathcal{S}_{\bar{v}}$, each robot must hold every baton at some time within some finite time interval $[0, T]$. We prove by contradiction that, if there is a baton b_q of velocity 0, then there exists a robot whose trajectory will not intersect the trajectory of b_q. Thus, such a robot would not obtain the information about the velocity and the position of robot r_q.

Consider cycle counters $c_j(t)$ for each robot r_j, $0 \leq j \leq n - 1$, where the reference point is set to $\rho = r_q(0)$. Because $v_q = 0$, there is always a robot of $\mathcal{S}_{\bar{v}}$ that remains motionless at point ρ. In other words, each robot of $\mathcal{S}_{\bar{v}}$, in order to hold baton b_q has to move to position ρ and collide with the current robot at that position. Observe that it is not possible that all robots arrive at point ρ from the same direction around the ring. Indeed, in such a case the robot velocities would be all positive or all negative implying $\bar{v} \neq 0$. Consequently, observe that there must exist two time moments t_1, t_2 and two consecutive robots r_i and r_{i+1} (where index $i + 1$ is taken modulo n) such that one of these robots visited ρ at time t_1 while walking in one direction and the other robot visited ρ at time t_2 while walking in the opposite direction. Notice that $t_1 \neq t_2$, since we supposed no three robots meeting simultaneously, and ρ coincides with a stationary robot.

Suppose, that r_i arrived at ρ at time t_1 while walking clockwise and r_{i+1} arrived at ρ at time t_2 while walking counterclockwise. As robots are arranged in the counterclockwise order around the ring it follows that within the time interval $[t_1, t_2]$ each other robot has to walk counterclockwise through ρ (or walk more times counterclockwise than clockwise) increasing its cycle counter.

Let $\mathcal{S}' = (\mathcal{P}', V_{\bar{v}})$, where $\mathcal{P}' = (r_0'(t_1), \ldots, r_{n-1}'(t_1))$ and let c_j' be the respective cycle counter of robot r_j' for every $0 \leq j \leq n - 1$. Notice that during the interval of time $[0, t_2 - t_1]$ every robot of $\mathcal{S}'$ behaves exactly the same way as does every robot in $\mathcal{S}_{\bar{v}}$ in the interval of time $[t_1, t_2]$. Thus, at time $t^* = t_2 - t_1$ we have $c_j'(t^*) > 0$ for all $0 \leq j \leq n - 1$ which contradicts Lemma 5. This implies that there is at least one robot that does not learn the initial position of all robots.

The cases where r_{i+1}, rather than r_i, arrived at ρ at time t_1 and when the directions of r_i and r_{i+1} while walking through ρ are reversed, are symmetric.

Suppose now that no robot has the initial velocity $\bar{v} = 0$. Consider any robot r_i and the interval $I_2 = [r_i(0) - 1, r_i(0) + 1]$ of the infinite line L. By Corollary 1 robot r_i never leaves interval I_2 during its movement, hence its trajectory is bound to the vertical strip of width 2 (cf. Fig. 3). Consider any baton b_j. Suppose, by symmetry, that $v_j > 0$. Take the trajectory of a copy of baton b_j which origins from the left half of I_2, i.e. from the segment $[r_i(0) - 1, r_i(0)]$. This trajectory will go across the vertical strip of width 2 enclosing I_2 and leave it before time $\frac{2}{|v_j|}$, forcing the meeting of robot r_i and baton b_j. If $v_j < 0$ we need to take a copy of b_j starting at the right half of I_2, i.e. in $[r_i(0), r_i(0) + 1]$ and the argument is the same. The time of $\frac{2}{|v_j|}$ is maximized for j minimizing $|v_j|$. ∎

An example of an infeasible robot configuration is shown in Fig. 2, in which robot r_0 never learns the initial position of robot r_1.

4 The Localization Algorithm

In this section we present an algorithm to solve the localization problem. The algorithm is based on Lemma 3. According to Theorem 1, if robots have knowledge of the velocities of other robots on the ring, it is possible for them to detect the infeasibility of the system without even starting to move, or stop it when the variable *clock* reaches the value of $\frac{2}{min_{0 \leq i \leq n-1}|v_i - \bar{v}|}$. Otherwise, the algorithm is designed to run indefinitely. However, we can also assume that a central authority, perhaps having the knowledge of the velocities of the robots in the system, may modify robot's signal variable *move* to halt the execution. The present algorithm may report the position of the same robot more than once; this may be clearly avoided providing robots with linear-size memory to recall all previously output robots.

The main theorem ensures that at this time all robots have discovered all the initial positions if the system is feasible.

In algorithm RingLocalization we assume that a robot has at any time immediate access to its clock as well as to the information of its current velocity through the variables *clock* and *velocity*, respectively. So the value of these variables can not be modified by the robot and they get updated instantaneously as a collision happens. We can assume that the values of these variables correspond to the readings of robots' sensors. A robot uses auxiliary variables, namely *old_velocity* and *pos* for recalling the position and the velocity of the robot detected through its last collision.

Algorithm RingLocalization;
1.   **var** *pos* ← 0, *old_velocity* ← *velocity* : **real;** *move* ← **true** : **boolean;**
2.   **reset** *clock* to 0;
3.   **while** *move* **do**
4.   **walk until** collision;
5.   *pos* ← ((*velocity* − *old_velocity*) · *clock* + *pos*) mod 1;
6.   **output** ("Robot of velocity" *velocity* "detected at position" *pos*);
7.   *old_velocity* ← *velocity*;

Since variable *pos* clearly keeps track of $d_s = d^{(+)}(p_q, p_s)$, Theorem 1 and Lemma 3 imply the following result.

Theorem 2. *Let* $\mathcal{S} = (\mathcal{P}, V)$ *be a system of robots. Suppose that no robot has initial velocity* $\bar{v}$, *meaning* $v_i \neq \bar{v}$ *for all* $v_i \in V$, *and that the algorithm* RingLocalization *is executed by each robot for time* $\frac{2}{min_{0 \leq i \leq n-1}|v_i - \bar{v}|}$. *Then, every robot correctly reports the initial positions and directions of all robots on the ring with respect to its initial position.*

5   Conclusions

We characterized configurations of all feasible systems. Observe that without the knowledge of velocities, even if the number of robots in the system is known, it is impossible for a robot to decide at any time if the system is infeasible. Indeed, by Theorem 1 to any system $\mathcal{S}$ it is possible to add a new robot of velocity equal to the average $\bar{v}$, making $\mathcal{S}$ infeasible for at least some robot r_i of $\mathcal{S}$. Consequently, given arbitrarily large time T^* it is also possible to add to $\mathcal{S}$ a robot of velocity close to $\bar{v}$, so the system stays feasible but not within the time bound of T^*

Notice also, that already for two robots at small distance ϵ, starting in opposite directions with small velocities v_1 and $v_2 = -v_1$ it takes time $\frac{1-\epsilon}{2v_1}$ to get the first collision, so the worst-case time of localization algorithm proportional to $\frac{1}{min_{0 \leq i \leq n-1}|v_i - \bar{v}|}$ is unavoidable.

References

1. Czyzowicz, J., Gąsieniec, L., Kosowski, A., Kranakis, E., Ponce, O.M., Pacheco, E.: Position discovery for a system of bouncing robots. In: Aguilera, M.K. (ed.) DISC 2012. LNCS, vol. 7611, pp. 341–355. Springer, Heidelberg (2012)
2. Friedetzky, T., Gąsieniec, L., Gorry, T., Martin, R.: Observe and remain silent (communication-less agent location discovery). In: Rovan, B., Sassone, V., Widmayer, P. (eds.) MFCS 2012. LNCS, vol. 7464, pp. 407–418. Springer, Heidelberg (2012)
3. Kranakis, E., Krizanc, D., Markou, E.: The mobile agent rendezvous problem in the ring. Synthesis Lectures on Distributed Computing Theory 1(1), 1–122 (2010)
4. Suzuki, I., Yamashita, M.: Distributed anonymous mobile robots: Formation of geometric patterns. SIAM J. Comput. 28(4), 1347–1363 (1999)
5. Das, S., Flocchini, P., Santoro, N., Yamashita, M.: On the computational power of oblivious robots: forming a series of geometric patterns. In: PODC, pp. 267–276 (2010)
6. Angluin, D., Aspnes, J., Diamadi, Z., Fischer, M.J., Peralta, R.: Computation in networks of passively mobile finite-state sensors. Distributed Computing 18(4), 235–253 (2006)
7. Angluin, D., Aspnes, J., Eisenstat, D.: Stably computable predicates are semilinear. In: PODC, pp. 292–299 (2006)
8. Czyzowicz, J., Gasieniec, L., Kosowski, A., Kranakis, E.: Boundary patrolling by mobile agents with distinct maximal speeds. Algorithms–ESA 2011, 701–712 (2011)
9. Cohen, R., Peleg, D.: Local spreading algorithms for autonomous robot systems. Theor. Comput. Sci. 399(1-2), 71–82 (2008)
10. Murphy, T.: Dynamics of hard rods in one dimension. Journal of Statistical Physics 74(3), 889–901 (1994)
11. Sevryuk, M.: Estimate of the number of collisions of n elastic particles on a line. Theoretical and Mathematical Physics 96(1), 818–826 (1993)
12. Tonks, L.: The complete equation of state of one, two and three-dimensional gases of hard elastic spheres. Physical Review 50(10), 955 (1936)
13. Wylie, J., Yang, R., Zhang, Q.: Periodic orbits of inelastic particles on a ring. Physical Review E 86(2), 026601 (2012)
14. Jepsen, D.: Dynamics of a simple many-body system of hard rods. Journal of Mathematical Physics 6, 405 (1965)
15. Cooley, B., Newton, P.: Random number generation from chaotic impact collisions. Regular and Chaotic Dynamics 9(3), 199–212 (2004)
16. Cooley, B., Newton, P.: Iterated impact dynamics of n-beads on a ring. SIAM Rev. 47(2), 273–300 (2005)
17. Susca, S., Bullo, F.: Synchronization of beads on a ring. In: 46th IEEE Conference on Decision and Control, pp. 4845–4850 (2007)
18. Wang, H., Guo, Y.: Synchronization on a segment without localization: algorithm and applications. In: International Conference on Intelligent Robots and Systems, IROS, pp. 3441–3446 (2009)
19. Gregory, R.: Classical mechanics. Cambridge University Press (2006)

Fast Collaborative Graph Exploration[*]

Dariusz Dereniowski[1], Yann Disser[2], Adrian Kosowski[3],
Dominik Pająk[3], and Przemysław Uznański[3]

[1] Gdańsk University of Technology, Poland
[2] TU Berlin, Germany
[3] CEPAGE project, Inria Bordeaux Sud-Ouest, France

Abstract. We study the following scenario of online graph exploration. A team of k agents is initially located at a distinguished vertex r of an undirected graph. At every time step, each agent can traverse an edge of the graph. All vertices have unique identifiers, and upon entering a vertex, an agent obtains the list of identifiers of all its neighbors. We ask how many time steps are required to complete exploration, i.e., to make sure that every vertex has been visited by some agent.

We consider two communication models: one in which all agents have global knowledge of the state of the exploration, and one in which agents may only exchange information when simultaneously located at the same vertex. As our main result, we provide the first strategy which performs exploration of a graph with n vertices at a distance of at most D from r in time $O(D)$, using a team of agents of polynomial size $k = Dn^{1+\epsilon} < n^{2+\epsilon}$, for any $\epsilon > 0$. Our strategy works in the local communication model, without knowledge of global parameters such as n or D.

We also obtain almost-tight bounds on the asymptotic relation between exploration time and team size, for large k. For any constant $c > 1$, we show that in the global communication model, a team of $k = Dn^c$ agents can always complete exploration in $D(1 + \frac{1}{c-1} + o(1))$ time steps, whereas at least $D(1 + \frac{1}{c} - o(1))$ steps are sometimes required. In the local communication model, $D(1 + \frac{2}{c-1} + o(1))$ steps always suffice to complete exploration, and at least $D(1 + \frac{2}{c} - o(1))$ steps are sometimes required. This shows a clear separation between the global and local communication models.

1 Introduction

Exploring an undirected graph-like environment is relatively straight-forward for a single agent. Assuming the agent is able to distinguish which neighboring vertices it has previously visited, there is no better systematic traversal strategy than a simple depth-first search of the graph, which takes $2(n-1)$ moves in total for a graph with n vertices. The situation becomes more interesting if multiple agents want to collectively explore the graph starting from a common location. If arbitrarily many agents may be used, then

[*] This work was initiated while A. Kosowski was visiting Y. Disser at ETH Zurich. Supported by ANR project DISPLEXITY and by NCN under contract DEC-2011/02/A/ST6/00201. The authors are grateful to Shantanu Das for valuable discussions and comments on the manuscript. The full version of this paper is available online at: http://hal.inria.fr/hal-00802308.

F.V. Fomin et al. (Eds.): ICALP 2013, Part II, LNCS 7966, pp. 520–532, 2013.
© Springer-Verlag Berlin Heidelberg 2013

we can generously send n^D agents through the graph, where D is the distance from the starting vertex to the most distant vertex of the graph. At each step, we spread out the agents located at each node (almost) evenly among all the neighbors of the current vertex, and thus explore the graph in D steps.

While the cases with one agent and arbitrarily many agents are both easy to understand, it is much harder to analyze the spectrum in between these two extremes. Of course, we would like to explore graphs in as few steps as possible (i.e., close to D), while using a team of as few agents as possible. In this paper we study this trade-off between exploration time and team size. A trivial lower bound on the number of steps required for exploration with k agents is $\Omega(D + n/k)$: for example, in a tree, some agent has to reach the most distant node from r, and each edge of the tree has to be traversed by some agent. We look at the case of larger groups of agents, for which D is the dominant factor in this lower bound. This complements previous research on the topic for trees [6,8] and grids [17], which usually focused on the case of small groups of agents (when n/k is dominant).

Another important issue when considering collaborating agents concerns the model that is assumed for the communication between agents. We need to allow communication to a certain degree, as otherwise there is no benefit to using multiple agents for exploration [8]. We may, for example, allow agents to freely communicate with each other, independent of their whereabouts, or we may restrict the exchange of information to agents located at the same location. This paper also studies this tradeoff between global and local communication.

The Collaborative Online Graph Exploration Problem. We are given a graph $G = (V, E)$ rooted at some vertex r. The number of vertices of the graph is bounded by n. Initially, a set $\mathcal{A}$ of k agents is located at r. We assume that vertices have unique identifiers that admit a total ordering. In each step, an agent visiting vertex v receives a complete list of the identifiers of the nodes in $N(v)$, where $N(v)$ is the neighborhood of v. Time is discretized into steps, and in each step, an agent can either stay at its current vertex or slide along an edge to a neighboring vertex. Agents have unique identifiers, which allows agents located at the same node and having the same exploration history to differentiate their actions. We do not explicitly bound the memory resources of agents, enabling them in particular to construct a map of the previously visited subgraph, and to remember this information between time steps. An *exploration strategy* for G is a sequence of moves performed independently by the agents. A strategy explores the graph G in t time steps if for all $v \in V$ there exists time step $s \leq t$ and an agent $g \in \mathcal{A}$, such that g is located at v in step s. Our goal is to find an exploration strategy which minimizes the time it takes the explore a graph in the worst case, with respect to the shortest path distance D from r to the vertex furthest from r in the graph.

We distinguish between two communication models. In exploration *with global communication* we assume that, at the end of each step s, all agents have complete knowledge of the explored subgraph. In particular, in step s all agents know the number of edges incident to each vertex of the explored subgraph which lead to unexplored vertices, but they have no information on any subgraph consisting of unexplored vertices. In exploration *with local communication* two agents can exchange information only if they occupy the same vertex. Thus, each agent g has its own view on which vertices

Table 1. Our bounds for the time required to explore general graphs with using Dn^c agents. The same upper and lower bounds hold for trees. The lower bounds use graphs with $D = n^{o(1)}$.

Communication Model	Upper bound	Lower bound
Global communication:	$D \cdot (1 + \frac{1}{c-1} + o(1))$ Thm. 3	$D \cdot (1 + \frac{1}{c} - o(1))$ Thm. 5
Local communication :	$D \cdot (1 + \frac{2}{c-1} + o(1))$ Thm. 3	$D \cdot (1 + \frac{2}{c} - o(1))$ Thm. 5

were explored so far, constructed based only the knowledge that originates from the agent's own observations and from other agents that it has met.

Our results. Our main contribution is an exploration strategy for a team of polynomial size to explore graphs in an asymptotically optimal number of steps. More precisely, for any $\epsilon > 0$, the strategy can operate with $Dn^{1+\epsilon} < n^{2+\epsilon}$ agents and takes time $O(D)$. It works even under the local communication model and without prior knowledge of n or D.

We first restrict ourselves to the exploration of trees (Section 2). We show that with global communication trees can be explored in time $D \cdot (1 + 1/(c - 1) + o(1))$ for any $c > 1$, using a team of Dn^c agents. Our approach can be adapted to show that with local communication trees can be explored in time $D \cdot (1 + 2/(c - 1) + o(1))$ for any $c > 1$, using the same number of agents. We then carry the results for trees over to the exploration of general graphs (Section 3). We obtain precisely the same asymptotic bounds for the number of time steps needed to explore graphs with Dn^c agents as for the case of trees, under both communication models.

Finally, we provide lower bounds for collaborative graph exploration that almost match our positive results (Section 4). More precisely, we show that, in the worst case and for any $c > 1$, exploring a graph with Dn^c agents takes at least $D \cdot (1 + 1/c - o(1))$ time steps in the global communication model, and at least $D \cdot (1 + 2/c - o(1))$ time steps in the local communication model. Table 1 summarizes our upper and corresponding lower bounds.

Related Work. Collaborative online graph exploration has been intensively studied for the special case of trees. In [8], a strategy is given which explores any tree with a team of k agents in $O(D + n/\log k)$ time steps, using a communication model with whiteboards at each vertex that can be used to exchange information. This corresponds to a competitive ratio of $O(k/\log k)$ with respect to the optimum exploration time of $\Theta(D + n/k)$ when the graph is known. In [13] authors show that the competitive ratio of the strategy presented in [8] is precisely $k/\log k$. Another DFS-based algorithm, given in [2], has an exploration time of $O(n/k + D^{k-1})$ time steps, which provides an improvement only for graphs of small diameter and small teams of agents, $k = O(\log_D n)$. For a special subclass of trees called sparse trees, [6] introduces online strategies with a competitive ratio of $O(D^{1-1/p})$, where p is the density of the tree as defined in that work. The best currently known lower bound is much lower: in [7], it is shown that any deterministic exploration strategy with $k < \sqrt{n}$ has a competitive ratio of $\Omega(\log k/\log\log k)$, even in the global communication model. A stronger lower bound of $\Omega(k/\log k)$ holds for so-called greedy algorithms [13]. Both for deterministic and randomized strategies, the competitive ratio is known to be at least $2 - 1/k$, when $k < \sqrt{n}$ [8]. None of these lower bounds concern larger teams of agents. In [16] a lower bound of $\Omega(D^{1/(2c+1)})$ on competitive ratio is shown to hold for a team of $k = n^c$ agents, but this lower bound only concerns so-called rebalancing algorithms which keep all agents at the same height in the tree throughout the exploration process.

The same model for online exploration is studied in [17], where a strategy is proposed for exploring graphs which can be represented as a $D \times D$ grid with a certain number of disjoint rectangular holes. The authors show that such graphs can be explored with a team of k agents in time $O(D \log^2 D + n \log D/k)$, i.e., with a competitive ratio of $O(\log^2 D)$. By adapting the approach for trees from [7], they also show lower bounds on the competitive ratio in this class of graphs of $\Omega(\log k/\log\log k)$ for deterministic strategies and $\Omega(\sqrt{\log k}/\log\log k)$ for randomized strategies. These lower bounds also hold in the global communication model.

Collaborative exploration has also been studied with different optimization objectives. An exploration strategy for trees with global communication is given in [7], achieving a competitive ratio of $(4 - 2/k)$ for the objective of minimizing the maximum number of edges traversed by an agent. In [5] a corresponding lower bound of $3/2$ is provided.

Our problem can be seen as an online version of the k Traveling Salesmen Problem (k-TSP) [9]. Online variants of TSP (for a single agent) have been studied in various contexts. For example, the geometric setting of exploring grid graphs with and without holes is considered by [10,11,14,15,17], where a variety of competitive algorithms with constant competitive ratios is provided. A related setting is studied in [4], where an agent has to explore a graph while being attached to the starting point by a rope of restricted length. A similar setting is considered in [1], in which each agent has to return regularly to the starting point, for example for refueling. Online exploration of polygons is considered in [3,12].

2   Tree Exploration

We start our considerations by designing exploration strategies for the special case when the explored graph is a tree T rooted at a vertex r. For any exploration strategy, the set of all encountered vertices (i.e., all visited vertices and their neighbors) at the beginning of step $s = 1, 2, 3, \ldots$ forms a connected subtree of T, rooted at r and denoted by $T^{(s)}$. In particular, $T^{(1)}$ is the vertex r together with its children, which have not yet been visited. For $v \in V(T)$ we write $T^{(s)}(v)$ to denote the subtree of $T^{(s)}$ rooted at v. We denote by $L(T^{(s)}, v)$ the number of leaves of the tree $T^{(s)}(v)$. Note that $L(T^{(s)}, v) \le L(T^{(s+1)}, v)$ because each leaf in $T^{(s)}(v)$ is either a leaf of the tree $T^{(s+1)}$ or the root of a subtree containing at least one vertex. If v is an unencountered vertex at the beginning of step s, i.e., its parent was not yet visited, we define $L(T^{(s)}, v) = 1$.

2.1 Tree Exploration with Global Communication

We are ready to give the procedure TEG (*Tree Exploration with Global Communication*). The pseudocode uses the command "move$^{(s)}$", describing the move to be performed by each agent, specifying the destination at which the agent appears at the start of time step $s + 1$. Since the agents can communicate globally, the procedure can centrally coordinate the movements of each agent. For simplicity we assume that x agents spawn in r in each time step, for some given value of x. Then, the total number of agents used after l steps is simply lx.

Procedure TEG (tree T with root r, integer x) **at time step** s:
   Place x new agents at r.
   for each $v \in V(T^{(s)})$ which is not a leaf **do**: { determine moves of the agents located at v }
      Let $\mathcal{A}_v^{(s)}$ be the set of agents currently located at v.
      Denote by $v_1, v_2, \ldots, v_d$ the set of children of v.
      Let $i^* := \arg\max_i\{L(T^{(s)}, v_i)\}$. { v_{i^*} is the child of v with the largest value of L }
      Partition $\mathcal{A}_v^{(s)}$ into disjoint sets $\mathcal{A}_{v_1}, \mathcal{A}_{v_2}, \ldots, \mathcal{A}_{v_d}$, such that:
      (i) $|\mathcal{A}_{v_i}| = \left\lfloor \frac{|\mathcal{A}_v^{(s)}| \cdot L(T^{(s)}, v_i)}{L(T^{(s)}, v)} \right\rfloor$, for $i \in \{1, 2, \ldots, d\} \setminus \{i^*\}$,
      (ii) $|\mathcal{A}_{v_{i^*}}| = |\mathcal{A}_v^{(s)}| - \sum_{i \in \{1,2,\ldots,d\} \setminus \{i^*\}} |\mathcal{A}_{v_i}|$.

      for each $i \in \{1, 2, \ldots, d\}$ **do for each** agent $g \in \mathcal{A}_{v_i}$ **do move**$^{(s)}$ g to vertex v_i.
   end for
end procedure TEG.

    The following lemma provides a characterization of the tradeoff between exploration time and the number of agents x released at every round in procedure TEG. In the following, all logarithms are with base 2 unless a different base is explicitly given.

Lemma 1. *In the global communication model, procedure* TEG *with parameter x explores any rooted tree T in at most* $D \cdot (1 + \frac{1}{\log_n x - 1 - \log_n (2 \log x)})$ *time steps, for* $x > 6(n \log n + 1)$.

Proof. Fix any leaf f of the tree T. We want to prove that procedure TEG visits the leaf f after at most $D \cdot (1 + \frac{1}{\log_n x - 1 - \log_n (2 \log x)})$ time steps. Take the path $\mathcal{F} = (f_0, f_1, f_2, \ldots, f_{D_f})$ from r to f in T, where $r = f_0, f = f_{D_f}$, and $D_f \leq D$. We define the *wave* of agents w_s starting from r at time s and traversing the path $\mathcal{F}$ as the maximum sequence of the non-empty sets of agents which leave the root in step s and traverse edges of $\mathcal{F}$ in successive time steps, i.e., $w_s = (\mathcal{A}_{f_0}^{(s)}, \mathcal{A}_{f_1}^{(s+1)}, \ldots)$, where we use the notation from procedure TEG. The size of wave w_s in step $s+t$ is defined to be $|\mathcal{A}_{f_t}^{(s)}|$, i.e., the number of exploring agents located at vertex f_t at the beginning of time step $s + t$; initially, every wave has size $|\mathcal{A}_{f_0}^{(s)}| = x$. Note that each agent in $\mathcal{A}_{f_i}^{(s+i)}, 0 \leq i < D_f$, is located at r at the start of time step s. We denote the number of leaves in the subtree of $T^{(i)}$ rooted at f_j by $\lambda_j^{(i)} = L(T^{(i)}, f_j)$. Recall that if f_j is not yet discovered in step i, by definition of the function L, we have $\lambda_j^{(i)} = 1$. In general, $1 \leq \lambda_j^{(i)} \leq n$. We define

$$\alpha_i = \frac{x}{2} \frac{\lambda_1^{(i)}}{\lambda_0^{(i)}} \frac{\lambda_2^{(i+1)}}{\lambda_1^{(i+1)}} \cdots \frac{\lambda_{D_f}^{(i+D_f-1)}}{\lambda_{D_f-1}^{(i+D_f-1)}},$$

and define α_i^* as the number of agents of the i-th wave that reach the leaf f, i.e., the size of the i-th wave in step $i + D_f$. If $\alpha_1^* = \alpha_2^* = \cdots = \alpha_{i-1}^* = 0$ and $\alpha_i^* \geq 1$ for some time step i, then we say that leaf f is explored by the i-th wave. Before we proceed with the analysis, we show the following auxiliary claim.

Claim (*). Let i be a time step for which $\alpha_i \geq \log x$. Then, $\alpha_i^* \geq \alpha_i$, and thus α_i is a lower bound on the number of agents reaching f in step $i + D_f$.

Proof (of the claim). We define $c_j = \lambda_{j+1}^{(i+j)}/\lambda_j^{(i+j)}$ for $j = 0, \ldots, D_f - 1$. For $i \geq 1$ we have $\alpha_i = x/2 \prod_{j=0}^{D_f-1} c_j$. Since $c_j \leq 1$ for all j and since $\alpha_i \geq \log x$, there exist at most $\log x$ different j such that $c_j \leq 1/2$. Denote the set of all such j by $\mathcal{J}$, with $|\mathcal{J}| \leq \log x$. Also, denote the size of wave w_i in step $i + s$ by a_s (for $s = 0, 1, 2, \ldots$), in particular $a_0 = x$.

Consider some index s for which $c_s > 1/2$. We have $\lambda_{s+1}^{(i+s)}/\lambda_s^{(i+s)} > 1/2$, thus more than half of all leaves of the tree $T^{(i+s)}(f_s)$ also belong to the tree $T^{(i+s)}(f_{s+1})$. But then, in time step $i + s + 1$, agents are sent from f_s to f_{s+1} according to the definition in expression (ii) in procedure TEG. Thus, we can lower-bound the size of wave w_i in step $i + s + 1$ by $a_{s+1} \geq a_s c_s$. Otherwise, if $c_s \leq 1/2$ (i.e., if $s \in \mathcal{J}$), then agents are sent according the definition in expression (i) in procedure TEG, and hence $a_{s+1} \geq \lfloor a_s c_s \rfloor$. Note that these bounds also hold if there are no agents left in the wave, i.e., $a_s = a_{s+1} = 0$. Thus, we have:

$$a_{s+1} \geq a_s c_s - \delta_s, \quad \text{where } \delta_s = \begin{cases} 1, & \text{if } s \in \mathcal{J}, \\ 0, & \text{otherwise.} \end{cases}$$

In this way we expand the expression for $\alpha_i^* = a_{D_f}$:

$$\alpha_i^* = a_{D_f} \geq a_{D_f-1} c_{D_f-1} - \delta_{D_f-1} \geq \ldots \geq (\ldots((a_0 c_0 - \delta_0)c_1 - \delta_1)c_2 - \ldots)c_{D_f-1} - \delta_{D_f-1} =$$

$$= x \prod_{j=0}^{D_f-1} c_j - \sum_{j=0}^{D_f-1} \left(\delta_j \prod_{p=j+1}^{D_f-1} c_j \right) \geq 2\alpha_i - \sum_{j=0}^{D_f-1} \delta_j \geq 2\alpha_i - |\mathcal{J}| \geq 2\alpha_i - \log x.$$

Since by assumption $\alpha_i \geq \log x$, we obtain $\alpha_i^* \geq 2\alpha_i - \log x \geq \alpha_i$, which completes the proof of the claim.

We now show that if the number of waves a in the execution of the procedure is sufficiently large, then there exists an index $i \leq a$, such that $\alpha_i \geq \log x$. Thus, taking into account Claim (*), leaf f is explored at the latest by the a-th wave.

Take a waves and consider the product $\prod_{i=1}^a \alpha_i$. Note that $\lambda_{D_f}^{(s)} = 1$ for every s. Thus, simplifying the product of all α_i by shortening repeating terms in numerators and denominators, and using $1 \leq \lambda_j^{(i)} \leq n$, we get

$$\prod_{i=1}^a \alpha_i = (\tfrac{x}{2})^a \prod_{i=1}^a \prod_{j=0}^{D_f-1} \frac{\lambda_{j+1}^{(i+j)}}{\lambda_j^{(i+j)}} = (\tfrac{x}{2})^a \frac{\prod_{i=1}^a \prod_{j=0}^{D_f-1} \lambda_{j+1}^{(i+j)}}{\prod_{i=1}^a \prod_{j=0}^{D_f-1} \lambda_j^{(i+j)}} = (\tfrac{x}{2})^a \frac{\prod_{i'=0}^{a-1} \prod_{j'=1}^{D_f} \lambda_{j'}^{(i'+j')}}{\prod_{i=1}^a \prod_{j=0}^{D_f-1} \lambda_j^{(i+j)}} =$$

$$= (\tfrac{x}{2})^a \frac{\left(\prod_{j'=1}^{D_f} \lambda_{j'}^{(j')} \right) \left(\prod_{i'=1}^{a-1} \prod_{j'=1}^{D_f-1} \lambda_{j'}^{(i'+j')} \right) \left(\prod_{i'=1}^{a-1} \lambda_{D_f}^{(i'+D_f)} \right)}{\left(\prod_{i=1}^a \lambda_0^{(i)} \right) \left(\prod_{i=1}^{a-1} \prod_{j=1}^{D_f-1} \lambda_j^{(i+j)} \right) \left(\prod_{j=1}^{D_f-1} \lambda_j^{(a+j)} \right)} \geq \frac{(x/2)^a}{n^a n^{D_f-1}} \geq \frac{(x/2)^a}{n^{a+D}}.$$

$$(1)$$

We want to find a, such that $\prod_{i=1}^{a} \alpha_i \geq (\log x)^a$. Taking into account (1), it is sufficient to find a satisfying

$$\frac{(x/2)^a}{n^{a+D}} \geq (\log x)^a,$$

which for sufficiently large x (we take $x > 6(n \log n + 1)$) can be equivalently transformed by taking logarithms and arithmetic to the form:

$$a \geq \frac{D}{\log_n x - 1 - \log_n (2 \log x)}.$$

Hence, for $a = \lceil \frac{D}{\log_n x - 1 - \log_n (2 \log x)} \rceil$, we have that there exists some i such that $\alpha_i \geq \log x$. For the same i we have $\alpha_i^* \geq \log x$, by Claim (*). Thus, a waves are sufficient to explore the path $\mathcal{F}$. This analysis can be done for any leaf f, thus it is enough to send a waves in order to explore the graph G. Considering that a wave w_i is completed by the end of step $D+i-1$, the exploration takes at most $D+a-1$ time steps in total. Thus, the exploration takes at most $D \cdot (1 + \frac{1}{\log_n x - 1 - \log_n (2 \log x)})$ time steps. $\square$

We remark that in the above Lemma, the total number of agents used throughout all steps of procedure TEG is $x \cdot D \cdot (1 + \frac{1}{\log_n x - 1 - \log_n (2 \log x)})$. For any $c > 1$, by appropriately setting $x = \Theta(n^c)$, we directly obtain the following theorem.

Theorem 1. *For any fixed $c > 1$ and known n, the online tree exploration problem with global communication can be solved in at most $D \cdot \left(1 + \frac{1}{c-1} + o(1)\right)$ time steps using a team of $k \geq Dn^c$ agents.* $\square$

2.2   Tree Exploration with Local Communication

In this section we propose a strategy for tree exploration under the local communication model. In the implementation of the algorithm we assume that whenever two agents meet, they exchange all information they possess about the tree. Thus, after the meeting, the knowledge about the explored vertices and their neighborhoods, is a union of the knowledge of the two agents before the meeting. Since agents exchange information only if they occupy the same vertex, at any time s, the explored tree $T^{(s)}$ may only partially be known to each agent, with different agents possibly knowing different subtrees of $T^{(s)}$.

In order to obtain a procedure for the local communication model, we modify procedure TEG from the previous section. Observe that in procedure TEG, agents never move towards the root of the tree, hence, in the local communication model, agents cannot exchange information with other agents located closer to the root. The new strategy is given by the procedure TEL (*Tree Exploration with Local Communication*).

In procedure TEL, all agents are associated with a state flag which may be set either to the value "exploring" or "notifying". Agents in the "exploring" state act similarly as in global exploration, with the requirement that they always move to a vertex in groups of 2 or more agents. Every time a group of "exploring" agents visits a new vertex, it detaches two of its agents, changes their state to "notifying", and sends them back along the path leading back to the root. These agents notify every agent they encounter on their way about the discovery of the new vertices. Although information about the discovery may be delayed, in every step s, all agents at vertex v know the entire subtree $T^{(s')}(v)$ which was explored until some previous time step $s' \leq s$. The state flag also has a third state, "discarded", which is assigned to agents no longer used in the exploration process.

The formulation of procedure TEL is not given from the perspective of individual agents, however, based on its description, the decision on what move to make in the current step can be made by each individual agent. The correctness of the definition of the procedure relies on the subsequent lemma, which guarantees that for a certain value s' the tree $T^{(s')}(v)$ is known to all agents at v.

Procedure TEL (tree T with root r, integer x) **at time step** s:

Place x new agents at r in state "exploring".

for each $v \in V(T^{(s)})$ which is not a leaf **do**: { determine moves of the agents located at v }

    if $v \neq r$ **then for each** agent g at v in state "notifying" **do move**$^{(s)}$ g to the parent of v.

    if v contains at least two agents in state "exploring" **and** agents at v do not have

        information of any agent which visited v before step s **then**:

        { send two new notifying agents back to the root from newly explored vertex v }

        Select two agents g^*, g^{**} at v in state "exploring".

        Change state to "notifying" for agents g^* and g^{**}.

        move$^{(s)}$ g^* to the parent of v. { g^{**} will move to the parent one step later }

    end if

    Let $\mathcal{A}_v^{(s)}$ be the set of all remaining agents in state "exploring" located at v.

    Denote by $v_1, v_2, \ldots, v_d$ all children of v, and by δ the distance from r to v.

    $s' := \lfloor \frac{\delta + s}{2} \rfloor$. { s' is a time in the past such that $T^{(s')}(v)$ is known to the agents at v }

    Let $i^* := \arg \max_i \{ L(T^{(s')}, v_i) \}$. { v_{i^*} is the child of v with the largest value of L }

    Partition $\mathcal{A}_v^{(s)}$ into disjoint sets $\mathcal{A}_{v_1}, \mathcal{A}_{v_2}, \ldots, \mathcal{A}_{v_d}$, such that:

      (i) $|\mathcal{A}_{v_i}| = \left\lfloor \frac{|\mathcal{A}_v^{(s)}| \cdot L(T^{(s')}, v_i)}{L(T^{(s')}, v)} \right\rfloor$, for $i \in \{1, 2, \ldots, d\} \setminus \{i^*\}$,

      (ii) $|\mathcal{A}_{v_{i^*}}| = |\mathcal{A}_v^{(s)}| - \sum_{i \in \{1, 2, \ldots, d\} \setminus \{i^*\}} |\mathcal{A}_{v_i}|$.

    for each $i \in \{1, \ldots, d\}$ **do if** $|\mathcal{A}_{v_i}| \geq 2$ **then for each** agent $g \in \mathcal{A}_{v_i}$ **do move**$^{(s)}$ g to v_i.

    for each $i \in \{1, \ldots, d\}$ **do if** $|\mathcal{A}_{v_i}| = 1$ **then** change state to "discarded" for agent in $\mathcal{A}_{v_i}$.

end for

for each $v \in V(T^{(s)})$ which is a leaf **do move**$^{(s)}$ all agents located at v to the parent of v.

end procedure TEL.

Lemma 2. *Let T be a tree rooted at some vertex r and let v be a vertex with distance δ to r. After running procedure TEL until time step s, all agents which are located at vertex v at the start of time step s know the tree $T^{(s')}(v)$, for $s' = \lfloor \frac{\delta + s}{2} \rfloor$.*

(Some proofs are omitted from this extended abstract.)

Lemma 3. *In the local communication model, procedure TEL with parameter x explores any rooted tree T in at most $D \cdot \left(1 + \frac{2 + 1/\log n}{\log_n x - 1 - \log_n(4 \log x)}\right)$ time steps, for $x > 17(n \log n + 1)$.*

Proof (sketch). As in the proof of Lemma 1, we consider any leaf f and the path $\mathcal{F} = (f_0, f_1, \ldots, f_{D_f})$ from r to f. As before, we denote the number of leaves in the subtree of $T^{(i)}$ rooted at f_j by $\lambda_j^{(i)} = L(T^{(i)}, f_j)$. Recall that if f_j is not yet discovered in step i, we have $L(T^{(i)}, f_j) = 1$. We adopt the definition of a wave from Lemma 1. We define the values α_i differently, however, to take into account the fact that the procedure relies on a delayed exploration tree, and that some waves lose agents as a result of deploying

notifying agents: $\alpha_i = \dfrac{x}{4} \dfrac{\lambda_1^{\left(\left\lfloor \frac{i}{2} \right\rfloor\right)}}{\lambda_0^{\left(\left\lfloor \frac{i}{2} \right\rfloor\right)}} \dfrac{\lambda_2^{\left(\left\lfloor \frac{i}{2} \right\rfloor + 1\right)}}{\lambda_1^{\left(\left\lfloor \frac{i}{2} \right\rfloor + 1\right)}} \cdots \dfrac{\lambda_{D_f}^{\left(\left\lfloor \frac{i}{2} \right\rfloor + D_f - 1\right)}}{\lambda_{D_f-1}^{\left(\left\lfloor \frac{i}{2} \right\rfloor + D_f - 1\right)}}$.

We call a wave that discovered at least $\lceil \log x \rceil$ new nodes (or equivalently, a wave whose agents were the first to visit at least $\lceil \log x \rceil$ nodes of the tree) a *discovery wave*. Thus, there are at most $\left\lfloor \frac{D_f}{\lceil \log x \rceil} \right\rfloor \leq \left\lfloor \frac{D}{\log x} \right\rfloor$ discovery waves along the considered path. Observe that if a wave is not a discovery wave, then the number of notifying agents it sends out is at most $2 \log x$.

We define by α_i^* the number of agents of the i-th wave that reach leaf f. We first prove that the following analogue of Claim (*) from the proof of Lemma 1 holds for non-discovery waves (we leave out the details from this extended abstract).

Claim (**). Let i be a time step for which w_i is not a discovery wave and $\alpha_i \geq \log x$. Then, $\alpha_i^* \geq \alpha_i$, and thus α_i is a lower bound on the number of agents reaching f in step $i + D_f$.

Finally, we prove that if the number of waves a in the execution of the procedure is sufficiently large, i.e. $a \geq D \cdot \left(\frac{2 + 1/\log n}{\log_n x - 1 - \log_n (4 \log x)} \right)$, there exists an index $i \leq a$, such that wave w_i is not a discovery wave and $\alpha_i \geq \log x$. Exploration is then completed when the last wave reaches leaves, i.e. in $D + a - 1$ steps, which completes the proof. □

Acting as in the previous Subsection, from Lemma 3 we obtain a strategy for online exploration of trees in the model with local communication.

Theorem 2. *For any fixed $c > 1$, the online tree exploration problem can be solved in the model with local communication and knowledge of n using a team of $k \geq Dn^c$ agents in at most $D \left(1 + \frac{2}{c-1} + o(1)\right)$ time steps.* □

3   General Graph Exploration

In this section we develop strategies for exploration of general graphs, both with global communication and with local communication. These algorithms are obtained by modifying the tree-exploration procedures given in the previous section.

Given a graph $G = (V, E)$ with root vertex r, we call $P = (v_0, v_1, v_2, \ldots, v_m)$ with $r = v_0$, $v_i \in V$, and $\{v_i, v_{i+1}\} \in E$ a *walk* of length $\ell(P) = m$. Note that a walk may contain a vertex more than once. We introduce the notation $P[j]$ to denote v_j, i.e., the j-th vertex of P after the root, and $P[0, j]$ to denote the walk $(v_0, v_1, \ldots, v_j)$, for $j \leq m$. The last vertex of path P is denoted by $end(P) = P[\ell(P)]$. The concatenation of a vertex u to path P, where $u \in N(end(P))$ is defined as the path $P' \equiv P + u$ of length $\ell(P) + 1$ with $P'[0, \ell(P)] = P$ and $end(P') = u$.

Let $\mathcal{P}$ be the set of walks P in G having length $0 \le \ell(P) < n$. We introduce a linear order on walks in $\mathcal{P}$ such that for two walks P_1 and P_2, we say that $P_1 < P_2$ if $\ell(P_1) < \ell(P_2)$, or $\ell(P_1) = \ell(P_2)$ and there exists an index $j < \ell(P_1)$ such that $P_1([0,j]) = P_2([0,j])$ and $P_1([j+1]) < P_2([j+1])$. The comparison of vertices from V is understood as comparison of their identifiers in G.

We now define the tree T with vertex set $\mathcal{P}$, root $(r) \in \mathcal{P}$, such that vertex P' is a child of vertex P if and only if $P' = P + u$, for some $u \in N(end(P))$. We first show that agents can simulate the exploration of T while in fact moving around graph G. Intuitively, while an agent is following a path from the root to the leaves of T, its location in T corresponds to the walk taken by this agent in G.

Lemma 4. *A team of agents can simulate the virtual exploration of tree T starting from root (r), while physically moving around graph G starting from vertex r. The simulation satisfies the following conditions:*

(1) *An agent virtually occupying a vertex P of T is physically located at a vertex $end(P)$ in G.*

(2) *Upon entering a vertex P of T in the virtual exploration, the agent obtains the identifiers of all children of P in T.*

(3) *A virtual move along an edge of T can be performed in a single time step, by moving the agent to an adjacent location in G.*

(4) *Agents occupying the same virtual location P in T can communicate locally, i.e., they are physically located at the same vertex of G.*

We remark that the number of vertices of tree T is exponential in n. Hence, our goal is to perform the simulation with only a subset of the vertices of T. For a vertex $v \in V$, let $P_{\min}(v) \in \mathcal{P}$ be the minimum (with respect to the linear order on $\mathcal{P}$) walk ending at v. We observe that, by property (1) in Lemma 4, if, for all $v \in V$, the vertex $P_{\min}(v)$ of T has been visited by at least one agent in the virtual exploration of T, the physical exploration of G is completed. We define $\mathcal{P}_{\min} = \{P_{\min}(v) : v \in V\}$, and show that all vertices of $\mathcal{P}_{\min}$ are visited relatively quickly if we employ the procedure TEG (or TEL) for T, subject to a simple modification. In the original algorithm, we divided the agents descending to the children of the vertex according to the number of leaves of the discovered subtrees. We introduce an alternate definition of the function $L(T^{(s)}, v)$, so as to take into account only the number of vertices in $T^{(s)}$ corresponding to walks which are smallest among all walks in $T^{(s)}$ sharing the same end-vertex.

Lemma 5. *Let $T^{(s)} \subseteq T$ be a subtree of T rooted at (r). For $P \in V(T^{(s)})$, let $L(T^{(s)}, P)$ be the number of vertices v of G, for which the subtree of $T^{(s)}$ rooted at P contains a vertex representing the smallest walk contained in $T^{(s)}$ which ends at v:*

$$L(T^{(s)}, P) = \left| V(T^{(s)}(P)) \cap \bigcup_{v \in V} \left\{ \min\{P' \in V(T^{(s)}) : end(P') = v\} \right\} \right|,$$

and for $P \in \mathcal{P} \setminus V(T^{(s)})$, let $L(T^{(s)}, P) = 1$. Subject to this definition of L, procedure TEG with parameter $x > 6(n \log n + 1)$ (procedure TEL with parameter $x > 17(n \log n + 1)$) applied to tree T starting from root (r) visits all vertices from $\mathcal{P}_{\min}$

within $D \cdot (1 + \frac{1}{\log_n x - 1 - \log_n (2 \log x)})$ *(respectively,* $D \cdot (1 + \frac{2 + 1/\log n}{\log_n x - 1 - \log_n (4 \log x)}))$ *time steps.*

Proof. The set $\mathcal{P}_{\min}$ spans a subtree $T_{\min} = T[\mathcal{P}_{\min}]$ in T, rooted at (r). We can perform an analysis analogous to that used in the Proofs of Lemmas 1 and 3, evaluating sizes of waves of agents along paths in the subtree $T_{\min}$. We observe that for any $P \in \mathcal{P}_{\min}$ which is not a leaf in $T_{\min}$, we always have $L(T^{(s)}, P) \geq 1$. Moreover, we have $L(T^{(s)}, P) \leq |V(T^{(s)}(P))|$, and so $L(T^{(s)}, P) \leq n$. Since these two bounds were the only required properties of the functions L in the Proofs of Lemmas 1 and 3, the analysis from these proofs applies within the tree $T_{\min}$ without any changes. It follows that each vertex of $\mathcal{P}_{\min}$ is reached by the exploration algorithm within $D \cdot (1 + \frac{1}{\log_n x - 1 - \log_n (2 \log x)})$ time steps in case of global communication, and within $D \cdot (1 + \frac{2 + 1/\log n}{\log_n x - 1 - \log_n (4 \log x)})$ time steps in case of local communication.   □

We recall that by Lemma 4, one step of exploration of tree T can be simulated by a single step of an agent running on graph G. Thus, appropriately choosing $x = \Theta(n^c)$ in Lemma 5, we obtain our main theorem for general graphs.

Theorem 3. *For any $c > 1$, the online graph exploration problem with knowledge of n can be solved using a team of $k \geq Dn^c$ agents:*

- *in at most $D \cdot \left(1 + \frac{1}{c-1} + o(1)\right)$ time steps in the global communication model.*
- *in at most $D \cdot \left(1 + \frac{2}{c-1} + o(1)\right)$ time steps in the local communication model.*

For the case when we do not assume knowledge of (an upper bound on) n, we provide a variant of the above theorem which also completes exploration in $O(D)$ steps, with a slightly larger multiplicative constant.

Theorem 4. *For any $c > 1$, there exists an algorithm for the local communication model, which explores a rooted graph of unknown order n and unknown diameter D using a team of k agents, such that its exploration time is $O(D)$ if $k \geq Dn^c$.*

We remark that by choosing $x = \Theta(n \log n)$ in Lemma 2, we can also explore a graph using $k = \Theta(Dn \log n)$ agents in time $\Theta(D \log n)$, with local communication. This bound is the limit of our approach in terms of the smallest allowed team of agents.

4   Lower Bounds

In this section, we show lower bounds for exploration with Dn^c agents, complementary to the positive results given by Theorem 3. The graphs that produce the lower bound are a special class of trees. The same class of trees appeared in the lower bound from [8] for the competitive ratio of tree exploration algorithms with small teams of agents. In our scenario, we obtain different lower bounds depending on whether communication is local or global.

Theorem 5. *For all $n > 1$ and for every increasing function f, such that $\log f(n) = o(\log n)$, and every constant $c > 0$, there exists a family of trees $\mathcal{T}_{n,D}$, each with n vertices and height $D = \Theta(f(n))$, such that*

(i) *for every exploration strategy with global communication that uses Dn^c agents there exists a tree in $\mathcal{T}_{n,D}$ such that number of time steps required for its exploration is at least $D\left(1 + \frac{1}{c} - o(1)\right)$,*

(ii) *for every exploration strategy with local communication that uses Dn^c agents there exists a tree in $\mathcal{T}_{n,D}$ such that number of time steps required for exploration is at least $D\left(1 + \frac{2}{c} - o(1)\right)$.*

When looking at the problem of minimizing the size of the team of agents, our work (Theorem 4) shows that it is possible to achieve asymptotically-optimal online exploration time of $O(D)$ using a team of $k \leq Dn^{1+\epsilon}$ agents, for any $\epsilon > 0$. For graphs of small diameter, $D = n^{o(1)}$, we can thus explore the graph in $O(D)$ time steps using $k \leq n^{1+\epsilon}$ agents. This result almost matches the lower bound on team size of $k = \Omega(n^{1-o(1)})$ for the case of graphs of small diameter, which follows from the trivial lower bound $\Omega(D + n/k)$ on exploration time (cf. e.g. [8]). The question of establishing precisely what team size k is necessary and sufficient for performing exploration in $O(D)$ steps in a graph of larger diameter remains open.

References

1. Awerbuch, B., Betke, M., Rivest, R.L., Singh, M.: Piecemeal graph exploration by a mobile robot. Information and Computation 152(2), 155–172 (1999)
2. Brass, P., Cabrera-Mora, F., Gasparri, A., Xiao, J.: Multirobot tree and graph exploration. IEEE Transactions on Robotics 27(4), 707–717 (2011)
3. Czyzowicz, J., Ilcinkas, D., Labourel, A., Pelc, A.: Worst-case optimal exploration of terrains with obstacles. Information and Computation 225, 16–28 (2013)
4. Duncan, C.A., Kobourov, S.G., Kumar, V.S.A.: Optimal constrained graph exploration. ACM Transactions on Algorithms 2(3), 380–402 (2006)
5. Dynia, M., Korzeniowski, M., Schindelhauer, C.: Power-aware collective tree exploration. In: Grass, W., Sick, B., Waldschmidt, K. (eds.) ARCS 2006. LNCS, vol. 3894, pp. 341–351. Springer, Heidelberg (2006)
6. Dynia, M., Kutyłowski, J., Meyer auf der Heide, F., Schindelhauer, C.: Smart robot teams exploring sparse trees. In: Královič, R., Urzyczyn, P. (eds.) MFCS 2006. LNCS, vol. 4162, pp. 327–338. Springer, Heidelberg (2006)
7. Dynia, M., Łopuszański, J., Schindelhauer, C.: Why robots need maps. In: Prencipe, G., Zaks, S. (eds.) SIROCCO 2007. LNCS, vol. 4474, pp. 41–50. Springer, Heidelberg (2007)
8. Fraigniaud, P., Gąsieniec, L., Kowalski, D.R., Pelc, A.: Collective tree exploration. Networks 48(3), 166–177 (2006)
9. Frederickson, G.N., Hecht, M.S., Kim, C.E.: Approximation algorithms for some routing problems. SIAM Journal on Computing 7(2), 178–193 (1978)
10. Gabriely, Y., Rimon, E.: Competitive on-line coverage of grid environments by a mobile robot. Computational Geometry 24(3), 197–224 (2003)
11. Herrmann, D., Kamphans, T., Langetepe, E.: Exploring simple triangular and hexagonal grid polygons online. CoRR, abs/1012.5253 (2010)
12. Higashikawa, Y., Katoh, N.: Online exploration of all vertices in a simple polygon. In: Proc. 6th Frontiers in Algorithmics Workshop and the 8th Int. Conf. on Algorithmic Aspects of Information and Management (FAW-AAIM), pp. 315–326 (2012)
13. Higashikawa, Y., Katoh, N., Langerman, S., Tanigawa, S.-I.: Online graph exploration algorithms for cycles and trees by multiple searchers. Journal of Combinatorial Optimization (2013)

14. Icking, C., Kamphans, T., Klein, R., Langetepe, E.: Exploring an unknown cellular environment. In: Proc. 16th European Workshop on Computational Geometry (EuroCG), pp. 140–143 (2000)
15. Kolenderska, A., Kosowski, A., Małafiejski, M., Żyliński, P.: An improved strategy for exploring a grid polygon. In: Kutten, S., Žerovnik, J. (eds.) SIROCCO 2009. LNCS, vol. 5869, pp. 222–236. Springer, Heidelberg (2010)
16. Łopuszański, J.: Tree exploration. Tech-report, Institute of Computer Science, University of Wrocław, Poland (2007) (in Polish)
17. Ortolf, C., Schindelhauer, C.: Online multi-robot exploration of grid graphs with rectangular obstacles. In: Proc. 24th ACM Symp. on Parallelism in Algorithms and Architectures (SPAA), pp. 27–36 (2012)

Deterministic Polynomial Approach in the Plane

Yoann Dieudonné[1] and Andrzej Pelc[2,*]

[1] MIS, Université de Picardie Jules Verne, France
[2] Département d'informatique, Université du Québec en Outaouais,
Gatineau, Québec, Canada

Abstract. Two mobile agents with range of vision 1 start at arbitrary
points in the plane and have to accomplish the task of *approach*, which
consists in getting at distance at most one from each other, i.e., in getting
within each other's range of vision. An adversary chooses the initial positions of the agents, their possibly different starting times, and assigns
a different positive integer label and a possibly different speed to each
of them. Each agent is equipped with a compass showing the cardinal
directions, with a measure of length and a clock. Each agent knows its
label and speed but not those of the other agent and it does not know
the initial position of the other agent relative to its own. Agents do not
have any global system of coordinates and they cannot communicate.
Our main result is a deterministic algorithm to accomplish the task of
approach, working in time polynomial in the unknown initial distance
between the agents, in the length of the smaller label and in the inverse
of the larger speed. The distance travelled by each agent until approach is
polynomial in the first two parameters and does not depend on the third.
The problem of approach in the plane reduces to a network problem: that
of rendezvous in an infinite grid.

1 Introduction

Among numerous tasks performed by mobile agents one of the most basic and well
studied is that of meeting (or rendezvous) of two agents [4,30]. Agents are mobile
entities equipped with computational power and they may model humans, animals,
mobile robots, or software agents in communication networks. Applications of rendezvous are ubiquitous. People may want to meet in an unknown town, rescuers
have to find a lost tourist in the mountains, while animals meet to mate or to give
food to their offsprings. In human-made environments mobile robots meet to exchange collected samples or to divide between them the task of future exploration
of a contaminated terrain, while software agents meet to share data collected from
nodes of a network or to distribute between them the task of collective network
maintenance and checking for faulty components. The basic task of meeting of two
agents is a building block for gathering many agents. In all these cases it is important to achieve the meeting in an efficient way.

* Supported in part by NSERC discovery grant and by the Research Chair in Distributed Computing of the Université du Québec en Outaouais.

F.V. Fomin et al. (Eds.): ICALP 2013, Part II, LNCS 7966, pp. 533–544, 2013.

The Model and the Problem. Agents are usually modeled as points moving in a graph representing the network, or in the plane. In the first case the meeting of two agents is defined as both agents being at the same time in the same node of the graph [15,31] or in the same point inside an edge [11,14]. This is possible to achieve even when agents have radius of vision 0, i.e., when they cannot sense the other agent prior to the meeting. As observed in [13], if agents freely circulate in the plane and can start in arbitrary unknown points of it, bringing them simultaneously to the same point is impossible with radius of vision 0. A natural assumption in this case is that agents have some positive radius of vision and the task is to bring them within this distance, so they can see each other. Once this is achieved, agents are in contact, so they can get even closer and exchange information or objects. It should be noted that the expression *radius of vision* does not need to be interpreted optically. It is a distance at which the agents can mutually sense each other optically, audibly (e.g. by emitting sounds in the dark), chemically (animals smelling each other), or even by touch (real agents are not points but have positive size, so if a point is chosen inside each of them, there is some positive distance s, such that they will touch before these chosen points get at distance s). Since in this paper we study agents moving in the plane, we are interested in the above described task of bringing the points representing them at some pre-defined positive distance. Without loss of generality we assume that this distance is 1 and we call *approach* the task of bringing the points representing the two agents at distance at most 1.

An adversary chooses the initial positions of the agents, which are two arbitrary points of the plane, it chooses their possibly different starting times, and it assigns a different label and a possibly different speed to each of them. At all times each of the agents moves at the assigned speed or stays idle. Labels are positive integers. Each agent is equipped with a compass showing the cardinal directions, with a measure of length, and with a clock. Clocks of the agents are not necessarily synchronized. Hence an agent can perform basic actions of the form: "go North/East/South/West at a given distance" and "stay idle for a given amount of time". In fact, these will be the only actions performed by agents in our solution. Each agent knows its label, having a clock and a measure of length it can calculate its speed, but it has no information about the other agent: it does not know the initial position of the other agent relative to its own, the distance separating them, it does not know the speed of the other agent or its label. Agents do not have any global system of coordinates and, prior to accomplishing the approach, they cannot communicate. The *cost* of an algorithm accomplishing the task of approach is the total distance travelled by both agents, and the *time* of an algorithm is counted from the start of the later agent.

Our Results. Our main result is a deterministic algorithm to accomplish the task of approach, working in time polynomial in the unknown initial distance between the agents, in the length of (the binary representation of) the shorter label and in the inverse of the larger speed. (Hence it is also polynomial in the distance, the length of the other label and the inverse of the other speed.) The cost of the algorithm is polynomial in the first two parameters and does not depend on the third. The problem of approach in the plane reduces to a network problem: that of rendezvous in an infinite grid. Due to lack of space, all the proofs will appear in the journal version of the paper.

Discussion and Open Problems. In this paper we are only interested in deterministic solutions to the approach problem. Randomized solutions, based on random walks on a grid, are well known [29]. Let us first discuss the assumptions concerning the equipment of the agents. From an application point of view the three tools provided to the agents, i.e., a compass, a unit of length and a clock, do not seem unrealistic when agents are humans or robots. Nevertheless it is an interesting question if these tools are really necessary. It is clear that an agent needs some kind of compass just to be able to change direction of its walk. However it remains open if our result remains true if the compasses of the agents are subject to some level of inaccuracy. The same remark concerns the measure of length: without it an agent could not carry out any moving plan, but it remains open if the result is still valid if agents have different, possibly completely unrelated units of length. Probably the most interesting question concerns the necessity of equipping the agents with a clock. In our solution clocks play a vital role, as the algorithm is based on interleaving patterns of moves with prescribed waiting periods. However, the possibility of designing a polynomial algorithm for the task of approach without any waiting periods, with each agent always traveling at its steady speed prescribed by the adversary, is not excluded. Such an algorithm could possibly work without relying on any clock.

We count the execution time of the algorithm from the starting time of the later agent. Notice that counting time from the start of the earlier agent does not make sense. Indeed, the adversary can assign an extremely small speed to the earlier agent, give a large speed to the later one and start it only when the earlier agent traversed half of the initial distance between them. The time of traversing this distance by the earlier agent can be arbitrarily large with respect to the inverse of the speed of the later agent, and approach must use at least this time, if the initial distance is larger than 2.

Notice that assigning different labels to agents is the only way to break symmetry between them in a deterministic way, and hence to ensure deterministic approach. Anonymous (identical) agents would make identical moves, and hence could never approach if started simultaneously with the same speed.

Concerning the complexity of our algorithm, our goal in this paper is only to keep both the time and the cost polynomial. We do not make any attempt at optimizing the obtained polynomials. Some improvements may have been obtained by using more complicated but slightly more efficient procedures or by performing tighter analysis. However getting optimal time and cost seems to be a very challenging open problem.

Next, it is interesting to ponder the degree of asynchrony allowed in the navigation of the agents. Some asynchrony is included in our model, by allowing the adversary to assign arbitrary, possibly different, mutually unknown speeds to both agents. Nevertheless we assume that when agents move, they move at constant speed. A higher level of asynchrony was assumed in [7,11,13,14,22]: the adversary could change the speed of each agent arbitrarily or halt the agent for an arbitrary finite time. In such a model it is of course impossible to limit the time of approach, so our main result could not remain valid. However, it is still perhaps possible to preserve our other result: design an algorithm in the scenario of arbitrarily varying speeds of agents, in which the distance travelled by each agent until approach is polynomial in the unknown initial distance between the agents and in the length of the shorter label. We leave this as an open problem.

Finally, let us consider the issue of the memory size of the agents. In our model we do not impose any restriction on it, treating agents, from the computational point of view, as Turing machines. However, it is easy to see that the execution of our algorithm requires memory of $O(\log D + \log L + \log(1/v))$ bits, where D is the initial distance between the agents, L is the agent's label and v is its speed. As for the corresponding lower bound, $\Omega(\log L)$ bits of memory are necessary to store the label of the agent, and it can be shown that $\Omega(\log D)$ bits of memory are necessary for approach even in the easier scenario where agents are on a line instead of the plane. By contrast, the lower bound $\Omega(\log(1/v))$ is much less clear. Indeed, if a wait-free solution not using the clock is possible, it could perhaps be implemented by agents whose memory size does not depend on their speed. Also this question remains open.

Related Work. The literature on rendezvous can be broadly divided according to whether the agents move in a randomized or in a deterministic way. An extensive survey of randomized rendezvous in various scenarios can be found in [4], cf. also [2,3,5,6,26]. In the sequel we briefly discuss the literature on deterministic rendezvous that is more closely related to our scenario. This literature is naturally divided according to the way of modeling the environment: agents can either move in a graph representing a network, or in the plane. Deterministic rendezvous in networks has been surveyed in [30].

In most papers on rendezvous in networks a synchronous scenario was assumed, in which agents navigate in the graph in synchronous rounds. Rendezvous with agents equipped with tokens used to mark nodes was considered, e.g., in [27]. Rendezvous of two agents that cannot mark nodes but have unique labels was discussed in [15,25,31]. These papers are concerned with the time of synchronous rendezvous in arbitrary graphs. In [15] the authors show a rendezvous algorithm polynomial in the size of the graph, in the length of the shorter label and in the delay between the starting time of the agents. In [25,31] rendezvous time is polynomial in the first two of these parameters and independent of the delay. Memory required by two anonymous agents to achieve deterministic rendezvous has been studied in [20,21] for trees and in [12] for general graphs.

Rendezvous of more than two agents, often called gathering, has been studied, e.g., in [16,17,28,32]. In [16] agents were anonymous, while in [32] the authors considered gathering many agents with unique labels. Gathering many labeled agents in the presence of Byzantine agents was studied in [17]. Gathering many agents in the plane has been studied in [8,9,19] under the assumption that agents are memoryless, but they can observe other agents and make navigation decisions based on these observations. Fault-tolerant aspects of this problem were investigated, e.g., in [1,10]. On the other hand, gathering memoryless agents in a ring, assuming that agents can see the entire ring and positions of agents in it, was studied in [23,24].

Asynchronous rendezvous of two agents in a network has been studied in [7,11,13,14,18,22] in a model when the adversary can arbitrarily change the speed of each agent or halt the agent for an arbitrary finite time. As mentioned previously, in this model time cannot be bounded, hence the authors concentrated on the cost of rendezvous, measured as the total number of edge traversals executed by both agents. In [14] the authors investigated the cost of rendezvous in the infinite line and in the ring. They also proposed a rendezvous algorithm for an arbitrary graph with a known upper bound on the size of the graph. This assumption was subsequently removed in [13], but both in [14] and in [13] the cost of rendezvous was exponential in the size of the graph and in the larger label. In [22] asynchronous rendezvous was studied for anonymous agents and the cost was again exponential. The result from [13] implies a solution to the problem of approach in the plane at cost *exponential* in the initial distance between agents and in the larger of the labels.

The first asynchronous rendezvous algorithms at cost polynomial in the initial distance of the agents were presented in [7,11]. In these papers the authors worked in infinite multidimensional grids and their result implies a solution to the problem of approach in the plane at cost polynomial in the initial distance of the agents. However, they used the powerful assumption that each agent knows its starting position in a global system of coordinates. It should be stressed that the assumptions and the results in these papers are incomparable to the assumptions and results in the present paper. In [7,11] the authors allow the adversary to arbitrarily control and change the speed of each agent, and hence cannot control the time of rendezvous and do not use clocks. To get polynomial cost (indeed, their cost is close to optimal) they use the assumption of known starting position in an absolute system of coordinates. By contrast, we assume arbitrary, possibly different and unknown but constant speeds of each agent, use clocks and different integer labels of agents but not their positions (in fact our agents are completely ignorant of where they are) and obtain an algorithm of polynomial time and cost (in the previously described parameters).

In a recent paper [18] we designed a rendezvous algorithm working for an arbitrary finite graph in the above asynchronous model. The algorithm has cost polynomial in the size of the graph and in the length of the smaller label. Again, the assumptions and the results are incomparable to those of the present paper. First, in [18], as in [7,11,13], time cannot be controlled and cost in [18] is polynomial in the size of the graph. More importantly, it is unlikely that the methods from [18], tailored for arbitrary finite graphs, could be used even to obtain our present result about cost. Indeed, it is easy to see that making cost polynomial in the initial distance is not possible in arbitrary graphs, as witnessed by the case of the clique: the adversary can hold one agent at a node and make the other agent traverse $\Theta(n)$ edges before rendezvous (even at steady speed), in spite of the initial distance 1. Also in [18] agents walk in the same finite graph, which is not the case in our present scenario.

2 Preliminaries

It follows from [13] that the problem of approach can be reduced to that of rendezvous in an infinite grid, in which every node u is adjacent to 4 nodes at Euclidean distance 1 from it, and located North, East, South and West from node u. We call this grid a *basic grid*. Rendezvous in this grid means simultaneously bringing two agents starting at arbitrary nodes of the grid to the same node or to the same point inside some edge.

Hence in the rest of the paper we will consider rendezvous in a basic grid, instead of the task of approach. Instructions in a rendezvous algorithm are: "go North/East/South/West at distance 1" and "stay idle for a given amount of time". Before executing our rendezvous algorithm in a basic grid, an agent performs two preprocessing procedures. The first is the procedure of transforming the label of the agent and works as follows. Let $L = (b_0 b_1 \ldots b_{r-1})$ be the binary representation of the label of the agent. We define its transformed label L^* as the binary sequence $b_0 b_0 b_1 b_1 \ldots b_{r-1} b_{r-1} 01)$. Notice that the length of the transformed label L^* is $2r + 2$, where r is the length of label L. Moreover, transformed labels are never prefixes of each other and they must differ at some position different from the first. This is why original labels are transformed.

The second preprocessing procedure performed by an agent is computing the inverse of its speed: the agent measures the time θ it takes it to traverse a distance of length 1. This time, called the *basic time* of the agent, will be used to establish the length of waiting periods in the execution of the algorithm. In fact, measuring θ can be done when the agent traverses the first edge of the basic grid indicated by the algorithm.

We denote by Δ be the initial distance between agents in the basic grid. Notice that $D \leq \Delta \leq \sqrt{2}D$, where D is the initial Euclidean distance between the agents. Let λ be the length of the shorter of the transformed labels of agents, and let τ be the shorter basic time.

3 Algorithm

Patterns. We first describe several patterns of moves that will be used by our algorithm. All the patterns are routes in the basic grid, and distances between nodes are measured also in the basic grid, i.e., in the Manhattan metric. We use N (resp. E,S,W) to denote the instruction "make an edge traversal by port North (resp. East, South, West)". We define the *reverse path* to the path $v_1, \ldots, v_k$ of the agent as the path $v_k, v_{k-1}, \ldots, v_1$. We also define a sub-path of the path $v_1, \ldots, v_k$ as $v_i, v_{i+1}, \ldots, v_{j-1}, v_j$, for some $1 \leq i < j \leq k$.

Pattern $BALL(v, s)$, for a node v and an integer $s \geq 1$, visits all nodes of the grid at distance at most s from v and traverses all edges of the grid between such nodes.

Moreover, executing this pattern the agent is always at distance at most s from v. Let $S(v, i)$ be the set of nodes at distance exactly i from v. Pattern $BALL(v, s)$ is executed in s phases, each of which starts and ends at v. Phase 1 is the unit cross with center v corresponding to the sequence NSEWSNWE. Suppose that phase $i - 1$, for $i > 1$, has been executed. Let $v_1, \ldots, v_q$ be nodes of $S(v, i - 1)$ with v_1 situated North of v, and all other nodes of $S(v, i - 1)$ ordered clockwise. Phase i consists of q stages $\sigma_1, \ldots, \sigma_q$. Stage σ_1 consists of going from v to v_1 using the shortest path and performing the unit cross NSEWSNWE with center v_1. Stage σ_j, for $1 < j < q$ consists of going from v_{j-1} to v_j using the unique path of length 2 with midpoint in $S(v, i - 2)$ and performing the unit cross NSEWSNWE with center v_j. Stage σ_q consists of going from v_{q-1} to v_q, performing the unit cross NSEWSNWE with center v_q and going back to v by the lexicographically smallest shortest path (coded as a sequence of letters N, E, S, W).

Pattern $SUPERBALL(v, s)$, for a node v and an integer $s \geq 0$, consists of performing the sequence of patterns $BALL(v, 1)$, $BALL(v, 2), \ldots, BALL(v, s)$, followed by the reverse path of this sequence of patterns.

For the subsequent patterns we will use the following notation. For $i = 1, \ldots, s$, let $w(i, 1)$, $w(i, 2)$, $\ldots$,$w(i, q(i))$ be the enumeration of all nodes u at distance at most i from v in the lexicographic order of the lexicographically smallest shortest path from v to u.

Pattern $FLOWER(v, s, k)$ is executed in phases $1, 2, \ldots, ks$. Each phase consists of two parts. For $i \leq s$, part 1 of phase i consists of $q(i)$ stages. Stage j consists of going from v to $w(i, j)$ by the lexicographically smallest shortest path $\pi(i, j)$, then executing $SUPERBALL(w(i, j), i)$ and then backtracking to v using the path reverse to $\pi(i, j)$. For $i > s$, part 1 of phase i is the same as part 1 of phase s, except that $SUPERBALL(w(s, j), s)$ is replaced by $SUPERBALL(w(s, j), i)$. Part 2 of every phase is backtracking using the reverse path to that used in part 1.

Pattern $BOUQUET(v, s, k)$ is executed in epochs $1, 2, \ldots, ks$. Each epoch consists of two parts. For $i \leq s$, part 1 of epoch i consists of $q(i)$ stages. Stage j consists of going from v to $w(i, j)$ by the lexicographically smallest shortest path $\pi(i, j)$, then executing phase 1, phase 2, ..., phase i of $FLOWER(w(i, j), s, k)$ and then backtracking to v using the path reverse to $\pi(i, j)$. For $i > s$, part 1 of phase i is the same as part 1 of phase s, except that the execution of phase 1, phase 2, ..., phase s of $FLOWER(w(s, j), s, k)$ is replaced by the execution of phase 1, phase 2, ..., phase s, phase $s + 1$, ..., phase i of $FLOWER(w(s, j), s, k)$. Part 2 of every epoch is backtracking using the reverse path to that used in part 1.

Pattern $CATCH - BOUQUET(v, s, k)$ is executed in $q(s)$ stages. Stage j consists of going from v to $w(s, j)$ by the lexicographically smallest shortest path $\pi(s, j)$, then executing pattern $BOUQUET(w(s, j), s, k)$ and then backtracking to v using the path reverse to $\pi(s, j)$.

Pattern $BORDER(v, s, n)$ consists of executing n times $SUPERBALL(v, s)$.

Apart from the above patterns of moves, our algorithm will use procedures $WAIT_0(v, s, k)$, $WAIT_1(v, s, k)$, $WAIT_2(v, s, k)$, and $WAIT_3(v, s, k)$. Each of these procedures consists of waiting at the initial position v of the agent for a prescribed period of time. We will specify these periods of waiting later on.

The Main Idea. The main idea of our rendezvous algorithm in the basic grid is the following. In order to guarantee rendezvous, symmetry in the actions of the agents must be broken. Since agents have different transformed labels, this can be done by designing the algorithm so that each agent processes consecutive bits of its transformed label, acting differently when the current bit is 0 and when it is 1. The aim is to force rendezvous when each agent processes the bit corresponding to the position where their transformed labels differ. This approach requires to overcome two major difficulties. The first is that due to the possibly different starting times and different speeds, agents may execute corresponding bits of their transformed labels at different times. This problem is solved in our algorithm by carefully scheduling patterns $BORDER$ and waiting times, in order to synchronize the agents. Patterns $BORDER$ and waiting times have the following role in this synchronization effort. While a pattern $BORDER$ executed by one agent pushes the other agent to proceed in its execution, or otherwise rendezvous is accomplished, waiting periods slow down the executing agent. The joint application of these two algorithmic ingredients guarantees that agents will at some point execute almost simultaneously the bit on which they differ. The second difficulty is to orchestrate rendezvous after the first difficulty has been overcome, i.e., when each agent executes this bit. This is done by combining waiting periods with patterns that are included in one another for some parameters. Our algorithm is designed in such a way that the execution of bit 0 consists of executing a pattern $FLOWER$ followed by a waiting period followed by pattern $CATCH - BOUQUET$, while the execution of bit 1 consists of executing a pattern $BOUQUET$ followed by a waiting period. According to our algorithm, either $FLOWER$ will be included in $BOUQUET$, or $BOUQUET$ will be included in $CATCH - BOUQUET$. If one agent executes a pattern P' included in the pattern P'' executed simultaneously by the other agent, one agent must "catch" the other, i.e., rendezvous must occur. The main role of the waiting periods associated with these patterns is to slow down the agent executing the pattern included in the other. Indeed, these waiting periods ensure that the agent executing pattern P' does not complete it too early and start some other action. It can be shown that this synchronization occurs soon enough in the execution to guarantee rendezvous at polynomial time and cost.

Description of the Algorithm. We are now ready to present a detailed description of our rendezvous algorithm in the basic grid, executed by an agent with transformed label $L^* = (c_0 c_1 \ldots c_{\ell-1})$ of length ℓ and with basic time θ. The agent starts at node v. For technical reasons we define $c_j = 0$ for all $j \geq \ell$. The main "repeat" loop is executed until rendezvous is accomplished. At this time both agents stop.

For any pattern P we will use the notation $\mathcal{C}[P]$ to denote the number of edge traversals in the execution of P.

Algorithm Meeting
$s := 1; i := 0$
repeat
    $j := 0$
    while $j \leq s - 1$ **do**
        if $c_j = 0$ **then**
            execute $FLOWER(v, s, 4i + 1)$
            execute $WAIT_0(v, s, 4i + 1)$
            execute $CATCH - BOUQUET(v, s, 4i + 1)$
        else
            execute $BOUQUET(v, s, 4i + 1)$
            execute $WAIT_1(v, s, 4i + 1)$
        endif
        $j := j + 1; i := i + 1; N := 3 \cdot C[CATCH - BOUQUET(v, s, 4i + 1)]$
        if $j \leq s - 1$ **then**
            execute $BORDER(v, (4i + 1)s, N)$
            execute $WAIT_2(v, s, 4i + 1)$
        else
            $M := 2s \cdot C[BORDER(v, (4i + 1)s, N)]$
            execute $BORDER(v, (4i + 1)s, M)$
            execute $WAIT_3(v, s, 4i + 1)$
        endif
    endwhile
    $s := s + 1$

It remains to give the lengths of the waiting periods used by our algorithm. To this end we introduce the following terminology. We first define fences and walls. A *wall* is a pattern $BORDER$ executed in the algorithm immediately prior to a waiting period $WAIT_3$. A *fence* is any other pattern $BORDER$. We next define *pieces* as follows. Notice that any execution of the algorithm can be viewed as a concatenation of chunks of the following form: some sequence of instructions Q immediately followed by a wall, immediately followed by a waiting period $WAIT_3$. We define the first piece Q_1 as the sequence of instructions before the first wall, and the i-th piece Q_i, for $i > 1$, as the sequence of instructions between the end of the $(i-1)$th $WAIT_3$ and the beginning of the ith wall. We next define *segments*. Consider any piece. It can be viewed as a concatenation of chunks of the following form: some sequence of instructions S immediately followed by a fence, immediately followed by a waiting period $WAIT_2$. We define the first segment S_1 of a piece as the sequence of instructions before the first fence of this piece, and the i-th segment S_i of the piece, for $i > 1$, as the sequence of instructions between the end of the $(i - 1)$th $WAIT_2$ in the piece and the beginning of the ith fence in it. Notice that segments correspond to bits of the transformed label of the agent: these are sequences of instructions executed in the statement "**if** $c_j = 0$ **then** ... **else** ...". A pattern $FLOWER, BOUQUET$ or $CATCH - BOUQUET$ will be called an *atom* of its segment. Now we are ready to give the lengths of waiting periods $WAIT_0(v, s, 4i + 1)$, $WAIT_1(v, s, 4i + 1)$, $WAIT_2(v, s, 4i + 1)$, and $WAIT_3(v, s, 4i + 1)$.

Consider the waiting period $WAIT_0(v, s, 4i + 1)$.

Let $\tau_0 = \theta \cdot C[FLOWER(v, s, 4i + 1)]$ be the time spent by the agent to perform $FLOWER(v, s, 4i + 1)$. The length of the waiting period $WAIT_0(v, s, 4i + 1)$ is defined as $\tau_0 \cdot C[BOUQUET(v, 2s, 4i' + 1)]$. Notice that if $WAIT_0(v, s, 4i + 1)$ is located in the m-th segment of the s-th piece, then its length upper-bounds the time of executing the first atom of the m-th segment of the $(2s)$-th piece – assuming that this segment corresponds to bit 1 – by an agent with basic time τ_0 (because this atom is $BOUQUET(v, 2s, 4i' + 1)$). This property is essential for the proof of correctness.

Consider the waiting period $WAIT_1(v, s, 4i + 1)$ located in the m-th segment of the s-th piece. Let $\tau_1 = \theta \cdot C[BOUQUET(v, s, 4i + 1)]$ be the time spent by the agent to perform $BOUQUET(v, s, 4i + 1)$. Let $i' = i + s^2 + s(s - 1)/2$. The length of the waiting period $WAIT_1(v, s, 4i + 1)$ is defined as the time of executing the m-th segment of the $(2s)$-th piece – assuming that this segment corresponds to bit 0 – by an agent by basic time τ_1.

Consider the waiting period $WAIT_2(v, s, 4i + 1)$ located immediately before the m-th segment of the s-th piece. Let τ_2 be the sum of times spent by the agent with basic time θ to perform the following chunks: the $(m - 1)$-th segment of the s-th piece, the $(m - 1)$-th fence of the s-th piece, and the first atom of the m-th segment of the s-th piece. The length of the waiting period $WAIT_2(v, s, 4i + 1)$ is defined as the time to perform the $(m - 1)$-th segment of the $(2s)$-th piece – assuming that this segment corresponds to bit 1 – together with the $(m - 1)$-th fence of the $(2s)$-th piece, by an agent with basic time τ_2.

Consider the waiting period $WAIT_3(v, s, 4i + 1)$. This period is located immediately before the $(s + 1)$-th piece. Let τ_3 be the sum of times spent by the agent with basic time θ to perform the following chunks: the s-th piece, the s-th wall, and the first atom of the first segment of the $(s + 1)$-th piece. The length of the waiting period $WAIT_3(v, s, 4i + 1)$ is defined as the time to perform the $(2s)$-th wall by an agent with basic time τ_3.

The lengths of the waiting periods having been defined, the description of our algorithm is now complete.

4 Correctness and complexity

In this section we formulate results stating that Algorithm Meeting accomplishes rendezvous in the basic grid, that its execution time is polynomial in Δ, λ and τ, and that its cost is polynomial in Δ and λ. The proofs of these results will appear in the journal version of the paper.

The following theorem implies that Algorithm Meeting is correct.

Theorem 1. *The rendezvous of agents executing Algorithm Meeting must occur before the first time when one of them completes the $(2(\Delta + \lambda) + 1)$-th piece.*

Our next result estimates the complexity of Algorithm Meeting.

Theorem 2. *Let Δ be the initial distance between agents in the basic grid. Let λ be the length of the shorter of the transformed labels of agents. Let τ be the shorter of the basic times of the agents. Then the execution time of Algorithm Meeting is polynomial in Δ, λ and τ, and its cost is polynomial in Δ and λ.*

Since Δ is linear in the initial Euclidean distance between agents, and the length of the transformed label of an agent is linear in the length of the original label, in view of the reduction described in Section 2 we have the following corollary concerning the task of approach in the plane.

Corollary 1. *Let D be the initial Euclidean distance between agents in the plane, let γ be the length of the binary representation of the shorter label and let τ be the inverse of the larger speed of the agents. Then deterministic approach between agents is possible in time polynomial in D, γ and τ, and at cost polynomial in D and γ.*

References

1. Agmon, N., Peleg, D.: Fault-tolerant gathering algorithms for autonomous mobile robots. SIAM J. Comput. 36, 56–82 (2006)
2. Alpern, S.: The rendezvous search problem. SIAM J. on Control and Optimization 33, 673–683 (1995)
3. Alpern, S.: Rendezvous search on labelled networks. Naval Research Logistics 49, 256–274 (2002)
4. Alpern, S., Gal, S.: The theory of search games and rendezvous. Int. Series in Operations research and Management Science. Kluwer Academic Publisher (2002)
5. Alpern, J., Baston, V., Essegaier, S.: Rendezvous search on a graph. Journal of Applied Probability 36, 223–231 (1999)
6. Anderson, E., Weber, R.: The rendezvous problem on discrete locations. Journal of Applied Probability 28, 839–851 (1990)
7. Bampas, E., Czyzowicz, J., Gąsieniec, L., Ilcinkas, D., Labourel, A.: Almost optimal asynchronous rendezvous in infinite multidimensional grids. In: Lynch, N.A., Shvartsman, A.A. (eds.) DISC 2010. LNCS, vol. 6343, pp. 297–311. Springer, Heidelberg (2010)
8. Cieliebak, M., Flocchini, P., Prencipe, G., Santoro, N.: Solving the robots gathering problem. In: Baeten, J.C.M., Lenstra, J.K., Parrow, J., Woeginger, G.J. (eds.) ICALP 2003. LNCS, vol. 2719, pp. 1181–1196. Springer, Heidelberg (2003)
9. Cohen, R., Peleg, D.: Convergence properties of the gravitational algorithm in asynchronous robot systems. SIAM J. Comput. 34, 1516–1528 (2005)
10. Cohen, R., Peleg, D.: Convergence of autonomous mobile robots with inaccurate sensors and movements. SIAM J. Comput. 38, 276–302 (2008)
11. Collins, A., Czyzowicz, J., Gąsieniec, L., Labourel, A.: Tell me where I am so I can meet you sooner. In: Abramsky, S., Gavoille, C., Kirchner, C., Meyer auf der Heide, F., Spirakis, P.G. (eds.) ICALP 2010. LNCS, vol. 6199, pp. 502–514. Springer, Heidelberg (2010)
12. Czyzowicz, J., Kosowski, A., Pelc, A.: How to meet when you forget: Log-space rendezvous in arbitrary graphs. Distributed Computing 25, 165–178 (2012)
13. J. Czyzowicz, A. Labourel, A. Pelc, How to meet asynchronously (almost) everywhere. ACM Transactions on Algorithms 8, article 37 (2012)

14. De Marco, G., Gargano, L., Kranakis, E., Krizanc, D., Pelc, A., Vaccaro, U.: Asynchronous deterministic rendezvous in graphs. Theoretical Computer Science 355, 315–326 (2006)
15. Dessmark, A., Fraigniaud, P., Kowalski, D., Pelc, A.: Deterministic rendezvous in graphs. Algorithmica 46, 69–96 (2006)
16. Dieudonné, Y., Pelc, A.: Anonymous meeting in networks. In: Proc. 24rd Annual ACM-SIAM Symposium on Discrete Algorithms (SODA 2013), pp. 737–747 (2013)
17. Dieudonné, Y., Pelc, A., Peleg, D.: Gathering despite mischief. In: Proc. 23rd Annual ACM-SIAM Symposium on Discrete Algorithms (SODA 2012), pp. 527–540 (2012)
18. Dieudonné, Y., Pelc, A., Villain, V.: How to meet asynchronously at polynomial cost. In: Proc. 32nd Annual ACM Symposium on Principles of Distributed Computing, PODC 2013 (to appear, 2013)
19. Flocchini, P., Prencipe, G., Santoro, N., Widmayer, P.: Gathering of asynchronous oblivious robots with limited visibility. In: Ferreira, A., Reichel, H. (eds.) STACS 2001. LNCS, vol. 2010, pp. 247–258. Springer, Heidelberg (2001)
20. Fraigniaud, P., Pelc, A.: Deterministic rendezvous in trees with little memory. In: Taubenfeld, G. (ed.) DISC 2008. LNCS, vol. 5218, pp. 242–256. Springer, Heidelberg (2008)
21. Fraigniaud, P., Pelc, A.: Delays induce an exponential memory gap for rendezvous in trees. In: Proc. 22nd Ann. ACM Symposium on Parallel Algorithms and Architectures (SPAA 2010), pp. 224–232 (2010)
22. Guilbault, S., Pelc, A.: Asynchronous rendezvous of anonymous agents in arbitrary graphs. In: Fernàndez Anta, A., Lipari, G., Roy, M. (eds.) OPODIS 2011. LNCS, vol. 7109, pp. 421–434. Springer, Heidelberg (2011)
23. Klasing, R., Kosowski, A., Navarra, A.: Taking advantage of symmetries: Gathering of many asynchronous oblivious robots on a ring. Theoretical Computer Science 411, 3235–3246 (2010)
24. Klasing, R., Markou, E., Pelc, A.: Gathering asynchronous oblivious mobile robots in a ring. Theoretical Computer Science 390, 27–39 (2008)
25. Kowalski, D., Malinowski, A.: How to meet in anonymous network. In: Flocchini, P., Gąsieniec, L. (eds.) SIROCCO 2006. LNCS, vol. 4056, pp. 44–58. Springer, Heidelberg (2006)
26. Kranakis, E., Krizanc, D., Morin, P.: Randomized rendez-vous with limited memory. In: Laber, E.S., Bornstein, C., Nogueira, L.T., Faria, L. (eds.) LATIN 2008. LNCS, vol. 4957, pp. 605–616. Springer, Heidelberg (2008)
27. Kranakis, E., Krizanc, D., Santoro, N., Sawchuk, C.: Mobile agent rendezvous in a ring. In: Proc. 23rd Int. Conference on Distributed Computing Systems (ICDCS 2003), pp. 592–599. IEEE (2003)
28. Lim, W., Alpern, S.: Minimax rendezvous on the line. SIAM J. on Control and Optimization 34, 1650–1665 (1996)
29. Mitzenmacher, M., Upfal, E.: Probability and computing: randomized algorithms and probabilistic analysis. Cambridge University Press (2005)
30. Pelc, A.: Deterministic rendezvous in networks: A comprehensive survey. Networks 59, 331–347 (2012)
31. Ta-Shma, A., Zwick, U.: Deterministic rendezvous, treasure hunts and strongly universal exploration sequences. In: Proc. 18th ACM-SIAM Symposium on Discrete Algorithms (SODA), pp. 599–608 (2007)
32. Yu, X., Yung, M.: Agent rendezvous: a dynamic symmetry-breaking problem. In: Meyer auf der Heide, F., Monien, B. (eds.) ICALP 1996. LNCS, vol. 1099, pp. 610–621. Springer, Heidelberg (1996)

Outsourced Pattern Matching

Sebastian Faust[1,*], Carmit Hazay[2], and Daniele Venturi[3,**]

[1] Security and Cryptography Laboratory, EPFL, Switzerland
[2] Faculty of Engineering, Bar-Ilan Univeristy, Israel
[3] Department of Computer Science, Aarhus University, Denmark

Abstract. In secure delegatable computation, computationally weak devices (or clients) wish to outsource their computation and data to an *untrusted* server in the cloud. While most earlier work considers the general question of how to securely outsource *any* computation to the cloud server, we focus on *concrete* and *important* functionalities and give the first protocol for the *pattern matching* problem in the cloud. Loosely speaking, this problem considers a text T that is outsourced to the cloud S by a client C_T. In a query phase, clients $C_1, \ldots, C_l$ run an efficient protocol with the server S and the client C_T in order to learn the positions at which a pattern of length m matches the text (and nothing beyond that). This is called the *outsourced pattern matching* problem and is highly motivated in the context of delegatable computing since it offers storage alternatives for massive databases that contain confidential data (e.g., health related data about patient history). Our constructions offer *simulation-based security* in the presence of semi-honest and malicious adversaries (in the random oracle model) and limit the communication in the query phase to $O(m)$ bits plus the number of occurrences — which is optimal. In contrast to generic solutions for delegatable computation, our schemes do not rely on fully homomorphic encryption but instead uses novel ideas for solving pattern matching, based on efficiently solvable instances of the subset sum problem.

1 Introduction

The problem of securely outsourcing computation to an *untrusted* server gained momentum with the recent penetration of *cloud computing* services. In cloud computing, clients can lease computing services on demand rather than maintaining their own infrastructure. While such an approach naturally has numerous advantages in cost and functionality, the outsourcing mechanism crucially needs to enforce privacy of the outsourced data and integrity of the computation. Cryptographic solutions for these challenges have been put forward with the concept of *secure delagatable computation* [1,6,11,2,8].

* Supported in part by the BEAT project 7th Framework Research Programme of the European Union, grant agreement number: 284989.
** Supported from the Danish National Research Foundation, the National Science Foundation of China (under the grant 61061130540), the Danish Council for Independent Research (under the DFF Starting Grant 10-081612) and also from the CFEM research center within which part of this work was performed.

F.V. Fomin et al. (Eds.): ICALP 2013, Part II, LNCS 7966, pp. 545–556, 2013.
© Springer-Verlag Berlin Heidelberg 2013

In secure delegatable computation, computationally weak devices (or clients) wish to outsource their computation and data to an *untrusted* server. The ultimate goal in this setting is to design efficient protocols that minimize the computational overhead of the clients and instead rely on the extended resources of the server. Of course, the amount of work invested by the client in order to verify the correctness of the computation shall be *substantially* smaller than running the computation by itself. Indeed, if this was not the case then the client could carry out the computation itself. Another ambitious goal of delegatable computation is to design protocols that minimize the *communication* between the cloud and the client.

Most recent works in the area of delegatable computation propose solutions to securely outsource *any functionality* to an untrusted server [1,6,11,2]. Such generic solutions often suffer from rather poor efficiency and high communication overhead due to the use of fully homomorphic encryption [12]. An exception is the randomized encoding technique used by [1] which instead relies on garbled circuits. Furthermore, these solution concepts typically examine a restricted scenario where a *single client* outsources its computation to an *external untrusted server*. Only few recent works study the setting with *multiple clients* that mutually distrust each other and wish to securely outsource a joint computation on their inputs with reduced costs, e.g., [15,17]. Of course, also in this more complex setting recent constructions build up on fully homomorphic encryption.

To move towards more practical schemes, we may focus on particularly efficient constructions for specific important functionalities. This approach has the potential to avoid the use of fully homomorphic encryption by exploiting the structure of the particular problem we intend to solve. Some recent works have considered this question [3,22,20]. While these schemes are more efficient than the generic constructions mentioned above, they typically only achieve very limited privacy or do not support multiple distrusting clients. In this paper, we follow this line of work and provide the first protocols for *pattern matching* in the cloud. In contrast to most earlier works, our constructions achieve a high-level of security, while avoiding the use of FHE and minimizing the amount of communication between the parties. We emphasize that even with the power of fully homomorphic encryption it is not clear how to get down to communication complexity that is linear in the number of matches in two rounds.[1]

Pattern Matching in the Cloud. The problem of pattern matching considers a text T of length n and a pattern of length m with the goal to find all the locations where the pattern matches the text. In a secure pattern matching protocol, one party holds the text whereas the other party holds the pattern and attempts to learn all the locations of the pattern in the text (and only that), while the party holding the text learns nothing about the pattern. Unfortunately, such protocols are not directly applicable in the cloud setting, mostly because the

---

[1] A one-round solution based on FHE would need a circuit that tolerates the maximal number of matches — which in the worst case is proportional to the length of the text.

communication overhead per search query grows linearly with the text length. Moreover, the text holder delegates its work to an external untrusted server and cannot control the content of the server's responses.

In the outsourced setting we consider a set of clients $C_T, (C_1, \ldots, C_l)$ that interact with a server S in the following way. (**1**) In a *setup phase* client C_T uploads a preprocessed text to an external server S. This phase is run only once and may be costly in terms of computation and communication. (**2**) In a *query phase* clients $C_1, \ldots, C_l$ query the text by searching patterns and learn the matched text locations. The main two goals of our approach are as follows:

1. *Simulation-based security:* We model outsourced pattern matching by a strong simulation-based security definition (cf. Section 2). Namely, we define a new reactive outsourced functionality $\mathcal{F}_{\mathrm{OPM}}$ that ensures the secrecy and integrity of the outsourced text and patterns. For instance, a semi-honest server does not gain any information about the text and patterns, except of what it can infer from the answers to the search queries. If the server is maliciously corrupted the functionality implies the correctness of the queries' replies as well. As in the standard secure computation setting, simulation-based modeling is simpler and stronger than game-based definitions.

2. *Sublinear communication complexity during query phase:* We consider an *amortized* model, where the communication and computational costs of the clients are reduced with the number of queries. More concretely, while in the setup phase communication and computation is linear in the length of the text, we want that during the query phase the overall communication and the work put by the clients is *linear in the number of matches* (which is optimal). Of course, we also require the server running in polynomial-time. Clearly, such strong efficiency requirement comes at a price as it allows the server to learn the number of matches. We model this additional information by giving the server *some leakage* for each pattern query which will be described in detail below.

1.1 Our Contribution

To simplify notation we will always only talk about a single client C that interacts with C_T and S in the query phase.

Modeling Outsourced Pattern Matching. We give a specification of an ideal execution with a trusted party by defining a reactive outsourced pattern matching functionality $\mathcal{F}_{\mathrm{OPM}}$. This functionality works in two phases: In the preprocessing phase client C_T uploads its preprocessed text $\widetilde{T}$ to the server. Next, in an iterative query phase, upon receiving a search query p the functionality asks for the approvals of client C_T (as it may also refuse for this query in the real execution), and the server (as in case of being corrupted it may abort the execution). To model the additional leakage that is required to minimize communication we ask the functionality to forward to the server the matched positions in the text upon receiving an approval from C_T. Our functionality returns all matched positions but can be modified so that only the first few matched positions are returned.[2]

Difficulties with Simulating $\mathcal{F}_{\mathrm{OPM}}$. The main challenge in designing a simulator for this functionality is in case when the server is corrupted. In this case the simulator must *commit* to some text in a way that later allows him (when taking the role of the server, given some trapdoor) to reply to pattern queries in a consistent way. More precisely, when the simulator commits to a preprocessed text, the leakage that the corrupted server obtains (namely, the positions where the pattern matches the text) has to be consistent with the information that it later sees during the query phases. This implies that the simulator must have flexibility when it later matches the committed text to the trapdoors. This difficulty does not arise in the classic two-party setting since there the simulator always plays against a party that contributes an input to the computation which it can first extract, whereas here the server is just a tool to run a computation. Due to this inherent difficulty the text must be *encoded* in a way, that given a search query p and a list of text positions $(i_1, \ldots, i_t)$, one can produce a trapdoor for p in such a way that the "search" in the preprocessed text, using this trapdoor, yields $(i_1, \ldots, i_t)$. We note that alternative solutions that permute the text to prevent the server from learning the matched positions, necessarily require that the server does not collude with the clients. In contrast, our solutions allow such strong collusion between the clients and the server.

Solutions Based on Searchable/Non-Committing Encryption. To better motivate our solution, let us consider a toy example first. Assume we encrypt each substring of length m in T using searchable encryption [4], which allows running a search over an encrypted text by producing a trapdoor for the searched word (or a pattern p). Given the trapdoor, the server can check each ciphertext and return the text positions in which the verification succeeds. The first problem that arises with this approach is that searchable encryption does not ensure the privacy of the searched patterns. While this issue may be addressed by tweaking existing constructions of searchable encryption, a more severe problem is that the simulator must commit in advance to (searchable) encryptions of a text that later allow to "find" p at positions that are consistent with the leakage. In other words: all the plaintexts in the specified positions must be associated with the keyword p ahead of time. Of course, as the simulator does not know the actual text T it cannot produce such a consistent preprocessed text. An alternative solution may be given by combining searchable encryption with techniques from non-committing encryptions [5]. Note that it is unclear how to combine these two tools even in the random oracle model.

---

[2] This definition is more applicable for search engines where the first few results are typically more relevant, whereas the former variant is more applicable for a DNA search where it is important to find all matched positions. For simplicity we only consider the first variant, our solutions support both variants.

Semi-Honest Outsourced Pattern Matching from Subset Sum. Our first construction for outsourced pattern matching is secure against semi-honest adversaries. In this construction client C_T generates a vector of random values, conditioned on that the sum of elements in all positions that match the pattern equals some specified value that will be explained below. Namely, C_T builds an instance $\widetilde{T}$ for the *subset sum problem*, where given a trapdoor R the goal is to find whether there exists a subset in $\widetilde{T}$ that sums to R. More formally, the subset sum problem is parameterized by two integers ℓ and M. An instance of the problem is generated by picking random vectors $\widetilde{T} \leftarrow \mathbb{Z}_M^\ell$, $\mathbf{s} \leftarrow \{0,1\}^\ell$ and outputting $(\widetilde{T}, R = \widetilde{T} \cdot \mathbf{s} \bmod M)$. The problem is to find $\mathbf{s}$ given $\widetilde{T}$ and a trapdoor R. Looking ahead, we will have such a trapdoor R_p for each pattern p of length m, such that if p matches T then with overwhelming probability there will be a unique solution to the subset sum instance $(\widetilde{T}, R_p)$. This unique solution is placed at exactly the positions where the pattern appears in the text. The client C that wishes to search for a pattern p obtains this trapdoor from C_T and will hand it to the server. Consequently, we are interested in easy instances of the subset sum problem since we require the server to solve it for each query. This is in contrast to prior cryptographic constructions, e.g., [18] that design cryptographic schemes based on the hardness of this problem. We therefore consider low-density instances which can be solved in polynomial time by a reduction to a short vector in a lattice [16,10,7].

We further note that the security of the scheme relies heavily on the unpredictability of the trapdoor. Namely, in order to ensure that the server cannot guess the trapdoor for some pattern p (and thus solve the subset problem and find the matched locations), we require that the trapdoor is unpredictable. We therefore employ a pseudorandom function (PRF) F on the pattern and fix this value as the trapdoor, where the key k for the PRF is picked by C_T and the two clients C_T and C communicate via a secure two-party protocol to compute the evaluation of the PRF.

Efficiency Considerations. The scheme described above does not yet satisfy the desired properties outlined in the previous paragraphs and has a very limited usage in practice. Recall that the server is asked to solve subset sum instances of the form $(\widetilde{T}, R_p)$, where $\widetilde{T}$ is a vector of length $\ell = n - m + 1$ with elements from $\mathbb{Z}_M$ for some integer M. In order to ensure correctness we must guarantee that given a subset sum instance, each trapdoor has a unique solution with high probability. In other words, the collision probability, which equals $2^\ell / M$ (stated also in [13]), should be negligible. Fixing $M = 2^{\kappa + n}$ for a security parameter κ, ensures this for a large enough κ, say whenever $\kappa \geq 80$. On the other hand, we need the subset sum problem to be solvable in polynomial time. A simple calculation (see Eq. (1)), yields in this case a value of $\ell \approx \sqrt{\kappa}$. This poses an inherent limitation on the length of the text to be preprocessed. For instance, even using a high value of $\kappa \approx 10^4$ (yielding approximately subset sum elements of size 10 KByte) limits the length of the text to only 100 bits. This scheme also requires quadratic communication complexity in the text length during the setup phase since client C_T sends $O(n^2 + \kappa n)$ bits.

An Improved Solution Using Packaging. To overcome this limitation, we employ an important extension of our construction based on *packaging.* First, the text is partitioned into smaller pieces of length $2m$ which are handled separately by the protocol, where m is some practical upper bound on the pattern length. Moreover, every two consecutive blocks are overlapping in m positions, so that we do not miss any match in the original text. Even though this approach introduces some overhead, yielding a text T' of overall length $2n$, note that now Eq. (1) yields $\ell = 2m - m + 1 = m + 1 < \sqrt{\kappa}$, which is an upper bound on the length of the pattern (and not on the length of the text as before). Namely, we remove the limitation on the text length and consider much shorter blocks lengths for the subset sum algorithm. As a result, the communication complexity in the setup phase is $O(mn + \kappa n)$, whereas the communication complexity in the query phase is $O(\kappa m)$. For short queries (which is typically the case), these measures meet the appealing properties we are sought after.

This comes at a price though since we now need to avoid using in each block the same trapdoor for some pattern p, as repetitions allow the server to extract potential valid trapdoors (that have not been queried yet) and figure out information about the text. We solve this problem by requiring from the function outputting the trapdoors to have some form of "programmability" (which allows to simulate the answers to all queries consistently). Specifically, we implement this function using the random oracle methodology on top of the PRF, so that a trapdoor now is computed by $\mathcal{H}(\mathsf{F}(k,p)\|b)$, for b being the block number. Now, the simulator can program the oracle to match with the positions where the pattern appears in each block. Note that using just the random oracle (without the PRF) is not sufficient as well, since an adversary that controls the server and has access to the random oracle can apply it on p as well.

Malicious Outsourced Pattern Matching. We extend our construction to the malicious setting as well, tolerating malicious attacks. Our proof ensures that the server returns the correct answers by employing Merkle commitments and zero-knowledge (ZK) sets. Informally speaking, Merkle commitments are succinct commitment schemes for which the commitment size is independent of the length of the committed value (or set). This tool is very useful in ensuring correctness, since now, upon committing to $\widetilde{T}$, the server decommits the solution to the subset sum trapdoor and client C can simply verify that the decommitted values correspond to the trapdoor. Nevertheless, this solution does not cover the case of a mismatch. Therefore, a corrupted server can always return a "no-match" massage. In order to resolve this technicality we borrow techniques from ZK sets arguments [19], used for proving whether an element is in a specified set without disclosing any further information. Next, proving security against a corrupted C is a straightforward extension of the semi-honest proof using the modifications we made above and the fact that the protocol for implementing the oblivious PRF evaluation is secure against malicious adversaries as well.

The case of a corrupted C_T is more challenging since we first need to extract the text T, but also verify C_T's computations with respect to the random oracle when it produces $\widetilde{T}$. The only proof technique that we are aware of for proving correctness when using a random oracle is cut-and-choose (e.g., as done in [14]), which inflates the communication complexity by an additional statistical parameter. Instead, we do not require that the server can verify immediately the correctness of the outsourced text, but only ensure that if C_T cheats with respect to some query p, then it will be caught during the query phase whenever p is queried. The crux of our protocol is that the simulator does not need to verify all computations at once, but only the computations with respect to the asked queries. This enables us to avoid the costly cut-and-choose technique since verification is done using a novel technique of *derandomizing* C_T's computations. We notice that this requires us to slightly adjust the description of our idealized functionality. For space reasons, we defer the details to the full version [9] and focus on the semi-honest case here.

We remark that all the solutions described above can be combined together into a single protocol which is secure even in the case of a collusion between S and client C. When a collusion between S and client C_T occurs we cannot guarantee either privacy or correctness since the simulator cannot extract the text, as the preprocessed protocol is "run" between the two corrupted parties. We stress that collusion does not imply that security collapses into the standard two-party setting.

2   Modeling Outsourced Pattern Matching

The outsourced pattern matching consists of two phases. In the *setup phase* a client C_T uploads a (preprocessed) text $\tilde{T}$ to an external server S. This phase is run only once. In the *query phase* client C queries the text by searching patterns and learn the matched text locations. We formalize security using the ideal/real paradigm. Denote by T_j the substring of length m that starts at text location j. The pattern matching ideal functionality in the outsourced setting is depicted in Fig. 1. We write $|T|$ for the bit length of T and assume that client C asks a number of queries p_i ($i \in [\lambda]$, $\lambda \in \mathbb{N}$).

The Definition. Formally, denote by $\mathbf{IDEAL}_{\mathcal{F}_{OPM}, \mathsf{Sim}(z)}(\kappa, (-, T, (p_1, \ldots, p_\lambda)))$ the output of an ideal adversary Sim, server S and clients C_T, C in the above ideal execution of $\mathcal{F}_{OPM}$ upon inputs $(-, (T, (p_1, \ldots, p_\lambda)))$ and auxiliary input z given to Sim.

We implement functionality $\mathcal{F}_{OPM}$ via a protocol $\pi = (\pi_{Pre}, \pi_{Query}, \pi_{Opm})$ consisting of three two-party protocols, specified as follows. Protocol π_{Pre} is run in the preprocessing phase by C_T to preprocess text T and forward the outcome $\tilde{T}$ to S. During the query phase, protocol π_{Query} is run between C_T and C (holding a pattern p); this protocol outputs a trapdoor R_p that depends on p and will enable the server to search the preprocessed text. Lastly, protocol π_{Opm} is run by S upon input the preprocessed text and trapdoor R_p (forwarded by C); this protocol returns C the matched text positions (if any). We denote by $\mathbf{REAL}_{\pi, \mathsf{Adv}(z)}(\kappa, (-, T, (p_1, \ldots, p_\lambda)))$ the output of adversary Adv, server S and clients C_T, C in a real execution of $\pi = (\pi_{Pre}, \pi_{Query}, \pi_{Opm})$ upon inputs $(-, (T, (p_1, \ldots, p_\lambda)))$ and auxiliary input z given to Adv.

Functionality $\mathcal{F}_{\text{OPM}}$

Let $m, \lambda \in \mathbb{N}$. Functionality $\mathcal{F}_{\text{OPM}}$ sets the table $\mathcal{B}$ initially to the empty and proceeds as follows, running with clients C_T and C, server S and adversary Sim.

1. Upon receiving a message (text, T, m) from C_T, send (preprocess, $|T|, m$) to S and Sim, and record (text, T).
2. Upon receiving a message (query, p_i) from client C (for $i \in [\lambda]$), where message (text, $\cdot$) has been recorded and $|p_i| = m$, it checks if the table $\mathcal{B}$ already contains an entry of the form $(p_i, \cdot)$. If this is not the case then it picks the next available identifier id from $\{0,1\}^*$ and adds (p_i, id) to $\mathcal{B}$. It sends (query, C) to C_T and Sim.
   (a) Upon receiving (approve, C) from client C_T, read (p_i, id) from $\mathcal{B}$ and send (query, $C, (i_1, \ldots, i_t), \text{id}$) to server S, for all text positions $\{i_j\}_{j \in [t]}$ such that $T_{i_j} = p_i$. Otherwise, if no (approve, C) message has been received from C_T, send $\perp$ to C and abort.
   (b) Upon receiving (approve, C) from Sim, read (p_i, id) from $\mathcal{B}$ and send (query, $p_i, (i_1, \ldots, i_t), \text{id}$) to client C. Otherwise, send $\perp$ to client C.

Fig. 1. The outsourced pattern matching functionality

Definition 1 (Security of outsourced pattern matching). *We say that π securely implements $\mathcal{F}_{\text{OPM}}$, if for any PPT real adversary* Adv *there exists a PPT simulator* Sim *such that for any tuple of inputs $(T, (p_1, \ldots, p_\lambda))$ and auxiliary input z,*

$$\{\mathbf{IDEAL}_{\mathcal{F}_{\text{OPM}}, \text{Sim}(z)}(\kappa, (-, T, (p_1, \ldots, p_\lambda)))\}_{\kappa \in \mathbb{N}}$$
$$\stackrel{c}{\approx} \{\mathbf{REAL}_{\pi, \text{Adv}(z)}(\kappa, (-, T, (p_1, \ldots, p_\lambda)))\}_{\kappa \in \mathbb{N}}.$$

The schemes described in the next sections, implement the ideal functionality $\mathcal{F}_{\text{OPM}}$ in the random oracle model.

3   A Scheme with Passive Security

In this section we present our implementation of the outsourced pattern matching functionality $\mathcal{F}_{\text{OPM}}$ that is formalized in Fig. 1, and prove its security against semi-honest adversaries. A scheme with security against malicious adversaries is described in the full version of this paper [9], building upon the protocol in this section. Recall first that in the outsourced variant of the pattern matching problem, client C_T preprocesses the text T and then stores it on the server S in such a way that the preprocessed text can be used later to answer search queries submitted by client C. The challenge is to find a way to hide the text (in order to obtain privacy), while enabling the server to carry out searches on the hidden text whenever it is in possession of an appropriate trapdoor.

Protocol $\pi_{\mathsf{SH}} = (\pi_{\mathsf{Pre}}, \pi_{\mathsf{Query}}, \pi_{\mathsf{Opm}})$

Let $\kappa \in \mathbb{N}$ be the security parameter and let M, m, n, μ be integers, where for simplicity we assume that n is a multiple of $2m$. Further, let $\mathcal{H} : \{0,1\}^\mu \to \mathbb{Z}_M$ be a random oracle and $\mathsf{F} : \{0,1\}^\kappa \times \{0,1\}^m \to \{0,1\}^\mu$ be a PRF. Protocol π_{SH} involves a client C_T holding a text $T \in \{0,1\}^n$, a client C querying for patterns $p \in \{0,1\}^m$, and a server S. The interaction between the parties is specified below.

Setup phase, π_{Pre}. The protocol is invoked between client C_T and server S. Given input T and integer m, client C_T picks a random key $k \in \{0,1\}^\kappa$ and prepares first the text T for the packaging by writing it as

$$T' := (B_1, \ldots, B_u) = ((T[1], \ldots, T[2m]),$$
$$(T[m+1], \ldots, T[3m]), \ldots, (T[n-2m+1], \ldots, T[n])),$$

where $u = n/m - 1$. Next, for each block B_b and each of the $m + 1$ patterns $p \in \{0,1\}^m$ that appear in B_b we proceed as follows (suppose there are at most t matches of p in B_b).

1. Client C_T evaluates $R_p := \mathcal{H}(\mathsf{F}(k,p)\|b)$, samples $a_1, \ldots, a_{t-1} \in \mathbb{Z}_M$ at random and then fixes a_t such that $a_t = R_p - \sum_{j=1}^{t-1} a_j \bmod M$.

2. Set $\widetilde{B}_b[v_j] = a_j$ for all $j \in [t]$ and $v_j \in [m+1]$. Note that here we denote by $\{v_j\}_{j \in [t]}$ $(v_j \in [m+1])$ the set of indexes corresponding to the positions where p occurs in B_b. Later in the proof we will be more precise and explicitly denote to which block v_j belongs by using explicitly the notation v_{j_b}.

Finally, we outsource the text $\widetilde{T} = (\widetilde{B}_1, \ldots, \widetilde{B}_u)$ to S.

Query phase, π_{Query}. Upon issuing a query $p \in \{0,1\}^m$ by client C, clients C_T and C engage in an execution of protocol π_{Query} which implements the oblivious PRF functionality $(k,p) \mapsto (-, \mathsf{F}(k,p))$. Upon completion, C learns $\mathsf{F}(k,p)$.

Oblivious pattern matching phase, π_{Opm}. This protocol is executed between server S (holding $\widetilde{T}$) and client C (holding $\mathsf{F}(k,p)$). Upon receiving $\mathsf{F}(k,p)$ from C, the server proceeds as follows for each block $\widetilde{B}_b$. It interprets $(\mathcal{H}(\mathsf{F}(k,p)\|b), \widetilde{B}_b)$ as a subset sum instance and computes $\mathbf{s}$ as the solution of $\widetilde{B}_b \cdot \mathbf{s} = \mathcal{H}(\mathsf{F}(k,p)\|b)$. Let $\{v_j\}_{j \in [t]}$ denote the set of indexes such that $\mathbf{s}[v_j] = 1$, then the server S returns the set of indexes $\{\varphi(b, v_j)\}_{b \in [u], j \in [t]}$ to the client C.

Fig. 2. Semi-honest outsourced pattern matching

We consider a new approach and reduce the pattern matching problem to the subset sum problem. Namely, consider a text T of length n, and assume we want to allow to search for patterns of length m. For some integer $M \in \mathbb{N}$, we assign to each distinct pattern p that appears in T a random element $R_p \in \mathbb{Z}_M$. Letting $\ell = n - m + 1$, the preprocessed text $\widetilde{T}$ is a vector in $\mathbb{Z}_M^\ell$ with elements specified as follows. For each pattern p that appears t times in T, we sample random values $a_1, \ldots, a_t \in \mathbb{Z}_M$ such that $R_p = \sum_{j=1}^t a_j$. Denote with $i_j \in [\ell]$ the jth position in T where p appears and set $\widetilde{T}[i_j] = a_j$. Notice that for each pattern p, there exists a vector $\mathbf{s} \in \{0,1\}^\ell$ such that $R_p = \widetilde{T} \cdot \mathbf{s}$. Hence, the positions in $\widetilde{T}$ where pattern p matches are identified by a vector $\mathbf{s}$ and can be viewed as the solution for the subset sum problem instance $(R_p, \widetilde{T})$.

Roughly, our protocol works as follows. During protocol π_{Pre}, we let the client C_T generate the preprocessed text $\widetilde{T}$ as described above, and send the result to the server S. Later, when a client C wants to learn at which positions a pattern p matches in the text, clients C and C_T run protocol π_{Query}; at the end of this protocol, C learns the trapdoor R_p corresponding to p. Hence, during π_{Opm}, client C sends this trapdoor to S, which can solve the subset sum problem instance $(R_p, \widetilde{T})$. The solution to this problem corresponds to the matches of p, which are forwarded to the client C. To avoid that C_T needs to store all trapdoors, we rely on a PRF to generate the trapdoors itself. More precisely, instead of sampling the trapdoors R_p uniformly at random, we set $R_p := \mathsf{F}(k, p)$, where F is a PRF. Thus, during the query phase C and C_T run an execution of an oblivious PRF protocol; where C learns the output of the PRF, i.e., the trapdoor R_p.

Efficiency. Although the protocol described above provides a first basic solution for the outsourced pattern matching, it suffers from a strong restriction as only very short texts are supported. (On the positive side, the above scheme does not rely on a random oracle.) The server S is asked to solve subset sum instances of the form $(\widetilde{T}, R_p)$, where $\widetilde{T}$ is a vector of length $\ell = n - m + 1$ with elements from $\mathbb{Z}_M$ for some integer M. To achieve correctness, we require that each subset sum instance has a unique solution with high probability. In order to satisfy this property, one needs to set the parameters such that the value $2^\ell / M$ is negligible. Fixing $M = 2^{\kappa + \ell}$ achieves a reasonable correctness level.

On the other hand, we need to let S solve subset sum instances efficiently. The hardness of subset sum depends on the ratio between ℓ and $\log M$, which is usually referred to as the *density* Δ of the subset sum instance. In particular both instances with $\Delta < 1/\ell$ (so called *low-density* instances) and $\Delta > \ell / \log^2 \ell$ (so called *high-density* instances) can be solved in polynomial time. Note that, however, the constraint on the ratio $2^\ell / M$ immediately rules out algorithms for high-density subset sum (e.g., algorithms based on dynamic programming, since they usually need to process a matrix of dimension M). On the other hand, for low-density instances, an easy calculation shows that $\ell + \kappa > \ell^2$, so that we need to choose κ, ℓ in such a way that

$$\ell < \frac{1}{2}\left(\sqrt{1 + 4\kappa} - 1\right). \tag{1}$$

The above analysis yields a value of $\ell \approx \sqrt{\kappa}$. This poses an inherent limitation on the length of the text. For instance, even using $\kappa \approx 10^4$ (yielding approximately subset sum elements of size 10 KByte) limits the length of the text to only 100 bits.

Packaging. To overcome this severe limitation, we partition the text into smaller pieces each of length $2m$, where each such piece is handled as a separate instance of the protocol. More specifically, for a text $T = (T[1], \dots, T[n])$ let $(T[1], \dots, T[2m])$, $(T[m+1], \dots, T[3m]), \dots$ be blocks, each of length $2m$, such that every two consecutive blocks overlap in m bits. Then, for each pattern p that appears in the text the client C_T computes an individual trapdoor for each block where the pattern p appears. In other words, suppose that pattern p appears in block B_b then we compute the trapdoor for this block (and pattern p) as $\mathcal{H}(\mathsf{F}(k,p)||b)$. Here, $\mathcal{H}$ is a cryptographic hash function that will be modeled as a random oracle in our proofs. Given the trapdoors, we apply the preprocessing algorithm to each block individually.

The sub-protocols π_{Query} and π_{Opm} work as described above with a small change. In π_{Query} client C learns the output of the PRF $\mathsf{F}(k,p)$ instead of the actual trapdoors and in π_{Opm} client C forwards directly the result $\mathsf{F}(k,p)$ to S. The server can then compute the actual trapdoor using the random oracle. This is needed to keep the communication complexity of the protocol low. Note that in this case if we let $\{v_{j_b}\}_{j_b \in [t_b]}$ be the set of indices corresponding to the positions where p occurs *in a given block* B_b, the server needs to map these positions to the corresponding positions in T (and this has to be done for each of the blocks where p matches). It is easy to see that such a mapping from a position v_{j_b} in block B_b to the corresponding position in the text T can be computed as $\varphi(b, v_j) = (b-1)m + v_j$. The entire protocol is shown in Fig. 2.

Note that now each of the preprocessed blocks $\widetilde{B}_b$ consist of $\ell = m+1$ elements in $\mathbb{Z}_M$. The advantage is that the blocks are reasonably short which yields subset sum instances of the form $(\widetilde{B}_b, R_p)$. Combined with Eq. (1) this yields a value of $\ell = 2m - m + 1 = m + 1 < \sqrt{\kappa}$, which is an upper bound on the length of the pattern (and not on the length of the text as before). By combining many blocks we can support texts of *any length* polynomial in the security parameter. Finally, we emphasize that the communication/computational complexities of π_{Query} depends on the underlying oblivious PRF evaluation. This in particular only depends on m (due to the algebraic structure of the [21] PRF). Using improved PRFs can further reduce the communication complexity. On the other hand, the communication complexity of π_{Opm} is dominated by the number of matches of p in T which is optimal.

We state the following result. The proof can be found in the full version [9].

Theorem 1. *Let $\kappa \in \mathbb{N}$ be the security parameter. For integers n, m we set $\lambda = \mathsf{poly}(\kappa), \mu = \mathsf{poly}(\kappa), u = n/m - 1, \ell = (m+1)u$ and $M = 2^{m+\kappa+1}$. We furthermore require that κ is such that $2^{m+1}/M$ is negligible (in κ). Assume $\mathcal{H} : \{0,1\}^\mu \to \mathbb{Z}_M$ is a random oracle and $\mathsf{F} : \{0,1\}^\kappa \times \{0,1\}^m \to \{0,1\}^\mu$ is a pseudorandom function. Then, protocol π_{SH} from Fig. 2 securely implements the $\mathcal{F}_{\mathrm{OPM}}$ functionality in the presence of semi-honest adversaries.*

References

1. Applebaum, B., Ishai, Y., Kushilevitz, E.: From secrecy to soundness: Efficient verification via secure computation. In: Abramsky, S., Gavoille, C., Kirchner, C., Meyer auf der Heide, F., Spirakis, P.G. (eds.) ICALP 2010. LNCS, vol. 6198, pp. 152–163. Springer, Heidelberg (2010)
2. Asharov, G., Jain, A., López-Alt, A., Tromer, E., Vaikuntanathan, V., Wichs, D.: Multiparty computation with low communication, computation and interaction via threshold FHE. In: Pointcheval, D., Johansson, T. (eds.) EUROCRYPT 2012. LNCS, vol. 7237, pp. 483–501. Springer, Heidelberg (2012)

3. Benabbas, S., Gennaro, R., Vahlis, Y.: Verifiable delegation of computation over large datasets. In: Rogaway, P. (ed.) CRYPTO 2011. LNCS, vol. 6841, pp. 111–131. Springer, Heidelberg (2011)
4. Boneh, D., Di Crescenzo, G., Ostrovsky, R., Persiano, G.: Public key encryption with keyword search. In: Cachin, C., Camenisch, J.L. (eds.) EUROCRYPT 2004. LNCS, vol. 3027, pp. 506–522. Springer, Heidelberg (2004)
5. Canetti, R., Feige, U., Goldreich, O., Naor, M.: Adaptively secure multi-party computation. In: STOC, pp. 639–648 (1996)
6. Chung, K.-M., Kalai, Y., Vadhan, S.: Improved delegation of computation using fully homomorphic encryption. In: Rabin, T. (ed.) CRYPTO 2010. LNCS, vol. 6223, pp. 483–501. Springer, Heidelberg (2010)
7. Coster, M.J., Joux, A., LaMacchia, B.A., Odlyzko, A.M., Schnorr, C.-P., Stern, J.: Improved low-density subset sum algorithms. Computational Complexity 2, 111–128 (1992)
8. Damgård, I., Faust, S., Hazay, C.: Secure two-party computation with low communication. In: Cramer, R. (ed.) TCC 2012. LNCS, vol. 7194, pp. 54–74. Springer, Heidelberg (2012)
9. Faust, S., Hazay, C., Venturi, D.: Outsourced pattern matching. Cryptology ePrint Archive, Report 2013/XX, http://eprint.iacr.org/
10. Frieze, A.M.: On the lagarias-odlyzko algorithm for the subset sum problem. SIAM J. Comput. 15(2), 536–539 (1986)
11. Gennaro, R., Gentry, C., Parno, B.: Non-interactive verifiable computing: Outsourcing computation to untrusted workers. In: Rabin, T. (ed.) CRYPTO 2010. LNCS, vol. 6223, pp. 465–482. Springer, Heidelberg (2010)
12. Gentry, C.: Fully homomorphic encryption using ideal lattices. In: STOC, pp. 169–178 (2009)
13. Impagliazzo, R., Naor, M.: Efficient cryptographic schemes provably as secure as subset sum. J. Cryptology 9(4), 199–216 (1996)
14. Ishai, Y., Kilian, J., Nissim, K., Petrank, E.: Extending oblivious transfers efficiently. In: Boneh, D. (ed.) CRYPTO 2003. LNCS, vol. 2729, pp. 145–161. Springer, Heidelberg (2003)
15. Kamara, S., Mohassel, P., Raykova, M.: Outsourcing multi-party computation. IACR Cryptology ePrint Archive, 2011:272 (2011)
16. Lagarias, J.C., Odlyzko, A.M.: Solving low-density subset sum problems. J. ACM 32(1), 229–246 (1985)
17. López-Alt, A., Tromer, E., Vaikuntanathan, V.: On-the-fly multiparty computation on the cloud via multikey fully homomorphic encryption. In: STOC, pp. 1219–1234 (2012)
18. Lyubashevsky, V., Palacio, A., Segev, G.: Public-key cryptographic primitives provably as secure as subset sum. In: Micciancio, D. (ed.) TCC 2010. LNCS, vol. 5978, pp. 382–400. Springer, Heidelberg (2010)
19. Micali, S., Rabin, M.O., Kilian, J.: Zero-knowledge sets. In: FOCS, pp. 80–91 (2003)
20. Mohassel, P.: Efficient and secure delegation of linear algebra. IACR Cryptology ePrint Archive 2011:605 (2011)
21. Naor, M., Reingold, O.: Number-theoretic constructions of efficient pseudo-random functions. In: FOCS, pp. 458–467 (1997)
22. Papamanthou, C., Tamassia, R., Triandopoulos, N.: Optimal verification of operations on dynamic sets. In: Rogaway, P. (ed.) CRYPTO 2011. LNCS, vol. 6841, pp. 91–110. Springer, Heidelberg (2011)

Learning a Ring Cheaply and Fast[*]

Emanuele G. Fusco[1], Andrzej Pelc[2],[**], and Rossella Petreschi[1]

[1] Computer Science Department, Sapienza, University of Rome, 00198 Rome, Italy
{fusco,petreschi}@di.uniroma1.it
[2] Département d'informatique, Université du Québec en Outaouais,
Gatineau, Québec J8X 3X7, Canada
pelc@uqo.ca

Abstract. We consider the task of learning a ring in a distributed way: each node of an unknown ring has to construct a labeled map of it. Nodes are equipped with unique labels. Communication proceeds in synchronous rounds. In every round every node can send arbitrary messages to its neighbors and perform arbitrary local computations. We study tradeoffs between the time (number of rounds) and the cost (number of messages) of completing this task in a deterministic way: for a given time T we seek bounds on the smallest number of messages needed for learning the ring in time T. Our bounds depend on the diameter D of the ring and on the *delay* $\theta = T - D$ above the least possible time D in which this task can be performed. We prove a lower bound $\Omega(D^2/\theta)$ on the number of messages used by any algorithm with delay θ, and we design a class of algorithms that give an almost matching upper bound: for any positive constant $0 < \varepsilon < 1$ there is an algorithm working with delay $\theta \leq D$ and using $O(D^2(\log^* D)/\theta^{1-\varepsilon})$ messages.

Keywords: labeled ring, message complexity, time, tradeoff.

1 Introduction

The Model and the Problem. Constructing a labeled map of a network is one of the most demanding distributed tasks that nodes can accomplish in a network. Each node has a distinct label and in the beginning each node knows only its own label. Moreover, ports at each node of degree d are arbitrarily numbered $0, \ldots, d - 1$. The goal is for each node to get an isomorphic copy of the graph underlying the network, including node labels and port numbers. Once nodes acquire this map, any other distributed task, such as leader election [9,13], minimum weight spanning tree construction [2], renaming [1], etc. can be performed by nodes using only local computations. Thus constructing a labeled map converts all distributed network problems to centralized ones, in the sense

---

[*] This research was done during the visit of Andrzej Pelc at Sapienza, University of Rome, partially supported by a visiting fellowship from this university.
[**] Partially supported by NSERC discovery grant and by the Research Chair in Distributed Computing at the Université du Québec en Outaouais.

F.V. Fomin et al. (Eds.): ICALP 2013, Part II, LNCS 7966, pp. 557–568, 2013.

that nodes can solve them simulating a central monitor. We are interested in the efficiency of deterministic algorithms for labeled map construction.

In this paper we use the extensively studied $\mathcal{LOCAL}$ model of communication [12]. In this model, communication proceeds in synchronous rounds and all nodes start simultaneously. In each round each node can exchange arbitrary messages with all its neighbors and perform arbitrary local computations. The time of completing a task is the number of rounds it takes. Our goal is to investigate tradeoffs between the *time* of constructing a labeled map and its *cost*, i.e., the number of messages needed to perform this task. To see extreme examples of such a tradeoff, consider the map construction task on an n-node ring. The fastest way to complete this task is in time D, where $D = \lfloor n/2 \rfloor$ is the diameter of the ring. This can be achieved by flooding, but the number of messages used is then $\Theta(n^2)$. On the other hand, cost $\Theta(n)$ (which is optimal) can be achieved by a version of the time slicing algorithm [11], but then time may become very large and depends on the labels of the nodes.

The general problem of tradeoffs between time and cost of labeled map construction can be formulated as follows.

> For a given time T, what is the smallest number of messages needed for constructing a labeled map by each node in time T?

For trees this problem is trivial: leaves of an n-node tree initiate the communication process and information about ever larger subtrees gets first to the central node (or central pair of adjacent nodes) and then back to all leaves, using time equal to the diameter of the tree and $O(n)$ messages, both of which are optimal. However, as soon as there are cycles in the network, there is no canonical place to start information exchange on each cycle and proceeding fast seems to force many messages to be sent in parallel, which in turn intuitively implies large cost. This phenomenon is present already in the simplest such network, i.e., the ring. Indeed, our study shows that meaningful tradeoffs between time and cost of labeled map construction already occur in rings.

We consider rings whose nodes have unique labels that are binary strings of length polynomial in the size of the ring. (Our results are valid also for much longer labels, but these can be dismissed for practicality reasons.) In the beginning, every node knows only its own label, the allowed time T and the diameter D of the ring. Equivalently, we provide each node with its label, with the diameter D and with the *delay* $\theta = T - D$, which is the extra time allowed on top of the minimum time D in which labeled map construction can be achieved, knowing D a priori.

Knowing its own label is an obvious assumption. Without any additional knowledge, nodes would have to assume the least possible time and hence do flooding at quadratic cost. Instead of providing nodes with D and θ, we could have provided them only with the allowed delay over the least possible time of learning the ring *without* a priori knowledge of the diameter. This would not affect our asymptotic bounds. However, it would result in more cumbersome formulations because, without knowing D a priori, the optimal time of labeled map construction varies between D and $D+1$, depending on whether the ring is of even or odd size. We are interested in achieving map construction with small delay: in particular, we assume $\theta \leq D$.

We assume that messages are of arbitrary size, but in our algorithms they need only to be sufficiently large to contain already acquired information about the n-node ring, i.e., strings of up to n labels and port numbers. This is a natural assumption for the task of labeled map construction whose output has large size, similarly as is done, e.g., in gossiping [7]. This should be contrasted with such tasks as leader election [9], distributed minimum weight spanning tree construction [2], or distributed coloring [8], where each node has to output only a small amount of information, and considered messages are often of small size.

Our Results. We prove almost tight upper and lower bounds on the minimum cost (number of messages) needed to deterministically perform labeled map construction on a ring in a given time. Our bounds depend on the diameter D of the ring and on the *delay* $\theta = T - D$ above the least possible time D in which this task can be performed. We prove a lower bound $\Omega(D^2/\theta)$ on the cost of any algorithm with delay θ, and we design a class of algorithms that give an almost matching upper bound: for any positive constant $0 < \varepsilon < 1$ there is an algorithm working with delay $\theta \leq D$ and using $O(D^2(\log^* D)/\theta^{1-\varepsilon})$ messages. We also provide tradeoffs between time and cost of labeled map construction for a more general class of graphs, when the delay is larger.

Due to the lack of space, several proofs are omitted.

Related Work. The task of constructing a map of a network has been studied mostly for anonymous networks, both in the context of message passing systems [14] and using a mobile agent exploring a network [3]. The goal was to determine the feasibility of map construction (also called topology recognition) and to find fast algorithms performing this task. For networks with unique labels, map construction is of course always feasible and can be done in time equal to the diameter of the network plus one (in the $\mathcal{LOCAL}$ model), which is optimal.

Tradeoffs between the time and the number of messages have been studied for various network problems, including leader election [6,9,13] weak unison [10], and gossiping [5]. It should be noticed that if the requirement concerning time is loose, i.e., concerns only the order of magnitude, then there are no tradeoffs to speak of for labeled map construction. It follows from [2] that minimum weight spanning tree construction can be done in time $O(n)$ and at cost $O(m+n\log n)$ in any network with n nodes and m edges, both of which are known to be optimal. This implies the same complexities for constructing a labeled map. However, our results show that the task of labeled map construction is very sensitive to time: time vs. cost tradeoffs occur for the ring between the time spans D and $2D$.

To the best of our knowledge, the problem of time vs. cost tradeoffs for labeled map construction has never been studied before.

2 The Lower Bound

The main result of this section is a lower bound $\Omega(D^2/\theta)$ on the cost of any labeled map construction algorithm working with delay θ on a ring with diameter D.

We prove the lower bound on the class of oriented rings of even size. (Restricting the class on which the lower bound is proved only increases the strength of the result.) We formalize orientation by assigning port numbers 0 and 1 in the clockwise order at each node. For every node v, let $\ell(v)$ be its label.

We first define the *history* $H(v,t)$ of node v at time t. Intuitively $H(v,t)$ represents the entire knowledge that node v can acquire by time t. Since we want to prove a lower bound on cost, it is enough to assume that whenever a node v sends a message to a neighbor in round $t+1$, the content of this message is its entire history $H(v,t)$. We define histories of all nodes by simultaneous induction on t. Define $H(v,0)$ as the one-element sequence $\langle \ell(v) \rangle$. In the inductive definition, we will use two symbols, s_0 and s_1, corresponding to the lack of message (silence) on port 0 and 1, respectively. Assume that histories of all nodes are defined until round t. We define $H(v,t+1)$ as:

- $\langle H(v,t), s_0, s_1 \rangle$, if v did not get any message in round $t+1$;
- $\langle H(v,t), s_0, H(u,t) \rangle$, if v did not get any message in round $t+1$ on port 0 but received a message on port 1 from its clockwise neighbor u in that round;
- $\langle H(v,t), H(w,t), s_1 \rangle$, if v did not get any message in round $t+1$ on port 1 but received a message on port 0 from its counterclockwise neighbor w in that round;
- $\langle H(v,t), H(w,t), H(u,t) \rangle$, if v received a message on port 0 from its counterclockwise neighbor w and a message on port 1 from its clockwise neighbor u, in round $t+1$.

We define a *communication pattern* until round t for the set E of all edges of the ring as a function $f : E \times \{1,\ldots,t\} \longrightarrow \{0,1\}$, where $f(e,i) = 0$, if and only if no message is sent on edge e in round i. Executing a map construction algorithm A on a given ring determines a communication pattern, which in turn determines histories $H(v,t)$, for all nodes v and all rounds t.

For any path $\pi_k = \langle u_0 \ldots u_k \rangle$ between nodes u_0 and u_k we define, by induction on k, the *communication delay* $\delta(\pi_k, f)$ induced on π_k by the communication pattern f. For $k = 1$, $\delta(\pi_1, f) = d$, if and only if, $f(\{u_0, u_1\}, i+1) = 0$, for all $i < d$, and $f(\{u_0, u_1\}, d+1) = 1$. In particular, if $f(\{u_0, u_1\}, 1) = 1$ then $\delta(\pi_1, f) = 0$. Suppose that $\delta(\pi_{k-1}, f)$ has been defined. We define $\delta(\pi_k, f) = \delta(\pi_{k-1}, f) + d$, if and only if, $f(\{u_{k-1}, u_k\}, \delta(\pi_{k-1}, f) + k + i) = 0$, for all $i < d$, and $f(\{u_{k-1}, u_k\}, \delta(\pi_{k-1}, f) + k + d) = 1$. In particular, if $f(\{u_{k-1}, u_k\}, \delta(\pi_{k-1}, f) + k) = 1$ then $\delta(\pi_k, f) = \delta(\pi_{k-1}, f)$. Intuitively the communication delay on a path between u and v indicates the additional time, with respect to the length of this path, that it would take node v to acquire any information about node u, along this path, if no information could be coded by silence. In fact some information can be coded by silence, and analyzing this phenomenon is the main conceptual difficulty of our lower bound proof. In particular, we will show that if map construction has to be performed quickly, then the number of configurations that can be coded by silence is small with respect to the total number of possible instances, and hence many messages have to be used for some of them.

We define the communication delay induced by a communication pattern f between a node x and its antipodal node $\bar{x}$ as the minimum of the delays induced by f on the two paths connecting x and $\overline{x}$. By $N(v, i)$ we denote the neighborhood of v with radius i, i.e., the set of nodes at distance at most i from v, including v itself. We also use $N_{\leftarrow}(v, i)$ and $N_{\rightarrow}(v, i)$ to denote the part of the neighborhood $N(v, i)$ clockwise (respectively counterclockwise) from v, including v itself.

The next lemma will be used to provide a necessary condition for correctness of a map construction algorithm working with a given delay. This condition will be crucial in proving the lower bound on the cost of such algorithms.

Lemma 1. *Let A be a labeled map construction algorithm. Let R and R' be two rings of size $2D$, such that R' is obtained from R by changing the label $\ell(x)$ of a single node x to label $\ell'(x)$. Let $\overline{x}$ be the antipodal node of node x. Assume that A determines the same communication pattern f on R and R'. Let τ be the delay induced by f between x and $\overline{x}$. Then the history $H(\overline{x}, D + \tau - 1)$ is the same in R and R'.*

Theorem 1. *Any labeled map construction algorithm A working with delay θ on the class of rings of diameter D has cost $\Omega(D^2/\theta)$.*

Proof. Let A be a labeled map construction algorithm working with delay θ on the class of rings of diameter D. Consider an oriented ring R of size $2D$. We will assign labels to nodes in R in such a way that A uses at least $D\lfloor D/(\theta + 1)\rfloor$ messages.

If, for some node x there exist two labels $\ell(x)$ and $\ell'(x)$ such that, for some labeling of the remaining nodes, the communication pattern f determined by A on both resulting rings is the same, and the delay induced by f between x and its antipodal node $\overline{x}$ is larger than θ, then algorithm A is incorrect, by Lemma 1. Indeed, the history $H(\overline{x}, D + \theta)$ would be the same in both labeled rings, and node $\overline{x}$ would fail in the correct construction of the labeled map for one of them.

Hence, for any node x, there can be only one label $\ell(x)$ for any communication pattern inducing communication delay larger than θ between x and $\overline{x}$. Let X be the set of all such labels. Since there are at most $D + \theta \leq 2D$ rounds of communication, there are at most $(2D)^{2D}$ distinct communication patterns. Hence $|X| \leq (2D)^{2D}$. Let $x_0, x_1, \ldots, x_{2D-1}$ be the clockwise enumeration of all nodes in the ring. Assign the lexicographically smallest label $\ell_i \notin X \cup \{\ell_j : j < i\}$ to node x_i. Recall that by our assumption, labels are binary sequences of length polynomial in D. In fact for the purpose of this proof it is enough to work with sequences of length bounded by D^2. Indeed, $(2D)^{2D} \in o(2^{D^2})$, hence there are enough available labels outside of X for the construction of our labeled ring.

We will show that algorithm A uses at least $D\lfloor D/(\theta + 1)\rfloor$ messages for the above labeling of ring R. Let g be the communication pattern induced by algorithm A on this labeled ring. Let $y_0, y_1, \ldots, y_{\lfloor 2D/(\theta+1)\rfloor - 1}$ be nodes of the ring R such that y_{i+1} is at clockwise distance $\theta + 1$ from y_i. Assume, without loss of generality, that for at least $\lfloor D/(\theta+1)\rfloor$ nodes y_j, the communication delay on the clockwise path between y_j and its antipodal node $\overline{y_j}$ is at most θ. Let Y be the set of these nodes y_j. For any node $y = x_h$ in Y, we define the set Z_y of size D as follows. (All additions of indices are modulo $2D$.) Elements of Z_y are ordered pairs composed of an edge and a round number, of the form $(\{x_{h+p}, x_{h+p+1}\}, r_p)$, for $0 \leq p < D$, where $r_p = \delta(\langle x_h \ldots x_{h+p+1}\rangle, g) + p + 1$. By the definition of communication delay induced on a path, we have $g(\{x_{h+p}, x_{h+p+1}\}, r_p) = 1$, for all $0 \leq p < D$.

We now show that sets Z_y are pairwise disjoint. Pick two nodes from Y: a node $y = x_h$ and a node $y' = x_{h+d}$ at clockwise distance $d < D$ from x_h. Consider a node x_{h+p}, with $d \leq p < D$. Consider two pairs, $(\{x_{h+p}, x_{h+p+1}\}, r_p) \in Z_{x_h}$ and $(\{x_{h+p}, x_{h+p+1}\}, r_{p-d}) \in Z_{x_{h+d}}$. Since $\delta(\langle x_{h+d} \ldots x_{h+p+1}\rangle, g) \leq \theta$, we have $r_{p-d} = p - d + \delta(\langle x_{h+d} \ldots x_{h+p+1}\rangle, g) + 1 \leq p - d + \theta + 1$. By definition of Y, we have $d > \theta$, hence $r_p = p + \delta(\langle x_h \ldots x_{h+p+1}\rangle, g) + 1 \geq p + 1 > p - d + \theta + 1 \geq r_{p-d}$. This implies that $r_p \neq r_{p-d}$ and hence sets Z_y and $Z_{y'}$ are disjoint. Notice that if y' is at distance D from y, then y' is the antipodal node of y and hence Z_y and $Z_{y'}$ are disjoint because the edges in their elements are different. It follows that all sets Z_y are pairwise disjoint, hence $\cup_{y \in Y} Z_y$ has at least $D\lfloor D/(\theta+1)\rfloor$ elements. Since each element corresponds to at least one message sent, we conclude that the algorithm uses at least $D\lfloor D/(\theta + 1)\rfloor \in \Omega(D^2/\theta)$ messages.   □

3   The Algorithm

The general idea of our labeled map construction algorithm is to spend the allowed delay θ in a preprocessing phase that deactivates some nodes, using the residual time D for a phase devoted to information spreading. This results in a reduction of the overall cost of the algorithm, with respect to flooding, since non-active nodes are only responsible for relaying messages originated at nodes that remained active after the preprocessing phase. Hence, this approach requires to deactivate as many nodes as possible. However, within delay θ, we cannot afford to deactivate sequences of consecutive nodes of length larger than 2θ. Indeed, deactivating such long sequences would imply that the label of some non-active node is unknown to all active nodes, which would make the time of the information spreading phase exceed the remaining D rounds. We reconcile these opposite requirements by defining local rules that allow us to deactivate almost half of the currently active nodes, without deactivating two consecutive ones. This process is then iterated as many times as possible within delay θ.

The preprocessing phase of our algorithm is divided into stages, each of which is in turn composed of multiple steps. In the first stage, all nodes are *active*. Nodes that become *non-active* at the end of a stage will never become active again. In order to simplify the description of the algorithm, we will use the concept of *residual* ring. In such a ring, the set of nodes is a subset of the original set of nodes, and edges correspond to paths of consecutive removed nodes. In particular, stage i is executed on the residual ring R_i composed of nodes that remained active at the end of the previous stage. Communication between consecutive nodes R_i is simulated by a multi-hop communication in the original ring, where non-active nodes relay messages of active nodes. Each simulated message exchange during stage i is allotted 2^{i-1} rounds.

Steps inside stage i are devoted to the election of (i, j)-*leaders*, where j is the number of the step. At the beginning of the first step of stage i, $(i, 0)$-leaders are all still active nodes. Step j of stage i is executed on the residual ring $R_{i,j}$ composed of $(i, j-1)$-leaders from the step $j-1$. Multi-hop communication between two consecutive nodes in $R_{i,j}$ is allotted $2^{i-1}4^{j-1}$ rounds.

Whenever a node v (active or not) sends or relays a message to its neighbor w, it appends to the message its label and the port number, at v, corresponding to the edge $\{v, w\}$. In order to simplify the description of the algorithm, we omit these message parts. We use log to denote logarithms to base two.

We first introduce three procedures that will be used as parts of our algorithm. The first procedure is due to Cole and Vishkin [4] and Goldberg et al. [8]. It colors every ring with at most three colors, so that adjacent nodes have distinct colors. We call it Procedure RTC as an abbreviation of ring three coloring.

Procedure RTC
Input: i, j.
The procedure starts from a ring whose nodes have unique labels of k bits and produces a coloring of the ring using at most 3 colors in time $O(\log^* k)$. Let $\{1, 2, 3\}$ be the set of these colors. Let $\alpha \log^* k$, where α is a positive constant, be an upper bound on the duration of this procedure, when labels are of k bits. The procedure with input i, j is executed on the residual ring $R_{i,j}$.   ◇

The second procedure elects (i, j)-leaders in the ring $R_{i,j}$.

Procedure Elect
Input: i, j.
Each node u sends its color $c(u) \in \{1, 2, 3\}$ to its neighbors in $R_{i,j}$.
Let v and w be the neighbors of u in $R_{i,j}$.
Node u becomes an (i, j)-leader, if and only if $c(u) > c(v)$ and $c(u) > c(w)$.   ◇

The third procedure is used to deactivate a subset of active nodes at the end of each stage.

Procedure Deactivate
Input: i, ε.
Each $(i, \lceil \log(8/\varepsilon) \rceil)$-leader u sends its color $c(u) \in \{1, 2, 3\}$ to both its neighbors in R_i.
All nodes in R_i that are not $(i, \lceil \log(8/\varepsilon) \rceil)$-leaders, upon receiving a message containing a sequence of colors from a neighbor in R_i, add their color to the message and relay it to the other neighbor in R_i.

Let l and r be two consecutive $(i, \lceil \log(8/\varepsilon) \rceil)$-leaders. Let S be the sequence of consecutive active nodes between l and r. Each node in the sequence S, upon discovering the sequence of colors in S and its position in the sequence, proceeds according to the following rules.

- If S is of odd length, i.e., $S = \langle la_1 \ldots a_{k-1}a_k b_{k-1} \ldots b_1 r \rangle$, nodes a_t and b_t become non-active, for all odd values of t. This means that every second node is deactivated, starting from both ends.
- If S is of even length, i.e., $S = \langle la_k \ldots a_1 b_1 \ldots b_k r \rangle$, nodes a_t and b_t become non-active, for all even values of t. This means that every second node is deactivated, starting from the neighbors of the two central nodes.   ◇

We are now ready to provide a detailed description of our labeled map construction algorithm. For each task that cannot be carried out locally, we allot a specific number of rounds to maintain synchronization between the execution of a given part of the algorithm by different nodes. In the analysis we will show that the allotted times are always sufficient.

Algorithm RingLearning
Input: D, θ, and ε.
Phase 1 – preprocessing
set all nodes as active – (locally);
for $i \leftarrow 1$ **to** $\lfloor \log \theta \rfloor - 2\lceil \log(8/\varepsilon) \rceil - \lceil \log(\alpha(\log^* D + 3)) \rceil$ //STAGE
    construct the residual ring R_i of active nodes – (locally);
    elect all nodes in R_i as $(i, 0)$-leaders – (locally);
    for $j \leftarrow 1$ **to** $\lceil \log(8/\varepsilon) \rceil$ //STEP
        construct the residual ring $R_{i,j}$ of $(i, j - 1)$-leaders – (locally);
        assign color $c(u)$ to all nodes u in $R_{i,j}$ with procedure RTC(i, j);
            (allotted time $2^{i-1}4^{j-1}\alpha(\log^* D + 1)$)
        elect (i, j)-leaders with procedure Elect(i, j);
            (allotted time $2^{i-1}4^{j-1}$)
    run procedure Deactivate(i, ε) in R_i;
        (allotted time $2^{i-1}4^{\lceil \log(8/\varepsilon) \rceil}$)
Phase 2 – information spreading
in round $\theta + 1$ each node that is still active constructs locally a labeled map of the part of the original ring consisting of nodes from which it received messages during Phase 1, and sends this map to its neighbors;
both active and non-active nodes that receive a message from one neighbor, send it to the other neighbor;
at time $D + \theta$, all nodes have the labeled map of the ring and stop. ◇

We now prove the correctness of Algorithm RingLearning and analyze it by estimating its cost for a given delay θ. The first two lemmas show that the time 2^{i-1} allotted for multi-hop communication between consecutive active nodes in stage i, and the time $2^{i-1}4^{j-1}$ allotted for multi-hop communication between consecutive $(i, j - 1)$-leaders in step j of stage i, are sufficient to perform the respective tasks.

Lemma 2. *The distance between two consecutive (i, j)-leaders is at most $2^{i-1}4^j$.*

The next two lemmas will be used to prove the correctness of Algorithm RingLearning.

Lemma 3. *All calls to procedures RTC, Elect, and Deactivate can be carried out within times allotted in Algorithm RingLearning.*

Proof. In view of Lemma 2, time $2^{i-1}4^{j-1}$ is sufficient to perform a message exchange between consecutive $(i, j - 1)$-leaders in stage i.

Let L be the length of the binary strings that are labels of nodes. Since L is polynomial in D, the execution of Procedure RTC(i, j) in the residual ring $R_{i,j}$

is completed in time at most $2^{i-1}4^{j-1}\alpha\log^* L \leq 2^{i-1}4^{j-1}\alpha(\log^* D + 1)$. Hence the allotted time is sufficient.

Running Procedure $\mathtt{Elect}(i,j)$ requires time $2^{i-1}4^{j-1}$ to allow each $(i,j-1)$-leader to learn the new color of its neighboring $(i,j-1)$-leaders. Hence the allotted time is sufficient.

Running Procedure $\mathtt{Deactivate}(i,\varepsilon)$ on the residual ring R_i takes time $2^{i-1}4^{\lceil\log(8/\varepsilon)\rceil}$. Indeed, within this time, all nodes between two consecutive $(i,\lceil\log(8/\varepsilon)\rceil)$-leaders learn labels of all nodes between them and decide locally if they should be deactivated. Hence the allotted time is sufficient.   □

Lemma 4. $\lfloor\log\theta\rfloor - 2\lceil\log(8/\varepsilon)\rceil - \lceil\log(\alpha(\log^* D + 3))\rceil$ *stages can be completed in time θ.*

We are now ready to prove the correctness of our algorithm.

Theorem 2. *Upon completion of Algorithm* $\mathtt{RingLearning}$ *all nodes of the ring correctly construct its labeled map.*

Proof. The correctness of Procedure $\mathtt{RTC}$ follows from [8], provided that enough time is allotted for its completion. Elections of (i,j)-leaders are carried out according to the largest color rule by Procedure $\mathtt{Elect}(i,j)$, provided that each node knows the colors assigned to its neighbors in $R_{i,j}$. Decisions to become non-active can be carried out locally by each node, according to the appropriate rule from Procedure $\mathtt{Deactivate}$, provided that nodes of each sequence S between two $(i,\lceil\log(8/\varepsilon)\rceil)$-leaders know the entire sequence. By Lemma 3 the times allotted to all three procedures are sufficient to satisfy the above conditions.

Due to Lemma 4, all nodes stop executing the preprocessing phase within round θ, hence D more rounds are available for the information spreading phase. At the end of stage i each $(i,\lceil\log(8/\varepsilon)\rceil)$-leader knows the sequences of node labels and port numbers connecting it to both closest $(i,\lceil\log(8/\varepsilon)\rceil)$-leaders. Hence, at the beginning of the information spreading phase, the union of the sequences known to all active nodes covers the entire ring, and consecutive sequences overlap. This in turn implies that, after D rounds of the information spreading phase, all nodes get the complete labeled map of the ring.   □

The next three lemmas are used to analyze the cost of Algorithm $\mathtt{RingLearning}$, running with delay θ.

Lemma 5. *At the end of stage i there are at most $n((\varepsilon/2+1)/2)^i$ active nodes in a ring of size n.*

Lemma 6. *The cost of the preprocessing phase of Algorithm* $\mathtt{RingLearning}$ *with input parameters D, θ, and ε, where $0 < \varepsilon < 1$, is $O(D\log^* D\,\theta^{\log(1+\varepsilon)}\log\theta/\varepsilon^2)$.*

Proof. As shown in the proof of Lemma 4, the time used for stage i is at most $2^{i-1}4^s\alpha(\log^* D + 3)$, where $s = \lceil\log(8/\varepsilon)\rceil$ is the number of steps in each stage. In view of Lemma 5, during stage i there are at most $n((\varepsilon/2+1)/2)^{i-1}$ active nodes in a ring of size n. Hence the cost of stage i is at most

$$2^{i-1}4^s\alpha(\log^* D + 3) \cdot n \left(\frac{\varepsilon/2 + 1}{2}\right)^{i-1}.$$

Since the number of stages is less than $\log \theta$, the overall cost of the preprocessing phase is less than

$$\sum_{i=1}^{\lfloor \log \theta \rfloor} 2^{i-1}4^s\alpha(\log^* D + 3) \cdot n \left(\frac{\varepsilon/2 + 1}{2}\right)^{i-1}.$$

Bounding each summand with the last one which is the largest we obtain

$$\sum_{i=1}^{\lfloor \log \theta \rfloor} 2^{i-1}4^s\alpha(\log^* D+3) \cdot n \left(\frac{\varepsilon/2 + 1}{2}\right)^{i-1} \le \alpha n(\log^* D+3)\theta^{\log(1+\varepsilon/2)} \log \theta \left\lceil \frac{8}{\varepsilon} \right\rceil^2,$$

which is $O(D \log^* D \, \theta^{\log(1+\varepsilon)} \log \theta / \varepsilon^2)$.   □

Lemma 7. *The cost of the information spreading phase of Algorithm* RingLearning *with input parameters* D, θ, *and* ε, *where* $0 < \varepsilon < 1$, *is* $O(D^2 \log^* D/(\varepsilon^2 \theta^{1-\varepsilon}))$.

Theorem 3. *The cost of Algorithm* RingLearning, *executed in time* $D + \theta$ *in a ring of diameter* D, *is* $O(D^2 \log^* D/\theta^{1-\varepsilon})$, *for any constant parameter* $0 < \varepsilon < 1$ *and any* $\theta \le D$.

Proof. Lemmas 6 and 7 imply that the cost of Algorithm RingLearning, executed with parameters D, θ, and ε, in a ring of diameter D, is of the order $O(D \log^* D \, \theta^{\log(1+\varepsilon/2)} \log \theta + D^2 \log^* D/(\theta^{1-\varepsilon}))$, for any constant $0 < \varepsilon < 1$. Since $\log(1 + \varepsilon/2) - \varepsilon$ is negative for all $\varepsilon > 0$, and $\theta \le D$, we have $\theta^{1+\log(1+\varepsilon/2)-\varepsilon} \log \theta < D$, for sufficiently large D. Hence $\frac{D}{\theta^{1-\varepsilon}} > \theta^{\log(1+\varepsilon/2)} \log \theta$, which implies

$$O\left(D \log^* D \, \theta^{\log(1+\varepsilon/2)} \log \theta + \frac{D^2 \log^* D}{\theta^{1-\varepsilon}}\right) = O\left(\frac{D^2 \log^* D}{\theta^{1-\varepsilon}}\right).$$

     □

4   Discussion and Open Problems

We proved almost matching upper and lower bounds for the tradeoffs between time and cost of the labeled map construction task in the class of rings. Can these tradeoffs be generalized to a larger class of networks? Since lower bounds are stronger when established on a more restricted class of graphs, the challenge would be to extend our algorithms, that provide an almost matching upper bound, to more general networks.

   First observe that our approach could not be extended directly. Indeed, we rely on repeated stages consisting of coloring with few colors and of node deactivation.

A subsequent stage works on the residual network of active nodes from the previous stage. In the case of rings, the residual network remains a ring and hence coloring with few colors can be done again. As soon as we move to networks of degree higher than two, the maximum degree of the residual network can grow exponentially in the number of stages, and thus the technique of fast coloring with few colors cannot be applied repeatedly. However, allowing delays larger than D, but still linear in D, permits to use a different approach that is successful on a larger class of networks.

Consider the class of networks in which neighborhoods of nodes grow polynomially in the radius. More precisely, let $N_r(v)$ be the set of all nodes within distance at most r from v. We will say that a network has *polynomially growing neighborhoods*, if there exists a constant $c \geq 1$ (called the growth parameter) such that $|N_r(v)| \in \Theta(r^c)$ for all nodes v. Notice that the class of networks with polynomially growing neighborhoods is fairly large, as it includes, e.g., all multidimensional grids and tori, as well as rings. On the other hand, all such networks have bounded maximum degree.

Consider the following *doubling algorithm*, working in two phases. The preprocessing phase of the algorithm is a generalization of the leader election algorithm for rings from [9]. In the beginning all nodes are active. Each node v that is active at the beginning of stage $i \geq 0$ has the largest label in the neighborhood $N_{2^i}(v)$. In stage i, every active node sends its label at distance 2^{i+1} and it remains active at the end of this stage, if it does not receive any larger label. We devote a given amount of time τ to the preprocessing phase. The rest of the algorithm is the information spreading phase, in which each node that is still active constructs independently a BFS spanning tree of the network in time D. Information exchange in each BFS tree is then initiated by its leaves and completed in additional time $2D$. (Hence BFS trees constructed by active nodes are used redundantly, but - as will be seen - the total cost can still be controlled.) Upon completion of information spreading, each node has a labeled map of the whole network.

We now analyze the cost of the above doubling algorithm.

Proposition 1. *The cost of the doubling algorithm, executed in time $3D + \tau$ on a network with polynomially growing neighborhoods, of diameter D and size n, is in $O(n \log \tau + nD^\beta / \tau^\beta)$, for some constant $\beta > 1$.*

In particular, for $\tau \in \Theta(D)$, i.e., when the total available time is $(3 + \eta)D$ for some constant $\eta > 0$, the total cost is $O(n \log D)$.

We close the paper with two open problems. The above tradeoffs are valid for fairly large running times (above $3D$). This means that the tradeoff curve remains flat for a long period of time. It is thus natural to ask for tradeoffs between cost and time for delays below D, i.e., for overall time below $2D$. Can such tradeoffs be established for some other classes of networks (such as bounded degree networks or even just grids and tori), similarly as we did for rings?

Finally, notice that for rings the information spreading phase can be performed in time $2D$ (instead of $3D$) by letting each active node initiate two sequences of messages (one clockwise, and the other counterclockwise), each containing labels of all already visited nodes. Moreover, the overall cost of the doubling algorithm, executed in time $2D + \tau$ on a ring of diameter D and size n, is $O(n \log \tau + nD/\tau) = O(D \log \tau + D^2/\tau)$. This should be compared to the cost of Algorithm RingLearning, that can be as small as $O(D^{1+\varepsilon} \log^* D)$ for total time $2D$ and any constant $\varepsilon > 0$. The cost of the doubling algorithm becomes asymptotically smaller when the overall time is larger than $2D + D^{1-\varepsilon}/\log^* D$. Closing the small gap between our bounds on the time vs. cost tradeoffs for labeled map construction on rings is another open problem.

References

1. Attiya, H., Bar-Noy, A., Dolev, D., Koller, D., Peleg, D., Reischuk, R.: Renaming in an asynchronous environment. Journal of the ACM 37, 524–548 (1990)
2. Awerbuch, B.: Optimal distributed algorithms for minimum weight spanning tree, counting, leader election and related problems. In: Proc. 19th Annual ACM Symposium on Theory of Computing (STOC 1987), pp. 230–240 (1987)
3. Chalopin, J., Das, S., Kosowski, A.: Constructing a map of an anonymous graph: Applications of universal sequences. In: Proc. 14th International Conference on Principles of Distributed Systems (OPODIS 2010), pp. 119–134 (2010)
4. Cole, R., Vishkin, U.: Deterministic coin tossing with applications to optimal parallel list ranking. Information and Control 70, 32–53 (1986)
5. Czumaj, A., Gasieniec, L., Pelc, A.: Time and cost trade-offs in gossiping. SIAM Journal on Discrete Mathematics 11, 400–413 (1998)
6. Fredrickson, G.N., Lynch, N.A.: Electing a leader in a synchronous ring. Journal of the ACM 34, 98–115 (1987)
7. Gasieniec, L., Pagourtzis, A., Potapov, I., Radzik, T.: Deterministic communication in radio networks with large labels. Algorithmica 47, 97–117 (2007)
8. Goldberg, A.V., Plotkin, S.A., Shannon, G.E.: Parallel symmetry- breaking in sparse graphs. SIAM Journal on Discrete Mathematics 1, 434–446 (1988)
9. Hirschberg, D.S., Sinclair, J.B.: Decentralized extrema-finding in circular configurations of processes. Communications of the ACM 23, 627–628 (1980)
10. Israeli, A., Kranakis, E., Krizanc, D., Santoro, N.: Time-message trade-offs for the weak unison problem. Nordic Journal of Computing 4, 317–341 (1997)
11. Lynch, N.L.: Distributed algorithms. Morgan Kaufmann Publ. Inc., San Francisco (1996)
12. Peleg, D.: Distributed Computing, A Locality-Sensitive Approach, Philadelphia. SIAM Monographs on Discrete Mathematics and Applications (2000)
13. Peterson, G.L.: An $O(n \log n)$ unidirectional distributed algorithm for the circular extrema problem. ACM Transactions on Programming Languages and Systems 4, 758–762 (1982)
14. Yamashita, M., Kameda, T.: Computing on anonymous networks: Part I - characterizing the solvable cases. IEEE Trans. Parallel and Distributed Systems 7, 69–89 (1996)

Competitive Auctions for Markets with Positive Externalities

Nick Gravin[1] and Pinyan Lu[2]

[1] Division of Mathematical Sciences, School of Physical and Mathematical Sciences,
Nanyang Technological University, Singapore
ngravin@pmail.ntu.edu.sg
[2] Microsoft Research Asia, China
pinyanl@microsoft.com

Abstract. In digital goods auctions, the auctioneer sells an item in unlimited supply to a set of potential buyers. The objective is to design a truthful auction that maximizes the auctioneer's total profit. Motivated by the observation that the buyers' valuation of the good might be interconnected through a social network, we study digital goods auctions with positive externalities among buyers. This defines a multi-parameter auction design problem where the private valuation of every buyer is a function of the set of other winning buyers. The main contribution of this paper is a truthful competitive mechanism for subadditive valuations. Our competitive result is with respect to a new solution benchmark $\mathcal{F}^{(3)}$. On the other hand, we show a surprising impossibility result if comparing to the stronger benchmark $\mathcal{F}^{(2)}$, where the latter has been used quite successfully in digital goods auctions without externalities [16].

1 Introduction

In economics, the term externality is used to describe situations in which private costs or benefits to the producers or purchasers of a good or service differ from the total social costs or benefits entailed in its production and consumption. In this context a benefit is called a positive externality, while a cost is referred to as a negative one. One needs not to go far to find examples of positive external influence in digital and communications markets, when a customer's decision to buy a good or purchase a service strongly relies on its popularity among his/her friends or generally among other customers, e.g. instant messenger and cell phone users will want a product that allows them to talk easily and cheaply with their friends. Another good example is social network, where a user is more likely to appreciate membership in a network if many of his/her friends are already using it. There exist a number of applications, like the very popular Farm Ville in online social network Facebook, where a user would have more fun when participating with friends. In fact, quite a few such applications explicitly reward players with a large number of friends.

On the other hand, negative external effects occur when a potential buyer, e.g. a big company, incurs a great loss if a subject it fights for, like a small firm or

F.V. Fomin et al. (Eds.): ICALP 2013, Part II, LNCS 7966, pp. 569–580, 2013.

company, goes to its direct competitor. Another well-studied example related to computer science is the allocation of advertisement slots [1, 13–15, 17, 23], where every customer would like to see a smaller number of competitors' advertisements on a web page that contains his/her own advert. One may also face mixed externalities as in the case of selling nuclear weapons [21], where countries would like to see their allies win the auction rather than their foes.

We investigate the problem of *mechanism design* for auctions with positive externalities. We study a scenario where an auctioneer sells the good, of no more than a single copy in the hands of each customer. We define a model for externalities among the buyers in the sealed-bid auction with an unlimited supply of the good. This types of auctions arise naturally in digital markets, where making a copy of the good (e.g. cd with songs or games, or extra copy of online application) has a negligible cost compared to the final price and can be done at any time the seller chooses.

A similar agenda has been introduced in the paper [18], where authors consider a Bayesian framework and study positive externalities in the social networks with single-parameter bidders and submodular valuations. The model in the most general form can be described by a number of bidders n, each with a non-negative private valuation function $v_i(S)$ depending on the possible winning set S. This is a natural multi-parameter mechanism design model that may be considered a generalization of the classical auctions with unlimited supply, i.e. auctions where the amount of items being sold is greater than the number of buyers.

Traditionally the main question arising in such situations is how to maximize the seller's revenue. In literature on the classical auctions without any externalities many diverse approaches to this question have been developed. In the current work we pick a classical approach and benchmark (cf. [16]), namely the best-uniform-price benchmark called $\mathcal{F}$, which is different from Bayesian framework. There one seeks to maximize the ratio of the mechanism's revenue to the revenue of $\mathcal{F}$ taken in the worst case over all possible bids. In particular a mechanism is called competitive if such a ratio is bounded by some uniform constant for each possible bid. However, it was shown that there is no competitive truthful mechanism w.r.t. $\mathcal{F}$, and therefore to get around this problem, a slightly modified benchmark $\mathcal{F}^{(2)}$ [16] was proposed. The only difference of $\mathcal{F}^{(2)}$ to $\mathcal{F}$ is in one additional requirement that at least two buyers should be in a winning set. Thus $\mathcal{F}^{(2)}$ becomes a standard benchmark in analyzing digital auctions [11, 12, 16, 20]. Similarly to $\mathcal{F}^{(2)}$ one may define benchmark $\mathcal{F}^{(k)}$ for any fixed constant k. It turns out that the same benchmarks can be naturally adopted to the case of positive externalities. Surprisingly $\mathcal{F}^{(2)}$ fails to serve as a benchmark in social networks with positive externalities, i.e. no competitive mechanism exists w.r.t. $\mathcal{F}^{(2)}$. Therefore, we go further and consider the next natural candidate for the benchmark, which is $\mathcal{F}^{(3)}$.

The main contribution of this paper is an universally truthful competitive mechanism for the general multi-parameter model with subadditive valuations (substantially broader class than submodular) w.r.t. $\mathcal{F}^{(3)}$ benchmark. We complement this result with a proof that no truthful mechanism can achieve constant ratio w.r.t. $\mathcal{F}^{(2)}$. In order to do so we introduce a restricted model with a single private parameter which in some respects resembles the one considered in [18]; further for this restricted model we give a simple geometric characterization of all truthful mechanisms and based on the characterization then show that there is no competitive truthful mechanism w.r.t. $\mathcal{F}^{(2)}$.

Our model is the so-called multi-parameter or multi-dimensional model (see [25]), as utility of every agent may not be described by a single real number for all possible outcomes of the mechanism. Mechanism design in this case is known to be harder than in the single-parameter domains.

1.1 Related Work

Many studies on externalities in the direction of pricing and marketing strategies over social networks have been conducted over the past few years. In many ways, they have been caused by the development of social-networks on the Internet, which has allowed companies to collect information about each user and user relationships.

Earlier works have generally been focused on the influence maximization problems (see Chapter 24 of [24]). For instance, Kempe *et al.* [22] study the algorithmic question of searching a set of nodes in a social network of highest influence. From the economic literature one could name such papers as [26], which studies the effect of network topology on a monopolist's profits and [10], which studies a multi-round pricing game, where a seller may lower his price in an attempt to attract low value buyers. These works take no heed of algorithmic motivation.

There are several more recent papers [2, 7, 9, 19] studying the question of revenue maximization as well as work studying the posted price mechanisms [3, 5, 8, 19].

We could not continue without mentioning a beautiful line of research on revenue maximization for classical auctions, where the objective is to maximize the seller's revenue compared to a benchmark in the worst case. We cite here only some papers that are most relevant to our setting [4, 11, 12, 16, 20]. With respect to the refined best-uniform-price benchmark $\mathcal{F}^{(2)}$ a number of mechanisms with constant competitive ratio were obtained; each subsequent paper improving the competitive ratio of the previous one [11, 12, 16, 20]. The best known current mechanism is due to Hartline and McGrew [20] and has a competitive ratio of 3.25. On the other hand a lower bound of 2.42 has been proven in [16] by Goldberg *et.al.*. The question of closing the gap still remains open.

2 Preliminaries

We suppose that in a marketplace n agents are present, the set of which we denote by $[n]$. Each agent i has a private valuation function v_i, which is a non-negative real number for each possible winner set $S \subset [n]$. The seller organizes a single round sealed bid auction, where agents submit their valuations $b_i(S)$ to an auctioneer for all possible winner sets S, and the auctioneer then chooses agents who will obtain the good and vector of prices to charge each of them. The auctioneer is interested in maximizing his/her revenue.

For every $i \in [n]$ we impose the following mild requirements on v_i.

1. $v_i(S) \geq 0$.
2. $v_i(S) = 0$ if $i \notin S$.
3. $v_i(S)$ is a monotone sub-additive function of S, i.e.
   (a) $v_i(S) \leq v_i(R)$ if $S \subseteq R \subseteq [n]$.
   (b) $v_i(S \cup R) \leq v_i(S) + v_i(R)$, for each $i \in S, R \subseteq [n]$

We should note here that the sub-additivity requirement is only for those subsets that include the agent i. This is a natural assumption since $v_i(S) = 0$ if $i \notin S$.

2.1   Mechanism Design

Each agent in turn would like to get a positive utility that is as high as possible and may lie strategically about his/her valuation. The utility $u_i(S)$ of an agent i for a winning set S is simply the difference of his valuation $v_i(S)$ and the price p_i the auctioneer charges i. Thus one of the desired properties for the auction is the well known concept of truthfulness or incentive compatibility, i.e. the condition that every agent maximizes his utility by truth telling.

It is worth mentioning that our model is that of multi-parameter mechanism design and, moreover, that collecting the whole bunch of values $v_i(S)$ for every $i \in [n]$ and $S \subset [n]$ would require an exponential amount of bits in n and thus is inefficient. However, in the field of mechanism design there is a way to get around such a problem of exponential input size with the broadly recognized concept of black box value queries. The latter simply means that the auctioneer, instead of getting the whole collection of bids instantly, may ask during the mechanism execution every agent i only for a small part of his input, i.e. a number of questions about valuation of i for certain sets. We note that the agent still may lie in the response to each such query. We denote the bid of i by $b_i(S)$ to distinguish it from the actual valuation $v_i(S)$. Thus if we are interested in designing a computationally efficient mechanism, we can only ask in total a polynomial in n number of queries.

Throughout the paper, with $\mathcal{M}$ we denote a mechanism with allocation rule $\mathcal{A}$ and payment rule $\mathcal{P}$. Allocation algorithm $\mathcal{A}$ may ask queries about valuations of any agent for any possible set of winners. Thus $\mathcal{A}$ has an oracle black box access to the collection of bid functions $b_i(S)$. For each agent i in the winning set S the payment algorithm decides a price p_i to charge. The utility of agent i is then $u_i = v_i(S) - p_i$ if $i \in S$ and 0 otherwise. To emphasize the fact that agents may report untruthfully we will use $u_i(b_i)$ notation for the utility function in the general case and $u_i(v_i)$ in the case of truth telling. We assume voluntary participation for the agents, that is $u_i \geq 0$ for each i who reports the truth.

2.2   Revenue Maximization and Possible Benchmarks

We discuss here the problem of revenue maximization from the seller's point of view. The revenue of the auctioneer is simply the total payment $\sum_{i \in S} p_i$ of all buyers in the winning set. We assume that the seller incurs no additional cost for making a copy of the good. This assumption is essential for our model, since unlike the classical digital auction case there is no simple reduction of the settings with a positive price per issuing the item to the settings with zero price.

The best revenue the seller can hope for is $\sum_{i \in [n]} v_i([n])$. However, it is not realistic when the seller does not know agents' valuation functions. We follow the tradition of previous literature [11, 12, 16, 20] of algorithmic mechanism design on competitive auctions with limited or unlimited supply and consider the best revenue uniform price benchmark, which is defined as maximal revenue that the auctioneer can get for a fixed uniform price for the good. In the literature on classical competitive auctions this benchmark was called $\mathcal{F}$ and is formally defined as follows.

Definition 1 ($\mathcal{F}$ without Externalities). *For the vector of agent's bids* **b**

$$\mathcal{F}(\mathbf{b}) = \max_{c \geq 0, S \subset [n]} \left(c \cdot |S| \middle| \forall i \in S \;\; b_i \geq c \right).$$

This definition generalizes naturally to our model with externalities and is defined rigorously as follows.

Definition 2 ($\mathcal{F}$ with Externalities). *For the collection of agents' bid functions* **b**.

$$\mathcal{F}(\mathbf{b}) = \max_{c \geq 0, S \subset [n]} \left(c \cdot |S| \middle| \forall i \in S \;\; b_i(S) \geq c \right).$$

The important point in considering $\mathcal{F}$ in the setting of classical auctions is that the auctioneer, when he/she is given in advance the best uniform price, can run a truthful mechanism with corresponding revenue. It turns out that the same mechanism works neatly for our model. Specifically, a seller who is given the price c in advance can begin with the set of all agents and drop one by one those agents with negative utility ($b_i(S) - c < 0$); once there are left no agents to delete, the auctioneer sells the item to all surviving buyers at the given price c.

In these circumstances, a natural problem arising for the auctioneer is to devise a truthful mechanism which has a good approximation ratio of the mechanism's revenue to the revenue of the benchmark at any possible bid vector **b**. Such a ratio is usually called the *competitive* ratio of a mechanism. However, it was shown (cf. [16]) that no truthful mechanism can guarantee any constant competitive ratio w.r.t. $\mathcal{F}$. Specifically, the unbounded ratio appears in the instances where the benchmark buys only one item at the highest price. To overcome this obstacle, a slightly modified benchmark $\mathcal{F}^{(2)}$ has been proposed and a number of competitive mechanisms w.r.t. $\mathcal{F}^{(2)}$ were obtained [11, 12, 16, 20]. The only difference of $\mathcal{F}^{(2)}$ from $\mathcal{F}$ is one additional requirement that at least two buyers should be in the winning set. Similarly, for any $k \geq 2$ we may define $\mathcal{F}^{(k)}$.

Definition 3.

$$\mathcal{F}^{(k)}(\mathbf{b}) = \max_{c \geq 0, S \subset [n]} \left(c \cdot |S| \middle| |S| \geq k, \;\; \forall i \in S \;\; b_i(S) \geq c \right).$$

However, in case of our model the benchmark $\mathcal{F}^{(2)}$ does not imply the existence of a constant approximation truthful mechanism. In order to illustrate that later in Section 4 we will introduce a couple of new models which differ from the original one in certain additional restrictions on the domain of agent's bids. We further give a complete characterization of truthful mechanisms for these new restricted settings substantially exploiting the fact that every agent's bidding language is single-parameter. Later, we use that characterization to argue that no truthful mechanism can achieve constant approximation with respect to $\mathcal{F}^{(2)}$ benchmark even for these cases. On the positive side, and quite surprisingly, we can furnish our work in the next section with the truthful mechanism which has a constant approximation ratio w.r.t. $\mathcal{F}^{(3)}$ benchmark for the general case of multi-parameter bidding.

3   Competitive Mechanism

Here we give a competitive truthful mechanism, that is a mechanism which guarantees that the auctioneer gets a constant fraction of the revenue he could get for the best fixed price benchmark assuming that all agents bid truthfully. We call it PROMOTION-TESTING-SELLING MECHANISM. In the mechanism we give the good to certain agents *for free*, that is without requiring any payment. The general scheme of the mechanism is as follows.

PROMOTION-TESTING-SELLING MECHANISM

1. Put every agent at random into one of the sets A, B, C.
2. Denote $r_A(C)$ and $r_B(C)$ the largest fixed price revenues one can extract from C given that, respectfully, either A, or B got the good for free.
3. Let $r(C) = \max\{r_A(C), r_B(C)\}$.
4. Sell items to agents in A for free.
5. Apply COST SHARING MECHANISM($r(C)$, B, A) to extract revenue $r(C)$ from set B given that A got the good for free.

Bidders in A receive items for free and increase the demand of agents from B. One may say that they "advertise" the goods and resemble the promotion that occurs when selling to participants. The agents in C play the role of the "testing" group, the only service of which is to determine the right price. Note that we take no agents of the testing group into the winning set, therefore, they have nothing to gain for bidding untruthfully. The agents of B appear to be the source of the mechanism's revenue, which is being extracted from B by a cost sharing mechanism as follows.

We note here that a more "natural" mechanism is simply to set that $r(C) = r_A(C)$ rather than $\max\{r_A(C), r_B(C)\}$. But unfortunately, we have a counter example to show that this simpler mechanism cannot guarantee a constant approximation ratio compared to our benchmark.

COST SHARING MECHANISM(r,X,Y)

1. $S \leftarrow X$.
2. Repeat until $T = \emptyset$:
   - $T \leftarrow \{i | i \in S$ and $b_i(S \cup Y) < \frac{r}{|S|}\}$.
   - $S \leftarrow S \setminus T$.
3. If $S \neq \emptyset$ sell items to everyone in S at $\frac{r}{|S|}$ price.

Lemma 4. PROMOTION-TESTING-SELLING MECHANISM *is universally truthful.*

Proof. The partitioning of the set $[n]$ into A, B, C does not depend on the agent bids. When the partition is fixed, our mechanism becomes deterministic. Therefore, we are only left to prove the truthfulness for that deterministic part. Let us do so by going through the proof separately for each set A, B and C.

- Bids of agents in A do not affect the outcome of the mechanism. Therefore, they have no incentive to lie.
- No agent from C could profit from bidding untruthfully, since her utility will be zero regardless of the bid.
- Let us note that the COST SHARING MECHANISM is applied to the agents in B and the value of r does not depend on their bids, since both $r_A(C)$ and $r_B(C)$ are retracted from C irrespectively of bids from A and B. Also let us note that at each step of the cost sharing mechanism the possible payment $\frac{r}{|S|}$ is rising, and meanwhile the valuation function, because of monotonicity condition, is going down. Hence, manipulation of a bid does not help any agent to survive in the winning set and receive positive utility, if by bidding truthfully he/she has been dropped from the winning set. Mis-reporting a bid could not help an agent to alter the surviving set and at the same time remain a winner. These two observations conclude the proof of truthfulness for B.

Therefore, from now on we may assume that $b_i(S) = v_i(S)$.

Theorem 5. PROMOTION-TESTING-SELLING MECHANISM *is universally truthful and has an expected revenue of at least* $\frac{\mathcal{F}^{(3)}}{324}$.

Proof. We are left to prove the lower bound on the competitive ratio of our mechanism, as we have shown the truthfulness in Lemma 4.

For the purpose of analysis, we separate the random part of our mechanism into two phases. In the first phase, we divide agents randomly into three groups S_1, S_2, S_3 and in the second one, we label the groups at random by A, B and C. Note that the combination of these two phases produces exactly the same distribution over partitions as in the mechanism.

Let S be the set of winners in the optimal $\mathcal{F}^{(3)}$ solution and the best fixed price be p^*. For $1 \leq i \neq j \leq 3$ we may compute r_{ij} the largest revenue for a fixed price that one can extract from set S_i given S_j is "advertising" the good, that is agents in S_j get the good for free and thus increase the valuations of agents from S_i though contribute nothing directly to the revenue.

First, let us note that the cost-sharing part of our mechanism will extract one of these r_{ij} from at least one of the six possible labels for every sample of the dividing phase (in general cost-sharing mechanism may extract 0 revenue, e.g. if the target revenue is set too high). Indeed, let i_0 and j_0 be the indexes for which $r_{i_0 j_0}$ achieves maximum over all r_{ij} and let $k_0 = \{1, 2, 3\} \setminus \{i_0, j_0\}$. Then the cost-sharing mechanism will retract the revenue $r(C) = max(r_A(C), r_B(C))$ on the labeling with $S_{j_0} = A$, $S_{i_0} = B$ and $S_{k_0} = C$. It turns out, as we will prove in the following lemma, that one can get a lower bound on this revenue within a constant factor of $r_{\mathcal{F}}(C)$; the revenue we got from the agents of C in the benchmark $\mathcal{F}^{(3)}$.

Lemma 6. $r(C) \geq \frac{r_{\mathcal{F}}(C)}{4}$.

Proof. Let $S_c = S \cap C$. Thus, by the definition of $\mathcal{F}^{(3)}$, we have $r_{\mathcal{F}}(C) = |S_c| \cdot p^*$ and for all $i \in S_c$, $v_i(S) \geq p^*$.

We define a subset T of S_c as a final result of the following procedure.

1. $T \leftarrow \emptyset$ and $X \leftarrow \{i | i \in S_c$ and $v_i(A \cup \{i\}) \geq \frac{p^*}{2}\}$.
2. While $X \neq \emptyset$
   - $T \leftarrow T \cup X$,
   - $X \leftarrow \{i | i \in S_c$ and $v_i(A \cup T \cup \{i\}) \geq \frac{p^*}{2}\}$

For any agent of T we have $v_i(A \cup T) \geq \frac{p^*}{2}$ because the valuation function is monotone. Now if $|T| \geq \frac{|S_c|}{2}$, we get the desired lower bound. Indeed,

$$r(C) \geq r_A(C) \geq \frac{|S_c|}{2} \cdot \frac{p^*}{2} = \frac{|S_c| \cdot p^*}{4} = \frac{r_{\mathcal{F}}(C)}{4}.$$

Otherwise, let $W = S_c \setminus T$. Then we have $|W| \geq \frac{|S_c|}{2}$. For an agent $i \in W$ it holds true that $v_i(A \cup T \cup \{i\}) < \frac{p^*}{2}$, since otherwise we should include i into T. However, since i wins in the optimal $\mathcal{F}^{(3)}$ solution, we have $v_i(S) \geq p^*$. The former two inequalities together with the subadditivity of $v_i(\cdot)$ $(v_i(S \setminus (A \cup T)) + v_i(A \cup T \cup \{i\}) \geq v_i(S))$ allow us to conclude that $v_i(S \setminus (A \cup T)) \geq \frac{p^*}{2}$ for each $i \in W$. Hence, we get $v_i(B \cup W) \geq \frac{p^*}{2}$ for each $i \in W$, since $S \setminus (A \cup T) \subseteq B \cup W$. Therefore, we are done with the lemma's proof, since

$$r(C) \geq r_B(C) \geq |W| \cdot \frac{p^*}{2} \geq \frac{|S_c| \cdot p^*}{4} = \frac{r_{\mathcal{F}}(C)}{4}.$$

Let k_1, k_2, k_3 be the number of winners of the optimal $\mathcal{F}^{(3)}$ solution, respectively, in S_1, S_2, S_3.

For any fixed partition S_1, S_2, S_3 of the dividing phase by applying Lemma 6, we get that the expected revenue of our mechanism over a distribution of six permutations in the second phase should be at least

$$\frac{1}{6} \cdot \frac{1}{4} \min\{k_1, k_2, k_3\} \cdot p^*.$$

In order to conclude the proof of the theorem we are only left to estimate the expected value of $\min\{k_1, k_2, k_3\}$ from below by some constant factor of $|S|$. The next lemma will do this for us.

Lemma 7. *Let $m \geq 3$ items independently at random be put in one of the three boxes and let a, b and c be the random variables denoting the number of items in these boxes. Then $\mathbb{E}[\min\{a, b, c\}] \geq \frac{2}{27}m$.*

By definition of the benchmark $F^{(3)}$ we have $m = k_1 + k_2 + k_3 \geq 3$ and thus we can apply Lemma 7. Combining every bound we have so far on the expected revenue of our mechanism we conclude the proof with the following lower bound.

$$\frac{1}{6} \cdot \frac{1}{4} \mathbb{E}\left[\min\{k_1, k_2, k_3\}\right] \cdot p^* \geq \frac{1}{24} \cdot \frac{2}{27} \cdot p^* \cdot m = \frac{F^{(3)}}{324}.$$

4   Restricted Single-Parameter Valuations

Here we introduce a couple of special restricted cases of the general setting with a single parameter bidding language. For these models we only specify restrictions on the valuation functions. In each case we assume that t_i is a single private parameter for agent i that he submits as a bid and $w_i(S)$ and $w_i'(S)$ are fixed publicly known functions for each possible winning set S. The models then are described as follows.

- *Additive* valuation $v_i(t_i, S) = t_i + w_i(S)$.
- *Scalar* valuation $v_i(t_i, S) = t_i \cdot w_i(S)$.
- *Linear* valuation $v_i(t_i, S) = t_i w_i(S) + w_i'(S)$, i.e. combination of previous two.

Note that we still require $w_i(S) = w_i'(S) = 0$ if $i \notin S$. These settings are now single parameter domains, which is the most well studied and understood case in mechanism design.

4.1   A Characterization

The basic question of mechanism design is to describe truthful mechanisms in terms of simple geometric conditions. Given a vector of n bids, $\mathbf{b} = (b_1, \ldots, b_n)$, let b_{-i} denote the vector, where b_i is replaced with a '?'. It is well known that truthfulness implies a *monotonicity* condition stating that if an agent i wins for the bid vector $\mathbf{b} = (b_{-i}, b_i)$ then she should win for any bid vector (b_{-i}, b_i') with $b_i' \geq b_i$. In single-dimensional domains monotonicity turns out to be a sufficient condition for truthfulness [6], where prices are determined by the threshold functions.

In our model, valuation of an agent may vary for different winning sets and thus may depend on his/her bid. Nevertheless, any truthful mechanism still has to have a bid-independent allocation rule, although now it is not sufficient for the truthfulness. However, in the case of linear valuation functions we are capable of giving a complete characterization.

Theorem 8. *In the model with linear valuation functions $v_i(t_i, S) = t_i \cdot w_i(S) + w_i'(S)$ an allocation rule $\mathcal{A}$ may be truthfully implemented if and only if it satisfies the following conditions:*

1. *$\mathcal{A}$ is bid-independent, that is for each agent i, bid vector $\mathbf{b} = (b_{-i}, b_i)$ with $i \in \mathcal{A}(\mathbf{b})$ and any $b_i' \geq b_i$, it holds that $i \in \mathcal{A}(b_{-i}, b_i')$.*
2. *$\mathcal{A}$ encourages asymptotically higher bids, i.e. for any fixed b_{-i} and $b_i' \geq b_i$, it holds that $w_i(\mathcal{A}(b_{-i}, b_i')) \geq w_i(\mathcal{A}(b_{-i}, b_i))$.*

Here we prove that these conditions are indeed necessary. The sufficiency part of the theorem is deferred to the full paper version, where we prove the characterization for a slightly more general family of single parameter valuation functions.

Proof. The necessity of the first monotonicity condition was known, so we prove here that the second condition is also necessary. In the truthful mechanism, an agent's payment should not depend on his/her bid, if by changing it the mechanism does not shift the allocated set. We denote by p the payment of agent i for winner set $\mathcal{A}(b_{-i}, b_i)$ and by p' the payment of agent i for winner set $\mathcal{A}(b_{-i}, b_i')$. If the agent's true value is b_i, by truthfulness, we have

$$b_i \cdot w_i(\mathcal{A}(b_{-i}, b_i)) + w_i'(\mathcal{A}(b_{-i}, b_i)) - p \geq b_i \cdot w_i(\mathcal{A}(b_{-i}, b_i')) + w_i'(\mathcal{A}(b_{-i}, b_i')) - p'.$$

And if the agent's true value is b_i', we have

$$b_i' \cdot w_i(\mathcal{A}(b_{-i}, b_i')) + w_i'(\mathcal{A}(b_{-i}, b_i')) - p' \geq b_i' \cdot w_i(\mathcal{A}(b_{-i}, b_i)) + w_i'(\mathcal{A}(b_{-i}, b_i)) - p.$$

Adding these two inequalities and using the fact that $b_i' \geq b_i$, we have

$$w_i(\mathcal{A}(b_{-i}, b_i')) \geq w_i(\mathcal{A}(b_{-i}, b_i)).$$

4.2   From $\mathcal{F}^{(2)}$ to $\mathcal{F}^{(3)}$

Here we show that the usage of $\mathcal{F}^{(2)}$ as a benchmark may lead to an unbounded approximation ratio even for the restricted single parameter scalar valuations. This justifies why we used a slightly modified benchmark $\mathcal{F}^{(3)}$ in Section 3.

Theorem 9. *There is no universally truthful mechanism that can achieve a constant approximation ratio w.r.t. $\mathcal{F}^{(2)}$.*

Proof. Consider the example of two people, in which every bidder evaluates the outcome, where both agents get items much higher than the outcome, where only one agent gets the item. That is $v_1(x, \{1\}) = x$, $v_2(y, \{2\}) = y$ and $v_1(x, \{1, 2\}) = Mx$, $v_2(y, \{1, 2\}) = My$ for a large constant M. We note that these are single parameter scalar valuations. We also note that these valuation functions are indeed subadditive according to our definition. The subadditive requirement is only for the subsets that includes the current agent and, in fact, any valuation function for two agents is subadditive by our definition.

We will show that any universally truthful mechanism $\mathcal{M}_\mathcal{D}$ with a distribution $\mathcal{D}$ over truthful mechanisms cannot achieve an approximation ratio better than

M. Each truthful mechanism $\mathcal{M}$ in $\mathcal{D}$ either sells items to both bidders for some pair of bids (b_1, b_2), or for all pairs of bids sells not more than one item. In the first case, by our characterization of truthful mechanisms (see theorem 8), $\mathcal{M}$ should also sell two items for the bids (x, b_2) and (b_1, y), where $x \geq b_1$ and $y \geq b_2$. Therefore, $\mathcal{M}$ has to sell two items for any bid (x, y) with $x \geq b_1$ and $y \geq b_2$. Let us denote the first and the second group of mechanisms in $\mathcal{D}$ by $\mathcal{G}_1$ and $\mathcal{G}_2$ respectively.

For any small ϵ we may pick sufficiently large x_0, such that at least $1-\epsilon$ fraction of $\mathcal{G}_1$ mechanisms in $\mathcal{D}$ are selling two items for the bids $(x = \frac{x_0}{2M}, y = \frac{y_0}{2M})$. Note that

- revenue of $\mathcal{F}^{(2)}$ for the bids (x_0, x_0) is $2Mx_0$,
- revenue of any $\mathcal{M}$ in $\mathcal{G}_2$ for the bids (x_0, x_0) is not greater than x_0,
- revenue of more than $1 - \epsilon$ fraction of $\mathcal{G}_1$ mechanisms in $\mathcal{D}$ is not greater than $2M\frac{x_0}{2M} = x_0$.
- revenue of the remaining ϵ fraction of $\mathcal{G}_1$ mechanisms is not greater than $2Mx_0$.

Thus we can upper bound the revenue of $\mathcal{M}_\mathcal{D}$ by $x_0(1 - \epsilon) + 2Mx_0\epsilon$ while the revenue of $\mathcal{F}^{(2)}$ is $2Mx_0$. By choosing sufficiently large M and small ϵ we get an arbitrarily large approximation ratio.

Remark 10. In fact, the same inapproximability results w.r.t. $\mathcal{F}^{(2)}$ holds for a weaker notion of truthfulness, namely truthfulness in expectation.

References

1. Aggarwal, G., Feldman, J., Muthukrishnan, S.M., Pál, M.: Sponsored search auctions with markovian users. In: Papadimitriou, C., Zhang, S. (eds.) WINE 2008. LNCS, vol. 5385, pp. 621–628. Springer, Heidelberg (2008)
2. Akhlaghpour, H., Ghodsi, M., Haghpanah, N., Mahini, H., Mirrokni, V.S., Nikzad, A.: Optimal Iterative Pricing over Social Networks. In: Proceedings of the Fifth Workshop on Ad Auctions (2009)
3. Akhlaghpour, H., Ghodsi, M., Haghpanah, N., Mirrokni, V.S., Mahini, H., Nikzad, A.: Optimal iterative pricing over social networks (Extended abstract). In: Saberi, A. (ed.) WINE 2010. LNCS, vol. 6484, pp. 415–423. Springer, Heidelberg (2010)
4. Alaei, S., Malekian, A., Srinivasan, A.: On random sampling auctions for digital goods. In: EC, pp. 187–196 (2009)
5. Anari, N., Ehsani, S., Ghodsi, M., Haghpanah, N., Immorlica, N., Mahini, H., Mirrokni, V.S.: Equilibrium pricing with positive externalities (Extended abstract). In: Saberi, A. (ed.) WINE 2010. LNCS, vol. 6484, pp. 424–431. Springer, Heidelberg (2010)
6. Archer, A., Tardos, É.: Truthful mechanisms for one-parameter agents. In: FOCS, pp. 482–491 (2001)
7. Arthur, D., Motwani, R., Sharma, A., Xu, Y.: Pricing strategies for viral marketing on Social Networks, pp. 101–112. Springer (2009)
8. Candogan, O., Bimpikis, K., Ozdaglar, A.: Optimal pricing in the presence of local network effects. In: Saberi, A. (ed.) WINE 2010. LNCS, vol. 6484, pp. 118–132. Springer, Heidelberg (2010)

9. Chen, W., Lu, P., Sun, X., Tang, B., Wang, Y., Zhu, Z.A.: Optimal pricing in social networks with incomplete information. In: Chen, N., Elkind, E., Koutsoupias, E. (eds.) WINE 2011. LNCS, vol. 7090, pp. 49–60. Springer, Heidelberg (2011)

10. Domingos, P., Richardson, M.: Mining the network value of customers. In: ACM SIGKDD, pp. 57–66. ACM Press, New York (2001)

11. Feige, U., Flaxman, A.D., Hartline, J.D., Kleinberg, R.D.: On the competitive ratio of the random sampling auction. In: Deng, X., Ye, Y. (eds.) WINE 2005. LNCS, vol. 3828, pp. 878–886. Springer, Heidelberg (2005)

12. Fiat, A., Goldberg, A.V., Hartline, J.D., Karlin, A.R.: Competitive generalized auctions. In: STOC, pp. 72–81 (2002)

13. Ghosh, A., Mahdian, M.: Externalities in online advertising. In: WWW, pp. 161–168. ACM (2008)

14. Ghosh, A., Sayedi, A.: Expressive auctions for externalities in online advertising. In: WWW, pp. 371–380. ACM (2010)

15. Giotis, I., Karlin, A.R.: On the equilibria and efficiency of the GSP mechanism in keyword auctions with externalities. In: Papadimitriou, C., Zhang, S. (eds.) WINE 2008. LNCS, vol. 5385, pp. 629–638. Springer, Heidelberg (2008)

16. Goldberg, A.V., Hartline, J.D., Karlin, A.R., Saks, M., Wright, A.: Competitive auctions. Games and Economic Behavior 55(2), 242–269 (2006)

17. Gomes, R., Immorlica, N., Markakis, E.: Externalities in keyword auctions: An empirical and theoretical assessment. In: Leonardi, S. (ed.) WINE 2009. LNCS, vol. 5929, pp. 172–183. Springer, Heidelberg (2009)

18. Haghpanah, N., Immorlica, N., Mirrokni, V.S., Munagala, K.: Optimal auctions with positive network externalities. In: EC, pp. 11–20 (2011)

19. Hartline, J., Mirrokni, V., Sundararajan, M.: Optimal marketing strategies over social networks. In: WWW, pp. 189–198. ACM (2008)

20. Hartline, J.D., McGrew, R.: From optimal limited to unlimited supply auctions. In: EC, pp. 175–182. ACM (2005)

21. Jehiel, P., Moldovanu, B., Stacchetti, E.: How (not) to sell nuclear weapons. American Economic Review 86(4), 814–829 (1996)

22. Kempe, D., Kleinberg, J.M., Tardos, É.: Influential nodes in a diffusion model for social networks. In: Caires, L., Italiano, G.F., Monteiro, L., Palamidessi, C., Yung, M. (eds.) ICALP 2005. LNCS, vol. 3580, pp. 1127–1138. Springer, Heidelberg (2005)

23. Kempe, D., Mahdian, M.: A cascade model for externalities in sponsored search. In: Papadimitriou, C., Zhang, S. (eds.) WINE 2008. LNCS, vol. 5385, pp. 585–596. Springer, Heidelberg (2008)

24. Kleinberg, J.: Cascading behavior in networks: algorithmic and economic issues. Cambridge University Press (2007)

25. Nisan, N., Roughgarden, T., Tardos, É., Vazirani, V.V.: Algorithmic game theory. Cambridge University Press (2007)

26. Sääskilahti, P.: Monopoly pricing of social goods, vol. 3526. University Library of Munich, Germany (2007)

Efficient Computation of Balanced Structures

David G. Harris[1,*], Ehab Morsy[2], Gopal Pandurangan[3,**],
Peter Robinson[4,***], and Aravind Srinivasan[5,*]

[1] Department of Applied Mathematics, University of Maryland,
College Park, MD 20742
davidgharris29@hotmail.com

[2] Division of Mathematical Sciences, Nanyang Technological University,
Singapore 637371 and Department of Mathematics, Suez Canal University,
Ismailia 22541, Egypt
ehabmorsy@gmail.com

[3] Division of Mathematical Sciences, Nanyang Technological University,
Singapore 637371 and Department of Computer Science,
Brown University, Providence, RI 02912
gopalpandurangan@gmail.com

[4] Division of Mathematical Sciences, Nanyang Technological University,
Singapore 637371
peter.robinson@ntu.edu.sg

[5] Department of Computer Science and Institute for Advanced Computer Studies,
University of Maryland, College Park, MD 20742
srin@cs.umd.edu

Abstract. Basic graph structures such as maximal independent sets (MIS's) have spurred much theoretical research in distributed algorithms, and have several applications in networking and distributed computing as well. However, the extant (distributed) algorithms for these problems do not necessarily guarantee fault-tolerance or load-balance properties: For example, in a star-graph, the central vertex, as well as the set of leaves, are both MIS's, with the latter being much more fault-tolerant and balanced — existing distributed algorithms do not handle this distinction. We propose and study "low-average degree" or "balanced" versions of such structures. Interestingly, in sharp contrast to, say, MIS's, it can be shown that checking whether a structure is balanced, will take substantial time. Nevertheless, we are able to develop good sequential and distributed algorithms for such "balanced" versions. We also complement our algorithms with several lower bounds.

* Supported in part by NSF Award CNS-1010789.
** Supported in part by Nanyang Technological University grant M58110000, Singapore Ministry of Education (MOE) Academic Research Fund (AcRF) Tier 2 grant MOE2010-T2-2-082, and MOE AcRF Tier 1 grant MOE2012-T1-001-094, and by a grant from the United States-Israel Binational Science Foundation (BSF).
*** Supported in part by Nanyang Technological University grant M58110000 and Singapore Ministry of Education (MOE) Academic Research Fund (AcRF) Tier 2 grant MOE2010-T2-2-082.

F.V. Fomin et al. (Eds.): ICALP 2013, Part II, LNCS 7966, pp. 581–593, 2013.

1   Introduction

Fundamental graph-theoretic structures such as maximal independent set (MIS) and minimal dominating set (MDS) and their efficient distributed computation are very important, especially in the context of distributed computing and networks where they have many applications [8]. MIS, for example, is a basic building block in distributed computing and is useful in basic tasks such as monitoring, scheduling, routing, clustering, etc. (e.g., [7,9]). Extensive research has gone into designing fast distributed algorithms for these problems since the early eighties (e.g., see [5,10] and the references therein) . We now know that problems such as MIS are quite *local*, i.e., they admit distributed algorithms that run in a small number of *rounds* (typically logarithmic in the network size). However, one main drawback of these algorithms is that there is no guarantee on the *quality* of the structure output. For example, the classical MIS algorithm of Luby [6] computes an MIS in $O(\log n)$ rounds (throughout, n stands for number of nodes in the network) with high probability, but does not give any guarantees on the properties of the output MIS. (Another $O(\log n)$ round parallel algorithm was independently found by Alon, Babai, and Itai [1].) In this paper, we initiate a systematic study of "balanced" versions of these structures, i.e., the *average degree* of the nodes belonging to the structure (the degrees of nodes in the structure are with respect to the original subgraph, and not with respect to the subgraph induced by the structure) is *small*, in particular, compared to the average degree of the graph (note that, in general, the best possible balance we can achieve is the average degree of the graph, as in a regular graph). For example, as we define later, a balanced MIS (BMIS) is an MIS where the average degree of nodes belonging to the MIS is small.

We note that the *maximum* independent set (which is a well-studied NP-complete problem [2]) in a graph G is not necessarily a BMIS in G. Consider the graph G that contains a complete graph K_p (assume p is even), and a complete bipartite graph $K_{A,B}$ with $|A| = 2$ and $|B| = 3$. Each vertex in A is connected to a different half of the set of vertices in K_p (i.e., one vertex of A is connected to one half of vertices of K_p and the second vertex of A is connected to the other half of K_p), and each vertex in B is connected to all vertices in K_p. Clearly, B is the maximum independent set in G and has average degree $p + 2$, while A is a BMIS in G since its average degree is $p/2 + 3$. Thus BMIS is a different problem compared to the maximum independent set problem (which is not the focus of this paper).

There are two key motivations for studying balanced structures. The first is from an application viewpoint. In distributed networks, especially in resource-constrained networks such as ad hoc networks, sensor and mobile networks, it is important to design structures that favor load balancing of tasks among nodes (belonging to the structure). This is crucial in extending the lifetime of the network (see e.g., [11] and the references therein). For example, in a typical application, an MIS (or an MDS) can be used to form clusters with low diameter, with the nodes in the MIS being the "clusterheads" [7]. Each clusterhead is responsible for monitoring the nodes that are adjacent to it. Having an MIS with low degree is useful in a resource/energy-constrained setting since the number of nodes monitored *per* node in the MIS will be low (on average). This can lead to better load balancing, and consequently less resource or energy consumption per node, which is crucial for ad hoc and sensor networks, and help in extending the lifetime of such networks while also leading to better fault-tolerance. For example, in an n-node star graph, the above requirements imply that it is better for the leaf nodes to form the MIS rather than the central node alone. In fact, the average degree of the MIS formed by the leaf nodes (which is 1) is within a constant factor of the average degree of a star (which is close to 2), whereas the average degree of the MIS consisting of the central node alone (which is $n - 1$) is much larger.

Another potential application of balanced structures is in the context of dynamic networks where one would like to *maintain* structures such as MIS efficiently even when nodes or links (edges) fail or change with time. For example, this is a feature of ad hoc and mobile networks where links keep changing either due to mobility or failures. BMIS can be a good candidate for maintaining an MIS efficiently in an incremental fashion: since the degrees of nodes in the MIS are balanced, this will lead to less overhead per insertion or deletion.

The second key motivation of our work is understanding the complexity of local computation of globally optimal (or near optimal) fundamental structures. The correctness of structures such as MIS or MDS can be verified *strictly locally* by a distributed algorithm. In the case of MIS, for example, each node can check the MIS property by communicating only with its neighbors; if there is a violation at least one node will raise an alarm. On the other hand, it is not difficult to show that the correctness of balanced structures such as BMIS cannot be locally verified (in the above sense) as the BMIS refers to a "global" property: nodes have to check the average degree property, in addition to the MIS property. In fact, one can show that at least D rounds (D being the network diameter) would be needed to check whether a structure is a BMIS. Moreover, we prove that BMIS is an NP-hard problem and hence the optimality of the structure is not easy to check even in a centralized setting. A key issue that we address in this paper is whether one can compute near-optimal local (distributed) solutions to balanced global structures such as BMIS. A main result of this paper is that despite the global nature, we can design efficient distributed algorithms that output high quality balanced structures.

Our work is also a step towards understanding the algorithmic complexity of balanced problems. While every MIS is an MDS, they differ significantly in their balanced versions. In particular, we show that there exist graphs for which no MIS is a good BMDS. Hence we need a different approach to compute a good BMDS as compared to a good BMIS. Even for BMIS, we show that while one can (for example) use Luby's algorithm [6] to efficiently compute an MIS, the same approach fails to compute a good quality BMIS. We present new algorithms for computing such balanced structures.

1.1 Problems Addressed and our Results

We consider an undirected simple graph $G = (V, E)$ with n nodes and m edges. We denote the average degree of G by $\delta = \frac{2m}{n}$. More generally, given any subset

$S \subseteq V$, we define the average degree of S, denoted by δ_S, as the total degree of the vertices of S divided by the number of vertices in S, i.e., $\delta_S = \frac{\sum_{v \in S} d_v}{|S|}$, where d_v is the degree of node v in G. To simplify the problem, we assume that G has no isolated vertices, i.e. $d_v \geqslant 1$ for all $v \in V$. (This assumption can be easily removed).

We consider the following fundamental graph structures: A **Maximal Independent Set (MIS)** is an inclusion-maximal vertex subset $S \subseteq V$ such that no two vertices in S are neighbors. A **Minimal Dominating Set (MDS)** is an inclusion-minimal vertex subset $S \subseteq V$ such that every vertex in G is either in S or is a neighbor of a vertex in S. A **Minimal Vertex Cover (MVC)** is an inclusion-minimal vertex subset $S \subseteq V$ such that every edge in G has at least one endpoint in S.

This paper is concerned with the "balanced" versions of the above problems which are optimization versions of the above binary problems.

1. **Balanced Maximal Independent Set (BMIS):** Given an undirected graph G, a BMIS is an MIS S in G that minimizes the average degree of S. In other words, the BMIS has the minimum average degree among all MIS's in G.

2. **Balanced Minimal Dominating Set (BMDS):** Given an undirected graph G, a BMDS is an MDS D in G that minimizes the average degree of D.

3. **Balanced Minimal Vertex Cover (BMVC):** Given an undirected graph G, a BMVC is an MVC C in G that minimizes the average degree of C.

Our Results. We first note that the trivial lower bound is δ for all balanced problems which follows from the example of a regular graph (where all nodes have the same degree). Hence, in general, the average degree of a balanced structure cannot be guaranteed to be less than δ. On the other hand, there exist graphs where the average degree of the BMIS is significantly smaller than δ (e.g., consider a graph on $2n$ nodes, out of which n nodes form a complete graph and each of these vertices is also connected by a single edge to one each of the rest of the n nodes). This leads us to two basic questions: (i) In every given graph G, does there always exist a BMIS whose average degree is at most δ? and (ii) Can question (i) be answered for a specific graph G in polynomial time? We answer both questions in the negative.

We show that unlike MIS, its balanced version, BMIS, is NP-hard. In particular, in the full paper [4] we show that the following *decision version* of the problem is NP-complete: *"Given a graph G, is there an MIS in G with average degree at most δ?"* In fact we show that the optimization version BMIS is quite *hard to approximate in polynomial time*: it cannot be approximated in polynomial time to within a factor of $\Omega(\sqrt{n})$ (cf. [4]).

Henceforth, we focus on obtaining solutions for BMIS that are good *compared* to the average degree of the graph. We show that we can obtain near-tight solutions that compare well with δ. The following are our main results:

Theorem 1. *There is a (centralized) algorithm that selects an MIS of average degree at most $\delta^2/8 + O(\delta)$ and runs in $O(nm^3 \log n)$ time with high probability.*[1]

---

[1] "With high probability" (w.h.p.) means with probability $\geqslant 1 - 1/n^{\Omega(1)}$.

To show the above theorem (the full proof is in [4]) we show that Luby's MIS algorithm[6] returns an MIS with average degree at most $\delta^2/8 + O(\delta)$, albeit *with inverse polynomially small* probability. This can be easily turned into a *centralized* algorithm by repeating this algorithm a polynomial number of times till the desired bound is obtained. (However, this does not give a fast distributed algorithm.) The above algorithm is nearly optimal with respect to the average degree of the MIS, as we show an almost matching lower bound (this also answers the question (i) posed above in the negative):

Theorem 2. *For any real number $\alpha > 1$, there is a graph G with average degree $\leqslant \alpha$, but in which every MIS has average degree $\geqslant \alpha^2/8 + 3\alpha/4 + 5/8$.*

We next consider *distributed* approximation algorithms for BMIS and show that we can output near-optimal solutions fast, i.e., solutions that are close to the lower bound. We consider the following standard model for our distributed algorithms where the given graph G represents a system of n nodes (each node has a distinct ID). Each node runs an instance of the distributed algorithm and the computation advances in synchronous *rounds*, where, in each round, nodes can communicate with their neighbors in G by sending messages of size $O(\log n)$. A node initially has only *local knowledge* limited to itself and its neighbors (it may however know n, the network size). We assume that local computation (performed by the node itself) is free as long it is polynomial in the network size. Each node u has local access to a special bit (initially 0) that indicates whether u is part of the output set. Our focus is on the *time complexity*, i.e., the number of rounds of the distributed computation.

We present two distributed algorithms for BMIS (cf. Section 2.1), the second algorithm gives a better bound on the average degree at the cost of (somewhat) increased run time. However, both algorithms are fast, i.e., run in polylogarithmic rounds.

Theorem 3. *Consider a graph $G = (V, E)$ with average degree δ.*
1. *There is a distributed algorithm that runs in $O(\log n \log \log n)$ rounds and with high probability outputs an MIS with average degree $O(\delta^2)$.*
2. *There is a distributed algorithm that runs in $\log^{2+o(1)} n$ rounds and with high probability outputs an MIS with average degree $(1 + o(1))(\delta^2/4 + \delta)$.*

Note that in general, due to the lower bound (cf. Theorem 2), the bounds provided by algorithms of the above theorem are optimal up to constant factors

We next present results on BMDS. Since an MIS is also an MDS, an algorithm for MIS can also be used to output an MDS. However, this can lead to a bad approximation guarantee, since there are graphs for which *every* MIS has a bad average degree compared to some MDS. This follows from the graph family used in the proof of Theorem 2: while the average degree of *every* MIS (of any graph in the family) is $\Omega(\delta^2)$, there exists an MDS with average degree only $O(\delta)$. Because an MIS is also an MDS, the results of Theorem 3 also hold for BMDS. Our next theorem shows that much better guarantees are possible for BMDS.

Theorem 4. *Any graph G with average degree δ has a minimal dominating set with average degree at most $O(\frac{\delta \log \delta}{\log \log \delta})$. Furthermore, there is a sequential randomized algorithm for finding such an MDS in polynomial time w.h.p.*

The next theorem shows that the bound of Theorem 4 is optimal (in general), up to constant factors:

Theorem 5. *For any real number $\alpha > 0$, there are graphs with average degree $\leqslant \alpha$, but for which any MDS has an average degree of $\Omega(\frac{\alpha \log \alpha}{\log \log \alpha})$.*

Finally, we show the following result for the BMVC problem which shows that there cannot be any bounded approximation algorithm for the problem:

Theorem 6. *For any real number $\alpha > 2$, there are graphs for which the average degree is at most α, but for which the average degree of any MVC approaches to infinity.*

2   Balanced Maximal Independent Set

We first prove Theorem 2 which shows that there are graphs G for which the degree of every MIS is much larger than the degree of G itself. More importantly, the theorem gives a lower bound on the quality of BMIS in general: one cannot guarantee an MIS whose average degree is less than $\frac{\delta^2}{8} + \Theta(\delta)$.

Proof of Theorem 2. Consider the graph consisting of a copies of K_b, as well one copy of $K_{c,c}$, where $b = \lfloor \frac{3+\alpha}{2} \rfloor$ and $c = \lfloor \frac{1}{2}\sqrt{2ab(\alpha - b + 1) + \alpha^2} + \alpha \rfloor$.

The resulting graph has average degree $\frac{ab(b-1)+2c^2}{ab+2c} \leqslant \alpha$. Every MIS of this graph contains one vertex from each K_b, as well as one half of the vertices of $K_{c,c}$, for an average degree of $\frac{ab+c^2}{a+c}$. As a tends to infinity, such average degree increasingly approaches $\frac{(3+\alpha-b)b}{2} \geqslant \alpha^2/8 + 3\alpha/4 + 5/8$.

2.1   Distributed Algorithms for BMIS

This section is devoted for designing different distributed algorithms for BMIS.[2] In particular we will prove Theorem 3 (Parts 1, and 2). The proposed algorithms do not require any global information of the original graph other than n.

Proof of Part 1 of Theorem 3. We propose a distributed algorithm that constructs an MIS I of G such that the following two properties hold with high probability: (a) I has average degree at most $O(\delta^2)$, and (b) I is constructed within $O(\log n \times \log \log n)$ rounds.

This algorithm does not require any global information of the original graph, other than the network size n. The algorithm depends on a parameter ϕ, which is held constant. For any constant $c > 0$, one can choose ϕ appropriately so that the distributed algorithms succeeds with probability $1 - n^{-c}$; the parameter ϕ

---

[2] Omitted proofs are included in the full paper [4].

First Phase – Luby's algorithm on $G_{\phi\sqrt{n/\log n}}$:
(Recall that G_x denotes the subgraph of vertices of degree $\leqslant x$.)
1. Each vertex v in $G_{\phi\sqrt{n/\log n}}$ marks itself independently with probability $1/(2d_v)$.
2. If two adjacent nodes are marked, unmark the one with higher degree (breaking ties arbitrarily).
3. Add any marked nodes to the independent set I.

Second Phase – Extending the MIS:
1. For $i = 0, 1, \ldots, \frac{1}{2}\lceil \log_2 \log n - \log_2 \phi \rceil$, repeat the following:
2. Let $x_i = 2^i \phi \sqrt{n/\log n}$. Run Luby's MIS algorithm for $\phi \log n$ iterations to extend the current independent set I to an MIS of the graph G_{x_i}.

Third Phase:
1. Using Luby's algorithm, extend the current independent set I to an MIS of G.

Algorithm 1. Distributed Algorithm for Approximating BMIS

will only affect the running time by a constant factor. This is the strongest form in which an algorithm can be said to succeed with high probability.

The algorithm has three phases, which are intended to address the cases where $\delta \leqslant O(\sqrt{n/\log n})$, $\Theta(\sqrt{n/\log n}) \leqslant \delta \leqslant \Theta(\sqrt{n})$, and $\delta \geqslant \Omega(\sqrt{n})$ respectively. The first phase runs Luby's algorithm for MIS on the vertices with degree $\leqslant \sqrt{n/\log n}$. The next phase gradually extends the resulting independent set by finding MIS's of the subgraphs consisting of successively larger degrees. Finally, using Luby's algorithm, this is extended to an MIS of G itself. It is easy to see that this leads to an MIS of G, and the resulting algorithm runs in $O(\log n \times \log \log n)$ rounds. We will also show that if we run only Phases I and III of this algorithm, then we can obtain an MIS of degree $O(\delta^2 \log \delta)$ in time $O(\log n)$.

We introduce the following definition which will be used throughout the proof. For any real number s, we let G_s denote the subgraph of G induced on the vertices of degree $\leqslant s$. This notation is used in describing Algorithm 1.

The following basic principle will be used in a variety of places in this proof:

Proposition 1. *Suppose a graph G has n vertices and average degree δ. Suppose $s > 1$. Then the subgraph $G_{s\delta}$ contains at least $n(1 - 1/s)$ vertices.*

We now show that this algorithm has good behavior in the first two parameter regimes. The third regime $\delta = \Omega(\sqrt{n})$ is trivial.

Lemma 1 (First Phase). *Suppose $\delta \leqslant \frac{\phi}{2}\sqrt{n/\log n}$. Then with probability $1 - n^{-\Omega(1/\phi)}$, the independent set produced at the end of the first phase, contains $\Omega(n/\delta)$ vertices. In particular, the final MIS produced has average degree $O(\delta^2)$.*

Proof. Let n', δ' denote the number of vertices and average degree of the graph $G_{\phi\sqrt{n/\log n}}$. Note $\delta' \leqslant \delta$. By Proposition 1 we have $n' \geqslant n/2$.

For each vertex $v \in G_{\phi\sqrt{n/\log n}}$ let X_v be the random variable indicating that v was marked, and X'_v the random variable indicating that v was accepted

1: Let $\phi > 1$ be a fixed parameter. Initialize $I = \emptyset$.
2: **for** $i = 0, \ldots, \lceil \log_\phi n \rceil$ **do**
3:    Using any distributed MIS algorithm, extend I to an MIS of the graph G_{ϕ^i}.
4: Return the final MIS I.

Algorithm 2. Greedy Distributed Approximation Algorithm for BMIS

into I (i.e. it did not conflict with a higher-degree vertex). Let Y denote the size of this independent set, i.e., $Y = \sum_{v \in G_{\phi\sqrt{n/\log n}}} X'_v$. As shown in [6], we have $E[Y] = \Omega(n'/\delta') = \Omega(n/\delta)$. Now, we want to show that Y is concentrated around its mean. As described in [3], this sum can be viewed as a "read-k family"; each variable X'_v is a Boolean function of the underlying independent variables X, and each variable X_v affects at most $k = O(\phi\sqrt{n/\log n})$ of the Boolean functions. Hence this sum obeys a similar concentration bound to the Chernoff bound, albeit the exponent is divided by k. In particular, the probability that Y deviates below a constant factor from its mean is given by $P(Y \leqslant (1-x)E[Y]) \leqslant e^{-\frac{E[Y]x^2}{2k}} \leqslant e^{-\Omega(\frac{\log n}{\phi})} = n^{-\Omega(1/\phi)}$. Hence, with high probability, the total number of vertices returned from the first phase is $\Omega(n/\delta)$ as desired.   □

Lemma 2 (Second Phase). *Suppose $\frac{\phi}{2}\sqrt{n/\log n} < \delta \leqslant \frac{1}{2}\sqrt{n}$. Then with probability $1 - n^{-\Omega(1/\phi)}$, the independent set produced at the end of the second phase, contains $\Omega(n/\delta)$ vertices. In particular, the final MIS produced has average degree $O(\delta^2)$.*

Proof. Let n', δ' represent the number of vertices and average degree of $G_{\sqrt{n}}$. By Proposition 1 we must have $n' = \Omega(n)$ and $\delta' \leqslant \delta$.

If $\delta' \leqslant \frac{\phi}{2}\sqrt{n/\log n}$, then by Proposition 1 there would be $\Omega(n)$ vertices of G with degree $\leqslant \frac{\phi}{2}\sqrt{n/\log n}$. By Lemma 1, phase 1 would then produce an independent set with $\Omega(n/\sqrt{n/\log n}) = \Omega(n/\delta^2)$ vertices.

So suppose $\delta' \geqslant \frac{\phi}{2}\sqrt{n/\log n}$. Now, as i increases, x_i is multiplied by a factor of 2 as it increases from $\phi\sqrt{n/\log n}$ to $\sqrt{n}$. In particular, there is some value of i which has $2\delta' \leqslant x_i \leqslant 4\delta'$. At this point, the standard analysis shows that $\phi \log n$ iterations of Luby's algorithm produces, with probability $1 - n^{-\Omega(1/\phi)}$, an MIS of the graph G_{x_i}. By Proposition 1, G_{x_i} contains $\Omega(n')$ vertices. Furthermore, *any* MIS of G' must contain $\Omega(n'/\delta')$ vertices; the reason for this is that the maximum degree of any vertex in G_{x_i} is $O(\delta')$, and it is necessary to select $\Omega(n'/\delta')$ simply to ensure that every vertex is covered by the MIS.

Now, at stage i we produce an MIS of G_{x_i} which contains $\Omega(n/\delta)$ vertices. This is eventually extended to an MIS of G with $\Omega(n/\delta)$ vertices.   □

Proof of Part 2 of Theorem 3. The greedy algorithm for BMIS is very simple. We label the vertices in order of increasing degree (breaking ties arbitrarily). Each vertex is added to the IS (Initially, IS=$\emptyset$), unless it was adjacent to a earlier vertex already selected. This is a simple deterministic algorithm which requires time $O(m)$.

Theorem 7. *The greedy algorithm produces an MIS of degree at most $\frac{\delta^2}{4} + \delta$.*

Proof. Order the vertices in order of increasing degree $d_1 \leqslant d_2 \leqslant \ldots \leqslant d_n$. Define the indicator variable x_v to be 1 if $v \in I$ and 0 otherwise, where I is the MIS produced. For any pair of vertices u and v with $d_u \geqslant d_v$, we also define the indicator y_{vu} to be 1 if $v \in I$ and there is an edge from v to u. (It may seem strange to include the variable y_{vv}, as we always have $y_{vv} = 0$ in the intended solution, but this will be crucial in our proof, which is based on LP relaxation.)

As the greedy algorithm selects v iff no earlier vertex was adjacent to it, we have $x_v = 1$ if and only if $y_{1v} = y_{2v} = \cdots = y_{v-1,v} = 0$. In particular, x_v satisfies the linear constraint $x_v \geqslant 1 - y_{1v} - y_{2v} - \cdots - y_{vv}$. The variables x, y also clearly satisfy the linear constraints $\forall v: 0 \leqslant x_v \leqslant 1$, $\forall v \leqslant u: 0 \leqslant y_{vu}$, and $\forall v: \sum_u y_{vu} \leqslant d_v x_v$. which we refer to as the *core constraints*. The final MIS contains $\sum x_v$ vertices and $\sum_v d_v x_v$ edges, and hence the average degree of the resulting MIS is $\delta_I = \sum_v d_v x_v / \sum_v x_v$.

We wish to find an upper bound on the ratio $R = \frac{\sum_v d_v x_v}{\sum_v x_v}$. The variables x, y satisfy many other linear and non-linear constraints, and in particular are forced to be integral. However, we will show that the core constraints are sufficient to bound R. The way we will prove this is to explicitly construct a solution x, y which satisfies the core constraints and maximizes R subject to them, and then show that the resulting x, y still satisfies $R \leqslant \frac{\delta^2}{4} + \delta$.

Let x, y be real vectors which maximizes R among all real vectors satisfying the core constraints, and among all such vectors, which minimize $\sum_{u>v} y_{vu}$. Suppose $y_{vu} > 0$ for some $u > v$. If $x_u = 1$, then we simply decrement y_{vu} by ϵ. The constraint $x_u \geqslant 1 - y_{1u} - \cdots - y_{uu}$ clearly remains satisfied as $x_u = 1$, and all other constraints are unaffected. The objective function is also unchanged. However, this reduces $\sum_{u>v} y_{vu}$, contradicting maximality of x, y.

Suppose $y_{vu} > 0$ for some $u > v$, and $x_v < 1$ strictly. Note that $y_{vu} \leqslant d_v x_v$, so we must have $x_v > 0$ strictly. For some sufficiently small ϵ, we change x, y as follows: $y'_{vu} = y_{vu} - \epsilon$, $y'_{vv} = y_{vv} + \frac{\epsilon}{d_v+1}$, $x'_v = x_v - \frac{\epsilon}{d_v+1}$, $x'_u = x_u + \frac{\epsilon}{d_u+1}$, and $y'_{uu} = y_{uu} + \frac{\epsilon d_u}{d_u+1}$. All other values remain unchanged. We claim that the constraints on x, y are still preserved. Furthermore, the numerator of R increases the denominator decreases; hence $R' \geqslant R$. This contradicts the maximality of x, y.

In summary, we can assume $y_{vu} = 0$ for all $u > v$. In this case, the core constraints on v become simply $1 - y_{vv} \leqslant x_v \leqslant 1$ and $y_{vv} \leqslant d_v x_v$. It is a simple exercise to maximize R subject to these constraints (every vertex operates completely independently), and the maximum is achieved when $x_v = \frac{1}{d_v+1}$ for $d_v \leqslant t$, and $x_v = 1$ for $d_v > t$. In this case, the objective function $R(x)$ satisfies $R \leqslant \frac{\sum_{d_v \leqslant t} \frac{d_v}{d_v+1} + \sum_{d_v>t} d_v}{\sum_{d_v \leqslant t} \frac{1}{d_v+1} + \sum_{d_v>t} 1}$ Let δ_S, δ_B denote the average degrees of the vertices of degree $\leqslant t, > t$ respectively, and let n_S, n_B represent the number of such vertices. Then by concavity, we have

$$R \leqslant \frac{n_S \frac{\delta_S}{\delta_S+1} + n_B \delta_B}{\frac{n_S}{\delta_S+1} + n_B} \leqslant \frac{\delta(\delta_B - \delta_S) + \delta_B \delta_S(\delta - \delta_S)}{\delta_S(\delta - \delta_S) + (\delta_B - \delta_S)}$$

1: Mark each vertex of degree $> 2\delta$ independently with prob. $\frac{\log t}{t}$ where $t = \frac{2\delta \log \delta}{\log \log \delta}$.
2: Mark every vertex of degree $\leqslant 2\delta$.
3: If any vertex v is not marked, and none of the neighbors of v are marked, then mark v.
4: Let M denote the set of marked vertices at this point. M forms a dominating set of G, but is not necessarily minimal. Using any algorithm, select a minimal dominating set $M' \subseteq M$.
5: Check if $\delta_{M'} \leqslant t$. If so, return M'. Otherwise, return FAIL.

Algorithm 3. Approximation Algorithm for BMDS

Routine calculus shows that this achieves its maximum value at $\delta_B = \infty, \delta_S = \delta/2$, yielding $R \leqslant \delta^2/4 + \delta$ as claimed. □

This greedy algorithm can be converted, with only a little loss, to a distributed algorithm as shown in Algorithm 2. This algorithm is basically the sequential greedy algorithm, except we are quantizing the degrees to multiples of some parameter ϕ. Allowing $\phi \to 1$ sufficiently slowly, we obtain an algorithm which requries $\log^{2+o(1)} n$ rounds and returns an MIS of degree $(1 + o(1))(\delta^2/4 + \delta)$ w.h.p. As we have seen in Theorem 2, this is within a factor of 2 of the lowest degree possible.

3 Balanced Minimal Dominating Set

For arbitrary graphs, we turn our attention to designing algorithms for finding approximate solutions to BMDS. Since any MIS in a given graph G is also an MDS in G, all algorithms designed for BMIS also return an BMDS in G of the same average degree. Thus, we have the same bounds (and distributed algorithms) corresponding to those in Section 2. However, for BMDS, better bounds are possible. Given a graph with average degree δ, we will show a polynomial-time algorithm that finds an MDS of average degree $O(\frac{\delta \log \delta}{\log \log \delta})$. We will also construct a family of graphs G for which every MDS has average degree $\Omega(\frac{\delta \log \delta}{\log \log \delta})$.

Proof of Theorem 4. For a target degree t, and any set of vertices V_0, we define $S_t^{V_0} = \sum_{v \in V_0}(d_v - t)$. Our goal is to find an MDS X with $S_t^X \leqslant 0$, for some $t = O(\delta \log \delta / \log \log \delta)$.

Let $x = 2\delta$ and divide the vertices into three classes: A, the set of vertices of degree $\leqslant x$; B, the set of vertices of degree $> x$, which have at least one neighbor in A; and C, the set of vertices of degree $> x$, all of whose neighbors are in B or C. Mark each vertex in $B \cup C$ with probability $p = \frac{\log t}{t}$. Next, form the set $Y \subseteq B \cup C$, by inserting all marked vertices in $B \cup C$ and vertices in C with no marked neighbors. Clearly Y dominates C, and $A \cup Y$ dominates G. Now, select two subsets $A' \subseteq A$ and $Y' \subseteq Y$ such that $X = A' \cup Y'$ is an MDS of G.

We first examine $S_t^{Y'}$. Any vertex of G with degree $\leqslant t$ contributes at most 0 to $S_t^{Y'}$. Otherwise, suppose v has degree $\geqslant t$. If $v \in B$, it is selected for Y with probability at most $\frac{\log t}{t}$. If $v \in C$, all its neighbors are marked with probability $\frac{\log t}{t}$, so it is selected for Y with probability at most $\frac{\log t}{t} + (1 - \frac{\log t}{t})t \leqslant 2\frac{\log t}{t}$. Hence the expected contribution of such vertex to $S_t^{Y'}$ is at most $2d_v\frac{\log t}{t}$. Summing over all such vertices, we have $E[S_t^{Y'}] \leqslant 2|B \cup C|\delta_{B \cup C}\frac{\log t}{t}$, where $\delta_{B \cup C}$ denote the average degree of vertices in $B \cup C$.

Now, some of the vertices in A are dominated by B-vertices of Y'. Let A_0 be the set of vertices *not dominated* by Y'. These vertices can only be dominated by vertices of A', so we must have $|A'|(\delta_{A'} + 1) \geqslant |A_0|$. Subject to the conditions $|A'|(\delta_{A'} + 1) \geqslant |A_0|$ and $\delta_{A'} \leqslant x$, we have $S_t^{A'} \leqslant \frac{|A_0|(x-t)}{x+1}$. The ultimate MDS may contain vertices in $A - A_0$ as well; however, as $x \leqslant t$, these will have a negative contribution to S_t, and hence they will only help up in showing an upper bound on S_t.

Consider the expected value of $E[|A_0|]$. A vertex $v \in A$ lies in A_0 if none of its neighbors are marked (this is not a necessary condition), and vertices are marked independently with probability p. Hence $E[|A_0|] \geqslant \sum_{v \in A}(1-p)^{d_v} \geqslant |A|(1-p)^{\delta_A}$. Putting all this together, we have that the final MDS $X = A' \cup Y'$ satisfies $E[S_t^X] \leqslant 2p|B \cup C|\delta_{B \cup C} + |A|\frac{x-t}{x+1}(1-p)^{\delta_A}$. For δ sufficiently large, p approaches zero, so that that $(1-p)^{\delta_A} \leqslant e^{-2p\delta_A}$.

We know that $|A| + |B| + |C| = n$, and $|A|\delta_A + (|B| + |C|)\delta_{B \cup C} \leqslant n\delta$. Eliminating $|A|, |B|, |C|$ we have

$$E[S_t^X] \leqslant n\left(\frac{2\delta_{B \cup C}(\delta - \delta_A)\log t}{t(\delta_{B \cup C} - \delta_A)} - \frac{(\delta_{B \cup C} - \delta)(t - 2\delta)t^{-\frac{\delta_A}{t}}}{(2\delta + 1)(\delta_{B \cup C} - \delta_A)}\right)$$

Routine calculus shows that, for t sufficiently large, this achieves its maximum value at $\delta_{B \cup C} \to \infty, \delta_A = \frac{t \log\left(\frac{t - 2\delta}{4\delta + 2}\right)}{\log t}$, yielding

$$E[S_t^X] \leqslant \frac{2\delta \log t}{t} - 2\log\left(\frac{t - 2\delta}{4\delta + 2}\right) - 2.$$

For $t = 2\delta \log \delta / \log\log \delta$, the RHS approaches to $-\infty$ as $\delta \to \infty$. This implies that $E[S_t^X] \leqslant 0$, so there is a positive probability of selecting an MDS of average degree $\leqslant t$.

Note that for $t = 2\delta \log \delta / \log\log \delta$, we have $S_t^X \geqslant -\delta n \log^{O(1)} n$ and $E[S_t^X] \leqslant -\Omega(1)$. By Markov's inequality, the random variable S_t^X is negative with probability $\geqslant \delta^{-1}n^{-1}\log^{-O(1)} n$. Hence, after $\delta n \log^{O(1)} n$ iterations of this sampling process, we find an MDS of degree $O(\frac{\delta \log \delta}{\log\log \delta})$ with high probability. The total work expended is $m^2 \log^{O(1)} n$. This is summarized as Algorithm 3.

We next prove Theorem 5 which shows that this bound $O(\frac{\delta \log \delta}{\log\log \delta})$ is optimal, up to constant factors.

Proof of Theorem 5. We will construct a graph of average degree $\delta = O(\alpha)$, all of whose MDS's have degree $\Omega(\frac{\alpha \log \alpha}{\log \log \alpha})$. To simplify the proof, we will ignore rounding issues. As all the quantities tend to infinity with α, such rounding issues are negligible for α sufficiently large.

Define $k = \log_2(\alpha \log \alpha / \log \log \alpha)$. We define a random process which constructs a graph with three types of vertices, which we denote A, B, C (these play the same role as in the proof of Theorem 4). The vertices in A, B are organized into clusters of related vertices. For class A, there are $l = 4^k/\alpha^2$ clusters of size α. For class B, there are r clusters of size $2^k/r$, for some $r = \Theta(k)$ (the constant will be specified later). There are $2^k - 1$ vertices in class C. These are not organized into clusters but are considered individually. We index these vertices by the non-zero k-dimensional binary vectors over the finite field $GF(2)$. That is, C corresponds to $C = GF(2)^k - \mathbf{0}$.

We add the following edges to the graph (some of these edges are deterministic, some are random):

1. From each A-vertex to the other vertices in the same A-cluster.
2. From each B-vertex to all the other B-vertices, even those outside its cluster.
3. For each B-cluster b, we choose a random non-zero binary vector v_b in $GF(2)^k$. For each vertex in C, indexed by vector w, we construct an edge from all the B-vertices in the cluster b to the vertex w iff $v_b.w = 1$. The dot product here is taken over the field $GF(2)$.
4. For each A-cluster a, we select $\alpha/(2^k/r)$ B-clusters uniformly at random, with replacement. We add an edge from every vertex in the A-cluster a to every vertex in the selected B-clusters.

This graph has degree $O(\alpha)$. In the full paper [4], we show that with high probability, every MDS of G has degree $\Omega(\frac{\alpha \log \alpha}{\log \log \alpha})$.

4   Conclusion

We initiate the study (graph-theoretic, algorithmic, and distributed) of the balanced versions of some fundamental graph-theoretic structures. As discussed in Section 1, the study of balanced structures can be useful in providing fault-tolerant, load-balanced MISs and MDSs. We develop reasonably-close upper and lower bounds for many of these problems. Furthermore, for the BMIS problem, we are able to develop fast (local) distributed algorithms that achieves an approximation close to the best possible in general. A main open problem that is left open is whether one can do the same for the BMDS problem. We view our results also as a step in understanding the complexity of local computation of these structures whose optimality itself cannot be verified locally.

Acknowledgement. We would like to thank the anonymous referees for the helpful comments.

References

1. Alon, N., Babai, L., Itai, A.: A fast and simple randomized parallel algorithm for the maximal independent set problem. J. Algorithms 7(4), 567–583 (1986)
2. Garey, M.R., Johnson, D.S.: Computers and Intractability: A Guide to the Theory of NP-Completeness. W. H. Freeman (1979)
3. Gavinsky, D., Lovett, S., Saks, M., Srinivasan, S.: A tail bound for read-k families of functions. arXiv preprint arXiv:1205.1478 (2012)
4. Harris, D.G., Morsy, E., Pandurangan, G., Robinson, P., Srinivasan, A.: Efficient Computation of Balanced Structures, http://static.monoid.at/balanced.pdf
5. Kuhn, F., Moscibroda, T., Wattenhofer, R.: Local computation: Lower and upper bounds. CoRR, abs/1011.5470 (2010)
6. Luby, M.: A simple parallel algorithm for the maximal independent set problem. SIAM J. Comput. 15(4), 1036–1053 (1986)
7. Moscibroda, T.: Clustering. In: Algorithms for Sensor and Ad Hoc Networks, pp. 37–60 (2007)
8. Peleg, D.: Distributed Computing: A Locality-Sensitive Approach. SIAM (2000)
9. Rajaraman, R.: Topology control and routing in ad hoc networks: a survey. SIGACT News 33(2), 60–73 (2002)
10. Suomela, J.: Survey of local algorithms. ACM Comput. Surv. 45(2) (2013)
11. Zhang, H., Shen, H.: Balancing energy consumption to maximize network lifetime in data-gathering sensor networks. IEEE TPDS 20(10), 1526–1539 (2009)

A Refined Complexity Analysis
of Degree Anonymization in Graphs

Sepp Hartung[1], André Nichterlein[1], Rolf Niedermeier[1], and Ondřej Suchý[2]

[1] Institut für Softwaretechnik und Theoretische Informatik, TU Berlin
{sepp.hartung,andre.nichterlein,rolf.niedermeier}@tu-berlin.de
[2] Faculty of Information Technology, Czech Technical University in Prague
ondrej.suchy@fit.cvut.cz

Abstract. Motivated by a strongly growing interest in graph anonymization in the data mining and databases communities, we study the NP-hard problem of making a graph k-anonymous by adding as few edges as possible. Herein, a graph is k-anonymous if for every vertex in the graph there are at least $k - 1$ other vertices of the same degree. Our algorithmic results shed light on the performance quality of a popular heuristic due to Liu and Terzi [ACM SIGMOD 2008]; in particular, we show that the heuristic provides optimal solutions in case that many edges need to be added. Based on this, we develop a polynomial-time data reduction, yielding a polynomial-size problem kernel for the problem parameterized by the maximum vertex degree. This result is in a sense tight since we also show that the problem is already NP-hard for H-index three, implying NP-hardness for smaller parameters such as average degree and degeneracy.

1 Introduction

For many scientific disciplines, including the understanding of the spread of diseases in a globalized world or power consumption habits with impact on fighting global warming, the availability of (anonymized) social network data becomes more and more important. In a landmark paper, Liu and Terzi [16] introduced the following simple graph-theoretic model for identity anonymization on (social) networks. Herein, they transferred the k-anonymity concept known for tabular data in databases [9] to graphs.

DEGREE ANONYMITY [16]
Input: An undirected graph $G = (V, E)$ and two positive integers k and s.
Question: Is there an edge set E' over V with $|E'| \leq s$ such that $G' = (V, E \cup E')$ is k-anonymous , that is, for every vertex $v \in V$ there are at least $k - 1$ other vertices in G' having the same degree?

Liu and Terzi [16] assume in this model that an adversary (who wants to deanonymize the network) knows only the degree of the vertex of a target individual;

F.V. Fomin et al. (Eds.): ICALP 2013, Part II, LNCS 7966, pp. 594–606, 2013.

this is a modest adversarial model. Clearly, there are stronger adversarial models which (in many cases very realistically) assume that the adversary has more knowledge, making it possible to breach privacy provided by a "k-anonymized graph" [20]. Moreover, it has been argued that graph anonymization has fundamental theoretical barriers which prevent a fully effective solution [1]. DEGREE ANONYMITY, however, provides the perhaps most basic and still practically relevant model for graph anonymization; it is the subject of active research [4, 5, 18].

Graph anonymization problems are typically NP-hard. Thus, almost all algorithms proposed in this field are heuristic in nature, this also being true for algorithms for DEGREE ANONYMITY [16, 18]. Indeed, as the field of graph anonymization is young and under strong development, there is very little research on its theoretical foundations, particularly concerning computational complexity and algorithms with provable performance guarantees [6].

Our Contributions. Our central result is to show that DEGREE ANONYMITY has a polynomial-size problem kernel when parameterized by the maximum vertex degree Δ of the input graph. In other words, we prove that there is a polynomial-time algorithm that transforms any input instance of DEGREE ANONYMITY into an equivalent instance with at most $O(\Delta^7)$ vertices. Indeed, we encounter a "win-win" situation when proving this result: We show that Liu and Terzi's heuristic strategy [16] finds an optimal solution when the size s of a minimum solution is larger than $2\Delta^4$. As a consequence, we can bound s in $O(\Delta^4)$ and, hence, a polynomial kernel we provide for the combined parameter (Δ, s) is also a polynomial kernel only for Δ. Furthermore, our kernelization has the useful property (e. g. for approximations) that each solution derived for the kernel instance one-to-one corresponds to a solution of the original instance. While this kernelization directly implies fixed-parameter tractability for DEGREE ANONYMITY parameterized by Δ, we also develop a further improved fixed-parameter algorithm.

In addition, we prove that the polynomial kernel for Δ is tight in the sense that even for constant values of the "stronger" parameter (that is, provably smaller) H-index[1], DEGREE ANONYMITY becomes NP-hard. The same proof also yields NP-hardness in 3-colorable graphs. Further, from a parameterized perspective, we show that DEGREE ANONYMITY is W[1]-hard when parameterized by the solution size s (the number of added edges), even when $k = 2$. In other words, there is no hope for tractability even when the level k of anonymity is low and the graph needs only few edge additions (meaning little perturbation) to achieve k-anonymity.

Why is the parameter "maximum vertex degree Δ" of specific interest? First, note that from a parameterized complexity perspective it seems to be a "tight" parameterization in the sense that for the only little "stronger" parameter H-index our results already show NP-hardness for H-index three (also implying hardness e.g. for the parameters degeneracy and average degree). Social networks typically have few vertices with high degree and many vertices of small degree.

---

[1] The H-index of a graph G is the maximum integer h such that G has at least h vertices with degree at least h. Thus G has at most h vertices of degree larger than h.

Leskovec and Horvitz [15] studied a huge instant-messaging network (180 million nodes) with maximum degree bounded by 600. For the DBLP co-author graph generated in February 2012 with more than 715,000 vertices we measured a maximum degree of 804 and an H-index of 208, so there are not more than 208 vertices with degree larger than 208. Thus, a plausible strategy might be to only anonymize vertices of "small" degree and to remove high-degree vertices for the anonymization process because it might be overly expensive to anonymize these high-degree vertices and since they might be well-known (that is, not anonymous) anyway. Indeed, high-degree vertices can be interpreted as outliers [2], potentially making their removal plausible.

Related Work. The most important reference is Liu and Terzi's work [16] where the basic model was introduced, sophisticated (heuristic) algorithms (also using algorithms to determine the realizability of degree sequences) have been developed and validated on experimental data. Somewhat more general models have been considered by Zhou and Pei [25] (studying the neighborhood of vertices instead of only the degree) and by Chester et al. [5] (anonymizing a *subset* of the vertices of the input). Chester et al. [4] investigate the variant of adding vertices instead of edges. Building on Liu and Terzi's work, Lu et al. [18] propose a "more efficient and more effective" algorithm for DEGREE ANONYMITY. Again, this algorithm is heuristic in nature. Today, the field of graph anonymization has grown tremendously with numerous surveys and research directions. We only mention some directly related work.

There are many other, often more complicated models for graph anonymization. Weaknesses DEGREE ANONYMITY (mainly depending on the assumed adversary model where for many practical situations the adversary may e.g. have an auxiliary network that helps in de-anonymizing) of and other models have been pointed out [1, 20, 24]. In conclusion, given the generality of background knowledge an adversary may or may not have, graph anonymization remains a chimerical target [18] and, thus, a universally best model is not available.

Finally, from a (parameterized) computational complexity perspective, the closest work we are aware of is due to Mathieson and Szeider [19] who provide a study on editing graphs to satisfy degree constraints. In their basic model, each vertex is equipped with a degree list and the task is to edit the graph such that each vertex achieves a degree contained in its degree list. They study the editing operations edge addition, edge deletion, and vertex deletion and provide numerous parameterized tractability and intractability results. Interestingly, on the technical side they also rely on the computation of general factors in graphs (as we do) and they also study kernelization, where they leave as most challenging open problem to extend their kernelization results to cases that include vertex deletion and edge addition, emphasizing that the presence of edge additions makes their approach inapplicable.

Due to the lack of space, many technical details are deferred to a full version of the paper.

2 Preliminaries

Parameterized complexity. The concept of parameterized complexity was pioneered by Downey and Fellows [7] (see also [8, 21] for more recent textbooks). A parameterized problem is called *fixed-parameter tractable* if there is an algorithm that decides any instance (I, k), consisting of the "classical" instance I and a parameter $k \in \mathbb{N}_0$, in $f(k) \cdot |I|^{O(1)}$ time, for some computable function f solely depending on k. A core tool in the development of fixed-parameter algorithms is polynomial-time kernelization [3, 12]. Here, the goal is to transform a given problem instance (I, k) in polynomial time into an equivalent instance (I', k'), the so-called *kernel*, such that $k' \leq g(k)$ and $|I'| \leq g(k)$ for some function g. If g is a polynomial, then it is called a *polynomial kernel*. A parameterized problem that is classified as W[1]-hard (using so-called parameterized reductions) is unlikely to admit a fixed-parameter algorithm. There is good complexity-theoretic reason to believe that W[1]-hard problems are not fixed-parameter tractable.

Graphs and Anonymization. We use standard graph-theoretic notation. All graphs studied in this paper are simple, i.e., there are no self-loops and no multi-edges. For a given graph $G = (V, E)$ with vertex set V and edge set E we set $n = |V|$ and $m = |E|$. Furthermore, by $\deg_G(v)$ we denote the degree of a vertex $v \in V$ in G and Δ_G denotes the maximum degree of any vertex of G. For $0 \leq a \leq \Delta_G$ let $D_G(a) = \{v \in V \mid \deg_G(v) = a\}$ be the *block* of degree a, that is, the set of all vertices with degree a in G. Thus, being k-anonymous is equivalent to each block being of size either zero or at least k. The complement graph of G is denoted by $\overline{G} = (V, \overline{E})$, $\overline{E} = \{\{u, v\} \mid u, v \in V, \{u, v\} \notin E\}$. The subgraph of G induced by a vertex subset $V' \subseteq V$ is denoted by $G[V']$. For an edge subset $E' \subseteq E$, $V(E')$ denotes the set of all endpoints of edges in E' and $G[E'] = (V(E'), E')$. For a set of edges S with endpoints in a graph G, we denote by $G + S$ the graph that results by inserting all edges in S into G and we call S an edge insertion set for G. Thus, DEGREE ANONYMITY is the question whether there is an edge insertion set S of size at most s such that $G + S$ is k-anonymous. In this case S is called *k-insertion set* for G. We omit subscripts if the graph is clear from the context.

3 Hardness Results

In this section we provide two polynomial-time many-to-one reductions yielding three (parameterized) hardness results.

Theorem 1. DEGREE ANONYMITY *is NP-hard on 3-colorable graphs and on graphs with H-index three.*

Proof (Sketch). We give a reduction from the NP-hard INDEPENDENT SET problem, where given a graph $G = (V, E)$ and a positive integer h, the question is whether there is a size-h independent set, that is, a vertex subset of pairwise nonadjacent vertices. We assume without loss of generality that in the given INDEPENDENT SET instance (G, h) it holds that $|V| \geq 2h + 1$. We construct an equivalent instance $(G' = (V', E'), k, s)$ for DEGREE ANONYMITY as follows. We start with a copy G' of G, denoting with $v' \in V'$ the copy of the vertex $v \in V$. Then, for each vertex $v \in V$ we add to G' degree-one vertices adjacent to v' such that v' has degree Δ_G in G'. Next we add a star with $\Delta_G + h - 1$ leaves and denote its central vertex c. We conclude the construction by setting $k = h + 1$ and $s = \binom{h}{2}$.

INDEPENDENT SET is NP-hard on 3-colorable graphs [23, Lemma 6] and on graphs with maximum degree three [11, GT20]. Observe that if G is 3-colorable, then G' is also 3-colorable. Furthermore, if G has maximum degree three, then only the central vertex c might have degree larger than three, implying that the H-index of G' is three. □

DEGREE ANONYMITY is W[1]-hard with respect to the standard parameterization, that is, by the size of edges s that are allowed to add:

Theorem 2. DEGREE ANONYMITY *is W[1]-hard parameterized by the number of inserted edges s, even if $k = 2$.*

4   Polynomial Kernel for the Maximum Degree Δ

In this main section we provide a polynomial kernel with respect to the parameter maximum degree Δ (Theorem 4). Our proof has two main ingredients: first we show in Section 4.2 a polynomial kernel with respect to the combined parameter (Δ, s); second we show in Section 4.3 that a slightly modified variant of Liu and Terzi's heuristic [16] exactly solves any instance having a minimum-size k-insertion set of size at least $(\Delta^2 + 4\Delta + 3)^2$. Hence, either we can solve a given instance in polynomial time or we can upper-bound s by $(\Delta^2 + 4\Delta + 3)^2$, implying that the kernel polynomial in (Δ, s) is indeed polynomial only in Δ. We begin by presenting the main technical tool used in our work, the so-called f-FACTOR problem.

4.1 f-Factor Problem

DEGREE ANONYMITY has a close connection to the polynomial-time solvable f-FACTOR problem [17, Chapter 10]: Given a graph $G = (V, E)$ and a function $f : V \to \mathbb{N}_0$, does there exist an f-*factor*, that is, a subgraph $G' = (V, E')$ of G such that $\deg_{G'}(v) = f(v)$ for all vertices? One can reformulate DEGREE ANONYMITY using f-FACTOR as follows: Given an instance (G, k, s), the question is whether there is a function $f : V \to \mathbb{N}_0$ such that the complement graph $\overline{G}$ contains an f-factor, $\sum_{v \in V} f(v) \leq 2s$ (every edge is counted twice in the sum of degrees), and for all $v \in V$ it holds that $|\{u \in V \mid \deg_G(u) + f(u) = \deg_G(v) + f(v)\}| \geq k$ (the k-anonymity requirement). The following lemma guarantees the existence of an f-factor in graphs fulfilling certain requirements on the maximum degree and the number of vertices.

Lemma 1 ([14]). *Let $G = (V, E)$ be a graph with minimum vertex degree δ and let $a \leq b$ be two positive integers. Suppose further that*

$$\delta \geq \frac{b}{a+b}|V| \text{ and } |V| > \frac{a+b}{a}(b+a-3).$$

Then, for any function $f : V \to \{a, a+1, ..., b\}$ where $\sum_{v \in V} f(v)$ is even, G has an f-factor.

As we are interested in an f-factor in the complement graph of our input graph G, we use Lemma 1 with $a = 1$, $b = \Delta + 2$, and minimum degree $\delta \geq n - \Delta - 1$. This directly leads to the following.

Corollary 1. *Let $G = (V, E)$ be a graph with n vertices, minimum degree $n - \Delta - 1$, $\Delta \geq 1$, and let $f : V \to \{1, \ldots, \Delta + 2\}$ be a function such that $\sum_{v \in V} f(v)$ is even. If $n \geq \Delta^2 + 4\Delta + 3$, then G has an f-factor.*

4.2 Polynomial Kernel for (Δ, s)

Our kernelization algorithm is based on the following observation. For a given graph G, consider for some $1 \leq i \leq \Delta_G$ the block $D_G(i)$, that is, the set of all vertices of degree i. If $D_G(i)$ contains many vertices, then the vertices are "interchangeable":

Observation 1. *Let $G = (V, E)$ be a graph, let S be a k-insertion set for G, and let $v \in V(S) \cap D_G(i)$ be a vertex such that $|D_G(i)| > (\Delta + 2)s$. Then there exists a vertex $u \in D_G(i) \setminus V(S)$ such that replacing in every edge of S the vertex v by u results in a k-insertion set for G.*

Proof. Since $|S| \leq s$, the vertex v can be incident to at most s edges in S. Denoting the set of these edges by S^v, obviously one can replace v by $u \in D_G(i)$ if u is non-adjacent to all vertices in $V(S^v) \setminus \{v\}$ (this allows to insert all edges) and $u \notin V(S)$ (the size of all blocks in $G + S$ does not change). However, as $V(S)$ contains at most $2s$ vertices from $D_G(i)$ and each of the at most s vertices in $V(S^v) \setminus \{v\}$ has at most Δ_G neighbors in G, it follows that such a vertex $u \in D_G(i)$ exists if $|D_G(i)| > (\Delta + 2)s$. $\qquad\square$

By Observation 1, in our kernel we only need to keep at most $(\Delta + 2)s$ vertices in each block: If in an optimal k-insertion set S there is a vertex $v \in V(S)$ that we did not keep, then by Observation 1 we can replace v by some vertex we kept. There are two major problems that need to be fixed to obtain a kernel: First, when removing vertices from the graph, the degrees of the remaining vertices change. Second, k might be "large" and, thus, removing vertices (during kernelization) in one block may breach the k-anonymity constraint. To overcome the first problem we insert some "dummy-vertices" which are guaranteed not to be contained in any k-insertion set. However, to solve the second problem we need to adjust the parameter k as well as the number of vertices that we keep from each block.

Algorithm 1. The pseudocode or the algorithm producing a polynomial kernel with respect to (Δ, s).

1: **procedure** PRODUCEPOLYKERNEL$(G = (V, E), k, s)$
2:      **if** $|V| \leq \Delta(\beta + 4s)$ **then**   // β is defined as $\beta = (\Delta + 4)s + 1$
3:          **return** (G, k, s)

4:      $k' \leftarrow \min\{k, \beta\}$; $A \leftarrow \emptyset$
5:      **for** $i \leftarrow 1$ **to** Δ **do**
6:          **if** $2s < |D_G(i)| < k - 2s$ **then**
7:              **return** trivial no-instance   // insufficient budget for $D_G(i)$
8:          **if** $k \leq \beta$ **then**   // determine retained vertices
9:              $x \leftarrow \min\{|D_G(i)|, \beta + 4s\}$
10:         **else if** $|D_G(i)| \leq 2s$ **then**
11:             $x \leftarrow |D_G(i)|$
12:         **else**
13:             $x \leftarrow k' + \min\{4s, (|D_G(i)| - k)\}$ // observe that $k' = \beta$.
14:         add x vertices from $D_G(i)$ to A
15:     $G' = G[A]$
16:     **for each** $v \in A$ **do**   // add vertices to preserve degree of retained vertices
17:         add to G' $\deg_G(v) - \deg_{G'}(v)$ many degree-one vertices adjacent to v
18:     denote with P the set of vertices added in Line 17
19:     by adding matched pairs of vertices, ensure that $|P| \geq \max\{4\Delta + 4s + 4, k'\}$
20:     **if** $\Delta + s + 1$ is even **then**
21:         $G^F = (P, E^F) \leftarrow (\Delta + s + 1)$-factor in $\overline{G'[P]}$
22:     **else**
23:         $G^F = (P, E^F) \leftarrow (\Delta + s + 2)$-factor in $\overline{G'[P]}$
24:     $G' \leftarrow G' + E^F$
25:     **return** (G', k', s)

We now explain the kernelization algorithm in detail (see Algorithm 1 for the pseudocode). Let (G, s, k) be an instance of DEGREE ANONYMITY. For brevity we set $\beta = (\Delta + 4)s + 1$. We compute in polynomial time an equivalent instance (G', s, k') with at most $O(\Delta^3 s)$ vertices: First set $k' = \min\{k, \beta\}$ (Line 4). We arbitrarily select from each block $D_G(i)$ a certain number x of vertices and collect all these vertices into the set A (Line 14). To cope with the above mentioned second problem, the "certain number" is defined in a case distinction on the value of k (see Lines 5 to 14). Intuitively, if k is large then we distinguish between "small" blocks of size at most $2s$ and "large" blocks of size at least $k - 2s$. Obviously, if there is a block which is neither small nor large, then the instance is a no-instance (see Line 7). Thus, in the problem kernel we keep for small blocks the "distance to size zero" and for large blocks the "distance to size k". Furthermore, in order to separate between small and large blocks it is sufficient that $k' > 4s$. However, to guarantee that Observation 1 is applicable, the case distinction is a little bit more complicated, see Lines 5 to 14.

We start building G' by first copying $G[A]$ into it (Line 15). Next, adding a pendant vertex to v means that we add a new vertex to G' and make it adjacent to v. For each $v \in A$ we add pendant vertices to v to ensure that $\deg_{G'}(v) = \deg_G(v)$ (Line 17). The vertices of A stay untouched in the following. Denote the set of all pendant vertices by P. Next, we add enough pairwise adjacent vertices to P to ensure that $|P| \geq \max\{k', 4\Delta + 4s + 4\}$ (Line 19). Hence, $|P| \leq \max\{|A| \cdot \Delta, k', 4\Delta + 4s + 4\} + 1$. To avoid that vertices in P help to anonymize the vertices in A we "shift" the degree of the vertices in P (see Lines 20 to 24): We add edges between the vertices in P to ensure that the degree of all vertices in P is $\Delta + s + 2$ (when $\Delta + s + 1$ is even) or $\Delta + s + 3$ (when $\Delta + s + 2$ is even). For the ease of notation let χ denote the new degree of the vertices in P. Observe that before adding edges all vertices in P have degree one in G'. Thus, the minimum degree in $\overline{G'[P]}$ is $|P| - 2$. Furthermore, for each $v \in P$ we denote by $f(v)$ the number of incident edges v requires to have the described degree. It follows that $f(v)$ is even and hence $\sum_{v \in P} f(v)$ is even. Hence setting $a = b = \chi$ fulfills all conditions of Lemma 1. Thus, the required f-factor exists and can be found in $O(|P|\sqrt{|P|(\Delta + s)})$ time [10]. This completes the description of the kernelization algorithm.

The key point of the correctness of the kernelization is to show that without loss of generality, no k-insertion set S for G' of size $|S| \leq s$ affects any vertex in P. This is ensured by "shifting" the degree of all vertices in P by $s + 1$ (or $s + 2$), implying that none of the vertices in A can "reach" the degree of any vertex in P by adding at most s edges. Hence each block either is a subset of A or of P. We now prove that we may assume that an edge insertion set does not affect any vertex in P. All what we need to prove this is the fact that A contains at least $\beta + 4s$ vertices from at least one block in G. Observe that this is ensured by the check in Line 2.

Lemma 2. *If there is a k-insertion set S for G' with $|S| \leq s$, then there is also a k-insertion set S' for G' with $|S'| = |S|$ such that $V(S') \cap P = \emptyset$.*

Based on Lemma 2 we now prove the correctness of our kernelization algorithm.

Theorem 3. DEGREE ANONYMITY *admits a kernel with $O(\Delta^3 s)$ vertices.*

Proof. The polynomial kernel is computed by Algorithm 1. As f-FACTOR can be solved in polynomial time [10], Algorithm 1 runs in polynomial time. The correctness of the kernelization algorithm is deferred to a full version. It remains to show the size of the kernel. To this end, observe that each block in A has size at most $\beta + 4s$ (see Lines 9, 11, and 13). Thus, $|A| = O(\Delta\beta) = O(\Delta^2 s)$. Furthermore, the set P contains at most $\max\{\Delta|A|, k', 4s + 4\Delta + 1\}$ vertices (see Lines 17 to 19). Thus, $|P| = O(\Delta^3 s)$ and, hence, the reduced instance contains $O(\Delta^3 s)$ vertices.   □

4.3   A Polynomial-Time Algorithm for "Large" Solution Instances

In this section we show that if a minimum size k-insertion set S is "large" compared to Δ, then one can solve the instance in polynomial time (Lemma 5). Towards this, we first show that a large solution influences the degree of many vertices. Then the main idea is that if it influences the degree of "many" vertices from the same block, say $D_G(i)$, then by Observation 1 the corresponding vertices can be arbitrarily "interchanged". Thus it is not important to know which vertex from $D_G(i)$ has to be "moved" up to a certain degree by adding edges, because Observation 1 ensures that we can greedily find one. This, however, implies that the actual structure of the input graph (which forbids to insert certain edges since they are already present) no longer matters. Hence, we solve DE-GREE ANONYMITY without taking the graph structure into account. Thereby, if we can k-anonymize the degree sequence corresponding to G (the sequence of degrees of G) such that "many" degrees have to be adjusted, then by Corollary 1 we can conclude that $\overline{G}$ contains an f-factor where $f(v)$ captures the difference between the degree of v in G and the anonymized degree sequence. The f-factor can be found in polynomial time [10] and, hence, a k-insertion set can be found in polynomial time. We now formalize this idea.

We first show that a "large" minimum-size k-insertion set increases the maximum degree by at most two.

Lemma 3. *Let $G = (V, E)$ be a graph and let S be a minimum-size k-insertion set. If $|V(S)| \geq \Delta^2 + 4\Delta + 3$, then the maximum degree in $G + S$ is at most $\Delta + 2$.*

Proof. Let G be a graph with maximum degree Δ and k an integer. Let S be a minimum-size edge set such that $G + S$ is k-anonymous and $|V(S)| \geq \Delta^2 + 4\Delta + 3$. Now assume towards a contradiction that the maximum degree in $G + S$ is at least $\Delta + 3$. We show that there exists an edge set S' such that $G + S'$ is k-anonymous, $|S'| < |S|$, and $G + S'$ has maximum degree at most $\Delta + 2$.

First we introduce some notation. Let f be a function $f : V \to \mathbb{N}_0$ defined as $f(v) = \deg_{G+S}(v) - \deg_G(v)$ for all $v \in V$. Furthermore, denote with X the set of all vertices having degree more than $\Delta + 2$ in $G + S$, that is, $X = \{v \in V \mid f(v) + \deg_G(v) \geq \Delta + 3\}$. Observe that $G[S]$ is an f-factor of $\overline{G}$ and $2|S| = \sum_{v \in V} f(v)$. We now define a new function $f' : V \to \mathbb{N}_0$ such that $\overline{G}$ contains an f'-factor denoted by $G' = (V, S')$ where the edge set S' has the properties as described in the previous paragraph.

We define f' for all $v \in V$ as follows:

$$f'(v) = \begin{cases} f(v) & \text{if } v \notin X, \\ \Delta - \deg_G(v) + 1 & \text{if } v \in X \text{ and } f(v) + \deg_G(v) - \Delta - 1 \text{ is even}, \\ \Delta - \deg_G(v) + 2 & \text{else.} \end{cases}$$

First observe that $\deg_G(v) + f'(v) \leq \Delta + 2$ for all $v \in V$. Furthermore, observe that $f'(v) = f(v)$ for all $v \in V \setminus X$ and for all $v \in X$ it holds that $f'(v) < f(v)$ and $f(v) - f'(v)$ is even. Thus, $\sum_{v \in V} f(v) > \sum_{v \in V} f'(v)$ and $\sum_{v \in V} f'(v)$ is even. It remains to show that (i) $\overline{G}$ contains an f'-factor $G' = (V, S')$ and (ii) $G + S'$ is k-anonymous.

To prove (i) let $\widetilde{V} = \{v \in V \mid f'(v) > 0\}$ and observe that if $f(v) > 0$, then, by definition of f', we have $f'(v) > 0$ and hence $\widetilde{V} = V(S)$. Furthermore, let $\widetilde{G} = \overline{G[\widetilde{V}]}$. Observe that $\widetilde{G}$ has minimum degree $|\widetilde{V}| - \Delta - 1$ and $|\widetilde{V}| =$

$|V(S)| \geq \Delta^2 + 4\Delta + 3$. Thus, the conditions of Corollary 1 are satisfied and hence $\widetilde{G}$ contains an $f'|_{\widetilde{V}}$-factor $\widetilde{G}' = (\widetilde{V}, S')$. By definition of $\widetilde{V}$ it follows that $G' = (V, S')$ is an f'-factor of $\overline{G}$. Thus, it remains to show (ii).

Assume towards a contradiction that $G + S'$ is not k-anonymous, that is, there exists some vertex $v \in V$ such that $1 \leq |D_{G+S'}(\deg_{G+S'}(v))| < k$. Let $d = \deg_{G+S}(v)$ and $d' = \deg_{G+S'}(v)$. Observe that $d' = \deg_G(v) + f'(v)$. Thus, if $v \notin X$, then by definition of f' it holds that $d' = \deg_G(v) + f(v) = d \leq \Delta + 2$. Hence, for all vertices $u \in D_{G+S}(d')$ it follows that $u \notin X$. Thus, $D_{G+S}(d') \subseteq D_{G+S'}(d')$ and since $G + S$ is k-anonymous we have $|D_{G+S'}(d')| \geq k$, a contradiction. On the other hand, if $v \in X$, that is, $d > \Delta + 2$, then $|D_{G+S}(d)| \geq k$ since $G + S$ is k-anonymous. Furthermore, by the definitions of $D_{G+S}(d)$, f, and X we have for all $u \in D_{G+S}(d)$ that $\deg_G(u) + f(u) = d$, $u \in X$, and, thus, $f'(u) + \deg_G(u) = d'$. Therefore, $D_{G+S}(d) \subseteq D_{G+S'}(d')$ and $|D_{G+S'}(d')| \geq k$, a contradiction. $\square$

We now formalize the anonymization of degree sequences. A multiset of positive integers $\mathcal{D} = d_1, \ldots, d_n$, d_i that corresponds to the degrees of all vertices in a graph is called *degree sequence*. A degree sequence $\mathcal{D}$ is k-anonymous if each number in $\mathcal{D}$ occurs at least k times in $\mathcal{D}$. Clearly, the degree sequence of a k-anonymous graph G is k-anonymous. Moreover, if a graph G can be transformed by at most s edge insertions into a k-anonymous graph, then the degree sequence of G can be transformed into a k-anonymous degree sequence by increasing the integers by no more than $2s$ in total (clearly, in the other direction this fails in general because of the graph structure). As we are only interested in a degree sequence corresponding to a graph of a DEGREE ANONYMITY instance where s is large, by Lemma 3 we can require the integers in a k-anonymous degree sequence to be at upper-bounded by $\Delta + 2$.

k-DEGREE SEQUENCE ANONYMITY (k-DSA)

Input: Two positive integers k and s and a degree sequence $\mathcal{D} = d_1, \ldots, d_n$ with $d_1 \leq d_2 \leq \ldots \leq d_n$ and $\Delta = d_n$.

Question: Is there a k-anonymous degree sequence $\mathcal{D}' = d_1', \ldots, d_n'$ with $d_i \leq d_i'$ and $\max_{1 \leq i \leq n} d_i' \leq \Delta + 2$ such that $\sum_{i=1}^{n} d_i' - d_i = 2s$?

By slightly modifying a dynamic programming-based heuristic introduced by Liu and Terzi [16], we next prove that k-DEGREE SEQUENCE ANONYMITY is polynomial-time solvable. Note that Liu and Terzi [16] use their heuristic to solve DEGREE ANONYMITY by first solving the problem on the degree sequence of the input graph G and then trying to "realize" (adding the corresponding edges to G) the produced k-anonymous degree sequence.

Lemma 4. k-DEGREE SEQUENCE ANONYMITY *can be solved in polynomial time.*

We now have all ingredients to solve DEGREE ANONYMITY in polynomial time in case it has a "large" minimum-size k-insertion set. More formally, let (G, k, s) be an instance and let $\mathcal{D}$ be the degree sequence of G, then first find the largest $i \leq s$ such that $(\mathcal{D}, k, i)$ is a yes-instance for k-DEGREE SEQUENCE ANONYMITY. If i is "large", then we prove that we can transfer the solution to G. In all other cases, since any k-insertion set for G of size $j \leq s$ directly implies that $(\mathcal{D}, k, j)$ is a yes-instance for k-DEGREE SEQUENCE ANONYMITY, it follows that we can bound the parameter s by a function in Δ.

Lemma 5. *Let (G, k, s) be an instance of DEGREE ANONYMITY. Either one can decide the instance in polynomial time or (G, k, s) is a yes-instance if and only if $(G, k, \min\{(\Delta^2 + 4\Delta + 3)^2, s\})$ is a yes-instance.*

By Lemma 5 it follows that in polynomial time we can either find a solution or we have $s < (\Delta^2 + 4\Delta + 3)^2$. By Theorem 3 this implies our main result.

Theorem 4. DEGREE ANONYMITY *admits an $O(\Delta^7)$-vertex kernel.*

5 Fixed-Parameter Algorithm

We provide a direct combinatorial algorithm for the combined parameter (Δ, s). Roughly speaking, for fixed k-insertion set S the algorithm branches into all suitable structures of $G[S]$, that is, graphs of at most $2s$ vertices with vertex labels from $\{1, \ldots, \Delta\}$. Then the algorithm checks whether the respective structure occurs as a subgraph in $\overline{G}$ such that the labels on the vertices match the degree of the corresponding vertex in G.

Theorem 5. DEGREE ANONYMITY *can be solved in $(6s^2\Delta^3)^{2s} \cdot s^2 \cdot n^{O(1)}$ time.*

Note that due to the upper bound $s < (\Delta^2 + 4\Delta + 3)^2$ (see Lemma 5) and the polynomial kernel for the parameter Δ (see Theorem 4), Theorem 5 also provides an algorithm running in $\Delta^{O(\Delta^4)} + n^{O(1)}$ time.

6 Conclusion

One of the grand challenges of theoretical research on computationally hard problems is to gain a better understanding of when and why heuristic algorithms work [13]. In this theoretical study, we contributed to a better theoretical understanding of a basic problem in graph anonymization, on the one side partially explaining the quality of a successful heuristic approach [16] and on the other side providing a first step towards a provably efficient algorithm for relevant special cases (bounded-degree graphs). Our work just being one of the first steps in the so far underdeveloped field of studying the computational complexity of graph anonymization [6], there are numerous challenges for future research. For instance, our focus was on classification results rather than engineering the upper bounds, a natural next step to do. Second, it would be interesting to perform a data-driven analysis of parameter values on real-world networks in order to gain parameterizations that can be exploited in a broad-band multivariate complexity analysis [22] of DEGREE ANONYMITY. Finally, with DEGREE ANONYMITY we focused on a very basic problem of graph anonymization; there are numerous other models (partially mentioned in the introductory section) that ask for similar studies.

References

[1] Aggarwal, C.C., Li, Y., Yu, P.S.: On the hardness of graph anonymization. In: Proc. 11th IEEE ICDM, pp. 1002–1007. IEEE (2011)

[2] Aggarwal, G., Feder, T., Kenthapadi, K., Khuller, S., Panigrahy, R., Thomas, D., Zhu, A.: Achieving anonymity via clustering. ACM Transactions on Algorithms 6(3), 1–19 (2010)

[3] Bodlaender, H.L.: Kernelization: New upper and lower bound techniques. In: Chen, J., Fomin, F.V. (eds.) IWPEC 2009. LNCS, vol. 5917, pp. 17–37. Springer, Heidelberg (2009)

[4] Chester, S., Kapron, B.M., Ramesh, G., Srivastava, G., Thomo, A., Venkatesh, S.: k-Anonymization of social networks by vertex addition. In: Proc. 15th ADBIS (2). CEUR Workshop Proceedings, vol. 789, pp. 107–116 (2011), CEUS-WS.org

[5] Chester, S., Gaertner, J., Stege, U., Venkatesh, S.: Anonymizing subsets of social networks with degree constrained subgraphs. In: Proc. ASONAM, pp. 418–422. IEEE Computer Society (2012)

[6] Chester, S., Kapron, B., Srivastava, G., Venkatesh, S.: Complexity of social network anonymization. Social Network Analysis and Mining (2012) (online available)

[7] Downey, R.G., Fellows, M.R.: Parameterized Complexity. Springer (1999)

[8] Flum, J., Grohe, M.: Parameterized Complexity Theory. Springer (2006)

[9] Fung, B.C.M., Wang, K., Chen, R., Yu, P.S.: Privacy-preserving data publishing: A survey of recent developments. ACM Computing Surveys 42(4), 14:1–14:53 (2010)

[10] Gabow, H.N.: An efficient reduction technique for degree-constrained subgraph and bidirected network flow problems. In: Proc. 15th STOC, pp. 448–456. ACM (1983)

[11] Garey, M.R., Johnson, D.S.: Computers and Intractability: A Guide to the Theory of NP-Completeness. Freeman (1979)

[12] Guo, J., Niedermeier, R.: Invitation to data reduction and problem kernelization. SIGACT News 38(1), 31–45 (2007)

[13] Karp, R.M.: Heuristic algorithms in computational molecular biology. J. Comput. Syst. Sci. 77(1), 122–128 (2011)

[14] Katerinis, P., Tsikopoulos, N.: Minimum degree and f-factors in graphs. New Zealand J. Math. 29(1), 33–40 (2000)

[15] Leskovec, J., Horvitz, E.: Planetary-scale views on a large instant-messaging network. In: Proc. 17th WWW, pp. 915–924. ACM (2008)

[16] Liu, K., Terzi, E.: Towards identity anonymization on graphs. In: Proc. ACM SIGMOD 2008, pp. 93–106. ACM (2008)

[17] Lovász, L., Plummer, M.D.: Matching Theory. Annals of Discrete Mathematics, vol. 29. North-Holland (1986)

[18] Lu, X., Song, Y., Bressan, S.: Fast identity anonymization on graphs. In: Liddle, S.W., Schewe, K.-D., Tjoa, A.M., Zhou, X. (eds.) DEXA 2012, Part I. LNCS, vol. 7446, pp. 281–295. Springer, Heidelberg (2012)

[19] Mathieson, L., Szeider, S.: Editing graphs to satisfy degree constraints: A parameterized approach. J. Comput. Syst. Sci. 78(1), 179–191 (2012)

[20] Narayanan, A., Shmatikov, V.: De-anonymizing social networks. In: Proc. 30th IEEE SP, pp. 173–187. IEEE (2009)

[21] Niedermeier, R.: Invitation to Fixed-Parameter Algorithms. Oxford University Press (2006)

[22] Niedermeier, R.: Reflections on multivariate algorithmics and problem parameterization. In: Proc. 27th STACS. LIPIcs, vol. 5, pp. 17–32. Schloss Dagstuhl–Leibniz-Zentrum für Informatik (2010)

[23] Phillips, C., Warnow, T.J.: The asymmetric median tree—a new model for building consensus trees. Discrete Appl. Math. 71(1-3), 311–335 (1996)

[24] Sala, A., Zhao, X., Wilson, C., Zheng, H., Zhao, B.Y.: Sharing graphs using differentially private graph models. In: Proc. 11th ACM SIGCOMM, pp. 81–98. ACM (2011)

[25] Zhou, B., Pei, J.: The k-anonymity and l-diversity approaches for privacy preservation in social networks against neighborhood attacks. Knowl. Inf. Syst. 28(1), 47–77 (2011)

Sublinear-Time Maintenance of Breadth-First Spanning Tree in Partially Dynamic Networks[*]

Monika Henzinger[1,**], Sebastian Krinninger[1,**], and Danupon Nanongkai[2,***]

[1] University of Vienna, Fakultät für Informatik, Austria
[2] Nanyang Technological University, Singapore

Abstract. We study the problem of maintaining a *breadth-first spanning tree* (BFS tree) in *partially dynamic* distributed networks modeling a sequence of either failures or additions of communication links (but not both). We show $(1 + \epsilon)$-approximation algorithms whose amortized time (over some number of link changes) is *sublinear* in D, the *maximum diameter* of the network. This breaks the $\Theta(D)$ time bound of recomputing "from scratch".

Our technique also leads to a $(1 + \epsilon)$-approximate incremental algorithm for single-source shortest paths (SSSP) in the sequential (usual RAM) model. Prior to our work, the state of the art was the classic *exact* algorithm of [9] that is optimal under some assumptions [27]. Our result is the first to show that, in the incremental setting, this bound can be beaten in certain cases if a small approximation is allowed.

1 Introduction

Complex networks are among the most ubiquitous models of interconnections between a multiplicity of individual entities, such as computers in a data center, human beings in society, and neurons in the human brain. The connections between these entities are constantly changing; new computers are gradually added to data centers, or humans regularly make new friends. These changes are usually *local* as they are known only to the entities involved. Despite their locality, they could affect the network *globally*; a single link failure could result in several routing path losses or destroy the network connectivity. To maintain its robustness, the network has to quickly respond to changes and repair its infrastructure. The study of such tasks has been the subject of several active areas of research, including dynamic, self-healing, and self-stabilizing networks.

One important infrastructure in distributed networks is the *breadth-first spanning (BFS) tree* [23,25]. It can be used, for instance, to approximate the network diameter and to provide a communication backbone for broadcast of information

---

[*] Full version available at `http://eprints.cs.univie.ac.at/3703`
[**] The research leading to these results has received funding from the European Union's Seventh Framework Programme (FP7/2007-2013) under grant agreement no. 317532.
[***] Work partially done while at University of Vienna, Austria.

through the network, routing, and control. In this paper, we study the problem of maintaining a BFS tree on dynamic distributed networks. Our main interest is repairing a BFS tree as fast as possible after each topology change.

Model. We model the communication network by the CONGEST model [25], one of the major models of (locality-sensitive) distributed computation. Consider a synchronous network of processors modeled by an undirected unweighted graph G, where nodes model the processors and edges model the bounded-bandwidth links between the processors. We let $V(G)$ and $E(G)$ denote the set of nodes and edges of G, respectively. For any node u and v, we let $d_G(u,v)$ be the distance between u and v in G. The processors (henceforth, nodes) are assumed to have unique IDs of $O(\log n)$ bits and infinite computational power. Each node has limited topological knowledge; in particular, it only knows the IDs of its neighbors and knows *no* other topological information. The communication is synchronous and occurs in discrete pulses, called *rounds*. All the nodes wake up simultaneously at the beginning of each round. In each round each node u is allowed to send an arbitrary message of $O(\log n)$ bits through each edge (u,v) that is adjacent to u, and the message will reach v at the end of the current round. There are several measures to analyze the performance of such algorithms, a fundamental one being the *running time*, defined as the worst-case number of rounds of distributed communication.

We model dynamic networks by a sequence of *attack* and *recovery* stages following the initial *preprocessing*. The dynamic network starts with a preprocessing on the initial network denoted by N_0, where nodes communicate on N_0 for some number of rounds. Once the preprocessing is finished, we begin the first attack stage where we assume that an adversary, who sees the current network N_0 and the states of all nodes, inserts and deletes an arbitrary number of edges in N_0. We denote the resulting network by N_1. This is followed by the first recovery stage where we allow nodes to communicate on N_1. After the nodes have finished communicating, the second attack stage starts, followed by the second recovery stage, and so on. We assume that N_t is connected for every stage t. For any algorithm, we let the *total update time* be the total number of rounds needed by nodes to communicate during all recovery stages. Let the *amortized update time* be the total time divided by q which is defined to be the number of edges inserted and deleted. Important parameters in analyzing the running time are n, the number of nodes (which remains the same throughout all changes) and D, the *maximum diameter*, defined to be the maximum diameter among all networks in $\{N_0, N_1, \ldots\}$. Note that $D \leq n$ since we assume that the network remains connected throughout. Following the convention from the area of (sequential) dynamic graph algorithms, we say that a dynamic network is *fully dynamic* if both insertions and deletions can occur in the attack stages. Otherwise, it is *partially dynamic*. Specifically, if only edge insertions can occur, it is an *incremental dynamic network*. If only edge deletions can occur, it is *decremental*.

Our model highlights two aspects of dynamic networks: (1) how quick a network can recover its infrastructure after changes (2) how edge failures and additions affect the network. These aspects have been studied earlier but we are not aware of any previous model identical to ours. To highlight these aspects, there are also a few assumptions inherent in our model. First, it is assumed that the network remains static in each recovery stage. This assumption is often used (e.g. [18,13,21,24]) and helps emphasizing the running time aspect of dynamic networks. Second, our model assumes that only edges can change. While there are studies that assume that nodes can change as well (e.g. [18,13]), this assumption is common and practical; see, e.g., [1,4,22,8] and references therein. Our amortized update time is also similar in spirit to the amortized communication complexity heavily studied earlier (e.g. [4]). Finally, the results in this paper are on partially dynamic networks. While fully dynamic algorithms are more desirable, we believe that the partially dynamic setting is worth studying, for two reasons. The first reason, which is our main motivation, comes from an experience in the study of sequential dynamic algorithms, where insights from the partially dynamic setting often lead to improved fully dynamic algorithms. Moreover, partially dynamic algorithms can be useful in cases where one type of changes occurs much more frequently than the other type.

Problem. We are interested in maintaining an approximate BFS tree. For any $\alpha \geq 1$, an α-*approximate BFS tree* of graph G with respect to a given root s is a spanning tree T such that for every node v, $d_T(s, v) \leq \alpha d_G(s, v)$ (note that, clearly, $d_T(s, v) \geq d_G(s, v)$). If $\alpha = 1$, then T is an (exact) BFS tree. The goal of our problem is to maintain an approximate BFS tree T_t at the end of each recovery stage t in the sense that every node v knows its approximate distance to the preconfigured root s in N_t and, for each neighbor u of v, v knows if u is its parent or child in T_t.

Results. Clearly, maintaining a BFS tree by recomputing it from scratch in every recovery stage requires $\Theta(D)$ time. Our main results are partially dynamic algorithms that break this time bound over a long run. They can maintain, for any constant $0 < \epsilon \leq 1$, a $(1 + \epsilon)$-approximate BFS tree in time that is *sublinear in D* when amortized over $\omega(\frac{n \log D}{D})$ edge changes. To be precise, the amortized update time over q edge changes is $O((1 + \frac{n^{1/3} D^{2/3}}{\epsilon^{2/3} q^{1/3}}) \log D + n/q)$ in the incremental setting and $O((\frac{n^{1/7} D^{6/7}}{q^{1/7}}) \log D + \frac{n}{\epsilon^{7/2} q})$ in the decremental one. For the particular case of $q = \Omega(n)$, we get amortized update times of $O(D^{2/3} \log D)$ and $O(D^{6/7} \log D)$ for the incremental and decremental cases, respectively. Our algorithms do not require any prior knowledge about the dynamic network, e.g., D and q. We note that, while there is no previous literature on this problem, one can parallelize the algorithm of Even and Shiloach [9] (also see [17,27]) to obtain an amortized update time of $O(nD/q + 1)$ over q changes in both the incremental and the decremental setting. This bound is sublinear in D when $q = \omega(n)$. Our algorithms give a sublinear time guarantee for a smaller number of changes, especially in applications where D is large. Consider, for example, an application where we want to maintain a BFS tree of the network under link failures until the network diameter is larger than, say $n/10$ (at this point, the network will alert an administrator). In this case, our algorithms guarantee a sublinear amortized update time after a polylogarithmic number of failures while previous algorithms cannot do so.

In the sequential (usual RAM) model, our technique also gives an incremental $(1 + \epsilon)$-approximation algorithm for the single-source shortest paths (SSSP) problem with an amortized update time of $O(mn^{2/5}/q^{3/5})$ per insertion and $O(1)$ query time, where m is the number of edges in the final graph, and q is the number of edge insertions. Prior to this result, only the classic exact algorithm of Even and Shiloach [9] from the 80s, with $O(mn/q)$ amortized update time, was known. No further progress has been made in the last three decades, and Roditty and Zwick [27] provided an explanation for this by showing that the algorithm of [9] is likely to be the fastest combinatorial exact algorithm, assuming that there is no faster combinatorial algorithm for Boolean matrix multiplication. Very recently, Bernstein and Roditty [5] showed that, in the decremental setting, this bound can be broken if a small approximation is allowed. Our result is the first one of the same spirit in the *incremental* setting; i.e., we brake the bound of Even and Shiloach for the case $q = o(n^{3/2})$.

Related Work. The problem of computing on dynamic networks is a classic problem in the area of distributed computing, studied from as early as the 70s; see, e.g. [4] and references therein. The main motivation is that dynamic networks better capture real networks, which experience failures and additions of new links. There is a large number of models of dynamic networks in the literature, each emphasizing different aspects of the problem. Our model closely follows the model of the sequential setting and, as discussed earlier, highlights the amortized update time aspect. It is closely related to the model in [20] where the main goal is to optimize the amortized update time using static algorithms in the recovery stages. The model in [20] is still slightly different from us in terms of allowed changes. For example, the model in [20] considers weighted networks and allows small weight changes but no topological changes; moreover, the message size can be unbounded (i.e., the static algorithm in the recovery stage operates under the so-called LOCAL model). Another related model the *controlled dynamic model* (e.g. [19,2]) where the topological changes do not happen instantaneously but are delayed until getting a permit to do so from the resource controller. Our algorithms can be used in this model as we can delay the changes until each recovery stage is finished. Our model is similar to, and can be thought of as a combination of, two types of models: those in, e.g., [18,13,21,24] whose main interest is to determine how fast a network can recover from changes using static algorithms in the recovery stages, and those in, e.g., [4,1,8], which focus on the amortized cost per edge change. Variations of partially dynamic distributed networks have also been considered (e.g. [14,26,7,6]).

The problem of constructing a BFS tree has been studied intensively in various distributed settings for decades (see [25,23] and references therein). The studies were also extended to more sophisticated structures such as minimum spanning trees (e.g. [11]) and Steiner trees [16]. These studies usually focus on *static* networks, i.e., they assume that the network never changes and want to construct a BFS tree once, from scratch. While we are not aware of any results on maintaining a BFS tree on dynamic networks, there are a few related results. Much previous attention (e.g. [4]) has been paid on the problem of *maintaining a spanning tree*. In a seminal paper by Awerbuch et al. [4], it was shown that the amortized message complexity of maintaining a spanning tree can be significantly smaller than the cost of the previous approach of recomputing from scratch [1]. Our result is in the same spirit as [4] in breaking the cost of recomputing from scratch. An attempt to maintain spanning trees of small diameter has also motivated a problem called *best swap*. The goal is to replace a failed edge in the spanning tree by a new edge in such a way that the diameter is minimized. This problem has recently gained considerable attention in both sequential (e.g. [15,3]) and distributed (e.g. [12,10]) settings.

In the sequential dynamic graph algorithms literature, a problem similar to ours is the single-source shortest paths (SSSP) problem on undirected graphs. This problem has been studied in partially dynamic settings and has applications to other problems, such as all-pairs shortest paths and reachability. As we have mentioned earlier, the classic bound of [9], which might be optimal [27], has recently been improved by a decremental approximation algorithm [5], and we achieve a similar result in the incremental setting.

2   Main Technical Idea

All our algorithms are based on a simple idea of *lazy updating*. Implementing this idea on different models requires modifications to cope with difficulties and to maximize efficiency. In this section, we explain the main idea by sketching a simple algorithm and its analysis for the incremental setting in the sequential and the distributed model. We start with an algorithm that has *additive error*: Let κ and δ be parameters. For every recovery stage t, we maintain a tree T_t such that $d_{T_t}(s, v) \leq d_{N_t}(s, v) \leq d_{T_t}(s, v) + \kappa\delta$ for every node v. We will do this by recomputing a BFS tree from scratch for $O(q/\kappa + nD/\delta^2)$ times.

During the preprocessing, our algorithm constructs a BFS tree of N_0, denoted by T_0. This means that every node u knows its parent and children in T_0 and the value of $d_{T_0}(s, u)$. Suppose that, in the first attack stage, an edge is inserted, say (u, v) where $d_{N_0}(s, u) \leq d_{N_0}(s, v)$. As a result, the distances from v to s might decrease, i.e. $d_{N_1}(s, v) < d_{N_0}(s, v)$. In this case, the distances from s to some other nodes (e.g. the children of v in T_0) could decrease as well, and we may wish to recompute the BFS tree. Our approach is to do this *lazily*: We recompute the BFS tree only when the distance from v to s decreases by at least δ; otherwise, we simply do nothing! In the latter case, we say that v *is lazy*. Additionally, we regularly "clean up" by recomputing the BFS tree after every κ insertions.

To prove an additive error of $\kappa\delta$, observe that errors occur for this single insertion only when v is lazy. Intuitively, this causes an additive error of δ since we could have decreased the distance of v and other nodes by at most δ, but we did not. This argument can be extended to show that if we have i lazy nodes, then the additive error will be at most $i\delta$. Since we do the cleanup every κ insertions, the additive error will be at most $\kappa\delta$ as claimed.

For the number of BFS tree recomputations, first observe that the cleanup clearly contributes $O(q/\kappa)$ recomputations in total, over q insertions. Moreover, a recomputation also could be caused by some node v, whose distance to s decreases by at least δ. Since every time a node v causes a recomputation, its distance decreases by at least δ, and $d_{N_0}(s, v) \leq D$, v will cause the recomputation at most D/δ times. This naive argument shows that there are nD/δ recomputations (caused by n different nodes) in total. This analysis is, however, *not* enough for our purpose. A tighter analysis, which is crucial to all our algorithms relies on the observation that when v causes a recomputation, the distance from v's neighbor, say v', to s also decreases by at least $\delta-1$. Similarly, the distance of v''s neighbor to s decreases by at least $\delta - 2$, and so on. This leads to the conclusion that one recomputation corresponds to $(\delta+(\delta-1)+(\delta-2)+\ldots) = \Omega(\delta^2)$ distance decreases. Thus, the number of recomputations is at most nD/δ^2. Combining the two bounds, we get that the number of BFS tree computations is $O(q/\kappa+nD/\delta^2)$ as claimed. We get the total time in sequential and distributed models by multiplying this number by m, the final number of edges, and D (time for BFS tree computation), respectively.

To convert the additive error into a multiplicative error of $(1 + \epsilon)$, we execute the above algorithm only for nodes whose distances to s are greater than $\kappa\delta/\epsilon$. For other nodes, we can use the algorithm of Even and Shiloach [9] to maintain a BFS tree of depth $\kappa\delta/\epsilon$. This requires an additional time of $O(m\kappa\delta/\epsilon)$ in the sequential model and $O(n\kappa\delta/\epsilon)$ in the distributed model.

By setting κ and δ appropriately, the above algorithm immediately gives us the claimed time bound for the sequential model. For incremental distributed networks, we need one more idea called *layering*, where we use different values of δ and κ depending on the distance of each node to s. In the decremental setting, the situation is much more difficult, mainly because it is nearly impossible for a node v to determine how much its distance to s has increased after a deletion. Moreover, unlike the incremental case, nodes cannot simply "do nothing" when an edge is deleted. We have to cope with this using several other ideas, e.g., constructing an imaginary tree (where edges sometimes represent paths).

3   Incremental Algorithm

We now present a framework for an incremental algorithm that allows up to q edge insertions and provides an additive approximation of the distances to a distinguished node s. Subsequently we will explain how to use this algorithm to get $(1 + \epsilon)$-approximations in the RAM model and the distributed model, respectively. For simplicity we assume that the initial graph is connected. The algorithm can be modified to work without this assumption within the same running time. We defer the details to a full version of the paper.

The algorithm (see Algorithm 1) works in *phases*. At the beginning of every phase we compute a BFS tree T_0 of the current graph, say G_0. Every time an edge (u, v) is inserted, the distances of some nodes to s in G might decrease. Our algorithm tries to be *as lazy as possible*. That is, when the decrease does not exceed some parameter δ, our algorithm keeps its tree T_0 and accepts an *additive error* of δ for every node. When the decrease exceeds δ, our algorithm starts a new phase and recomputes the BFS tree. It also start a new phase after every κ edge insertions to keep the additive error limited to $\kappa\delta$. The algorithm will answer a query for the distance from a node x to s by returning $d_{G_0}(x, s)$, the distance from x to s at the beginning of the current phase. It can also return the path from x to s in T_0 of length $d_{G_0}(x, s)$. Besides δ and κ, the algorithm has a third parameter X which indicates up to which distance from s the BFS tree will be computed. In the following we denote by G_0 the state of the graph at the beginning of the current phase and by G we denote the current state of the graph after all insertions. It is easy to see that the algorithm gives the desired additive

---

Algorithm 1. Incremental algorithm

1: **procedure** INSERT(u, v)
2:     $k \leftarrow k + 1$
3:     **if** $k = \kappa$ **then** INITIALIZE( )
4:     **if** $d_{G_0}(u, s) > d_{G_0}(v, s) + \delta$ **then** INITIALIZE( )
5: **procedure** INITIALIZE( )   *(Start new phase)*
6:     $k = 0$
7:     Compute BFS tree T of depth X rooted at s and current distances $d_{G_0}(\cdot, s)$

---

approximation by considering the shortest path of a node x to the root s in the current graph G. By the main rule in Line 4 of the algorithm, the inequality $d_{G_0}(u, s) \le d_{G_0}(v, s) + \delta$ holds for every edge (u, v) that was inserted since the beginning of the current phase (otherwise a new phase would have been started). Since at most κ edges have been inserted, the additive error is at most $\kappa\delta$.

Lemma 1. *For every node x such that $d_{G_0}(x, s) \le X$, Algorithm 1 maintains the invariant $d_G(x, s) \le d_{G_0}(x, s) \le d_G(x, s) + \kappa\delta$.*

If we insert an edge (u, v) such that the inequality $d_{G_0}(u, s) \le d_{G_0}(v, s) + \delta$ does not hold, we cannot guarantee the additive error anymore. Nevertheless the algorithm makes a lot of progress in some sense: After the edge insertion, u is linked to v whose initial distance to s was significantly smaller than the one from u to s. This implies that the distance from u to s has decreased by at least δ since the beginning of the current phase.

Lemma 2. *If an edge (u, v) is inserted such that $d_{G_0}(u, s) > d_{G_0}(v, s) + \delta$, then $d_{G_0}(u, s) \ge d_G(u, s) + \delta$.*

Since we consider undirected, unweighted graphs, a large decrease in distance for one node also implies a large decrease in distance for many other nodes.

Lemma 3. *Let G and G' be unweighted, undirected graphs such that G is connected, $V(G) = V(G')$, and $E(G) \subseteq E(G')$. If there is a node y such that $d_G(y, s) \ge d_{G'}(y, s) + \delta$, then $\sum_{x \in V(G)} d_G(x, s) \ge \sum_{x \in V(G')} d_{G'}(x, s) + \Omega(\delta^2)$.*

This is the key observation for the efficiency of our algorithm as it limits the number of times a new phase starts, which is the expensive part of our algorithm.

Lemma 4. *If $\kappa \leq q$ and $\delta \leq X$, then the total running time of Algorithm 1 is $O(T_{BFS}(X) \cdot q/\kappa + T_{BFS}(X) \cdot nX/\delta^2 + q)$ where $T_{BFS}(X)$ is the time needed for computing a BFS tree up to depth X.*

3.1   RAM Model

In the RAM model we use a standard approach for turning the additive $\kappa\delta$-approximation of Algorithm 1 into a multiplicative $(1 + \epsilon)$-approximation. We use the Even-Shiloach algorithm for maintaining the *exact* distances to s up to depth $\kappa\delta/\epsilon$. Algorithm 1 now only has to be approximately correct for every node x such that $d_G(x, s) \geq \kappa\delta/\epsilon$, and indeed $d_G(x, s) + \kappa\delta \leq d_G(x, s) + \epsilon d_G(x, s) = (1 + \epsilon)d_G(x, s)$. The Even-Shiloach tree adds $O(m\kappa\delta/\epsilon)$ to the running time of Lemma 4 (where computing a BFS tree takes time $O(m)$). We choose the parameters κ and δ in a way that balances the dominant terms in the running time.

Theorem 5. *In the RAM model, there is an incremental $(1 + \epsilon)$-approximate SSSP algorithm for inserting up to q edges that has a total update time of $O(mn^{2/5}q^{2/5}/\epsilon)$ where m is the number of edges in the final graph. It answers distance and path queries in optimal worst-case time.*

3.2   Distributed Model

In the distributed model, we use a different approach for obtaining the $(1 + \epsilon)$-approximation. We run $\log D$ parallel instances of Algorithm 1, where instance i provides a $(1 + \epsilon)$-approximation for nodes in the distance range from 2^i to 2^{i+1}. For every such range, we can determine the targeted additive approximation A that guarantees a $(1 + \epsilon)$-approximation in that range. We set the parameters κ and δ in a way that gives $\kappa\delta = A$ and minimizes the running time of Algorithm 1. Here, this approach is more efficient than the two-layered approach that uses a single Even-Shiloach tree for small distances. The fact that the time for computing a BFS tree of depth X is proportional to X in the distributed model nicely fits into the running time of the multi-layer approach.

Theorem 6. *In the distributed model, there is an incremental algorithm for maintaining a $(1 + \epsilon)$-approximate BFS tree under up to q edge insertions that has a total running time of $O((q^{2/3}n^{1/3}D^{2/3}/\epsilon^{2/3} + q) \log D + n)$, where D is the initial diameter.*

4   Decremental Algorithm

In the decremental setting we use an algorithm of the same flavor as in the incremental setting (see Algorithm 2). However, the update procedure is more complicated because it is not obvious which edge should be used to repair the tree after a deletion. Our solution exploits the fact that in the distributed model it is relatively cheap to examine the local neighborhood of every node. As in

---

Algorithm 2. Decremental algorithm

 1: **procedure** DELETE(u, v)
 2:     $k \leftarrow k + 1$
 3:     **if** $k = \kappa$ **then** INITIALIZE( )
 4:     Delete edge (u, v) from F
 5:     Compute tree T from F and $d_{G_0}(\cdot, s)$: $T \leftarrow$ UPDATETREE($F, d_{G_0}(\cdot, s)$)
 6:     **if** UPDATETREE reports distance increase by at least δ **then** INITIALIZE( )

 7: **procedure** INITIALIZE( )   *(Start new phase)*
 8:     $k = 0$
 9:     Compute BFS tree T of depth X rooted at s. Set $F \leftarrow T$
10:     Compute current distances $d_{G_0}(\cdot, s)$

11: **procedure** UPDATETREE($F, d_{G_0}(\cdot, s)$)
12:     *At any time:* $U = \{u \in V \mid u$ *has no outgoing edge in F and* $u \neq s\}$
13:     **while** $U \neq \emptyset$ **do**
14:         **for all** $u \in U$ **do** u marks every node x such that $d_F^w(x, u) \leq 3\kappa\delta$
15:         **for all** $u \in U$ **do**   *(Search process)*
16:             u tries to find a node v by breadth-first search such that
17:             (1) $d_G(u, v) + d_{G_0}(v, s) \leq d_{G_0}(u, s) + \delta$
18:             (2) v is not marked
19:             (3) $d_G(u, v) \leq (\kappa + 1)\delta + 1$
20:             **if** such a node v could be found **then**
21:                 Add edge (u, v) of weight $d_F^w(u, v) = d_G(u, v)$ to F
22:         **if** no such node v could be found for any $u \in U$ **then**
23:             **return** "distance increase by at least δ"
24:     **return** F

---

the incremental setting, the algorithm has the parameters κ, δ, and X. The tree update procedure of Algorithm 2 either computes a (weighted) tree T that approximates the real distances with additive error $\kappa\delta$, or it reports a distance increase by at least δ since the beginning of the current phase. Let T_0 denote the BFS tree computed at the beginning of the current phase and let F be the forest resulting from removing those edges from T_0 that have already been deleted in the current phase. After every edge deletion, the tree update procedure tries to rebuild a tree T by starting from F. Every node u that had a parent in T_0 but has no parent in F tries to find a "good" node v to reconnect to. This process is repeated until F is a full tree again. Algorithm 2 imposes three conditions (Lines 17-19) on a "good" node v. Condition (1) guarantees that the error introduced by each reconnection is at most δ, (2) avoids that the reconnections introduce any cycles, and (3) states that the node v should be found relatively close to u. This is the key to efficiently find such a node.

4.1   Analysis of Tree Update Procedure

For the analysis of the tree update procedure of Algorithm 2, we assume that the edges in F are directed. When we compute a BFS tree, we consider all edges as directed towards the root. The *weighted, directed* distance from x to y in F

is denoted by $d_F^w(x, y)$. We assume for the analysis that F initially contains *all* nodes. By T we denote the graph returned by the algorithm. By Condition (1), every reconnection made by the tree update procedure adds an additive error of δ. In total there are at most κ reconnections (one per previous edge deletion) and therefore the total additive error introduced is $\kappa\delta$.

Lemma 7. *After every iteration, we have, for all nodes x and y such that $d_F^w(x, y) < \infty$, $d_F^w(x, y) + d_{G_0}(y, s) \leq d_{G_0}(x, s) + \kappa\delta$.*

By Condition (2) we avoid that the reconnection process introduces any cycle. Ideally, a node $u \in U$ should reconnect to a node v that is in the subtree of the root s. We could achieve this if every node in U marked its whole subtree in F. However, this would be too inefficient. Instead, marking the subtree up to a limited depth $(3\kappa\delta)$ is sufficient.

Lemma 8. *After every iteration, the graph F is a forest.*

We can show that the algorithm makes progress in every iteration. There is always at least one node for which a "good" reconnection is possible that fulfills the conditions of the algorithm. Even more, if such a node does not exist, then there is a node whose distance to s has increased by at least δ since the beginning of the current phase.

Lemma 9. *In every iteration, if $d_G(x, s) \leq d_{G_0}(x, s) + \delta$ for every node $x \in U$, then for every node $u \in U$ with minimal $d_{G_0}(u, s)$, there is a node $v \in V$ such that (1) $d_G(u, v) + d_{G_0}(v, s) \leq d_{G_0}(u, s) + \delta$, (2) v is not marked, and (3) $d_G(u, v) \leq (\kappa + 1)\delta + 1$.*

The marking and the search process both take time $O(\kappa\delta)$. Since there is at least one reconnection in every iteration (unless the algorithm reports a distance increase), there are at most κ iterations that take time $O(\kappa\delta)$ each.

Lemma 10. *The tree update procedure of Algorithm 2 either reports "distance increase" and guarantees that there is a node x such that $d_G(x, s) > d_{G_0}(x, s) + \delta$, or it computes a tree T such that for every node x we have $d_{G_0}(x, s) \leq d_G(x, s) \leq d_T^w(x, s) \leq d_{G_0}(x, s) + \kappa\delta$. It runs in time $O(\kappa^2\delta)$.*

In the following we clarify some implementation issues of the tree update procedure in the distributed model.

Weighted Edges. The tree computed by the algorithm contains weighted edges. Such an edge e corresponds to a path P of the same distance in the network. We implement weighted edges by a routing table for every node v that stores the next node on P if a message is sent over v as part of the weighted edge e.

Avoiding Congestion. The marking can be done in parallel without congestion because the trees in the forest F do not overlap. We avoid congestion during the search as follows. If a node receives more than one message from its neighbors, we always give priority to search requests originating from the node u with the lowest distance $d_{G_0}(u, s)$ to s in G_0. By Lemma 9, we know that the search of at least one of the nodes u with minimal $d_{G_0}(u)$ will always be successful.

Reporting Deletions. The nodes that do not have a parent in F before the update procedure starts do not necessarily know that a new edge deletion has happened. Such a node only has to become active and do the marking and the search if there is a change in its neighborhood of distance $3\kappa\delta$, otherwise it can still use the weighted edge in the tree T that it previously used because the three conditions imposed by the algorithm will still be fulfilled. After the deletion of an edge (x, y), the nodes x and y can inform all nodes at distance $3\kappa\delta$ in time $O(\kappa\delta)$, which is within our projected running time.

4.2 Analysis of Decremental Distributed Algorithm

The tree update procedure provides an additive approximation of the shortest paths. It also provides a means for detecting that the distance of some node to s has increased by at least δ since the beginning of the current phase. Using the distance increase argument of Lemma 3, we can provide a running time analysis for the decremental algorithm that is very similar to the incremental algorithm.

Lemma 11. *If $q \leq \kappa$ and $\delta \leq X$, then the total running time of Algorithm 2 is $O(qX/\kappa + nX^2/\delta^2 + q\kappa^2\delta)$. It maintains the following invariant: If $d_{G_0}(x, s) \leq X$, then $d_{G_0}(x, s) \leq d_G(x, s) \leq d_T^w(x, s) \leq d_{G_0}(x, s) + \kappa\delta$.*

We use a similar approach as in the incremental setting to get the $(1 + \epsilon)$-approximation. We run i parallel instances of the algorithm where each instance covers the distance range from 2^i to 2^{i+1}. By a careful choice of the parameters κ and δ for each instance we can guarantee a $(1 + \epsilon)$-approximation.

Theorem 12. *In the decremental distributed dynamic network, we can maintain a $(1 + \epsilon)$-approximate BFS tree over q edge deletions in a total running time of $O((q^{6/7} n^{1/7} D^{6/7}) \log D + (n + q)/\epsilon^{7/2})$, where D is the maximum diameter.*

5 Conclusion and Open Problems

In this paper, we showed that maintaining a breadth-first search spanning tree can be done in an amortized update time that is sublinear in D in partially dynamic networks. Many problems remain open. For example, can we get a similar result for the case of *fully dynamic* networks? How about *weighted* networks (even partially dynamic ones)? Can we also get a sublinear time bound for the *all-pairs shortest paths* problem? Moreover, in addition to the sublinear time complexity achieved in this paper, it is also interesting to obtain algorithms with small bounds on message complexity and memory.

Acknowledgements. We thank the reviewers of ICALP 2013 for pointing to related papers and to an error in an example given in the previous version.

References

1. Afek, Y., Awerbuch, B., Gafni, E.: Applying static network protocols to dynamic networks. In: FOCS, pp. 358–370 (1987)
2. Afek, Y., Awerbuch, B., Plotkin, S.A., Saks, M.E.: Local management of a global resource in a communication network. J. ACM 43(1), 1–19 (1996)
3. Alstrup, S., Holm, J., de Lichtenberg, K., Thorup, M.: Maintaining information in fully dynamic trees with top trees. ACM Transactions on Algorithms 1(2), 243–264 (2005), announced at ICALP 1997 and SWAT 2000
4. Awerbuch, B., Cidon, I., Kutten, S.: Optimal maintenance of a spanning tree. J. ACM 55(4) (2008); announced at FOCS 1990
5. Bernstein, A., Roditty, L.: Improved dynamic algorithms for maintaining approximate shortest paths under deletions. In: SODA, pp. 1355–1365 (2011)
6. Cicerone, S., D'Angelo, G., Stefano, G.D., Frigioni, D.: Partially dynamic efficient algorithms for distributed shortest paths. Theor. Comput. Sci. 411(7-9), 1013–1037 (2010)
7. Cicerone, S., D'Angelo, G., Stefano, G.D., Frigioni, D., Petricola, A.: Partially dynamic algorithms for distributed shortest paths and their experimental evaluation. JCP 2(9), 16–26 (2007)
8. Elkin, M.: A near-optimal distributed fully dynamic algorithm for maintaining sparse spanners. In: PODC, pp. 185–194 (2007)
9. Even, S., Shiloach, Y.: An on-line edge-deletion problem. J. ACM 28(1), 1–4 (1981)
10. Flocchini, P., Enriques, A.M., Pagli, L., Prencipe, G., Santoro, N.: Point-of-failure shortest-path rerouting: Computing the optimal swap edges distributively. IEICE Transactions 89-D(2), 700–708 (2006)
11. Garay, J., Kutten, S., Peleg, D.: A sublinear time distributed algorithm for minimum-weight spanning trees. SIAM J. on Computing 27, 302–316 (1998); announced at FOCS 1993
12. Gfeller, B., Santoro, N., Widmayer, P.: A distributed algorithm for finding all best swap edges of a minimum-diameter spanning tree. IEEE Trans. Dependable Sec. Comput. 8(1), 1–12 (2011), announced at DISC 2007
13. Hayes, T.P., Saia, J., Trehan, A.: The forgiving graph: a distributed data structure for low stretch under adversarial attack. Distributed Computing 25(4), 261–278 (2012); announced at PODC 2009
14. Italiano, G.F.: Distributed algorithms for updating shortest paths. In: WDAG(DISC), pp. 200–211 (1991)
15. Italiano, G.F., Ramaswami, R.: Maintaining spanning trees of small diameter. Algorithmica 22(3), 275–304 (1998)
16. Khan, M., Kuhn, F., Malkhi, D., Pandurangan, G., Talwar, K.: Efficient distributed approximation algorithms via probabilistic tree embeddings. Distributed Computing 25(3), 189–205 (2012); announced at PODC 2008
17. King, V.: Fully dynamic algorithms for maintaining all-pairs shortest paths and transitive closure in digraphs. In: FOCS, pp. 81–91 (1999)
18. Korman, A.: Improved compact routing schemes for dynamic trees. In: PODC, pp. 185–194 (2008)
19. Korman, A., Kutten, S.: Controller and estimator for dynamic networks. Inf. Comput. 223, 43–66 (2013)

20. Korman, A., Peleg, D.: Dynamic routing schemes for graphs with low local density. ACM Transactions on Algorithms 4(4) (2008)
21. Krizanc, D., Luccio, F.L., Raman, R.: Compact routing schemes for dynamic ring networks. Theory Comput. Syst. 37(5), 585–607 (2004)
22. Kuhn, F., Lynch, N.A., Oshman, R.: Distributed computation in dynamic networks. In: STOC, pp. 513–522 (2010)
23. Lynch, N.A.: Distributed Algorithms. Morgan Kaufmann Publishers, San Francisco (1996)
24. Malpani, N., Welch, J.L., Vaidya, N.H.: Leader election algorithms for mobile ad hoc networks. In: DIAL-M, pp. 96–103 (2000)
25. Peleg, D.: Distributed computing: a locality-sensitive approach. SIAM, Philadelphia (2000)
26. Ramarao, K.V.S., Venkatesan, S.: On finding and updating shortest paths distributively. J. Algorithms 13(2), 235–257 (1992)
27. Roditty, L., Zwick, U.: On dynamic shortest paths problems. Algorithmica 61(2), 389–401 (2011); announced at ESA 2004

Locally Stable Marriage with Strict Preferences*

Martin Hoefer[1] and Lisa Wagner[2]

[1] Max-Planck-Institut für Informatik and Saarland University, Germany
mhoefer@mpi-inf.mpg.de
[2] Dept. of Computer Science, RWTH Aachen University, Germany
lwagner@cs.rwth-aachen.de

Abstract. We study two-sided matching markets with locality of information and control. Each male (female) agent has an arbitrary strict preference list over all female (male) agents. In addition, each agent is a node in a fixed network. Agents learn about possible partners dynamically based on their current network neighborhood. We consider convergence of dynamics to locally stable matchings that are stable with respect to their imposed information structure in the network. While existence of such states is guaranteed, we show that reachability becomes NP-hard to decide. This holds even when the network exists only among one side. In contrast, if only one side has no network and agents remember a previous match every round, reachability is guaranteed and random dynamics converge with probability 1. We characterize this positive result in various ways. For instance, it holds for random memory and for memory with the most recent partner, but not for memory with the best partner. Also, it is crucial which partition of the agents has memory. Finally, we conclude with results on approximating maximum locally stable matchings.

1 Introduction

Matching problems form the basis of many assignment and allocation tasks encountered in computer science, operations research, and economics. A prominent and popular approach in all these areas is *stable matching*, as it captures aspects like distributed control and rationality of participants that arise in many assignment problems today. A variety of allocation problems in markets can be analyzed within the context of two-sided stable matching, e.g., the assignment of jobs to workers [2,5], organs to patients [18], or general buyers to sellers. In addition, stable marriage problems have been successfully used to study distributed resource allocation problems in networks [9].

In this paper, we consider a game-theoretic model for matching with distributed control and information. Agents are rational agents embedded in a (social) network and strive to find a partner for a joint relationship or activity,

* Supported by DFG Cluster of Excellence MMCI and grant Ho 3831/3-1.
An extended full version of this paper can be found at
http://arxiv.org/abs/1207.1265

F.V. Fomin et al. (Eds.): ICALP 2013, Part II, LNCS 7966, pp. 620–631, 2013.
© Springer-Verlag Berlin Heidelberg 2013

e.g., to do sports, write a research paper, exchange data etc. Such problems are of central interest in economics and sociology, and they act as fundamental coordination tasks in distributed computer networks. Our model extends the stable marriage problem, in which we have sets U and W of men and women. Each man (woman) can match to at most one woman (man) and has a complete strict preference list over all women (men). Given a matching M, a *blocking pair* is a man-woman pair such that both strictly improve by matching to each other. A matching without blocking pair is a *stable matching*.

A central assumption in stable marriage is that every agent knows all agents it can match to. In reality, however, agents often have limited information about their matching possibilities. For instance, in a large society we would not expect a man to match up with any other woman immediately. Instead, there exist restrictions in terms of knowledge and information that allow some pairs to match up directly, while others would have to get to know each other before being able to start a relationship. We incorporate this aspect by assuming that agents are embedded in a fixed network of *links*. Links represent an enduring knowledge relation that is not primarily under the control of the agents. Depending on the interpretation, links could represent, e.g., family, neighbor, colleague or teammate relations. Each agent strives to build one *matching edge* to a partner. The set of links and edges defines a dynamic information structure based on *triadic closure*, a standard idea in social network theory: If two agents have a common friend, they are likely to meet and learn about each other. Translated into our model this implies that each agent can match only to partners in its 2-hop neighborhood of the network of matching edges and links. Then, a *local blocking pair* is a blocking pair of agents that are at hop distance at most 2 in the network. Consequently, a *locally stable matching* is a matching without local blocking pairs. Local blocking pairs are a subset of blocking pairs. In turn, every stable matching is a locally stable matching, because it allows no (local or global) blocking pairs. Thus, one might be tempted to think that locally stable matchings are easier to find and/or reach using distributed dynamics than ordinary stable matchings. In contrast, we show in this paper that locally stable matchings have a rich structure and can behave quite differently than ordinary stable matchings. Our study of locally stable matching with arbitrary strict preferences significantly extends recent work on the special case of correlated or weighted matching [11], in which preferences are correlated via matching edge benefits.

Contribution We concentrate on the important case of two-sided markets, in which a (locally) stable matching is always guaranteed to exist. Our primary interest is to characterize convergence properties of iterative round-based dynamics with distributed control, in which in each round a local blocking pair is resolved. We focus on the REACHABILITY problem: Given an instance and an initial matching, is there a sequence of local blocking pair resolutions leading to a locally stable matching? In Section 3 we see that there are cases, in which a locally stable matching might never be reached. This is in strong contrast to the case of weighted matching, in which it is easy to show convergence of every sequence of local blocking pair resolutions with a potential function. In fact, it is NP-hard to decide REACHABILITY, even if the network exists only among one partition of agents. Moreover, there exist games and initial matchings such that *every* sequence of local blocking pairs terminating in a locally stable matching is exponentially long. Hence, REACHABILITY might even be outside NP. If we need to decide REACHABILITY for a given initial matching *and a specific locally stable matching to be reached*, the problem is even NP-hard for correlated matching.

Our NP-hardness results hold even if the network exists only among one partition. In Section 4, we concentrate on a more general class of games in which links exist in one partition and between partitions (i.e., one partition has no internal links). This is a natural assumption when considering objects that do not generate knowledge about each other, e.g., when matching resources to networked nodes or users, where initially resources are only known to a subset of users. Here we characterize the impact of memory on distributed dynamics. For *recency memory*, each agent remembers in every round the *most recent partner* that is different from the current one. With recency memory, REACHABILITY is always true, and for every initial matching there exists a sequence of polynomially many local or remembered blocking pairs leading to a locally stable matching. In contrast, for *quality memory* where all agents remember their *best partner* REACHABILITY stays NP-hard. This formally supports the intuition that recency memory is more powerful than quality memory, as the latter yields agents that are "hung up" on preferred but unavailable partners. This provides a novel distinction between recency and quality memory.

Our positive results for recency memory in Section 4 imply that if we pick local blocking pairs uniformly at random in each step, we achieve convergence with probability 1. The proof relies only on the memory of one partition. In contrast, if only the other partition has memory, we obtain NP-hardness of REACHABILITY. Convergence with probability 1 can also be guaranteed for *random memory* if in each round each agent remembers one of his previous matches chosen uniformly at random. The latter result holds even when links exist among or between both partitions. However, using known results on stable marriage with full information [1], convergence time can be exponential with high probability, independently of any memory.

In contrast to ordinary stable matchings, two locally stable matchings can have very different size. This motivates our search for maximum locally stable matchings in Section 5. While a simple 2-approximation algorithm exists, we can show a non-approximability result of $1.5 - \varepsilon$ under the unique games conjecture. For spatial reasons most of the proofs are omitted but can be found in the full version.

Related Work Locally stable matchings were introduced by Arcaute and Vassilvitskii [2] in a two-sided job-market model, in which links exist only among one partition. The paper uses strong uniformity assumptions on the preferences and addresses the lattice structure for stable matchings and a local Gale-Shapley algorithm. More recently, we studied locally stable matching with correlated preferences in the roommates problem, where arbitrary pairs of agents can be matched [11]. Using a potential function argument, REACHABILITY is always true and convergence guaranteed. Moreover, for every initial matching there is a polynomial sequence of local blocking pairs that leads to a locally stable matching. The expected convergence time of random dynamics, however, is exponential. If we restrict to resolution of pairs with maximum benefit, then for random memory the expected convergence time becomes polynomial, but for recency or quality memory convergence time remains exponential, even if the memory is of polynomial size.

For an introduction to stable marriage and some of its variants we refer the reader to several books in the area [10, 19]. There is a significant literature on dynamics, especially in economics, which is too broad to survey here. These works usually do not address issues like computational complexity or worst-case bounds. We focus on a subset of prominent analytical works related to our scenario. For the stable marriage problem, it is known that better-response dynamics, in which agents sequentially deviate to blocking pairs, can cycle [16]. On the other hand, REACHABILITY is always true, and for every initial matching there exists a sequence of polynomially many steps to a stable matching [20]. If blocking pairs are chosen uniformly at random at each step, convergence time is exponential [1] in the worst case.

In the roommates problem, in which every pair of agents can be matched, stable matchings can be absent. Deciding existence and computing stable matchings if they exist can be done in polynomial time [14]. In addition, if a stable matching exists, then REACHABILITY is always true [8]. A similar statement can be made even more generally for relaxed concepts like P-stable matchings that always exist [12]. Ergodic sets of the underlying Markov chain have been studied [13] and related to random dynamics [15]. In addition, for computing (variants of) stable matchings via iterative entry dynamics see [4–6].

The problem of computing a maximum locally stable matchings has recently been considered in [7]. In addition to characterizations for special cases, a NP-hardness result is shown and non-approximability of $(21/19 - \varepsilon)$ unless P= NP. Computing maximum stable matchings with ties and incomplete lists has generated a significant amount of research interest over the past decade. The currently best results are a 1.5-approximation algorithm [17] and $(4/3 - \varepsilon)$-hardness under the unique games conjecture [21].

2   Preliminaries

A *network matching game* (or *network game*) consists of a (social) *network* $N = (V, L)$, where V is a set of vertices representing *agents* and $L \subseteq \{\{u, v\} | u, v \in V, u \neq v\}$ is a set of fixed *links*. A set $E \subseteq \{\{u, v\} | u, v \in V, u \neq v\}$ defines the *potential matching edges*. A *state* is a matching $M \subseteq E$ such that for each $v \in V$ we have $|\{e \mid e \in M, v \in e\}| \leq 1$. An edge $e = \{u, v\} \in M$ provides utilities $b_u(e), b_v(e) > 0$ for u and v, respectively. If for every $e \in E$ we have $b_u(e) = b_v(e) = b(e) > 0$, we speak of *correlated preferences* or a *correlated network game*. Otherwise, we will assume that each agent has a total order $\succ$ over its possible matching partners, and for every agent the utility of matching edges is given according to this ranking. Throughout the paper we focus on the *two-sided* or *bipartite* case, which is often referred to as the *stable marriage problem*, where V is divided into two disjoint sets U and W such that $E \subseteq \{\{u, w\} \mid u \in U, w \in W\}$. Note that this does not imply that N is bipartite. If further the vertices of U are isolated in N, we speak of a *job-market game* for consistency with [2, 11].

To describe stability in network matching games, we assume agents u and v are *accessible* in state M if they have a distance of at most 2 in the graph $G = (V, L \cup M)$. A state M has a *local blocking pair* $e = \{u, v\} \in E$ if u and v are accessible and are each either unmatched in M or matched through an edge e' such that e' serves a strictly smaller utility than e. Thus, in a local blocking pair both agents can strictly increase their utility by generating e (and possibly dismissing some other edge thereby). A state M that has no local blocking pair is a *locally stable matching*.

Most of our analysis concerns iterative round-based dynamics, where in each round we pick one local blocking pair, add it to M, and remove all edges that conflict with this new edge. We call one such step a *local improvement step*. With *random dynamics* we refer to the process when in each step the local blocking pair is chosen uniformly at random from the ones available. A local blocking pair $\{u, v\}$ that is resolved and not a link as well must be connected by some distance-2 path (u, w, v) in M before the step. This path can consist of two links, or of exactly one link and one matching edge. In the latter case, let w.l.o.g. $\{u, w\}$ be the matching edge. As u can have only one matching edge, the local improvement step will delete $\{u, w\}$ to create $\{u, v\}$. For simplicity, we will refer to this fact as "an edge moving from $\{u, w\}$ to $\{u, v\}$" or "u's edge moving from w to v".

In subsequent sections we will assume that agents have memory that allows to "remember" one matching partner from a former round. In this case, a pair $\{u, v\}$ of agents becomes accessible not only by a distance-2 path in G, but also when u appears in the memory of v. Hence, in this case a local blocking pair can be based solely on access through memory. For *random memory*, we assume that in every round each agent remembers a previous matching partner chosen uniformly at random. For *recency memory*, each agent remembers the last matching partner that is different from the current partner. For *quality memory*, each agent remembers the best previous matching partner.

3   Complexity of REACHABILITY and Duration

Complexity In contrast to ordinary stable marriage, there exist small examples that show REACHABILITY is not always true for locally stable matchings (see [11, Thm. 5] or our circling gadget in the full version). Here we consider complexity of this problem and show NP-hardness results when agents have strict preferences. This is in contrast to correlated network games, where REACHABILITY is true and for every initial matching there is always a sequence to a locally stable matching of polynomial length [11]. Still, we here show that given a particular matching to reach, deciding REACHABILITY becomes also NP-hard, even for correlated job-market games. We present the proof of the latter result in detail, as it provides the basic idea for the omitted NP-hardness proofs as well.

Theorem 1. *It is* NP-*hard to decide* REACHABILITY *from the initial matching* $M = \emptyset$ *to a given locally stable matching in a correlated network game.*

Proof. We use a reduction from 3SAT . Given a 3SAT formula with k variables $x_1, \ldots, x_k$ and l clauses $C_1, \ldots, C_l$, where clause C_j contains the literals $l1_j, l2_j$ and $l3_j$, we have

$$U = \{u_{x_i} | i = 1 \ldots k\} \cup \{u_{C_j} | j = 1 \ldots l\} \cup \{b_h | h = 1 \ldots k + l - 1\},$$
$$W = \{v_{x_i}, x_i, \overline{x}_i | i = 1 \ldots k\} \cup \{v_{C_j} | j = 1 \ldots l\} \cup \{a, a_1\}.$$

For the social links see the picture shown below.

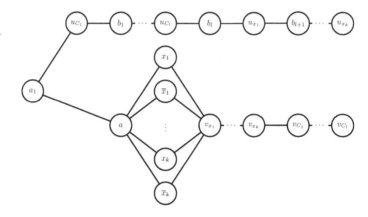

We do not restrict the set of matching edges, but assume that every edge not appearing in the list below has benefit $\epsilon \ll 1$ (resulting in them being irrelevant for the dynamics). The other benefits are given as follows.

$u \in U$	$w \in W$	$b(\{u, w\})$	
u_{C_j}	a	j	$j = 1, \ldots, l$
u_{x_i}	a	$i + l$	$i = 1, \ldots, k$
b_h	a	$h + \frac{1}{2}$	$h = 1, \ldots, k + l - 1$
u_{C_j}	$l1_j / l2_j / l3_j$	$k + l + 1$	$j = 1, \ldots, l$
u_{x_i}	$x_i / \overline{x}_i$	$k + l + 1$	$i = 1, \ldots, k$
u_{C_j}	v_{x_i}	$k + l + 1 + i$	$i = 1, \ldots, k, j = 1, \ldots, l$
u_{x_i}	$v_{x_{i'}}$	$k + l + 1 + i'$	$i = 1, \ldots, k, i' = 1, \ldots, i$
u_{C_j}	$v_{C_{j'}}$	$2k + l + 1 + j'$	$j = 1, \ldots, l, j' = 1, \ldots, j$

Our goal is to reach $M^* = \{\{u_s, v_s\} | s \in \{x_1, \ldots, x_k\} \cup \{C_1, \ldots, C_l\}\}$.

First, note that additional matching edges can only be introduced at $\{u_{C_1}, a\}$. Furthermore, once a vertex u_y, $y \in \{x_1, \ldots, x_k\} \cup \{C_1, \ldots, C_l\}$, is matched to a vertex other than a, it blocks the introduction of any edge for a vertex lying behind u_y on the path from u_{C_1} to u_{x_k}. Also, the vertices b_h prevent that an edge is moved on from one u-vertex to another after it has left a. Thus, at the time when an edge to a clause u-vertex is created that still exists in the final matching (but is connected to some v_{C_j} then), the edges for all variable u-vertices must have been created already.

Assume that the 3SAT formula is satisfiable. Then we first create a matching edge at $\{u_{C_j}, a\}$, move it over the u-and b-vertices to u_{x_k}, and then move it into the branching to the one of x_k or $\overline{x}_k$ that negates its value in the satisfying assignment. Similarly, one after the other (in descending order), we create a matching edge at a for each of the variable u-vertices and move it into the branching to the variable vertex that negates its value in the satisfying assignment. As every clause is fulfilled, at least one of the three vertices that yield an improvement for the clause u-vertex from a is not blocked by a matching edge to a variable u-vertex. Then, the edges to clause u-vertices can bypass the existing edges (again, one after the other in descending order) and reach their positions in M^*. After that, the variable-edges can leave the branching and move to their final position in the same order as before.

Now assume that we can reach M^* from $\emptyset$. We note that the edges to clause u-vertices have to overtake the edges to variable u-vertices somewhere on the way to reach their final position. The only place to do so is in the branching leading over the x_i and $\overline{x}_i$. Thus all variable-edges have to wait at some x_i or $\overline{x}_i$ until the clause-edges have passed. But from a, vertex u_{x_i} is only willing to switch to x_i or $\overline{x}_i$. Thus, every vertex blocks out a different variable (either in its true or in its false value). Similarly, a vertex u_{C_j} will only move further from a if it can reach one of its literals. Hence, if all clauses can bypass the variables, then for every clause there was one of its literals left open for passage. Thus, if we set each variable to the value that yields the passage for clause-edges in the branching, we obtain a satisfying assignment.   □

Corollary 1. *It is* NP-*hard to decide* REACHABILITY *to a given locally stable matching in a correlated job-market game.*

Theorem 2. *It is* NP-*hard to decide* REACHABILITY *from the initial matching* $M = \emptyset$ *to an arbitrary locally stable matching in a bipartite network game.*

Length of Sequences We now consider the number of improvement steps required to reach locally stable matchings. In general, there is a network game and an initial matching such that we need an exponential number of steps before reaching any locally stable matching. This is again in contrast to the correlated case, where every reachable stable matching can be reached in a polynomial number of steps. We present the latter result and defer the much more technical proof of the general lower bound to the full version.

Theorem 3. *For every network game with correlated preferences, every locally stable matching* $M^* \in E$ *and initial matching* $M_0 \in E$ *such that* M^* *can be reached from* M_0 *through local improvement steps, there exists a sequence of at most* $O(|E|^3)$ *local improvement steps leading from* M_0 *to* M^*.

Proof. Consider an arbitrary sequence between M_0 and M^*. We will explore which steps in the sequence are necessary and which parts can be omitted. We rank all edges by their benefit (allowing multiple edges to have the same rank) such that $r(e) > r(e')$ iff $b(e) > b(e')$ and set $r_{max} = max\{r(e)|e \in E\}$. Recall from Section 2 that we can account edges in the way that every edge e has at most one direct predecessor e' in the sequence, which was necessary to build e. Because e was a local blocking pair, we know $r(e') < r(e)$. Thus, every edge e has at most r_{max} predecessors. Our proof is based on two crucial observations:

(1) An edge can only be deleted by a stronger edge, that is, every chain of one edge deleting the next is limited in length by r_{max}.
(2) If an edge is created, then possibly moved, and finally deleted without deleting an edge on its way, this edge would not have to be introduced in the first place.

Suppose our initial matching is the empty matching, then every edge in the locally stable matching has to be created and by (repeated application of) (2) we only need to create and move edges that are needed for the final matching. Thus we have $|M^*|$ edges, which each made at most r_{max} steps.

Now if we start with an arbitrary matching, the sequence might be forced to delete some edges that cannot be used for the final matching. Each of these edges generates a chain of edges deleting each other throughout the sequence, but (1) tells us that this chain is limited as well as the number of steps each of these edges has to make. The only remaining issue is what happens to edges "accidentally" deleted during this procedure. Again, we can use (2) to argue that there is no reason to rebuild such an edge just to delete it again. Thus, such deletions can happen only once for every edge we had in M_0 (not necessarily on the position it had in M_0). It does not do any harm if it happens to an edge of one of the deletion-chains, as it would just end as desired. For the edges remaining in $|M^*|$ the same bounds holds as before. Thus, we have an overall bound of $|M_0| \cdot r_{max} \cdot r_{max} + |M^*| \cdot r_{max} \in O(|E|^3)$ steps, where the first term results from the deletion chains and the second one from the edges surviving in the final matching.   □

Theorem 4. *There is a network game with strict preferences such that a locally stable matching can be reached by a sequence of local improvement steps from the initial matching $M = \emptyset$, but every such sequence has length $2^{\Omega(|V|)}$.*

4   Memory

Given the impossibility results in the last section, we now focus on the impact of memory. As a direct initial result, no memory can yield reachability of a *given* locally stable matching, even in a correlated job-market game.

Corollary 2. *It is* NP-*hard to decide* REACHABILITY *to a given locally stable matching in a correlated job-market game with any kind of memory.*

Let us instead concentrate on the impact of memory on reaching an arbitrary locally stable matching. For our treatment we will focus on the case in which the network links $L \subseteq (W \times W) \cup (U \times W)$. We assume that in every step, every agent remembers one previous matching partner.

Quality Memory. With quality memory, each agent remembers the best matching partner he ever had before. While this seems quite a natural choice and appears like a smart strategy, it can be easily fooled by starting with a much-liked partner, who soon after matches with someone more preferred and never becomes available again. This way the memory becomes useless which leaves us with the same dynamics as before.

Proposition 1. *There is a network game with strict preferences, links $L \subseteq (W \times W) \cup (U \times W)$, quality memory and initial matching $M = \emptyset$ such that no locally stable matching can be reached with local improvement steps from M. This even holds if every agent remembers the best k previous matches.*

Theorem 5. *It is NP-hard to decide REACHABILITY to an arbitrary locally stable matching in a network game with quality memory.*

Recency Memory. With recency memory, each agent remembers the last partner he has been matched to. This is again quite a very natural choice as it expresses the human character of remembering the latest events best. Interestingly, here we actually can ensure that a locally stable matching can be reached.

Theorem 6. *For every network game with strict preferences, links $L \subseteq (U \times W) \cup (W \times W)$, recency memory and every initial matching, there is a sequence of $O(|U|^2|W|^2)$ many local improvement steps to a locally stable matching.*

Proof. Our basic approach is to construct the sequence in two phases similarly as in [1]. In the first phase, we let the matched vertices from U improve, but ignore the unmatched ones. In the second phase, we make sure that vertices from W have improved after every round.

   Preparation phase: As long as there is at least one $u \in U$ with u matched and u part of a blocking pair, allow u to switch to the better partner.

   The preparation phase terminates after at most $|U| \cdot |W|$ steps, as in every round one matched $u \in U$ strictly improves in terms of preference. This can happen at most $|W|$ times for each matched u. In addition, the number of matched vertices from U only decreases.

   Memory phase: As long as there is a $u \in U$ with u part of a blocking pair, pick u and execute a sequence of local improvement steps involving u until u is not part of any blocking pair anymore. For every edge $e = \{u', w\}$ with $u' \neq u$ that was deleted during the sequence, recreate e from the memory of u'.

   We claim that if we start the memory phase after the preparation phase, at the end of every round we have the following invariants: The vertices from W that have been matched before are still matched, they do not have a worse partner than before, and at least one of them is matched strictly better than before. Also, only unmatched vertices from U are involved in local blocking pairs.

Obviously, at the end of the preparation phase the only U-vertices in local blocking pairs are unmatched, i.e., initially only unmatched U-vertices are part of local blocking pairs. Let u be the vertex chosen in the following round of the memory phase. At first we consider the outcome for $w \in W$. If w is the vertex matched to u in the end, then w clearly has improved. Otherwise w gets matched to its former partner (if it had one) through memory and thus has the same utility as before. In particular, every w that represents an improvement to some u' but was blocked by a higher ranked vertex still remains blocked. Together with the fact that u plays local improvement steps until it is not part of a local blocking pair anymore, this guarantees that all matched U-vertices cannot improve at the end of the round. As one W-vertex improves in every round, we have at most $|U| \cdot |W|$ rounds in the memory phase, where every round consists of at most $|W|$ steps by u and at most $|U| - 1$ edges reproduced from memory. □

The existence of sequences to (locally) stable matchings also implies that random dynamics converge in the long run with probability 1 [8,12,20]. In general, we cannot expect fast convergence here, as there are instances where random dynamics yield an exponential sequence with high probability even if all information is given – e.g., reinterpret the instance from [1] with $L = U \times W$, then every agent knows every possible partner and memory has no effect.

Observe that the previous proof relies only on the recency memory of partition U. Hence, the existence of short sequences holds even if only agents from U have memory. In contrast, if only agents from W have recency memory, the previous NP-hardness constructions can be extended.

Theorem 7. *It is* NP-*hard to decide* REACHABILITY *to an arbitrary locally stable matching in a network game with links $L \subseteq (U \times W) \cup (W \times W)$ and recency memory only for agents in W.*

Random Memory. Finally, with random memory, each agent remembers a partner chosen uniformly at random in each step. We consider random memory and reaching a locally stable matching from every starting state even in general network games. While we cannot expect fast convergence, we can show that random memory helps with reachability:

Theorem 8. *For every network game with random memory, random dynamics converge to a locally stable matching with probability 1.*

5   Maximum Locally Stable Matchings

As the size of locally stable matchings can vary significantly – up to the point where the empty matching as well as a matching that includes every vertex is locally stable – it is desirable to target locally stable matchings of maximal size. We address the computational complexity of finding maximum locally stable matchings by relating it to the independent set problem.

Theorem 9. *For every graph $G = (V, E)$ there is a job-market game that admits a maximum locally stable matching of size $|V| + k$ if and only if G holds a maximum independent set of size k.*

Proof. Given a graph $G = (V, E)$, $|V| = n$, we construct the job-market game with network $N = (V' = U \cup W, L)$. For every $v \in V$ we have $u_{v,1}, u_{v,2} \in U$ and $w_{v,1}, w_{v,2} \in W$. We have the links $\{w_{v,1}, w_{v',2}\}$ and $\{w_{v',2}, w_{v,2}\}$ if $v' \in N(v)$. We allow matching edges $\{u_{v,1}, w_{v,1}\}$, $\{u_{v,1}, w_{v',2}\}$ for $v' \in N(v)$, $\{u_{v,1}, w_{v,2}\}$ and $\{u_{v,2}, w_{v,2}\}$. Each $u_{v,1}$ prefers $w_{v,2}$ to every $w_{v',2}$, $v' \in N(v)$, and every $w_{v',2}$ to $w_{v,1}$. The preferences between the different neighbors can be chosen arbitrarily. Each $w_{v,2}$ prefers $u_{v,1}$ to every $u_{v',1}$, $v' \in N(v)$, and every $u_{v',2}$ to $u_{v,2}$. Again the neighbors can be ordered arbitrarily. Vertices $w_{v,1}$ and $u_{v,2}$ have only one possible matching partner.

We claim that G has a maximum independent set of size k iff N has a locally stable matching of size $n + k$.

Let S be a maximum independent set in G. Then $M = \{\{u_{v,1}, w_{v,2}\}| \ v \in V \setminus S\} \cup \{\{u_{v,1}, w_{v,1}\}, \{u_{v,2}, w_{v,2}\} \mid v \in S\}$ is a locally stable matching as the edges $\{u_{v,1}, w_{v,2}\}$ are always stable. For the other vertices the independent set property tells us that for $v \in S$ all vertices $v' \in N(S)$ generate stable edges $\{u_{v',1}, w_{v',2}\}$ that keep $u_{v,1}$ from switching to $w_{v',2}$. Thus $\{u_{v,1}, w_{v,1}\}$ is stable and $w_{v,2}$ cannot see $u_{v,1}$ which stabilizes $\{u_{v,2}, w_{v,2}\}$.

Now let M be a maximum locally stable matching for the job-market game. Further we chose M such that every $u_{v,1}$ is matched, which is possible as replacing a matching partner of $w_{v,2}$ by (the unmatched) $u_{v,1}$ will not generate instabilities or lower the size of M. We note that no $u_{v,1}$ is matched to some $w_{v',2}$ with $v \neq v'$ as from there $u_{v,1}$ and $w_{v,2}$ can see each other and thus constitute a blocking pair. Then, for $S = \{v|u_{v,2} \in M\}$, $|S| = |M| - n$ and S is an independent set, as every $u_{v,2}$ can only be matched to its vertex $w_{v,2}$, which means that $u_{v,1}$ must be matched to $w_{v,1}$. But this edge is only stable if every $w_{v',2}$, $v' \in N(v)$, is blocked by $u_{v',1}$. Hence for every $v \in S$ $N(v) \cap S = \emptyset$.   □

This result allows us to transfer hardness of approximation results for independent set to locally stable matching.

Corollary 3. *Finding a maximum locally stable matching is NP-complete. Under the unique games conjuncture the problem cannot be approximated within $1.5 - \varepsilon$, for any constant ε.*

In fact, our reduction applies in the setting of the job-market game, where one side has no network at all. This shows that even under quite strong restrictions the hardness of approximation holds. In contrast, it is easy to obtain a 2-approximation in every network game that admits a globally stable matching.

Proposition 2. *If a (globally) stable matching exists, every such stable matching is a 2-approximation for the maximum locally stable matching.*

References

1. Ackermann, H., Goldberg, P., Mirrokni, V., Röglin, H., Vöcking, B.: Uncoordinated two-sided matching markets. SIAM J. Comput. 40(1), 92–106 (2011)
2. Arcaute, E., Vassilvitskii, S.: Social networks and stable matchings in the job market. In: Leonardi, S. (ed.) WINE 2009. LNCS, vol. 5929, pp. 220–231. Springer, Heidelberg (2009)
3. Austrin, P., Khot, S., Safra, M.: Inapproximability of vertex cover and independent set in bounded degree graphs. Theory of Computing 7(1), 27–43 (2011)
4. Biró, P., Cechlárová, K., Fleiner, T.: The dynamics of stable matchings and half-matchings for the stable marriage and roommates problems. Int. J. Game Theory 36(3-4), 333–352 (2008)
5. Blum, Y., Roth, A., Rothblum, U.: Vacancy chains and equilibration in senior-level labor markets. J. Econom. Theory 76, 362–411 (1997)
6. Blum, Y., Rothblum, U.: "Timing is everything" and martial bliss. J. Econom. Theory 103, 429–442 (2002)
7. Cheng, C., McDermid, E.: Maximum locally stable matchings. In: Proc. 2nd Intl. Workshop Matching under Preferences (MATCH-UP), pp. 51–62 (2012)
8. Diamantoudi, E., Miyagawa, E., Xue, L.: Random paths to stability in the roommates problem. Games Econom. Behav. 48(1), 18–28 (2004)
9. Goemans, M., Li, L., Mirrokni, V., Thottan, M.: Market sharing games applied to content distribution in ad-hoc networks. IEEE J. Sel. Area Comm. 24(5), 1020–1033 (2006)
10. Gusfield, D., Irving, R.: The Stable Marriage Problem: Structure and Algorithms. MIT Press (1989)
11. Hoefer, M.: Local matching dynamics in social networks. Inf. Comput. 222, 20–35 (2013)
12. Inarra, E., Larrea, C., Moris, E.: Random paths to P-stability in the roommates problem. Int. J. Game Theory 36(3-4), 461–471 (2008)
13. Inarra, E., Larrea, C., Moris, E.: The stability of the roommate problem revisited. Core Discussion Paper 2010/7 (2010)
14. Irving, R.: An efficient algorithm for the "stable roommates" problem. J. Algorithms 6(4), 577–595 (1985)
15. Klaus, B., Klijn, F., Walzl, M.: Stochastic stability for rommate markets. J. Econom. Theory 145, 2218–2240 (2010)
16. Knuth, D.: Marriages stables et leurs relations avec d'autres problemes combinatoires. Les Presses de l'Université de Montréal (1976)
17. McDermid, E.: A 3/2-approximation algorithm for general stable marriage. In: Albers, S., Marchetti-Spaccamela, A., Matias, Y., Nikoletseas, S., Thomas, W. (eds.) ICALP 2009, Part I. LNCS, vol. 5555, pp. 689–700. Springer, Heidelberg (2009)
18. Roth, A., Sönmezc, T., Ünver, M.U.: Pairwise kidney exchange. J. Econom. Theory 125(2), 151–188 (2005)
19. Roth, A., Sotomayor, M.O.: Two-sided Matching: A study in game-theoretic modeling and analysis. Cambridge University Press (1990)
20. Roth, A., Vate, J.V.: Random paths to stability in two-sided matching. Econometrica 58(6), 1475–1480 (1990)
21. Yanagisawa, H.: Approximation algorithms for stable marriage problems. PhD thesis, Kyoto University, Graduate School of Informatics (2007)

Distributed Deterministic Broadcasting in Wireless Networks of Weak Devices*

Tomasz Jurdzinski[1], Dariusz R. Kowalski[2], and Grzegorz Stachowiak[1]

[1] Institute of Computer Science, University of Wrocław, Poland
[2] Department of Computer Science, University of Liverpool, United Kingdom

Abstract. Many futuristic technologies, such as Internet of Things or nano-communication, assume that a large number of simple devices of very limited energy and computational power will be able to communicate efficiently via wireless medium. Motivated by this, we study broadcasting in the model of ad-hoc wireless networks of weak devices with uniform transmission powers. We compare two settings: with and without local knowledge about immediate neighborhood. In the latter setting, we prove $\Omega(n \log n)$-round lower bound and develop an algorithm matching this formula. This result could be made more accurate with respect to network density, or more precisely, the maximum node degree Δ in the communication graph. If Δ is known to the nodes, it is possible to broadcast in $O(D\Delta \log^2 n)$ rounds, which is almost optimal in the class of networks parametrized by D and Δ due to the lower bound $\Omega(D\Delta)$. In the setting with local knowledge, we design a scalable and almost optimal algorithm accomplishing broadcast in $O(D \log^2 n)$ communication rounds, where n is the number of nodes and D is the eccentricity of a network. This can be improved to $O(D \log g)$ if network granularity g is known to the nodes. Our results imply that the cost of "local communication" is a dominating component in the complexity of wireless broadcasting by weak devices, unlike in traditional models with non-weak devices in which well-scalable solutions can be obtained even without local knowledge.

1 Introduction

1.1 The Model

We consider a wireless network consisting of n *stations*, also called *nodes*, deployed into an Euclidean plane and communicating by a wireless medium. The *Euclidean metric* on the plane is denoted $\text{dist}(\cdot, \cdot)$. Each station v has its *transmission power* P_v, which is a positive real number. There are three fixed model parameters: path loss $\alpha > 2$, threshold $\beta \geq 1$, and ambient noise $\mathcal{N} > 0$.

---

* The full version of the paper is available at [13]. This work was supported by the Polish National Science Centre grant DEC-2012/06/M/ST6/00459.

F.V. Fomin et al. (Eds.): ICALP 2013, Part II, LNCS 7966, pp. 632–644, 2013.

The $SINR(v, u, \mathcal{T})$ ratio, for given stations u, v and a set of (transmitting) stations $\mathcal{T}$, is defined as follows:

$$SINR(v, u, \mathcal{T}) = \frac{P_v \text{dist}(v, u)^{-\alpha}}{\mathcal{N} + \sum_{w \in \mathcal{T} \setminus \{v\}} P_w \text{dist}(w, u)^{-\alpha}} \tag{1}$$

In the *weak devices model* considered in this work, a station u successfully receives a message from a station v in a round if $v \in \mathcal{T}$, $u \notin \mathcal{T}$, and:

a) $P_v \text{dist}^{-\alpha}(v, u) \geq (1 + \varepsilon)\beta\mathcal{N}$, and
b) $SINR(v, u, \mathcal{T}) \geq \beta$,

where $\mathcal{T}$ is the set of stations transmitting at that time and $\varepsilon > 0$ is a fixed *signal sensitivity parameter* of the model.[1]

Ranges and Uniformity. The *communication range* r_v of a station v is the radius of the ball in which a message transmitted by the station is heard, provided no other station transmits at the same time. A network is *uniform*, when transmission powers P_v and thus ranges of all stations r_v are equal, or *nonuniform* otherwise. In this paper, only uniform networks are considered, i.e., $P_v = P$ and $r = r_v = (P/(\mathcal{N}\beta(1 + \varepsilon)))^{1/\alpha}$. The *range area* of a station v is defined to be the ball of radius r centered in v.

Communication Graph and Graph Notation. The *communication graph* $G(V, E)$, also called the *reachability graph*, of a given network consists of all network nodes and edges (v, u) such that u is in the range area of v. Note that the communication graph is symmetric for uniform networks. By a *neighborhood* of a node u we mean the set (and positions) of all neighbors of u in G, i.e., the set $\{w \mid (w, u) \in E(G)\}$. The *graph distance* from v to w is equal to the length of a shortest path from v to w in the communication graph, where the length of a path is equal to the number of its edges. The *eccentricity* of a node is the maximum graph distance from this node to any other node (note that the eccentricity is of order of the diameter). By Δ we denote the maximum degree of a node in the communication graph.

Synchronization. It is assumed that algorithms work synchronously in rounds, each station can either act as a sender or as a receiver during a round. We do not assume global clock ticking.

Carrier Sensing. We consider the model *without carrier sensing*, that is, a station u has no other feedback from the wireless channel than receiving or not receiving a message in a round t.

---

[1] This model is motivated by the fact that it is too costly for weak devices to have receivers doing signal acquisition continuously, c.f., [7]. Therefore, in many systems they rather wait for an energy spike, c.f., condition (a), and once they see it, they start sampling and correlating to synchronize and acquire a potential packet preamble [19]. Once synchronized, they can detect signals, c.f., condition (b).

Knowledge of Stations. Each station has its unique ID from the set $[N]$,[2] where N is polynomial in n. Stations also know their locations, and parameters n, N. Some subroutines use the granularity g, defined as r divided by the minimum distance between any two stations (c.f., [5]). We distinguish between networks *without local knowledge (ad hoc)*, where stations do not know anything about the topology of the network, and networks *with local knowledge*, in which each station knows locations and IDs of its neighbors in the communication graph.

Broadcasting Problem and Complexity Parameters. In the broadcast problem, there is one distinguished node, called the *source*, which initially holds a piece of information (also called a source message or a broadcast message). The goal is to disseminate this message to all other nodes. The complexity measure is the worst-case time to accomplish the broadcast task, taken over all connected networks with specified parameters. Time, also called the *round complexity*, denotes the number of communication rounds in the execution of a protocol: from the round when the source is activated with its source message till the broadcast task is accomplished. For the sake of complexity formulas, we consider the following parameters: n, N, D, and g.

Messages and Initialization of Stations Other than Source. We assume that a single message sent in the execution of any algorithm can carry the broadcast message and at most polynomial, in the size of the network, number of control bits. (For the purpose of our algorithms, it is sufficient that positions of stations on the plane are stored with accuracy requiring $O(\log n)$ bits; therefore, we assume that each message contains the position of its sender.) A station other than the source starts executing the broadcast protocol after the first successful receipt of the source message; it is often called a *non-spontaneous wake-up model.*

1.2   Our Results

In this paper we present the first study of deterministic distributed broadcasting in wireless networks of weak devices with uniform transmission powers, deployed in the two dimensional Euclidean space. We distinguish between the two settings: with and without local knowledge about the neighbors in the communication graph. In the latter model, we developed an algorithm accomplishing broadcast in $O(n \log n)$ rounds, which matches the lower bound (Sections 2.1 and 2.3, resp.). Then, an algorithm accomplishing broadcast in time $O(D\Delta \log^2 n)$ is presented, where D is the eccentricity of the source and Δ is the largest degree of a node in the communication graph (Section 2.2). This algorithm is close to the lower bound $\Omega(D\Delta)$, see Section 2.3. Our solution for networks with local knowledge works in $O(D \log^2 n)$ rounds (Section 3), which provides only a small $O(\log^2 n)$ overhead over the straightforward lower bound of $\Omega(D)$, and is faster, in the worst case, than any algorithm designed for networks without local knowledge of eccentricity $D = o(n/\log n)$ or maximal degree $\Delta = \omega(1)$. It also implies that the cost of learning neighborhoods by stations in wireless network

---

[2] We denote $[i] = \{1, 2, \ldots, i\}$, $[i, j] = \{i, i+1, \ldots, j\}$ for $i, j \in \mathbb{N}$.

is much higher, by factor around $\min\{n/D, \Delta\}$, than the cost of broadcast itself (i.e., broadcast performed when such neighborhoods would be provided). If the granularity g is known, a complexity $O(D \log g)$ can be achieved by a variation of the algorithm mentioned above.

Our results rely on novel techniques which simultaneously exploit specific properties of conflict resolution in the SINR model (see e.g., [1]) and several algorithmic techniques developed for a different radio network model. In particular, we show how to efficiently combine a novel SINR-based communication technique, ensuring several simultaneous point-to-point communications inside the range area of one station (which is unfeasible to achieve in the radio network model), with strongly selective families and methods based on geometric grids developed in the context of radio networks. As a result, we are able to transform algorithms relying on the knowledge of network's granularity into algorithms of asymptotically similar performance (up to a $\log n$ factor) that do not require such knowledge; this is in particular demonstrated in the leader election algorithms.

Details of some algorithms and technical proofs can be found in the full version of the paper [13].

1.3   Previous and Related Results

To the best of our knowledge, this is the first theoretical study of the problem of distributed deterministic broadcasting in ad hoc wireless networks of weak devices. In what follows, we list most relevant results in the SINR-based model and in the older, but still related, radio network model.

SINR Models. In the model of (uniform) weak devices, distributed algorithms for building a backbone structure in $O(\Delta \text{ polylog } n)$ rounds were constructed in [11]. Unlike in our broadcast problem, in [11] it was assumed that all nodes simultaneously start building the backbone. That result combined with the results of this work implicates that there is an extra cost payed for the lack of initial synchronization. If devices are not weak (i.e., not restricted by the fact that the signal must be sufficiently strong in order to be noticed), broadcasting can be done in $O(D \log^2 n)$, as proved in [14]. Combined with results in this paper, it proves a complexity gap between the two models: weak and non-weak devices.

Under the SINR-based models in ad hoc setting, a few other problems were also studied, such as deterministic data aggregation [10] and *local* broadcasting [20], in which nodes have to inform only their neighbors in the corresponding reachability graph. The considered setting allowed power control by algorithms, in which, in order to avoid collisions, stations could transmit with any power smaller than the maximal one. Randomized solutions for contention resolution [15] and local broadcasting [8] were also obtained.

There is a vast amount of work on centralized algorithms under the SINR model. The most studied problems include connectivity, capacity maximization, and link scheduling types of problems; for recent results and references we refer the reader to the survey [9]. Multiple Access Channel properties were also recently studied under the SINR model, c.f., [18].

Radio Network Model. In this model, a transmitted message is successfully heard if there are no other simultaneous transmissions from the neighbors of the receiver in the communication graph. This model does not take into account the real strength of the received signals, and also the signals from outside of the close proximity. In the geometric ad hoc setting, Dessmark and Pelc [4] were the first who studied this problem. They analyzed the impact of local knowledge, defined as the range within which stations can discover the nearby stations. Emek et al. [5] designed a broadcast algorithm working in time $O(Dg)$ in Unit Disc Graphs (UDG) radio networks with eccentricity D and granularity g. Later, Emek et al. [6] developed a matching lower bound $\Omega(Dg)$. In the *graph-based* model of radio networks, in which stations are not explicitly deployed in a metric space, the fastest $O(n \log(n/D))$-round deterministic algorithm was developed by Kowalski [16], and almost matching lower bound was given by Kowalski and Pelc [17], who also studied fast randomized solutions (in parallel with [3]). The above results hold without assuming local knowledge. With local knowledge, Jurdzinski and Kowalski [12] showed a lower bound $\Omega(\sqrt{Dn \log n})$ on the number of rounds and an algorithm of complexity $O(D\sqrt{n} \log^6 n)$.

1.4   Technical Preliminaries

In the broadcast problem, a round counter could be easily maintained by already informed nodes by passing it along the network with the source message, so in all algorithms we in fact assume having a global clock. For simplicity of analysis, we assume that every message sent during the execution of our broadcast protocols contains the broadcast message; in practice, further optimization of a message content could be done in order to reduce the total number of transmitted bits in real executions. In a given round t we say that a station v transmits *c-successfully* in round t if v transmits a message in round t and this message is heard by each station u in the Euclidean distance at most c from v. We say that a station v transmits *successfully* in round t if it transmits r-successfully, i.e., each of its neighbors in the communication graph can hear its message. Finally, v transmits *successfully* to u in round t if v transmits a message and u receives this message in round t. We say that a station that received the broadcast message is *informed*.

Grids. Given a parameter $c > 0$, we define a partition of the 2-dimensional space into square boxes of size $c \times c$ by the grid G_c, in such a way that: all boxes are aligned with the coordinate axes, point $(0,0)$ is a grid point, each box includes its left side without the top endpoint and its bottom side without the right endpoint and does not include its right and top sides. We say that (i,j) are the coordinates of the box with its bottom left corner located at $(c \cdot i, c \cdot j)$, for $i,j \in \mathbb{Z}$. A box with coordinates $(i,j) \in \mathbb{Z}^2$ is denoted $C(i,j)$. As observed in [4,5], the *grid* $G_{r/\sqrt{2}}$ is very useful in the design of the algorithms for UDG (unit disk graph) radio networks, where r is equal to the range of each station. This follows from the fact that $r/\sqrt{2}$ is the largest parameter of a grid such that each station in a box is in the range of every other station in that box. We fix $\gamma = r/\sqrt{2}$ and call G_γ the *pivotal grid*. If not stated otherwise, our considerations will refer to (boxes of) G_γ. The boxes C, C' of the pivotal grid are *neighbors* in a network if there are stations $v \in C$ and $v' \in C'$ such that the edge (v, v') belongs to the communication graph. We define the set DIR $\subset [-2, 2]^2$ such that $(d_1, d_2) \in$ DIR iff it is possible that boxes $C(i, j)$ and $C(i + d_1, j + d_2)$ are neighbors.

Schedules. A (general) *broadcast schedule* S of length T wrt $N \in \mathbb{N}$ is a mapping from $[N]$ to binary sequences of length T. A station with identifier $v \in [N]$ *follows* the schedule S of length T in a fixed period of time consisting of T rounds, when v transmits a message in round t of that period iff the position $t \mod T$ of $S(v)$ is equal to 1. For the tuples (i_1, j_1), (i_2, j_2) the relation $(i_1, j_1) \equiv (i_2, j_2) \mod d$ for $d \in \mathbb{N}$ denotes that $(|i_1 - i_2| \mod d) = 0$ and $(|j_1 - j_2| \mod d) = 0$. A set of stations A on the plane is δ-*diluted* wrt G_c, for $\delta \in \mathbb{N} \setminus \{0\}$, if for any two stations $v_1, v_2 \in A$ with grid coordinates (i_1, j_1) and (i_2, j_2), respectively, the relationship $(i_1, j_1) \equiv (i_2, j_2) \mod d$ holds. We say that δ-*dilution* is applied to a schedule S if each round of an execution of S is replaced with δ^2 rounds parameterized by $(i, j) \in [0, \delta - 1]^2$ such that a station $v \in C(a, b)$ can transmit a message only in the rounds (i, j) such that $(i, j) \equiv (a, b) \mod \delta$.

Proposition 1. *For each $\alpha > 2$ and $\varepsilon > 0$, there exists a constant d_0 such that the following properties hold. Assume that a set of n stations A is d-diluted wrt the grid G_x, where $x = \gamma/c$, $c \in \mathbb{N}$, $c > 1$ and $d \geq d_0$. Moreover, at most one station from A is located in each box of G_x. Then, if all stations from A transmit simultaneously, each of them transmits $\frac{2r}{c}$-successfully.*

Proposition 2. *For each $\alpha > 2$ and $\varepsilon > 0$, there exists a constant d satisfying the following property. Let A be a set of stations such that $\min_{u,v \in A} \{dist(u, v)\} = x \cdot \sqrt{2}$, where $x \leq \gamma$. If a station $u \in C(i, j)$ for a box $C(i, j)$ of G_x is transmitting in a round t and no other station in any box $C(i', j')$ of G_x such that $\max\{|i - i'|, |j - j'|\} \leq d$ is transmitting at that round, then v can hear the message from u at round t.*

Selective families. A family $S = (S_0, \ldots, S_{s-1})$ of subsets of $[N]$ is a (N, k)-*ssf* *(strongly-selective family)* of length s if, for every non empty subset Z of $[N]$ such that $|Z| \leq k$ and for every element $z \in Z$, there is a set S_i in S such that $S_i \cap Z = \{z\}$. It is known that there exists (N, k)-ssf of size $O(k^2 \log N)$ for every $k \leq N$, c.f., [2]. We identify a family of sets $S = (S_0, \ldots, S_{s-1})$ with the broadcast schedule S' such that the ith bit of $S'(v)$ is equal to 1 iff $v \in S_i$.

2   Algorithms without Local Knowledge

2.1   Size Dependent Algorithm

In this section we consider networks in which a station knows only n, N, its own ID and its coordinates in the Euclidean space. We develop an algorithm SIZEUBR, which executes repeatedly two interleaved threads.

The **first thread** keeps combining stations into groups such that eventually, for any box C of the pivotal grid, all stations located in C form one group. Moreover, each group has the leader, and eventually each station should be aware of (i) which group it belongs to, (ii) which station is the leader of that group, and (iii) which stations belong to that group. Upon waking up, each station forms a group with a single element (itself), and the groups increase gradually by merging. The merging process builds upon the following observation. Let σ be the smallest distance between two stations and let u, v be the closest stations. Thus, there is at most one station in each box of the grid $G_{\sigma/\sqrt{2}}$. Then, if u transmits a message and no other station in distance $d \cdot \sigma$, for some constant d, transmits at the same time, then v can hear that message (see Prop. 2). Using a $(N, (2d+1)^2)$-strongly-selective family as a broadcast schedule S on the set of leaders of groups, c.f., [16], one can assure that such a situation occurs in each $O(\log N)$ rounds. If u can hear v and v can hear u during such a schedule, the groups of u and v can be merged. In order to coordinate the merging process, we implicitly build a matching among pairs (u, v) such that u can hear v and v can hear u during execution of S.

The **second thread** is supposed to guarantee that the broadcast message is transmitted from boxes containing informed stations to their neighbors. Each station determines its temporary ID (TID) as the rank of its ID in the set of IDs in its group. Using these TIDs, the stations apply round-robin strategy. Thus, if each group corresponds to all stations in the appropriate box, transmissions are successful (see Prop. 1), and thus they guarantee that neighbors of a box containing informed stations will also contain informed stations.

The main problem with implementation of these ideas is that, as long as there are many groups inside a box, transmissions in the second thread may cause unwanted interferences. Another problem is that the set of stations attending the protocol changes gradually, when new stations become informed and can join the execution of the protocol. These issues are managed by measuring the progress of a protocol using amortized analysis. The details of the implementation and analysis can be found in the full version of the paper.

Theorem 1. *Algorithm* SizeUBr *performs broadcasting in each n-node network in $O(n \log n)$ rounds, in the setting without local knowledge.*

2.2   Degree Dependent Algorithm

In this section we present the algorithm GenBroadcast whose complexity is optimized with respect to maximal degree of the communication graph. The core of this algorithm is a leader election procedure which, given a set of stations V, chooses exactly one station (the leader) in each box C of the pivotal grid containing at least one element from V. This procedure works in $O(\log n \cdot \log N) = O(\log^2 n)$ rounds and it is executed repeatedly during GenBroadcast. The set of stations attending a particular leader election execution consists of all stations which received the broadcast message and have not been chosen leaders of their boxes in previous executions of the leader election procedure. Moreover, at

Algorithm 1. LeaderElection(V, n)

1: For each $v \in V$: $cand(v) \leftarrow true$;
2: **for** $i = 1, \ldots, \log n + 1$ **do**   ▷ **Elimination**
3:     **for** $j, k \in [0, 2]$ **do**
4:         Execute S twice on the set:
5:         $\{w \in V \mid cand(w) = true$ and $w \in C(j', k')$
6:             such that $(j' \bmod 2, k' \bmod 2) = (j, k)\}$;
7:         Each $w \in V$ determines and stores X_w during the first execution of S, and
8:         X_v, for each $v \in X_w$, during the second execution of S;
9:         **for** each $v \in V$ **do**
10:             $u \leftarrow \min(X_v)$;
11:             **if** $X_v = \emptyset$ or $v > \min(X_u \cup \{u\})$ **then** $cand(v) \leftarrow false$; $ph(v) \leftarrow i$;
12: For each $v \in V$: $state(v) \leftarrow active$;   ▷ **Selection**
13: **for** $i = \log n, (\log n) - 1, \ldots, 2, 1$ **do**
14:     $V_i \leftarrow$ GranLeaderElection($\{v \in V \mid$
15:         $ph(v) = i, state(v) = active\}, 1/n$);   ▷ V_i – leaders
16:     Each element $v \in V_i$ sets $state(v) \leftarrow leader$ and
17:         transmits successfully using constant dilution (see Prop. 1);
18:     Simultaneously, for each $v \in V$ which can hear $u \in$ box(v): $state(v) \leftarrow passive$.

---

the end of each execution of the leader election procedure, each leader chosen in that execution transmits a message successfully — this can be done in a constant number of rounds, by using d-dilution with appropriate constant d (c.f., Prop. 1). In this way, each station receives the source message after $O(D\Delta \log^2 n)$ rounds. (Note that there are at most Δ stations in a box of the pivotal grid.)

In the following, we describe the leader election algorithm — its pseudo-code is presented as Algorithm 1. We are given a set of stations V of size at most n. The set V is not known to stations, each station knows merely whether it belongs to V or it does not belong to V. In the algorithm, we use (N, e)-ssf S of size $s = O(\log N)$, where $e = (2d + 1)^2$ and d is the constant depending merely on the parameters of the model, the same as in Section 2.1 (see also Prop. 2). Let X_v, for a given execution of S be the set of stations which belong to box(v) and v can hear them during that execution.

The following proposition combines properties of ssf with Prop. 2.

Proposition 3. *For each $\alpha > 2$ and $\varepsilon > 0$, there exists a constant k satisfying the following property. Let W be a 3-diluted (wrt the pivotal grid) set of stations and let C be a box of the pivotal grid. If $\min_{u, v \in C \cap W} = x \leq r/n$ and dist(u, v) = x for some $u, v \in W$ such that box(u) = box(v) = C, then v can hear the message from u during an execution of a (N, k)-ssf on W.*

The leader election algorithm consists of two stages. The first stage gradually eliminates elements from the set of candidates for the leaders of boxes in consecutive executions of the ssf S in the first for loop. Therefore, we call this stage *Elimination*. Let *phase* l of Elimination stage denote the executions of S for $i = l$. Each station v "eliminated" in phase l has assigned the value $ph(v) = l$. Let $V(l) = \{v \mid ph(v) > l\}$ and $V_C(l) = \{v \mid ph(v) > l$ and box(v) = $C\}$ for $l \in \mathbb{N}$

and C being a box of the pivotal grid. That is, $V_C(l)$ is the set of stations from C which are *not* eliminated until phase l. The key property of the sets $V_C(l)$ is that $|V_C(l+1)| \le |V_C(l)|/2$ and the granularity of $V_C(l_C^\star)$ is *smaller* than n for each box C and $l \in \mathbb{N}$, where $l_C^\star$ is the largest $l \in \mathbb{N}$ such that $V_C(l)$ is not empty. Therefore, we can choose the leader of each box C by applying (simultaneously in each box) the granularity dependent leader election algorithm GranLeader-Election, described later in Section 3.2 on the set $V_C(l_C^\star)$ and with upper bound n on granularity. Note that we can elect the leaders in $O(\log N) = O(\log n)$ rounds in this way. However, the stations in C do not necessary know the value of $l_C^\star$. Therefore, the second stage (called *Selection*) applies the granularity dependent leader election on $V(\log n)$, $V(\log n - 1)$, $V(\log n - 2)$ and so on. When the leader of a box C is chosen, all stations in C become silent (state *passive* in line 18), i.e., they do not attend the following executions of GranLeaderElection. It is important that a station becomes silent after the leader of its box is chosen, since granularity of $V_C(i)$ might be larger than n for $i < l_C^\star$. Activity of stations from such a box C for $i < l_C^\star$ during the Selection stage could cause large interferences preventing other boxes from electing leaders.

Recall that each leader broadcasts successfully at the end of the execution of LeaderElection in which it is elected. If each station attends consecutive leader election executions until it becomes a leader in its box, the broadcasting message is transmitted from a box C to all its neighbors in $O(\Delta \log^2 n)$ rounds, since there are at most Δ station in each box of the pivotal grid. Therefore, we obtain a GeneralBroadcast algorithm providing the following result.

Theorem 2. *Algorithm GeneralBroadcast completes broadcast in $O(D\Delta \log^2 n)$ rounds in any network without local knowledge.*

2.3   Lower Bounds

Theorem 3. *There exist: (i) an infinite family of n-node networks requiring $\Omega(n \log n)$ rounds to accomplish deterministic broadcast, and (ii) for every $D\Delta = O(n)$, an infinite family of n-node networks of diameter D and maximum degree Δ requiring $\Omega(D\Delta)$ rounds to accomplish deterministic broadcast.*

Proof (Sketch). We describe a family of networks $\mathcal{F}$ such that broadcasting requires time $\Omega(D \log N)$. By L_i we denote the set of stations in distance i from the source in the communication graph. Each element of $\mathcal{F}$ is formed as a sequential composition of D networks $V_1, \ldots, V_D$ of eccentricity 3 each, such that:

- the source s is connected with two nodes v_1, v_2 in L_1 with arbitrary IDs and fixed positions;
- v_1, v_2 are connected with w, the only element of L_2, and satisfy the condition:

$$P \cdot \text{dist}(v_1, w)^{-\alpha} = P \cdot \text{dist}(v_2, w)^{-\alpha} - \mathcal{N}/2 . \tag{2}$$

Sequential composition of networks $V_1, \ldots, V_D$ stands for identifying the element w of network component V_i with the source s of network component V_{i+1}.

Note that if v_1 and v_2 transmit simultaneously in a network component V_i, the message is *not* received by w. Using simple counting argument, one can force such choice of IDs of v_1 and v_2 that $\Omega(\log N)$ rounds are necessary until a round in which exactly one of v_1, v_2 transmits successfully a message to w under the SINR model of weak devices. Since $D = \Theta(n)$ in the above construction and $\log n = \Theta(\log N)$, the bound $\Omega(n \log N)$ holds.

The above proof can be extended to obtain lower bounds $\Omega(D\Delta)$, by considering the following class of network components V_i: the source s, located in the origin point $(0,0)$, is the only element of L_0; L_1 consists of Δ nodes $v_0, \ldots, v_{\Delta-1}$, where the position of v_i is $(\gamma \cdot \frac{i}{\Delta}, \gamma)$ for $0 \leq i \leq \Delta - 1$; and L_2 contains only one node w_j with coordinates $(\gamma \cdot \frac{j}{\Delta}, \gamma + r)$, i.e., w_j can receive a message only from v_j. $\square$

This result can also be transformed to the case of randomized algorithms. We sketch an idea of these transformations by considering networks from the family $\mathcal{F}$ described in the proof of Theorem 3. Recall that each element of the layer L_1 should transmit as the only element of L_1 in order to guarantee that the only element of L_2 is informed, regardless of its location. However, by simple counting arguments, the expectation of the number of steps after which some of elements of L_1 transmit as the only one is $\Omega(\log n)$ or $\Omega(\Delta)$, respectively.

3   Algorithms for Networks with Local Knowledge

In this section we assume that each station knows n, N as well as IDs and locations of all stations in its range area. We start with presenting a generic algorithmic scheme. Next, we describe an algorithm for networks with additionally known granularity bound g. Finally we provide a solution for the general setting when granularity g is not known in advance.

3.1   Generic Algorithmic Scheme

In the first round the source sends the broadcast message. Then, we repeat the generic procedure Inter-Box-Bdcst, whose ith repetition is aimed at transmitting the broadcast message from boxes of the pivotal grid containing at least one station that has received the broadcast message in the previous execution of Inter-Box-Bdcst (or from the source) to boxes which are their neighbors. The specific implementation of procedure Inter-Box-Bdcst depends on the considered setting.

Each station v is in state $s(v)$, which may be equal to one of the following three values: asleep, active, or idle. At the beginning, the source sends the source message and all stations of its box in the pivotal grid set their states to active, while all the remaining stations are in the asleep state. The states of stations change only at the end of Inter-Box-Bdcst, according to the following rules:

Rule 1: All stations in state active change their state to idle.
Rule 2: A station u changes its state from asleep to active if it has received the broadcast message from a station v in the current execution of Inter-Box-Bdcst

such that either v was in state active or v belongs to the same box of the pivotal grid as u. That is, let C be a box of the pivotal grid, let $u \in C$ be in state asleep at the beginning of Inter-Box-Bdcst. The only possibility that u receives a message and it does not change its state from asleep to active at the end of Inter-Box-Bdcst is that each message received by u is sent by a station v which is in state asleep when it sends the message and $v \notin C$.

The intended properties of an execution of Inter-Box-Bdcst are:

(I) For each box C of the pivotal grid, states of all stations in C are equal.
(P) The broadcast message is (successfully) sent from each box C containing stations in state *active* to all stations located in boxes which are neighbors of C.

The following proposition easily follows from the above stated properties.

Proposition 4. *If (I) and (P) are satisfied, the source message is transmitted to the whole network in $O(D \cdot T)$ rounds, where T is the number of rounds in a single execution of Inter-Box-Bdcst.*

3.2   A Granularity-Dependent Algorithm

First, we develop a broadcasting algorithm with known granularity g of a network. The main ingredient of this protocol is a leader election algorithm, called GranLeaderElection(A, g), which, given a set of stations A chooses the leader in each box of the pivotal grid containing stations from A (at the beginning, each station knows only whether it belongs to A or not). The idea of the leader election procedure is as follows. Granularity g implies that each station is the leader of a box of G_x, where $x = \gamma/h$ for $h = \min(2^i \,|\, 2^i \geq g)$. Then, leaders of boxes of $G_{2^i x}$ are chosen among leaders of boxes of $G_{2^{i-1} x}$ for $i = 1, 2, \ldots, \lceil \log h \rceil$ in constant number of rounds with help of Prop. 1. Thus, leaders in boxes of the pivotal grid can be chosen in $O(\log g)$ rounds.

Given the above (local) leader election procedure, the procedure Inter-Box-Bdcst is implemented as follows. For each direction $(d_1, d_2) \in$ DIR, the leaders are elected in all boxes among station in state active which have neighbors in the direction (d_1, d_2). Then, these leaders send messages successfully using dilution (see Prop. 1). Moreover, since each station knows all stations in its box, the station with smallest ID among newly informed in each box sends the broadcast message which is delivered to all stations from its box. In this way, the procedure Inter-Box-Bdcst satisfying the invariants (I) and (P) working in time $O(\log g)$ is obtained which gives the broadcasting algorithm working in time $O(D \log g)$.

Theorem 4. *Algorithm* GRANUBR *accomplishes broadcast in any n-node network of diameter D and granularity g in $O(D \log g)$, in the setting with local knowledge.*

3.3   General Algorithm

In this section we develop Algorithm DiamUBr, which also builds on the generic scheme from Section 3.1. Procedure Inter-Box-Bdcst required by the generic algorithm is implemented as in Section 3.2, the only difference is that GranLeader-Election with complexity $O(\log g)$ is replaced with the procedure LeaderElection from Section 2.2 (Alg. 1). By Prop. 4, and by the round complexity $O(\log^2 n)$ of algorithm LeaderElection, we obtain the following result.

Theorem 5. *Algorithm DiamUBr completes broadcast in any n-node network of diameter D in $O(D \log^2 n)$ rounds, in the setting with local knowledge.*

References

1. Avin, C., Emek, Y., Kantor, E., Lotker, Z., Peleg, D., Roditty, L.: Sinr diagrams: towards algorithmically usable sinr models of wireless networks. In: PODC, pp. 200–209 (2009)
2. Clementi, A.E.F., Monti, A., Silvestri, R.: Selective families, superimposed codes, and broadcasting on unknown radio networks. In: SODA, pp. 709–718 (2001)
3. Czumaj, A., Rytter, W.: Broadcasting algorithms in radio networks with unknown topology. In: FOCS, pp. 492–501 (2003)
4. Dessmark, A., Pelc, A.: Broadcasting in geometric radio networks. J. Discrete Algorithms 5(1), 187–201 (2007)
5. Emek, Y., Gasieniec, L., Kantor, E., Pelc, A., Peleg, D., Su, C.: Broadcasting in udg radio networks with unknown topology. Distributed Computing 21(5), 331–351 (2009)
6. Emek, Y., Kantor, E., Peleg, D.: On the effect of the deployment setting on broadcasting in euclidean radio networks. In: PODC, pp. 223–232 (2008)
7. Goldsmith, A.J., Wicker, S.B.: Design challenges for energy-constrained ad hoc wireless networks. IEEE Wireless Communications 9(4), 8–27 (2002)
8. Goussevskaia, O., Moscibroda, T., Wattenhofer, R.: Local broadcasting in the physical interference model. In: DIALM-POMC, pp. 35–44 (2008)
9. Goussevskaia, O., Pignolet, Y.A., Wattenhofer, R.: Efficiency of wireless networks: Approximation algorithms for the physical interference model. Foundations and Trends in Networking 4(3), 313–420 (2010)
10. Hobbs, N., Wang, Y., Hua, Q.-S., Yu, D., Lau, F.C.M.: Deterministic distributed data aggregation under the SINR model. In: Agrawal, M., Cooper, S.B., Li, A. (eds.) TAMC 2012. LNCS, vol. 7287, pp. 385–399. Springer, Heidelberg (2012)
11. Jurdzinski, T., Kowalski, D.R.: Distributed backbone structure for algorithms in the SINR model of wireless networks. In: Aguilera, M.K. (ed.) DISC 2012. LNCS, vol. 7611, pp. 106–120. Springer, Heidelberg (2012)
12. Jurdzinski, T., Kowalski, D.R.: On the complexity of distributed broadcasting and MDS construction in radio networks. In: Baldoni, R., Flocchini, P., Binoy, R. (eds.) OPODIS 2012. LNCS, vol. 7702, pp. 209–223. Springer, Heidelberg (2012)
13. Jurdzinski, T., Kowalski, D.R., Stachowiak, G.: Distributed deterministic broadcasting in wireless networks of weak devices under the sinr model. CoRR, abs/1210.1804 (2012)
14. Jurdzinski, T., Kowalski, D.R., Stachowiak, G.: Distributed deterministic broadcasting in uniform-power ad hoc wireless networks. CoRR, abs/1302.4059 (2013)

15. Kesselheim, T., Vöcking, B.: Distributed contention resolution in wireless networks. In: Lynch, N.A., Shvartsman, A.A. (eds.) DISC 2010. LNCS, vol. 6343, pp. 163–178. Springer, Heidelberg (2010)
16. Kowalski, D.R.: On selection problem in radio networks. In: PODC, pp. 158–166 (2005)
17. Kowalski, D.R., Pelc, A.: Broadcasting in undirected ad hoc radio networks. Distributed Computing 18(1), 43–57 (2005)
18. Richa, A., Scheideler, C., Schmid, S., Zhang, J.: Towards jamming-resistant and competitive medium access in the sinr model. In: Proc. 3rd ACM Workshop on Wireless of the Students, by the Students, for the Students, S3 2011, pp. 33–36 (2011)
19. Schmid, S., Wattenhofer, R.: Algorithmic models for sensor networks. In: IPDPS. IEEE (2006)
20. Yu, D., Wang, Y., Hua, Q.-S., Lau, F.C.M.: Distributed local broadcasting algorithms in the physical interference model. In: DCOSS, pp. 1–8 (2011)

Secure Equality and Greater-Than Tests with Sublinear Online Complexity

Helger Lipmaa[1] and Tomas Toft[2]

[1] Institute of CS, University of Tartu, Estonia
[2] Dept. of CS, Aarhus University, Denmark

Abstract. Secure multiparty computation (MPC) allows multiple parties to evaluate functions without disclosing the private inputs. Secure comparisons (testing equality and greater-than) are important primitives required by many MPC applications. We propose two equality tests for ℓ-bit values with $O(1)$ online communication that require $O(\ell)$ respectively $O(\kappa)$ total work, where κ is a correctness parameter.

Combining these with ideas of Toft [16], we obtain (i) a greater-than protocol with sublinear online complexity in the arithmetic black-box model ($O(c)$ rounds and $O(c \cdot \ell^{1/c})$ work online, with $c = \log \ell$ resulting in logarithmic online work). In difference to Toft, we do not assume two mutually incorruptible parties, but $O(\ell)$ offline work is required, and (ii) two greater-than protocols with the same online complexity as the above, but with overall complexity reduced to $O(\log \ell(\kappa + \log\log \ell))$ and $O(c \cdot \ell^{1/c}(\kappa + \log \ell))$; these require two mutually incorruptible parties, but are highly competitive with respect to online complexity when compared to existing protocols.

Keywords: Additively homomorphic encryption, arithmetic black box, secure comparison, secure equality test.

1   Introduction

Secure multiparty computation (MPC) considers the following problem: n parties hold inputs, $x_1, \ldots, x_n$, for a function, f; they wish to evaluate f without disclosing their inputs to each other or third-parties. Numerous solutions to this problem exist; many provide secure arithmetic over a field or ring, e.g., $\mathbb{Z}_M$ for an appropriate M, by relying either on secret sharing or additively homomorphic encryption. The overall structure of those solutions is similar, thus the details of the constructions may be abstracted away and MPC-protocols can be constructed based on secure arithmetic. This idea was formalized as the arithmetic black-box (ABB) by Damgård and Nielsen [7]. For a longer discussion of MPC and the ABB, see Sect. 2.

Secure $\mathbb{Z}_M$-arithmetic may be used to emulate integer computation when inputs/outputs are less than M (which typically can be chosen quite freely). However, other operations may be needed. Secure comparison – equality testing (Eq) and greater-than testing (GT) – are two important problems in the (MPC)

F.V. Fomin et al. (Eds.): ICALP 2013, Part II, LNCS 7966, pp. 645–656, 2013.

Table 1. A comparison of sublinear GT protocols for bitlength ℓ

Result	Online rounds	Online work	Overall work	Correctness
Adversary structure with two mutually incorruptible parties				
[16]	$O(c)$	$O(c \cdot \ell^{1/c}(\kappa + \log \ell))$	$O(c \cdot \ell^{1/c}(\kappa + \log \ell))$	Statistical
[16]	$O(\log \ell)$	$O(\log \ell(\kappa + \log\log \ell))$	$O(\log \ell(\kappa + \log\log \ell))$	Statistical
This paper	$O(c)$	$O(c \cdot \ell^{1/c})$	$O(c \cdot \ell^{1/c}(\kappa + \log \ell))$	Statistical
This paper	$O(\log \ell)$	$O(\log \ell)$	$O(\log \ell(\kappa + \log\log \ell)))$	Statistical
Arbitrary adversary structure				
[18]	$O(1)$	$O(\sqrt{\ell/\log\ell})$	$O(\ell)$	Perfect
This paper	$O(c)$	$O(c \cdot \ell^{1/c})$	$O(\ell)$	Perfect
This paper	$O(\log \ell)$	$O(\log \ell)$	$O(\ell)$	Perfect

literature. They are required for tasks as diverse as auctions, data-mining, and benchmarking. A prime example is the first real-world MPC execution [4], which required both integer additions and GT tests.

In this paper, we introduce two new Eq tests and improve over state of the art GT testing in the ABB model. The main focus is online efficiency, i.e., parties may generate joint randomness in advance (e.g, while setting up an auction) to increase efficiency once the inputs have been supplied (bids have been given).

Related Work. Secure comparison and its applications is a very active topic with too many papers to mention all. Damgård et al. [6] proposed the first constant-rounds protocols which required $O(\ell \log \ell)$ secure multiplications. Nishide and Ohta [13] improved this to $O(\ell)$ work for GT and $O(\kappa)$ work for equality where κ is a correctness parameter.

Until recently, all GT tests had a complexity (at least) linear in the bitlength, ℓ, of the inputs, but in [16], Toft proposed the first sublinear constructions. These utilized proofs of boundedness and required the presence of 2 mutually incorruptible parties, i.e., one of two named parties was required to be honest. This is naturally satisfied in the two-party case ($n = 2$), but the multiparty case is left with either a corruption threshold of 1 or a non-standard adversary structure. In [18], Yu proposed a sublinear, constant-rounds protocol in the ABB model based on *sign modules*. His protocol requires $O(\sqrt{\ell/\log \ell})$ operations online and works for an ABB over a finite field, i.e., prime M. It does not appear that the ideas work with composite M such as is needed by Paillier encryption. See Table 1 for an overview of existing sublinear GT tests.

Contribution. We propose a collection of actively secure protocols. We first introduce two new protocols for equality testing of ℓ-bit values. The first is perfectly correct with $O(1)$ ABB-operations online and $O(\ell)$ ABB-operations overall. The second reduces overall communication to $O(\kappa)$ at the cost of imperfect correctness, i.e., κ is a correctness parameter; it also requires two mutually incorruptible parties. Both improve online complexity dramatically over previous work. Additionally, we use these in combination with ideas of [16] to obtain new GT tests for ℓ-bit values in the ABB model. We end up with multiple variations.

First, ABB-protocols with $O(\log \ell)$ work and rounds (respectively $O(c \cdot \ell^{1/c})$ work in $O(c)$ rounds for constant c) online; $O(\ell)$ work overall. Second, we reduce overall work to $O(\log \ell \cdot (\kappa + \log\log \ell))$ $(O(c \cdot \ell^{1/c}(\kappa + \log \ell))$ respectively) at the cost of requiring two mutually incorruptible parties. All constructions require proofs of boundedness to prevent active attacks. In contrast to [18], we do not utilize sign modules, hence our protocols work for Paillier encryption-based MPC as well. In that setting our GT tests are the first with sublinear online complexity and arbitrary adversary structure.

2 Preliminaries

The Arithmetic Black-Box. Many MPC protocols work by having parties supply their inputs "secretly," e.g., using secret sharing, which allows a value to be split between parties such that it remains unknown unless sufficiently many agree. A homomorphic scheme allows parties to compute sums, while secure multiplication requires interaction. Once the desired result has been computed, it is straightforward to output it by reconstructing. The arithmetic black-box of [7] captures this type of behaviour, making it a convenient model for presenting MPC protocols. This allows protocol construction with focus on the task at hand rather than "irrelevant details" such as the specifics and security guarantees of the underlying cryptographic primitives.

Formally, the arithmetic black-box is an ideal functionality, $\mathcal{F}_{\text{ABB}}$, and protocols are constructed in a hybrid model where access to this functionality is given. $\mathcal{F}_{\text{ABB}}$ can be thought of as a (virtual) trusted third party, who provides storage of elements of a ring, $\mathbb{Z}_M$, as well as arithmetic computation on stored values. Here, M will be either a prime or an RSA-modulus, i.e., the product of two large, odd primes. We provide an intuitive presentation of the ABB here; see [7] or [17] for a formal definition. Full simulation-based proofs are possible; due to space constraints we merely sketch privacy proofs.

Secure storage (input/output) can be thought of as secret sharing and we use the notation of Damgård et al. [6], writing ABB-values in square brackets, $[\![x]\!]$. ABB-arithmetic is written using *"plaintext space,"* infix operators, e.g., $[\![x \cdot y + z]\!] \leftarrow [\![x]\!] \cdot [\![y]\!] + [\![z]\!]$. Naturally, such operations eventually refer to protocols between $P_1, \ldots, P_n$, e.g., the protocols of Ben-Or et al. [3].

The complexity of a protocol in the $\mathcal{F}_{\text{ABB}}$ hybrid model is the number of basic operations performed, input/output and arithmetic. We assume that these operations may be executed concurrently, i.e., that executing the underlying cryptographic protocols concurrently does not violate security. Round complexity is defined as the number of sequential sets of concurrent operations performed (basic operations typically require a constant number of rounds; in this case constant-rounds in the ABB model implies constant-rounds in the actual protocol). Finally, we focus on communication complexity and therefore consider addition costless; typical ABB realizations are based on additively homomorphic primitives, and this is a standard choice. Additionally, ABB-computatation occurs in two phases: (i) random values are generated within $\mathcal{F}_{\mathrm{ABB}}$ before the inputs are known (preprocessing or offline phase), and (ii) when the inputs are available within $\mathcal{F}_{\mathrm{ABB}}$, the result is computed (online phase). Focus is predominantly on the efficiency of the online phase.

Known ABB Constructions and Additional Primitives. The following known primitives are needed in the proposed constructions. These are exclusively needed as part of the preprocessing phase; in practice it may be preferable to utilize simpler (non-constant-rounds) solutions.

- RandElem: Generates a uniformly random, secret element of $\mathbb{Z}_M$ stored within the ABB. Considered as 1 multiplication and 1 rounds, [6].
- RandBit: Generates a uniformly random, secret bit stored within the ABB. $O(1)$ multiplications in $O(1)$ rounds, [6].[1]
- RandBits: Generates a uniformly random $\mathbb{Z}_M$-value r and its binary representation $r = \sum 2^i r_i$, $r_i \in \{0,1\}$ stored as elements of $\mathbb{Z}_M$, $O(\log M)$ multiplications in $O(1)$ rounds, [6].
- RandInv: Generates a uniformly random element in $\mathbb{Z}_M^*$ along with its inverse; $O(1)$ multiplications in $O(1)$ rounds, [6].
- $\mathtt{prefix}_\times$: Prefix product takes a vector of invertible, secret values, $[\![r_1]\!]$, $[\![r_2]\!]$, ..., $[\![r_m]\!]$, and computes the prefix-product, i.e., $[\![\prod_{i=1}^j r_i]\!]$ for $1 \le j \le m$, using $O(m)$ ABB operations in $O(1)$ rounds [2,6].

We also require that the ABB can verify that an input is of bounded size, e.g., $[\![x]\!] < 2^\ell$. x is known by the inputter, P_i, so this corresponds to executing a *proof of boundedness*. A communication-efficiently solution ($\Theta(1)$ group elements) can be obtained using the sum-of-four-squares technique, [10]: P_i supplies an integer input (decomposed into squares) which is converted to a $\mathbb{Z}_M$ element; this can be done using *integer commitments* (for encryption) or *linear integer secret sharing scheme* [15] (for Shamir sharing). An alternative is to use the constant-communication non-interactive zero-knowledge argument of [5]; there, P_i commits to a vector of digits of x and uses the techniques of [8,11] to prove that the *encrypted* x belongs to the given range.

Disclose-If-Equal. In a disclose-if-equal (DIE) protocol between Alice and Bob, Alice gets to know Bob's secret β exactly if she encrypted x (where x is a value known to Bob only, or possibly to both). Otherwise, Alice should obtain a (pseudo)random plaintext. See [1,9] for original definitions.

If the plaintext space is $\mathbb{Z}_M$ for a prime M (as it is in the case of the secret sharing setting), one can use the following simple protocol inspired by [1] (here, Enc_{pk} means encryption by public key pk and Dec_{sk} means decryption by the corresponding secret key sk): (1) Alice sends $q \leftarrow Enc_{pk}(\alpha)$ to Bob. (2) If the ciphertext is invalid, Bob returns $\perp$. Otherwise, he returns $a \leftarrow (q \cdot Enc_{pk}(-x))^r \cdot$

---

[1] When M is an RSA-modulus, complexity is linear in the number of parties, for simplicity we assume that this is constant. (This is only used in preprocessing.)

$Enc_{pk}(\beta)$, where $r \leftarrow \mathbb{Z}_M$. (3) Alice computes $Dec_{sk}(a) = r(\alpha - x) + \beta$, which is equal to β when $\alpha = x$. Clearly, this protocol is perfectly complete, and encryption, decryption, and exponent-arithmetic can be replaced by ABB operations.

If M is not a prime but has sufficiently large prime factors (like in the case of existing additively homomorphic public-key cryptosystems), then the resulting DIE protocol, proposed by Laur and Lipmaa in [9], is somewhat more complicated. Let ℓ be the bitlength of β. Let $T \leftarrow \lfloor 2^{-\ell} \cdot M \rfloor$. Let $spf(M)$ be the smallest prime factor of the plaintext group order M. We assume $\ell \leq \frac{1}{2} \log_2 M + \log_2 \varepsilon$, where $\varepsilon \leq 2^{-80}$ is the hiding parameter. Here we assume that Bob knows the public key and Alice knows the secret key and the parties use an additively homomorphic public-key cryptosystem like the one by Paillier [14]. (1) Alice sends $q \leftarrow Enc_{pk}(\alpha)$ to Bob. (2) If the ciphertext is invalid, Bob returns $\perp$. Otherwise, he returns $a \leftarrow (q \cdot Enc_{pk}(-x))^r \cdot Enc_{pk}(\beta + 2^\ell \cdot t)$, where $r \leftarrow \mathbb{Z}_M$ and $t \leftarrow \mathbb{Z}_T$. (3) Alice computes $Dec_{sk}(a) \mod 2^\ell$.

As shown in [9], this protocol is $(1 - \varepsilon)$-semisimulatable [12] (that is, game-based computationally private against a malicious server, and simulation-based statistically private against a malicious client) as long as $2^{\ell-1}/spf(M)$ is bounded by ε. That is, if $x \neq \alpha$ then the distribution of $U(\mathbb{Z}_M) \cdot (\alpha - x) + 2^\ell \cdot U(\mathbb{Z}_T)$ is ε-far from the uniform distribution $U(\mathbb{Z}_M)$ on $\mathbb{Z}_M$. Since in the case of Paillier, $spf(M) \approx \sqrt{M}$, we need that $\ell - 1 - \frac{1}{2} \cdot \log_2 M \leq \log_2 \varepsilon$ or $\ell < \frac{1}{2} \cdot \log_2 M + \log_2 \varepsilon$, as mentioned. The idea behind including the additional term $2^\ell \cdot t$ in the Laur-Lipmaa protocol is that if M is composite, then $Dec_{pk}((q \cdot Enc_{pk}(-x))^{U(\mathbb{Z}_M)}) = U(\mathbb{Z}_M) \cdot (\alpha - x)$ can be a random element of a nontrivial subgroup of $\mathbb{Z}_M$ and thus far from random in $\mathbb{Z}_M$; adding $2^\ell \cdot U(\mathbb{Z}_T)$ guarantees that the result is almost uniform in $\mathbb{Z}_M$.

3 Secure Equality Tests

It is well-known that equality testing can be implemented using a zero-test (given additively homomorphic primitives) as $x = y \Leftrightarrow x - y = 0$; w.l.o.g., we focus on testing whether x equals 0 and present two new, secure protocols.

The first zero-test is based on the Hamming distance between a mask and the masked value. Complexity is linear in the bit-length, but only $O(1)$ ABB multiplications and outputs are needed online. Hence, when a preprocessing phase is present, this is highly efficient. Additionally, we present a variation allowing comparison of ℓ-bit numbers with $O(\ell)$ preprocessing and $O(1)$ work online, when $2^{\ell+k+\log n} \ll M$ for statistical security parameter k, and n parties.

The second approach is based on DIE and reduces the problem from arbitrary size inputs to κ-bit inputs, where κ is a correctness parameter, e.g., 80. This simpler problem may then be solved, e.g., using the Hamming-based approach.

3.1 Equality from Hamming Distance

Let $\ell_M = \lceil \log_2 M \rceil$ be the bitlength of M. The protocol, denoted eq_H, is seen as Protocol 1. It is a variation of [13] with a highly optimized online phase. (Though phrased differently, Nishide and Ohta [13] did essentially the same thing).

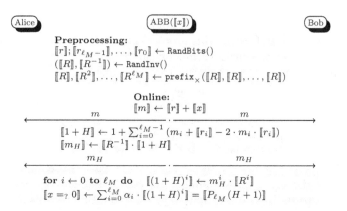

Protocol 1. eq_H, secure zero-testing based on Hamming distance

Correctness. Picking a uniformly random, unknown $[\![r]\!]$ and revealing $m = [\![x]\!] + [\![r]\!]$ allows testing $x = 0$ by testing whether $m = r$. If $[\![r]\!]$ in generated in binary, we can compute the Hamming distance $[\![H]\!] = \sum_{i=0}^{\ell_M-1} [\![r_i]\!] \oplus m_i = \sum_{i=0}^{\ell_M-1} (m_i + [\![r_i]\!] - 2 \cdot m_i \cdot [\![r_i]\!])$, and test if $H = 0$. Since $H \leq \ell_M$, the latter is simpler than the general zero-test. Let $P_{\ell_M}(x) = \sum_{i=0}^{\ell_M} \alpha_i \cdot x^i$ denote the (at most) ℓ_M-degree polynomial that maps 1 to 1 and $x \in \{2, 3, \ldots, \ell_M + 1\}$ to 0.[2] Evaluating P_{ℓ_M} at $1 + H$ determines $H + 1 = 1 \Leftrightarrow m = r \Leftrightarrow x = 0$.

To avoid $\Omega(\ell_M)$ online multiplications when computing the $\ell_M + 1$ powers of $[\![1 + H]\!]$, the following trick is used: A uniformly random value, $[\![R]\!] \in \mathbb{Z}_M^*$ is chosen in advance, and its exponents $[\![R^0]\!], [\![R^1]\!], \ldots, [\![R^{\ell_M}]\!]$ and inverse, $[\![R^{-1}]\!]$, are computed in the offline phase. In the online phase, $m_H = [\![R^{-1}]\!] \cdot [\![1 + H]\!]$ is computed and revealed, and the powers of $[\![1 + H]\!]$ are computed from the powers of $[\![R]\!]$ and the powers of m_H, which can be done locally by all parties: $m_H^i \cdot [\![R^i]\!] = (R^{-1}(1 + H))^i \cdot [\![R^i]\!] = [\![(1 + H)^i]\!]$.

Privacy. Two values are revealed in eq_H, m and m_H. Since r is uniformly random in $\mathbb{Z}_M$, then so is $m = x + r$. Similarly, since $1 + \ell_M$ is smaller[3] than the smallest prime factor of M we have $1 + H \in \mathbb{Z}_M^*$. Thus $(1 + H) \cdot R^{-1}$ is uniformly random in $\mathbb{Z}_M^*$ as R is uniformly random. Simulation in the $\mathcal{F}_{\mathrm{ABB}}$-hybrid model consists of providing "fake" m and m_H distributed as the real ones.

Complexity. The preprocessing phase consists of generating $[\![r]\!]$ along with its bits, $[\![r_{i-1}]\!]$ as well as $[\![R]\!]$, $[\![R^{-1}]\!]$, and $[\![R^i]\!]$ for $i \in \{1, \ldots, \ell_M\}$. Overall this amounts to $O(\ell_M)$ work. Online, only 1 ABB-multiplication (to compute m_H) and 2 outputs are needed. Computing the Hamming distance and evaluating P_{ℓ_M} are costless.

---

[2] P_{ℓ_M} exists both when M is a prime or an RSA-modulus and the coefficients, α_i, can be computed using Lagrange interpolation. For technical reasons, the input to P_{ℓ_M} must belong to $\mathbb{Z}_M^*$, this is ensured by adding 1.

[3] Always the case since M is either a prime or the product of two large primes.

Bounded Inputs. If the input is of bounded size, $[\![x]\!] < 2^\ell$, and $2^{\ell+k+\log n} \ll M$ where k is a statistical security parameter, the following variation is possible: Each party P_j inputs a uniformly random k-bit value, $r^{(j)}$, and the n parties jointly generate ℓ random bits, $[\![r_i]\!]$, using RandBit. The ABB then computes $[\![r]\!] \leftarrow \sum_{j=1}^n [\![r^{(j)}]\!] \cdot 2^\ell + (\sum_{i=0}^{\ell-1} 2^i [\![r_i]\!])$. Here, r statistically masks x: $m \bmod 2^\ell$ is uniformly random, while a single $r^{(j)}$ masks the ℓ'th carrybit of the addition, $x + r$, i.e., $\lfloor m/2^\ell \rfloor$ is statistically indistinguishable from a sum of uniformly random k-bit values plus the r_i of malicious parties. Testing equality between $r \bmod 2^\ell$ and $m \bmod 2^\ell$ is sufficient; note that this zero-test allows equality testing even when the difference between the inputs is negative.

Theorem 1. *Given two ℓ-bit values $[\![x]\!]$ and $[\![y]\!]$ stored in an n-party arithmetic black-box for $\mathbb{Z}_M$ augmented with a proof of boundedness, equality may be computed with 2 outputs and 1 ABB-multiplication in the online phase and $O(\ell)$ operations overall. This is the case both when $\ell = \ell_M$ as well as when $2^{\ell+k+\log n} \ll M$, where k is a statistical security parameter.*

3.2   Equality from DIE

We utilize the DIE protocol in the ABB model to construct a statistically correct zero test (and hence an equality test) in the presence of mutually incorruptible parties, denoted Alice and Bob. Complexity linear in the correctness parameter, κ, i.e., it is only useful when the input is of greater bitlength, say $\ell = 1000$ and $\kappa = 80$. For the sake of concreteness, we describe the case where M is composite.

The idea is to transform $[\![x]\!]$, $x \in \{0,1\}^\ell$, to $[\![y]\!]$, where $y = 0$ when $x = 0$, and y is $(1 - \varepsilon)$-close to uniformly random, for an exponentially small ε, when $x \neq 0$. Note that here we use the security parameter κ as the bitlength in the DIE protocol. (See Sect. 2 for the explanation of $\varepsilon = 2^{\kappa-1}/spf(M)$.) The value y is then used to "mask" $t \cdot 2^\kappa + \beta$, i.e., disclose it when $x = 0$ and hide it otherwise. The value revealed to Alice is always statistically close to uniformly random, hence reducing it modulo 2^κ and testing equality with β provides a zero test with a probability of failure of $2^{-\kappa}$. Details are seen as Protocol 2, where eq denotes the equality test from Sect. 3.1 but for κ-bit inputs. We focus on the case when M is an RSA-modulus and limit the description to the two-party case. The main benefit of this combined protocol is that by combining it with the equality test above replaces the $O(\ell)$ offline computation/communication with $O(\kappa)$ offline computation/communication. As a drawback, it requires two mutually incorruptible parties and has only has statistical (not perfect) correctness.

Correctness. When $x = 0$, we have $m = t \cdot 2^\kappa + \beta$ and therefore $\tilde{m} = \beta$. Thus, the final equality test correctly determines equality with 0. When $x \neq 0$, $[\![x]\!] \cdot [\![r]\!]$ is $(1 - \varepsilon)$-close to uniformly random since $[\![r]\!]$ is generated using RandElem and therefore guaranteed to be uniformly random. This implies that m is statistically close to uniformly random, independently of $t \cdot 2^\kappa + \beta$. Thus, m reveals statistically almost no information about β. We remark that the ABB must verify not only

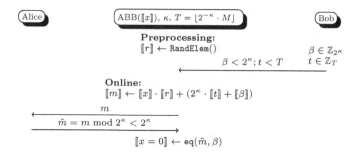

Protocol 2. $\mathsf{eq}_{\mathrm{DIE}}$, secure zero-testing based on disclose-if-equal

$\tilde{m} < 2^\kappa$, but also that $\tilde{m} = m \bmod 2^\kappa$. This can be done by providing not only $\tilde{m} = m \bmod 2^\kappa < 2^\kappa$, but also $\lfloor m/2^\kappa \rfloor < \lfloor M/2^\kappa \rfloor$, and verifying that $m = \lfloor m/2^\kappa \rfloor \cdot 2^\kappa + \tilde{m}$ (e.g., by outputting the difference).

Privacy. A corrupt Bob receives no outputs from the ABB, hence simulation is trivial: do nothing. For a corrupt Alice, note that the only value leaving the ABB is m, hence this is the only possible information leak. Since Bob is honest, $t \cdot 2^\kappa + \beta$ is chosen correctly, thus, no matter the value of x, m will be statistically close to uniformly random – either due to Bob's random choice or the addition of $x \cdot r$. Hence, simulation will consist of a uniformly random element.

Complexity. The protocol consists of one random element generation, three inputs, and one output plus the invocation of eq. Using eq_H of the previous subsection, implies $O(\kappa)$ work overall and $O(1)$ (but a slightly worse constant) work online. We state the following theorem:

Theorem 2. *Given two ℓ-bit values $[\![x]\!]$ and $[\![y]\!]$ stored in an n-party arithmetic black-box for $\mathbb{Z}_M$ augmented with a proof of boundedness, equality may be computed with 3 outputs and 2 ABB-multiplication in the online phase and $O(\kappa)$ operations overall when two mutually incorruptible parties are present.*

4 Greater-Than with Sublinear online Complexity

Toft [16] recently introduced the first sublinear GT protocols, i.e., protocols computing $[\![x \geq y]\!]$ from $[\![x]\!]$ and $[\![y]\!]$. Utilizing eq_H and $\mathsf{eq}_{\mathrm{DIE}}$ from Sect. 3, we propose two different (and orthogonal) improvements: (i) We can eliminate the need for two mutually incorruptible parties; this comes at the cost of linear preprocessing, or (ii) We improve efficiency when two mutually incorruptible parties exist by an order of magnitude. Similarly to [16] we assume $2^{\ell+k+\log n} \ll M$, where k is a statistical security parameter and n the number of parties.

The overall idea behind Toft's construction is to perform a GT-test through $\log \ell$ equality tests: If the $\ell/2$ most significant bits of $[\![x]\!]$ and $[\![y]\!]$ differ then

$\boxed{\text{Alice}}$ $\qquad\qquad$ $\boxed{\text{ABB}(\llbracket x\rrbracket, \llbracket y\rrbracket, \ell)}$ $\qquad\qquad$ $\boxed{\text{Bob}}$

if $\ell = 1$ then return
$(\llbracket e_1^{(\ell)}\rrbracket, \ldots, \llbracket e_{S_e}^{(\ell)}\rrbracket) \leftarrow \text{eq}_{(\ell/2),\text{preproc}};$
for $i \leftarrow 0$ to $\ell - 1$ do $\llbracket r_i\rrbracket \leftarrow \text{RandBit}$
$\llbracket r_\perp^{(\ell)}\rrbracket \leftarrow \sum_{i=0}^{\ell/2-1} 2^i \llbracket r_i\rrbracket;$
$\llbracket r_\top^{(\ell)}\rrbracket \leftarrow \sum_{i=0}^{\ell/2-1} 2^i \llbracket r_{i+\ell/2}\rrbracket;$

$r^{(A,\ell)} \leftarrow \mathbb{Z}_{2^k}$ $\qquad\qquad\qquad\qquad\qquad\qquad\qquad\qquad\qquad$ $r^{(B,\ell)} \leftarrow \mathbb{Z}_{2^k}$

$\overset{r^{(A,\ell)} < 2^k}{\underset{\phantom{r^{(A,\ell)} < 2^k}}{\longrightarrow}} \cdot \overset{r^{(B,\ell)} < 2^k}{\underset{\phantom{r^{(B,\ell)} < 2^k}}{\longleftarrow}}$

$\llbracket R^{(\ell)}\rrbracket \leftarrow 2^\ell(\llbracket r^{(A,\ell)}\rrbracket + \llbracket r^{(B,\ell)}\rrbracket) + 2^{\ell/2}\llbracket r_\top^{(\ell)}\rrbracket + \llbracket r_\perp^{(\ell)}\rrbracket$
$(\llbracket g_1^{(\ell)}\rrbracket, \ldots, \llbracket g_{S_g}^{(\ell)}\rrbracket) \leftarrow \text{gt}_{(\ell/2),\text{log,preproc}}$

Protocol 3. $\text{gt}_{(\ell),\text{log,preproc}}$: Preprocessing for the secure, ℓ-bit GT test, $\text{gt}_{(\ell),\text{log}}$

ignore the $\ell/2$ least significant ones; if they are equal then continue with the $\ell/2$ least significant ones. (This description is not correct, but provides sufficient intuition at this point.)

4.1 Sublinear Online Communication in the ABB Model

The main idea of the construction, $\text{gt}_{(\ell),\text{log}} = (\text{gt}_{(\ell),\text{log,preproc}}, \text{gt}_{(\ell),\text{log,online}})$, for comparing ℓ-bit numbers is to take the two mutually incorruptible parties of [16] and implement one using the ABB and executing the other "publicly." The core task of the ABB-party is then to generate appropriately distributed random values. Letting $\text{eq}_{(\ell)} = (\text{eq}_{(\ell),\text{preproc}}, \text{eq}_{(\ell),\text{online}})$ denote an equality test for ℓ-bit numbers (and its offline and online phases), preprocessing consists of invoking $\text{eq}_{(2^j),\text{preproc}}$ as well as generating $\log \ell$ random values $\llbracket R^{(2^j)}\rrbracket \leftarrow 2^{(2^j)}(\sum_{i=1}^n \llbracket r^{(i,2^j)}\rrbracket) + 2^{(2^{j-1})}\llbracket r_\top^{(2^j)}\rrbracket + \llbracket r_\perp^{(2^j)}\rrbracket$ for $j \in \{1, \ldots, \log \ell\}$, where $\llbracket r^{(i,2^j)}\rrbracket$ is a uniformly random k-bit number supplied by P_i and $\llbracket r_\top^{(2^j)}\rrbracket, \llbracket r_\perp^{(2^j)}\rrbracket$ are uniformly random 2^{j-1}-bit values unknown to all. Details are seen as Protocol 3.

The online phase of the construction is seen as Protocol 4 and explained in the correctness argument below. For clarity, we include preprocessed values implicitly in invocations of subprotocols.

Correctness. Correctness is immediate for single bit inputs: $1 - y + x \cdot y$ is 1 exactly when $x \geq y$. For $\ell > 1$, the goal is to transform the comparison of ℓ-bit integers to a comparison of $\ell/2$-bit integers. Observe that $\lfloor z/2^\ell \rfloor$ equals the desired result and that this can be computed as $2^{-\ell}(\llbracket z\rrbracket - \llbracket z \bmod 2^\ell\rrbracket)$. Further, since $2^{\ell+k+\log n} \ll M$, we have $z \bmod 2^\ell \equiv m - r \bmod 2^\ell$. We reduce m and $\llbracket r\rrbracket$ before the subtraction, which ensures that the result lies between -2^ℓ and 2^ℓ. The correct result is obtained by adding 2^ℓ when this is negative, i.e., when $\llbracket r \bmod 2^\ell\rrbracket > m \bmod 2^\ell$. The latter implies $\llbracket f\rrbracket = 1$ since we recursively compare the $\ell/2$ most- or least-significant bits of $\llbracket r \bmod 2^\ell\rrbracket$ and $m \bmod 2^\ell$ depending on whether the $\ell/2$ most significant bits differed.

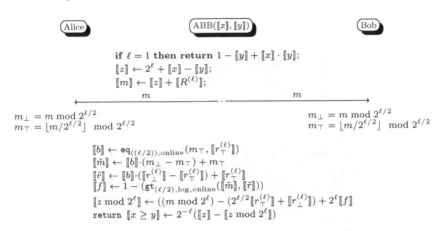

Protocol 4. $\mathsf{gt}_{(\ell),\log,\text{online}}$: Online phase of the secure, ℓ-bit GT test, $\mathsf{gt}_{(\ell),\log}$

Privacy. In each recursive call, $m = z + r$ is revealed, but this is statistically indistinguishable from a random value distributed as r – as above r statistically masks z as the bit-length is (at least) k bits longer; for honest ith party P_i, $2^\ell \cdot r^{(i)} + 2^{\ell/2} \cdot r_\top + r_\perp$ is uniformly random.

Complexity. Preprocessing requires $O(\ell)$ work: Though there is a logarithmic number of rounds, each one deals with a problem of half size. Hence, the combined $r_\top$, $r_\perp$ and random masks for $\mathsf{eq}_{H,(\cdot)}$ are only $O(\ell)$ bits overall. Round complexity is $O(1)$, as the iterations can be preprocessed in parallel.

Each online iteration (for $j \in \{1,\dots,2^\ell\}$) requires an output (m) and an execution of $\mathsf{eq}_{(2^j),\text{online}}$. Additionally, an ABB-multiplication is used to copy the most significant differing halves (if these exist). The remaining computation is purely local or in the form of ABB-additions. Thus, the overall complexity is $O(\log \ell)$ given that $\mathsf{eq}_{(\cdot),\text{online}}$ requires a constant number of ABB operations.

Implementing eq as $\mathsf{eq}_{H,(\cdot)}$, the above results in a protocol with $3 \log \ell$ outputs and $2 \log \ell + 1$ ABB-multiplications online; three outputs and two ABB-multiplications per iteration and a single secure ABB-multiplication in the final, single-bit comparison. We state the following theorem:

Theorem 3. *Given two ℓ-bit values $[x]$ and $[y]$ stored in an n-party arithmetic black-box for $\mathbb{Z}_M$ augmented with a proof of boundedness, greater-than may be computed with $3 \log \ell$ outputs and $2 \log \ell + 1$ ABB-multiplications in the online phase when $2^{\ell+k+\log n} \ll M$, where k is a statistical security parameter.*

We may adapt the constant-rounds protocol of [16] to the present setting. Sketching the solution, let c be a (constant) integer and split $m \bmod 2^\ell$ into $\ell^{1/c}$ strings of $\ell^{1-1/c}$ length. The most significant differing strings may be determined using $O(\ell^{1/c})$ equality tests and arithmetic; these are then compared recursively. Overall this requires c iterations and $O(c \cdot \ell^{1/c})$ equality tests and ABB-multiplications/outputs.

Theorem 4. *Given two ℓ-bit values $[\![x]\!]$ and $[\![y]\!]$ stored in an n-party arithmetic black-box for $\mathbb{Z}_M$ augmented with a proof of boundedness, greater-than may be computed with $O(c \cdot \ell^{1/c})$ ABB operations in $O(c)$ rounds in the online phase when $2^{\ell+k+\log n} \ll M$, where k is a security parameter.*

4.2   Sublinear, DIE-Based Greater-Than

$\mathsf{eq}_{\mathrm{DIE},(\cdot)}$ is much more efficient than the equality test used in [16]. Thus, combining this with Toft's original protocol[4] improves practical efficiency and reduces the theoretical online complexity – $O(\log \ell)$ rounds and work online and[5] $O(\log \ell(\kappa + \log\log \ell))$ ABB-operations overall. The constant-rounds protocol may also be combined with $\mathsf{eq}_{\mathrm{DIE}}$ resulting in an $O(c)$ rounds protocol with $O(c \cdot \ell^{1/c})$ work online and $O(\ell^{1/c}(\kappa + \log \ell))$ work overall. We state the following theorems:

Theorem 5. *Given two ℓ-bit values $[\![x]\!]$ and $[\![y]\!]$ stored in an n-party arithmetic black-box for $\mathbb{Z}_M$ augmented with a proof of boundedness, GT may be computed in the presence of two mutually incorruptible parties with $4 \log \ell$ outputs and $3 \log \ell + 1$ ABB-multiplications in the online phase and $O(\log \ell(\kappa + \log\log \ell))$ operations overall when $2^{\ell+k+\log n} \ll M$, where k is a statistical security parameter.*

Theorem 6. *Given two ℓ-bit values $[\![x]\!]$ and $[\![y]\!]$ stored in an n-party arithmetic black-box for $\mathbb{Z}_M$ augmented with a proof of boundedness, greater-than may be computed in the presence of two mutually incorruptible parties with $O(c \cdot \ell^{1/c})$ ABB-operations in $O(c)$ rounds in the online phase and $O(\ell^{1/c}(\kappa + \log \ell))$ operations overall when $2^{\ell+k+\log n} \ll M$, where k is a statistical security parameter.*

Acknowledgements. The first author was supported by the Estonian Research Council, and European Union through the European Regional Development Fund. The second author was supported by COBE financed by "The Danish Agency for Science, technology and Innovation." Additional support from the Danish National Research Foundation and The National Science Foundation of China (under the grant 61061130540) for the Sino-Danish Center for the Theory of Interactive Computation.

References

1. Aiello, W., Ishai, Y., Reingold, O.: Priced Oblivious Transfer: How to Sell Digital Goods. In: Pfitzmann, B. (ed.) EUROCRYPT 2001. LNCS, vol. 2045, pp. 119–135. Springer, Heidelberg (2001)
2. Bar-Ilan, J., Beaver, D.: Non-Cryptographic Fault-Tolerant Computing in a Constant Number of Rounds of Interaction. In: Rudnicki, P. (ed.) PODC 1989, pp. 201–209. ACM Press (1989)

---

[4] The key difference from Protocol 4 is that Bob selects r, while only Alice learns m.
[5] We add $\log\log \ell$ to κ to compensate for a non-constant number of equality tests.

3. Ben-Or, M., Goldwasser, S., Wigderson, A.: Completeness Theorems for Non-Cryptographic Fault-Tolerant Distributed Computation. In: STOC 1988, pp. 1–10. ACM Press (1988)
4. Bogetoft, P., Christensen, D.L., Damgård, I., Geisler, M., Jakobsen, T., Krøigaard, M., Nielsen, J.D., Nielsen, J.B., Nielsen, K., Pagter, J., Schwartzbach, M., Toft, T.: Secure Multiparty Computation Goes Live. In: Dingledine, R., Golle, P. (eds.) FC 2009. LNCS, vol. 5628, pp. 325–343. Springer, Heidelberg (2009)
5. Chaabouni, R., Lipmaa, H., Zhang, B.: A Non-interactive Range Proof with Constant Communication. In: Keromytis, A.D. (ed.) FC 2012. LNCS, vol. 7397, pp. 179–199. Springer, Heidelberg (2012)
6. Damgård, I.B., Fitzi, M., Kiltz, E., Nielsen, J.B., Toft, T.: Unconditionally Secure Constant-Rounds Multi-party Computation for Equality, Comparison, Bits and Exponentiation. In: Halevi, S., Rabin, T. (eds.) TCC 2006. LNCS, vol. 3876, pp. 285–304. Springer, Heidelberg (2006)
7. Damgård, I.B., Nielsen, J.B.: Universally Composable Efficient Multiparty Computation from Threshold Homomorphic Encryption. In: Boneh, D. (ed.) CRYPTO 2003. LNCS, vol. 2729, pp. 247–264. Springer, Heidelberg (2003)
8. Groth, J.: Short Pairing-Based Non-interactive Zero-Knowledge Arguments. In: Abe, M. (ed.) ASIACRYPT 2010. LNCS, vol. 6477, pp. 321–340. Springer, Heidelberg (2010)
9. Laur, S., Lipmaa, H.: A New Protocol for Conditional Disclosure of Secrets and Its Applications. In: Katz, J., Yung, M. (eds.) ACNS 2007. LNCS, vol. 4521, pp. 207–225. Springer, Heidelberg (2007)
10. Lipmaa, H.: On Diophantine Complexity and Statistical Zero-Knowledge Arguments. In: Laih, C.-S. (ed.) ASIACRYPT 2003. LNCS, vol. 2894, pp. 398–415. Springer, Heidelberg (2003)
11. Lipmaa, H.: Progression-Free Sets and Sublinear Pairing-Based Non-Interactive Zero-Knowledge Arguments. In: Cramer, R. (ed.) TCC 2012. LNCS, vol. 7194, pp. 169–189. Springer, Heidelberg (2012)
12. Naor, M., Pinkas, B.: Oblivious Transfer and Polynomial Evaluation. In: STOC 1999, pp. 245–254. ACM Press (1999)
13. Nishide, T., Ohta, K.: Multiparty Computation for Interval, Equality, and Comparison Without Bit-Decomposition Protocol. In: Okamoto, T., Wang, X. (eds.) PKC 2007. LNCS, vol. 4450, pp. 343–360. Springer, Heidelberg (2007)
14. Paillier, P.: Public-Key Cryptosystems Based on Composite Degree Residuosity Classes. In: Stern, J. (ed.) EUROCRYPT 1999. LNCS, vol. 1592, pp. 223–238. Springer, Heidelberg (1999)
15. Thorbek, R.: Linear Integer Secret Sharing. Ph.D. thesis, Aarhus University (2009)
16. Toft, T.: Sub-linear, Secure Comparison with Two Non-colluding Parties. In: Catalano, D., Fazio, N., Gennaro, R., Nicolosi, A. (eds.) PKC 2011. LNCS, vol. 6571, pp. 174–191. Springer, Heidelberg (2011)
17. Toft, T.: Primitives and Applications for Multiparty Computation. Ph.D. thesis, Aarhus University (2007)
18. Yu, C.H.: Sign Modules in Secure Arithmetic Circuits. Tech. Rep. 2011/539, IACR (October 1, 2011), http://eprint.iacr.org/2011/539 (checked in February 2013)

Temporal Network Optimization
Subject to Connectivity Constraints*

George B. Mertzios[1], Othon Michail[2],
Ioannis Chatzigiannakis[2], and Paul G. Spirakis[2,3]

[1] School of Engineering and Computing Sciences, Durham University, UK
[2] Computer Technology Institute & Press "Diophantus" (CTI), Patras, Greece
[3] Department of Computer Science, University of Liverpool, UK
george.mertzios@durham.ac.uk, {michailo,ichatz,spirakis}@cti.gr

Abstract. In this work we consider *temporal networks*, i.e. networks defined by a *labeling* λ assigning to each edge of an *underlying graph G* a set of *discrete* time-labels. The labels of an edge, which are natural numbers, indicate the discrete time moments at which the edge is available. We focus on *path problems* of temporal networks. In particular, we consider *time-respecting* paths, i.e. paths whose edges are assigned by λ a strictly increasing sequence of labels. We begin by giving two efficient algorithms for computing shortest time-respecting paths on a temporal network. We then prove that there is a *natural analogue of Menger's theorem* holding for arbitrary temporal networks. Finally, we propose two *cost minimization parameters* for temporal network design. One is the *temporality* of G, in which the goal is to minimize the maximum number of labels of an edge, and the other is the *temporal cost* of G, in which the goal is to minimize the total number of labels used. Optimization of these parameters is performed subject to some *connectivity constraint*. We prove several lower and upper bounds for the temporality and the temporal cost of some very basic graph families such as rings, directed acyclic graphs, and trees.

1 Introduction

A *temporal* (or *dynamic*) *network* is, loosely speaking, a network that changes with time. This notion encloses a great variety of both modern and traditional networks such as information and communication networks, social networks, transportation networks, and several physical systems.

In this work, embarking from the foundational work of Kempe *et al.* [KKK00], we consider *discrete time*, that is, we consider networks in which changes occur at discrete moments in time, e.g. days. This choice is not only a very natural

---

* Supported in part by (i) the project FOCUS implemented under the "ARISTEIA" Action of the OP "Education and Lifelong Learning" and co-funded by the EU (ESF) and Greek National Resources, (ii) the FET EU IP project MULTIPLEX under contract no 317532, and (iii) the EPSRC Grant EP/G043434/1. Full version: http://ru1.cti.gr/aigaion/?page=publication&kind=single&ID=977

F.V. Fomin et al. (Eds.): ICALP 2013, Part II, LNCS 7966, pp. 657–668, 2013.

abstraction of many real systems but also gives to the resulting models a purely combinatorial flavor. In particular, we consider those networks that can be described via an underlying graph G and a labeling λ assigning to each edge of G a (possibly empty) set of discrete labels. Note that this is a generalization of the single-label-per-edge model used in [KKK00], as we allow many time-labels to appear on an edge. These labels are drawn from the natural numbers and indicate the discrete moments in time at which the corresponding connection is available. For example, in the case of a communication network, availability of a communication link at some time t may mean that a communication protocol is allowed to transmit a data packet over that link at time t.

In this work, we initiate the study of the following fundamental network design problem: "*Given an underlying (di)graph G, assign labels to the edges of G so that the resulting temporal graph $\lambda(G)$ minimizes some parameter while satisfying some connectivity property*". In particular, we consider two cost optimization parameters for a given graph G. The first one, called *temporality* of G, measures the maximum number of labels that an edge of G has been assigned. The second one, called *temporal cost* of G, measures the total number of labels that have been assigned to all edges of G (i.e. if $|\lambda(e)|$ denotes the number of labels assigned to edge e, we are interested in $\sum_{e \in E} |\lambda(e)|$). Each of these two cost measures can be minimized subject to some particular connectivity property $\mathcal{P}$ that the temporal graph $\lambda(G)$ has to satisfy. In this work, we consider two very basic connectivity properties. The first one, that we call the *all paths* property, requires the temporal graph to preserve every simple path of its underlying graph, where by "preserve a path of G" we mean that the labeling should provide at least one strictly increasing sequence of labels on the edges of that path (we also call such a path *time-respecting*).

For an illustration, consider a directed ring $u_1, u_2, \ldots, u_n$. We want to determine the temporality of the ring subject to the *all paths* property, that is, we want to find a labeling λ that preserves every simple path of the ring and at the same time minimizes the maximum number of labels of an edge. Consider the paths $P_1 = (u_1, \ldots, u_n)$ and $P_2 = (u_{n-1}, u_n, u_1, u_2)$. It is immediate to observe that an increasing sequence of labels on the edges of path P_1 implies a decreasing pair of labels on edges (u_{n-1}, u_n) and (u_1, u_2). On the other hand, path P_2 uses first (u_{n-1}, u_n) and then (u_1, u_2) thus it requires an increasing pair of labels on these edges. It follows that in order to preserve both P_1 and P_2 we have to use a second label on at least one of these two edges, thus the temporality is at least 2. Next, consider the labeling that assigns to each edge (u_i, u_{i+1}) the labels $\{i, n+i\}$, where $1 \leq i \leq n$ and $u_{n+1} = u_1$. It is not hard to see that this labeling preserves all simple paths of the ring. Since the maximum number of labels that it assigns to an edge is 2, we conclude that the temporality is also at most 2. In summary, the temporality of preserving all simple paths of a directed ring is 2.

The other connectivity property that we define, called the *reach* property, requires the temporal graph to preserve a path from node u to node v whenever v is reachable from u in the underlying graph. Furthermore, the minimization of each of our two cost measures can be affected by some problem-specific constraints on the labels that we are allowed to use. We consider here one of the most natural constraints, namely an upper bound of the *age* of the constructed labeling λ, where the age of a labeling λ is defined to be equal to the maximum label of λ minus its minimum label plus 1. Now the goal is to minimize the cost parameter, e.g. the temporality, satisfy the connectivity property, e.g. *all paths*, and additionally guarantee that the age does not exceed some given natural k. Returning to the ring example, it is not hard to see, that if we additionally restrict the age to be at most $n-1$ then we can no longer preserve all paths of a ring using at most 2 labels per edge. In fact, we must now necessarily use the worst possible number of labels, i.e. $n-1$ on every edge.

Minimizing such parameters may be crucial as, in most real networks, making a connection available and maintaining its availability does not come for free. At the same time, such a study is important from a purely graph-theoretic perspective as it gives some first insight into the structure of specific families of temporal graphs (e.g. *no temporal ring exists with fewer than $n+1$ labels*). Finally, we believe that our results are a first step towards answering the following fundamental question: "*To what extent can algorithmic and structural results of graph theory be carried over to temporal graphs?*". For example, is there an analogue of Menger's theorem for temporal graphs? One of the results of the present work is an affirmative answer to the latter question.

1.1   Related Work

Single-label Temporal Graphs and Menger's Theorem. The model of temporal graphs that we consider in this work is a direct extension of the single-label model studied in [Ber96] and [KKK00] to allow for many labels per edge. In [KKK00], Kempe *et al.*, among other things, proved that there is no analogue of Menger's theorem, at least in its original formulation, for arbitrary single-label temporal networks. In this work, we go a step ahead showing that if one reformulates Menger's theorem in a way that takes time into acount then a very natural temporal analogue of Menger's theorem is obtained. Furthermore, in the present work, we consider a path as *time-respecting* if its edges have strictly increasing labels and not non-decreasing as in the above papers.

Continuous Availabilities (Intervals). Some authors have naturally assumed that an edge may be available for continuous time-intervals. The techniques used there are quite different than those needed in the discrete case [XFJ03, FT98].

Distributed Computing on Dynamic Networks. A notable set of recent works has studied (distributed) computation in *worst-case* dynamic networks in which the topology may change arbitrarily from round to round (see e.g. [KLO10, MCS12]). Population protocols [AAD+06] and variants [MCS11a] are collections of passively mobile finite-state agents that compute something useful in the limit. Another interesting direction assumes random network dynamicity and the interest is on determining "good" properties of the dynamic network that hold with high probability and on designing protocols for distributed tasks [CMM+08, AKL08]. For introductory texts cf. [CFQS12, MCS11b, Sch02].

Distance Labeling. A distance labeling of a graph G is an assignment of unique labels to the vertices of G so that the distance between any two vertices can be inferred from their labels alone [GPPR01, KKKP04]. There, the labeling parameter to be minimized is the binary length of an appropriate distance encoding, which is different from our cost parameters.

1.2   Contribution

In Section 2, we formally define the model of temporal graphs under consideration and provide all further necessary definitions. In Section 3, we give two efficient algorithms for computing shortest time-respecting paths. Then in Section 4 we present an analogue of Menger's theorem which we prove valid for arbitrary temporal graphs. In the full paper, we also apply our Menger's analogue to substantially simplify the proof of a recent result on distributed token gathering. In Section 5, we formally define the temporality and temporal cost optimization metrics for temporal graphs. In Section 5.1, we provide several upper and lower bounds for the temporality of some fundamental graph families such as rings, directed acyclic graphs (DAGs), and trees, as well as an interesting trade-off between the temporality and the age of rings. Furthermore, we provide in Section 5.2 a generic method for computing a lower bound of the temporality of an arbitrary graph w.r.t. the *all paths* property, and we illustrate its usefulness in cliques and planar graphs. Finally, we consider in Section 5.3 the temporal cost of a digraph G w.r.t. the *reach* property, when additionally the age of the resulting labeling $\lambda(G)$ is restricted to be the smallest possible. We prove that this problem is **APX**-hard. To prove our claim, we first prove (which may be of interest in its own right) that the Max-XOR(3) problem is **APX**-hard via a PTAS reduction from Max-XOR. In Max-XOR(3) problem, we are given a 2-CNF formula ϕ, every literal of which appears in at most 3 clauses, and we want to compute the greatest number of clauses of ϕ that can be simultaneously XOR-satisfied. Then we provide a PTAS reduction from Max-XOR(3) to our temporal cost minimization problem. On the positive side, we provide an $(r(G)/n)$-factor approximation algorithm for the latter problem, where $r(G)$ denotes the total number of reachabilities in G.

2   Preliminaries

Given a (di)graph $G = (V, E)$, a *labeling* of G is a mapping $\lambda : E \rightarrow 2^{\mathbb{N}}$, that is, a labeling assigns to each edge of G a (possibly empty) set of natural numbers, called *labels*.

Definition 1. *Let G be a (di)graph and λ be a labeling of G. Then $\lambda(G)$ is the temporal graph (or dynamic graph) of G with respect to λ. Furthermore, G is the underlying graph of $\lambda(G)$.*

We denote by $\lambda(E)$ the multiset of all labels assigned to the underlying graph by the labeling λ and by $|\lambda| = |\lambda(E)|$ their cardinality (i.e. $|\lambda| = \sum_{e \in E} |\lambda(e)|$). We also denote by $\lambda_{\min} = \min\{l \in \lambda(E)\}$ the minimum label and by $\lambda_{\max} = \max\{l \in \lambda(E)\}$ the maximum label assigned by λ. We define the *age* of a temporal graph $\lambda(G)$ as $\alpha(\lambda) = \lambda_{\max} - \lambda_{\min} + 1$. Note that in case $\lambda_{\min} = 1$ then we have $\alpha(\lambda) = \lambda_{\max}$. For every graph G we denote by $\mathcal{L}_G$ the set of all possible labelings λ of G. Furthermore, for every $k \in \mathbb{N}$, we define $\mathcal{L}_{G,k} = \{\lambda \in \mathcal{L}_G : \alpha(\lambda) \le k\}$.

For every time $r \in \mathbb{N}$, we define the *rth instance of a temporal graph* $\lambda(G)$ as the static graph $\lambda(G,r) = (V, E(r))$, where $E(r) = \{e \in E : r \in \lambda(e)\}$ is the (possibly empty) set of all edges of the underlying graph G that are assigned label r by labeling λ. A temporal graph $\lambda(G)$ may be also viewed as a *sequence of static graphs* $(G_1, G_2, \ldots, G_{\alpha(\lambda)})$, where $G_i = \lambda(G, \lambda_{\min} + i - 1)$ for all $1 \le i \le \alpha(\lambda)$. Another, often convenient, representation of a temporal graph is the following.

Definition 2. *The* static expansion *of a temporal graph $\lambda(G)$ is a DAG $H = (S, A)$ defined as follows. If $V = \{u_1, u_2, \ldots, u_n\}$ then $S = \{u_{ij} : \lambda_{\min} - 1 \le i \le \lambda_{\max}, 1 \le j \le n\}$ and $A = \{(u_{(i-1)j}, u_{ij'}) : \text{if } j = j' \text{ or } (u_j, u'_j) \in E(i) \text{ for some } \lambda_{\min} \le i \le \lambda_{\max}\}$.*

A *journey* (or *time-respecting path*) J of a temporal graph $\lambda(G)$ is a path $(e_1, e_2, \ldots, e_k)$ of the underlying graph $G = (V, E)$, where $e_i \in E$, together with labels $l_1 < l_2 < \ldots < l_k$ such that $l_i \in \lambda(e_i)$ for all $1 \le i \le k$. In words, a journey is a path that uses strictly increasing edge-labels. If labeling λ defines a journey on some path P of G then we also say that λ *preserves* P. A natural notation for a journey is $(e_1, l_1), (e_2, l_2), \ldots, (e_k, l_k)$ where each (e_i, l_i) is called a *time-edge*. A (u, v)-journey J is called *foremost from time* $t \in \mathbb{N}$ if $l_1 \ge t$ and l_k is minimized. We say that a journey J *leaves from node u* (*arrives at*, resp.) *at time t* if (u, v, t) $((v, u, t)$, resp.) is a time-edge of J. Two journeys are called *out-disjoint* (*in-disjoint*, respectively) if they never leave from (arrive at, resp.) the same node at the same time. If, in addition to the labeling λ, a positive weight $w(e) > 0$ is assigned to every edge $e \in E$, then we get a *weighted* temporal graph. If this is the case, then a journey J is called *shortest* if it minimizes the sum of the weights of its edges.

Throughout the text, unless otherwise stated, we denote by n the number of nodes of (di)graphs and by $d(G)$ the diameter of a (di)graph G, that is, the length of the longest shortest path between any two nodes of G. Finally, by δ_u we denote the degree of a node $u \in V(G)$ (in case of an undirected graph G).

3   Journey Problems

Theorem 1. *Let $\lambda(G)$ be a temporal graph, $s \in V$ be a source node, and t_{start} a time s.t. $\lambda_{\min} \le t_{start} \le \lambda_{\max}$. There is an algorithm that correctly computes for all $w \in V \setminus \{s\}$ a foremost (s, w)-journey from time t_{start}. The running time of the algorithm is $O(n\alpha^3(\lambda) + |\lambda|)$.*

Theorem 2. *Let $\lambda(G)$ be a weighted temporal graph and let $s, t \in V$. Assume also that $|\lambda(e)| = 1$ for all $e \in E$. Then, we can compute a shortest journey J between s and t in $\lambda(G)$ (or report that no such journey exists) in $O(m \log m + \sum_{v \in V} \delta_v^2) = O(n^3)$ time, where $m = |E|$.*

4   A Menger's Analogue for Temporal Graphs

In this section, we prove that, in contrast to an important negative result from [KKK00], there is a natural analogue of Menger's theorem that is valid for all temporal networks. In the full paper, we also apply our theorem to substantially simplify the proof of a recent token gathering result.

When we say that we remove *node departure time* (u, t) we mean that we remove *all edges leaving u at time t*, i.e. we remove the set $\{(u, v) \in E : t \in \lambda(u, v)\}$. So, when we ask how many node departure times are needed to separate two nodes s and v we mean how many node departure times must be selected so that after the removal of all the corresponding time-edges the resulting temporal graph has no (s, v)-journey.

Theorem 3 (Menger's Temporal Analogue). *Take any temporal graph $\lambda(G)$, where $G = (V, E)$, with two distinguished nodes s and v. The maximum number of out-disjoint journeys from s to v is equal to the minimum number of node departure times needed to separate s from v.*

Proof. Assume, in order to simplify notation, that $\lambda_{\min} = 1$. Take the static expansion $H = (S, A)$ of $\lambda(G)$. Let $\{u_{i1}\}$ and $\{u_{in}\}$ represent s and v over time, respectively (first and last colums, respectively), where $0 \leq i \leq \lambda_{\max}$. We extend H as follows. For each u_{ij}, $0 \leq i \leq \lambda_{\max} - 1$, with at least 2 outgoing edges to nodes different than $u_{(i+1)j}$, e.g. to nodes $u_{(i+1)j_1}, u_{(i+1)j_2}, \ldots, u_{(i+1)j_k}$, we add a new node w_{ij} and the edges (u_{ij}, w_{ij}) and $(w_{ij}, u_{(i+1)j_1}), (w_{ij}, u_{(i+1)j_2}), \ldots, (w_{ij}, u_{(i+1)j_k})$. We also define an edge capacity function $c : A \rightarrow \{1, \lambda_{\max}\}$ as follows. All edges of the form $(u_{ij}, u_{(i+1)j})$ take capacity $\lambda_{\max}$ and all other edges take capacity 1. We are interested in the maximum flow from u_{01} to $u_{\lambda_{\max}n}$. As this is simply a usual static flow network, the max-flow min-cut theorem applies stating that the maximum flow from u_{01} to $u_{\lambda_{\max}n}$ is equal to the minimum of the capacity of a cut separating u_{01} from $u_{\lambda_{\max}n}$. Finally, observe that (i) the maximum number of out-disjoint journeys from s to v is equal to the maximum flow from u_{01} to $u_{\lambda_{\max}n}$ and (ii) the minimum number of node departure times needed to separate s from v is equal to the minimum of the capacity of a cut separating u_{01} from $u_{\lambda_{\max}n}$.   □

5   Minimum Cost Temporal Connectivity

In this section, we introduce (in Definition 3) the *temporality* and *temporal cost* measures. These measures can be minimized subject to some particular connectivity property $\mathcal{P}$ that the labeled graph $\lambda(G)$ has to satisfy. For simplicity of notation, we consider the connectivity property $\mathcal{P}$ as a subset of the set $\mathcal{L}_G$ of all possible labelings λ on the (di)graph G. Furthermore, the minimization of each of these two cost measures can be affected by some problem-specific constraints on the labels that we are allowed to use. We consider one of the most natural constraints, namely an upper bound on the *age* of the constructed labeling.

Definition 3. *Let* $G = (V, E)$ *be a (di)graph,* $\alpha_{\max} \in \mathbb{N}$, *and* $\mathcal{P}$ *be a connectivity property. Then the* temporality *of* $(G, \mathcal{P}, \alpha_{\max})$ *is*

$$\tau(G, \mathcal{P}, \alpha_{\max}) = \min_{\lambda \in \mathcal{P} \cap \mathcal{L}_{G, \alpha_{\max}}} \max_{e \in E} |\lambda(e)|$$

and the temporal cost *of* $(G, \mathcal{P}, \alpha_{\max})$ *is*

$$\kappa(G, \mathcal{P}, \alpha_{\max}) = \min_{\lambda \in \mathcal{P} \cap \mathcal{L}_{G, \alpha_{\max}}} \sum_{e \in E} |\lambda(e)|$$

Furthermore $\tau(G, \mathcal{P}) = \tau(G, \mathcal{P}, \infty)$ *and* $\kappa(G, \mathcal{P}) = \kappa(G, \mathcal{P}, \infty)$.

Note that Definition 3 can be stated for an arbitrary property $\mathcal{P}$ of the labeled graph $\lambda(G)$ (e.g. some proper coloring-preserving property). Nevertheless, we only consider here $\mathcal{P}$ to be a connectivity property of $\lambda(G)$. In particular, we investigate the following two connectivity properties $\mathcal{P}$:

- $all\text{-}paths(G) = \{\lambda \in \mathcal{L}_G : \text{for all simple paths } P \text{ of } G, \lambda \text{ preserves } P\}$,
- $reach(G) = \{\lambda \in \mathcal{L}_G : \text{for all } u, v \in V \text{ where } v \text{ is reachable from } v \text{ in } G, \lambda$ preserves at least one simple path from u to $v\}$.

5.1 Basic Properties of Temporality Parameters

5.1.1 Preserving All Paths

We begin with some simple observations on $\tau(G, all\ paths)$. Recall that given a (di)graph G our goal is to label G so that all simple paths of G are preserved by using as few labels per edge as possible. First note that if $p(G)$ is the length of the longest path in G then $\tau(G, all\ paths) \leq p(G)$ for all graphs G: just give to every edge the labels $\{1, 2, \ldots, p(G)\}$.

A topological sort of a digraph G is a linear ordering of its nodes such that if G contains an edge (u, v) then u appears before v in the ordering. It is well known that a digraph G can be topologically sorted iff is a DAG.

Proposition 1. *If* G *is a DAG then* $\tau(G, all\ paths) = 1$.

Proof. Take a topological sort $u_1, u_2, \ldots, u_n$ of G. Give to every edge (u_i, u_j), where $i < j$, label i. $\square$

5.1.2 Preserving All Reachabilities

Now, instead of preserving all paths, we impose the apparently simpler requirement of preserving just a single path between every reachability pair $u, v \in V$. We claim that it is sufficient to understand how $\tau(G, reach)$, behaves on strongly connected digraphs. Let $\mathcal{C}(G)$ be the set of all strongly connected components of a digraph G. The following lemma proves that, w.r.t. the $reach$ property, the temporality of any digraph G is upper bounded by the maximum temporality of its components.

Lemma 1. $\tau(G, reach) \leq \max_{C \in \mathcal{C}(G)} \tau(C, reach)$ *for every digraph* G.

Lemma 1 implies that any upper bound on the temporality of preserving the reachabilities of strongly connected digraphs can be used as an upper bound on the temporality of preserving the reachabilities of general digraphs. An interesting question is whether there is some bound on $\tau(G, reach)$ either for all digraphs or for specific families of digraphs. By using Lemma 1, it can be proved that indeed there is a very satisfactory generic upper bound.

Theorem 4. $\tau(G, reach) \leq 2$ *for all digraphs* G.

5.1.3   Restricting the Age

Now notice that for all G we have $\tau(G, reach, d(G)) \leq d(G)$; recall that $d(G)$ denotes the diameter of (di)graph G. Indeed it suffices to label each edge by $\{1, 2, \ldots, d(G)\}$. Thus, a clique G has trivially $\tau(G, reach, d(G)) = 1$ as $d(G) = 1$ and we can only have large $\tau(G, reach, d(G))$ in graphs with large diameter. For example, a directed ring G of size n has $\tau(G, reach, d(G)) = n - 1$. Indeed, assume that from some edge e, label $1 \leq i \leq n - 1$ is missing. It is easy to see that there is some shortest path between two nodes of the ring that in order to arrive by time $n - 1$ must use edge e at time i. As this label is missing, it uses label $i+1$, thus it arrives by time n which is greater than the diameter. On a ring we can preserve the diameter only if all edges have the labels $\{1, 2, \ldots, n - 1\}$.

On the other hand, there are graphs with large diameter in which $\tau(G, reach, d(G))$ is small. This may also be the case even if G is strongly connected. For example, consider the graph with nodes $u_1, u_2, \ldots, u_n$ and edges (u_i, u_{i+1}) and (u_{i+1}, u_i) for all $1 \leq i \leq n - 1$. In words, we have a directed line from u_1 to u_n and an inverse one from u_n to u_1. The diameter here is $n - 1$ (e.g. the shortest path from u_1 to u_n) but $\tau(G, reach, d(G)) = 1$: simply label one path $1, 2, \ldots, n - 1$ and label the inverse one $1, 2, \ldots, n - 1$ again, i.e. give to edges (u_i, u_{i+1}) and (u_{n-i+1}, u_{n-i+2}) label i. Now consider an undirected tree T.

Theorem 5. *If* T *is an undirected tree then* $\tau(T, all\ paths, d(T)) \leq 2$.

We next present an interesting trade-off between the temporality and the age of a directed ring.

Theorem 6. *If* G *is a directed ring and* $\alpha = (n-1)+k$, *where* $1 \leq k \leq n-1$, *then* $\tau(G, all\ paths, \alpha) = \Theta(n/k)$ *and in particular* $\lfloor \frac{n-1}{k+1} \rfloor + 1 \leq \tau(G, all\ paths, \alpha) \leq \lceil \frac{n}{k+1} \rceil + 1$. *Moreover,* $\tau(G, all\ paths, n - 1) = n - 1$ *(i.e. when* $k = 0$*).*

5.2   A Generic Method for Lower Bounding Temporality

We show here that there are graphs G for which $\tau(G, all\ paths) = \Omega(p(G))$ (recall that $p(G)$ denotes the length of the longest path in G), that is graphs in which the optimum labeling, w.r.t. temporality, is very close to the trivial labeling $\lambda(e) = \{1, 2, \ldots, p(G)\}$, for all $e \in E$.

Definition 4. *Call a set* $K = \{e_1, e_2, \ldots, e_k\} \subseteq E(G)$ *of edges of a digraph* G *an* edge-kernel *if for every permutation* $\pi = (e_{i_1}, e_{i_2}, \ldots, e_{i_k})$ *of* K *there is a simple path of* G *that visits all edges of* K *in the ordering defined by* π.

The following theorem states that an edge-kernel of size k needs at least k labels on some edge(s).

Theorem 7 (Edge-Kernel Lower Bound). *If a digraph G contains an edge-kernel of size k then $\tau(G, \text{all paths}) \geq k$.*

The usefulness of Theorem 7 is that it allows us to establish a lower bound k on the temporality of a graph G by only proving the existence of an edge-kernel of size k in G. We now apply this to complete digraphs and planar graphs.

Lemma 2. *If G is a complete digraph of order n then it has an edge-kernel of size $\lfloor n/2 \rfloor$.*

Now Theorem 7 implies that if G is a complete digraph then $\lfloor n/2 \rfloor \leq \tau(G, \text{all paths}) \leq n - 1$.

Lemma 3. *There exist planar graphs having edge-kernels of size $\Omega(n^{\frac{1}{3}})$.*

5.3 Computing the Cost

5.3.1 Hardness of Approximation

Consider a boolean formula ϕ in conjunctive normal form with two literals in every clause (2-CNF). Let τ be a truth assignment of the variables of ϕ and $\alpha = (\ell_1 \vee \ell_2)$ be a clause of ϕ. Then α is *XOR-satisfied* in τ, if one of the literals $\{\ell_1, \ell_2\}$ of the clause α is true in τ and the other one is false in τ. The number of clauses of ϕ that are XOR-satisfied in τ is denoted by $|t(\phi)|$. The formula ϕ is *XOR-satisfiable* if there exists a truth assignment τ of ϕ such that every clause of ϕ is XOR-satisfied in τ. The *Max-XOR* problem is the following maximization problem: given a 2-CNF formula ϕ, compute the greatest number of clauses of ϕ that can be simultaneously XOR-satisfied in a truth assignment τ, i.e. compute the greatest value for $|t(\phi)|$. The *Max-XOR(k)* problem is the special case of the the Max-XOR problem, where every literal of the input formula ϕ appears in at most k clauses of ϕ. Max-XOR is known to be **APX**-hard, i.e. it does not admit a PTAS unless $\mathbf{P} = \mathbf{NP}$ [KMSV99, CKS01]. In the next lemma we prove that Max-XOR(3) remains **APX**-hard by providing a PTAS reduction from Max-XOR.

Lemma 4. *The Max-XOR(3) problem is **APX**-hard.*

Now we provide a reduction from the Max-XOR(3) problem to the problem of computing $\kappa(G, reach, d(G))$. Let ϕ be an instance formula of Max-XOR(3) with n variables $x_1, x_2, \ldots, x_n$ and m clauses. Since every variable x_i appears in ϕ (either as x_i or as $\overline{x_i}$) in at most 3 clauses, it follows that $m \leq \frac{3}{2}n$. We will construct from ϕ a graph G_ϕ having length of a directed cycle at most 2. Then, as we prove in Theorem 8, $\kappa(G_\phi, reach, d(G_\phi)) \leq 39n - 4m - 2k$ if and only if there exists a truth assignment τ of ϕ with $|t(\phi)| \geq k$, i.e. τ XOR-satisfies at least k clauses of ϕ. Since ϕ is an instance of Max-XOR(3), we can replace every clause $(\overline{x_i} \vee \overline{x_j})$ by the clause $(x_i \vee x_j)$ in ϕ, since $(\overline{x_i} \vee \overline{x_j}) = (x_i \vee x_j)$ in XOR. Furthermore, whenever $(\overline{x_i} \vee x_j)$ is a clause of ϕ, where $i < j$, we can replace this clause by $(x_i \vee \overline{x_j})$, since $(\overline{x_i} \vee x_j) = (x_i \vee \overline{x_j})$ in XOR. Thus, we can assume w.l.o.g. that every clause of ϕ is either of the form $(x_i \vee x_j)$ or $(x_i \vee \overline{x_j})$, $i < j$.

For every $i = 1, 2, \ldots, n$ we construct the graph $G_{\phi,i}$ of Figure 1. Note that the diameter of $G_{\phi,i}$ is $d(G_{\phi,i}) = 9$ and the maximum length of a directed cycle in $G_{\phi,i}$ is 2. In this figure, we call the induced subgraph of $G_{\phi,i}$ on the 13 vertices $\{s^{x_i}, u_1^{x_i}, \ldots, u_6^{x_i}, v_1^{x_i}, \ldots, v_6^{x_i}\}$ the *trunk* of $G_{\phi,i}$. Furthermore, for every $p \in \{1, 2, 3\}$, we call the induced subgraph of $G_{\phi,i}$ on the 5 vertices $\{u_{7,p}^{x_i}, u_{8,p}^{x_i}, v_{7,p}^{x_i}, v_{8,p}^{x_i}, t_p^{x_i},\}$ the *pth branch* of $G_{\phi,i}$. Finally, we call the edges $u_6^{x_i} u_{7,p}^{x_i}$ and $v_6^{x_i} v_{7,p}^{x_i}$ the *transition edges* of the *pth* branch of $G_{\phi,i}$. Furthermore, for every $i = 1, 2, \ldots, n$, let $r_i \leq 3$ be the number of clauses in which variable x_i appears in ϕ. For every $1 \leq p \leq r_i$, we assign the *pth* appearance of the variable x_i (either as x_i or as $\overline{x_i}$) in a clause of ϕ to the *pth* branch of $G_{\phi,i}$.

Consider now a clause $\alpha = (\ell_i \vee \ell_j)$ of ϕ, where $i < j$. Then, by our assumptions on ϕ, it follows that $\ell_i = x_i$ and $\ell_j \in \{x_j, \overline{x_j}\}$. Assume that the literal ℓ_i (resp. ℓ_j) of the clause α corresponds to the *pth* (resp. to the *qth*) appearance of the variable x_i (resp. x_j) in ϕ. Then we identify the vertices of the *pth* branch of $G_{\phi,i}$ with the vertices of the *qth* branch of $G_{\phi,j}$ as follows. If $\ell_j = x_j$ then we identify the vertices $u_{7,p}^{x_i}, u_{8,p}^{x_i}, v_{7,p}^{x_i}, v_{8,p}^{x_i}, t_p^{x_i}$ with the vertices $v_{7,q}^{x_j}, v_{8,q}^{x_j}, u_{7,q}^{x_j}, u_{8,q}^{x_j}, t_q^{x_j}$, respectively. Otherwise, if $\ell_j = \overline{x_j}$ then we identify the vertices $u_{7,p}^{x_i}, u_{8,p}^{x_i}, v_{7,p}^{x_i}, v_{8,p}^{x_i}, t_p^{x_i}$ with the vertices $u_{7,q}^{x_j}, u_{8,q}^{x_j}, v_{7,q}^{x_j}, v_{8,q}^{x_j}, t_q^{x_j}$, respectively. This completes the construction of the graph G_ϕ. Note that, similarly to the graphs $G_{\phi,i}$, $1 \leq i \leq n$, the diameter of G_ϕ is $d(G_\phi) = 9$ and the maximum length of a directed cycle in G_ϕ is 2. Furthermore, note that for each of the m clauses of ϕ, one branch of a gadget $G_{\phi,i}$ coincides with one branch of a gadget $G_{\phi,j}$, where $1 \leq i < j \leq n$, while every $G_{\phi,i}$ has three branches. Therefore G_ϕ has exactly $3n - 2m$ branches which belong to only one gadget $G_{\phi,i}$, and m branches that belong to two gadgets $G_{\phi,i}, G_{\phi,j}$.

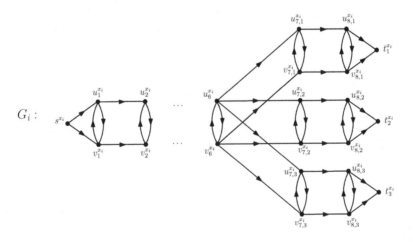

Fig. 1. The gadget $G_{\phi,i}$ for the variable x_i

Theorem 8. *There exists a truth assignment* τ *of* ϕ *with* $|t(\phi)| \geq k$ *if and only if* $\kappa(G_\phi, reach, d(G_\phi)) \leq 39n - 4m - 2k$.

Using Theorem 8, we are now ready to prove the main theorem of this section.

Theorem 9 (Hardness of Approximating the Temporal Cost). *The problem of computing* $\kappa(G, reach, d(G))$ *is* **APX**-*hard, even when the maximum length of a directed cycle in* G *is* 2.

Proof. Denote now by $\text{OPT}_{\text{Max-XOR(3)}}(\phi)$ the greatest number of clauses that can be simultaneously XOR-satisfied by a truth assignment of ϕ. Then Theorem 8 implies that

$$\kappa(G_\phi, reach, d(G_\phi)) \leq 39n - 4m - 2 \cdot \text{OPT}_{\text{Max-XOR(3)}}(\phi)$$

Note that a random assignment XOR-satisfies each clause of ϕ with probability $\frac{1}{2}$, and thus we can easily compute (even deterministically) an assignment τ that XOR-satisfies $\frac{m}{2}$ clauses of ϕ. Therefore $\text{OPT}_{\text{Max-XOR(3)}}(\phi) \geq \frac{m}{2}$, and thus, since every variable x_i appears in at least one clause of ϕ, it follows that $n \leq m \leq 2 \cdot \text{OPT}_{\text{Max-XOR(3)}}(\phi_1)$.

Assume that there is a PTAS for computing $\kappa(G, reach, d(G))$. Then, for every $\varepsilon > 0$ we can compute in polynomial time a labeling λ for the graph G_ϕ, such that $|\lambda| \leq (1 + \varepsilon) \cdot \kappa(G_\phi, reach, d(G_\phi))$.

Given such a labeling λ we can compute by the sufficiency part ($\Leftarrow$) of the proof of Theorem 8 a truth assignment τ of ϕ such that $39n - 4m - 2|t(\phi)| \leq |\lambda|$, i.e. $2|t(\phi)| \geq 39n - 4m - |\lambda|$.

Therefore it follows by all the above that $2|t(\phi)| \geq 39n - 4m - (1 + \varepsilon) \cdot \kappa(G_\phi, reach, d(G_\phi)) \geq 39n - 4m - (1 + \varepsilon) \cdot (39n - 4m - 2 \cdot \text{OPT}_{\text{Max-XOR(3)}}(\phi)) = \varepsilon (4m - 39n) + 2(1 + \varepsilon) \cdot \text{OPT}_{\text{Max-XOR(3)}}(\phi) \geq -35\varepsilon m + (2 + 2\varepsilon) \cdot \text{OPT}_{\text{Max-XOR(3)}}(\phi) \geq -35\varepsilon \cdot 2\text{OPT}_{\text{Max-XOR(3)}}(\phi) + (2 + 2\varepsilon) \cdot \text{OPT}_{\text{Max-XOR(3)}}(\phi) = (2 - 68\varepsilon) \cdot \text{OPT}_{\text{Max-XOR(3)}}(\phi)$ and thus

$$|t(\phi)| \geq (1 - 34\varepsilon) \cdot \text{OPT}_{\text{Max-XOR(3)}}(\phi).$$

That is, assuming a PTAS for computing $\kappa(G, reach, d(G))$, we obtain a PTAS for the Max-XOR(3) problem, which is a contradiction by Lemma 4. Therefore computing $\kappa(G, reach, d(G))$ is **APX**-hard. Finally, notice that the constructed graph G_ϕ has maximum length of a directed cycle at most 2. $\qquad\square$

5.3.2   Approximating the Cost

In this section, we provide an approximation algorithm for computing $\kappa(G, reach, d(G))$, which complements the hardness result of Theorem 9. Given a digraph G define, for every $u \in V$, u's reachability number $r(u) = |\{v \in V : v \text{ is reachable from } u\}|$ and $r(G) = \sum_{u \in V} r(u)$, that is $r(G)$ is the total number of reachabilities in G.

Theorem 10. *There is an* $\frac{r(G)}{n-1}$-*factor approximation algorithm for computing* $\kappa(G, reach, d(G))$ *on any weakly connected digraph* G.

References

[AAD+06]   Angluin, D., Aspnes, J., Diamadi, Z., Fischer, M.J., Peralta, R.: Computation in networks of passively mobile finite-state sensors. In: Distributed Computing, pp. 235–253 (March 2006)

[AKL08]   Avin, C., Koucký, M., Lotker, Z.: How to explore a fast-changing world (Cover time of a simple random walk on evolving graphs). In: Aceto, L., Damgård, I., Goldberg, L.A., Halldórsson, M.M., Ingólfsdóttir, A., Walukiewicz, I. (eds.) ICALP 2008, Part I. LNCS, vol. 5125, pp. 121–132. Springer, Heidelberg (2008)

[Ber96]   Berman, K.A.: Vulnerability of scheduled networks and a generalization of Menger's theorem. Networks 28(3), 125–134 (1996)

[CFQS12]   Casteigts, A., Flocchini, P., Quattrociocchi, W., Santoro, N.: Time-varying graphs and dynamic networks. IJPEDS 27(5), 387–408 (2012)

[CKS01]   Creignou, N., Khanna, S., Sudan, M.: Complexity classifications of boolean constraint satisfaction problems. SIAM Monographs on Discrete Mathematics and Applications (2001)

[CMM+08]   Clementi, A.E., Macci, C., Monti, A., Pasquale, F., Silvestri, R.: Flooding time in edge-markovian dynamic graphs. In: Proc. of the 27th ACM Symp. on Principles of Distributed Computing (PODC), pp. 213–222 (2008)

[FT98]   Fleischer, L., Tardos, É.: Efficient continuous-time dynamic network flow algorithms. Operations Research Letters 23(3), 71–80 (1998)

[GPPR01]   Gavoille, C., Peleg, D., Pérennes, S., Raz, R.: Distance labeling in graphs. In: Proc. of the 12th annual ACM-SIAM Symposium on Discrete Algorithms (SODA), Philadelphia, PA, USA, pp. 210–219 (2001)

[KKK00]   Kempe, D., Kleinberg, J., Kumar, A.: Connectivity and inference problems for temporal networks. In: Proceedings of the 32nd Annual ACM Symposium on Theory of Computing (STOC), pp. 504–513 (2000)

[KKKP04]   Katz, M., Katz, N.A., Korman, A., Peleg, D.: Labeling schemes for flow and connectivity. SIAM Journal on Computing 34(1), 23–40 (2004)

[KLO10]   Kuhn, F., Lynch, N., Oshman, R.: Distributed computation in dynamic networks. In: Proceedings of the 42nd ACM Symposium on Theory of Computing (STOC), pp. 513–522. ACM, New York (2010)

[KMSV99]   Khanna, S., Motwani, R., Sudan, M., Vazirani, U.: On syntactic versus computational views of approximability. SIAM Journal on Computing 28(1), 64–191 (1999)

[MCS11a]   Michail, O., Chatzigiannakis, I., Spirakis, P.G.: Mediated population protocols. Theoretical Computer Science 412(22), 2434–2450 (2011)

[MCS11b]   Michail, O., Chatzigiannakis, I., Spirakis, P.G.: New Models for Population Protocols. In: Lynch, N.A. (ed.) Synthesis Lectures on Distributed Computing Theory. Morgan & Claypool (2011)

[MCS12]   Michail, O., Chatzigiannakis, I., Spirakis, P.G.: Causality, influence, and computation in possibly disconnected synchronous dynamic networks. In: Baldoni, R., Flocchini, P., Binoy, R. (eds.) OPODIS 2012. LNCS, vol. 7702, pp. 269–283. Springer, Heidelberg (2012)

[Sch02]   Scheideler, C.: Models and techniques for communication in dynamic networks. In: Alt, H., Ferreira, A. (eds.) STACS 2002. LNCS, vol. 2285, pp. 27–49. Springer, Heidelberg (2002)

[XFJ03]   Xuan, B., Ferreira, A., Jarry, A.: Computing shortest, fastest, and foremost journeys in dynamic networks. International Journal of Foundations of Computer Science 14(02), 267–285 (2003)

Strong Bounds for Evolution in Networks*

George B. Mertzios[1] and Paul G. Spirakis[2,3]

[1] School of Engineering and Computing Sciences, Durham University, UK
[2] Department of Computer Science, University of Liverpool, UK
[3] Computer Technology Institute and University of Patras, Greece
george.mertzios@durham.ac.uk, spirakis@cti.gr

Abstract. This work extends what is known so far for a basic model of evolutionary antagonism in undirected networks (graphs). More specifically, this work studies the generalized Moran process, as introduced by Lieberman, Hauert, and Nowak [Nature, 433:312-316, 2005], where the individuals of a population reside on the vertices of an undirected connected graph. The initial population has a single *mutant* of a *fitness* value r (typically $r > 1$), residing at some vertex v of the graph, while every other vertex is initially occupied by an individual of fitness 1. At every step of this process, an individual (i.e. vertex) is randomly chosen for reproduction with probability proportional to its fitness, and then it places a copy of itself on a random neighbor, thus replacing the individual that was residing there. The main quantity of interest is the *fixation probability*, i.e. the probability that eventually the whole graph is occupied by descendants of the mutant. In this work we concentrate on the fixation probability when the mutant is initially on a specific vertex v, thus refining the older notion of Lieberman et al. which studied the fixation probability when the initial mutant is placed at a random vertex. We then aim at finding graphs that have many "strong starts" (or many "weak starts") for the mutant. Thus we introduce a parameterized notion of *selective amplifiers* (resp. *selective suppressors*) of evolution. We prove the existence of *strong* selective amplifiers (i.e. for $h(n) = \Theta(n)$ vertices v the fixation probability of v is at least $1 - \frac{c(r)}{n}$ for a function $c(r)$ that depends only on r), and the existence of quite strong selective suppressors. Regarding the traditional notion of fixation probability from a random start, we provide strong upper and lower bounds: first we demonstrate the non-existence of "strong universal" amplifiers, and second we prove the *Thermal Theorem* which states that for any undirected graph, when the mutant starts at vertex v, the fixation probability at least $(r-1)/(r + \frac{\deg v}{\deg_{\min}})$. This theorem (which extends the "Isothermal Theorem" of Lieberman et al. for regular graphs) implies an almost tight lower bound for the usual notion of fixation probability. Our proof techniques are original and are based on new domination arguments which may be of general interest in Markov Processes that are of the general birth-death type.

---

* This work was partially supported by (i) the FET EU IP Project MULTIPLEX (Contract no 317532), (ii) the ERC EU Grant ALGAME (Agreement no 321171), and (iii) the EPSRC Grant EP/G043434/1. The full version of this paper is available at http://arxiv.org/abs/1211.2384

F.V. Fomin et al. (Eds.): ICALP 2013, Part II, LNCS 7966, pp. 669–680, 2013.

1   Introduction

Population and evolutionary dynamics have been extensively studied [2, 6, 7, 15, 21, 24, 25], mainly on the assumption that the evolving population is homogeneous, i.e. it has no spatial structure. One of the main models in this area is the Moran Process [19], where the initial population contains a single *mutant* with fitness $r > 0$, with all other individuals having fitness 1. At every step of this process, an individual is chosen for reproduction with probability proportional to its fitness. This individual then replaces a second individual, which is chosen uniformly at random, with a copy of itself. Such dynamics as the above have been extensively studied also in the context of strategic interaction in evolutionary game theory [11–14, 23].

In a recent article, Lieberman, Hauert, and Nowak [16] (see also [20]) introduced a generalization of the Moran process, where the individuals of the population are placed on the vertices of a connected graph (which is, in general, directed) such that the edges of the graph determine competitive interaction. In the generalized Moran process, the initial population again consists of a single mutant of fitness r, placed on a vertex that is chosen uniformly at random, with each other vertex occupied by a non-mutant of fitness 1. An individual is chosen for reproduction exactly as in the standard Moran process, but now the second individual to be replaced is chosen among its neighbors in the graph uniformly at random (or according to some weights of the edges) [16, 20]. If the underlying graph is the complete graph, then this process becomes the standard Moran process on a homogeneous population [16, 20]. Several similar models describing infections and particle interactions have been also studied in the past, including the SIR and SIS epidemics [10, Chapter 21], the voter and antivoter models and the exclusion process [1, 9, 17]. However such models do not consider the issue of different fitness of the individuals.

The central question that emerges in the generalized Moran process is how the population structure affects evolutionary dynamics [16, 20]. In the present work we consider the generalized Moran process on arbitrary finite, undirected, and connected graphs. On such graphs, the generalized Moran process terminates almost surely, reaching either *fixation* of the graph (all vertices are occupied by copies of the mutant) or *extinction* of the mutants (no copy of the mutant remains). The *fixation probability* of a graph G for a mutant of fitness r, is the probability that eventually fixation is reached when the mutant is initially placed at a random vertex of G, and is denoted by $f_r(G)$. The fixation probability can, in principle, be determined using standard Markov Chain techniques. But doing so for a general graph on n vertices requires solving a linear system of 2^n linear equations. Such a task is not computationally feasible, even numerically. As a result of this, most previous work on computing fixation probabilities in the generalized Moran process was either restricted to graphs of small size [6] or to graph classes which have a high degree of symmetry, reducing thus the size of the corresponding linear system (e.g. paths, cycles, stars, and cliques [3–5]). Experimental results on the fixation probability of random graphs derived from grids can be found in [22].

A recent result [8] shows how to construct fully polynomial randomized approximation schemes (FPRAS) for the probability of reaching fixation (when $r \geq 1$) or extinction (for all $r > 0$). The result of [8] uses a Monte Carlo estimator, i.e. it runs the generalized Moran process several times[1], while each run terminates in polynomial time with high probability [8]. Note that improved lower and upper bounds on the fixation probability immediately lead to a better estimator here. Ontil now, the only known general bounds for the fixation probability on connected undirected graphs, are that $f_r(G) \geq \frac{1}{n}$ and $f_r(G) \leq 1 - \frac{1}{n+r}$.

Lieberman et al. [16,20] proved the *Isothermal Theorem*, stating that (in the case of undirected graphs) the fixation probability of a regular graph (i.e. of a graph with overall the same vertex degree) is equal to that of the complete graph (i.e. the homogeneous population of the standard Moran process), which equals to $(1 - \frac{1}{r})/(1 - \frac{1}{r^n})$, where n is the size of the population. Intuitively, in the Isothermal Theorem, every vertex of the graph has a *temperature* which determines how often this vertex is being replaced by other individuals during the generalized Moran process. The complete graph (or equivalently, any regular graph) serves as a benchmark for measuring the fixation probability of an arbitrary graph G: if $f_r(G)$ is larger (resp. smaller) than that of the complete graph then G is called an *amplifier* (resp. a *suppressor*) [16,20]. Until now only graphs with similar (i.e. a little larger or smaller) fixation probability than regular graphs have been identified [3–5,16,18], while no class of strong amplifiers/suppressors is known so far.

Our Contribution. The structure of the graph, on which the population resides, plays a crucial role in the course of evolutionary dynamics. Human societies or social networks are never homogeneous, while certain individuals in central positions may be more influential than others [20]. Motivated by this, we introduce in this paper a new notion of measuring the success of an advantageous mutant in a structured population, by counting the number of initial placements of the mutant in a graph that guarantee fixation of the graph with large probability. This provides a refinement of the notion of fixation probability. Specifically, we do not any more consider the fixation probability as the probability of reaching fixation when the mutant is placed at a random vertex, but we rather consider the probability $f_r(v)$ of reaching fixation when a mutant with fitness $r > 1$ is introduced at a specific vertex v of the graph; $f_r(v)$ is termed the *fixation probability of vertex v*. Using this notion, the fixation probability $f_r(G)$ of a graph $G = (V, E)$ with n vertices is $f_r(G) = \frac{1}{n} \sum_{v \in V} f_r(v)$.

We aim in finding graphs that have many "strong starts" (or many "weak starts") of the mutant. Thus we introduce the notions of $(h(n), g(n))$-*selective amplifiers* (resp. $(h(n), g(n))$-*selective suppressors*), which include those graphs with n vertices for which there exist at least $h(n)$ vertices v with $f_r(v) \geq 1 - \frac{c(r)}{g(n)}$ (resp. $f_r(v) \leq \frac{c(r)}{g(n)}$) for an appropriate function $c(r)$ of r. We contrast this new

---

[1] For approximating the probability to reach fixation (resp. extinction), one needs a number of runs which is about the inverse of the best known lower (resp. upper) bound of the fixation probability.

notion of $(h(n), g(n))$-selective amplifiers (resp. suppressors) with the notion of $g(n)$-*universal amplifiers* (resp. *suppressors*) which include those graphs G with n vertices for which $f_r(G) \geq 1 - \frac{c(r)}{g(n)}$ (resp. $f_r(G) \leq \frac{c(r)}{g(n)}$) for an appropriate function $c(r)$ of r. For a detailed presentation and a rigorous definition of these notions we refer to Section 2.

Using these new notions, we prove that there exist strong selective amplifiers, namely $(\Theta(n), n)$-selective amplifiers (called the *urchin graphs*). Furthermore we prove that there exist also quite strong selective suppressors, namely $(\frac{n}{\phi(n)+1}, \frac{n}{\phi(n)})$-selective suppressors (called the $\phi(n)$-*urchin graphs*) for *any* function $\phi(n) = \omega(1)$ with $\phi(n) \leq \sqrt{n}$.

Regarding the traditional measure of the fixation probability $f_r(G)$ of undirected graphs G, we provide upper and lower bounds that are much stronger than the bounds $\frac{1}{n}$ and $1 - \frac{1}{n+r}$ that were known so far [8]. More specifically, first of all we demonstrate the nonexistence of "strong" universal amplifiers by showing that for any graph G with n vertices, the fixation probability $f_r(G)$ is strictly less than $1 - \frac{c(r)}{n^{3/4+\varepsilon}}$, for any $\varepsilon > 0$. This is in a wide contrast with what happens in directed graphs, as Lieberman et al. [16] provided directed graphs with arbitrarily large fixation probability (see also [20]).

On the other hand, we provide our lower bound in the *Thermal Theorem*, which states that for any vertex v of an arbitrary undirected graph G, the fixation probability $f_r(v)$ of v is at least $(r-1)/(r + \frac{\deg v}{\deg_{\min}})$ for any $r > 1$, where $\deg v$ is the degree of v in G (i.e. the number of its neighbors) and $\deg_{\min}$ (resp. $\deg_{\max}$) is the minimum (resp. maximum) degree in G. This result extends the Isothermal Theorem for regular graphs [16]. In particular, we consider here a different notion of *temperature* for a vertex than [16]: the temperature of vertex v is $\frac{1}{\deg v}$. As it turns out, a "hot" vertex (i.e. with hight temperature) affects more often its neighbors than a "cold" vertex (with low temperature). The Thermal Theorem, which takes into account the vertex v on which the mutant is introduced, provides immediately our lower bound $(r-1)/(r + \frac{\deg_{\max}}{\deg_{\min}})$ for the fixation probability $f_r(G)$ of any undirected graph G. The latter lower bound is almost tight, as it implies that $f_r(G) \geq \frac{r-1}{r+1}$ for a regular graph G, while the Isothermal Theorem implies that the fixation probability of a regular graph G tends to $\frac{r-1}{r}$ as the size of G increases. Note that our new upper/lower bounds for the fixation probability lead to better time complexity of the FPRAS proposed in [8], as the Monte Carlo technique proposed in [8] now needs to simulate the Moran process a less number of times (to estimate fixation or extinction).

Our techniques are original and of a constructive combinatorics flavor. For the class of strong selective amplifiers (the urchin graphs) we introduce a novel decomposition of the Markov chain $\mathcal{M}$ of the generalized Moran process into $n-1$ smaller chains $\mathcal{M}_1, \mathcal{M}_2, \ldots, \mathcal{M}_{n-1}$, and then we decompose each $\mathcal{M}_k$ into two even smaller chains $\mathcal{M}_k^1, \mathcal{M}_k^2$. Then we exploit a new way of composing these smaller chains (and returning to the original one) that is carefully done to maintain the needed domination properties. For the proof of the lower bound in the Thermal Theorem, we first introduce a new and simpler weighted process that bounds fixation probability from below (the generalized Moran process is a special case of this new process). Then we add appropriate dummy states to its (exponentially large) Markov chain, and finally we iteratively modify the resulting chain by maintaining the needed monotonicity properties. Eventually this results to the desired lower bound of the Thermal Theorem. Finally, our proof for the non-existence of strong universal amplifiers is done by contradiction, partitioning appropriately the vertex set of the graph and discovering an appropriate independent set that leads to the contradiction.

2   Preliminaries

Throughout the paper we consider only finite, connected, undirected graphs $G = (V, E)$. Our results apply to connected graphs as, otherwise, the fixation probability is necessarily zero. The edge $e \in E$ between two vertices $u, v \in V$ is denoted by $e = uv$. For a vertex subset $X \subseteq V$, we write $X + y$ and $X - y$ for $X \cup \{y\}$ and $X \cap \{y\}$, respectively. Furthermore, throughout r denotes the fitness of the mutant, while the value r is considered to be independent of the size n of the network, i.e. we assume that r is constant. For simplicity of presentation, we call a vertex v "infected" if a copy of the mutant is placed on v. For every vertex subset $S \subseteq V$ we denote by $f_r(S)$ the fixation probability of the set S, i.e. the probability that, starting with exactly $|S|$ copies of the mutant placed on the vertices of S, the generalized Moran process will eventually reach fixation. By the definition of the generalized Moran process $f_r(\emptyset) = 0$ and $f_r(V) = 1$, while for $S \notin \{\emptyset, V\}$,

$$f_r(S) = \frac{\sum_{xy \in E} \left(\frac{r}{\deg x} f(S + y) + \frac{1}{\deg y} f(S - x) \right)}{\sum_{xy \in E} \left(\frac{r}{\deg x} + \frac{1}{\deg y} \right)}$$

Therefore, eliminating self-loops in the above Markov process,

$$f_r(S) = \frac{\sum_{xy \in E, x \in S, y \notin S} \left(\frac{r}{\deg x} f_r(S + y) + \frac{1}{\deg y} f_r(S - x) \right)}{\sum_{xy \in E, x \in S, y \notin S} \left(\frac{r}{\deg x} + \frac{1}{\deg y} \right)} \quad (1)$$

In the next definition we introduce the notions of *universal* and *selective* amplifiers.

Definition 1. *Let $\mathcal{G}$ be an infinite class of undirected graphs. If there exists an $n_0 \in \mathbb{N}$, an $r_0 \geq 1$, and some function $c(r)$, such that for every graph $G \in \mathcal{G}$ with $n \geq n_0$ vertices and for every $r > r_0$:*

- *$f_r(G) \geq 1 - \frac{c(r)}{g(n)}$, then $\mathcal{G}$ is a class of $g(n)$-universal amplifiers,*
- *there exists a subset S of at least $h(n)$ vertices of G, such that $f_r(v) \geq 1 - \frac{c(r)}{g(n)}$ for every vertex $v \in S$, then $\mathcal{G}$ is a class of $(h(n), g(n))$-selective amplifiers.*

Moreover, $\mathcal{G}$ is a class of strong universal *(resp.* strong selective*) amplifiers if $\mathcal{G}$ is a class of n-universal (resp. $(\Theta(n), n)$-selective) amplifiers.*

Similarly to Definition 1, we introduce the notions of *universal* and *selective* suppressors.

Definition 2. *Let $\mathcal{G}$ be an infinite class of undirected graphs. If there exist functions $c(r)$ and $n_0(r)$, such that for every $r > 1$ and for every graph $G \in \mathcal{G}$ with $n \geq n_0(r)$ vertices:*

- *$f_r(G) \leq \frac{c(r)}{g(n)}$, then $\mathcal{G}$ is a class of $g(n)$-universal suppressors,*
- *there exists a subset S of at least $h(n)$ vertices of G, such that $f_r(v) \leq \frac{c(r)}{g(n)}$ for every vertex $v \in S$, then $\mathcal{G}$ is a class of $(h(n), g(n))$-selective suppressors.*

Moreover, $\mathcal{G}$ is a class of strong universal *(resp.* strong selective*) suppressors if $\mathcal{G}$ is a class of n-universal (resp. $(\Theta(n), n)$-selective) suppressors.*

Note that $n_0 = n_0(r)$ in Definition 2, while in Definition 1 n_0 is not a function of r. The reason for this is that, since we consider the fitness value r to be constant, the size n of G needs to be sufficiently large with respect to r in order for G to act as a suppressor. Indeed, if we let r grow arbitrarily, e.g. if $r = n^2$, then for *any* graph G with n vertices the fixation probability $f_r(v)$ tends to 1 as n grows. The next lemma follows by Definitions 1 and 2.

Lemma 1. *If $\mathcal{G}$ is a class of $g(n)$-universal amplifiers (resp. suppressors), then $\mathcal{G}$ is a class of $(\Theta(n), g(n))$-selective amplifiers (resp. suppressors).*

The most natural question that arises by Definitions 1 and 2 is whether there exists any class of strong selective amplifiers/suppressors, as well as for which functions $h(n)$ and $g(n)$ there exist classes of $g(n)$-universal amplifiers/suppressors and classes of $(h(n), g(n))$-selective amplifiers/suppressors. In Section 3 and 4 we provide our results on amplifiers and suppressors, respectively.

3   Amplifier Bounds

In this section we prove that there exist no strong universal amplifiers (Section 3.1), although there exists a class of strong selective amplifiers (Section 3.2).

3.1   Non-existence of Strong Universal Amplifiers

Theorem 1. *For any function $g(n) = \Omega(n^{\frac{3}{4}+\varepsilon})$ for some $\varepsilon > 0$, there exists no graph class $\mathcal{G}$ of $g(n)$-universal amplifiers for any $r > r_0 = 1$.*

Proof (sketch). The proof is done by contradiction. It involves a surprising partition of the vertices of the graph into three sets V_1, V_2, V_3, where V_1 and V_2 are independent sets, and $N(v) \subseteq V_3$ for every $v \in V_1 \cup V_2$. For the detailed proof we refer to the full paper in the Appendix.

Corollary 1. *There exists no infinite class $\mathcal{G}$ of undirected graphs which are strong universal amplifiers.*

3.2   A Class of Strong Selective Amplifiers

In this section we present the first class $\mathcal{G} = \{G_n : n \geq 1\}$ of strong selective amplifiers, which we call the *urchin* graphs. Namely, the graph G_n has $2n$ vertices, consisting of a clique with n vertices, an independent set of n vertices, and a perfect matching between the clique and the independent set, as it is illustrated in Figure 1(a). For every graph G_n, we refer for simplicity to a vertex of the clique of G_n as a *clique vertex* of G_n, and to a vertex of the independent set of G_n as a *nose* of G_n, respectively. We prove in this section that the class $\mathcal{G}$ of urchin graphs are strong selective amplifiers. Namely, we prove that, whenever $r > r_0 = 5$, the fixation probability of any nose v of any graph G_n is $f_r(v) \geq 1 - \frac{c(r)}{n}$, where $c(r)$ is a function that depends only on the mutant fitness r.

Let v be a clique vertex (resp. a nose) and u be its adjacent nose (resp. clique vertex). If v is infected and u is not infected, then v is called an *isolated clique vertex* (resp. *isolated nose*), otherwise v is called a *covered clique vertex* (resp. *covered nose*). Let $k \in \{0, 1, \ldots, n\}$, $i \in \{0, 1, 2, \ldots, n-k\}$, and $x \in \{0, 1, 2, \ldots, k\}$. Denote by $Q_{i,x}^k$ the state of G_n with exactly i isolated clique vertices, x isolated noses, and $k - x$ covered noses. An example of the state $Q_{i,x}^k$ is illustrated in Figure 1. Furthermore, for every $k, i \in \{0, 1, \ldots, n\}$, we define the state P_i^k of G_n as follows. If $i \leq k$, then P_i^k is the state with exactly i covered noses and $k - i$ isolated noses. If $i > k$, then P_i^k is the state with exactly k covered noses and $i - k$ isolated clique vertices. Note that $Q_{i,0}^k = P_{k+i}^k$ and $Q_{0,x}^k = P_{k-x}^k$, for every $k \in \{0, 1, \ldots, n\}$, $i \in \{0, 1, 2, \ldots, n-k\}$, and $x \in \{0, 1, 2, \ldots, k\}$. Two examples of the state P_i^k, for the cases where $i \leq k$ and $i > k$, are shown in Figure 1.

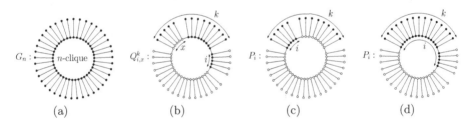

(a)　　　　　　(b)　　　　　　(c)　　　　　　(d)

Fig. 1. (a) The "urchin" graph G_n. Furthermore, the state (b) $Q_{i,x}^k$ and the state P_i^k, where (c) $i \leq k$, and (d) $i > k$.

Let $k \in \{1, 2, \ldots, n-1\}$. For all appropriate values of i and x, we denote by $q_{i,x}^k$ (resp. p_i^k) the probability that, starting at state $Q_{i,x}^k$ (resp. P_i^k) we eventually arrive to a state with $k + 1$ infected noses before we arrive to a state with $k - 1$ infected noses.

Lemma 2. *Let* $1 \leq k \leq n-1$. *Then* $q_{i,x}^k > q_{i-1,x-1}^k$, *for every* $i \in \{1, 2, \ldots, n-k\}$ *and every* $x \in \{1, 2, \ldots, k\}$.

Corollary 2. *Let* $k \in \{1, 2, \ldots, n-1\}$, $i \in \{0, 1, \ldots, n-k\}$, *and* $x \in \{0, 1, \ldots, k\}$. *Then* $q_{i,x}^k > p_{k+i-x}^k$.

Note by Corollary 2 that, in order to compute a lower bound for the fixation probability $f_r(v)$ of a nose v of the graph G_n, we can assume that, whenever we have k infected noses and i infected clique vertices, we are at state P_i^k. That is, in the Markov chain of the generalized Moran process, we replace any transition to a state $Q_{i,x}^k$ with a transition to state P_{k+i-x}^k. Denote this relaxed Markov chain by $\mathcal{M}$; we will compute a lower bound of the fixation probability of state P_0^1 in the Markov chain $\mathcal{M}$ (cf. Theorem 2).

In order to analyze $\mathcal{M}$, we decompose it first into the $n-1$ smaller Markov chains $\mathcal{M}_1, \mathcal{M}_2, \ldots, \mathcal{M}_{n-1}$, as follows. For every $k \in \{1, 2, \ldots, n-1\}$, the Markov chain $\mathcal{M}_k$ captures all transitions of $\mathcal{M}$ between states with k infected noses. We denote by F_{k-1} (resp. F_{k+1}) an *arbitrary* state with $k-1$ (resp. $k+1$) infected noses. Moreover, we consider F_{k-1} and F_{k+1} as absorbing states of $\mathcal{M}_k$. Since we want to compute a lower bound of the fixation probability, whenever we arrive at state F_{k+1} (resp. at state F_{k-1}), we assume that we have the smallest number of infected clique vertices with $k+1$ (resp. with $k-1$) infected noses. That is, whenever $\mathcal{M}_k$ reaches state F_{k+1}, we assume that $\mathcal{M}$ has reached state P_{k+1}^{k+1} (and thus we move to the Markov chain $\mathcal{M}_{k+1}$). Similarly, whenever $\mathcal{M}_k$ reaches state F_{k-1}, we assume that $\mathcal{M}$ has reached state P_0^{k-1} (and thus we move to the Markov chain $\mathcal{M}_{k-1}$).

A Decomposition of $\mathcal{M}_k$ into Two Markov Chains. In order to analyze the Markov chain $\mathcal{M}_k$, where $k \in \{1, 2, \ldots, n-1\}$, we decompose it into two smaller Markov chains $\{\mathcal{M}_k^1, \mathcal{M}_k^2\}$.

In $\mathcal{M}_k^1$, we consider the state P_{k+1}^k absorbing. For every $i \in \{0, 1, \ldots, k\}$ denote by h_i^k the probability that, starting at state P_i^k in $\mathcal{M}_k^1$, we eventually reach state P_{k+1}^k before we reach state F_{k-1}. In this Markov chain $\mathcal{M}_k^1$, every transition probability between two states is equal to the corresponding transition probabilities in $\mathcal{M}_k$.

In $\mathcal{M}_k^2$, we denote by s_i^k, where $i \in \{k, k+1, \ldots, n\}$, the probability that starting at state P_i^k we eventually reach state F_{k+1} before we reach state F_{k-1}. In this Markov chain $\mathcal{M}_k^2$, the transition probability from state P_k^k to state P_{k+1}^k (resp. to state F_{k-1}) is equal to h_k^k (resp. $1 - h_k^k$), while all other transition probabilities between two states in $\mathcal{M}_k^2$ are the same as the corresponding transition probabilities in $\mathcal{M}_k$.

Urchin Graphs are Strong Selective Amplifiers. We now conclude our analysis by combining the results of Section 3.2 on the two Markov chains $\mathcal{M}_1$ and $\mathcal{M}_2$. In the Markov chain $\mathcal{M}$, the transition from state P_k^k to the states P_k^k, P_0^{k-1} is done through the Markov chain $\mathcal{M}_1$, and the transition from state P_k^k to the states P_{k+1}^{k+1}, P_0^{k-1} is done through the Markov chain $\mathcal{M}_2$, respectively.

In the Markov chain $\mathcal{M}$, the transition probability from state P_k^k to state P_{k+1}^{k+1} (resp. P_0^{k-1}) is s_k^k (resp. $1 - s_k^k$). Recall that s_k^k is the probability that, starting at P_k^k in $\mathcal{M}_2$ (and thus also in $\mathcal{M}$), we reach state F_{k+1} before we reach F_{k-1}. Furthermore, the transition probability from state P_0^k to state P_k^k is equal to the probability that, starting at P_0^k in $\mathcal{M}_1$, we reach P_k^k before we reach F_{k-1}. Note that this probability is larger than h_0^k. Therefore, in order to compute a lower bound of the fixation probability of a nose in G_n, we can assume that in $\mathcal{M}$ the transition probability from state P_0^k to P_k^k (resp. P_0^{k-1}) is h_0^k (resp. $1 - h_0^k$).

Note that for every $k \in \{2, \dots, n-1\}$ the infected vertices of state P_0^k is a strict subset of the infected vertices of state P_k^k. Therefore, in order to compute a lower bound of the fixation probability of state P_0^1 in $\mathcal{M}$, we can relax $\mathcal{M}$ by changing every transition from state P_{k-1}^{k-1} to state P_k^k to a transition from state P_{k-1}^{k-1} to state P_0^k, where $k \in \{2, \dots, n-1\}$. After eliminating the states P_k^k in $\mathcal{M}'$, where $k \in \{1, 2, \dots, n-1\}$, we obtain an equivalent birth-death process $\mathcal{B}_n$. Denote by p_1 the fixation probability of state P_0^1 in $\mathcal{B}_n$, i.e. p_1 is the probability that, starting at state P_0^1 in $\mathcal{B}_n$, we eventually arrive to state P_n^n. For the next theorem we use the lower bounds of Section 3.2.

Theorem 2. *For any $r > 5$ and for sufficiently large n, the fixation probability p_1 of state P_0^1 in $\mathcal{B}_n$ is $p_1 \geq 1 - \frac{c(r)}{n}$, for some appropriate function $c(r)$ of r.*

We are now ready to provide our main result in this section.

Theorem 3. *The class $\mathcal{G} = \{G_n : n \geq 1\}$ of urchin graphs is a class of strong selective amplifiers.*

4   Suppressor Bounds

In this section we prove our lower bound for the fixation probability of an arbitrary undirected graph, namely the *Thermal Theorem* (Section 4.1), which generalizes the analysis of the fixation probability of regular graphs [16]. Furthermore we present for every function $\phi(n)$, where $\phi(n) = \omega(1)$ and $\phi(n) \leq \sqrt{n}$, a class of $(\frac{n}{\phi(n)+1}, \frac{n}{\phi(n)})$-selective suppressors in Section 4.2.

4.1   The Thermal Theorem

Consider a graph $G = (V, E)$ and a fitness value $r > 1$. Denote by $\mathcal{M}_r(G)$ the generalized Moran process on G with fitness r. Then, for every subset $S \notin \{\emptyset, V\}$ of its vertices, the fixation probability $f_r(S)$ of S in $\mathcal{M}_r(G)$ is given by (1), where $f_r(\emptyset) = 0$ and $f_r(V) = 1$. That is, the fixation probabilities $f_r(S)$, where $S \notin \{\emptyset, V\}$, are the solution of the linear system (1) with boundary conditions $f_r(\emptyset) = 0$ and $f_r(V) = 1$.

Suppose that at some iteration of the generalized Moran process the set S of vertices are infected and that the edge $xy \in E$ (where $x \in S$ and $y \notin S$) is activated, i.e. either x infects y or y disinfects x. Then (1) implies that the probability that x infects y is higher if $\frac{1}{\deg x}$ is large; similarly, the probability that y disinfects x is higher if $\frac{1}{\deg y}$ is large. Therefore, in a fashion similar to [16], we call for every vertex $v \in V$ the quantity $\frac{1}{\deg v}$ the *temperature* of v: a "hot" vertex (i.e. with high temperature) affects more often its neighbors than a "cold" vertex (i.e. with low temperature). It follows now by (1) that for every set $S \notin \{\emptyset, V\}$ there exists at least one pair $x(S), y(S)$ of vertices with $x(S) \in S$, $y(S) \notin S$, and $x(S)y(S) \in E$ such that

$$f_r(S) \geq \frac{\frac{r}{\deg x(S)} f_r(S + y(S)) + \frac{1}{\deg y(S)} f_r(S - x(S))}{\frac{r}{\deg x(S)} + \frac{1}{\deg y(S)}} \tag{2}$$

Thus, solving the linear system that is obtained from (2) by replacing inequalities with equalities, we obtain a lower bound for the fixation probabilities $f_r(S)$, where $S \notin \{\emptyset, V\}$. In the next definition we introduce a weighted generalization of this linear system, which is a crucial tool for our analysis in obtaining the Thermal Theorem.

Definition 3. (the linear system L_0)   Let $G = (V, E)$ be an undirected graph and $r > 1$. Let every vertex $v \in V$ have weight (temperature) $d_v > 0$. The linear system L_0 on the variables $p_r(S)$, where $S \subseteq V$, is given by the following equations whenever $S \notin \{\emptyset, V\}$:

$$p_r(S) = \frac{r d_{x(S)} p_r(S + y(S)) + d_{y(S)} p_r(S - x(S))}{r d_{x(S)} + d_{y(S)}} \tag{3}$$

with boundary conditions $p_r(\emptyset) = 0$ and $p_r(V) = 1$.

With a slight abuse of notation, whenever $S = \{u_1, u_2, \ldots, u_k\}$, we denote $p_r(u_1, u_2, \ldots, u_k) = p_r(S)$.

Observation 1. The linear system L_0 in Definition 3 corresponds naturally to the Markov chain $\mathcal{M}_0$ with one state for every subset $S \subseteq V$, where the states $\emptyset$ and V are absorbing, and every non-absorbing state S has exactly two transitions to the states $S + y(S)$ and $S - x(S)$ with transition probabilities $q_S = \frac{r d_{x(S)}}{r d_{x(S)} + d_{y(S)}}$ and $1 - q_S$, respectively.

Observation 2. Let $G = (V, E)$ be a graph and $r > 1$. For every vertex $x \in V$ let $d_x = \frac{1}{\deg x}$ be the temperature of x. Then $f_r(S) \geq p_r(S)$ for every $S \subseteq V$, where the values $p_r(S)$ are the solution of the linear system L_0.

Before we provide the Thermal Theorem (Theorem 4), we first prove an auxiliary result in the next lemma which generalizes the Isothermal Theorem of [16] for regular graphs, i.e. for graphs with the same number of neighbors for every vertex.

Lemma 3. Let $G = (V, E)$ be a graph with n vertices, $r > 1$, and d_u be the same for all vertices $u \in V$. Then $p_r(u) = \frac{1 - \frac{1}{r}}{1 - \frac{1}{r^n}} \geq 1 - \frac{1}{r}$ for every vertex $u \in V$.

We are now ready to provide our main result in this section which provides a lower bound for the fixation probability on arbitrary graphs, parameterized by the maximum ratio between two different temperatures in the graph.

Theorem 4 (Thermal Theorem). *Let $G = (V, E)$ be a connected undirected graph and $r > 1$. Then $f_r(v) \geq \frac{r-1}{r + \frac{\deg v}{\deg_{\min}}}$ for every $v \in V$.*

The lower bound for the fixation probability in Theorem 4 is almost tight. Indeed, if a graph $G = (V, E)$ with n vertices is regular, i.e. if $\deg u = \deg v$ for every $u, v \in V$, then $f_r(G) = \frac{1 - \frac{1}{r}}{1 - \frac{1}{r^n}}$ by Lemma 3 (cf. also the Isothermal Theorem in [16]), and thus $f_r(G) \cong \frac{r-1}{r}$ for large enough n. On the other hand, Theorem 4 implies for a regular graph G that $f_r(G) \geq \frac{r-1}{r+1}$.

4.2   A Class of Selective Suppressors

In this section we present for every function $\phi(n)$, where $\phi(n) = \omega(1)$ and $\phi(n) \leq \sqrt{n}$, the class $\mathcal{G}_{\phi(n)} = \{G_{\phi(n),n} : n \geq 1\}$ of $(\frac{n}{\phi(n)+1}, \frac{n}{\phi(n)})$-selective suppressors. We call these graphs $\phi(n)$-*urchin graphs*, since for $\phi(n) = 1$ they coincide with the class of urchin graphs in Section 3.2. For every n, the graph $G_{\phi(n),n} = (V_{\phi(n),n}, E_{\phi(n),n})$ has n vertices. Its vertex set $V_{\phi(n),n}$ can be partitioned into two sets $V^1_{\phi(n),n}$ and $V^2_{\phi(n),n}$, where $|V^1_{\phi(n),n}| = \frac{n}{\phi(n)+1}$ and $|V^2_{\phi(n),n}| = \frac{\phi(n)}{\phi(n)+1}n$, such that $V^1_{\phi(n),n}$ induces a clique and $V^2_{\phi(n),n}$ induces an independent set in $G_{\phi(n),n}$. Furthermore, every vertex $u \in V^2_{\phi(n),n}$ has $\phi(n)$ neighbors in $V^1_{\phi(n),n}$, and every vertex $v \in V^1_{\phi(n),n}$ has $\phi^2(n)$ neighbors in $V^2_{\phi(n),n}$. Therefore $\deg v = n + \phi^2(n) - 1$ for every $v \in V^1_{\phi(n),n}$ and $\deg u = \phi(n)$ for every $u \in V^2_{\phi(n),n}$.

Theorem 5. *For every function $\phi(n)$, where $\phi(n) = \omega(1)$ and $\phi(n) \leq \sqrt{n}$, the class $\mathcal{G}_{\phi(n)} = \{G_{\phi(n),n} : n \geq 1\}$ of $\phi(n)$-urchin graphs is a class of $(\frac{n}{\phi(n)+1}, \frac{n}{\phi(n)})$-selective suppressors.*

References

1. Aldous, D., Fill, J.: Reversible Markov Chains and Random Walks on Graphs. Monograph in preparation,
   http://www.stat.berkeley.edu/aldous/RWG/book.html
2. Antal, T., Scheuring, I.: Fixation of strategies for an evolutionary game in finite populations. Bulletin of Math. Biology 68, 1923–1944 (2006)
3. Broom, M., Hadjichrysanthou, C., Rychtar, J.: Evolutionary games on graphs and the speed of the evolutionary process. Proceedings of the Royal Society A 466(2117), 1327–1346 (2010)
4. Broom, M., Hadjichrysanthou, C., Rychtar, J.: Two results on evolutionary processes on general non-directed graphs. Proceedings of the Royal Society A 466(2121), 2795–2798 (2010)
5. Broom, M., Rychtar, J.: An analysis of the fixation probability of a mutant on special classes of non-directed graphs. Proceedings of the Royal Society A 464(2098), 2609–2627 (2008)
6. Broom, M., Rychtar, J., Stadler, B.: Evolutionary dynamics on small order graphs. Journal of Interdisciplinary Mathematics 12, 129–140 (2009)

7. Sasaki, A., Taylor, C., Fudenberg, D., Nowak, M.A.: Evolutionary game dynamics in finite populations. Bulletin of Math. Biology 66(6), 1621–1644 (2004)
8. Diáz, J., Goldberg, L., Mertzios, G., Richerby, D., Serna, M., Spirakis, P.: Approximating fixation probabilities in the generalized moran process. In: Proceedings of the ACM-SIAM Symposium on Discrete Algorithms (SODA), pp. 954–960 (2012)
9. Durrett, R.: Lecture notes on particle systems and percolation. Wadsworth Publishing Company (1988)
10. Easley, D., Kleinberg, J.: Networks, Crowds, and Markets: Reasoning about a Highly Connected World. Cambridge University Press (2010)
11. Gintis, H.: Game theory evolving: A problem-centered introduction to modeling strategic interaction. Princeton University Press (2000)
12. Hofbauer, J., Sigmund, K.: Evolutionary Games and Population Dynamics. Cambridge University Press (1998)
13. Imhof, L.A.: The long-run behavior of the stochastic replicator dynamics. Annals of applied probability 15(1B), 1019–1045 (2005)
14. Kandori, M., Mailath, G.J., Rob, R.: Learning, mutation, and long run equilibria in games. Econometrica 61(1), 29–56 (1993)
15. Karlin, S., Taylor, H.: A First Course in Stochastic Processes, 2nd edn. Academic Press, NY (1975)
16. Lieberman, E., Hauert, C., Nowak, M.A.: Evolutionary dynamics on graphs. Nature 433, 312–316 (2005)
17. Liggett, T.M.: Interacting Particle Systems. Springer (1985)
18. Mertzios, G.B., Nikoletseas, S., Raptopoulos, C., Spirakis, P.G.: Natural models for evolution on networks. In: Chen, N., Elkind, E., Koutsoupias, E. (eds.) Internet and Network Economics. LNCS, vol. 7090, pp. 290–301. Springer, Heidelberg (2011)
19. Moran, P.A.P.: Random processes in genetics. Proceedings of the Cambridge Philosophical Society 54, 60–71 (1958)
20. Nowak, M.A.: Evolutionary Dynamics: Exploring the Equations of Life. Harvard University Press (2006)
21. Ohtsuki, H., Nowak, M.A.: Evolutionary games on cycles. Proceedings of the Royal Society B: Biological Sciences 273, 2249–2256 (2006)
22. Rychtář, J., Stadler, B.: Evolutionary dynamics on small-world networks. International Journal of Computational and Mathematical Sciences 2(1), 1–4 (2008)
23. Sandholm, W.H.: Population games and evolutionary dynamics. MIT Press (2011)
24. Taylor, C., Iwasa, Y., Nowak, M.A.: A symmetry of fixation times in evoultionary dynamics. Journal of Theoretical Biology 243(2), 245–251 (2006)
25. Traulsen, A., Hauert, C.: Stochastic evolutionary game dynamics. In: Reviews of Nonlinear Dynamics and Complexity, vol. 2. Wiley, NY (2008)

Fast Distributed Coloring Algorithms
for Triangle-Free Graphs*

Seth Pettie and Hsin-Hao Su

University of Michigan

Abstract. *Vertex coloring* is a central concept in graph theory and an important symmetry-breaking primitive in distributed computing. Whereas degree-Δ graphs may require palettes of $\Delta+1$ colors in the worst case, it is well known that the chromatic number of many natural graph classes can be much smaller. In this paper we give new distributed algorithms to find (Δ/k)-coloring in graphs of girth 4 (triangle-free graphs), girth 5, and trees, where k is at most $(\frac{1}{4} - o(1)) \ln \Delta$ in triangle-free graphs and at most $(1 - o(1)) \ln \Delta$ in girth-5 graphs and trees, and $o(1)$ is a function of Δ. Specifically, for Δ sufficiently large we can find such a coloring in $O(k + \log^* n)$ time. Moreover, for *any* Δ we can compute such colorings in roughly logarithmic time for triangle-free and girth-5 graphs, and in $O(\log \Delta + \log_\Delta \log n)$ time on trees. As a byproduct, our algorithm shows that the chromatic number of triangle-free graphs is at most $(4 + o(1))\frac{\Delta}{\ln \Delta}$, which improves on Jamall's recent bound of $(67 + o(1))\frac{\Delta}{\ln \Delta}$. Also, we show that $(\Delta + 1)$-coloring for triangle-free graphs can be obtained in sublogarithmic time for any Δ.

1 Introduction

A proper t-coloring of a graph $G = (V, E)$ is an assignment from V to $\{1, \ldots, t\}$ (colors) such that no edge is monochromatic, or equivalently, each color class is an independent set. The *chromatic number* $\chi(G)$ is the minimum number of colors needed to properly color G. Let Δ be the maximum degree of the graph. It is easy to see that sometimes $\Delta + 1$ colors are necessary, e.g., on an odd cycle or a $(\Delta+1)$-clique. Brooks' celebrated theorem [9] states that these are the *only* such examples and that every other graph can be Δ-colored. Vizing [31] asked whether Brooks' Theorem can be improved for triangle-free graphs. In the 1970s Borodin and Kostochka [8], Catlin [10], and Lawrence [21] independently proved that $\chi(G) \leq \frac{3}{4}(\Delta + 2)$ for triangle-free G, and Kostochka (see [17]) improved this bound to $\chi(G) \leq \frac{2}{3}(\Delta + 2)$.

Existential Bounds. Better asymptotic bounds were achieved in the 1990s by using an iterated approach, often called the "Rödl Nibble". The idea is to color a very small fraction of the graph in a sequence of rounds, where after each

---

* This work is supported by NSF CAREER grant no. CCF-0746673, NSF grant no. CCF-1217338, and a grant from the US-Israel Binational Science Foundation.

F.V. Fomin et al. (Eds.): ICALP 2013, Part II, LNCS 7966, pp. 681–693, 2013.

round some property is guaranteed to hold with some small non-zero probability. Kim [18] proved that in any girth-5 graph G, $\chi(G) \leq (1+o(1))\frac{\Delta}{\ln \Delta}$. This bound is optimal to within a factor-2 under *any* lower bound on girth. (Constructions of Kostochka and Masurova [19] and Bollobás [7] show that there is a graph G of arbitrarily large girth and $\chi(G) > \frac{\Delta}{2\ln \Delta}$.) Building on [18], Johansson (see [23]) proved that $\chi(G) = O(\frac{\Delta}{\ln \Delta})$ for any triangle-free (girth-4) graph G.[1] In relatively recent work Jamall [14] proved that the chromatic number of triangle-free graphs is at most $(67+o(1))\frac{\Delta}{\ln \Delta}$.

Algorithms. We assume the $\mathcal{LOCAL}$ model [26] of distributed computation.[2] Grable and Panconesi [12] gave a distributed algorithm that Δ/k-colors a girth-5 graph in $O(\log n)$ time, where $\Delta > \log^{1+\epsilon'} n$ and $k \leq \epsilon \ln \Delta$ for any $\epsilon' > 0$ and some $\epsilon < 1$ depending on ϵ'.[3] Jamall [15] showed a sequential algorithm for $O(\Delta/\ln \Delta)$-coloring a triangle-free graph in $O(n\Delta^2 \ln \Delta)$ time, for any $\epsilon' > 0$ and $\Delta > \log^{1+\epsilon'} n$.

Note that there are *two* gaps between the existential [14,18,23] and algorithmic results [12,15]. The algorithmic results use a constant factor more colors than necessary (compared to the existential bounds) and they only work when $\Delta \geq \log^{1+\Omega(1)} n$ is sufficiently large, whereas the existential bounds hold for all Δ.

New Results. We give new distributed algorithms for (Δ/k)-coloring triangle-free graphs that simultaneously improve on both the existential and algorithmic results of [12,14,15,23]. Our algorithms run in $\log^{1+o(1)} n$ time for *all* Δ and in $O(k+\log^* n)$ time for Δ sufficiently large. Moreover, we prove that the chromatic number of triangle-free graphs is $(4+o(1))\frac{\Delta}{\ln \Delta}$.

Theorem 1. *Fix a constant $\epsilon' > 0$. Let Δ be the maximum degree of a triangle-free graph G, assumed to be at least some $\Delta_{\epsilon'}$ depending on ϵ'. Let $k \geq 1$ be a parameter such that $2\epsilon' \leq 1 - \frac{4k}{\ln \Delta}$. Then G can be (Δ/k)-colored, in time $O(k + \log^* \Delta)$ if $\Delta^{1-\frac{4k}{\ln \Delta}-\epsilon'} = \Omega(\ln n)$, and, for any Δ, in time on the order of*

$$\min \left(e^{O(\sqrt{\ln \ln n})}, \Delta + \log^* n \right) \cdot (k + \log^* \Delta) \cdot \frac{\log n}{\Delta^{1-\frac{4k}{\ln \Delta}-\epsilon'}} = \log^{1+o(1)} n$$

The first time bound comes from an $O(k + \log^* \Delta)$-round procedure, each round of which succeeds with probability $1 - 1/\text{poly}(n)$. However, as Δ decreases the probability of failure tends to 1. To enforce that each step succeeds with high

---

[1] We are not aware of any extant copy of Johansson's manuscript. It is often cited as a DIMACS Technical Report, though no such report exists. Molloy and Reed [23] reproduced a variant of Johansson's proof showing that $\chi(G) \leq 160\frac{\Delta}{\ln \Delta}$ for triangle-free G.

[2] In short, vertices host processors which operate is synchronized rounds; vertices can communicate one arbitrarily large message across each edge in each round; local computation is free; *time* is measured by the number of rounds.

[3] They claimed that their algorithm could also be extended to triangle-free graphs. Jamall [15] pointed out a flaw in their argument.

probability we use a version of the Local Lemma algorithm of Moser and Tardos [24] optimized for the parameters of our problem.[4]

By choosing $k = \ln \Delta/(4 + \epsilon)$ and $\epsilon' = \epsilon/(2(4 + \epsilon))$, we obtain new bounds on the chromatic number of triangle-free graphs.

Corollary 1. *For any $\epsilon > 0$ and Δ sufficiently large (as a function of ϵ), $\chi(G) \leq (4 + \epsilon)\frac{\Delta}{\ln \Delta}$. Consequently, the chromatic number of triangle-free graphs is $(4 + o(1))\frac{\Delta}{\ln \Delta}$, where the $o(1)$ is a function of Δ.*

Our result also extends to girth-5 graphs with $\Delta^{1-\frac{4k}{\ln \Delta}-\epsilon'}$ replaced by $\Delta^{1-\frac{k}{\ln \Delta}-\epsilon'}$, which allows us to $(1 + \epsilon)\Delta/\ln \Delta$-color such graphs. Our algorithm can clearly be applied to trees (girth ∞). Elkin [11] noted that with Bollobás's construction [7], Linial's lower bound [22] on coloring trees can be strengthened to show that it is impossible to $o(\Delta/\ln \Delta)$-color a tree in $o(\log_\Delta n)$ time. We prove that it *is* possible to $(1 + o(1))\Delta/\ln \Delta$-color a tree in $O(\log \Delta + \log_\Delta \log n)$ time. Also, we show that $(\Delta + 1)$-coloring for triangle-free graphs can be obtained in $\exp(O(\sqrt{\log \log n}))$ time.

Technical Overview. In the iterated approaches of [12, 14, 18, 23] each vertex u maintains a *palette*, which consists of the colors that have not been selected by its neighbors. To obtain a t-coloring, each palette consists of colors $\{1, \ldots, t\}$ initially. In each round, each u tries to assign itself a color (or colors) from its palette, using randomization to resolve the conflicts between itself and the neighbors. The *c-degree* of u is defined to be the number of its neighbors whose palettes contain c. In Kim's algorithm [18] for girth-5 graphs, the properties maintained for each round are that the c-degrees are upper bounded and the palette sizes are lower bounded. In girth-5 graphs the neighborhoods of the neighbors of u only intersect at u and therefore have a negligible influence on each other, that is, whether c remains in one neighbor's palette has little influence on a different neighbor of u. Due to this independence one can bound the c-degree after an iteration using standard concentration inequalities. In triangle-free graphs, however, there is no guarantee of independence. If two neighbors of u have identical neighborhoods, then after one iteration they will either both keep or both lose c from their palettes. In other words, the c-degree of u is a random variable that may not have any significant concentration around its mean. Rather than bound c-degrees, Johansson [23] bounded the entropy of the remaining palettes so that each color is picked nearly uniformly in each round. Jamall [14] claimed that although each c-degree does not concentrate, the *average* c-degree (over each c in the palette) does concentrate. Moreover, it suffices to consider only those colors within a constant factor of the average in subsequent iterations.

Our (Δ/k)-coloring algorithm performs the same coloring procedure in each round, though the behavior of the algorithm has two qualitatively distinct phases.

---

[4] Note that for many reasonable parameters (e.g., $k = O(1), \Delta = \log^{1-\delta} n$), the running time is *sub*logarithmic.

In the first $O(k)$ rounds the c-degrees, palette sizes, and probability of remaining uncolored are very well behaved. Once the available palette is close to the number of uncolored neighbors the probability of remaining uncolored begins to decrease drastically in each successive round, and after $O(\log^* n)$ rounds all vertices are colored, w.h.p.

Our analysis is similar to that of Jamall [14] in that we focus on bounding the average of the c-degrees. However, our proof needs to take a different approach, for two reasons. First, to obtain an efficient *distributed* algorithm we need to obtain a tighter bound on the probability of failure in the last $O(\log^* n)$ rounds, where the c-degrees shrink faster than a constant factor per round. Second, there is a small flaw in Jamall's application of Azuma's inequality in Lemma 12 in [14], the corresponding Lemma 17 in [15], and the corresponding lemmas in [16]. It is probably possible to correct the flaw, though we manage to circumvent this difficulty altogether. See the full version for a discussion of this issue.

The second phase presents different challenges. The natural way to bound c-degrees using Chernoff-type inequalities gives error probabilities that are exponential *in the c-degree*, which is fine if it is $\Omega(\log n)$ but becomes too large as the c-degrees are reduced in each coloring round. At a certain threshold we switch to a different analysis (along the lines of Schneider and Wattenhofer [30]) that allows us to bound c-degrees with high probability in the *palette* size, which, again, is fine if it is $\Omega(\log n)$.

In both phases, if we cannot obtain small error probabilities (via concentration inequalities and a union bound) we revert to a distributed implementation of the Moser-Tardos Lovász Local Lemma algorithm [24]. We show that for certain parameters the symmetric LLL can be made to run in *sub*logarithmic time. For the extensions to trees and the $(\Delta + 1)$-coloring algorithm for triangle-free graphs, we adopt the ideas from [5,6,29] to reduce the graph into several smaller components and color each of them separately by deterministic algorithms [4,25], which will run faster as the size of each subproblem is smaller.

Organization. Section 2 presents the general framework for the analysis. Section 3 describes the algorithms and discusses what parameters to plug into the framework. Section 4 describes the extension to graphs of girth 5, trees, and the $(\Delta + 1)$-coloring algorithm for triangle-free graphs.

2   The Framework

Every vertex maintains a *palette* that consists of all colors not previously chosen by its neighbors. The coloring is performed in rounds, where each vertex chooses zero or more colors in each round. Let G_i be the graph induced by the uncolored vertices after round i, so $G = G_0$. Let $N_i(u)$ be u's neighbors in G_i and let $P_i(u)$ be its palette after round i. The *c-neighbors* $N_{i,c}(u)$ consist of those $v \in N_i(u)$ with $c \in P_i(v)$. Call $|N_i(u)|$ the *degree* of u and $|N_{i,c}(u)|$ the *c-degree* of u after round i. This notation is extended to sets of vertices in a natural way, e.g., $N_i(N_i(u))$ is the set of neighbors of neighbors of u in G_i.

Algorithm 2 describes the iterative coloring procedure. In each round, each vertex u selects a set $S_i(u)$ of colors by including each $c \in P_{i-1}(u)$ independently with probability π_i to be determined later. If some $c \in S_i(u)$ is not selected by any neighbor of u then u can safely color itself c. In order to remove dependencies between various random variables we exclude colors from u's palette more aggressively than is necessary. First, we exclude any color *selected* by a neighbor, that is, $S_i(N_{i-1}(u))$ does not appear in $P_i(u)$. The probability that a color c is *not* selected by a neighbor is $(1 - \pi_i)^{|N_{i-1,c}(u)|}$. Suppose that this quantity is at least some threshold β_i for all c. We force c to be kept with probability *precisely* β_i by putting c in a keep-set $K_i(u)$ with probability $\beta_i/(1 - \pi_i)^{|N_{i-1,c}(u)|}$. The probability that $c \in K_i(u)\backslash S_i(N_{i-1}(u))$ is therefore β_i, assuming $\beta_i/(1 - \pi_i)^{|N_{i-1,c}(u)|}$ is a valid probability; if it is not then c is *ignored*. Let $\widehat{P}_i(u)$ be what remains of u's palette. Algorithm 2 has two variants. In Variant B, $P_i(u)$ is exactly $\widehat{P}_i(u)$ whereas in Variant A $P_i(u)$ is the subset of $\widehat{P}_i(u)$ whose c-degrees are sufficiently low, less than $2t_i$, where t_i is a parameter that will be explained below.

Include each $c \in P_{i-1}(u)$ in $S_i(u)$ independently with probability π_i.
For each c, calculate $r_c = \beta_i/(1 - \pi_i)^{|N_{i-1,c}(u)|}$.
If $r_c \leq 1$, include $c \in P_{i-1}(u)$ in $K_i(u)$ independently with probability r_c.
return $(S_i(u), K_i(u))$.

Algorithm 1. Select(u, π_i, β_i)

---

repeat
    Round $i = 1, 2, 3, \ldots$.
    for each $u \in G_{i-1}$ **do**
        $(S_i(u), K_i(u)) \leftarrow \text{Select}(u, \pi_i, \beta_i)$
        Set $\widehat{P}_i(u) \leftarrow K_i(u) \setminus S_i(N_{i-1}(u))$
        if $S_i(u) \cap \widehat{P}_i(u) \neq \emptyset$ **then** color u with any color in $S_i(u) \cap \widehat{P}_i(u)$ **end if**
        (Variant A) $P_i(u) \leftarrow \{c \in \widehat{P}_i(u) \mid |N_{i,c}(u)| \leq 2t_i\}$
        (Variant B) $P_i(u) \leftarrow \widehat{P}_i(u)$
    end for
    $G_i \leftarrow G_{i-1} \setminus \{\text{colored vertices}\}$
until the termination condition occurs

Algorithm 2. Coloring-Algorithm$(G_0, \{\pi_i\}, \{\beta_i\})$

The algorithm is parameterized by the sampling probabilities $\{\pi_i\}$, the ideal c-degrees $\{t_i\}$ and the ideal probability $\{\beta_i\}$ of retaining a color. The $\{\beta_i\}$ define how the ideal palette sizes $\{p_i\}$ degrade. Of course, the *actual* palette sizes and c-degrees after i rounds will drift from their ideal values, so we will need to reason about approximations of these quantities. We will specify the initial parameters and the terminating conditions when applying both variants in Section 3.

2.1   Analysis A

Given $\{\pi_i\}$, p_0, t_0, and δ, the parameters for Variant A are derived below.

$$\beta_i = (1 - \pi_i)^{2t_{i-1}} \qquad\qquad \alpha_i = (1 - \pi_i)^{(1-(1+\delta)^{i-1}/2)p_i'}$$
$$p_i = \beta_i p_{i-1} \qquad\qquad\qquad t_i = \max(\alpha_i \beta_i t_{i-1}, T) \qquad\qquad (1)$$
$$p_i' = (1 - \delta/8)^i p_i \qquad\qquad t_i' = (1 + \delta)^i t_i$$

Let us take a brief tour of the parameters. The sampling probability π_i will be inversely proportional to t_{i-1}, the ideal c-degree at end of round $i - 1$. (The exact expression for π_i depends on ϵ'.) Since we filter out colors with more than twice the ideal c-degree, the probability that a color is not selected by any neighbor is at least $(1 - \pi_i)^{2t_{i-1}} = \beta_i$. Note that since $\pi_i = \Theta(1/t_{i-1})$ we have $\beta_i = \Theta(1)$. Thus, we can force all colors to be retained in the palette with probability precisely β_i, making the ideal palette size $p_i = \beta_i p_{i-1}$. Remember that a c-neighbor stays a c-neighbor if it remains uncolored *and* it does not remove c from its palette. The latter event happens with probability β_i. We use α_i as an upper bound on the probability that a vertex remains uncolored, so the ideal c-degree should be $t_i = \alpha_i \beta_i t_{i-1}$. To account for deviations from the ideal we let p_i' and t_i' be approximate versions of p_i and t_i, defined in terms of a small error control parameter $\delta > 0$. Furthermore, certain high probability bounds will fail to hold if t_i becomes too small, so we will not let it go below a threshold T.

When the graph has girth 5, the concentration bounds allow us to show that $|P_i(u)| \geq p_i'$ and $|N_{i,c}(u)| \leq t_i'$ with certain probabilities. As pointed out by Jamall [14,15], $|N_{i,c}(u)|$ does not concentrate in triangle-free graphs. He showed that the average c-degree, $\bar{n}_i(u) = \sum_{c \in P_i(u)} |N_{i,c}(u)|/|P_i(u)|$, concentrates and will be bounded above by t_i' with a certain probability. Since $\bar{n}_i(u)$ concentrates, it is possible to bound the fraction of colors filtered for having c-degrees larger than $2t_i$.

Let $\lambda_i(u) = \min(1, |P_i(u)|/p_i')$. Since $P_i(u)$ is supposed to be at least p_i', if we do not filter out colors, $1 - \lambda_i(u)$ can be viewed as the fraction that has been filtered. In the following we state an induction hypotheses equivalent to Jamall's [14].

$$D_i(u) \leq t_i', \text{ where } D_i(u) = \lambda_i(u)\bar{n}_i(u) + (1 - \lambda_i(u))2t_i$$

$D_i(u)$ can be interpreted as the average of the c-degrees of $P_i(u)$ with $p_i' - |P_i(u)|$ dummy colors whose c-degrees are exactly $2t_i$. Notice that $D_i(u) \leq t_i'$ also implies $1 - \lambda_i(u) \leq (1 + \delta)^i/2$, because $(1 - \lambda_i(u))2t_i \leq D_i(u) \leq t_i'$. Therefore:

$$|P_i(u)| \geq (1 - (1 + \delta)^i/2)p_i'$$

Recall $P_i(u)$ is the palette consisting of colors c for which $|N_{i,c}(u)| \leq 2t_i$.

The main theorem for this section shows the inductive hypothesis holds with a certain probability. See the full version for the proof.

Theorem 2. *Suppose that $D_{i-1}(x) \leq t_{i-1}'$ for all $x \in G_{i-1}$, then for a given $u \in G_{i-1}$, $D_i(u) \leq t_i'$ holds with probability at least $1 - \Delta e^{-\Omega(\delta^2 T)} - (\Delta^2 + 2)e^{-\Omega(\delta^2 p_i')}$.*

2.2   Analysis B

Analysis A has a limitation for smaller c-degrees, since the probability guarantee becomes smaller as t_i goes down. Therefore, Analysis A only works well for $t_i \geq T$, where T is a threshold for certain probability guarantees. For example, if we want Theorem 2 to hold with high probability in n, then we must have $T \gg \log n$.

To get a good probability guarantee below T, we will use an idea by Schneider and Wattenhofer [30]. They took advantage of the trials done for each color inside the palette, rather than just considering the trials on whether each neighbor is colored or not. We demonstrate this idea in the proof of Theorem 3 in the full version. The probability guarantee in the analysis will not depend on the current c-degree but on the *initial* c-degree and the current palette size.

The parameters for Variant B are chosen based on an initial lower bound on the palette size p_0, upper bound on the c-degree t_0, and error control parameter δ. The selection probability is chosen to be $\pi_i = 1/(t_{i-1}+1)$ and the probability a color remains in a palette $\beta_i = (1 - \pi_i)^{t_{i-1}}$. The ideal palette size and its relaxation are $p_i = \beta_i p_{i-1}$ and $p'_i = (1 - \delta)^i p_i$, and the ideal c-degree $t_i = \max(\alpha_i t_{i-1}, 1)$. One can show the probability of remaining uncolored is upper bounded by $\alpha_i = 5t_0/p'_i$,

Let $E_i(u)$ denote the event that $|P_i(u)| \geq p'_i$ and $|N_{i,c}(u)| < t_i$ for all $c \in P_i(u)$. Although a vertex could lose its c-neighbor if the c-neighbor becomes colored or loses c in its palette, in this analysis, we only use the former to bound its c-degree. Also, if $E_{i-1}(u)$ is true, then $\Pr(c \notin S_i(N_{i-1}(u))) > \beta_i$ for all $c \in P_{i-1}(u)$. Thus in $\text{Select}(u, \pi_i, \beta_i)$, we will not ignore any colors in the palette. Each color remains in the palette with probability exactly β_i.

The following theorem shows the inductive hypothesis holds with a certain probability. See the full version for the proof.

Theorem 3. *If $E_{i-1}(x)$ holds for all $x \in G_{i-1}$, then for a given $u \in G_{i-1}$, $E_i(u)$ holds with probability at least $1 - \Delta e^{-\Omega(t_0)} - (\Delta^2 + 1)e^{-\Omega(\delta^2 p'_i)}$*

3   The Coloring Algorithms

The algorithm in Theorem 1 consists of two phases. Phase I uses Analysis A and Phase II uses Analysis B. First, we will give the parameters for both phases. Then, we will present the distributed algorithm that makes the induction hypothesis in Theorem 2 ($D_i(u) \leq t'_i$) and Theorem 3 ($E_i(u)$) hold for all $u \in G_i$ with high probability in n for every round i.

Let $\epsilon_1 = 1 - \frac{4k}{\ln \Delta} - \frac{2\epsilon'}{3}$ and $\epsilon_2 = 1 - \frac{4k}{\ln \Delta} - \frac{\epsilon'}{3}$. We will show that upon reaching the terminating condition of Phase I (which will be defined later), we will have $|P_i(u)| \geq \Delta^{\epsilon_2}$ for all $u \in G_i$ and $|N_{i,c}(u)| < \Delta^{\epsilon_1}$ for all $u \in G_i$ and all $c \in P_i(u)$. At this point, for a non-constructive version, we can simply apply the results about list coloring constants [13, 27, 28] to get a proper coloring, since at this point there is an $\omega(1)$ gap between $|N_{i,c}(u)|$ and $|P_i(u)|$ for every $u \in G_i$. One can turn the result of [27] into a distributed algorithm with the aid of Moser-Tardos

Lovász Local Lemma algorithm to amplify the success probability. However, to obtain an efficient distributed algorithm we use Analysis B in Phase II.

Since our result holds for large enough Δ, we can assume whenever necessary that Δ is sufficiently large. The asymptotic notation will be with respect to Δ.

3.1 Parameters for Phase I

In this phase, we use Analysis A with the following parameters and terminating condition: $\pi_i = \frac{1}{2Kt_{i-1}+1}$, where $K = 4/\epsilon'$ is a constant, $p_0 = \Delta/k$, $t_0 = \Delta$ and $\delta = 1/\log^2 \Delta$. This phase ends after the round when $t_i \leq T \overset{\text{def}}{=} \Delta^{\epsilon_1}/3$.

First, we consider the algorithm for at most the first $O(\log \Delta)$ rounds. For these rounds, we can assume the error $(1 + \delta)^i \leq \left(1 + \frac{1}{\log^2 \Delta}\right)^{O(\log \Delta)} \leq e^{O(1/\log \Delta)} = 1 + o(1)$ and similarly $(1 - \delta/8)^i \geq \left(1 - \frac{1}{\log^2 \Delta + 1}\right)^{O(\log \Delta)} \geq e^{-O(1/\log \Delta)} = 1 - o(1)$. We will show the algorithm reaches the terminating condition during these rounds, where the error is under control.

The probability a color is retained, $\beta_i = (1 - \pi_i)^{2t_{i-1}} \geq e^{-1/K}$, is bounded below by a constant. The probability a vertex remains uncolored is at most $\alpha_i = (1 - \pi_i)^{(1-(1+\delta)^{i-1}/2)p'_i} \leq e^{-(1-o(1))Cp_{i-1}/t_{i-1}}$, where $C = 1/(4Ke^{1/K})$.

Let $s_i = t_i/p_i$ be the ratio between the ideal c-degree and the ideal palette size. Initially, $s_0 = k$ and $s_i = \alpha_i s_{i-1} \leq s_{i-1}e^{-(1-o(1))(C/s_{i-1})}$. Initially, s_i decreases roughly linearly by C for each round until the ratio $s_i \approx C$ is a constant. Then, s_i decreases rapidly in the order of iterated exponentiation. Therefore, it takes roughly $O(k + \log^* \Delta)$ rounds to reach the terminating condition where $t_i \leq T$. Our goal is to show upon reaching the terminating condition, the palette size bound p_i is greater than T by some amount, in particular, $p_i \geq 30e^{3/\epsilon'}\Delta^{\epsilon_2}$. See the full version for the proof of the following Lemma.

Lemma 1. *Phase I terminates in $(4 + o(1))Ke^{1/K}k + O(\log^* \Delta)$ rounds, where $K = 4/\epsilon'$. Moreover, $p_i \geq 30e^{3/\epsilon'}\Delta^{\epsilon_2}$ for every round i in this phase.*

Thus, if the induction hypothesis $D_i(u) \leq t'_i$ holds for every $u \in G_i$ for every round i during this phase, we will have $|P_i(u)| \geq (1 - (1 + \delta)^i/2)p'_i \geq 10e^{3/\epsilon'}\Delta^{\epsilon_2}$ for all $u \in G_i$ and $|N_{i,c}(u)| \leq 2t_i < \Delta^{\epsilon_1}$ for all $u \in G_i$ and all $c \in P_i(u)$ in the end.

3.2 Parameters for Phase II

In Phase II, we will use Analysis B with the following parameters and terminating condition: $p_0 = 10e^{3/\epsilon'}\Delta^{\epsilon_2}$, $t_0 = \Delta^{\epsilon_1}$ and $\delta = 1/\log^2 \Delta$. This phase terminates after $\frac{3}{\epsilon'}$ rounds.

First note that the number of rounds $\frac{3}{\epsilon'}$ is a constant. We show $p'_i \geq 5\Delta^{\epsilon_2}$ for each round $1 \leq i \leq \frac{3}{\epsilon'}$, so there is always a sufficient large gap between the current palette size and the initial c-degree, which implies the shrinking factor of the c-degrees is $\alpha_i = 5t_0/p'_i \leq \Delta^{-\epsilon'/3}$. Since p_i shrinks by at most a $\beta_i \geq e^{-1}$ factor every round, $p'_i \geq (1 - \delta)^i \prod_{j=1}^i \beta_j p_0 \geq ((1 - \delta)e^{-1})^i 10e^{3/\epsilon'}\Delta^{\epsilon_2} \geq 5\Delta^{\epsilon_2}$.

Now since $\alpha_i \leq \Delta^{-\epsilon'/3}$, after $\frac{3}{\epsilon'}$ rounds, $t_i \leq t_0 \prod_{j=1}^{i} \alpha_j \leq \Delta \left(\Delta^{-\epsilon'/3} \right)^{\frac{3}{\epsilon'}} \leq 1$. The c-degree bound, $t_{\epsilon'/3}$, becomes 1. Recall that the induction hypothesis $E_i(u)$ is the event that $|P_i(u)| \geq p_i'$ and $|N_{i,c}(u)| < t_i$. If $E_i(u)$ holds for every $u \in G_i$ for every round i during this phase, then in the end, every uncolored vertex has no c-neighbors, as implied by $|N_{i,c}(u)| < t_i \leq 1$. This means these vertices can be colored with anything in their palettes, which are non-empty.

3.3 The Distributed Coloring Algorithm

We will show a distributed algorithm that makes the induction hypothesis in Phase I and Phase II hold with high probability in n.

Fix the round i and assume the induction hypothesis holds for all $x \in G_{i-1}$. For $u \in G_{i-1}$, define $A(u)$ to be the bad event that the induction hypothesis fails at u (i.e. $D_i(u) > t_i'$ in Phase I or $E_i(u)$ fails in Phase II). Let $p = e^{-\Delta^{1 - \frac{4k}{\ln \Delta} - \epsilon'}} / (e\Delta^4)$. By Theorem 2 and Theorem 3, $\Pr(A(u))$ is at most:

$$\Delta e^{-\Omega(\delta^2 T)} + (\Delta^2 + 2)e^{-\Omega(\delta^2 p_i')} \quad \text{or} \quad \Delta e^{-\Omega(t_0)} + (\Delta^2 + 1)e^{-\Omega(\delta^2 p_i')}$$

Since $T = \Delta^{\epsilon_1}/3, t_0 = \Delta^{\epsilon_1}, p_i' \geq \Delta^{\epsilon_2}, \Pr(A(u)) \leq p$ for large enough Δ.

If $\Delta^{1 - \frac{4k}{\ln \Delta} - \epsilon'} > c \log n$, then $p < 1/n^c$. By the union bound over all $u \in G_{i-1}$, the probability that any of the $A(u)$ fails is at most $1/n^{c-1}$. The induction hypothesis holds for all $u \in G_i \subseteq G_{i-1}$ with high probability. In this case, $O(k + \log^* \Delta)$ rounds suffice, because each round succeeds with high probability.

On the other hand, if $\Delta^{1 - \frac{4k}{\ln \Delta} - \epsilon'} < c \log n$, then we apply Moser and Tardos' resampling algorithm to make $A(u)$ simultaneously hold for all u with high probability. At round i, the bad event $A(u)$ depends on the random variables which are generated by $\text{Select}(v, \pi_i, \beta_i)$ for v within distance 2 in G_{i-1}. Therefore, the dependency graph $G_{i-1}^{\leq 4}$ consists of edges (u, v) such that $dist_{G_{i-1}}(u, v) \leq 4$. Each event $A(u)$ shares variables with at most $d < \Delta^4$ other events. The Lovász Local Lemma [1] implies that if $ep(d+1) \leq 1$, then the probability that all $A(u)$ simultaneously hold is guaranteed to be non-zero. Moser and Tardos showed how to boost this probability by resampling. In each round of resampling, their algorithm finds an MIS I in the dependency graph induced by the set of bad events B and then resamples the random variables that I depends on. In our case, it corresponds to finding an MIS I in $G_{i-1}^{\leq 4}[B]$, where $B = \{u \in G_{i-1} \mid A(u) \text{ fails}\}$. Then, we redo $\text{Select}(v, \pi_i, \beta_i)$ for $v \in G$ within distance 2 from I to resample the random variables that I depends on. By plugging in the parameters for the symmetric case, their proof shows if $ep(d+1) \leq 1 - \epsilon$, then the probability any of the bad events occur after t rounds of resampling is at most $(1 - \epsilon)^t n/d$. Thus, $O(\log n / \log(\frac{1}{1-\epsilon}))$ rounds will be sufficient for all $A(u)$ to hold with high probability in n.[5]

---

[5] In the statement of Theorem 1.3 in [24], they used $1/\epsilon$ as an approximation for $\log(\frac{1}{1-\epsilon})$. However, this difference can be significant in our case, when $1 - \epsilon$ is very small.

As shown in previous sections, $p \le e^{-\Delta^{1-\frac{4k}{\ln \Delta}-\epsilon'}}/(e\Delta^4)$. We can let $1-\epsilon = ep(d+1) \le e^{-\Delta^{1-\frac{4k}{\ln \Delta}-\epsilon'}}$. Therefore, $O(\log n/\Delta^{1-\frac{4k}{\ln \Delta}-\epsilon'})$ resampling rounds will be sufficient. Also, an MIS can be found in $O(\Delta + \log^* n)$ time [3, 20], or $e^{O(\sqrt{\log \log n})}$ since $\Delta \le (c \log n)^{1/(1-\frac{4k}{\ln \Delta}-\epsilon')} \le (c\log n)^{1/\epsilon'} \le \log^{O(1)} n$ [5]. Each of the $O(k+\log^* \Delta)$ rounds is delayed by $O(\log n/\Delta^{1-\frac{4k}{\ln \Delta}-\epsilon'})$ resampling rounds, which are futher delayed by the rounds needed to find an MIS. Therefore, the total number of rounds is

$$O\left((k+\log^* \Delta) \cdot \frac{\log n}{\Delta^{1-\frac{4k}{\ln \Delta}-\epsilon'}} \cdot \min\left(\exp\left(O\left(\sqrt{\log \log n}\right)\right), \Delta + \log^* n\right)\right)$$

Note that this is always at most $\log^{1+o(1)} n$.

4   Extensions

4.1   Graphs of Girth at Least 5

For graphs of girth at least 5, existential results [18, 23] show that there exists $(1 + o(1))\Delta/\ln \Delta$-coloring. We state a matching algorithmic result. The proof will be included in the full version.

Theorem 4. *Fix a constant $\epsilon' > 0$. Let Δ be the maximum degree of a girth-5 graph G, assumed to be at least some $\Delta_{\epsilon'}$ depending on ϵ'. Let $k \ge 1$ be a parameter such that $2\epsilon' \le 1 - \frac{k}{\ln \Delta}$. Then G can be (Δ/k)-colored, in time $O(k + \log^* \Delta)$ if $\Delta^{1-\frac{k}{\ln \Delta}-\epsilon'} = \Omega(\ln n)$, and, for any Δ, in time on the order of*

$$\min\left(e^{O(\sqrt{\ln \ln n})}, \Delta + \log^* n\right) \cdot (k + \log^* \Delta) \cdot \frac{\log n}{\Delta^{1-\frac{k}{\ln \Delta}-\epsilon'}} = \log^{1+o(1)} n$$

4.2   Trees

Trees are graphs of infinity girth. According to Theorem 4, it is possible to get a (Δ/k)-coloring in $O(k + \log^* \Delta)$ time if $\Delta^{1-\frac{k}{\ln \Delta}-\epsilon'} = \Omega(\log n)$. If $\Delta^{1-\frac{k}{\ln \Delta}-\epsilon'} = O(\log n)$, we will show that using additional $O(q)$ colors, it is possible to get a $(\Delta/k + O(q))$-coloring in $O\left(k + \log^* n + \frac{\log \log n}{\log q}\right)$ time. By choosing $q = \sqrt{\Delta}$, we can find a $(1 + o(1))\Delta/\ln \Delta$-coloring in $O(\log \Delta + \log_\Delta \log n)$ rounds.

The algorithm is the same with the framework, except that at the end of each round we delete the *bad* vertices, which are the vertices that fail to satisfy the induction hypothesis. The remaining vertices must satisfy the induction hypothesis, and then we will continue the next round on these vertices. Using the idea from [5,6,29], we can show that after $O(k+\log^* \Delta)$ rounds of the algorithm, the size of each component formed by the bad vertices is at most $O\left(\Delta^4 \log n\right)$ with high probability. See the full version for the proof.

Barenboim and Elkin's deterministic algorithm [4] obtains $O(q)$-coloring in $O\left(\frac{\log n}{\log q} + \log^* n\right)$ time for trees (arboricity $= 1$). We then apply their algorithm on each component formed by bad vertices. Since the size of each component is at most $O(\Delta^4 \log n)$, their algorithm will run in $O\left(\frac{\log\log n + \log \Delta}{\log q} + \log^* n\right)$ time, using the additional $O(q)$ colors. Note that this running time is actually $O\left(\frac{\log\log n}{\log q} + \log^* n\right)$, since $\Delta = \log^{O(1)} n$.

4.3 $(\Delta + 1)$-Color Triangle-Free Graphs in Sublogarithmic Time

We show that $(\Delta + 1)$-coloring in triangle-free graphs can be obtained in $\exp(O(\sqrt{\log\log n}))$ rounds for any Δ. Let $k = 1$ and $\epsilon' = 1/4$. By Theorem 1, there exists a constant Δ_0 such that for all $\Delta \geq \Delta_0$, if $\Delta^{1/2} \geq \log n$, then a $(\Delta+1)$-coloring can be found in $O(\log^* \Delta)$ time. If $\Delta < \Delta_0$, then $(\Delta+1)$ can be solved in $O(\Delta+\log^* n) = O(\log^* n)$ rounds [3,20]. Otherwise, if $\Delta_0 \leq \Delta < \log^2 n$, then we can apply the same technique for trees to bound the size of each bad component by $O(\Delta^4 \log n) = \text{polylog}(n)$, whose vertices failed to satisfy the induction hypothesis in the $O(\log^* \Delta)$ rounds. Panconesi and Srinivasan's deterministic network decomposition algorithm [25] obtains $(\Delta + 1)$-coloring in $\exp(O(\sqrt{\log n}))$ for graphs with n vertices. In fact, their decomposition can also obtain a proper coloring as long as the graph can be greedily colored (e.g. the palette size is more than the degree for each vertex). Therefore, by applying their algorithm, each bad component can be properly colored in $\exp(O(\sqrt{\log\log n}))$ rounds.

5   Conclusion

The time bounds of Theorem 1 show an interesting discontinuity. When Δ is large we can cap the error at $1/\text{poly}(n)$ by using standard concentration inequalities and a union bound. When Δ is small we can use the Moser-Tardos LLL algorithm to reduce the failure probability again to $1/\text{poly}(n)$. Thus, the distributed complexity of our coloring algorithm is tied to the distributed complexity of the constructive Lovász Local Lemma.

We showed that $\chi(G) \leq (4+o(1))\Delta/\ln \Delta$ for triangle-free graphs G. It would be interesting to see if it is possible to reduce the palette size to $(1+o(1))\Delta/\ln \Delta$, matching Kim's [18] bound for girth-5 graphs.

Alon et al. [2] and Vu [32] extended Johansson's result [23] for triangle-free graphs to obtain an $O(\Delta/\log f)$-coloring for locally sparse graphs (the latter also works for list coloring), in which no neighborhood of any vertex spans more than Δ^2/f edges. It would be interesting to extend our result to locally sparse graphs and other sparse graph classes.

References

1. Alon, N., Spencer, J.H.: The Probabilistic Method. Wiley Series in Discrete Mathematics and Optimization. Wiley (2011)
2. Alon, N., Krivelevich, M., Sudakov, B.: Coloring graphs with sparse neighborhoods. Journal of Combinatorial Theory, Series B 77(1), 73–82 (1999)

3. Barenboim, L., Elkin, M.: Distributed $(\Delta + 1)$-coloring in linear (in Δ) time. In: STOC 2009, pp. 111–120. ACM, New York (2009)
4. Barenboim, L., Elkin, M.: Sublogarithmic distributed MIS algorithm for sparse graphs using Nash-Williams decomposition. Distrib. Comput. 22, 363–379 (2010)
5. Barenboim, L., Elkin, M., Pettie, S., Schneider, J.: The locality of distributed symmetry breaking. In: FOCS 2012, pp. 321–330 (October 2012)
6. Beck, J.: An algorithmic approach to the lovász local lemma. Random Structures & Algorithms 2(4), 343–365 (1991)
7. Bollobás, B.: Chromatic number, girth and maximal degree. Discrete Mathematics 24(3), 311–314 (1978)
8. Borodin, O.V., Kostochka, A.V.: On an upper bound of a graph's chromatic number, depending on the graph's degree and density. Journal of Combinatorial Theory, Series B 23(2-3), 247–250 (1977)
9. Brooks, R.L.: On colouring the nodes of a network. Mathematical Proceedings of the Cambridge Philosophical Society 37(02), 194–197 (1941)
10. Catlin, P.A.: A bound on the chromatic number of a graph. Discrete Math. 22(1), 81–83 (1978)
11. Elkin, M.: Personal communication
12. Grable, D.A., Panconesi, A.: Fast distributed algorithms for Brooks-Vizing colorings. Journal of Algorithms 37(1), 85–120 (2000)
13. Haxell, P.E.: A note on vertex list colouring. Comb. Probab. Comput. 10(4), 345–347 (2001)
14. Jamall, M.S.: A Brooks' Theorem for Triangle-Free Graphs. ArXiv e-prints (2011)
15. Jamall, M.S.: A Coloring Algorithm for Triangle-Free Graphs. ArXiv e-prints (2011)
16. Jamall, M.S.: Coloring Triangle-Free Graphs and Network Games. Dissertation. University of California, San Diego (2011)
17. Jensen, T.R., Toft, B.: Graph coloring problems. Wiley-Interscience series in discrete mathematics and optimization. Wiley (1995)
18. Kim, J.H.: On brooks' theorem for sparse graphs. Combinatorics. Probability and Computing 4, 97–132 (1995)
19. Kostochka, A.V., Mazuronva, N.P.: An inequality in the theory of graph coloring. Metody Diskret. Analiz. 30, 23–29 (1977)
20. Kuhn, F.: Weak graph colorings: distributed algorithms and applications. In: SPAA 2009, pp. 138–144. ACM, New York (2009)
21. Lawrence, J.: Covering the vertex set of a graph with subgraphs of smaller degree. Discrete Mathematics 21(1), 61–68 (1978)
22. Linial, N.: Locality in distributed graph algorithms. SIAM J. Comput. 21(1), 193–201 (1992)
23. Molloy, M., Reed, B.: Graph Colouring and the Probabilistic Method. Algorithms and Combinatorics. Springer (2001)
24. Moser, R.A., Tardos, G.: A constructive proof of the general lovász local lemma. J. ACM 57(2), 11:1–11:15 (2010)
25. Panconesi, A., Srinivasan, A.: On the complexity of distributed network decomposition. Journal of Algorithms 20(2), 356–374 (1996)
26. Peleg, D.: Distributed Computing: A Locality-Sensitive Approach. Monographs on Discrete Mathematics and Applications. SIAM (2000)
27. Reed, B.: The list colouring constants. Journal of Graph Theory 31(2), 149–153 (1999)

28. Reed, B., Sudakov, B.: Asymptotically the list colouring constants are 1. J. Comb. Theory Ser. B 86(1), 27–37 (2002)
29. Rubinfeld, R., Tamir, G., Vardi, S., Xie, N.: Fast local computation algorithms. In: ICS 2011, pp. 223–238 (2011)
30. Schneider, J., Wattenhofer, R.: A new technique for distributed symmetry breaking. In: PODC 2010, pp. 257–266. ACM, New York (2010)
31. Vizing, V.G.: Some unsolved problems in graph theory. Uspekhi Mat. Nauk 23(6(144)), 117–134 (1968)
32. Van Vu, H.: A general upper bound on the list chromatic number of locally sparse graphs. Comb. Probab. Comput. 11(1), 103–111 (2002)

Author Index